Shallow Foundations

Shallow Foundations
Discussions and Problem Solving

Tharwat M. Baban
B.Sc. (Hons.); M.S. (Berkeley, Calif.)
University Professor and
Geotechnical Consultant
(Retired)

WILEY Blackwell

This edition first published 2016

Registered Office
John Wiley & Sons, Ltd, The Atrium, Southern Gate, Chichester, West Sussex, PO19 8SQ, United Kingdom

Editorial Offices
9600 Garsington Road, Oxford, OX4 2DQ, United Kingdom
The Atrium, Southern Gate, Chichester, West Sussex, PO19 8SQ, United Kingdom

For details of our global editorial offices, for customer services and for information about how to apply for permission to reuse the copyright material in this book please see our website at www.wiley.com

Library of Congress Cataloging-in-Publication Data applied for.

Hardback ISBN: 9781119056119

A catalogue record for this book is available from the British Library.

Wiley also publishes its books in a variety of electronic formats. Some content that appears in print may not be available in electronic books.

Cover image: Courtesy of the Author

Set in 9.5/11.5pt Minion by SPi Global, Pondicherry, India

Printed in Singapore by C.O.S. Printers Pte Ltd

1 2016

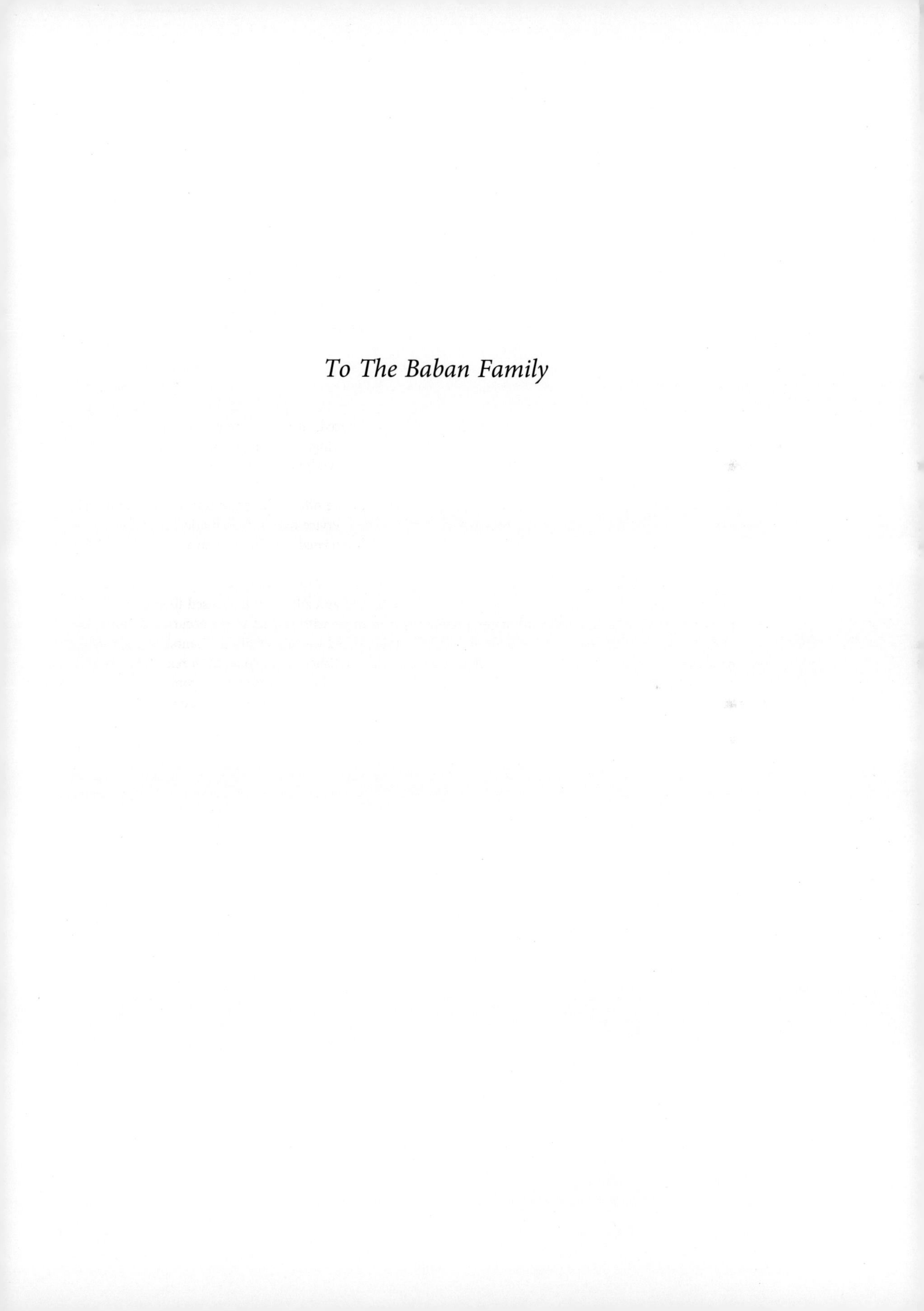

To The Baban Family

Contents

Preface

This book is intended primarily to introduce civil engineers, especially geotechnical engineers and all civil engineering students reading the specialist subjects of soil mechanics and geotechnical engineering, to the fundamental concepts and application of shallow foundation analysis and design. Also, the furnished material can be considered as an essential reference work for practising civil engineers, consulting engineers and government authorities. The primary focus of this book is on interfacing structural elements with the underlying soil, which is, in the author's opinion, where the major focus of shallow foundation engineering lies.

The book is not intended to be a specific text book on soil mechanics or geotechnical engineering. Therefore, there is no part of the text alone that could be used as a core syllabus for a certain course. However, it is the author's opinion that more than 70% of the book is core material at the advanced undergraduate levels. It is expected that civil engineering students will find the text helpful in better understanding the fundamental concepts and their implications for the analysis and design of shallow foundations. The author tried to present the material such that separable topics and subtopics are covered in separate sections, with clear and unambiguous titles and subtitles. Thus, it would not be difficult for a university lecturer to draw up a personalised reading schedule, appropriate to his or her own course. It is hoped that the book can establish itself as an effective reference and a useful text in most of the engineering colleges and technical institutions.

Generally, the given material is of an advanced level and, therefore, it is assumed that the reader has a good understanding of basic statics and the mechanics of materials and has studied the basic principles of soil mechanics, lateral earth pressures and reinforced concrete. SI units are used throughout all the chapters and, therefore, the reader also needs to have sufficient background knowledge regarding the use of these units.

The book would be very beneficial to the reader, since it provides essential data for the design of shallow foundations under ordinary circumstances. The necessary background concepts and theories are generally presented clearly in concise forms of formulas or charts, and their applications are highlighted through solving a relatively large number of realistic problems. Moreover, the worked problems are of the types usually faced by civil engineers in practice and, therefore, the obtained information will be most valuable.

Generally, the subject matter is introduced here by first discussing the particular topic and then solving a number of pertinent objective problems that highlight the relevant theories, concepts and analysis methods. A list of crucial references is given at the end of each chapter. Thus, each chapter consists of three parts: *discussions*, *problem solving* and *references*.

The "discussion" part is presented in a clear and concise but precise manner, keeping in view the avoidance of unnecessary details. In some chapters, where the topics are of special difficulty, full guiding explanations are given; where the subject of study is simpler, less detailed treatment is provided.

The "problem solving" part gives a relatively comprehensive range of worked out problems to consolidate an understanding of the principles and illustrate their applications in simple practical situations. A total of 180 worked problems have been provided. The author's academic and professional career has proved to him that geotechnical engineers and civil engineering students need to be well acquainted with the correct and effective use of the theoretical and empirical principles and formulas they have learned. An effective way to lessen the deficiency may be through solving, as much as possible, a variety of problems of the type or nearly similar to those engineers face in practice. For these reasons the author considers the "problem solving" part, on which the book is partially based, as a vital portion of the text.

The "references" part that comes directly at the end of each chapter enriches the discussions part with valuable sources of information and increases its reliability. Moreover, the furnished references will be very beneficial to any ambitious fresh civil engineer or undergraduate student who may wish later to undertake higher studies in the subject.

The text comprises six chapters. The chapters are devoted mostly to the geotechnical and structural aspects of shallow foundation design. A brief overview of each chapter follows:

- *Chapter 1* deals with site investigation in relation to the analysis and design of shallow foundations. Unlike the other chapters, this chapter requires various topics, field tests in particular, to be discussed separately. Therefore, only a general and relatively brief overview of the overall subject matter, consistent with the chapter title, is given in the main "discussion" part. Discussion individual topics is given in the "problem solving" part directly below the relevant problem statement. Solutions of 27 problems have been provided. These solutions and those of the other five chapters have fully worked out calculations.
- *Chapter 2* presents introductory discussions and explanations of various topics pertaining to shallow foundations, their analysis and design. It discusses type and depth of shallow foundations, performance requirements, sulfate and organic acid attack on concrete, distribution of contact pressures and settlements and vertical stress increase in soils due to foundation loads. Solutions of 21 problems are presented.
- *Chapter 3* concerns settlements due to foundation loads. The chapter discusses various types of settlements of foundations on both coarse- and fine-grained soils, methods of settlement estimation, methods of estimating and accelerating consolidation settlement, settlement due to secondary compression (creep), estimation of settlements over the construction period and settlement of foundations on rock. Solutions of 56 problems are introduced.
- *Chapter 4* deals with the bearing capacity of shallow foundations. The chapter discusses most of the significant aspects of the subject matter, among them: bearing capacity failure mechanism, bearing capacity equations, bearing capacity of footings with concentric and eccentric vertical loads, bearing capacity of footings with inclined loads, effects of water table and other factors on bearing capacity, uplift capacity of shallow foundations, bearing capacity of foundations on layered soils and on slopes, and bearing capacity of rock. Solutions of 40 problems are provided.
- *Chapter 5* deals with the structural design of different types of shallow foundations. Structural design of plain concrete spread footings, pedestals, pile caps and the foundations of earth-retaining concrete walls are also included. The discussion part of the chapter covers most of the major aspects of the subject matter such as: selection of materials, design loads, structural action of isolated and continuous footings, eccentrically loaded footings, modulus of subgrade reaction and beams on elastic foundations, rigid and flexible design methods and so on. Design calculations of 28 typical problems are presented in a step by step order. All the structural designs conform to the Building Code Requirements for Structural Concrete (ACI 318) and Commentary, USA.
- *Chapter 6* deals with Eurocode Standards in relation to the design of spread foundations. The discussion part of the chapter covers certain important topics such as: Eurocode background and applications, basis of design and requirements, principles of limit states design, design approaches, partial factors and load combinations, geotechnical and structural designs of spread foundations and so on. The problem solving part of the chapter provides design calculations of eight typical

problems, as an attempt to introduce the concerned engineer to application of the design rules stipulated by Eurocodes 2 and 7 (or EN 1992 and 1997) rather than the geotechnical and structural related issues.

It is well known that not all civil engineers are acquainted with all internationally recognised codes, such as the ACI Code and Eurocodes, at the same time. Therefore, this chapter is especially written for those civil engineers, specially geotechnical engineers, who are unfamiliar with the technical rules and requirements of the Eurocode Standards (Structural Eurocode). The implementation of the Eurocode is extended to all the European Union countries and there are firm steps toward their international adoption.

It must be clear that, despite every care taken to ensure correctness or accuracy, some errors might have crept in. The author will be grateful to the readers for bringing such errors, if any, to his notice. Also, suggestions for the improvement of the text will be gratefully acknowledged.

Tharwat M. Baban

Acknowledgements

The author wishes to express his gratitude to those writers and researchers whose findings are quoted and to acknowledge his dependence on those authors whose works provided sources of material; among them are:

Giuseppe Scarpelli, Technical University of Marche Region, Ancona, Italy
Trevor L.L. Orr, Trinity College, Dublin, Ireland
Andrew J. Bond, Geocentrix Ltd, Banstead, UK
Andrew Harris, Kingston University, London, UK
Jenny Burridge, The Concrete Center, London, UK

The author is truly grateful to Dr. Paul Sayer, who was kind enough to suggest the inclusion of Chapter 6.

The author wishes to thank all the members of his family, especially his wife Mrs. Sawsan Baban, for their encouragement, assistance and patience in preparing the manuscript of this book.

Finally, the author likes to thank John Wiley and Sons, Ltd and their production staff for their cooperation and assistance in the final development and production of this book.

CHAPTER 1

Site Investigation in Relation to Analysis and Design of Foundations

1.1 General

The stability and safety of a structure depend upon the proper performance of its foundation. Hence, the first step in the successful design of any structure is that of achieving a proper foundation design.

Soil mechanics is the basis of foundation design since all engineered constructions rest on the earth. Before the established principles of soil mechanics can be properly applied, it is necessary to have a knowledge of the distribution, types and engineering properties of subsurface materials. Therefore, an adequate site investigation is an essential preliminary to enable a safe and economic design and to avoid any difficulties during construction. A careful site investigation can minimise the need for overdesign and reduce the risks of underdesign. A designer who is well equipped with the necessary reliable information can use a lower factor of safety, thereby achieving a more economical design. With enough information available, construction troubles can be decreased and, therefore, construction costs are decreased too.

A site investigation usually costs a small percentage of total construction costs. According to Bowles (2001), elimination of the site exploration, which usually ranges from about 0.5 to 1.0% of total construction costs, only to find after construction has started that the foundation must be redesigned is certainly a false economy. However, a geotechnical engineer planning a subsurface exploration program for a specific job must keep in mind the relative costs of the exploration versus the total construction costs. It is understood that there is no hard and fast procedure for an economical planning a site investigation programme. Each condition must be weighed with good judgment and relative economy.

Nowadays, it is doubtful that any major structures are designed without site exploration being undertaken. Sometimes, for small jobs, it may be more economical to make the foundation design on conservative values rather than making elaborate borings and tests, especially, when the condition of the adjacent structures is an indication that the site is satisfactory. However, generally, design of structures without site investigation is not recommended by civil engineers.

The cheapest and most common method of subsurface exploration is by drilling boreholes. Test pits are too expensive for general exploration, but they may be used for more careful examination if found to be needed.

Shallow Foundations: Discussions and Problem Solving, First Edition. Tharwat M. Baban.

1.2 Site Investigation

A successful investigation of a site for an important structure will generally falls under the following five headings:

(1) Reconnaissance
(2) Subsurface exploration
(3) Laboratory tests
(4) Compiling information
(5) Geotechnical report.

1.2.1 Reconnaissance

Office Reconnaissance: This phase of reconnaissance comprises the following duties:

- Review of plans, boring logs and construction records of existing structures in the area.
- Study of the preliminary plans and designs of the proposed structure, including the approximate magnitude of the loads to be transmitted to the supporting material.
- Review of other backlogs of information already compiled on the same general area and similar structures.
- Review of other information pertaining to the site area obtained from such sources as different types of maps (i.e., topographic, geologic and agricultural maps), photographs, records of adjacent bridge if exists, underground utility constructions and well drilling logs.
- Formulation of a boring plan should be made during the latter phases of the office reconnaissance. This prepared boring plan should be reviewed during the field reconnaissance. The objective should be the development of a maximum of subsurface information through the use of a minimum number of boreholes. Spacing, number and depth of boreholes will be discussed later in conjunction with the Solution of Problem 1.4.

Field Reconnaissance: This phase of reconnaissance should commence with a visit to the site of the proposed structure. It should always be made by a Soils or Foundation Engineer who will complete the Geotechnical Report. Whenever possible, it is desirable that this engineer be accompanied by the driller or the driller foreman. Notes on items to be observed are as follows:

- Surface Soils: Surface soils are easily revealed through the use of a shovel or post-hole diggers. These soils may sometimes be identified as belonging to some particular formation, and usually they indicate the underlying material.
- Gullies, Excavations and Slopes: Any cut or hole near the proposed structure site is a subsurface window, and for its depth it will provide more information than borehole since it may be examined in detail.
- Surface and Subsurface Water: The presence of either surface or subsurface water is an important factor in both preparation of boring plans and foundation design. All surface flows should be noted, and all opportunities should be taken to observe the groundwater level.
- Study of Existing Structures: The existing structures within an area are valuable sources of information. A very close examination of them with regard to their performance, type of foundation, apparent settlement, load, location and age will yield a wealth of data.
- Topography: To some extent, topography is indicative of subsurface conditions. In narrow, steep stream beds, rock is likely to be near the surface with little overlying stream-deposited soil. On the other hand, wide, flat valleys indicate deep soil deposits.

- Information required by the Drill Crew. The drill crew needs to know how to get to the site, where to drill, what equipment to take, and what difficulties to expect. Generally, the following types of information are usually needed:
- Information regarding verification of the boring plan which was already prepared during the office reconnaissance phase. The proposed locations of boreholes should be checked for accessibility. Desirable deletions, additions, and relocations should be made as are necessary to better suit the crew's capabilities and to add completeness to the subsurface information.
- Type of drilling and equipment needed. Notes should be made as to which type of drilling is best suited to the site (i.e., rotary, auger, etc.).
- Reference points and bench marks. The reconnaissance should determine if reference points and bench marks are in place adjacent to the site and properly referenced on the plans.
- Utilities. Underground and overhead utilities located at the site should be accurately shown on the plans or their locations should be staked on the ground.

- Geophysical Survey: The field reconnaissance may require a geophysical survey of the site. The use of geophysical methods provides information on the depths to the *soil* and *rock layers*, the homogeneity of the layer and the type of soil or rock present. This information can be used to supplement the boring plan.
- Field Reconnaissance Report: A concise and informative field reconnaissance report, in which all decisions concerning the boring plan and the drill crew are delineated, should be prepared. It can be facilitated by the use of a special check list or form.

1.2.2 Subsurface Exploration

General: After the reconnaissance has been completed, the results should be given to the foreman of the drill crew in order to carry out the foundation investigation at the site. Briefly, the subsurface or foundation exploration consists of making the borings and collecting the pertinent samples (i.e., drilling and sampling), performing the required field tests in conjunction with the drilling, coring, and identification of materials. Each job site must be studied and explored according to its subsurface conditions and the size and type of the proposed structure. The civil engineer or the experienced geologist in charge of the exploration task should endeavour to furnish complete data such that a reliable study of practical foundation types can be made.

Before the arrival of the drill crew at the exploration site, enough survey control should have been previously carried out with reference to at least one bench mark already established at the site. The borehole locations should be staked in conformity with the boring plan. The stakes could indicate the borehole number and the existing ground surface elevation.

Drilling: It is defined as that process which advances the borehole. There are various methods of drilling or boring, namely: auger drilling, rotary drilling, wash boring, drilling by continuous sampling, percussion drilling and rock coring. Most of these methods are best suited for some particular problem or type of information sought. It is doubtful if an organisation (authority responsible for site investigation) would adopt any one method for all of its work unless the work was limited to one particular area. The same argument is true with respect to the various types of equipment used in drilling and coring.

Sampling: It is defined as that process wherein samples of the subsurface materials are obtained. As there are various methods of drilling and various equipment types, also there are different types of sampling, namely: split-barrel or spoon sampling, push barrel or thin-walled Shelby tube and stationary piston type sampling, wet barrel or double wall and dry barrel or single wall sampling, retractable plug sampling and rock coring.

Samples: The samples obtained during subsurface exploration should always represent the material encountered, that is, representative samples. These samples are disturbed, semi-disturbed or undisturbed. Usually, disturbed and semi-disturbed samples are used for identification and classification of the material. Undisturbed samples are used for determining the engineering properties such as strength, compressibility, permeability and so on, density and natural moisture content of the material. Disturbed samples are those in which the material (soil or rock) structure has not been maintained, and they are used in those tests in which structure is not important. On the other hand, relatively undisturbed samples have structures maintained enough that they could be used in engineering properties determination of the material. The degree of disturbance depends upon several factors such as type of materials being sampled, sampler or core barrel used, the drilling equipment, methods of transporting and preserving samples and driller skill. Extended exposure of the sample material to the atmosphere will change relatively undisturbed sample into an unusable state; therefore, the methods of obtaining and maintaining samples cannot be overemphasised.

Field Tests: there are various types of tests performed in the field, in conjunction with the drillings, in order to determine soil properties in situ. These tests are: dynamic penetration tests, such as standard (SPT) cone penetration and driven probe and driven casing; static cone penetration (CPT); in-place vane shear; plate-load (may not be in conjunction with the drillings); pressuremeter; flat-plate dilatometer; and other tests made in field laboratories, such as classification tests and unconfined strength test. Rock quality designation (RQD) test is performed on rock core samples. *Note:* Discussion of a particular field test will be given in the "*Problem Solving*" part of this chapter directly below the relevant problem statement.

Field Boring Log and Borehole Logging: The log is a record which should contain all of the information obtained from a boring whether or not it may seem important at the time of drilling. The process of recording the information in a special field log form is "logging". It is important to record the maximum amount of accurate information. This record is the "field" boring log. The importance of good logging and field notes cannot be overemphasised, and it is most necessary for the logger (who may be a soil engineer, a geologist, a trained technician or a trained drill crew foreman) to realise that a good field description must be recorded. The field boring log is the major portion of the factual data used in the analysis of foundation conditions.

Groundwater Table: The location of the groundwater level is an important factor in foundation analysis and design, and emphasis should be placed upon proper determination and reporting of this data.

In order to determine the elevation of groundwater it is suggested that at least two boreholes be left open for the duration of the subsurface exploration and periodically checked as to water level. These two boreholes should have their final check made no earlier than 24 h after the completion of exploration. The depth to the water level should be recorded on the boring log each time a reading is made, along with the time lapse since completion of the boring. When there is significant difference between the two borings checked, or when the logger deems it otherwise necessary, the water level in other boreholes should be checked.

Note: It is obvious that details of all the various methods and descriptions of the above mentioned drilling, samples and sampling, field tests, field boring log and borehole logging are too large in bulk for inclusion in the discussions. However, those interested in further information and details, should see various standards, practice codes and manuals, such as AASHTO Manual on Subsurface Investigations (1988), ASTM, BSI and Eurocode standards.

1.2.3 Laboratory Tests

Economical foundation design requires the use of the physical properties of the foundation material (soil or rock). The physical properties may be determined by in situ tests, load tests and laboratory tests. Results of laboratory tests, in addition to their use in foundations design, are used to predict the foundation behavior based on the experience of similar tested materials and their performance in the field. The two main reasons for making these laboratory tests are first to verify classification and second to determine engineering properties. An adequate amount of laboratory testing should be conducted to simulate the most sever design criteria. Generally, the amount of testing performed will depend on the subsurface conditions, laboratory facilities and type of the proposed structure.

Laboratory tests for foundations design will generally fall into four categories: classification, strength, compressibility and swelling and soil collapsibility. Other tests, such as permeability and compaction tests, may be required, especially when the proposed structure is a bridge or a dam.

Note: Significance, apparatus and procedure of various types of laboratory tests can be found in geotechnical books and recognised laboratory standard manuals (ASTM, AASHTO, BSI, etc.). The detailed description of these items is too bulky for inclusion in these discussions. The author assumes that the reader will have access to the latest volumes of the standards just mentioned. Nevertheless, Table 1.1 presents a summary list of ASTM and AASHTO tests frequently used for the laboratory testing of soils.

Table 1.1 ASTM and AASHTO standards for frequently used laboratory testing of soils.

		Test designation	
Test category	**Name of test**	**ASTM**	**AASHTO**
Visual identification	Practice for identification of soils (visual-manual procedure)	D 2488	—
	Practice for description of frozen soils (visual-manual procedure)	D 4083	—
Index properties	Test method for determination of water (moisture) content of soil by direct heating method	D 4959	T 265
	Test method for specific gravity of soils	D 854	T 100
	Method for particle-size analysis of soils	D 422	T 88
	Test method for amount of material in soils finer than the no. 200 sieve	D1140	—
	Test method for liquid limit, plastic limit and plasticity index of soils	D 4318	T 89
			T 90
	Test method for laboratory compaction characteristics of soil using standard effort (600 kN.m/m^3)	D 698	T 99
	Test method for laboratory compaction characteristics of soil using modified effort (2700 kN.m/m^3)	D 1557	T 180
Corrosivity	Test method for pH of peat material	D 2976	—
	Test method for pH of soils	D 2972	—
	Test method for pH of soil for use in corrosion testing	G 51	T 289
	Test method for sulfate content	D 4230	T 290
	Test method for resistivity	D 1125	T 288
		G 57	
	Test method for chloride content	D 512	T 291
	Test method for moisture, ash and organic matter of peat and other organic soils	D 2974	T194
	Test method for classification of soils for engineering purposes	D 2487	M 145
		D 3282	
Strength Properties	Unconfined compressive strength of cohesive soil	D 2166	T 208
	Unconsolidated, undrained compressive strength of clay and silt soils in triaxial compression	D 2850	T 296

(Continued)

Table 1.1 (*Continued*)

Test category	Name of test	Test designation	
		ASTM	**AASHTO**
	Consolidated undrained triaxial compression test on cohesive soils	D 4767	T 297
	Direct shear test of soils for unconsolidated drained conditions	D 3080	T 236
	Modulus and damping of soils by the resonant-column method (small-strain properties)	D 4015	—
	Test method for laboratory miniature vane shear test for saturated fine-grained clayey soil	D 4648	—
	Test method for bearing ratio of soils in place	D 4429	—
Strength Properties	California bearing ratio (CBR) of laboratory-compacted soils	D 1883	—
	Test method for resilient modulus of soils	—	T 294
	Method for resistance R-value and expansion pressure of compacted soils	D 2844	T 190
Permeability	Test method for permeability of granular soils (constant head)	D 2434	T 215
	Test method for measurement of hydraulic conductivity of saturated porous materials using flexible wall perameters	D 5084	—
Compression Properties	Method for one-dimensional consolidation characteristics of soils (oedometer test)	D 2435	T 216
	Test method for one-dimensional swell or settlement potential of cohesive soils	D 4546	T 258
	Test method for measurement of collapse potential of soils	D 5333	—

1.2.4 Compiling Information

General: Having carried the investigation through the steps of reconnaissance, subsurface exploration and laboratory testing, the next step is obviously the compilation of all the information.

Prior to preparing finished (final) logs, all samples should be checked by the Engineer in charge. This should ideally be performed immediately after the borings are completed and even while the boring program is still in progress. Significant soil characteristics that may have been omitted from the field logs may be identified, and the Engineer thus alerted to a potential foundation problem that might otherwise have gone undetected.

Finished Boring Log: It is important to differentiate clearly between the "field" boring log and the "finished" boring log. The field log is a factual record of events during boring operation, whereas the finished log is a graphical representation. The field log gives a wide range of information in notes or tabular form. The finished log is drawn from the data given in the field log as well as from results of visual inspection of samples; it represents a graphical picture of subsurface conditions. Information obtained from laboratory test results along with other necessary information are also utilised in its preparation. Also, various data such as results of field tests, depth location of different samples and groundwater level are superimposed upon it. It is important that the *soil* or *rock* in each stratum be clearly described and classified. A typical finished boring log is shown in Figure 1.1.

Soil Profile: In many cases it may be advantageous to plot a soil profile along various longitudinal or transverse lines. This should be done by plotting the boreholes in their true location, but with vertical scale exaggerated, connecting the similar strata by lines and shading similar areas by means of an identifying cross-section mark. The groundwater level should also be shown on the plot. Thus a probable representation of the subsurface conditions between the boreholes can be given, although a pocket or more of different formation may exist between any two adjacent boreholes. A representative example of an interpreted subsurface profile is shown in Figure 1.2.

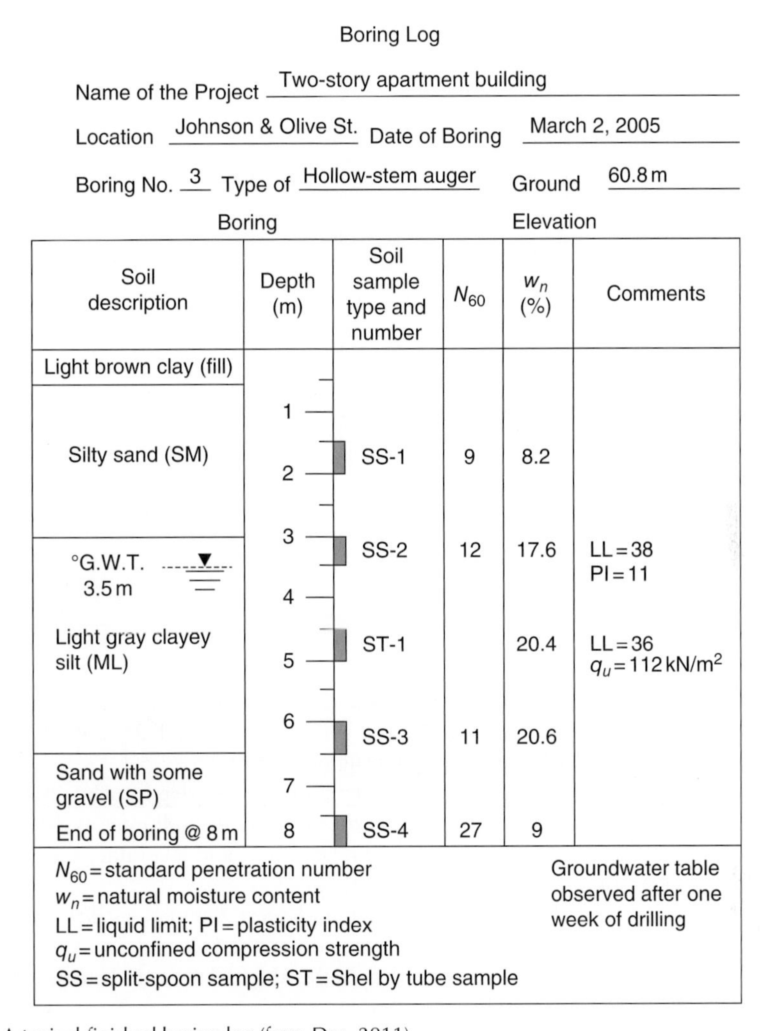

Boring Log

Name of the Project: Two-story apartment building

Location: Johnson & Olive St. Date of Boring: March 2, 2005

Boring No. 3 Type of Boring: Hollow-stem auger Ground Elevation: 60.8 m

Soil description	Depth (m)	Soil sample type and number	N_{60}	w_n (%)	Comments
Light brown clay (fill)					
Silty sand (SM)	1, 2	SS-1	9	8.2	
°G.W.T. 3.5 m; Light gray clayey silt (ML)	3	SS-2	12	17.6	LL = 38, PI = 11
	4, 5	ST-1		20.4	LL = 36, q_u = 112 kN/m^2
	6	SS-3	11	20.6	
Sand with some gravel (SP)	7				
End of boring @ 8 m	8	SS-4	27	9	

N_{60} = standard penetration number
w_n = natural moisture content
LL = liquid limit; PI = plasticity index
q_u = unconfined compression strength
SS = split-spoon sample; ST = Shel by tube sample

°Groundwater table observed after one week of drilling

Figure 1.1 A typical finished boring log (from Das, 2011).

1.2.5 Final Geotechnical Report

After information compiling has been successfully completed, a final written Geotechnical Report should be prepared and presented to the designer for use in the foundations design. Additionally, this report will furnish the Resident Engineer with data regarding anticipated construction problems, as well as serving to establish a firm basis for the contractor to estimate costs, unless the organisation's policies or regulations restrict the release of such information. A good policy may be of releasing such information, which should be as accurate as possible, provided that it is expressly and formally understood that the organisation will not be liable for any damages or losses incurred as a result of reliance upon it in the bidding or in the construction operations.

The Geotechnical Engineer who writes the Report should avoid including extraneous data which are of no use to the Designer or Resident Engineer. Also, his recommendations should be brief, concise and, where possible, definite. The Report should include:

- Authorisation of the site investigation and the job contract number and date.
- A description of the investigation scope.
- A description of the proposed structure for which the subsurface exploration has been conducted.

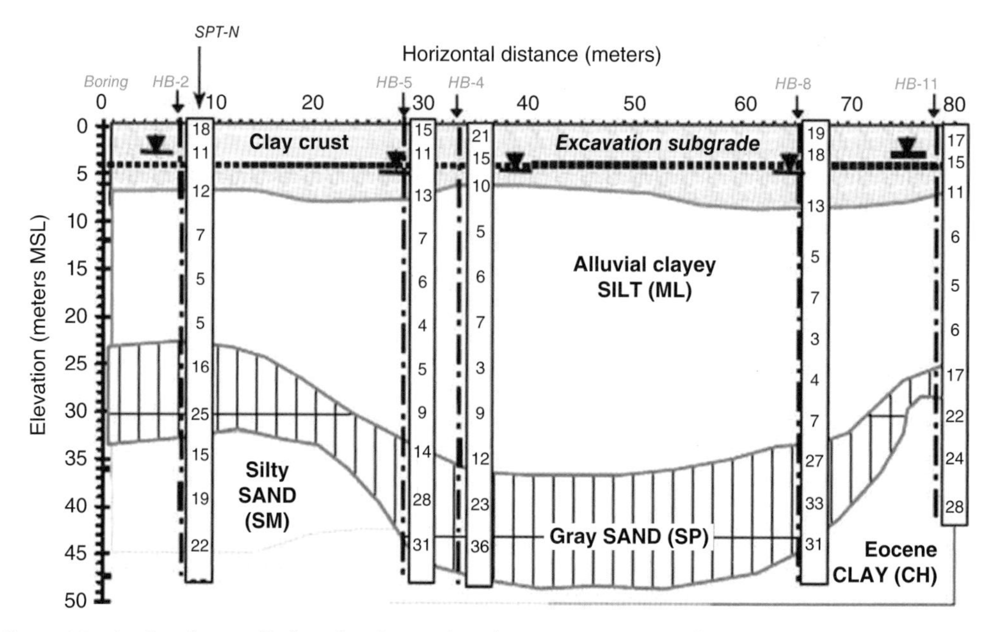

Figure 1.2 A subsurface profile based on boring data showing cross- sectional view (from FHWA, 2002).

- A description of the location of the site, including any structures nearby, drainage conditions, the nature of vegetation on the site and surrounding it, and any other features unique to the site.
- A description of the geological setting of the site.
- Details of the field exploration including number and depths of boreholes, types of borings involved and so on. A general description of the subsurface conditions, as determined from soil specimens and rock cores, and from results of related field and laboratory tests.
- A description of groundwater conditions.
- Brief description of suitable types of foundations, foundations depth and bearing capacity analysis.
- Recommendations and conclusions regarding the type and depth of foundations, allowable design bearing pressures for the supporting soil and rock layers, and solutions for any anticipated design and construction problems.
- The following presentations (mostly graphical) also need to be attached (as appendices) to the Report:
  - A site location map.
  - A plan view of the boreholes location with respect to the proposed location of each structure.
  - Finished boring logs.
  - Summary of the laboratory test results.
  - Other special graphical presentations.

Problem Solving

Problem 1.1

A rock stratum was cored for a length of 1.00 m (i.e., length of core advance). Total length of the recovered core was 0.75 m. Total length of the recovered pieces which are 100 mm or larger was 0.60 m. Determine:

(a) The rock recovery ratio (*RRR*) and the rock quality.
(b) The rock quality designation (*RQD*) and the rock quality.
(c) An approximate value for the reduced field modulus of elasticity (E_f) of the rock if its laboratory modulus of elasticity (E_{lab}) is 20 000 MPa.

Discussion:
General evaluation or classification of rock quality may be determined by calculating rock recovery ratio (*RRR*) and rock quality designation (*RQD*), where

$$RRR = \frac{\text{Total length of core recovered}}{\text{Length of core advance}} \quad (1.1)$$

$$RQD = \frac{\Sigma \text{Lengths of intact pieces of core} > 10\,\text{mm}}{\text{Length of core advance}} \quad (1.2)$$

Value of *RRR* near to 1.0 usually indicates good-quality rock, while in badly fractured or soft rock the *RRR* value may be 0.5 or less. Table 1.2 presents a general relationship between *RQD* and in situ rock quality (Deere, 1963). An approximate relationship between *RQD* and ratio of field and laboratory modulus of elasticity (E_f/E_{lab}) was added to Table 1.2 (Bowles, 2001). This relationship can be used as a guide only.

Table 1.2 Correlation for *RQD* and in situ rock quality.

RQD	Rock quality	E_f/E_{lab}*
<0.25	Very poor	0.15
0.25–0.5	Poor	0.2
0.50–0.75	Fair	0.25
0.75–0.9	Good	0.3–0.7
>0.9	Excellent	0.7–1.0

* Approximately for field/laboratory ratio of compression strengths also.

Solution:

(a) Equation (1.1): $RRR = \frac{0.75}{1.00} = 0.75 \rightarrow$ *Fair rock quality*

(b) Equation (1.2): $RQD = \frac{0.60}{1.00} = 0.60 \rightarrow$ *Fair rock quality* (see Table 1.2)

(c) $RQD = 0.6 \rightarrow \frac{E_f}{E_{lab}} = 0.25$ (see Table 1.2)

$$E_f = 20000 \times 0.25 = 5000\,MPa$$

Problem 1.2

(a) What are the factors on which disturbance of a soil sample depends?
(b) The cutting edge of a sampling tube has outside diameter D_o = 50.8 mm and inside diameter D_i = 47.6 mm, while the sampling tube has D_{ot} = 50.03 mm and D_{it} = 48.50 mm. Is the tube sampler well designed?

(*Continued*)

Discussion:

Disturbance of a soil sample, during the sampling process, depends on many factors such as rate of penetration of the sampling tube, method of applying the penetrating force (by pushing or driving), the presence of gravel, the area ratio A_r, the inside and outside clearances (see Figure 1.3) and proper working of the check valve inside the sampler head.

In obtaining undisturbed soil samples, the *sample disturbance* can be greatly reduced by: pushing the sampler with a moderate rate of penetration, using stainless steel tube with inside protective coating as required by ASTM D 1587, frequent cleaning the sampler head so that the check valve works properly, and using sampler of $A_r < 10\%$ with its inside clearance $C_i = 0.5–3.0\%$ and outside clearance $C_o = 0–2\%$. The factors A_r, C_i and C_o are usually expressed as:

$$A_r = \frac{D_o^2 - D_i^2}{D_i^2} \times 100 \qquad C_i = \frac{D_{it} - D_i}{D_i} \times 100 \qquad C_o = \frac{D_o - D_{ot}}{D_{ot}} \times 100$$

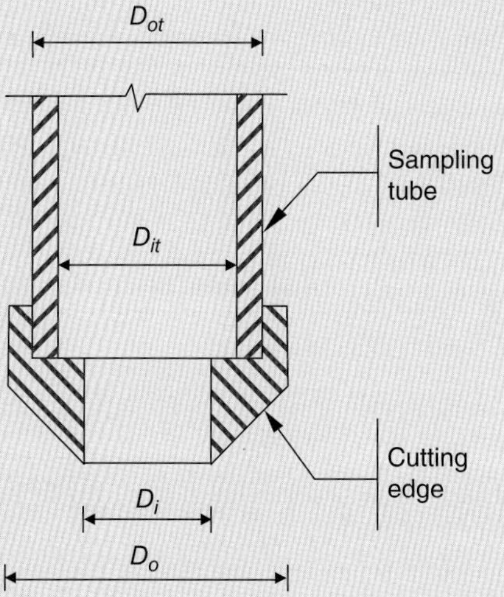

Figure 1.3 Sampling tube and the inside and outside diameters.

Solution:

(a) The answer is included in the discussion.

(b)

$$A_r = \frac{D_o^2 - D_i^2}{D_i^2} \times 100 = \frac{50.8^2 - 47.63^2}{47.63^2} \times 100 = 13.75\%$$

$$C_i = \frac{D_{it} - D_i}{D_i} \times 100 = \frac{48.50 - 47.63}{47.63} \times 100 = 1.83\%$$

$$C_o = \frac{D_o - D_{ot}}{D_{ot}} \times 100 = \frac{50.80 - 50.03}{50.03} \times 100 = 1.54\%$$

The A_r value is not satisfactory because it is greater than 10%; however, it is close to the limit. The C_i and C_o values are both satisfactory because they are less than 3% and 2%, respectively. Therefore, one may judge that *the sampler to some extent (approximately) is well designed.*

Problem 1.3

A thin-walled tube sampler was pushed into soft clay at the bottom of a borehole a distance of 600 mm. When the tube was recovered, a measurement down inside the tube indicated a recovered sample length of 585 mm. What is the soil recovery ratio *SRR*, and what (if anything) happened to the sample? If D_o and the D_i (outside and inside diameters) of the sampling tube were 76.2 and 73.0 mm respectively, what is the probable sample quality?

Discussion:
Similar to the rock recovery ratio, there is also soil recovery ratio, *SRR*, which can be used in estimating degree of disturbance of soil samples, where

$$SRR = \frac{\text{Actual length of recovered soil sample}}{\text{Total length of sampler in soil}} \tag{1.3}$$

A recovery ratio of 1 indicates, theoretically, that the sample did not become compressed during sampling. A recovery ratio less than 1 indicates compressed sample in a disturbed state. Also, when a recovery ratio is greater than 1, the sample is disturbed due to loosening from rearrangement of stone fragments, roots, removal of overburden, or other factors.

Solution:
Equation (1.3): $SRR = \frac{585}{600} = 0.98 < 1 \rightarrow$ *slightly compressed sample*

The sample is compressed from friction on the tube and, or from pressure of entrapped air above the sample due to incapable head check valve.

$$A_r = \frac{D_o^2 - D_i^2}{D_i^2} \times 100 = \frac{76.2^2 - 73.0^2}{73.0^2} \times 100 = 8.96\,\% < 10\%$$

$$C_i = \frac{D_{it} - D_i}{D_i} \times 100 = \frac{73.0 - 73.0}{73.0} \times 100 = 0\% < (0.5\text{–}3.0\%)$$

$$C_o = \frac{D_o - D_{ot}}{D_{ot}} \times 100 = \frac{76.2 - 76.2}{76.2} \times 100 = 0\%$$

These results and that of *SRR* indicate that *the sample is nearly undisturbed.*

Problem 1.4

A five story office building is to be built in a deep, moderately uniform fine-grained soil deposit. The bedrock has a depth of over 75 m. The foundation level will be located at a depth of 1 m below ground surface. At the foundation level, the unfactored uniformly distributed load (q) from the building structure is 70 kPa (or 1434 psf). The building will be 50 m wide and 85 m long. Appropriate values for any missing data may be assumed. Estimate:

(a) The required number, spacing and location of exploratory boreholes.
(b) The required depth of the boreholes.

Discussion:
A boring plan should include number, spacing, location and depth of borings in addition to the sampling intervals. There are many variables involved in formulation a boring plan such as general knowledge of subsurface conditions, knowledge of proposed structure, economy by scheduling an inexcessive number and depth of boreholes, and personal experience, preference and judgment of the responsible geotechnical engineer.

Boring spacing: Required number and location configuration of borings (or boreholes) is defined by their spacing. Boring spacing depends mainly on the uniformity of soil strata, the type of structure and loading conditions. Erratic subsurface conditions require closely spaced borings, whereas uniform conditions require a maximum spacing. Structures sensitive to settlements and structures subjected to heavy loads require extensive

(*Continued*)

subsurface knowledge; therefore borings should be closely spaced. Knowledge of the regional conditions as well as the type and nature of the proposed structure is the best guide to the necessary boreholes spacing.

It is a difficult subject to pinpoint in exact figures the required number of boreholes and their tentative spacing for the many reasons stated above; therefore, there are no binding rules on the number and spacing of boreholes.

For buildings taller than three stories or 12.2 m, the BOCA (1996) National Building Code requires at least one boring for every 230 m^2 of built-over area, while others prefer boreholes drilled at or near all the corners and also at important locations. For buildings a minimum of three borings, where the surface is level and the first two borings indicate regular stratification, may be adequate. Five borings are generally preferable (four borings at building corners and one at centre), especially if the site is not level. On the other hand, a single boring may be sufficient for small and less important building, an antenna or industrial process tower base with the hole drilled at centre.

There are tables which give suggested spacing of boreholes (Table 1.3) or required area for each borehole (Table 1.4). These tables should be used as general guidelines only.

Table 1.3 Suggested spacing of boreholes.

Type of project	Boreholes spacing, m
One story building	25–30
Multistory building	15–25
Residential subdivision planning	60–100
Highway	150–300*
Earth dam	25–50
Bridge**	—

* It is reduced to even 30 m for erratic subsurface conditions.
** Minimum of one borehole at location of each pier or abutment.

Table 1.4 Suggested required area for each borehole.

Subsurface conditions	Structure foot print area for each borehole, m^2
Poor quality and/or erratic	100–300
Average	200–400
High quality and uniform	300–1000

Furthermore, the following are two key clauses of BSI publications regarding the spacing of borings:

BSI 5930: 1999 Code of Practice for Site Investigations – Clause 12.6

"Although no hard and fast rules can be laid down, a relatively close spacing between points of exploration, e.g. 10 m to 30 m, are often appropriate for structures. For structures small in plan area, exploration should be made at a minimum of three points, unless other reliable information is available in the immediate vicinity."

BSI EN 1997-2: 2007 Ground Investigation and Testing – Clause B.3

"The following spacing of investigation points should be used as guidance:

- For high-rise and industrial structures, a grid pattern with points at 15 m to 40 m distance,
- For large-area structures, a grid pattern with points at not more than 60 m distance,
- For linear structures, a spacing of 20 m to 200 m,
- For special structures, two to six investigation points per foundation."

Depth of Borings: As it was the case with the required spacing and number of borings and due to nearly the same main reasons, here also, there is no absolute and unique rule to determine the required depth of exploratory borings. Borings should reach depths adequate to explore the nature of the subsurface soils, mainly encountered in the influence stress zones, including all strata likely to contribute significantly to settlement. There are many empirical and semi-empirical criteria, rules and equations established by individual researchers, engineering agencies and societies, to estimate the required minimum depth of borings. Some of these are as follows:

ASCE Criteria (1972): "Unless bedrock is encountered first, borings should be carried to such depth that the net increase in soil stress σ_z under the weight of the structure is less than 10 percent of the average load of the structure q, or less than 5 percent of the effective overburden stress σ'_O in the soil at that depth, whichever is the lesser depth." These two criteria are shown in the scheme below.

Criterion 1: $\sigma_z = 0.10\ q \rightarrow$ find z_1
Criterion 2: $\sigma_z = 0.05\ \sigma'_O \rightarrow$ find z_2
Use z_1 or z_2, whichever is smaller.
Boring depth = $z_b = (z_1 \text{ or } z_2) + D$
D = the anticipative foundation depth

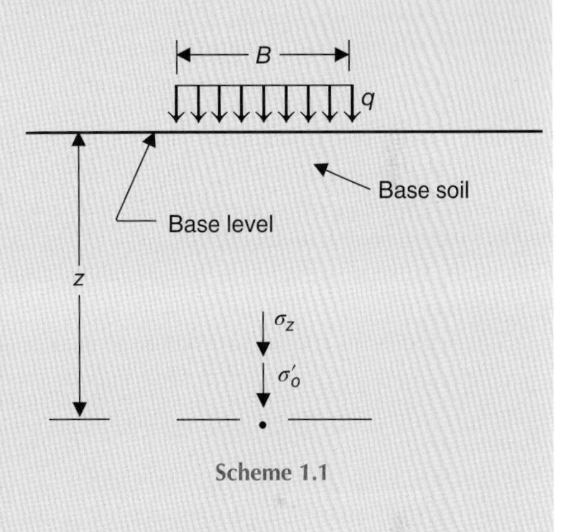

Scheme 1.1

Smith (1970) used Criterion 2 in the form: $\sigma_z = M\ \sigma'_O$; in which M is the compressibility criterion equals 10% for fine-grained soils and 20% for course-grained soils. He used the Boussinesq principles of stress distribution to determine the influence depth z, and established a set of curves relating the influence depth to the equivalent square dimension of a uniformly loaded rectangular or circular area. Also, he derived equations for estimating minimum depth of borings for embedded footing systems, mats or rafts and pile groups.

Baban (1992) established the same relationships using the ASCE criteria and Westergaard principles of stress distribution; with an assumed average value for Poisson's ratio of the base soil equals 0.3. He found that for embedded footing systems, mats or rafts and pile groups of width B larger than 20 m, Criterion 2 always controls the minimum boring depth. The equations derived by Baban are as follows:

According to the ASCE Criterion 1,

$$Z = \left\{0.5\left[\left(0.9B^2 + 0.9L^2\right)^2 + 122B^2L^2\right]^{1/2} - 0.4\left(B^2 + L^2\right)\right\}^{1/2} \tag{1.4}$$

According to the ASCE Criterion 2,

$$\tan\left(\frac{0.025\,\pi\,\gamma'_A z}{q}\right) = \left[\frac{BL}{\left(1.144B^2Z^2 + 1.144L^2Z^2 + 1.309Z^4\right)^{1/2}}\right] \tag{1.5}$$

$$Z_b = Z + D$$

γ'_A = average effective unit weight of soil within depth Z in lb/ft^3, q is in lb/ft^2, and Z, B and L are in feet.

Sowers and Sowers (1970) suggested a rough estimate of minimum boring depth (unless bedrock is encountered) required for building structures according to number of stories S, as follows:

$$\text{For light steel or narrow concrete buildings, } z_b = 3\,S^{0.7} + D \tag{1.6}$$

$$\text{For heavy steel or wide concrete buildings, } z_b = 6\,S^{0.7} + D \tag{1.7}$$

(*Continued*)

Coduto (2001) adapted from Sowers (1979) equations for z_b considering both the number of stories and the subsurface conditions, as presented in Table 1.5.

Table 1.5 Minimum boring depth for different subsurface conditions.

Subsurface conditions	Minimum boring depth, m
Poor	$z_b = 6\ S^{0.7} + D$
Average	$z_b = 5\ S^{0.7} + D$
Good	$z_b = 3\ S^{0.7} + D$

It is noteworthy there is a general rule of thumb, often adopted in practice, requires minimum boring depth equals 2× the least lateral plan dimensions of the building or 10 m below the lowest building elevation. This is because the foundation or footing dimensions are seldom known in advance of borings. According to Bowles (2001), where 2× width is not practical as, say, for a one-story warehouse or department store, boring depths of 6 to 15 m may be adequate. On the other hand, for important (or high-rise) structures that have small plan dimensions, it is common to extend one or more of the borings to competent (hard) soil or to bedrock regardless of depth.

When the foundations are taken up to rock, it should be insured that large boulders are not mistaken as bedrock. The minimum depth of core boring into the bedrock should be 3 m to establish it as a rock.

Solution:

(a) Try the suggested guidelines as follows:

- According to BSI requirements, for large-area structures, a grid pattern with boring spacing not more than 60 m should be used. A spacing range of 10–30 m is appropriate.
- According to BOCA, minimum of one boring shall be used for every 230 m^2. The building area is 50 × 85 = 4250 m^2 which requires about 20 boreholes (considered relatively large number). This number of boreholes requires a grid of four rows and five columns with about 16.5 m vertical and 21.0 m horizontal centre to centre spacing (considered relatively small spacing). Moreover, using 20 boreholes is uneconomical.
- Others may suggest five boreholes, but undoubtedly this requires too large a borehole spacing.
- According to Table 1.3, for multistory buildings, borehole spacing of 15–25 m is suggested.
- According to Table 1.4, for uniform subsurface soil condition, 300–1000 m^2 for every borehole is suggested.

If 10 boreholes are used, the area per borehole will be 425 m^2; considered a reasonable figure. These boreholes may be distributed in a triangular pattern, with three rows in the long direction. Each of the two side rows gets three boreholes, the middle row gets the remaining four boreholes. The boreholes of each row will have equal spacing. Let the centre of the exterior boreholes be located 0.65 m inside the building area. In this way, the centre to centre borehole spacing will be 27.9 m. This borehole spacing is considered adequate. Hence:

Use 10 boreholes.

The boreholes will be distributed and spacing as shown in Figure 1.4.

(b) Depth of boreholes: Try all the available criteria and established relationships. In order to accomplish this it is necessary to assume reasonable values for certain missing data, as follows:

- The ground water table is so deeply located that it would not be reached during subsurface exploration.
- The average effective unit weight of the subsurface soil equals 18 kN/m^3 (or 112.4 pcf).
- The expected type of support would be shallow foundation such as column footing system, mat and raft foundation; rests on soil (not piles).

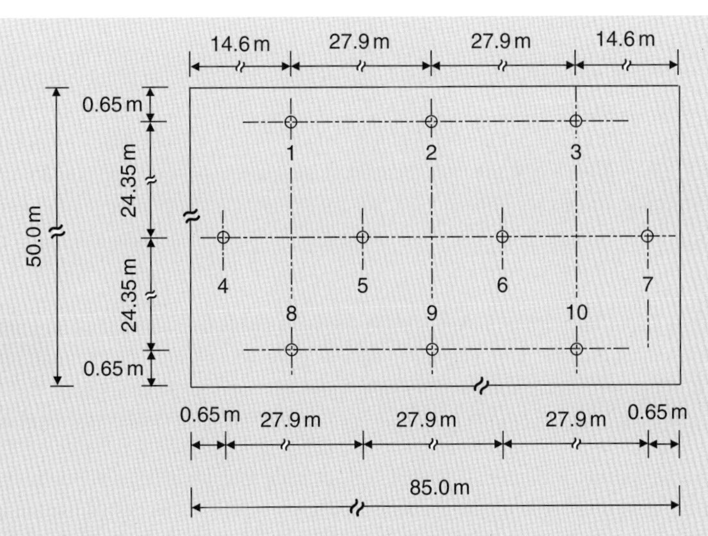

Figure 1.4 Borehole spacing.

According to Smith (1970):

$$\text{For } M = 0.1, q/(\gamma' M) = 1434/(112.4 \times 0.1) = 127.6 \text{ ft}$$

Equivalent width $B_e = A^{1/2} = (50 \times 3.28 \times 85 \times 3.28)^{1/2} = 213.8$ ft, and for $L/B = 85/50 = 1.7$, from Smith curves find $D = z = 100$ ft. Hence, the minimum borehole depth is

$$z_b = \text{foundation depth} + z$$

$$= 1 + \left(\frac{100}{3.28}\right) = 31.49 \text{ m; say } 32\,m$$

According to Baban (1992):
The ASCE Criteria 2 controls because $B = 50 > 20$ m

$$\frac{\gamma'}{q} = \frac{112.4}{1434} = \left(\frac{1}{12.76}\right)\text{ft}^{-1}, \ \frac{L}{B} = \frac{85}{50} = 1.7, B = 50 \times 3.28 = 164 \text{ ft}$$

Using the applicable equation or from table and interpolating, find $z = 125$ ft (computed from the equation)

$$z_b = \text{foundation depth} + z$$

$$= \left(\frac{125}{3.28}\right) + 1 = 39.11 \text{ m; say } 39\,m$$

According to Coduto (2001) and Sowers (1970):

$$z_b = 6S^{0.7} + D = 6 \times 5^{0.7} + 1$$

$$= 6 \times 3.09 + 1 = 19.54 \text{ m, } say\ 20.0\,m$$

According to the 2 × width rule of thumb, $z_b = 2 \times 50 = 100$ m; it is unrealistic and uneconomical.
Refer to Figure 1.4. The responsible Geotechnical Engineer may recommend the following borehole depths:

- All four boreholes nearest to the corners shall be drilled to 20 m depth (boreholes 1, 3, 8 and 10).
- The two boreholes, each at centre of an exterior row, shall be drilled to 25 m depth (boreholes 2 and 9).
- The two exterior boreholes of the middle row shall be drilled to 25 m depth (boreholes 4 and 7). The remaining two middle boreholes shall be drilled to 35 m depth (boreholes 5 and 6).

Problem 1.5

Field vane shear tests (FVST) were conducted in a layer of organic clay (not peat). The rectangular vane dimensions were 63.5 mm width × 127 mm height. At the 2 m depth, the torque T required to cause failure was 51 N.m. The liquid limit LL and plastic limit PL of the clay were 50 and 20, respectively. The effective unit weight γ' of the clay is 18 kN/m^3. Estimate the design undrained vane shear strength s_u, the preconsolidation pressure σ'_c and the ratio OCR of the clay.

Discussion:

The in situ or field vane shear test (FVST – ASTM D-2573) is used to estimate the in situ *undrained shear strength* s_u (or undrained cohesion c_u) of very soft to soft clay, silty clay (muck) and clayey silt soils. It is practically used when sensitive fine-grained soils are encountered. However, these materials must be free of gravel, large shell particles, and roots in order to avoid inaccurate results and probable damage to the vane.

Briefly, the test apparatus consists of a small diameter shaft with four tapered or rectangular thin blades or fins (Figure 1.5), and suitable extension rods and fasteners used to connect the shear vane with a torque measuring device at the ground surface, as shown in Figure 1.6a. There are various forms of the equipment; the device shown in the figure is somewhat an *antique* version, belongs to The Bureau of Reclamation, USA; (Gibbs and Holtz, 1960). Also, there are different vane sizes (the diameter D and height H dimensions) used for soils of different consistencies; the softer the soil, the larger the vane diameter should be. A standard vane has the H/D ratio equals 2, where D is the overall vane width. Typical vane dimensions being 150 mm by 75 mm and 100 mm by 50 mm.

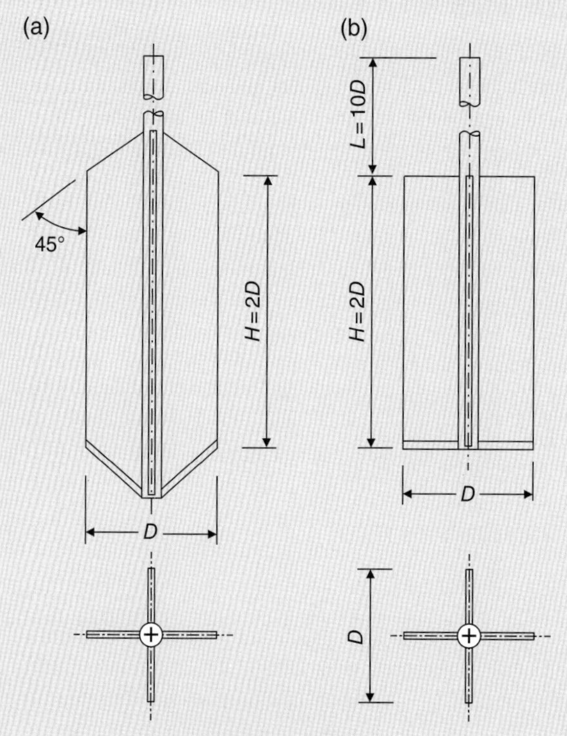

Figure 1.5 Field vane shapes: (a) tapered vane, (b) rectangular vane.

The test is conducted by pushing the shear vane into the soil at bottom of a borehole, without disturbing the soil appreciably, until the vane top is about 15 cm below the hole bottom. After a short time lapse of 5–10 min, the torque is applied at the top of the extension rod to rotate the fins at a standard rate of 0.1 degree per second. This rotation will induce failure in the soil surrounding the vanes. The shear surface has the same vein dimensions *D* and *H*. The maximum torque *T* required to cause failure is measured and recorded. Steps of the test procedure are schematically illustrated in Figure 1.6b. *Note:* Refer to ASTM D-2573 for more details of the test apparatus and procedure.

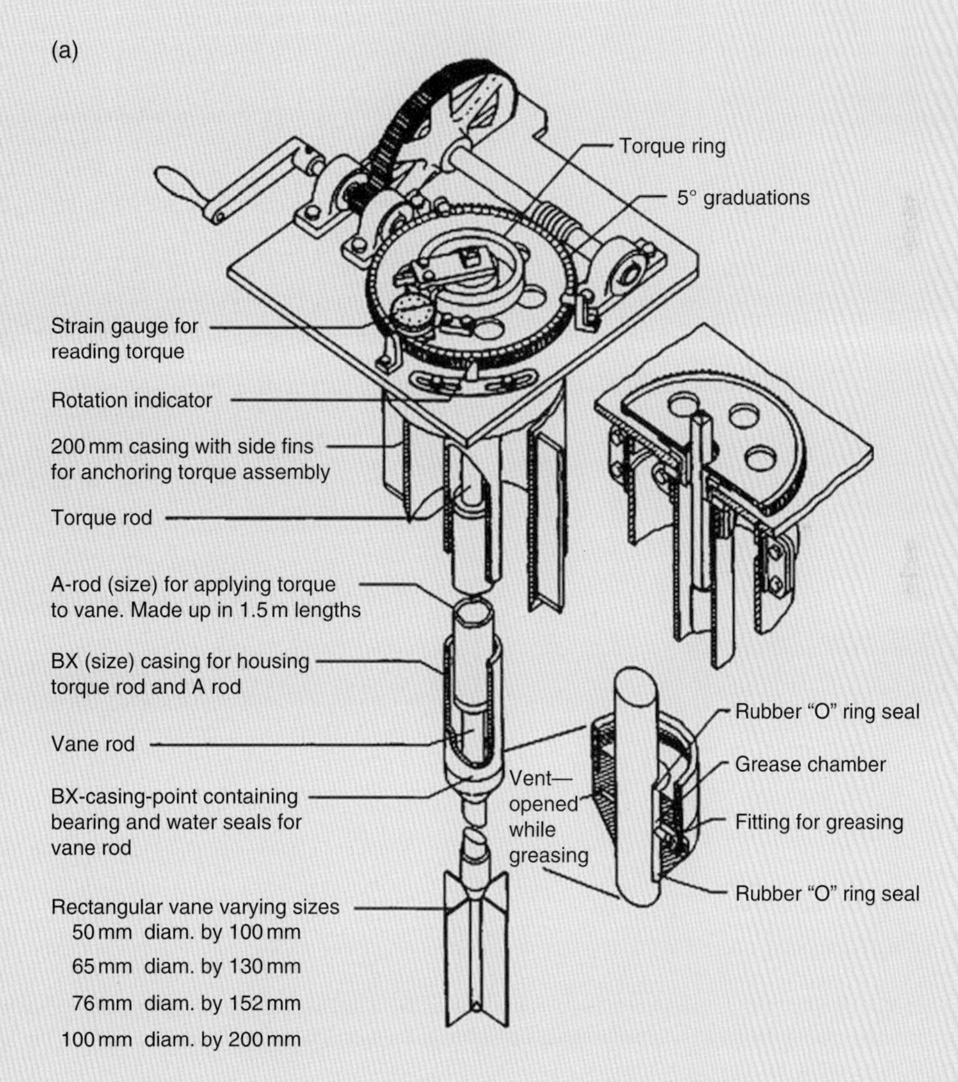

Figure 1.6 (a) Field vane shear apparatus (after Gibbs and Holtz 1.6; USBR). (b) Field vane shear test procedure.

(*Continued*)

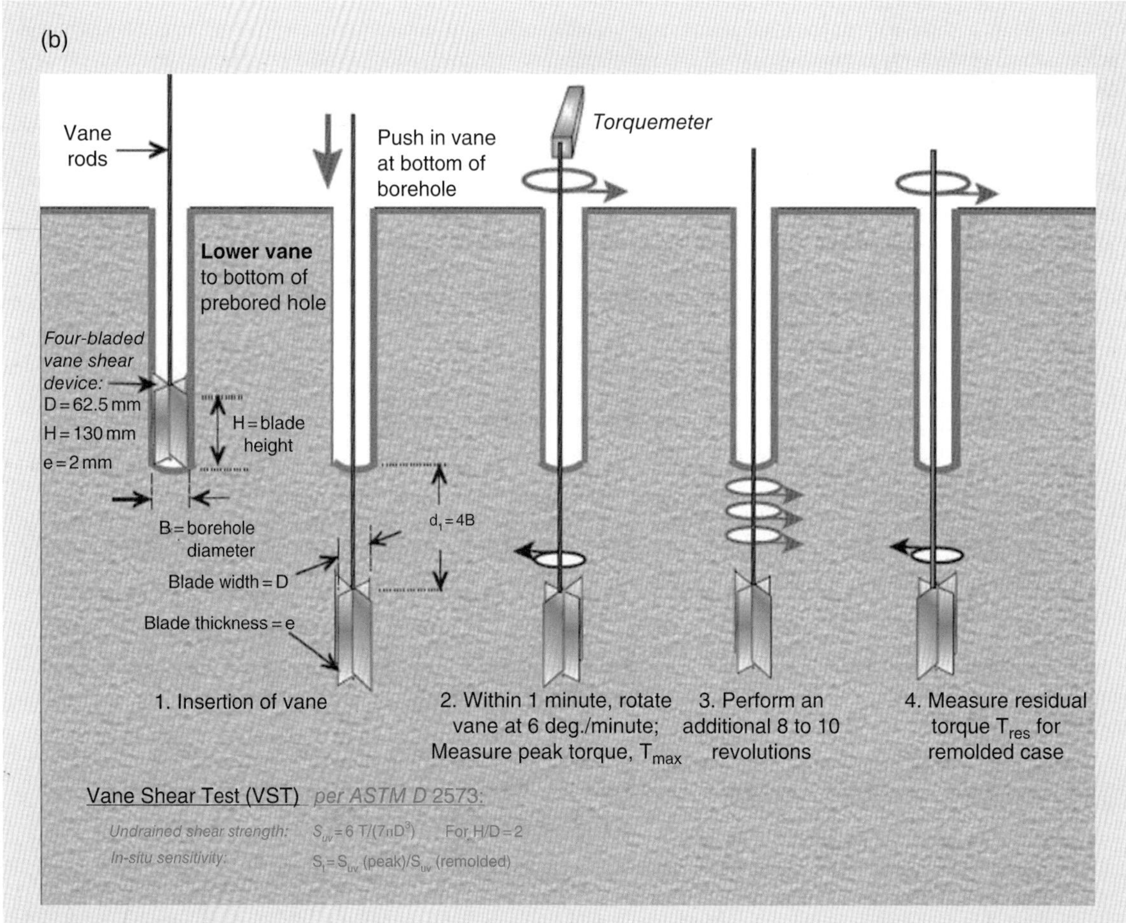

Figure 1.6 (*Continued*)

The following equations relate the torque T at failure to the undrained vane shear strength $s_{u,v}$ (or $c_{u,v}$) and vane dimensions:

$$T = \pi s_{u,v}\left(\frac{D^2H}{2} + \frac{D^3}{6}\right) \tag{1.8}$$

$$T = 7\pi s_{u,v}\left(\frac{D^3}{6}\right) \text{ for } \frac{H}{D} = 2, \qquad s_{u,v} = \frac{6T}{7\pi D^3} \tag{1.9}$$

Note: In using these equations, all the units should be consistent; D and H are in metres, $s_{u,v}$ is in kN/m^2 and T is in kN.m.

Researchers found that the $s_{u,v}$ values obtained from the FVST are too high for design purposes; its use reduces the factor of safety considerably. It is recommended to use the empirical correction factor λ with $s_{u,v}$ in order to obtain appropriate value for design undrained shear strength. Hence,

$$\text{Design } s_{u,v} = \lambda \times s_{u,v} \tag{1.10}$$

The correction factor λ may be obtained from curves or calculated using any of the following equations:

Bjerrum (1972):

$$\lambda = 1.7 - 0.54\log[PI(\%)] \tag{1.11}$$

PI = plasticity index
Morris and Williams (1994):

$$\lambda = 7.01\, e^{-0.08[LL(\%)]} + 0.57 \quad (1.12)$$

$$\lambda = 1.18\, e^{-0.08(PI)} + 0.57 \quad (\text{for PI} > 5) \quad (1.13)$$

For organic soils other than peat, an additional correction factor of 0.85 is recommended be used with the corrected $s_{u,v}$ (Terzaghi, Peck, and Mesri, 1996). However, this correction may not be necessary when the Morris and Williams correction factors are used.

Researchers derived empirical equations for estimating the effective *preconsolidation pressure* σ'_c and the *overconsolidation ratio OCR* (or σ'_c/σ'_O) of natural clays using results of the field vane shear test. Mayne and Mitchell (1988) found the following correlations:

$$\sigma'_c = 7.04\left[c_{u,v\,(field)}\right]^{0.83}, \quad c_{u,v\,(field)} \text{ and } \sigma'_c \text{ are in kN/m}^2 \quad (1.14)$$

$$OCR = \beta \frac{c_{u,v\,(field)}}{\sigma'_O}, \quad c_{u,v\,(field)} \text{ and } \sigma'_O \text{ are in kN/m}^2 \quad (1.15)$$

$$\beta = 22(PI)^{-0.48}$$

$$\beta = \frac{222}{\omega(\%)}, \quad \omega(\%) = \text{moisture content} \quad \text{(Hansbo, 1957)} \quad (1.16)$$

$$\beta = \frac{1}{0.08 + 0.0055(PI)} \quad \text{(Larsson, 1980)} \quad (1.17)$$

Solution:

T = 51 N.m = 51 × 10^{-3} kN.m;
D = 63.5 mm = 63.5 × 10^{-3} m
H = 127 mm = 127 × 10^{-3} m; H/D = 2

Equation (1.9):
$$s_{u,v} = \frac{6T}{7\pi D^3} = \frac{6 \times 51 \times 10^{-3}}{7 \times \pi (63.5 \times 10-3)^3} = 54.32\,\text{kN/m}^2$$

Equation (1.11):
$$\lambda = 1.7 - 0.54\log[PI(\%)]$$
$$= 1.7 - 0.54\log(50-20) = 0.902$$

As the soil is organic (not peat), an additional correction factor of 0.85 is recommended.

$$\text{Design}\, s_{u,v} = 0.85 \times \lambda \times s_{u,v} = 0.85 \times 0.902 \times 54.32 = \mathit{41.65\,kN/m^2}$$

Equation (1.14):
$$\sigma'_c = 7.04\left[c_{u,v\,(field)}\right]^{0.83}$$
$$= 7.04(54.32)^{0.83} = \mathit{193.9\,kN/m^2}$$

Equation (1.15):
$$OCR = \beta \frac{c_{u,v\,(field)}}{\sigma'_O}$$
$$\beta = 22(PI)^{-0.48} = 22(50-20)^{-0.48} = 4.3$$
$$OCR = \beta \frac{c_{u,v\,(field)}}{\sigma'_O} = 4.3 \times \frac{54.32}{2 \times 18} = 6.5 > \left(\frac{193.9}{2 \times 18} = 5.4!?\right)$$

(Continued)

Equation (1.17): $\beta = \dfrac{1}{0.08 + 0.0055(PI)} = \dfrac{1}{0.08 + 0.0055(50-20)} = 4.08$

$$OCR = \beta \frac{c_{u,v\,(field)}}{\sigma'_O} = 4.08 \times \frac{54.32}{2 \times 18} = 6.2 > \left(\frac{193.9}{2 \times 18} = 5.4!? \right)$$

Problem 1.6

At the site of a proposed structure the variation of the SPT number (N_{60}) with depth in a deposit of normally consolidated sand is as given below:

Depth, m:	1.5	3.0	4.5	6.0	7.5	9.0
N_{60}:	6	8	9	8	13	14

The ground water table (W.T) is located at a depth of 6.5 m. The dry unit weight γ_{dry} of the sand above W.T is 18 kN/m^3, and its saturated unit weight γ_{sat} below W.T is 20.2 kN/m^3.

(a) Correct the SPT numbers.
(b) Select an appropriate SPT corrected number (N'_{60}) for use in designs. Assume the given depths are located within the influence zone (zone of major stressing).

Discussion:
One of the most common in situ tests conducted during subsurface exploration is the standard penetration test SPT. When a borehole is extended to a predetermined depth, the drill tools are removed and a standard split-spoon sampler connected to a drill rod is lowered to the bottom of the hole. The sampler is 50 mm in external diameter, 35 mm in internal diameter and about 650 mm in length (Figure 1.7a, b). The sampler is driven into the soil by means of blows from a standard 65 kg hammer which freely drops a standard 760 mm distance to an anvil at top of the drill rod (Figure 1.8). The number of blows N required for the spoon penetration of three 150 mm intervals is recorded. The number of blows required for the last two intervals are added to give the field standard penetration number, recorded as N_{field} at the particular depth. This number is generally referred to as N-value. If 50 blows are reached before a penetration of 300 mm, no further blows should be applied but the actual penetration should be recorded. The test depth intervals are, generally, 1–2 m. At conclusion of a test the sampler is withdrawn and the soil extracted. If the test is to be carried out in gravelly soils the driving shoe is replaced by a solid 60° cone and the test is usually called a *dynamic cone penetration test.* There is evidence that slightly higher results are obtained in the same material when the normal driving shoe is replaced by the 60° cone.

Note: For more details of the test refer to ASTM D-1586.

The SPT is very useful for determining the relative density D_r and the effective angle of shearing resistance Ø′ of cohesionless soils. To a lesser degree, it can also be used to estimate the undrained shear strength S_u (or C_u) of cohesive soils. The SPT N-value is used to estimate allowable bearing capacity of granular soils. It may be useful to point out herein, that the SPT N-value of soil deposits which contain large boulders and gravel may be erratic and un-reliable. In loose coarse gravel deposits, the spoon sampler tends to slide into the large voids and gives low values of N. Excessively large values of N may be expected if the sampler is blocked by a large stone or gravel. According to Bowles (2001), the penetration test in gravel or gravelly soils requires careful interpretation, since pushing a piece of gravel can greatly change the blow count. Generally if one uses a statistical average of the blow count in the stratum from several borings, either excessively high or very low (pushing gravel or creating a void space) values will be averaged out so that approximately the correct blow count can be estimated for design.

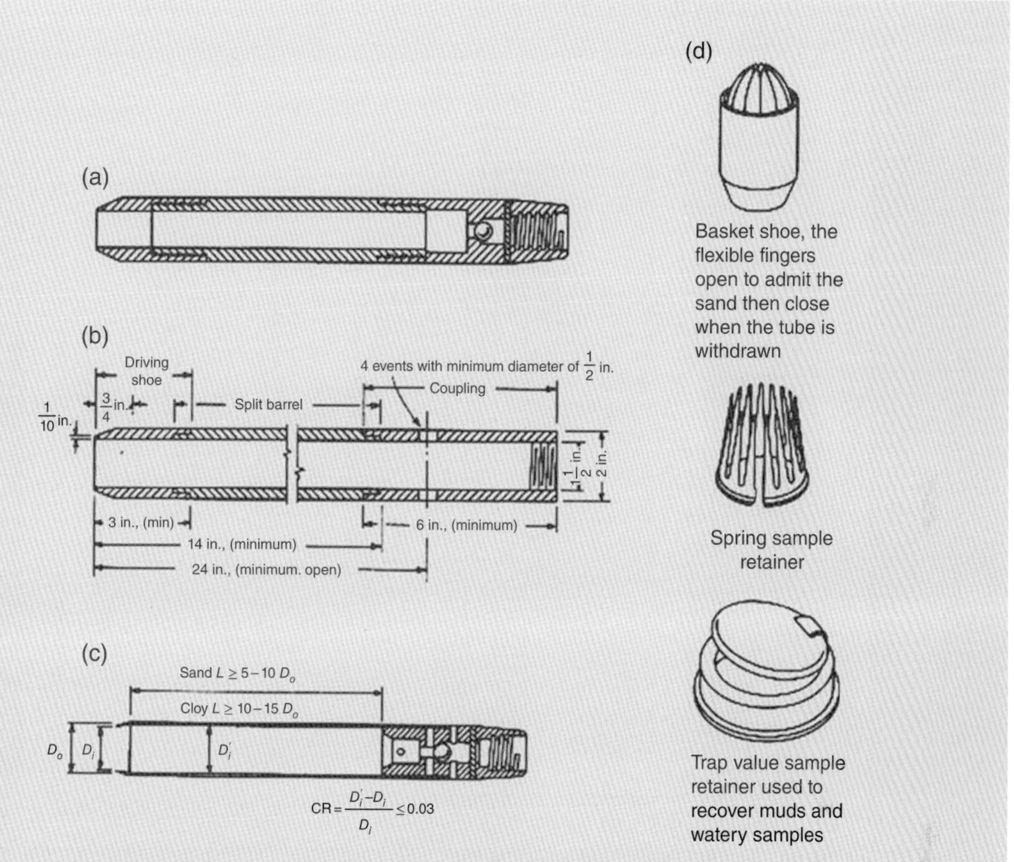

Figure 1.7 (a) Standard split-spoon sampler; (b, d) dimensions and inserts of the standard split-spoon sampler, respectively; (c) Shelby tube (from Bowles, 2001).

Corrections of SPT number:
The recorded N_{field} should be adjusted and corrected for the effects of some factors such as effective overburden pressure (C_N), hammer efficiency (η_h), borehole size (η_b), sampler (η_s) and rod length (η_r).

Due to various types of driving hammers with different efficiencies, the hammer energy ratio E_r is referenced to a standard or basic energy ratio, E_{rb}, such that η_h be equal to ratio of the particular hammer energy to the standard E_{rb}, which is equal to either 60 or 70%.

When N_{field} is adjusted to η_h, η_b, η_s and η_r corrections, it is generally referred to as N_{60} or N_{70} according to the E_{rb} value, as follows:

$$N_{60} = \frac{E_r \cdot \eta_b \cdot \eta_s \cdot \eta_r \cdot N_{field}}{60} \tag{1.18}$$

$$N_{70} = \frac{E_r \cdot \eta_b \cdot \eta_s \cdot \eta_r \cdot N_{field}}{70} \tag{1.19}$$

Table 1.6 gives values of E_r for different types of SPT hammers. Table 1.7 gives values for borehole, sampler and rod correction factors.

When N_{60} and N_{70} are corrected for the effective overburden pressure, they are written as N'_{60} and N'_{70}, where

$$N'_{60} = C_N\, N_{60}$$

$$N'_{70} = C_N\, N_{70}$$

(*Continued*)

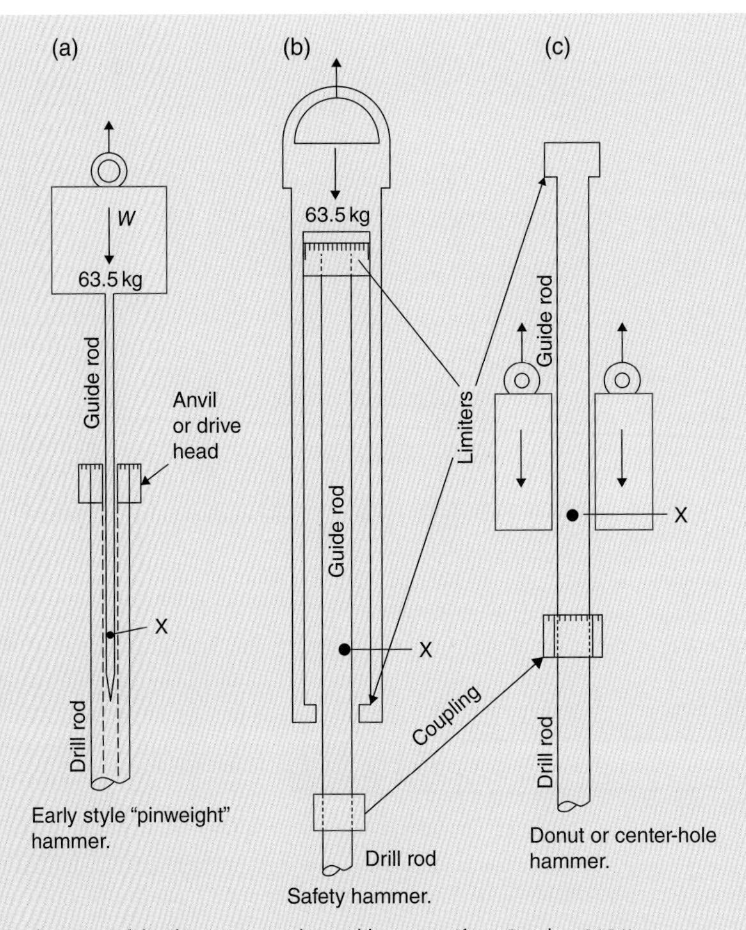

Figure 1.8 Schematic diagrams of the three commonly used hammers (from Bowles, 2001).

Table 1.6 SPT hammer efficiencies (adapted from Clayton, 1990).

Country	Hammer type	Hammer release mechanism	Efficiency, E_r
Argentina	Donut	Cathead	45
Brazil	Pin weight	Hand dropped	72
China	Automatic	Trip	60
	Donut	Hand dropped	55
	Donut	Cathead	50
Colombia	Donut	Cathead	50
Japan	Donut	Tombi triggers	78–85
	Donut	Cathead, two turns + special release	65–67
UK	Automatic	Trip	73
US	Safety	Cathead, two turns	55–60
	Donut	Cathead, two turns	45
Venezuela	Donut	Cathead	43

Table 1.7 Borehole, sampler and rod correction factors (adapted from Skempton, 1986).

Factor	Equipment variables	Value
η_b	65–115 mm (2.5–4.5 in)	1.0
	150 mm (6 in)	1.05
	200 mm (8 in)	1.15
η_s	Standard sampler	1.0
	with liner for dense sand and clay	0.8
	with liner for loose sand	0.9
η_r	< 4 m (< 13 ft)	0.75
	4–6 m (13–20 ft)	0.85
	6–10 m (20–30 ft)	0.95
	> 10 m (> 30 ft)	1.0

C_N = effective overburden pressure factor

$$C_N \leq 2$$

Terzaghi and Peck (1967) have recommended the following additional correction where the soil is fine sand or silty sand below the water table:

$$N'_{field} = 15 + 0.5(N_{field} - 15), \text{ for } N_{field} > 15 \tag{1.20}$$

Note: This correction is applied first and then the overburden and other corrections are applied.

The effective overburden pressure factor C_N, may be determined from different empirical relationships, such as: Liao and Whitman (1986):

$$C_N = \sqrt{\frac{p_a}{\sigma'_O}}, \text{ where } \sigma'_O = \text{effective overburden pressure, kPa} \tag{1.21}$$

Skempton (1986):

$$C_N = \frac{2}{1 + \left(\frac{\sigma'_O}{P_a}\right)}, \text{ for normally consolidated fine sand} \tag{1.22}$$

$$C_N = \frac{3}{2 + \left(\frac{\sigma'_O}{P_a}\right)}, \text{ for normally consolidated coarse sand} \tag{1.23}$$

$$C_N = \frac{1.7}{0.7 + \left(\frac{\sigma'_O}{P_a}\right)}, \text{ for overconsolidated sand} \tag{1.24}$$

(*Continued*)

Seed et al (1975):

$$C_N = 1 - 1.25\log\left(\frac{\sigma'_O}{P_a}\right) \tag{1.25}$$

In the above C_N equations, note that σ'_O is the effective overburden pressure and P_a = atmospheric pressure ($\cong$100 kN/m^2).

Figure 1.9 may give the most realistic relationship.

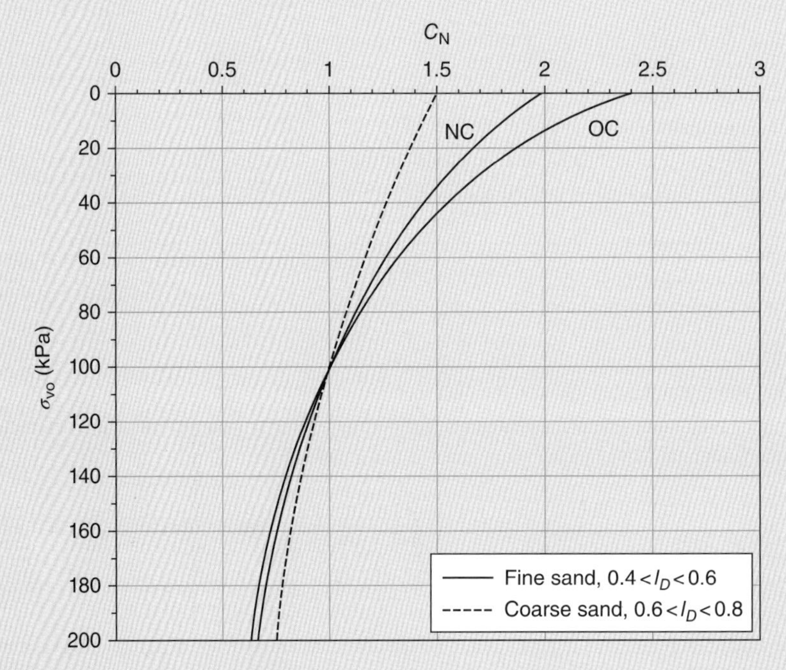

Figure 1.9 Variation of correction factor C_N with effective overburden pressure for coarse-grained soils (from Knappett and Craig, 2012).

For cohesionless soils, the SPT number can be correlated with the relative density D_r and effective angle of shearing resistance Ø′. Some of the published empirical relationships are:

- Marcuson and Bieganousky (1977):

$$D_r\% = 11.7 + 0.76\left(222N_{60} + 1600 - 53\sigma'_o - 50C_u^2\right)^{1/2} \tag{1.26}$$

- Kulhawy and Mayne (1990):

$$D_r\% = 12.2 + 0.75\left(222N_{60} + 2311 - 711OCR - \frac{779\sigma'_o}{P_a} - 50C_u^2\right)^{1/2} \tag{1.27}$$

- Cubrinovisky and Ishihara (1999):

$$D_r\% = \left[\frac{N_{60}\left(0.23 + \dfrac{0.06}{D_{50}}\right)^{1.7}}{9}\left(\frac{1}{\dfrac{\sigma'_o}{P_a}}\right)\right]^{1/2} \tag{1.28}$$

- Kulhawy and Mayne (1990):

$$D_r = \sqrt{\frac{N'_{60}}{C_P C_A C_{OCR}}} \times 100 \tag{1.29}$$

Where:
$C_p = 60 + 25 \log D_{50}$ = Grain size correction factor

$C_A = 1.2 + 0.05 \log\left(\dfrac{t}{100}\right)$ = Aging correction factor

t = age of soil in year (time since deposition)
$C_{OCR} = (OCR)^{0.18}$ = Overconsolidation correction factor

N'_{60} = corrected SPT N-value

σ'_o = effective overburden pressure, kN/m^2

P_a = atmospheric pressure ($\approx 100\ kPa$)
D_{50} = sieve size through which 50 % of the soil will pass (mm)
C_u = uniformity coefficient of the sand

$$OCR = \text{overconsolidation ratio} = \frac{\text{preconsolidation pressure, } \sigma'_C}{\text{effective overburden pressur, } \sigma'_O}$$

Note: When there is shortage in data to compute the C_p, C_A and C_{OCR} parameters; estimated approximate values may be assumed. For example, a value for D_{50} may be estimated from visual examination of the soil with reference to particle size classification tables. If there is no reliable source available to estimate the age of a sand deposit; a value of $t = 1000$ years may be assumed since the aging factor is not very sensitive to the t value. Also, for the overconsolidation correction factor, values of about one in loose sands to about four in dense sands would be adequate (Coduto, 2001).

Table 1.8 gives approximate relation between D_r, N'_{60}, Ø,′ unit weight γ and denseness description for cohesionless soils.

Table 1.8 Correlation for D_r, N'_{60}, Ø′, γ, soil denseness.

N'_{60}	D_r, %	Ø′	γ, kN/m^3	Denseness
0–5	5–15	25–32°	11–16	Very loose
5–10	15–30	27–35°	14–18	Loose
10–30	30–60	30–40°	17–20	Medium
30–50	60–85	35–45°	17–22	Dense
> 50	> 85	> 45°	20–23	Very dense

(*Continued*)

Peck et al. (1974) gives a correlation between N'_{60} and $\varnothing'$ in a graphical form which can be approximated (Wolff, 1989) as

$$\varnothing' = 27.1 + 0.3N'_{60} - 0.00054\left(N'_{60}\right)^2 \tag{1.30}$$

Kulhawy and *Mayne* (1990):

$$\varnothing' = \tan^{-1}\left[\frac{N_{60}}{12.2 + 20.3\left(\frac{\sigma'_O}{P_a}\right)}\right]^{0.34} \tag{1.31}$$

Hatanac and *Ushida* (1996):

$$\varnothing' = \sqrt{20N'_{60}} + 20 \tag{1.32}$$

The SPT number N_{60} is correlated to the modulus of elasticity (E_s) of sands as follows:
Kulhawy and *Mayne* (1990):

$$\frac{E_s}{P_a} = \alpha N_{60} \tag{1.33}$$

Where: P_a = atmospheric pressure (same unit as E_s)
α = 5, 10 and 15 for sands with fines, clean normally consolidated sand, and clean overconsolidated sand, respectively.

The SPT number is also correlated to unconfined compression strength q_u and undrained shear strength s_u of fine-grained soils as given in Table 1.9 and the following equations. It is important to point out that the obtained values of q_u and s_u should be used as a guide only; local cohesive soil samples should be tested to verify that the given relationships or correlations are valid.

Table 1.9 Correlations for N'_{60}, q_u and consistency of fine-grained soils.

N'_{60}	Consistency	q_u, kN/m^2
0–2	Very soft	0–25
2–5	Soft	25–50
5–10	Medium stiff	50–100
10–20	Stiff	100–200
20–30	Very stiff	200–400
> 30	Hard	> 400

Stroud and Butler (1975):

$$s_u = kN_{60}\ \text{(for insensitive overconsolidated clay)} \tag{1.34}$$

where $k = 4.5\,\text{kPa}$ for $PI > 30$

$k = 4 - 6\,\text{kPa}$ for $PI = 30 \sim 15$

Hara et al (1971):

$$c_u = 29(N_{60})^{0.72}, \text{ kPa} \tag{1.35}$$

Mayne and Kemper (1988):

$$OCR = 0.193\left(\frac{N_{60}}{\sigma'_O}\right)^{0.689} \tag{1.36}$$

where σ'_O = effective overburden pressure in MN/m^2

Szechy and Varga (1978) gave the correlations presented in Table 1.10.

Table 1.10 Correlations for N_{60}, CI, q_u and consistency.

N_{60}	Consistency	CI	q_u, MN/m^2
< 2	Very soft	< 0.5	< 25
2–8	Soft to medium	0.5–0.75	25–80
8–15	Stiff	0.75–1.0	80–150
15–30	Very stiff	1.0–1.5	150–400
> 30	Hard	> 1.5	> 400

Where:

CI = consistency index $= \dfrac{LL - \omega}{LL - PI}$

ω = natural moisture content
LL = liquid limit
PL = plastic limit
N_{60} = SPT N-value corrected for field procedures only

Note: As mentioned earlier, any correlation or relationship between N_{60}, s_u, c_u, OCR, CI and q_u for cohesive soils is approximate and should be used as a guide only.

Solution:
(a)

Depth, m:	1.5	3.0	4.5	6.0	7.5	9.0
N_{60}:	6	8	9	8	13	14

$$N'_{60} = C_N N_{60}$$

Equation (1.22): $C_N = \dfrac{2}{1 + \left(\dfrac{\sigma'_O}{P_a}\right)}$ $\quad P_a \cong 100\,\text{kPa}$

(*Continued*)

At 1.5 m depth, $\sigma'_O = 1.5 \times 18 = 25\,\text{kPa} \rightarrow C_N = 1.6 \rightarrow N'_{60} = 10$

At 3.0 m depth, $\sigma'_O = 3.0 \times 18 = 54\,\text{kPa} \rightarrow C_N = 1.3 \rightarrow N'_{60} = 10$

At 4.5 m depth, $\sigma'_O = 4.5 \times 18 = 81\,\text{kPa} \rightarrow C_N = 1.1 \rightarrow N'_{60} = 10$

At 6.0 m depth, $\sigma'_O = 6.0 \times 18 = 108\,\text{kPa} \rightarrow C_N = 0.96 \rightarrow N'_{60} = 8$

At 7.5 m depth, $\sigma'_O = 6.5 \times 18 + 1(20.2 - 10) = 127\,\text{kPa} \rightarrow C_N = 0.88 \rightarrow N'_{60} = 11$

At 9.0 m depth, $\sigma'_O = 6.5 \times 18 + 2.5(20.2 - 10) = 143\,\text{kPa} \rightarrow C_N = 0.82 \rightarrow N'_{60} = 12$

Depth, m:	1.5	3.0	4.5	6.0	7.5	9.0
N'_{60}:	*10*	*10*	*10*	*8*	*11*	*12*

(b) The design N'_{60} is the average of all N'_{60} values of SPT conducted at the depth intervals encountered within the influence zone. For example, for a spread footing the zone of interest is from about one-half footing width B above the estimated foundation level to a depth of about $2B$ below that level. For shallow foundations, weighted average using depth increment $\times$ N may be preferable to an ordinary arithmetic average; that is, $N_{average} = \sum(N \times z_i)/\sum z_i$ and not $\sum N_i/i$.

For deep foundations, the simple ordinary average may be adequate unless the stratum is very thick. In case of thick stratum, it may be better to subdivide the stratum into several strata and average the N values for each subdivision (Bowles, 2001).

Depth, m	N'_{60}	$N'_{60} \times$ **Depth**
1.5	10	15
3.0	10	30
4.5	10	45
6.0	8	48
7.5	11	83
9.0	12	108

$$\sum z_i = 31.5 \qquad \sum N'_{60} \times z_i = 329$$

Average $N'_{60} = 329/31.5 = 10.44 \rightarrow$ use 10 (only use integers)
Design $N'_{60} = 10$

Problem 1.7

A standard penetration test was performed in a 150 mm diameter borehole at a depth of 9.5 m below the ground surface. The driller used a UK-style automatic trip hammer, a standard SPT sampler and a 10-m drill rod. The actual blow count N_{field} was 19. The soil is normally consolidated fine sand with a unit weight of 18.0 kN/m^3 and D_{50} = 0.4 mm. The ground water table is at a depth of 15 m. Compute (a) N_{60}, (b) N'_{60}, (c) D_r%, (d) Ø′, and (e) denseness of the fine sand.

Solution:

(a) From Table 1.6, for the UK automatic trip hammer, $E_r = 73\%$
From Table 1.7, for 150 mm diameter borehole, $\eta_b = 1.05$
From Table 1.7, for a standard sampler, $\eta_s = 1.0$
From Table 1.7, for a 10-m drill rod, $\eta_r = 0.95$

Equation (1.18):
$$N_{60} = \frac{E_r \cdot \eta_b \cdot \eta_s \cdot \eta_r \cdot N_{field}}{60} = \frac{73 \times 1.05 \times 1 \times 0.95 \times 19}{60} = 23$$

(b)

$$N'_{60} = C_N\, N_{60}$$

Assume using Equation (1.18):
$$C_N = \sqrt{\frac{p_a}{\sigma'_O}} = \sqrt{\frac{100}{9.5 \times 18}} = 0.76$$

$$N'_{60} = 0.76 \times 23 = 17$$

(c) Equation (1.28):

$$D_r\% = \left[\frac{N_{60}\left(0.23 + \frac{0.06}{D_{50}}\right)^{1.7}}{9}\left(\frac{1}{\frac{\sigma'_O}{P_a}}\right)\right]^{½}$$

$$= \left[\frac{23\left(0.23 + \frac{0.06}{0.4}\right)^{1.7}}{9}\left(\frac{1}{\frac{9.5 \times 18}{100}}\right)\right]^{½}(100) = 54\%$$

(d) Equation (1.31):

$$Ø' = \tan^{-1}\left[\frac{N_{60}}{12.2 + 20.3\left(\frac{\sigma'_O}{P_a}\right)}\right]^{0.34}$$

$$= \tan^{-1}\left[\frac{23}{12.2 + 20.3\left(\frac{9.5 \times 18}{100}\right)}\right]^{0.34} = \tan^{-1}(0.785) \rightarrow Ø' = 38^\circ$$

(e) Table 1.8: for $N'_{60} = 17$, $D_r = 54\%$, $Ø' = 38\%$ and $\gamma = 18\ kN/m^3$, the soil may be classified as *medium dense sand.*

Problem 1.8

A 15 cm diameter borehole has been drilled through a fine sand deposit to a depth of 7 m. At this depth, the SPT, using a US safety hammer and a standard SPT sampler, gave $N_{field} = 23$. Boring then continued to greater depths, encountering water table at a depth of 11 m. Compute N'_{60}, $Ø'$, and D_r at the test location, and use the data to estimate the denseness of the sand.

Solution:

The measured N_{field} of 23, with an estimated allowance of about ± 20% for the correction factors, suggests medium-dense sand (Table 1.8). Knowing that fine sand may be considered poorly graded, a value for γ equals 18 kN/m^3 may be assumed appropriate (Table 1.8).

(Continued)

$$\text{At 7 m depth, } \sigma'_O = 7 \times 18 = 126\ \text{kN/m}^2$$

Equation (1.25):

$$C_N = 1 - 1.25\log\left(\frac{\sigma'_O}{P_a}\right)$$

$$= 1 - 1.25\log\left(\frac{126}{100}\right) = 0.87$$

From Table 1.6, for the US safety hammer, E_r = 55–65%

$$\text{For } E_{rb} = 60\%, \eta_h = \frac{(55+60)/2}{60} = 0.96$$

$$N_{60} = \eta_h \times \eta_b \times \eta_s \eta_r \times N_{field}$$

$$N_{60} = 0.96 \times 1.05 \times 1 \times 0.95 \times 23 = 22$$

$$N'_{60} = C_N\, N_{60} = 0.87 \times 22 = 19$$

Equation (1.32):

$$\emptyset' = \sqrt{20N'_{60}} + 20 = \sqrt{20 \times 19} + 20 = 39°$$

Equation (1.29):

$$D_r = \sqrt{\frac{N'_{60}}{C_P C_A C_{OCR}}} \times 100$$

$$C_p = 60 + 25\ \log D_{50}$$

Grain size of fine sand ranges from 0.075 to 0.425 mm (ASTM D 2487) Estimate D_{50} = 0.2 mm; hence, C_P = 60 + 25 log 0.2 = 42.5

$$C_A = 1.2 + 0.05\log\left(\frac{t}{100}\right)$$

Since there is shortage in data and there is no reliable source available, assume t = 1000 years; hence, $C_A = 1.2 + 0.05\log\left(\frac{1000}{100}\right) = 1.25$

$$C_{OCR} = (OCR)^{0.18}$$

Also, for the same reasons just mentioned, assume OCR = 2.5; hence, $C_{OCR} = (2.5)^{0.18} = 1.18$

$$D_r = \sqrt{\frac{19}{42.5 \times 1.25 \times 1.18}} \times 100 = 55\%$$

Problem 1.9

Two plate-load tests on a sandy soil were performed using plates 0.3 × 0.3 m and 0.6 × 0.6 m. For a 20 mm settlement, the loads were 30 and 72 kN, respectively. Find dimension B of a square footing required to carry a 200 kN column load with an allowable settlement of 20 mm, (a) assume extrapolating plate-load test results to the footing size is approximately justified and the bearing load V increases linearly with B (for the given settlement), (b) use Housel's method.

Discussion:

A method to obtain load-settlement relationship and to estimate allowable bearing capacity of foundations is by conducting *field load tests*. One of these tests is the Plate Load Test or PLT. To carry out a PLT, a pit of the size 5 B_{plate} × 5 B_{plate} is excavated to a depth equals to the estimated foundation depth D_f. The dimension B_{plate} is the width or diameter of the rigid steel plate. The size of the plate is usually 0.3-m square, and its thickness is 25 mm. Sometimes, large size plates of 0.6-m square are used. A central hole of the same plate size is excavated in the pit such that its depth $D_h = D_f\left(\frac{B_h}{B_f}\right)$, where B_f is the pit width, and B_h is the hole size. Usually, for shallow foundation depths, excavation of the central hole is optional.

For conducting the test, the plate is placed in the central hole in a horizontal position. The vertical load is applied by means of a suitable hydraulic jack. The reaction to the jack is provided by means of a reaction beam, a truss, a loaded platform and so on (Figure 1.10). A seating load of 7 kPa is first applied, which is released after sometimes. Then the load is applied in successive increments until failure of the ground in shear is attained or, more usually, until the bearing pressure on the plate reaches some multiple, say two or three, of the bearing pressure proposed for the full-scale foundations. The magnitude and rate of settlements are observed and recorded.

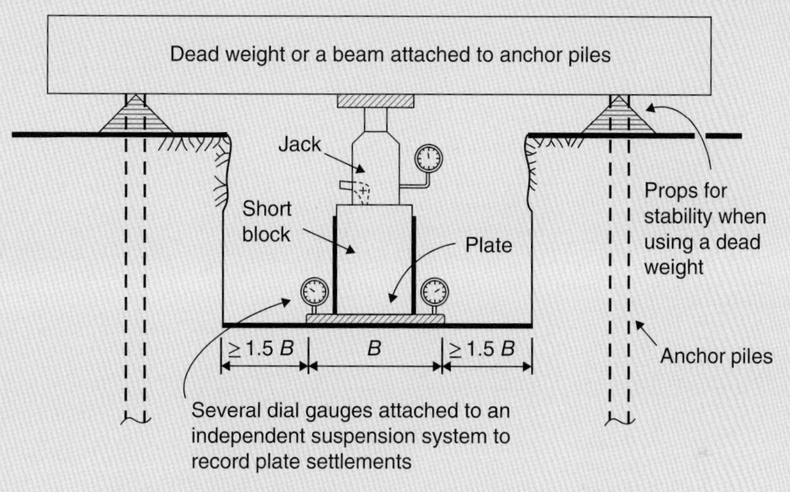

Figure 1.10 Plate-load testing (reproduced from Bowles, 2001).

Pavement subgrades are tested using circular plates of relatively large diameter (Fig. 1.11). Some laboratories use a steel plate of 30-inch (76.2 cm) diameter and 5/8 inch (15.9 mm) thick, and stiffened with plates of 26-inch (66.0 cm) and 22-inch (55.9 cm) diameter, of the same thickness, placed on top of it.

Figure 1.11 Plate load test in the field (Courtesy of Bjara M. Das, Henderson, NV).

(*Continued*)

Note: For more detailed discussion on the test tools and apparatus, preparation, procedure and data recording, refer to ASTM D-1194.

Figure 1.12a shows a typical plot of settlement versus log time (as for the consolidation test), and Figure 1.12b shows a plot of load versus settlement. When the slope of the time-settlement curve for a certain load increment is approximately horizontal, the maximum settlement for that load can be obtained as a point on the load-settlement curve. When the load-settlement curve approaches the vertical, one interpolates the ultimate load intensity q_{ult}. Sometimes, however, q_{ult} is obtained as that value corresponding to a specified settlement (as, say, 25 mm).

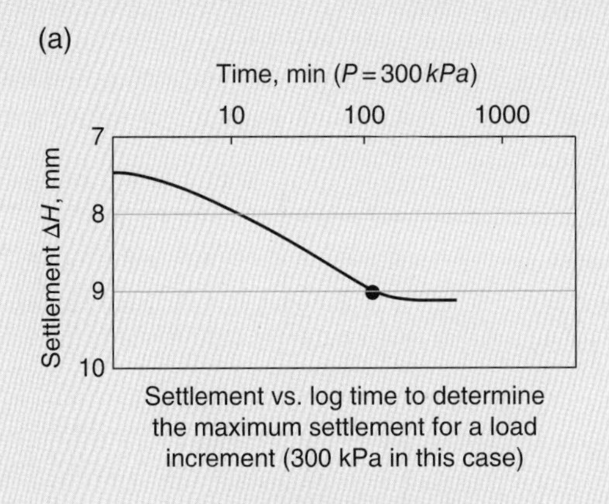

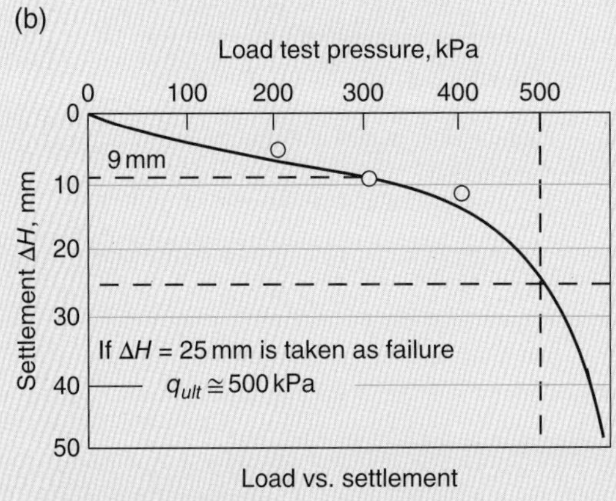

Figure 1.12 Plate load test data (reproduced from Bowles, 2001). (a) Settlement versus load time to determine the maximum settlement for a load increment (300 kPa in this case). (b) Load versus settlement.

The following points are related to results of a plate-load test:

(1) Extrapolating plate-load test results to full-size footings is not standard. For clay soils it might be said that q_{ult} is independent of footing size, giving:

$$q_{ult,\,footing} = q_{ult,\,plate} \tag{1.37}$$

According to Housel, size of a footing to carry a given load for a given safe settlement may be established by using data from at least two PLT in the following equation

$$V = qA + sP \tag{1.38}$$

where V = total load on a bearing area A
A = contact area of footing or plate
P = perimeter of footing or plate

q = bearing pressure beneath A
P = perimeter shear

For sand soils, practically, extrapolation may be justified using

$$q_{ult,\,footing} = q_{ult,\,plate}\left(\frac{B_{footing}}{B_{plate}}\right) \tag{1.39}$$

However, the use of this equation is not recommended unless the ratio $B_{footing}/B_{plate}$ is not much more than about three. When the ratio is 6 to 15 or more, the extrapolation from a plate-load test is little more than a guess that could be obtained at least as reliably using SPT or cone penetration test (CPT) correlations (Bowles, 2001).

(2) For clayey soils (cohesive soils) and for the same q_{ult} of the footing and plate, the following empirical approximate relationship regarding footing settlement may be used:

$$S_{footing} = S_{plate}\left(\frac{B_{footing}}{B_{plate}}\right) \tag{1.40}$$

However, since a plate-load test is of short duration, consolidation settlements usually cannot be predicted. If the test is performed on a material overlying a saturated compressible stratum, the test may give highly misleading information. As a precautionary measure against this event, it is good practice to have borings performed at the plate-load test site. *Note:* For clayey soils, settlement is normally determined from laboratory consolidation test results and not from results of plate-load test.

For sandy soils (cohesionless soils) and for the same q_{ult} of the footing and plate, the following empirical relationship regarding footing settlement may be used:

$$S_{footing} = S_{plate}\left[\frac{B_{footing}\left(B_{plate}+0.3\right)}{B_{plate}\left(B_{footing}+0.3\right)}\right]^2 \tag{1.41}$$

In this equation, both $B_{footing}$ and B_{plate} are in meters.

(3) In designing a shallow foundation for an allowable settlement, a trial and error procedure is usually adopted. First, a value for $B_{footing}$ is assumed and q_o is calculated using $q_o = Q/A_{footing}$, where Q is the structure load (such as a column load). Then, for the computed value of q_o, from the load-settlement curve a value for S_{plate} is estimated. The value of $S_{footing}$ is computed from the above equations and compared with the allowable settlement value. The procedure is repeated till the computed settlement value is equal to the allowable settlement.

(4) The plate load test can also be used for determination of the stress-strain modulus (or deformation modulus) E_s of the supporting soil, used in the known immediate settlement equation

$$\Delta H = qB\frac{1-\mu^2}{E_s}I \tag{1.42}$$

Where: ΔH = settlement
q = uniform load per unit area
B = least lateral dimension of the loaded area
I = influence factor
μ = Poisson's ratio of the soil obtained from tables

(*Continued*)

The expression $\left(\frac{1-\mu^2}{E_s}I\right)$ equals to $\left(\frac{\Delta H}{qB}\right)$ which is the slope of an assumed straight line plotted with coordinates ΔH and $q\,B$. It is determined from results of two or more plate-load tests, using plates of the same shape. The main features of Equation (1.42) and its applications will be outlined in Chapter 3.

(5) The *influence* depth of the plate (about 2 B_{plate}) is much smaller than that of the real footing, so the test reflects only the load carrying capacity of the near surface soils, whereas, usually, soils type and property vary with depth. This can introduce large errors. Several complete foundation failures occurred in spite of the use of plate-load tests (Terzaghi and Peck, 1967; Mesri, 1996). Obviously, for the reasons just mentioned, plate-load test is not used for design of deep foundations. Because of the development of better methods of testing and analysis, current engineering practice rarely uses plate-load tests for design of foundations. However, these tests are still useful for other design problems, such as those involving wheel loads on pavement subgrades, where the service loads act over smaller area.

Solution:

(a) The linear proportionality requires

$$V = aB + b$$

$$\text{From the first PLT}: 30 = a(0.3) + b$$
$$\text{From the second PLT}: 72 = a(0.6) + b$$
$$\text{Therefore, } a = \frac{42}{0.3} = 140 \text{ and } b = 30 - 140(0.3) = -12$$

For full-size footing:

$$V = 140\ B - 12$$
$$V = \text{column load} = 200\ \text{kN}$$
$$200 = 140\ B - 12$$

Therefore, $B = \frac{212}{140} = 1.514\ \text{m}$

Try 1.5 m × 1.5 m square footing

(b) Equation 1.38 : $V = qA + sP$

From the first PLT : $30 = q(0.3)^2 + s(4 \times 0.3) = 0.09q + 1.2\ s$

From the second PLT : $72 = q(0.6)^2 + s(4 \times 0.6) = 0.36q + 2.4\ s$ or $36 = 0.18q + 1.2\ s$

Therefore, $q = \frac{36-30}{0.09} = 66.67$ and $s = \frac{30 - 66.67(0.09)}{1.2} = 20$

For full-size footing:

$$200 = 66.67B^2 + 20 \times 4B \text{ or } B^2 + 1.2B - 3 = 0$$

$$B = \frac{-1.2 \pm \sqrt{1.2^2 - 4 \times 1(-3)}}{2 \times 1} = \frac{-1.2 + 3.67}{2} = 1.23\ \text{m}$$

Try 1.3 m × 1.3 m square footing.

Obviously, the assumed linear proportionality used in solution (a) is in error, whereas the Housel's equation used in solution (b) is based on research results; expected to be more correct. Therefore, one may decide on using *1.3 m × 1.3 m square footing.*

Problem 1.10

A standard plate-load test was conducted on a clay soil using a plate of 0.6 × 0.6 m. Under the ultimate vertical load V of 96.8 kN the plate settlement was 5 mm. If the clay supports a square footing, what will be its net safe bearing capacity *net* q_{safe} with a safety factor $SF = 3$? What size of square footing is required to carry a column load Q of 550 kN and what will be the settlement?

Solution:
In a standard PLT, actually, $q_{ult,plate}$ is *net* q_{ult} rather than *gross* q_{ult} since width of the test pit is as large as about five times the plate width.

Equation (1.37):

$$q_{ult,\,footing} = q_{ult,plate} \text{ or}$$

$$net\, q_{ult,\,footing} = net\, q_{ult,\,plate}$$

$$net\, q_{ult,\,plate} = \frac{V}{A_{plate}} = \frac{96.8}{0.6 \times 0.6} = 268.9\,\text{kPa. Hence}$$

$$net\, q_{ult,\,footing} = 268.9\,\text{kPa}$$

$$net\, q_{safe} = \frac{net\, q_{ult}}{SF} = \frac{268.9}{3} = 89.6\,kPa$$

The footing area

$$A_{footing} = Q/net\, q_{safe} = \frac{550}{89.6} = 6.14\,\text{m}^2$$

Use a square footing 2.5 m × 2.5 m

Equation (1.40):

$$S_{footing} = S_{plate}\left(\frac{B_{footing}}{B_{plate}}\right) = 0.005 \times \frac{2.5}{0.6} = 0.021\,\text{m}$$

The footing settlement = 21 mm.

Problem 1.11

A standard plate-load test was performed with a plate of 0.3 × 0.3 m at a depth of 1.0 m below the ground surface in a highly cohesive soil with Ø = 0°. The water table was located at a depth of 5 m below the ground surface. Failure occurred at a load of 4500 kg. The foundation level will be located at the same depth of the test. The total unit weight of the cohesive soil above water table $\gamma = 19\ \text{kN/m}^3$. Using the Terzaghi general bearing capacity equation (see any geotechnical text book), what would be the net ultimate bearing capacity for a 1.5 m wide continuous footing?

Solution:
Terzaghi general bearing capacity equation:

$$Gross\, q_{ult} = c\,N_c s_c + \gamma'\,D_f N_q + 0.5\gamma'\,B\,N_\gamma s_\gamma$$

Where c = soil cohesion; N_c, N_q and N_γ = Terzaghi bearing capacity factors; s_c and s_γ = shape factors; B = footing base width.

$$\text{For } Ø = 0° \text{ condition}: N_c = 5.7;\ N_q = 1;\ N_\gamma = 0$$

$$\text{For square footings}: s_c = 1.3;\ s_c = 0.8$$

$$\text{For continuous footings}: s_c = 1;\ s_\gamma = 1$$

(*Continued*)

Water table is located at 4 m depth below the foundation level which is deeper than B, hence, $\gamma' = \gamma = 19$ kN/m3

$$net\ q_{ult} = gross\ q_{ult} - \gamma' D_f$$
$$= c \times 5.7 \times 1.3 + 19 \times 1 \times 1 + 0 - 19 \times 1$$
$$= 7.41\,c$$

Equation (1.37): $q_{ult,footing} = q_{ult,plate}$ or

$$net\ q_{ult,\text{footing}} = net\ q_{ult,plate}$$

$$net\ q_{ult,\ plate} = \frac{4500}{0.3 \times 0.3} = 50\,000\ \text{kg/m}^2$$

$$7.41c = 50\,000\ \text{kg/m}^2 \rightarrow c = 6748\ \text{kg/m}^2$$

For continuous footings,

$$net\ q_{ult} = c \times 5.7 \times 1$$
$$= 6748 \times 5.7 = 38\,462\ \text{kg/m}^2$$
$$= 384.62\ \text{kN/m}^2$$

Use $net\ q_{ult} = 385\ kN/m^2$.

Problem 1.12

Two plate load tests on a cohesionless soil yielded the following data:

Plate size, m	Load V, kN	s_{plate}, mm
0.6 × 0.6	40	10
0.9 × 0.9	40	3

(a) Determine the stress-strain modulus E_s of the supporting soil. Assume the soil Poisson's ratio $\mu = 0.3$.
(b) What will be the differential settlement of two footings; one of the size 2.5 × 2.5 m, supporting a column load of 700 kN; the other of the size 3.0 × 3.0 m, supporting a column load of 1000 kN? Assume the footings are so apart that there will be no overlap of stresses take place in the supporting soil.

Solution:

(a) Equation (1.42):

$$\Delta H = q B \frac{1 - \mu^2}{E_s} I$$

From the first PLT, let $q_1 = q_{ult}$:

$$q_1 = \frac{40}{0.6 \times 0.6} = 111.1\ \text{kN/m}^2,\ \text{and}\ q_1 B_1 = 111.1 \times 0.6 = 66.7\ \text{kN/m}$$

From the second PLT, let $q_2 = q_{ult}$:

$$q_2 = \frac{40}{0.9 \times 0.9} = 49.4\ \text{kN/m}^2,\ \text{and}\ q_2 B_2 = 49.4 \times 0.9 = 44.5\ \text{kN/m}$$

From a plot of ΔH versus $q\ B$ or by calculation:

$$\frac{1-\mu^2}{E_s}I = \frac{\Delta H}{qB} = \frac{\Delta H1 - \Delta H2}{q_1B_1 - q_2B_2} = \frac{(0.010-0.003)}{(66.7-44.5)}$$

$$= \frac{0.007}{22.2} = 3.15\times10^{-4}$$

From tables (available in any geotechnical or foundation engineering text book), for a point at centre of a *rigid* square plate or footing, obtain $I = 0.82$ (approximately),

$$\frac{1-\mu^2}{E_s}I = \frac{1-0.3^2}{E_s}\times0.82 = \frac{0.75}{E_s}$$

$$\frac{0.75}{E_s} = 3.15\times10^{-4}$$

$$E_s = \frac{0.75}{3.15\times10^{-4}} = 2381\ kPa$$

(b) Equation (1.41):

$$S_{footing} = S_{plate}\left[\frac{B_{footing}\left(B_{plate}+0.3\right)}{B_{plate}\left(B_{footing}+0.3\right)}\right]^2$$

$$\text{First footing: load intensity} = 700/(2.5\times2.5) = 112.0\ \text{kN/m}^2$$

$$\text{Second footing: load intensity} = 1000/(3\times3) = 111.1\ \text{kN/m}^2$$

$$\text{The load intensity of the first PLT} = 111.1\ \text{kN/m}^2$$

$$\text{The load intensity of the second PLT} = 49.4\ \text{kN/m}^2$$

Use the results of the first PLT, since its load intensity is equal or very close to that of the footings.

$$\text{Settlement of the first footing} = 0.01\left[\frac{2.5(0.6+0.3)}{0.6(2.5+0.3)}\right]^2 = 0.018\ \text{m}$$

$$\text{Settlement of the second footing} = 0.01\left[\frac{3(0.6+0.3)}{0.6(3+0.3)}\right]^2 = 0.019\ \text{m}$$

The differential settlement of the footings is $0.019-0.018 = 0.001$ m

The differential settlement = 1 mm.

Problem 1.13

A static cone penetration test was conducted in a deposit of normally consolidated and moderately compressible dry sand. The test results were as given below:

Depth (m):	1.50	3.00	4.50	6.00	7.50	9.00
Point resistance q_c (MN/m^2):	2.05	4.23	6.01	8.18	9.97	12.42

(Continued)

Assume the dry unit weight of the sand deposit = 16 kN/m^3. Estimate its average peak friction angle $Ø'$ and relative density D_r.

Discussion:

The static cone penetration test or cone penetrometer test, CPT, also known as the Dutch cone penetration test, has found wide application in lieu of the SPT, especially in European countries.

The test is particularly used for soft clays, soft silts and fine to medium sand deposits. It is not well adapted to gravel deposits or to stiff and hard cohesive soil deposits. Briefly, in its original version, the test consists in pushing down a standard steel cone, usually a 60° cone of a 1000 mm^2 base area, into ground at a steady rate of 10–20 mm/s for a depth of 40 mm each time (each stroke) over stages (intervals) of 200 mm; the resistance in units of pressure is recorded for each stroke. The cone is pushed by applying thrust and not by driving. Data usually recorded as point or tip bearing resistance q_c, side friction resistance q_s and total penetration resistance q_T. Pore pressures, vertical alignment, and temperature may also be taken if allowed by the equipment configuration. A CPT allows nearly continuous testing at many sites, which is often valuable. If the soil is stratified, the test may be performed in parallel with a drilling machine. In this case the hole is drilled to a soft material, a CPT is done, boring recommences and so on. This test is rather popular for sites where there are deep deposits of transported soil such as in flood plains, river deltas and along coastlines.

Although many different penetrometer styles and configurations have been used, however, the most widely used are the mechanical Dutch cone (Figure 1.13), and the electric friction cone (Figure 1.14). The operation of the two

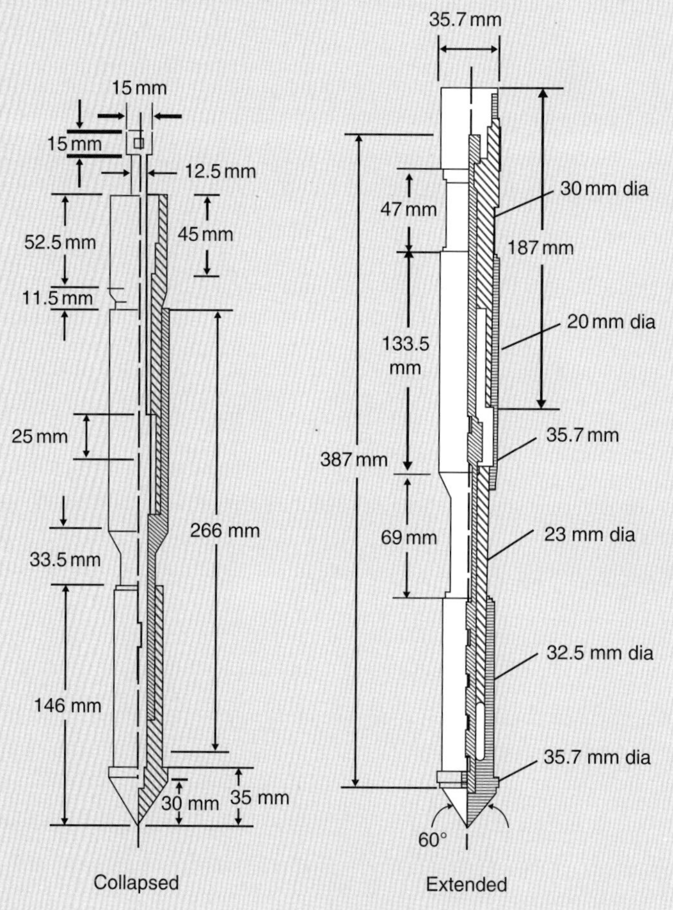

Figure 1.13 A mechanical friction cone penetrometer (from ASTM, 2001).

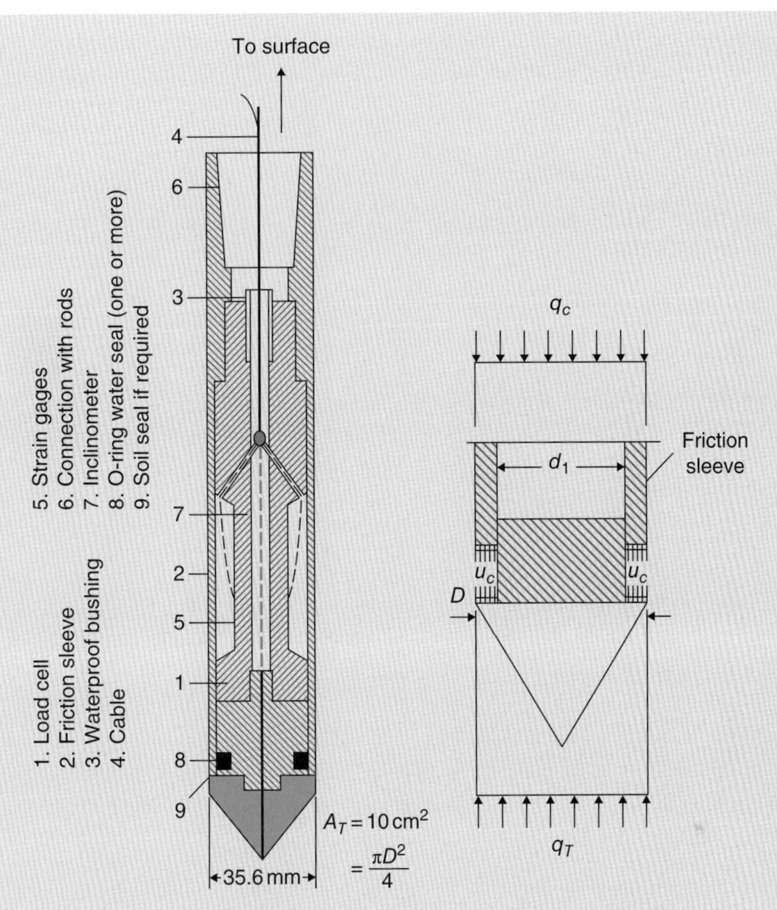

Figure 1.14 An electric friction cone penetrometer (from Bowles, 2001).

types differs in that the mechanical cone is advanced in stages and measures q_c and q_s at intervals of 20 cm, whereas the electric cone is able to measure q_c and q_s continuously with depth. In either case, the CPT defines the soil profile with much greater resolution than does the SPT. Another advantage of the CPT is that the disturbance to the soil is minimal.

In conducting a CPT, using a developed mechanical cone, initially the inner rod is pushed downwards a distance of 40 mm, causing the cone only to penetrate the soil, and the cone point resistance q_c is recorded. The outer shaft is now advanced to the cone base, and the side (or skin) friction resistance q_s is recorded. Now the cone and the engaged friction sleeve are advanced in combination to obtain the total penetration resistance q_T, which should be approximately the sum of the point and side resistances just measured.

In using the electric friction cone, the cone penetration resistance is measured by means of a load cell inside the body of the instrument and can thus be recorded continuously as the penetrometer is pushed into the soil. The results are normally plotted automatically, against depth, by means of a chart recorder. The friction sleeve is mechanically separate from the conical point; side resistance is measured by means of a second load cell. Cone resistance and side resistance can thus be measured independently. A full description of the test apparatus, procedure and results interpretation is given by Meigh (1987). *Note:* This test has been standardised by ASTM as D-3441.

The CPT data are used to classify a soil, to establish the allowable bearing capacity of shallow foundation elements, or to design piles. The CPT data processing and evaluating should always be handled by a specialised

(*Continued*)

geotechnical engineer with enough experience because the data can be so erratic that may need good engineering judgment.

The measured point resistance q_c and side friction q_s are used to compute the friction ratio f_r as

$$f_r = \frac{q_s}{q_c} \times 100\,(\%) \tag{1.43}$$

The friction ratio f_r may also be estimated using D_{50} as in the following empirical equations: Anagnostopoulos et al (2003),

$$f_r(\%) = 1.45 - 1.36 \log D_{50} \text{ (using electric cone)} \tag{1.44}$$

$$f_r(\%) = 0.781 - 1.611 \log D_{50} \text{ (using mechanical cone)} \tag{1.45}$$

Note: In developing these equations, the D_{50} of soils ranged from 0.01 mm to about 10 mm.

The friction ratio f_r is primarily used for soil classification as illustrated in the charts of Figures 1.15 and 1.16.

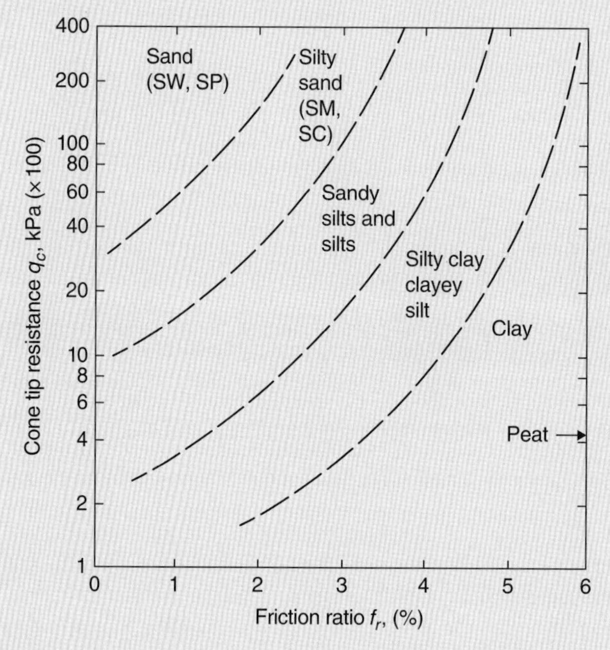

Figure 1.15 Soil classification chart for standard electric friction or mechanical cone (from Bowles, 2001, after Robertson and Campanella, 1983).

The friction ratio may be used to give an estimate for soil sensitivity S_t as follows (Robertson and Campanella, 1983):

$$S_t \approx \frac{10}{f_r} \quad (f_r \text{ is in } \%) \tag{1.46}$$

CPT correlations for cohesive soils:

The cone bearing resistance q_c (tip bearing resistance) and the undrained shear strength s_u (or c_u) are correlated through the following empirical equation (Mayne and Kemper, 1988):

$$s_u = \frac{q_c - \sigma_o}{N_k} \tag{1.47}$$

$\sigma_o = \gamma Z$ = overburden pressure where the q_c is measured. It is in the same units of q_c and same type of pressure (i.e., if q_c is an effective pressure, σ'_o shall be used).

Zone	Soil behaviour type
1	Sensitive fine gained
2	Organic material
3	Clay
4	Silty clay to clay
5	Clayey silt to silty clay
6	Sandy silt to clayey silt
7	Silty sand to sandy silt
8	Sand to silt
9	Sand
10	Gravelly sand to sand
11	Very stiff fine grained*
12	Sand to clayey sand*

*Overconsolidated or cemented

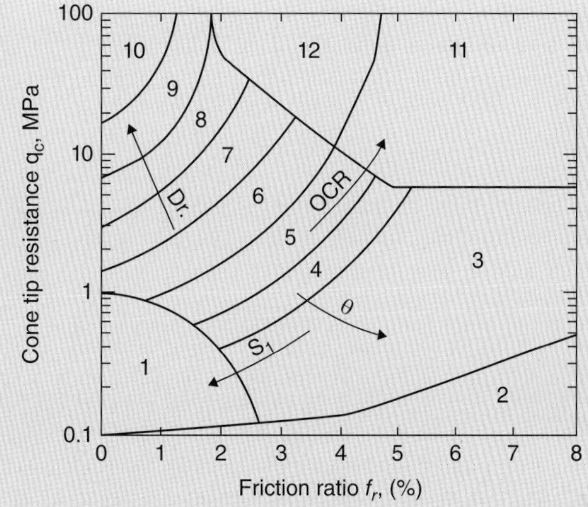

Figure 1.16 Soil behaviour type chart based on q_c and f_r (reproduced from paper by P.K. Robertson, 2010; it is the original Robertson and Campanella chart, 1986).

N_k = Cone factor or bearing capacity factor, constant for a certain soil; depends mainly on soil plasticity index (PI) and sensitivity (S_t), and on the cone penetrometer type. Its range is 5–30. The recommended values for mechanical and electric cones are 20 and 15, respectively. Anagnostopoulos et al. (2003) determined these two values equal 18.9 and 17.2, respectively. They also showed that c_u equals 0.79 q_s and q_s for mechanical and electric cones, respectively. For normally consolidated clays of $S_t < 4$ and $PI < 30$, a value of $N_k = 18$ may be satisfactory.

The preconsolidation pressure σ_c' and overconsolidation ratio OCR are correlated as follows (Mayne and Kemper, 1988):

$$\sigma_c' = 0.243(q_c)^{0.96} \qquad (q_c \text{ and } \sigma_c' \text{ are in MN/m}^2) \tag{1.48}$$

$$OCR = 0.37\left(\frac{q_c - \sigma_o}{\sigma_o'}\right)^{1.01} \tag{1.49}$$

CPT correlations for cohesionless soils:

Lancellotta (1983) and Jamiolkowsk et al (1985) showed that for the normally consolidated sand, the relative density D_r and the cone resistance q_c can be correlated as:

$$D_r(\%) = 66 \times \log\frac{q_c}{\sqrt{\sigma_o'}} - 98 \tag{1.50}$$

where the q_c and σ_O' are in t/m^2

According to Kulhawy and Mayne (1990), the above equation can be rewritten as

$$D_r(\%) = 68\left[\log\left(\frac{q_c}{\sqrt{p_a\sigma_o'}}\right) - 1\right] \tag{1.51}$$

$$P_a = \text{atmospheric pressure} \ (\approx 100\,\text{kPa})$$

Figure 1.17 represents the empirical relationship recommended by Baldi et al. (1982), and Robertson and Campanella (1983), for normally consolidated quartz sand.

(Continued)

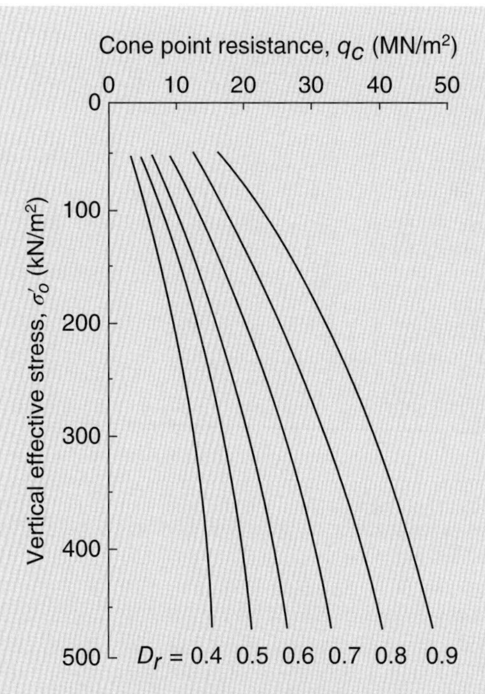

Figure 1.17 Correlation between q_c, D_r and σ'_o for normally consolidated quartz sand (after Baldi et al, 1982; Robertson and Campanella, 1983).

Also, D_r, OCR and q_c are correlated by Kulhawy and Mayne (1990) as

$$D_r = \sqrt{\left(\frac{1}{305 Q_c OCR^{1.8}}\right)\left[\frac{\frac{q_c}{p_a}}{\left(\frac{\sigma'_o}{p_a}\right)^{0.5}}\right]} \tag{1.52}$$

Where Q_c = compressibility factor; using 0.91, 1.0 and 1.09 for high, moderate and low compressibility of sand, respectively.

For normally consolidated quartz sand, the effective friction angle $\varnothing'$, σ'_o and q_c can be correlated and expressed as (Kulhawy and Mayne, 1990):

$$\varnothing' = \tan^{-1}\left[0.1 + 0.38\log\left(\frac{q_c}{\sigma'_o}\right)\right] \tag{1.53}$$

Ricceri et al. (2002) suggested a similar correlation for ML and SP-SM soil types as

$$\varnothing' = \tan^{-1}\left[0.38 + 0.27\log\left(\frac{q_c}{\sigma'_o}\right)\right] \tag{1.54}$$

An approximate correlation (Bowles, 1996) for $\varnothing'$ is

$$\varnothing' = 29^\circ + \sqrt{q_c} + 5^\circ \text{ for gravel}; -5^\circ \text{ for silty sand, where } q_c \text{ is in MPa}$$

Lee et al. (2004) established a relationship between the horizontal effective stress σ'_h, $\text{Ø}'$ and q_c as follows:

$$\text{Ø}' = 15.575\left(\frac{q_c}{\sigma'_h}\right)^{0.1714} \tag{1.55}$$

A number of correlations have been proposed to estimate the SPT N-values from CPT results in both cohesive and cohesionless soils. These correlations, generally, use a form of $q_c = k\ N$. Unfortunately, all the available correlations cannot be used with much confidence. According to Meyerhof, $q_c \approx 4N_{55}$, where q_c is in kg/cm^2. Figure 1.18 gives a relatively reliable correlation between mean grain size D_{50} (grain size at which 50% of the soil is finer, mm) and q_c/N ratio. Also, Table 1.11 gives approximate range values of q_c/N'_{60} ratio for different soils, using q_c in MPa.

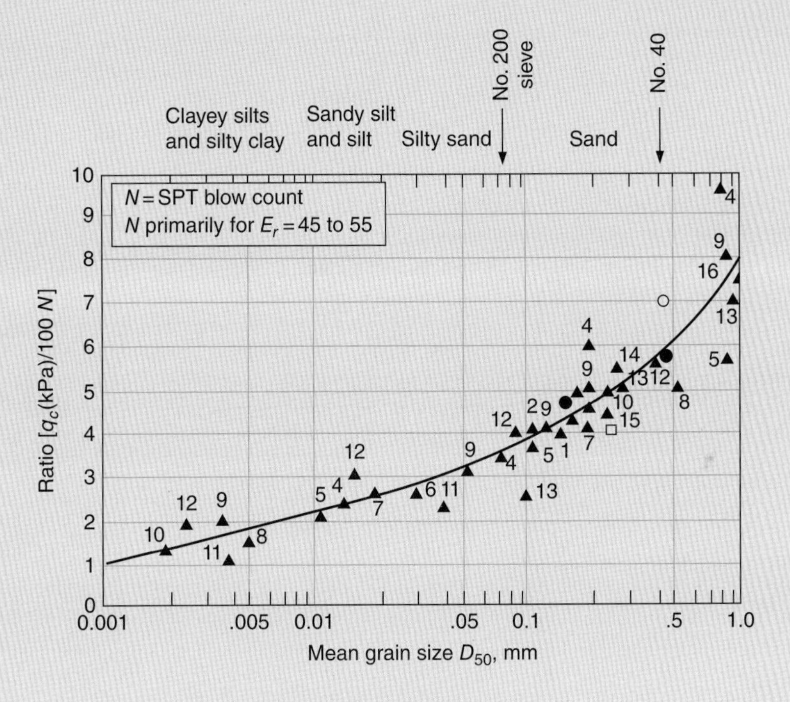

Example: From sieve analysis $D_{50} = 0.5$ mm

From in situ CPT $q_{c,av} = 60$ kg/cm^2 = 6000 kPa

Required: Estimate N_{55}

Solution: Enter chart at $D_{50} = 0.5$ project vertically to curve and then horizontally to read

$q_c/100\ N_{55} = 6.2$

$$N_{55} = \frac{6000}{100 \times 6.2} = 9.6 \rightarrow 10 \text{ blows (0.3 m penetration)}$$

Check using Meyerhof equation $q_c = 4N_{55}$ (approximately):

$$N_{55} = \frac{q_c}{4} = \frac{60}{4} = 15$$

Figure 1.18 A plot of correlation between q_c/N ratios and mean grain size (after Robertson and Campanella, 1983; Ismael and Jeragh, 1986; reference numbers correspond to references in original sources).

(Continued)

Table 1.11 Approximate range values of q_c/N'_{60} ratio for different soils.

Soil type	q_c/N'_{60}
Silts, sandy silts and slightly cohesive silt-sand mixtures	0.1–0.2
Clean fine to medium sands and slightly silty sands	0.3–0.4
Coarse sands and sands with little gravel	0.5–0.7
Sandy gravels and gravels	0.8–1.0

Anagnostopoulos et al. (2003) proposed the following correlation:

$$\frac{\left(\frac{q_c}{p_a}\right)}{N_{60}} = 7.6429 D_{50}^{0.26} \tag{1.56}$$

where p_a = atmospheric pressure (≅100 kPa); q_c in kPa.

Schmertmann (1970) established empirical correlations between E_s and the cone resistance q_c. Because CPT can provide a continuous plot of q_c versus depth, it is possible to model E_s as a function of depth, which is especially useful. Table 1.12 presents design range values of E_s/q_c for sands, adapted from Schmertmann et al. (1978), Robertson and Campanella (1989), and other sources.

Table 1.12 Design range values of E_s/q_c for sands (from Coduto, 2001).

Soil type	USCS Group Symbol	E_s/q_c
Young, normally consolidated clean silica sands*	SW or SP	2.5–3.5
Aged, normally consolidated clean silica sands**	SW or SP	3.5–6.0
Overconsolidated clean silica sands	SW or SP	6.0–10.0
Normally consolidated silty or clayey sand	SM or SC	1.5
Overconsolidated silty or clayey sand	SM or SC	3.0

* Age < 100 years;
** Age > 100 years

Solution:

Equation (1.53):

$$\emptyset' = \tan^{-1}\left[0.1 + 0.38\log\left(\frac{q_c}{\sigma'_o}\right)\right]$$

$$\sigma'_O = 16\,z, \text{where } z = \text{depth, m}$$

$$\emptyset' = \tan^{-1}\left[0.1 + 0.38\log\left(\frac{q_c}{16\,z}\right)\right]$$

Equation (1.52):

$$D_r = \sqrt{\left(\frac{1}{305 Q_c OCR^{1.8}}\right)\left[\frac{\frac{q_c}{p_a}}{\left(\frac{\sigma'_o}{pa}\right)^{0.5}}\right]}$$

$$p_a = 100\,\text{kN/m}^2;\ \sigma'_O = 16\,z$$

For normally consolidated sand $OCR = 1$

For moderately compressible sand $Q_c = 1$

$$D_r = \sqrt{\left(\frac{1}{305}\right)\left[\frac{\frac{q_c}{100}}{\left(\frac{16\,z}{100}\right)^{0.5}}\right]} = \sqrt{\frac{q_c}{12200\,z^{0.5}}}$$

Obtain $Ø'$ and D_r values for different values of q_c at the given depths, as shown in the table below:

Z, m	q_c, kN/m^2	σ'_O, kN/m^3	$Ø'$, degree	D_r, %
1.5	2050	24	39.8	37.0
3.0	4230	48	40.0	44.7
4.5	6010	72	39.7	48.2
6.0	8180	96	39.8	52.3
7.5	9970	120	39.7	54.6
9.0	12420	144	39.9	58.3

$$\text{Average } Ø' = \sum Ø'/6 = 238.9/6 = 39.8°$$
$$\text{Average } D_r = \sum D_r/6 = 295.1/6 = 49.2\%$$

Note: To obtain the average values for $Ø'$ and D_r, one may prefer to use the same procedure as that followed for N'_{60} in Solution (b) of Problem 1.6. If that procedure is used, the average values will be 39.8° and 52.5%.

Problem 1.14

A cone penetration test was conducted, using an electric-friction cone, in a clay deposit. The cone penetration resistance, at 6 m depth, was 0.8 MN/m^2. Sensitivity of the clay $S_t < 4$, and its $PI < 30$. Ground water table was located at 2 m depth. The unit weight γ of the clay above and below water table was 18 and 20 kN/m^3, respectively. Find the over consolidation ratio OCR, and the undrained cohesion c_u.

Solution:

Equation (1.49):

$$OCR = 0.37\left(\frac{q_c - \sigma_o}{\sigma'_o}\right)^{1.01}$$

$$q_c = 0.8\,\text{MN/m}^2 = 800\,\text{kN/m}^2$$
$$\sigma_o = 2 \times 18 + 4 \times 20 = 116\,\text{kN/m}^2$$
$$\sigma'_o = 2 \times 18 + 4(20 - 10) = 76\,\text{kN/m}^2$$

(*Continued*)

$$OCR = 0.37\left(\frac{800-116}{76}\right)^{1.01} = 3.4$$

Equation (1.49):

$$s_u = \frac{q_c - \sigma_o}{N_k} = c_u$$

For electric friction penetrometers a value of $N_k = 15$ may be used. However, because the clay has $S_t < 4$ and $PI < 30$, and in order to be more on the safe side, assume using $N_k = 18$.

$$c_u = \frac{800-116}{18} = 38\,kN/m^2$$

Problem 1.15

Figure 1.19 shows an electric cone penetration record at site of a proposed structure. The average soil unit weight $\gamma = 19$ kN/m^3. Classify the soil encountered at the depth interval of 10–12 m. Also estimate the internal friction angle $\text{Ø}'$, the relative density D_r and the undrained shear strength S_u of the classified soil, as applicable (according to the soil type). Assume the soil is normally consolidated and of low compressibility.

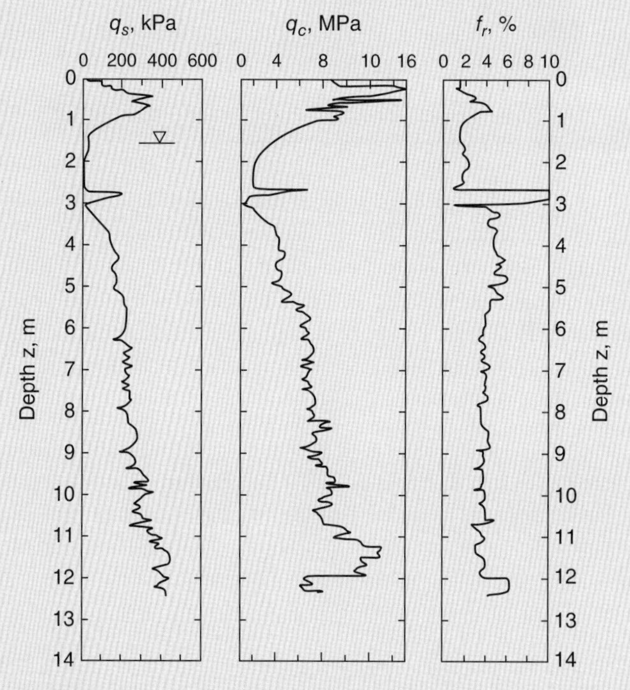

Figure 1.19 Cone penetration record.

Solution:

Estimate the average value for q_c = 10 MPa = 10 000 kN/m^2

Estimate the average value for f_r = 4

Use Figure 1.15 and classify the soil as *sandy silt*

Consider the soil as cohesionless (approximately); estimate $\text{Ø}'$ and D_r:

Equation (1.53): $$\varnothing' = \tan^{-1}\left[0.1 + 0.38\log\left(\frac{q_c}{\sigma'_o}\right)\right]$$

$$\sigma'_O = 11 \times 19 = 209\ \text{kN/m}^2$$

$$\varnothing' = \tan^{-1}\left[0.1 + 0.38\log\left(\frac{10000}{209}\right)\right] = 36.44^\circ$$

Equation (1.52): $$D_r = \sqrt{\left(\frac{1}{305Q_cOCR^{1.8}}\right)\left[\frac{\frac{q_c}{p_a}}{\left(\frac{\sigma'_o}{p_a}\right)^{0.5}}\right]}$$

$p_a = 100\ \text{kPa};\ \sigma'_O = 209\ \text{kPa}$

For normally consolidated sand, $OCR = 1$

For sand of low compressibility, $Q_c = 1.09$

$$D_r = \sqrt{\left(\frac{1}{305 \times 1.09 \times 1}\right)\left[\frac{\frac{10000}{100}}{(209/100)^{0.5}}\right]} = \sqrt{0.208} = 0.456 = 45.6\%$$

Problem 1.16

Using the data given in Problem 1.15, estimate the equivalent SPT N-value of the classified soil.

Solution:
The soil was classified as medium dense sandy silt. However, the soil grain size curve is not available. Visual examination of the soil is not possible since there is no representative sample available. Therefore, the soil D_{50} should be estimated from soil classification with reference to grain size range for silt and sand. According to ASTM particle size classification, the lower limit of sand grain size is 0.075 mm and that of silt is 0.006 mm.

Assume $D_{50} = 0.06$ mm.

From Figure 1.18, obtain $(q_c/100N) = 3.4$

For $q_c = 10\ \text{MPa} = 10\,000\ \text{kN/m}^2$, compute $N_{55} = 29$

According to Meyerhof: $$N \approx \frac{q_c\,(\text{kg/cm}^2)}{4} \approx \frac{(10000/100)}{4} \approx \underline{25}$$

According to Equation (1.56): $$\frac{(q_c/q_a)}{N_{60}} = 7.6429\,D_{50}^{0.26}$$

$$\frac{(10000/100)}{N_{60}} = 7.6429(0.06)^{0.26} = 3.68,\ N_{60} = 27$$

Since the three results are somewhat close, it may be reasonable to use their average. *Use N = 27.*

Problem 1.17

A pressuremeter test (PMT) was conducted in a soft saturated clay.
Given: corrected $V_o = 535\,\text{cm}^3$, $v_o = 46\,\text{cm}^3$, $p_o = 42.4\,\text{kN/m}^2$, $p_f = 326.5\,\text{kN/m}^2$, and $v_f = 180\,\text{cm}^3$.
Assuming Poisson's ratio $\mu_s = 0.5$ and using Figure 1.22, calculate the pressuremeter modulus E_{sp}.

Discussion:
A PMT is a cylindrical device designed to apply a uniform radial pressure to the sides of a borehole at the required depth below ground surface. Although it can be used in different types of soil, however, its best applications are in relatively fine-grained sedimentary soil deposits. The original PMT was developed in the year 1956 by Louis Menard in an attempt to overcome the problem of sampling disturbance and to insure that the in-place soil structure is adequately represented. There are two different basic types: (1) The Menard PMT, which is lowered into a pre-formed borehole (Figure 1.20a); (2) The self-boring PMT, which forms its own borehole and thus causes much less disturbance to the soil prior to testing (Figure 1.20b).

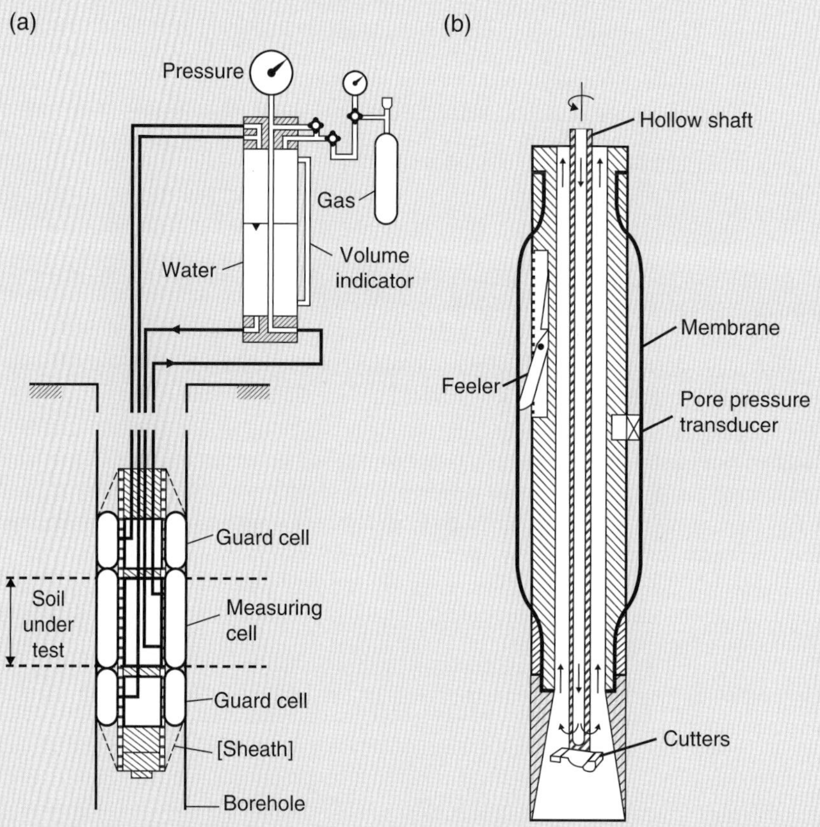

Figure 1.20 Basic features of (a) Menard pressuremeter and (b) self-boring pressuremeter (from Knappett and Craig, 2012).

The device basically consists of a cylindrical rubber cell, usually of 58 mm in diameter and 535 cm^3 in volume, and two guard cells of the same diameter arranged coaxially. The probe length is usually 420 mm. The device is lowered into a carefully prepared (slightly oversize) borehole to the required depth. As recommended by ASTM, for diameters of 44, 58 and 74 mm, the required borehole diameter ranges are 45–53 mm, 60–70 mm and 76–89 mm, respectively. The central measuring cell is expanded against the borehole wall by means of water pressure, measurements of the applied pressure and the corresponding increase in volume being recorded. Pressure is applied to the water by compressed gas (usually nitrogen) in a controlled gas cylinder at the surface. The increase in volume of the pressure cell is determined from the movement of the water-gas interface in the other control cylinder (Figure 1.20a). Readings being normally taken at 15, 30, 60 and 120 s after a pressure increment has been applied. The process is continued until the total volume of the expanded cavity becomes as large as twice the volume of the original cavity, or, the pressure limit of the device is reached. A new test begins after the probe has been deflated and advanced to another depth. A version of Menard pressuremeter test is schematically shown in Figure 1.21. *Note:* Refer to ASTM D 4719 for more details of the test.

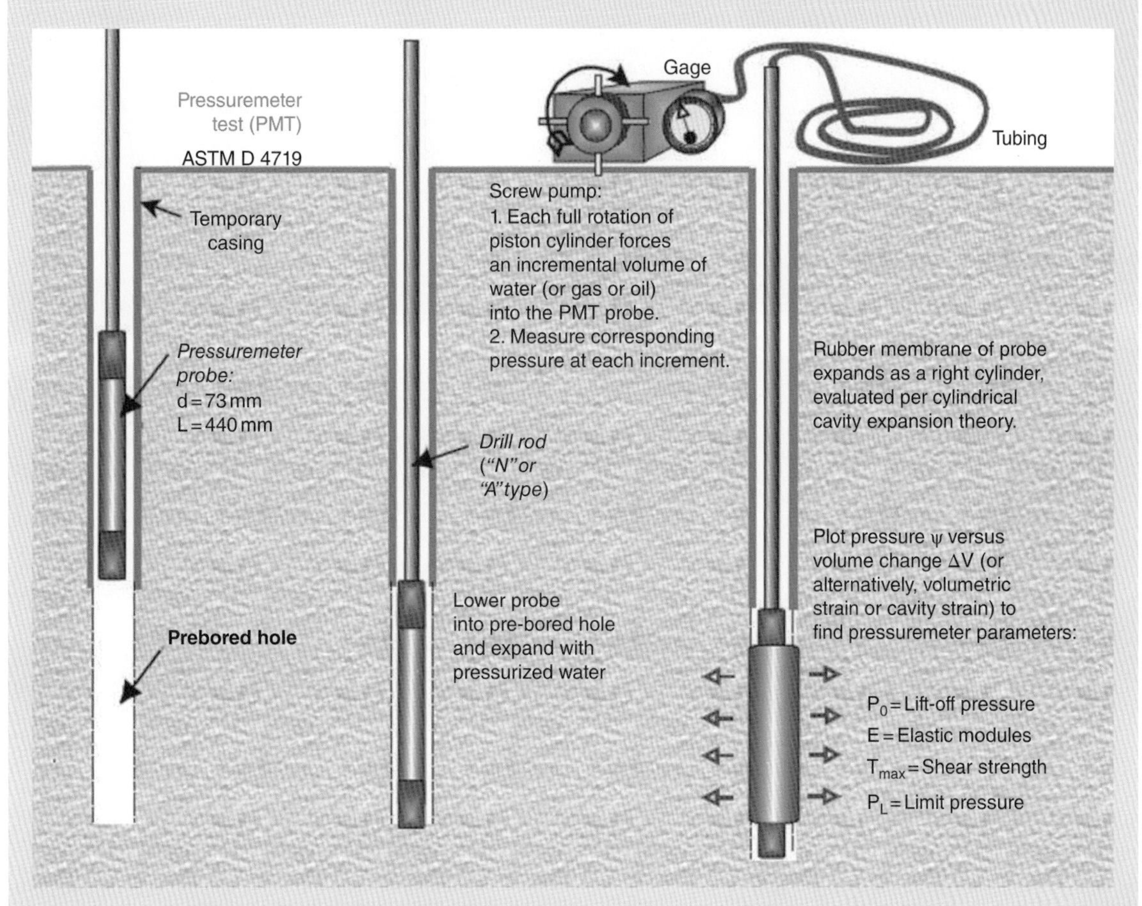

Figure 1.21 A version of the Menard pressuremeter test schematic (from FHWA, 2002).

Generally, a pressuremeter is designed for maximum pressures in the ranges 2.5–10.0 MPa in soils and 10–20 MPa in very stiff soils and weak rocks. Corrections must be made to the measured pressure, volume change and cavity deformation in order to account for: (a) the head difference between the water level in the cylinder and the test level in the borehole (or between the pressure transducer and the element), (b) the pressure required to stretch the rubber cell (i.e. for stiffness of the membrane) and (c) the expansion of the control cylinder and tubing under pressure. The two other guard cells are expanded under the same pressure as in the measuring cell but using compressed gas. The increase in volume of the guard cells is not measured. The function of the guard cells is to eliminate end effects, insuring a state of plane strain.

(*Continued*)

The results of a Menard pressuremeter test are represented by a graphical form of corrected pressure (P) against total volume (V), as shown in Figure 1.22. Zone I of the figure represents the reloading stage during which the soil around the borehole is pushed back into the original state (the at-rest state before drilling the borehole). At this state the cell initial volume (V_o) has been increased to ($V_o + v_o$), and the cell pressure (p_o) approximately equals to the in situ lateral earth pressure (p_h), depending on procedure and insertion disturbance. Zone II represents an assumed elastic zone in which the cell volume versus cell pressure is practically linear. In this stage, within the linear section of the $P - V$ plot, the *shear modulus* of the soil may be considered equal to slope (dp/dV) multiplied by volume V. Pressure (p_f) represents the yield (or creep) pressure. Zone III is the plastic (or creep) zone, and the pressure (p_l) represents the *limit pressure.*

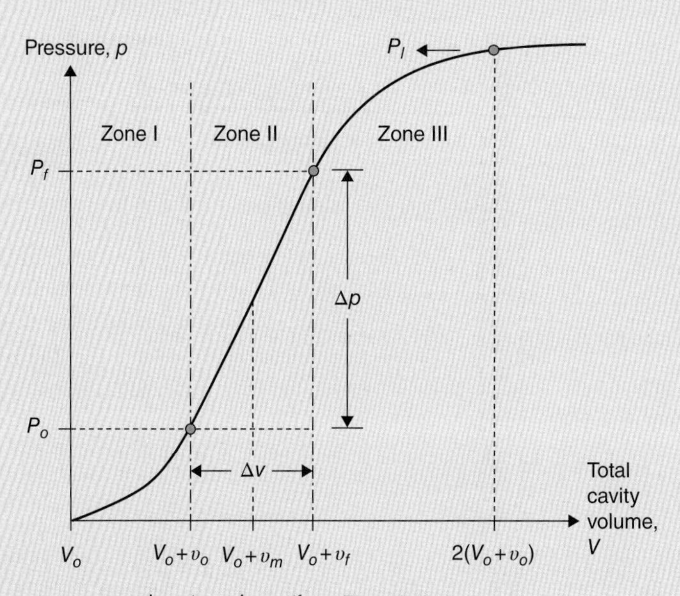

Figure 1.22 Plot of pressure versus total cavity volume (from Das, 2011).

Results of a pressuremeter test can be used in solving the following equations:

$$E_{sp} = 2(1+\mu_s)(V_o + v_m)\left(\frac{\Delta p}{\Delta v}\right) \tag{1.57}$$

where:

E_{sp} = pressuremeter modulus

$v_m = \dfrac{v_o + v_f}{2}$ (for a pre-drilled borehole using Menard pressuremeter)

$v_m = \dfrac{v_f}{2}$ (for a self-boring pressuremeter)

$$\Delta p = p_f - p_o,\ \Delta_v = v_f - v_o$$

μ_s = Poisson's ratio of soil; Menard recommended $\mu_s = 0.33$, but other values can be used.

$\dfrac{\Delta p}{\Delta v}$ = slope of the linear section of the $P - V$ plot (Figure 1.22).

According to Ohya et al. (1982):

$$E_{sp}(\text{kN/m}^2) = 1930\,N_{60}^{0.63} \qquad \text{(for clays)} \tag{1.58}$$

$$E_{sp}(\text{kN/m}^2) = 908N_{60}^{0.66} \qquad \text{(for sands)} \tag{1.59}$$

$$E_{sp} = 2(1+\mu_s)G_{sp} \tag{1.60}$$

$$G_{sp} = (V_o + v_m)\left(\frac{\Delta p}{\Delta v}\right) \tag{1.61}$$

where:

G_{sp} = pressuremeter shear modulus

$$\frac{\Delta V}{V} = 1-(1+\varepsilon_c)^{-2} \tag{1.62}$$

$\frac{\Delta V}{V}$ = volumetric strain of soil

where ε_c = circumferential strain

= (increase in cavity radius Δr/radius r_o)

$$K_o = \frac{p_h}{\sigma_o} \approx \frac{p_o}{\sigma_o} \tag{1.63}$$

where K_o = at-rest earth-pressure coefficient.

σ_o = total vertical stress

According to Kulhawy and Mayne (1990):

$$\sigma'_c = 0.45\,p_l \tag{1.64}$$

where σ'_c = Preconsolidation pressure

p_l = limit pressure (Figure 1.22)

According to Baguelin et al. (1978):

$$c_u = \frac{p_l - p_o}{N_p} \tag{1.65}$$

where $N_p = 1 + \ln\left(\frac{E_{sp}}{3c_u}\right)$; c_u = the undrained shear strength of soil. Typical values of N_p vary between 5 and 12.

In the case of saturated clays, it is possible to obtain the value of c_u by iteration from the following expression:

$$p_l - p_o = c_u\left[\ln\left(\frac{G_{sp}}{c_u}\right) + 1\right] \tag{1.66}$$

Note: Experience indicates that c_u values obtained from pressuremeter tests are probably higher by about 50% than the values obtained from other tests.

A comprehensive review of the use of pressuremeter, including examples of test results and their applications in design, has been given by Mair and Wood (1987) and Clarke (1995). Also, an analysis for the interpretation

(*Continued*)

of pressuremeter tests in sands has been given by Hughes, Wroth and Windle (1977). Those who are interested in the pressuremeter test should refer to E. Winter (1982) for the test and calibration details and to Briaud and Gambin (1984), and Briaud (1989) for borehole preparation (which is extremely critical).

Solution:

Equation (1.57): $$E_{sp} = 2(1+\mu_s)(V_o + v_m)\left(\frac{\Delta p}{\Delta v}\right)$$

$$v_m = \frac{v_o + v_f}{2} = (46 + 180)/2 = 113\ \text{cm}^3$$

$$\Delta p = p_f - p_o = 326.5 - 42.4 = 284.1\ \text{kPa}$$

$$\Delta V = v_f - v_o = 180 - 46 = 134\ \text{cm}^3$$

$$E_{sp} = 2(1+0.5)(535\times10^{-6} + 113\times10^{-6})\left(\frac{284.1}{134\times10^{-6}}\right) = 4122\ kPa$$

Problem 1.18

Refer to Problem 1.17. The test was conducted at a depth of 4 m below the existing ground surface. Saturated γ of the clay was 20 kN/m^3. The test results gave the limit pressure $p_l = 400$ kN/m^2. Determine: σ'_c, G_{sp}, K_o and a design value for c_u. Also, find an approximate value for N_{60}.

Solution:

Equation (1.64): $$\sigma'_c = 0.45\,p_l = 0.45\times400 = 180\ kN/m^2$$

Equation (1.61): $$G_{sp} = (V_o + v_m)\left(\frac{\Delta p}{\Delta v}\right)$$

$$= (535\times10^{-6} + 113\times10^{-6})\left(\frac{284.1}{134\times10^{-6}}\right)$$

$$= 648\times2.12 = 1374\ kN/m^2.$$

Equation (1.63): $$K_o = \frac{p_h}{\sigma_o} \approx \frac{p_o}{\sigma_o} = \frac{42.4}{20\times4} = 0.53$$

Equation (1.65): $$c_u = \frac{p_l - p_o}{N_p}$$

Typical values for N_p vary in the range 5 to 12; assume $N_p = 8.5$.

$$c_u = \frac{400-42.4}{8.5} = 42.1\ \text{kN/m}^2$$

Equation (1.66): $$p_l - p_o = c_u\left[\ln\left(\frac{G_{sp}}{c_u}\right)+1\right]$$

$400 - 42.4 = c_u\left[\ln\left(\frac{1374}{c_u}\right)+1\right]$. Using iteration, find c_u = 98.4 kN/m^2

However, it is recommended to reduce this value by 50%; hence, c_u = 49.2 kN/m^2. Average of the two results is 45.7 kN/m^2. Hence,

$$\text{Use } c_u = 45\ kN/m^2$$

Equation (1.58): For clays, $E_{sp}\,(\text{kN/m}^2) = 1930\,N_{60}^{0.63}$
Equation (1.60): $E_{sp} = 2(1+\mu_s)G_{sp} = 2(1+0.5)(1374) = 4122\ \text{kN/m}^2$

$$N_{60}^{0.63} = \frac{4122}{1930} = 2.136$$

$N_{60} = (2.136)^{\frac{1}{0.63}} = (2.136)^{1.59} = 3.34$. Hence, $N_{60} = 3$

Problem 1.19

Plot the following corrected pressuremeter test data and estimate p_h, K_o and E_{sp}. Assume average $\gamma = 17.65\ \text{kN/m}^3$, $\mu_s = 0.3$ and the test depth = 2.60 m. What is the "limiting pressure"?

V, cm^3:	55	88	110	130	175	195	230	300	400	500
P, kPa:	10	30	110	192	290	325	390	430	460	475

Solution:
Figure 1.23 shows the plot of the pressure P versus volume V.

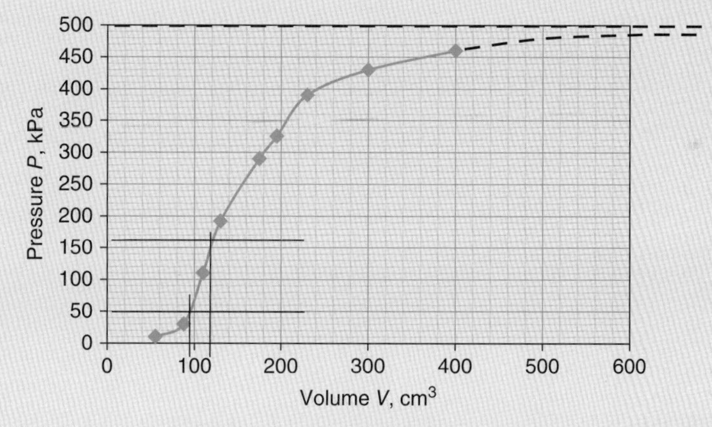

Figure 1.23 Plot of pressure versus volume.

From the plot, obtain:

$$p_o = 50\ \text{kPa}; V_o + v_o = 95\ \text{cm}^3;\ p_f = 160\ \text{kPa};\ V_o + v_f = 120\ \text{cm}^3$$
$$p_h \approx p_o;\ p_h \approx 50\ kPa.$$

Equation (1.63):
$$K_o = \frac{p_h}{\sigma_o} \approx \frac{p_o}{\sigma_o} = \frac{50}{2.60 \times 17.65} = 1.09$$

$$\Delta p = p_f - p_o = 160 - 50 = 110\ \text{kPa}$$
$$\Delta V = (V_o + v_f) - (V_o + v_o) = 120 - 95 = 25\ \text{cm}^3$$
$$V_o + v_m = V_o + v_o + \frac{\Delta V}{2} = 95 + \frac{25}{2} = 107.5\ \text{cm}^3$$

(Continued)

Equation (1.57):

$$E_{sp} = 2(1+\mu_s)(V_o + v_m)\left(\frac{\Delta p}{\Delta v}\right)$$

$$E_{sp} = 2(1+0.3)(107.5\times10^{-6})\left(\frac{110}{25\times10^{-6}}\right) = 1230\,kPa$$

Extrapolate the plot in Figure 1.23 and estimate the limiting pressure ≅ *495 kPa.*

Problem 1.20

Plot the following corrected pressuremeter test data. The test depth is at 4 m below the ground surface. Estimate E_{sp}, σ_3, K_o, G_{sp}, p_l, σ_c' and c_u. Use: $\mu_s = 0.5$, $\gamma = 19.81$ kN/m^2. In estimating c_u assume $N_p = 6$.

V, cm^3:	40	70	88	132	180	250	330	500	600	700
P, kPa:	10	20	38	178	305	410	460	480	490	493

Solution:
Figure 1.24 shows the plot of the pressure P versus volume V.

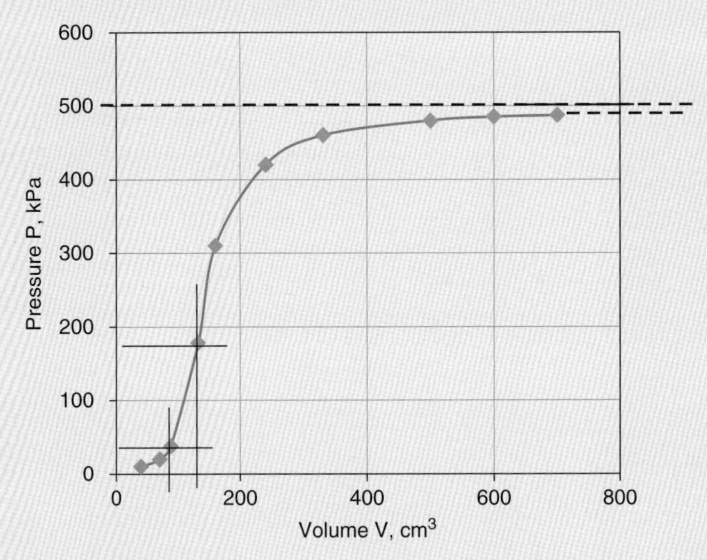

Figure 1.24 Plot of pressure versus volume.

From the plot, obtain:

$$p_o = 38\,\text{kPa};\ V_o + v_o = 88\,\text{cm}^3;\ p_f = 178\,\text{kPa};\ V_o + v_f = 132\,\text{cm}^3$$

$$\sigma_3 = p_h \approx p_o;\ \sigma_3 \approx 38\,kPa$$

Equation (1.63):

$$K_o = \frac{p_h}{\sigma_o} = \frac{\sigma_3}{\sigma_o} = \frac{38}{4\times19.81} = 0.48$$

$$\Delta p = p_f - p_o = 178 - 38 = 140\,\text{kPa}$$
$$\Delta V = (V_o + v_f) - (V_o + v_o) = 132 - 88 = 44\,\text{cm}^3$$
$$V_o + v_m = V_o + v_o + \frac{\Delta v}{2} = 88 + \frac{44}{2} = 110\ \text{cm}^3$$

Equation (1.57): $$E_{sp} = 2(1+\mu_s)(V_o + v_m)\left(\frac{\Delta p}{\Delta v}\right)$$

$$E_{sp} = 2(1+0.5)(110\times10^{-6})\left(\frac{140}{44\times10^{-6}}\right) = 1050\,kPa$$

Equation (1.61): $$G_{sp} = (V_o + v_m)\left(\frac{\Delta p}{\Delta v}\right) = (110\times10^{-6})\left(\frac{140}{44\times10^{-6}}\right) = 350\,kPa$$

Extrapolate the plot in Figure 1.24 and estimate the limiting pressure ≅ *500 kPa*

Equation (1.64): $$\sigma'_c = 0.45\,p_l = 0.45\times500 = 225\,kN/m^2$$

Equation (1.65): $$c_u = \frac{p_l - p_o}{N_p} = \frac{500-38}{6} = 77\,kPa$$

Problem 1.21

A dilatometer test (DMT) was conducted in a clay deposit. The water table was located at a depth of 3 m below the ground surface. At 8 m depth the contact pressure (p_1) was 280 kPa and the expansion stress (p_2) was 350 kPa. Assume $\sigma'_o = 95\,\text{kPa}$ at the 8 m depth and $\mu = 0.35$. Determine (a) Coefficient of at-rest earth pressure K_o, (b) Overconsolidation ratio *OCR* and (c) Modulus of elasticity E_s.

Discussion:

A DMT is an in situ test carried out in order to assess the in place stresses and compressibility of soils. The test uses a device called *dilatometer* developed during the late 1970s in Italy by Silvano Marchetti. It is also known as a *flat dilatometer* or a *Marchetti dilatometer*. The device consists of a spade-shaped flat plate (tapered blade) measuring 240 mm (length) × 95 mm (width) × 15 mm (thickness) with an expandable steel-faced membrane (pressure cell) 60 mm in diameter on one face of the probe (Figure 1.25a). Figure 1.26 shows the Marchetti dilatometer along with its control panel and accessories. The dilatometer probe is inserted down by pushing or driving (if necessary). The CPT pushing equipment (or some other suitable device) can be used for pressing the dilatometer (Figure 1.25b).

In soils where the expected SPT-N value is greater than 35–40, it can be driven or pushed from the bottom of a predrilled borehole using the SPT drilling and testing equipment. After insertion of the probe to the depth of interest z, the DMT is conducted as in the following steps:

(1) Apply nitrogen gas pressure to the expandable membrane so as to move it 0.05 mm into the soil and record the required pressure (termed "liftoff" pressure) as p_1. The operator gets a signal at liftoff.

(2) Increase the probe pressure until the membrane expands $\Delta d = 1.1$ mm into the adjacent soil and record this pressure as p_2. This pressure is recorded after the operator receives a signal again.

(3) Decrease the pressure and take a reading when the membrane has returned to the liftoff position. Record this pressure as p_3.

(4) The probe is then pushed to the next depth position, which is normally from 150 to 300 mm further down into the ground, and the test is repeated. The probing is continued until the desired depth is reached.

According to Schmertmann (1988), the CPT and DMT are complementary tests. The CPT is a good way to evaluate soil strength, whereas the DMT assess the compressibility and in situ stresses. These three kinds of information form the basis for most foundation engineering analysis. In addition, the dilatometer tapered blade is most easily pressed into the ground using a conventional CPT rig, so it is a simple matter to conduct both CPT and DMT while mobilising only a minimum of equipment.

(Continued)

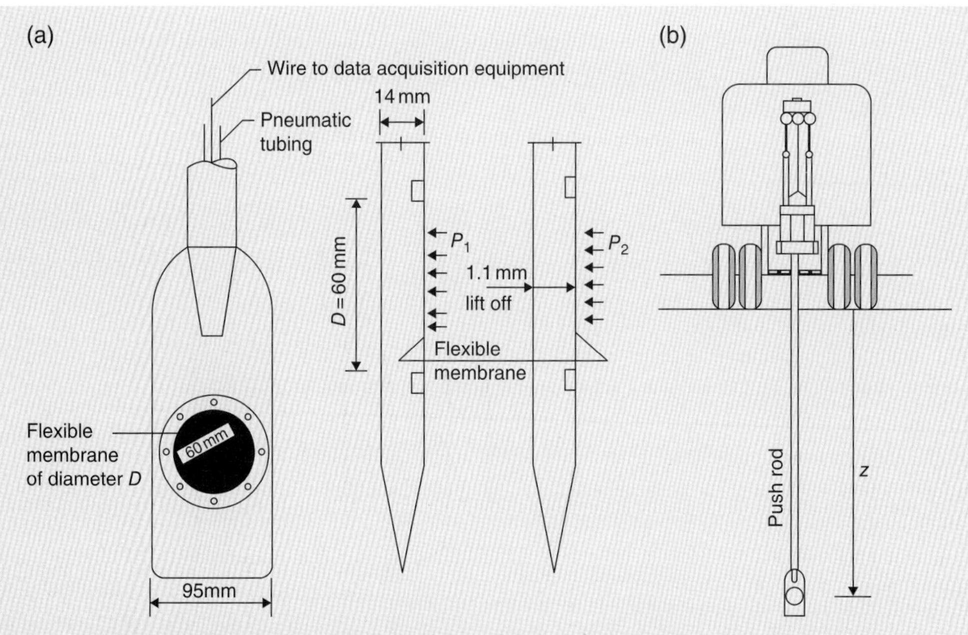

Figure 1.25 Schematic diagrams of a flat-plate dilatometer and the equipment inserter. (a) Marchetti dilatometer (after Marchetti, 1980). (b) The dilatometer pushed to depth z for test.

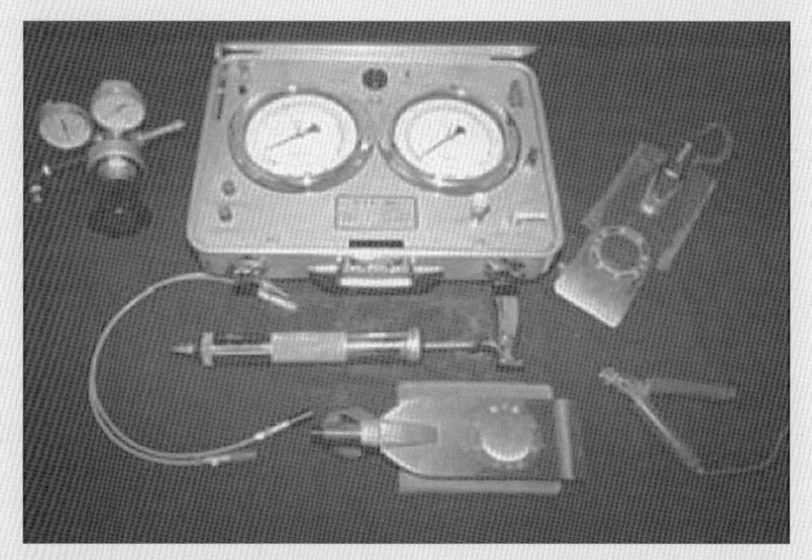

Figure 1.26 Dilatometer along with control panel and accessories (Courtesy of N.Sivakugan, James Kook University, Australia).

According to both Marchetti (1980) and Schmertmann (1986) the DMT can be used to obtain the full range of soil E_D, K_o, OCR, s_u, Ø and m_v parameters for both strength and compressibility analyses.

A given test data are reduced to obtain the following parameters:

(1) Dilatometer modulus E_D. According to Marchetti (1980),

$$\Delta d = \frac{2D(P_2 - P_1)}{\pi}\left(\frac{1-\mu^2}{E_s}\right) \tag{1.67}$$

where Δd = 1.1 mm, D = 60 mm (membrane diameter)

$$E_D = \frac{E_s}{1-\mu^2} = 34.7(P_2 - P_1) \tag{1.68}$$

where E_s, P_1 and P_2 are in kN/m^2

(2) The lateral stress index K_D is defined as

$$K_D = \frac{p_1 - u}{\sigma'_o} \tag{1.69}$$

(3) The material or deposit index I_D is defined as

$$I_D = \frac{p_2 - p_1}{p_2 - u} \tag{1.70}$$

The pore pressure u may be computed as the static pressure from groundwater table, which must be known or estimated. The effective overburden pressure $\sigma'_o = \gamma' z$ must be computed by estimating the soil unit weight or taking tube samples for a more direct determination.

The lateral stress index K_D is related to the coefficient of at-rest earth pressure K_o and therefore indirectly to the overconsolidation ratio OCR. Determination of K_o is approximate since the probe blade of finite thickness has been inserted into the soil. Figure 1.27 may be used to estimate K_o from K_D.

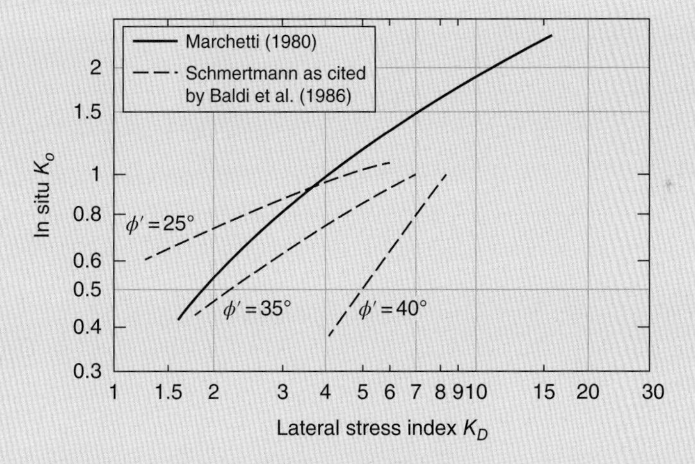

Figure 1.27 Correlations between K_D and K_o (after Baldi et al. 1986).

A typical data set of a DMT results might be as follows (Bowles, 2001):

Depth z, m	Rod push, kg	p_1, bar	p_2, bar	u, bar (100 kPa)
2.10	1400	2.97	14.53	0.21
2.40	1250	1.69	8.75	0.24
2.70	980	1.25	7.65	0.27

Here the depths shown are from 2.1 to 2.7 m. The probe push ranged from 1400 to 980 kg (the soil became softer) and, as should be obvious, values of p_2 are greater than p_1. With the groundwater table at the ground surface the static pore pressure is directly computed as u = 9.81z/100 bars.

(*Continued*)

The results of a dilatometer test are usually presented in form of graphs showing variation of p_1, p_2, K_D and I_D with depth.

According to Marchetti (1980):

$$K_o = \left(\frac{K_D}{B_D}\right)^{\alpha} - C_D \tag{1.71}$$

Where:

B_D	α	C_D	According to:
1.50	0.47	0.6	Marchetti ($K_D < 8$)
1.25–??	0.44 – 0.46	0 – 0.6	Others
7.40	0.54	0.0	—
2.00	0.47	0.6	(for sensitive clay)

$$OCR = (n K_D)^m \tag{1.72}$$

Where:

n	m	According to
0.5	1.56	Marchetti ($K_D < 8$; $I_D < 1.2$)
0.225–??	1.30 – 1.75	Others

$$\frac{c_u}{\sigma'_o} = 0.22 \,(\text{for normally consolidated clay}) \tag{1.73}$$

$$\left(\frac{c_u}{\sigma'_o}\right)_{OC} = \left(\frac{c_u}{\sigma'_o}\right)_{NC} (0.5K_D)^{1,25} \tag{1.74}$$

where OC = overconsolidated soil

NC = normally consolidated soil.

$$E_s = \left(1 - \mu_s^2\right) E_D \tag{1.75}$$

There are other relevant correlations using the results of DMT as follows:

Kamei and Iwasaki (1995):

$$c_u = 0.35\, \sigma'_o \,(0.47K_D)^{1.14} \tag{1.76}$$

Ricceri et al. (2002):

For ML and SP-SM soils,

$$\varnothing'_{ult} = 31 + \frac{K_D}{0.236 + 0.066K_D} \tag{1.77}$$

$$\emptyset'_{ult} = 28 + 14.6 \log K_D - 2.1(\log K_D)^2 \quad (1.78)$$

The material index I_D and the stress index K_D together are related to the soil type and to the soil consistency or density, as illustrated in Figure 1.28.

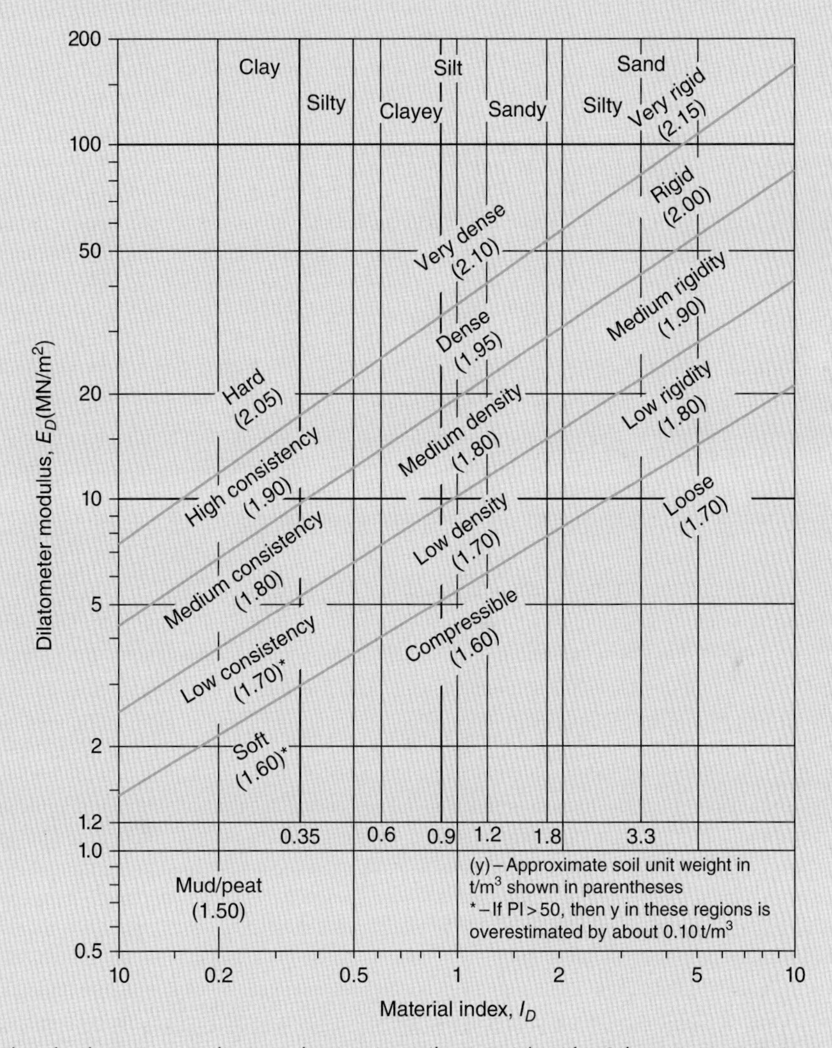

Figure 1.28 Chart for determining soil type and estimating soil unit weight (After Schmertmann, 1986).

Solution:

(a) Equation (1.71): $K_o = \left(\frac{K_D}{B_D}\right)^{\alpha} - C_D$

Equation (1.69): $K_D = \frac{p_1 - u}{\sigma'_o} = \frac{280 - 5 \times 10}{95} = 2.42$

According to Marchetti (1980):

$B_D = 1.5,\ \alpha = 0.47,\ C_D = 0.6,$ for $K_D < 8$

$$K_o = \left(\frac{2.42}{1.5}\right)^{0.47} - 0.6 = 0.65$$

(Continued)

(b) Equation (1.72): $OCR = (n\,K_D)^m$
According to Marchetti (1980):

$$n = 0.5,\ m = 1.56\ \text{for}\ K_D < 8;\ I_D < 1.2$$

$$OCR = (0.5 \times 2.42)^{1.56} = 1.35$$

(c) Equation (1.68):

$$E_D = \frac{E_s}{1-\mu^2} = 34.7(P_2 - P_1)$$

$$\frac{E_s}{1-\mu^2} = 34.7(350 - 280) = 2429\ \text{kPa}$$

$$E_s = (1-\mu^2)E_D = (1 - 0.35^2)(2429) = 2131\ kPa$$

Problem 1.22

A dilatometer test was conducted in a sand deposit at a depth of 6 m. The water table was located at a depth of 2 m below the ground surface. Given: sand γ_d = 14.5 kN/m^3, γ_{sat} = 19.8 kN/m^3 and contact stress p_1 = 260 kN/m^2. Estimate the soil friction angles $Ø'$ and $Ø'_{ult}$.

Solution:

Equation (1.69):

$$K_D = \frac{p_1 - u}{\sigma'_o} = \frac{260 - 4 \times 10}{2 \times 14.5 + 4(19.8 - 10)} = 3.23$$

Equation (1.77):

$$Ø' = 31 + \frac{K_D}{0.236 + 0.066K_D} = 31 + \frac{3.23}{0.236 + 0.066 \times 3.23} = 38.2°$$

Equation (1.78):

$$Ø'_{ult} = 28 + 14.6\log K_D - 2.1(\log K_D)^2$$

$$= 28 + 14.6\log 3.23 - 2.1(\log 3.23)^2$$

$$= 28 + 7.434 - 0.545\quad Ø'_{ult} = 34.9°$$

Problem 1.23

Below is a corrected DMT data set. For the individually assigned depth value, estimate E_D, K_D, I_D, K_o, and soil description. In case the soil classification suggests fine-grained soil, an estimate of c_u is required.

Depth z, m	Rod push, kg	p_1, bar	p_2, bar	u, bar*
2.10	1400	2.97	14.53	0.21
2.40	1250	1.69	8.75	0.24
2.70	980	1.25	7.65	0.27

*1 bar ≈ 100 kPa.

Solution:

DMT at 2.1 m depth: Both γ and μ are not known; they must be assumed. Therefore, Assume $\gamma_{sat} = 19\ kN/m^3$ and $\mu = 0.35$. From the u values it is clear that the groundwater table is located at the ground surface. Use 1 bar = 100 kN/m^2.

Equation (1.68):
$$E_D = \frac{E_S}{1-\mu^2} = 34.7(P_2 - P_1)$$
$$= 34.7(1453 - 297) = 40\,113\ kPa$$

Equation (1.69):
$$K_D = \frac{p_1 - u}{\sigma'_o} = \frac{297 - 21}{2.1(19 - 10)} = 14.6$$

Equation (1.70):
$$I_D = \frac{p_2 - p_1}{p_2 - u} = \frac{1453 - 297}{1453 - 21} = 0.81$$

In the Marchetti equation of K_o [i.e. Equation (1.71)] the assigned values for B_D, α and C_D factors may not be applicable because K_D is greater than 8. In such a case we may assign *other's* values for the factors and use Equation (1.71), or estimate K_o from the Schmertmann curves of Figure 1.27. Let us try both possibilities:

(1) Using other values for B_D, α and C_D factors [refer to the table of the factors which belong to Equation (1.71)]. Assume average values as follows:

$$B_D = \frac{1.25 + 7.4 + 2.0}{3} = 3.55; \quad \alpha = \frac{0.45 + 0.54 + 0.47}{3} = 0.49; \quad C_D = \frac{0.3 + 0 + 0.6}{3} = 0.3$$

Equation (1.71):
$$K_o = \left(\frac{K_D}{B_D}\right)^{\alpha} - C_D = \left(\frac{14.6}{3.55}\right)^{0.49} - 0.3 = 1.7$$

(2) Using Figure 1.27. For $K_D = 14.60$ and a Ø value between 35° and 40°, the Schmertmann curves give a value for $K_o \cong 1.5$.

Using the average of the estimated two values, obtain $K_o = 1.6$.

The soil is considered overconsolidated since the upper limit for K_o in normally consolidated soils is $\cong 1$. Using the Schmertmann chart of Figure 1.28, for the estimated E_D and I_D values, the soil is very dense clayey silt. Therefore, the soil may be described as *very dense overconsolidated clayey silt.*

DMT at 2.4 m depth:

Equation (1.68):
$$E_D = \frac{E_S}{1-\mu^2} = 34.7(P_2 - P_1) = 34.7(875 - 169)$$
$$= 24\,498\ kPa$$

Equation (1.69):
$$K_D = \frac{p_1 - u}{\sigma'_o} = \frac{169 - 24}{2.4(19 - 10)} = 6.71$$

Equation (1.70):
$$I_D = \frac{p_2 - p_1}{p_2 - u} = \frac{875 - 169}{875 - 24} = 0.83$$

Equation (1.71):
$$K_o = \left(\frac{K_D}{B_D}\right)^{\alpha} - C_D; \; K_D = 6.71 < 8$$
$$= \left(\frac{6.71}{1.5}\right)^{0.47} - 0.6 = 1.42 > 1 \rightarrow \text{overconsolidated soil.}$$

Using the Schmertmann chart of Figure 1.28, for the estimated E_D and I_D values, the soil is dense clayey silt. Therefore, the soil may be described as *dense overconsolidated clayey silt.*

DMT at 2.7 m depth:

Equation (1.68):
$$E_D = \frac{E_S}{1-\mu^2} = 34.7 = (P_2 - P_1) = 34.7(765 - 125)$$
$$= 22\,208\ kPa$$

(Continued)

Equation (1.69): $K_D = \dfrac{p_1 - u}{\sigma'_o} = \dfrac{125-27}{2.7(19-10)} = 4.03$

Equation (1.70): $I_D = \dfrac{p_2 - p_1}{p_2 - u} = \dfrac{765-125}{765-27} = 0.87$

Equation (1.71): $K_o = \left(\dfrac{K_D}{B_D}\right)^{\alpha} - C_D;\ K_D = 4.03 < 8$

$$= \left(\frac{4.03}{1.5}\right)^{0.47} - 0.6 = 1.0 \rightarrow \text{slightly overconsolidated soil.}$$

Using the Schmertmann chart of Figure 1.23, for the estimated E_D and I_D values, the soil is dense clayey silt. Therefore, the soil may be described as *dense slightly overconsolidated clayey silt.*

Summary of results:

z, m	E_D	K_D	I_D	K_o	Soil description
2.10	40113	14.60	0.81	1.80	very dense overconsolidated clayey silt
2.40	24498	6.71	0.83	1.42	dense overconsolidated clayey silt
2.70	22208	4.03	0.87	1.00	dense slightly overconsolidated clayey silt

Problem 1.24

The results of a refraction survey at a site are given in Table 1.13. Determine the thickness and the *P*-wave velocity of the materials encountered.

Table 1.13 Refraction survey site results.

Distance from the source of disturbance (m)	Time of first arrival of *P*-waves (s × 10^{-3})
10.0	50.0
20.0	100.0
24.0	118.0
30.49	148.2
45.73	174.2
60.98	202.8
76.22	228.6
91.46	256.7

Discussion:

In the last five decades, even earlier, the trend for increased knowledge of subsurface conditions has placed emphasis upon methods which save time and money in obtaining this information. This is why *geophysical methods* have been developed. Among the several developed and used, the *seismic refraction* and *electrical resistivity* methods have received the widest use. The basic philosophy of geophysical methods of exploration is that changes in subsurface conditions near the surface of the earth can be detected by the nature of the physical characteristics of the material. Generally speaking, these methods are particularly suited for large sites and for areas inaccessible to other forms of equipment when the geology of these areas is well known, that is, when the materials to be encountered are known but the depths at which they lie are to be determined. The main uses of seismic refraction and resistivity methods in foundation exploration are:

(1) To determine the depths and homogeneity of the *soil* and *rock layers*. The methods are accurate enough; however, classification and identification of the soil and rock type cannot always be accurately accomplished. The engineering properties of soils, to some extent, can be detected; however, the results would not be accurate enough for direct use in the foundations design.
(2) To detect subsurface anomalies such as faults, caves, cavities and sinkholes. Caves and sinkholes, however, can best be located by resistivity methods.
(3) To aid in planning the initial boring plan. The boring plan will require less change and a better estimate of the cost and time required for detailed exploration can be made if a geophysical investigation is conducted first.
(4) To complement borings, so a more complete soil profile can be drawn. Closely spaced borings are costly and may not find trouble areas that may often be easily detected by a geophysical investigation. Complementary use of geophysical data and boring can save time and money during all phases of design and construction.

Seismic Refraction Method (ASTM D 5777): In this method impact or shock waves are generated either by the detonation of explosives or by striking a metal plate with a large hammer at a certain location on the ground surface, such as point *A* in Figure 1.29. The equipment consists of one or more sensitive vibration transducers, called geophones, installed at a number of points in a straight line with increasing distances from the source of wave generation, and an extremely accurate time-measuring device called seismograph. The length of the line of points should be three to five times the required depth of investigation. A circuit between the detonator or hammer and the seismograph starts the timing mechanism at the instant of detonation or impact. The geophone is also connected electrically to the seismograph. When the first wave reaches the geophone the timing mechanism stops and the time interval is recorded in milliseconds.

Seismic waves bend (refract) with different velocities when crossing different strata interface. Using this wave property and the principles of wave propagation in layered systems, different strata thickness can be determined.

An impact on the ground surface creates different types of stress wave, such as the compression or *P*-wave and shear or *S*-wave. The *P*-waves travel faster than the *S*-waves; hence the first arrival of disturbance waves will be related to *P*-waves in various layers. *P*-wave and *S*-wave velocities range in common geotechnical materials are given in Tables 1.14 and 1.15, respectively.

The velocity of *P*-waves in a medium is

$$v_P = \sqrt{\frac{\lambda + 2G_s'}{\rho}} = \sqrt{\frac{E_s(1-\mu_s)}{\frac{\gamma}{g}(1-2\mu_s)(1+\mu_s)}} \tag{1.79}$$

Where λ = Lame's constant $= 2\mu_s G_s'/(1-2\mu_s)$
ρ = mass density of medium = γ/g
G_s' = dynamic shear modulus of medium $= E_s/2(1+\mu_s)$
g = acceleration due to gravity (9.81 m/s^2)
E_s = modulus of elasticity of medium
γ = unit weight of medium
μ_s = Poisson's ratio of medium

The velocity v_p and thickness Z of various layers are determined as follows:

- Times of first wave arrival t_1, t_2, t_3 ... for distances x_1, x_2, x_3 ... from impact source are recorded.
- A plot of time t versus distance x is prepared (Figure 1.30).
- Slope of the lines l_1, l_2, l_3 ... are determined using slope of $l_1 = \frac{1}{v_1}$, slope of $l_2 = \frac{1}{v_2}$, slope of $l_3 = \frac{1}{v_3}$, ...; velocities v_1, v_2, v_3, ..., are the *P*-wave velocities in layers I, II, III; ..., respectively (Figure 1.29).

(*Continued*)

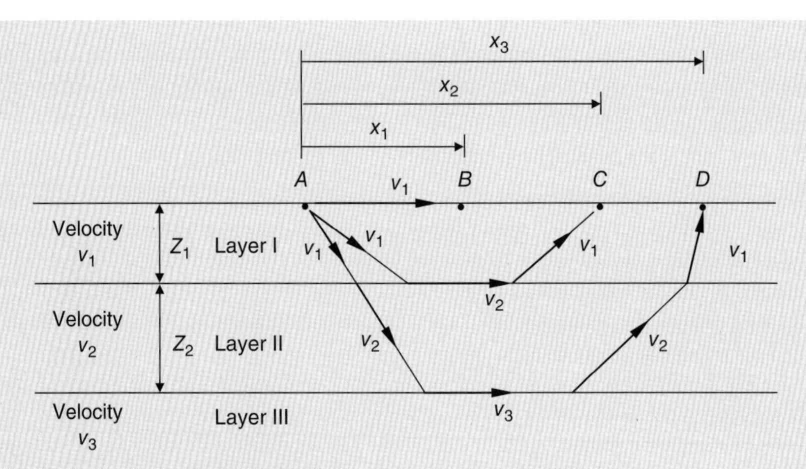

Figure 1.29 Refraction of seismic waves in subsurface layers.

- Thickness Z_1 of the layer I is determined using

$$Z_1 = \frac{1}{2}(x_c)\left[\frac{v_2 - v_1}{v_2 + v_1}\right]^{\frac{1}{2}} \tag{1.80}$$

where x_c is a distance, as shown in Figure 1.30.

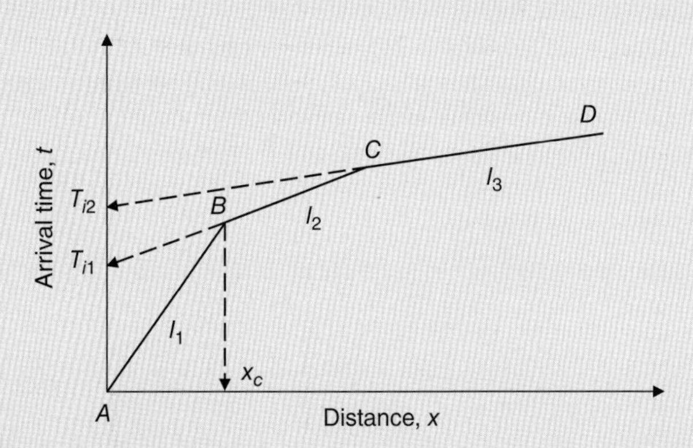

Figure 1.30 A graph of time *t* versus distance *x*.

- Thickness Z_2 of the layer II is determined as

$$Z_2 = \frac{1}{2}\left[T_{i2} - 2Z_1\frac{\sqrt{v_3^2 - v_1^2}}{v_3 v_1}\right]\frac{v_3 v_2}{\sqrt{v_3^2 - v_2^2}} \tag{1.81}$$

where T_{i2} is the time intersept of line l_3 extended backwards, as shown in Figure 1.30.

The *P*-wave velocities in various strata indicate the types of soil and rock that are present below the ground surface. Table 1.14 gives *P*-wave velocities range in different types of soil and rock at shallow depths.

Table 1.14 Range of *P*-wave velocity in various soils and rocks.

Type of material	*P*-wave velocity (m/s)
Soil	
Sand, dry silt and fine-grained topsoil	200–1000
Alluvium	500–2000
Compacted clays, clayey gravel and dense clayey sand	1000–2500
Loess	250–750
Rock	
Slate and shale	2500–5000
Sandstone	1500–5000
Granite	4000–6000
Sound limestone	5000–10 000

Seismic Refraction Problems and Limitations: Most of the problems that arise in the use of seismic refraction are due to field conditions. Seismic refraction theory assumes the following conditions exist:

(1) The soil and rock are to be considered homogeneous, isotropic and elastic material.
(2) No decrease in velocity with depth occurs; that is $v_1 < v_2 < v_3$. ...
(3) Contrast in elastic properties of adjacent layers exists.

Soils, however, are neither homogeneous nor isotropic. Anomalies or discontinuities such as cavities, faults, boulders, and sinkholes can introduce errors if careful attention is not given to the interpretation of the seismic data.

Most soil and rock deposits increase in density with depth which is in favor of the second assumption. If not, the velocity of the shock waves will not increase with depth and errors will be introduced. When velocities of adjacent strata are within about 60–90 m/s of each other, determination of depth to the interface is virtually impossible from seismic data. Layers of clay or shale (of low velocity) underlying limestone (of high velocity) will not be recognisable on a time distance plot. The thickness of the harder material will appear to include the thickness of the softer layer. This condition, known as the "hidden layer" problem, can be corrected to some extent. If a hidden layer problem is suspected, the maximum errors in depth determination can be detected using nomographs.

Complications also arise if the elastic properties of the adjacent strata do not contrast enough to define the interface. A layer of hard clay lying on soft shale or sandstone is an example of this problem.

When a soil is saturated below water table, the *P*-wave velocity may be deceptive. *P*-waves can travel with velocity of about 1500 m/s through water. For dry loose soil the velocity may be well below 1500 m/s. if the presence of the groundwater has not been detected, the *P*-wave velocity may be erroneously interpreted to indicate a stronger material (e.g. sandstone) than is actually present in the site.

In general, geophysical interpretation should always be verified by the results obtained from borings.

Cross-hole Seismic Method: The cross-hole seismic technique is an in situ seismic test conducted so that shear wave velocity v_s and dynamic shear modulus G_s' of soil or rock can be determined effectively (Stokoe and Woods, 1972).

In this method (Bowles, 2001), two boreholes a known distance apart are drilled to some depth, preferably on each side of the proposed foundation location so that the shear wave can be measured between the two boreholes and across the base zone. At a depth of about *B* a sensor device is located in the side or bottom of one hole and a shock-producing device (or small blast) in the other (Figure 1.31a). A trigger is supplied with the shock so that the

(*Continued*)

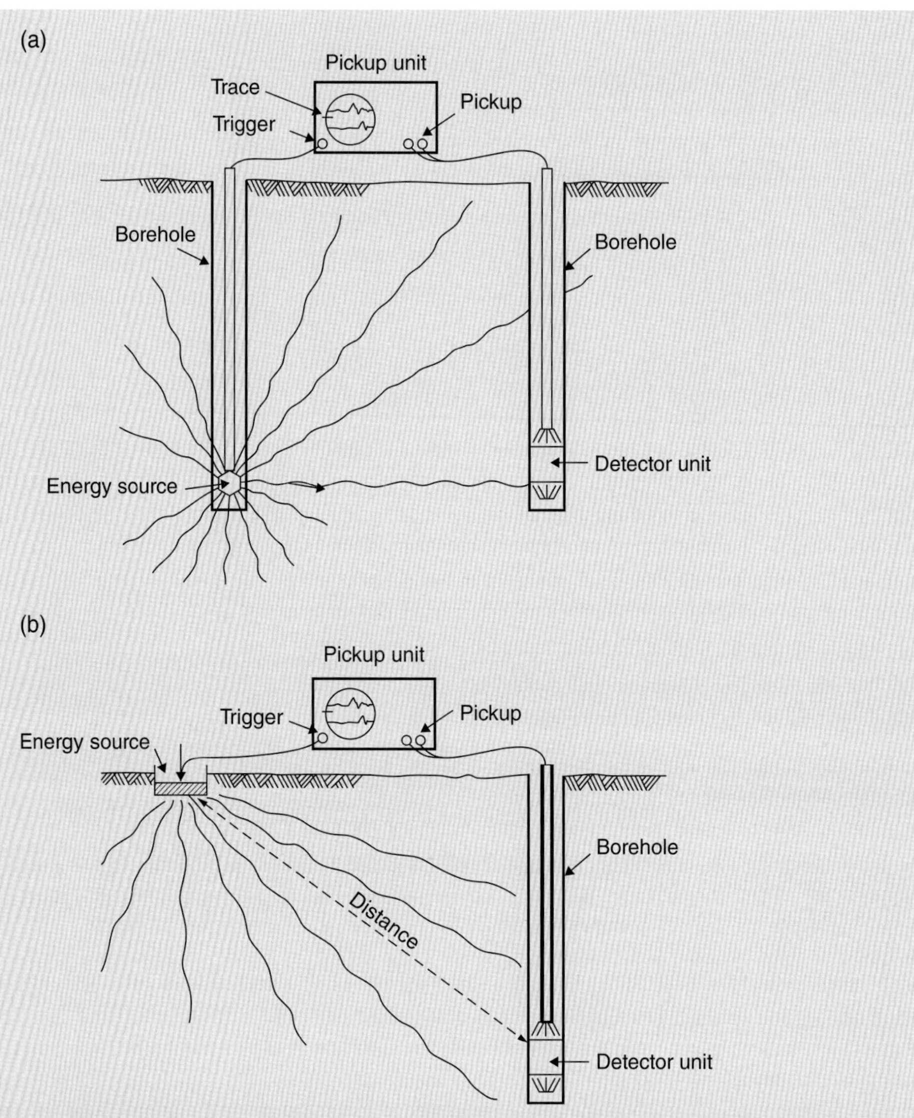

Figure 1.31 Two seismic methods for obtaining dynamic shear modulus G_s'. (a) Cross-hole method, (b) Down-hole method (from Bowles, 2001).

time for the induced shear wave can be observed at the pickup unit. The time of travel T_h of the known distance D_h between the two holes gives the shear wave velocity v_s as

$$v_s = \frac{D_h}{T_h} \tag{1.82}$$

The *dynamic shear modulus* of the material at the depth at which the test is taken can be determined from the relation

$$G_s' = \frac{v_s^2 \gamma}{g} \quad \text{or} \quad v_s = \sqrt{\frac{G_s'}{(\gamma/g)}} \tag{1.83}$$

Down-hole Seismic Method: The down-hole method is similar to the cross-hole but has the advantage of only requiring one boring, as shown in Figure 1.31b. In this method, the hole is drilled, and a shock device is located a known distance away. A shock detector is located at some known depth in the hole and a shock applied. As with the cross-hole method, we can measure the time T_h for arrival of the shear wave and, by computing the diagonal side of the triangle, obtain the travel distance D_h. The detector device is then placed at a greater depth and the test repeated until a reasonably average value of v_s is obtained.

Table 1.15 *S*-wave velocities range in common geotechnical materials at shallow depths.

Type of material	*S*-wave velocity (m/s)
Hard rocks (e.g. metamorphic)	1400+
Firm to hard rocks (e.g. igneous, conglomerates, competent sedimentary)	700–1400
Gravelly soils and soft rocks (e.g. sandstone, shale, soils with > 20% gravel)	375–700
Stiff clays and sandy soils	200–375
Soft soils (e.g. loose submerged fills and soft clays)	100–200
Very soft soils (e.g. marshland, reclaimed soil)	50–100

Hoar and Stokoe (1978) discuss in some detail precautionary measures to take in making either of the above mentioned two tests so that the results can justify the test effort (Bowles, 2001).

Electrical Resistivity Method: There are two types of the electrical resistivity methods:

(1) The electrical resistivity sounding method. It is used when the variation of resistivity of the subsurface materials with depth is required at a predetermined location; resulting in rough estimates of types and depths of strata. In other words, the method can indicate subsurface variation when a hard layer overlies a soft layer or vice versa. It can also be used to locate the groundwater table.

(2) The electrical resistivity mapping (profiling) method. Here, the resistivity soundings are conducted along different profile lines across the area such that contours of resistivity variation with depth can be plotted.

The methods depend on differences in the electrical resistance of different types of soil and rock. Resistivity of a material depends upon type of material, its water content and the concentration of dissolved salts in the pore water. The mineral particles of a soil are poor conductors of electric current; they are of high resistivity. Resistivity of a soil decreases (conductivity increases) as both the water content and pore water salts concentration increase. Approximate ranges of resistivity value for different types of soil and rock are given in Table 1.16.

Table 1.16 Representative values of resistivity.

Material	Resistivity (ohm.m)
Sand	500–1500
Clays, saturated silt	0–100
Clayey sand	200–500
Gravel	1500–40 000
Weathered rock	1500–2500
Sound rock	> 1500

(Continued)

Resistivity of any conducting material having a length l and cross section area A is given by the equation:

$$\rho = \frac{RA}{l} \tag{1.84}$$

where ρ = electrical resistivity (ohm.m)

$R = \text{electrical resistance (ohm)} = \frac{\text{voltage drop } E}{\text{current } I}$ (Ohm's law)

The in situ common electrical sounding procedure involves driving four electrodes, which are usually in the form of metal spikes, into the ground at equal distances L apart in a straight line (Figure 1.32a). The two outer electrodes are known as current electrodes, and the inner electrodes are called potential electrodes. The current I, usually 50–100 mA, from a battery, flows through the soil between the two current electrodes producing an electrical field within the soil. The potential (voltage) drop E is then measured between the two potential electrodes. The mean (apparent) resistivity ρ is given by the equation:

$$\rho = \frac{2\pi L E}{I} \tag{1.85}$$

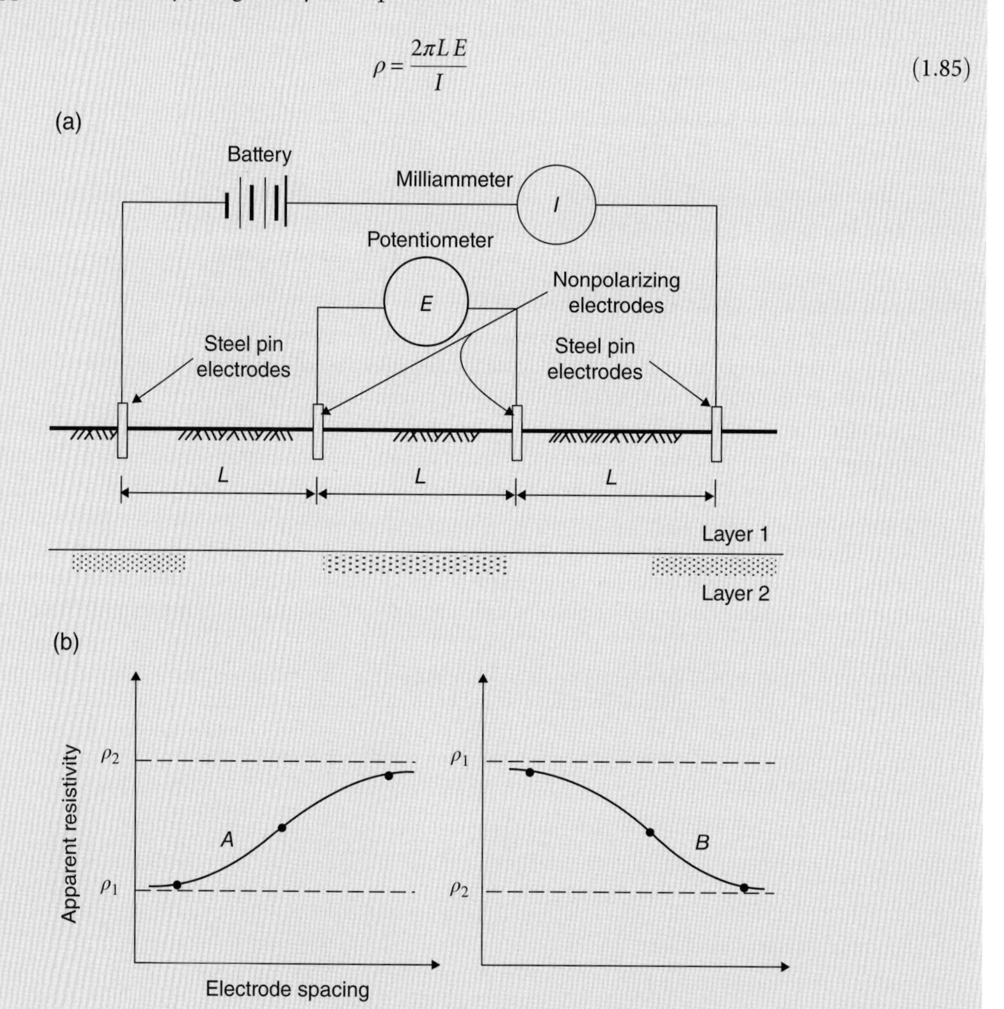

Figure 1.32 (a) Electrical resistivity method schematically illustrated; (b) apparent resistivity versus electrode spacing (redrawn from Knappett and Craig, 2012).

Equation (1.85) gives the mean resistivity up to a depth equals the distance L below the ground surface, as the depth of current penetration below the ground surface is approximately equal to the spacing L of the two potential electrodes.

A series of readings are taken, the (equal) spacing of the electrodes being increased for each successive reading; however the central position of the four electrodes remains unchanged. As L is increased, the resistivity ρ is influenced by a greater depth of soil. The distance L is thus gradually increases to a distance equal to the required depth of exploration. The maximum depth is usually about 30 m or so. If ρ decreases with increasing L a stratum of lower resistivity, such as a saturated clay layer, is beginning to influence the readings. If ρ increases with increasing L it can be concluded that an underlying stratum of higher resistivity, such as a gravel layer, is beginning to influence the readings.

The mean resistivity ρ is plotted against the electrode spacing L, preferably on log-log paper. Characteristic curves for a two-layer structure are illustrated in Figure 1.32b. For curve A the resistivity of layer 1 is lower than that of layer 2; for curve B the resistivity of layer 1 is higher than that of layer 2. The curves become asymptotic to lines representing the true resistivity ρ_1 and ρ_2 of the respective layers. Approximate layer thickness can be obtained by comparing the observed curve of ρ versus L with a set of standard curves, or, by using other methods of interpretation, such as the method illustrated by the graph of Figure 1.33, which relates the sum of the apparent resistivity $\Sigma\rho$ to the electrode spacing. In this figure, the slopes ρ_1 and ρ_2 are actual resistivity of layers 1 and 2, respectively.

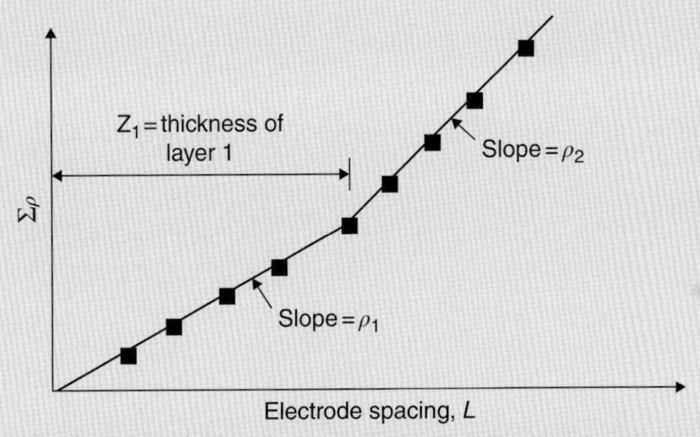

Figure 1.33 An approximate method for determining resistivity and thickness of subsurface layers.

Resistivity Problems and Limitations: The theoretical basis for electrical resistivity assumes:

(1) The soil or rock is homogeneous and isotropic.
(2) Uniform resistivity exists.
(3) The soil layers are parallel to each other and to ground surface.

Unfortunately, the assumed soil conditions seldom exist in the field because soil and rock are not homogeneous and isotropic. Therefore the user must be aware of the effects on data caused by anomalies, such as faults, folded layers, caves, sinkholes or intermixing of soil and rock in the layers. Also, buried pipelines, underground cables, tunnels or any buried metallic materials will cause anomalous readings.

Nonparallel layers cause bending or warping of the current flow lines causing erroneous resistivity readings. If the ground surface and the interfaces are not parallel the exact depth below centre of the fixed location (centre of the spread) may not be found due to varying layer thickness.

Materials differ in their electrical resistivity. The resistivity of materials does not change due to texture only, but also because of the changes in the moisture and electrolytic content. As a result, considerable overlapping of the resistivity of various materials will exist. Overlapping makes interpretation of resistivity data very difficult in some cases.

(Continued)

It is noteworthy: (1) there is no masked layer problem with electrical resistivity methods as there is with seismic methods, (2) an increase in material density is not necessary, as is the case with seismic refraction, (3) as the resistivity measures change in texture rather than change in strength and consolidation; it is generally used for locating coarse-grained soil deposits within fine-grained soil deposits, (4) the resistivity methods are not considered as reliable as the seismic methods, (5) as it is the case with seismic methods, the resistivity data should also be supplemented by few borings, (6) since seismic and resistivity methods measure different properties, they can be used effectively to complement each other.

Solution:

Plot the times of first arrival of *P* waves against the distance of geophone from the source of disturbance, as shown in the scheme below.

From the plot:

Slope of segment $ab = \frac{1}{v_1} = \frac{(148.2 - 50.0) \times 10^{-3}}{30.49 - 10} = \frac{0.098}{20.49}$; hence,

$$v_1 = 209\ m/s$$

Slope of segment $bc = \frac{1}{v_2} = \frac{(228.6 - 174.2) \times 10^{-3}}{76.22 - 45.73} = \frac{0.054}{30.49}$; hence

$$v_2 = 565\ m/s$$

From the plot: $x_c = 30.5$ m

Equation (1.80): $Z_1 = \frac{1}{2}(x_c)\left[\frac{v_2 - v_1}{v_2 + v_1}\right]^{\frac{1}{2}} = \frac{1}{2} \times 30.5\left[\frac{565 - 209}{565 + 209}\right]^{\frac{1}{2}} = 10.34\,\text{m}$

Thickness of the top layer = *10.34 m*

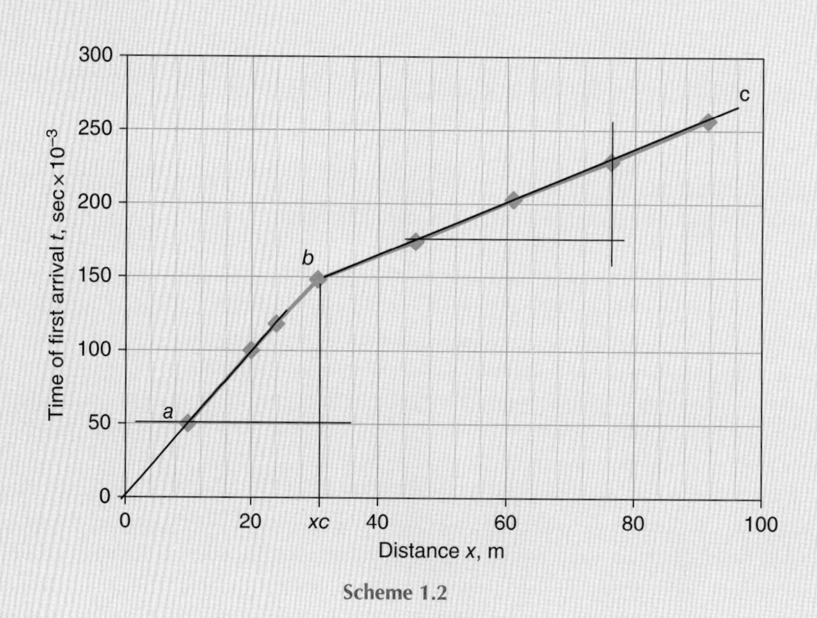

Scheme 1.2

Problem 1.25

The plot in the scheme below represents the results of a seismic refraction survey.

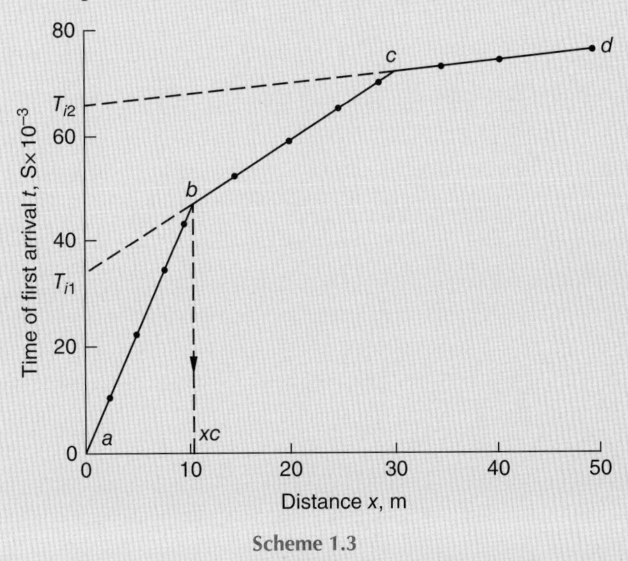

Scheme 1.3

Determine the P-wave velocities and the thickness of the material encountered in the third stratum.

Solution:
From the plot:

Slope of segment $ab = \dfrac{1}{v_1} = \dfrac{47 \times 10^{-3}}{10.5}$, $v_1 = 223.4\,m/s$ (*first layer*)

Slope of segment $ab = \dfrac{1}{v_2} = \dfrac{25 \times 10^{-3}}{19.5}$, $v_2 = 780.0\,m/s$ (*second layer*)

Slope of segment $cd = \dfrac{1}{v_3} = \dfrac{4.5 \times 10^{-3}}{20}$, $v_3 = 4444.4\,m/s$ (*third layer*)

Comparing these velocity values with those given in Table 1.14, we find that the third layer is *rock*.
From the plot: $x_c = 10.5$ m

Equation (1.80):
$$Z_1 = \frac{1}{2}(x_c)\left[\frac{v_2 - v_1}{v_2 + v_1}\right]^{\frac{1}{2}}$$

$$\text{Thickness of the first layer} = Z_1 = \frac{1}{2}x_c\left[\frac{v_2 - v_1}{v_2 + v_1}\right]^{\frac{1}{2}}$$

$$= \frac{1}{2} \times 10.5\left[\frac{780.0 - 223.4}{780.0 + 223.4}\right]^{\frac{1}{2}} = 3.91\,m$$

Equation (1.81):
$$Z_2 = \frac{1}{2}\left[T_{i2} - 2Z_1\frac{\sqrt{v_3^2 - v_1^2}}{v_3 v_1}\right]\frac{v_3 v_2}{\sqrt{v_3^2 - v_2^2}}$$

From the plot: $T_{i2} = 66 \times 10^{-3}$

(*Continued*)

Thickness of the second layer is

$$Z_2 = \frac{1}{2}\left[66 \times 10^{-3} - 2(3.91)\frac{\sqrt{4444.4^2 - 223.4^2}}{4444.4 \times 223.4}\right]\frac{4444.4 \times 780}{\sqrt{4444.4^2 - 780^2}} = 12.29\,m$$

Surface of the rock layer is located at the depth $D = Z_1 + Z_2 = 16.2\ m$

Problem 1.26

Using the data of Problem 1.25 determine the dynamic shear modulus G'_s and E_s of the material encountered in the top layer. Assume the unit weight γ of the material equals 16 kN/m^3, and the Poisson's ratio $\mu_s = 0.35$.

Solution:

Equation (1.79):

$$v_P = \sqrt{\frac{\lambda + 2G'_s}{\rho}}$$

$$\lambda = \text{Lame's constant} = \frac{G'_s}{(1 - 2\mu_s)};\ \rho = \frac{\gamma}{g} = \frac{\gamma}{9.81}$$

$$v_p = \sqrt{\frac{G'_s}{\frac{\gamma}{9.81}(1 - 2\mu_s)} + \frac{2G'_s}{\frac{\gamma}{9.81}}} = \sqrt{\frac{G'_s}{\frac{16}{9.81}(1 - 2 \times 0.35)} + \frac{2G'_s}{\frac{16}{9.81}}} = \sqrt{3.27G'_s}$$

$$v_p = 223.4\,\text{m/s (from Solution of Problem 1.25)}$$

$$223.4 = \sqrt{3.27G'_s}$$

$$G'_s = \frac{223.4^2}{3,\ 27} = 15\,262\,kN/m^2$$

$$G'_s = \frac{E_s}{2(1 + \mu_s)}$$

$$E_s = 2G'_s(1 + \mu_s) = 2 \times\ 15262\,(1 + 0.35) = 41\,207\,kN/m^2$$

Problem 1.27

A seismic refraction test yielded the following data:
$v_1 = 610$ m/s; $v_2 = 4268$ m/s; $x_c = 91.5$ m
The profile was a two-layer system and no v_3 was detected.
Determine the thickness of the top layer.

Solution:

Equation (1.80):

$$Z_1 = \frac{1}{2}(x_c)\left[\frac{v_2 - v_1}{v_2 + v_1}\right]^{\frac{1}{2}}$$

Thickness of the top layer

$$Z_1 = \frac{1}{2}x_c\left[\frac{v_2 - v_1}{v_2 + v_1}\right]^{\frac{1}{2}}$$

$$= \frac{1}{2} \times 91.5\left[\frac{4268 - 610}{4268 + 610}\right]^{\frac{1}{2}} = 40\,m$$

References

AASHTO (1988), *Manual on Subsurface Investigations,* American Association of State Highway and transportation officials, Developed by the Subcommittee on Materials, Washington, D.C.

ASCE (1972), "Subsurface Investigation for Design and Construction of Foundations of Buildings," ASCE *Journal of the Soil Mechanics and Foundations Division*, Vol. **98** Nos SM5, SM6, SM7 and SM8.

ASTM (2009), *Annual Book of ASTM Standards*, Vol. 04.08, American Society for Testing and Materials, West Conshohocken, PA.

Anagnostopoulos, A., Koukis, G., Sabtakakis, N. and Tsiambaos, G. (2003), "Empirical Correlations of Soil Parameters Based on Cone Penetration Tests (CPT) for Greek Soils," *Geotechnical and Geological Engineering*, Vol. **21**, No. 4, pp. 377–387.

Baban, Tharwat M. (1992), "The Required Depth of Soil Exploration for a Building Foundation." *The Scientific Journal of Salahaddin University*, Vol. **5**, No. 3, pp. 77–87.

Baguelin, F., Jezequel, J. F. and Shields, D. H. (1978), *The Pressuremeter and Foundation Engineering*, Trans Tech Publications, Clousthal, Germany.

Baldi, G., Bellotti, R., Gionna, V. and Jamiolkowski, M. (1982), "Design Parameters for Sands from CPT," *Proceedings, Second European Symposium on Penetration Testing, Amsterdam*, Vol. **2**, pp. 425–438.

Baldi, G., et al. (1986), "Flat Dilatometer Tests in Calibration Chambers," *14th PSC*, ASCE, pp. 431–446.

Bjerrum, L. (1972), "Embankments on Soft Ground," *Proceedings of the Specialty Conference, ASCE*, Vol. **2**, pp. 1–54.

BOCA (1996), *National Building Code*, Building Officials and Code Administrators International, Inc., Country Club Hills, IL.

Bowles, J. E. (2001), *Foundation Analysis and Design,* 5th edn, McGraw-Hill, New York.

Briaud, J. L. (1989), "The Pressuremeter Test for Highway Applications," *Report FDHWA-IP-008,* Federal Highway Administration, Washington, D.C., 148.

Briaud, J. L. and M. Gambin (1984), "Suggested Practice for Drilling Boreholes for Pressuremeter Testing," *GTJ*, ASTM, Vol. **7**, No. 1, March, pp. 36–40.

BSI (1999), *BS 5930: 1999 Code of Practice for Site Investigations* – Clause 12.6, British Standards Institution, London.

BSI (2007), *BS EN 1997-2: 2007 Ground Investigation and Testing* – Clause B.3, British Standards Institution, London.

BSI (2006), *Geotechnical Investigation and Testing – Sampling Methods and Groundwater Measurements. Technical Principles for Execution BS EN ISO 22475-1:2006,* British Standards Institution, London.

Chandler, R. J. (1988), "The In situ Measurement of the Undrained Shear Strength of Clays Using the Field Vane," *ASTM STP No.* **1014**, pp. 13–44.

Clarke, B. G. (1995), *Pressuremeters in Geotechnical Design,* International Thomson Publishing/BiTech Publishers, Vancouver.

Clayton, C. R. I. (1990), "SPT Energy Transmission: Theory, Measurement and Significance," *Ground Engineering,* Vol. **23**, No. 10, pp. 35–43.

Coduto, Donald P. (2001), *Foundation Design: Principles and Practices,* 2nd edn, Prentice-Hall, Inc., New Jersey.

Craig, R. F. (2004), *Soil Mechanics,* 7th edn, Chapman and Hall, London.

Cubrinovski, M. and Ishihara, K. (1999), "Empirical Correlations between SPT N-Values and Relative Density for Sandy Soils," *Soils and Foundations,* Vol. **39**, No. 5, pp. 61–92.

Das, Braja M. (2011), *Principles of Foundation Engineering,* 7th edn, CENGAGE Learning, United States.

Deere, D. U. (1963), "Technical Description of Rock Cores for Engineering Purposes," *Felsmechanik und Ingenieurgeologie,* Vol. **1**, No. 1, pp. 16–22.

FHWA (2002), *Subsurface Investigations-Geotechnical Site Characterization,* Publication No. FHWA NHI-01-031, Federal Highway Administration, Washington, D.C.

Gibbs, H. J. and Holtz, W. G. (1960), "Shear Strength of Cohesive Soils," *First PSC ASCE*, pp. 33–162.

Hansbo, S. (1957), *A New Approach to the Determination of the Shear Strength of Clay by the Fall Cone Test,* Swedish Geotechnical Institute, Report No. **114**.

Hara, A., Ohata, T. and Niwa, M. (1971), "Shear Modulus and Shear Strength of Cohesive Soils," *Soils and Foundations,* Vol. **14**, No. 3, pp. 1–12.

Hatanaka, M. and Ushida, A. (1996), "Empirical Correlation between Penetration Resistance and Internal Friction Angle of Sandy Soils," *Soils and Foundations,* Vol. **36**, No. 4, pp. 1–10.

Hoar, R. J. and K. H. Stokoe II (1978), "Generation and Measurement of Shear Waves In situ," *ASTM STP No.* **645**, pp. 3–29.

Housel, W. S. (1929), "a Practical Method for the Selection of Foundations Based on Fundamental Research in Soil Mechanics," *University of Michigan Engineering Research Bull.* **13**, Ann Arbour, Mich., October.

Hughes, J. M. O., Worth, C. P. and Windle, D. (1977), "Pressuremeter Tests in Sands," *Geotechnique,* **27**, pp. 455–477.

Ismael, N. F. and A. M. Jeragh (1986) "Static Cone Tests and Settlement of Calcareous Desert Sands," *CGJ*, Vol. **23**, No. 3, Aug, pp. 297–303.

Jamiolkowski, M., Ladd, C. C., Germaine, J. T. et al. (1985), "New Developments in Field and Laboratory Testing of Soils," *11th ICSMFE*, Vol. **1**, pp. 57–153.

Kamie, T. and Iwasaki, K. (1995), "Evaluation of Undrained Shear Strength of Cohesive Soils Using a Flat Dilatometer," *Soils and Foundations,* Vol. **35**, No. 2, pp. 111–116.

Knappett, J. A. and Craig, R. F. (2012), *Soil Mechanics*, 8th edn, Spon Press, Abingdon, United Kingdom.

Kulhawy, F. H. and Mayne, P. W. (1990), *Manual on Engineering Soil Properties for Foundation Design*, Electric Power Research Institute, Palo Alto, California.

Lancellotta, R. (1983), *Analisi di Affidabilita in Ingegneria Geotecnica*, Atti Instituto Scienza Construzioni, No. 625, Politecnico di Torino.

Larson, R. (1980), "Undrained Shear Strength in Stability Calculation of Embankments and Foundations on Clay," *Canadian Geotechnical Journal*, Vol. **17**, pp. 591–602.

Lee, L., Salgado R. and Carraro, A. H. (2004), "Stiffness Degradation and Shear Strength of Silty Sand," *Canadian Geotechnical Journal*, Vol. **41**, No. 5, pp. 831–843.

Liao, S. S. C. and Whitman, R. V. (1986), "Overburden Correction Factors for SPT in Sand," *Journal of Geotechnical Engineering Division*, ASCE, Vol. **112**, No. 3, March, pp. 373–377.

Mair, R. J. and Wood, D. M. (1987), *Pressuremeter Testing: Methods and Interpretation*, Construction Industry Research and Information Association/Butterworths, London.

Marchetti, S. (1980), "In Situ Test by Flat Dilatometer," *Journal of Geotechnical Engineering Division*, ASCE, Vol. **106**, GT. 3, March, pp. 299–321.

Marcuson, W. F., III and Bieganousky, W. A. (1980), "SPT and Relative Density in Coarse Sands," *Journal of Geotechnical Engineering Division*, ASCE, Vol. **103**, No. 11, pp. 1295–1309.

Mayne, P. W. and Mitchell, J. K. (1988), "Profiling of Over consolidation Ratio in Clays by Field Vane," *Canadian Geotechnical Journal*, Vol. **25**, No. 1, pp. 150–158.

Mayne, P. W. and Kemper, J. B. (1988), "Profiling OCR in Stiff Clays by CPT and SPT," *Geotechnical Testing Journal*, ASTM, Vol. **11**, No. 2, pp. 139–147.

Meigh, A. C. (1987), *Cone Penetrometer Testing*, Construction Industry Research and Information Association/Butterworths, London.

Menard, L. (1956), "An Apparatus for Measuring the Strength of Soils in Place," MSc Thesis, University of Illinois, Illinois.

Morris, P. M. and Williams, D. T. (1994), "Effective Stress Vane Shear Strength Correction Factor Correlations," *Canadian Geotechnical Journal*, Vol. **31**, No. 3, pp. 335–342.

Ohya, S., Imai, T. and Matsubara, M. (1982), "Relationships between *N* Value by SPT and LLT Pressuremeter Results," *Proceedings, Second European Symposium on Penetration Testing,* **Vol. 1**, Amsterdam, pp. 125–130.

Peck, R. B., Hanson, W. E. and Thornburn T. H. (1974), *Foundation Engineering*, 2nd edn, John Wiley & Sons, Inc., New York.

Ramaswamy, S. D., Daulah, I. U. and Hasan, Z. (1982), "Pressuremeter Correlations with Standard Penetration and Cone Penetration Tests," *2nd ESOPT*, Vol. **1**, pp. 137–142.

Ricceri, G., Simonini, P. and Cola, S. (2002), "Applicability of Piezocone and Dilatometer to Characterize the Soils of the Venice Lagoon," *Geotechnical and Geological Engineering*, Vol. **20**, No. 2, pp. 89–121.

Robertson, P. K. (2010), "Soil Behaviour Type from the CPT: an update," *Second International Symposium on Cone Penetration Testing,* CPT10, Technical Paper No. 2-56, Huntington Beach, CA, USA, www.cpt10.com. Or, Soil behaviour type from the CPT: an update. P.K. Robertson Gregg Drilling and Testing Inc., Signal Hill, California, USA.

Robertson, P. K. and Campanella, R G. (1986), "Interpretation of Cone Penetration Tests. Part I: Sand and Part II: Clay," *Canadian Geotechnical Journal,* Vol. **20**, No. 4, pp. 718–745.

Robertson, P. K. and Campanella, R. G. (1983), "SPT-CPT Correlations," *Journal of Geotechnical Engineering Division*, ASCE, Vol. **109**, No. 11, pp. 1449–1459.

Robertson, P. K. and Campanella, R. G. (1989), *Guidelines for Geotechnical Design Using the Cone Penetrometer Test and CPT with Pore Pressure Measurement,* 4th edn, Hogentogler and Co., Columbia, MD.

Schmertmann, J. H. (1970), "Static Cone to Compute Static Settlement over Sand," *Journal of Soil Mechanics and Foundations Division*, ASCE, Vol. **96**, SM 3, pp. 1011–1043.

Schmertmann, J. H. (1978), *Guidelines for Cone Penetration Test: Performance and Design,* Report FHWA-TS-78-209, Federal Highway Administration, Washington, D.C.

Schmertmann, J. H. (1986), "Suggested Method for Performing the Flat Dilatometer Test," *Geotechnical Testing Journal,* ASTM, Vol. **9**, No. 2, pp. 93–101.

Seed, H. B., Arango, I. and Chan, C. K. (1975), *Evaluation of Soil Liquefaction Potential during Earthquakes,* Report No. EERC 75-28, Earthquake Engineering Research Centre, University of California, Berkeley.

Skempton, A. W. (1986), "Standard Penetration Test Procedures and the Effects in Sands of Overburden Pressure, Relative Density, Particle Size, Aging and Over consolidation," *Geotechnique*, Vol. **36**, No. 3, pp. 425–447.

Smith, R. E. (1970), "Guide for Depth of Foundation Exploration," *Journal of Soil Mechanics and Foundations Division*, ASCE, Vol. **96**, SM 2, March, pp. 377–384.

Sowers, G. B. and Sowers, G. F. (1970), *Introductory Soil Mechanics and Foundations,* 3rd edn, Macmillan, New York.

Sowers, G. F. (1979), *Introductory Soil Mechanics and Foundations,* 4th edn, Macmillan, New York.

Stokoe, K. H. and Woods, R. D. (1972), "In Situ Shear Wave Velocity by Cross-Hole Method," *Journal of Soil Mechanics and Foundations Division,* ASCE, Vol. **98**, SM 5, pp. 443–460.

Stroud, M. A. and Butler, F. G. (1975), "The Standard Penetration Test and the Engineering Properties of Glacial Materials," *Proceedings of a Symposium on the Engineering Behaviour of Glacial Materials,* pp. 124–135, Midland Geotechnical Society, Birmingham.

Szechy, K. and Varga, L. (1978), *Foundation Engineering – Soil Exploration and Spread Foundation,* Akademial Klado, Budapest, Hungary.

Terzaghi, K., Peck, R. B. and Mesri, G. (1996), *Soil Mechanics in Engineering Practice,* 3rd edn, John Wiley & Sons, Inc., New York.

Terzaghi, K., Peck, R. B. (1967), *Soil Mechanics in Engineering Practice,* 2nd edn, John Wiley & Sons, Inc., New York.

Winter, E. (1982), "Suggested Practice for Pressuremeter Testing in Soils," *Geotechnical Testing Journal,* ASTM, Vol. **5**, No. 3/4, Sept, pp. 85–88.

Wolff, T. F. (1989), "Pile Capacity Prediction Using Parameter Functions," in *Predicted and Observed Axial Behaviour of Piles, Results of a Pile Prediction Symposium,* ASCE Geotechnical Special Publication No. 23, pp. 96–106.

CHAPTER 2
Shallow Foundations – Introductory Chapter

2.1 General

In civil engineering, the *foundation* may be defined as that part of a structure which transmits loads directly to the underlying soil and/or rock safely. This is also sometimes called the *substructure*, since the *superstructure* brings load onto the foundation, or substructure. The term *foundation soil* is commonly used to describe the underlying material within the adjacent zone which will be affected by the substructure and its loads. This is also sometimes called the *supporting soil*, or *base material*, since the total structure will rest on it.

Generally, foundations may be broadly classified into two categories: (1) *Shallow foundations*, (2) *Deep foundations*.

A shallow foundation transmits loads to the near-surface strata; it is positioned at a shallow depth. According to Terzaghi and Peck (1967), a foundation is shallow if the least dimension (usually the width B) of the structural base is equal to or greater than the foundation depth D_f. This criterion is reasonable for normal shallow foundations, but it is unsatisfactory for narrow or very wide foundations. Some investigators have suggested that D_f can be as great as 3–4B (Das, 2011). Here, D_f is the vertical distance from the ground surface down to a level at which the bottom of the structural base is located. The level at this depth is usually known as the *foundation level*.

In practice, all the spread footing foundations are usually shallow foundations. Also, the mat and raft foundations are mostly shallow foundations. Here, "spread" is a generic word can be applied to any structural base that spreads its structure load on a large area. It is function of the base to spread the load laterally to the underlying strata so that the stress intensity is reduced to a value that the foundation soil or rock can safely carry. In practice, however, a spread footing is particularly used to mean an individual column footing.

Before the early twentieth century, almost all the spread footings were made of steel grillage or masonry. The steel grillage footings included several layers of railroad tracks or I-beams placed in both directions. The masonry footings were dimension-stone footings built of stones cut and dressed to specific sizes fit together with joints of very small gap, or, they were rubble-stone footings built of random size material joined with mortar. These footings prevailed until the advent of reinforced concrete in the early twentieth century. Compared to steel grillage or masonry footings, reinforced concrete footings are very strong, economical, durable and easy to build. Also, they are thinner and smaller in plan

Shallow Foundations: Discussions and Problem Solving, First Edition. Tharwat M. Baban.

dimensions than the old masonry footings, so they do not require large excavations and do not intrude into basements. Therefore, nearly all shallow foundations are now made of reinforced concrete.

2.2 Types of Shallow Foundations

There are different types of shallow foundations. They are different due to the differences in their geometrical shape, design and construction. They are of the following types:

(1) *Strip or continuous footing.* Load bearing walls are supported by continuous-strip footings (also known as wall footings) as shown in Figure 2.1. Transverse reinforcing steel bars are provided at the footing bottom to satisfy the bending requirements of the footing projection. Longitudinal steel bars are required to satisfy shrinkage requirements. Longitudinal steel will, in general, be more effective in the top of the footing than in the bottom.

A strip footing is also used to support a row of columns which are so closely spaced that their individual footings overlap or nearly touch each other. In this case it is more economical to excavate and concrete a strip foundation than to work in a large number of individual pits. In fact it is often thought to be more economical to provide a strip footing whenever the distance between the adjacent individual footings is less than their dimensions. The individual footings are formed by inserting vertical joints in a continuous strip of concrete. Also, footings of this type, if the foundation soil conditions permit, are more economical in reinforcing steel than continuous combined (beam) footings.

(2) *Spread footing (also, individual, single, isolated or independent column footing).* This is an independent footing provided to support an individual column. The footing may be circular, square or rectangular slab of uniform thickness or may be stepped or sloped, used to spread the load over a large area (Figure 2.2). An advantage of this type of footing foundation is that the footing size can be adjusted to the same or different contact soil pressures. Also, the footings can be located at different foundation levels so as to distribute the structure's load advantageously over the site, holding the differential settlement within certain limits. It is often thought that this type of foundation becomes more economical than the continuous-strip foundation when the required foundation depth is greater than 1.5 m.

(3) *Combined footing.* When a reinforced concrete slab supports a line of two or more columns, it is called combined footing. An ordinary combined footing which supports two columns may be rectangular or trapezoidal in plan (Figure 2.3). When the slab supports more than two columns it is called continuous combined footing. In this case, more often the slab has a longitudinal pedestal beam (rib), and the footing is designed as a continuous inverted T-beam (Figure 2.4b). Combined footings are used when the columns are so close to each other that their individual footings would touch each other or overlap. They are also used when the foundation soil is erratic and of relatively low bearing capacity. A combined footing may be used when a column is located near or right next to a property limit; it often supports the edge column and an interior column to avoid using an undesirable eccentrically loaded spread footing adjacent to the property line (Figure 2.4a).

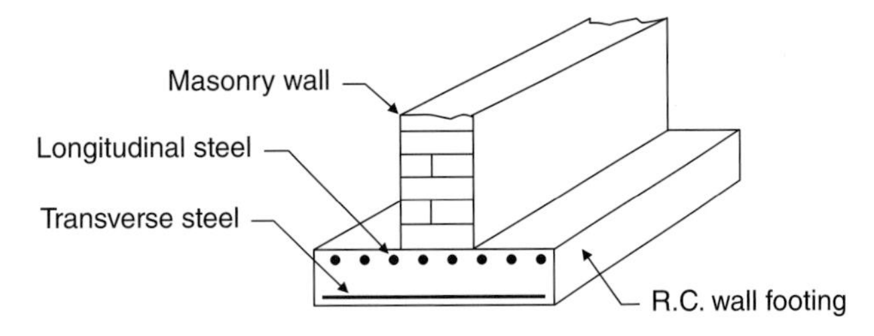

Figure 2.1 Strip or continuous wall footing.

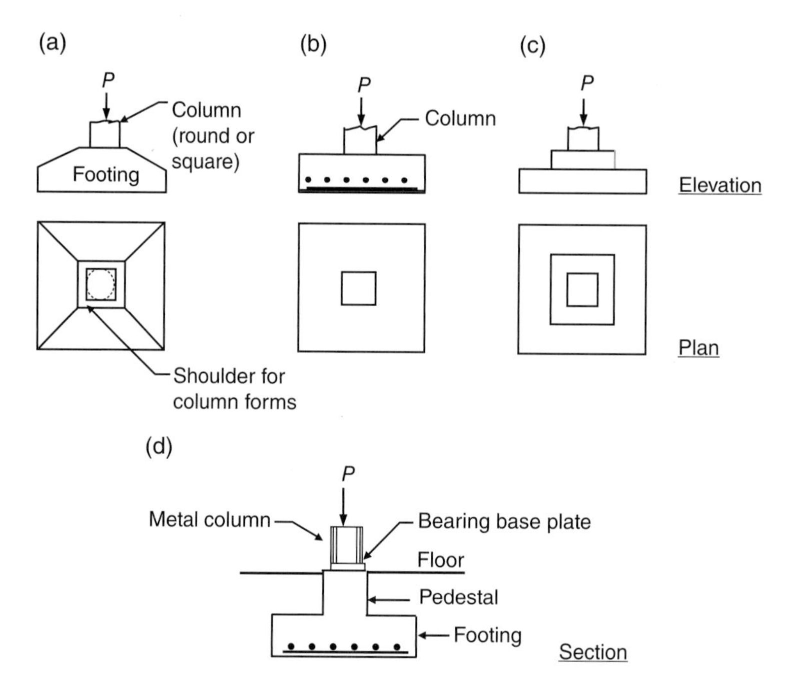

Figure 2.2 Typical spread column footings: (a) sloped footing, (b) footing of uniform thickness, (c) stepped footing and (d) footing with pedestal.

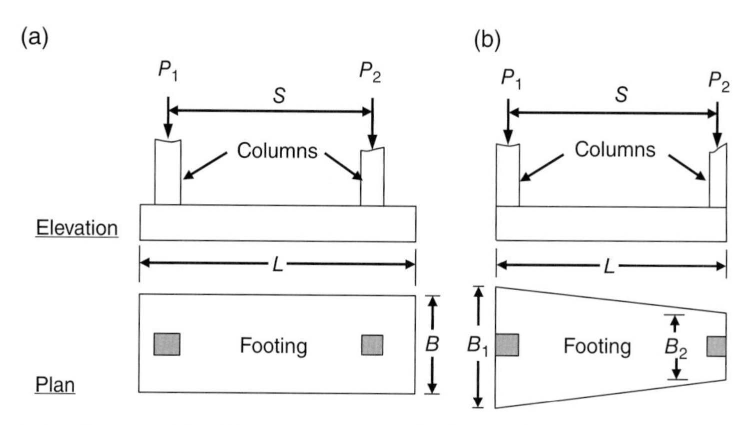

Figure 2.3 Typical ordinary combined footings: (a) rectangular footing; (b) trapezoid footing.

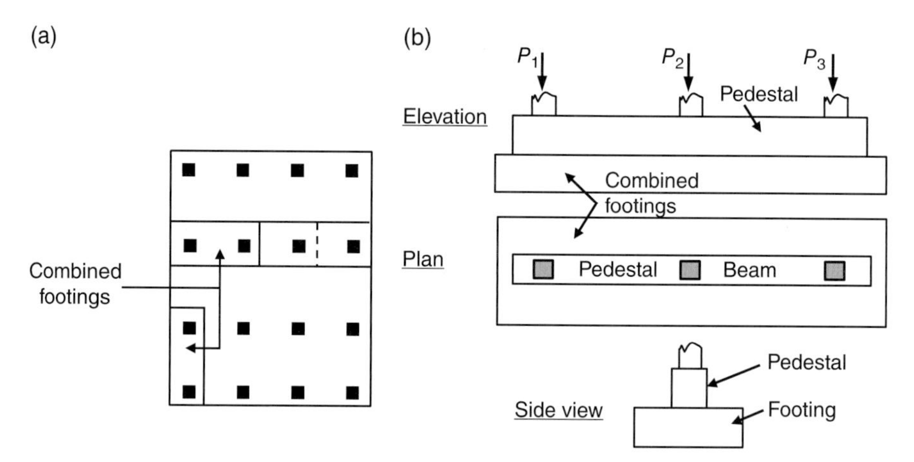

Figure 2.4 (a) Typical combined footings, (b) Combined footing with a pedestal beam designed as an inverted T-beam to reduce footing mass.

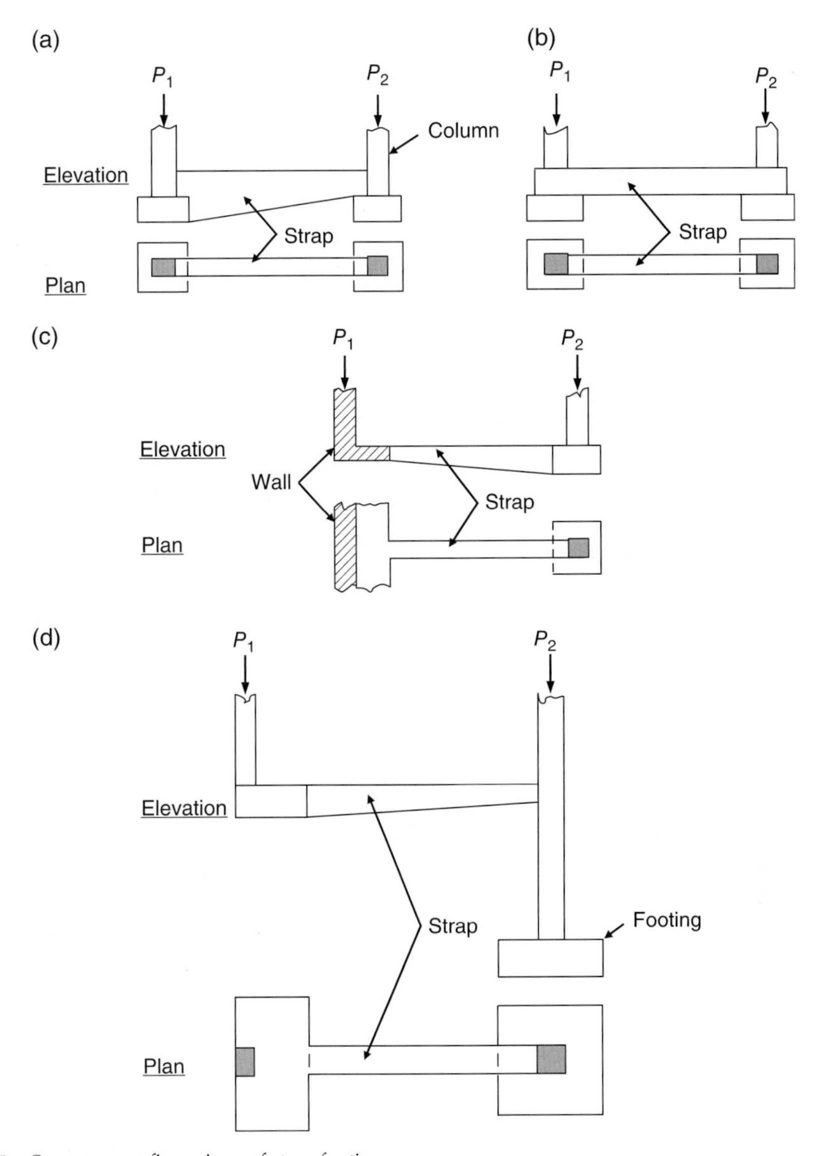

Figure 2.5 Common configurations of strap footing.

(4) *Strap footing.* A strap footing (also known as a cantilever footing) consists of two individual footings connected to each other with a structural rigid beam called *strap*; so it may be considered as a special form of the ordinary combined footing. Figure 2.5 shows the common arrangements of strap footings. The strap should be out of contact with soil so that there are no soil reaction and the srap could act as a rigid beam. Strap footings may be used in place of the combined footings to achieve the same purposes. However, this type of foundation would be economical than rectangular or trapezoidal combined footing foundations only when the allowable soil bearing capacity is high and the distance between the columns are relatively large. Similar to the continuous combined footing, strap footing can be continuous supporting more than two columns in one row; designed as continuous strap footing.

(5) *Ring spread footing.* This is a continuous-strip footing that has been wrapped into a circle (Figure 2.6). This type of footing is commonly used to support above-ground and elevated storage tanks, transmission towers, TV antennas and various process tower structures. Regarding the

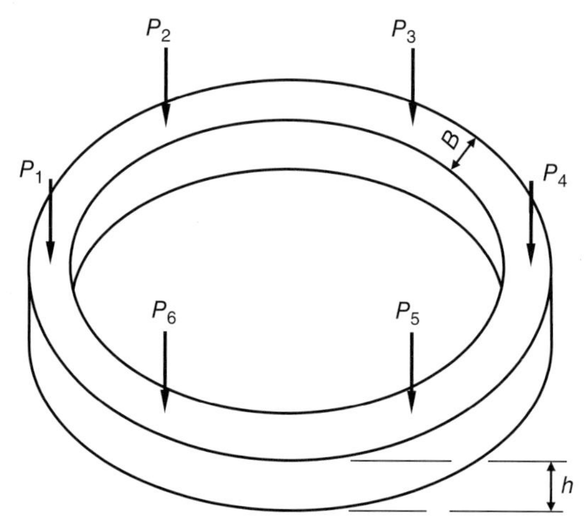

Figure 2.6 Ring footing supporting six columns.

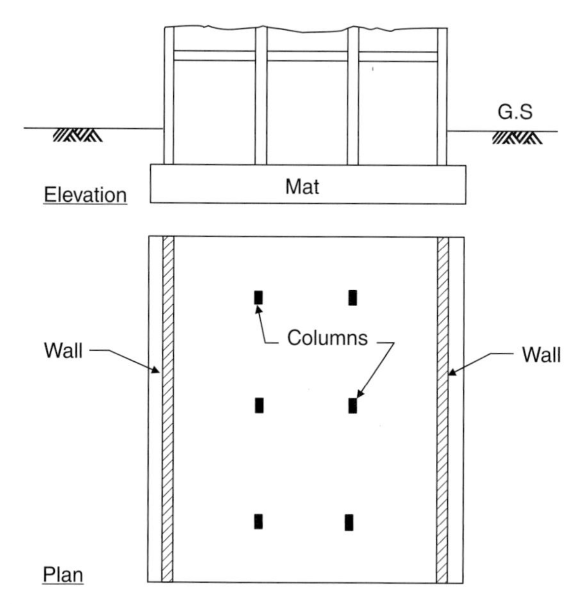

Figure 2.7 A mat foundation.

above-ground storage tanks, the contents weight of these tanks is spread uniformly across the total base area, and this weight is probably greater than that of the tank itself. Accordingly, the geotechnical analyses of these tanks usually treat them as circular mat foundations with the diameters equal to diameter of the tank. Obviously, this case is quite different from that with an elevated storage tank which is usually supported by a number of columns distributed in a circular line. Here, the footing may be treated as a circular (ring) continuous combined footing.

(6) *Mat foundation.* A simple mat foundation consists of a single reinforced concrete slab, relatively thick, used to support two or more columns both ways, and/or walls of a structure or a part of it (Figure 2.7).

A mat foundation may be used where the base soil has a low bearing capacity, and where there is a large variation in the loads on the individual columns. It is commonly used when the column loads are so large that more than 50% of the construction area is covered by the conventional spread footings. It is common to use mat foundation when the structure contains basement (especially deep basement) in order to spread the column loads to a more uniform pressure distribution, to decrease danger of differential settlement, and to provide the floor slab for the basement. Moreover, a particular advantage is to provide a water barrier for basements at or below the ground water table, since the basement walls can be constructed integral with the mat.

A mat foundation may be supported by piles to control buoyancy due to high ground water level, or, where the base soil is susceptible to large settlement.

(7) *Raft foundation.* A raft foundation consists of a relatively thin reinforced concrete slab cast integrally with reinforced concrete beams either above or below the slab in both directions (Figure 2.8). Sometimes raft and mat are used synonymously since they are desired to achieve the same goal. Raft foundations are usually used on compressible soil in order to distribute the building load over the entire building area. Sometimes stiffening walls or basement walls are used as part of the raft, making a heavy large *cellular* construction capable of supporting a heavy structure (Figure 2.9).

The raft with the beams below the slab (Figure 2.8b) has the advantages of ease of construction (if the groundwater level is not a problem) and providing a level surface slab which can form the ground floor or the basement floor. However, it is necessary to construct the beams in trenches which can cause difficulties in in soft or loose soil that requires continuous support sheeting and

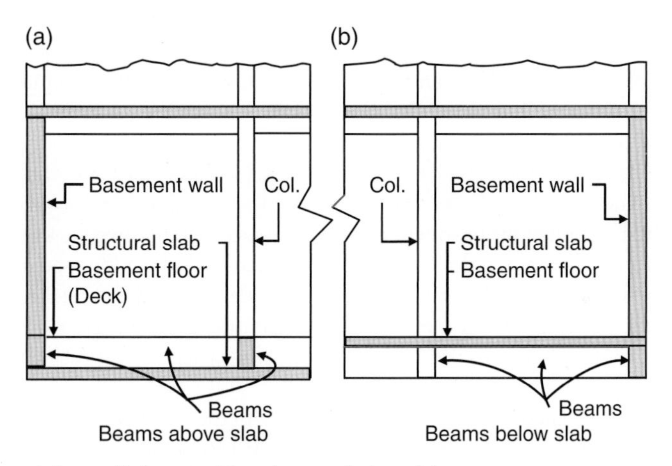

Figure 2.8 Raft foundations with beams either above or below slab.

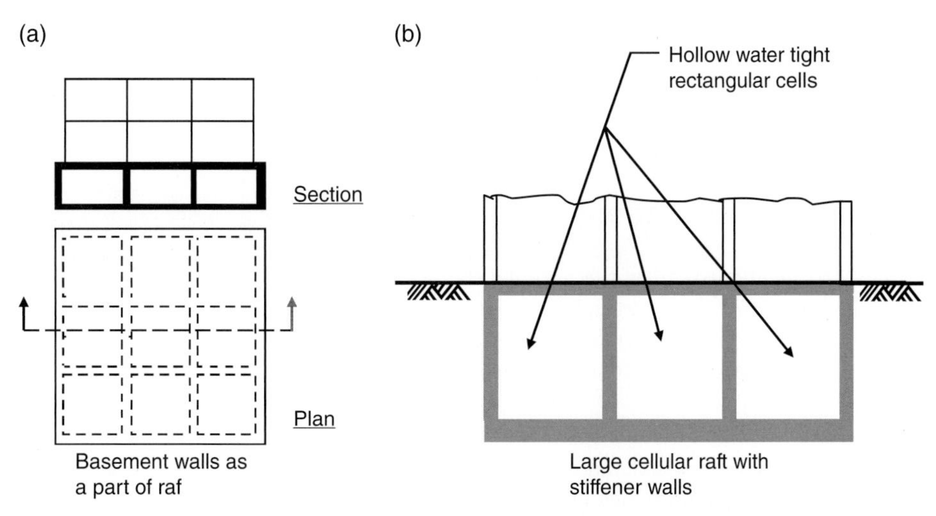

Figure 2.9 Cellular raft foundations.

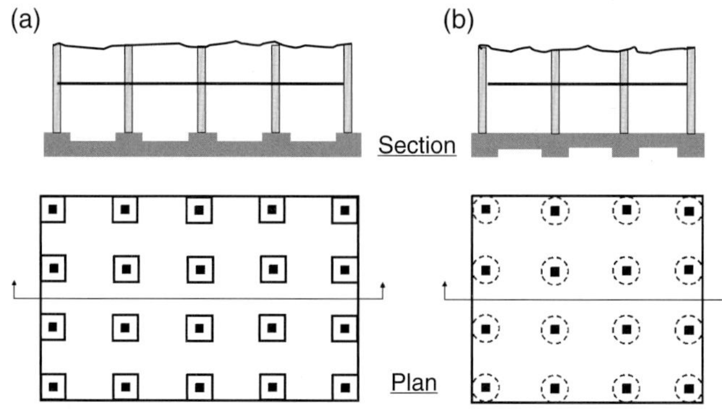

Figure 2.10 Raft foundations consist of a flat plate (slab) thickened under columns, or with pedestals.

strutting. Construction may also be difficult when the trenches are excavated in water-bearing soil requiring additional space in the excavation for subsoil drainage and sumps. Another disadvantage is in the design because the loads from the foundation slabs are transmitted to the beams by tension.

The raft with beams formed above the slab (Figure 2.8a) has an advantage in design because the load from the slab is transferred to the beams in a conventional manner, producing compression and diagonal tension due to shear. Also, it insures that the beams are constructed in clean dry conditions above the foundation slab. Where excavation for the foundation slab has to be undertaken in water-bearing soil it is easier to deal with water in a large open excavation than in the confines of narrow trenches. However, this type of raft foundation requires the provision of an upper slab (deck) to form the ground or the basement floor of the structure. This involves the construction and removal of soffit form work for the deck slab, or the alternative of filling the spaces between the beams with granular material to provide a surface on which the deck slab can be cast. Precast concrete slabs can be used for the top decking, but these require the addition concrete screed to receive the floor finish.

Some designers prefer a raft consists of a flat plate (slab) thickened under columns, or with pedestals (Figure 2.10), than a raft consists of slab and beams.

(8) *Floating foundation.* A raft foundation with the weight of the excavated material Q_m from the basement approximately equal to the superimposed total weight of the structure Q_s is often called *floating foundation* or *compensated foundation.* When this condition exists the foundation is usually referred to as *fully compensated foundation.* On the other hand when $Q_m < Q_s$ the foundation is referred to as *partially compensated foundation.* The use of a floating foundation is probably the most effective and practical method of controlling total settlement, especially, when structures are to be built on very soft soils. Another advantage of using this technique of floatation is that the factor of safety against bearing capacity failure will be greatly increased.

2.3 Depth of Foundations

As mentioned earlier, the vertical distance between the ground surface (G.S.) and the foundation level (bottom of the structural base) is called *foundation depth* (sometimes the *footing depth*), usually represented by the symbol D_f. Foundation depth is one of the basic factors that control the allowable bearing capacity of a foundation material. In all cases, the foundation depth must not be shallower than depth of:

(a) Overburden soil and/or rock as a requisite of the allowable bearing capacity.
(b) Debris and garbage fill materials.
(c) Peat and muck.

(d) Top soil consisting of loose and organic materials.
(e) Frost penetration.
(f) Zone of high volume change.
(g) Scour.

The materials mentioned in (b) and (c) are considered unsuitable foundation materials and must be removed from the construction area. If they are too deep, the material directly under the footing is removed and replaced with lean concrete, or replaced with compacted clean granular soil (such as clean sub-base material) in an area relatively larger than the footing area (Figure 2.11).

Generally, soils encountered within the zone of frost penetration remains in an unstable condition because of alternate seasonal freezing and thawing. Consequently, a footing located within the zone is lifted during cold weather due to freezing of the soil porewater, and it settles during worm weather due to thawing of the frozen water. Therefore, frost action can cause damage to the foundation and to the superstructure unless D_f is greater than the depth of frost penetration. Gravel and coarse sand above water level containing less than about 3% fine sand, silt, or clay particles cannot hold any water and consequently are not subject to frost damage. Other soils are subjected to frost heave within the depth of frost penetration.

Some clay soils of relatively high plasticity, usually called *expansive soils*, undergo excessive volume change (usually seasonal) by shrinking upon drying and swelling upon wetting. Consequently, footings rest on such soils will move up and down. This process can cause severe damage to the total structure unless D_f is greater than the depth within which large volume change takes place.

Bridge piers, abutments, bases for retaining walls and footings for other structures adjacent to or located in flowing water must be located at a depth such that erosion or *scour* does not undercut the soil and cause a failure.

The difference in foundation depths b of adjacent footings should not be so large as to produce undesirable overlapping of stresses in the foundation soils. This is generally avoided by maintaining the maximum difference in foundation levels b equal to $m/2$ for footings rest on soil and m for footings on rock, where m is the horizontal distance between two adjacent footings (Figure 2.12). These requirements should be satisfied to prevent *shear failure* and *excessive settlement* take place in the supporting soils.

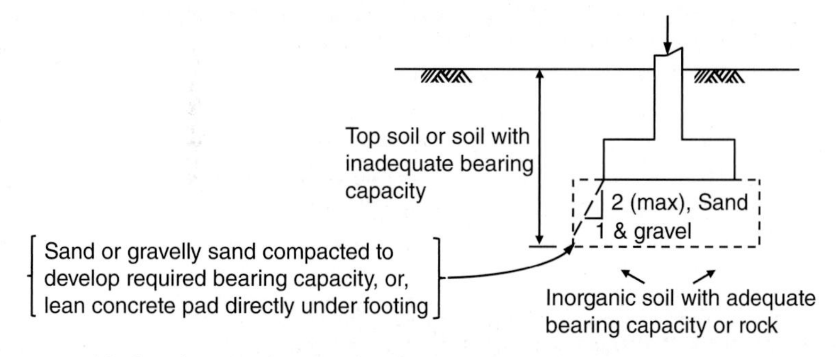

Figure 2.11 Unsuitable foundation soil replaced with sand and gravel or lean concrete pad.

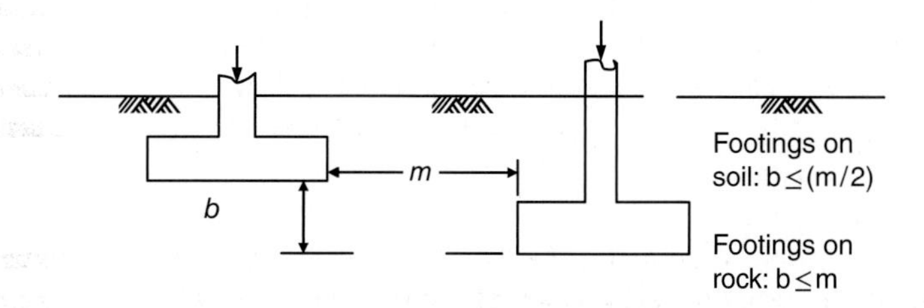

Figure 2.12 Suggested limits for difference in foundation depth of adjacent footings.

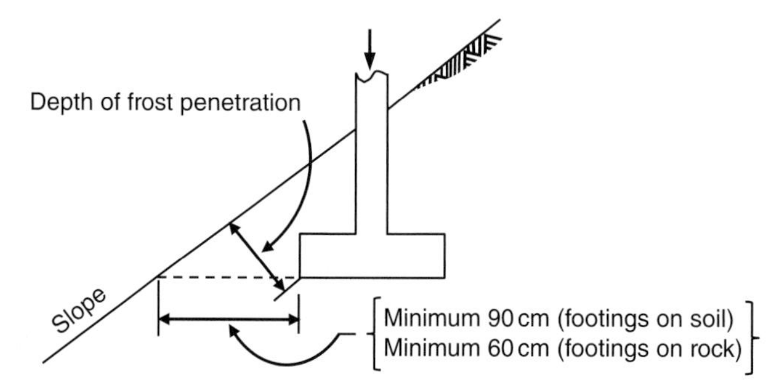

Figure 2.13 Foundation depth requirements for footings on slope.

If a new footing is lower than an existing footing, the new excavation may cause soil flow laterally from beneath the existing footing. This may result in settlement cracks in the existing structure. In this case, the safe depth difference b may be approximately determined for a c – Ø soil using the following equation:

$$b_{safe} \cong \frac{2c}{(SF)(\gamma\sqrt{K})} - \frac{q_o}{(SF)\gamma} \tag{2.1}$$

where c = soil cohesion
SF = suitable factor of safety
q_o = footing (or base) contact pressure
γ = average unit weight of soil
K = coefficient of lateral earth pressure, $K_a \leq K \leq K_p$

For cohesionless soils such as sand, the equation reveals that one cannot excavate to a depth greater than that of the existing foundation, that is $b = 0$.

Footings on sloping ground should have sufficient edge distance of about 1 m minimum to provide protection against erosion (Figure 2.13).

Underground defects or utilities, such as cavities, old mine tunnels, sewer lines, water mains and underground cables, may affect the foundation depth. Sometimes the solution may require shifting the utilities, relocating the foundations or even an abandonment of the site.

It is not good practice to place footings on the ground surface (i.e. $D_f = 0$) even in localities where the requirements of bearing capacity, frost penetration, seasonal volume change and so on, mentioned earlier, are not obstacles because of the possibility of surface erosion. Generally the minimum depth of foundation should be 50 cm below the natural ground surface.

2.4 Foundation Performance Requirements

2.4.1 General

It is important to remember that the performance of foundations is based on an interface between the loadings from the structure and the supporting ground or strata. Foundations must be designed both to interface with the soil at a safe stress level and to limit settlements to an acceptable amount. In all cases the most economical solution will be selected, provided that it satisfies the performance requirements.

The factors related to the ground conditions are the allowable bearing capacity and location of the supporting strata, the soil composition, the ground level and gradients and the groundwater level. The allowable bearing capacity of the supporting ground is one of the key elements in the selection of appropriate foundations for all types of structures. The ground of a satisfactory bearing strength may be located at a considerable depth below the surface that may require using a deep foundation rather than a shallow foundation form which is unlikely to be efficient or cost effective. It is not uncommon to find buildings being constructed on sites made up of filled ground which is of greater variability in composition than the natural ground strata even they are ununiformed or heterogeneous deposits. With respect to ground surface; it is quite rare for building sites to be truly flat and level. Where the groundwater is present, it is common to lower its level below the construction zone either permanently or for the duration of the construction work. If the groundwater rises above the foundation level, the foundation will be subject to uplift or flotation, which would have to be taken into account. Also, there is a potential problem of ground subsidence in the area surrounding the construction site if there is significant lowering of groundwater level. Obviously, solutions to such problems increase the construction cost.

The factors related to the structure loads are the type and application, source and duration of the loads. There are different types of loads such as axial compressive or tensile, shear, moment and torsional loads. Knowing nature of these design loads can in turn help in selecting the type of foundation to be used. The nature of framed buildings is such that the loads are likely to be concentrated at the point of application, that is the column bases. Hence the use of pads (spread bases) and piles tends to be most common. However, there may be situations where there are also uniformly distributed loads, such as from masonry cladding for example. These must also be dealt with and a combination of foundation solutions may applied in a given situation.

In design of foundations for a structure the designer should already have appropriate answers to questions such as what functions the proposed structure will accomplish? What are the required design criteria? What would be the acceptable performance? As long as failure is an unacceptable difference between expected and observed performance, the foundations like the other engineering products will have varying degrees of performance which are not the same for all structures.

In view of the above discussions, the performance requirements of foundations concern the followings:

(1) Strength
(2) Serviceability
(3) Constructibility
(4) Economy

2.4.2 Strength Requirements

Foundations are required to be strong enough that catastrophic failures not be allowed to take place. There are two types of strength requirements: the geotechnical and the structural strength requirements.

Geotechnical strength requirements are those that directly related to the supporting soil or rock strata. In footing foundations geotechnical strength is expressed as *bearing capacity* considering *shear* in the supporting material; the failure due to shear is usually referred to as *bearing capacity failure* (Figure 2.14). In the design of foundations there must be a *sufficient factor of safety* against such failures. Geotechnical strength analysis is usually performed using *working (unfactored) loads*, that is using allowable stress design (ASD) methods. However, using the load and resistance factor design (LRFD) methods are not uncommon.

Structural strength requirements are those that directly related to the structural foundation components. The foundation elements are designed to avoid structural failures, similar to the other structural analyses. The strength analyses are made using either LRFD or ASD methods, depending on the type of structure and its foundation, the structural material, and the governing design code. It is important to realise that foundations which are loaded beyond there structural capacity will, in principle, fail catastrophically.

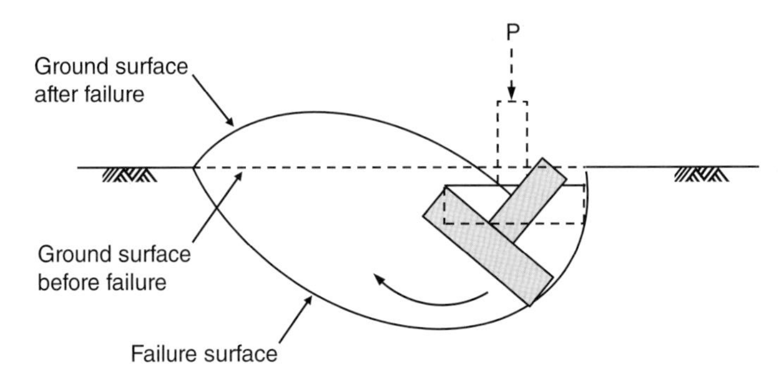

Figure 2.14 A bearing capacity failure beneath a spread footing foundation. The soil has failed in shear, causing the foundation to collapse.

2.4.3 Serviceability Requirements

The second performance requirement is *serviceability*. A foundation should have adequate serviceability performance when subjected to the service loads during its design life time. That is it should not excessively: (a) settle under vertical downward loads, move laterally due to lateral loads, (b) move upward due to heave of the supporting soil or uplift pressure or (c) tilt because of uneven settlement or heave occurring under a part of the structure. Also, it should not excessively move due to other factors such as changes in groundwater level, seasonal changes (dry and wet periods), internal erosion (piping), soil creep, adjacent excavations and buildings and vibrations. Moreover, the foundation concrete must be durable; resistant to the various physical and chemical processes, such as sulfate and acid attacks, that cause deterioration. In practice, all foundations are subjected to one or more of these undesirable events or processes; however, there are tolerable limits laid down in various building codes and specifications in order to control their effects and to achieve acceptable performance.

The most important seviceability requirement is the one that concerns *settlement*. The variability of soil in combination with unanticipated loads can result in settlement problems over which the designer may have little control. However, a relatively low number of modern buildings collapse from excessive settlements, but it is not uncommon for a partial collapse or a localised failure in a structural member to occur. More common occurrences are unsightly wall and floor cracks, uneven floors (sags and slopes), sticking doors and windows and the like.

2.4.4 Constructibility Requirements

The third performance requirement is *constructibility*. According to the Construction Industry Institute (CII) at the University of Texas in Austin, USA, constructibility may be defined as "the optimum use of construction knowledge and experience in planning, design, procurement and field operations to achieve the overall project objectives". In some universities research has been developed for new management methods and techniques to improve the construction industry. In addition, local and regional groups of construction users have been formed, resulting in increased awareness of the benefits to be gained through improved constructibility programs. These benefits include improvements in quality and reliability, as well as savings in time and money.

The foundation should be buildable with available construction personnel and, as much as possible, without having to use extraordinary methods or equipment. A proper design requires insuring that a structural member can be constructed without degrading the specified material quality of the product. For example, problems of constructibility of concrete structures occur most often because of the attempt to design slimmer columns or beams of smaller cross-section area. These designs, although satisfying the governing design codes, reduce space for placing concrete, and can create problems in obtaining good vibration and proper compaction. In case there is no enough previous experience required for construction of a certain kind of projects, it may be necessary that all concerned parties carefully work together to achieve the desired result.

2.4.5 Economy Requirements

The last performance requirement is *economy*. In general, it is common practice to be conservative in design of foundations. This is because of the uncertainties in soil properties, loads and the uncertainty in the nature and distribution of the load transfer between foundations and the ground (soil–structure interaction).

Due to the fact that foundations being the most important part of the total structure but the most difficult to access if problems later develop, a conservative design or even an overdesign has a better return on investment here than in other parts of the structure or the project. However, gross over-conservatism is not warranted. An overly conservative design can be very expensive to build, especially with large projects where the foundation is a greater portion of the total project cost. Being excessively conservative is an ethics problem, unless the client is made aware of the several alternatives and accepts the more conservative recommendation as being in his or her interests. It is necessary that a designer always try to produce designs that are both safe and cost-effective. Achieving the optimum balance between safety and cost is a part of good engineering. It is important to remember, designs that minimise the required quantity of construction materials do not necessarily minimise the cost. In some cases, designs that use more mterials may be easier to build, and thus have a lower overall cost.

As a summary, foundation design tends to be more conservative than other structural designs because the unknowns are not as well quantified; consequences of catastrophic foundation failure are much greater, as the entire structure fails with the foundation; reduction in weight may not be beneficial in foundation design, depending upon the circumstance.

2.5 Sulfate and Organic Acid Attack on Concrete

2.5.1 Sulfate Attack

Concrete in foundations may have to withstand attack by water soluble sulfate salts existing in the ground or in chemical wastes. Calcium sulfate occurs naturally in soils, usually clays, as crystalline gypsum. Sodium and magnesium sulfates are usually occur to a lesser degree, but as they are more soluble than the calcium sulfate they are potentially more dangerous. The severity of attack on concrete foundations depends mainly on the type and quality of the concrete, the concentration of sulfates (SO_4), the level of and fluctuations in the groundwater table and the climatic conditions.

Sulfates in solution can react with cement to form insoluble calcium sulfate and calcium sulfoaluminate crystals. The latter compound is highly hydrated and contains 31 molecules of water of hydration. The internal stresses in the concrete, created by the expansion accompanying the formation of calcium sulfoaluminate, are sufficient to cause disruption of the concrete at the surface. This disruption exposes fresh areas to attack again and if there is flow of groundwater bringing fresh sulfates to the affected area, the rate of disruption can be very rapid causing mechanical failure of the concrete as whole.

In areas that were formerly used for agricultural purposes, some fertilisers contain a high concentration of sulfates that may cause problems to concrete foundations of structures built in such areas. The same is true for some industrial wastes.

Attack takes place when there is groundwater; and for the disruption of concrete to continue there must be replenishment of the sulfates. If the groundwater is always static the attack does not penetrate beyond the outer skin of concrete. Thus there is little risk of serious attack on structures buried in clay soils provided that there is no flow of groundwater such as might occur along a loosely backfilled trench. There is risk of attack in clayey soils in certain climatic conditions, for example when hot,

Table 2.1 Suggested maximum water–cement ratio, sulfate concentration and cement type.

Water-soluble sulfate in soil (% by weight)	Sulfates in water (ppm)	Sulfate attack hazard	Cement type	Maximum water–cement ratio
0.0–0.1	0–150	Negligible	–	–
0.1–0.2	150–1500	Moderate	II	0.5
0.2–2.0	1500–10000	Severe	V	0.45
>2.0	>10000	Very severe	V plus pozzolan	0.45

dry conditions cause an upward flow of water by capillarity from sulfate-bearing waters below foundation level. Similar conditions can occur when water is drawn up to the ground floor of a building due to the drying action of domestic heating or furnaces.

In case the laboratory tests indicate that the soil or groundwater has high sulfate content, for example more than 0.1% by weight in soil or more than 150 ppm in water, the foundation concrete may be protected against sulfate attack by using one or more of the following methods:

(1) Reduce the water:cement ratio. This reduces the permeability of concrete; thus retarding the chemical reactions and disruption. This is one of the most effective methods of resisting sulfate attack. Table 2.1 presents suggested maximum water:cement ratio with sulfate concentration and cement type. The table is adapted from Kosmatka and Panarese, 1988, and Portland Cement Association (PCA), 1991.

(2) Increase the cement content. The permeability of concrete is also reduced when the cement content is increased. A minimum cement content of 335 kg/m^3 is recommended.

(3) Use *sulfate resisting cement*. In American practice, type I cement (ASTM C 150-71) is similar in chemical and physical properties to British ordinary Portland cement. The American type II cement has a moderate sulfate resistance, and type V has a high sulfate resistance equivalent to that of British sulfate resisting cement (BS 4027). Pozzolan additives to the type V cement also help.

(4) Coat the concrete with an asphalt emulsion. This alternative method is usually used for retaining walls or buried concrete pipes, but not practical for foundations.

(5) Use well-compacted dense impermeable concrete made with sulfate resisting cement or in severe conditions use a protective membrane formed from a heavy coat of hot bitumen or from polythene sheeting. This method provides the best form of protection for high sulfate concentrations in ordinary foundation work.

2.5.2 Organic Acid Attack

The acidic nature of organic matter constituents tends to give an acid reaction to the water in soil which in turn may have a corrosive effect on materials buried in the soil. In certain marsh peats, oxidation of pyrite or marcasite can produce free sulfuric acid which is highly aggressive to concrete. The presence of free sulfuric acid is indicated by pH values lower than 4.3 and high sulfate content. However, the pH value of the groundwater provides a rather crude measure of the potential aggressiveness to concrete of naturally occurring organic acids. In European countries reliance is generally placed on obtaining a dense impermeable concrete as a means of resisting attack by organic acids rather than the use of special cements. Researchers found that ordinary Portland or rapid-hardening Portland cement combined with ground-granulated slag or of ash provides better resistance to acid attack than concrete made only with ordinary Portland cement.

Where the acids are organic and derived from natural sources the reader may find Table 2.2 more convenient to use; it gives precautions for foundations against attack by organic acids in peaty or

Table 2.2 Precautions for foundations against attack by organic acids in peaty or marshy soils (from Tomlinson, 2001).

pH value	Precautions for foundations above groundwater table in any soil and below groundwater level in impervious clay	Precautions for foundations in contact with flowing groundwater in permeable soil
6.0	None necessary	None necessary
6.0–4.5	Use ordinary Portland cement at ≥ 370 kg/m^3, max. W/C ratio = 0.5	Use ordinary Portland cement at 380 kg/m^3, or sulfate-resisting cement at ≥ 350 kg/m^3, maximum W/C ratio = 0.5
4.5	Use ordinary Portland cement at ≥ 400 kg/m^3, or sulfate – resisting cement at ≥ 390 kg/m^3, maximum W/C ratio = 0.5.	Use super-sulfate cement or ordinary Portland cement at ≥ 400 kg/m^3, or sulfate-resisting cement at ≥ 390 kg/m^3, max. W/C ratio = 0.5; plus protection by external sheathing.

Note: The cement contents recommended in the table are suitable for medium-workability concrete (50–75 mm slump).

marshy soils. This table has been based on practice in Germany and the Netherlands where extensive deposits of marshy soil are present.

2.6 Pressures under Foundations

2.6.1 Contact Pressure and Contact Settlement

The vertical pressure at contact surface between the foundation bottom and its base material (soil or rock) is termed *contact pressure*; and the induced settlement is termed *contact settlement*.

Contact pressure distribution and contact settlement beneath symmetrically loaded shallow foundations are functions of the type of base material, the consistency or density of soil or rock quality and the relative rigidity of the base material and foundation. For the limiting cases of rigidity, that is perfectly flexible or perfectly rigid foundations, the distribution of contact pressures and settlements for cohesionless soils (sands) and cohesive soils (clays) are as follows:

Consider a *flexible footing* carrying a *uniformly distributed load on the surface of a cohesionless soil* (Figure 2.15). Uniform contact pressure distribution exists, whereas, the contact settlement is not uniform. The outer edge of the footing undergoes a relatively large settlement compared to that at the centre, as shown by the dashed curve. Below the centre of the footing the settlement is small because the soil develops strength and rigidity as soon as it is loaded by the footing.

For a *rigid footing resting on a surface of cohesionless soil* the settlement must be uniform. In cohesionless soils, for uniform settlement to occur there must be a relatively large pressure under centre of the footing and no pressure at the edge (Figure 2.16a). If the average pressure is relatively small or if the footing width is large, the non-uniform pressure distribution is somewhat flatter over the central portion of the footing (Figure 2.16b).

For *rigid footings placed in a cohesionless soil deposit below the ground surface* there is some strength below the edge of the footing, therefore, the pressure is not zero at the edges (Figure 2.16c). For *very deep rigid footings on sand* the contact pressure distribution may be more like that discussed below for rigid footings in cohesive soils.

Consider a *uniformly loaded flexible footing on highly cohesive soil.* The contact pressure distribution is uniform, but the resulting contact compressive strain (contact settlement) distribution beneath the footing shows a dish pattern with maximum settlement at the centre (Figure 2.17). This is because the greater stress below the centre of the footing in the subsurface horizontal layers, such as the one shown, must cause a greater compressive strain at this location.

For a *rigid footing on highly cohesive soil* the contact settlement must be uniform, whereas the contact pressure distribution is not; it is much larger at the edges than that at the centre (Figure 2.18).

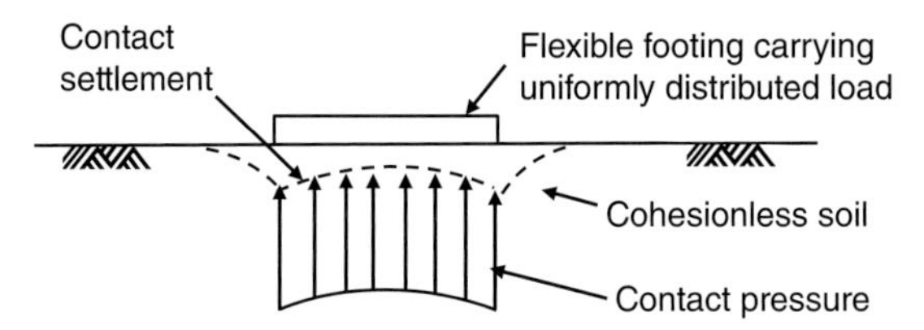

Figure 2.15 Contact pressure distribution and contact settlement beneath a uniformly loaded flexible footing rests on cohesionless soil.

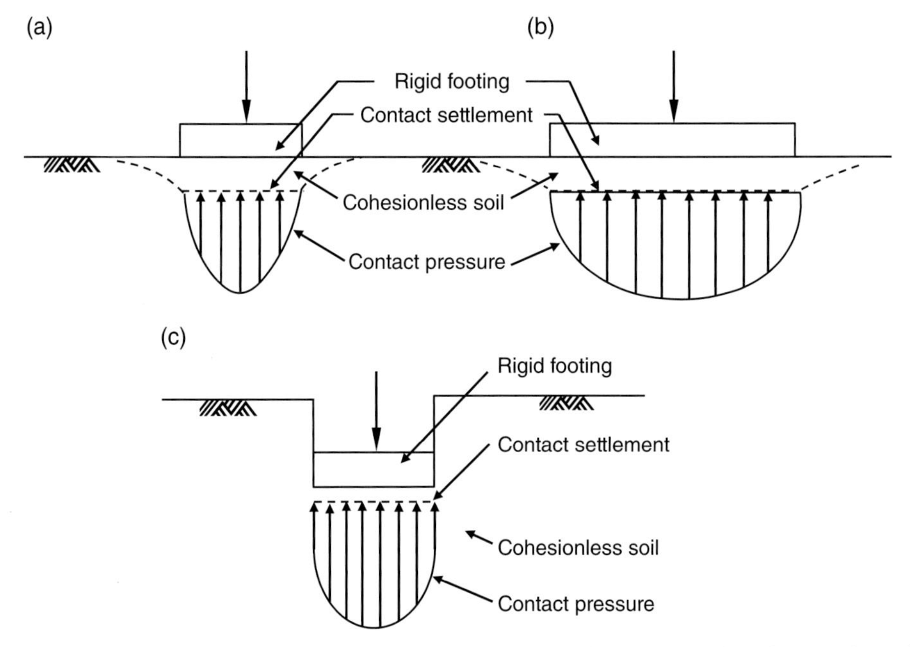

Figure 2.16 Contact settlement and pressure distribution beneath rigid footing placed (a), (b) on surface of cohesionless soil (c) in cohesionless soil.

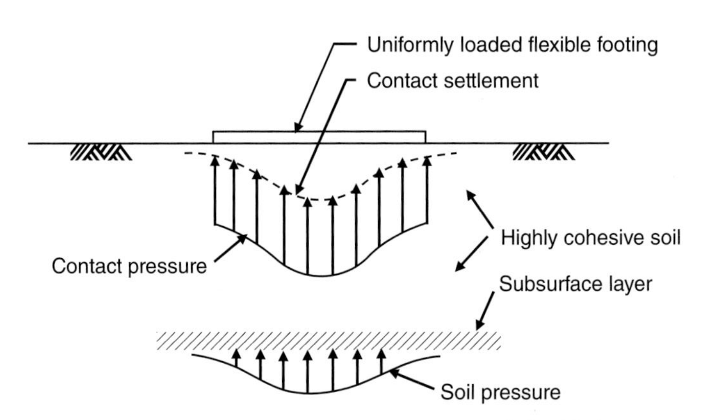

Figure 2.17 Contact pressure distribution and contact settlement beneath flexible footing on surface of highly cohesive soil.

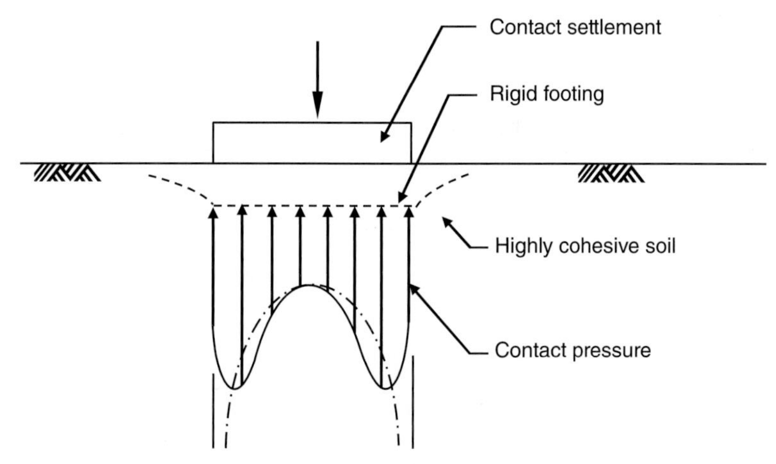

Figure 2.18 Contact pressure distribution and contact settlement beneath a rigid footing on surface of highly cohesive soil.

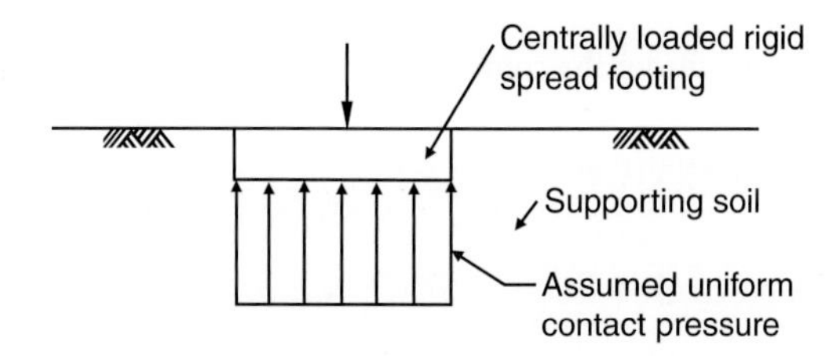

Figure 2.19 Assumed uniform contact pressure distribution beneath a spread footing.

According to the theory of elasticity, for an elastic material of infinite strength the contact pressure distribution is as indicated in the Figure by the long dash-dot curve which shows an infinite stress at the edge of the footing. Actually an infinite stress cannot occur, but the stress at the edges may be much larger than that at the centre.

In practice, spread footings are usually of intermediate to very high rigidity, so the actual contact pressure distribution is not uniform. Bearing capacity and settlement analyses as well as structural design of footings based on such a distribution would be very complex. Therefore, it is common practice to use the uniformly distributed contact pressure beneath centrally loaded spread footings, as shown in Figure 2.19. Results of some field measurements indicate this simplification in design of spread footings does not cause serious error and the assumption can be considered adequate. However, after a design has been prepared on this basis, it is suggested that the designer review it and strength it at locations where the actual distribution gives greater stresses than are given by the assumed distribution. In Figure 2.18, for example, the bending moment in the footing is much larger for the distribution shown than it is for a case of uniform pressure; additional reinforcing steel is needed to carry this greater moment.

With *mat* or *raft foundations* the problems of contact pressure distribution and deformation are quite different from those with spread footings. Usually, mat foundations have a much smaller thickness to width ratio, and thus are more flexible than spread footings. Therefore, the assumption of rigidity is no longer valid. Also, the assumption of linear contact pressure distribution would be erroneous unless the supporting material is mud or peat or soft soil. At locations of heavy loads, such as those of columns and bearing walls, a mat settles more than at locations with relatively light loads because the pressure beneath the heavily loaded zones are greater. Figure 2.20a shows contact pressure distribution beneath a mat rests on strong bedrock; the column loads are transmitted to the rock on a relatively

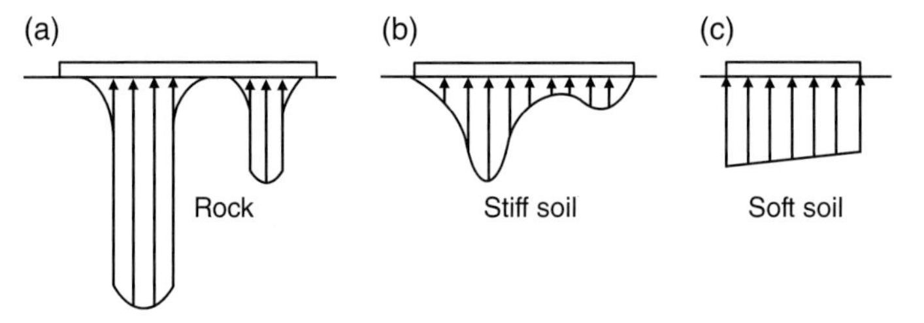

Figure 2.20 Distribution of contact pressure under a mat foundation.

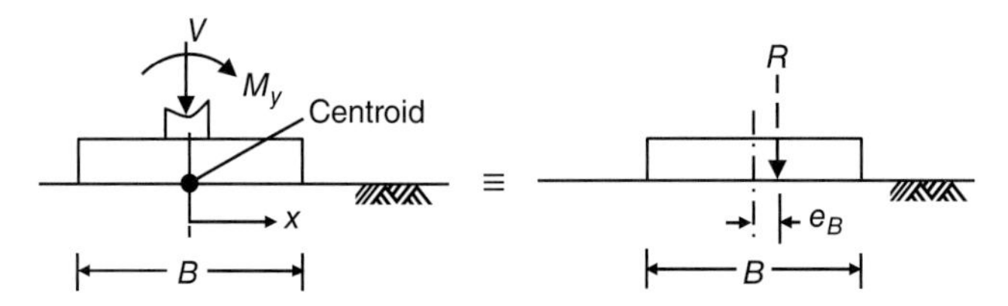

Figure 2.21 Eccentrically loaded spread footing.

small area directly under the columns. When a mat rests on stiff soils, the column loads are distributed to the supporting soil in larger areas (Figure 2.20b). The pressure beneath a mat foundation rests on soft soil approaches planar distribution (Figure 2.20c).

2.6.2 Contact Pressure under Eccentrically Loaded Spread Footings

In certain cases, as with a footing of a retaining wall, foundations are subjected to eccentric loading; the resultant of all the loads and moments applied off the centre of the foundation base area. Usually, this loading condition results from an eccentric vertical load (or a resultant of vertical loads) alone or from a concentric load plus moments in one or more directions. In general, any combinations of vertical loads and moments can be represented by a vertical load shifted to a fictitious location with an eccentricity, e, relative to the centroid of the base area, as shown in Figure 2.21.

This eccentricity may be calculated as follows:

For column footings and rigid mats subjected to eccentric vertical loads without moments,

$$e = \frac{V \times e_v}{V + W_f} \tag{2.2}$$

For continuous footings subjected to eccentric vertical loads without moments,

$$e = \frac{\left(\frac{V}{L}\right)e_v}{(V/L) + (W_f/L)} \tag{2.3}$$

For column footings and rigid mats subjected to concentric vertical loads with moments,

$$e = \frac{M}{V + W_f} \tag{2.4}$$

For continuous footings subjected to concentric vertical loads with moments,

$$e = \frac{\frac{M}{L}}{(V/L) + (W_f/L)} \tag{2.5}$$

where: e = eccentricity of the resultant vertical load
V = applied vertical load
e_v = eccentricity of applied vertical load
M = applied total moment
W_f = weight of foundation
L = length of continuous footing.

Eccentric loads produce non-uniform contact pressure distribution under footings. For practical design purposes, the contact pressure distribution is usually assumed linear, as shown in (Figure 2.22a). The contact pressure is calculated by assuming linearly elastic action in compression, across the contact between the footing and the soil. Eccentricity of the resultant load R larger than $B/6$ or $L/6$ (Figure 2.23) will cause a portion of a footing to lift off the soil (i.e. soil in tension), since the soil-footing interface cannot resist tension. The dimensions B and L are the actual width and length of the base area, respectively. For a rectangular footing, the distances $B/6$ and $L/6$ are called the *kern distance.*

The resultant load applied within the *kern*, the shaded area in Figure 2.22b, will cause compression over the entire area beneath the footing, and the contact pressure q under rigid spread foundations can be calculated using the following common flexural equation:

$$q = \frac{R}{A} \pm \frac{M_y}{I_y} x \pm \frac{M_x}{I_x} y \tag{2.6}$$

In this equation, I_x and I_y are the moments of inertia of the base about x and y axis, respectively. The x and y distances define location of q with respect to centroid of the base area. The moments M_x and M_y are about x and y axis, respectively.

For rectangular footings of B and L base dimensions Equation (2.6) may be written as:

$$q = \frac{R}{BL}\left(1 \pm \frac{6e_x}{B} \pm \frac{6e_y}{L}\right) \tag{2.7}$$

where $e_x = e_B$; $e_y = e_L$. They are eccentricities of R in x and y directions. This equation is applicable only when:

$$\frac{6e_B}{B} + \frac{6e_L}{L} \leq 1.0 \tag{2.8}$$

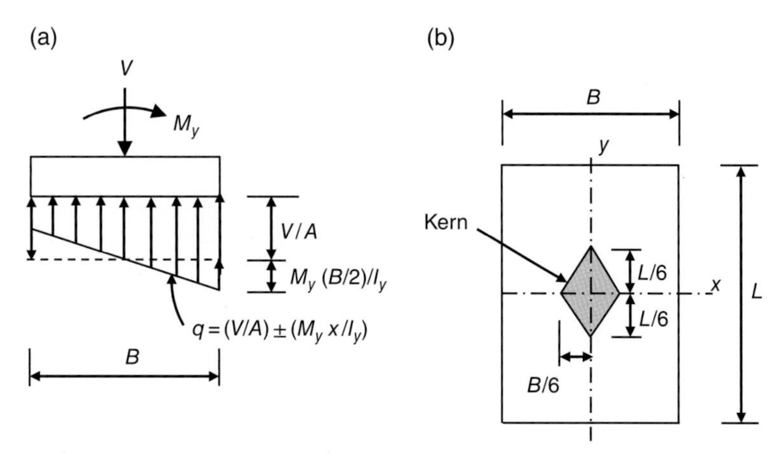

Figure 2.22 Contact pressure under eccentrically loaded rectangular spread footing with kern distances $B/6$ and $L/6$ indicated.

If the applied resultant load falls outside the kern area, Equation (2.6) will not be satisfied. However, the soil reaction resultant remains equal and opposite in direction to the resultant R, as shown in Figure 2.23. Generally, such a pressure distribution would not be acceptable because it makes insufficient use of the footing concrete and tends to overload the supporting soil. Therefore, foundations with eccentric loads and/or moments must have R always fall within the kern. This criterion maintains compressive stresses along the entire base area.

The kern distance for circular footings is $r/4$ where r is radius of the circular base area, as shown in Figure 2.24.

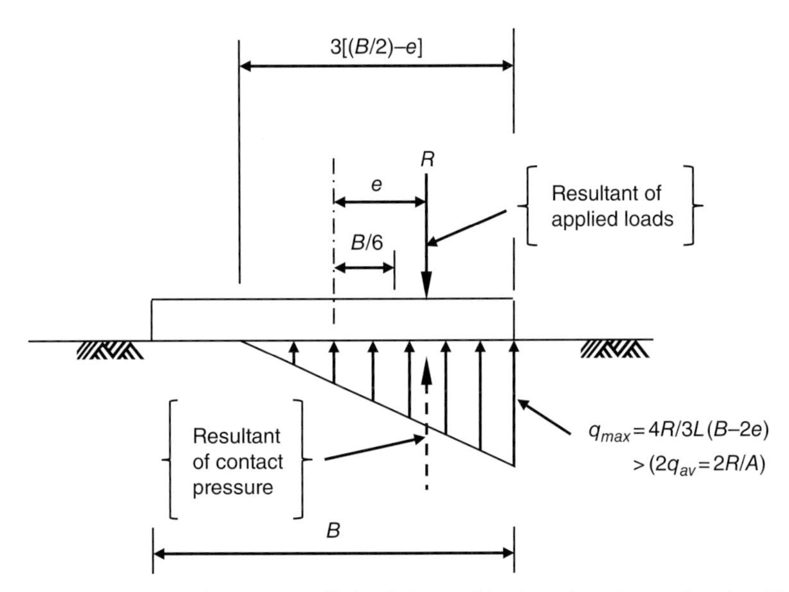

Figure 2.23 Contact pressure under eccentrically loaded spread footing where the resultant load falls outside the kern.

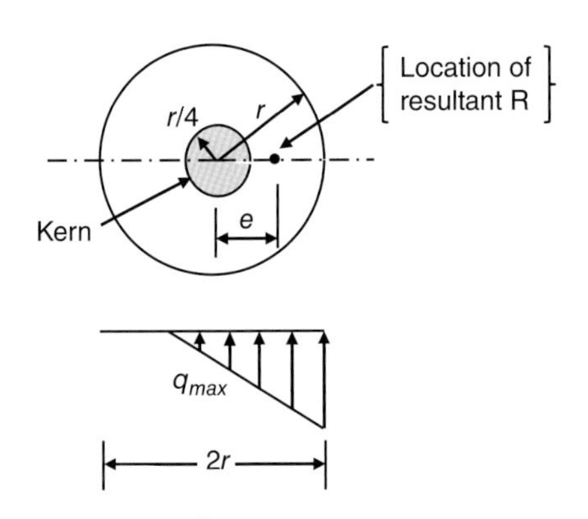

For $e \leq r/4$: $q = (R/A)\ [1 \pm (4e/r)]$, $A = \pi r^2$

For $e > \frac{r}{4}$: $q_{max} = k\left(\frac{R}{A}\right)$

Values of k are tabulated below:

e/r:	0.25	0.30	0.35	0.40	0.45	0.50	0.55	0.60	0.65	0.70	0.75	0.80	0.90
k:	2.00	2.20	2.43	2.70	3.10	3.55	4.22	4.92	5.90	7.20	9.20	13.0	80.0

Figure 2.24 Contact pressure under eccentrically loaded circular spread footings.

2.7 Vertical Stresses in a Soil Mass due to Foundation Loads

2.7.1 General

Knowledge of stresses at a point in a soil mass is required for the settlement analysis of foundations, the stability analysis of the soil mass, and the determination of earth pressures. Stresses in a soil deposit are due to weight of the soil itself, may be called *geostatic stresses*, and due to the external applied loads, such as the foundation loads, maybe called *induced stresses*. In soil engineering problems the geostatic stresses are significant; unlike many other civil engineering problems wherein the stresses due to self-weight are relatively small, such as in design of a steel superstructure. In a loaded soil mass, both the normal stress σ and the shear stress τ may exist. The normal horizontal stresses are usually represented by variables σ_x and σ_y, and the normal vertical stress by σ_z. Normal stresses, in reality, are nearly always compressive. Geotechnical engineers use compressive stresses positive while tensile stresses negative.

Regarding the geostatic stresses, foundation engineers are more interested in the vertical compressive geostatic stress. The vertical total stress σ_z and effective stress σ_z' at a depth z below the ground surface may be computed as $\sigma_z = \gamma z$ and $\sigma_z' = \gamma z - u = \gamma' z$, where u is the soil pore water pressure, γ is the average total unit weight of the soil and γ' is the effective unit weight. Generally, the unit weight of a natural soil deposit increases with depth due to the weight of soil above. Therefore, the unit weight of soil cannot be taken as constant. For this reason, the geostatic vertical stress may be expressed as

$$\sigma_z = \int_0^z \gamma d_z \tag{2.9}$$

For layered soils with variable unit weight, the vertical stress is given by

$$\sigma_z = \gamma_1 \Delta z_1 + \gamma_2 \Delta z_2 + \cdots + \gamma_n \Delta z_n = \Sigma \gamma \, \Delta z \tag{2.10}$$

In the following subsections the induced vertical stresses due to external applied loads, such as foundation loads, will be discussed.

2.7.2 Vertical Stress Due to a Concentrated Load

Normal and shear stresses at a point within a semi-infinite homogeneous, isotropic and elastic mass, due to a concentrated point load on the mass surface, where determined by Boussinesq (1883). Accordingly, the vertical stress increase σ_z at any point A due to a surface point load V (Figure 2.25) is given by

$$\sigma_z = \frac{3V}{2\pi z^2} \frac{1}{\left[1 + (r/z)^2\right]^{5/2}} = \frac{3V}{2\pi z^2}(I) \tag{2.11}$$

$$I = \frac{1}{\left[1 + (r/z)^2\right]^{5/2}} = \text{Influence factor}$$

Equation (2.11) is universally known as Boussinesq's equation.

The form of variation of σ_z with z and r is illustrated in Figure 2.26. The left-hand side of the figure shows the variation of σ_z with z on the vertical through the point of application of the load V (i.e. for $r = 0$). The right-hand side of the figure shows the variation of σ_z with three different values of z. It must be realised that these vertical stresses do not include the vertical stress due to self-weight of the overburden soil.

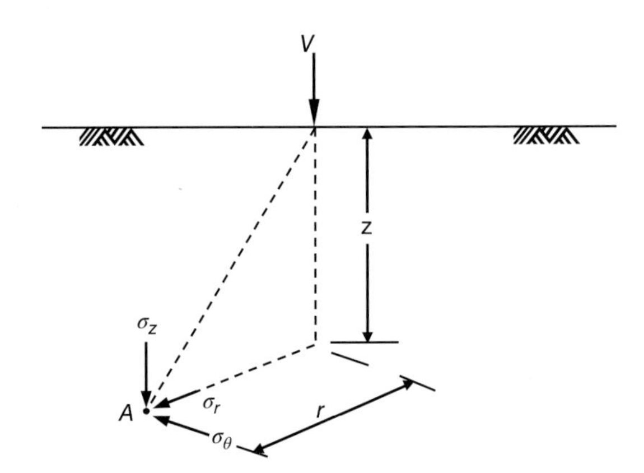

Figure 2.25 Stresses at point A of depth z below ground surface acted on by a point load.

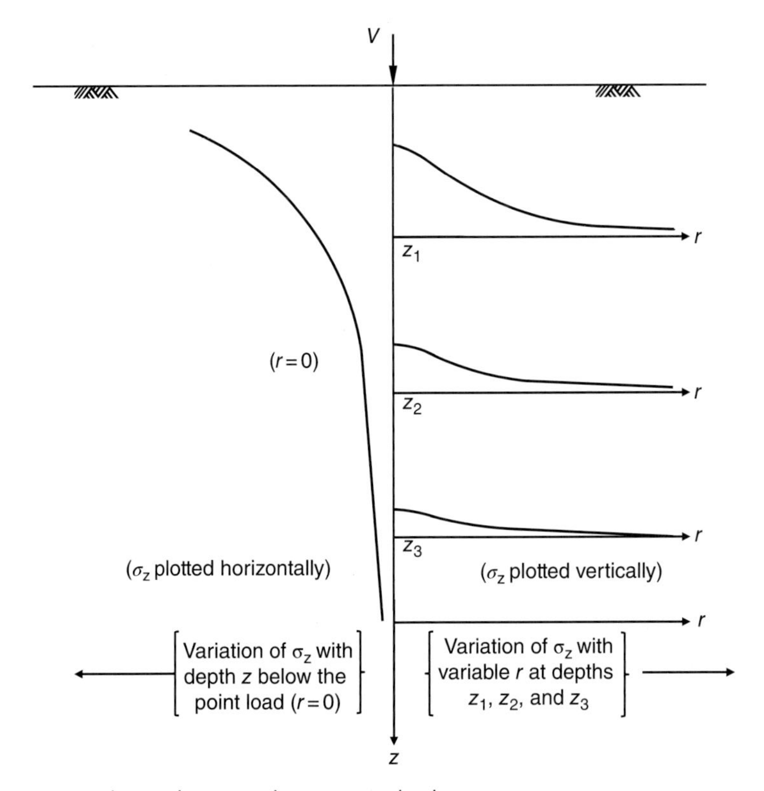

Figure 2.26 Variation of vertical stress σ_z due to a point load.

2.7.3 Vertical Stress Due to a Line Load

The Boussinesq's solution described in Section 2.7.2 can be used to obtain the vertical stresses in a soil mass due to a vertical line load. Figure 2.27 shows a line load of infinite length having intensity V per unit length on the surface of a semi-infinite soil mass. The vertical stress increase at point A inside the soil mass is

$$\sigma_z = \frac{2V}{\pi} \frac{z^3}{\left(x^2 + z^2\right)^2} \tag{2.12}$$

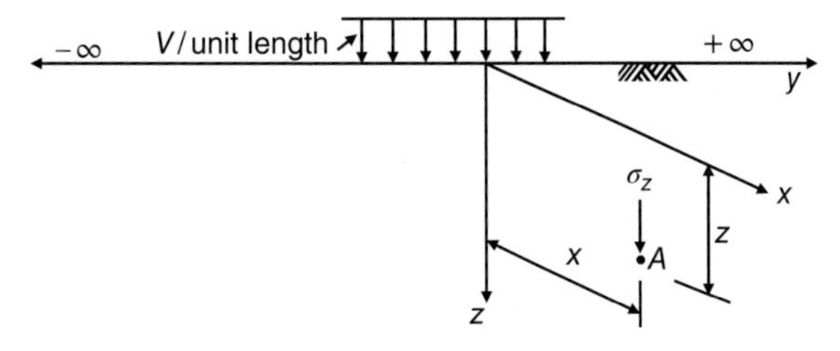

Figure 2.27 Vertical stress due to a line load of infinite length.

2.7.4 Vertical Stress Due to a Uniformly Loaded Strip Area

Referring to Figure 2.28a, the vertical stress at point A due to a uniform load per unit area q on a flexible strip area of width B and infinite length are given in terms of the angles α and β, as follows:

$$\sigma_z = \frac{q}{\pi}[\alpha + \sin\alpha\,\cos(\alpha + 2\beta)] \tag{2.13}$$

The vertical stress at points below the centre of the strip, such as point A of Figure 2.28b, is given as

$$\sigma_z = \frac{q}{\pi}(\alpha + \sin\alpha) \tag{2.14}$$

The vertical stress at point A of Figure 2.28c due to pressure increasing linearly from zero to q on a strip area of width B and infinite length is given in terms of the angles α and β as follows:

$$\sigma_z = \frac{q}{\pi}\left(\frac{x}{B}\theta - \frac{1}{2}\sin 2\beta\right) \tag{2.15}$$

When the point is located within B under the loaded area, β is negative.

2.7.5 Vertical Stress Due to a Uniformly Loaded Circular Area

Referring to Figure 2.29, the stresses at a point due to a uniformly loaded circular area with intensity q can be obtained by dividing the loaded area into many small elements, each with an area of dA and a small concentrated load at its centre equal to

$$dV = q\,dA$$

From the Boussinesq's equation of vertical stress due to a concentrated point load, the vertical stress at point A due to dV is

$$d\sigma_Z = \frac{3(dV)}{2\pi z^2}\frac{1}{\left[1 + (r/z)^2\right]^{5/2}} = \frac{3q}{2\pi z^2}\frac{1}{\left[1 + (r/z)^2\right]^{5/2}}dA$$

The stress increase σ_z can be found by integrating this equation over the loaded area. If we consider the vertical stress at any depth z underneath the centre of the loaded area of radius R, we will have

$$\sigma_Z = \frac{3q}{2\pi}\int_0^{2\pi}\int_0^R \frac{1}{z^2\left[1 + (r/z)^2\right]^{5/2}} r.\,d\theta.dr$$

where $r.d\theta.dr = dA$

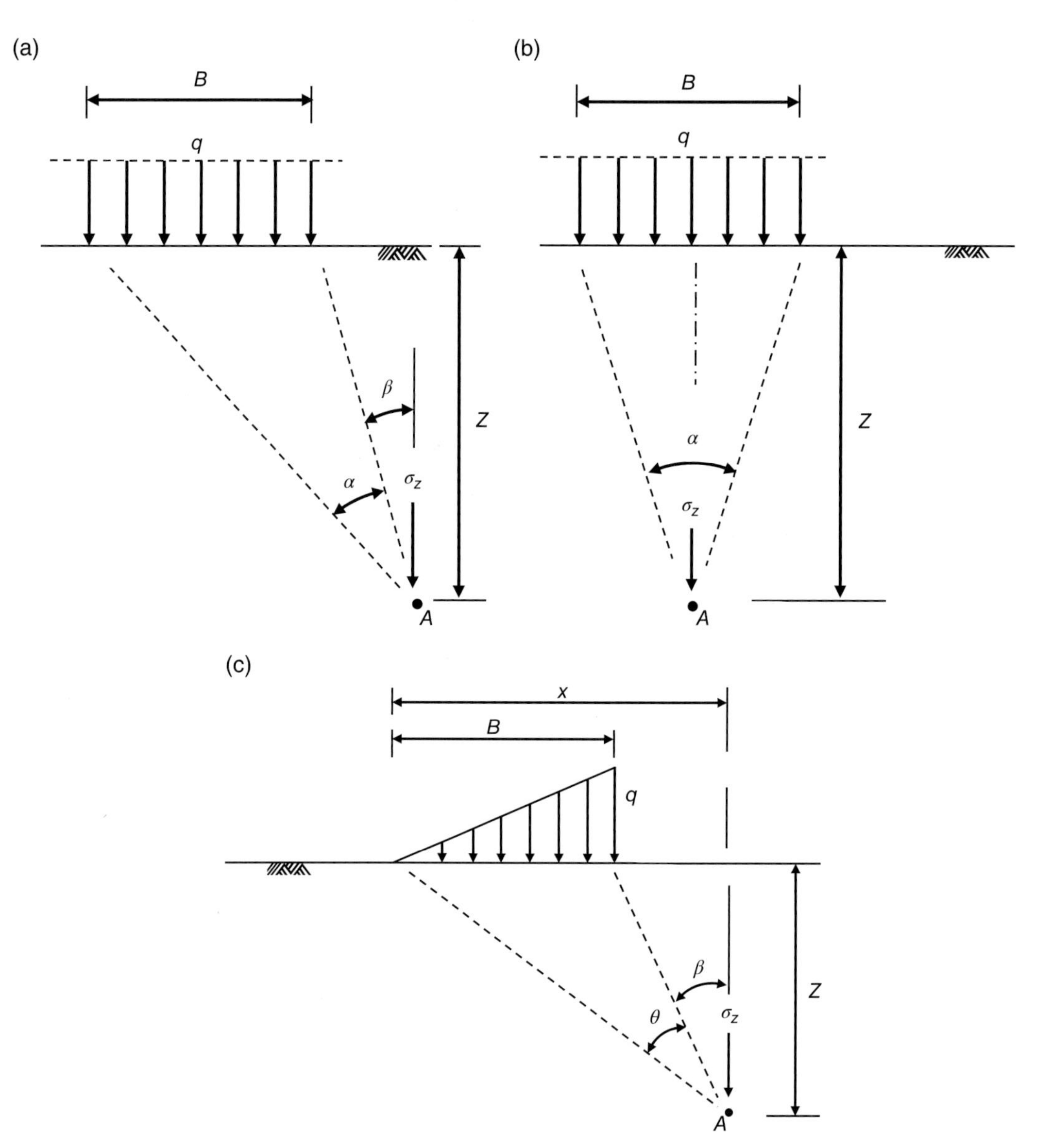

Figure 2.28 (a, b) Vertical stresses due to a uniformly loaded strip area of infinite length. (c) Vertical stresses due to a triangularly loaded trip area of infinite length.

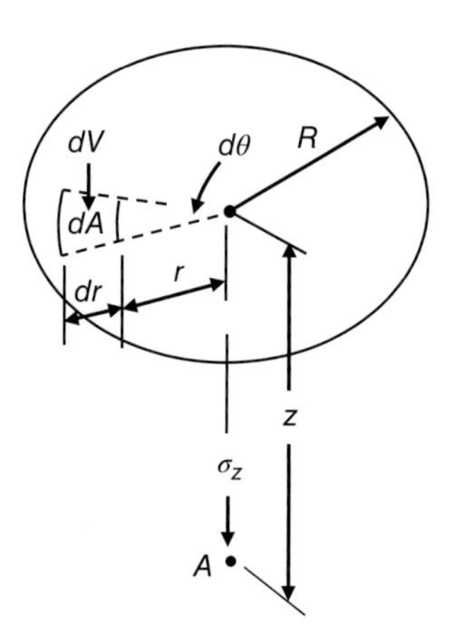

Figure 2.29 Vertical stresses below center of a uniformly loaded circular area.

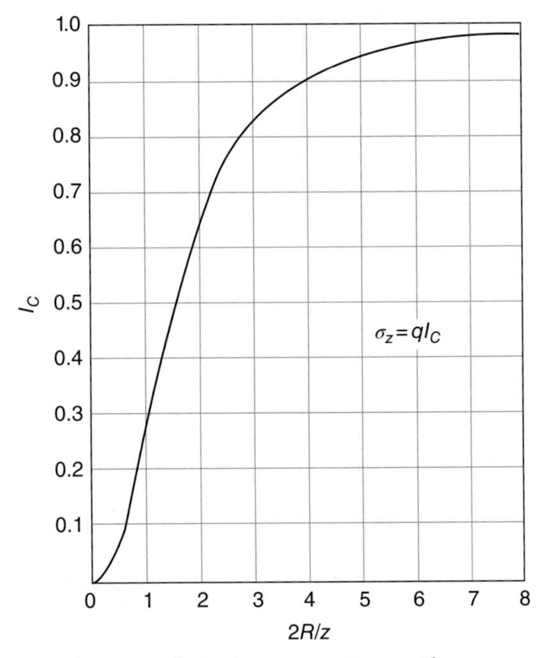

Figure 2.30 Vertical stresses under centre of circular area carrying a uniform pressure.

Performing the integrations and inserting the limits, we obtain

$$\sigma_z = q\left\{1 - \frac{1}{\left[1 + (R/z)^2\right]^{3/2}}\right\} = qI_c \tag{2.16}$$

This equation can be used to directly obtain the vertical stress at depth z under centre of a uniformly loaded flexible circular area of radius R. The bracket part of the equation is usually known as influence factor I_c which is function of the ratio R/z. For selected values of R/z, values of I_c can easily be calculated and tabulated, or, may be obtained in terms of the ratio $2R/z$ from the plot of Figure 2.30.

2.7.6 Vertical Stress Due to a Uniformly Loaded Rectangular Footing

The vertical stress increase at depth z under a *corner* of a *uniformly loaded flexible rectangular area* (Figure 2.31) can be obtained using the integration technique of the Boussinesq's equation.

Consider an elemental area $dA = dx\,dy$ on the loaded rectangular area (Figure 2.31). If the load per unit area is q, the total load on the elemental area is

$$dV = q\,dx\,dy$$

This elemental load dV may be treated as a point load which causes a vertical stress increase $d\sigma_z$ at point A. Boussinesq's equation of point load can be written as

$$\sigma_z = \frac{3V}{2\pi}\frac{z^3}{\left(r^2 + z^2\right)^{5/2}}$$

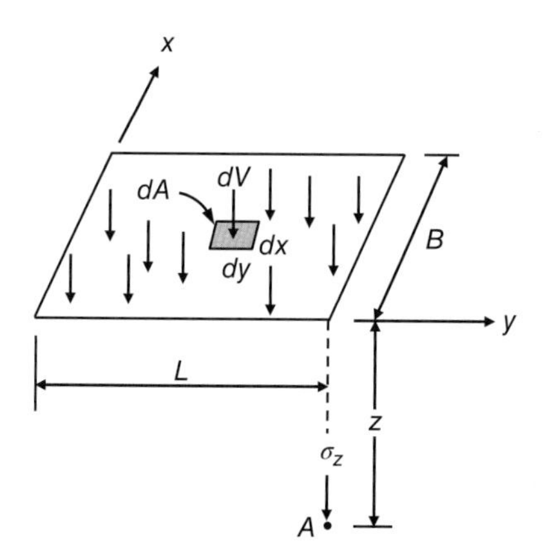

Figure 2.31 Vertical stresses σ_z at depth z below a corner of a uniformly loaded rectangular area.

Using $r^2 = x^2 + y^2$; the equation can be rewritten for the elemental point load as

$$d\sigma_z = \frac{3q(dx\,dy)z^3}{2\pi(x^2+y^2+z^2)^{5/2}}$$

The value of σ_z at point A caused by the entire loaded rectangular area of dimensions B and L may now be obtained by integrating the preceding equation:

$$\sigma_z = \int_{y=0}^{L}\int_{x=0}^{B} \frac{3q(dx\,dy)z^3}{2\pi(x^2+y^2+z^2)^{5/2}} = qI_r \tag{2.17}$$

The influence factor I_r is function of two variables m and n, where

$$m = B/z;\ n = L/z;$$

It is of interest to note that for any magnitude of z the vertical stress σ_z depends only on the ratios m and n and the surface load intensity q.

The variations of I_r values with m and n, required in computing σ_z below *corner* of a uniformly loaded rectangular area, are given in Table 2.3. The table is usually known as *Newmark Table*, since the earliest use of the integration technique of the Boussinesq's equation has been attributed to Newmark (1935).

Values of I_r in terms of m and n are also given in the chart of Figure 2.32 (note that I_r is I_{qr} in the chart); due to Fadum (1948).

Equation (2.17) can also be used to find the vertical stress at a point which is not located below the corner (Figure 2.33). The rectangular area, carrying a uniform pressure q, is subdivided into four rectangles such that each rectangle has a corner at the point where the vertical stress is required. The principle of superposition is used to determine the vertical stress at the point.

Figure 2.33a shows location of any point P on the uniformly loaded rectangular area *EFGH*. This rectangle is subdivided into four smaller rectangles *EIPL, IFJP, PJGK* and *LPKH*. These rectangles have their influence factors I_1, I_2, I_3 and I_4, respectively. The vertical stress at point A of depth z below point P due to the uniformly loaded rectangular area *EFGH* is

$$\sigma_z = q\,(I_1 + I_2 + I_3 + I_4)$$

Table 2.3 Values of the influence factor I_r (Equation 2.17) to compute vertical stresses beneath the *corner* of a uniformly loaded rectangular area.

	m									
n	0.1	0.2	0.3	0.4	0.5	0.6	0.7	0.8	0.9	1.0
0.1	0.005	0.009	0.013	0.017	0.020	0.022	0.024	0.026	0.027	0.028
0.2	0.009	0.018	0.026	0.033	0.039	0.043	0.047	0.050	0.053	0.055
0.3	0.013	0.026	0.037	0.047	0.056	0.063	0.069	0.073	0.077	0.079
0.4	0.017	0.033	0.047	0.060	0.071	0.080	0.087	0.093	0.098	0.101
0.5	0.020	0.039	0.056	0.071	0.084	0.095	0.103	0.110	0.116	0.120
0.6	0.022	0.043	0.063	0.080	0.095	0.107	0.117	0.125	0.131	0.136
0.7	0.024	0.047	0.069	0.087	0.103	0.117	0.128	0.137	0.144	0.149
0.8	0.026	0.050	0.073	0.093	0.110	0.125	0.137	0.146	0.154	0.160
0.9	0.027	0.053	0.077	0.093	0.116	0.131	0.144	0.154	0.162	0.168
1.0	0.028	0.055	0.079	0.101	0.120	0.136	0.149	0.160	0.168	0.175
1.1	0.029	0.056	0.082	0.104	0.124	0.140	0.154	0.165	0.174	0.181
1.2	0.029	0.057	0.083	0.106	0.126	0.143	0.157	0.168	0.178	0.185
1.3	0.030	0.058	0.085	0.108	0.128	0.146	0.160	0.171	0.181	0.189
1.4	0.030	0.059	0.086	0.109	0.130	0.147	0.162	0.174	0.184	0.191
1.5	0.030	0.059	0.086	0.110	0.131	0.149	0.164	0.176	0.186	0.194
2.0	0.031	0.061	0.089	0.113	0.135	0.153	0.169	0.181	0.192	0.200
2.5	0.031	0.062	0.089	0.114	0.136	0.155	0.170	0.183	0.194	0.202
3.0	0.031	0.062	0.090	0.115	0.137	0.155	0.171	0.184	0.195	0.203
5.0	0.032	0.062	0.090	0.115	0.137	0.156	0.172	0.185	0.196	0.204
10.0	0.032	0.062	0.090	0.115	0.137	0.156	0.172	0.185	0.196	0.205

	m									
n	1.1	1.2	1.3	1.4	1.5	2.0	2.5	3.0	5.0	10.0
0.1	0.029	0.029	0.030	0.030	0.030	0.031	0.031	0.031	0.032	0.032
0.2	0.056	0.057	0.058	0.059	0.059	0.061	0.062	0.062	0.062	0.062
0.3	0.082	0.083	0.085	0.086	0.086	0.089	0.089	0.090	0.090	0.090
0.4	0.104	0.106	0.108	0.109	0.110	0.113	0.114	0.115	0.115	0.115
0.5	0.124	0.126	0.128	0.130	0.131	0.135	0.136	0.137	0.137	0.137
0.6	0.140	0.143	0.146	0.147	0.149	0.153	0.155	0.155	0.156	0.156
0.7	0.154	0.157	0.160	0.162	0.164	0.169	0.170	0.171	0.172	0.172
0.8	0.165	0.168	0.171	0.174	0.176	0.181	0.183	0.184	0.185	0.185
0.9	0.174	0.178	0.181	0.184	0.186	0.192	0.194	0.195	0.196	0.196
1.0	0.181	0.185	0.189	0.191	0.194	0.200	0.202	0.203	0.204	0.205
1.1	0.186	0.191	0.195	0.198	0.200	0.207	0.209	0.211	0.212	0.212
1.2	0.191	0.196	0.200	0.203	0.205	0.212	0.215	0.216	0.217	0.218

(*Continued*)

Table 2.3 (*Continued*)

n	m = 1.1	1.2	1.3	1.4	1.5	2.0	2.5	3.0	5.0	10.0
1.3	0.195	0.200	0.204	0.207	0.209	0.217	0.220	0.221	0.222	0.223
1.4	0.198	0.203	0.207	0.210	0.213	0.221	0.224	0.225	0.226	0.227
1.5	0.200	0.205	0.209	0.213	0.216	0.224	0.227	0.228	0.230	0.230
2.0	0.207	0.212	0.217	0.221	0.224	0.232	0.236	0.238	0.240	0.240
2.5	0.209	0.215	0.220	0.224	0.227	0.236	0.240	0.242	0.244	0.244
3.0	0.211	0.216	0.221	0.225	0.228	0.238	0.242	0.244	0.246	0.247
5.0	0.212	0.217	0.222	0.226	0.230	0.240	0.244	0.246	0.249	0.249
10.0	0.212	0.218	0.223	0.227	0.230	0.240	0.244	0.247	0.249	0.250

Note: The factors m and n are interchangeable.

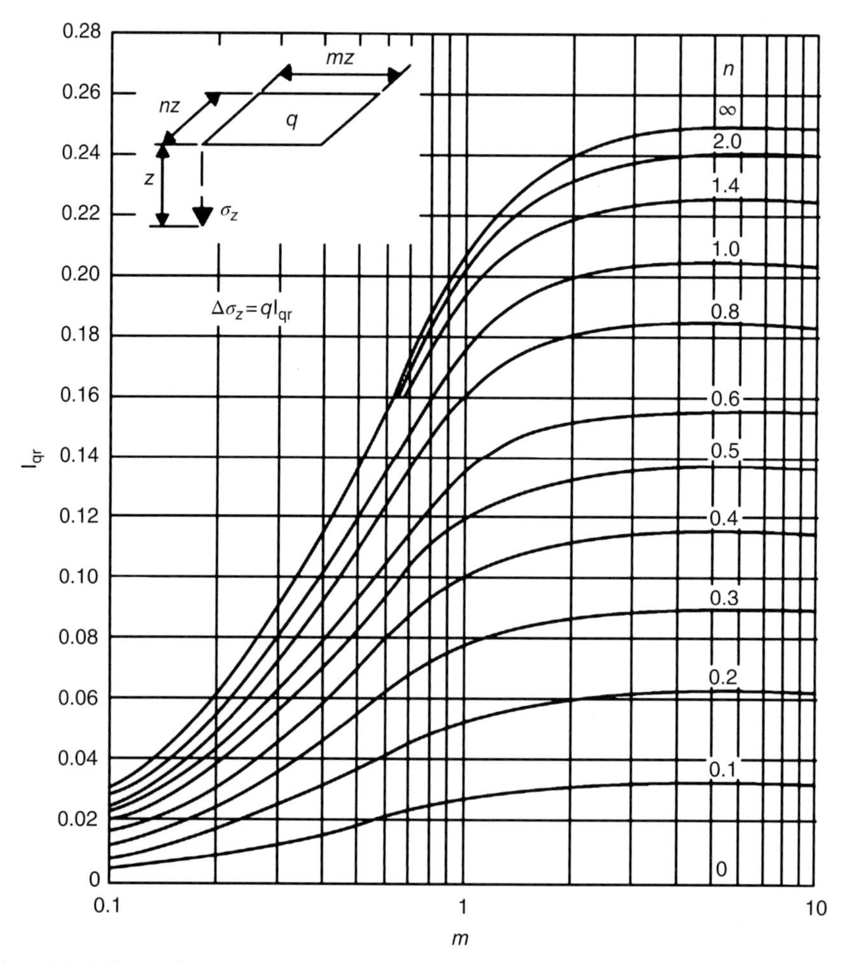

Figure 2.32 Values of the influence factor for computing vertical stress under corner of rectangular area carrying a uniform pressure *(reproduced from Knappett and Craig, 2012; after R.E. Fadum, 1948).*

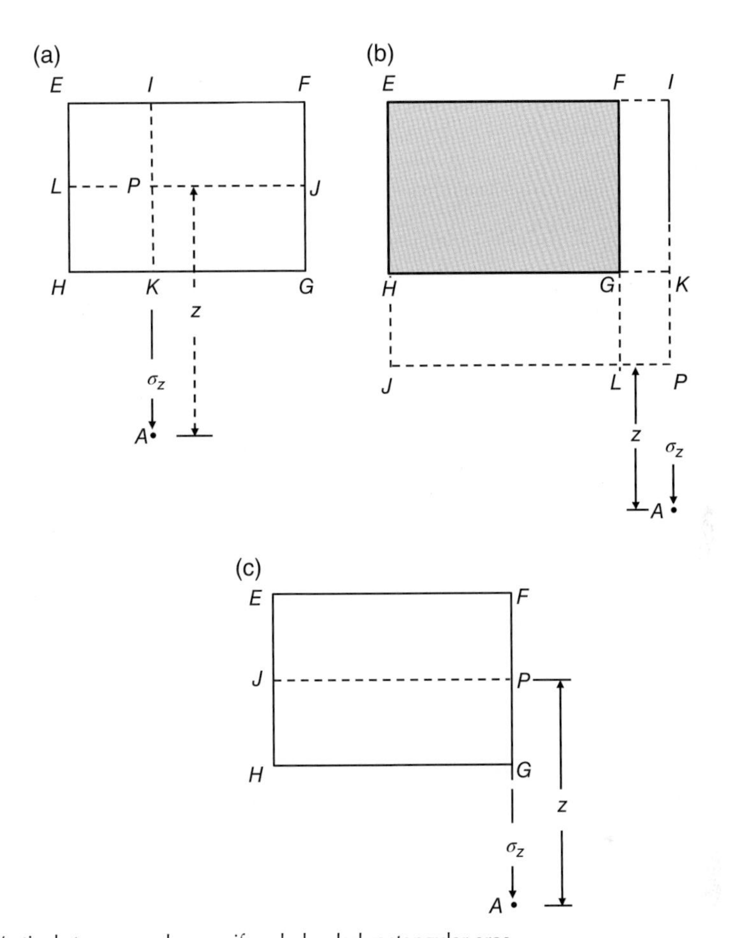

Figure 2.33 Vertical stresses under a uniformly loaded rectangular area.

Figure 2.33b shows location of any point *P* outside the loaded area *EFGH*. In this case, a large rectangle *EIPJ* is drawn such that point *P* is one of its corners. Therefore, the vertical stress at point *A* of depth *z* below point *P* due to the uniformly loaded rectangular area *EFGH* is

$$\sigma_z = q\,(I_1 - I_2 - I_3 + I_4)$$

The influence factors I_1, I_2, I_3 and I_4 belong to the rectangles *EIPJ, FIPL, HKPJ* and *GKPL*, respectively.

Figure 2.33c shows any point *P* located on an edge of the loaded area *EFGH*. In order to find the vertical stress at point *A* of depth *z* below point *P*, the area is divided into two smaller rectangles *EFPJ* and *JPGH*. Hence,

$$\sigma_z = q\,(I_1 + I_2)$$

The influence factors I_1 and I_2 belong to the rectangles *EFPJ* and *JPGH*, respectively.

There is an approximate method, referred to as the $2V:1H$ *method*, used to estimate vertical stresses caused by foundation loads (Figure 2.34). However, since the method gives only rough estimates of stress values, it is not very much used at present.

According to this method, the foundation load spreads out within the supporting soil on a slope two vertical to one horizontal.

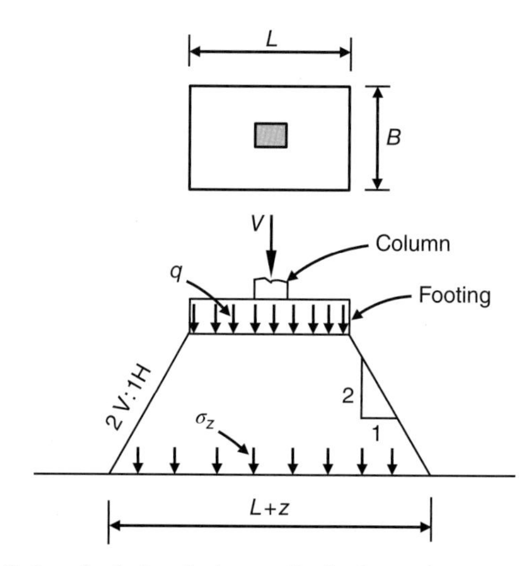

Figure 2.34 Approximate 2: 1 method of vertical stress distribution under a rectangular footing.

Thus, a vertical load V, distributed as a uniform pressure q over a horizontal area $B \times L$ (loaded area), causes an increase in pressure σ_z, at depth z below the loaded area, over an area with dimensions $(B + z)$ $(L + z)$. Hence,

$$\sigma_z = \frac{V}{(B + z)(L + z)} \quad \text{or} \quad \sigma_z = q\frac{B \times L}{(B + z)(L + z)} \tag{2.18}$$

According to Bowles (2001), the $2V : 1H$ method compares reasonably well with more theoretical methods for values of z equals B to about $4B$ but should not be used for z values between 0 and B value.

2.7.7 Newmark's Chart Method of Determining Vertical Stresses

Equation 2.16 can be rearranged and written in the following form:

$$\frac{R}{z} = \sqrt{\left(1 - \frac{\sigma_z}{q}\right)^{-2/3} - 1}$$

As it is clear, the ratios R/z and σ_z/q are non-dimensional quantities. This equation reveals that the R/z ratio is the relative size of a uniformly loaded circular area which gives a unique pressure ratio σ_z/q on a soil element of depth z in the soil mass.

Values of R/z that correspond to various pressure ratios may be computed and tabulated, as shown below:

σ_z/q:	0.0	0.1	0.2	0.3	0.4	0.5	0.6	0.7	0.8	0.9	1.0
R/z:	0.0	0.27	0.4	0.518	0.637	0.766	0.918	1.11	1.387	1.908	∞

Using these R/z and σ_z/q values, Newmark (1942) developed influence charts, such as the chart of Figure 2.35, which can be used to determine vertical pressure σ_z below any point x on a uniformly loaded flexible area of any shape.

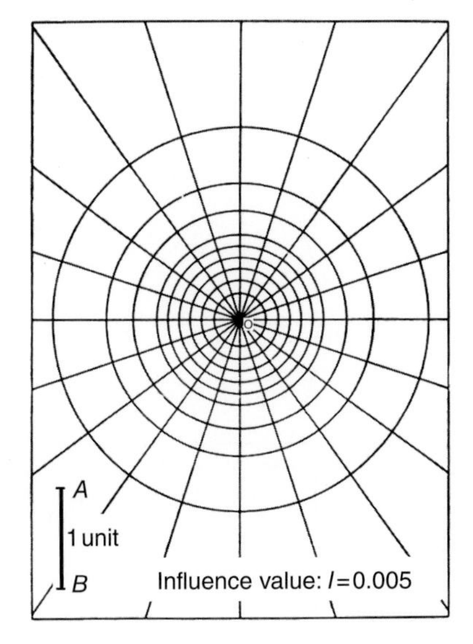

Figure 2.35 Influence chart for computing vertical stresses based on Boussinesq's theory *(after Newmark, 1942)*.

The influence chart is useful in cases where access to a computer is not practical and there are several footings with different contact pressures or where the footing is irregular-shaped and σ_z is desired for a number of points.

An influence chart consists of concentric circles and radial lines (Figure 2.35). The radii of the circles are equal to the R/z values corresponding to $\sigma_z/q = 0, 0.1, 0.2 \ldots 1.0$ (note that for $\sigma_z/q = 0$, $R/z = 0$, and for $\sigma_z/q = 1$, $R/z = \infty$, so nine circles are shown). The unit length for plotting the circles is the length of the scale line AB. The circles are divided by several equally spaced radial lines. The influence value I of the chart is given by $1/N$, where N equals to number of elements in the chart. In the chart of Figure 2.35, there are 200 elements; hence, the influence value $I = 0.005$.

Steps of the procedure for obtaining vertical stress σ_z at a point of depth z below a uniformly loaded area of any shape are as follows:

- Determine the depth z below the uniformly loaded area at which the vertical stress is required.
- Plot the plan of the loaded area on a transparent paper to a scale such that the length of the scale line AB represents the depth z at which the vertical stress σ_z is required. On the drawn plan locate the point, say point x, below which σ_z is required.
- Place the plan on top of the influence chart with point x located at the origin of the chart.
- Count the number of elements M occupied by the plan of the loaded area.
- The vertical stress at the point under consideration is given by

$$\sigma_z = I\,Mq \tag{2.19}$$

where I and M are dimensionless quantities.

2.7.8 Pressure Bulbs Method of Determining Vertical Stresses

Based on Boussinesq's equation, the vertical stresses under continuous (strip), rectangular, square and circular footings have been computed. For example, the results due to uniformly loaded square and continuous areas are shown in Figures 2.36 and 2.37. The magnitudes of vertical stresses at various points in a soil mass are given in terms of the uniform contact pressures q_o in Figure 2.36 and q in Figure 2.37. The vertical stress is designated q in Figure 2.36. The *pressure bulbs* are isobars (lines

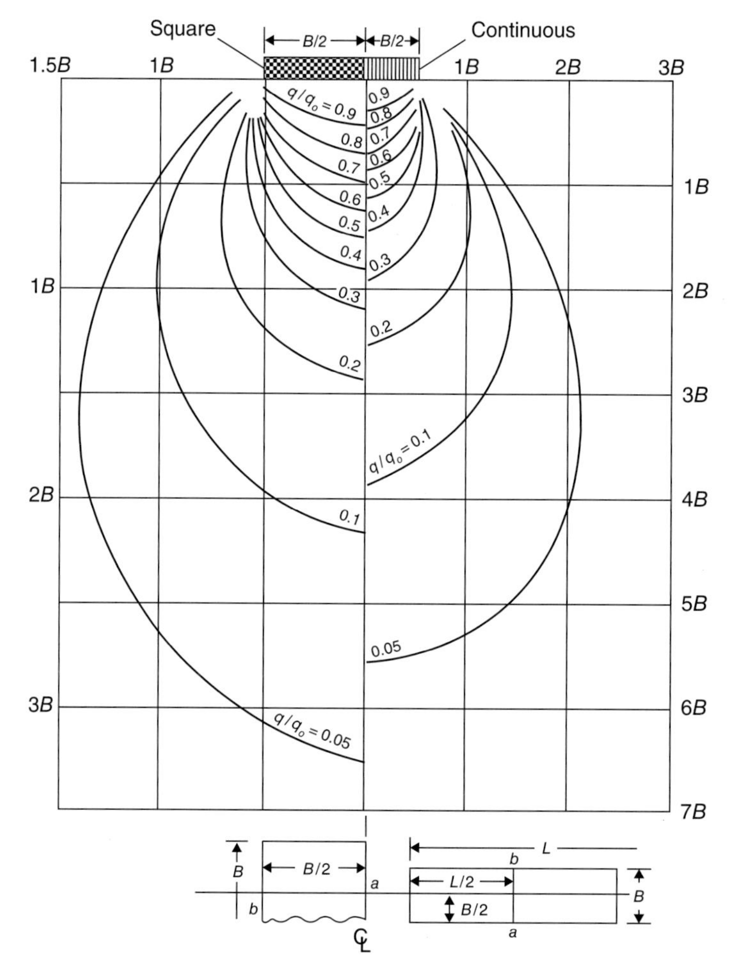

Figure 2.36 Pressure isobars (also called pressure bulbs) based on the Boussinesq's equation for square and long footings. Applicable only along line *ab* from centre to edge of base *(From Bowles, 2001)*.

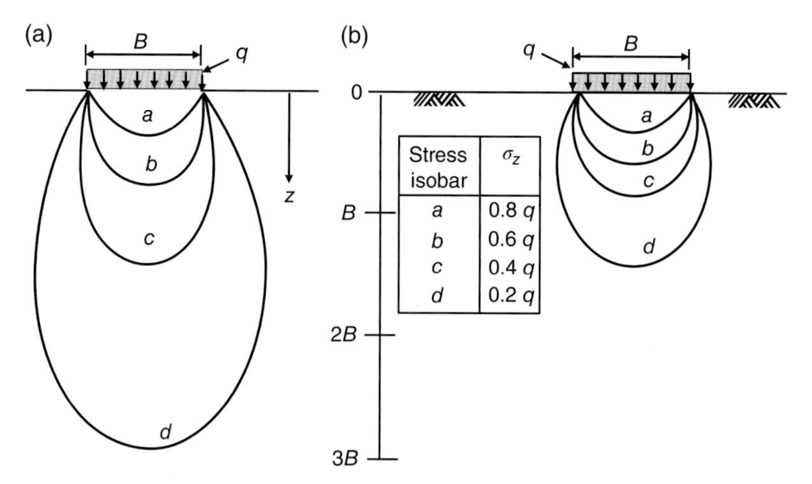

Stress isobar	σ_z
a	0.8 q
b	0.6 q
c	0.4 q
d	0.2 q

Figure 2.37 Contours of equal vertical stress: (a) under strip area, (b) under square area *(redrawn from Knappett and Craig, 2012)*.

of constant vertical stress) obtained by constructing vertical stress profiles at selected points across the footing width B and interpolating points of vertical stress intensity.

All concepts of the size of the pressure bulb depend on an arbitrary choice of the magnitude of a stress at which other stress values are considered to pass from appreciable or significant to inappreciable or negligible. If vertical stresses are considered to be of inappreciable magnitude when they are smaller than 20% of the uniform contact pressure at the surface of loading, the isobar (stress contour) labelled 0.2 in Figure 2.36 or 0.2 q in Figure 2.37 may be said to define the outline of the pressure bulb. However, it is generally considered that the zone inside the isobar labelled 0.1 is the *bulb of pressure*. The zones outside this bulb of pressure are assumed to have negligible vertical stresses. On this basis the depth of the bulb is between 1.5 and 2.0 times the diameter of the uniformly loaded circular surface area. It is, of course, an arbitrary choice, but in general bulb of pressure is considered to have a depth roughly 1.5 times the breadth of the loaded area.

Knowledge of isobars is useful for determining the effect of the foundation load on the vertical stresses at various points in the supporting material. Consequently, the bulb of pressure has a significant effect on the settlement of structures. Also, it is of extreme importance in subsurface exploration regarding the depth to which preliminary borings should penetrate.

2.7.9 Average Vertical Stress Due to a Loaded Rectangular Area

The average vertical stress, with limits of depth $z = 0$ to $z = H$, below the *corner* of a uniformly loaded rectangular area (Figure 2.38), can be evaluated as

$$\sigma_{z,av} = \frac{1}{H}\int_0^H (qI)\,dz = qI_a \tag{2.20a}$$

where $I_a = f(m,n)$, $m = \dfrac{B}{H}$, and $n = \dfrac{L}{H}$

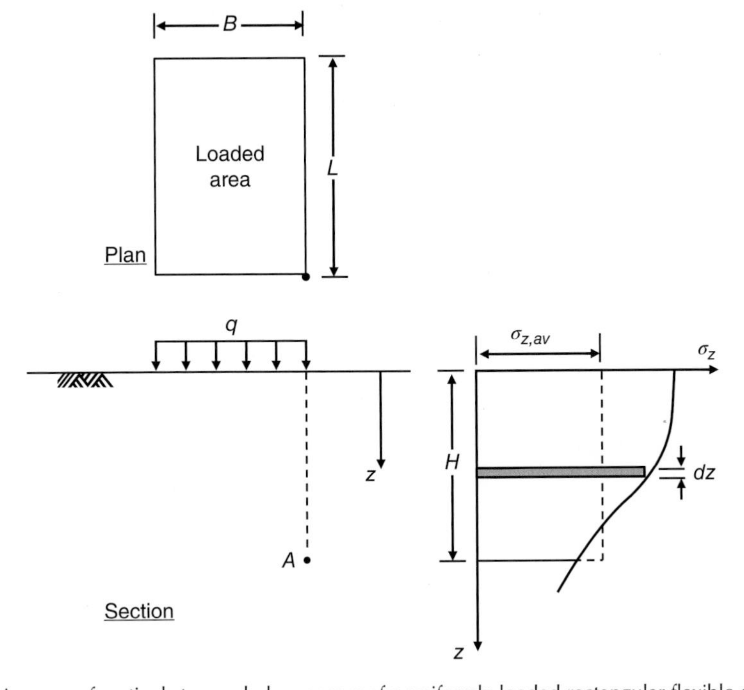

Figure 2.38 Average of vertical stresses below corner of a uniformly loaded rectangular flexible area.

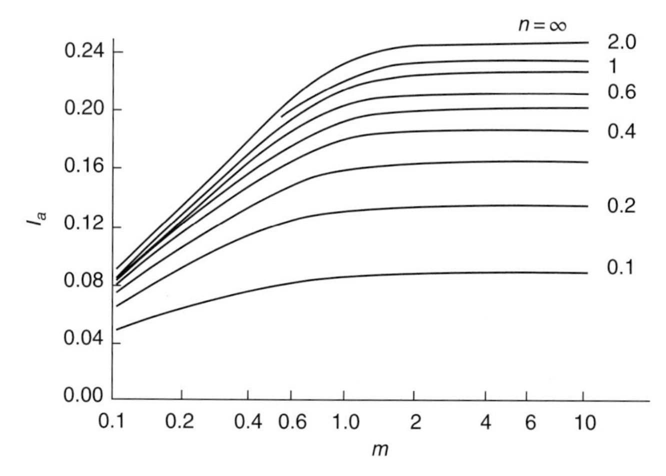

Figure 2.39 Griffiths influence factor, I_a.

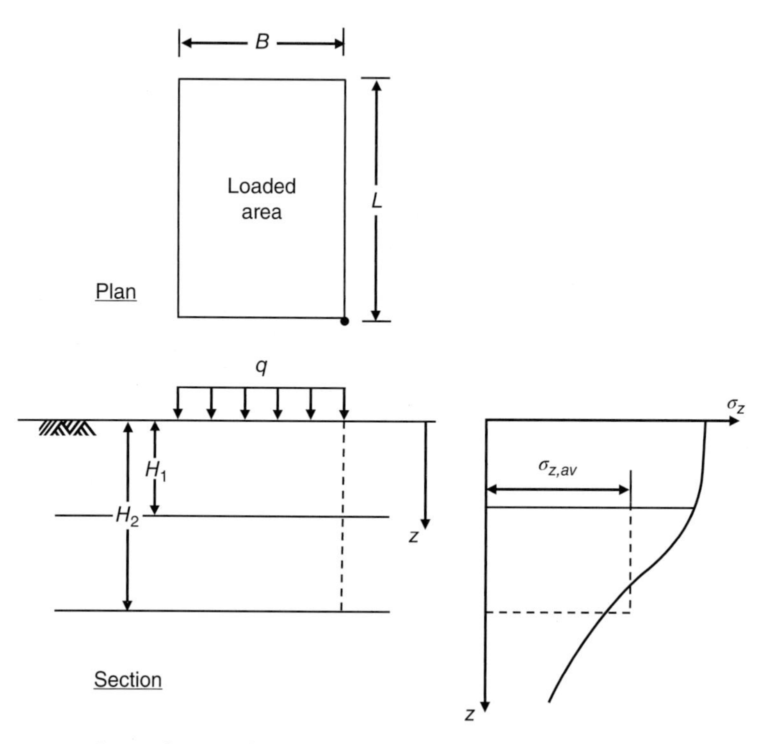

Figure 2.40 Average of vertical stresses between $z = H_1$ and $z = H_2$ below the corner of a uniformly loaded rectangular flexible area.

The variation of I_a with m and n is shown in Figure 2.39, as proposed by Griffiths (1984).

Considering a given layer between $z = H_1$ and $z = H_2$, as shown in Figure 2.40, the average vertical stress below the *corner* of a uniformly loaded rectangular flexible area can be determined as (Griffiths, 1984)

$$\sigma_{av} = q\left[\frac{H_2 I_{a(H_2)} - H_1 I_{a(H_1)}}{H_2 - H_1}\right] \tag{2.20b}$$

Where $I_{a(H_2)} = I_a = f\left(m = \frac{B}{H_2}, n = \frac{L}{H_2}\right)$ for $z = 0$ *to* $z = H_2$, and

$$I_{a(H_1)} = I_a = f\left(m = \frac{B}{H_1}, n = \frac{L}{H_1}\right) \text{ for } z = 0 \text{ to } z = H_1$$

Another simple and nearly accurate method for determination of an average vertical stress σ_{av} in a layer is by the use of the following *Simpson's rule*, provided that the vertical stress values at top, middle and bottom of the layer are known:

$$\sigma_{av} = \frac{1}{6}(\sigma_{zt} + 4\sigma_{zm} + \sigma_{zb})$$

Where σ_{zt}, σ_{zm} and σ_{zb} are vertical stresses at top, middle and bottom of the layer.

2.7.10 Westergaard's Equations

An actual soil mass may not be as homogeneous and isotropic as Boussinesq's solution assumes. Sedimentary soil deposits are generally anisotropic.

Typical clay strata usually have partings or thin lenses of coarser and more rigid materials sandwiched within them. These lenses or thin sheets are the cause of a greatly increased resistance to lateral strain.

Westergaard (1938) derived equations, for computing vertical stress σ_z, based on conditions which are nearly analogous to extreme conditions exist in reality. Therefore, Westergaard's elastic solution can give a better estimate of vertical stress in actual sedimentary soil deposits and in a soil mass consists of layered strata of fine and coarse materials.

Westergaard's equation for the vertical stress caused by a surface point load Q is

$$\sigma_z = \frac{Q\frac{1}{2\pi}\sqrt{\frac{1-2\mu}{2-2\mu}}}{z^2\left[\frac{(1-2\mu)}{(2-2\mu)} + \left(\frac{r}{z}\right)^2\right]^{3/2}} \tag{2.21}$$

As it is seen, the Westergaard's equation, unlike that of the Boussinesq's, includes Poisson's ratio μ; the other terms are the same as defined in the Boussinesq equation. Also, as done for the Boussinesq equation we can write this equation as

$$\sigma_z = \frac{Q}{z^2} I_w \tag{2.22}$$

Where, I_w represents *Westergaard influence coefficient*. For $\mu = 0.30$ we obtain the following values:

r/z	0.0	0.1	0.2	0.3	0.4	0.5	0.75	1.0	1.5	2.0
I_w	0.557	0.529	0.458	0.369	0.286	0.217	0.109	0.058	0.021	0.01

According to Taylor (1948), for cases of point loads with r/z less than about 0.8, the Westergaard's equation, assuming $\mu = 0$, gives values of vertical stresses which are approximately equal to two thirds of the values given by the Boussinesq equation. Also, it gives results as reasonable as any for use in connection with soil analyses, since it gives the flattest curve of vertical stress variation (or influence factor) with depth and a flat curve is the logical shape for a case of large lateral restraint.

For the assumed condition of $\mu = 0$, Equation (2.21) becomes

$$\sigma_z = \frac{Q \quad \frac{1}{\pi}}{z^2\left[1+2\left(\frac{r}{z}\right)^2\right]^{3/2}} = \frac{Q}{z^2}(I_W) \tag{2.23}$$

Similarly as for the Boussinesq equation, Westergaard's equation, Equation (2.21), can also be integrated over uniformly loaded circular and rectangular areas to obtain equations for σ_z, as written below:

For a circular area

$$\sigma_z = q\left(1-\sqrt{\frac{a}{\left(\frac{r}{z}\right)^2 + a}}\right) = q(I_{Wc}) \quad \left(a = \frac{1-2\mu}{2-2\mu}\right) \tag{2.24}$$

For a rectangular area; σ_z below corner of the area

$$\sigma_z = \frac{q}{2\pi}\cot^{-1}\left[a\left(\frac{1}{m^2}+\frac{1}{n^2}\right)+a^2\frac{1}{m^2n^2}\right]^{1/2} = \frac{q}{2\pi}(I_{Wr}) \tag{2.25a}$$

$$\sigma_z = \frac{q}{2\pi}\cot^{-1}\left[\frac{1}{2m^2}+\frac{1}{2n^2}+\frac{1}{4m^2n^2}\right]^{1/2} \quad (\text{for } \mu = 0) \tag{2.25b}$$

The factors m and n are as previously defined with the Newmark table or Fadum chart.

Also, just like Newmark's charts, there are influence charts based on Westergaard's equation, for a given value of μ, available in geotechnical text books. They are used in the same manner as that for the Boussinesq (Newmark) charts.

According to Taylor (1948), it has been found that estimates of settlements obtained by use of the Boussinesq equations for estimation of stresses, which is a procedure widely in use, are in the great majority of cases larger than the observed settlements. This may be somewhat of an indication that the Boussinesq equations give stress values which are relatively large, although it also could be the result of other assumptions that are used in settlement estimates.

For these reasons the Westergaard equations tend to be accepted by many engineers as somewhat preferable to the Boussinesq equations for use in settlement predictions of sedimentary soil deposits.

Problem Solving

Problem 2.1

Compute the effective contact pressure q beneath the following foundations:

(a) An individual column footing is 0.5 m × 2.0 m × 2.0 m. The column load is 500 kN. Unit weights of concrete and the overburden soil are 24 and 18 kN/m^3, respectively. The foundation depth is 2.5 m, and the water table is located at a depth of 2.2 m below the ground surface.

(b) A continuous concrete footing is 0.75 m wide and 0.50 m thick, supports a wall load of 100 kN/m. The foundation level is at a depth of 0.5 m below the ground surface. The soil has a unit weight of 17 kN/m^3. The ground water table is at a depth of 8 m below the ground surface.

(c) A mat foundation of a building structure is 1.2 m × 40 m × 80 m. The sum of the column and wall loads is 800 MN. The foundation depth is 10 m, and the water table exists at 5 m depth below the ground surface. The unit weight γ of the soil above and below water table are 17 and 20 kN/m^3 respectively.

Also, compute the net increase in the vertical effective stress immediately below the mat centre.

Solution:

(a)

$$W_f = 2 \times 2 \times 0.5 \times 24 = 48\,\text{kN}$$

Weight of overburden (backfill) soil

$$= 2 \times 2 \times 2 \times 18 = 144\,\text{kN}$$

The upward pore water pressure

$$= 10(2.5 - 2.2) = 3\,\text{kN/m}^2$$

The effective contact pressure is

$$q = \frac{(500 + 48 + 144)}{2 \times 2} - 3 = 170\,kPa$$

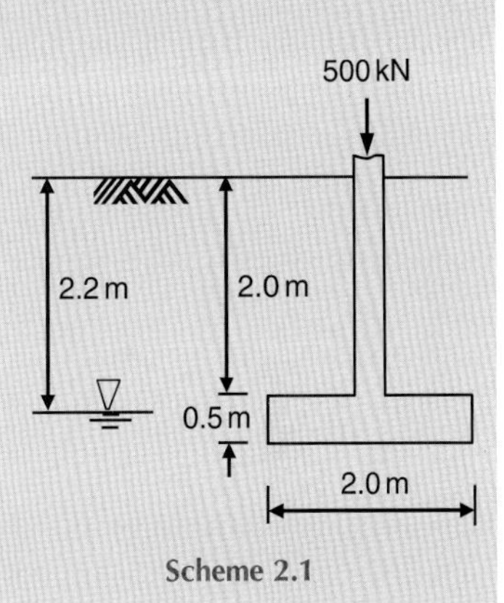

Scheme 2.1

(b)

$$W_f = 1 \times 0.75 \times 0.5 \times 24 = 9\,\text{kN}$$

Here, since the foundation depth equals the footing thickness there will be no backfill soil.

Also, since the water table exists below the foundation level, there will be no uplift pressure. Hence, the effective contact pressure is

$$q = \frac{(100 + 9)}{1 \times 0.75} = 145.3\,kPa$$

(Continued)

(c)

$$W_f = 1.2 \times 40 \times 80 \times 24 = 92160\,\text{kN}$$

The effective contact pressure is

$$q = [(800000 + 92160)/(40 \times 80)] - 5 \times 10 = 228.8\,kPa$$

The vertical effective stress (i.e. the effective overburden pressure) at the foundation level due to the effective weight of the overburden soil is

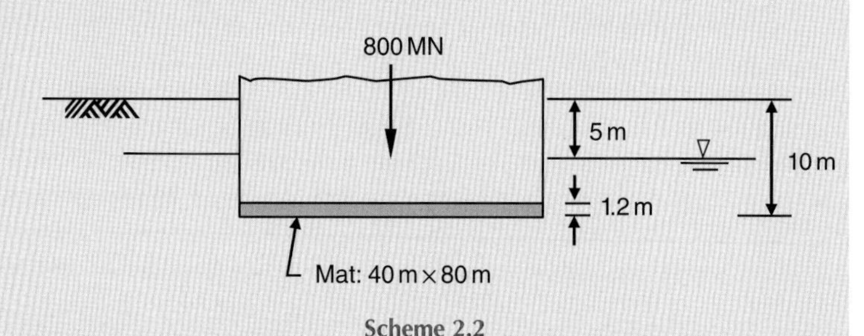

Scheme 2.2

$$\sigma'_o = 5 \times 17 + 5\,(20 - 10) = 135\ \text{kPa}$$

Hence, the net increase in the vertical effective stress immediately below the mat centre due to the foundation load is

$$228.8 - 135.0 = 93.8\,kPa$$

Problem 2.2

A continuous footing is 1.5 m wide, subjected to a concentric vertical load V = 200 kN/m and moment M_y = 40 m.kN/m acting laterally across the footing, as shown in the scheme. The ground water table is at a great depth. Unit weight of concrete = 24 kN/m^3. Determine whether the resultant force on the footing base acts within the middle third and compute the maximum and minimum contact pressures.

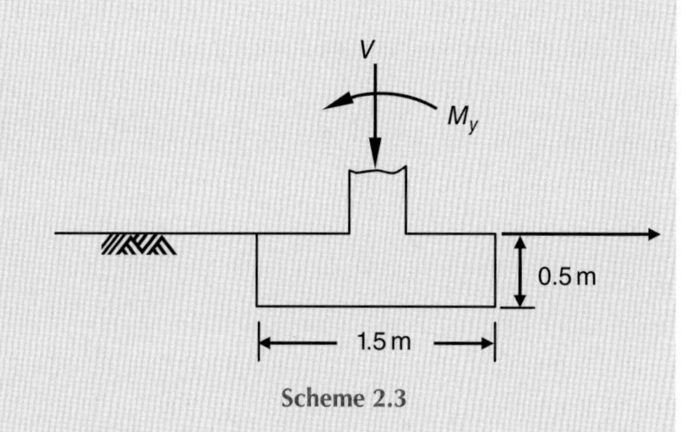

Scheme 2.3

Solution:

$$W_f = 0.5 \times 1.5 \times 1.0 \times 24 = 18\,\text{kN/m}$$

Equation (2.5):

$$e_x = \frac{\dfrac{M_y}{L}}{\dfrac{V}{L} + \dfrac{W_f}{L}} = \frac{40}{200 + 18} = 0.183\,\text{m}$$

$\frac{B}{6} = \frac{1.5}{6} = 0.25\,\text{m} > e$. Therefore, *the resultant force acts within the middle third.*

The ground water table has no effect on the contact pressure, since it exists at a great depth below the foundation level.

Equation(2.7) :

$$q = \frac{R}{BL}\left(1 \pm \frac{6e_x}{B} \pm \frac{6e_y}{L}\right) = \frac{200+18}{1.5\times 1}\left(1 \pm \frac{6\times 0.183}{1.5} \pm 0\right)$$

$$q_{max} = 145.33 + 106.56 = \mathit{251.89\,kPa}$$

$$q_{min} = 145.33 - 106.56 = \mathit{38.77\,kPa}$$

Problem 2.3

Six cylindrical grain silos are to be supported by a mat foundation, as shown in the scheme below. Weight of the mat is 90 MN, and empty weight of each silo is 29 MN. Each silo can hold up 110 MN. Each silo will be filled independently. For the various critical loading conditions find whether the resultant force falls within the kern. In case the resultant falls within the kern compute the maximum and minimum contact pressures and show the contact pressure diagram, and if it does not, revise the plan dimensions of the mat so that Equation (2.8) is satisfied.

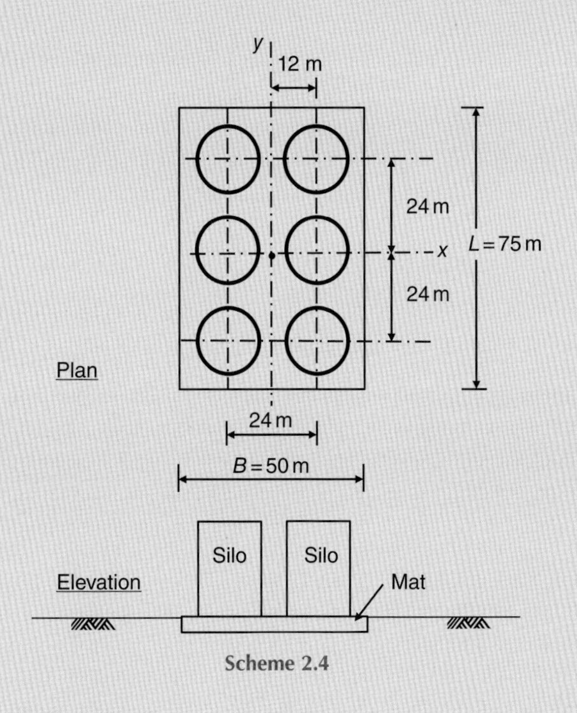

Scheme 2.4

(Continued)

Solution:

(1) *Check one-way eccentricity*

(a) One-way eccentricity in L or y direction occurs when two adjacent (in B or x direction) corner silos are full of grains and the other four silos are empty.

Weight of grains in the two silos cause a moment about x axis equals

$$M_x = 2 \times 110 \times 24 = 5280 \text{ MN.m}$$

$$R = 6 \times 29 + 2 \times 110 + 90 = 484 \text{ MN}$$

$$e_L = e_y = \frac{M_x}{R} = \frac{5280}{484} = 10.91 \text{ m}; \; e_B = e_x = 0$$

$\frac{L}{6} = \frac{75}{6} = 12.5 > e_L \rightarrow$ *the resultant R falls within the kern*

or,

Equation (2.8): $\frac{6e_B}{B} + \frac{6e_L}{L} \leq 1$

$\frac{6e_B}{B} + \frac{6e_L}{L} = 0 + \frac{6 \times 10.91}{75} = 0.87 < 1 \rightarrow$ *the resultant R falls within the kern.*

Equation (2.7): $q = \frac{R}{BL}\left(1 \pm \frac{6e_x}{B} \pm \frac{6e_y}{L}\right) = \frac{484}{50 \times 75}\left(1 \pm 0 \pm \frac{6 \times 10.91}{75}\right)$

Use the plus sign (+) for soil in compression, and the minus sign (−) for soil in tension.

$$q_{\max} = \frac{484}{50 \times 75}\left(1 + \frac{6 \times 10.91}{75}\right) = 0.24175 \text{ MN/m}^2 = 241.75 \, kPa$$

$$q_{\min} = \frac{484}{50 \times 75}\left(1 - \frac{6 \times 10.91}{75}\right) = 0.01642 \text{ MN/m}^2 = 16.42 \, kPa$$

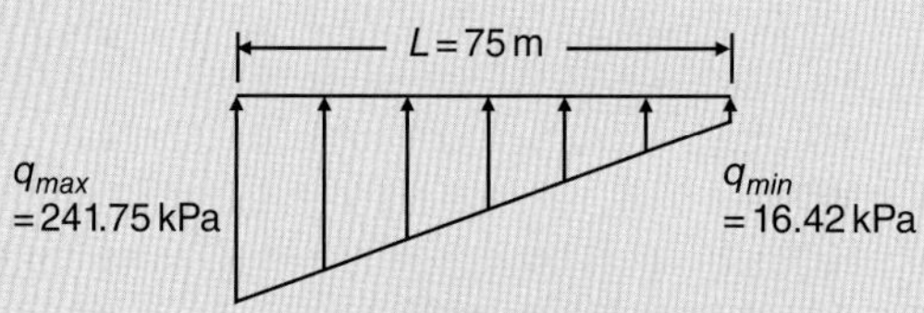

L = 75 m

q_{max} = 241.75 kPa

q_{min} = 16.42 kPa

Scheme 2.5

(b) One-way eccentricity in B or x direction occurs when three silos in one row are full, and the other three silos are empty.

Weight of grains in the three silos cause a moment about y axis equals

$$M_y = 3 \times 110 \times 12 = 3960 \text{ MN.m}$$

$$R = 6 \times 29 + 3 \times 110 + 90 = 594 \text{ MN}$$

$$e_B = e_x = \frac{M_y}{R} = \frac{3960}{594} = 6.67 \text{ m}; \; e_L = e_y = 0$$

$\frac{6e_B}{B} + \frac{6e_L}{L} = \frac{6 \times 6.67}{50} + 0 = 0.80 < 1 \rightarrow$ *the resultant R falls within the kern.*

$$q_{max} = \frac{R}{A}\left(1 + \frac{6e_B}{B}\right)$$

$$= \frac{594}{50 \times 75}\left(1 + \frac{6 \times 6.67}{50}\right) = 0.28518 \ \text{MN/m}^2 = 285.18\,kPa$$

$$q_{min} = \frac{R}{A}\left(1 - \frac{6e_B}{B}\right)$$

$$= \frac{594}{50 \times 75}\left(1 - \frac{6 \times 6.67}{50}\right) = 0.03168 \ \text{MN/m}^2 = 31.68\,kPa$$

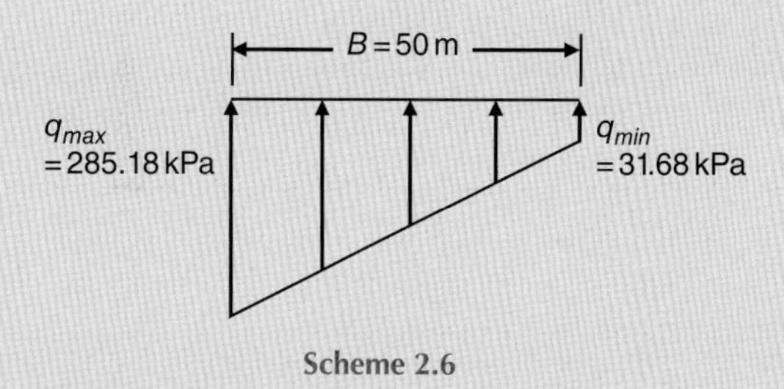

Scheme 2.6

(c) One-way eccentricity in B or x direction occurs when one of the two silos, located on x axis, is full and the other five silos are empty.

Weight of grains in the silo cause a moment about y axis equals

$$M_y = 1 \times 110 \times 12 = 1320\,\text{MN.m}$$

$$R = 6 \times 29 + 1 \times 110 + 90 = 374\,\text{MN}$$

$$e_B = e_x = \frac{M_y}{R} = \frac{1320}{374} = 3.53\,\text{m};\ e_L = e_y = 0$$

$$\frac{6e_B}{B} + \frac{6e_L}{L} = \frac{6 \times 3.53}{50} + 0 = 0.42 < 1 \rightarrow \textit{the resultant R falls within the kern.}$$

$$q_{max} = \frac{R}{A}\left(1 + \frac{6e_B}{B}\right)$$

$$= \frac{374}{50 \times 75}\left(1 + \frac{6 \times 3.53}{50}\right) = 0.1416 \ \text{MN/m}^2 = 141.62\,kPa$$

$$q_{min} = \frac{R}{A}\left(1 - \frac{6e_B}{B}\right)$$

$$= \frac{374}{50 \times 75}\left(1 - \frac{6 \times 3.53}{50}\right) = 0.05749 \ \text{MN/m}^2 = 57.49\,kPa$$

(d) One-way eccentricity in B or x direction occurs when the two corner silos, located on a line parallel to y axis, are full and the other four silos are empty. Since computing eccentricity and pressures are similar to those presented in (b), this will be left to the reader.

(Continued)

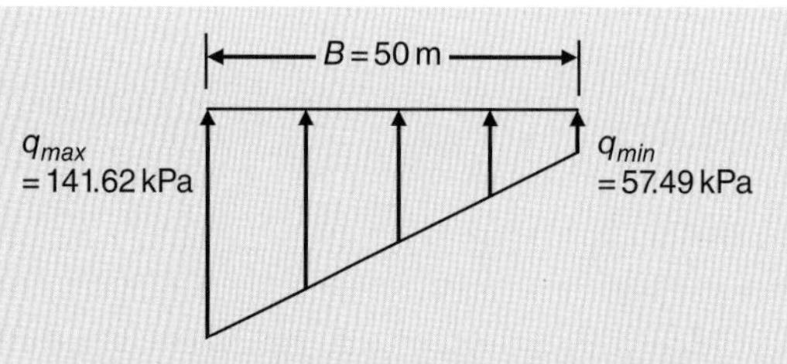

Scheme 2.7 (belongs to (c) in previous page)

(2) *Check two-way eccentricity*

(a) Eccentricities e_x and e_y occur simultaneously when one corner silo is full, and the other five silos are empty.

$$M_x = 1 \times 110 \times 24 = 2640 \text{ MN.m}$$

$$M_y = 1 \times 110 \times 12 = 1320 \text{ MN.m}$$

$$R = 6 \times 29 + 1 \times 110 + 90 = 374 \text{ MN}$$

$$e_B = e_x = \frac{M_y}{R} = \frac{1320}{374} = 3.53 \text{ m}; \; e_L = e_y = \frac{M_x}{R} = \frac{2640}{374} = 7.06 \text{ m}$$

$\frac{6e_B}{B} + \frac{6e_L}{L} = \frac{6 \times 3.53}{50} + \frac{6 \times 7.06}{75} = 0.424 + 0.565 = 0.989 \text{ m} < 1 \rightarrow$ *the resultant R falls within the kern but very close to the limit.*

It may not be necessary now to compute the contact pressures for this case because in the remaining critical cases the resultant may not be located within the kern and a revise of mat dimensions may be required.

(b) Eccentricities e_x and e_y occur simultaneously when two adjacent silos, located on a line parallel to y axis, are full and the other four silos are empty.

$$M_x = 1 \times 110 \times 24 = 2640 \text{ MN.m}$$

$$M_y = 2 \times 110 \times 12 = 2640 \text{ MN.m}$$

$$R = 6 \times 29 + 2 \times 110 + 90 = 484 \text{ MN}$$

$$e_B = e_x = \frac{M_y}{R} = \frac{2640}{484} = 5.45 \text{ m}; \; e_L = e_y = \frac{M_x}{R} = \frac{2640}{484} = 5.45 \text{ m}$$

$\frac{6e_B}{B} + \frac{6e_L}{L} = \frac{6 \times 5.45}{50} + \frac{6 \times 5.45}{75} = 0.654 + 0.436 = 1.09 \text{ m} > 1 \rightarrow$ *the resultant R does not fall within the kern but very close to limit.*

Therefore, the mat dimensions should be revised.

However, it would be better to examine the following case too before doing any change in dimensions.

(c) Eccentricities e_x and e_y occur simultaneously when two adjacent corner silos and another adjacent silo are full and the other three silos are empty.

$$M_x = 2 \times 110 \times 24 = 5280 \text{ MN.m}$$

$$M_y = 1 \times 110 \times 12 = 1320 \text{ MN.m}$$

$$R = 6 \times 29 + 3 \times 110 + 90 = 594 \text{ MN}$$

$$e_B = e_x = \frac{M_y}{R} = \frac{1320}{594} = 2.22\ \text{m},\ e_L = e_y = \frac{M_x}{R} = \frac{5280}{594} = 8.89\ \text{m}$$

$\frac{6e_B}{B} + \frac{6e_L}{L} = \frac{6 \times 2.22}{50} + \frac{6 \times 8.89}{75} = 0.266 + 0.711 = 0.977\ \text{m} < 1 \rightarrow$ *the resultant R falls within the kern but very close to the limit.*

The above results reveal that the mat dimensions need to be revised, as indicated in case (2b). Try a mat area 56 m × 80 m provided that centre to centre of the silos and their distances from the *x* and *y* axis remain unchanged (i.e. the mat projections are increased only). Weight of the mat becomes 107.52 MN.

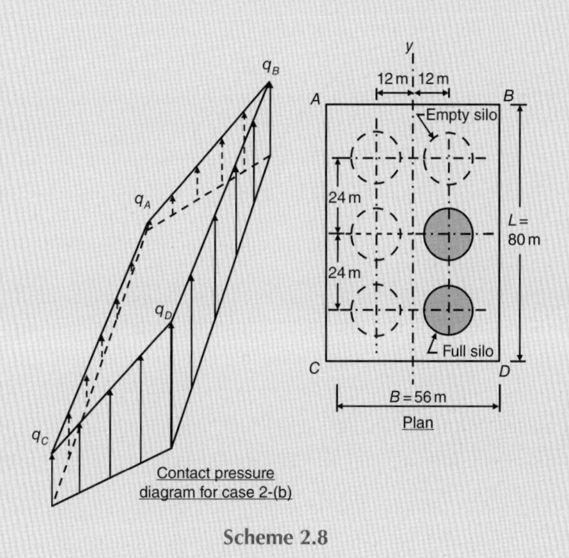

Scheme 2.8

Consider case (2b) again:

$$M_x = 1 \times 110 \times 24 = 2640\ \text{MN.m}$$

$$M_y = 2 \times 110 \times 12 = 2640\ \text{MN.m}$$

$$R = 6 \times 29 + 2 \times 110 + 107.52 = 501.52\ \text{MN}$$

$$e_B = e_x = \frac{M_y}{R} = \frac{2640}{501.52} = 5.26\ \text{m};\ e_L = e_y = \frac{M_x}{R} = \frac{2640}{501.52} = 5.26\ \text{m}$$

$\frac{6e_B}{B} + \frac{6e_L}{L} = \frac{6 \times 5.26}{56} + \frac{6 \times 5.26}{80} = 0.564 + 0.395 = 0.96\ \text{m}$; hence, *the resultant R falls within the kern.*

Equation (2.7):

$$q = \frac{R}{BL}\left(1 \pm \frac{6e_x}{B} \pm \frac{6e_y}{L}\right)$$

$$q_A = \frac{R}{BL}\left(1 - \frac{6e_x}{B} - \frac{6e_y}{L}\right) = \frac{501.52}{56 \times 80}\left(1 - \frac{6 \times 5.26}{56} - \frac{6 \times 5.26}{80}\right)$$

$$q_A = 0.112\,(1 - 0.564 - 0.395) = 0.0046\ \text{MN/m}^2 = \mathit{4.6\ kPa}$$

$$q_B = 0.112\,(1 + 0.564 - 0.395) = 0.1309\ \text{MN/m}^2 = \mathit{130.9\ kPa}$$

$$q_C = 0.112\,(1 - 0.564 + 0.395) = 0.0046\ \text{MN/m}^2 = \mathit{93.1\ kPa}$$

$$q_D = 0.112\,(1 + 0.564 + 0.395) = 0.0046\ \text{MN/m}^2 = \mathit{219.4\ kPa}$$

Problem 2.4

The square footing shown in the scheme below is subjected to a concentric column load $V = 300$ kN and moment $M = 50$ kN.m acting clockwise about the diagonal axis AD. The footing is 0.6 m thick. The foundation depth equals 0.6 m and the ground water table is encountered at a great depth. Determine whether the resultant force falls within the kern of the footing base area, and compute contact pressure at corner points A, B, C and D.

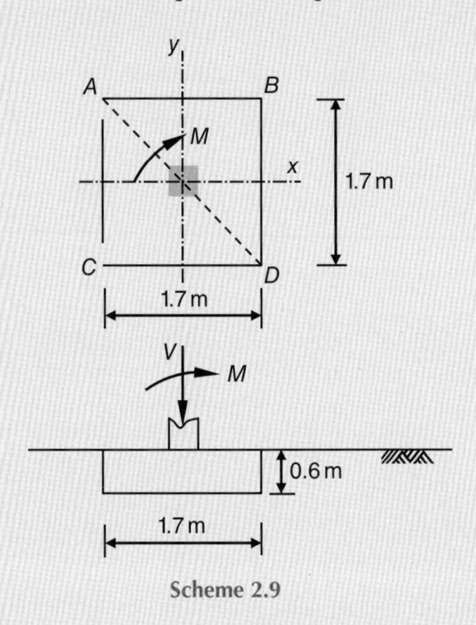

Scheme 2.9

Solution:
The resultant force is

$$R = 300 + 0.6 \times 1.7 \times 1.7 \times 24 = 341.62\,\text{kN}$$

$e = \dfrac{50}{341.62} = 0.15\,\text{m}$. The components of this eccentricity are

$$e_x = e(\cos 45^o) = 0.15 \times 0.707 = 0.11\text{ m} < (1.7/6)$$
$$e_y = e(\sin 45^o) = 0.15 \times 0.707 = 0.11\text{ m} < (1.7/6)$$

$\dfrac{6e_B}{B} + \dfrac{6e_L}{L} = \dfrac{6 \times 0.11}{1.7} + \dfrac{6 \times 0.11}{1.7} = 0.39 + 0.39 = 0.78 < 1$. Hence, *the resultant R falls within the kern.*

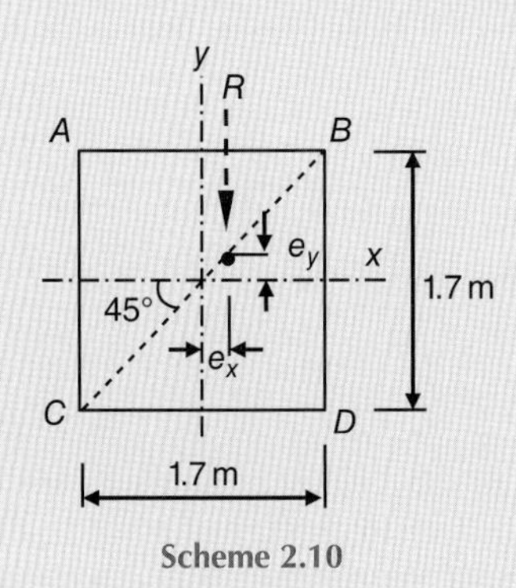

Scheme 2.10

$$q = \frac{R}{BL}\left(1 \pm \frac{6e_x}{B} \pm \frac{6e_y}{L}\right)$$

$$q_A = \frac{R}{BL}\left(1 - \frac{6e_x}{B} + \frac{6e_y}{L}\right) = \frac{341.62}{1.7 \times 1.7}\left(1 - \frac{6 \times 0.11}{1.7} + \frac{6 \times 0.11}{1.7}\right)$$

$$= 118.21(1 - 0.39 + 0.39) = \mathit{118.21\ kPa}$$

$$q_B = 118.21(1 + 0.39 + 0.39) = \mathit{210.41\ kPa}$$

$$q_C = 118.21(1 - 0.39 - 0.39) = \mathit{26.01\ kPa}$$

$$q_D = 118.21(1 + 0.39 - 0.39) = \mathit{118.21\ kPa}$$

Problem 2.5

(a) Two columns A and B are to be supported on a trapezoidal combined footing, as shown in the scheme below. The footing is 0.7 m thick. The top of the footing is flush with the ground surface, and the ground water table is located at a great depth. The vertical dead load on columns A and B is 500 and 1400 kN, respectively. Determine dimension B_2 so that the resultant of the column loads acts through the centroid of the footing.

(b) In addition to the dead loads, columns A and B also can carry vertical live loads of up to 800 and 1200 kN, respectively. The live load on each column is independent of that on the other column; that is, one could be carrying the full live load while the other does not. Using the dimensions obtained in (a), find the worst possible combination of the columns live load.

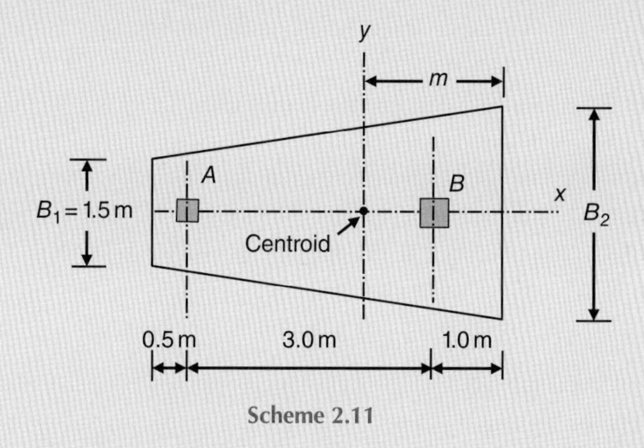

Scheme 2.11

Solution:

(a) Centroid of the trapezoid is located at distance m from B_2, as shown. The trapezoid rule gives

$$m = \frac{L}{3}\left(\frac{2B_1 + B_2}{B_1 + B_2}\right) = \frac{4.5}{3}\left(\frac{2 \times 1.5 + B_2}{B_1 + B_2}\right) = \frac{4.5 + 1.5B_2}{B_2 + 1.5} \quad \cdots\cdots\cdots(1)$$

$$W_f = 24 \times 0.7 \times 4.5\left(\frac{1.5 + B_2}{2}\right) = 56.7 + 37.8B_2$$

$$R = W_f + 500 + 1400 = 56.7 + 37.8B_2 + 1900 = 1956.7 + 37.8B_2$$

(Continued)

Moment of R about B_2 equals to $\sum$ moment of components:

$$m(1956.7+37.8B_2)=m(56.7+37.8B_2)+4\times 500+1\times 1400$$

$$m=\frac{3400}{1900}=1.79\,\text{m}\cdots\cdots(2).$$

Hence,

$$1.79=\frac{4.5+1.5B_2}{B_2+1.5}\rightarrow 1.79B_2+2.69=4.5+1.5B_2\rightarrow 0.29B_2=1.81\rightarrow$$

$$B_2=\frac{1.81}{0.29}=6.241\text{ m}\rightarrow \text{Use } B_2=6.3\,m$$

$$m=\frac{4.5+1.5B_2}{B_2+1.5}=\frac{4.5+1.5\times 6.3}{6.3+1.5}=1.79\,\text{m}\rightarrow \text{unchanged.}$$

(b) In order to find the worst possible combination of the columns live load, it is necessary to find which combination produces the greatest maximum contact pressure q_{max}. Therefore the following possible loading conditions may be examined:

(1) The two column live loads are present at the same time.

$$A=(4.5)\left(\frac{1.5+6.3}{2}\right)=17.55\text{ m}^2,\ W_f=0.7\times 17.55\times 24=294.84\text{ kN}$$

$$R=294.84+(500+800)+(1400+1200)=4194.84\,\text{kN}$$

Moment of R about B_2 equals to $\sum$ moment of components:

$$4194.84x'=294.84\times 1.79+1300\times 4+2600\times 1=8327.76$$

$$x'=\frac{8327.76}{4194.84}=1.99\,\text{m}>m;\ e_x=x'-m=1.99-1.79=0.2$$

Equation (2.6):

$$q=\frac{R}{A}\pm\frac{M_y}{I_y}x\pm\frac{M_x}{I_x}y$$

$$M_x=0;\ x_{\max}=L-m=4.5-1.79=2.71\text{ m.}$$

For a trapezoid area

$$I_y=\frac{B_1^2+4B_1B_2+B_2^2}{36(B_1+B_2)}L^3$$

$$I_y=\frac{1.5^2+4\times 1.5\times 6.3+6.3^2}{36(1.5+6.3)}(4.5^3)=25.88\text{ m}^4$$

$$M_y=Re_x=4194.84\times 0.2=838.97\,\text{kN.m}$$

$$q_{\max}=\frac{4194.84}{17.55}+\frac{838.97\times 2.71}{25.88}=239.02+87.85=326.87\,kPa$$

It may be noticed that $q_{\min}$ is also positive. Therefore, the supporting soil is totally in compression.

(2) Live load of column A is present only.

$$R = 294.84 + (500 + 800) + (1400) = 2994.84\ \text{Kn}$$

Moment of R about B_2 equals to $\sum$ moment of components:

$$2994.84x' = 294.84 \times 1.79 + 1300 \times 4 + 1400 \times 1 = 7127.76$$

$$x' = \frac{7127.76}{2994.84} = 2.38\ \text{m} > m;\ e_x = x' - m = 2.38 - 1.79 = 0.59\ \text{m}$$

$$q_{max} = \frac{R}{A} + \frac{M_y x_{max}}{I_y},\ x_{max} = L - m = 4.5 - 1.79 = 2.71\ \text{m}$$

$$M_y = Re_x = 2994.84 \times 0.59 = 1766.96\ \text{kN.m}$$

$$q_{max} = \frac{2994.84}{17.55} + \frac{1766.96 \times 2.71}{25.88} = 170.65 + 185.03 = 355.68\ \text{kPa}$$

However, in this case, q_{min} is negative which indicates tension between the soil and a part of the footing bottom. Because soil cannot take any tension, separation between the footing and the soil underlying it will take place. This condition tends to overload the base soil, that is to increase q_{max}. Therefore, the maximum contact pressure needs to be recomputed, using the same concept as outlined for rectangular footings (see Figure 2.23).

The soil reaction resultant R_s remains equal and opposite in direction to the resultant R, as shown below. Let L' represent length of the base area under which the soil is in compression, and m' equals the distance of its centroid from side B, as shown in the scheme below.

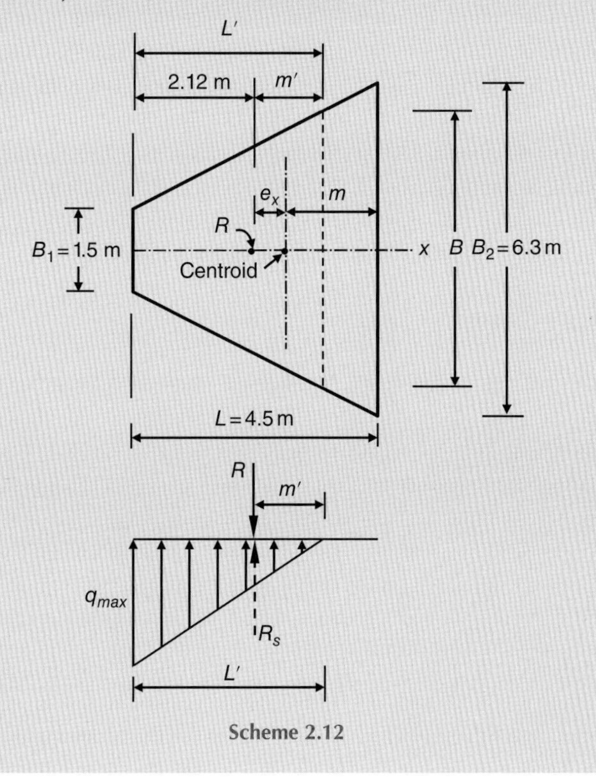

Scheme 2.12

(*Continued*)

$$L' - (L - x') = m' = \frac{L'}{3}\left(\frac{2B_1 + B}{B_1 + B}\right)$$

$$L' - (4.5 - 2.38) = \frac{L'}{3}\left(\frac{2 \times 1.5 + B}{1.5 + B}\right)$$

$$L' - 2.12 = \frac{L'}{3}\left(\frac{2 \times 1.5 + B}{1.5 + B}\right) \ldots\ldots\ldots\ldots(1)$$

$$B = B_1 + 2\left[L'\left(\frac{B_2 - B_1}{2}L\right)\right] = 1.5 + L'\left(\frac{6.3 - 1.5}{4.5}\right)$$

$$B = 1.5 + 1.07L' \ldots\ldots\ldots\ldots(2)$$

Hence,

$$L' - 2.12 = \frac{L'}{3}\left(\frac{2 \times 1.5 + 1.5 + 1.07L'}{1.5 + 1.5 + 1.07L'}\right) = \frac{L'}{3}\left(\frac{4.5 + 1.07L'}{3 + 1.07L'}\right)$$

$$3L' + 1.07(L')^2 - 6.36 - 2.27L' = 1.5L' + 0.36(L')^2$$

$$(L')^2 - 1.08(L') - 8.96 = 0$$

$$L' = \frac{-(-1.08) \pm \sqrt{(-1.08)^2 - 4(1)(-8.96)}}{2(1)} = \frac{1.08 \pm 6.08}{2} = 3.58\,\text{m}$$

$$R_s = R = 2994.84\,\text{kN}$$

$$R_s = \frac{q_{max} L'}{2}\left(\frac{B_1 + B}{2}\right) = \frac{q_{max}(3.58)}{2}\left(\frac{1.5 + 1.5 + 1.07 \times 3.58}{2}\right) = 6.11\,q_{max}$$

$$q_{max} = \frac{2994.84}{6.11} = \mathit{490.15\,kPa} > 355.68\,\text{kPa}$$

(3) Live load of column B is present only.

$$R = 294.84 + (500) + (1400 + 1200) = 3394.84\,\text{kN}$$

Moment of R about B_2 equals to $\sum$ moment of components:

$$3394.84\,x' = 294.84 \times 1.79 + 500 \times 4 + 2600 \times 1 = 5127.76$$

$$x' = \frac{5127.76}{3394.84} = 1.51\,\text{m} < m;\ e_x = m - x' = 1.79 - 1.51 = 0.28$$

$$q_{max} = \frac{R}{A} + \frac{M_y x_{max}}{I_y},\ x_{max} = m = 1.79\,\text{m}$$

$$M_y = Re_x = 3394.84 \times 0.28 = 950.56\,\text{kN.m}$$

$$q_{max} = \frac{3394.84}{17.55} + \frac{950.56 \times 1.79}{25.88} = 193.44 + 65.75 = \mathit{259.19\,kPa}$$

Therefore, the worst possible combination of the columns live load occurs when the live load of column A is present only.

Problem 2.6

A water tower has a circular foundation of 10 m diameter. If the total weight of the tower, including the foundation, is 20 MN, calculate the vertical stress increase at a depth of 2.5 m below the foundation level.

Solution:
The base area $A = (22/7)10^2/4 = 78.57\ \text{m}^2$

The uniformly distributed load $q = 20/78.57 = 0.255\ \text{MN/m}^2$

Equation (2.16):

$$\sigma_z = q\left\{1 - \frac{1}{\left[1 + (R/z)^2\right]^{3/2}}\right\}$$

$$= 0.255\left\{1 - \frac{1}{\left[1 + (5/2.5)^2\right]^{3/2}}\right\}$$

$$= 0.255 \times 0.911 = 0.2319\ \text{MN/m}^2 = \mathit{231.9\,kPa}$$

Problem 2.7

A monument weighing 15 MN is erected on the ground surface. Considering the load as a concentrated one, determine the vertical pressure directly under the monument at a depth of 8 m below the ground surface.

Solution:

Equation (2.11):

$$\sigma_z = \frac{3V}{2\pi z^2}\frac{1}{\left[1 + (r/z)^2\right]^{5/2}} = \frac{3 \times 15}{2\pi(8)^2}\frac{1}{\left[1 + (0/z)^2\right]^{5/2}}$$

$$\sigma_z = \frac{3 \times 15}{2\pi(8)^2} = 0.1119\ \text{MN/m}^2 = \mathit{111.9\,kPa}$$

Problem 2.8

A square foundation 5 × 5 m is to carry a load of 4 MN. Calculate the vertical stress increase at a depth of 5 m below the centre of the foundation, using (a) The Newmark table, (b) The approximate 2 *V*: 1 *H* ratio method

Solution:
(a) The area is divided into four squares, each of $B = L = 5/2 = 2.5$ m.

$$B/z = L/z = 2.5/5 = 0.5$$

From Table 2.3 find $I = 0.084$

$$\sigma_z = q(4I) = \frac{4}{5 \times 5} \times 4 \times 0.084 = 0.05376\ \text{MN/m}^2 = \mathit{53.76\,kPa}$$

(Continued)

(b) Using the approximate 2 V: 1 H ratio method

$$\sigma_z = \frac{V}{(B+z)(L+z)} = \frac{4}{(5+5)(5+5)} = 0.040\,\text{MN/m}^2 = 40\,\text{kPa}$$

Problem 2.9

Two columns A and B situated 6 m apart. Column A transfers a load of 500 kN, and column B transfers a load of 250 kN. Determine the resultant vertical stress increase at points vertically below the columns on a horizontal plane 2 m below the ground surface.

Solution:
Consider the point below column A:

Equation (2.11):

$$\sigma_z = \frac{3V}{2\pi z^2}\frac{1}{\left[1+(r/z)^2\right]^{5/2}}$$

Vertical stress increase due to load of column A

$$\sigma_z = \frac{3\times 500}{2\pi 2^2}\frac{1}{\left[1+(0/2)^2\right]^{5/2}} = 59.67\,\text{kPa}$$

Vertical stress increase due to load of column B

$$\sigma_z = \frac{3V}{2\pi z^2}\frac{1}{\left[1+(r/z)^2\right]^{5/2}} = \frac{3\times 250}{2\pi 2^2}\frac{1}{\left[1+(6/2)^2\right]^{5/2}} = 29.83\times\frac{1}{10^{5/2}} = 0.09\,\text{kPa}$$

The resultant vertical stress increase = 59.67 + 0.09 = *59.76 kPa*
Consider the point below column B:

Equation (2.11):

$$\sigma_z = \frac{3V}{2\pi z^2}\frac{1}{\left[1+(r/z)^2\right]^{5/2}}$$

Vertical stress increase due to load of column B

$$\sigma_z = \frac{3\times 250}{2\pi 2^2}\frac{1}{\left[1+(0/2)^2\right]^{5/2}} = 29.83\,\text{kPa}$$

Vertical stress increase due to load of column A

$$\sigma_z = \frac{3\times 500}{2\pi 2^2}\frac{1}{\left[1+(6/2)^2\right]^{5/2}} = 59.67\times\frac{1}{10^{5/2}} = 0.19\,\text{kPa}$$

The resultant vertical stress increase = 29.83 + 0.19 = *30.02 kPa*

Problem 2.10

An excavation 3 × 6 m for a foundation is to be made to a depth of 2.5 m below ground level in a soil of bulk unit weight = 20 kN/m^3. What effect this excavation will have on the vertical pressure at a depth of 6 m measured from the ground surface vertically below the centre of foundation? The influence factor for $m = 0.43$ and $n = 0.86$ is 0.10.

Solution:
The area is divided into four rectangles, each of $B = 1.5$ m and $L = 3$ m.

$$m = B/z = 1.5/(6-2.5) = 0.43$$
$$n = L/z = 3/(6-2.5) = 0.86$$

From Table 2.3 find $I = 0.1$

The decrease in vertical stress is
$\sigma_z = q(4I) = (20 \times 2.5)(4 \times 0.1) = 20$ kPa. Therefore, the vertical stress at 6 m depth will be *decreased by 20 kPa.*

Problem 2.11

The contact pressure due to a mat foundation 10 × 20 m is 250 kPa. Determine the vertical stress increase at a depth of 5 m below P at the foundation level, as shown in the scheme below. Use the Fadum chart of Figure 2.32.

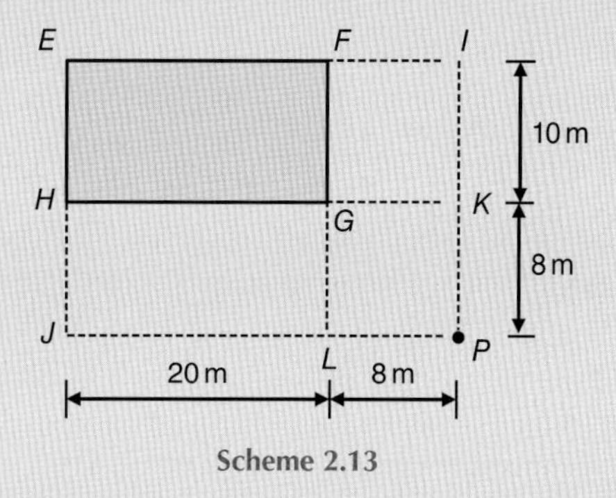

Scheme 2.13

Solution:
A large rectangle EIPJ is drawn such that point p is one of its corners.

$$\sigma_z = q(I_1 - I_2 - I_3 + I_4)$$

I_1 = influence factor for rectangle EIPJ; $m = \frac{18}{5} = 3.6$, $n = \frac{28}{5} = 5.6$

I_2 = influence factor for rectangle FIPL; $m = \frac{8}{5} = 1.6$, $n = \frac{18}{5} = 3.6$

I_3 = influence factor for rectangle HKPJ; $m = \frac{8}{5} = 1.6$, $n = \frac{28}{5} = 5.6$

I_4 = influence factor for rectangle GKPL; $m = \frac{8}{5} = 1.6$, $n = \frac{8}{5} = 1.6$

(*Continued*)

From the Fadum chart:

$$I_1 = 0.248,\ I_2 = 0.233,\ I_3 = 0.233,\ I_4 = 0.223$$

$$\sigma_z = 250\,(0.248 - 0.233 - 0.233 + 0.223) = \mathit{1.25\,kPa}$$

Problem 2.12

A column footing foundation is shown in the schemes (plan and profile) below. Determine the average increase in pressure within the clay layer below the centre of the footing using (a) The Griffiths equation, (b) The Newmark method based on the Boussinesq solution, (c) The Westergaard solution; $\mu = 0.45$, and (d) The approximate $2V$: $1H$ ratio method.

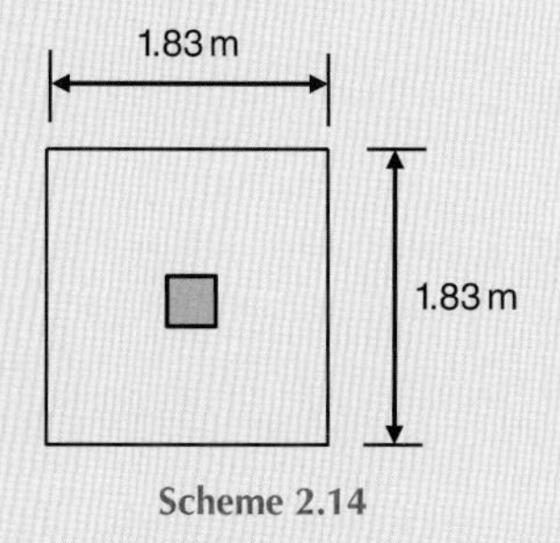

Scheme 2.14

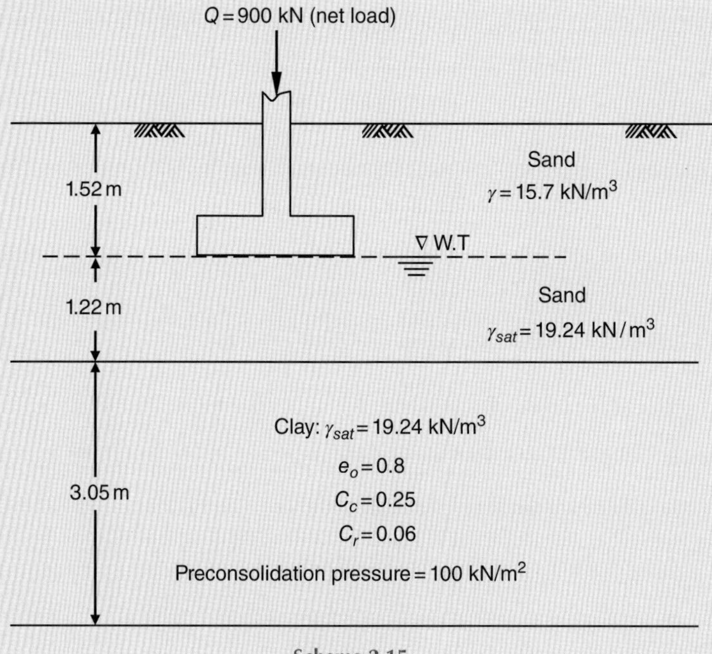

Scheme 2.15

Solution:

(a) Using the Griffiths equation

Equation (2.20b): $$\sigma_{av} = q\left[\frac{H_2 I_{a(H_2)} - H_1 I_{a(H_1)}}{H_2 - H_1}\right]$$

The loaded area can be divided into four squares, each measuring 0.915 × 0.915 m.

$$H_1 = 1.22\text{ m}; H_2 = 1.22 + 3.05 = 4.27\text{ m}$$

$$I_{a(H_2)}: m = \frac{B}{H_2} = \frac{0.915}{4.27} = 0.214,\ n = \frac{L}{H_2} = \frac{0.915}{4.27} = 0.214$$

From the Griffiths chart estimate $I_{a(H_2)} = 0.1$

$$I_{a(H_1)}: m = \frac{B}{H_1} = \frac{0.915}{1.22} = 0.75,\ n = \frac{L}{H_1} = \frac{0.915}{1.22} = 0.75$$

From the Griffiths chart estimate $I_{a(H_1)} = 0.204$

$$\sigma_{av} = \frac{900}{1.83 \times 1.83}\left[\frac{4.27 \times 0.10 - 1.22 \times 0.204}{4.27 - 1.22}\right] = 268.74 \times 0.059 = 15.86\,kPa$$

The average increase in pressure in the clay layer below the centre of the foundation is

$$4\sigma_{av} = 4 \times 15.86 = 63.44\,kPa$$

(b) Using the Newmark method based on the Boussinesq solution

$$\sigma_z = q(4I)$$

The average increase in pressure σ_{av} may be approximated by the use of Simpson's rule

$$\sigma_{av} = \frac{1}{6}(\sigma_{zt} + 4\sigma_{zm} + \sigma_{zb})$$

where σ_{zt}, σ_{zm} and σ_{zb} are vertical stresses at top, middle and bottom of the clay layer, respectively.
At top:

Table 2.3: $$m = \frac{B}{z} = \frac{0.915}{1.22} = 0.75,\ n = \frac{L}{z} = \frac{0.915}{1.22} = 0.75 \rightarrow I = 0.137$$

$$\sigma_{zt} = \frac{900}{1.83 \times 1.83}(4 \times 0.137) = 147.27\text{ kPa}$$

At middle:

Table 2.3: $$m = \frac{B}{z} = \frac{0.915}{2.75} = 0.33,\ n = \frac{L}{z} = \frac{0.915}{2.75} = 0.33 \rightarrow I = 0.043$$

$$\sigma_{zm} = \frac{900}{1.83 \times 1.83}(4 \times 0.043) = 46.22\text{ kPa}$$

(Continued)

At bottom:

Table 2.3: $m = \frac{B}{z} = \frac{0.915}{4.27} = 0.21,\ n = \frac{L}{z} = \frac{0.915}{4.27} = 0.21 \rightarrow I = 0.021$

$$\sigma_{zb} = \frac{900}{1.83 \times 1.83}(4 \times 0.021) = 22.57\,\text{kPa}$$

$$\sigma_{av} = \frac{1}{6}(147.27 + 4 \times 46.22 + 22.57) = \mathit{59.12\,kPa}$$

(c) Using the Westergaard equation with $\mu = 0.45$ for saturated clay

Equation [2.25 – (a)]: $\sigma_z = \frac{q}{2\pi} \cot^{-1}\left[a\left(\frac{1}{m^2} + \frac{1}{n^2}\right) + a^2 \frac{1}{m^2 n^2}\right]\frac{1}{2}$

$$a = \frac{1 - 2\mu}{2 - 2\mu} = \frac{1 - 2 \times 0.45}{2 - 2 \times 0.45} = 0.09$$

Follow the same procedure used in (b) above:
At top:

$$m = \frac{B}{z} = \frac{0.915}{1.22} = 0.75,\ n = \frac{L}{z} = \frac{0.915}{1.22} = 0.75$$

$$\sigma_{zt} = 4\left\{\frac{900}{1.83 \times 1.83 \times 2 \times \pi} \cot^{-1}\begin{bmatrix} 0.09\left(\frac{1}{m^2} + \frac{1}{n^2}\right) \\ +0.09^2 \frac{1}{m^2 n^2} \end{bmatrix}^{1/2}\right\}$$

$$= 4\left\{42.75 \cot^{-1}\left[0.09\left(\frac{1}{0.75^2} + \frac{1}{0.75^2}\right) + 0.09^2 \frac{1}{0.75^2 \times 0.75^2}\right]^{1/2}\right\}$$

$$= 4(42.75 \cot^{-1} 0.588) = 4(42.75 \times 0.33\,\pi) = 177.65\,\text{kPa}$$

At middle:

$$m = \frac{B}{z} = \frac{0.915}{2.75} = 0.33,\ n = \frac{L}{z} = \frac{0.915}{2.75} = 0.33$$

$$m = \frac{B}{z} = \frac{0.915}{2.75} = 0.33,\ n = \frac{L}{z} = \frac{0.915}{2.75} = 0.33$$

$$\sigma_{zm} = 4(42.75 \cot^{-1} 0.921) = 4(42.75 \times 0.26\,\pi) = 139.72\,\text{kPa}$$

At bottom:

$$m = \frac{B}{z} = \frac{0.915}{4.27} = 0.21,\ n = \frac{L}{z} = \frac{0.915}{4.27} = 0.21$$

$$\sigma_{zb} = 4\left\{42.75 \cot^{-1}\left[0.09\left(\frac{1}{0.21^2} + \frac{1}{0.21^2}\right) + 0.09^2 \frac{1}{0.21^2 \times 0.21^2}\right]^{1/2}\right\}$$

$$\sigma_{zb} = 4(42.75 \cot^{-1} 0.921) = 4(42.75 \times 0.11\,\pi) = 59.12\,\text{kPa}$$

$$\sigma_{av} = \frac{1}{6}(177.65 + 4 \times 139.72 + 59.12) = 132.6\,\mathit{kPa}$$

(d) Using the approximate $2V:1H$ ratio method

Equation (2.18): $$\sigma_z = \frac{V}{(B+z)(L+z)}$$

$$V = Q = 900\ \text{kPa}$$

At top:

$$\sigma_{zt} = \frac{900}{(1.83+1.22)(1.83+1.22)} = 96.75\ \text{kPa}$$

At middle:

$$\sigma_{zm} = \frac{900}{(1.83+2.75)(1.83+2.75)} = 42.91\ \text{kPa}$$

At bottom:

$$\sigma_{zb} = \frac{900}{(1.83+4.27)(1.83+4.27)} = 24.19\ \text{kPa}$$

$$\sigma_{av} = \frac{1}{6}(\sigma_{zt} + 4\sigma_{zm} + \sigma_{zb})$$

$$\sigma_{av} = \frac{1}{6}(96.75 + 4\times 42.91 + 24.19) = 48.76\ kPa$$

Problem 2.13

An embankment is shown in the scheme below. Determine the vertical stress increase under the embankment at points A_1 and A_2.

Solution:
The embankment section may be divided into a rectangle at the middle and two triangles on sides of the embankment, as shown below. Superposition enables the vertical stress increase to be obtained at points A_1 and A_2 using Equations (2.13), (2.14) and (2.15) as applicable.

Vertical stress increase at point A_1:

$$\tan\frac{\alpha}{2} = \frac{2.5}{5} = 0.5 \rightarrow \frac{\alpha}{2} = 26.6^\circ \rightarrow \alpha = 53.2^\circ;\ \beta = \frac{\alpha}{2} = 26.6^\circ$$

$$\tan(\theta+\beta) = \frac{14+2.5}{5} = 3.3 \rightarrow (\theta+\beta) = 73.1^\circ \rightarrow \theta = 46.5^\circ$$

Equation (2.14): $$\sigma_z = \frac{q}{\pi}(\alpha + \sin\alpha)$$

Equation (2.15): $$\sigma_z = \frac{q}{\pi}\left(\frac{x}{B}\theta - \frac{1}{2}\sin 2\beta\right)$$

$$\Sigma\sigma_z = \left[\frac{7\times 17.5}{\pi}\left(\frac{53.2}{180}\pi + \sin 53.2^\circ\right)\right] + 2\left[\frac{7\times 17.5}{\pi}\left(\frac{16.5}{14}\times\frac{46.5}{180}\pi - \frac{1}{2}\sin 53.2^\circ\right)\right]$$

$$= 67.42 + 43.42 = 110.84\ kPa$$

(Continued)

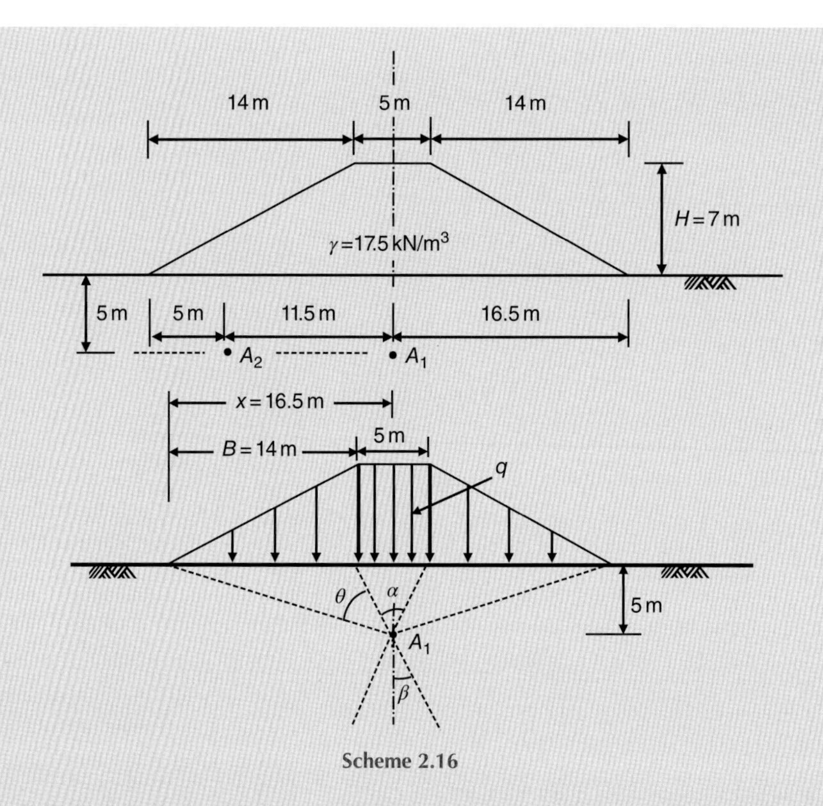

Scheme 2.16

Vertical stress increase at point A_2:

Vertical stress increase at point A_2 equals to sum of the vertical stresses due to loads of the rectangle at the middle and the two triangles on the sides, computed as shown below.

(1) Vertical stress due to load of the triangle on the right side of the embankment.

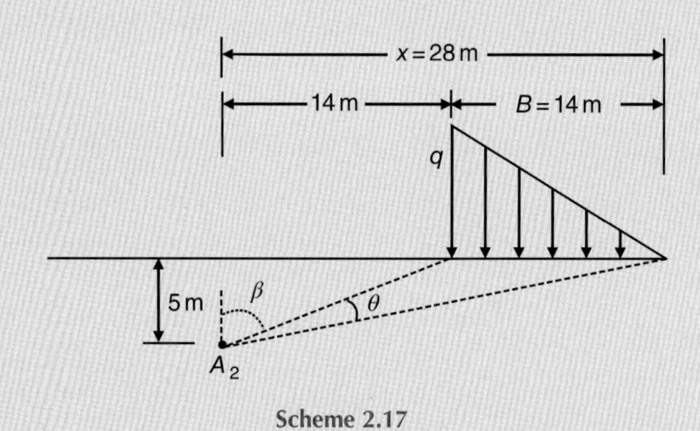

Scheme 2.17

$$\tan\beta = \frac{14}{5} = 2.8 \rightarrow \beta = 70.4°, \tan(\theta + \beta) = \frac{28}{5} = 5.6 \rightarrow$$

$$\theta + \beta = 79.84° \rightarrow \theta = 9.44°$$

Equation (2.15): $\sigma_z = \frac{q}{\pi}\left(\frac{x}{B}\theta - \frac{1}{2}\sin 2\beta\right) = \frac{7\times 17.5}{\pi}\left[\frac{28}{14}\times\frac{9.44}{180}\pi - \frac{1}{2}\sin 140.8^\circ\right]$

$$\sigma_{z1} = \sigma_z = 39(0.33 - 0.316) = 0.55\text{ kPa}$$

(2) Vertical stress due to load of the triangle on the left side of the embankment.

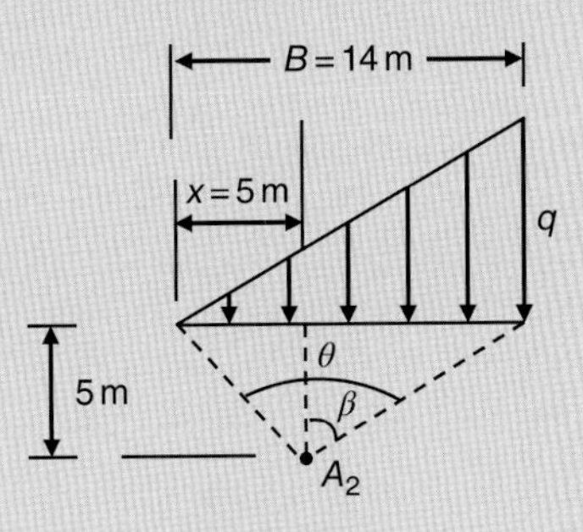

Scheme 2.18

$$\tan\beta = \frac{9}{5} = 1.8 \rightarrow \beta = 61^\circ,\ \theta = 45 + 61 = 106^\circ$$

Note that, in this case, angle β should be treated negative, since point A_2 is located within the distance B.

Equation (2.15): $\sigma_z = \frac{q}{\pi}\left[\frac{x}{B}\theta - \frac{1}{2}\sin(-2\beta)\right]$

$$= \frac{7\times 17.5}{\pi}\left[\frac{5}{14}\times\frac{106}{180}\pi - \frac{1}{2}\sin(-122^\circ)\right]$$

$$\sigma_{z2} = \sigma_z = 39(0.661 + 0.424) = 42.32\text{ kPa}$$

(3) Vertical stress due to load of the rectangle at middle of the embankment.

$$\tan(\alpha+\beta) = (14/5) = 2.8 \rightarrow \alpha+\beta = 70.4^\circ$$

$$\tan\beta = (9/5) = 1.8 \rightarrow \beta = 61^\circ;\ \alpha = 9.4^\circ$$

Equation (2.13): $\sigma_z = \frac{q}{\pi}[\alpha + \sin\alpha\cos(\alpha+2\beta)]$

$$\sigma_{z3} = \sigma_z = 39\left[\frac{9.4\pi}{180} + \sin 9.4^\circ\cos 131.4^\circ\right]$$

$$\sigma_{z3} = 39(0.164 - 0.108) = 2.19\text{ kPa}$$

$$\sum\sigma_z = \sigma_{z1} + \sigma_{z2} + \sigma_{z3}$$

$$= 0.55 + 42.32 + 2.19 = \mathit{45.06\ kPa}$$

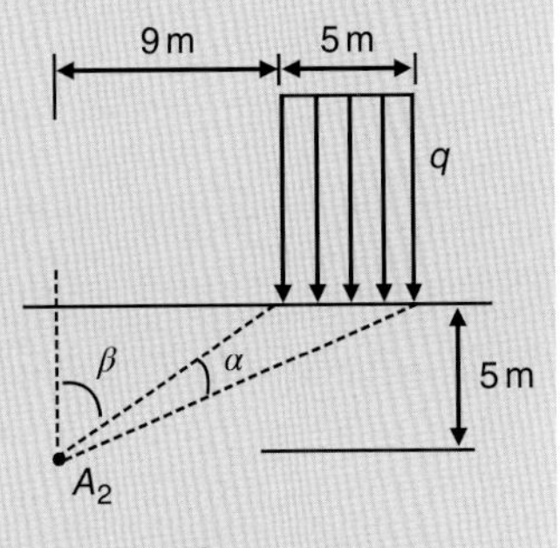

Scheme 2.19

Problem 2.14

There is a line load of 120 kN/m acting on the ground surface along y axis, as shown below. Determine the vertical stress at a point which has x and z coordinates at 2.0 m and 3.5 m, respectively.

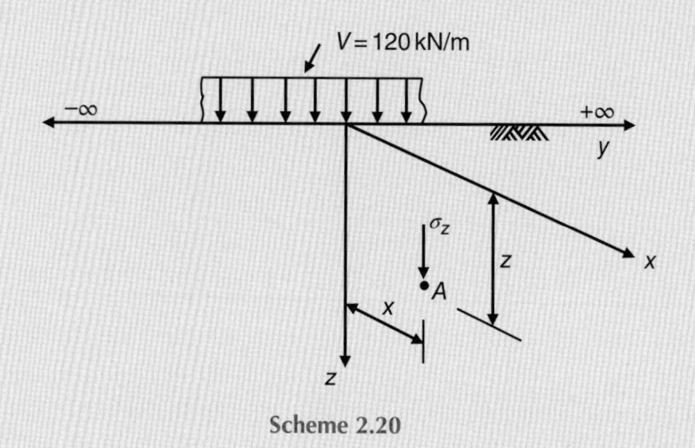

Scheme 2.20

Solution:

Equation (2.12):

$$\sigma_z = \frac{2V}{\pi} \frac{z^3}{(x^2 + z^2)^2}$$

$$\sigma_z = \frac{2 \times 120}{\pi} \times \frac{3.5^3}{(2^2 + 3.5^2)^2} = 12.4\,kPa/m\,length$$

Problem 2.15

Determine the vertical stress at point P which is 3 m below and at a radial distance of 3 m from the vertical point load of 100 kN. Use:

(a) Boussinesq solution
(b) Westergaard solution with $\mu = 0$
(c) Westergaard solution with $\mu = 0.4$

Solution:
(a)

Equation (2.11):

$$\sigma_z = \frac{3V}{2\pi z^2} \frac{1}{\left[1 + (r/z)^2\right]^{5/2}} = \frac{3 \times 100}{2\pi \times 3^2} \frac{1}{\left[1 + (3/3)^2\right]^{5/2}} = 0.94\,kPa$$

(b)

Equation (2.23):

$$\sigma_z = \frac{Q \quad \frac{1}{\pi}}{z^2\left[1 + 2\left(\frac{r}{z}\right)^2\right]^{3/2}} = \frac{100 \times \frac{1}{\pi}}{3^2\left[1 + 2\left(\frac{3}{3}\right)^2\right]^{3/2}} = 0.68\,kPa$$

(c)

$$\sigma_z = \frac{Q\dfrac{1}{2\pi}\sqrt{\dfrac{1-2\mu}{2-2\mu}}}{z^2\left[(1-2\mu)/(2-2\mu)+(r/z)^2\right]^{3/2}} = \frac{100\times\dfrac{1}{2\pi}\sqrt{\dfrac{1-2\times0.4}{2-2\times0.4}}}{3^2\left[(1-2\times0.4)/(2-2\times0.4)+(3/3)^2\right]^{3/2}} = 0.573\,kPa$$

Problem 2.16

The scheme below shows an embankment load on a silty clay layer of soil. Determine the vertical stress increase at points A, B and C, located at a depth of 5 m below the ground surface.

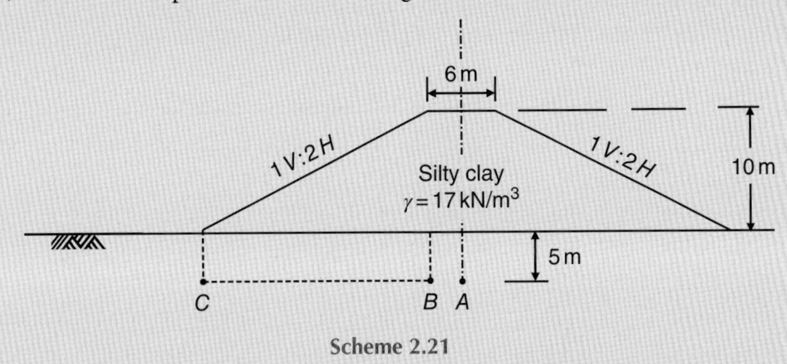

Scheme 2.21

Solution:

The embankment section may be divided into a rectangle at the middle and two triangles on sides of the embankment, as shown below. Superposition enables the vertical stress increase to be obtained at points A, B and C using Equations (2.13), (2.14) and (2.15) as applicable.

Vertical stress increase at point A:

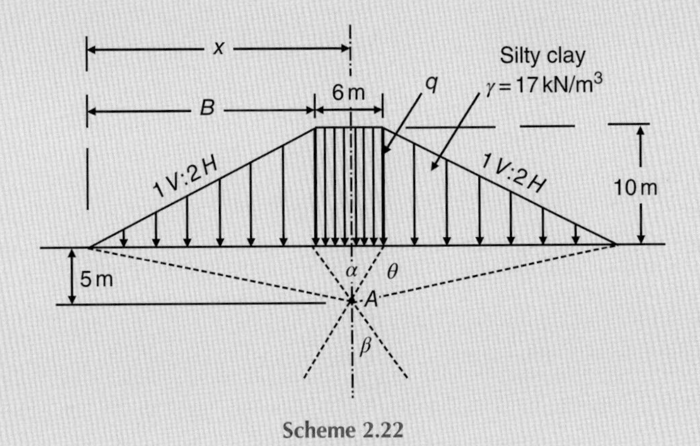

Scheme 2.22

$$\tan\frac{\alpha}{2} = \frac{3}{5} = 0.6 \rightarrow \frac{\alpha}{2} = 31^\circ \rightarrow \alpha = 62^\circ;\ \beta = \frac{\alpha}{2} = 31^\circ$$

$$\tan(\theta+\beta) = \frac{23}{5} = 4.6 \rightarrow (\theta+\beta) = 77.75^\circ \rightarrow \theta = 46.75^\circ$$

(*Continued*)

Equation (2.14): $$\sigma_z = \frac{q}{\pi}(\alpha + \sin\alpha)$$

Equation (2.15): $$\sigma_z = \frac{q}{\pi}\left(\frac{x}{B}\theta - \frac{1}{2}\sin 2\beta\right)$$

$$\Sigma\sigma_z = \left[\frac{10\times 17}{\pi}\left(\frac{62}{180}\pi + \sin 62\right)\right] + 2\left[\frac{10\times 17}{\pi}\left(\frac{23}{20}\times\frac{46.75}{180}\pi - \frac{1}{2}\sin 62\right)\right]$$

$$= 106.31 + 53.79 = \mathit{160.1\ kPa}$$

Vertical stress increase at point *B*:

Vertical stress increase at point *B* equals to sum of the vertical stresses due to loads of the rectangle at the middle and the two triangles on the sides, computed as shown below.

(1) Vertical stress due to load of the triangle on the right side of the embankment.

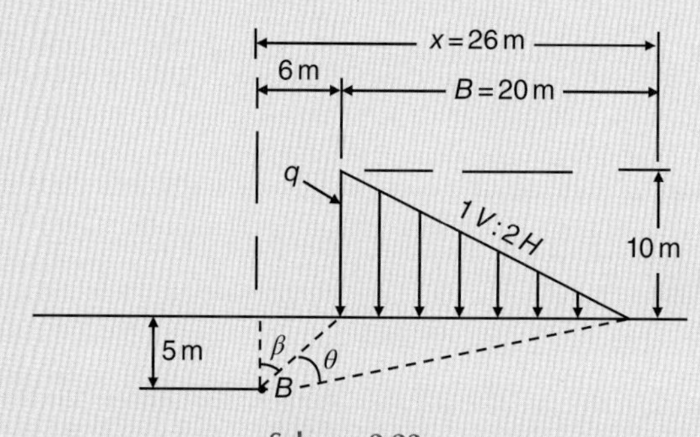

Scheme 2.23

$$\tan\beta = \frac{6}{5} = 1.2 \rightarrow \beta = 50.3°, \tan(\theta+\beta) = \frac{26}{5} = 5.2 \rightarrow$$

$$\theta + \beta = 79.12^o \rightarrow \theta = 28.82°$$

Equation (2.15): $$\sigma_z = \frac{q}{\pi}\left(\frac{x}{B}\theta - \frac{1}{2}\sin 2\beta\right)$$

$$\sigma_{z1} = \sigma_z = \frac{10\times 17}{\pi}\left[\frac{26}{20}\times\frac{28.82}{180}\pi - \frac{1}{2}\sin 100.6°\right]$$

$$= 54.09(0.654 - 0.491) = 8.82\,\text{kPa}$$

(2) Vertical stress due to load of the triangle on the left side of the embankment.

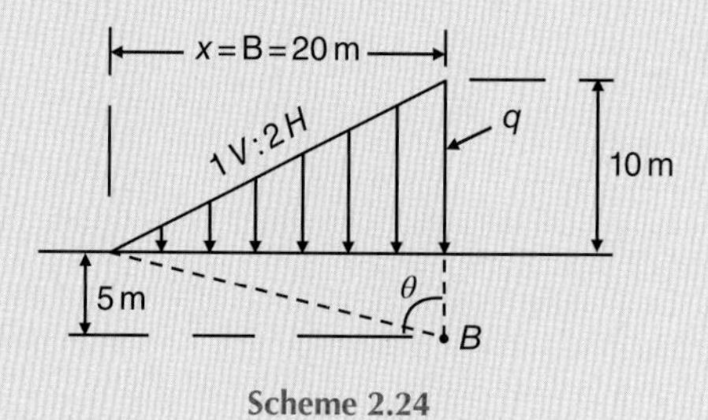

Scheme 2.24

$$\beta = 0°,\ \tan\theta = \frac{20}{5} = 4 \rightarrow \theta = 76°$$

$$\text{Equation (2.15)}: \ \sigma_z = \frac{q}{\pi}\left(\frac{x}{B}\theta - \frac{1}{2}\sin 2\beta\right)$$

$$\sigma_{z2} = \sigma_z = \frac{q}{\pi}\left[\frac{x}{B}\theta\right] = 54.09\left[\frac{20}{20} \times \frac{76}{180}\pi\right] = 71.78\,\text{kPa}$$

(3) Vertical stress due to load of the rectangle at middle of the embankment.

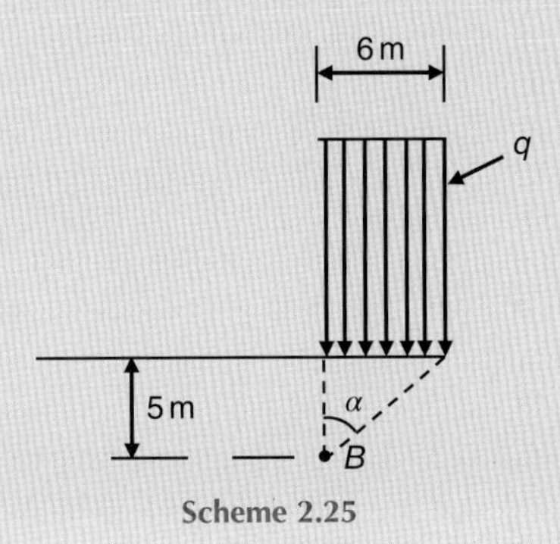

Scheme 2.25

$$\beta = 0°,\ \tan(\alpha + \beta) = \tan\alpha = \frac{6}{5} = 1.2 \rightarrow \alpha = 50.3°$$

Equation (2.13):
$$\sigma_z = \frac{q}{\pi}[\alpha + \sin\alpha\ \cos(\alpha + 2\beta)]$$

$$\sigma_{z3} = \sigma_z = 54.09\left[\frac{50.3 \times \pi}{180} + \sin 50.3°\ \cos 50.3°\right]$$
$$= 54.09(0.878 + 0.491) = 74.05\,\text{kPa}$$

$$\Sigma\sigma_z = \sigma_{z1} + \sigma_{z2} + \sigma_{z3}$$
$$= 8.82 + 71.78 + 74.05 = \mathit{154.65\,kPa}$$

Vertical stress increase at point *C*:
The determination of vertical stress at point *C* proceeds in the same manner as that for vertical stress at point A_2 in Problem 2.14. The necessary computations are left for the reader.

Problem 2.17

Three point loads 10, 7.5 and 9 MN, act in line 5 m apart near the surface of a soil mass, as shown below. Calculate the vertical stress at a depth of 4 m vertically below the centre load.

Solution:
Vertical stress due to a point load is

Equation (2.11):
$$\sigma_z = \frac{3V}{2\pi z^2}\frac{1}{\left[1 + (r/z)^2\right]^{5/2}}$$

(Continued)

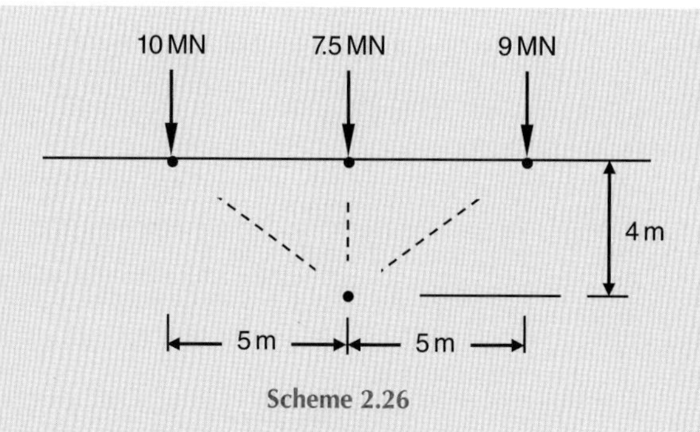

Scheme 2.26

Vertical stress due to the 10 MN load plus the 9 MN load is

$$\sigma_z = \frac{3}{2\pi \times 4^2} \frac{1}{\left[1+(5/4)^2\right]^{5/2}} (10+9) = (0.00284)(19) = 0.054\,\text{MN/m}^2$$

Vertical stress due to the 7.5 MN load is

$$\sigma_z = \frac{3 \times 7.5}{2\pi \times 4^2} \frac{1}{\left[1+(0/4)^2\right]^{5/2}} = 0.224\,\text{MN/m}^2$$

$$\text{Total}\,\sigma_z = 0.054 + 0.224 = 0.278\,\text{MN/m}^2 = 278\,kPa$$

Problem 2.18

A strip footing 2.5 m wide carries a uniform pressure of 300 kPa on the surface of a deposit of sand, as shown in the scheme below. The water table is at the surface. The sand has a saturated unit weight = 20 kN/m^3. Determine the effective vertical stress at a point 3 m below the centre of the footing before and after the application of the pressure.

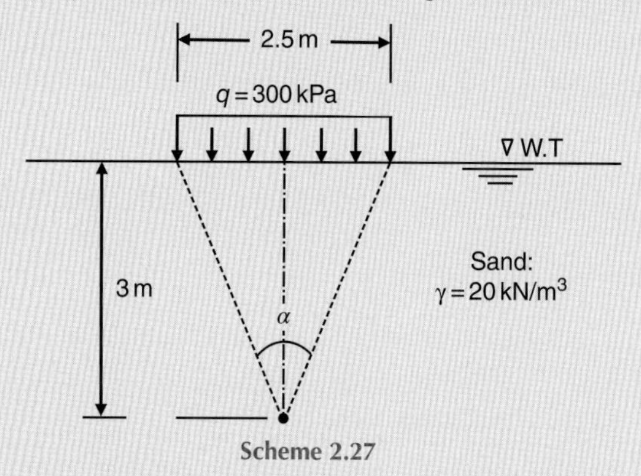

Scheme 2.27

Solution:

Before loading:

Effective unit weight of soil $= \gamma' = \gamma_{sat} - \gamma_w$

The effective unit weight of the sand $= \gamma' = 20 - 10 = 10\,\text{kN/m}^3$

The effective vertical stress $= \sigma'_z = 3 \times \gamma' = 3 \times 10 = 30\,kPa$

After loading:

$$\tan\frac{\alpha}{2} = \frac{2.5/2}{3} = 0.417 \rightarrow \frac{\alpha}{2} = 22.65° \rightarrow \alpha = 45.3°$$

Equation (2.14):
$$\sigma_z = \frac{q}{\pi}(\alpha + \sin\alpha) = \frac{300}{\pi}\left(\frac{45.3}{180}\pi + \sin 45.3°\right)$$
$$= 143.35\,\text{kPa}$$

The effective vertical stress $= \sigma'_z + \sigma_z = 143.35 + 30.00 = \mathit{173.35\,kPa}$

Problem 2.19

Loads P and $2P$ of two columns are to be applied to the surface of a 6 m thick layer of dense sand which overlies a layer of clay. Determine the maximum spacing of the columns if the settlement of the heavier column is not to be greater than 1.5 times that of the other column. Assume the settlement is due to the clay alone and that its compressibility characteristics are the same for each load.

Solution:
Settlement of a clay layer may be assumed proportional to the effective vertical stress causing it provided that compressibility characteristics of the clay as well as the layer thickness remain unchanged. This is readily noticed in the known equation of consolidation settlement: $s_c = m_v\sigma'_z H$; in which m_v is the coefficient of volume compressibility of the clay and H is the layer thickness.

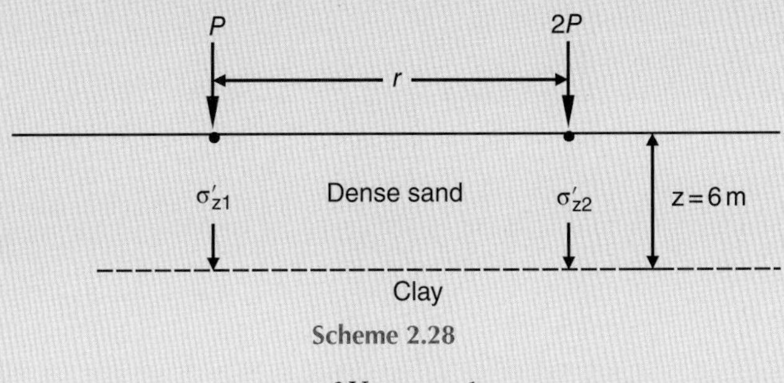

Scheme 2.28

Equation (2.11):
$$\sigma_z = \frac{3V}{2\pi z^2}\frac{1}{\left[1+(r/z)^2\right]^{5/2}}$$

Let $V = P$, and assume $x = [1 + (r/z)^2]^{5/2}$. Also, let the vertical effective stress increase below centre of load P equals σ'_{z1} and that below centre of load $2P$ equals σ'_{z2}, as shown in the scheme above.

$$\sigma'_{z1} = \frac{3P}{2\pi z^2} + \frac{6P}{2\pi z^2}\frac{1}{x} = \frac{3P}{2\pi z^2}\left[1 + \frac{2}{x}\right]$$

$$\sigma'_{z2} = \frac{6P}{2\pi z^2} + \frac{3P}{2\pi z^2}\frac{1}{x} = \frac{3P}{2\pi z^2}\left[2 + \frac{1}{x}\right]$$

$$s_c = m_v\sigma'_z H;\ \frac{s_{c2}}{s_{c1}} \le \frac{1.5}{1} \rightarrow \frac{m_v\sigma'_{z2}H}{m_v\sigma'_{z1}H} \le \frac{1.5}{1} \rightarrow \frac{\sigma'_{z2}}{\sigma'_{z1}} \le \frac{1.5}{1}$$

(Continued)

$$\text{Use } \frac{\sigma'_{z2}}{\sigma'_{z1}} = \frac{1.5}{1}$$

$$\frac{\sigma'_{z2}}{\sigma'_{z1}} = \frac{\frac{3P}{2\pi z^2}\left[2+\frac{1}{x}\right]}{\frac{3P}{2\pi z^2}\left[1+\frac{2}{x}\right]} = \frac{\left[2+\frac{1}{x}\right]}{\left[1+\frac{2}{x}\right]} = \frac{2x+1}{x+2} = 1.5$$

$$1.5x + 3 = 2x + 1 \rightarrow x = 4 = \left[1+(r/z)^2\right]^{5/2}$$

$$4 = \left[1+(r/z)^2\right]^{5/2} = \left[1+(r/6)^2\right]^{5/2}$$

$$1.74 = 1 + (r/6)^2 \rightarrow 0.74 = (r/6)^2 \rightarrow$$

$$r^2 = 26.64 \rightarrow r = 5.16\,\text{m}$$

Therefore, the maximum spacing of the columns = *5.16 m*

Problem 2.20

Centres of two columns *A* and *B* are 3 m apart. Column *A* is supported by a footing 1.2 × 1.2 m and column *B* by a footing 1 × 1 m. Column *A* is founded at a depth of 2.5 m below ground level and column *B* at 1.5 m. The contact pressure on the soil under each footing is 425 kPa. Find the increase in vertical stress at a depth of 9 m vertically below the centres of *A* and *B* and the point midway between them. Use the Newmark method.

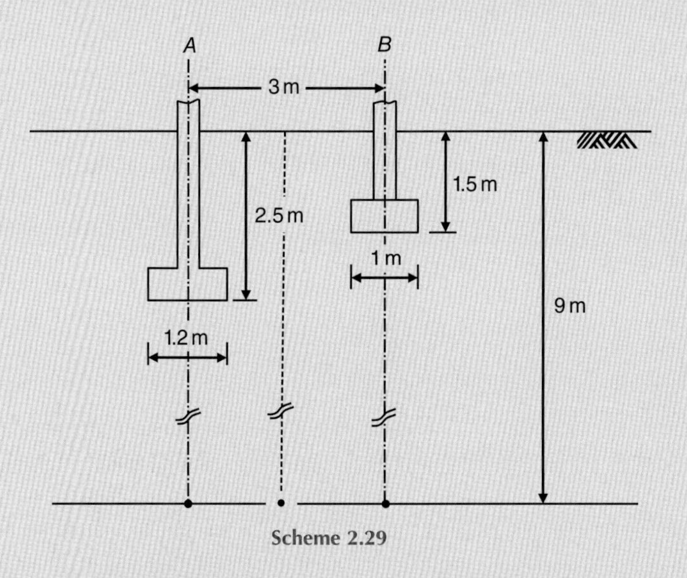

Scheme 2.29

Solution:

Vertical stress increase below centre of column *A*:

(1) Vertical stress increase due to load of column *A* is

$$\sigma_{A1} = q(4I)$$
$$q = 425\,\text{kPa}$$

$$B/z = L/z = 0.6/6.5 \approx 0.1 \rightarrow I = 0.0047$$

$$\sigma_{A1} = q(4I) = 425 \times 4 \times 0.0047 = 7.99\,\text{kPa}$$

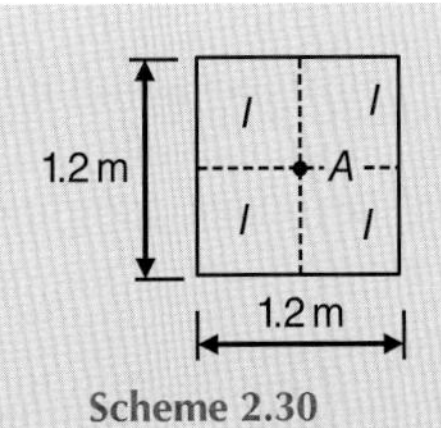

Scheme 2.30

(2) Vertical stress increase due to load of column B is obtained using the scheme shown below.

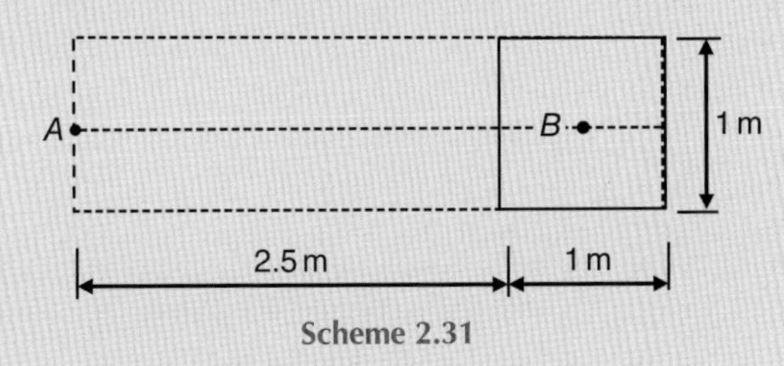

Scheme 2.31

For the two long rectangles: $\frac{B}{Z} = \frac{0.5}{7.5} = 0.067$ and $\frac{L}{Z} = \frac{3.5}{7.5} = 0.467 \rightarrow$

$$I_{long} = 0.012$$

For the two short rectangles: $\frac{B}{Z} = \frac{0.5}{7.5} = 0.067$ and $\frac{L}{Z} = \frac{2.5}{7.5} = 0.333 \rightarrow$

$$I_{short} = 0.01$$

$$\sigma_{A2} = q\left(2I_{long} - 2I_{short}\right) = 425(2 \times 0.012 - 2 \times 0.01) = 1.7\,\text{kPa}$$

Vertical stress increase below centre of column A is

$$\sigma_A = \sigma_{A1} + \sigma_{A2} = 7.99 + 1.7 = 9.67\ kPa$$

Vertical stress increase below centre of column B :

(1) Vertical stress increase due to load of column B is

$$\sigma_{B1} = q(4I)$$

$$q = 425\,\text{kPa}$$

$$B/z = L/z = 0.5/7.5 = 0.067 \rightarrow I = 0.003$$

$$\sigma_{A1} = q(4I) = 425 \times 4 \times 0.003 = 5.1\,\text{kPa}$$

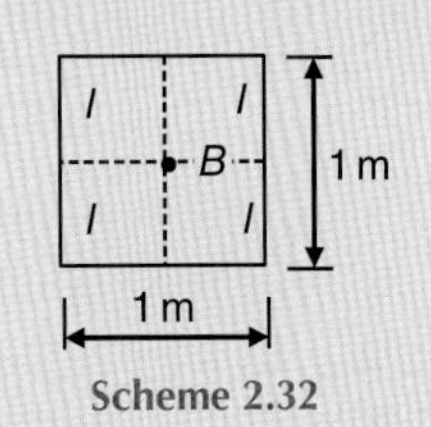

Scheme 2.32

(2) Vertical stress increase due to load of column A is obtained using the scheme shown below.

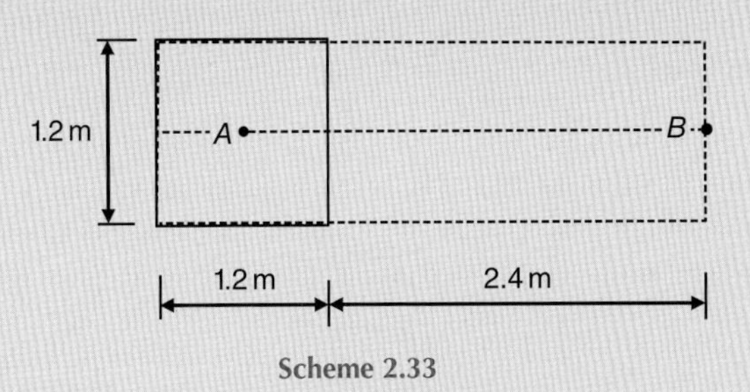

Scheme 2.33

(*Continued*)

For the two long rectangles: $\frac{B}{Z}=\frac{0.6}{6.5}=0.092$ and $\frac{L}{Z}=\frac{3.6}{6.5}=0.467 \rightarrow$

$$I_{long}=0.016$$

For the two short rectangles: $\frac{B}{Z}=\frac{0.6}{6.5}=0.092$ and $\frac{L}{Z}=\frac{2.4}{6.5}=0.369 \rightarrow$

$$I_{short}=0.0125$$

$$\sigma_{B2}=q\left(2I_{long}-2I_{short}\right)=425(2\times 0.016-2\times 0.0125)=2.98\,\text{kPa}$$

Vertical stress increase below centre of column B is

$$\sigma_B=\sigma_{B1}+\sigma_{B2}=5.1+2.98=\mathit{8.08\,kPa}$$

Vertical stress increase below the point midway between centres of A and B:

(1) Vertical stress increase due to load of column A is obtained using the scheme shown below.

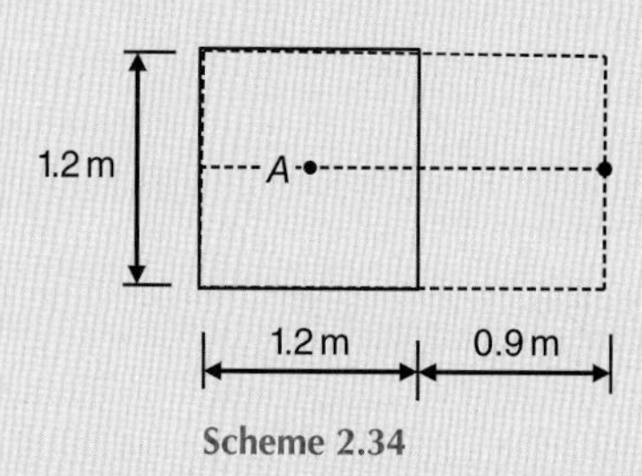

Scheme 2.34

For the two long rectangles: $\frac{B}{Z}=\frac{0.6}{6.5}=0.092$ and $\frac{L}{Z}=\frac{2.1}{6.5}=0.323 \rightarrow$

$$I_{long}=0.013$$

For the two short rectangles: $\frac{B}{Z}=\frac{0.6}{6.5}=0.092$ and $\frac{L}{Z}=\frac{0.9}{6.5}=0.138 \rightarrow$

$$I_{short}=0.006$$

$$\sigma_A=q\left(2I_{long}-2I_{short}\right)=425(2\times 0.013-2\times 0.006)=5.95\,\text{kPa}$$

(2) Vertical stress increase due to load of column B is obtained using the scheme shown below.

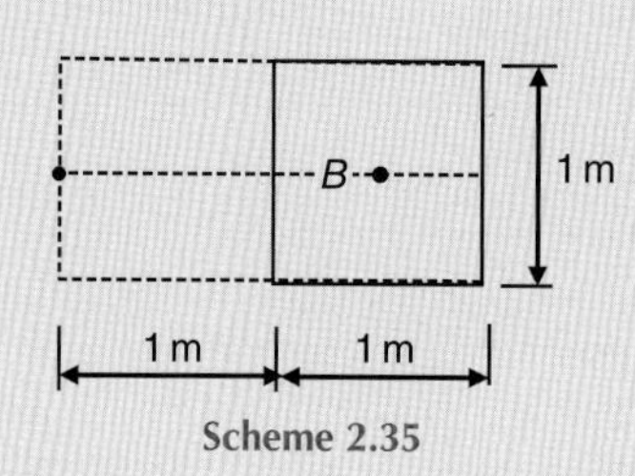

Scheme 2.35

For the two long rectangles: $\frac{B}{Z}=\frac{0.5}{7.5}=0.067$ and $\frac{L}{Z}=\frac{2}{7.5}=0.267 \rightarrow$

$$I_{long} = 0.008$$

For the two short rectangles: $\frac{B}{Z}=\frac{0.5}{7.5}=0.067$ and $\frac{L}{Z}=\frac{1}{7.5}=0.133 \rightarrow$

$$I_{short} = 0.004$$

$$\sigma_B = q\left(2I_{long} - 2I_{short}\right) = 425(2\times 0.008 - 2\times 0.004) = 3.4\,\text{kPa}$$

Increase in vertical stress below the point midway between centres of A and B is

$$\sigma_z = \sigma_A + \sigma_B = 5.95 + 3.4 = 9.35\,kPa$$

Problem 2.21

A cantilever retaining wall is shown in the scheme below. Height of the stem is 4 m, and there is a line load V of 150 kN/m length on the surface of the backfill acting at a distance of 2 m from top of the wall as shown. Calculate (a) the increase in vertical stress at points A and B due to the load V, and plot a linear pressure distribution between the points, (b) the total thrust on the structure.

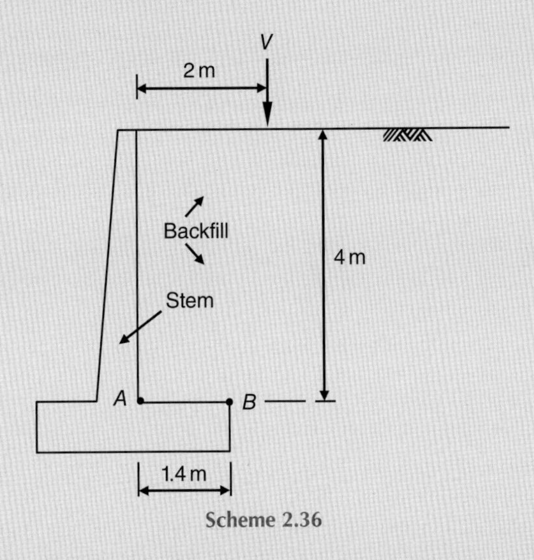

Scheme 2.36

Solution:

(a)

Equation (2.12):
$$\sigma_z = \frac{2V}{\pi}\frac{z^3}{\left(x^2+z^2\right)^2}$$

$$\sigma_{zA} = \frac{2\times 150}{\pi}\times\frac{4^3}{\left(2^2+4^2\right)^2} = 15.27\,kPa$$

$$\sigma_{zB} = \frac{2\times 150}{\pi}\times\frac{4^3}{\left(0.6^2+4^2\right)^2} = 22.82\,kPa$$

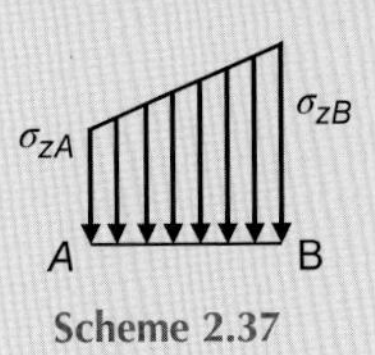

Scheme 2.37

(*Continued*)

(b) Similar to Equation (2.12), the following equation (not presented in Section 2.7.3, since it was beyond the scope of the subject matter) gives horizontal stress increase:

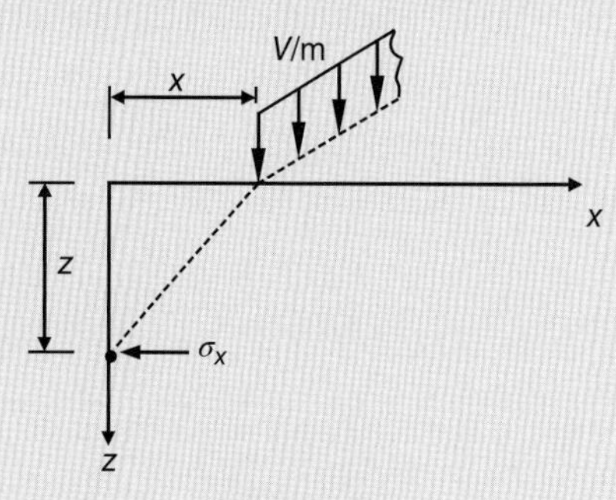

Scheme 2.38

$$\sigma_x = \frac{2V}{\pi}\frac{zx^2}{(x^2+z^2)^2}$$

However, in this case, since the relatively rigid retaining wall will tend to interfere with the lateral strain due to the load V, the lateral pressure p_x shall be twice as great as σ_x. For this loading condition, the lateral pressure is given by:

$$p_x = \frac{4V}{\pi}\frac{zx^2}{(x^2+z^2)^2}$$

In terms of the dimensions given in the scheme below, this equation becomes

$$p_x = \frac{4V}{\pi h}\frac{m^2 n}{(m^2+n^2)^2}$$

The total thrust on the stem due to the line load V is given by:

$$P_x = \int_0^1 p_x h\, dn = \frac{2V}{\pi}\frac{1}{m^2+1}$$

$$x = mh = 2 \rightarrow m = \frac{2}{h} = \frac{2}{4} = 0.5$$

$$P_x = \frac{2\times 150}{\pi}\frac{1}{0.5^2+1} = 76.36\, kN/m$$

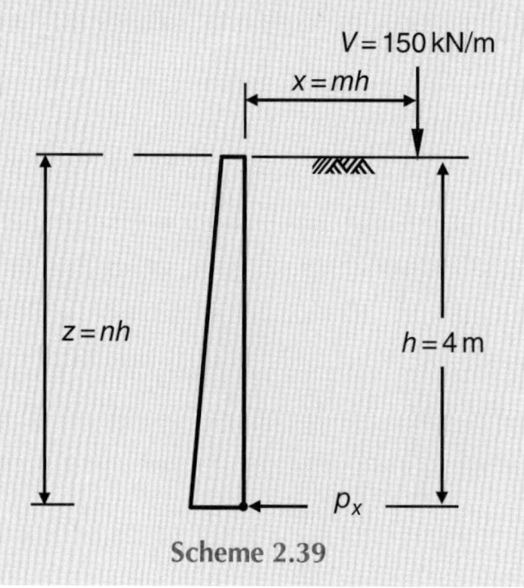

Scheme 2.39

References

Bowles, J. E. (2001), *Foundation Analysis and Design*, 5th edn, McGraw-Hill, New York.

Boussinesq, M. J. (1883), *Application Des Potentials à l'Étude de l'Équilibre et du Movvement Des Solides Elastiques*, Gauthier-Villars, Paris (in French).

Coduto, Donald P. (2001), *Foundation Design: Principles and Practices*, 2nd edn, Prentice-Hall, Inc., New Jersey.

Das, Braja M. (2011), *Principles of Foundation Engineering*, 7th edn, CENGAGE Learning, United States.

Fadum, R. E. (1948), "Influence Values for Estimating Stresses in Elastic Foundations", *Proceedings of 2nd International Conference SFME*, Rotterdam, Vol. **3**, pp. 77–84.

Griffiths, D. V. (1984), "A Chart for Estimating the Average Vertical Stress Increase in an Elastic Foundation below a Uniformly Rectangular Loaded Area", *Canadian Geotechnical Journal*, Vol. **21**, No. 4, pp. 710–713.

Knappett, J. A., and Craig, R. F. (2012), *Soil Mechanics*, 8th edn, Spon, Abingdon, United Kingdom.

Kosmatca, S. H., and Panarese, W. C. (1988), *Design and Control of Concrete Mixtures*, 13th edn, Portland-Cement Association, Skokie, Illinois.

Newmark, N. M. (1935), *Simplified Computation of Vertical Pressures in Elastic Foundations*, Engineering Experiment Station Circular No. 24, University of Illinois, Urbana, Illinois.

Newmark, N. M. (1942), *Influence Charts for Computation of Stresses in Elastic Foundations*, University of Illinois Bulletin No. 338, University of Illinois, Urbana, Illinois.

PCA (1991), "Durability of Concrete in Sulfate-Rich Soils", *Concrete Technology Today*, Vol. **12**, No. 3, pp. 6–8, Portland Cement Association, New York.

Taylor, D. W. (1948), *Fundamentals of Soil Mechanics*, 1st edn, John Wiley and Sons, Inc., New York.

Terzaghi, K., and Peck, R. B. (1967), *Soil Mechanics in Engineering Practice*, 2nd edn, John Wiley and Sons, Inc., New York.

Tomlinson, M. J. (2001), *Foundation Design and Construction*, 7th edn, Pearson Education, Harlow, United Kingdom.

Westegaard, H. M. (1938), "A Problem of Elasticity Suggested by a Problem in Soil Mechanics: Soft Material Reinforced by a Numerous Strong Horizontal Sheets", *Contributions to Mechanics of Solids*, Stephen Timoshenko 60th Anniversary Volume, The Macmillan Company, New York.

CHAPTER 3

Shallow Foundations – Settlement

3.1 General

The vertical downward displacements at the ground surface or the vertical downward displacement of a structure are often called *settlement*. It is usually caused by direct application of structural loads on the foundation which in turn cause compression of the supporting material (soil or rock). However, in addition to the settlement under loads, foundation settlement may also occur due to some or combination of other causes, as follows:

(1) Seasonal swelling and shrinking of expansive soils.
(2) Ground water lowering or a falling ground water table. Prolonged lowering of water level in fine-grained soils may introduce settlement due to consolidation. Repeated lowering and rising of water level in loose granular soils tends to compact the soil and cause settlement. Pumping water or draining water by pipes or tiles from granular soils without adequate filter material as protection may in a period of time wash and carry a sufficient amount of fine particles away from the soil and cause settlement.
(3) Underground erosion. It may cause formation of cavities in the subsoil which when collapse settlement occurs.
(4) Changes in the vicinity. If there are changes adjacent to the property such as recently placed fill, excavation, construction of a new structure, underground tunnelling or mining and so on, settlement may occur due to increase in the stresses.
(5) Vibrations and shocks. Vibrations due to pile driving or oscillating machineries as well as shocks due to blasting or earthquake cause settlements, especially in granular soils.
(6) Ground movement on earth slopes. If surface erosion, slow creep or landslides occur, there may be settlement problems.

Theoretically, no damage is done to the superstructure if the foundations settle uniformly. However, settlement exceeding a certain limit may cause trouble in utilities such as water pipe lines, sewers, telephone lines; also trouble in the surface drainage configuration and in access from streets. Sometimes buildings must join existing structures, and it is required that floors of the two buildings be at the same level. If the new building settles excessively, the floors will no longer remain at the same level, causing serious serviceability problems. In addition to all these

Shallow Foundations: Discussions and Problem Solving, First Edition. Tharwat M. Baban.

unfavourable incidents, excessive settlement may cause aesthetic problems long before there is any threat to structural integrity or serviceability.

More serious troubles may occur due to excessive *differential settlements*. Differential settlement can be computed as the difference in settlement between two adjacent points. It has been found from actual observations of various existing buildings that differential settlement seldom exceeds 75% of the maximum total settlement. Actually, in most constructions, the subsoil is not homogeneous and the load carried by various shallow foundations of a given structure can vary widely. Consequently, settlements of varying degrees in different parts of the structure are expected.

In the design of foundations, there are certain settlement requirements must be satisfied. The requirements are usually stated in terms of the allowable total settlement S_{Ta} and the allowable total differential settlement ΔS_{Ta}, as follows:

$$S_T \leq S_{Ta} \tag{3.1)-(a}$$

$$\Delta S_T \leq \Delta S_{Ta} \tag{3.1)-(b}$$

where S_T = total settlement, and ΔS_T = total differential settlement.

Figure 3.1 defines the necessary settlement parameters. Line *AE* represents base line of various footings for a given structure. Center of the footings at *A*, *B*, *C*, *D* and *E* have gone through varying degrees of settlement. The total settlement at *A* is AA' and that at *B* is BB' and so on. Definitions of the various settlement parameters are as follows:

S_T = Total settlement of a given point
ΔS_T = Difference in total settlement between any two points
α = Gradient between two successive points
β = Angular Distortion $= \dfrac{\Delta S_T}{L}$
ω = Tilt or tilt angle
Δ = Relative deflection (i.e. movement from a straight line joining two reference points)
$\dfrac{\Delta}{L}$ = Deflection ratio.

MacDonald and Skempton (1955) made a study of settlements in 98 buildings, mostly older structures of load-bearing wall, steel and reinforced concrete construction. As a result, they proposed acceptable limiting values for the maximum allowable differential settlement ΔS_{Ta}, total settlement S_{Ta} and angular distortion β, to be used for building purposes, as presented in Table 3.1 (Skempton and MacDonald, 1956).

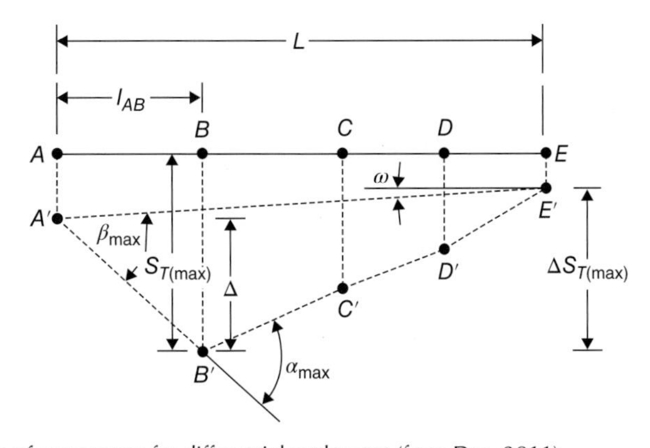

Figure 3.1 Definition of parameters for differential settlement (from Das, 2011).

Table 3.1 Tolerable settlements, mm.

Criterion	Isolated foundation	Rafts
Angular distortion (cracking), β_{max}	1/300	
Greatest differential settlement, $\Delta S_{T(max)}$		
Clays	45 (35)*	
Sands	32 (25)	
Maximum total settlement, $S_{T(max)}$		
Clays	75	75–125 (65–100)
Sands	50	50–75 (35–65)

[a] Recommended values are in the parentheses.

Table 3.2 A composite guide for estimating differential settlement.

Construction and/or material	Maximum δ/L
Masonry (centre sag)	1/250–1/700
(edge sag)	1/500–1/1000
Masonry and steel	1/500
Steel with metal siding	1/250
Tall structures	<1/300
Storage tanks (centre – to – edge)	<1/300

Also, one might use Table 3.2, a composite from several sources, as a guide in estimating differential settlement (Bowles, 2001). Here, L = column spacing and δ = differential settlement between two adjacent columns.

Recommendations of the European Committee for Standardisation (1994) regarding the limiting values for the serviceability and maximum accepted foundation movements are presented in Table 3.3.

Values of acceptable deflection ratios from The Soviet Code of Practice for buildings on both unfrozen and frozen ground are given in Table 3.4 (Mikhejef et al., 1961; Polshin and Tokar, 1957).

There are empirical criteria have been established by many researchers for limiting the movement of structures in order to prevent or minimise cracking or other forms of structure damage. These criteria are shown in Table 3.5. It will be noted from this table that the critical factor for framed buildings and reinforced load-bearing walls is the relative rotation (or angular distortion), whereas the deflection ratio is the criterion for unreinforced load-bearing walls which may fail by sagging or hogging.

In Table 3.5, the limiting values for framed buildings are for structural members of average dimensions. Values may be much less for exceptionally large and stiff beams or columns for which the limiting values of angular distortion should be obtained by structural analysis.

According to Coduto (2001), Table 3.6 presents a synthesis of the abovementioned studies, expressed in terms of the allowable angular distortion β_a; compiled from Wahls (1994), AASHTO (1996) and

Table 3.3 Recommendations of the European Committee for Standardisation of Differential Settlement Parameters (from Das, 2011).

Item	Parameter	Magnitude	Comments
Limiting values for serviceability (European Committee for Standardization, 1994a)	S_T	25 mm	Isolated shallow foundation
		50 mm	Raft foundation
	ΔS_T	5 mm	Frames with rigid cladding
		10 mm	Frames with flexible cladding
		20 mm	Open frames
	β	1/500	—
Maximum acceptable foundation movement (European Committee for Standardization, 1994b)	S_T	50	Isolated shallow foundation
	ΔS_T	20	Isolated shallow foundation
	β	≈1/500	—

Table 3.4 Allowable deflection ratios for structures (from Bowles, 2001).

	Structure on:		
Structure	**Sand or hard clay**	**Plastic clay**	**Average maximum settlement, mm**
Crane runway	0.003	0.003	
Steel and concrete frames	0.002	0.002	100
End rows of brick-clad frame	0.0007	0.001	150
Where strain does not occur	0.005	0.005	
Multistory brick wall			25 $L/H \geq 2.5$
L/H to 3[a]	0.0003	0.0004	100 $L/H \leq 1.5$
Multistory brick wall			
L/H over 5	0.0005	0.0007	
One-story mill buildings	0.001	0.001	
Smokestacks, water towers, ring foundations	0.004	0.004	300

[a] L/H = ratio of the length to the height of a building.

other sources. These values already include a factor of safety of at least 1.5. They may be used to compute the allowable differential settlement as follows:

$$\Delta S_{Ta} = \beta_a L \tag{3.2}$$

Where:
ΔS_{Ta} = allowable differential settlement
β_a = allowable angular distortion (from the following table)
L = column spacing (horizontal distance between columns)

Note: Be sure to consider local practice and precedent when developing design values of ΔS_{Ta}.

Table 3.5 Limiting values of distortion and deflection ratio of buildings (from Tomlinson, 2001).

Type of structure	Type of damage	Limiting values: Values of relative rotation (angular distortion) — Skempton and MacDonald	Mayerhof	Polshin and Tokar	Bjerrum
Framed buildings and reinforced load bearing walls	Structural damage	1/15	1/250	1/200	1/150
	Cracking in walls and partitions	1/300 (but 1/500 recommended)	1/500	1/500 (0.7/1000 to 1/1000 for end bays)	1/500
		Values of deflection ratio Δ/L: Mayerhof	Polshin and Tokar	Burland and Wroth	
Unreinforced load-bearing walls	Cracking by sagging	0.4/1000	$L/H = 3$: 0.3 to 0.4×10^{-3}	At $L/H = 1$: 0.4×10^{-3} At $L/H = 5$: 0.8×10^{-3}	
	Cracking by hogging	–	–	At $L/H = 1$: 0.2×10^{-3} At $L/H = 5$: 0.4×10^{-3}	

Table 3.6 Allowable angular distortions β_a (compiled from Wahls, 1994; AASHTO, 1996; and other sources).

Tyle of structure	β_a
Steel tanks	1/25
Bridges with simply-supported spans	1/125
Bridges with continuous spans	1/250
Buildings which are very tolerant of differential settlement, such as industrial buildings with corrugated steel siding and no sensitive interior finish	1/250
Typical commercial and residential buildings	1/500
Overhead traveling crane rails	1/500
Buildings which are especially intolerant of differential settlements, such as that with sensitive wall or floor finishes	1/1000
Machinery[a]	1/1500
Buildings with unreinforced masonry load bearing walls Length/height≤ 3	1/2500
Length/height≥ 5	1/1250

[a] Large machines, such as turbines or large punch presses, often have their own foundation, separate from that of the building that houses them. It often is appropriate to discuss allowable differential settlement issues with the machine manufacturer.

Total settlement (S_T) under loads of a structural foundation is generally made up of three components. The *immediate settlement* (S_i) takes place immediately after application of the loading or within a short period of time (not more than several days) as a result of elastic deformation of the soil without change in water content. It may predominate in all coarse-grained soils with a large coefficient of permeability (not less than 0.001 m/s) and in unsaturated (degree of saturation less than 90%) fine-grained soils. The *consolidation settlement* (S_c) takes place as a result of volume reduction of the soil caused by extrusion of some of the porewater from the soil. It predominates in saturated or nearly saturated fine-grained soils unless the soil is very organic. Analysis of consolidation settlement requires estimate of both the settlement and how long a time it will take for most of the settlement to occur. *Secondary compression or creep* (S_s) caused by the viscous resistance of the soil to continuing readjustment of the soil particles into a closer (or denser) state under the compressive load. This phenomenon is associated with both immediate and consolidation-type settlements, although it is usually not of much significance with immediate settlements. Secondary compression may be the larger component of total settlement in some soils, particularly in soils with a large organic content.

Thus, the total settlement (S_T) is given by

$$S_T = S_i + S_c + S_s \tag{3.3}$$

3.2 Immediate Settlement

Immediate settlement or *elastic* settlement of a shallow foundation can be estimated by using the theory of elasticity.

The settlement under the *corner* of a uniformly loaded *flexible* rectangular base of dimensions $B \times L$, or a round base converted to an equivalent square, on the *surface* of an elastic half-space can be computed as

$$S_i = q \times B \times \frac{1-\mu^2}{E_s} \times m \times I_S \tag{3.4}$$

According to Fox (1948), the settlement is reduced when the loaded area (the foundation base) is placed at some depth in the ground. He suggested a factor I_F be used with the above equation as follows:

$$S_i = q \times B \times \frac{1-\mu^2}{E_s} \times m \times I_S \times I_F \tag{3.5}$$

Where:

q = Contact pressure intensity (uniformly distributed load), in units of E_s
B = Least lateral dimension of the contributing base area, in units of S_i
L = Length of the contributing base area in units of S_i
E_s = *Average* modulus of elasticity of soil for a depth equals to H
H = Thickness of the soil layer or $4B$ whichever is smaller
I_F = Influence factor (Fox, 1948) depends on Poisson's ratio of soil μ, $\frac{B}{L}$ or $\frac{L}{B}$ and $\frac{D}{B}$; it can be obtained from Figure 3.2 or from Table 3.7 (approximately).
m = Number of corners contributing to S_i
I_S = Influence factor (Steinbrenner) $= I_1 + \frac{1-2\mu}{1-\mu} I_2$
I_1 and I_2 = Factors obtained from equations by Steinbrenner or from Table 3.8, depend on M and N
$M = \frac{L}{B}$, $N = \frac{H}{B}$, H = Thickness of the soil layer, in units of B

For S_i at *centre* use $\frac{B}{2}$ instead of B, $\frac{L}{2}$ instead of L, and $m = 4$.
For S_i at *corner* use $B = B, L = L$, and $m = 1$.

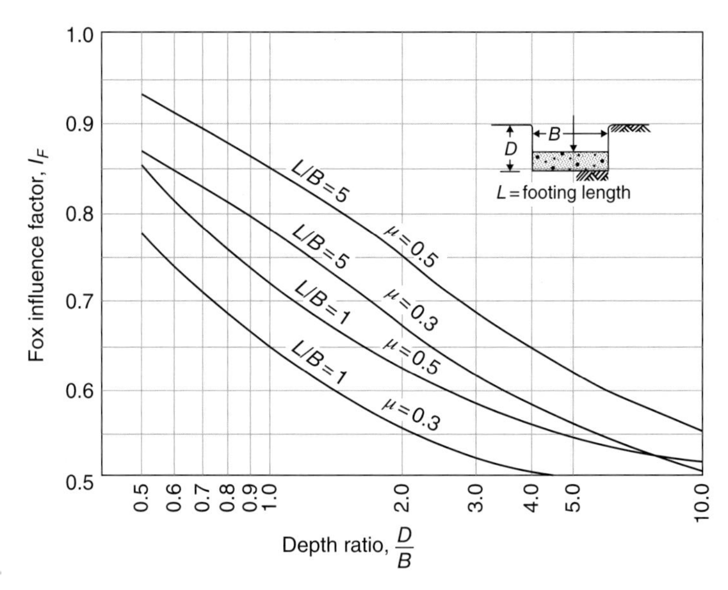

Figure 3.2 Influence factor I_F for a foundation base at depth D. Use the actual base width B for the $\frac{D}{B}$ ratio.

Table 3.7 Variation of I_F with $\frac{B}{L}$, $\frac{D}{B}$ and μ.

		B/L		
μ	***D/B***	**0.2**	**0.5**	**1.0**
0.3	0.2	0.95	0.93	0.90
	0.4	0.90	0.86	0.81
	0.6	0.85	0.80	0.74
	1.0	0.78	0.71	0.65
0.4	0.2	0.97	0.96	0.93
	0.4	0.93	0.89	0.85
	0.6	0.89	0.84	0.78
	1.0	0.82	0.75	0.69
0.5	0.2	0.99	0.98	0.96
	0.4	0.95	0.93	0.89
	0.6	0.92	0.87	0.82
	1.0	0.85	0.79	0.72

For S_i at *middle of side B* use $\frac{B}{2}$ instead of B, $L = L$, and $m = 2$.

For S_i *at middle of side L* where $\frac{L}{2}$ is larger than B, use $B = B$, $\frac{L}{2}$ for L, and $m = 2$; otherwise, use B for L, $\frac{L}{2}$ for B, and $m = 2$.

Table 3.8 Values of I_1 and I_2 to compute I_S factor (from Bowles, 2001).

N	M = 1.0	1.0	1.2	1.3	1.4	1.5	1.6	1.7	1.8	1.9	2.0
0.2	I_1 = 0.009	0.008	0.008	0.008	0.008	0.008	0.007	0.007	0.007	0.007	0.007
	I_2 = 0.041	0.042	0.042	0.042	0.042	0.042	0.043	0.043	0.043	0.043	0.043
0.4	0.033	0.032	0.031	0.030	0.029	0.028	0.028	0.027	0.027	0.027	0.027
	0.066	0.068	0.069	0.070	0.070	0.071	0.071	0.072	0.072	0.073	0.073
0.6	0.066	0.064	0.063	0.061	0.060	0.059	0.058	0.057	0.056	0.056	0.055
	0.079	0.081	0.083	0.085	0.087	0.088	0.089	0.090	0.091	0.091	0.092
0.8	0.104	0.102	0.100	0.098	0.096	0.095	0.093	0.092	0.091	0.090	0.089
	0.083	0.087	0.090	0.093	0.095	0.097	0.098	0.100	0.101	0.102	0.103
1.0	0.142	0.140	0.138	0.136	0.134	0.132	0.130	0.129	0.127	0.126	0.125
	0.083	0.088	0.091	0.095	0.098	0.100	0.102	0.104	0.106	0.108	0.109
1.5	0.224	0.224	0.224	0.223	0.222	0.220	0.219	0.217	0.216	0.214	0.213
	0.075	0.080	0.084	0.089	0.093	0.096	0.099	0.102	0.105	0.108	0.110
2.0	0.285	0.288	0.290	0.292	0.292	0.292	0.292	0.292	0.291	0.290	0.289
	0.064	0.069	0.074	0.078	0.083	0.086	0.090	0.094	0.097	0.100	0.102
3.0	0.363	0.372	0.379	0.384	0.389	0.393	0.396	0.398	0.400	0.401	0.402
	0.048	0.052	0.056	0.060	0.064	0.068	0.071	0.075	0.078	0.081	0.084
4.0	0.408	0.421	0.431	0.440	0.448	0.455	0.460	0.465	0.469	0.473	0.476
	0.037	0.041	0.044	0.048	0.051	0.054	0.057	0.060	0.063	0.066	0.069
5.0	0.437	0.452	0.465	0.477	0.487	0.496	0.503	0.510	0.516	0.522	0.526
	0.031	0.034	0.036	0.039	0.042	0.045	0.048	0.050	0.053	0.055	0.058
6.0	0.457	0.474	0.489	0.502	0.514	0.524	0.534	0.542	0.550	0.557	0.563
	0.026	0.028	0.031	0.033	0.036	0.038	0.040	0.043	0.045	0.047	0.050
7.0	0.471	0.490	0.506	0.520	0.533	0.545	0.556	0.566	0.575	0.583	0.590
	0.022	0.024	0.027	0.029	0.031	0.033	0.035	0.037	0.039	0.041	0.043
8.0	0.482	0.502	0.519	0.534	0.549	0.561	0.573	0.584	0.594	0.602	0.611
	0.020	0.022	0.023	0.025	0.027	0.029	0.031	0.033	0.035	0.036	0.038
9.0	0.491	0.511	0.529	0.545	0.560	0.574	0.587	0.598	0.600	0.618	0.627
	0.017	0.019	0.021	0.023	0.024	0.026	0.028	0.029	0.031	0.033	0.034
10.0	0.498	0.519	0.537	0.554	0.570	0.584	0.597	0.610	0.621	0.631	0.641
	0.016	0.017	0.019	0.020	0.022	0.023	0.025	0.027	0.028	0.030	0.031
20.0	0.529	0.553	0.575	0.595	0.614	0.631	0.647	0.662	0.677	0.690	0.702
	0.008	0.009	0.010	0.010	0.011	0.012	0.013	0.013	0.014	0.015	0.016
500.0	**0.560**	0.587	0.612	0.635	0.656	0.677	0.696	0.714	0.731	0.748	0.763
	0.000	0.000	0.000	0.000	0.000	0.000	0.001	0.001	0.001	0.001	0.001

(Continued)

Table 3.8 (*Continued*)

N	*M* = 2.5	4.0	5.0	6.0	7.0	8.0	9.0	10.0	25.0	50.0	100.0
0.2	I_1 = 0.007	0.006	0.006	0.006	0.006	0.006	0.006	0.006	0.006	0.006	0.006
	I_2 = 0.043	0.044	0.044	0.044	0.044	0.044	0.044	0.044	0.044	0.044	0.044
0.4	0.026	0.024	0.024	0.024	0.024	0.024	0.024	0.024	0.024	0.024	0.024
	0.074	0.075	0.075	0.075	0.076	0.076	0.076	0.076	0.076	0.076	0.076
0.6	0.053	0.051	0.050	0.050	0.050	0.049	0.049	0.049	0.049	0.049	0.049
	0.094	0.097	0.097	0.098	0.098	0.098	0.098	0.098	0.098	0.098	0.098
0.8	0.086	0.082	0.081	0.080	0.080	0.080	0.079	0.079	0.079	0.079	0.079
	0.107	0.111	0.112	0.113	0.113	0.113	0.113	0.114	0.114	0.114	0.114
1.0	0.121	0.115	0.113	0.112	0.112	0.112	0.111	0.111	0.110	0.110	0.110
	0.114	0.120	0.122	0.123	0.123	0.124	0.124	0.124	0.125	0.125	0.125
1.5	0.207	0.197	0.194	0.192	0.191	0.190	0.190	0.189	0.188	0.188	0.188
	0.118	0.130	0.134	0.136	0.137	0.138	0.138	0.139	0.140	0.140	0.140
2.0	0.284	0.271	0.267	0.264	0.262	0.261	0.260	0.259	0.257	0.256	0.256
	0.114	0.131	0.136	0.139	0.141	0.143	0.144	0.145	0.147	0.147	0.148
3.0	0.402	0.392	0.386	0.382	0.378	0.376	0.374	0.373	0.368	0.367	0.367
	0.097	0.122	0.131	0.137	0.141	0.144	0.145	0.147	0.152	0.153	0.154
4.0	0.484	0.484	0.479	0.474	0.470	0.466	0.464	0.462	0.453	0.451	0.451
	0.082	0.110	0.121	0.129	0.135	0.139	0.142	0.145	0.154	0.155	0.156
5.0	0.553	0.554	0.552	0.548	0.543	0.540	0.536	0.534	0.522	0.519	0.519
	0.070	0.098	0.111	0.120	0.128	0.133	0.137	0.140	0.154	0.156	0.157
6.0	0.585	0.609	0.610	0.608	0.604	0.601	0.598	0.595	0.579	0.576	0.575
	0.060	0.087	0.101	0.111	0.120	0.126	0.131	0.135	0.153	0.157	0.157
7.0	0.618	0.653	0.658	0.658	0.656	0.653	0.650	0.647	0.628	0.624	0.623
	0.053	0.078	0.092	0.103	0.112	0.119	0.125	0.129	0.152	0.157	0.158
8.0	0.643	0.688	0.697	0.700	0.700	0.698	0.695	0.692	0.672	0.666	0.665
	0.047	0.071	0.084	0.095	0.104	0.112	0.118	0.124	0.151	0.156	0.158
9.0	0.663	0.716	0.730	0.736	0.737	0.736	0.735	0.732	0.710	0.704	0.702
	0.042	0.064	0.077	0.088	0.097	0.105	0.112	0.118	0.149	0.156	0.158
10.0	0.679	0.740	0.758	0.766	0.770	0.770	0.770	0.768	0.745	0.738	0.735
	0.038	0.059	0.071	0.082	0.091	0.099	0.106	0.112	0.147	0.156	0.158
20.0	0.756	0.856	0.896	0.925	0.945	0.959	0.969	0.977	0.982	0.965	0.957
	0.020	0.031	0.039	0.046	0.053	0.059	0.065	0.071	0.124	0.148	0.156
500.0	0.832	0.977	1.046	1.102	1.150	1.191	1.227	1.259	1.532	1.721	1.879
	0.001	0.001	0.002	0.002	0.002	0.003	0.003	0.003	0.008	0.016	0.031

Equation (3.5) is strictly applicable to *flexible* bases on the half-space. Here, the half-space may consist of either cohesionless materials of any water content or unsaturated cohesive soils, with no or very small organic content.

In practice most foundations are not perfectly rigid. Even thick footings deflect under the superstructure loads.

It was found that if the footing is rigid the settlement will be uniform and the I_S factor will be reduced by about 7%. Hence, S_i of a *rigid* base can be estimated as

$$S_{i(\text{rigid})} = 0.93 S_{i(\text{flexible})} \tag{3.6}$$

It may be more correct to use the *weighted average* of E_s than the simple ordinary average. The weighted average E_s in the depth $z = H$ can be computed (where for n layers, $H = \sum_i^n H_i$ as

$$E_{s,\,av} = \frac{H_1 E_{s1} + H_2 E_{s2} + \ldots + H_n E_{sn}}{H} \tag{3.7}$$

Another equation may be considered as an *improved* equation compared to Equation (3.5) for calculation of the elastic settlement was presented by Mayne and Poulos (1999). The equation takes into account the increase in the modulus of elasticity of the soil with depth, the rigidity of the base, the depth of embedment of the foundation and the location of rigid layers at a limited depth. The equation gives the elastic settlement below *centre* of a uniformly loaded rectangular base of the equivalent diameter B_e, as

$$S_i = \frac{q_o B_e I_G I_F I_E}{E_o}\left(1 - \mu_s^2\right) \tag{3.8}$$

Where:

q_o = Contact pressure intensity in units of E_o
E_o = Soil modulus of elasticity considered at the foundation level
I_G = Influence factor for the variation of E_s with depth, $= f\left(\beta = \frac{E_o}{kB_e}, \frac{H}{B_e}\right)$, obtained from Figure 3.4
I_F = Foundation rigidity correction factor, obtained from Figure 3.5
I_E = Foundation embedment correction factor, obtained from Figure 3.6
μ_s = Poisson's ratio of soil
$E_s = (E_o + kz)$, as shown in Figure 3.3
E_f = Modulus of elasticity of the foundation material

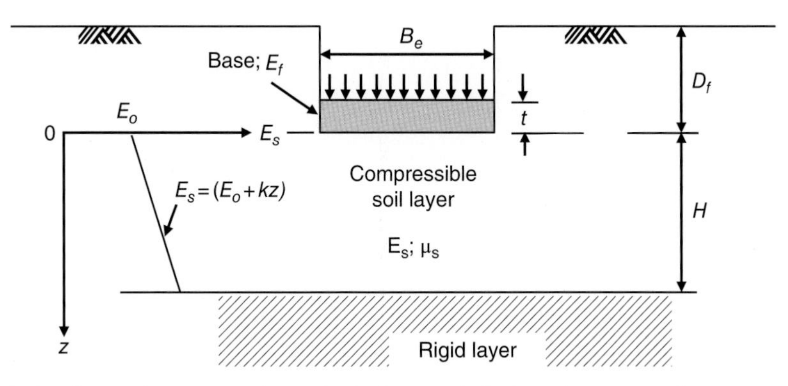

Figure 3.3 Defined general parameters for use in the improved elastic settlement equation.

B_e = Equivalent diameter; for a rectangular base: $B_e = \sqrt{\dfrac{4BL}{\pi}}$

D_f = Foundation depth
t = Base thickness

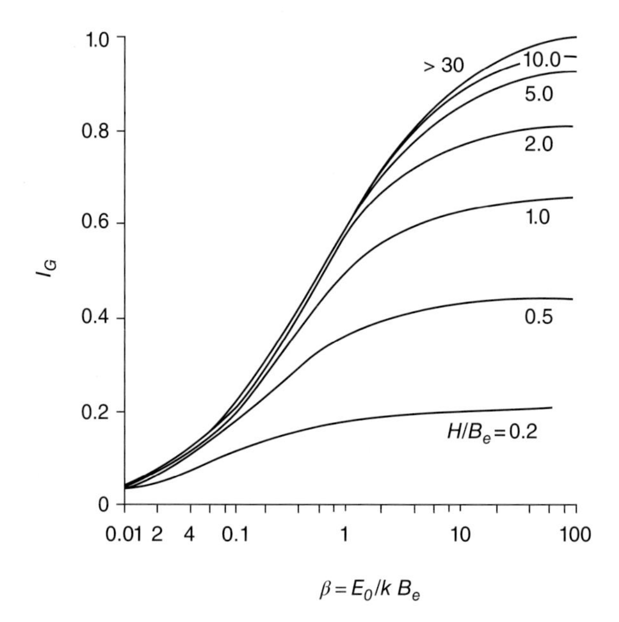

Figure 3.4 Variation of I_G with β.

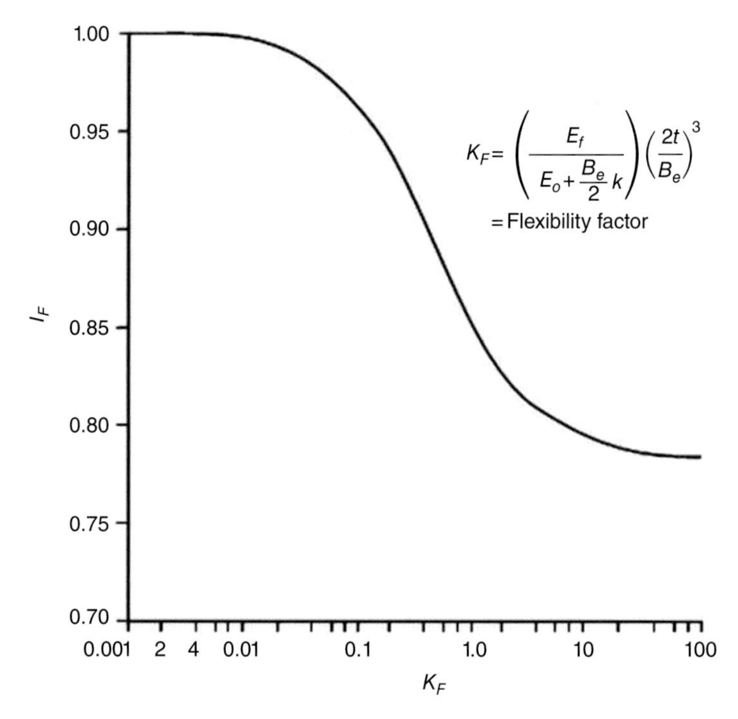

Figure 3.5 Variation of factor I_F with flexibility factor K_F.

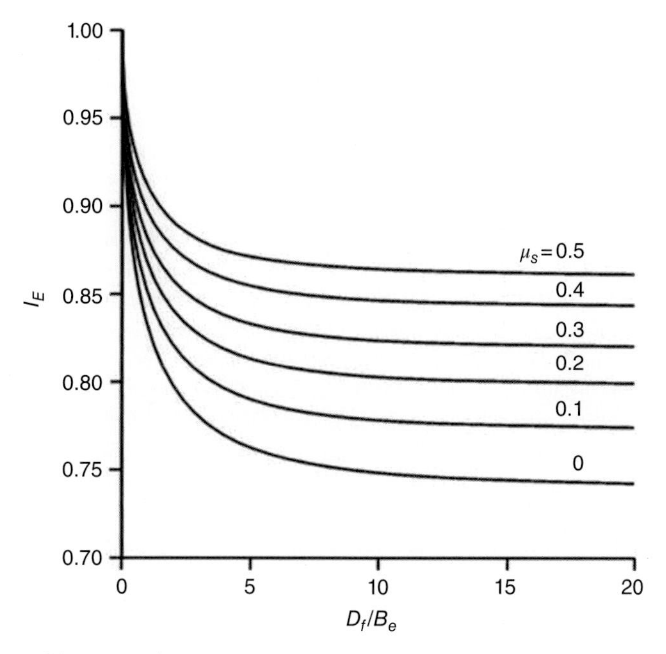

Figure 3.6 Variation of factor I_E with D_f/B_e.

The foundation rigidity correction factor can be expressed as

$$I_F = \frac{\pi}{4} + \frac{1}{4.6 + 10\left\{E_f/[E_o + (B_e k/2)]\right\}(2t/B_e)^3} \tag{3.9}$$

The foundation embedment correction factor can be expressed as

$$I_E = 1 - \frac{1}{3.5[\exp(1.22\mu_s - 0.4)]\left[(B_e/D_f) + 1.6\right]} \tag{3.10}$$

Figure 3.3 shows a foundation with an equivalent diameter B_e, located at a depth D_f below the ground surface. A rigid layer is located at a depth H below the bottom of the foundation. Variation of E_s with depth is also shown.

3.3 Settlement of Foundations on Coarse-grained Soils

3.3.1 General

Settlements of foundations on sands, gravel deposits and granular fill materials take place almost immediately as the foundation loading is imposed on them. Therefore, Equation (3.5) or (3.8) can be used with enough confidence provided that the *elastic parameters* E_s and μ are computed to a reasonable degree of accuracy. This is because these parameters especially the E_s, are mainly responsible for magnitude of the soil strain ε occurs under the imposed load. Since the elastic settlement is simply $S_i = \int_0^H \varepsilon\, dh = \sum_{i=1}^{n} \varepsilon_i H_i$, any method that accurately gives the strains in the identified influence depth H would give an accurate evaluation of the settlement S_i. However, because of difficulty of obtaining undisturbed samples of sand and gravel soils, there is no practicable laboratory test procedure for determining their elastic parameters or consolidation characteristics. For these reasons geotechnical engineers resort to empirical or semi-empirical methods for estimation of foundation settlements, based on the results of in situ PLT, SPT, CPT, pressuremeter test and dilatometer test.

Due to the many limitations the PLT suffers from (as indicated in the discussion of Problem 1.9), the test has little real value. Sutherland (1974) concluded that there is no reliable method for extrapolating the settlement of a standard plate to the settlement of an actual foundation at the same location.

The pressuremeter and the dilatometer tests tend to obtain more direct measurement of E_s. However, the value of E_s obtained from these tests is generally the horizontal value, whereas the vertical value is usually needed for settlement. According to Bowles (2001), most soils are anisotropic, so the horizontal E_s value may considerably different from the vertical E_s value. Overconsolidation may also alter the vertical and horizontal values of a stress–strain modulus. Therefore, in computing S_i the geotechnical engineer should use results of these tests with caution.

As already noted, because the laboratory values of E_s are difficult and expensive to obtain and are generally not very good anyway owing to sampling disturbance and due to the limitations of the other in situ tests just mentioned above, the standard penetration test (SPT) and cone penetration test (CPT) have been widely used to obtain the stress–strain modulus E_s. For this purpose there are many empirical equations and/or correlations have been developed by researchers [see Equations (1.33), (1.57) and (3.32), Tables 1.12 and 3.10]. Also, a comprehensive study of the subject has been done by Bowles (2001); the outcome was a table, such as Table 3.9, which gives empirical equations for *stress-strain modulus* E_s by several test methods. However, Bowles suggested that the value to use should be based on local experience with that equation giving the best fit for that locality. Bowles, in his table, indicated that the SPT-*N* values should be estimated as N_{55} and not N_{70}. As suggested by Bowles, the value ranges

Table 3.9 Equations for stress-strain modulus E_s by several test methods (reproduced from Bowles, 2001). *Note:* The *N* -values should be estimated as N_{55}; and E_s should be used in kPa for SPT and units of q_c for CPT.

Soil	SPT	CPT
Sand (normally consolidated)	$E_s = 500(N+15)$	$E_s = (2 \text{ to } 4)q_e$
	$= 7000\sqrt{N}$	$= 800\sqrt{q_c}$
	$= 6000N$	— — —
	— — —	
	‡$E_s = (15000 \text{ to } 22000)\cdot\ln N$	$E_s = 1.2(3D_e^2+2)q_e$
Sand (saturated)	$E_s = 250(N+15)$	*$E_s = (1+D_e^2)q_e$
		$E_s = Fq_e$
		$e = 1.0 \quad F = 3.5$
		$e = 0.6 \quad F = 7.0$
Sandt, all (norm. concol.)	¶E_s = (2600 to 2900)N	
Sand (overconsolidated)	†E_s = 40 000 + 1050N	$E_s = (6 \text{ to } 30)q_c$
	$E_{p(\text{OCR})} = E_{s,\text{nc}}\sqrt{\text{OCR}}$	
Gravelly sand	$E_s = 1200(N+6)$	
	$= 600(N+6) \; N \le 15$	
	$= 600(N+6) + 2000 \; N > 15$	
Clayey sand	$E_s = 320(N+15)$	$E_s = (3 \text{ to } 6)q_c$
Silts, sandy silt, or clayey silt	$E_s = 300(N+6)$	$E_s = (1 \text{ to } 2)q_c$

Table 3.9 (*Continued*)

Soil	SPT	CPT
	If $q_c < 2500$ kPa use $2500 < q_c < 5000$ use	$^{\$}E'_s = 2.5q_e$
		$E'_s = 4q_c + 5000$
	where	
	$E'_s = \text{constrained modulus} = \dfrac{E_s(1-\mu)}{(1+\mu)(1-2\mu)} = \dfrac{1}{m_e}$	
Soft clay or clayey silt		$E_s = (3 \text{ to } 8)q_c$
Clay and silt	$I_p > 30$ or *organic*	$E_s = (100 \text{ to } 500)s_u$
Silty or sandy clay	$I_r < 30$ or *stiff*	$E_s = (500 \text{ to } 1500)s_u$
		Again, $E_{i,OCR} = E_{s,nc}\sqrt{\text{OCR}}$
		Use smaller s_u – coefficient for highly plastic clay

Of general application in clays is

$$E_s = Ksw \text{ (units of } s_u) \qquad (a)$$

where K is defined as

$$K = 4200 - 142.54I_P + 1.73I_P^2 - 0.0071I_P3 \qquad (b)$$

and I_p = plasticity index in percent. Use $20\% \leq I_p \leq 100\%$ and round K to the nearest multiple of 10.

Another equation of general application is

$$E_s = 9400 - 8900I_p + 11600I_c - 8800S \text{ (kPa)} \qquad (c)$$

I_P, I_c, S = previously defined above and/or in chapter 2

* Vesic (1970); †based on plot of D'Appolonia et al. (1970); ‡USSR (may not be standard blow count N); ¶Japanese Design Standards (lower value for structures); $Senneset et al. (1988).

for static stress-strain modulus E_s presented in Table 3.10 and the values or value ranges of Poisson's ratio μ presented in Tables 3.11 and 3.12 should also be used as references or guides.

3.3.2 Estimation of Settlements from SPT

(1) *Using bearing capacity equations.*
Mayerhof (1956, 1974) published equations for computing the net allowable bearing capacity (*net* q_a) for 25 mm settlement. Considering the accumulation of field observations and the stated opinions of the author and others, Bowles (1977) adjusted the Mayerhof equations for an approximate 50% increase in *net* q_a to obtain the following:

$$net\ q_a = \frac{N'_{55}}{F_1} K_d W_r \qquad \text{for footings;} \quad B \leq F_4 \qquad (3.11)$$

Table 3.10 Value ranges* for static stress–strain modulus E_s for selected soils.

Soil	E_s, MPa
Clay	
Very soft	2–15
Soft	5–25
Medium	15–50
Hard	50–100
Sandy	25–250
Glacial till	
Loose	10–150
Dense	150–720
Very dense	500–1440
Loess	15–60
Sand	
Silty	5–20
Loose	10–25
Dense	50–81
Sand and gravel	
Loose	50–150
Dense	100–200
Shale	150–5000
Silt	2–20

*The value ranges are too large to use an "average" value for design.

Table 3.11 Values or value ranges for Poisson's ratio μ.

Type of soil	μ
Clay, saturated	0.4–0.5
Clay, unsaturated	0.1–0.3
Sandy clay	0.2–0.3
Silt	0.3–0.35
Sand, gravelly sand	−0.1–1.00
commonly used	0.3–0.4
Rock	0.1–0.4 [a]
Loess	0.1–0.3
Ice	0.36
Concrete	0.15
Steel	0.33

[a] Depends somewhat on type of rock

Table 3.12 Commonly used values ranges for μ.

μ	Soil type
0.4–0.5	Most clay soils
0.45–0.50	Saturated clay soils
0.3–0.4	Cohesionless—medium and dense
0.2–0.35	Cohesionless—loose to medium

$$net\ q_a = \frac{N'_{55}}{F_2}\left(\frac{B+F_3}{B}\right)^2 K_d W_r \qquad \text{for footings;}\ B > F_4 \tag{3.12}$$

$$net\ q_a = \frac{N'_{55}}{F_2} K_d W_r \qquad \text{for mats} \tag{3.13}$$

where:

K_d = depth factor = $\left[1 + 0.33\left(\frac{D}{B}\right)\right] \le 1.33$

W_r = water reduction factor due to ground water table (W.T)

$\cong 0.5 + \frac{0.5\, z_w}{B}$; z_w = depth of W.T below foundation level $\le B$

D = depth of foundation

B = width of footing

F_1, F_2, F_3 and F_4 = factors depend on the SPT hammer energy ratio E_r, as given below:

N'	F_1	F_2	F_3	F_4
N'_{55}	0.05	0.08	0.3	1.2
N'_{70}	0.04	0.06	0.3	1.2

Note that, for complete saturation (submergence) conditions, that is when W.T is above the foundation level or when $z_w = 0$, $W_r \cong 0.5$; and $W_r \cong 1.0$ for $z_w \ge B$.

In these equations N'_{55} is the statistical average corrected value (refer to Solution (b) of Problem 1.6) for the footing influence zone of about 0.5 B above the foundation level to at least 2 B below.

In the above three equations the allowable soil pressure is for an assumed 25-mm settlement. In general, for cohesionless soils, it is possible to assume that settlement S is proportional to net soil pressure *net q*. Based on this assumption, the settlement S_i caused by any given net soil pressure $netq_{(S_i)}$ is

$$S_i = S_o \times \frac{netq_{(S_i)}}{netq_{a,\,(S_o)}} \tag{3.14}$$

where S_o = 25 mm, S_i = required settlement in mm.

(2) *Using Burland and Burbidge empirical relationship*

Burland and Burbidge (1985) established the following empirical relationship, based on SPT results, for estimating settlement S_i of foundations on sands and/or gravels:

$$\frac{S_i}{B_R} = \alpha_1\alpha_2\alpha_3 \left[\frac{1.25\left(\frac{L}{B}\right)}{0.25 + (L/B)}\right]^2 \left(\frac{B}{B_R}\right)^{0.7} \left(\frac{q'}{p_a}\right) \tag{3.15}$$

where:
α_1 = a constant
α_2 = compressibility index
α_3 = correction for depth of influence
B_R = reference width = 0.3 m
B = width of actual foundation
L = length of actual foundation
p_a = atmospheric pressure = 100 kN/m^2
q' = contact pressure in kN/m^2

Values of q', α_1, α_2 and α_3 are given in Table 3.13.
Equation (3.15) may be applied as follows:

(a) Obtain the *field* penetration number N_{60} with depth at the foundation location. The following adjustments of each N_{60} -value may be necessary, depending on field conditions:
For gravel or sandy gravel, the adjusted N_{60} is

$$N_{60(a)} = 1.25\,N_{60} \tag{3.16}$$

For fine sand or silty sand below the ground water table and $N_{60} > 15$, the adjusted N_{60} is

$$N_{60(a)} = 15 + 0.5\,(N_{60} - 15) \tag{3.17}$$

(b) Determine the *influence depth* z' as follows:
Case I. If N_{60} or $[N_{60(a)}]$ is nearly constant with depth, calculate z' from:

$$\frac{z'}{B_R} = 1.4\left(\frac{B}{B_R}\right)^{0.75} \tag{3.18}$$

Case II. If N_{60} or $[N_{60(a)}]$ is increasing with depth, use Equation (3.18) to calculate z'
Case III. If N_{60} or $[N_{60(a)}]$ is decreasing with depth, calculate $z' = 2B$ or to the bottom of soft soil layer measured from the bottom of the foundation, whichever is smaller.

Table 3.13 Values of q', α_1, α_2 and α_3.

Soil type	q'	α_1	α_2	α_3
Normally consolidated sand	*net q*	0.14	$\frac{1.71}{[\bar{N}_{60} \text{ or } \bar{N}_{60(a)}]^{1.4}}$	$\alpha_3 = \frac{H}{Z'}\left(2 - \frac{H}{Z'}\right)$ for $H \le z'$
Overconsolidated sand for *net* $q \le \sigma'_c$	*net q*	0.047	$\frac{0.57}{[\bar{N}_{60} \text{ or } \bar{N}_{60(a)}]^{1.4}}$	$\alpha_3 = 1$ for $H > z'$
where σ'_c = preconsolidation pressure				where H = depth of compressible layer
Overconsolidated sand for *net* $q > \sigma'_c$	*net* $q - 0.67\,\sigma'_c$	0.14	$\frac{0.57}{[\bar{N}_{60} \text{ or } \bar{N}_{60(a)}]^{1.4}}$	z' = influence depth

Note: $\bar{N}_{60}$ or $[\bar{N}_{60(a)}]$ = the statistical average of N_{60} or $[N_{60(a)}]$ values within the depth of stress influence.

3.3.3 Estimation of Settlements from CPT

(1) *The Buisman–DeBeer method*

According to this method, the constant of compressibility (C) of the sand is proportional to static cone penetration resistance (q_c), as in the following empirical equation:

$$C = 1.5\frac{q_c}{\sigma'_o} \tag{3.19}$$

where:

σ'_o is the effective overburden pressure at the depth of measurement.

Settlement of a sand layer of thickness H can be estimated using the equation:

$$S_i = \frac{H}{C}\ln\frac{\sigma'_o + \sigma_z}{\sigma'_o} \tag{3.20}$$

where:

σ_z is the increase in vertical stress at the centre of the layer.

Considering an elemental layer of small thicknesses (dz) with C assumed constant; the settlement is computed by means of the equation:

$$S_i = \int_{z=0}^{z=H} \frac{1}{C}\left\{\ln\left(\frac{\sigma'_o + \sigma_z}{\sigma'_o}\right)\right\}dz \tag{3.21}$$

or approximately:

$$S_i = \sum_0^H \frac{2.3\sigma'_0}{1.5q_c}\Delta z \log\frac{\sigma'_o + \sigma_z}{\sigma'_o}$$

$$S_i = \sum_0^H 1.53\frac{\sigma'_o}{q_c}\Delta z \log\frac{\sigma'_o + \sigma_z}{\sigma'_o} \tag{3.22}$$

In practice the thickness H is divided into suitable layers (thickness Δz) provided that within each of which the value of q_c is assumed constant. In deep deposits the summation may be terminated at the depth at which the stress increment σ_z becomes less than 10% of the effective overburden pressure σ'_0. According to Craig (2004), the Buisman–DeBeer method is strictly applicable only to normally consolidated sands. In the case of overconsolidated sands the method will give settlements which are too high. Based on a study of case records, Mayerhof (1965) recommended that the foundation pressure producing the allowable settlement by the Buisman–DeBeer method should be increased by 50%. This is approximately equivalent to using the following equation for the constant of compressibility:

$$C = 1.9\frac{q_c}{\sigma'_o} \tag{3.23}$$

(2) *The Schmertmann method*

A semi-empirical equation to compute settlement of granular soils was proposed by Schmertmann (1978). The equation is based on a simplified distribution of vertical strain ε_z under the centre of a shallow foundation, expressed in the form of a *strain influence factor* I_z. According to this method, the settlement is

$$S_i = C_1 C_2 C_3 q_{net} \sum_{z=0}^{z=z_2} \frac{I_z}{E_s} \Delta z \tag{3.24}$$

where:
C_1 = depth factor $= 1 - 0.5\sigma'_o / q_{net}$

C_2 = correction factor for creep in soil $= 1 + 0.2 \log \frac{t}{0.1}$

C_3 = shape factor $= 1.03 - \frac{0.03L}{B} \geq 0.73$
  = 1 for square and circular foundations
q_{net} = net contact pressure $= q_{total} - \sigma'_o$ in kN/m^2
σ'_o = effective overburden pressure at foundation level in kN/m^2
q_{total} = total contact pressure in kN/m^2
E_s = *equivalent* modulus of elasticity (linear function), in kN/m^2
Δz = thickness of a soil layer in m
t = time in years
z_2 = final depth below foundation where I_z is insignificant or zero.

The assumed distribution of I_z with depth is shown in Figure 3.7; depth is expressed in terms of the width B of the footing. This is a simplified distribution, based on both theoretical and experimental results, in which it is assumed that strains become insignificant at a depth of $2B$ below square or circular footings and at a depth of $4B$ below continuous ($L/B \geq 10$) footings. It should be noted that the maximum vertical strain does not occur immediately below the footing as is the case with vertical stress.

As already noted in Section 3.3.1, there are several direct correlations between E_s and q_c, such as those presented in Table 1.11 or given by Equation 1.57, which may be used to obtain a suitable value for E_s. Schmertmann et al. (1978) recommended $E_s = 2.5\,q_c$ for square or circular foundations and $E_s = 3.5\,q_c$ for long strip foundations with $L/B \geq 10$.

The measured q_c/depth profile, to a depth z_2 below the footing, is divided into suitable layers (thickness Δz) within each of which the value of q_c is assumed constant. The value of I_z at the centre of each

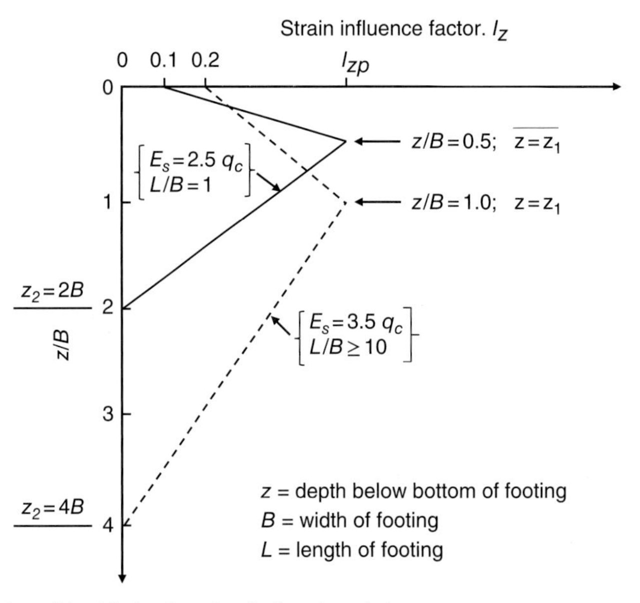

Figure 3.7 Distribution of I_z with depth under shallow foundations.

layer is determined by superimposing the distribution of I_z in Figure 3.7. It being assumed that this distribution of I_z is independent of the sand heterogeneity. The sum of the settlement of the layers equals S_i.

Figure 3.7 is drawn such that: for square or circular foundations value of I_z at depth $z = 0$ (i.e. foundation level), $z = 0.5B$ and $z = 2B$ are respectively equal to 0.1, peak value of I_z and 0; for continuous foundations with $L/B \geq 10$ value of I_z at depth $z = 0$, $z = B$ and $z = 4B$ are respectively equal to 0.2, peak value of I_z and 0; for rectangular foundations with $1 < L/B < 10$ interpolation can be made. The peak strain influence factor I_{zp} is calculated as

$$I_{zp} = 0.5 + 0.1\sqrt{\frac{q_{net}}{\sigma'_{zp}}} \tag{3.25}$$

where σ'_{zp} initial vertical effective stress at depth of I_{zp}.

The exact value of I_z at any given depth z below the foundation level may be most easily computed using equations of the straight lines in Figure 3.7, as follows:

Considering square and circular foundations:

$$\text{For } z = 0 \text{ to } \frac{B}{2} \qquad I_z = 0.1 + \left(\frac{z}{B}\right)(2I_{zp} - 0.2) \tag{3.26}$$

$$\text{For } z = \frac{B}{2} \text{ to } 2B \qquad I_z = 0.667 I_{zp}\left(2 - \frac{z}{B}\right) \tag{3.27}$$

Considering continuous foundations ($L/B \geq 10$):

$$\text{For } z = 0 \text{ to } B \qquad I_z = 0.2 + \left(\frac{z}{B}\right)(I_{zp} - 0.2) \tag{3.28}$$

$$\text{For } z = B \text{ to } 4B \qquad I_z = 0.333 I_{zp}\left(4 - \frac{z}{B}\right) \tag{3.29}$$

Considering rectangular foundations ($1 < L/B < 10$):

$$I_z = I_{zs} + 0.111(I_{zc} - I_{zs})\left(\frac{L}{B} - 1\right) \tag{3.30}$$

where:

z = depth below foundation to centre of layer
$I_{zc} = I_z$ for a continuous foundation
$I_{zs} = I_z$ for a square foundation

When only minimal subsurface data is available, as it is often the case with SPT and the soil appears to be fairly homogeneous, it may be possible to consider E_s constant with depth between the bottom of foundation and the depth of influence ($\Delta z = 2B = H$) for square and circular footings and ($\Delta z = 4B = H$) for continuous footings. Consequently, it may be possible to use $I_{z,av}$ instead of the I_z in Equation 3.24; considered constant within the influence depth.

For square and circular foundations ($\Delta z = 2B = H$):

$$I_{z,\,av} = 0.0125 + 0.5\,I_{zp}$$

$$S_i = C_1 C_2 C_3 q_{net}\left[\frac{B\left(I_{zp} + 0.025\right)}{E_s}\right] \tag{3.31-(a)}$$

For continuous footings ($\Delta z = 4B = H$):

$$I_{z,\,av} = 0.025 + 0.5\, I_{zp}$$

$$S_i = C_1 C_2 C_3 q_{net} \left[\frac{B\,(2\,I_{zp} + 0.1)}{E_s} \right] \qquad (3.31)\text{-(b)}$$

It is useful to understand clearly that Schmertmann's method was developed primarily for ordinary spread footings, so the various empirical data used to calibrate the method have been developed with this type of foundations in mind. However, in principle, the method also may be used with mat foundations. In using the Schmertmann's method with mats, an overestimated settlement is obtained because their depth of influence is much greater and the equivalent modulus values at these depths is larger than predicted from correlations based on results of tests which are usually performed at relatively shallow depths.

According to Coduto (2001); when applying Schmertmann's method to mat foundations, it is best to progressively increase the E_s values with depth, such that E_s at 30 m is about three times that predicted, as just mentioned above.

It is noteworthy that the Schmertmann's method also may be used with E_s values based on the standard penetration test. However, generally, these values may not be as precise as those obtained from the cone penetration test. For those projects in which the soil conditions are satisfactory and the loads are relatively small, the SPT data is considered adequate for use with the Schmertmann's method.

There are several direct correlations between E_s and SPT N-values have been developed; see Table 3.9 and Equation (1.33). The correlations between q_c and N given in Tables 1.11, 1.12 and Figure 1.16 may be used, indirectly, to obtain E_s values.

Also, the following relationship gives approximate values of E_s:

$$E_s = \beta_o \sqrt{OCR} + \beta_1 N_{60} \qquad (3.32)$$

where:

E_s = equivalent modulus of elasticity in kPa
N_{60} = SPT N–value corrected for field procedures only
OCR = overconsolidation ratio. Use $OCR = 1$ unless there is clear evidence of overconsolidation
β_o, β_1 = correlation factors, as given below:

Soil type	β_o	β_1
Clean sands (SW and SP)	5000	1200
Silty sands and clayey sands (SM and SC)	2500	600

3.4 Settlement of Foundations on Fine-grained Soils

3.4.1 General

Before starting any settlement analysis it is necessary to study carefully the particular *soil report*, especially, those items concerning description and engineering properties of soils encountered in each stratum. In the case of thick clay strata, it must not be assumed that the compressibility is constant throughout the depth of the strata. Clays usually show progressively decreasing compressibility and

Table 3.14 Approximate thicknesses of soil layers for manual computation of consolidation settlement of shallow foundations.

Layer Number	Approximate Layer Thickness	
	Square Footing	Continuous Footing
1	$B/2$	B
2	B	$2B$
3	$2B$	$4B$

increase in modulus of elasticity (deformation modulus) with increasing depth. For large and important structures it is worthwhile to make settlement analyses for the highest compressibility and maximum depth of compressible strata and the lowest compressibility with the minimum depth of strata and then to compare the two analyses to obtain an idea of the differential settlement if these two extremes of conditions occur over the area of the structure. Since strain varies non-linearly with depth, analyses that use a large number of thin layers produce a more precise results than those that use a few thick layers. However, unless computer is used in settlement computations of a large number of thin layers, it would be too tedious to do the computations by hand, so manual computations normally use less number of layers. According to Coduto (2001), for most soils, the guidelines in Table 3.14 should produce reasonable results.

When considering long-term consolidation settlement, it is essential that the foundation loading used in the analysis should be realistic and representative of the *sustained* loading over the time period under consideration. This is a different procedure from that used when calculating safe bearing capacity. In the latter case the most severe loading conditions are allowed for, with full provision for maximum imposed loading. The imposed loading used in a settlement analysis is an *average value* representing the continuous load over the time period being considered.

Wind loading is only considered in settlement analyses for high structures where it represents a considerable proportion of the total loads. If so, the wind loads, expressed as equivalent static loads, representing the average of continuous wind over the full period are allowed for. However, many well-known building codes, such as ICBO, BOCA and ICC, allow 33% greater allowable bearing capacity for short term loads (such as wind and seismic loads) which are *included* in the design total load combinations. Usually, geotechnical engineers do not use this increase in the allowable bearing capacity for foundations supported on soft clayey soils.

In the case of deep compressible soils the lowest level considered in the settlement analysis is the point where the increase in vertical stress σ_z resulting from the net foundation pressure q_n is equal or less than 20% of the effective overburden pressure σ'_o, as shown in Figure 3.8. However, some authorities prefer 10% instead of the 20%. In the case of soil layers of limited thicknesses, the lowest level considered is the bottom of the layer or the depth $z = 4B$ below the foundation level, whichever is giving a shallower depth.

As already noted in Section 2.6.1, the distribution of contact pressure and settlement beneath a shallow foundation are functions of the type of the supporting soil and the relative rigidity of the foundation and soil. In a ground steel storage tank the weight of the stored liquid is supported directly on the plate steel floor. Usually this type of floor is so thin that could be considered to be perfectly flexible. The settlement beneath the floor centre and edge would not be the same and their difference (i.e. differential settlement) could be computed. On the other hand, a spread footing is much more rigid than the plate steel floor. This increase in rigidity causes the settlement to be nearly uniform or uniform beneath the

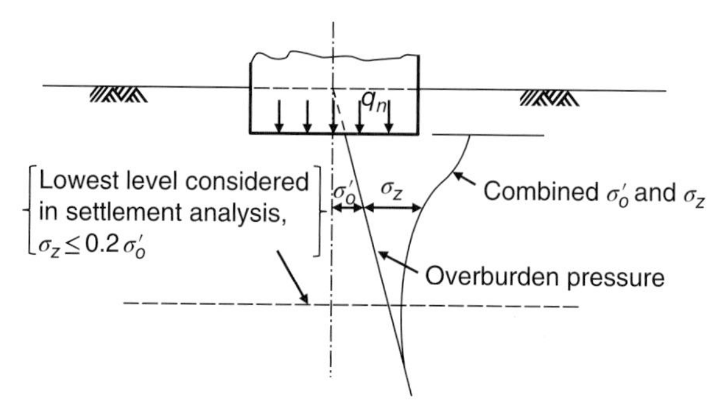

Figure 3.8 Vertical pressure and stress distribution for deep clay layer (redrawn from Tomlinson, 2001).

Table 3.15 Values of rigidity factor r for computation of total settlement S_t at the centre of a shallow foundation (adapted from Coduto, 2001).

Foundation rigidity	r – Values for S_t at center of foundation
Perfectly flexible (i.e. steel tanks)	1.00
Intermediate (i.e. mat foundations)	0.85–1.00, typically about 0.90
Perfectly rigid (i.e. spread footings)	0.85

footing. Another possibility is associated with mat foundations. A mat is more rigid than a plate steel floor and less rigid than a spread footing. Obviously, there will be differential settlement between centre and the edge, but not as much as that of a plate steel floor.

For these reasons, in order to account for the degree of rigidity of foundations, results of settlement computations may be multiplied by a *rigidity* factor r. Table 3.15 presents values of r for various conditions. When r is used as1, which is preferred by many geotechnical engineers, the design will be conservative. This practice may not have serious impact on construction costs of small or moderate-size structures and can be considered acceptable. The use of $r < 1$ is justified when the subsurface conditions have been well defined and the provided data are complete and reliable to make a more precise analysis.

3.4.2 Immediate Settlement of Fine-grained Soils

The immediate settlement, that is the elastic settlement, beneath the corner or centre of a flexible uniformly loaded rectangular area, can be calculated from Equations (3.5) and (3.8), respectively, as discussed in Section 3.2.

In saturated clays or clayey silts the immediate settlement occurs under undrained conditions; hence, the *undrained* modulus E_u is required. It is the general practice to obtain drained and undrained E-values of fine-grained soils from laboratory tests on undisturbed samples taken from boreholes. The undrained modulus E_u of clays can be determined from relationships with the undrained shear strength, as presented in Table 3.9. Also, the value of E_u can be determined from results of pressuremeter tests or other in situ tests.

In the case of an extensive, homogeneous deposit of saturated clay, it is a reasonable approximation to assume that E_u is constant throughout the deposit.

Researchers demonstrated that for certain soils, such as normally consolidated clays, there is a significant departure from linear stress- strain behaviour within the range of working stress, that is local

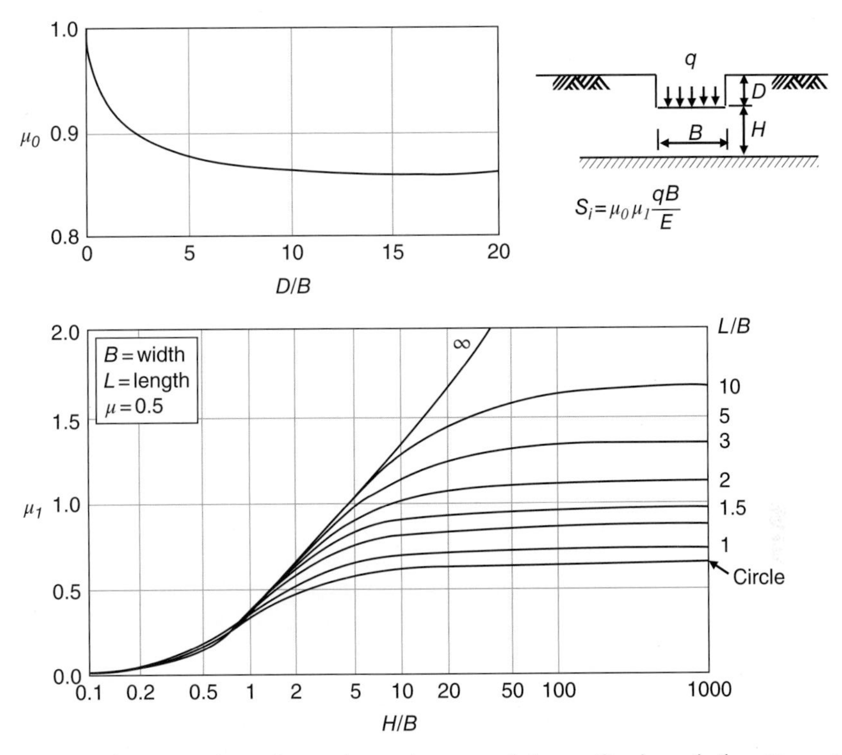

Figure 3.9 Factors for average immediate settlement in saturated clay or silty clay soils (from Knappett and Craig, 2012; after Christian and Carrier, 1978).

yielding will occur within this range and the immediate settlement will be underestimated; hence, geotechnical engineers should be aware of that.

In principle the settlement under fully drained conditions could be estimated using the settlement elastic equations if the value of *drained* modulus E_d and the value of Poisson's ratio μ for the soil skeleton could be determined. If the results of drained triaxial tests are not available, E_d for overconsolidated clays can be obtained approximately from the relationship $E_d = 0.6\,E_u$. Alternatively if values of the coefficient of volume compressibility m_v from oedometer tests are available, the relationship $E_d = 1/m_v$ may be used.

In practice, in most of cases, the soil deposit will be of limited thickness and will be underlain by a hard stratum. Christian and Carrier (1978) proposed the following equation for determining *average* vertical displacement S_i (in mm) under a uniformly loaded *flexible* area carrying a uniform pressure q:

$$S_i = \mu_o \mu_1 \frac{qB}{E} \tag{3.33}$$

where μ_o depends on the depth of foundation and μ_1 depends on the layer thickness and the shape of the loaded area. The width B is in metres; q is in kN/m^2 and E is in MN/m^2.

Values of the coefficients μ_o and μ_1 for Poisson's ratio μ equal to 0.5 are given in Figure 3.9. It may be noted that the rigidity factor is not included in the equation. The principle of superposition can be used in cases of a number of soil layers each having a different value of E. The equation is used mainly to estimate the immediate settlement of foundations on saturated clays; such settlement occurs under undrained conditions, the suitable value of μ being 0.5 and therefore the value of undrained modulus E_u should be used in the equation.

According to Tomlinson (2001), due to the difficulty of obtaining representative values of the deformation modulus of clay, either by correlation with the undrained shear strength or directly from field or laboratory tests, it may be preferable to determine the immediate settlement from relationships established by Burland et al. (1977), as follows:

For stiff over-consolidated clays:
Immediate settlement = S_i = 0.5 to 0.6 S_{oed}
Consolidation settlement = S_c = 0.5 to 0.4 S_{oed}

Final settlement = S_{oed} = settlement calculated from results of an oedometer test (i.e. one-dimensional consolidation test).

For soft normally-consolidated clays:
Immediate settlement = S_i = 0.1 S_{oed}
Consolidation settlement = S_c = S_{oed}
Final settlement = 1.1 S_{oed}

3.4.3 Consolidation Settlement

(1) *Lateral strain is neglected – Terzaghi method*
Laboratory one-dimensional consolidation tests (ASTM Test Designation D-2435) on representative undisturbed saturated fine-grained soil specimens can be conducted to determine the consolidation settlement caused by various incremental loadings.

In order to estimate consolidation settlement, the value of coefficient of volume compressibility m_v or the values of compression index C_c, expansion (swelling) index C_e, recompression index C_r and effective preconsolidation pressure σ_c' are required.

Consider a layer of saturated clay of thickness *H*: due to construction the total vertical stress in an elemental layer of thickness *dz* at depth *z* is increased by σ_z, as shown in Figure 3.10. It is assumed that the condition of *zero lateral strain* applies within the clay layer; which is the same condition with the specimen under the *oedometer test* due to the confining ring. After the completion of consolidation an equal increase in effective vertical stress $\sigma_z' = \sigma_1' - \sigma_o'$ will have taken place corresponding to a stress increase σ_o' to σ_1' and a reduction in void ratio from e_o to e_1 on the $e-\sigma'$ curve. The reduction in volume ΔV per unit initial volume V_o of clay can be written in terms of void ratio as

$$\frac{\Delta V}{V_o} = \frac{e_o - e_1}{1 + e_o}$$

The lateral strain is assumed equal to zero; hence, the reduction in volume per unit volume is equal to the reduction in thickness per unit thickness, that is the settlement per unit depth.

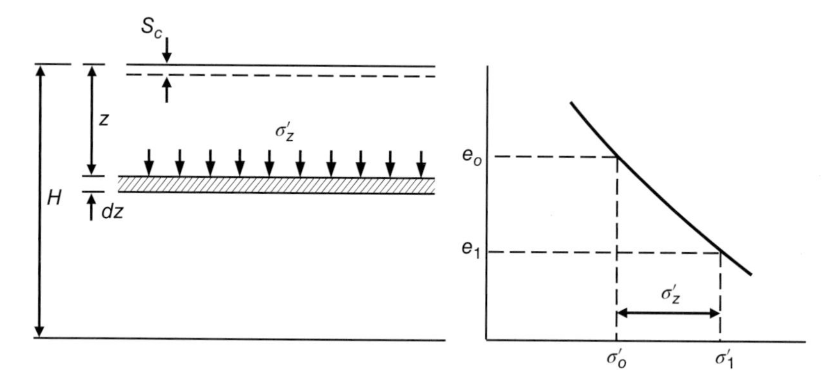

Figure 3.10 Consolidation settlements.

Therefore, by proportion, the consolidation settlement dS_c of the elemental layer of thickness d_z will be given by:

$$dS_c = \frac{e_o - e_1}{1 + e_o} dz = \left(\frac{e_o - e_1}{\sigma'_1 - \sigma'_o}\right)\left(\frac{\sigma'_1 - \sigma'_o}{1 + e_o}\right) dz$$

The coefficient of volume compressibility m_v is defined as the volume change ΔV per unit volume V_o per unit increase in effective stress σ'_z, written as

$$m_v = \frac{e_o - e_1}{1 + e_o} \times \frac{1}{\sigma'_1 - \sigma'_o} = \frac{1}{H_o}\left(\frac{H_o - H_1}{\sigma'_1 - \sigma'_o}\right)$$

Therefore, $dS_c = m_v\, \sigma'_z\, dz$ and the settlement of the layer of thickness H is given by

$$S_c = \int_{z=0}^{z=H} m_v\, \sigma'_z\, dz$$

If m_v and σ'_z are assumed constant with depth, then

$$S_c = m_v\, \sigma'_z H \tag{3.34}$$

or

$$S_c = \frac{\Delta e}{1 + e_o} H \tag{3.35}$$

It is noteworthy that Equation (3.34) or (3.35) is general and can be used for *normally* consolidated, *overconsolidated* and *underconsolidated* clays.

According to the Terzaghi theory of consolidation, the void ratio versus log of pressure relationship for *normally* consolidated clays is a straight line. The slope of this straight line is the compression index $C_c = \dfrac{\Delta e}{\log \sigma'_1 - \log \sigma'_o}$, as shown in Figure 3.11. Therefore, $\Delta e = C_c \log(\sigma'_1/\sigma'_o)$.

Hence, Equation (3.35) can be written as

$$S_c = \frac{C_c \log(\sigma'_1/\sigma'_o)}{1 + e_o} H$$

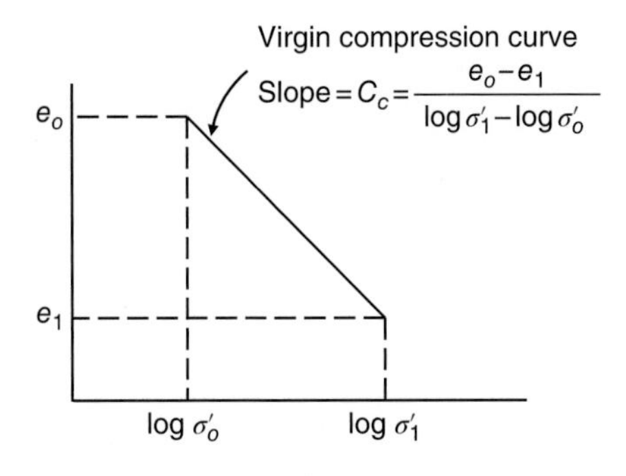

Figure 3.11 Relationship of pressure-versus-void ratio for normally consolidated clay, according to the Terzaghi theory of consolidation.

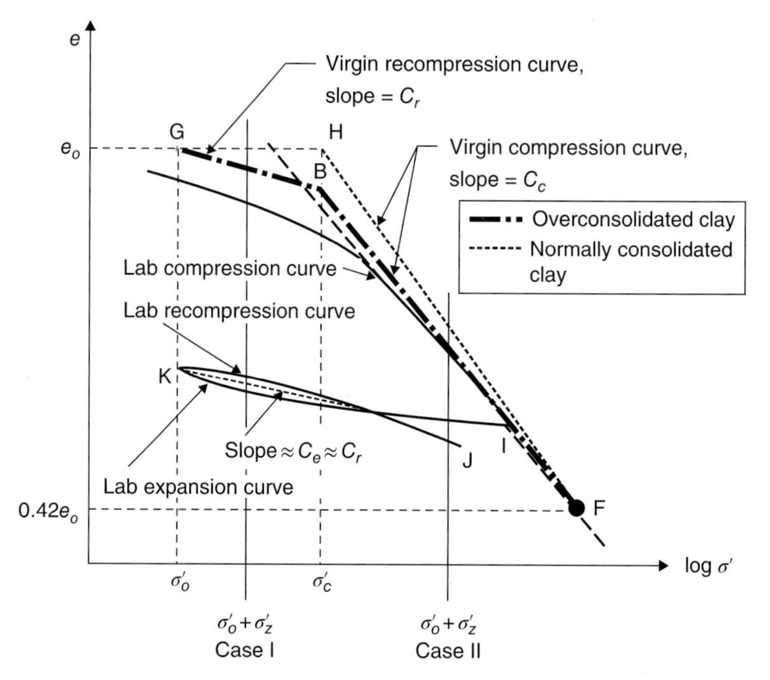

Figure 3.12 In situ and laboratory consolidation curves for normally and overconsolidated clays.

or

$$S_c = \frac{C_c H}{1+e_o} \log \frac{\sigma'_o + \sigma'_z}{\sigma'_o} \quad \left(\text{for } \sigma'_o \approx \sigma'_c;\ OCR \approx 1\right) \tag{3.36-(a)}$$

According to Terzaghi and Peck (1967), for *normally* consolidated undisturbed clays with low or moderate sensitivity

$$C_c \cong 0.009(LL - 10) \tag{3.36-(b)}$$

Figure 3.12 shows the virgin compression curves for both normally-consolidated and overconsolidated clays. It also shows the expansion and recompression curves. The slope of the expansion curve KI is the expansion (or swelling) index C_e, while, the slope of the recompression curve KJ or that of the curve GB is the recompression index C_r. The dotted line HF is the virgin compression curve for normally consolidated clays, that is when $\sigma'_o \approx \sigma'_c$. Lines GB and BF are respectively the virgin recompression and compression curves for overconsolidated clays.

Derivation of the consolidation settlement equations for *overconsolidated* clays proceeds in the same manner as that for the normally consolidated clays, as follows:

Case I: $\left[\sigma'_o < \left(\sigma'_1 = \sigma'_o + \sigma'_z\right) \le \sigma'_c\ ; OCR > 1\right]$

$$S_c = \frac{C_r H}{1+e_o} \log \frac{\sigma'_o + \sigma'_z}{\sigma'_o} \tag{3.37}$$

Case II: $\left[\sigma'_o < \sigma'_c < \left(\sigma'_1 = \sigma'_o + \sigma'_z\right); OCR > 1\right]$

$$S_c = \frac{C_r H}{1+e_o} \log \frac{\sigma'_c}{\sigma'_o} + \frac{C_c H}{1+e_o} \log \frac{\sigma'_o + \sigma'_z}{\sigma'_c} \tag{3.38}$$

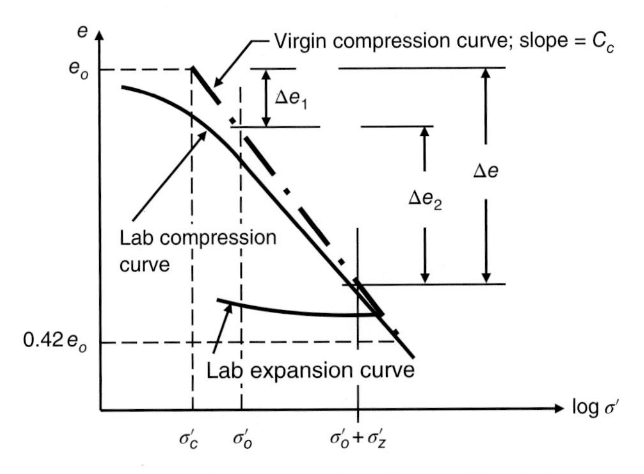

Figure 3.13 In situ and laboratory consolidation curves for underconsolidated clays.

If a clay deposit has not reached equilibrium under the overburden loads, i.e. still consolidating, it is said to be *underconsolidated* clay. Based on the definition of normally consolidated clays, the *OCR* of underconsolidated clays must be less than 1. This condition normally occurs in areas of recent land fill and newly transported clay deposits, particularly via water, which tend initially to produce somewhat loose deposits with large void ratios. Following the same technique used in defining the virgin compression curves for normally and overconsolidated clays; the same type of curves may be obtained for underconsolidated clays too, as shown in Figure 3.13.

The settlement equation for underconsolidated clays can be written as

$$S_c = \frac{C_c H}{1 + e_o} \log \frac{\sigma'_o + \sigma'_z}{\sigma'_c} \quad \left(\sigma'_1 > \sigma'_o > \sigma'_c; OCR < 1\right) \tag{3.39}$$

Figure 3.14 shows the Casagrande (1936) empirical construction to obtain from the $e - \log \sigma'$ curve the maximum effective vertical stress that has acted on an overconsolidated clay deposit in the past; referred to as the *preconsolidation pressure*, σ'_c. Steps of the construction are as follows:

- Extend back the straight part BC of the curve to a point E.
- Determine the point D of maximum curvature on the recompression part AB of the curve.
- Draw the tangent to the curve at D and bisect the angle between the tangent and the horizontal through D.
- The vertical through the point of intersection of the bisector and EC gives the approximate value of the preconsolidation pressure.

An indication that a clay deposit is overconsolidated is when the natural moisture content of the clay is nearer to its plastic limit *PL* than to the liquid limit *LL*. Whenever possible the preconsolidation pressure for overconsolidated clays should not be exceeded in construction. Settlement will not usually be great if the effective vertical stress remains below σ'_c; only if σ'_c is exceeded will settlement be large.

The laboratory σ'_c represents the preconsolidation stress only at the sample depth. To estimate σ'_c at a desired depth shallower than the sample depth in the same strata (i.e. in soil strata with the same geologic origin), the difference in effective overburden pressure $\Delta\sigma'_o$ at the two depths (the sample depth and the desired depth) is subtracted from the laboratory σ'_c and, for a deeper depth in the same strata, the difference is added.

Any of the two expressions $OCR = \frac{\sigma'_c}{\sigma'_o}$ and $\sigma'_m = \sigma'_c - \sigma'_o$ may be used to classify clay soils with

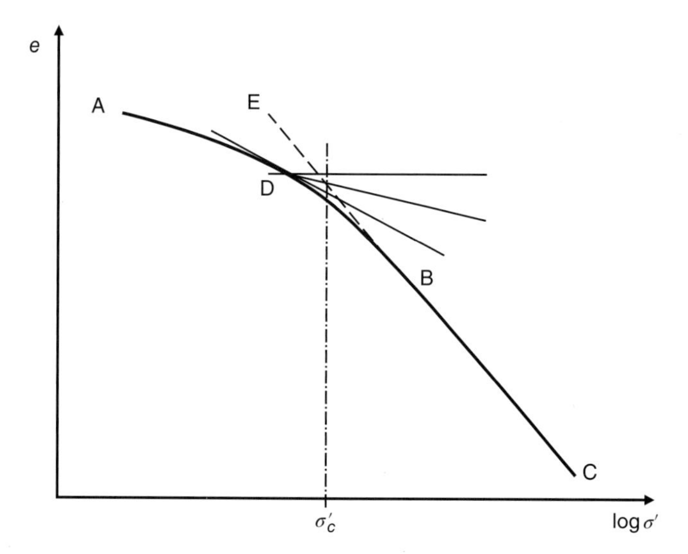

Figure 3.14 Determination of preconsolidation pressure.

Table 3.16 Types of overconsolidated clay soils.

OCR	σ'_m (kPa)	Classification
1–3	0–100	Lightly overconsolidated
3–6	100–400	Moderately overconsolidated
>6	>400	Heavily overconsolidated

respect to their degree of overconsolidation as indicated in Table 3.16, where OCR = the *overconsolidation ratio* and σ'_m = the *overconsolidation margin.*

(2) *Lateral strain is considered – Skempton–Bjerrum method*
In reality the condition of zero lateral strain is satisfied approximately in the cases of thin clay layers and of layers under loaded areas which are large compared with the layer thickness. In practice, however, there are many situations where significant *lateral strain* will occur and the initial excess pore water pressure will depend on the in-situ stress conditions. In these cases, there will be an immediate settlement, under undrained conditions, in addition to the consolidation settlement. Immediate settlement is zero if the lateral strain is zero, as assumed in the one-dimensional method of calculating settlement. In the Skempton–Bjerrum method the *final* settlement S (excluding settlement due to creep or secondary compression) of a foundation on clay equals to the immediate settlement S_i plus the consolidation settlement S_c. However, Skempton and Bjerrum (1957) accounted for differences in the way excess pore water pressures are generated when the soil experiences lateral strain. This is reflected in the three-dimensional adjustment coefficient K (also known as settlement coefficient or settlement ratio), using

$$S_c = K\, S_{oed} \tag{3.40}$$

An equation for the coefficient K can be derived as follows:

From soil mechanics, if there is no change in static pore water pressure, the *initial* value of excess pore water pressure u_i at a point in a *fully saturated* clay layer is given as

$$\begin{aligned} u_i &= \Delta\sigma_3 + A(\Delta\sigma_1 - \Delta\sigma_3) \\ &= \Delta\sigma_1 \left[A + \frac{\Delta\sigma_3}{\Delta\sigma_1}(1 - A) \right] \end{aligned} \tag{3.41}$$

where A is a pore pressure coefficient. Value of A for a fully saturated soil can be determined from measurements of pore water pressure during the application of principal stress difference $\Delta\sigma_1$ under undrained conditions in a triaxial compression test (see ASTM D2850).

By the Skempton–Bjerrum method, consolidation settlement is expressed as

$$\begin{aligned} S_c &= \int_0^H m_v u_i dz, \\ &= \int_0^H m_v \Delta\sigma_1 \left[A + \frac{\Delta\sigma_3}{\Delta\sigma_1}(1 - A) \right] dz, \end{aligned}$$

where H is the thickness of the clay layer. By the one-dimensional consolidation method, settlement calculated from results of an oedometer test is

$$S_{oed} = \int_0^H m_v \Delta\sigma_1 dz \quad (\text{where } \Delta\sigma_1 = \sigma'_z)$$

A settlement coefficient K is introduced, such that $S_c = K\,S_{oed}$, where

$$K = \frac{\int_0^H m_v \Delta\sigma_1 \left[A + \frac{\Delta\sigma_3}{\Delta\sigma_1}(1 - A) \right] dz}{\int_0^H m_v \Delta\sigma_1 dz} \tag{3.42}$$

Values of the settlement coefficient K, for circular footing of diameter B and strip footing of width B, in terms of A and the ratio H/B are given in Figure 3.15. For square and rectangular foundations B will be the diameter of the equivalent circle.

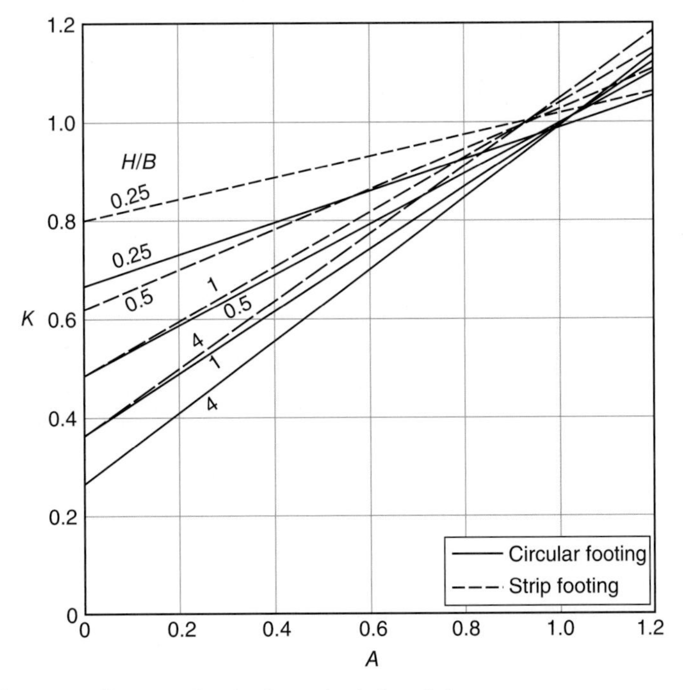

Figure 3.15 Settlement coefficients K for circular and strip foundations.

Table 3.17 Values of K_{OC} for round footings.

OCR	K_{OC}		
	B/H = 4.0	B/H = 1.0	B/H = 0.2
1	1	1	1
2	0.986	0.957	0.929
3	0.972	0.914	0.842
4	0.964	0.871	0.771
5	0.950	0.829	0.707
6	0.943	0.800	0.643
7	0.929	0.757	0.586
8	0.914	0.729	0.529
9	0.900	0.700	0.493
10	0.886	0.671	0.457
11	0.871	0.643	0.429
12	0.864	0.629	0.414
13	0.857	0.614	0.400
14	0.850	0.607	0.386
15	0.843	0.600	0.371
16	0.843	0.600	0.357

According to Leonards (1976), the settlement ratio K_{OC} for *circular* foundation on overconsolidated clay is a function of overconsolidation ratio OCR and the ratio B/H and the consolidation settlement is:

$$S_c = K_{OC}\, S_{oed} \tag{3.43}$$

According to Das (2011), the interpolated values of K_{OC} from the work of Leonards (1976) are as given in Table 3.17.

According to Craig (2004), values of K are typically within the following ranges:

Soft, sensitive clays 1.0 to 1.2
Normally consolidated clays 0.7 to 1.0
Lightly overconsolidated clays 0.5 to 0.7
Heavily overconsolidated clays 0.2 to 0.5

3.4.4 Estimation of the Rate of Consolidation Settlement

Sometimes it is necessary to know the rate at which the foundations will settle during the long process of consolidation of a clay layer. In practical problems it is the *average* degree of consolidation U over the depth of the layer as a whole that is of interest; the consolidation settlement at time t being given by the product of U and the final settlement.

In *one-dimensional consolidation*, the dimensionless time factor

$$T_v = \frac{c_v t}{d^2} \tag{3.44}$$

Hence,

$$t = \frac{T_v d^2}{c_v} \quad \text{or} \quad t = \frac{10^{-7} \times T_v \times d^2}{3.154 \times c_v} \tag{3.45}$$

where

t = time in years
c_v = average coefficient of consolidation over the range of pressure involved, obtained from oedometer test, in m^2/sec
$d = H$ (thickness of compressible layer in metres) for drainage at top or at bottom only, whereas, $d = \frac{H}{2}$ for drainage at top and bottom.

The following empirical equations, for the condition of constant initial porewater pressure, give almost exact relationships between the average degrees of consolidation U and time factor T_v:

$$T_v = \frac{\pi}{4} U^2 \quad (\text{for } U < 60\%) \tag{3.46}$$

$$T_v = 1.781 - 0.933 \log(100 - U\%) \quad (\text{for } U > 60\%) \tag{3.47}$$

Sivaram and Swamee (1977) also developed an empirical relationship between T_v and U, for the condition of constant initial porewater pressure, which is valid for U varying from 0 to 100%. The equation is of the form

$$T_v = \frac{\left(\frac{\pi}{4}\right)\left(\frac{U\%}{100}\right)^2}{\left[1 - \left(\frac{U\%}{100}\right)^{5.6}\right]^{0.357}} \tag{3.48}$$

The total settlement at any time t is given by

$$S_t = S_i + US_c \tag{3.49}$$

3.4.5 Method of Accelerating the Rate of Consolidation Settlement

The use of preload fills and other means to precompress soils in advance of construction of permanent facilities is a relatively inexpensive but effective method for improving poor foundation soils. The benefits of *precompression* are to increase shear strength and to decrease post construction settlements to tolerable values. Temporary surcharge loading makes it possible to practically eliminate, in advance of construction of structures or paving of roadways or runways, subsoil settlement that would otherwise occur subsequent to completion of construction. However, many compressible subsoil deposits are sufficiently thick or impermeable so that consolidation occurs slowly and preload fills required to precompress the soil within the time available may become so high that they become uneconomical or, because of possible foundation instability, require large and costly berms. When this happens, it may be wise to install artificial internal drainage channels in the poor subsoil to accelerate the rate of consolidation. One method of doing this consists of installing *vertical drains*.

The slow rate of consolidation in saturated clays of low permeability is accelerated by means of vertical drains which shorten the drainage path within the clay. Consolidation is then due mainly to horizontal radial drainage, resulting in the faster dissipation of excess pore water pressure; vertical drainage becomes of minor importance. In theory the final magnitude of consolidation settlement is the same, only the *rate* of settlement being affected. The traditional vertical drains are *sand drains* (Figure 3.16). These drains are constructed by drilling holes through the clay layer (or layers) in the field at regular intervals. The holes are then backfilled with suitably graded sand. This can be achieved by several means, such as: (a) rotary drilling and then backfilling with sand, (b) drilling by continuous-flight auger with a hollow stem and backfilling with sand (through the hollow stem) and (c) driving hollow steel piles. The soil inside the pile is then jetted out, after which backfilling with sand is done. Typical diameters of sand drains are 200–400 mm and drains have been installed to depths of over 30 m. The sand should be capable of allowing the efficient drainage of water without permitting fine soil particles to be washed in.

In the case of an embankment constructed over a highly compressible clay layer (Figure 3.17), vertical drains installed in the clay would enable the embankment to be brought into service much sooner and there would be a quicker increase in the shear strength of the clay. A degree of consolidation of the order of 80% would be desirable at the end of construction. Any advantages, of course, must be set against the additional cost of the installation.

Prefabricated drains (PVDs) are also used and are generally cheaper than backfilled drains for a given area of treatment. One such type consists of a filter stocking, generally of woven polypropylene, filled with sand, a typical diameter being 65 mm; compressed air is used to ensure that the stocking is completely filled with sand. This type of drain is very flexible and is usually unaffected by lateral

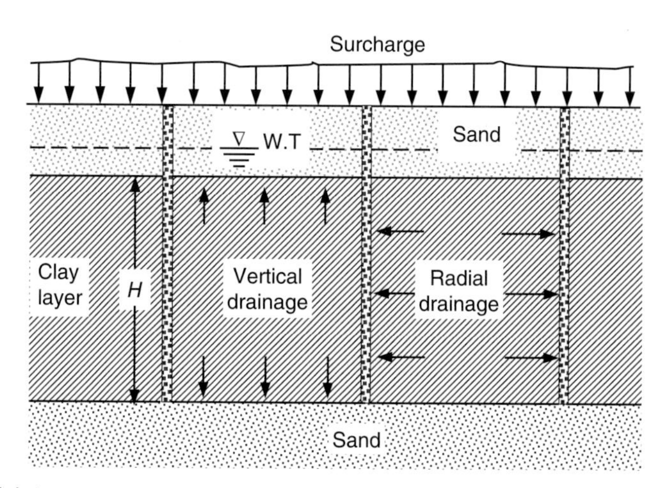

Figure 3.16 Sand drains.

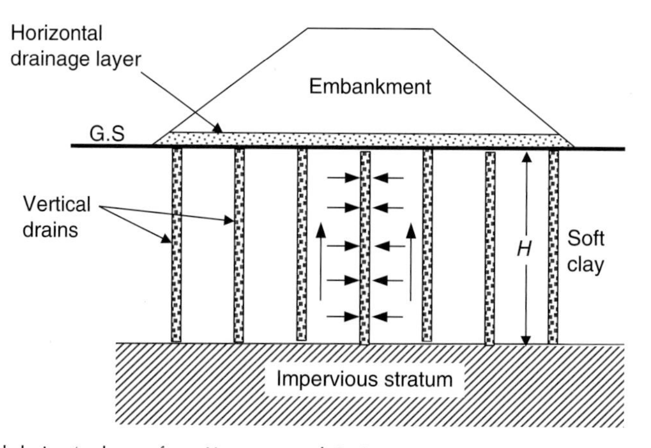

Figure 3.17 Vertical drains (redrawn from Knappett and Craig, 2012).

ground movements. Another type of PVDs is the *band* drain, consisting of a flat plastic core with drainage channels, enclosed by a thin layer of geotextile filter fabric (Figure 3.18); the fabric must have sufficient strength to prevent it from being squeezed into the channels. The main function of the fabric is to prevent the passage of fine soil particles which might clog the channels in the core. Typical dimensions of a band drain are 100×4 mm and the equivalent diameter is generally assumed to be the perimeter divided by π. The PVDs are installed either by insertion into pre-bored holes or by placing them inside a mandrel or casing which is then driven or vibrated into the ground (Figure 3.19); thus, drilling

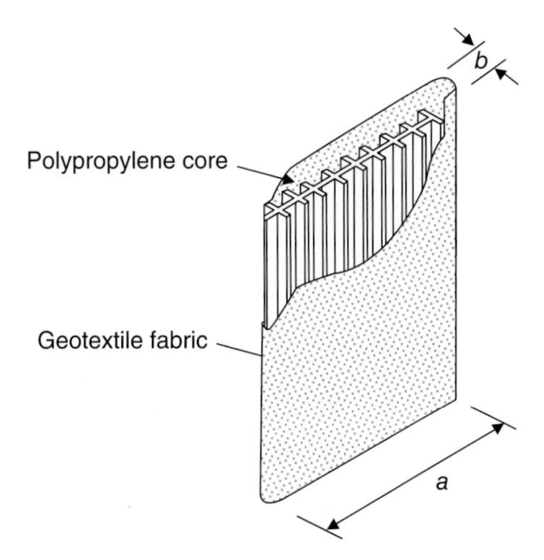

Figure 3.18 Prefabricated vertical drain.

Figure 3.19 Installation of PVDs in the field (from Das, 2011).

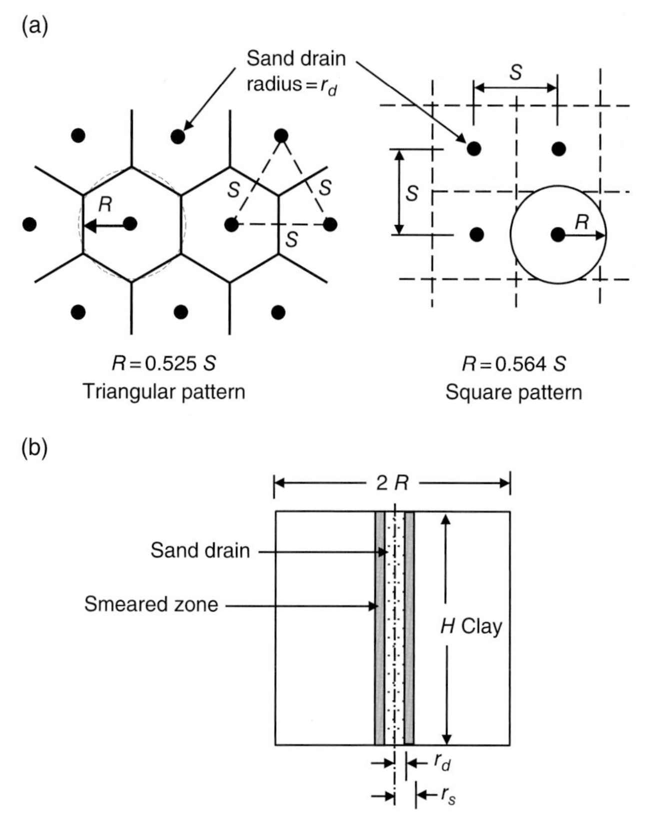

Figure 3.20 Vertical sand drains: (a) sand drain patterns, (b) vertical section of a sand drain with smeared zone indicated.

would not be required and installation would be much faster, which are considered advantages of PVDs over vertical sand drains.

The spacing of the drains is the most important design consideration because the object in using the vertical drains is to reduce the length of the drainage path. The drains are usually spaced in either a square or a triangular pattern, as shown in Figure 3.20a.

The centre to centre spacing of the drains must obviously be less than the thickness of the clay layer and there is no point in using vertical drains in relatively thin clay layers. It is clear to understanding that a successful design requires the coefficients of consolidation in both the horizontal and vertical directions c_h and c_v respectively to be known fairly accurately. It is noteworthy that the ratio c_h/c_v is normally between 1 and 2; the higher the ratio the more beneficial the drain installation will be. The values of the coefficients for the clay adjacent to the drains may be significantly reduced due to remoulding during installation (especially if a mandrel is used), an effect known as *smear* (Figure 3.20b). The smear effect can be taken into account by assuming a reduced value of c_h or by using a reduced drain diameter in the design computations.

In the case of vertical sand drains both radial and vertical drainage contribute to the average degree of consolidation U. The three-dimensional form of the consolidation equation in polar coordinates, with different soil properties in the horizontal and vertical directions, is

$$\frac{\partial u_e}{\partial t} = c_h \left(\frac{\partial^2 u_e}{\partial r^2} + \frac{1}{r}\frac{\partial u_e}{\partial r} \right) + c_v \frac{\partial^2 u_e}{\partial z^2} \tag{3.50}$$

The vertical prismatic blocks of soil surrounding the drains are replaced by equivalent cylindrical blocks, of radius R having the same cross-sectional area (Figure 3.20a). The solution to Equation (3.50) can be written in two parts: $U_v = f(T_v)$ and $U_r = f(T_r)$, where

U_v = average degree of consolidation due to vertical drainage
U_v = average degree of consolidation due to radial drainage drainage (radial drainage)
T_v = time factor for consolidation due to vertical drainage
T_r = time factor for consolidation due to radial drainage

$$T_v = c_v t / d^2 \tag{3.51}$$

$$T_r = c_h\, t / 4R^2 \tag{3.52}$$

Equation (3.52) confirms the fact that the smaller the value of R, i.e. the closer the spacing of the drains, the quicker the consolidation process due to radial drainage proceeds. The solution for radial drainage, due to Barron (1948), is given in Figure 3.21. The relationship between U_r and T_r depends on the ratio $n = R/r_\text{d}$, where R is the radius of the equivalent cylindrical block and r_d is the radius of the drain.

For a given surcharge and time duration, the *average* degree of consolidation U due to drainage in the vertical and radial directions is:

$$U = 1 - (1 - U_v)(1 - U_r) \tag{3.53}$$

According to Olson (1977), if the surcharge is applied during a certain period of time t_c, that is in the form of *ramp* loading (not instantaneous loading) as shown below, then:

(a) For radial drainage, instead of using the solution of Figure 3.21 the following equations are to be used:

$$U_r = \frac{T_r}{T_{rc}} - \frac{1}{AT_{rc}}[1 - \exp(-AT_r)] \quad (\text{for } T_r \le T_{rc}) \tag{3.54}$$

$$U_r = 1 - \frac{1}{AT_{rc}}[\exp(AT_{rc}) - 1]\exp(-AT_r) \quad (\text{for } T_r \ge T_{rc}) \tag{3.55}$$

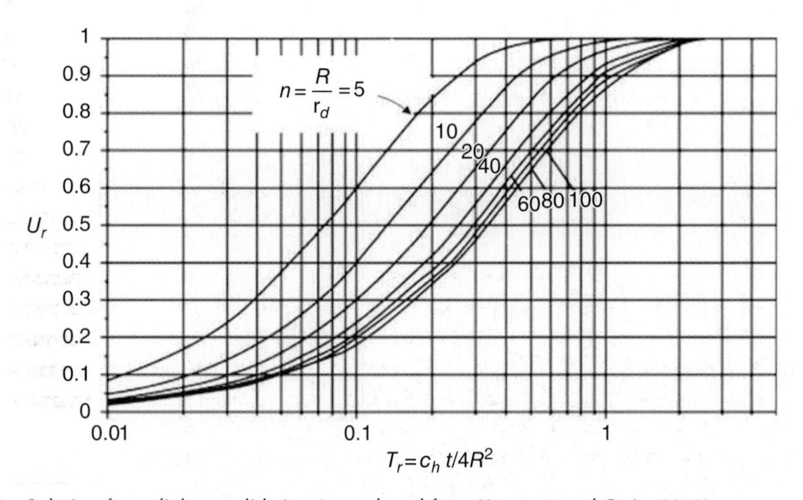

Figure 3.21 Solution for radial consolidation (reproduced from Knappett and Craig, 2012).

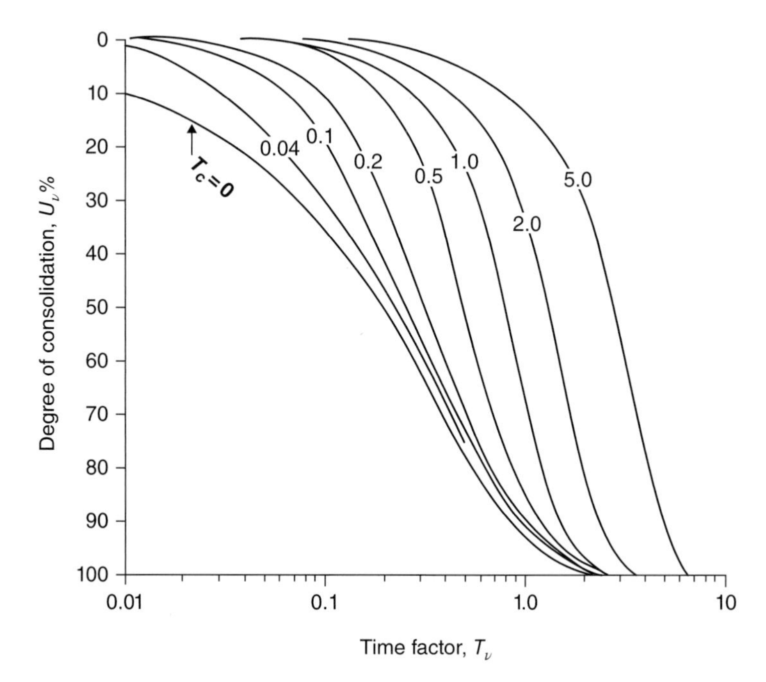

Figure 3.22 Variation of U_v with T_v and T_c (after Olson, 1977).

where

$$T_{rc} = \frac{c_h\, t_c}{4R^2},\ T_r = \frac{c_h\, t}{4R^2}$$

$$A = \frac{2}{[n^2/(n^2-1)]\ln(n) - [(3n^2-1)/4n^2]}$$

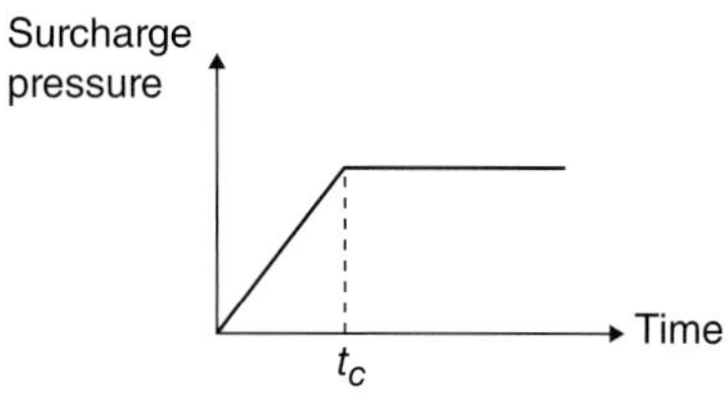

Surcharge pressure

Time

t_c

Scheme 3.1

(b) For vertical drainage, instead of using Equations (3.46), (3.47) and (3.48), the curves of Figure 3.22 are to be used. This figure gives the variation of U (%) with T_v and T_c (Olson, 1977). Note that

$$T_c = \frac{c_v\, t_c}{d^2};\ T_v = \frac{c_v\, t}{d^2}$$

It is necessary to realise that the effect of *smear* is not included in all the equations presented above; that is they are applicable for no-smear cases. For sand drain problems in which smear effect is to be considered, see Solution of Problems 3.45 and 3.46.

3.4.6 Estimation of Settlements over the Construction Period

Before the actual construction of a structure starts, necessary excavations will be carried out which cause a reduction in net load, resulting in swelling the foundation clay soil. Structural loads are applied to the soil over a period of time and not instantaneously. Actual settlement will not begin until the applied load exceeds the weight of the excavated soil. Terzaghi proposed an empirical method of correcting the instantaneous time–settlement curve to allow for the construction period (Figure 3.23).

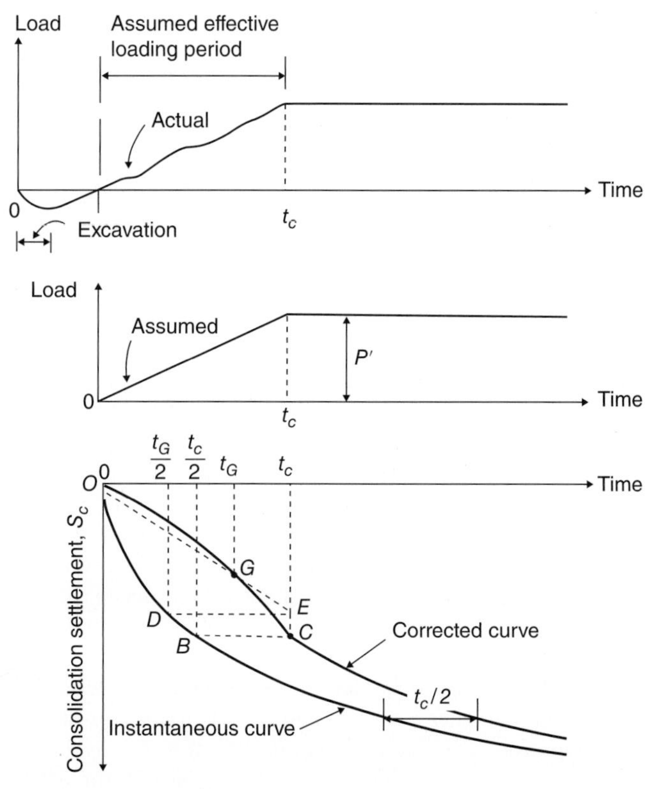

Figure 3.23 Corrected consolidation settlement curve during the construction period.

Assume t_c represents the effective construction period and P' represents the net load which is the gross load less the weight of the excavated soil. The period t_c is measured from the time when P' is zero and it is assumed that the net load is applied uniformly over the time t_c, as shown in Figure 3.23. Also, it is assumed that the degree of consolidation at time t_c is the same as if the load P' had been acting as a constant load for the period $t_c/2$. The settlement curve due to the instantaneously applied net load is first plotted as shown by the lower curve. The first point C on the corrected curve is obtained by intersection of a perpendicular dropped from a point where t_c is located on the time abscissa with the horizontal line BC, where B is the point of intersection of another perpendicular, dropped from point of $t_c/2$ with the instantaneous curve. Thus the settlement at any time during the construction period is equal to that occurring for instantaneous loading at half that time; however, since the load then acting is not the total net load P', the settlement value so obtained must be reduced in the proportion of that load to the total load. Therefore, to obtain any other point G on the corrected curve for a period t_G, a perpendicular is dropped from $t_G/2$ to intersect the instantaneous loading curve at D. A horizontal line DE is drawn; then the intersection of OE with the perpendicular from the point t_G gives the intermediate point G on the corrected curve for time t_G.

During the period after completion of construction, the corrected curve will be the instantaneous curve offset by $t_c/2$, as shown in Figure 3.23. The corrected total settlement curve can be obtained by adding the immediate settlement to the corrected consolidation settlement.

3.4.7 Secondary Compression

Figure 3.24 shows variation of e with $\log t$ under a given load increment. The final part of the experimental void ratio–log time curve represents the curve of the secondary compression, which is practically linear. According to the Terzaghi theory of consolidation, the primary consolidation is due entirely

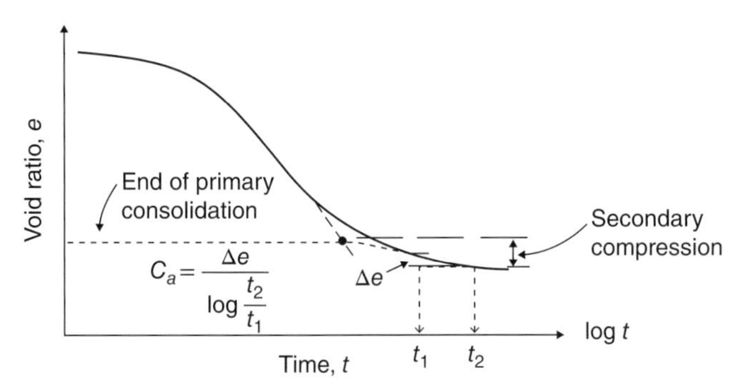

Figure 3.24 Variation of e with log *t* under a given load increment and definition of secondary compression index.

to the dissipation of the excess porewater pressure, with permeability alone governing the time dependency of the consolidation process. However, practice observations as well as experimental results show that compression does not cease when the excess porewater pressure has dissipated to zero but continues at a gradually decreasing rate under constant effective stress.

Secondary compression (or *creep*) is thought to be due to the gradual readjustment of the clay particles into a more stable configuration following the structural disturbance caused by the decrease in void ratio, especially if the clay is laterally confined. Another factor is the gradual lateral displacements which take place in thick clay layers. Some authorities relate secondary compression to the clay *adsorbed water* (the water molecules that are held to the clay particles). Researchers measured the secondary compression in a clay soil after the water was replaced by a non-polar liquid, such as CCl_4 and found its value was considerably less than that in the undisturbed clay. Hence, the adsorbed water may exert an important influence on the viscoelastic behaviour of the clay–particle contacts. The *rate* of secondary compression is thought to be controlled by the highly viscous film of adsorbed water surrounding the clay mineral particles in the soil. A very low viscous flow of adsorbed water takes place from the zones of film contact, allowing the soil particles to move closer together. The viscosity of the film increases as the particles move closer, resulting in a decrease in the rate of compression of the soil. According to Knappett and Craig (2012), it is presumed that primary consolidation and secondary compression proceed simultaneously from the time of loading.

The secondary compression index C_α can be defined by the slope of the final part of the curve of Figure 3.24; written as

$$C_\alpha = \frac{\Delta e}{\log t_2 - \log t_1} = \frac{\Delta e}{\log(t_2/t_1)} \tag{3.56}$$

where

Δe = change in void ratio
t_1, t_2 = time

From the results of laboratory tests and field observations, Simons (1974) derived the following equation for clays:

$$C_\alpha = 0.00018 \times \text{natural moisture content (in percent)} \tag{3.57}$$

The magnitude of the secondary compression can be calculated as

$$S_s = C'_\alpha H \log(t_2/t_1) \tag{3.58}$$

where $C'_\alpha = \dfrac{C_\alpha}{(1+e_p)}$

e_p = void ratio at the end of primary consolidation
H = thickness of clay layer

The magnitude of secondary compression in a given time is generally greater in normally consolidated clays than in overconsolidated clays.

Mesri (1973) correlated C'_α with the natural moisture content w of several soils, from which it appears that

$$C'_\alpha = 0.0001\,w \tag{3.59}$$

where w = natural moisture content, in percent.

For most overconsolidated clays, the range of C'_α values are 0.0005 to 0.001. Also, Mesri and Godlewski (1977) compiled the magnitude of C_α/C_c for a number of soils (C_c = compression index). Accordingly, the following relationships can be given:

$$\text{For inorganic clays and silts}: \; C_\alpha/C_c = 0.04 \mp 0.01 \tag{3.60}$$

$$\text{For organic clays and silts}: \; C_\alpha/C_c = 0.05 \mp 0.01 \tag{3.61}$$

$$\text{For peats}: \; C_\alpha/C_c = 0.075 \mp 0.01 \tag{3.62}$$

For a particular soil under oedometer test, the magnitude of secondary compression over a given time, as a percentage of the total compression, increases as the ratio of pressure increment to initial effective pressure decreases; the magnitude of secondary compression also increases as the thickness of the oedometer specimen decreases and as temperature increases. The secondary compression characteristics of an oedometer specimen cannot normally be extrapolated to the case of a full-scale foundation (Knappett and Craig, 2012).

3.5 Settlement of Foundations on Rock

Closely jointed rock formations and weak to moderately weak weathered rocks possess a degree of compressibility such that it is necessary to make estimates of settlement wherever heavily loaded structures are founded on these formations.

Settlement of a rock mass may be estimated using equations from elastic theory similar to Equation (3.4). Meigh (1976) proposed the following settlement equation:

$$\rho_i = q\left(\frac{B}{E_f}\right) I'_p F_B F_D \tag{3.63}$$

where ρ_i = immediate settlement beneath corner of a uniformly loaded area
q = net foundation pressurer
B = width of foundation
E_f = deformation modulus of rock at the foundation level
I'_p = influence factor (Figure 3.25)
F_B = correction factor for roughness of the base (Figure 3.26a)
F_D = correction factor for depth of embedment in rock (Figure 3.26b)

In the case of a *flexible* rectangular foundation, the loaded area is divided into four equal rectangles and the settlement computed at the corner of each rectangle. The settlement at the centre of the foundation is then four times the corner settlement. In the case of relatively *rigid* foundation, the settlements

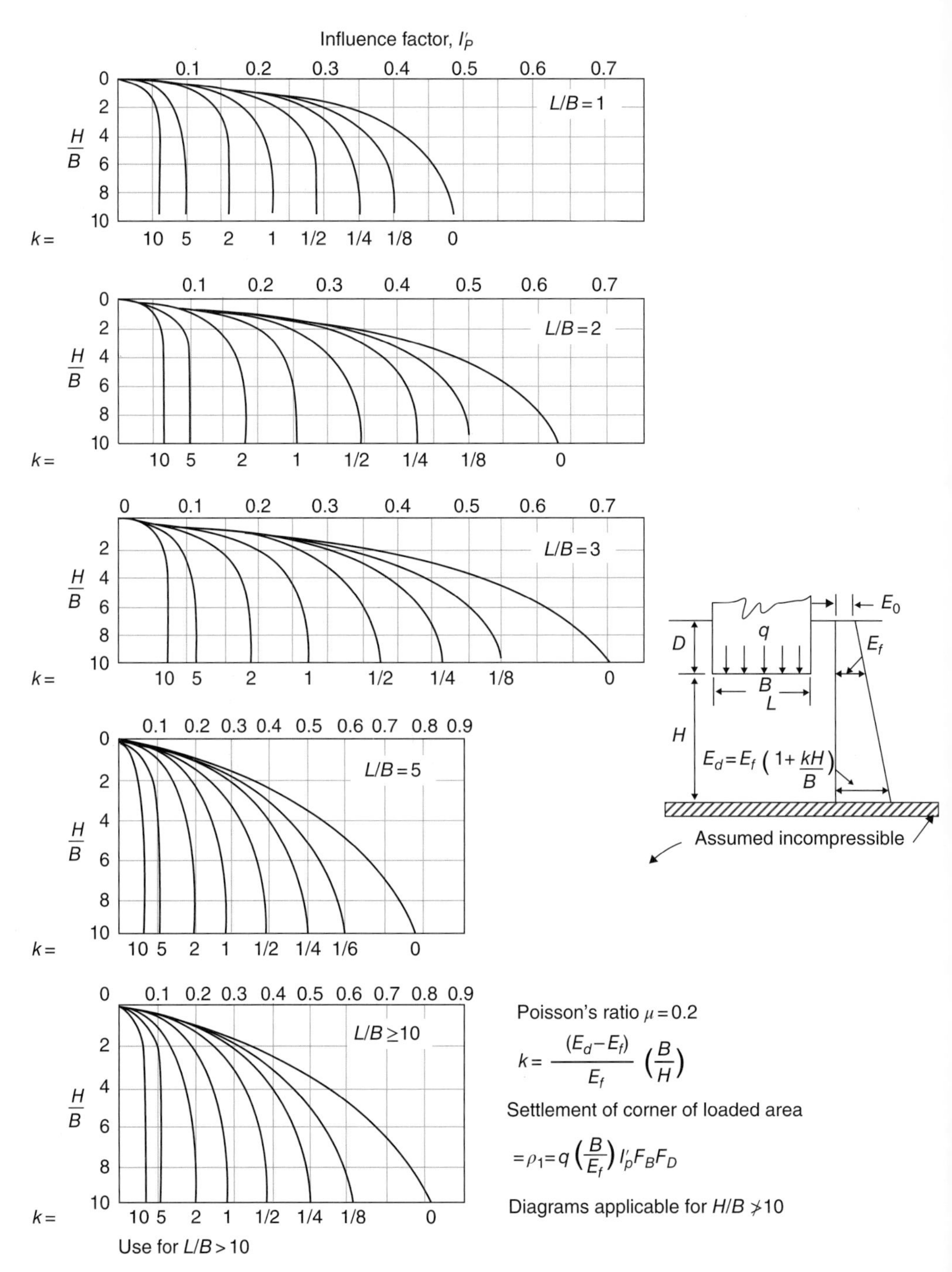

Figure 3.25 Values of influence factor I'_p for deformation modulus increasing linearly with depth and Poisson's ratio of 0.2 (from Tomlinson, 2001).

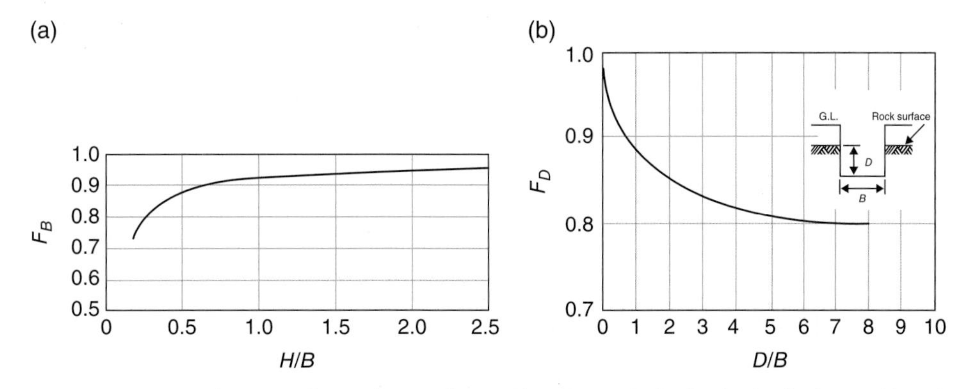

Figure 3.26 Correction factors: (a) for roughness of base of foundation; (b) for depth of embedment of foundation below surface of rock (from Tomlinson, 2001).

at centre, at the midpoint of a longer side and at a corner of the loaded area are calculated. Then the average settlement of the rigid foundation is given by (Tomlinson, 2001),

$$\rho_{i,\text{ ave}} = 1/3\left(\rho_{i,\text{ center}} + \rho_{i,\text{ midpoint long side}} + \rho_{i,\text{ corner}}\right) \tag{3.64}$$

Meigh obtained curves for influence factors for various values of the constant k using a value for Poisson's ratio equals to 0.2, as shown in Figure 3.25. The value of k is computed from the following equation:

$$k = \left(\frac{E_d - E_f}{E_f}\right)\left(\frac{B}{z}\right) \tag{3.65}$$

For a rock mass of thickness H below the foundation level

$$k = \left(\frac{E_d - E_f}{E_f}\right)\left(\frac{B}{H}\right)$$

Measured values of E against depth z are plotted and a straight line through the plotted points is drawn (Figure 3.25). Thus values of deformation modulus such as E_d and E_f are obtained, which are used in Equation (3.65) to obtain value of k.

$$\text{Correction factor } F_B = \frac{\text{Settlement for rough base}}{\text{Settlement for smooth base}}$$

The most important of the procedure is to obtain reliable variation of E values with depth. According to Tomlinson (2001), where the results of unconfined compression test on core samples are in sufficient numbers to be representative of the variation in strength of the rock over the depth stressed by the foundation loading, the deformation modulus E_m of the rock mass can be obtained from the relationship

$$E_m = (j)(M_r)(q_{uc}) \tag{3.66}$$

Table 3.18 Values of mass factor, j.

Quality classification[a]	*RQD* (%)	Fracture frequency per metre	Velocity index[b] $(V_f/V_L)^2$	Mass factor (j)
Very poor	0–25	15	0–0.2	0.2
Poor	25–50	15–8	0.2–0.4	0.2
Fair	50–75	8–5	0.4–0.6	0.2–0.8
Good	75–90	5–1	0.6–0.8	0.5–0.8
Excellent	90–100	1	0.8–1.0	0.8–1.0

[a] As BS 5930.
[b] V_f = wave velocity in field, V_L = wave velocity in laboratory.

Table 3.19 Modulus ratios M_r for different groups of rock.

Rock group	Description	M_r
Group 1	Pure limestones and dolomites	600
	Carbonate sandstones	
Group 2	Igneous	
	Oolite and marly limestones	
	Well-cemented sandstone	
	Indurated carbonate mudstones	
	Metamorphic rocks, including slates and schists (flat cleavage/foliation)	300
Group 3	Very marly limestones	
	Poorly cemented sandstones	
	Cemented mudstones and shales	
	Slates and schists (steep cleavage/foliation)	150
Group 4	Uncemented mudstones and shales	75

where j = rock mass factor related to the discontinuity spacing in the rock mass; its values are given by Hobbs (1974), as shown in Table 3.18,

M_r = modulus ratio, which is the ratio of E to q_{uc} of the intact rock,
q_{uc} = unconfined compressive strength.

Hobbs showed that rocks of various types could be grouped together, and for practical purposes a modulus ratio M_r could be assigned to each group as shown in Table 3.19.

The high porosity chalks such as that of south-east England and marls such as the Keuper Marl are special cases. Values of the mass modulus of deformation of chalk established by Hobbs (1974) and for Keuper Marl established by Chandler and Davis (1973) are shown in Table 3.20.

Table 3.20 Mass deformation modulus values for high porosity chalk and Keuper Marl.

Rock type	Weathering grade	Description	Deformation modulus (MN/m^2)
Chalk	I	Blocky, moderately weak, brittle	>400
	II	Blocky, weak, joints more than 200 mm apart and closed	90–400
	III	Rubbly to blocky, unweathered, joints 60–200 mm apart, open to 3 mm and sometimes filled with fragments	50–90
	IV	Rubbly, partly weathered, with bedding and jointing, joints 10–60 mm apart, open to 20 mm, often filled with weak remoulded chalk and fragments	35–50
	V	Structureless, remoulded, containing lumps of intact chalk	10–35
	VI	Extremely weak, structureless, containing small lumps of intact chalk	<10
Keuper Marl	I (unweathered)	Mudstone (often fissured)	>150
	II (slightly weathered)	Angular blocks of unweathered marl and mudstone with virtually no matrix	75–150
	III (Moderately weathered)	Matrix with frequent lithorelicts up to 25 mm	30–75
	IV (Highly to completely weathered)	Matrix with occasional claystone pellets less than 3 mm size (usually coarse sand) degrading to matrix only	Obtain from laboratory tests

Problem Solving

Problem 3.1

A steel-frame office building has a column spacing of 6 m. It is to be supported on spread footings founded on a clayey soil. What are the allowable total and differential settlements?

Solution:

Table 3.1 gives maximum allowable total settlement = 75 mm
Table 3.2 gives maximum allowable total settlement = 50mm
Table 3.2 gives allowable total settlement = 25 mm as limiting value for serviceability.
Use allowable total settlement = 25 *mm*

Tables 3.2, 3.4 and 3.5, relatively, give a small value for the allowable angular distortion $\beta = \frac{1}{500} = \frac{\Delta S_T}{L}$, which is a conservative value.

Hence, $\Delta S_T = \frac{L}{500} = \frac{6 \times 1000}{500} = 12\ \text{mm}$
Use allowable differential settlement = 12 *mm*

Problem 3.2

A rigid shallow foundation 1 × 2 m is shown in the scheme. Calculate the elastic settlement at the centre of the foundation.

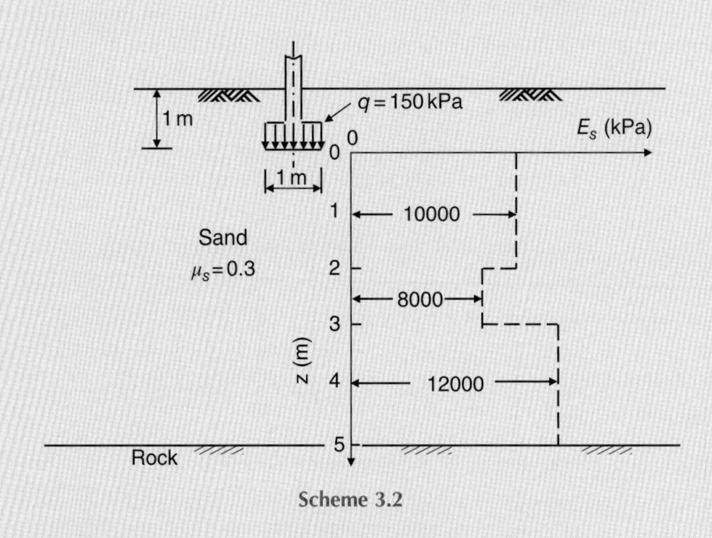

Scheme 3.2

Solution:

Equation (3.5):
$$S_i = q \times B \times \frac{1-\mu^2}{E_s} \times m \times I_S \times I_F$$

Equation(3.7):

$$E_{s,\,av} = \frac{H_1E_{s1} + H_2E_{s2} + \ldots + H_nE_{sn}}{H} = \frac{10\,000 \times 2 + 8000 \times 1 + 12\,000 \times 2}{5}$$
$$= 10\ 400\,\text{kPa}$$

For settlement at *centre* of the foundation

$$m = 4;\ B = \frac{1}{2}\ \text{m};\ L = \frac{2}{2} = 1\ \text{m};\ M = \frac{\frac{L}{2}}{(B/2)} = \frac{1}{(1/2)} = 2;\ N = \frac{H}{(B/2)} = \frac{5}{(1/2)} = 10$$

From Table 3.8 obtain $I_1 = 0.641$ and $I_2 = 0.031$. Hence,

$$I_S = I_1 + \frac{1-2\mu}{1-\mu} I_2 = 0.641 + \frac{1-2\times 0.3}{1-0.3} 0.031 = 0.659$$

$\frac{D}{B} = \frac{1}{1} = 1;\ \frac{B}{L} = \frac{1}{2} = 0.5;\ \mu = 0.3$. From Table 3.7 obtain $I_F = 0.71$

$$S_i = 150 \times \frac{1}{2} \times \frac{1-(0.3)^2}{10\ 400} \times 4 \times 0.659 \times 0.71 = 0.0123\ \text{m} = 12.3\ \text{mm}$$

Equation (3.6):
$$S_{i(rigid)} = 0.93 S_{i(flexible)}$$
$$S_i = 0.93 \times 12.3 = 11.4\,mm$$

Problem 3.3

Solve Problem 3.2 ignoring the given E_s-data and considering that the supporting soil is clean, normally consolidated sand with the following available data:

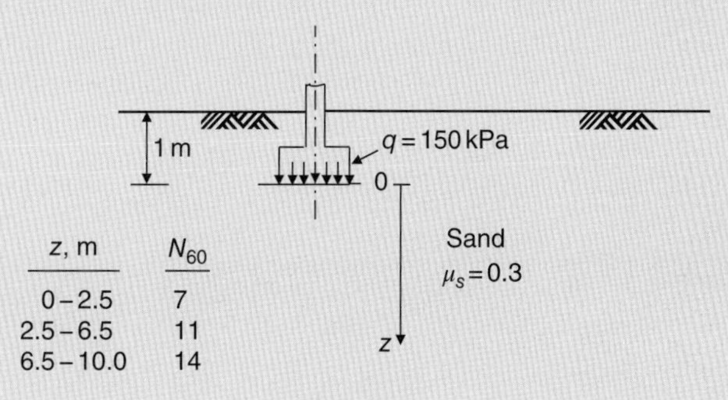

Scheme 3.3

Assume the rock layer exists at depth z = 10 m. Use Equation (1.33) for computing E_s values.

Solution:

Equation (3.5):
$$S_i = q \times B \times \frac{1-\mu^2}{E_s} \times m \times I_S \times I_F$$

Equation (1.33):
$$\frac{E_s}{P_a} = \alpha\, N_{60} \rightarrow E_s = \alpha\, P_a\, N_{60};\; P_a = 100 \text{ kPa}$$

For clean normally consolidated sand $\alpha = 10$

Thickness of the sand deposit = 10 m > (4B = 4 m); use $H = 4$ m

$$E_{s,\,av} = \frac{H_1 E_{s1} + H_2 E_{s2} + \ldots + H_n E_{sn}}{H} = \frac{10\times100\times7\times2.5 + 10\times100\times11\times1.5}{4} = 8500\,\text{kPa}$$

For settlement at *centre* of the foundation:

$$m = 4;\; B = \frac{1}{2}\text{ m};\; L = \frac{2}{2} = 1\text{ m};\; M = \frac{L}{2}(B/2) = \frac{1}{(1/2)} = 2;\; N = \frac{H}{(B/2)} = \frac{4}{(1/2)} = 8$$

From Table 3.8 obtain $I_1 = 0.611$ and $I_2 = 0.038$. Hence,

$$I_S = I_1 + \frac{1-2\mu}{1-\mu} I_2 = 0.611 + \frac{1-2\times0.3}{1-0.3}0.038 = 0.633$$

$\frac{D}{B} = \frac{1}{1} = 1;\; \frac{B}{L} = \frac{1}{2} = 0.5;\; \mu = 0.3$. From Table 3.7 obtain $I_F = 0.71$

$$S_i = 150 \times \frac{1}{2} \times \frac{1-(0.3)^2}{8500} \times 4 \times 0.633 \times 0.71 = 0.0144\,\text{m} = 14.4\,\text{mm}$$

Equation (3.6):
$$S_{i(rigid)} = 0.93 S_{i(flexible)}$$
$$S_i = 0.93 \times 14.4 = 13.4\,mm$$

Problem 3.4

Solve Problem 3.3 using the improved equation by Mayne and Poulos: Equation (3.8). Assume the footing thickness t = 0.30 m, and the modulus of elasticity of the foundation material $E_f = 15 \times 106$ kPa.

Solution:

Equation (3.8): $$S_i = \frac{q_o B_e I_G I_F I_E}{E_o}\left(1-\mu_s^2\right)$$

Obtain variation of E_s with depth z as follows:

Equation (1.33): $$\frac{E_s}{P_a} = \alpha N_{60} \rightarrow E_s = \alpha P_a N_{60};\ P_a = 100\,\text{kPa}$$

$$E_s = 10 \times 100 \times 7$$
$$= 7000\,\text{kPa, for the depth interval } 0 \sim 2.5\text{ m; at } z = 1.25\text{ m}$$
$$E_s = 10 \times 100 \times 11$$
$$= 11000\,\text{kPa, for the depth interval } 2.5 \sim 6.5\text{ m; at } z = 4.5\text{ m}$$
$$E_s = 10 \times 100 \times 14$$
$$= 14000\,\text{kPa, for the depth interval } 6.5 \sim 10.0\text{ m; at } z = 8.25\text{ m}$$

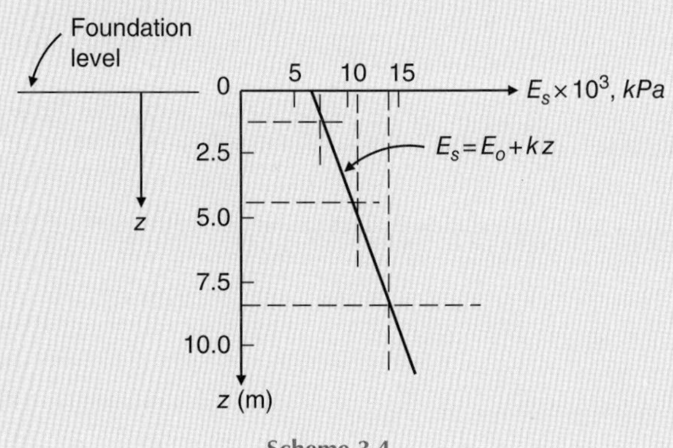

Scheme 3.4

From the plot obtain (approximately):

$$E_o = 6700\,\text{kPa}$$

$$k = \frac{14\,000 - 6700}{8.25} = 885\,\text{kN/m}$$

$$B_e = \sqrt{\frac{4BL}{\pi}} = \sqrt{\frac{4 \times 1 \times 2}{\pi}} = 1.6\ m$$

$$\beta = \frac{E_o}{kB_e} = \frac{6700}{885 \times 1.6} = 4.73;\ \frac{H}{B_e} = \frac{4B_e}{B_e} = 4$$

From Figure 3.4, for $\frac{H}{B_e} = 4;\ \beta = 4.73 \rightarrow I_G \approx 0.75$

Equation (3.9): $$I_F = \frac{\pi}{4} + \frac{1}{4.6 + 10\left[\dfrac{E_f}{E_o + (B_e k/2)}\right]\left(\dfrac{2t}{B_e}\right)^3}$$

$$= \frac{\pi}{4} + \frac{1}{4.6 + 10\left[\dfrac{15\times 10^6}{6700 + \{(1.6\times 885)/2\}}\right]\left(\dfrac{2\times 0.3}{1.6}\right)^3} = 0.787$$

Equation (3.10): $$I_E = 1 - \frac{1}{3.5[\exp(1.22\mu_s - 0.4)]\left[(B_e/D_f) + 1.6\right]}$$

$$= 1 - \frac{1}{3.5\left[\exp(1.22\times 0.3 - 0.4)\right][(1.6/1) + 1.6]} = 0.908$$

$$q_o = 150\,\text{kPa}$$

$$S_i = \frac{q_o B_e I_G I_F I_E}{E_o}\left(1 - \mu_s^2\right) = \frac{150\times 1.6\times 0.75\times 0.787\times 0.908}{6700}\left(1 - 0.3^2\right)$$

$$S_i = 0.0175\,\text{m} = 17.5\,mm$$

Problem 3.5

A rigid foundation 4 × 2 m carries a uniform pressure of 150 kN/m^2. It is located at a depth of 1 m in a layer of saturated clay 5 m thick, which is underlain by a second clay layer 8 m thick. The average values of E_u for the first and second layers are 40 and 75 MN/m^2, respectively. A hard stratum lies below the second layer. Using Equation (3.33), determine the average immediate settlement under the foundation.

Solution:

Equation (3.33): $$S_i = \mu_o \mu_1 \frac{qB}{E}$$

(a) Consider the upper clay layer, with $E_u = 40\,\text{MN/m}^2$:
D/B = ½ = 0.5 and therefore from Figure 3.9, $\mu_o = 0.94$
$\frac{H}{B} = \frac{4}{2} = 2$; $L/B = 4/2 = 2$. From Figure 3.9 obtain $\mu_1 = 0.60$

$$S_{i(a)} = 0.94\times 0.60\times \frac{150\times 2}{40} = 4.2\,\text{mm}$$

(b) Consider the two clay layers combined, with $E_u = 75\,\text{MN/m}^2$:
D/B = ½ = 0.5 and therefore from Figure 3.9, $\mu_o = 0.94$
$\frac{H}{B} = \frac{12}{2} = 6$; $L/B = 4/2 = 2$. From Figure 3.9 obtain $\mu_1 = 0.85$

$$S_{i(b)} = 0.94\times 0.85\times \frac{150\times 2}{75} = 3.2\,\text{mm}$$

(c) Consider the upper clay layer, with $E_u = 75\,\text{MN/m}^2$:
D/B = ½ = 0.5 and therefore from Figure 3.9, $\mu_o = 0.94$
$\frac{H}{B} = \frac{4}{2} = 2$; $L/B = 4/2 = 2$. From Figure 3.9 obtain $\mu_1 = 0.60$

(Continued)

$$S_{i(c)} = 0.94 \times 0.60 \times \frac{150 \times 2}{75} = 2.3\ \text{mm}$$

Hence, using the principle of superposition, the average settlement of the foundation, considering *flexible* loading area, is given by:

$$S_i = S_{i(a)} + S_{i(b)} - S_{i(c)} = 4.2 + 3.2 - 2.3 = 5.1\ \text{mm}.$$

Equation (3.6): $S_{i(rigid)} = 0.93 S_{i(flexible)}$

$S_i = 0.93 \times 5.1 = 4.7$; say 5 mm

Problem 3.6

Solve Problem 3.5 using equation Equation (3.5). Use $\mu = 0.5$ for saturated clays. Consider settlement at centre of the foundation only.

Solution:

Equation (3.5): $S_i = q \times B \times \dfrac{1-\mu^2}{E_s} \times m \times I_S \times I_F$

(a) Consider the upper clay layer, with $E_u = 40\ \text{MN/m}^2$:
For settlement at *centre* of the foundation:

$$m = 4;\ B = \frac{2}{2} = 1\ \text{m};\ L = \frac{4}{2} = 2\ \text{m};\ M = \frac{\frac{L}{2}}{(B/2)} = \frac{\frac{4}{2}}{(2/2)} = 2;\ N = \frac{H}{(B/2)} = \frac{4}{(2/2)} = 4$$

From Table 3.8 obtain $I_1 = 0.476$ and $I_2 = 0.069$. Hence,

$$I_S = I_1 + \frac{1-2\mu}{1-\mu} I_2 = 0.476 + \frac{1-2\times 0.5}{1-0.5} 0.069 = 0.476$$

$\dfrac{D}{B} = \dfrac{1}{2} = 0.5;\ \dfrac{B}{L} = \dfrac{2}{4} = 0.5;\ \mu = 0.5$. From Table 3.7 obtain $I_F = 0.90$

$$S_{i(a)} = 150 \times \frac{2}{2} \times \frac{1-(0.5)^2}{40\ 000} \times 4 \times 0.476 \times 0.9 = 0.0048\ \text{m} = 4.8\ \text{mm}$$

(b) Consider the two clay layers combined, with $E_u = 75\ \text{MN/m}^2$:
For settlement at *centre* of the foundation:

$$m = 4;\ B = \frac{2}{2} = 1\ \text{m};\ L = \frac{4}{2} = 2\ \text{m};\ M = \frac{\frac{L}{2}}{(B/2)} = \frac{\frac{4}{2}}{(2/2)} = 2;\ N = \frac{H}{(B/2)} = \frac{12}{(2/2)} = 12$$

From Table 3.8 obtain $I_1 = 0.653$ and $I_2 = 0.028$. Hence,

$$I_S = I_1 + \frac{1-2\mu}{1-\mu} I_2 = 0.653 + \frac{1-2\times 0.5}{1-0.5} 0.069 = 0.0.653$$

$\frac{D}{B}=\frac{1}{2}=0.5;\ \frac{B}{L}=\frac{2}{4}=0.5;\ \mu=0.5$. From Table 3.7 obtain $I_F=0.90$

$$S_{i(b)}=150\times\frac{2}{2}\times\frac{1-(0.5)^2}{75\ 000}\times4\times0.653\times0.9=0.0035\ \text{m}=3.5\ \text{mm}$$

(c) Consider the upper clay layer, with $E_u=75\ \text{MN/m}^2$:
For settlement at *centre* of the foundation:

$$m=4;B=\frac{2}{2}=1\ \text{m};L=\frac{4}{2}=2\ \text{m};M=\frac{\frac{L}{2}}{(B/2)}=\frac{\frac{4}{2}}{(2/2)}=2;N=\frac{H}{(B/2)}=\frac{4}{(2/2)}=4$$

From Table 3.8 obtain $I_1=0.476$ and $I_2=0.069$. Hence,

$$I_S=I_1+\frac{1-2\mu}{1-\mu}I_2=0.476+\frac{1-2\times0.5}{1-0.5}0.069=0.476$$

$\frac{D}{B}=\frac{1}{2}=0.5;\ \frac{B}{L}=\frac{2}{4}=0.5;\ \mu=0.5$. From Table 3.7 obtain $I_F=0.90$

$$S_{i(c)}=150\times\frac{2}{2}\times\frac{1-(0.5)^2}{75\ 000}\times4\times0.476\times0.9=0.0026\ \text{m}=2.6\ \text{mm}$$

Hence, using the principle of superposition, settlement at centre of the foundation, considering *flexible* loading area, is given by:

$$S_i=S_{i(a)}+S_{i(b)}-S_{i(c)}=4.8+3.5-2.6=5.7\ \text{mm}$$

Equation (3.6):

$$S_{i(\text{rigid})}=0.93S_{i(\text{flexible})}$$

$$S_i=0.93\times5.7=\mathit{5.3\ mm}$$

Another procedure is to calculate settlement of each layer alone and then adding them to obtain the total settlement, as follows:

(a) Settlement of the upper clay layer was calculated = 4.8 mm.
(b) Settlement of the lower clay layer:

Assume using the simple $2V:1H$ ratio method of stress distribution
At top of the layer: $q=150[(2\times4)/(2+4)(4+4)]=25\ \text{kN/m}^2$
For settlement at the *centre* of the foundation:

$$m=4;B=\frac{6}{2}=3\ \text{m};L=\frac{8}{2}=4\ \text{m};M=\frac{\frac{L}{2}}{(B/2)}=\frac{4}{3}=1.33$$

$$N=\frac{H}{(B/2)}=\frac{8}{3}=2.67$$

From Table 3.8 obtain $I_1=0.36$ and $I_2=0.074$. Hence,

$$I_S=I_1+\frac{1-2\mu}{1-\mu}I_2=0.36+\frac{1-2\times0.5}{1-0.5}0.074=0.36$$

(Continued)

$\frac{D}{B} = \frac{5}{6} = 0.83$; $\frac{B}{L} = \frac{6}{8} = 0.75$; $\mu = 0.5$. From Table 3.7 obtain $I_F = 0.79$

$$S_{i(b)} = 25 \times \frac{6}{2} \times \frac{1-(0.5)^2}{75\ 000} \times 4 \times 0.36 \times 0.79 = 0.0009\,\text{m} = 0.9\,\text{mm}$$

Hence, using the principle of superposition, settlement at centre of the foundation, considering *flexible* loading area, is given by

$$S_i = S_{i(a)} + S_{i(b)} = 4.8 + 0.9 = 5.7\,\text{mm}.$$

Equation (3.6):

$$S_{i(\text{rigid})} = 0.93 S_{i(\text{flexible})}$$

$$S_i = 0.93 \times 5.7 = 5.3\,mm$$

Problem 3.7

A bridge pier has a base 8.5 m long by 7.5 m wide, founded at a depth of 3.0 m. The base of the pier imposes net foundation contact pressure equals 220 kN/m^2 due to dead loading and 360 kN/m^2 due to combined dead and live loading. Borings showed dense sand and gravel with cobbles and boulders to a depth of 9 m, followed by very stiff overconsolidated clay to more than 25 m below ground level. Standard penetration tests gave an average N-value of 40 blows per 300 mm in the sand and gravel stratum. A number of oedometer tests were made on samples of the stiff clay; the following values for the coefficient of compressibility m_v corresponding to the respective increments of pressure (due to the two types of loading) were obtained as follows:

Net contact pressure q_n, kN/m^2	Depth of layer, m	m_v, m^2/kN
220	9–12	0.000 11
	12–18	0.000 03
360	9–12	0.000 20
	12–18	0.000 04

Triaxial tests on undisturbed samples of the clay (nearly saturated) gave minimum shear strength of 120 kN/m^2 and undrained modulus of deformation E_u of 40 MN/m^2. Calculate the immediate and long-term settlement of the bridge pier.

Solution:

(1) Calculating immediate settlements in sand and gravel stratum.
It is not known whether this stratum is overconsolidated; the preconsolidation pressure is not given, hence assume it is normally consolidated.
Assume using the Burland and Burbidge empirical relationship:

Equation (3.15):

$$\frac{S_i}{B_R} = \alpha_1\alpha_2\alpha_3 \left[\frac{1.25\left(\frac{L}{B}\right)}{0.25 + (L/B)}\right]^2 \left(\frac{B}{B_R}\right)^{0.7} \left(\frac{q'}{p_a}\right)$$

Use Equation (3.16) to adjust the N-value for the gravel content; and therefore,

$$N_{(a)} = 1.25 \times N = 1.25 \times 40 = 50$$

Equation 3.18: $$\frac{z'}{B_R} = 1.4\left(\frac{B}{B_R}\right)^{0.75}$$

$\frac{z'}{0.3} = 1.4\left(\frac{7.5}{0.3}\right)^{0.75}$; $z' = 4.7\,\text{m} < (H = 6\,\text{m})$; hence use $\alpha_3 = 1$

For normally consolidated sand and gravel, $\alpha_1 = 0.14$ and

$$\alpha_2 = 1.71/N_{(a)}^{1.4} = 1.71/50^{1.4} = 0.007$$

For *dead loading* only:

$$S_i = 0.3 \times 0.14 \times 0.007 \times 1 \left[\frac{1.25\left(\frac{8.5}{7.5}\right)}{0.25 + (8.5/7.5)}\right]^2 \left(\frac{7.5}{0.3}\right)^{0.7} \left(\frac{220}{100}\right) = 0.0065\,\text{m}$$

$S_i = 6.5\,mm$

For combined *dead and live* loading:

$$S_i = 6.5 \times \frac{360}{220} = 10.6\,\text{mm}$$

(2) Calculating immediate settlements in clay stratum
Determine the net average vertical stress at top of the clay layer using Fadum chart of Figure 2.32:
Below centre of the pier base at depth z = 6 m;

$$m = \frac{\frac{B}{2}}{z} = \frac{\frac{7.5}{2}}{6} = 0.63;\; n = \frac{\frac{L}{2}}{z} = \frac{\frac{8.5}{2}}{6} = 0.71;\; I = 0.12$$

$$\sigma_z = q(4I) = 220 \times 4 \times 0.12 = 105.6\,\text{kN/m}^2$$

Below middle of the long side at depth z = 6 m;

$$m = \frac{B}{z} = \frac{7.5}{6} = 1.25;\; n = \frac{\frac{L}{2}}{z} = \frac{\frac{8.5}{2}}{6} = 0.71;\; I = 0.158$$

$$\sigma_z = q(2I) = 220 \times 2 \times 0.158 = 69.5\,\text{kN/m}^2$$

Below middle of the short side at depth z = 6 m;

$$m = \frac{\frac{B}{2}}{z} = \frac{7.5/2}{6} = 0.63;\; n = \frac{L}{z} = \frac{8.5}{6} = 1.42;\; I = 0.156$$

$$\sigma_z = q(2I) = 220 \times 2 \times 0.156 = 68.6\,\text{kN/m}^2$$

Below corner of the pier base at depth z = 6 m;

$$m = \frac{B}{z} = \frac{7.5}{6} = 1.25;\; n = \frac{L}{z} = \frac{8.5}{6} = 1.42;\; I = 0.203$$

$$\sigma_z = q(I) = 220 \times 0.203 = 44.7\,\text{kN/m}^2$$

(*Continued*)

The average stress below middle of the long side is greater than that below middle of the short side; hence the average stress on *top* of the clay layer, for the dead loading only, may be calculated, approximately, as

$$q_{ave} = \sigma_{z,ave} = \frac{105.6 + \frac{69.5 + 44.7}{2}}{2} = 81.4\ \text{kN/m}^2$$

This pressure is distributed over an area $B' \times L' = \frac{220}{81.4} \times 7.5 \times 8.5$

$B' \times L' = 172.3\ m^2$; $\frac{B'}{L'} = \frac{7.5}{8.5} \rightarrow B' = \frac{7.5}{8.5} L'$; hence,

$L' = \sqrt{\frac{8.5 \times 172.3}{7.5}} = 14.0\ \text{m}$, and $B' = 12.3\ \text{m}$

For combined *dead and live loading*,

$q_{ave} = 81.4 \times \frac{360}{220} = 133.2\ \text{kN/m}^2$. It is distributed over the area $B' \times L'$

The pier base is nearly square. According to the pressure bulb of Figure 2.36 for a square area loaded with q, the pressure at a depth of $2B$ below the loaded area is about $0.1q$. Therefore, practically, we need not consider settlements below that depth. Hence, thickness of the clay layer to be used in computations is

$$H = 2 \times 7.5 - 6 = 9\ \text{m}$$

$$S_i = q \times B \times \frac{1 - \mu^2}{E_s} \times m \times I_S \times I_F$$

For settlement below the *centre* of the loaded area $B'L'$:

$$m = 4; B = \frac{B'}{2} = \frac{12.3}{2} = 6.15\ \text{m}; L = \frac{L'}{2} = \frac{14}{2} = 7\ \text{m}; M = \frac{\frac{L'}{2}}{(B'/2)} = 1.14$$

$N = \frac{H}{(B'/2)} = \frac{9}{6.15} = 1.46$. Assume $\mu = 0.5$

From Table 3.8 obtain $I_1 = 0.217$; $I_2 = 0.083$

$$I_S = I_1 + \frac{1 - 2\mu}{1 - \mu} I_2 = 0.217 + \frac{1 - 2 \times 0.5}{1 - 0.5} 0.083 = 0.217$$

$\frac{D}{B'} = \frac{9}{12.3} = 0.73$; $\frac{B'}{L'} = \frac{12.3}{14} = 0.88$. From Table 3.7 obtain $I_F = 0.8$

For *dead loading* only:

$$S_{i(clay)} = 81.4 \times \frac{12.3}{2} \times \frac{1 - (0.5)^2}{40\ 000} \times 4 \times 0.217 \times 0.8 = 0.0065\ \text{m}$$
$$= 6.5\ mm$$

For combined *dead and live loading*:

$$S_{i(clay)} = 6.5 \times \frac{133.2}{81.4} = 10.6\ mm$$

(3) Calculating consolidation settlements ($S_c = KS_{oed}$)
Consider the first 3 m clay layer:

Equation (2.20)-(a): $\sigma_{av} = qI_a$ (Below corner of the loaded area)

$$m = \frac{B'/2}{H} = \frac{12.3/2}{3} = 2.05;\ \text{and}\ n = \frac{L'/2}{H} = \frac{14/2}{3} = 2.33.$$

From Figure 2.38: $I_a = 0.248$
For *dead loading* only, below centre of the loaded area:

$$\sigma_{av} = 4qI_a = 4 \times 81.4 \times 0.248 = 80.7\,\text{kN/m}^2$$

For combined *dead and live loading*, below centre of the loaded area:

$$\sigma_{av} = 4qI_a = 4 \times 133.2 \times 0.248 = 132.1\,\text{kN/m}^2$$

Consider the remaining 6 m clay layer:

Equation (2.20)-(b): $\sigma_{av} = q\left[\dfrac{H_2 I_{a(H_2)} - H_1 I_{a(H_1)}}{H_2 - H_1}\right]$ (Below corner of the loaded area)

$$m = \frac{\frac{B'}{2}}{H_2} = \frac{\frac{12.3}{2}}{9} = 0.68\ \text{and}\ n = \frac{\frac{L'}{2}}{H_2} = \frac{\frac{14}{2}}{9} = 0.78;\ \text{from Figure 2.38}$$

$$I_{a(H_2)} = 0.2$$

$$m = \frac{\frac{B'}{2}}{H_1} = \frac{\frac{12.3}{2}}{3} = 2.05;\ \text{and}\ n = \frac{\frac{L'}{2}}{H_1} = \frac{\frac{14}{2}}{3} = 2.33.\ \text{From Figure 2.38:}$$

$$I_{a(H_1)} = 0.248$$

For *dead loading* only, below centre of the loaded area:

$$\sigma_{av} = 4 \times 81.4\left[\frac{9 \times 0.2 - 3 \times 0.248}{9-3}\right] = 57.3\,\text{kN/m}^2$$

For combined *dead and live loading*, below centre of the loaded area:

$$\sigma_{av} = 4 \times 133.2\left[\frac{9 \times 0.2 - 3 \times 0.248}{9-3}\right] = 93.8\,\text{kN/m}^2$$

$$S_{oed} = H \times m_v \times \sigma_{av}$$

$$S_c = K \times S_{oed}$$

(Continued)

Since the clay is overconsolidated, we can take $K = 0.5$.

Net contact pressure, kN/m^2	Depth of layer, m	Layer thickness, m	m_v m^2/kN	σ_{av} kN/m^2	S_{oed} mm	K	S_c mm
220	9–12	3	0.000 11	80.7	26.6	0.5	13.3
	12–18	6	0.000 03	57.3	10.3	0.5	5.2
360	9–12	3	0.000 20	132.1	79.3	0.5	39.7
	12–18	6	0.000 04	93.8	22.5	0.5	11.3

The total final settlement of the bridge pier is given by
S = immediate settlement in sand and gravel
+ Immediate settlement in clay
+ Consolidation settlement in clay
For the *dead loading* (*net* $q = 220\,\text{kN/m}^2$):

$$S = 6.5 + 6.5 + (13.3 + 5.2) = 31.5\ mm$$

For the *combined dead and live loadings* (*net* $q = 360\,\text{kN/m}^2$):

$$S = 10.6 + 10.6 + (39.7 + 11.3) = 72.2\ mm$$

The results may be interpreted such that a settlement of 13 mm will take place as the pier is constructed and this will increase to about 22 mm if and when the bridge sustains its maximum live load. The remaining settlement due to long-term consolidation of the clay under the dead load and sustained live load will take a long period of time to attain its final value of about 50 mm.

Problem 3.8

A compacted fill of 8.5 m height is to be placed over a large area. The soil profile under the filled area is shown in the scheme below. Oedometer tests on samples from points A and B produced the following results:

	C_c	C_r	e_o	$\sigma_c'(kPa)$
Sample A	0.25	0.08	0.66	101
Sample B	0.2	0.06	0.45	510

Compute S_{oed} and make an estimate of S_i and S_c.

Solution:
The loaded area is large and, therefore, B/z and L/z values will be large too (say three or above). Assume the same vertical pressure q will be transmitted to any point located in the concerned subsurface layers below centre of the filled area. Accordingly,

$$\sigma_z' = q = 20.3 \times 8.5 = 172.6\,\text{kPa}$$

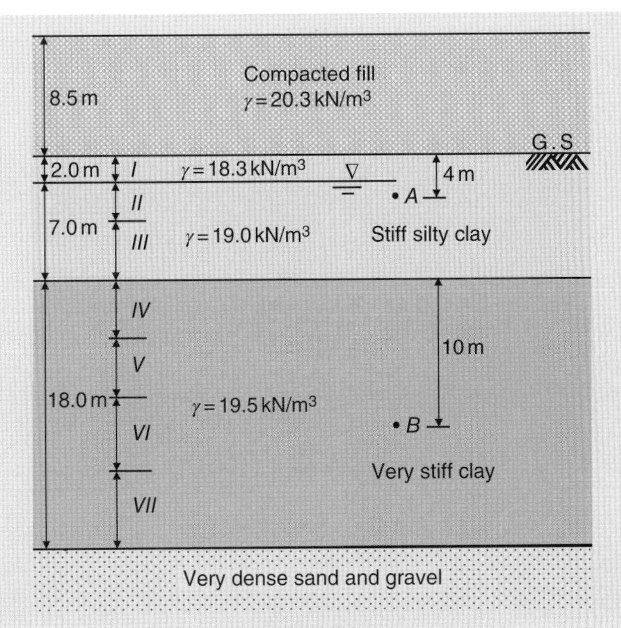

Scheme 3.5

At location of sample *A*:
$\sigma'_o = 2 \times 18.3 + 2(19 - 10) = 54.6\,\text{kPa} < \sigma'_c = 101\,\text{kPa}$. Hence, the soil is overconsolidated.

$$\sigma'_1 = \sigma'_o + \sigma'_z = 54.6 + 172.6 = 227.2\,\text{kPa} > \sigma'_c \rightarrow \text{Case II}$$

Equation (3.38):
$$S_c = \frac{C_r H}{1 + e_o}\log\frac{\sigma'_c}{\sigma'_o} + \frac{C_c H}{1 + e_o}\log\frac{\sigma'_o + \sigma'_z}{\sigma'_c}$$

At location of sample *B*:
$\sigma'_o = 2 \times 18.3 + 7(19 - 10) + 10(19.5 - 10) = 194.6\,\text{kPa} < \sigma'_c = 510\,\text{kPa}$. Hence, the soil is overconsolidated.

$$\sigma'_1 = \sigma'_o + \sigma'_z = 194.6 + 172.6 = 367.2\,\text{kPa} < \sigma'_c \rightarrow \text{Case I}$$

Equation (3.37):
$$S_c = \frac{C_r H}{1 + e_o}\log\frac{\sigma'_o + \sigma'_z}{\sigma'_o}$$

In order to compute settlements more accurately, the stiff silty clay is divided into three layers and the very stiff clay into four layers, as shown in the scheme above and as indicated in Table 3.21.

Calculate σ'_o, σ'_c and $(\sigma'_o + \sigma'_z)$ at midpoint of each layer and determine S_{oed} using Equations (3.37) and (3.38) as applicable.

For the Case II overconsolidation, it is required to determine the σ'_c value at midpoint of each layer, as follows:

Layer *I*; $\sigma'_c = 101 - [1 \times 18.3 + 2(19 - 10)] = 64.7\,\text{kPa}$
Layer *II*; $\sigma'_c = 101 - [0.5(19 - 10)] = 96.5\,\text{kPa}$
Layer *III*; $\sigma'_c = 101 + [3(19 - 10)] = 128.0\,\text{kPa}$

(Continued)

Table 3.21 Division of stiff silty clay and very stiff clay into layers.

Layer	H (m)	σ'_o (kPa)	σ'_c (kPa)	$\sigma'_o + \sigma'_z$ (kPa)	$\frac{C_r}{1+e_o}$	$\frac{C_c}{1+e_o}$	Eq.	S_{oed} mm
I	2.0	18.3	64.7	190.9	0.05	0.15	(3.38)	196
II	3.0	50.1	96.5	222.7	0.05	0.15	(3.38)	206
III	4.0	81.6	128.0	254.2	0.05	0.15	(3.38)	218
IV	4.0	118.6	--	291.2	0.04	0.14	(3.37)	62
V	4.0	156.6	--	329.2	0.04	0.14	(3.37)	52
VI	5.0	199.4	--	372.0	0.04	0.14	(3.37)	54
VII	5.0	246.9	--	419.5	0.04	0.14	(3.37)	46

Total $S_{oed} = 834\,mm$

The clays are stiff and overconsolidated; hence, according to Burland *et al.* (1977), the immediate settlement S_i and the consolidation settlement S_c may be estimated, approximately, as

$$S_i = 0.5\ S_{oed} = 0.5 \times 834 = 417\,mm$$

$$S_c = 0.5\ S_{oed} = 0.5 \times 834 = 417\,mm$$

Also, the S_c may be estimated, approximately, using Equation (3.42) as follows:

According to Craig (1995), value the settlement ratio K for lightly over-consolidated clays is typically between 0.5 and 0.7. Values of σ'_c and σ'_o indicate that the clays are lightly overconsolidated ($OCR < 3$). Hence, an average value of $K = 0.6$ may be used to estimate, approximately,

$$S_c = 0.6\ S_{oed} = 0.6 \times 834 = 500\,mm$$

$$S_i = 0.4 \times 834 = 334\,mm$$

Problem 3.9

A compacted fill of 3.0 m height is to be placed over a large area. The soil profile under the area is shown in the scheme below.

An oedometer test on a sample from point A produced the following results:

	C_c	C_r	e_o	$\sigma'_c(kPa)$
Sample A:	0.40	0.08	1.10	70.0

These results can be used in settlement calculations for the soft clay stratum as a whole.

Compute the total settlement below centre of the area due to the weight of the fill.

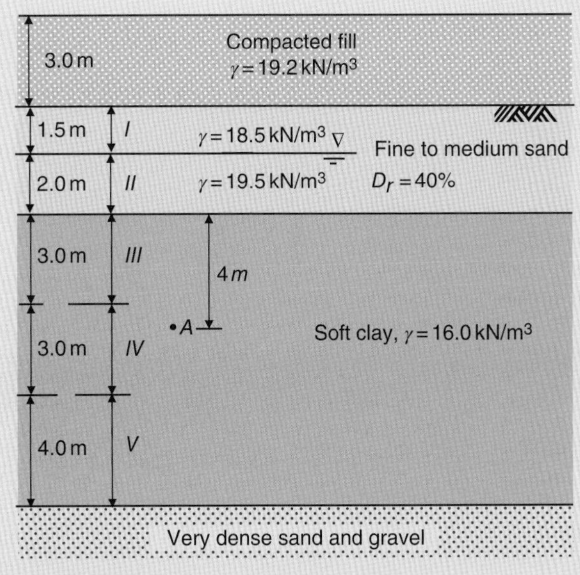

Scheme 3.6

Solution:
The loaded area is large and therefore, B/z and L/z values will be large too (say three or above). Assume the same vertical pressure q will be transmitted to any point located in the concerned subsurface layers below centre of the filled area. Accordingly,

$$\sigma_z' = q = 19.2 \times 3 = 57.6\,\text{kPa}$$

In order to compute the settlements more accurately, the fine to medium sand is divided into layers *I* and *II*, and the soft clay layers *III*, *IV* and *V*, as shown in the scheme above.

(1) Settlement in the fine to medium sand stratum.
Assume using the Buisman–DeBeer method:

Equation (3.20): $$S_i = \frac{H}{C}\ln\frac{\sigma_o' + \sigma_z}{\sigma_o'}$$

Equation (3.23): $C = 1.9\dfrac{q_c}{\sigma_o'}$ (Recommended by Meyerhof, 1965)

$$S_i = \frac{\sigma_o' H}{1.9 q_c}\ln\frac{\sigma_o' + \sigma_z}{\sigma_o'} = \frac{2.3\sigma_o' H}{1.9 q_c}\log\frac{\sigma_o' + \sigma_z}{\sigma_o'}.$$

Here, $\sigma_z = q = 57.6\,\text{kPa}$
An adequate value for q_c may be estimated from D_r and q_c correlations. For this purpose let us use the average of results using the relationships given by Equation (1.50) and Figure 1.17, as follows:

Equation (1.50): $$D_r(\%) = 66 \times \log\frac{q_c}{\sqrt{\sigma_o'}} - 98,$$

where q_c and σ_o' are in t/m^2

(*Continued*)

At midpoint of Layer *I*:

$$\sigma'_O = 0.75 \times 18.5 = 13.9\,\text{kPa} = 1.39\,\text{t/m}^2$$

$$40 = 66 \times \log\frac{q_c}{\sqrt{1.39}} - 98 \rightarrow \log(0.85q_c) = 2.09 \rightarrow$$

$$q_c = 143.53\,\text{t/m}^2 = 1435\,\text{kPa}$$

From Figure 1.17; $q_c = 1.333\frac{\text{MN}}{\text{m}^2} = 1333\,\text{kPa}$

Use $q_c = \frac{1435 + 1333}{2} = 1384\,\text{kPa}$

$$S_{i(\text{I})} = \frac{2.3 \times 1.5 \times 13.9}{1.9 \times 1384}\log\frac{13.9 + 57.6}{13.9} = 0.013\,\text{m}$$

At mid-point of Layer *II*:

$$\sigma'_O = 1.5 \times 18.5 + 1(19.5 - 10) = 37.25\,\text{kPa} = 3.73\,\text{t/m}^2$$

$$40 = 66 \times \log\frac{q_c}{\sqrt{3.73}} - 98 \rightarrow \log(0.52q_c) = 2.09 \rightarrow$$

$$q_c = 234.6\ \text{t/m}^2 = 2346\,\text{kPa}$$

From Figure 1.17, $q_c = 2.667\frac{\text{MN}}{\text{m}^2} = 2667\,\text{kPa}$

Use $q_c = \frac{2346 + 2667}{2} = 2507\,\text{kPa}$

$$S_{i(\text{II})} = \frac{2.3 \times 2 \times 37.25}{1.9 \times 2507}\log\frac{37.25 + 57.6}{37.25} = 0.015\,\text{m}$$

Settlement in the fine to medium sand stratum

$$S_{\text{sand}} = S_{i(I)} + S_{i(II)} = 0.013 + 0.15 = 0.028\,\text{m} = 28\ mm$$

(2) Settlement in the soft clay stratum:
At sample *A*: $\sigma'_o = 1.5 \times 18.5 + 2(19.5 - 10) + 4(16 - 10) = 70.75\,\text{kPa} \approx p'_c = 70.0\,\text{kPa}$. Hence, the clay is considered normally consolidated.

Calculate σ'_o and $(\sigma'_o + \sigma'_z)$ at midpoint of layers and determine S_{oed} using Equation (3.36)-(a)

Equation (3.36)-(a): $$S_c = \frac{C_c H}{1 + e_o}\log\frac{\sigma'_o + \sigma'_z}{\sigma'_o}$$

Layer	H (m)	σ'_o (kPa)	$(\sigma'_o + \sigma'_z)$ (kPa)	$\frac{C_c}{1+e_o}$	Eq.	S_{oed} (mm)
III	3	55.8	113.4	0.19	(3.36)	176
IV	3	73.8	131.4	0.19	(3.36)	143
V	4	94.8	152.4	0.19	(3.36)	157

$\Sigma\, S_{oed} = \underline{476\,\text{mm}}$

Scheme 3.7

The total settlement below centre of the area due to the weight of the fill = $S_{\text{total}} = S_{\text{sand}} + S_{oed} = 28 + 476 = 504\ mm$

Problem 3.10

The scheme below shows results of a CPT sounding performed at a certain site. The soil profile consists of normally consolidated sands with some interbedded silts. The groundwater table exists at a depth of 2.0 m below the ground surface. A long footing of 2.5 × 30.0 m is required to support a total load, including weights of the footing and backfill soil, equal to 197 kN/m^2 and will be founded at a depth of 2.0 m. Use Schmertmann's method to compute the settlement of this footing soon after construction and the settlement 50 years after construction.

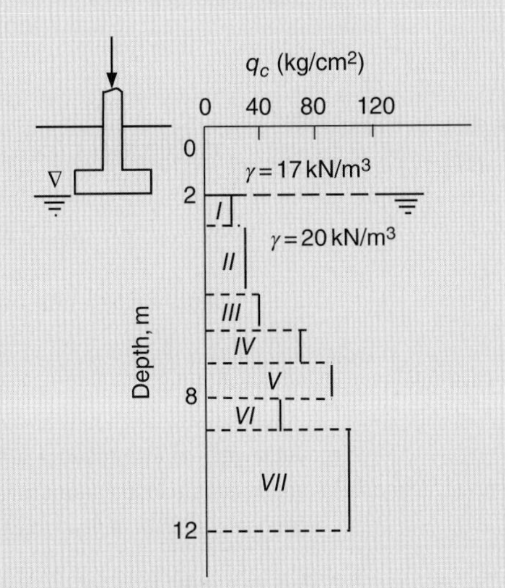

Layer	*I*	*II*	*III*	*IV*	*V*	*VI*	*VII*
Depth (m)	2–3	3–5	5–6	6–7	7–8	8–9	9–12
q_c (kg/cm^2)	20	30	41	68	90	58	108

Scheme 3.8

Solution:

Equation (3.24):

$$S_i = C_1 C_2 C_3 q_{net} \sum_{z=0}^{z=z_2} \frac{I_z}{E_s} \Delta z$$

According to Schmertmann, for long footings $\left(\frac{L}{B} > 10\right)$; $E_s = 3.5\, q_c$.

Values of E_s for each layer will be calculated (using 1 kPa = 0.01 kg/cm^2).
Influence depth = 4*B* below the foundation level. Hence, all the seven layers should be considered in the settlement computations.

Layer:	***I***	***II***	***III***	***IV***	***V***	***VI***	***VII***
E_s (kPa):	7000	10 500	14 350	23 800	31 500	20 300	37 800

(Continued)

Equation (3.25): $I_{zp} = 0.5 + 0.1\sqrt{\frac{q_{net}}{\sigma'_{zp}}}$

$$q_{net} = q - \sigma'_o = 197 - 2 \times 17 = 163\,\text{kPa}$$

For long footings, σ'_{zp} is calculated at depth B below foundation level; hence, $\sigma'_{zp} = 2 \times 17 + 2.5(20 - 10) = 59\,\text{kPa}$

$$I_{zp} = 0.5 + 0.1\sqrt{\frac{163}{59}} = 0.666$$

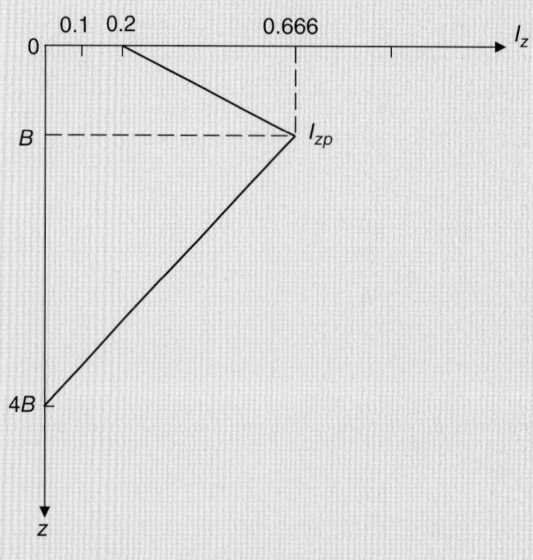

Scheme 3.9

Considering continuous foundations $(L/B \geq 10)$:

$$\text{For } z = 0 \text{ } to \text{ } B \quad I_z = 0.2 + \left(\frac{z}{B}\right)(I_{zp} - 0.2)$$

$$\text{For } z = B \text{ } to \text{ } 4B \quad I_z = 0.333 I_{zp}\left(4 - \frac{z}{B}\right)$$

Determine value of $\left(\frac{I_z}{E_s}\Delta z\right)$ at midpoint of each layer, as shown in Table 3.22

$$C_1 = 1 - \frac{0.5\sigma'_o}{q_{net}} = 1 - \frac{0.5(2 \times 17)}{163} = 0.896$$

$$C_2 = 1 + 0.2\log\frac{t}{0.1}$$

At t = 0.1 year $\rightarrow C_2 = 1$

$$C_3 = 1.03 - \frac{0.03L}{B} = 1.03 - \frac{0.03 \times 30}{2.5} = 0.67 < 0.73 \rightarrow \text{use } C_3 = 0.73$$

Table 3.22 Determining the value of $\left(\frac{I_z}{E_s}\Delta z\right)$ at the midpoint of each layer.

Layer	Es (kN/m²)	z (m)	I_z	Δz (m)	$I_z\ \Delta z/E_s$
I	7000	0.5	0.293	1.0	4.19×10^{-5}
II	10 500	2.0	0.573	2.0	10.91×10^{-5}
III	14 350	3.5	0.577	1.0	4.02×10^{-5}
IV	23 800	4.5	0.488	1.0	2.05×10^{-5}
V	31 500	5.5	0.399	1.0	1.27×10^{-5}
VI	20 300	6.5	0.310	1.0	1.53×10^{-5}
VII	37 800	8.5	0.133	3.0	1.06×10^{-5}

$\sum(I_z\ \Delta z/E_s) = 25.03 \times 10^{-5}$

$$S_i = C_1 C_2 C_3 q_{net} \sum_{z=0}^{z=z_2} \frac{I_z}{E_s}\Delta z = 0.896 \times 1 \times 0.73 \times 163 \times 25.03 \times 10^{-5}$$

$$S_i = 0.027\ \text{m} = 27\ mm$$

At $t = 50$ year: $C_2 = 1 + 0.2\log\frac{50}{0.1} = 1.54$

$$S_i = 27 \times 1.54 = 42\ mm$$

Problem 3.11

A 900-kN column load is to be supported on square footing founded at a depth of 1 m blow the ground surface. The foundation soil is a fairly homogeneous silty sand with an average N_{60} of 28 and $\gamma = 19\,\text{kN/m}^3$. Assume γ of the backfill material is approximately the same as that of concrete. The ground water table is at a depth of 15 m below the ground surface. The allowable total settlement is 20 mm. Determine the required footing width using the Schmertmann's method.

Solution:
Only a limited soil data is available and because the supporting soil is fairly homogeneous, it may be satisfactory to assume E_s constant with depth. Hence, Equation (3.31)-(a), derived from the Schmertmann's original equation, can be applied.

Equation (3.31)-(a): $$S_i = C_1 C_2 C_3 q_{net}\left[\frac{B\left(I_{zp} + 0.025\right)}{E_s}\right]$$

Estimate E_s from Equation (3.32):

$$E_s = \beta_o\ \sqrt{OCR} + \beta_1 N_{60} = 2500\ \sqrt{1} + 600 \times 28 = 19\,300\,\text{kPa}$$

(Continued)

As a first trial, assume $B = 2.0\,\text{m}$

$$q_{net} = q - \sigma'_o = \frac{900 + 2.0 \times 2.0 \times 1 \times 24}{2.0 \times 2.0} - 1 \times 19 = 230\,\text{kPa}$$

$$C_1 = 1 - \frac{0.5\sigma'_o}{q_{net}} = 1 - \frac{0.5 \times 1 \times 19}{230} = 0.959$$

Assume the total settlement will be completed after 50 years;

$$C_2 = 1 + 0.2\log\frac{t}{0.1} = 1 + 0.2\log\frac{50}{0.1} = 1.54$$

$$C_3 = 1.03 - \frac{0.03L}{B} = 1.03 - \frac{0.03 \times 2.5}{2.5} = 1$$

Equation (3.25): $$I_{zp} = 0.5 + 0.1\sqrt{\frac{q_{net}}{\sigma'_{zp}}}$$

For square footings, σ'_{zp} is calculated at depth $0.5B$ below foundation level; hence $\sigma'_{zp} = 1 \times 19 + 0.5 \times 2 \times 19 = 38\,\text{kPa}$

$$I_{zp} = 0.5 + 0.1\sqrt{\frac{q_{net}}{\sigma'_{zp}}} = 0.5 + 0.1\sqrt{\frac{230}{38}} = 0.746$$

$$S_i = 0.02 = 0.959 \times 1.54 \times 1 \times 230\left[\frac{B(0.746 + 0.025)}{19300}\right] = 0.0136B$$

$$B = 1.47\,\text{m} < 2.0\ m \rightarrow increase\ B$$

As a second trial, assume $B = 2.5\,\text{m}$

$$q_{net} = q - \sigma'_o = \frac{900 + 2.5 \times 2.5 \times 1 \times 24}{2.5 \times 2.5} - 1 \times 19 = 149\,\text{kPa}$$

$$C_1 = 1 - \frac{0.5\sigma'_o}{q_{net}} C_1 = 1 - \frac{0.5 \times 1 \times 19}{149} = 0.936$$

Assume the total settlement will be completed after 50 years;

$$C_2 = 1 + 0.2\log\frac{t}{0.1} = 1 + 0.2\log\frac{50}{0.1} = 1.54$$

$$C_3 = 1.03 - \frac{0.03L}{B} = C_3 = 1.03 - \frac{0.03 \times 2.5}{2.5} = 1$$

$$I_{zp} = 0.5 + 0.1\sqrt{\frac{q_{net}}{\sigma'_{zp}}} = 0.5 + 0.1\sqrt{\frac{149}{19\left(1 + \frac{2.5}{2}\right)}} = 0.69$$

$$S_i = 0.02 = 0.936 \times 1.54 \times 1 \times 149\left[\frac{B(0.69 + 0.025)}{19\,300}\right] = 0.008B$$

$$B = 2.5\,\text{m} = \text{the assumed value.}$$

Use the footing width $B = 2.5\,m$

Problem 3.12

Solve Problem 3.10 using the Buisman–DeBeer method.

Solution:

Equation (3.20): $$S_i = \frac{H}{C}\ln\frac{\sigma'_o + \sigma_z}{\sigma'_o}$$

Equation (3.23): $C = 1.9\frac{q_c}{\sigma'_o}$ (Recommended by Meyerhof, 1965)

$$S_i = \sum_0^H \frac{2.3\sigma'_o}{1.9q_c}\,\Delta z\log\frac{\sigma'_o + \sigma_z}{\sigma'_o}$$

Refer to Table 3.23. The depth z (below ground level) to the centre of each layer is obtained from the figure of Problem 10. The effective overburden pressure σ'_o and the stress increment σ_z at the centre of each layer are calculated. The σ_z-values are obtained using either Figure 2.32 or Table 2.3. Values of q_c in kg/cm^2 are converted to their corresponding values in kPa using 1 kPa = 0.01 kg/cm^2. The above settlement equations are applied to obtain the total settlement.

Table 3.23 Data for solving Problem 3.12.

Layer	z (m)	Δz (m)	σ'_o (kPa)	σ_z (kPa)	$\log\frac{\sigma'_o + \sigma_z}{\sigma'_o}$	q_c (kPa)	$\frac{2.3\sigma'_o}{1.9q_c}\Delta z$ (mm)	S_i (mm)
I	2.5	1.0	39	158.0	0.703	2000	23.6	16.6
II	4.0	2.0	54	104.3	0.467	3000	43.6	20.4
III	5.5	1.0	69	68.5	0.300	4100	20.4	6.1
IV	6.5	1.0	79	57.4	0.237	6800	14.1	3.3
V	7.5	1.0	89	45.6	0.180	9000	12.0	2.2
VI	8.5	1.0	99	38.5	0.143	5800	20.7	3.0
VII	10.5	3.0	119	31.3	0.101	10800	40.0	4.0
$\sum S_i = 55.6$								

The total settlement = *56 mm*

This result, compared with that obtained by Schmertmann's method (see Solution of Problem 3.10), is considered too conservative.

Problem 3.13

A column load of 500 kN is supported on a 1.5 × 2.0 m spread footing. The foundation depth equals 0.5m. Assume γ of the backfill material is approximately the same as γ of concrete. The soil profile consists of clean, well graded, normally consolidated sand with γ = 17 kN/m^3 and the following SPT results:

Depth (m):	1.0	2.0	3.0	4.0	5.0
N_{60}, (blow):	12	13	13	18	22

(Continued)

The ground water table is at a depth of 25 m. Using Schmertmann's method, compute the total settlement at t = 30 years.

Solution:

Equation (1.33): $$\frac{E_s}{P_a} = \alpha N_{60} \rightarrow E_s = \alpha P_a N_{60}; P_a = 100\,\text{kPa}$$

$$\alpha = 10\,\text{for normally consolidated sands} \rightarrow E_s = 1000\,N_{60}$$

Equation (3.32): $E_s = \beta_o\,\sqrt{OCR} + \beta_1 N_{60}$
$\beta_o = 5000$ and $\beta_1 = 1200$ (for clean normally consolidated sands)

$$E_s = 5000\,\sqrt{1} + 1200N_{60} = 5000\,+1200N_{60}$$

It may be a reasonable judgment to use the average of E_s values computed from these two equations, as presented below:

Depth (m):	1.0	2.0	3.0	4.0	5.0
E_s, (kPa):	15 700	16 800	16 800	22 300	26 700

$$L/B = 2/1.5 = 1.33$$

$$1 < L/B < 10.$$

Hence,

$$I_z = I_{zs} + 0.111(I_{zc} - I_{zs})\left(\frac{L}{B} - 1\right)$$

Considering square and circular foundations:

$$\text{For}\ z = 0\ \text{to}\ \frac{B}{2} \qquad I_{zs} = 0.1 + \left(\frac{z}{B}\right)(2I_{zp} - 0.2)$$

$$\text{For}\ z = \frac{B}{2}\ \text{to}\ 2B \qquad I_{zs} = 0.667I_{zp}\left(2 - \frac{z}{B}\right)$$

Considering continuous foundations ($L/B \geq 10$):

$$\text{For}\ z = 0\ \text{to}\ B \qquad I_{zc} = 0.2 + \left(\frac{z}{B}\right)(I_{zp} - 0.2)$$

$$\text{For}\ z = B\ \text{to}\ 4B \qquad I_{zc} = 0.333I_{zp}\left(4 - \frac{z}{B}\right)$$

$$I_{zp} = 0.5 + 0.1\sqrt{\frac{q_{net}}{\sigma'_{zp}}}$$

$$q_{net} = q - \sigma'_o = [500/(1.5 \times 2)] + 0.5 \times 24 - 0.5 \times 17 = 170.2\,\text{kPa}$$

For square footings: σ'_{zp} is calculated at depth $0.5B$ below foundation level; hence, $\sigma'_{zp} = 0.5 \times 17 + (1.5/2)(17) =$ 21.25 kPa

$$I_{zp} = 0.5 + 0.1\sqrt{\frac{170.2}{21.25}} = 0.783$$

For long footings: σ'_{zp} is calculated at depth B below foundation level; hence, $\sigma'_{zp} = 0.5 \times 17 + 1.5 \times 17 = 34$ kPa

$$I_{zp} = 0.5 + 0.1\sqrt{\frac{170.2}{34}} = 0.724$$

$$S_i = C_1 C_2 C_3 q_{net} \sum_{z=0}^{z=z_2} \frac{I_z}{E_s} \Delta z$$

Layer No.	E_S (kPa)	z (m)	I_{zs}	I_{zc}	I_z	Δz (m)	$I_z \Delta z/E_s$
1	15700	0.5	0.555	0.375	0.548	1	3.49×10^{-5}
2	16800	1.5	0.522	0.724	0.529	1	3.15×10^{-5}
3	16800	2.5	0.174	0.562	0.188	1	1.12×10^{-5}
4	22300	3.5	0	0.402	0.015	1	0.07×10^{-5}
5	26700	4.5	0	0.241	0.009	1	0.03×10^{-5}

$\Sigma\ (I_z \Delta z/E_s) = 7.86 \times 10^{-5}$

Scheme 3.10

$$C_1 = 1 - \frac{0.5\sigma'_o}{q_{\text{net}}} = 1 - \frac{0.5(0.5 \times 17)}{170.2} = 0.975$$

$$C_2 = 1 + 0.2\log\frac{t}{0.1} = 1 + 0.2\log\frac{30}{0.1} = 1.495$$

$$C_3 = 1.03 - \frac{0.03L}{B} = 1.03 - \frac{0.03 \times 2}{2.5} = 1.006 > 0.73$$

$$S_i = 0.975 \times 1.495 \times 1.006 \times 170.2 \times 7.86 \times 10^{-5}$$

$$S_i = 0.02\ \text{m} = 20\ \text{mm}$$

Total settlement at $t = 30$ years is *20 mm*

Values of ($I_z\ \Delta z/\ E_s$) indicate that the influence depth is, practically, equals to $2B$ (or 3 m) because the footing is too close to a square footing than to a long or continuous footing.

Problem 3.14

A raft 30 × 40 m supports a building. The net contact pressure equals 120 kPa; assumed uniformly distributed over the building area. The soil profile is shown in the scheme below. The laboratory test results gave an average value of m_v for the clay equals 0.32 m^2/MN. Determine the final settlement under the centre of the raft due to consolidation of the clay.

(*Continued*)

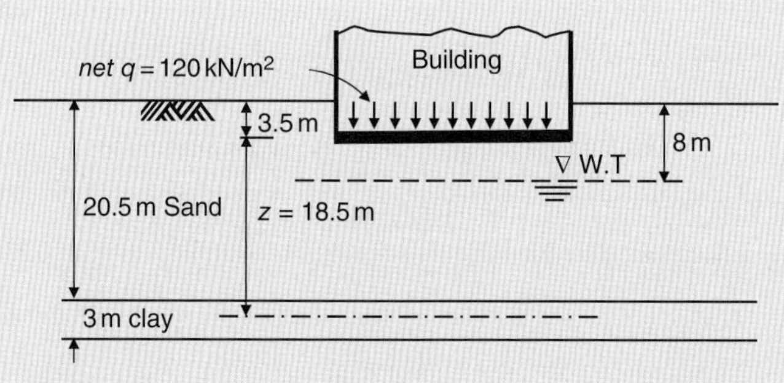

Scheme 3.11

Solution:
The clay layer thickness of 3 m is considered thin relative to the raft dimensions. Therefore, the condition of zero lateral strain is approximately satisfied and the consolidation process can be assumed one dimensional. Also, for this clay layer, it will be practically justified to consider the layer as a whole when consolidation settlement is computed.

Equation (3.34): $S_c = m_v \sigma'_z H$

σ'_z = The average stress increment in the clay layer due to the net contact pressure q; may be determined from Equation (2.20)-(b), as follows:

$$\sigma_{av} = q\left[\frac{H_2 I_{a(H_2)} - H_1 I_{a(H_1)}}{H_2 - H_1}\right] \text{ (Below corner of the loaded area)}$$

Refer to Figure 2.39:

$$I_{a(H_2)} : m = \frac{B/2}{H_2} = \frac{15}{20} = 0.75;\ n = \frac{L/2}{H_2} = \frac{20}{20} = 1 \rightarrow I_{a(H_2)} = 0.215$$

$$I_{a(H_1)} : m = \frac{B/2}{H_1} = \frac{15}{17} = 0.88;\ n = \frac{L}{H_1} = \frac{20}{17} = 1.18 \rightarrow I_{a(H_1)} = 0.223$$

Under the raft centre:

$$\sigma'_z = \sigma_{av} = 4 \times 120\left(\frac{20 \times 0.215 - 17 \times 0.223}{20 - 17}\right) = 81.44\,\text{kPa}$$

The final settlement due to consolidation of the clay layer is

$$S_c = \frac{0.32}{1000} \times 81.44 \times 3 = 0.078\,\text{m} = 78\ mm.$$

Problem 3.15

A net contact pressure equals 160 kN/m^2 is applied to a stiff clay layer 15 m thick. The footing is 6 m square founded at 2 m depth below ground surface. A layer of silty sand, 2 m thick, overlies the clay and a firm stratum lies immediately below the clay. Oedometer tests on specimens of the clay gave the value of $m_v = 0.13$ m^2/MN, and triaxial tests gave the value of pore pressure parameter $A = 0.35$. The undrained Young's modulus for the clay E_u was estimated equal 55 MN/m^2. Determine the total settlement under the footing centre.

Solution:

In order to compute the consolidation settlement more accurately, assume the clay is divided into five layers of equal thicknesses, as shown in the figure below.

The clay layer is relatively thick; there will be significant lateral strain (resulting in immediate settlement in the clay) and therefore, it will be more appropriate if the Skempton–Bjerrum method is used.

$$S_{\text{total}} = S_i + S_c$$

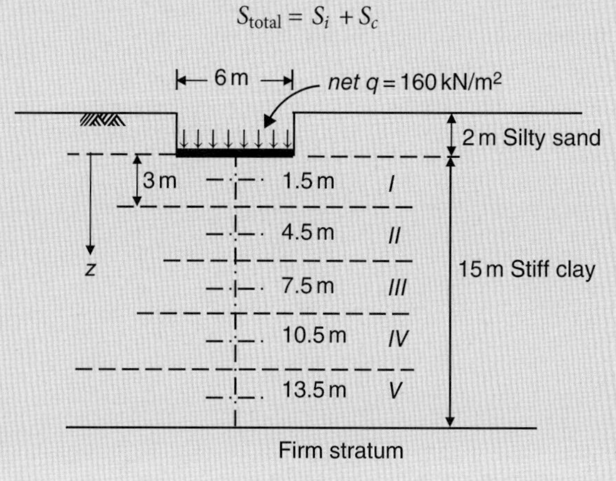

Scheme 3.12

(a) Immediate settlement.

Equation (3.33): $$S_i = \mu_o \mu_1 \frac{qB}{E_u}$$

$\frac{H}{B} = \frac{15}{6} = 2.5$; $\frac{D}{B} = \frac{2}{6} = 0.33$; $\frac{L}{B} = 1$; hence, from Figure 3.9:

$$\mu_o = 0.95 \text{ and } \mu_1 = 0.55$$

$$S_i = 0.95 \times 0.55 \times \frac{160 \times 6}{55 \times 1000} = 0.009 \text{ m} = 9 \text{ mm}$$

(b) Consolidation settlement.

Equation (3.40): $$S_c = K\, S_{oed}$$

$$\sum S_c = K \sum S_{oed}$$

For each layer:

$$S_{oed} = m_v \times \sigma'_z \times H = \frac{0.13}{1000} \times \sigma'_z \times 3 \times 1000 = 0.39\, \sigma'_z \text{ (mm)}$$

$\sigma'_z = q\,(4I)$; I is calculated from Table 2.3 or Figure 2.32.

The equivalent diameter of the base area = $\sqrt{\frac{4(6 \times 6)}{\pi}} = 6.77$ m

$$\frac{H}{B} = \frac{15}{6.77} = 2.2,\ A = 0.35$$

(Continued)

Layer	z (m)	I	σ'_z (kPa)	S_{oed} (mm)
I	1.5	0.233	149	58.1
II	4.5	0.121	78	30.4
III	7.5	0.060	38	14.8
IV	10.5	0.033	21	8.2
V	13.5	0.021	13	5.1

$\Sigma\ S_{oed} = 116.6$

Scheme 3.13

Figure 3.15 for $\dfrac{H}{B} = 2.2$ and $A = 0.35$, obtain $K = 0.55$

$$S_c = 0.55 \times 116.6 = 64\,\text{mm}$$

$$S_{\text{total}} = S_i + S_c = 9 + 64 = 73\ mm$$

Problem 3.16

A general soil profile at a certain site is shown in the scheme below. The lower sand layer is under artesian pressure, the piezometric level being 6 m above ground level. For the clay, $m_v = 0.94\,\text{m}^2/\text{MN}$ and $c_v = 1.4\,\text{m}^2/\text{year}$.

As a result of pumping from the artesian layer the piezometric level falls by 3 m over a period of 2 years. Draw the time-settlement curve due to consolidation of the clay for a period of 5 years from the start of pumping.

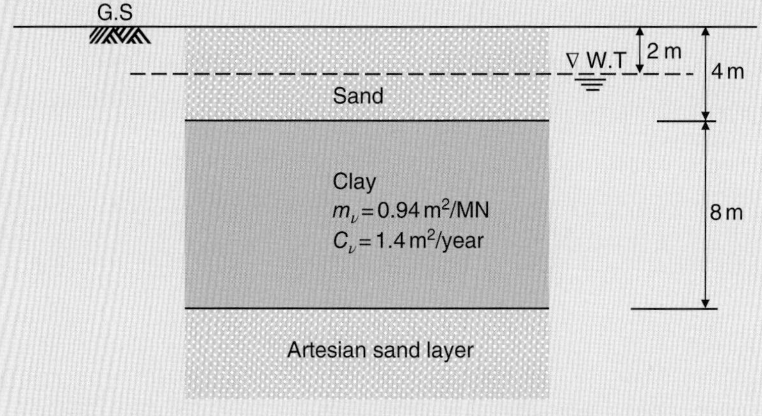

Scheme 3.14

Solution:

Consolidation of the clay layer occurs due to decrease in porewater pressure at the lower boundary of the clay as a consequence of pumping operation. As there is no change in total vertical stress; the effective vertical stress in the clay layer will increase with depth. The 3 m drop in piezometric level is equivalent to the decrease in the porewater pressure and increase in the effective vertical stress, which is $3\gamma_w$ at bottom of the clay layer, as shown below.

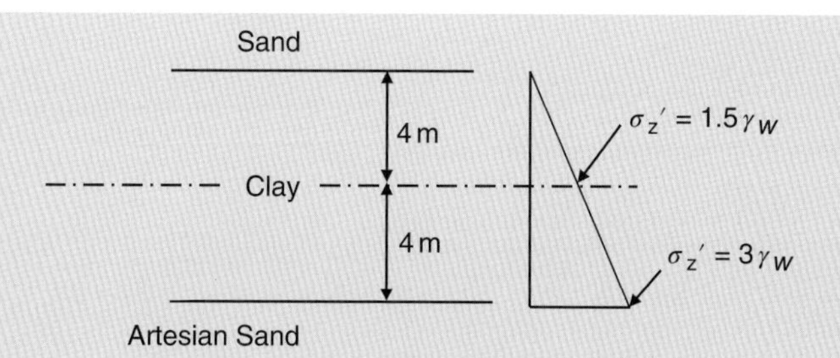

Scheme 3.15

σ_z' = increase in effective vertical stress

At the centre of the clay layer: $\sigma_z' = 1.5\gamma_w = 1.5 \times 10 = 15\,\text{kN/m}^2$

Equation (3.34):

$$S_c = m_v \times \sigma_z' \times H$$

$$S_c = \frac{0.94}{1000} \times 15 \times 8 = 0.113\,\text{m} = 113\,\text{mm}$$

Equation (3.44):

$$T_v = \frac{c_v t}{d^2}$$

$$T_v = \frac{c_v t}{d^2} = \frac{1.4 \times 5}{4^2} = 0.438$$

Equation (3.48):

$$T_v = \frac{\left(\frac{\pi}{4}\right)\left(\frac{U\%}{100}\right)^2}{\left[1-(U\%/100)^{5.6}\right]^{0.357}}$$

$$T_v = 0.438 = \frac{\left(\frac{\pi}{4}\right)\left(\frac{U\%}{100}\right)^2}{\left[1-(U\%/100)^{5.6}\right]^{0.357}}. \text{ Therefore, } U = 0.73$$

To obtain the time settlement curve, a series of values of U is selected up to 0.73 and the corresponding T_v calculated using either Equation (3.48) or Equations (3.46) and (3.47) as applicable. Then, the corresponding t values are calculated using Equation (3.44). The corresponding values of settlement (S_c) are given by the product of U and the calculated final settlement, as shown in Table 3.24.

Table 3.24 Data for the calculation of settlement values.

U	T_v	t (years)	S_c(mm)
0.10	0.008	0.09	11.3
0.20	0.031	0.35	22.8
0.30	0.070	0.79	33.9
0.40	0.126	1.42	45.2
0.50	0.196	2.21	56.5
0.60	0.285	3.22	67.8
0.73	0.438	5.00	82.5

(Continued)

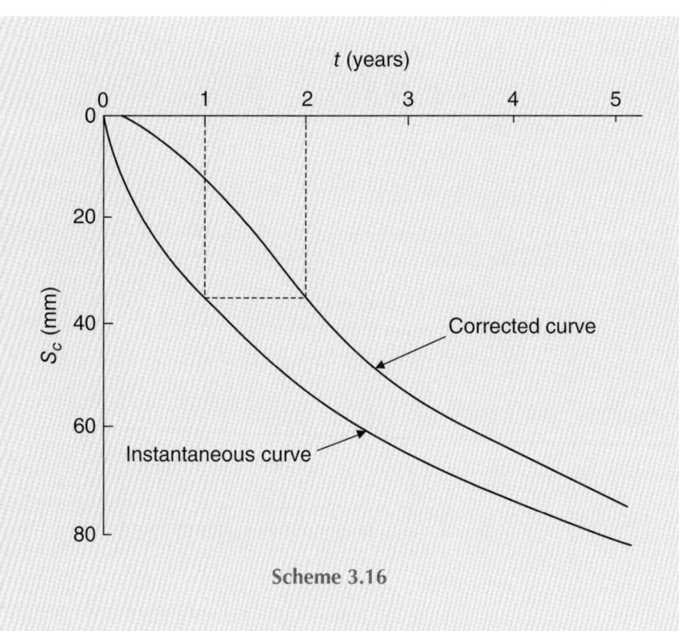

Scheme 3.16

Problem 3.17

At a certain site the soil profile consists of 8 m sand layer overlying a 6 m layer of normally consolidated clay, below which is an impermeable stratum. The water table exists at 2 m depth below the surface of the sand (ground surface), as shown in the scheme below.

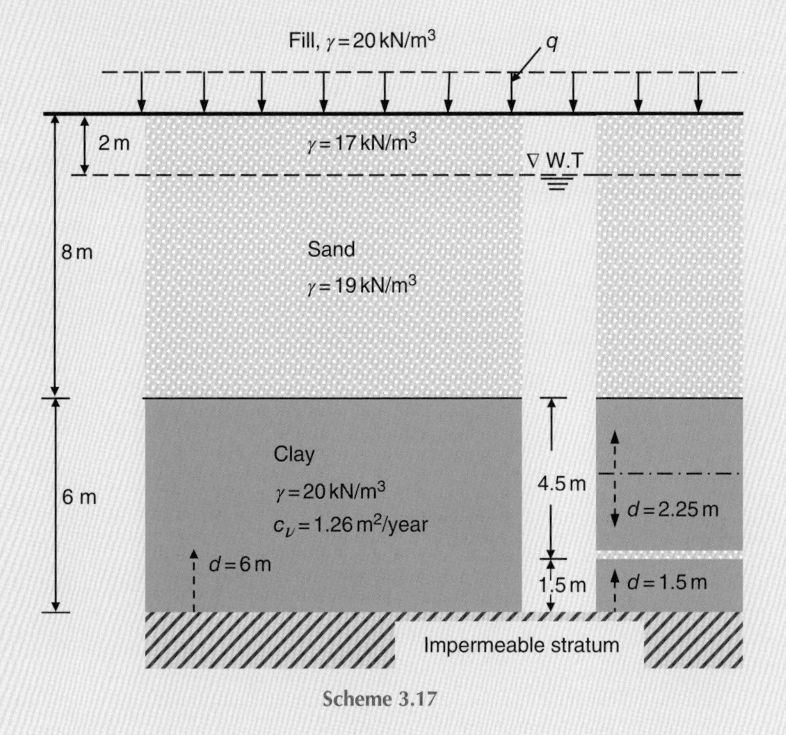

Scheme 3.17

The saturated unit weight of the sand = 19 kN/m^3 and that of the clay = 20 kN/m^3. The sand above water table has unit weight = 17 kN/m^3. For the clay, the relationship between void ratio and effective stress (in kN/m^2) can be represented by the equation $[e = 0.88 - 0.32\log(\sigma'/100)]$; and the coefficient of consolidation c_v = 1.26 m^2/year.

Over a period of 1 year a fill of 3 m depth is to be dumped on the surface over an extensive area. The fill has unit weight $\gamma = 20\,\text{kN/m}^3$.

(a) Calculate the final settlement of the area due to consolidation of the clay and the settlement after a period of 3 years from the start of dumping.

(b) If a very thin layer of sand, freely draining, exists 1.5 m above the bottom of the clay layer, what will be the final and 3 year settlements?

Solution:

(a) The fill covers a wide area; hence, the same fill pressure q is transmitted (approximately) to the concerned subsurface layers and the problem can be considered one-dimensional. Consider one-dimensional consolidation, taking the clay layer as a whole.

At the middle of the clay layer:

$$\sigma'_o = (17 \times 2) + 6(19 - 10) + 3(20 - 10) = 118\,\text{kN/m}^2$$

$$e_o = 0.88 - 0.32\log\left(\frac{118}{100}\right) = 0.857$$

From the given relationship it is clear that $C_c = 0.32$

Equation (3.36)-(a):
$$S_c = \frac{C_c H}{1 + e_o}\log\frac{\sigma'_o + \sigma'_z}{\sigma'_o}$$

Final consolidation settlement is

$$S_c = \frac{0.32 \times 6000}{1 + 0.857}\log\frac{118 + 3 \times 20}{118} = 185\,\text{mm}$$

The corrected value of time to allow for 1 year dumping period is

$$t = 3 - \frac{1}{2}(1) = 2.5\,\text{years}$$

The clay layer is half closed; hence, $d = 6\ m$

Equation (3.44):
$$T_v = \frac{c_v t}{d^2}$$
$$T_v = \frac{1.26 \times 2.5}{6^2} = 0.0875$$

Equation (3.48):
$$T_v = \frac{\left(\frac{\pi}{4}\right)\left(\frac{U\%}{100}\right)^2}{\left[1 - (U\%/100)^{5.6}\right]^{0.357}}$$
$$0.0875 = \frac{\left(\frac{\pi}{4}\right)\left(\frac{U\%}{100}\right)^2}{\left[1 - (U\%/100)^{5.6}\right]^{0.357}} \rightarrow U = 0.334$$

Settlement after 3 years = $U \times$ final consolidation settlement
Settlement after 3 years = 0.334 × 185 = *62 mm*

(Continued)

(b) Thickness of the drainage layer (i.e. the very thin layer of sand) can be ignored and from the point of view of drainage there is an open upper clay layer of thickness 4.5 m ($d = 2.25$ m) above a half closed lower clay layer of 1.5 m ($d = 1.5$ m), as shown in the scheme above.

Now, by proportion:

$T_{v1} = \dfrac{0.0875 \times 6^2}{2.25^2} = 0.622$. Equation (3.48) gives $U_1 = 0.825$ (upper layer; U)

$T_{v2} = \dfrac{0.0875 \times 6^2}{1.5^2} = 1.4$. Equation (3.48) gives $U_2 = 0.968$ (lower layer)

For each layer, S_c is proportional to HU. Therefore, if $\bar{U}$ is the overall degree of consolidation for the two layers combined:

$4.5U_1 + 1.5U_2 = 6 \times \bar{U}$

$4.5 \times 0.825 + 1.5 \times 0.968 = 6 \times \bar{U}$. Hence $\bar{U} = 0.86$ and the 3 year settlement is

$S_c = 0.86 \times 185 = 159.1$ mm; say *160 mm*

Problem 3.18

A 2.0 × 3.2 m flexible loaded area carries a uniformly distributed load = 210 kN/m^2, as shown in the scheme below. Estimate the elastic settlement below the centre of the loaded area. Assume that the foundation depth $D_f = 1.6$ m and $H = \infty$. Use Equation (3.5).

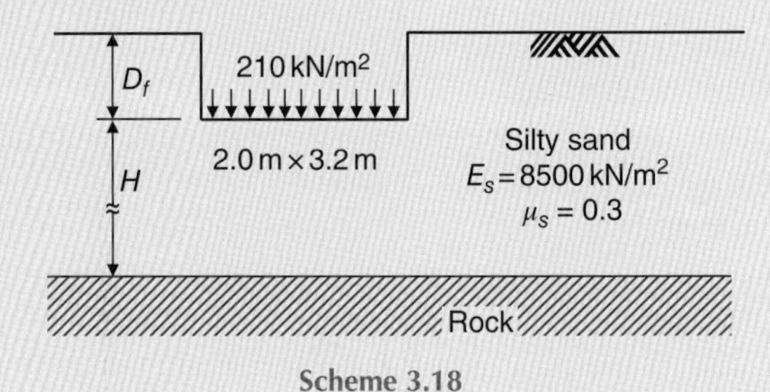

Scheme 3.18

Solution:

Elastic settlement S_i below the *centre* of the loaded area:

Equation (3.5): $$S_i = q \times B \times \frac{1-\mu^2}{E_s} \times m \times I_S \times I_F$$

$$m = 4;\quad \frac{B}{2} = \frac{2}{2} = 1 \text{ m};\quad \frac{L}{2} = \frac{3.2}{2} = 1.6 \text{ m};\quad M = \frac{\frac{L}{2}}{B/2} = \frac{1.6}{1} = 1.6;\quad B = \frac{2}{2} = 1$$

$$N = \frac{H}{B/2} = \frac{\infty}{1} = \infty$$

From Table 3.8 obtain $I_1 = 0.697$ and $I_2 = 0.0$; hence,

$$I_S = I_1 + \frac{1-2\mu}{1-\mu} I_2 = 0.697$$

$\dfrac{D}{B} = \dfrac{1.6}{2} = 0.8;\ \dfrac{B}{L} = \dfrac{2}{3.2} = 0.625,\ \mu = 0.3$. From Table 3.7: $I_F = 0.728$

$$S_i = 210 \times \frac{2}{2} \times \frac{1-(0.3)^2}{8500} \times 4 \times 0.697 \times 0.728 = 0.046 \text{ m} = 46 \textit{ mm}$$

Problem 3.19

Redo Problem 3.18, assuming that $D_f = 1.2$ m and $H = 4$ m.

Solution:

Elastic settlement S_i below the *centre* of the loaded area:

Equation (3.5):
$$S_i = q \times B \times \frac{1-\mu^2}{E_s} \times m \times I_S \times I_F$$

$$m = 4; \frac{B}{2} = \frac{2}{2} = 1\,\text{m}; \frac{L}{2} = \frac{3.2}{2} = 1.6\,\text{m}; M = \frac{\frac{L}{2}}{B/2} = \frac{1.6}{1} = 1.6; B = \frac{2}{2} = 1$$

$$N = \frac{H}{B/2} = \frac{4}{1} = 4$$

From Table 3.8 obtain $I_1 = 0.460$ and $I_2 = 0.057$; hence,

$$I_S = I_1 + \frac{1-2\mu}{1-\mu} I_2 = 0.460 + \left(\frac{1-2\times 0.3}{1-0.3}\right)(0.057) = 0.493$$

$\text{D/B} = \frac{1.2}{2} = 0.6; \frac{B}{L} = \frac{2}{3.2} = 0.625; \mu = 0.3$. From Table 3.7: $I_F = 0.778$

$$S_i = 210 \times \frac{2}{2} \times \frac{1-(0.3)^2}{8500} \times 4 \times 0.493 \times 0.778 = 0.0345\,\text{m} = 34.5\,mm$$

Problem 3.20

For a 3 × 3 m shallow foundation, supported by a layer of sand as shown in the scheme below, the following data are given:

$$D_f = 1.5\,\text{m}; t = 0.25\,\text{m}; E_o = 16\,000\,\text{kN/m}^2; E_f = 15 \times 10^6\,\text{kN/m}^2$$

$$k = 400\,(\text{kN/m}^2)/\text{m depth}; q_o = 150\,\text{kN/m}^2; \mu_s = 0.3; H = 20\,\text{m}$$

Calculate the elastic settlement.

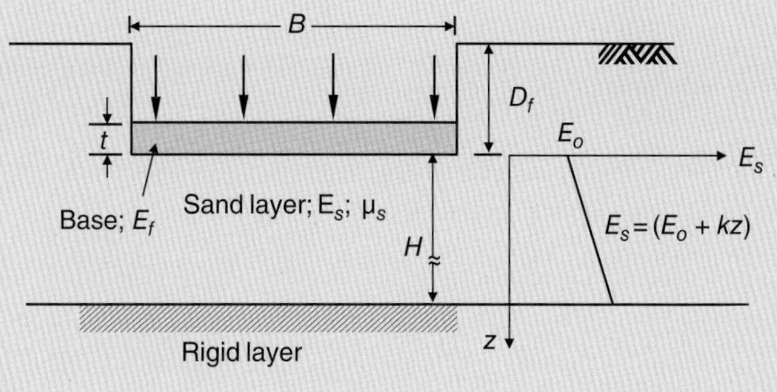

Scheme 3.19

(Continued)

Solution:

Equation (3.8): $S_i = \dfrac{q_o B_e I_G I_F I_E}{E_o}(1-\mu_s^2)$

$$B_e = \sqrt{\frac{4BL}{\pi}} = \sqrt{\frac{4\times3\times3}{\pi}} = 3.384\,\text{m}$$

$$\beta = \frac{E_o}{kB_e} = \frac{16\ 000}{400\times3.384} = 11.82;\ \frac{H}{B_e} = \frac{20}{3.384} = 5.91.$$

Figure 3.4 obtain $I_G = 0.85$

Equation (3.9):

$$I_F = \frac{\pi}{4} + \frac{1}{4.6 + 10\{E_f/[E_o + (B_e k/2)]\}(2t/B_e)^3}$$

$$= \frac{\pi}{4} + \frac{1}{4.6 + 10\{(15\times10^6)/[16\ 000 + (3.384\times400)/2]\}[(2\times0.25)/3.384]^3}$$

$$= 0.816$$

Equation (3.10):

$$I_E = 1 - \frac{1}{3.5\left[\exp(1.22\mu_s - 0.4)\right]\left[(B_e/D_f) + 1.6\right]}$$

$$I_E = 1 - \frac{1}{3.5\exp(1.22\times0.3-0.4)[(3.384/1.5)+1.6]} = 0.926$$

$$S_i = \frac{150\times3.384\times0.85\times0.816\times0.926}{16\ 000}(1-0.3^2) = 0.0183\,\text{m} = \mathit{18.3\ mm}$$

Problem 3.21

Estimate the consolidation settlement of the clay layer shown in the scheme in Problem 2.13, using the results of part (a) of the same problem.

Solution:
At the middle of the clay layer:

$$\sigma'_o = 1.52\times15.7 + 1.22(19.24-10) + \left(\frac{3.05}{2}\right)(19.24-10)$$

$$= 49.23\frac{\text{kN}}{\text{m}^2} < \sigma'_c = 100\,\text{kN/m}^2 \rightarrow \text{the clay is overconsolidated}$$

Solution of Problem 2.13, part (a), gave $\sigma'_z = 63.44\,\text{kN/m}^2$
$63.44 + 49.23 = 112.67\ \text{kN/m}^2 > \sigma'_c$. Hence, Equation (3.38) is applied.

Equation (3.38): $$S_c = \frac{C_r H}{1+e_o}\log\frac{\sigma'_c}{\sigma'_o} + \frac{C_c H}{1+e_o}\log\frac{\sigma'_o+\sigma'_z}{\sigma'_c}$$

$$S_c = \frac{0.06\times3.05}{1+0.8}\log\frac{100}{49.23} + \frac{0.25\times3.05}{1+0.8}\log\frac{49.23+63.44}{100} = 0.053\ \text{m} = \mathit{53\ mm}$$

Problem 3.22

A stratum of clay is 20 m thick and is just consolidated under its present overburden. The average saturated unit weight of the clay = $20\,\text{kN/m}^3$. Results of oedometer tests revealed that the coefficient of consolidation $c_v = 10\,\text{m}^2/\text{year}$ and the relationship between the effective pressure σ' (in kN/m^2) and the void ratio e of the clay could be expressed by the equation: $e = 0.8 - 0.3\log\dfrac{\sigma'}{100}$

At the top of the clay stratum the intergranular pressure and the hydrostatic pressure are 162 and $54\,\text{kN/m}^2$, respectively. To furnish a supply of water it is proposed to pump water from a sand stratum that underlies the clay. What ultimate settlement would occur at ground surface if the pressure in the water below the clay were to be permanently reduced by an amount $= 54\,\text{kN/m}^2$ by pumping and what would be the settlement after one year of pumping?

Solution:

The decrease in water pressure equals $54\,\text{kN/m}^2$ will cause the effective vertical stress (granular stress) be increased by the same amount.

At the centre of the clay layer the pressure increase is

$$\sigma'_Z = 54/2 = 27\,\text{kN/m}^2$$

The ultimate settlement would occur due to this pressure increment.

Since the clay is consolidated under its present overburden, it is considered as normally consolidated clay. Hence, Equation (3.36)-(a) is applied.

Equation (3.36)-(a): $$S_{oed} = \frac{C_c H}{1+e_o}\log\frac{\sigma'_o + \sigma'_z}{\sigma'_o} \quad (\text{for } \sigma'_o \approx \sigma'_c;\ OCR \approx 1)$$

$$\sigma'_o = 162 + 10(20-10) = 262\,\text{kN/m}^2$$

From the given equation it is clear that $C_c = 0.3$

$$e_o = 0.8 - 0.3\log\frac{262}{100} = 0.800 - 0.125 = 0.675$$

Ultimate settlement $$= S_{oed} = \frac{0.3 \times 20 \times 1000}{1+0.675}\log\frac{262+27}{262} = \mathit{153\ mm}$$

Equation (3.44): $$T_v = \frac{c_v t}{d^2}$$

The clay layer is open; hence, $d = 10$ m

$$T_v = \frac{10 \times 1}{10^2} = 0.1$$

Equation (3.48): $$T_v = \frac{\left(\frac{\pi}{4}\right)\left(\frac{U\%}{100}\right)^2}{\left[1-(U\%/100)^{5.6}\right]^{0.357}}$$

$$0.1 = \frac{\left(\frac{\pi}{4}\right)\left(\frac{U\%}{100}\right)^2}{\left[1-(U\%/100)^{5.6}\right]^{0.357}} \rightarrow U = 0.357$$

Settlement after 1 year of pumping $= U \times S_{oed} = 0.357 \times 153$

$= \mathit{55\ mm}$

(Continued)

Problem 3.23

The following results were obtained from an oedometer test on a specimen of saturated clay:

Pressure (kPa):	27	54	107	214	429	214	107	54
Void ratio:	1.243	1.217	1.144	1.068	0.994	1.001	1.012	1.024

A layer of this clay 8 m thick lies below a 4 m depth of sand, the water table being at the surface. The saturated unit weight for both soils $= 19\,\text{kN/m}^3$. A 4 m depth of fill, unit weight $= 21\,\text{kN/m}^3$, is placed on the sand over an extensive area. Determine the final settlement due to consolidation of the clay. If the fill were to be removed some time after the completion of consolidation, what heave would eventually take place due to swelling of the clay?

Solution:
A plot of e–$\log\sigma'$, using the given pressure–void ratio data, was drawn on a separate semi-logarithmic graph paper. From the plot the preconsolidation pressure σ'_c was approximately determined $= 73\,\text{kN/m}^2$.
At the centre of the clay layer, the effective overburden pressure is

$$\sigma'_o = 4(19-10) + 4(19-10) = 72\,\text{kN/m}^2 \approx \sigma'_c$$

Therefore, the clay is considered normally consolidated.
After the virgin compression curve and the expansion curve were found, approximately, the following $e - \log\sigma'$ relationships and indices could be obtained:

Compression: $e = 1.74 - 0.27\log\sigma'$; $C_c = 0.27$
Expansion: $e = 1.075 - 0.031\log\sigma'$; $C_e = 0.031$
Compression: $e_o = 1.74 - 0.27\log 72 = 1.24$

The fill covers a wide area; hence, the same fill pressure q is transmitted to the concerned subsurface layers and the problem can be considered one-dimensional. Consider one-dimensional consolidation, taking the clay layer as a whole:

The increase in the vertical effective stress at centre of the clay layer is

$$\sigma'_z = q = 4 \times 21 = 84\,\text{kN/m}^2$$

Equation (3.36)-(a):

$$S_{oed} = \frac{C_c H}{1+e_o}\log\frac{\sigma'_o + \sigma'_z}{\sigma'_o} \quad (\text{for } \sigma'_o \approx \sigma'_c;\ OCR \approx 1)$$

$$S_{oed} = \frac{0.27 \times 8 \times 1000}{1+1.24}\log\frac{72+84}{72} = 324\ mm$$

Expansion (heave):

$$\sigma'_o = 4 \times 21 + 4(19-10) + 4(19-10) = 156\ \text{kN/m}^2$$

$$e_o = 1.075 - 0.031\log 156 = 1.01$$

Removed pressure $\sigma'_z = q = 4 \times 21 = 84\,\text{kN/m}^2$ will be used as a negative pressure in the settlement equation. Hence,

$$\text{Heave (expansion)} = \frac{C_e H}{1+e_o}\log\frac{\sigma'_o-\sigma'_z}{\sigma'_o} = \frac{0.031\times 8\times 1000}{1+1.01}\log\frac{156-84}{156}$$

$$= -42\ mm \text{ (The negative sign indicates expansion)}$$

Problem 3.24

A compressible clay layer 5 m thick carries an overburden of pervious sand and rests on an impervious bed of rock. A structure with a large construction area, founded in the sand, causes the same increase in vertical effective stress throughout the clay layer. Results of an oedometer test on a sample of the clay, 19 mm thick, showed that the void ratio decreased from 0.85 to 0.83 under a corresponding increase in pressure and the consolidation was 75% complete after 30 min. The following values of time factor versus degree of consolidation are available:

U:	0.45	0.65	0.85
T_V:	0.15	0.30	0.60

Estimate the final settlement of the structure due to consolidation of the clay. Also, estimate the time elapsing before one-half of the final settlement has taken place.

Solution:

Equation (3.35): $\Delta H = \frac{\Delta e}{1+e_o}H$ = final settlement

Therefore, in an oedometer test: $dh = \frac{de}{1+e_o}h$

$$\text{Change in sample height} = dh = \frac{(0.85-0.83)}{1+0.85}\times 19 = 0.204\,\text{mm}$$

Therefore, the same clay under the corresponding increase in pressure in the field, by proportion, will settle 0.204(H/h)mm and the final settlement is

$$S = \frac{0.204\times 5\times 1000}{19} = 54\ mm$$

Refer to the plot of the given values of U and T_v shown below:
For $U = 0.75$ obtain $\sqrt{T_v} = 0.66$; $T_v = 0.436$

Equation (3.44): $$T_v = \frac{c_v t}{d^2}$$

$$c_v = \frac{T_v d^2}{t} = \frac{0.436\times(19/2)^2}{30\times 60} = 0.022\ \text{mm}^2/\text{s} = 0.022\times 10^{-6}\,\text{m}^2/\text{s}$$

For one-half of the final settlement, $U = 0.50$; $\sqrt{T_v} = 0.43$; $T_v = 0.185$

Equation (3.45): $$t = \frac{10^{-7}\times T_v\times d^2}{3.154\times c_v} = \frac{10^{-7}\times 0.185\times 5^2}{3.154\times 0.022\times 10^{-6}} = 6.67\ years$$

(Continued)

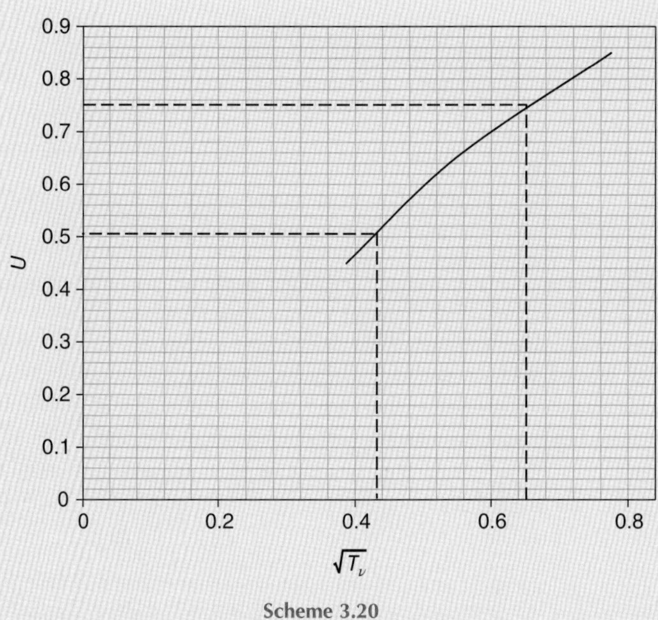

Scheme 3.20

Problem 3.25

In an oedometer test, with half-closed drainage condition, a specimen of saturated clay 19 mm thick reaches 50% consolidation in 20 min. How long would it take a layer of this clay 5 m thick to reach the same degree of consolidation under the same stress and drainage conditions? How long would it take the layer to reach 30% consolidation?

Solution:

For the same degree of consolidation U under the same stress and drainage conditions the time factor T_v is the same too.

Equation (3.44):
$$T_v = \frac{c_v t}{d^2}$$

Sample:
$$T_v = \frac{c_v t}{d^2} = \frac{c_v \times 20}{(19)^2}$$

$$T_v = \frac{c_v t}{d^2} = \frac{c_v \times t}{(5 \times 1000)^2}$$

Layer:
$$\frac{c_v \times 20}{(19)^2} = \frac{c_v \times t}{(5 \times 1000)^2}$$

$$t = \frac{20(5 \times 1000)^2}{(19)^2} \text{ (min)}$$

$$t = \frac{20(5 \times 1000)^2}{(19)^2} \times \frac{1}{365 \times 24 \times 60} = 2.63 \text{ } years$$

Equation (3.46):
$$T_v = \frac{\pi}{4} U^2 \quad (\text{for } U < 60\%)$$

For $U = 50\%$, $T_v = \frac{\pi}{4} \times 0.5^2 = \frac{c_v \times 2.63}{(5)^2}$

$$c_v = 1.867 \text{ m}^2/\text{year}$$

For 30% consolidation:

$$T_v = \frac{\pi}{4} \times 0.3^2 = \frac{c_v \times t}{(5)^2}$$

$$t = \frac{\frac{\pi}{4} \times 0.3^2 \times 5^2}{c_v} = \frac{\frac{\pi}{4} \times 0.3^2 \times 5^2}{1.867} = \mathit{0.95\ year}$$

Problem 3.26

Assuming the fill in Problem 3.23 is dumped very rapidly, what would be the value of excess porewater pressure at the centre of the clay layer after a period of three years? The layer is open and the value of $c_v = 2.4 \text{ m}^2/\text{year}$.

Solution:
The total stress increment, applied very rapidly ($t \cong 0$), equals to the pressure applied by the fill and will be carried initially entirely by the pore water, that is the initial value of excess pore water pressure $u_i = \Delta\sigma = 4 \times 21 = 84 \text{ kN/m}^2$. In particular, if u_i is constant throughout the clay layer, the excess porewater pressure u_e at any time $t > 0$ may be determined from the following equation based on the Terzaghi theory of one-dimensional consolidation:

$$u_e = \sum_{n=1}^{n=\infty} \frac{2u_i}{n\pi}(1 - \cos n\pi)\left(\sin \frac{n\pi z}{2H}\right) \exp\left(-\frac{n^2\pi^2 T_v}{4}\right)$$

where:

n = an odd integer; hence, $1 - \cos n\pi = 2$

$z = H$

$2H$ = thickness of the clay layer

Equation (3.44):

$$T_v = \frac{c_v t}{d^2} = \frac{2.4 \times 3}{(4)^2} = 0.45$$

Using $u_i = 84 \text{ kN/m}^2$, $n = 1$ and $T_v = 0.45$:

$$u_e = \frac{2 \times 84}{\pi}(2)\left(\sin \frac{\pi}{2}\right) \exp\left(-\frac{\pi^2 \times 0.45}{4}\right)$$

$$u_e = 106.91 \times 1 \times 0.329 = \mathit{35.2\ kN/m^2}$$

Problem 3.27

A 10 m depth of sand overlies a clay layer 8 m thick, below which is a further depth of sand. For the clay, $m_v = 0.83 \text{ m}^2/\text{MN}$ and $c_v = 4.4 \text{ m}^2/\text{year}$. Water table is at the surface level but is to be lowered permanently by 4 m, the initial lowering taking place over a period of 40 weeks. Calculate the final settlement due to consolidation of the clay, assuming no change in the weight of sand and the settlement 2 years after the start of lowering.

(Continued)

Solution:
Total increase in vertical effective stress equals to total decrease in the porewater pressure due to lowering of the water table, which is

$$4\gamma_w = 4 \times 10 = 40\,\text{kN/m}^2.$$

Assume taking the clay layer as a whole:
At the centre of the layer the increase in effective stress is

$$\sigma'_z = \frac{1}{2} \times 40 = 20\,\text{kN/m}^2$$

Equation (3.34): $S_c = m_v \sigma'_z H$

$$S_c = 8 \times \frac{0.83}{1000} \times 20 = 0.133\,\text{m} = 133\ mm$$

For calculation of settlement two years after the start of lowering, the time t would be equal to two years minus one-half of the total lowering period.

Equation (3.44): $$T_v = \frac{c_v t}{d^2} = \frac{4.4}{(4)^2}\left(2 - \frac{20 \times 7}{365}\right) = 0.444$$

Equation (3.48): $$T_v = \frac{\left(\frac{\pi}{4}\right)\left(\frac{U\%}{100}\right)^2}{\left[1 - \left(\frac{U\%}{100}\right)^{5.6}\right]^{0.357}}$$

$$0.444 = \frac{\left(\frac{\pi}{4}\right)\left(\frac{U\%}{100}\right)^2}{\left[1 - \left(\frac{U\%}{100}\right)^{5.6}\right]^{0.357}} \rightarrow U = 0.73$$

Settlement two years after the start of lowering the water table is

$$S = U \times S_c = 0.73 \times 133 = 97\ mm.$$

Problem 3.28

A raft foundation 60 × 40 m carrying a net pressure of 145 kN/m^2 is located at a depth of 4.5 m below the surface in a deposit of dense sandy gravel 22 m deep. The water table is at a depth of 7 m and below the sandy gravel is a layer of clay 5 m thick which, in turn, is underlain by dense sand. The average value of m_v for the clay is 0.22 m^2/MN. Determine the settlement below the centre of the raft, the corner of the raft and the centre of each edge of the raft, due to consolidation of the clay.

Solution:
Final settlement due to consolidation of the clay is given by

Equation (3.34): $$S_c = m_v \sigma'_z H$$

Calculation of σ_z' at depth z = 20 m (i.e. at middle of the clay layer), using Table 2.3 or Figure 2.32:

(a) Below the centre of the raft

$$\frac{B}{z} = \frac{20}{20} = 1; L/z = 30/20 = 1.5 \rightarrow I_r = 0.194$$
$$\sigma_z' = q(4I_r) = 145 \times 4 \times 0.194 = 112.52\ \text{kN/m}^2$$

(b) Below the corner of the raft

$$B/z = 40/20 = 2, L/z = 60/20 = 3 \rightarrow I_r = 0.238$$
$$\sigma_z' = q(I_r) = 145 \times 0.238 = 34.51\ \text{kN/m}^2$$

(c) Below the centre of a long edge of the raft

$$B/z = 30/20 = 1.5, L/z = 40/20 = 2 \rightarrow I_r = 0.224$$
$$\sigma_z' = q(2I_r) = 145 \times 2 \times 0.224 = 64.96\ \text{kN/m}^2$$

(d) Below centre of a short edge of the raft

$$B/z = 20/20 = 1, L/z = 60/20 = 3 \rightarrow I_r = 0.203$$
$$\sigma_z' = q(2I_r) = 145 \times 2 \times 0.203 = 58.87\ \text{kN/m}^2$$

Calculation of consolidation settlement:

(a) Below the raft centre

$$S_c = 5 \times \frac{0.22}{1000} \times 112.52 = 0.124\ \text{m} = \mathit{124\ mm}$$

(b) Below the raft corner

$$S_c = 5 \times \frac{0.22}{1000} \times 34.51 = 0.038\ \text{m} = \mathit{38\ mm}$$

(c) Below centre of a long edge

$$S_c = 5 \times \frac{0.22}{1000} \times 64.96 = 0.072\ \text{m} = \mathit{72\ mm}$$

(d) Below centre of a short edge

$$S_c = 5 \times \frac{0.22}{1000} \times 58.87 = 0.065\ \text{m} = \mathit{65\ mm}$$

Problem 3.29

An oil storage tank 35 m in diameter is located 2 m below the surface of a deposit of clay 32 m thick, which overlies a firm stratum, as shown in the scheme below. The net foundation pressures at the foundation level equal 105 kN/m^2. The average values of m_v and pore pressure coefficient A for the clay are 0.14 m^2/MN and 0.65, respectively. The undrained value of Young's modulus is estimated to be 40 MN/m^2. Determine the total settlement (excluding settlement due to secondary compression or creep) under the centre of the tank.

(*Continued*)

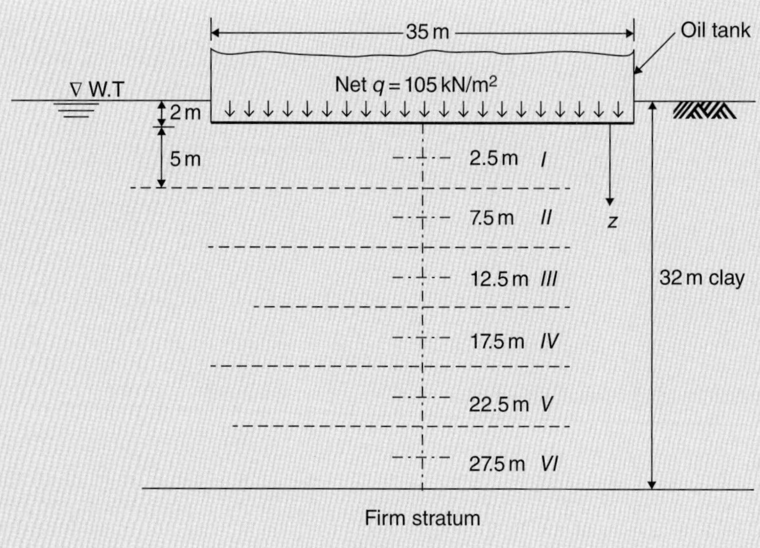

Scheme 3.21

Solution:

The clay deposit is considered a thick layer; there will be significant lateral strain (resulting in immediate settlement in the clay) and therefore it would be appropriate if the Skempton–Bjerrum method is used in the settlement computations. Accordingly, the total settlement (excluding settlement due to secondary compression or creep) is given by

Equation (3.3): $$S_T = S_i + S_c$$

Immediate settlement S_i may be calculated using Equation (3.33) and Figure 3.9, whereas the consolidation settlement S_c is calculated using Equation (3.40) and Figure 3.15.

In order to compute the consolidation settlement more accurately, assume the clay below the foundation level is divided into six layers of equal thicknesses, as shown in the scheme above:

(a) Immediate settlement

Equation (3.33): $$S_i = \mu_o \mu_1 \frac{qB}{E_u}$$

$$\frac{H}{B} = \frac{30}{35} = 0.857; \frac{D}{B} = \frac{2}{35} = 0.057$$

Figure 3.9: obtain $\mu_o = 0.99$ and $\mu_1 = 0.35$

$$S_i = 0.99 \times 0.35 \times \frac{105 \times 35}{40 \times 1000} = 0.032\,\text{m} = 32\,\text{mm}$$

(b) Consolidation settlement.

Equation (3.40): $$S_c = K \sum S_{oed}$$

For each layer:

$$S_{oed} = m_v \times \sigma'_z \times H = \frac{0.14}{1000} \times \sigma'_z \times 5 \times 1000 = 0.7\,\sigma'_z\ \text{mm}$$

Figure 2.30: $\sigma'_z = \sigma_z = q(I_c)$

Layer	z, (m)	I_c	σ'_z, (mPa)	S_{oed}, (mm)
I	2.5	1.00	105.0	73.5
II	7.5	0.94	98.7	69.1
III	12.5	0.82	86.1	60.3
IV	17.5	0.65	68.3	47.8
V	22.5	0.55	57.8	40.5
VI	27.5	0.40	42.0	29.4

$\Sigma S_{oed} = 320.6\,\text{mm}$

Scheme 3.22

$A = 0.65;\ \dfrac{H}{B} = 0.857$

Figure 3.15: obtain $K = 0.775$

$$S_c = 0.775 \times 320.6 = 248\ \text{mm}$$

Total settlement $S_T = S_i + S_c = 32 + 248 = \mathit{280\ mm}$

Problem 3.30

The foundation of a large building is at 3 m below ground level. The building contains one-story basement. The soil profile consists of topsoil and sand 7.5 m thick, overlies a clay layer 3 m thick. The water table is 6.25 m deep. Unit weights of the sand above and below water table are 20 and 22 kN/m^3, respectively. Saturated unit weight of the clay is 19.2 kN/m^3. The average value of m_v for the clay is 177×10^{-6} m^2/kN and the coefficient of consolidation c_v is 0.025 mm^2/s. Additional pressure on the clay due to weight of the building $= 200\,\text{kN/m}^2$ at the top of the stratum, decreasing uniformly with depth to 150 kN/m^2 at the bottom. Find:

(a) The original effective pressures at top and bottom of the clay stratum before excavation commences.
(b) The effective pressures at the top and bottom of the clay stratum after completion of the building.
(c) The magnitude of settlement of the building expected due to consolidation of the clay.
(d) The probable time in which 90 % of the settlement will occur. Assume half-closed drainage condition.

Solution:

(a) At top of the clay: $\sigma'_o = 6.25 \times 20 + 1.25(22-10) = \mathit{140\ kPa}$
At bottom of the clay: $\sigma'_o = 140 + 3(19.2-10) = \mathit{167.6\ kPa}$

(b) At top of the clay: $\sigma' = 140 + 200 - 3 \times 20 = \mathit{280\ kPa}$
At bottom of the clay: $\sigma' = 167.6 + 150 - 3 \times 20 = \mathit{257.6\ kPa}$

(c) The net increase in effective pressure at the centre of the clay layer $\sigma'_z = \dfrac{280+257.6}{2} - \dfrac{140+167.6}{2} =$ $268.8 - 153.8 = 115\,\text{kPa}$

or,

$$\sigma'_z = \frac{200+150}{2} - 3 \times 20 = 115\,\text{kPa}$$

Equation (3.34): $$S_c = m_v \times \sigma'_z \times H$$

$$= 177 \times 10^{-6} \times 115 \times 3 = 0.061\ \text{m} = \mathit{61\ mm}$$

(Continued)

(d) For 90 % of the settlement, the degree of consolidation $U = 0.90 > 0.60$; hence the time factor may be computed from

Equation (3.47):

$$T_v = 1.781 - 0.933\log(100 - U\,\%)$$

$$T_v = 1.781 - 0.933\log(100 - 90) = 0.85$$

Equation (3.45):

$$t = \frac{10^{-7} \times T_v \times d^2}{3.154 \times c_v}$$

$$t = \frac{10^{-7} \times 0.85 \times 3^2}{3.154 \times \dfrac{0.025}{1000^2}} = \frac{0.85 \times 9}{3.154 \times 0.25} = 9.7\ \textit{years}$$

Problem 3.31

Below the foundation of a structure there is a stratum of compressible clay 6 m thick with incompressible porous strata above and below. The average overburden pressure on the stratum before construction was 115 kPa and after completion of the structure the pressure increased to 210 kPa. Oedometer tests were carried out on a sample of clay initially 20 mm thick. Each pressure was allowed to act for 24 h and the decrease in thickness measured, the results being as follows:

Pressure, (kPa):	0	50	100	200	400
Thickness, mm:	20.0	19.8	19.4	19.0	18.6

Under a pressure of 100 kPa, 90 % of the total consolidation took place in 21 min.

Find (a) value of m_v for the increased stress range after completion of the structure, (b) the probable settlement of the structure and (c) the time in which 90% of this settlement may be expected to occur.

Solution:

(a)

$$m_v = \frac{e_o - e_1}{1 + e_o} \times \frac{1}{\sigma'_1 - \sigma'_o} = \frac{1}{H_o}\left(\frac{H_o - H_1}{\sigma'_1 - \sigma'_o}\right)$$

A plot of the given pressure versus thickness was drawn on a separate graph paper. From the plot, thickness of the oedometer sample under a pressure of 115 kPa was estimated at 19.34 mm and that under a pressure of 210 kPa was 18.97 mm. Hence,

$$m_v = \frac{1}{19.34}\left(\frac{19.34 - 18.98}{210 - 115}\right) = 1.97 \times 10^{-4}\, m^2/kN$$

(b)

$$\sigma'_z = 210 - 115 = 95\,\text{kPa}$$

Equation (3.34):

$$S_c = m_v \times \sigma'_z \times H$$

$$= 1.96 \times 10^{-4} \times 95 \times 6 \times 1000 = 112\ mm$$

(c) Equation (3.44):

$$T_v = \frac{c_v t}{d^2}$$

For the same 90 % of the total consolidation and pressure increment, the time factor of the clay stratum equals that of the test sample. Therefore, under a pressure of 100 kPa, we can write

$$\frac{c_v \times t}{3^2} = \frac{c_v \times 21}{(10/1000)^2}$$
$$t = 9 \times 21 \times 10\ 000 = 189 \times 10^4\ \text{min.}$$

Under a pressure of 95 kPa:

$$t = 189 \times 10^4 \times \frac{100}{95} = 198.95 \times 10^4\ \text{min}$$

The time in which 90% of the 112 mm settlement may be expected to occur is

$$t = \frac{198.95 \times 10^4}{365 \times 24 \times 60} = 3.8\ \textit{years}$$

Problem 3.32

On a flat site boreholes reveal that a 4 m layer of dense sand overlies a layer of compressible clay 8 m thick, below which stiffer clay extends to a considerable depth. The appropriate values of the coefficient of compressibility of the 8 m clay layer are:

Upper 4 m, $m_v = 0.746 \times 10^{-3}\text{m}^2/\text{kN}$
Lower 4 m, $m_v = 0.559 \times 10^{-3}\text{m}^2/\text{kN}$

It is proposed to store material on two adjacent paved areas as indicated in the scheme below. The pavement and the stored material together can be assumed to provide a uniform contact pressure q on the surface of the ground. Estimate the value of q in kPa if the differential settlement between points A and B is to be limited to 0.05 m. Use the pressure influence chart of Figure 2.32.

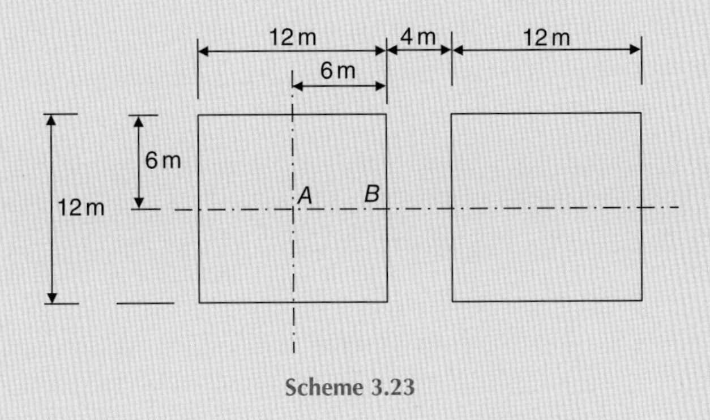

Scheme 3.23

(Continued)

Solution:

(A) Settlement at point *A*

Equation (3.34): $S_c = m_v \times \sigma'_z \times H$

(1) Settlement of the upper 4 m clay; $z = 6\,\text{m}$

Left area:

$\frac{B}{z} = \frac{L}{z} = \frac{6}{6} = 1$; Figure 2.31 gives $I = 0.175$ and $4I = 0.7$

$$\sigma'_z = q(4I) = 0.7\,q\,(\text{kPa})$$

Right area:

$$\frac{B}{z} = \frac{6}{6} = 1;\ \frac{L}{z} = \frac{22}{6} = 3.67 \rightarrow I = 0.204 \rightarrow 2I = 0.408$$

$$\frac{B}{z} = \frac{6}{6} = 1;\ L/z = 10/6 = 1.67 \rightarrow I = -0.195 \rightarrow 2I = -0.390$$

$$\sigma'_z = q(0.408 - 0.390) = 0.018\,q\,(\text{kPa})$$

$$S_c = \left(0.746 \times 10^{-3}\right)(0.7q + 0.018\,q)(4) = 2.143 \times 10^{-3}\,q\,(\text{m})$$

(2) Settlement of the lower 4 m clay, $z = 10$ m

Left area:

$\frac{B}{z} = \frac{L}{z} = \frac{6}{10} = 0.6$. Figure 2.31 gives $I = 0.107$ and $4I = 0.428$.

$$\sigma'_z = q(4I) = 0.428\,q\,(\text{kPa})$$

Right area:

$$\frac{B}{z} = \frac{6}{10} = 0.6; L/z = 22/10 = 2.2 \rightarrow I = 0.153 \rightarrow 2I = 0.306$$

$$\frac{B}{z} = \frac{6}{10} = 0.6; L/z = 10/10 = 1 \rightarrow I = -0.136 \rightarrow 2I = -0.272$$

$$\sigma'_z = (0.306 - 0.272)q = 0.034\,q\,(\text{kPa})$$

$$S_c = \left(0.559 \times 10^{-3}\right)(0.428\,q + 0.034\,q)(4) = 1.033 \times 10^{-3}\,q$$

Settlement at $A = 2.143 \times 10^{-3}\,q + 1.033 \times 10^{-3}\,q = 3.176 \times 10^{-3}\,q$

(B) Settlement at point *B*

Equation (3.34): $S_c = m_v \times \sigma'_z \times H$

(1) Settlement of the upper 4 m clay; $z = 6\,\text{m}$:

Left area:

$$\frac{B}{z} = \frac{6}{6} = 1;\ \frac{L}{z} = \frac{12}{6} = 2 \rightarrow I = 0.2, \text{and } 2I = 0.4$$

$$\sigma'_z = 0.4\,q\,(\text{kPa})$$

Right area:

$$\frac{B}{z} = \frac{6}{6} = 1;\ \frac{L}{z} = \frac{16}{6} = 2.67 \rightarrow I = 0.202; \text{ and } 2I = 0.404$$

$$\frac{B}{z}=\frac{4}{6}=0.667;\ \frac{L}{z}=\frac{6}{6}=1\rightarrow I=-0.145;\ \text{and}\ 2I=-0.29$$

$$\sigma'_z=(0.404-0.290)q=0.114\,q\ (\text{kPa})$$

$$S_c=\left(0.746\times10^{-3}\right)(0.404\,q+0.114\,q)(4)=1.546\times10^{-3}\,q$$

(2) Settlement of the lower 4 m clay, $z = 10$ m
Left area:

$$\frac{B}{z}=\frac{6}{10}=0.6;\ L/z=12/10=1.2\rightarrow I=0.143\ \text{and}\ 2I=0.286$$

$$\sigma'_z=0.286\,q\ \ (\text{kPa})$$

Right area:

$$\frac{B}{z}=\frac{6}{10}=0.6;\ \frac{L}{z}=\frac{16}{10}=1.6\rightarrow I=0.15,\ \text{and}\ 2I=0.3$$

$$\frac{B}{z}=\frac{4}{10}=0.4;\ L/z=6/10=0.60\rightarrow I=-0.08,\ \text{and}\ 2I=-0.16$$

$$\sigma'_z=(0.3-0.16)q=0.14\,q\ \ (\text{kPa})$$

$$S_c=(0.559\times10^{-3})(0.286\,q+0.14\,q)(4)=0.953\times10^{-3}\,q$$

Settlement at $B=1.546\times10^{-3}\,q+0.953\times10^{-3}\,q=2.499\times10^{-3}\,q$
The differential settlement between points A and $B=0.05$ m. Therefore, $0.05=3.176\times10^{-3}\,q-2.499\times10^{-3}\,q$

$$q=\frac{0.05}{0.677\times10^{-3}}=73.86\ kPa$$

Problem 3.33

Solve Problem 3.32 taking the 8 m clay layer as a whole.

Solution:

(A) Settlement at point A
Equation (3.34): $S_c=m_v\times\sigma'_z\times H$
Use average $m_v=\dfrac{0.746+0.559}{2}\times10^{-3}=0.653\times10^{-3}\ \text{m}^2/\text{kN}$

$$z=8\,\text{m}$$

Left area:
$\dfrac{B}{z}=\dfrac{L}{z}=\dfrac{6}{8}=0.75$; Figure 2.31 gives $I=0.14$ and $4I=0.56$

$$\sigma'_z=q(4I)=0.56\,q\ (\text{kPa})$$

(Continued)

Right area:

$$\frac{B}{z}=\frac{6}{8}=0.75;\ \frac{L}{z}=\frac{22}{8}=2.75 \rightarrow I=0.18;\ \text{and}\ 2I=0.36$$

$$\frac{B}{z}=\frac{6}{8}=0.75;\ \frac{L}{z}=\frac{10}{8}=1.1.25 \rightarrow I=-0.165 \rightarrow 2I=-0.33$$

$$\sigma'_z=q(0.36-0.33)=0.03\,q\ \ (\text{kPa})$$

$$S_c=(0.653\times10^{-3})(0.56q+0.03\,q)(8)=3.082\times10^{-3}q\ \ (\text{m})$$

(B) Settlement at point B

Equation (3.34): $$S_c=m_v\times\sigma'_z\times H$$

Use average $$m_v=\frac{0.746+0.559}{2}\times10^{-3}=0.653\times10^{-3}\ \text{m}^2/\text{kN}$$

$$z=8\,\text{m}$$

Left area:

$$\frac{B}{z}=\frac{6}{8}=0.75;\frac{L}{z}=\frac{12}{8}=1.5 \rightarrow I=0.17;\ \text{and}\ 2I=0.34$$

$$\sigma'_z=0.34\,q\ \ (\text{kPa})$$

Right area:

$$\frac{B}{z}=\frac{6}{8}=0.75;\frac{L}{z}=\frac{16}{8}=2 \rightarrow I=0.175;\ \text{and}\ 2I=0.35$$

$$\frac{B}{z}=\frac{4}{8}=0.5;\frac{L}{z}=\frac{6}{8}=0.75 \rightarrow I=-0.108;\ \text{and}\ 2I=-0.216$$

$$\sigma'_z=(0.35-0.216)q=0.134\,q\ \ (\text{kPa})$$

$$S_c=(0.653\times10^{-3})(0.34\,q+0.134\,q)(8)=2.476\times10^{-3}\,q$$

The differential settlement between points A and B = 0.05 m. Therefore,

$$0.05=3.082\times10^{-3}\,q-2.476\times10^{-3}\,q$$

$$q=\frac{0.05}{0.606\times10^{-3}}=82.51\ kPa$$

Problem 3.34

A building has a foundation base slab supported on a bed of compact sand. Under the sand there is a stratum of clay 4.5 m thick which in turn rests on impermeable shale. The ground water level is situated within the sand. The initial effective overburden pressure at the top of the clay is 85 kPa. Additional pressures due to the foundation loads are as follows:

Location	Additional pressure, kPa	
	Under centre of slab	Under corner of slab
Top of clay layer	85.0	45.0
Bottom of clay layer	25.0	10.5

The clay has an average unit weight $= 19.2\,\text{kN/m}^3$ and oedometer tests gave the following void ratio/effective pressure data:

e	0.93	0.91	0.88	0.85
σ', kPa	50	100	200	400

Estimate the final settlement under the centre and under a corner of the foundation base slab due to consolidation of the clay stratum.

Solution:

(A) Final settlement under centre of the base slab:

The initial effective overburden pressure at bottom of the clay layer is

$$85 + 4.5\,(19.2 - 10) = 126.4\,\text{kPa}$$

The average initial effective overburden pressure at centre of the clay layer is

$$\frac{85 + 126.4}{2} = 105.7\,\text{kPa}$$

The average increase in pressure at centre of the clay layer is

$$\frac{85 + 25}{2} = 55\,\text{kPa}$$

The total effective pressure at centre of the clay layer is

$$105.7 + 55 = 160.7\,\text{kPa}$$

The graph of e versus σ' is shown in the scheme below.

From this graph the initial void ratio $e_o = 0.908$ at the initial effective pressure $\sigma' = 105.7\,\text{kPa}$, and $\dfrac{de}{d\sigma'} = 3.5 \times 10^{-4}$

$$m_v = \left(\frac{1}{1 + e_o}\right)\frac{de}{d\sigma'} = \frac{1}{(1 + 0.908)} \times 3.5 \times 10^{-4} = 18.34 \times 10^{-5}\,\text{m}^2/\text{kN}$$

$$S_c = m_v \times \sigma'_z \times H = 18.34 \times 10^{-5} \times 55 \times 4.5 = 0.045\ \text{m} = 45\ mm$$

It is noteworthy that m_v may also be calculated approximately by assuming a straight line relationship between e and σ' over the current range of pressures, as follows:

(*Continued*)

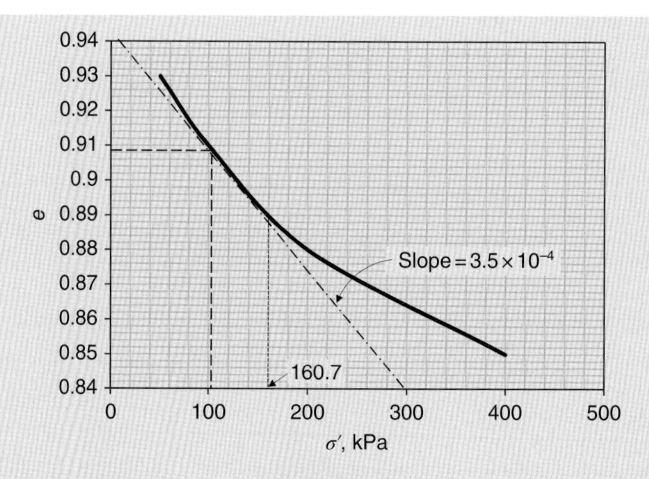

Scheme 3.24

$$m_v = \frac{1}{1+e_o}\left(\frac{e_o - e_1}{\sigma_1' - \sigma_o'}\right) = \frac{1}{1+0.908}\left(\frac{0.908-0.889}{160.7-1057}\right) = 18.11\times10^{-5}\,\text{m}^2/\text{kN}$$

(B) Final settlement under a corner of the base slab:
The average increase in pressure at centre of the clay layer is

$$\frac{45+10.5}{2} = 27.75\,\text{kPa}$$

The total effective pressure at centre of the clay layer = 133.45 kPa.
This pressure falls within the range of the pressures for which the computed m_v may be assumed constant. Therefore, the final settlement under a corner of the base slab is

$$S_c = m_v \times \sigma_z' \times H = 18.34\times10^{-5}\times27.75\times4.5\times1000 = 23\ mm$$

Problem 3.35

A soft to very soft silty clay layer 3 m thick exists between a layer of permeable clayey silty sand at top and a nearly impermeable layer of stiff silty clay at bottom. Oedometer tests were carried out on a sample of the soft clay 19 mm thick. From the plot of ΔH versus log time, the value of time t_{100} (the time at 100% primary consolidation) was 100 min. From the graph of e versus log σ' the compression index C_c was 0.32 and the void ratio e_p at the end of the primary consolidation was 0.72. Estimate the secondary compression of the silty clay layer that would occur from time of completion of the primary consolidation to 30 years of the load application in the field.

Solution:

Equation (3.58): $S_s = C_\alpha' H \log(t_2/t_1)$, where $C'\alpha = \dfrac{C_\alpha}{(1+e_p)}$.

Equation (3.60): estimate $\dfrac{C_\alpha}{C_c} = 0.04$; hence $C_\alpha = 0.04\times0.32 = 0.013$

$$C_\alpha' = \frac{0.013}{(1+0.72)} = 75.6\times10^{-4}$$

For the same clay soil:

$$T_v = \frac{c_v \times t_{\text{field}}}{d_{\text{field}}^{\;2}} = \frac{c_v \times t_{\text{lab}}}{d_{\text{lab}}^{\;2}};\ \text{hence}\ \ t_{\text{field}} = \frac{d_{\text{field}}^{\;2} \times t_{\text{lab}}}{d_{\text{lab}}^{\;2}} = \frac{3000^2 \times 100}{\left(\frac{19}{2}\right)^2}$$

$$t_{\text{field}} = 99.723 \times 10^5 \text{ min} = \frac{99.723 \times 10^5}{365 \times 24 \times 60} = 19 \text{ years}$$

$$t_1 = t_{\text{field}} = 19 \text{ years};\ t_2 = 30 \text{ years}$$

$$S_s = 75.6 \times 10^{-4} \times 3000 \times \log(30/19) = 4.5\ mm$$

However, it is very likely that the secondary compression will be larger than this small value, as some will occur during primary consolidation. Theoretically, at the end of the 19 years there is no excess pore pressure anywhere in the 3-m clay layer; however during this time period dissipation occurs from the top down, with secondary compression beginning before 19 years have elapsed in the upper regions.

Problem 3.36

A soil profile is shown in the scheme below. A uniformly distributed load q = 100 kPa is applied at the ground surface covering an extensive area. The clay is normally consolidated. Assume that the primary consolidation will be complete in 3.5 years. The secondary compression index $C_a = 0.011$. Estimate the total settlement 10 years after the load application due to consolidation and secondary compression of the clay layer.

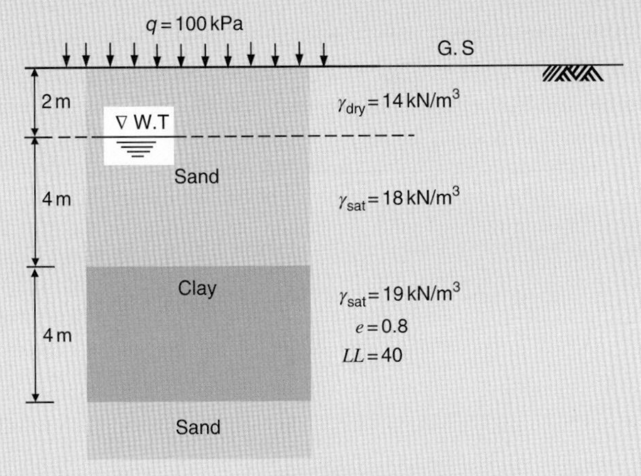

Scheme 3.25

Solution:

Equation (3.36)-(b): $$S_c = \frac{C_c H}{1 + e_o} \log \frac{\sigma'_o + \sigma'_z}{\sigma'_o}$$

The average effective overburden stress at the middle of the clay layer is

$$\sigma'_o = 2 \times 14 + 4(18 - 10) + 2(19 - 10) = 78 \text{ kN/m}^2$$

Equation (3.36)-(b): $C_c = 0.009(LL - 10) = 0.009(40 - 10) = 0.27$

The loaded area is a large area; hence, practically, $\sigma'_z = q = 100$ kPa

$$S_c = \frac{0.27 \times 4}{1 + 0.8} \log \frac{78 + 100}{78} = 0.215 \text{ m} = 215 \text{ mm}$$

(Continued)

Equation (3.58):

$$S_s = C'_\alpha H \log(t_2/t_1), \text{ where } C'_\alpha = \frac{C\alpha}{(1+e_p)}$$

$$C_c = \frac{\Delta e}{\log\frac{\sigma'_o + \sigma'_z}{\sigma'_o}}; \ \Delta e = e_o - e_p = C_c \log\frac{\sigma'_o + \sigma'_z}{\sigma'_o}; \text{ hence,}$$

$$e_p = e_0 - C_c \log\frac{\sigma'_o + \sigma'_z}{\sigma'_o} = 0.8 - 0.27\log\frac{78+100}{78} = 0.703; \text{ hence,}$$

$$C'_\alpha = \frac{0.011}{1+0.703} = 0.0065; \ \ t_1 = 3.5\,\text{years}; \ t_2 = 10\,\text{years}$$

$$S_s = 0.0065 \times 4\log(10/3.5) = 0.0118\,\text{m} = 12\,\text{mm}$$

Therefore, the total settlement 10 years after the load application due to consolidation and secondary compression of the clay layer is

$$215 + 12 = 227\ mm$$

Problem 3.37

A soil profile consists of two strata of normally consolidated homogeneous clay above an impervious layer and with a drainage layer of incompressible sand, $\gamma_{sat} = 19.6\,\text{kN/m}^3$, separating the strata, as shown in the scheme below. Undisturbed samples were taken from nearly middle of the clay layers, consolidation tests were carried out, $e - \log\sigma'$ curves were plotted and values of C_c were determined. Also, for one point on each curve (the straight part of the curve), the value of σ' and its corresponding *e* value were found. These data and values of specific gravity G_s are as tabulatedin the scheme:

Layer	G_s	C_c	e	σ', kPa
Upper clay layer	2.74	0.36	1.02	100
Lower clay layer	2.70	0.24	0.73	160

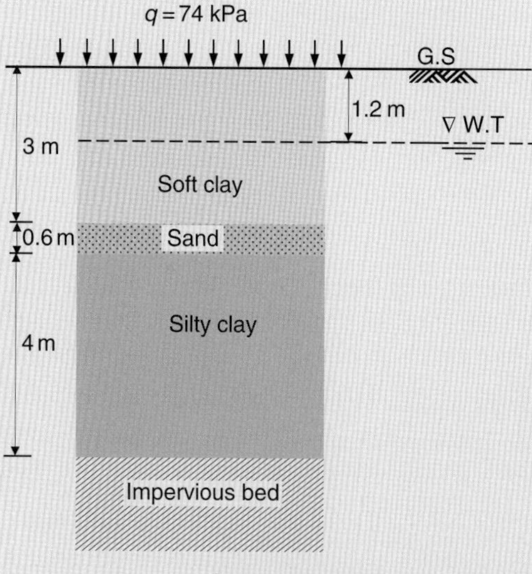

Scheme 3.26

A very large area is to be loaded on the surface with a uniform load of 74 kPa. Determine:

(a) The settlement due to consolidation of the clay strata under the added surface uniform load.
(b) The data required for drawing a time settlement curve using the following additional data:

Layer	**Height of specimen, cm**	**t_{50}, min**
Upper	2.54	4.6
Lower	1.62	3.1

Solution:

(a) The upper and lower clay strata are divided into convenient layers for the desired degree of accuracy. In the upper layer the water table is a convenient plane for division. The lower 4 m stratum is arbitrarily divided into two equal layers. The necessary calculations are carried out on the tabulated form in Table 3.25, in accordance with the following procedure:

(1) Number the layers and record them in column 1 of the table.
(2) Write the layers thickness H in column 2.
(3) Determine the effective overburden pressure σ'_o at mid-depth of each layer and record them in column 3. Practically, the first soft clay layer may be considered nearly saturated.

For layers I and II: $\gamma_{sat} = \dfrac{G_s + e}{1+e}\gamma_w = \dfrac{2.74 + 1.02}{1 + 1.02}(10) = 18.6\ \text{kN/m}^3$

For layers III and IV: $\gamma_{sat} = \dfrac{2.7 + 0.73}{1 + 0.73}(10) = 19.8\ \text{kN/m}^3$

For example, the effective overburden pressure σ'_o at the mid-depth of layer III is

$$\sigma'_o = 1.2 \times 18.6 + 1.8(18.6 - 10) + 0.6(19.6 - 10) + 1(19.8 - 10)$$
$$= 53.36\,\text{kPa}$$

(4) Record C_c values in column 4.

(5) Compute e_o values using the relationship $e_o = e + C_c \log\dfrac{\sigma'}{\sigma'_o}$ and record them in column 5. For example, at the mid-depth of layer III

$$e_o = 0.73 + 0.24\log\frac{160}{53.36} = 0.844$$

(6) Record values of $\dfrac{C_c H}{1+e_0}$ in column 6. Because $\dfrac{C_c}{1+e_0}$ is dimensionless, thickness H can be in any unit and the consolidation settlement S_c will be in the same unit.

(7) Determine and record in column 7 the effective pressure increase σ'_z produced at the mid-depth of each layer by the added load. In this problem, because the loaded area is large compared to the depth of the compressible layers, σ'_z may be considered as constant for the full depth and equals to the added uniform load q which is 74 kPa.

(Continued)

(8) Record in column 8 values of $(\sigma'_z + \sigma'_o)$

(9) Record in column 9 values of $\log\frac{\sigma'_o + \sigma'_z}{\sigma'_o}$

(10) Record in column 10 values of $S_c = \frac{C_c H}{1+e_o}\log\frac{\sigma'_o + \sigma'_z}{\sigma'_o}$ Add the S_c values to obtain the total settlement.

Table 3.25 Layers of the upper and lower clay strata and the calculations necessary to determine settlement due to consolidation of the clay strata.

1	2	3	4	5	6	7	8	9	10
Layer	H, m	σ'_o, kPa	C_c	e_o	$\frac{C_c H}{1+e_o}$, m	σ'_z, kPa	$\sigma'_o + \sigma'_z$	$\log\frac{\sigma'_o + \sigma'_z}{\sigma'_o}$	S_c, m
I	1.2	11.16	0.36	1.360	0.183	74	85.16	0.883	0.162
II	1.8	30.06	0.36	1.210	0.293	74	104.60	0.534	0.156
III	2.0	53.36	0.24	0.844	0.260	74	127.36	0.378	0.098
IV	2.0	72.96	0.24	0.812	0.265	74	146.96	0.304	0.081
									$\sum S_c = 0.497$ m

The settlement due to consolidation of the clay strata under the added surface uniform load is

$$\sum S_c = 0.497\ m = 497\ mm$$

(b) Considering the upper layer:

Equation (3.44): $$T_v = \frac{c_v t}{d^2}$$

$$c_v = \frac{T_v \times d^2}{t}$$

$d = H$ = thickness of compressible layer for drainage at top or at bottom, as it is the case with the lower silty clay layer, whereas $d = \frac{H}{2}$ for drainage at top and bottom at the same time, as it is the case with the upper soft clay layer and the laboratory specimens.

Height of the specimen which belongs to the upper layer = 2.54 cm and t_{50} = 4.6 min.

At time t_{50} the degree of consolidation $U = 50\%$. Hence, Equation (3.46) gives

$$T_v = \frac{\pi}{4}U^2 = \frac{\pi}{4}(0.5)^2 = 0.2$$

$$c_v = \frac{0.2 \times 1.27^2}{4.6} = 0.07\ \text{cm}^2/\text{min}$$

Now, for certain values of U, selected arbitrarily, the corresponding values of T_v are calculated from Equation (3.46) or (3.47), as applicable. Values of settlement and the corresponding values of time t are also calculated and tabulated, as shown in the scheme below. For example, using $U = 10\%$, Equation (3.46) gives $T_v = 0.008$. For the upper layer $d = \frac{H}{2} = 1.5$ m and $c_v = 0.07$ cm^2/min; hence,

$$t = \frac{0.008 \times (1.5 \times 100)^2}{0.07} = 2571\ \text{min}$$

Total settlement of the upper layer = 0.318 m; hence, settlement at time $t = 2571$ min is

$$S_c = U \times 0.318 = 0.1 \times 0.318 = 0.032\ \text{m}.$$

Considering the lower layer:
The computations just mentioned above are also required for the lower layer. For this case:

$$d = H = 4 \text{ m, specimen height} = 1.62 \text{ cm and } t_{50} = 3.1 \text{ min}$$

At time t_{50} the degree of consolidation $U = 50\%$; hence, Equation (3.46) gives

$$T_v = \frac{\pi}{4}U^2 = \frac{\pi}{4}(0.5)^2 = 0.2$$

$$c_v = \frac{0.2 \times 0.81^2}{3.1} = 0.0423 \text{ cm}^2/\text{min}$$

For example, using $U = 10\%$, Equation (3.46) gives $T_v = 0.008$. Therefore,

$$t = \frac{0.008 \times (4 \times 100)^2}{0.0423} = 30\ 260 \text{ min.}$$

Total settlement of the lower layer $= 0.179$ m; hence, settlement at time $t = 30\,260$ min is

$$S_c = U \times 0.179 = 0.1 \times 0.179 = 0.018 \text{ m.}$$

		Upper layer (drained 2 sides) $H = 3$ m, $S_c = 0.318$ m $\frac{d^2}{c_v} = 7.44$ month (1 month = 30 days)		Lower layer (draine 1 side) $H = 4$ m, $S_c = 0.179$ m $\frac{d^2}{c_v} = 87.56$ month (1 month = 30 days)	
$U\%$	T_v	Settlement, m $U \times 0.318$	t month	Settlement, m $U \times 0.179$	t month
10	0.008	0.032	0.060	0.018	0.70
30	0.071	0.095	0.528	0.054	6.22
50	0.197	0.159	1.466	0.090	17.25
70	0.405	0.223	3.013	0.125	35.46
90	0.848	0.286	6.309	0.161	74.25

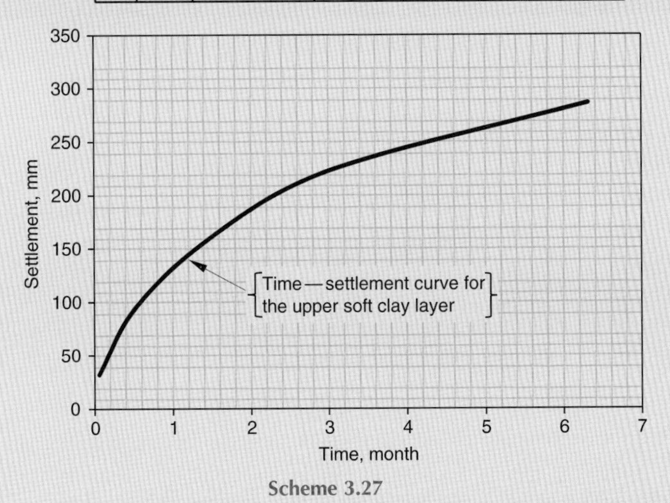

Scheme 3.27

The plot shown in the scheme represents a time–settlement curve which belongs to the upper soft clay layer.
A similar curve can be plotted for each of the lower silty clay layer and the two clay layers combined.

Problem 3.38

The plan of a group of independent square footings is shown in the scheme below. Each footing size is proportioned for a contact pressure of 100 kPa. Footing loads are assumed to be applied independently, so that differential settlements have no effect upon the applied loads. All the footings have the same foundation depth equals 3.6 m. Water table exists at a depth of 4.6 m below the ground surface. The scheme also shows the average soil profile of the site. The submerged unit weight γ' of the intervening drainage sand layer equals 10 kN/m^3. Results of the oedometer tests revealed that the clay strata are normally consolidated (i.e. $\sigma'_o \cong \sigma'_c$). The soil profile also shows data which are obtained from results of oedometer tests carried out on undisturbed samples of the clay soils. Estimate settlement of footing number five due to consolidation of the clay strata under the loads of the footings group.

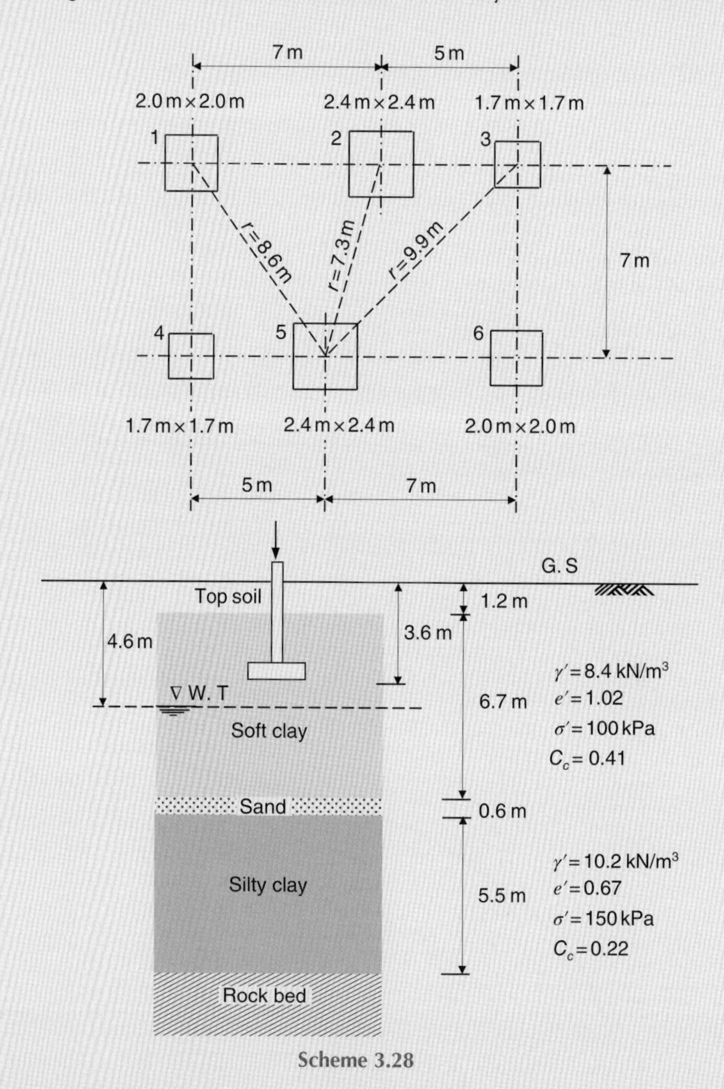

Scheme 3.28

Solution:

The computations for the total settlement analysis are organised in Table 3.26, following the same general procedure which was used in the solution of Problem 3.37. The soft clay below the foundation level is divided into three layers, whereas the silty clay below the intervening 0.6 m sand is divided into two layers, as indicated in the table. The increase in vertical stresses at the middle of the layers below centre of footing number five caused by the loads on the other footings in the group (i.e. footings numbered 1–4 & 6), is computed assuming the footings act as point loads at the foundation level. Because these footings are at some distance (i.e. the r distances) from the footing under consideration, this assumption does not produce an appreciable error. The increase in vertical stress caused by the load of footing number five is computed considering the load is uniformly distributed as pressure $q = 100$ kPa at its foundation level. In this stress computation the Fadum chart of Figure 2.32 will be used.

Results of the oedometer tests revealed that the clay strata are normally consolidated; hence the settlement due to consolidation of each clay layer is calculated using

Equation (3.36)-(a): $$S_c = \frac{C_c H}{1 + e_o}\log\frac{\sigma'_o + \sigma'_z}{\sigma'_o}$$

Table 3.26 Computations for the total settlement analysis.

1	2	3	4	5	6
Layer	**H, m**	**σ'_o, kPa**	**C_c**	**e_o**	**$\frac{C_c H}{1+e_o}$, m**
I	1.0	75.44	0.41	1.070	0.198
II	1.3	90.10	0.41	1.039	0.261
III	2.0	103.96	0.41	1.013	0.407
IV	2.5	131.11	0.22	0.683	0.327
V	3.0	159.16	0.22	0.664	0.397

	7			8				9			
Footing:	**No. 5**			**No. 1**				**No. 2**			
$4I$	q	σ'_z	z	r/z	V	σ'_z	z	r/z	V	σ'_z	
0.96	100	96.0	0.50	17.2	400	0.15	0.50	14.60	576	0.22	
0.72	100	72.0	1.15	7.48	400	0.03	1.15	6.35	576	0.04	
0.28	100	28.0	2.80	3.07	400	0.08	2.80	2.61	576	0.22	
0.08	100	8.0	5.65	1.52	400	0.31	5.65	1.29	576	0.73	
0.03	100	3.0	8.40	1.02	400	0.48	8.40	0.87	576	0.90	

	10				11				12			
Footing:	**No. 3**				**No. 4**				**No. 6**			
	z	r/z	V	σ'_z	z	r/z	V	σ'_z	z	r/z	V	σ'_z
	0.50	19.8	289	0.11	0.50	14.0	400	0.16	0.50	10.0	289	0.11
	1.15	8.61	289	0.02	1.15	6.09	400	0.03	1.15	4.35	289	0.07
	2.80	3.54	289	0.33	2.80	2.50	400	0.17	2.80	1.79	289	0.02
	5.65	1.75	289	0.13	5.65	1.24	400	0.58	5.65	0.88	289	1.00
	8.40	1.18	289	0.25	8.40	0.83	400	0.74	8.4	0.60	289	0.91

(Continued)

13	14	15	16	17
Layer	$\sum\sigma'_z$, **kPa**	$\sigma'_o + \sum\sigma'_z$, **kPa**	$\log\frac{\sigma'_o+\sum\sigma'_z}{\sigma'_o}$	S_c, **m**
I	96.75	172.19	0.358	0.071
II	72.19	162.29	0.256	0.067
III	28.82	132.78	0.106	0.043
IV	10.75	141.86	0.034	0.011
V	6.28	165.44	0.017	0.007
				$\sum S_c = 0.199$ m

Total settlement = *199 mm*

Problem 3.39

Assume the soft clay layer of Figure 3.17 is 12 m thick and construction of the embankment will increase the average total vertical stress in the clay by 70 kPa. Design requirement is that all except the last 20 mm of the settlement due to consolidation of the clay layer will have taken place after 6 months. For the clay:

$$c_v = 5\,\text{m}^2/\text{year};\ c_h = 8\,\text{m}^2/\text{year};\ m_v = 0.2\,\text{m}^2/\text{MN}.$$

Determine the spacing, in a square pattern, of 450 mm diameter sand drains to achieve the above requirement.

Solution:
Final settlement is calculated from

Equation (3.34):
$$S_c = m_v\sigma'_z H = \frac{0.2}{1000}\times 70\times 12 = 0.168\,\text{m} = 168\,\text{mm}$$

For t = 6 months, $U = \frac{168-20}{168} = 0.88$
Diameter of each sand drain is 0.45 m. Therefore, $r_d = 0.225\,\text{m}$
Radius of equivalent cylindrical block = $R = n\,r_d = 0.225\,n$
The layer is half-closed and therefore $d = H = 12\,\text{m}$

Equation (3.44):
$$T_v = \frac{c_v t}{d^2} = \frac{5\times\frac{6}{12}}{12^2} = 0.0174$$

Equation (3.48):
$$T_v = \frac{\left(\frac{\pi}{4}\right)\left(\frac{U_v\%}{100}\right)^2}{\left[1-(U_v\%/100)^{5.6}\right]^{0.357}} = 0.0174$$

$$U_v = 14.9\% = 0.149$$

Equation (3.53):
$$U = 1-(1-U_v)(1-U_r)$$
$$0.88 = 1-(1-0.149)(1-U_r).\quad \text{Hence, } U_r = 0.86$$

Equation (3.52):

$$T_r = \frac{c_h\, t}{4R^2} = \frac{8 \times \frac{6}{12}}{4(0.225n)^2}$$

$$n = \sqrt{\left(\frac{19.75}{T_r}\right)}$$

A trial and error solution is necessary to obtain the value of n. Start with a value of n corresponding to one of the curves in Figure 3.21 and obtain the value of T_r for $U_r = 0.86$ from that curve. Using this value of T_r the value of $\sqrt{\left(\frac{19.75}{T_r}\right)}$ is calculated and plotted against the selected value of n, as shown in the scheme below:

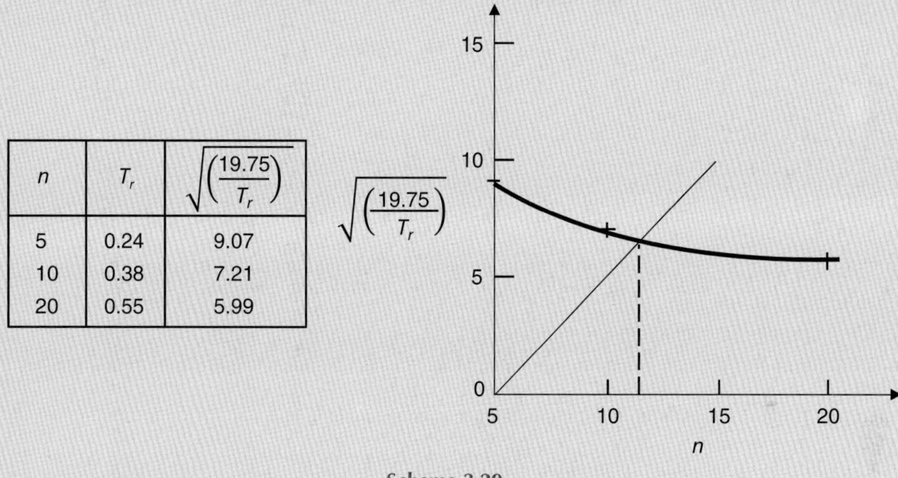

n	T_r	$\sqrt{\left(\frac{19.75}{T_r}\right)}$
5	0.24	9.07
10	0.38	7.21
20	0.55	5.99

Scheme 3.29

From the plot it is clear that $n = 11.53$.
Therefore, $R = 0.225 \times 11.53 = 2.594\,\text{m}$
Refer to Figure 3.20a. Spacing of drains in a square pattern is given by

$$S = \frac{R}{0.564}$$

$$= \frac{2.594}{0.564} = 4.6\ m$$

Problem 3.40

A half-closed clay layer is 8 m thick and it can be assumed that the coefficients of consolidation in both the vertical and horizontal directions are the same. Vertical sand drains 300 mm in diameter, spaced at 3 m centres in a square pattern, are to be used to increase the rate of consolidation of the clay under the increased vertical stress due to the construction of an embankment. Without sand drains the degree of consolidation at the time the embankment is due to come into use has been calculated as 25%. What degree of consolidation would be reached with the sand drains at the same time?

(*Continued*)

Solution:

Equation (3.46): $$T_v = \frac{\pi}{4}U_v^2 = \frac{\pi}{4}(0.25)^2 = 0.049$$

Equation (3.44): $$T_v = \frac{c_v t}{d^2}$$

$$c_v\ t = T_v \times d^2 = 0.049 \times 8^2 = 3.136\ \text{m}^2$$

$c_v = c_h$. Therefore, $c_h\ t = c_v\ t = 3.136\ \text{m}^2$

Equation (3.52): $$T_r = \frac{c_h\ t}{4R^2}$$

$$c_h\ t = T_r \times 4 \times R^2 = 3.136$$

Refer to Figure 3.20a. For sand drains in a square pattern

$$R = 0.564\,S = 0.564 \times 3 = 1.692\,\text{m}$$

$$R = n \times r_d = n \times \frac{300}{1000 \times 2} = 0.15\,n.\ \text{Hence,}\ n = \frac{1.692}{0.15} = 11.3$$

$T_r = \frac{3.136}{4 \times 1.692^2} = 0.274$. From Figure 3.21 obtain $U_r = 0.72$

Equation (3.53): $$U = 1 - (1 - U_v)(1 - U_r)$$

$$U = 1 - (1 - 0.25)(1 - 0.72) = 0.79$$

Therefore, the degree of consolidation that would be reached with the sand drains at the same time is

$$U = 79\%$$

Problem 3.41

A layer of saturated clay is 10 m thick, the lower boundary being impermeable; an embankment is to be constructed above the clay. Determine the time required for 90% consolidation of the clay layer. If 300 mm diameter sand drains at 4–m centres in a square pattern were installed in the clay, in what time would the same overall degree of consolidation be reached? The coefficients of consolidation in the vertical and horizontal directions respectively are 9.6 and 14.0 m^2/year.

Solution:

Equation (3.47): $$T_v = 1.781 - 0.933\log(100 - U\%)$$

$$T_v = 1.781 - 0.933\log(100 - 90) = 0.848$$

Equation (3.44): $$T_v = \frac{c_v t}{d^2}$$

Time required for 90% consolidation of the clay layer (without sand drains) is

$$t = T_v \times \frac{d^2}{c_v} = 0.848 \times \frac{10^2}{9.6} = 8.8\ \textit{years}$$

For sand drains in square pattern, $R = 0.564\,S = 0.564 \times 4 = 2.256\,\text{m}$

$$R = n \times r_d = 0.15\,n$$

$$n = \frac{2.256}{0.15} = 15.04$$

$$T_r = \frac{c_h\ t}{4R^2} = \frac{14 \times t}{4 \times 2.256^2} = 0.688\,t$$

For selected values of t determine the corresponding values of T_r and then from Figure 3.21 obtain the corresponding values of U_r.

$$T_v = \frac{c_v\, t}{d^2} = \frac{9.6 \times t}{10^2} = 0.096\, t$$

For selected values of t determine the corresponding values of T_v and then from Equation (3.48) obtain the corresponding values of U_v.

For the obtained values of U_v and U_r, compute values of U using

Equation (3.53): $U = 1 - (1 - U_v)(1 - U_r)$

All the computed values of T_r, U_r, T_v, U_v and U are shown in Table 3.27.

Table 3.27 All the computed values of T_r, U_r, T_v, U_v and U.

t,year	T_v	U_v	$(1 - U_v)$	T_r	U_r	$(1 - U_r)$	U
0.5	0.050	0.245	0.755	0.344	0.760	0.240	0.819
1.0	0.096	0.350	0.650	0.688	0.933	0.067	0.956
2.0	0.192	0.495	0.505	1.376	0.990	0.010	0.995

Draw a plot of t against U, as shown in the scheme below. Thus, from the plot, the time required for 90% consolidation of the clay layer, with sand drains, is

$$t = 0.67\, years$$

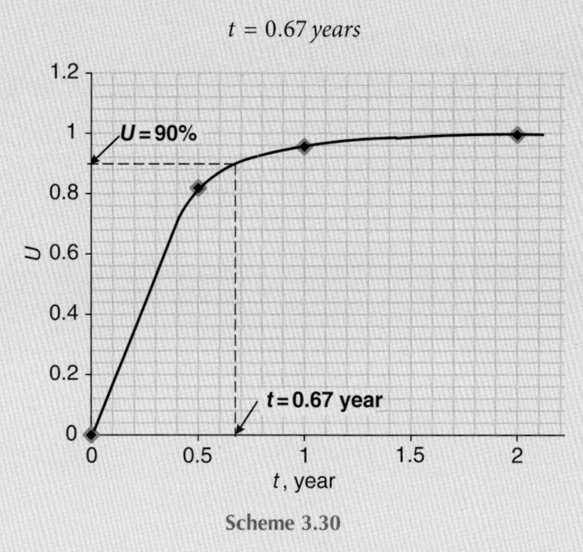

Scheme 3.30

Problem 3.42

During the construction of a highway bridge, the average permanent load on a normally consolidated clay layer, 6 m thick with open drainage at top and bottom, is expected to increase by about 115 kPa. The average effective overburden pressure at mid-depth of the layer is 210 kPa. The average values of C_c and e_o are 0.28 and 0.9, respectively. The consolidation coefficient is 0.36 m^2/month. Determine:

(a) The final consolidation settlement without precompression

(b) The added pressure $\Delta\sigma'_{(f)}$, due to weight of a temporary surcharge fill as a part of the total surcharge load, at the middle of the clay layer, needed to eliminate the consolidation settlement in nine months by precompression.

(Continued)

Solution:

(a) Equation (3.36)-(a):

$$S_c = \frac{C_c H}{1+e_o}\log\frac{\sigma'_o+\sigma'_z}{\sigma'_o}$$

$$= \frac{0.28\times 6}{1+0.9}\log\frac{210+115}{210} = 0.168 \text{ m} = \mathit{168\ mm}$$

(b) This technique of *precompression* is sometimes used to control or eliminate, as desired, consolidation settlement due to a structure load. Assume the permanent load of the structure causes increase in pressure, at the middle of the clay layer, equals $\Delta\sigma'_{(p)}$. Settlement S_c will take place if a total surcharge load, which causes increase in pressure $= \Delta\sigma'_{(p)} + \Delta\sigma'_{(f)}$, is acting for a sufficient period of time t. At that time, if the entire surcharge is removed and a permanent load of a structure is applied which causes increase in pressure equals $\Delta\sigma'_{(p)}$ no appreciable settlement will occur.

Equation (3.44): $$T_v = \frac{c_v t}{d^2} = \frac{0.36\times 9}{(6/2)^2} = 0.36$$

Equation (3.48): $$T_v = \frac{\left(\frac{\pi}{4}\right)\left(\frac{U\%}{100}\right)^2}{\left[1-(U\%/100)^{5.6}\right]^{0.357}} = 0.36. \text{ Compute } U = 66.5\%$$

According to this process of consolidation, the entire settlement due to $U = 100\%$ is computed as

$$S_c = \frac{100}{66.5}\times 0.168 = 0.253 \text{ m.}$$

Therefore, it is necessary σ'_z be increased so that the entire consolidation settlement is eliminated in nine months.

Equation (3.36)-(a):

$$S_c = \frac{C_c H}{1+e_o}\log\frac{\sigma'_o+\sigma'_z}{\sigma'_o}$$

$$0.253 = \frac{0.28\times 6}{1+0.9}\log\frac{210+\sigma'_z}{210}$$

$$\log\frac{210+\sigma'_z}{210} = \frac{0.253}{0.884} = 0.286$$

$$\frac{210+\sigma'_z}{210} = 1.93. \text{ Hence,}$$

$$\sigma'_z = 195.3 \text{ kPa}$$

$$\sigma'_z = \Delta\sigma'_{(p)} + \Delta\sigma'_{(f)}$$

$$195.3 = 115 + \Delta\sigma'_{(f)}$$

$$\Delta\sigma'_{(f)} = 195.3 - 115 = \mathit{80.3\ kPa}$$

Another method for determining (approximately) the surcharge pressure $\Delta\sigma'_{(f)}$ is by using the scheme below (Das, 2011), as follows:

Values of T_v and $U\%$ are computed as before.

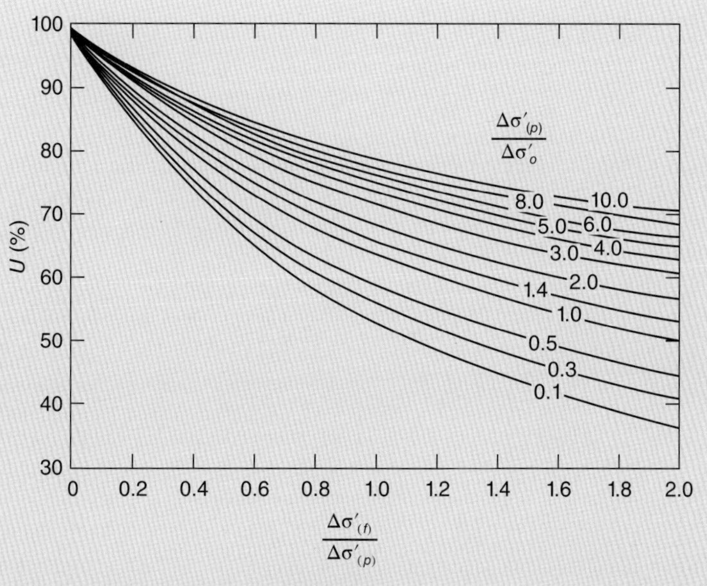

Scheme 3.31

$$\frac{\Delta\sigma'_{(p)}}{\Delta\sigma'_o} = \frac{115}{210} = 0.548$$

From the scheme, for $U = 66.5\%$ and $\left(\dfrac{\Delta\sigma'_{(p)}}{\Delta\sigma'_o}\right) = 0.548$, estimate

$$\frac{\Delta\sigma'_{(f)}}{\Delta\sigma'_{(p)}} = 0.7$$

Therefore, $\Delta\sigma'_{(f)} = 0.7 \times 115 = 80.5\ kPa$

Problem 3.43

Solve Problem 3.42 with the addition of some sand drains. Assume that $r_d = 0.1$ m, $R = 1.5$ m, $c_v = c_h$. Also, assume that this is a no-smear case. Consider that the surcharge is applied instantaneously.

Solution:

(a) Settlement due to consolidation of the normally consolidated clay will be *168 mm* as computed before.

(b) Values of T_v and U_v remain unchanged, that is $T_v = 0.36$ and $U_v = 66.5\%$.

$$n = \frac{R}{r_d} = \frac{1.5}{0.1} = 15 \text{ and } T_r = \frac{c_h\, t}{4R^2} = \frac{0.36 \times 9}{4 \times 1.5^2} = 0.36$$

From Figure 3.21, obtain $U_r = 0.77$

Equation (3.53):

$$U = 1 - (1 - U_v)(1 - U_r)$$
$$= 1 - (1 - 0.67)(1 - 0.77) = 0.924 = 92.4\%$$

(*Continued*)

The entire settlement due to $U = 100\%$ is

$$\frac{100}{92.4} \times 0.168 = 0.182 \text{ m}$$

Therefore, σ'_z should have an increased value so that the entire consolidation settlement is eliminated in nine months.

Equation (3.36)-(a): $\quad S_c = \frac{C_c H}{1+e_o} \log \frac{\sigma'_o + \sigma'_z}{\sigma'_o}$

$$0.182 = \frac{C_c H}{1+e_o} \log \frac{\sigma'_o + \sigma'_z}{\sigma'_o} = \frac{0.28 \times 6}{1+0.9} \log \frac{210+\sigma'_z}{210} \log \frac{210+\sigma'_z}{210} = \frac{0.182}{0.884} = 0.206.$$

Hence, $\frac{210+\sigma'_Z}{210} = 1.61$ and $\sigma'_Z = 128.1\,\text{kPa}$.

$\sigma'_Z = \Delta\sigma'_{(p)} + \Delta\sigma'_{(f)}$. Therefore,

$$\Delta\sigma'_{(f)} = 128.1 - 115 = \mathit{13.1\ kPa}.$$

or,

from the scheme in Problem 3.42: for $U = 92.4\%$ and $\left(\frac{\Delta\sigma'_{(p)}}{\Delta\sigma'_o}\right) = 0.548$, estimate $\left(\frac{\Delta\sigma'_{(f)}}{\Delta\sigma'_{(p)}}\right) = 0.115$. Therefore,

$$\Delta\sigma'_{(f)} = 0.115 \times 115 = 13.2\,\text{kPa}.$$

Problem 3.44

For a sand drain project the clay is normally consolidated and the entire surcharge is applied, as shown in the scheme below:

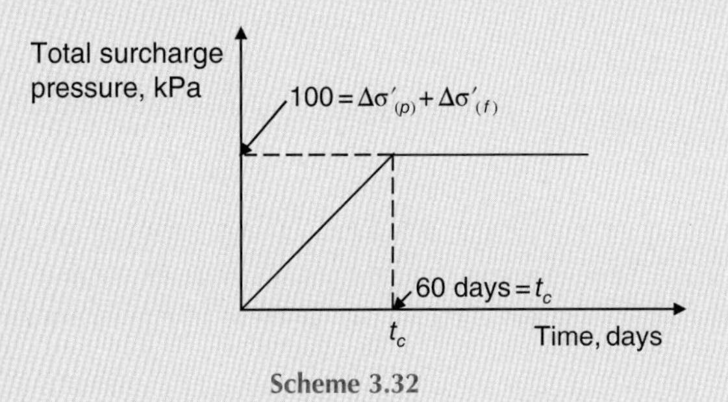

Scheme 3.32

The following data are given:

For the sand drains:
$r_d = 0.1\,\text{m}$, $R = 1.0$ m, $c_v = c_h$. Assume no-smear case.

For the clay: thickness $H = 4.6\,\text{m}$ (two-way drainage), $C_c = 0.31$, $e_o = 1.1$, $\sigma'_o = 50\,\text{kPa}$ at middle of clay layer, $c_v = 0.011\ \text{m}^2/\text{day}$

Calculate the degree of consolidation 30 days after the surcharge is first applied. Also, determine the consolidation settlement at that time due to the surcharge.

Solution:

Equation (3.44):

$$T_v = \frac{c_v\, t}{d^2} = \frac{0.011 \times 30}{(4.6/2)^2} = 0.062$$

$$T_c = \frac{c_v\, t_c}{d^2} = \frac{0.011 \times 60}{(4.6/2)^2} = 0.125$$

Using Figure 3.22, for $T_c = 0.125$ and $T_v = 0.062$, we obtain $U_v \approx 9\%$
Equation (3.54):

$$n = \frac{R}{r_d} = \frac{1}{0.1} = 10$$

$$T_{rc} = \frac{c_h\, t_c}{4R^2} = \frac{0.011\ \times 60}{4 \times 1^2} = 0.165$$

$$T_r = \frac{c_h\, t}{4R^2} = \frac{0.011\ \times 30}{4 \times 1^2} = 0.082 < T_{rc};\ \text{hence use}$$

$$U_r = \frac{T_r}{T_{rc}} - \frac{1}{AT_{rc}}[1 - \exp(-AT_r)];\ A = \frac{2}{[n^2/(n^2-1)]\ln(n) - [(3n^2-1)/4n^2]}$$

$$A = \frac{2}{[10^2/(10^2-1)]\ln(10) - [(3 \times 10^2 - 1)/(4 \times 10^2)]} = \frac{2}{1.576} = 1.269$$

$$U_r = \frac{0.082}{0.165} - \frac{1}{1.269 \times 0.165}[1 - \exp(-1.269 \times 0.082)] = 0.024 = 2.4\%$$

Equation (3.53):

$$U = 1 - (1 - U_v)(1 - U_r)$$

$$U = 1 - (1 - 0.090)(1 - 0.024) = 0.112 = 11.2\,\%$$

The final consolidation settlement $S_c = \dfrac{C_c H}{1 + e_o} \log \dfrac{\sigma'_o + \left(\Delta\sigma'_{(p)} + \Delta\sigma'_{(f)}\right)}{\sigma'_o}$

$$S_c = \frac{0.31 \times 4.6}{1 + 1.1} \log \frac{50 + (100)}{50} = 0.324\ \text{m}$$

The consolidation settlement after 30 days $= U \times S_c$

$= 0.112\ \times 0.324 = 0.036\ \text{m}$

$= 36\ mm$

Problem 3.45

For a sand drain project, the following are given:
Clay: Normally consolidated (one-way drainage)

$$H = 5.5\ \text{m}$$

$$C_c = 0.3; e_o = 0.76; c_v = 0.015\ \text{m}^2/\text{day}$$

$$\sigma'_o = 80\ \text{kPa, at middle of clay layer}$$

Sand drains: $r_d = 0.07$ m, $R = 1.25$ m, $r_s = r_d$, $c_r = c_v$.

(Continued)

A surcharge is applied as shown in the scheme below. Calculate the degree of consolidation and the consolidation settlement 50 days after the surcharge is first applied.

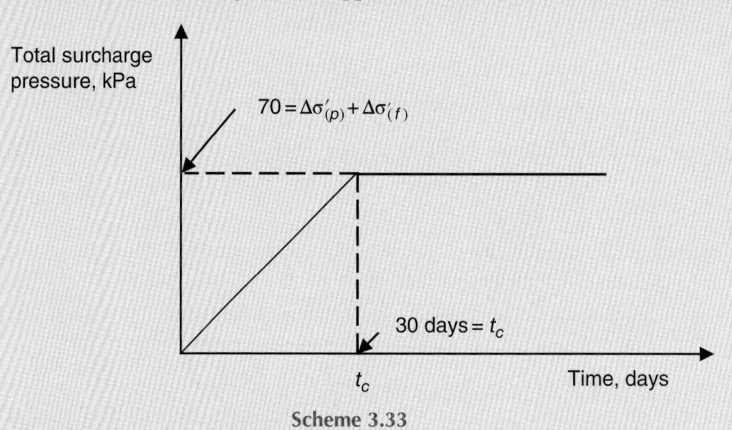

Scheme 3.33

Solution:

As the radius of the smeared zone is given in this Problem, it is possible to consider the smear effect in the solution; more realistic results would be achieved by doing so.

For the case in which smear effect is considered, the A factor in Equations (3.54) and (3.55) becomes

$$A = \frac{2}{\frac{n^2}{n^2 - S^2}\ln\left(\frac{n}{S}\right) - \frac{3}{4} + \frac{S^2}{4n^2} + \frac{k_h}{k_s}\left(\frac{n^2 - s^2}{n^2}\right)\ln S}$$

where: $n = \frac{R}{r_d}$; $S = \frac{r_s}{r_d}$

k_s = horizontal permeability in the smeared zone
k_h = horizontal permeability in the unsmeared zone

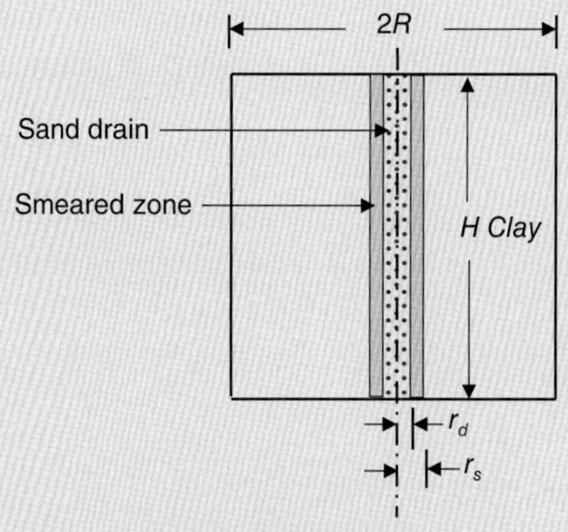

Scheme 3.34

$$T_c = \frac{c_v t_c}{d^2} = \frac{0.015 \times 30}{(5.5)^2} = 0.015$$

$$T_v = \frac{c_v t}{d^2} = \frac{0.015 \times 50}{(5.5)^2} = 0.025$$

For $T_c = 0.015$ and $T_v = 0.025$, Figure 3.22 gives

$$U_v \approx 13\%$$

$$n = \frac{R}{r_d} = \frac{1.25}{0.07} = 17.86$$

$$T_{rc} = \frac{c_h t_c}{4R^2} = \frac{0.015 \times 30}{4 \times 1.25^2} = 0.072$$

$T_r = \dfrac{c_h t}{4R^2} = \dfrac{0.015 \times 50}{4 \times 1.25^2} = 0.12 > T_{rc}$; hence use

Equation (3.55):

$$U_r = 1 - \frac{1}{AT_{rc}}[\exp(AT_{rc}) - 1]\exp(-AT_r)$$

$$A = \frac{2}{\dfrac{n^2}{n^2 - S^2}\ln\left(\dfrac{n}{S}\right) - \dfrac{3}{4} + \dfrac{S^2}{4n^2} + \dfrac{k_h}{k_s}\left(\dfrac{n^2 - s^2}{n^2}\right)\ln S}, \quad S = \frac{r_s}{r_d} = 1$$

$$A = \frac{2}{\dfrac{17.86^2}{17.86^2 - 1}\ln(17.86) - \dfrac{3}{4} + \dfrac{1}{4 \times 17.86^2} + (0)} = \frac{2}{2.89 - 0.75 + 0.0008} = 0.934$$

$$AT_{rc} = 0.934 \times 0.072 = 0.067; \quad AT_r = 0.934 \times 0.12 = 0.112$$

$$U_r = 1 - \frac{1}{0.067}[\exp(0.067) - 1]\exp(-0.112) = 0.075 = 7.5\%$$

Equation (3.53):

$$U = 1 - (1 - U_v)(1 - U_r)$$

$$U = 1 - (1 - 0.13)(1 - 0.075) = 0.195 = 19.5\%$$

The final consolidation settlement

$$S_c = \frac{CcH}{1 + e_o}\log\frac{\sigma'_o + \left(\Delta\sigma'_{(p)} + \Delta\sigma'_{(f)}\right)}{\sigma'_o}$$

$$S_c = \frac{0.3 \times 5.5}{1 + 0.76}\log\frac{80 + 70}{80} = 0.256 \text{ m}$$

The settlement after 50 days $= U \times S_c$

$$= 0.195 \times 0.256 \times 1000 = 50 \; mm$$

Problem 3.46

For a sand drain project, the following are given:

$$r_d = 0.2 \text{ m}, R = 2.5 \text{ m}, r_s = 0.3 \text{ m}, c_h = c_v = 0.3 \text{ m}^2/\text{month}, H = 6 \text{ m}, (k_h/k_s) = 2.$$

Determine:

(a) The degree of consolidation of the clay layer caused only by the sand drains after six months of application of the surcharge.

(*Continued*)

(b) The degree of consolidation of the clay layer caused by the combination of vertical drainage (drained at top and bottom) and radial drainage after six months of application of the surcharge.

Assume that the surcharge is applied instantaneously.

Solution:

(a) For the case in which *smear* effect is considered and the *surcharge* assumed *instantaneously* applied, according to Barron (1948),

$$U_r = 1 - \exp\left(\frac{-8T_r}{m}\right)$$

$$m = \frac{n^2}{n^2 - S^2}\ln\left(\frac{n}{S}\right) - \frac{3}{4} + \frac{S^2}{4n^2} + \frac{k_h}{k_s}\left(\frac{n^2 - s^2}{n^2}\right)\ln S;\ n = \frac{R}{r_d};\ S = \frac{r_s}{r_d}$$

$$n = \frac{2.5}{0.2} = 12.5;\ S = \frac{0.3}{0.2} = 1.5;\ T_r = \frac{c_h t}{4R^2} = \frac{0.3 \times 6}{4 \times 2.5^2} = 0.072$$

$$m = \frac{12.5^2}{12.5^2 - 1.5^2}\ln\left(\frac{12.5}{1.5}\right) - \frac{3}{4} + \frac{1.5^2}{4 \times 12.5^2} + 2\left(\frac{12.5^2 - 1.5^2}{12.5^2}\right)\ln 1.5$$

$$m = 2.151 - 0.75 + 0.004 + 0.799 = 2.204$$

$$U_r = 1 - \exp\left(\frac{-8 \times 0.072}{2.204}\right) = 1 - 0.77 = 0.23 = 23\%$$

(b) Equation (3.44):

$$T_v = \frac{c_v t}{d^2} = \frac{0.3 \times 6}{(6/2)^2} = 0.2$$

Equation (3.48):

$$T_v = \frac{\left(\frac{\pi}{4}\right)\left(\frac{U\%}{100}\right)^2}{\left[1 - (U\%/100)^{5.6}\right]^{0.357}}$$

$$0.2 = \frac{\left(\frac{\pi}{4}\right)\left(\frac{U\%}{100}\right)^2}{\left[1 - (U\%/100)^{5.6}\right]^{0.357}};\ \text{hence } U \approx 50\% = U_v$$

Equation (3.53):

$$U = 1 - (1 - U_v)(1 - U_r)$$

$$U = 1 - (1 - 0.5)(1 - 0.23) = 615 - 61.5\%$$

Problem 3.47

Redo part (a) of Problem 3.46 considering no-smear case.

Solution:

$$T_r = \frac{c_h t}{4R^2} = \frac{0.3 \times 6}{4 \times 2.5^2} = 0.072;\ n = \frac{2.5}{0.2} = 12.5$$

From Figure 3.21 obtain the corresponding value of $U_r = 0.28 = 28\%$

or,

$U_r = 1 - \exp\left(\frac{-8T_r}{m}\right)$. For no-smear case:

$$m = \left(\frac{n^2}{n^2-1}\right)\ln(n) - \frac{3n^2-1}{4n^2}$$

$$m = \frac{156.25}{156.25-1}\ln 12.5 - \frac{3\times 156.25-1}{4\times 156.25} = 1.794$$

$$U_r = 1-\exp\left(\frac{-8\times 0.072}{1.794}\right) = 0.275 = 27.5\%$$

Problem 3.48

For a large fill operation, the average permanent load at middle of the clay layer will increase by 75 kPa. The average effective overburden pressure on the clay layer before the fill operation was 110 kpa. For the clay the following are given:

Normally consolidated clay, drained at the top and bottom; $H = 8$ m; $C_c = 0.27$; $e_o = 1.02$; $c_v = 0.52\text{m}^2/\text{month}$. Determine: (a) the consolidation settlement of the clay layer caused by the addition of the permanent load, $\Delta\sigma'_{(p)}$, (b) the time required for 80% of consolidation settlement under the additional permanent load only, (c) the temporary surcharge pressure increase, $\Delta\sigma'_{(f)}$, which will be required to eliminate the entire consolidation settlement in 12 months by the precompression technique. Assume that the surcharge is applied instantaneously.

Solution:

(a) Equation (3.36)-(a):

$$S_c = \frac{C_c H}{1+e_o}\log\frac{\sigma'_o+\sigma'_z}{\sigma'_o}$$

$$\sigma'_z = \Delta\sigma'_{(p)} = 75 \text{ kPa}$$

$$S_c = \frac{C_c H}{1+e_o}\log\frac{\sigma'_o+\Delta\sigma'_z}{\sigma'_o} = \frac{0.27\times 8}{1+1.02}\log\frac{110+75}{110} = 0.241 \text{ m} = \mathit{241\ mm}$$

(b) For 80% consolidation settlement, $U = 80\%$

Assume using Equation (3.47):

$$T_v = 1.781 - 0.933\log(100-U\%)$$

$$T_v = 1.781 - 0.933\log(100-80) = 0.567$$

Equation (3.45):

$$t = \frac{T_v d^2}{c_v} = \frac{0.567\times\left(\frac{8}{2}\right)^2}{0.52} = \mathit{17.45\ month}$$

or,

Figure 3.22: for $T_c = 0$ and $U = 80\%$, obtain $T_v = 0.565$ Therefore,

$$t = \frac{0.565\times\left(\frac{8}{2}\right)^2}{0.52} = 17.44 \text{ months}$$

(Continued)

(c) Final consolidation settlement under the average permanent load $\Delta\sigma'_{(p)}$ is 241 mm.

Equation (3.44): $$T_v = \frac{c_v\,t}{d^2} = \frac{0.52 \times 12}{(8/2)^2} = 0.39$$

Use Figure 3.22 with $T_c = 0$ and $T_v = 0.39$ or either of Equations (3.47) and (3.48) and obtain $U \cong 70\%$. Therefore, settlement due to 100% consolidation is

$$\frac{100}{70} \times 241 = 344\,\text{mm} = 0.344\,\text{m}.$$

For this settlement to take place in 12 months by the precompression technique, the required load increase of the entire surcharge is

$\sigma'_z = \left[\Delta\sigma'_{(p)} + \Delta\sigma'_{(f)}\right]$. Therefore,

$$S_c = 0.344 = \frac{C_c H}{1+e_o}\log\frac{\sigma'_o + \left(\Delta\sigma'_{(p)} + \Delta\sigma'_{(f)}\right)}{\sigma'_o}$$

$$0.344 = \frac{0.27 \times 8}{1+1.02}\log\frac{110 + 75 + \Delta\sigma'_{(f)}}{110}$$

$$\log\frac{110 + 75 + \Delta\sigma'_{(f)}}{110} = 0.322. \text{ Hence, } \frac{110 + 75 + \Delta\sigma'_{(f)}}{110} = 2.1.$$

$$\Delta\sigma'_{(f)} = 46\ kPa$$

Another method for determining (approximately) the surcharge pressure $\Delta\sigma'_{(f)}$ is by using the scheme in Problem 3.42. Value of U is computed as before. Then, from the scheme, for values of $U = 70\%$ and $\left(\frac{\Delta\sigma'_{(p)}}{\sigma'_o}\right) = \left(\frac{75}{110}\right) = 0.682$, obtain $\left(\frac{\Delta\sigma'_{(f)}}{\Delta\sigma'_{(p)}}\right) = 0.63$. Therefore, $\Delta\sigma'_{(f)} = 0.63 \times 75 = 47$ kpa

Problem 3.49

The loading period for a structure was two years. Average settlement of 12 cm occurred five years after the structure's load was first applied due to consolidation of an underlying clay layer. It is known that the final settlement will exceed 30 cm. Estimate the settlement after 10 years from the time the loading was started.

Solution:
It is reasonable to measure the time t for the purpose of settlement calculations, from the middle of the loading period. Thus, the 12 cm settlement occurred during four years, that is t = 4 years and it is desired what settlement will occur at the required time t' = 9 years.

For a certain clay deposit:

$$S_c = U\,S_{c(\text{final})};\text{hence, } S'_c : S_c = U' : U$$

For $U < 60\%$: $$T_v = \frac{\pi}{4}U^2;\text{ hence, } \sqrt{T'_v} : \sqrt{T_v} = U' : U$$

Equation (3.44): $$T_v = \frac{c_v\,t}{d^2};\text{ hence, } \sqrt{T'_v} : \sqrt{T_v} = \sqrt{t'} : \sqrt{t}$$

Note that the terms with and without primes represent quantities for the required case and given case, respectively.

Now, we can write $S'_c : S_c = \sqrt{t'} : \sqrt{t}$ or $S'_c = 12 = \sqrt{9} : \sqrt{4}$

Therefore, the settlement after 10 years from the time the loading was started, $S'_c = 12 \times \frac{3}{2} = 18\ cm.$

Since the final settlement was given as greater than 30 cm the obtained 18 cm settlement probably corresponds to less than 60%, that is $U < 60\%$ and cannot be above 60% by a sufficient amount to invalidate the expression $T_v = \frac{\pi}{4}U^2$.

Problem 3.50

Building (A) settled 10 cm in three years due to consolidation of the underlying clay layer and it is known that the final settlement will be about 30 cm Building (B) and its underlying clay layer are, as far as the available data confirm, very similar to (A) except that the underlying clay layer is 20% thicker than the clay layer below (A). The available data also confirm that the increase in the average pressure at the middle of the two clay layers can be considered approximately alike. Estimate the final settlement of building (B) and also the settlement in three years.

Solution:

For a certain clay deposit:

$S_c = m_v \sigma'_z H$; hence, $S'_c : S_c = H' : H$

$H' = 1.2H$. Therefore, $S'_c : 30 = 1.2H : H$.

The final settlement of building(B) $= S'_c = 30 \times 1.2 = 36\ cm$

Equation (3.44):

$$T_v = \frac{c_v t}{d^2}$$

$$T_v = c_v\left(\frac{t}{H^2}\right);$$

$$T_v = c_v\left(\frac{t}{H^2}\right);\text{hence,}$$

$$T'_v : T_v = \frac{t'}{(H')^2} : \frac{t}{(H)^2}$$

$$T'_v : T_v = \frac{3}{(1.2H)^2} : \frac{3}{(H)^2} = \frac{1}{(1.2)^2}.$$

In this case the expression $T_v = \frac{\pi}{4}U^2$ is valid, that is $U < 60\%$; hence

$T'_v : T_v = (U')^2 : (U)^2$; hence, $U' : U = \frac{1}{1.2}$

$S_c = US_{c(\text{final})}$; $U = \frac{S_c}{S_{c(\text{final})}}$; hence, $U' : U = \frac{S_c'}{S'_{c(\text{final})}} : \frac{S_c}{S_{c(\text{final})}} = \frac{1}{1.2}$

$$\frac{S'_c S_{c(\text{final})}}{S_c S'_{c(\text{final})}} = \frac{1}{1.2}$$

$S_{c(\text{final})} = m_v \sigma'_z H$ and $S'_{c(\text{final})} = m_v \sigma'_z (1.2H)$; hence, $\frac{S_{c(\text{final})}}{S'_{c(\text{final})}} = \frac{1}{1.2}$

Therefore, $S'_c = S_c$. Thus at any time the settlement of buildings (A) and (B) are alike, as long as $U < 60\%$.

Therefore, settlement in three years = 10 cm.

Problem 3.51

The settlement analysis, based on two-way drainage of the underlying clay layer, for a proposed structure indicates 8 cm of settlement in four years and a final settlement of 25 cm. However, there is some indication that there may be no drainage at the bottom. For this second case, that is one-way drainage, determine the final settlement and the time required for the same 8 cm settlement.

Solution:
The general equation of final consolidation settlement $(S_c = m_v \sigma'_z H)$ does not include a term that depends on drainage conditions. Hence, the final settlement is independent of drainage conditions. Accordingly, for the second case, that is one-way drainage, the final settlement is *25 cm*.

Degrees of consolidation at 8 cm of final settlements are the same for both cases, since the final settlements are the same. Therefore, the time factors $\left(T_v = \frac{c_v t}{d^2}\right)$ are the same too. Hence,

$$\frac{c_v \times 4}{\left(\frac{d}{2}\right)^2} = \frac{c_v t}{d^2}$$

$$t = \frac{c_v \times d^2 \times 4 \times 4}{c_v \times d^2} = 16\,\text{years}$$

For the second case, that is one-way drainage, the time required for the same 8 cm settlement is *16 years*.

Problem 3.52

Consider the clay layer of Problem 3.51 with two-way drainage condition. One of the borings at the site showed thin sand strata at points about one-third and two-thirds of the layer. Assume that these two thin strata are completely drained but that other conditions remain unchanged. Determine the time required for 8 cm of settlement.

Solution:
Assume that the average increase in pressure σ'_z for the entire clay layer is constant throughout the height and that it may be used as σ'_z for each of the three partial heights. This assumption might lead to appreciable errors in estimates for the top and bottom thirds individually, but in estimates for the entire stratum the errors in the top third will approximately balance those in the bottom third.

For $U < 60\% : T_v = \frac{\pi}{4} U^2$; hence, $\sqrt{T'_v} : \sqrt{T_v} = U' : U = S'_c : S_c$

$$\sqrt{\frac{c_v t'}{d^2}} : \sqrt{\frac{c_v t}{d^2}} = S'_c : S_c$$

$$\frac{\sqrt{t'}}{\sqrt{t}} = \frac{S'_c}{S_c}.$$

The settlement in one year is computed as $S'_c = S_c \times \frac{\sqrt{t'}}{\sqrt{t}} = 8 \times \frac{1}{\sqrt{4}} = 4\,\text{cm}$

Therefore, the three sections (sublayers of the same thickness) together will settle $3 \times 4 = 12$ cm in one year and a check shows that this is less than 60% of the final 25 cm settlement.

$$\sqrt{t'} = \sqrt{t} \times \frac{S'_c}{S_c}.$$

$$\sqrt{t'} = \sqrt{1} \times \frac{8}{12}.$$

$$t' = \left(\sqrt{1} \times \frac{8}{12}\right)^2 = \frac{4}{9}\text{year.}$$

Therefore, the time required for 8 cm settlement = *0.444 years.*

Problem 3.53

The laboratory data for the clay layer of Problem 3.51, with two-way drainage condition, were from tests on a few samples. Subsequent tests on additional samples give what are believed to be more accurate determinations of the soil coefficients and show that the coefficient of volume compressibility m_v is 20% smaller and the coefficient of consolidation c_v is 30% smaller than originally obtained. Considering these changes into account, but otherwise using the same available data of the original two-way drainage condition, determine the final settlement and the time required for 8 cm settlement.

Solution:

Equation (3.44): $S_c = m_v \sigma'_z H$

$$S'_{c(\text{final})} = m'_v \sigma'_z H \quad \text{(after change)}$$

$$S_{c(\text{final})} = m_v \sigma'_z H \quad \text{(before change)}$$

$$m'_v = 0.8\, m_v$$

$$m'_v : m_v = S'_{c(\text{final})} : S_{c(\text{final})}$$

$$S'_{c(\text{final})} = S_{c(\text{final})} \times \frac{m'_v}{m_v} = 25 \times \frac{0.8\, m_v}{m_v} = 20 \text{ cm}$$

Taking the changes into account, the final settlement = *20 cm.*
Considering 8 cm settlement:

For $U < 60\%$:

$$T_v = \frac{\pi}{4} U^2 = \frac{c_v t}{(d/2)^2}$$

For $U = \frac{8}{25}$:

$$\frac{\pi}{4}\left(\frac{8}{25}\right)^2 = \frac{c_v \times 4}{(d/2)^2}; \text{ hence, } \left[\frac{c_v}{(d/2)^2}\right] = \frac{\pi}{16}\left(\frac{8}{25}\right)^2$$

For $U' = \frac{8}{20}$:

$$\frac{\pi}{4}\left(\frac{8}{20}\right)^2 = \frac{c'_v \times t}{(d/2)^2}; \; c'_v = 0.7 c_v; \text{ hence,}$$

(*Continued*)

$$\frac{\pi}{4}\left(\frac{8}{20}\right)^2 = \frac{0.7 \times c_v \times t}{(d/2)^2} = 0.7t\left[\frac{c_v}{(d/2)^2}\right];$$

$$t = \frac{\frac{\pi}{4}\left(\frac{8}{20}\right)^2}{0.7\left[\frac{c_v}{(d/2)^2}\right]} = \frac{\frac{\pi}{4}\left(\frac{8}{20}\right)^2}{0.7 \times \frac{\pi}{16}\left(\frac{8}{25}\right)^2} = \frac{4\left(\frac{8}{20}\right)^2}{0.7\left(\frac{8}{25}\right)^2} = \frac{4 \times 625}{0.7 \times 400} = 8.93\,\text{years}$$

Taking the changes into account, the time required for 8 cm of settlement = *8.93 years.*

Problem 3.54

The settlement analysis for a proposed building structure indicates 8 cm of settlement in four years and a final settlement of 25 cm. The underlying clay layer is drained at both its top and bottom surfaces. An alternate design of the structure uses a taller building (i.e. the same proposed building structure with increased number of stories only) which would increase the net building load by 25% On the basis of this heavier load, determine the final settlement, the settlement in four years and the time required for 8 cm settlement.

Solution:
Equation (3.44): $S_c = m_v \sigma'_z H$

$$S_{c(\text{final})} = m_v\, \sigma'_z H \quad \text{(before change)}$$

$$S'_{c(\text{final})} = m'_v\left(1.25\,\sigma'_z\right) H \quad \text{(after change)}$$

$$S'_{c(\text{final})} = \frac{1.25\sigma'_z \times 25}{\sigma'_z} = 31.25\ \text{cm}$$

The final settlement under the heavier building = *31.25 cm*
Equation (3.44):

$$T_v = \frac{c_v t}{d^2}$$

Since the quantity $\left(\frac{c_v}{d^2}\right)$ remains unchanged, for the same time t (i.e. $t' = t = 4$ years) the time factor T_v remains unchanged too (i.e. $T'_v = T_v$).
Equation (3.46):

$$T_v = \frac{\pi}{4}U^2 \ (\text{for}\, U < 60\%)$$

$$T_v = \frac{\pi}{4}U^2 \ (\text{before change})$$

$$T'_v = \frac{\pi}{4}U^2 \ (\text{after change})$$

Therefore, $U' = U$

$$S_c = U S_{c(\text{final})}$$

$U' = \frac{S_c'}{S'_{c(\text{final})}}$; hence, $U = \frac{S_c}{S_{c(\text{final})}} = \frac{8}{25} = U'$

$\frac{8}{25} = \frac{S_c}{32.25}$; hence, $S_c = \frac{31.25 \times 8}{25} = 10\,\text{cm}$

Under the heavier building, the settlement in four years = *10 cm*.

For $U = \frac{10}{31.25}$ and $t = 4\,\text{years}$:

$$T_v = \frac{\pi}{4}\left(\frac{10}{31.25}\right)^2 = 4\left(\frac{c_v}{d^2}\right);\ \text{hence}\ \left(\frac{c_v}{d^2}\right) = \left[\frac{\pi}{4\times 4}\left(\frac{10}{31.25}\right)^2\right]$$

For $U' = \frac{8}{31.25}$ and $t = t'$:

$$T_v' = \frac{\pi}{4}\left(\frac{8}{31.25}\right)^2 = t'\left(\frac{c_v}{d^2}\right) = t'\left[\frac{\pi}{4\times 4}\left(\frac{10}{31.25}\right)^2\right]$$

$$t' = \frac{\frac{\pi}{4}\left(\frac{8}{31.25}\right)^2}{\left[\frac{\pi}{4\times 4}\left(\frac{10}{31.25}\right)^2\right]} = \frac{4\times 8^2}{10^2} = 2.56\ \text{years}$$

The time required for 8 cm settlement, under the heavier building, is *2.56 years*.

Problem 3.55

A building consisting of 20 storeys is founded on a stiff raft 20 m wide by 40 m long at a depth of 2 m below ground level. The raft rests on a weathered rock becoming less weathered with increasing depth, until a relatively incompressible stratum is met at a depth of 45 m Deformation modulus, E_d, values obtained from plate-load tests made in a large-diameter borehole are shown in the scheme below. Estimate the settlement of the building for a net bearing pressure (net uniform contact pressure) of 250 kPa (a) considering the raft as a flexible base, (b) considering the raft as a rigid base.

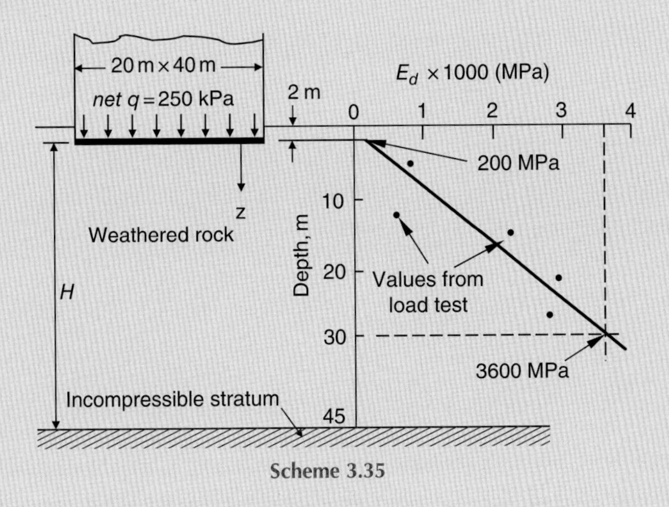

Scheme 3.35

(Continued)

Solution:

Variation of E_d with depth is obtained by plotting a straight line through the given points, shown in the figure. At foundation level E_f = 200 MPa and at a depth of 30 m below ground level, where z = 28 m, E_d = 3600 MPa.

Equation (3.65):

$$K=\left(\frac{E_d-E_f}{E_f}\right)\left(\frac{B}{z}\right)=\left(\frac{3600-200}{200}\right)\left(\frac{20}{28}\right)=12.14.$$

(a) Considering the raft as a flexible base.
The raft area is divided into four equal rectangles, each 10 × 20 m.

$H=45-2=43$ m; $\frac{H}{B}=\frac{43}{10}=4.3$; $\frac{L}{B}=\frac{20}{10}=2$; $k=12.14$. Therefore, from Figure 3.25: $I'_p\approx 0.05$

From Figure 3.26a, for $\frac{H}{B}=\frac{43}{20}=2.15$, the factor $F_B=0.96$

From Figure 3.26b, for $\frac{D}{B}=\frac{2}{20}=0.1$, the factor $F_D=1.0$

From Equation (3.63), immediate settlement at corner of each flexible rectangle is given by

$$\rho_i=q\left(\frac{B}{E_f}\right)I'_pF_BF_D=(250)\left(\frac{10}{200\times 1000}\right)(0.05)(0.96)(1)=6\times 1^{-4}\text{m}$$

The immediate settlement at the *centre* of the *flexible* raft is

$$\rho_i=4\left(6\times 1^{-4}\right)=2.4\times 10^{-3}\,\text{m}=2.4\,mm$$

(b) Considering the raft as a rigid base.
The average settlement of the rigid foundation is given by

Equation (3.64): $\rho_{i,\text{ ave}}=1/3\left(\rho_{i,\text{ centre}}+\rho_{i,\text{midpoint long side}}+\rho_{i,\text{ corner}}\right)$

$\rho_{i,corner}$:

$H=45-2=43$ m; $\frac{H}{B}=\frac{43}{20}=2.15$; $\frac{L}{B}=\frac{40}{20}=2$; $k=12.14$ Therefore, from Figure 3.25: $I'_p\approx 0.05$

From Figure 3.26a, for $\frac{H}{B}=\frac{43}{20}=2.15$, the factor $F_B=0.96$

From Figure 3.26b, for $\frac{D}{B}=\frac{2}{20}=0.1$, the factor $F_D=1.0$

From Equation (3.63), immediate settlement at corner of the flexible raft is given by

$$\rho_i=q\left(\frac{B}{E_f}\right)I'_pF_BF_D=(250)\left(\frac{20}{200\times 1000}\right)(0.05)(0.96)(1)=1.2\times 1^{-3}\text{m}$$

$\rho_{i,\text{ mid point long side}}$:

$$H = 45 - 2 = 43\text{m}; \frac{H}{B} = \frac{43}{20} = 2.15; \frac{L}{B} = \frac{20}{20} = 1; k = 12.14$$

Therefore, from Figure 3.25: $I'_P \approx 0.05$

From Figure 3.26a, for $\frac{H}{B} = \frac{43}{20} = 2.15$, the factor $F_B = 0.96$

From Figure 3.26b, for $\frac{D}{B} = \frac{2}{20} = 0.1$, the factor $F_D = 1.0$

From Equation (3.63), immediate settlement at the corner of each flexible rectangle is given by

$$\rho_i = q\left(\frac{B}{E_f}\right)I'_P F_B F_D = (250)\left(\frac{20}{200 \times 1000}\right)(0.05)(0.96)(1) = 1.2 \times 1^{-3}\text{m}$$

Immediate settlement at the midpoint of a long side of the flexible raft is

$$\rho_i = 2\left(1.2 \times 1^{-3}\right) = 2.4 \times 10^{-3} m$$

$\rho_{i,\text{ centre}}$:

From (a), $\rho_{i,\text{ centre}} = 2.4 \times 10^{-3}$ m

The average settlement of the *rigid* foundation is

$$\rho_{i,\text{ ave}} = \left(\frac{1}{3}\right)(2.4 \times 10^{-3} + 2.4 \times 10^{-3} + 1.2 \times 10^{-3}) = 2 \times 10^{-3}\text{ m}$$

$$= 2\ mm$$

There could, in addition, be some creep settlement in the weathered rock which might double the immediate settlement.

Problem 3.56

A column footing 1.8 m square is to be founded at a depth of 0.6 m on weak medium-bedded poorly cemented sandstone. The net foundation pressure (net contact pressure) is 1.55 MPa Examination of rock cores showed an average joint spacing of 250 mm Tests on the cores showed representative uniaxial compression strength of 1.75 MPa Make an estimation of the foundation settlement. Consider the footing as a rigid base.

Solution:
For the 250 cm joint spacing, the fracture frequency per metre (or joint spacing per metre) is (1000/250) = 4 From Table 3.18, for a fracture frequency of four per metre; a suitable value for the mass factor (j) is 0.6, Also, from Table 3.19, for poorly cemented sandstone the modulus ratio (M_r) is 150. Let q_{uc} of the intact rock equals the uniaxial compression strength, which is 1.75 MPa. The mass deformation modulus is given by Equation (3.66):

$$Em = jM_r q_{uc} = 0.6 \times 150 \times 1.75 = 157.5\text{MPa}$$

(*Continued*)

Assume the influence depth H of the rock below the foundation level is $4B = 4 \times 1.8 = 7.2\text{m}$
Divide the footing area into four equal squares each of 0.9 m width.

From Figure 3.25 for $\frac{L}{B} = \frac{0.9}{0.9} = 1$; $\frac{H}{B} = \frac{7.2}{0.9} = 8$ and $k = 0$ (using $E_m = E_d = E_f$), obtain $I'_p \approx 0.48$

From Figure 3.26a, for $\frac{H}{B} = \frac{7.2}{1.8} = 4$, the factor $F_B \approx 1$

From Figure 3.26b, for $\frac{D}{B} = \frac{0.6}{1.8} = 0.33$, the factor $F_D = 0.93$

From Equation (3.63), immediate settlement at corner of each flexible square is given by

$$\rho_i = q\left(\frac{B}{E_f}\right) I'_p F_B F_D = (1.55)\left(\frac{0.9}{157.5}\right)(0.48)(1)(0.93) = 3.954 \times 1^{-3}\text{m}$$

The immediate settlement at centre of the flexible raft is

$$4\left(3.954 \times 1^{-3}\right) = 15.82 \times 10^{-3}\text{m}$$

According to Equation (3.6) the rigidity factor is 0.93 and according to Table 3.15 the rigidity factor is 0.85. Assume the rigidity factor is 0.90. Accordingly, the average settlement of the foundation is

$$0.90 \times 15.82 \times 10^{-3} = 14.24 \times 10^{-3}\text{m} = 14.24\ \text{mm; say } 15\ mm$$

(Creep might increase this settlement to about 30 m in the long term.)

References

AASHTO (1996), *Standard Specifications for Highway Bridges, 16th edn*, American Association of State Highway and transportation officials, Washington, D.C.

Barron, R. A. (1948), "Consolidation of Fine-Grained Soils by Drain Wells", *Transactions ASCE*, **113**, pp. 718–742.

Bjerrum, L. (1963), "Discussion on Compressibility of soils", *Proceedings of the European Conference on Soil Mechanics and Foundation Engineering*, Wiesbaden, Vol. **2**, pp. 16–17.

Bjerrum, L. (1967), "Engineering Geology of Norwegian Normally Consolidated Marine Clays as Related to Settlement of Buildings", *Geotechnique*, **17**, pp. 83–118.

Bowles, J. E. (2001), *Foundation Analysis and Design*, 5th edn, McGraw-Hill, New York.

Bowles, J. E. (1977), *Foundation Analysis and Design*, 2nd edn, McGraw-Hill, New York.

Buisman, A. K. (1943), *Grondmechanica* (Soil Mechanics), 2nd edn, 281 pp., Waltman, Delft, Netherlands.

Burland, J. B. and Wroth, C. P. (1974), "Review Paper: Settlement of Buildings and Associated Damage", *Proceedings of the Conference on Settlement of Structures*, Pentech Press, Cambridge, pp. 611–654.

Burland, J, B. and Burbidge, M. C. (1985), "Settlement of Foundations on Sand and Gravel", *Proceedings, Institute of Civil Engineers*, Part I, Vol. **7**, pp. 1325–1381.

Burland, J, B., Broms, B. B. and De Mello, V. F. B. (1977), "Behaviour of Foundations and Structures", *Proceedings of the 9th ICSMFE, Tokyo*, Vol. **2**, Japanese Society SFME, pp. 495–538.

Casagrande, A. (1936), "The Determination of the Preconsolidation Load and Its Practical Significance", *Proceedings of the 1st ICSMFE*, Vol. **3**, pp. 60–64.

Chandler, R. J. and Davis, A. G. (1973), "Further Work on the Engineering Properties of Keuper Marl", *Construction Industry Research and Information Association Report No. 47*, England.

Christian, J. T. and Carrier, W. D. (1978), "Janbu, Bjerrum and Kjaernsli's chart reinterpreted", *Canadian Geotechnical Journal*, **15**, pp. 123–128, and discussion, **15**, pp. 436–437.

Coduto, Donald P. (2001), *Foundation Design: Principles and Practices*, 2nd edn, Prentice-Hall, Inc., New Jersey.

Craig, R. F. (2004), *Soil Mechanics*, 7th edn, Chapman and Hall, London.

D'Appolonia, D. J., D'Appolonia, E. D. and Brissetta, R. F. (1970), "Closure Settlement of Spread Footings on Sand", *Journal of Soil Mechanics and Foundations Division*, ASCE, Vol. **96**, SM 2, March, pp. 561–584.

Das, Braja M. (2011), *Principles of Foundation Engineering*, 7th edn, CENGAGE Learning, United States.

DeBeer, E. E. and Martens, A. (1957), "Method of Computation of an Upper Limit for the Influence of Heterogeneity of Sand Layers on the Settlement of Bridges", *Proceedings 4th International Conference on Soil Mechanics and Foundation Engineering, London*, Vol. **1**, pp. 275–281, Butterworths, London.

DeBeer, E. E. (1963), "The Scale Effect in the Transposition of the Results of Deep Sounding Tests on the Ultimate Bearing Capacity of Piles and Caisson Foundations", *Geotechnique*, **13**, pp. 39–75.

European Committee for Standardisation (1994a), *Basis of Design and Actions on Structures,* Eurocode 1, Brussels, Belgium.

European Committee for Standardisation (1994b), *Geotechnical Design, General Rules-Part 1,* Eurocode 7, Brussels, Belgium.

Fox, E. N. (1957), "The Mean Elastic Settlement of a Uniformly Loaded Area at a Depth below the Ground Surface", *Proceedings 2nd International Conference on Soil Mechanics and Foundation Engineering, London*, Vol. **1**, pp. 129–132, Rotterdam.

Hobbs, N, B. (1974), "General Report and State-of-the-Art Review", *Proceedings of the Conference on Settlement of Structures*, Pentech Press, Cambridge, pp. 579–609.

Knappett, J. A. and Craig, R. F. (2012), *Soil Mechanics*, 8th edn, Spon Press, Abingdon, Oxon, United Kingdom.

Leonards, G. A. (1976), *Estimating Consolidation Settlement of Shallow Foundations on Overconsolidated Clay*, Special Report No. 163, Transportation Research Board, Washington, D.C., pp. 13–16.

MacDonald, D. H. and Skempton, A. W. (1955), "A Survey of Comparisons between Calculated and Observed Settlements of Structures on Clay", *Conference on Correlation of Calculated and Observed Stresses and Displacements*, ICE, London, pp. 318–337.

Mayne, P. W. and Poulos, H. G. (1999), "Approximate Displacement Influence Factors for Elastic Shallow Foundations", *Journal of Geotechnical and Geoenvironmental Engineering*, ASCE, Vol. **125**, No. 6, pp. 453–460.

Meigh, A. C. (1976), "The Triassic Rocks, with Particular Reference to Predicted and Observed Performance of some Major Structures", *Geotechnique*, **26,** pp. 393–451, London.

Mesri, G. (1973), "Coefficient of Secondary Compression", *Journal of the Soil Mechanics and Foundations Division*, ASCE, Vol. **99**, No. SM1, pp. 122–137.

Mesri, G. and P. M. Godlewski (1977), "Time- and Stress-Compatibility Interrelationship", *Journal of Geotechnical Engineering Division*, ASCE, Vol. **103**, GT5, May, pp. 417–430.

Meyerhof, G. G. (1956), "Penetration Tests and Bearing Capacity of Cohesionless Soils", *Journal of Soil Mechanics and Foundations Division*, ASCE, Vol. **82**, SM1, pp. 1–19.

Meyerhof, G. G. (1965), "Shallow Foundations", *Proceedings ASCE*, **91**, No. SM2, pp. 21–31.

Meyerhof, G. G. (1974), "Ultimate Bearing Capacity of Footings on Sand layer Overlying Clay", *Canadian Geotechnical Journal*, Ottawa, Vol. **11**, No. 2, May, pp. 223–229.

Meyerhof, G. G. (1982), *The Bearing Capacity and Settlement of Foundations*, Tech-Press, Technical University of Nova Scotia, Halifax.

Mikhejef, V. V., Ushkalor, V. P., Tokar, R. A. et al. (1961), "Foundation Design in the USSR", *Fifth ICSMFE*, Vol. **1**, pp. 753–757.

Olson, R. E. (1977), "Consolidation under Time-Dependent Loading", *Journal of Geotechnical Engineering Division*, ASCE, Vol. **102**, No. GT1, pp. 55–60.

Polshin, D. E. and Tokar, R. A. (1957), "Maximum Allowable Non-Uniform Settlement of Structures", *Proceedings of the 4th ICSMFE, London*, Vol. **1**, pp. 402–405.

Schmertmann, J. H. and Hartman, J. P. (1978), "Improved Strain Influence Factor Diagrams", *Journal of the Soil Mechanics and Foundations Division*, ASCE, Vol. **104**, No. GT8, pp. 1131–1135.

Senneset, K., Sandven, R., Lunne, T. et al. (1988), "Piezocone Tests in Silty Soils", *Proceedings 1st ISOPT*, Vol. **2**, pp. 955–966.

Simons, N. A. (1974), "Review Paper: Normally Consolidated and lightly Overconsolidated Cohesive Materials", *Proceedings of the Conference on Settlements of Structures*, Pentech Press, Cambridge, pp. 500–530.

Sivaram, B. and Swamee, P. (1977), "A Computational Method for Consolidation Coefficient", *Soils and Foundations*, Tokyo, Vol. **17**, No. 2 pp. 48–52.

Skempton, A. W. and Bjerrum, L. (1957), "A Contribution to the Settlement Analysis of Foundations on Clay", *Geotechnique*, **7**, pp. 168–178.

Skempton, A. W. and MacDonald, D. H. (1956), "The Allowable Settlement of Buildings", *Proceedings of the Institute of Civil Engineering*, Part 3, **5**, pp. 727–784.

Sutherland, H. B. (1974), "Review paper: Granular Materials", *Proceedings of the Conference on Settlement of Structures*, Pentech Press, Cambridge, pp. 473–499.

Terzaghi, K., Peck, R. B. (1967), *Soil Mechanics in Engineering Practice*, 2nd edn, John Wiley, New York.

Tomlinson, M. J. (2001), *Foundation Design and Construction*, 7th edn, Pearson Education Ltd., Harlow, Essex.

Tomlinson, M. J. (1995), *Foundation Design and Construction*, 6th edn, Longman Scientific and Technical, Harlow, Essex.

Vesic, A. S. (1970), "Tests on Instrumented Piles, Ogeechee River Site", *Journal of Soil Mechanics and Foundations Division*, ASCE, Vol. **96**, SM 2, March, pp. 561–584.

Wahls, Harvey E. (1994), "Tolerable Deformations", *Vertical and Horizontal Deformations of Foundations and Embankments*, A. T. Yeung and G. Y, Felio, Eds., Vol. **2**, pp. 1611–1628, ASCE.

CHAPTER 4

Shallow Foundations – Bearing Capacity

4.1 General

As discussed in Section 2.4, the geotechnical strength, expressed as *bearing capacity*, is considered one of the most important performance aspects of shallow foundations. Therefore, the subject of bearing capacity is perhaps the most important of all subjects in soil engineering. Geotechnical strength requirements of a shallow foundation comprise two fundamental criteria which must be always satisfied:

(1) The shearing stresses transmitted to the supporting material must be smaller than the *shearing strength* by an amount sufficient to give an ample factor of safety, a value between 2.5 and 3.0 normally being specified.
(2) The settlement of the foundation should be tolerable and, in particular, differential settlement should not cause any unacceptable damage nor interfere with the function of the structure. This settlement criterion was discussed in Chapter 3.

This chapter will concern criterion (1), that is the bearing capacity requirements. In conjunction with this subject, sometimes the *shear failure* of the supporting soil is called *bearing capacity failure*, maybe because the observation of the behaviour of shallow foundations revealed that bearing capacity failure usually occurs due to shear failure of the supporting material.

Three principal *modes of shear failure*, shown in Figure 4.1, have been identified and described in literature: *general shear failure, local shear failure* and *punching shear failure.*

The mode of the *general shear failure* is characterised by the existence of a well-defined failure pattern, which consists of continuous slip surfaces from the edges of the footing to ground surface. Under the increased pressure towards the ultimate failure load, the state of plastic equilibrium is reached initially in the soil around the edges of the footing then gradually spreads downwards and outwards. Finally the state of plastic equilibrium is fully developed throughout the soil above the failure surfaces (Figure 4.1a). Bulging or heaving of the adjacent ground surface occurs on both sides of the footing, although the final slip movement (or soil collapse) occurs only on one side and it is often accompanied by tilting of the footing. The general shear failure is typical of soils of low compressibility such as dense or stiff soils. In this mode of failure, the pressure settlement curve comes to a vertical ultimate condition at a relatively small settlement, as illustrated by curve (*a*) in Figure 4.1.

Shallow Foundations: Discussions and Problem Solving, First Edition. Tharwat M. Baban.

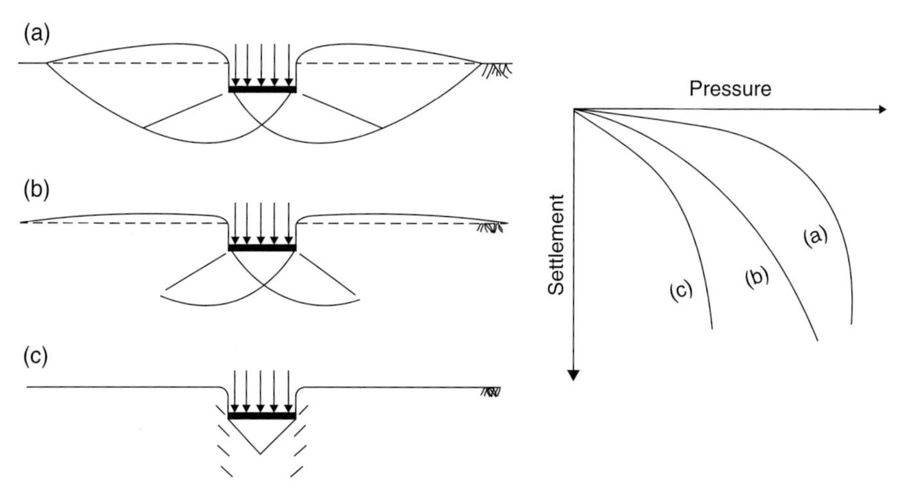

Figure 4.1 Modes of bearing capacity failure with typical pressure–settlement curves: (a) general shear failure, (b) local shear failure, (c) punching shear failure (reproduced from Knappett and Craig, 2012).

In contrast to the mode of general shear failure, the *punching shear failure* mode is characterised by a failure pattern that is not easily identified. Under the increasing pressure the downwards vertical movement of the footing is accompanied by compression of the soil immediately underneath; shear in vertical direction around the footing edges assists the progress of the footing penetration to continue (Figure 4.1c). There is little or no bulging of the ground surface away from the footing edges; both the vertical and horizontal equilibrium of the footing are maintained; hence, there is neither visible collapse nor substantial tilting. The ultimate load at failure is not defined and relatively large settlements occur, as illustrated by curve (*c*) in Figure 4.1. Punching shear failure depends on the compressibility of the supporting soil and the depth of foundation relative to its breadth. It occurs in very loose sands, thin crust of strong soil underlain by a very weak soil, in weak clays loaded under slow drained conditions, or in a soil of low compressibility if the footing is located at considerable depth.

The mode of *local shear failure* is characterised by a failure pattern which is clearly defined only immediately below the foundation (Figure 4.1b). There is significant compression of the soil under the footing and only partial development of the state of plastic equilibrium. Therefore the failure surfaces end somewhere in the soil mass without reaching the ground surface. However, there is a visible tendency toward soil bulging on the sides of the footing. There would be no catastrophic collapse or tilting of the footing. As indicated by curve (*b*) in Figure 4.1, local shear failure is characterised by the occurrence of relatively large settlements and the ultimate load at failure is not clearly defined. This type of shear failure mode is usually associated with soils of high compressibility. According to Vesic (1973), the local shear failure retains some characteristics of both general and punching shear failure modes, representing truly a transitional mode.

4.2 Basic Definitions

There are various terms, especially those which are directly related to pressures intensity, involved in bearing capacity analysis which requires clear understanding without confusion. Therefore, it may be useful at this stage to write down their definitions and symbols, as follows:

(1) *Total overburden pressure* σ_o is the intensity of total pressure or total stress due to the weights of both soil and soil water, on any horizontal plane at and below foundation level before construction operations come into action.

(2) *Effective overburden pressure* σ'_o is the intensity of the intergranular pressure, or effective normal stress, on any horizontal plane at and below foundation level before construction operations come into action. It is the total overburden pressure σ_o less the pore-water pressure u. Therefore,

$$\sigma'_o = \sigma_o - u$$

(3) *Gross foundation pressure* (or gross contact pressure, or, gross loading intensity), *gross* q, is the intensity of total pressure directly beneath the foundation (at the foundation level) after the structure has been erected and fully loaded. It is inclusive of the load of the foundation substructure, the loading from the superstructure and the total loading from any backfilled soil.

(4) *Gross effective foundation pressure, gross* q' is the gross foundation pressure less the uplift pressure due to height h of ground water table (W.T) above the foundation level, expressed as

$$gross\ q' = gross\ q - h\gamma_w$$

(5) *Net foundation pressure* (or net contact pressure, or, net loading intensity), *net* q, is the excess pressure or the difference between the gross foundation pressure *gross* q and the total overburden pressure σ_o directly beneath the foundation, expressed as

$$net\ q = gross\ q - \sigma_o \tag{4.1}$$

(6) *Net effective foundation pressure, net* q', is the effective excess pressure or the difference between the gross effective foundation pressure *gross* q' and the effective overburden pressure σ'_o directly beneath the foundation, expressed as

$$net\ q' = gross\ q' - \sigma'_o \tag{4.2}$$

The term *net* q' is used for calculating the net increase in effective stresses at any depth below the foundation level.

(7) *Gross ultimate bearing capacity* (or gross ultimate soil pressure), *gross* q_{ult}, is the minimum value of the gross effective contact pressure at which the supporting material fails in shear. It is often written as q_{ult} only.

(8) *Net ultimate bearing capacity* (or net ultimate soil pressure), *net* q_{ult}, is the minimum value of the excess effective contact pressure at which the supporting material fails in shear, expressed as

$$net\ q_{ult} = gross\ q_{ult} - \sigma'_o \tag{4.3}$$

(9) *Net safe bearing capacity, net* q_s is the net ultimate bearing capacity divided by a suitable safety factor *SF*, expressed as

$$net\ q_s = \frac{net\ q_{ult}}{SF} \tag{4.4}$$

(10) *Gross safe bearing capacity, net* q_s is the net safe bearing capacity plus the effective overburden pressure, expressed as

$$gross\ q_s = \frac{net\ q_{ult}}{SF} + \sigma'_o \tag{4.5}$$

Note that the σ'_o term is not divided by the safety factor and may not be logical to do so because its value is usually available in full. However, some authors define the *gross* q_s as the gross ultimate bearing capacity divided by a suitable safety factor, expressed as

$$gross\ q_s = \frac{gross\ q_{ult}}{SF} = \frac{net\ q_{ult} + \sigma'_o}{SF} = \frac{net\ q_{ult}}{SF} + \frac{\sigma'_o}{SF} \tag{4.6}$$

Table 4.1 Presumed bearing values (BS 8004: 1986).

Soil type	Bearing value (kN/m^2)	Remarks
Dense gravel or dense sand and gravel	>600	Width of foundation (B) not less than 1 m. Water table at least B below base of foundation
Medium-dense gravel or medium-dense sand and gravel	200–600	
Loose gravel or loose sand and gravel	<200	
Dense sand	>300	
Medium-dense sand	100–300	
Loose sand	<100	
Very stiff boulder clays and hard clays	300–600	Susceptible to long-term consolidation settlement
Stiff clays	150–300	
Firm clays	75–150	
Soft clays and silts	<75	
Very soft clays and silts	–	

(11) *Presumptive bearing value* is the net loading intensity considered appropriate to the particular type of soil for preliminary design purposes. The particular value is based either on local experience or by calculation from strength tests or field loading tests using a factor of safety against shear failure but without consideration of settlement. As an example, Table 4.1 gives presumed bearing values for different types of soils (BS 8004: 1986).

(12) *Net safe settlement pressure, net* $q_{s,\rho}$, is the maximum net effective pressure which the soil can carry without exceeding the particular allowable settlement.

(13) *Net allowable bearing capacity* or *allowable bearing pressure* (or: *net allowable soil pressure*; *design bearing capacity* or soil pressure) *net* q_a or *net* q_{design}. It is the maximum allowable net loading intensity that the supporting material can carry, taking into consideration the bearing capacity, the estimated amount and rate of settlement that will occur and the ability of the structure to accommodate the settlement. Hence, it is a function both of the site and of the structural conditions. It is either *net* q_s or *net* $q_{s,\rho}$, whichever is smaller; used in design of shallow foundations.

4.3 Gross and Net Foundation Pressures

The gross and net foundation pressures (or gross and net contact pressures) at the foundation level can be determined as follows:

(A) *The footing foundation is backfilled*
Refer to Figure 4.2a. The concrete footing is subjected to a superimposed load (a column load or a wall load per unit length) V. The footing is backfilled up to the ground surface with a material of unit weight γ. Footing base area $= A$, thickness $= D_c$ and unit weight $= \gamma_c$.

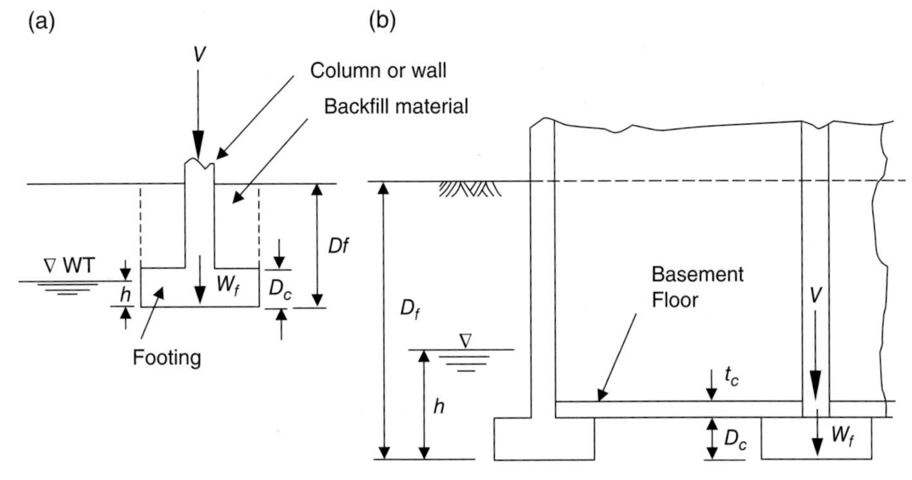

Figure 4.2 Footings subjected to superimposed loads.

The gross foundation pressure is given by

$$gross\ q = \frac{V + W_f}{A} \tag{4.7)-(a}$$

or

$$gross\ q = \frac{V}{A} + D_c\gamma_c + \gamma\left(D_f - D_c\right) \tag{4.7)-(b}$$

where W_f = total weights of the footing and the backfill material.

Assuming unit weight γ of the overburden soil above W.T is the same as that of the backfill material and the saturated unit weight of the overburden soil (below W.T) is γ_{sat}, then from Equation (4.1), the net foundation pressure is given by

$$net\ q = \left(\frac{V + W_f}{A}\right) - \sigma_o \tag{4.8)-(a}$$

or

$$\begin{aligned} net\ q &= \frac{V}{A} + D_c\gamma_c + \gamma\left(D_f - D_c\right) - \left[\gamma\left(D_f - h\right) + \gamma_{sat}h\right] \\ &= \frac{V}{A} - h(\gamma_{\text{sat}} - \gamma) + D_c(\gamma_c - \gamma) \end{aligned}$$

$$h(\gamma_{sat} - \gamma) = \gamma_w h.$$

Hence,

$$net\ q = \frac{V}{A} - \gamma_w h + D_c(\gamma_c - \gamma) \le net\ q_a \tag{4.8)-(b}$$

where: $\gamma_w h$ = uplift water pressure

From Equation (4.2), the net effective foundation pressure is given by

$$net\ q' = \left(\frac{V + W_f}{A} - \gamma_w h\right) - \sigma_o'$$

$$net\ q' = \frac{V}{A} + \gamma D_f - \gamma D_c + \gamma_c D_c - \gamma_w h - \left[\gamma\left(D_f - h\right) + h\left(\gamma_{sat} - \gamma_w\right)\right]$$

$$= \frac{V}{A} + \gamma D_f - \gamma D_c + \gamma_c D_c - \gamma_w h - \gamma D_f + \gamma h - \gamma_{sat} h + \gamma_w h \quad (4.9)\text{-(a)}$$

$$= \frac{V}{A} - h(\gamma_{sat} - \gamma) + D_c(\gamma_c - \gamma)$$

$$h(\gamma_{sat} - \gamma) = \gamma_w h.$$

Hence,

$$net\ q' = \frac{V}{A} - \gamma_w h + D_c(\gamma_c - \gamma) \quad \leq net\ q_a \quad (4.9)\text{-(b)}$$

It is noteworthy that both Equations (4.8)-(b) and (4.9)-(b) give the same result, indicating that it makes no difference whether the net foundation pressure is computed on the basis of *total* stress or *effective* stress approach.Equation (4.9)-(b) indicates that if $\gamma \cong \gamma_c$ and W.T is located at a depth greater than D_f, the net effective foundation pressure can be written as

$$net\ q' = \frac{V}{A} \quad (4.10)$$

For a safe design, the net effective foundation pressure $net\ q'$ should be less than or equal to the net allowable bearing capacity $net\ q_a$. Therefore,

$$net\ q' \leq net\ q_a \quad (4.11)$$

(B) *The footing foundation is not backfilled.*

Refer to Figure 4.2b. In this case the interior footings are directly below the basement floor, not backfilled and with no overburden soil above. Also, one side of the exterior footings are under the same loading conditions. Therefore, the overburden pressure should not be deducted from q when $net\ q$ or $net\ q'$ is calculated.

The gross foundation pressure is given by

$$gross\ q = \frac{V + W_f}{A} = \frac{V}{A} + D_c\gamma_c + t_c\gamma_c$$

$$net\ q = gross\ q = \frac{V}{A} + D_c\gamma_c + t_c\gamma_c \quad (4.12)$$

$$gross\ q' = \left(\frac{V}{A} - \gamma_w h\right) + D_c\gamma_c + t_c\gamma_c$$

$$net\ q' = gross\ q' = \left(\frac{V}{A} - \gamma_w h\right) + D_c\gamma_c + t_c\gamma_c \leq net\ q_a \quad (4.13)$$

where t_c = thickness of the floor slab

Comparing Equation (4.13) with Equation (4.9)-(b), it is observed that the load carrying capacity of a foundation is increased if it is not backfilled.

(C) *Mat or raft foundation under a basement.*
Assume the basement footings and the floor slabs in Figure 4.2b are replaced by a mat with thickness D_c and the building load is Q.

$$gross\ q = \frac{Q + W_f}{A}$$

Equation (4.7)-(a):

$$gross\ q = \frac{Q}{A} + \frac{A(D_c\gamma_c)}{A} = \frac{Q}{A} + D_c\gamma_c$$

Equation (4.1): $net\ q = gross\ q - \sigma_o$

$$net\ q = \frac{Q}{A} + D_c\gamma_c - \sigma_o$$

Assume γ' represents unit weight of the overburden soil above W.T, that is the effective unit weight.

$$net\ q = \frac{Q}{A} + D_c\gamma_c - \left[\gamma'(D_f - h) + \gamma_{sat}h\right] \tag{4.14}$$

Equation (4.2): $net\ q' = gross\ q' - \sigma'_o$

$$\begin{aligned} net\ q' &= \left(\frac{Q}{A} - \gamma_w h\right) + D_c\gamma_c - \sigma'_o \\ &= \left(\frac{Q}{A} - \gamma_w h\right) + D_c\gamma_c - \left[\gamma'(D_f - h) + h(\gamma_{sat} - \gamma_w)\right] \\ &= \frac{Q}{A} - \gamma_w h + D_c\gamma_c - \gamma' D_f + \gamma' h - \gamma_{sat}h + \gamma_w h \\ &= \frac{Q}{A} - h(\gamma_{sat} - \gamma') + D_c\gamma_c - \gamma' D_f \end{aligned} \tag{4.15}$$

$h(\gamma_{sat} - \gamma') = \gamma_w h$; hence,

$$net\ q' = \left[\left(\frac{Q}{A} - \gamma_w h\right) + D_c\gamma_c\right] - \gamma' D_f \leq net\ q_a$$

or

$$\left\{\begin{array}{l}\text{Net effective}\\ \text{foundation}\\ \text{pressure}\end{array}\right\} = \left\{\begin{array}{l}\text{Total applied effective}\\ \text{pressure due to the total}\\ \text{building loads including}\\ \text{weight of foundation}\end{array}\right\} - \left\{\begin{array}{l}\text{Effective}\\ \text{overburden}\\ \text{pressure}\end{array}\right\} \leq net\ q_a \tag{4.16}$$

Here, also, it makes no difference whether the net foundation pressure is computed on the basis of *total* stress or *effective* stress approach, as it is clear from Equations (4.14) and (4.15)

Equation (4.16) indicates that the net effective foundation pressure would reduce to zero if the total applied effective pressure $\left[\left(\frac{Q}{A} - \gamma_w h\right) + D_c\gamma_c\right]$ equals the effective overburden pressure ($\gamma' D_f$). This is the same principle of floating foundations (or compensated foundations) which has been discussed in Section 2.2.

4.4 Bearing Capacity Failure Mechanism for Long (Strip or Continuous) Footings

Assume the strip footing (width $= B$) of Figure 4.3 carries a uniform pressure q on the *surface* of a mass of elastic, homogeneous and isotropic soil (i.e. ideal soil). The unit weight γ of the soil is assumed to be zero (the same assumption by Prandtl, 1921; and Reissner, 1924) but it possesses shear strength parameters c and $\varnothing$. When the soil fails in shear, that is when the pressure q becomes equal to q_{ult}, the footing will have been pushed downwards into the soil mass, producing a state of plastic equilibrium, in the form of an active Rankine zone, below the footing, the angles ABC and BAC being $(45° + \varnothing/2)$. The downward movement of the wedge ABC, that is zone I, forces the adjoining soil sideways, producing outward lateral forces on both sides of the wedge. Therefore, passive Rankine zones ADE and BGF, that is zone III, develop on both sides of zone I, the angles DEA and GFB being $(45° - \varnothing/2)$. The transition between the downward movement of zone I and lateral movement of zone III occur through *zones of radial shear* (also known as *slip fan zones*) ACD and BCG, that is zone II. The slip surfaces CD and CG are logarithmic spirals (or circular arcs if $\varnothing = 0$) to which BC and ED, or AC and FG, are tangential. The soil mass below the surface EDCGF remains in the state of elastic equilibrium, whereas the soil above exists in the state of plastic equilibrium.

It may be useful to mention herein that stability of zone I depends mainly on whether the footing base is smooth or rough. When the base is perfectly smooth, the shear pattern in this wedge is identical with the shear pattern for an active Rankine zone. When the base is rough, the soil underneath the footing cannot spread laterally due to friction and adhesion between the bottom of the footing and the soil. As a consequence, the soil within this zone acts as a part of the footing and remains in elastic equilibrium. The depth of this wedge remains practically unchanged and still the footing sinks. According to Terzaghi (1943), the boundary AC or BC rises at an angle $(\psi = \varnothing)$ to the horizontal, provided no sliding movement of the footing occur. However, most other theories use $\left(\psi = 45° + \dfrac{\varnothing}{2}\right)$, as just indicated.

Based on the mechanism described above and using the Prandtl plastic equilibrium theory, the following exact expression can be obtained for the ultimate bearing capacity of a strip footing on the surface of a weightless ideal soil:

$$q_{\text{ult}} = (2+\pi)c_u$$
$$q_{\text{ult}} = 5.14\,c_u \quad (\text{For undrained condition, } \varnothing = 0°) \tag{4.17}$$

Where c_u = undrained cohesion of soil (or undrained shear strength)

However, in the general case shear strength parameters are c and $\varnothing$; hence it is necessary to consider surcharge pressure q_o acting on the soil surface as shown in Figure 4.3; otherwise if $c = 0$ the bearing capacity of the weightless soil would be zero. For this case, Reissner (1924) found the following solution (based on the Prandtl plastic equilibrium theory):

$$q_{ult} = c\cot\varnothing\left[\exp(\pi\tan\varnothing)\tan^2\left(45° + \frac{\varnothing}{2}\right) - 1\right] + q_o\left[\exp(\pi\tan\varnothing)\tan^2\left(45° + \frac{\varnothing}{2}\right)\right] \tag{4.18}$$

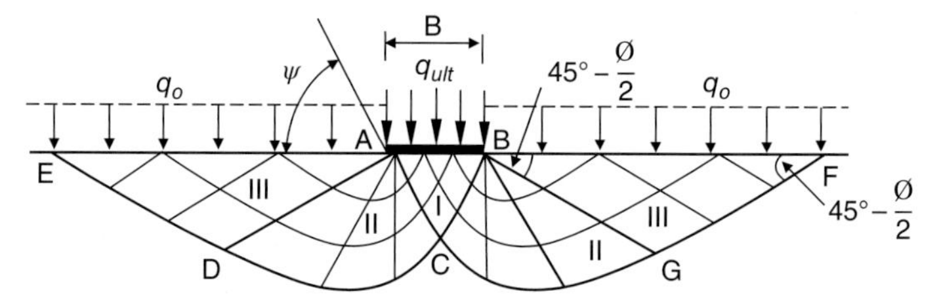

Figure 4.3 Bearing capacity failure under a strip footing.

We should not forget that Equation (4.18) was derived assuming weightless soil. Therefore, an additional term must be added to this equation to take into account the component of bearing capacity due to the self-weight of the soil. This third term cannot theoretically (or exactly) be determined mainly because of the inclination angle ψ which is unknown; hence the term can only be determined using approximate procedures (e.g. Terzaghi, 1943; Meyerhof, 1951, 1955; Terzaghi and Peck, 1966; Hansen, 1969; Vesic, 1973). According to Terzaghi, the ultimate bearing capacity (*gross* q_{ult}) of the soil under a shallow strip footing can be expressed by the following general equation (known as the Terzaghi general bearing capacity equation):

$$gross\, q_{ult} = cN_c + \gamma' D_f N_q + \frac{1}{2}\gamma' B N_\gamma \tag{4.19}$$

where N_c, N_q and N_γ are bearing capacity factors depending only on the value of $\emptyset$.

Contributions of the first, second and third terms in Equation (4.19) are due to the constant component of shear strength, the surcharge pressure and the self-weight of the soil, respectively. It is necessary to realise, however, that the superposition of the components of bearing capacity is not strictly correct for a plastic material; it leads to errors which are on the safe side, not exceeding 17–20% for $\emptyset = 30°$ to $40°$ and zero for $\emptyset = 0°$.

It may be useful to mention herein that there exists in literature a great variety of proposed solutions to the problem of ultimate bearing capacity, especially, those related to the bearing capacity factors. While variations in the proposed N_c and N_q values remain relatively insignificant, the differences in N_γ values, coming primarily from the sharp variation of this factor with the angle ψ, are substantial.

4.5 Bearing Capacity Equations

Various analytical methods have been established so that the problem of a two-dimensional solution for a strip footing is extended to a solution of three-dimensional problem for square, rectangular and circular footings. Accordingly, semi-empirical *shape*, *depth* and various *inclination* factors are applied to the ultimate bearing capacity solution for a strip footing.

Table 4.2 shows four general equations for estimating gross ultimate bearing capacity, currently in use, proposed by the four authors Terzaghi (1943), Meyerhof (1963), Hansen (1970) and Vesic (1973, 1975). The table also shows equations for computing the bearing capacity factors included in each equation. Table 4.3 gives values of bearing capacity factors for the Terzaghi bearing capacity equations. Table 4.4 gives values of bearing capacity factors for the Hansen, Meyerhof and Vesic equations. Table 4.5 shows equations for computing shape, depth and load inclination factors for use in the Meyerhof bearing capacity equations. Table 4.6 shows equations for computing shape and depth factors for use in either the Hansen or Vesic bearing capacity equations. Table 4.7 shows equations for computing load inclination, ground and base factors for use in the Hansen bearing capacity equations. Table 4.8 shows equations for computing load inclination, ground and base factors for use in the Vesic bearing capacity equations. According to Bowles (2001), as a result of his observations regarding the use of the various bearing capacity equations, one may suggest the following equation use:

- *Terzaghi* – best for very cohesive soils where $D/B \le 1$ or for a quick estimate of q_{ult} to compare with other methods. *Do not use* for footings with moments and/or horizontal forces or for tilted bases and/or sloping ground.
- *Hansen, Meyerhof, Vesic* – best for any situatin that applies, depending on user preference or familiarity with a particular method.
- *Hansen, Vesic* – best when base is tilted, when footing is on a slope or when $D/B > 1$.

Notes for Table 4.2–4.8:

(1) Note use of "effective" base dimensions B', L' in shape factors by Hansen but not by Vesic. $B' = B - 2\,e_B$, $L' = L - 2\,e_L$ and $A_f = B'L'$.

Table 4.2 Bearing capacity equations by the several authors indicated (*from Bowles, 2001*).

Terzaghi (1943).

$$q_{ult} = cN_c s_c + \bar{q}N_q + 0.5\gamma' B N_\gamma s_\gamma$$

$$N_q = \frac{a^2}{a\cos^2(45 + \phi/2)}$$

$$a = e^{(0.75\pi - \phi/2)\tan\phi}$$

$$N_c = (N_q - 1)\cot\phi$$

$$N_\gamma = \frac{\tan\phi}{2}\left(\frac{K_{p\gamma}}{\cos^2\phi} - 1\right)$$

For:	strip	round	square
s_c =	1.0	1.3	1.3
s_γ =	1.0	0.6	0.8

Meyerhof (1963).

Vertical load: $q_{ult} = cN_c s_c d_c + \bar{q}N_q s_q d_q + 0.5\gamma' B' N_\gamma s_\gamma d_\gamma$

Inclined load: $q_{ult} = cN_c d_c i_c + \bar{q}N_q d_q i_q + 0.5\gamma' B' N_\gamma d_\gamma i_\gamma$

$$N_q = e^{\pi\tan\phi}\tan^2\left(45 + \frac{\phi}{2}\right)$$

$$N_c = (N_q - 1)\cot\phi$$

$$N_\gamma = (N_q - 1)\tan(1.4\phi)$$

Hansen (1970).

General:† $q_{ult} = cN_c s_c d_c i_c g_c b_c + \bar{q}N_q s_q d_q i_q g_q b_q + 0.5\gamma' B' N_\gamma s_\gamma d_\gamma i_\gamma g_\gamma b_\gamma$

when $\phi = 0$

use $q_{ult} = 5.14 s_u (1 + s'_c + d'_c - i'_c - b'_c - g'_c) + \bar{q}$

N_q = same as Meyerhof above

N_c = same as Meyerhof above

$N_\gamma = 1.5(N_q - 1)\tan\phi$

Vesić (1973, 1975).

Use Hansen's equations above.

N_q = same as Meyerhof above

N_c = same as Meyerhof above

$N_\gamma = 2(N_q + 1)\tan\phi$

(2) In both Hansen and Vesic depth factors use actual dimensions B and L.
(3) Values of the s_i factors in the table above are consistent with vertical loads only. For inclined loads see Section 4.7.
(4) With a vertical load V accompanied by a horizontal load H_L and either $H_B = 0$ or $H_B > 0$ you may have to compute two sets of shape and depth factors as $s_{i,B}$, $s_{i,L}$, and $d_{i,B}$, $d_{i,L}$ as discussed in Section 4.7.
(5) Use H_i as either H_B or H_L or both if $H_L > 0$.
(6) Hansen (1970) did not give values of i_c for $\text{Ø} > 0$. The value given in the tables above is from Hansen (1961), which is also used by Vesic.
(7) Variable c_a = soil adhesion to the base, about 0.6 to $1.0 \times c$.

Table 4.3 Values of bearing capacity factors for use in the Terzaghi bearing capacity equations of Table 4.2.

Ø°	N_c	N_q	N_γ	$K_{P\gamma}$
0	5.7*	1.0	0.0	10.8
5	7.3	1.6	0.5	12.2
10	9.6	2.7	1.2	14.7
15	12.9	4.4	2.5	18.6
20	17.7	7.4	5.0	25.0
25	25.1	12.7	9.7	35.0
30	37.2	22.5	19.7	52.0
34	52.6	36.5	36.0	
35	57.8	41.4	42.4	82.0
40	95.7	81.3	100.4	141.0
45	172.3	173.3	297.5	298.0
48	258.3	287.9	780.1	
50	347.5	415.1	1153.2	800.0

Note: Values of N_γ for Ø of 0°, 34°, and 48° are original Terzaghi values used to back-compute $K_{p\gamma}$ (Bowles, 1996).
*$N_c = 1.5\,\pi + 1$. [See Terzaghi (1943), P. 127.]

Table 4.4 Values of bearing capacity factors for use in the Hansen, Meyerhof and Vesic bearing capacity equations of Table 4.2.

Ø°	N_c	N_q	$N_{\gamma(H)}$	$N_{\gamma(M)}$	$N_{\gamma(V)}$	N_q/N_c	$2\tan Ø(1 - \sin Ø)^2$
0	5.14[a]	1.0	0.0	0.0	0.0	0.195	0.000
5	6.49	1.6	0.1	0.1	0.4	0.242	0.146
10	8.34	2.5	0.4	0.4	1.2	0.296	0.241
15	10.97	3.9	1.2	1.1	2.6	0.359	0.294
20	14.83	6.4	2.9	2.9	5.4	0.431	0.315
25	20.71	10.7	6.8	6.8	10.9	0.514	0.311
26	22.25	11.8	7.9	8.0	12.5	0.533	0.308
28	25.79	14.7	10.9	11.2	16.7	0.570	0.299
30	30.13	18.4	15.1	15.7	22.4	0.610	0.289
32	35.47	23.2	20.8	22.0	30.2	0.653	0.276
34	42.14	29.4	28.7	31.1	41.0	0.698	0.262
36	50.55	37.7	40.0	44.4	56.2	0.746	0.247
38	61.31	48.9	56.1	64.0	77.9	0.797	0.231
40	75.25	64.1	79.4	93.6	109.3	0.852	0.214
45	133.73	134.7	200.5	262.3	271.3	1.007	0.172
50	266.50	318.5	567.4	871.7	761.3	1.195	0.131

Note: Values of N_c and N_q are the same for all three methods; subscripts identify author for N_γ.
[a] $N_c = \pi + 2$ as limit when $Ø \rightarrow 0°$

Table 4.5 Shape, depth, and load inclination factors, for use in the Meyerhof bearing capacity equations of Table 4.2.

Factor	Value	For
Shape	$s_c = 1 + 0.2K_P \frac{B}{L}$	Any ϕ
	$s_q = s_\gamma = 1 + 0.1K_P \frac{B}{L}$	$\phi > 10°$
	$s_q = s_\gamma = 1$	$\phi = 0$
Depth	$d_c = 1 + 0.2\sqrt{K_P}\frac{D}{B}$	Any ϕ
	$d_q = d_\gamma = 1 + 0.1\sqrt{K_P}\frac{D}{B}$	$\phi > 10$
	$d_q = d_\gamma = 1$	$\phi = 0$
Load-inclination (R, V, H, $\theta < \emptyset$)	$i_c = i_q = \left(1 - \frac{\theta°}{90°}\right)^2$	Any ϕ
	$i_\gamma = \left(1 - \frac{\theta°}{\phi°}\right)^2$	$\phi > 0$
	$i_\gamma = 0$ for $\theta > 0$	$\phi = 0$

Where: $K_P = \tan^2 (45 + \emptyset/2)$
θ = angle of resultant R measured from vertical without a sign; if $\theta = 0$ all $i_i = 1.0$
D = depth of foundation
B = width of footing
L = length of footing
s_i, d_i, i_i = shape, depth, and load – inclination factors, respectively

Table 4.6 Shape and depth factors for use in the Hansen and Vesic bearing capacity equations of Table 4.2. Use s'_c and d'_c when $\emptyset = 0$ only for Hansen equations. Subscripts H, V used for Hansen, Vesic, respectively.

Shape factors	Depth factors
$s'_{c(H)} = 0.2\frac{B'}{L'}$ $(\phi = 0°)$ $S_{c(H)} = 1.0 + \frac{N_q}{N_c}\cdot\frac{B'}{L'}$ $S_{c(V)} = 1.0 + \frac{N_q}{N_c}\cdot\frac{B}{L}$ $s_c = 1.0$ for strip	$d'_c = 0.4k$ $(\phi = 0°)$ $d_c = 1.0 + 0.4\,k$ $k = D/B$ for $D/B \le 1$ $k = \tan^{-1}(D/B)$ for $D/B > 1$ k in radians
$S_{q(H)} = 1.0 + \frac{B'}{L'}\sin\phi$ $S_{q(V)} = 1.0 + \frac{B}{L}\tan\phi$ for all ϕ	$d_q = 1 + 2\tan\phi(1 - \sin\phi)^2 k$ k defined above
$s_{\gamma(H)} = 1.0 - 0.4\frac{B'}{L'} \ge 0.6$ $s_{\gamma(V)} = 1.0 - 0.4\frac{B}{L} \ge 0.6$	$d_\gamma = 1.00$ for all ϕ

Table 4.7 Load inclination, ground and base factors for use in the Hansen bearing capacity equations of Table 4.2.

Load-inclination factors	$i'_c = 0.5 - 0.5 \times \sqrt{1 - \frac{H_i}{A_f C_a}}$ (for $\phi = 0$) $i_c = i_q - \frac{1 - i_q}{N_q - 1}$	$i_q = \left[1 - \frac{0.5 H_i}{V + A_f C_a \cot\phi}\right]^{a_1}$ $2 \le a_1 \le 5$	$i_\gamma = \left[1 - \frac{0.7 H_i}{V + A_f C_a \cot\phi}\right]^{a_2}$ $2 \le a_2 \le 5$ $i_\gamma = \left[1 - \frac{0.7 H_i}{V + A_f C_a \cot\phi}\right]^{a_2}$
Ground factors (base on slope)	$g'_c = \frac{B^\circ}{147^\circ}$ (for $\phi = 0$) $g_c = 1 - \frac{B^\circ}{147^\circ}$	$g_q = g_\gamma = (1 - 0.5 \cot\beta)^5$	
Base factors (tilted base)	$b'_c = \frac{\eta^\circ}{147^\circ}$ (for $\phi = 0$) $b_c = 1 - \frac{\eta^\circ}{147^\circ}$	$b_q = \exp(-1\eta \tan\phi)$ η in radians	$b_\gamma = \exp(-2.7\eta \tan\phi)$ η in radians

Table 4.8 Load inclination, ground and base factors for use in the Vesic bearing capacity equations of Table 4.2.

Load inclination factors	$i'_c = 1 - \frac{m H_i}{A_f C_a N_c}$ (for $\phi = 0$) $i_c = i_q - \frac{1 - i_q}{N_q - 1}$	$i_q = \left[1 - \frac{H_i}{V + A_f C_a \cot\phi}\right]^m$ $m = m_B = \frac{2 + B/L}{1 + b/L}$ $m = m_L = \frac{2 + B/L}{1 + B/L}$	$i_\gamma = \left[1 - \frac{H_i}{V + A_f C_a \cot\phi}\right]^{m+1}$
Ground factors (base on slope)	$g'_c = \frac{\beta}{5.14}$ (β in radians) $g_c = i_q - \frac{1 - i_q}{5.14 \cot\phi}$	$g_q = g_\gamma = (1 - \tan\beta)^2$	
Base factors (tilted base)	$b'_c = g'_c$ (for $\phi = 0$) $b_c = 1 - \frac{2\beta}{5.14 \tan\phi}$	$b_q = b_\gamma = (1 - \eta \tan\phi)^2$ (η in radians)	

(8) Variables Ø and c are angle of internal friction and cohesion of the base soil, respectively.

(9) Hansen suggests using:

$$Ø_{tr} \text{ (triaxial strain Ø) for } \frac{L}{B} \le 2$$

$$Ø_{ps} \text{ (plane strain Ø)} = 1.5\, Ø_{tr} - 17^\circ \text{ for } \frac{L}{B} > 2$$

$$Ø_{ps} = Ø_{tr} \text{ for } Ø_{tr} \le 34^\circ$$

(10) Variable δ = friction angle between footing and base soil $(0.5\, Ø \le \delta \le Ø)$

(11) Bowles (2001) suggests: $2 \le \alpha_1 \le 3$ and $3 \le \alpha_2 \le 4$

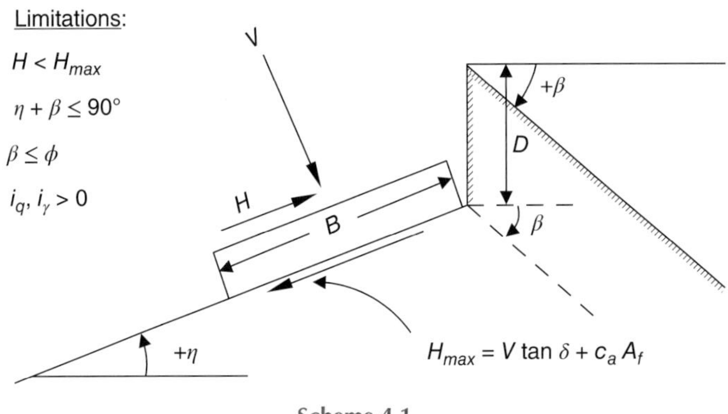

Scheme 4.1

(12) According to Vesic:
with $\varnothing = 0$ (and $\beta \neq 0$), use $N_\gamma = -2\sin(\pm\beta)$.
$m = m_B$ when $H_i = H_B$ (H parallel to B) and $m = m_L$ when $H_i = H_L$ (H parallel to L)
Use $m = \sqrt{m_B^2 + m_L^2}$ if you have both H_B and H_L.
Note use of B and L, not B' and L'
The H_i term ≤ 1.0 for computing i_q, i_γ (always).

(13) Vesic uses B' in the N_γ term even when $H_i = H_L$.

(14) Refer to the scheme 4.1 for identification of angles η and β, foundation depth D and location of H_i (parallel and at top of the base slab; usually also produces eccentricity). Especially note that V = force *normal* to the base and is not the resultant R from combining V and H_i.

4.6 Some Considerations Concerning the Use of Bearing Capacity Equations

(1) The requirement of moment equilibrium $\sum M = 0$ is not satisfied in the methods used in developing the bearing capacity equations; only the requirements $\sum F_H = \sum F_Y = 0$ are satisfied. However this error would not be serious since statics is obviously satisfied at ultimate loading.

(2) The common practice is, usually, to use conservative estimates for the soil parameters; hence, the bearing capacity equations tend to be conservative most of the time. Moreover, after obtaining a conservative q_{ult} this is reduced to the safe soil pressure q_s using a safety factor. Therefore, the probability is very high that the allowable soil pressure q_a is *safe*.

(3) Terzaghi developed his bearing capacity equations considering the mode of general shear failure in a dense or stiff soil and the mode of local shear failure in a loose or soft soil. According to Terzaghi, the same general bearing capacity equation of Table 4.2 is used for the mode of local shear failure provided that reduced values of the parameters c and $\varnothing$ are used, as follows:

$$c_{\text{local}} = \frac{2}{3}\, c_{\text{general}}$$

$$\tan \varnothing_{\text{local}} = \frac{2}{3} \tan \varnothing_{\text{general}}; \text{ or } \varnothing_{\text{local}} = \tan^{-1}\left(\frac{2}{3} \tan \varnothing_{\text{general}}\right)$$

Therefore, according to Terzaghi, for the mode of local shear failure c_{local} instead of c should be used in the N_c term and $\varnothing_{\text{local}}$ instead of $\varnothing$ should be used in computing the bearing capacity factors (using equations or Table 4.3).

(4) The researchers Vesic (1969) and De Beer (1965) found that, for very large values of B of the N_γ term, the limiting value of q_{ult} approaches that of deep foundations and suggest a reduction factor such as r_γ may be used with the N_γ term of all the bearing capacity equations. Bowles (1996) suggests:

$$r_\gamma = 1 - 0.25\log\left(\frac{B}{K}\right) \quad \text{for } B > 2\,\text{m} \tag{4.20}$$

where B = width or diameter of footing or mat foundation, in m, and $K = 2$

(5) It is not recommended using tables of N factors (Tables 4.3 and 4.4) that require interpolation over about 2°. For angles larger than 35° the factors change rapidly and by large amounts; interpolation can have a considerable error, so someone checking the work may not be able to verify q_{ult}.

(6) The N_c term (the shear strength term) predominates in cohesive soils.

(7) The N_q term (the surcharge term) predominates in cohesionless soils.

(8) The N_γ term (the weight term) provides some increase in bearing capacity for both (c-Ø) and cohesionless soils.

(9) One should avoid placing a footing on the surface of a cohesionless soil mass.

(10) It is recommended to compact loose cohesionless soils of relative density less than 50% to a higher density prior to placing footings in it.

(11) In case the soil beneath a footing is not homogeneous or is stratified, some judgment must be applied to determine the bearing capacity. Bearing capacity for footings on layered soils will be considered in Section 4.13.

(12) It is obvious that the Terzaghi bearing capacity equation is much easier to use than the others equation so that it has great appeal for many practitioners, particularly for bases with only vertical loads and $\frac{D}{B} \le 1$. The shape factors proposed by Terzaghi and Peck (1967) are still widely used in practice although they are considered to give conservative values of q_{ult}. The shape factors for strip, square and circular footings are as shown in Table 4.2. For a rectangular footing of breadth B and length L, the shape factors are obtained by linear interpolation between the values for a strip footing $\left(\frac{B}{L} = 0\right)$ and a square footing $\left(\frac{B}{L} = 1\right)$. Hence,

$$s_c = 1 + 0.3\frac{B}{L}, \text{ and } s_\gamma = 1 - 0.2\frac{B}{L}$$

(13) According to Vesic (1973), it is recommended that the depth factors d_i are not be used for shallow foundations ($D/B \le 1$) because of uncertainties in quality of the overburden. However, he did give the d_i values as shown in Table 4.6 despite this recommendation. Also, Vesic always used the actual B as B' in the N_γ term of the general bearing capacity equation of Table 4.2.

(14) In a study of bearing capacity theories, Skempton (1951) concluded that the bearing capacity factor N_c for saturated clays under undrained conditions ($Ø_u = 0$) is a function of the shape of the footing and the depth to breadth ratio, as shown in Figure 4.4. This figure also includes relationships suggested by Salgado et al. (2004), which are described by $N_c = \left[(2+\pi)(1+0.27)(d/B)^{0.5}\right]$. N_c for circular footings may be obtained by taking the square values. The factor for a rectangular footing of dimensions $B \times L$ (where $B < L$) is the value for a square footing multiplied by $(0.84 + 0.16\,B/L)$. In practice, N_c is normally limited at a value equals 9.0 for very deeply embedded square or circular foundations.

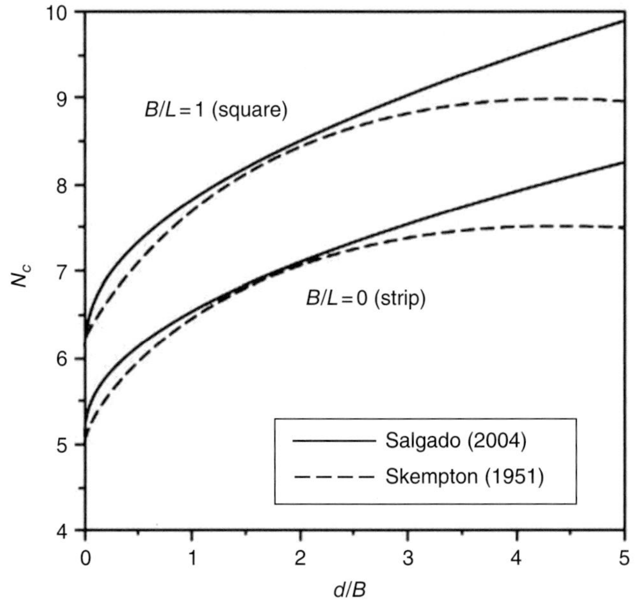

Figure 4.4 Bearing capacity factors N_c for embedded foundations in undrained soil (reproduced from Knappett and Craig, 2012).

4.7 Bearing Capacity of Footings with Inclined Loads

A footing may be subjected to inclined loading and such condition leads to a reduction in bearing capacity. This type of loading is common for footing of retaining walls loaded with both horizontal (lateral earth pressure) and vertical loading, for footings of many industrial process structures where horizontal wind loads are acting in combination with the gravity loads and for a number of other types of foundations which are subjected to horizontal and vertical loads simultaneously. The effect of inclined loading on bearing capacity can be taken into account by means of inclination factors, as shown in Tables 4.5, 4.7 and 4.8.

The Terzaghi equations have no direct provision for a reduction in cases where the load is inclined.

The Meyerhof inclination factors of Table 4.5 do not need explanation. However, he did not take into account the direction of the load horizontal component (H_B, H_L).

The Vesic inclination factors take into account the load direction (H_B, H_L) in computing the m exponents, as shown in Table 4.8. According to Vesic, the shape s factors are computed regardless of i factors.

The Hansen inclination and shape factors take into account the load direction (H_B, H_L) as follows:

(1) Compute the inclination factors using the equations given in Table 4.7 and using either the exponents given in that table or those suggested by Bowles mentioned before (see note 11 below Table 4.8). Accordingly, for:

$$H_B = 0;\ i'_{c,B} = 0 \text{ and } i_{c,B}, i_{q,B}, i_{\gamma,B} \text{ are all } 1.0$$
$$H_L = 0;\ i'_{c,L} = 0 \text{ and } i_{c,L}, i_{q,L}, i_{\gamma,L} \text{ are all } 1.0$$

(2) Use the computed inclination factors to compute Hansen shape factors, as follows:

$$s'_{c,B} = 0.2B'i_{c,B}/L' \qquad s'_{c,L} = 0.2L'i_{c,L}/B' \quad (\phi = 0 \text{ case})$$

$$s_{c,B} = 1.0 + \frac{N_q}{N_c}\cdot\frac{B'i_{c,B}}{L'} \qquad s_{c,L} = 1.0 + \frac{N_q}{N_c}\cdot\frac{L'i_{c,L}}{B'}$$

$$s_{q,B} = 1 + \sin\phi\cdot B'i_{q,B}/L' \quad s_{q,L} = 1 + \sin\phi\cdot L'i_{q,L}/B'$$

$$s_{\gamma,B} = 1 - 0.4B'i_{\gamma,B}/L'i_{\gamma,L} \quad s_{\gamma,L} = 1 - 0.4L'i_{\gamma,L}B'i_{\gamma,B}$$

$$\text{Limitation}: s_{\gamma,i} \geq 0.6 \ (\textit{if less than } 0.6 \text{ use } 0.60)$$

These factors are used in the following modifications of the Hansen general bearing capacity equation of Table 4.2:

$$\begin{aligned} q_{\text{ult}} &= c\,N_c s_{c,B} d_{c,B} i_{c,B} g_{c,B} b_{c,B} + \bar{q} N_q s_{q,B} d_{q,B} i_{q,B} g_{q,B} b_{q,B} \\ &\quad + 0.5\gamma B' N_\gamma s_{\gamma,B} d_{\gamma,B} i_{\gamma,B} g_{\gamma,B} b_{\gamma,B} \end{aligned} \tag{4.21}$$

or

$$\begin{aligned} q_{\text{ult}} &= c\,N_c s_{c,L} d_{c,L} i_{c,L} g_{c,L} b_{c,L} + \bar{q} N_q s_{q,L} d_{q,L} i_{q,L} g_{q,L} b_{q,L} \\ &\quad + 0.5\gamma L' N_\gamma s_{\gamma,L} d_{\gamma,L} i_{\gamma,L} g_{\gamma,L} b_{\gamma,L} \end{aligned} \tag{4.22}$$

4.8 Bearing Capacity of Footings with Eccentric Loads

A footing may be subjected to eccentric loading and such condition leads to a reduction in bearing capacity.

The estimate of ultimate bearing capacity for footings with eccentricity, using the general bearing capacity equations of Table 4.2, may be obtained by the following methods:

(1) *The* effective area *method (Meyerhof, 1953, 1963; Hansen, 1970).*
Refer to Figure 4.5. The procedure is as explained below:
(a) Determine the effective footing dimensions B' and L' using

$$B' = B - 2\,e_B;\ \text{where } e_B \text{ is the eccentricity parallel to } B$$

$$L' = L - 2\,e_L;\ \text{where } e_L \text{ is the eccentricity parallel to } L$$

The smaller of the two dimensions (i.e. B' and L') is the effective width of the footing used in the N_γ term of the bearing capacity equations. Effective area of the footing $A' = B'L'$.
The equivalent effective area A' and the effective width B′ for eccentrically loaded *circular* footings is given in a nondimensional form (Highter and Anders, 1985), as shown in the table of Figure 4.5b. Hence, the effective length $L' = A'/B'$. It may be noted that in the case of circular foundations under eccentric loading, the eccentricity is always one-way.
(b) Use the effective footing dimensions B' and L' in computing the shape factors.
(c) Use actual B and L dimensions in computing the depth factors.
(d) Use a general bearing capacity equation to obtain q_{ult}.
The ultimate load that the foundation can support is

$$Q_{\text{ult}} = q_{\text{ult}} \times A' = q_{\text{ult}}(B'L') \tag{4.23}$$

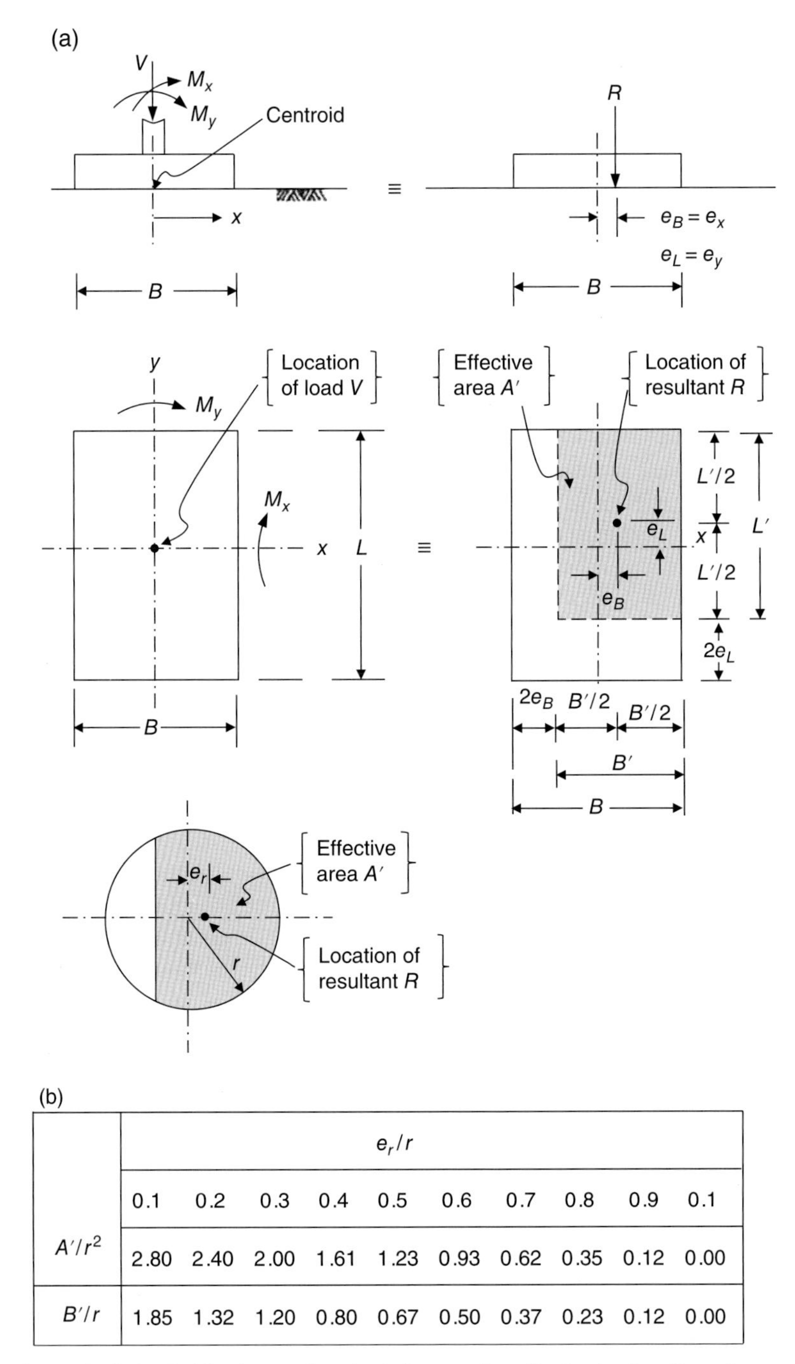

	e_r/r									
	0.1	0.2	0.3	0.4	0.5	0.6	0.7	0.8	0.9	0.1
A'/r^2	2.80	2.40	2.00	1.61	1.23	0.93	0.62	0.35	0.12	0.00
B'/r	1.85	1.32	1.20	0.80	0.67	0.50	0.37	0.23	0.12	0.00

Figure 4.5 Eccentrically loaded footings and method of computing effective footing dimensions.

The safety factor against bearing capacity failure is

$$SF = \frac{Q_{\text{ult}}}{R} \tag{4.24}$$

(2) *The* Meyerhof reduction factor *method.*
In this procedure the ultimate bearing capacity is computed for concentric loading condition using the *Meyerhof* bearing capacity equations of Table 4.2. This computed ultimate soil pressure, $q_{\text{ult}(c)}$, is then reduced with a reduction factor R_e to obtain the ultimate soil pressure for eccentric loading condition which is

$$q_{\text{ult}(e)} = q_{\text{ult}(c)} \times R_e \tag{4.25}$$

The original Meyerhof method gave reduction curves for obtaining R_e; however, Bowles (1982) converted the curves to the following suitable equations:

$$R_e = 1 - \frac{2e}{B} \quad \text{(cohesive soil)} \tag{4.26)-(a}$$

$$R_e = 1 - \left(\frac{e}{B}\right)^{\frac{1}{2}} \quad \text{(cohesionless soil and for } 0 < e/B < 0.3\text{)} \tag{4.26)-(b}$$

It should be evident that, for two-way eccentricity (e_x, e_y), two reduction factors are used for each type of soil. According to this method, the ultimate load that the foundation can support is

$$Q_{\text{ult}} = q_{\text{ult}(e)} \times A = q_{\text{ult}(e)}(BL) \tag{4.27}$$

(3) *Purkaystha and Char reduction factor method (for* granular *soils).*
Purkaystha and Char (1977) performed stability analysis of *strip* foundations under *eccentric vertical* loads supported by *sand,* using the method of slices. Their analysis resulted in the following eccentricity reduction factor,

$$R_k = 1 - \frac{q_{\text{ult(e)}}}{q_{\text{ult(c)}}} \tag{4.28}$$

$$R_k = a\left(\frac{e}{B}\right)^k \tag{4.29}$$

Where R_k = eccentricity reduction factor

$q_{\text{ult(e)}}$ = ultimate bearing capacity of strip foundations under eccenteric vertical loads
$q_{\text{ult(c)}}$ = ultimate bearing capacity of strip foundations under centeric vertical loads
a and k = functions of the embedment ratio D_f/B (Table 4.9)

By combining Equations (4.28) and (4.29),

$$q_{\text{ult(e)}} = q_{\text{ult(c)}}(1 - R_k) = q_{\text{ult(c)}}\left[1 - a\left(\frac{e}{B}\right)^k\right]$$
$$q_{\text{ult(e)}} = q_{\text{ult(c)}}\left[1 - a\left(\frac{e}{B}\right)^k\right] \tag{4.30}$$

Table 4.9 Variations of a and k, Equation (4.30).

D_f / B	a	k
0.00	1.862	0.730
0.25	1.811	0.785
0.50	1.754	0.800
1.00	1.820	0.888

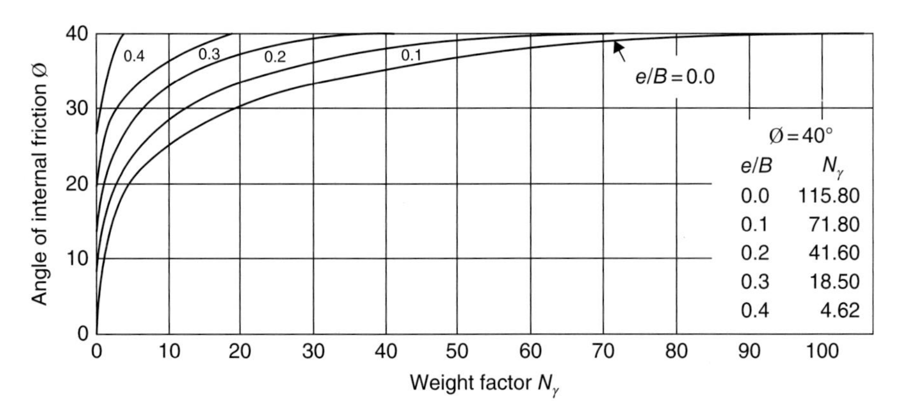

Figure 4.6 N_γ versus Ø.

Where

$$q_{\text{ult(c)}} = qN_qF_{qd} + \frac{1}{2}\gamma BN_\gamma F_{\gamma d} \tag{4.31}$$

The relationships for the depth factors F_{qd} and $F_{\gamma d}$ are given in Table 4.6.
The ultimate load per unit length of the strip foundation can be given as

$$Q_{\text{ult}} = Bq_{\text{ult(eccentric)}} \tag{4.32}$$

(4) *Prakash and Saran method*
According to Prakash and Saran (1971), the ultimate load per unit length of an *eccentrically* and *vertically* loaded *strip* (*continuous*) foundation can be estimated as

$$Q_{\text{ult}} = B\left[c'N_{c(e)} + qN_{q(e)} + \frac{1}{2}\gamma BN_{\gamma(e)}\right] \tag{4.33}$$

where $N_{\gamma(e)}$, $N_{q(e)}$ and $N_{c(e)}$ are bearing capacity factors depend on Ø, given in Figures 4.6, 4.7 and 4.8, respectively.

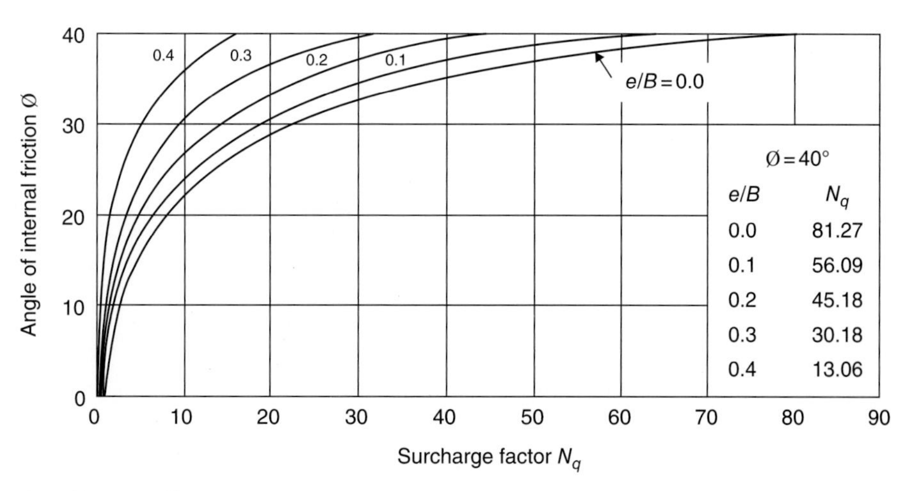

Figure 4.7 N_q versus Ø.

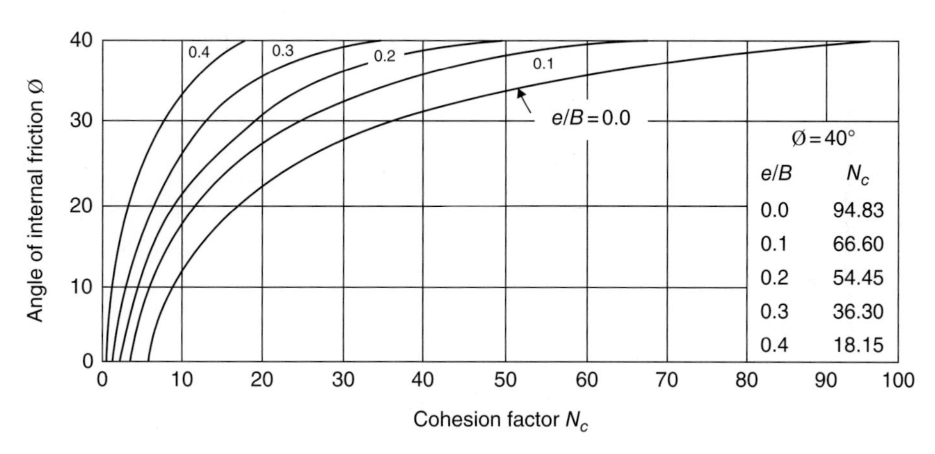

Figure 4.8 N_c versus Ø.

For an *eccentrically* and *vertically* loaded *rectangular* foundation, the ultimate load can be given as

$$Q_{\text{ult}} = BL\left[c'N_{c(e)}F_{cs(e)} + qN_{q(e)}F_{qs(e)} + \frac{1}{2}\gamma BN_{\gamma(e)}F_{\gamma s(e)}\right] \tag{4.34}$$

where $F_{cs\,(e)}$, $F_{qs(e)}$ and $F_{\gamma s(e)}$ are shape factors can be determined as

$$F_{cs(e)} = \left(1.2 - 0.025\frac{L}{B}\right) \geq 1$$

$$F_{qs(e)} = 1$$

$$F_{\gamma s(e)} = 1.0 + \left(\frac{2e}{B} - 0.68\right)\frac{B}{L} + \left[0.43 - \left(\frac{3}{2}\right)\left(\frac{e}{B}\right)\right]\left(\frac{B}{L}\right)^2 \quad \text{(For dense sands)}$$

$$F_{\gamma s(e)} = 1 \quad \text{(For loose sands)}$$

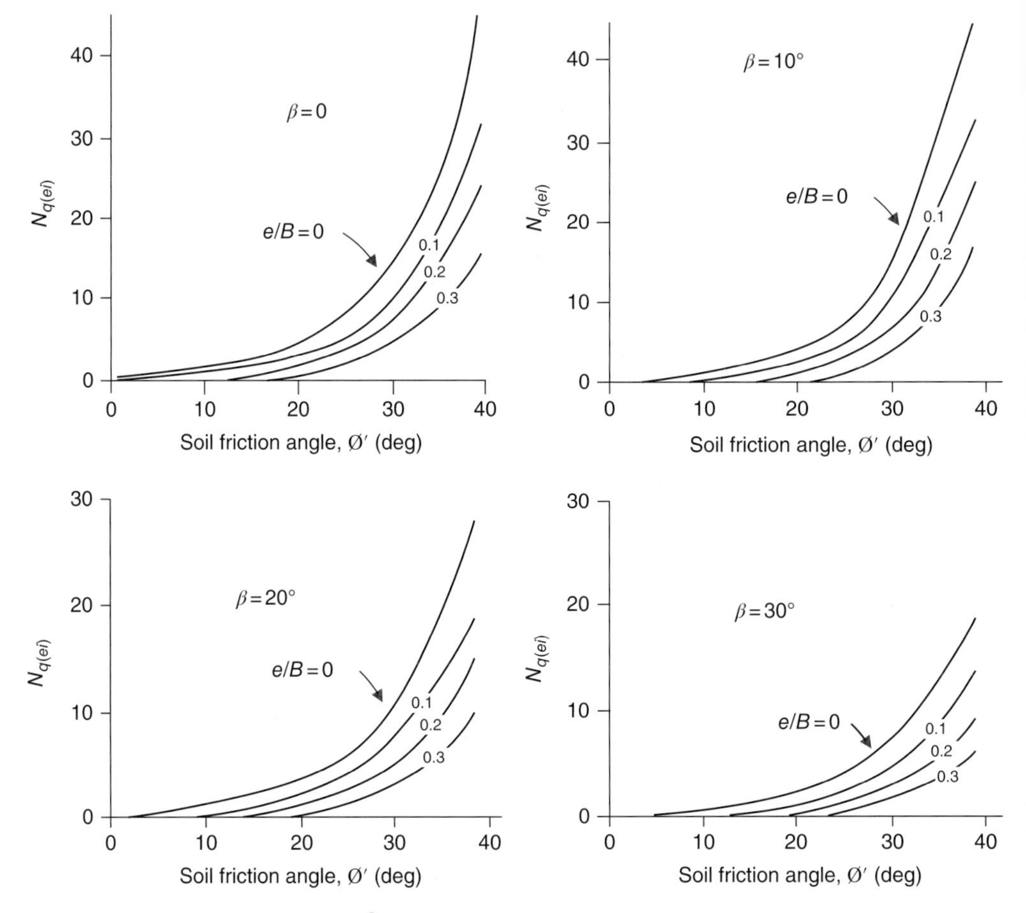

Figure 4.9 Variation of $N_{q(ei)}$ with Ø′, $\frac{e}{B}$ and β.

(5) *Saran and Agarwal Analysis*

Bearing capacity of a *continuous* foundation subjected to an *eccentric inclined* load was studied by Saran and Agarwal (1991). They proposed the following bearing capacity equation for a continuous foundation located at a depth D_f below the ground surface and subjected to an eccentric load (load eccentricity = e) which is inclined at an angle β to the vertical:

$$Q_{\text{ult}} = B\left[c'N_{c(ei)} + qN_{q(ei)} + \frac{1}{2}\gamma BN_{\gamma(ei)}\right] \tag{4.35}$$

The bearing capacity factors $N_{q(ei)}$, $N_{c(ei)}$ and $N_{\gamma(ei)}$ are obtained from Figures 4.9, 4.10 and 4.11, respectively. These figures are reproduced from Das (1911). Equation (4.35) gives ultimate load per unit length of a strip foundation.

4.9 Effect of Water Table on Bearing Capacity

The depth at which the ground water table (W.T) is located may have a significant effect on the bearing capacity of shallow foundations. This is mainly because the *effective* unit weights of both the base and surcharge soils are needed in the bearing capacity equations (such as those of Table 4.2). It is known

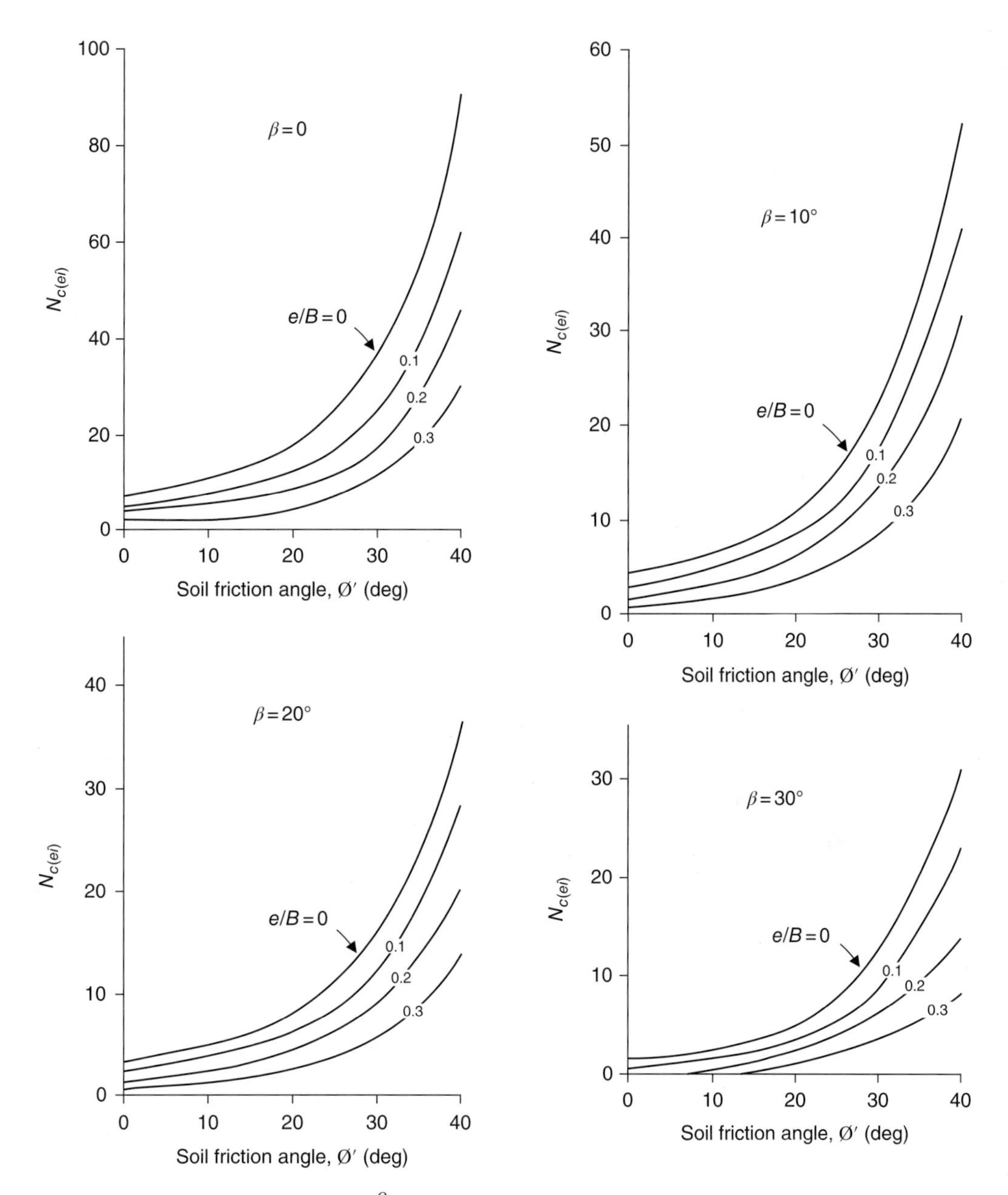

Figure 4.10 Variation of $N_{c(ei)}$ with $\varnothing'$, $\frac{e}{B}$ and β.

that the effective unit weight of a submerged soil will be reduced to about half of its unit weight above water table. Moreover, generally, the submergence of soils will cause the loss of all apparent cohesions, coming from capillary stresses or from weak cementation bonds. Thus, through submergence, all the three terms (cohesion, surcharge and weight) of the bearing capacity equation become considerably smaller. Therefore, it is essential that the bearing capacity analyses be made assuming the highest possible W.T at the particular location for the expected lifetime of the structure in question. It is necessary that the assessment of this highest possible level is made taking into consideration the probability of temporary high levels that could be expected in some locations during heavy rainstorms or floods,

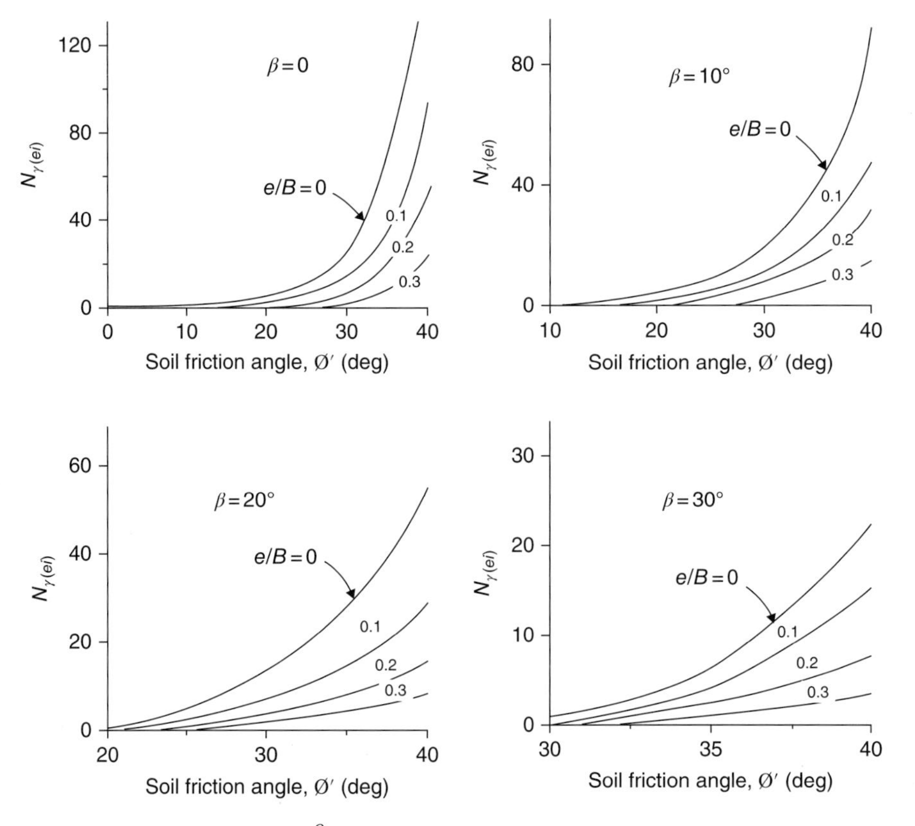

Figure 4.11 Variation of $N_{\gamma(ei)}$ with $Ø'$, $\frac{e}{B}$ and β.

although they may not appear in the official records. The effective unit weights needed in bearing capacity equations may be determined as follows:

(A) The effective unit weight in the weight term (i.e. N_γ term) of the bearing capacity equations of Table 4.2. There are three possible cases:

(1) The highest W.T exists at or above the foundation level, that is $z_w = 0$, where z_w is the depth to W.T below the foundation level. The base soil is submerged; hence, the effective unit weight γ' is the buoyant (submerged) unit weight γ_b. Therefore,

$$\gamma' = \gamma_b = \gamma_{sat} - \gamma_w \tag{4.36}$$

(2) The highest W.T exists within the depth $0 < z_w \le B$. The average effective unit weight may be determined as

$$\gamma' = (2H - z_w)\frac{z_w}{H^2}\gamma_{wet} + \frac{\gamma_b}{H^2}(H - z_w)^2 \tag{4.37}$$

where $H = 0.5\,B \tan\left(45 + \frac{Ø}{2}\right) \le B$

γ_{wet} = wet unit weight of soil within depth z_w

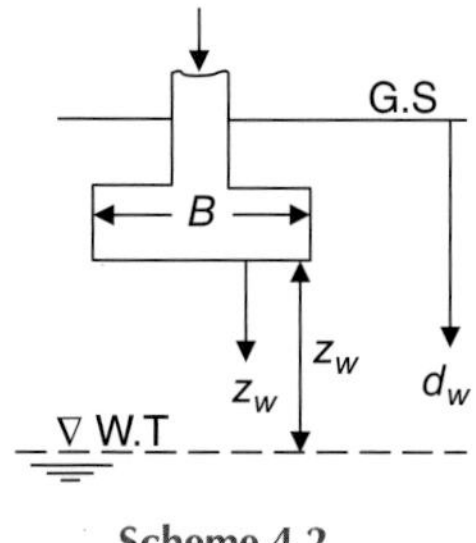

Scheme 4.2

Also, it may be determined as

$$\gamma' = \gamma_b + \left(\frac{z_w}{B}\right)(\gamma_{wet} - \gamma_b) \tag{4.38}$$

(3) The highest W.T exists permanently at a depth well below the foundation level, that is $z_w > B$. The average effective unit weight is determined as

$$\gamma' = \gamma_{wet} \tag{4.39}$$

(B) The effective unit weight needed for computing the effective surcharge (overburden) pressure q' in the surcharge term (i.e. N_q term) of the bearing capacity equations of Table 4.2. There are three possible cases:

(1) The highest W.T exists at or below the foundation level, that is $d_w \geq D_f$, where d_w is the depth to W.T below the ground surface. Hence, the average effective unit weight of the surcharge soil is determined as

$$\gamma' = \gamma_{wet} \tag{4.40)-(a}$$

and

$$q' = \gamma_{wet} D_f \tag{4.40)-(b}$$

(2) The highest W.T exists within the depth $0 < d_w < D_f$. In this case, $\gamma' = \gamma_{wet}$ is used for the soil within the depth d_w, and $\gamma_b = (\gamma_{sat} - \gamma_w)$ is used for the soil within the depth $(D_f - d_w)$. Hence,

$$q' = \gamma_{wet} d_w + \gamma_b (D_f - d_w) \tag{4.41}$$

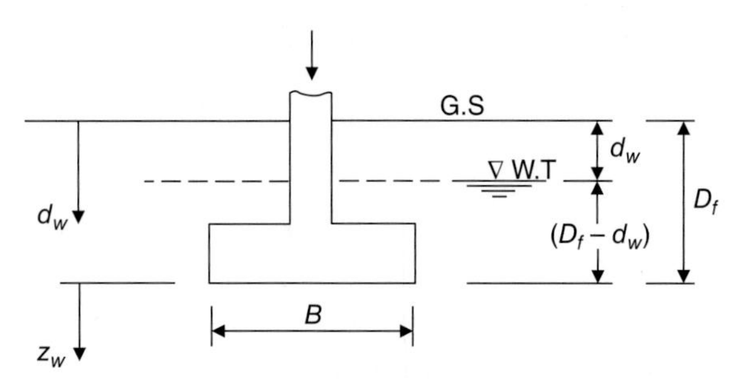

Scheme 4.3

(3) The highest W.T exists at or above the ground surface, i.e. $d_w = 0$. The surcharge soil is submerged; hence, the effective unit weight γ' is the buoyant (submerged) unit weight γ_b. Therefore,

$$\gamma' = \gamma_b = \gamma_{sat} - \gamma_w \tag{4.42-(a)}$$

$$q' = \gamma_b D_f \tag{4.42-(b)}$$

It may be useful to mention herein that all preceding considerations are based on the assumption that the seepage forces acting on the soil skeleton are negligible. In case there is significant ground-water seepage in any direction, it may have an effect on the bearing capacity. In addition to possible internal erosion of the soil (undermining, piping and similar phenomena), the seepage force adds a component to the body forces caused by gravity. When it is believed wise to consider seepage effects, a particular bearing capacity analysis would be necessary.

4.10 Influence of Soil Compressibility on Bearing Capacity

The bearing capacity equations of Table 4.2 are based on the assumption of incompressibility of soil and that they should be applied, strictly speaking, only to cases where the mode of general shear failure prevails. The other two failure modes are characteristic for compressible soils as previously explained in Section 4.1. Terzaghi (1943) accounted for the effect of soil compressibility, considering the mode of *local shear failure*, through using reduced values of cohesion c and friction angle $\emptyset$, as explained in Section 4.6. However, according to Vesič (1973), such an approach is not always satisfactory; and instead, he proposed using the following *compressibility factors* (ζ_{ic}), always smaller than 1, with the three terms of bearing capacity equations of Table 4.2:

For $I_r < I_{r(cr)}$:

$$\zeta_{cc} = 0.32 + 0.12\frac{B}{L} + 0.60\log I_r \quad \text{for } \emptyset = 0 \tag{4.43}$$

$$\zeta_{cc} = \zeta_{qc} - \frac{1-\zeta_{qc}}{N_c \tan\emptyset} \quad \text{for } \emptyset > 0 \tag{4.44}$$

$$\zeta_{qc} = \zeta_{\gamma c} = \exp\left\{\left(-4.4 + 0.6\frac{B}{L}\right)\tan\emptyset + \left[\frac{(3.07\sin\emptyset)(\log 2I_r)}{1+\sin\emptyset}\right]\right\} \tag{4.45}$$

For $I_r \geq I_{r(cr)}$:

$$\zeta_{cc} = \zeta_{qc} = \zeta_{\gamma c} = 1 \tag{4.46}$$

Where:

I_r = Rigidity index of soil at a depth approximately $B/2$ below the foundation level

$$I_r = \frac{G_s}{c + q'\tan\emptyset} \tag{4.47}$$

G_s = shear modulus of the soil
q' = effective overburden pressure at a depth of $(D_f + B/2)$
$I_{r(cr)}$ = Critical rigidity index of soil

Table 4.10 Values of critical rigidity index of soil.

	Critical rigidity index, $I_{r(cr)}$					
Ø (deg)	B/L = 0	B/L = 0.2	B/L = 0.4	B/L = 0.6	B/L = 0.8	B/L = 1.0
0	13.56	12.39	11.32	10.35	9.46	8.64
5	18.30	16.59	15.04	13.63	12.36	11.20
10	25.53	22.93	20.60	18.50	16.62	14.93
15	36.85	32.77	29.14	25.92	23.05	20.49
20	55.66	48.95	43.04	37.85	33.29	29.27
25	88.93	77.21	67.04	58.20	50.53	43.88
30	151.78	129.88	111.13	95.09	81.36	69.62
35	283.20	238.24	200.41	168.59	141.82	119.31
40	593.09	488.97	403.13	332.35	274.01	225.90
45	1440.94	1159.56	933.19	750.90	604.26	486.26

$$I_{r(cr)} = \frac{1}{2}\left\{ \exp\left[\left(3.30 - 0.45\frac{B}{L}\right)\cot\left(45 - \frac{\text{Ø}}{2}\right)\right]\right\} \tag{4.48}$$

The variations of $I_{r(cr)}$ with (B/L) for different values of Ø are given in Table 4.10.

4.11 Effect of Adjacent Footings on Bearing Capacity

Approximations for the spacing of footings to avoid interference between old and new footings were discussed in Section 2.3.

The analyses used in the derivation of the bearing capacity equations (Table 4.2) were concerned with the bearing capacity of isolated footings. All analyses based on the assumption that that the supporting soil is under the action of gravity forces alone and is not under the influence of any other footing. However, in engineering practice, there are conditions where footings are placed so close to each other that the influence zones of their supporting soils overlap.

This problem of stress overlap due to closely spaced, simultaneously loaded, footings has been investigated and studied by many researchers. For example, the results of experimental studies by Stuart (1962) and West and Stuart (1965) indicate that the effects of adjacent parallel strip footings in sand on bearing capacity may vary considerably with the angle of shearing resistance, Ø. With decreased Ø values the effects become small or negligible; however, for high Ø values they appear to be significant, especially if a footing is surrounded by others on both sides. According to Vesic (1973), these effects are considerably reduced as the value of (L/B) ratio approaches unity. Similarly, the compressibility of soils reduces and may eliminate completely the interference effects. There are practically no such effects in the case of punching shear failure. For these and other reasons, as Vesic suggested, it is not recommended to consider interference effects in bearing capacity computations. A designer should be aware, however, of the possibility of their existence in some special circumstances.

4.12 Bearing Capacity of Foundations on Slopes

4.12.1 General

In principle, it may be a good engineering practice to avoid constructing shallow foundations on or adjacent to sloping ground whenever possible. Among the reasons are:

(1) The foundation might be dislocated or undermined if a landslide is to take place.
(2) The reduction in lateral support makes bearing capacity failures more likely.
(3) The overburden and near-slope surface soils are more vulnerable to erosion by the action of water, wind and so on.
(4) The near-surface soils, particularly clay soils, may be slowly creeping downhill and this creep may cause the footings move slowly downslope.

There are, however, circumstances where foundations must be constructed on or adjacent to a slope. Examples include bridge abutments supported on approach embankments, foundations for electric transmission towers and some buildings. However, through adopting appropriate site investigation, design and construction methodologies, the adverse situations just mentioned above may be prevented from occurring.

Figure 4.12 shows footing setback as required by the Uniform Building Code (ICBO, 1997) and the International Building Code (ICC, 2000) for slopes steeper than $3H$: $1V$. The horizontal distance from the footing to the face of the slope should be at least $H/3$, but need not exceed 12 m. Fore slopes that are steeper than $1H$: $1V$, this setback distance should be measured from a line that extends from the toe of the slope at an angle of 45°. This criteria can be satisfied either by moving the footing away from the slope or by making it deeper. We must keep in mind, however, this setback criteria does not justify foundations on unstable slopes. Hence, appropriate slope stability analysis is required to verify the overall stability.

4.12.2 Solutions

(1) *Hansen (1970) and Vesic (1975) solutions.*
The ground factors g_i of Tables 4.7 and 4.8 account for the effect of sloping ground on bearing capacity. However, both the Hansen and Vesic methods appear too conservative, especially for slopes which exist in granular soils.
(2) *Meyerhof (1957) solution.*
Refer to Figure 4.13. The height of the slope is H and the slope makes an angle β with the horizontal. The edge of the foundation is located at a distance b from the top of the slope. At ultimate load (i.e. q_{ult}) the failure surface will be as shown in the figure.

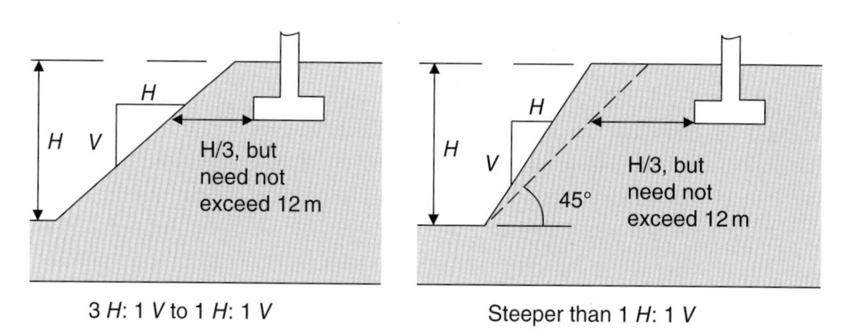

Figure 4.12 Setback distances of footings adjacent to slopes as required by the ICBO (1997) and the ICC (2000).

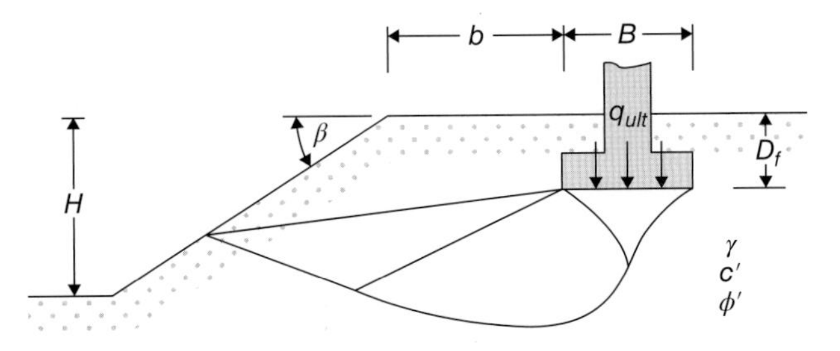

Figure 4.13 A shallow foundation adjacent to a slope (from Das, 2011).

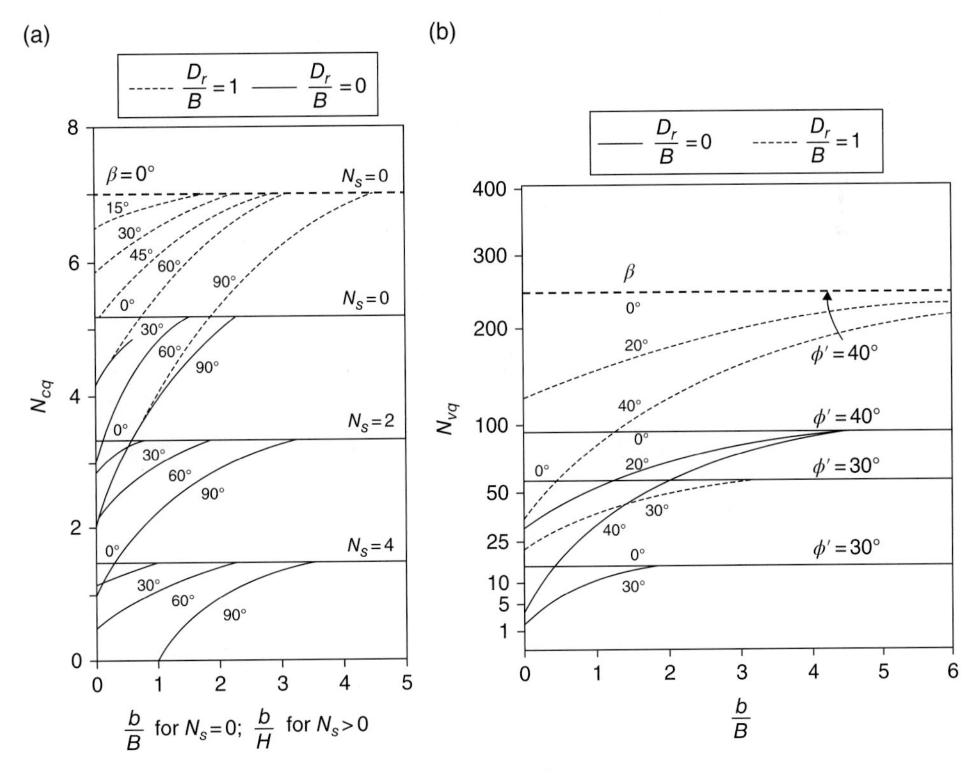

Figure 4.14 Meyerhof bearing capacity factors N_{cq} and $N_{\gamma q}$ (from Das, 2011).

Meyerhof derived the following theoretical equation for the gross ultimate bearing capacity of *continuous* foundations:

$$gross\, q_{\text{ult}} = c\, N_{cq} + \frac{1}{2}\gamma\, B\, N_{\gamma q} \tag{4.49}$$

The variations of N_{cq} and $N_{\gamma q}$ are shown in Figure 4.14a, b, respectively.

For purely granular soils ($c = 0$):

$$gross\, q_{\text{ult}} = \frac{1}{2}\gamma\, B\, N_{\gamma q} \tag{4.50}$$

For purely cohesive soils ($\varnothing = 0$):

$$gross\, q_{\text{ult}} = c\, N_{cq} \tag{4.51}$$

where c = undrained cohesion

In using N_{cq} of Figure 4.14a it is necessary to determine the *stability number*, N_s, where

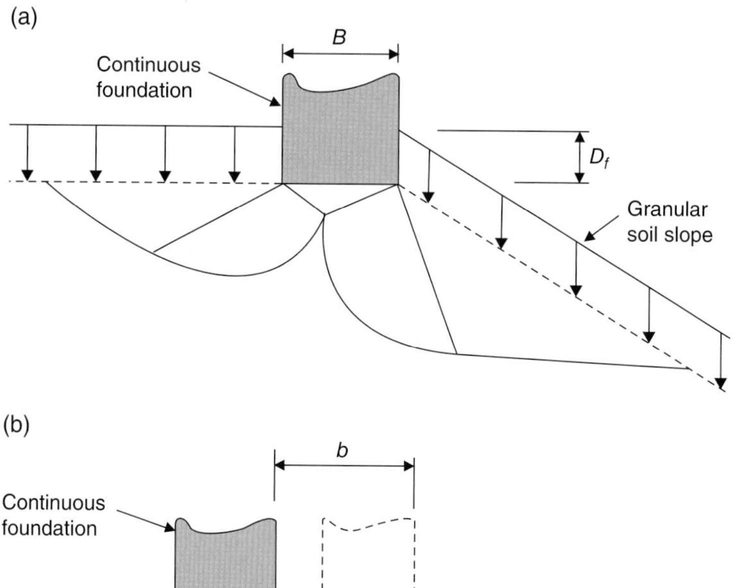

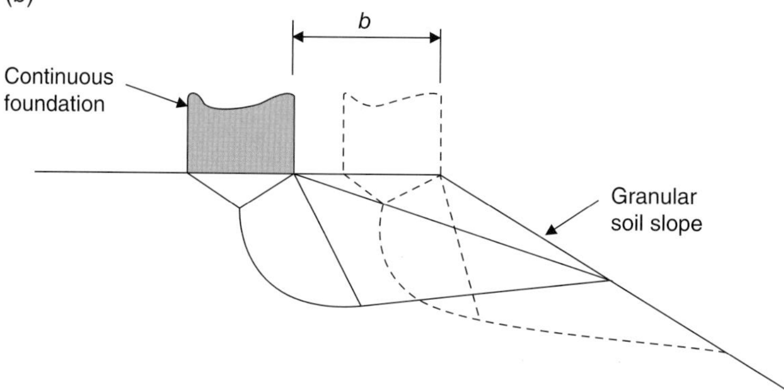

Figure 4.15 Schematic diagram of failure zones for: (a) embedment, $\frac{D_f}{B}>0$; (b) setback, $\frac{b}{B}>0$ (from Das, 2011).

$$N_s = \frac{\gamma H}{c} \tag{4.52}$$

If $B < H$, use the curves for $N_s = 0$. If $B \geq H$, use the curves for N_s

(3) *Stress characteristics solution.*
Graham, Andrews and Shields (1988), on the basis of the method of stress characteristics, developed a solution for the bearing capacity factor $N_{\gamma q}$ for *continuous* shallow foundations on top of *granular* soil slopes (Figure 4.15).

The variations of $N_{\gamma q}$ are shown in Figures 4.16, 4.17 and 4.18. Using the appropriate $N_{\gamma q}$ factor, the ultimate bearing capacity of continuous foundations is computed from Equation (4.50).

(4) *Bowles solution.*
According to Bowles (2001), the gross ultimate bearing capacity may be computed by any of the equations of Table 4.2 which contain ground factors; however, they give too conservative results. Bowles suggests using the Hansen equation modified to read as follows:

$$gross\ q_{\text{ult}} = c N_c' s_c i_c + \bar{q} N_q' s_q i_q + 0.5\gamma' B N_\gamma' s_\gamma i_\gamma \tag{4.53}$$

where N_c', N_q', and N_γ' are adjusted bearing capacity factors.

The depth d_i factors are not included in the foregoing equation since the depth effect is included in the computation of area ratios.

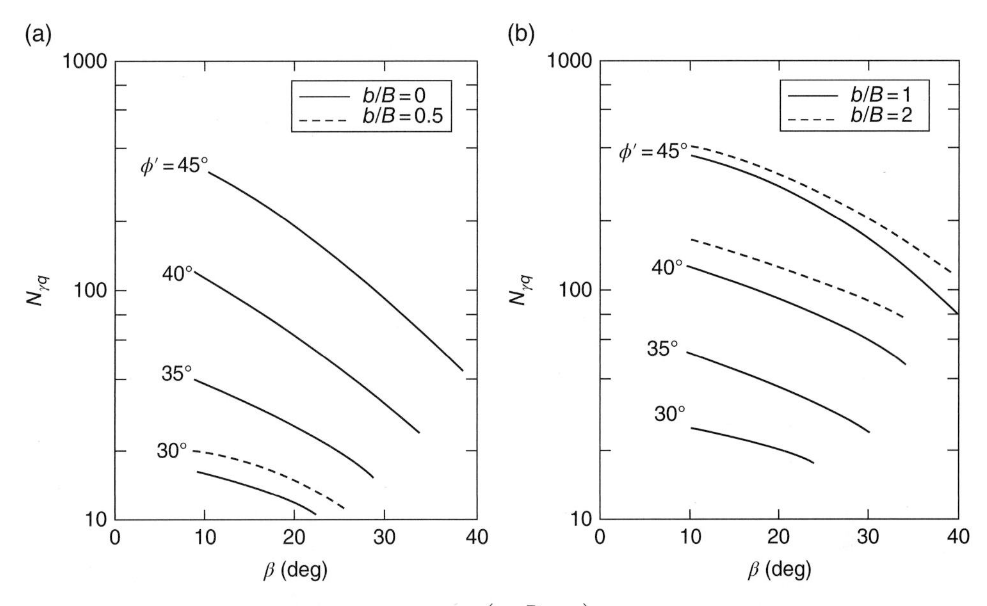

Figure 4.16 Graham et al.'s theoretical values of $N_{\gamma q}\left(\text{for } \frac{D_f}{B}=0\right)$; from: Graham, Andrews and Shields (1988).

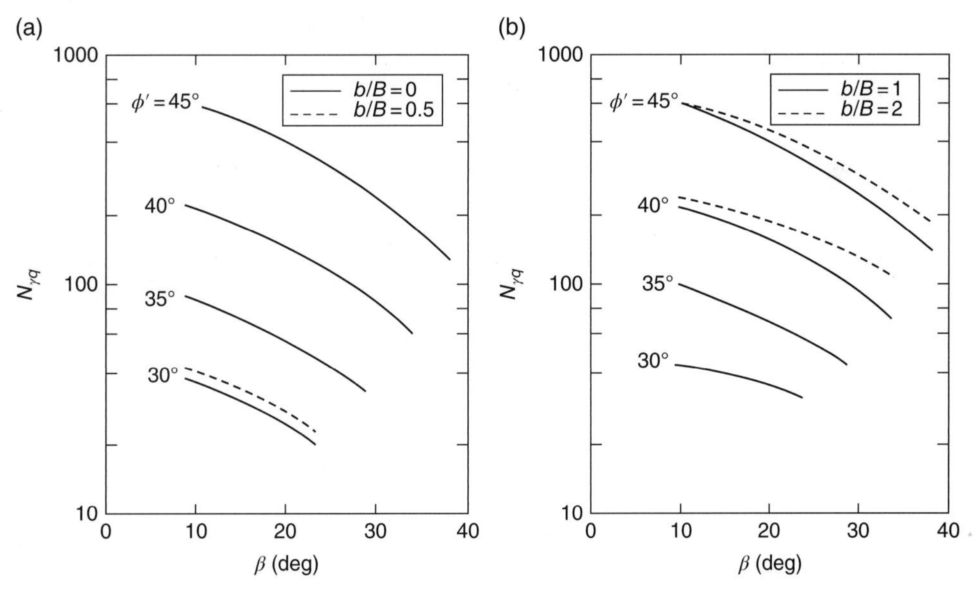

Figure 4.17 Graham et al.'s theoretical values of $N_{\gamma q}\left(\text{for } \frac{D_f}{B}=0.5\right)$; from: Graham, Andrews and Shields (1988).

The N_c' and N_q' factors are obtained from Table 4.11, developed by Bowles.

It will be conservative to use shape factors $s_c = s_q = 1$; but s_γ should be computed. It is recommended not to adjust $\varnothing_{tr}$ to $\varnothing_{ps}$, as there are considerable uncertainties in the stress state when there is loss of soil support on one side of the base, even for strip (or long) bases. The adjusted bearing capacity factor N_γ' is obtained as follows:

(1) For the ratio $\frac{b}{B} \geq 2$ (Figure 4.19b), use $N_\gamma' = N_{\gamma(H)}$, obtained from Table 4.4.

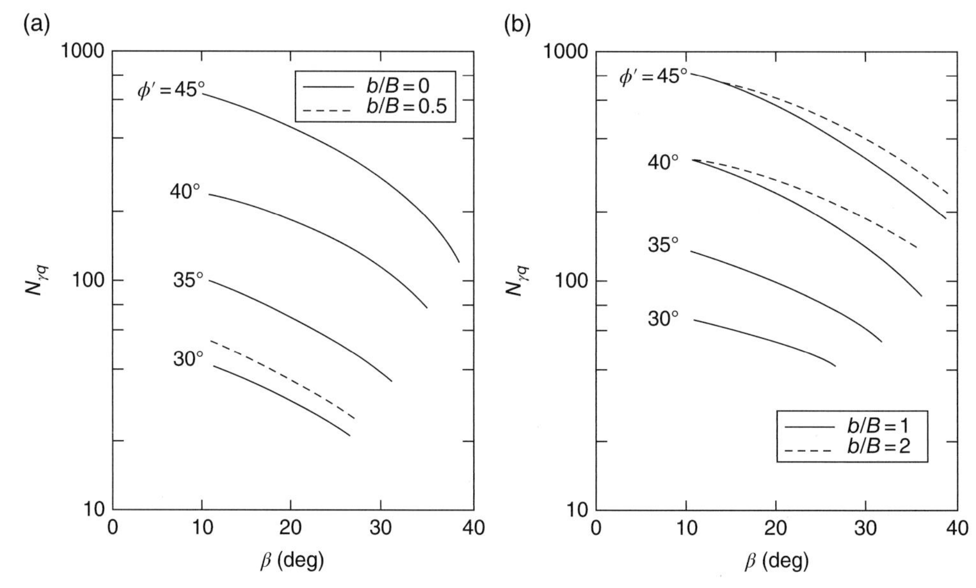

Figure 4.18 Graham et al.'s theoretical values of $N_{\gamma q}$ $\left(\text{for } \frac{D_f}{B} = 1\right)$; from: Graham, Andrews and Shields (1988).

(2) For the ratio $\frac{b}{B} < 2$, use $N'_\gamma = \text{adjusted}\, N_{\gamma(H)}$, as follows:

(a) Compute the Coulomb passive pressure coefficients for the slope angle β using $\beta = (-)$ for one computation and $(+)$ for the other. Use the friction angle $\delta = \emptyset$ and $\alpha = 90°$ for both computations. The Coulomb passive pressure coefficient K_p can be computed from equations or obtained from tables or curves, usually given in geotechnical text books. When $\beta = (+ \text{ or } 0)$ is used, the computed $K_p = K_{\max}$ on the base side away from the slope and when $\beta = (-)$ is used, the computed $K_p = K_{\min}$ (see Figure 4.19).

(b) Compute a *reduction factor* $R = K_{\min}/K_{\max}$

(c) Compute $N'_\gamma = \frac{N_{\gamma(H)}}{2} + \frac{N_{\gamma(H)}}{2}\left[R + \frac{b}{2B}(1 - R)\right]$. This equation is easily checked:

At $b = 0$ and foundation on slope; $N'_\gamma = \frac{N_{\gamma(H)}}{2} + \frac{N_{\gamma(H)}}{2}(R)$

At $b = 2B$ and foundation on top of slope and out of slope influence; $N'_\gamma = N_{\gamma(H)}$

4.13 Bearing Capacity of Footings on Layered Soils

4.13.1 General

All the preceding bearing capacity equations presented in Section 4.5 or referred to in the other sections have considered only the cases in which the supporting soil is homogeneous, with constant strength parameters and unit weight with depth and extends to a depth sufficient to fully enclose the rupture zone. However, in practice, many soil profiles exist which are not uniform; consist of different soil types in layers of variable thicknesses. In these cases, if the failure zone extends into the lower layer (or layers if they are very thin) the ultimate bearing capacity, undoubtedly, will be modified to some extent.

Table 4.11 Bearing capacity factors N'_c, N'_q for footings on or adjacent to a slope. Refer to Figure 4.19 for identification of variables.

	D/B = 0			*b/B* = 0		*D/B* = 0.75			*b/B* = 0		*D/B* = 1.50			*b/B* = 0	
β	Ø = 0°	10°	20°	30°	40°	0°	10°	20°	30°	40°	0°	10°	20°	30°	40°
0°	N'_c = 14	8.35	14.83	30.14	75.31	5.14	8.35	14.83	30.14	75.31	5.14	8.25	14.83	30.14	75.31
	N'_q = 03	2.47	6.40	18.40	64.20	1.03	2.47	6.40	18.40	64.20	1.03	2.47	6.40	18.40	64.20
10°	4.89	7.80	13.37	26.80	64.42	5.14	8.35	14.83	30.14	75.31	5.14	8.35	14.83	30.14	75.31
	1.03	2.47	6.40	18.40	64.20	0.92	1.95	4.43	11.16	33.94	1.03	2.47	5.85	14.13	40.81
20°	4.63	7.28	12.39	23.78	55.01	5.14	8.35	14.83	30.14	66.81	5.14	8.35	14.83	30.14	75.31
	1.03	2.47	6.40	18.40	64.20	0.94	1.90	4.11	9.84	28.21	1.03	2.47	5.65	12.93	35.14
25°	4.51	7.02	11.82	22.38	50.80	5.14	8.35	14.83	28.76	62.18	5.14	8.35	14.83	30.14	73.57
	1.03	2.47	6.40	18.40	64.20	0.92	1.82	3.85	9.00	25.09	1.03	2.47	5.39	12.04	31.80
30°	4.38	6.77	11.28	21.05	46.88	5.14	8.35	14.83	27.14	57.76	5.14	8.35	14.83	30.14	68.64
	1.03	2.47	6.40	18.40	64.20	0.88	1.70	3.54	8.08	21.91	1.03	2.47	5.04	10.99	28.33
60°	3.62	5.33	8.33	14.34	28.56	4.70	6.83	10.55	17.85	34.84	5.14	8.34	12.76	21.37	41.12
	1.03	2.47	6.40	18.40	64.20	0.37	0.63	1.17	2.36	5.52	0.62	1.04	1.83	3.52	7.80
	D/B = 0			*b/B* = 0.75		*D/B* = 0.75			*b/B* = 0.75		*D/B* = 1.50			*b/B* = 0.75	
β	Ø = 0°	10°	20°	30°	40°	0°	10°	20°	30°	40°	0°	10°	20°	30°	40°
10°	5.14	8.33	14.34	28.02	66.60	5.14	8.35	14.83	30.14	75.31	5.14	8.35	14.83	30.14	75.31
	1.03	2.47	6.40	18.40	64.20	1.03	2.34	5.34	13.47	40.83	1.03	2.47	6.40	15.79	45.45
20°	5.14	8.31	13.90	26.19	59.31	5.14	8.35	14.83	30.14	71.11	5.14	8.35	14.83	30.14	75.31
	1.03	2.47	6.40	18.40	64.20	1.03	2.47	6.04	14.39	40.88	1.03	2.47	6.40	16.31	43.96
25°	5.14	8.29	13.69	25.36	56.11	5.14	8.35	14.83	30.14	67.49	5.14	8.35	14.83	30.14	75.31
	1.03	2.47	6.40	18.40	64.20	1.03	2.47	6.27	14.56	40.06	1.03	2.47	6.40	16.20	42.35
30°	5.14	8.27	13.49	24.57	53.16	5.14	8.35	14.83	30.14	64.04	5.14	8.35	14.83	30.14	74.92
	1.03	2.47	6.40	18.40	64.20	1.03	2.47	6.40	14.52	38.72	1.03	2.47	6.40	15.85	40.23
60°	5.14	7.94	12.17	20.43	39.44	5.14	8.35	14.38	23.94	45.72	5.14	8.35	14.83	27.46	52.00
	1.03	2.47	6.40	18.40	64.20	1.03	2.47	5.14	10.05	22.56	1.03	2.47	4.97	9.41	20.33
	D/B = 0			*b/B* = 1.50		*D/B* = 0.75			*b/B* = 1.50		*D/B* = 1.50			*b/B* = 1.50	
10°	5.14	8.35	14.83	29.24	68.78	5.14	8.35	14.83	30.14	75.31	5.14	8.35	14.83	30.14	75.31
	1.03	2.47	6.40	18.40	64.20	1.03	2.47	6.01	15.39	47.09	1.03	2.47	6.40	17.26	49.77
20°	5.14	8.35	14.83	28.59	63.60	5.14	8.35	14.83	30.14	75.31	5.14	8.35	14.83	30.14	75.31
	1.03	2.47	6.40	18.40	64.20	1.03	2.47	6.40	18.40	53.21	1.03	2.47	6.40	18.40	52.58
25°	5.14	8.35	14.83	28.33	61.41	5.14	8.35	14.83	30.14	72.80	5.14	8.35	14.83	30.14	75.31
	1.03	2.47	6.40	18.40	64.20	1.03	2.47	6.40	18.40	55.20	1.03	2.47	6.40	18.40	52.97
30°	5.14	8.35	14.83	28.09	59.44	5.14	8.35	14.83	30.14	70.32	5.14	8.35	14.83	30.14	75.31
	1.03	2.47	6.40	18.40	64.20	1.03	2.47	6.40	18.40	56.41	1.03	2.47	6.40	18.40	52.63
60°	5.14	8.35	14.83	26.52	50.32	5.14	8.35	14.83	30.03	56.60	5.14	8.35	14.83	30.14	62.88
	1.03	2.47	6.40	18.40	64.20	1.03	2.47	6.40	18.40	46.18	1.03	2.47	6.40	16.72	36.17

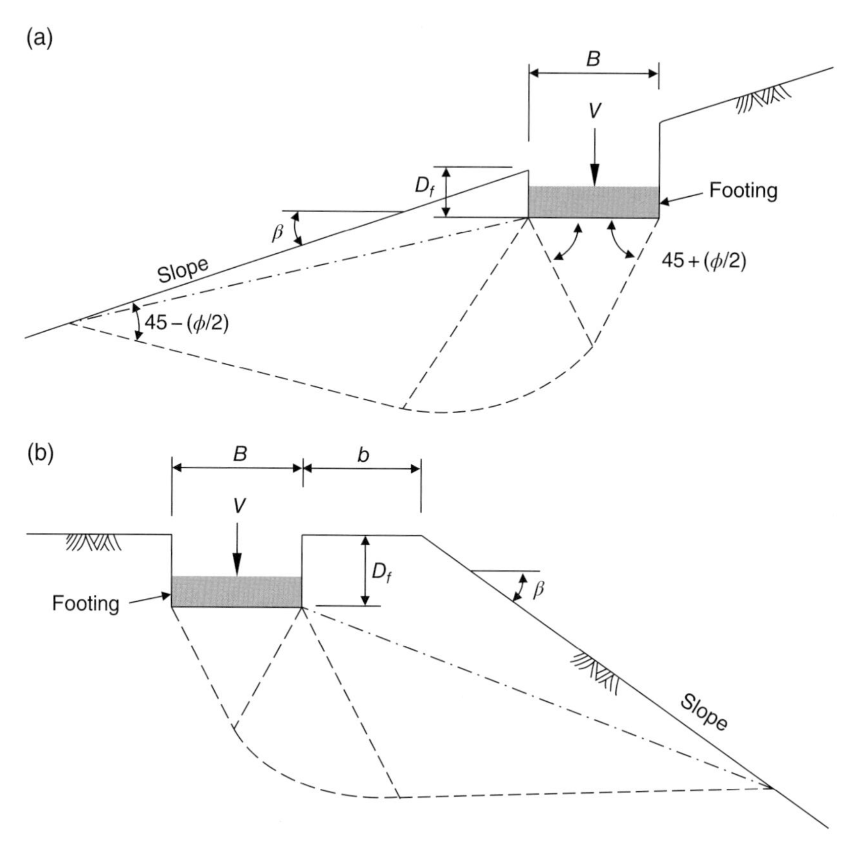

Figure 4.19 Foundation: (a) on a slope; (b) adjacent to a slope.

The easiest way to tackle this problem is estimating the bearing capacity using the *lowest values* of strength parameters (c, $Ø$) and unit weight (γ) in the zone between the foundation level and a depth B below that level (i.e. the zone in which bearing capacity failures occur), where B equals to the foundation width. This procedure may be too conservative, since the rupture zone comprises the stronger layers also. Therefore, this method may be adopted only when it is clear that the bearing capacity failure criterion does not control the geotechnical design even with a conservative analysis (i.e. the settlement criterion controls); thus, there would be no need to conduct a more detailed analysis. Another way to handle the problem of layered soils is using *weighted average values* of strength parameters and unit weight based on the relative thicknesses of each stratum in the zone in which bearing capacity failures occur. This method could be conservative or unconservative, but should provide acceptable results so long as the differences in the strength parameters are not too great (Coduto, 2001). A more detailed method, in which a series of trial slip surfaces under the foundation are considered, requires more effort to implement and would be appropriate only for critical projects on complex soil profiles. For these reasons, numerous solutions have been proposed to enable one to make an estimate of the ultimate bearing capacity. Some of these solutions will be presented in the following sections.

4.13.2 Ultimate Bearing Capacity: Stronger Soil Underlain by Weaker Soil

(A) *General case*

Meyerhof and Hanna (1978) and Meyerhof (1974) proposed semi-theoretical methods for estimating ultimate bearing capacity of a shallow *continuous* foundation supported by two-layered soils. The *stronger* soil layer which exists immediately beneath the foundation, underlain by a

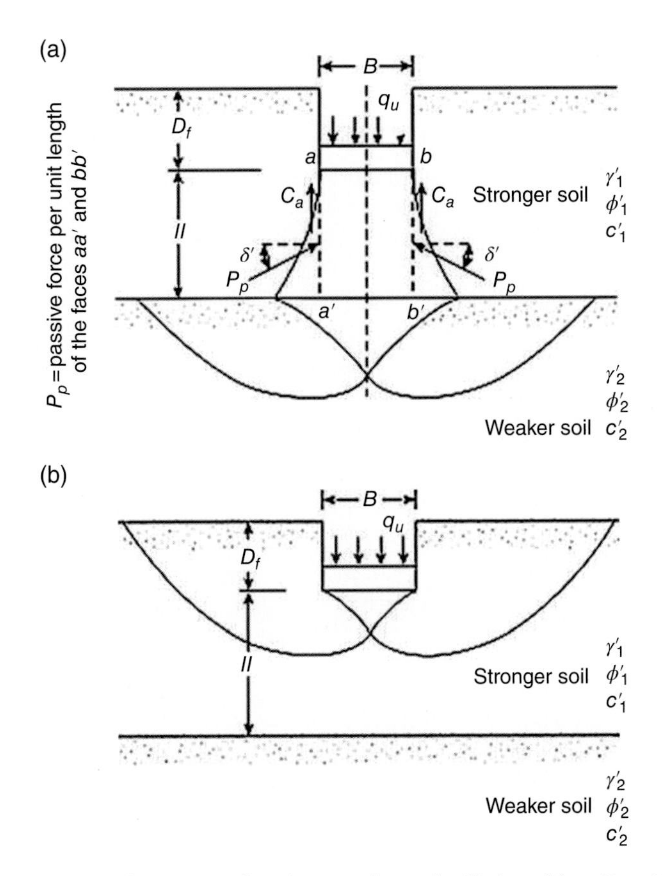

Figure 4.20 Bearing capacity of continuous foundation on layered soil(adapted from Das, 2011).

weaker soil layer, extends to a great depth, as shown in Figure 4.20. The physical parameters γ'_1, $Ø'_1, c'_1$ and γ'_2, $Ø'_2, c'_2$ belong to the stronger and weaker soil layers (top and bottom layers), respectively. In case the depth H is *small* compared with the foundation width B, the punching shear failure will take place in the top soil layer, followed by a general shear failure in the bottom soil layer, as shown in Figure 4.20a. In case, however, the depth H is large, then the general shear failure completely occurs in the top soil layer, as shown in Figure 4.20b. This condition is considered the upper limit of the ultimate bearing capacity.

Considering these two cases, the following equation for ultimate bearing capacity of a *continuous* footing could be derived (Das, 2011):

$$q_{\text{ult}} = \left\{ q_{\text{ult},b} + \left[\frac{2c'_a H}{B} + \gamma'_1 H^2 \left(1 + \frac{2D_f}{H}\right) \frac{K_s \tan Ø'_1}{B} - \gamma'_1 H \right] \right\} \leq q_{\text{ult},t} \tag{4.54}$$

Where

$$q_{\text{ult},t} = c'_1 N_{c(1)} + (\gamma'_1 D_f) N_{q(1)} + \frac{1}{2}\gamma'_1 B N_{\gamma(1)} \tag{4.55}$$

$$= \text{ultimate bearing capacity of the top (stronger) soil layer}$$

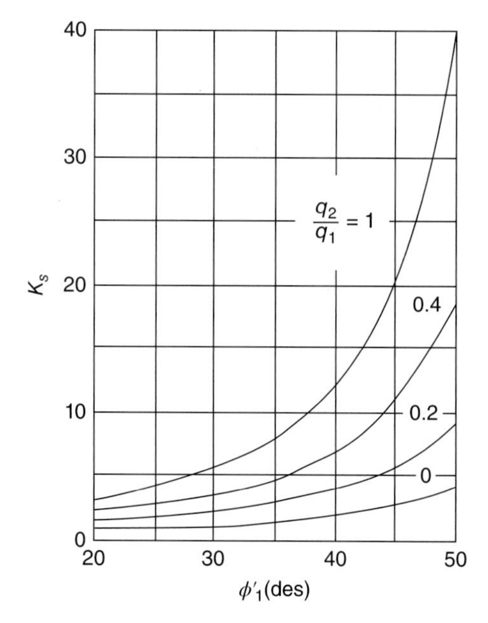

Figure 4.21 Meyerhof and Hanna's punching shear coefficient K_s.

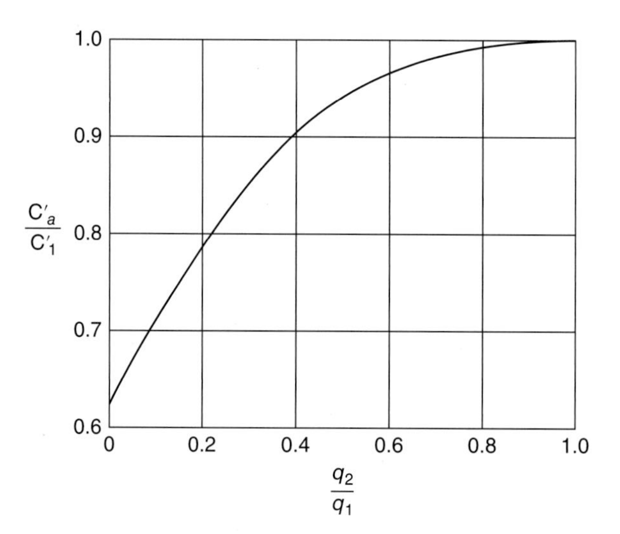

Figure 4.22 Variation of $\frac{c'_a}{c'_1}$ with $\frac{q_2}{q_1}$ based on the theory of Meyerhof and Hanna (1978).

$$q_{\text{ult},b} = c'_2 N_{c(2)} + \gamma'_1 (D_f + H) N_{q(2)} + \frac{1}{2}\gamma'_2 B N_{\gamma(2)} \tag{4.56}$$

$$q_1 = c'_1 N_{c(1)} + \frac{1}{2}\gamma'_1 B N_{\gamma(1)} \tag{4.57}$$

$$q_2 = c'_2 N_{c(2)} + \frac{1}{2}\gamma'_2 B N_{\gamma(2)} \tag{4.58}$$

K_s = punching shear coefficient; its variation with $\frac{q_2}{q_1}$ and $\varnothing'_1$ is shown in Figure 4.21..

c'_a = unit adhesion of stronger soil. The variation of $\frac{c'_a}{c'_1}$ with $\frac{q_2}{q_1}$ is shown in Figure 4.22.
C_a = adhesive force = $c'_a H$

For *rectangular* foundations, Equation 4.54 can be extended to the form

$$q_{\text{ult}} = \left\{ q_{\text{ult},b} + \left[\left(1+\frac{B}{L}\right)\left(\frac{2c'_a H}{B}\right) + \gamma'_1 H^2 \left(1+\frac{B}{L}\right)\left(1+\frac{2D_f}{H}\right)\left(\frac{K_s \tan \text{Ø}'_1}{B}\right) - \gamma'_1 H \right] \right\} \le q_{\text{ult},t} \quad (4.59)$$

Where

$$q_{\text{ult},t} = c'_1 N_{c(1)} F_{cs(1)} + (\gamma'_1 D_f) N_{q(1)} F_{qs(1)} + \frac{1}{2}\gamma'_1 B N_{\gamma(1)} F_{\gamma s(1)} \quad (4.60)$$

$$q_{\text{ult},b} = c'_2 N_{c(2)} F_{cs(2)} + \gamma'_1 (D_f + H) N_{q(2)} F_{qs(2)} + \frac{1}{2}\gamma'_2 B N_{\gamma(2)} F_{\gamma s(2)} \quad (4.61)$$

$F_{cs(1)}, F_{qs(1)}, F_{\gamma s(1)}$ = Shape factors with respect to the stronger soil layer

$F_{cs(2)}, F_{qs(2)}, F_{\gamma s(2)}$ = Shape factors with respect to the weaker soil layer

(B) *Special cases*

(1) *Strong sand* (top layer) *overlying saturated soft clay* ($\text{Ø}_2 = 0$)

From Equation (4.61):

$$q_{\text{ult},b} = \left(1 + 0.2\frac{B}{L}\right) 5.14 c_2 + \gamma'_1 (D_f + H) \quad (4.62)$$

From Equation (4.60):

$$q_{\text{ult},t} = (\gamma'_1 D_f) N_{q(1)} F_{qs(1)} + \frac{1}{2}\gamma'_1 B N_{\gamma(1)} F_{\gamma s(1)} \quad (4.63)$$

Hence, from Equation (4.59):

$$q_{\text{ult}} = \left[\left(1 + 0.2\frac{B}{L}\right) 5.14 c_2 + \gamma'_1 H^2 \left(1+\frac{B}{L}\right)\left(1+\frac{2D_f}{H}\right)\left(\frac{K_s \tan \text{Ø}'_1}{B}\right) + \gamma'_1 D_f \right] \le (\gamma'_1 D_f) N_{q(1)} F_{qs(1)} + \frac{1}{2}\gamma'_1 B N_{\gamma(1)} F_{\gamma s(1)} \quad (4.64)$$

where

c_2 = undrained cohesion of the soft clay

K_s is determined from Figure 4.21, using

$$\frac{q_2}{q_1} = \frac{c_2 N_{c(2)}}{\frac{1}{2}\gamma'_1 B N_{\gamma(1)}} = \frac{5.14 c_2}{0.5 \gamma'_1 B N_{\gamma(1)}} \quad (4.65)$$

(2) *Stronger sand* (top layer) *overlying weaker sand* ($c'_1 = 0, c'_2 = 0$)

From Equation (4.61):

$$q_{\text{ult},b} = \gamma'_1 (D_f + H) N_{q(2)} F_{qs(2)} + \frac{1}{2}\gamma'_2 B N_{\gamma(2)} F_{\gamma s(2)} \quad (4.66)\text{-(a)}$$

From Equation (4.60):

$$q_{\text{ult},t} = (\gamma'_1 D_f) N_{q(1)} F_{qs(1)} + \frac{1}{2}\gamma'_1 B N_{\gamma(1)} F_{\gamma s(1)} \quad (4.66)\text{-(b)}$$

Hence, from Equation (4.59):

$$q_{\text{ult}} = \left[\gamma_1'(D_f + H)N_{q(2)}F_{qs(2)} + \frac{1}{2}\gamma_2' BN_{\gamma(2)}F_{\gamma s(2)} + \gamma_1' H^2\left(1+\frac{B}{L}\right)\left(1+\frac{2D_f}{H}\right)\left(\frac{K_s \tan \text{Ø}_1'}{B}\right) - \gamma_1' H\right]$$
$$\leq (\gamma_1' Df)N_{q(1)}F_{qs(1)} + \frac{1}{2}\gamma_1' BN_{\gamma(1)}F_{\gamma s(1)} \quad (4.67)$$

K_s is determined from Figure 4.21, using

$$\frac{q_2}{q_1} = \frac{\frac{1}{2}\gamma_2' BN_{\gamma(2)}}{\frac{1}{2}\gamma_1' BN_{\gamma(1)}} = \frac{\gamma_2' BN_{\gamma(2)}}{\gamma_1' BN_{\gamma(1)}} \quad (4.68)$$

(3) *Stronger saturated clay* ($\text{Ø}_1 = 0$) *overlying weaker saturated clay* ($\text{Ø}_2 = 0$)
From Equation (4.61):

$$q_{\text{ult},b} = \left(1 + 0.2\frac{B}{L}\right)5.14c_2 + \gamma_1'(D_f + H)$$

From Equation (4.60):

$$q_{\text{ult},t} = \left(1 + 0.2\frac{B}{L}\right)5.14c_1 + \gamma_1' D_f \quad (4.69)$$

Hence, from Equation (4.59):

$$q_{\text{ult}} = \left[\left(1 + 0.2\frac{B}{L}\right)5.14c_2 + \left(1+\frac{B}{L}\right)\left(\frac{2c_a H}{B}\right) + \gamma_1' D_f\right]$$
$$\leq \left(1 + 0.2\frac{B}{L}\right)5.14c_1 + \gamma_1' D_f \quad (4.70)$$

where
c_1 and c_2 = undrained cohesion of the top and bottom clays,
c_a is determined from Fig. 4.22, using

$$\frac{q_2}{q_1} = \frac{5.14c_2}{5.14c_1} = \frac{c_2}{c_1} \quad (4.71)$$

4.13.3 Ultimate Bearing Capacity: Weaker Soil Underlain by Stronger Soil

Refer to Figure 4.23a. If the ratio $\frac{H}{B}$ is relatively small, the slip surface at ultimate load will pass through both soil layers, as shown in the left half of the figure. For large $\frac{H}{B}$ ratio, however, the failure surface will be completely located in the weaker soil layer which overlies the stronger soil, as shown in the right half of the figure. The ultimate bearing capacities q_1 and q_2 are the same as defined by Equations (4.57) and (4.58); however, the ratio $\frac{q_2}{q_1}$ should be greater than unity.

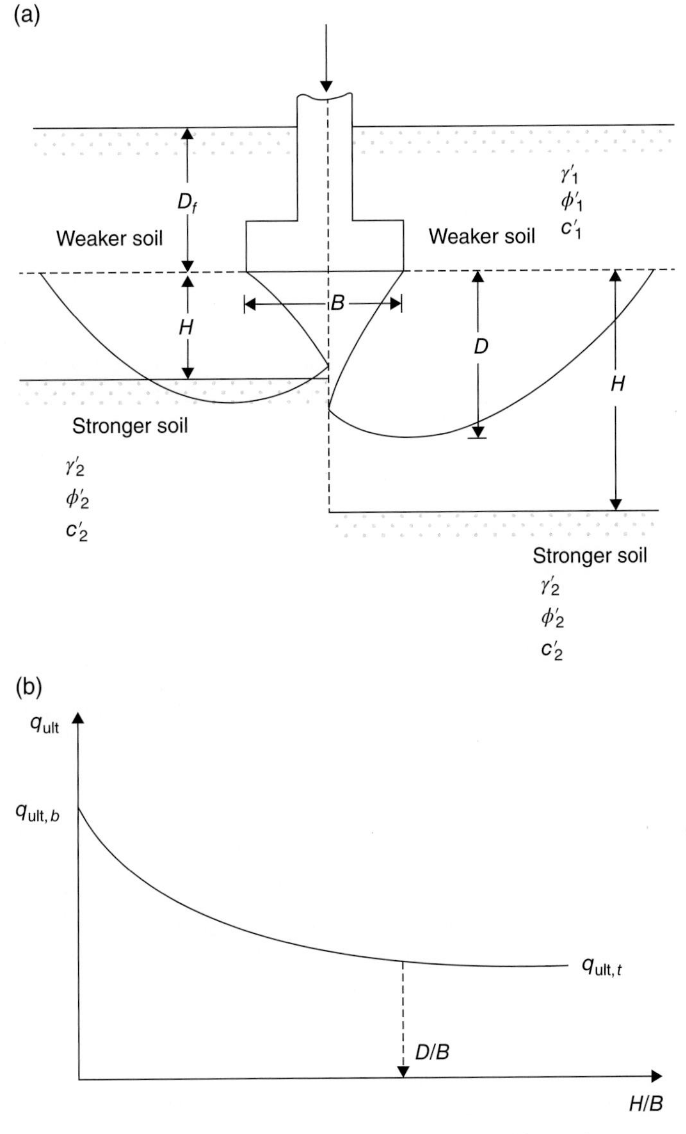

Figure 4.23 (a) Foundation on weaker soil layer underlain by stronger soil layer. (b) Nature of variation of q_{ult} with (H/B). From Das (2011).

For this case, the ultimate bearing capacity can be given by the following semi-theoretical general equation (Meyerhof, 1974; Meyerhof and Hanna, 1978):

$$q_{ult} = q_{ult,t} + (q_{ult,b} - q_{ult,t})\left(\frac{H}{D}\right)^2 \geq q_{ult,t} \tag{4.72}$$

Where:

$$q_{ult,t} = c_1 N_{c(1)} F_{cs(1)} + (\gamma'_1 D_f) N_{q(1)} F_{qs(1)} + \frac{1}{2}\gamma'_1 B N_{\gamma(1)} F_{\gamma s(1)} \tag{4.73}$$

$$= \text{ultimate bearing capacity of the top (weaker) soil layer}$$

$$q_{\text{ult},b} = c_2 N_{c(2)} F_{cs(2)} + (\gamma'_2 D_f) N_{q(2)} F_{qs(2)} + \frac{1}{2}\gamma'_2 B N_{\gamma(2)} F_{\gamma s(2)} \tag{4.74}$$

= ultimate bearing capacity of the bottom (stronger) soil layer

D = maximum depth of failure surface beneath the foundation in the thick bed of the upper (weaker) soil layer.

Meyerhof and Hanna (1978) suggested that:

$D \approx B$ for loose sand and clay
$D \approx 2B$ for dense sand

Equations (4.72) to (4.74) indicate that $q_{\text{ult},b} \geq q_{ult} \geq q_{\text{ult},t}$, as shown in Figure 4.23b.

Equation (4.72) is general; can be applied to special cases as needed following the same procedure delineated in Section 4.13.2

4.14 Safety Factor

One of the main performance requirements of any structure is strength. With respect to foundations, as explained in Section 2.4, this performance requirement comprises both the geotechnical strength and the structural strength. In general, structures are designed on the basis of determining the service loads and obtaining a suitable *ratio* of material strength to these loads, termed either a *safety factor* (SF; in geotechnical design) or *load factor* (in structural design). These factors are used to compensate for the many uncertainties usually associated with analysis and design of structures, particularly, foundations. Thus, reliable designs could be developed. In design, the selected safety factor defines the designer's estimate of the best compromise between cost and reliability. Therefore, in selection of an appropriate safety factor the designer should consider many factors or parameters, including: importance of structure, probability of failure and its consequences; uncertainties in materials properties, applied loads and analytical methods; cost–benefit ratio of additional conservatism in the design.

In the analysis and design of foundations, especially the geotechnical part, there are more uncertainties than those associated with analysis and design of superstructure elements. For this reason, safety factors in foundations are typically greater than those in the superstructure. These uncertainties and their causes may be stated as follows:

Insufficient knowledge of subsurface conditions
Deficient in accurate determination of the soil properties
Complexity of soil behavior and soil–structure interaction
Deficient in methods of analyses
Lack of control over environmental changes after construction

Coduto (2001) presents the necessary parameters and the typical values of safety factor which are usually considered in selecting an adequate design safety factor, as shown in Figure 4.24. Geotechnical engineers usually use SF between 2.5 and 3.5 for bearing capacity analyses of shallow foundations. Occasionally, however, they might use values as low as 2.0 or as high as 4.0.

It may be useful to mention herein that the true safety factor is probably greater than the design safety factor, for one or combination of the following reasons:

(1) Where settlement controls the final design, the safe bearing capacity will be reduced, which in turn increases the real safety factor.

(2) The shear strength data are usually interpreted conservatively, so the design strength parameters c and Ø implicitly contain another safety factor.

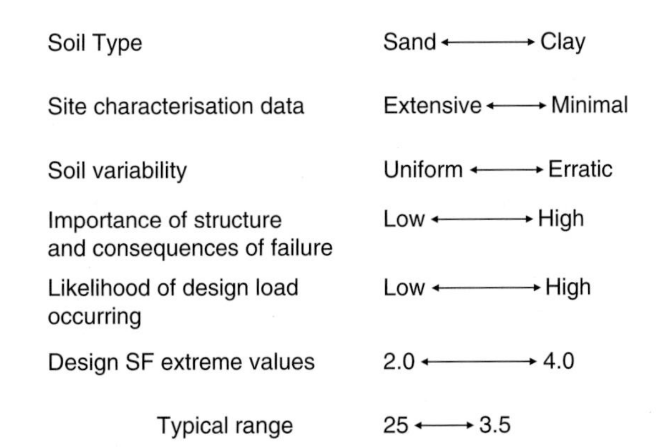

Figure 4.24 Factors considered in selecting design safety factor and typical values of SF (adapted from Coduto, 2001).

(3) The service loads are probably less than the design loads.
(4) Spread footings are commonly built somewhat larger than the plan dimensions.
(5) Where the same design safety factor for a spread footing (2.5–3.5) is also used for a mat foundation, the real safety factor of the mat will be higher because, normally, mats require a lower design safety factor (2.0–2.5).

4.15 Bearing Capacity from Results of In Situ Tests

(1) *From results of standard penetration test.*
The standard penetration test, SPT, has been described and explained in some detail in the discussion of solution of Problem 1.6. Also, the use of SPT results, that is N-values, in estimating settlement of *granular* soils was discussed in Section 3.3.2. In that discussion a set of allowable bearing capacity equations [Equations (3.11) to (3.13)], based on SPT results, have been presented. These equations give net allowable bearing capacity for a specified 25 mm settlement. It has been stated also that for cohesionless soils, it is possible to assume that settlement is proportional to net allowable soil pressure. Based on this assumption, a net allowable soil pressure, $netq_{(S_i)}$, for an allowable settlement other than 25 mm can be estimated using Equation (3.14).

(2) *From results of plate-load test.*
The plate-load test, PLT, was described and explained in some detail in the discussion of the Solution of Problem 1.9. Also, the use of PLT results, that is the time settlement curves (Figure 1.12a) for different load increments and the load settlement curve (Figure 1.12b), was explained in the discussion of the Solution of the same Problem. In that discussion a set of equations [Equations (1.37) to (1.39)], based on PLT results, were presented. These equations give ultimate bearing capacities, q_{ult}, for different types of soils. However, due to several factors which are explained in the same discussion and also in Section 3.3.1, the extrapolation of the PLT results to full-size footings is questionable except in very limited cases.

(3) *From results of cone penetration test.*
The static cone penetration test, CPT, was described and explained in some detail in the discussion of the Solution of Problem 1.13. Also, the CPT correlations for both cohesive and cohesionless soils (i.e. q_c with c_u or s_u and Ø correlations) have been provided. It is obvious that we can use these CPT correlations to obtain the necessary parameters so that the bearing capacity equations of Table 4.2 can be applied. Furthermore, correlations for the SPT-N values have been presented, as shown in Figure 1.18, Table 1.11 and Equation (1.56). The net allowable bearing capacity of sand can be estimated directly using Equations (3.11) to (3.13) in which the N'_{55} is replaced (approximately) by $\frac{q_c}{4}$ (Meyerhof, 1956), where q_c is in units of kg/cm^2.

According to Schmertmann (1978), the bearing capacity factors for use in the Terzaghi bearing capacity equation of Table 4.2 can be estimated as

$$0.8\,N_q \cong 0.8\,N_\gamma \cong q_c \quad \left(\text{for } \frac{\text{D}}{\text{B}} \le 1.5\right) \tag{4.75}$$

The q_c (kg/cm²) is averaged over the depth interval from about $B/2$ above to 1.1 B below the footing base.

For *cohesionless* soils one may use

$$q_{\text{ult}} = 28 - 0.0052\,(300 - q_c)^{1.5}\,(\text{kg/cm}^2) \quad \text{(Strip footings)} \tag{4.76}$$

$$q_{\text{ult}} = 48 - 0.009\,(300 - q_c)^{1.5}\,(\text{kg/cm}^2) \quad \text{(Square footings)} \tag{4.77}$$

For *clays* one may use

$$q_{\text{ult}} = 2 + 0.28\,q_c\,(\text{kg/cm}^2) \quad \text{(Strip footings)} \tag{4.78}$$

$$q_{\text{ult}} = 5 + 0.34\,q_c\,(\text{kg/cm}^2) \quad \text{(Square footings)} \tag{4.79}$$

According to Bowles (2001), Equations (4.76) through (4.79) are based on charts given by Schmertmann (1978) credited to an unpublished reference by Awakti.

4.16 Uplift Capacity of Shallow Foundations

Foundations under special circumstances, particularly in industrial applications, are subjected to external uplift or tension forces. Examples are: footings for the columns (legs) of an elevated water tank, bases for legs of power transmission towers, anchorages for the anchor cables of transmission towers and many other industrial equipment installations.

Balla (1961) considered the problem of round footings which develop tension resistance. He assumed circular failure surfaces and developed mathematical expressions. Later, Meyerhof and Adams (1968) studied the problem and proposed two conditions; one for shallow and the other for deep footings, as shown in Figure 4.25. They provided relationships to estimate the ultimate uplifting load for shallow circular and rectangular footings.

Refer to Figure 4.26. The ultimate load Q_u (or the ultimate tension T_u) can be expressed in the form of a nondimensional breakout factor F_q as follows:

$$F_q = \frac{Q_u}{A\,\gamma\,D_f} \tag{4.80}$$

where A = area of footing

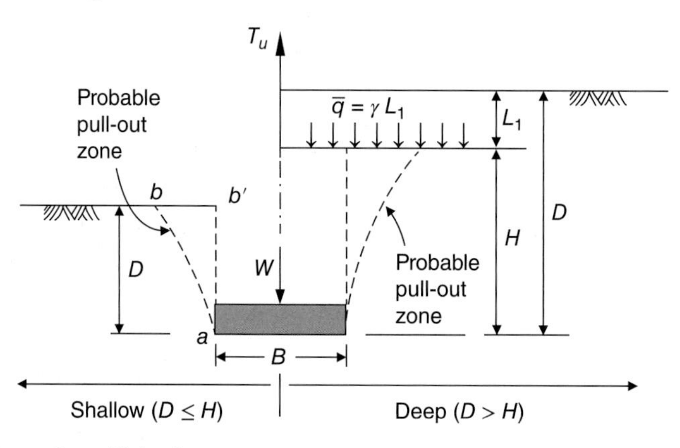

Figure 4.25 Footing under uplift load.

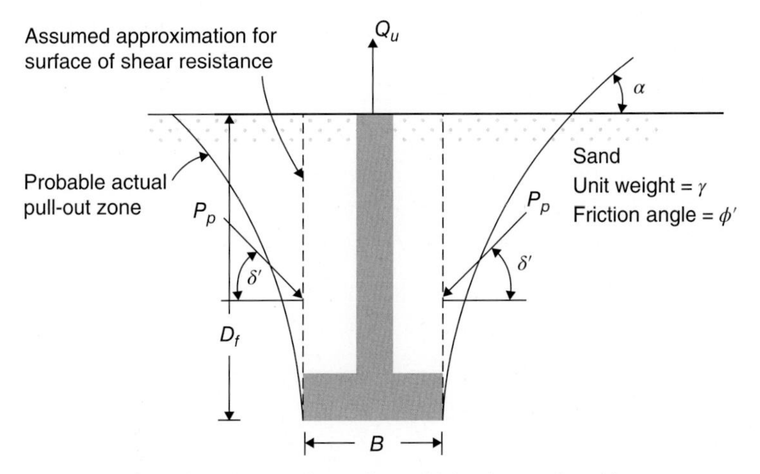

Figure 4.26 Shallow continuous foundation subjected to uplift load (reproduced from Das, 2011).

The factor F_q is a function of the soil friction angle $\text{Ø}'$ and the embedment ratio D_f/B. For a given soil friction angle, the breakout factor increases with increase in D_f/B and reaches its maximum value at D_f/B equals to the critical embedment ratio, $(D_f/B)_{cr}$. For shallow foundations the embedment ratio $D_f/B \le (D_f/B)_{cr}$, whereas for deep foundations $D_f/B > (D_f/B)_{cr}$. Using Equation (4.80) and the relationships developed by Meyerhof and Adams (1968), Das and Seely (1975) developed relationships for foundations subjected to uplift load as follows:

(A) *Uplift capacity of footings in granular soils* $(c' = 0)$
For shallow circular and square foundations:

$$F_q = 1 + 2\left[1 + m\left(\frac{D_f}{B}\right)\right]\left(\frac{D_f}{B}\right)K_u \tan \text{Ø}' \tag{4.81}$$

For shallow rectangular foundations:

$$F_q = 1 + \left\{\left[1 + 2m\left(\frac{D_f}{B}\right)\right]\left(\frac{B}{L}\right) + 1\right\}\left(\frac{D_f}{B}\right)K_u \tan \text{Ø}' \tag{4.82}$$

where

m = a coefficient which is a function of $\text{Ø}'$
K_u = nominal uplift coefficient

The variations of K_u, m and $(D_f/B)_{cr}$ for square and circular foundations are given in Table 4.12 (Meyerhof and Adams, 1968).
Das and Jones (1982) recommended the following critical embedment ratio for rectangular foundations:

$$\left(\frac{D_f}{B}\right)_{cr,\ rec} = \left(\frac{D_f}{B}\right)_{cr,\ sq}\left[0.133\left(\frac{L}{B}\right) + 0.867\right] \le 1.4\left(\frac{D_f}{B}\right)_{cr,\ sq} \tag{4.83}$$

Figure 4.27 shows variations of F_q with D_f/B and $\text{Ø}'$ for square and circular foundations. The variations have been calculated from Equation (4.81).

Table 4.12 Variations of K_u, m and $(D_f/B)_{cr}$ for circular and square foundations.

Ø′ (deg)	K_u	m	$(D_f/B)_{cr}$
20	0.856	0.05	2.5
25	0.888	0.10	3
30	0.920	0.15	4
35	0.936	0.25	5
40	0.960	0.35	7
45	0.960	0.50	9

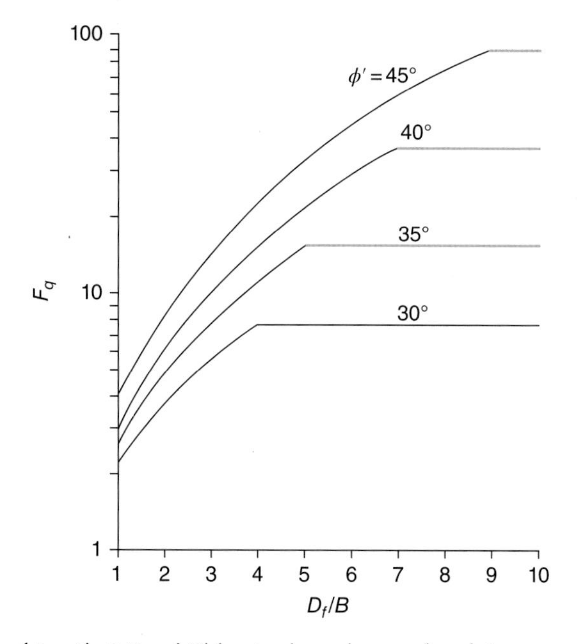

Figure 4.27 Variations of F_q with D_f/B and Ø′ for circular and square foundations.

According to Das (2011), the steps of the procedure to estimate the uplift capacity of foundations in granular soils are as below:

(1) Determine D_f, B, L and $Ø'$.

(2) Calculate D_f/B.

(3) Using Table 4.12 and Equation (4.83), calculate $(D_f/B)_{cr}$.

(4) If D_f/B is less than or equal to $(D_f/B)_{cr}$, it is a shallow foundation.

(5) If D_f/B is greater than $(D_f/B)_{cr}$, it is a deep foundation.

(6) For shallow foundations, use D_f/B calculated in Step 2 in Equation (4.81) or (4.82) to estimate F_q. Thus, $Q_u = F_q A \gamma D_f$.

(7) For deep foundations, substitute $(D_f/B)_{cr}$ for D_f/B in Equation (4.81) or (4.82) to obtain F_q from which the ultimate load Q_u may be obtained.

(B) *Uplift capacity of footings in cohesive soils* $(Ø' = 0)$

The ultimate load Q_u of a foundation in a cohesive soil, considering $Ø' = 0$, may be expressed as

$$Q_u = A\left(\gamma D_f + c_u F_c\right) \tag{4.84}$$

where

A = area of foundation

c_u = undrained shear strength of soil

F_c = breakout factor

The breakout factor increases with increase in D_f/B and reaches its maximum value of $F_c = F_c^*$ at D_f/B equals to the critical embedment ratio, $(D_f/B)_{cr}$. Das (1978, 1980) proposed

$$\left(\frac{D_f}{B}\right)_{cr} = 0.107\, c_u + 2.5 \le 7 \quad \text{(for square and circular footings)} \tag{4.85}$$

where

c_u = undrained shear strength of soil, in kN/m^2

$$\left(\frac{D_f}{B}\right)_{cr} = \left(\frac{D_f}{B}\right)_{cr,\ sq}\left[0.73 + 0.27\left(\frac{L}{B}\right)\right] \le 1.55\left(\frac{D_f}{B}\right)_{cr,\ sq} \tag{4.86}$$

$$\text{(for rectangular footings)}$$

where

L = length of footing

According to Das (2011), steps of the procedure to estimate the uplift capacity of foundations in cohesive soils are as below:

(1) Determine the undrained cohesion, c_u.

(2) Determine the $(D_f/B)_{cr}$ using Equations (4.85) and (4.86).

(3) Determine the $\frac{D_f}{B}$ ratio of the foundation. If $D_f/B \le (D_f/B)_{cr}$, it is a shallow foundation. However, if D_f/B is greater than $(D_f/B)_{cr}$, it is a deep foundation.

(4) For $D_f/B > (D_f/B)_{cr}$ (i.e. deep foundations)

$$F_c = F_c^* = 7.56 + 1.44\left(\frac{B}{L}\right)$$

$$Q_u = A\left\{\left[7.56 + 1.44\left(\frac{B}{L}\right)\right]c_u + \gamma D_f\right\} \tag{4.87}$$

(5) For $D_f/B \le (D_f/B)_{cr}$ (i.e. shallow foundations)

$$Q_u = A\left(B' F_c^* c_u + \gamma D_f\right) = A\left\{B'\left[7.56 + 1.44\left(\frac{B}{L}\right)\right]c_u + \gamma D_f\right\} \tag{4.88}$$

where:

$B' = \frac{F_c}{F_c^*}$, obtained from the average curve of Figure 4.28 which shows a plot of B' versus a non-dimensional factor α'.

$$\alpha' = \frac{D_f/B}{(D_f/B)_{cr}}$$

According to Das, the procedures outlined above give fairly good results for estimating the net ultimate uplift capacity of foundations and agree reasonably well with the theoretical solution of Merifield et al. (2003).

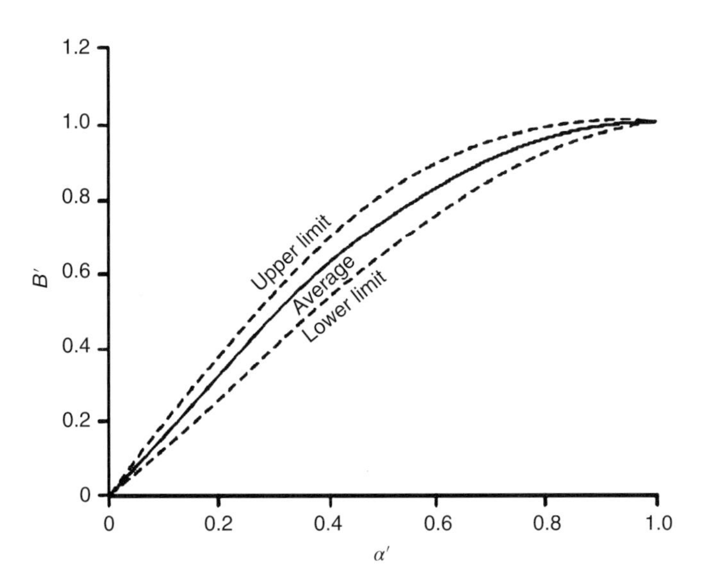

Figure 4.28 Plot of β' versus α'.

4.17 Some Comments and Considerations Concerning the Geotechnical Design of Shallow Foundations

(1) The geotechnical design of a shallow foundation comprises determination of the depth of foundation and the dimensions of the foundation base area through assessment of the results of bearing capacity and settlement analyses, as well as other considerations, so that the applied loads are transferred from the foundation to the ground safely.

The foundation depth was discussed in some detail in Section 2.3. Dimensions of the base area are determined such that the net effective foundation pressure *net* q', is always less or equal to the net allowable bearing capacity (or design soil pressure) *net* q_a; see Equation (4.11). Therefore, as it is obvious, the main and most important task in geotechnical design is how to determine *net* q_a reasonably so that the performance requirements of Section 2.4 are satisfied. However, the base dimensions are usually required in executing the bearing capacity and settlement computations. For example, the width B and length L dimensions are needed to compute shape, depth and load inclination factors which are contained in the ultimate bearing capacity equations of Table 4.2. Therefore, the analysis methods require trial process to obtain the design base dimensions.

(2) In selecting a design soil pressure, it is important to consider the following:

- If the supporting soil is very loose and saturated, it should be compacted and retested. Based on the available new data, an allowable bearing capacity can be selected.
- Usually, the allowable bearing capacity values for sand, except for narrow footings on loose saturated sand, are governed only by settlement considerations, because it can be taken for granted that the safety factor with respect to bearing capacity failure (base failure) is adequate (Terzaghi and Peck, 1967). On routine jobs the net allowable soil pressure on dry and moist sand can be determined from Equations (3.11) to (3.13) on the basis of the results of SPT or CPT. These Equations satisfy the condition that the maximum total and differential settlements are unlikely to exceed 25 and 20 mm, respectively. Obviously, results of computations using these equations may be verified, if desired, using other appropriate bearing capacity and settlement equations.
- The net ultimate bearing capacity for clay can be calculated on the basis of the equations given in Table 4.2 and the results of unconfined compression tests or undrained shear tests. The net

safe bearing capacity value for clay is usually determined using a safety factor with respect to a bearing capacity failure not less than three. After the net safe soil pressure has been selected, it is necessary to find out under this pressure whether the settlement will be tolerable. If the calculated settlement is tolerable, the net safe soil pressure is taken as the net allowable bearing capacity or design soil pressure. Otherwise, settlement controls and the net allowable bearing capacity will be selected accordingly. If the clay is normally consolidated, excessive differential settlements of about 50–100 mm or even 150 mm are commonly considered unavoidable. Attempts to reduce the settlement by reducing the net allowable soil pressures to values smaller than the net safe soil pressure are ineffective and wasteful (Terzaghi and Peck, 1967). Therefore, the designer must choose between two alternatives. Either he designs the footings using the net safe bearing capacity at the risk of large differential settlements, or else he provides the structure with another type of foundations (raft, pile or pier foundations). On the other hand, if the clay is overconsolidated (medium and stiff clays beneath a shallow overburden are usually precompressed), the differential settlement is likely to be tolerable. In doubtful cases the load test method may be used; however, if it is not expertly planned and executed, the results may be very misleading. The net allowable soil pressure on stiff fissured clays can be determined only by this method.

- Very loose or even loose saturated silt is unsuitable for supporting foundation loads. The allowable bearing capacity value of medium or dense silt of the rock-flour type (of nearly equidimensional quartz grains) can be estimated roughly by means of the methods proposed for sand. On the other hand, that of medium or stiff plastic silt (of mostly flake-shaped particles) can be approximated by the methods for clays.
- In case the area occupied by the footings exceeds half the total area covered by the building, it is commonly more economical and may be safer to provide the building with a mat or raft foundation (see Section 2.2).

(3) Reduction of differential settlement by adjusting footing size. The preceding equations of settlement (Chapter 3) indicated that the settlement of loaded areas with similar shape but different size increases at a given load intensity with increasing width B of the area. In case the footings of a structure differ greatly in size, the differential settlement due to this cause can be important; hence, some revision of the safety factor through change in the size of the footing may be necessary. In case the subsoil consists of sand, the differential settlement can be reduced by decreasing the size of the smallest footings, because even after the reduction the safety factor SF of these footings with respect to bearing capacity failure is likely to be adequate. If the subsoil consists of clay, the application of this procedure would reduce the value of SF for the smallest footings to less than three, which is not admissible. Therefore, the differential settlement of footing foundations on clay can be reduced only by increasing size of the largest footings beyond that required by the safe bearing capacity.

(4) Overall design steps of shallow foundations. The first step is to compute the effective load (the design load), according to a specified method (ASD or LRFD method), that will be transferred to the subsoil at the base of the foundation. The second step is to determine the allowable bearing capacity (the design soil pressure) for the supporting soil. The area of the foundation base is then obtained by dividing the design load by the design soil pressure. Finally, using a specified design method, the design loads are computed, the bending moments and shears in the base are determined and the structural design of the base is carried out.

(5) Design loads. There are two methods of expressing and working with design loads: the *allowable stress design* (ASD) method (also known as the *working stress design*, WSD, method) and the *load and resistance factor design* (LRFD) method (also known as the *ultimate strength design*, USD, method). The ASD method uses *unfactored (working) loads*, whereas the LRFD uses factored loads. All codes, except few ones, specify using the ASD method in *geotechnical* analysis and design, whereas the LRFD method is mostly specified for use in *structural* analysis and design.

As defined by the codes, the design load is the most critical combination of the various load sources. The ANSI/ASCE *Minimum Design Loads for Buildings and Other Structures* (ASCE, 1996) defines the ASD design load as the greatest of the following four load combinations [ANSI/ASCE 2.4.1]:

(a) D (Governs only when some of the loads act in opposite direction)
(b) $D+L+F+H+T+(L_r \text{ or } S \text{ or } R)$
(c) $D+L+(L_r \text{ or } S \text{ or } R)+(W \text{ or } E)$
(d) $D+(W \text{ or } E)$

where

D = Dead loads due to the weight of the structure, including permanently installed equipment.
L = Live loads; they are caused by the intended use and occupancy. These include loads from people, furniture, inventory, maintenance activities, moveable partitions, moveable equipment, vehicles and other similar sources.
L_r = Roof-live loads.
S = Snow loads, they are a special type of live load caused by accumulation of snow.
R = Rain loads; they are a special type of live load caused by accumulation of rain. Sometimes rain loads caused by ponding (the static accumulation of water on the roof) are considered separately.
H = Earth pressure loads due to the weight and lateral pressures from soil or rock, such as those acting on a retaining wall.
F = Fluid loads; they are caused by fluids with well-defined pressures and maximum heights, such as water in a storage tank.
E = Earthquake loads; they are the result of accelerations from earthquakes.
W = Wind loads; they are imparted by wind onto the structure.
T = Self-straining loads; they are due to temperature changes, shrinkage, moisture changes, creep, differential settlement and other similar processes.

There are codes which are using different load combinations for computing the ASD design load, so it is important to observe the applicable code for each project.

Foundation engineers normally have the authority to increase the design bearing capacity by one-third when they consider load combination (c) or (d), even if this increase is not specifically authorised by the prevailing code. An easy way to implement this criterion is by multiplying these load combinations by a factor of $0.75=1/1.33$ instead of increasing the static design bearing capacity. Thus, the design load combinations become

(a) D
(b) $D+L+F+H+T+(L_r \text{ or } S \text{ or } R)$
(c) $0.75[D+L+(L_r \text{ or } S \text{ or } R)+(W \text{ or } E)]$
(d) $0.75[D+(W \text{ or } E)]$

4.18 Bearing Capacity of Rock

Unless rocks are completely weathered to the consistency of sand, silt or clay, failure of foundations occurs in modes different from that of soils (Figure 4.29). In case where a rock is completely weathered as just described, it should be considered as a soil and the ultimate bearing capacity can be determined using the bearing capacity equations of Table 4.2. Critical conditions causing bearing capacity failure can occur where foundations rest on bedded and jointed unweathered or partly weathered rocks, as shown in Figure 4.29. Therefore, it would not be correct to assume that safe bearing pressures on rock are governed only from the considerations of permissible settlement. However, usually, settlement is more often of concern than is the bearing capacity.

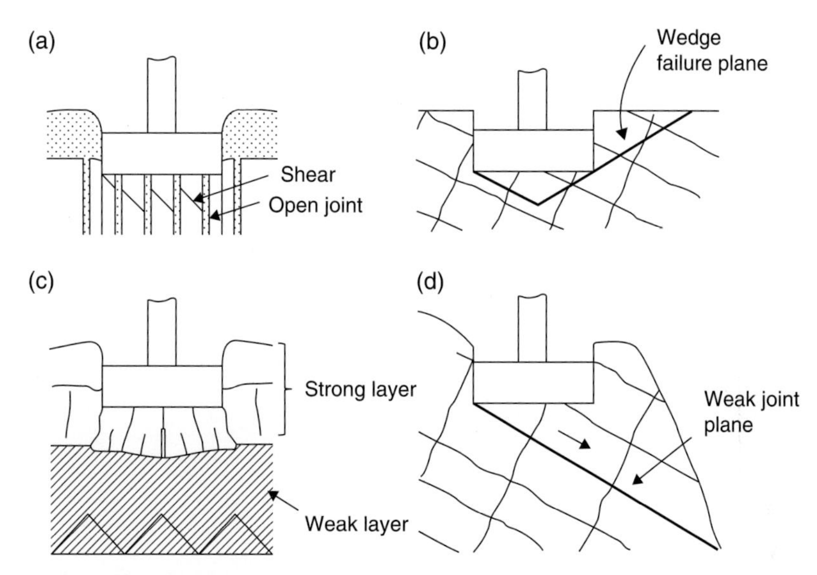

Figure 4.29 Failure of foundations on jointed rocks. (a) Shear failure of unsupported rock columns. (b) Wedge-type failure with closed joints (horizontal or inclined). (c) Splitting and punching with strong layer over weak layer. (d) Sliding on weak inclined joint. (From Tomlinson, 2001).

Table 4.13 Range of properties for selected rock groups; data from several sources (from Bowles, 2001).

Type or rock	Typical unit wt., kN/m^3	Modulus of elasticity E^a, MPa $\times 10^3$	Piosson's ratio, μ	Compressive strength, MPa
Basalt	28.0	17–103	0.27–0.32	170–415
Granite	26.4	14–83	0.26–0.30	70–276
Schist	26.0	7–83	0.18–0.22	35–105
Limestone	26.0	21–103	0.24–0.45	35–170
Porous limestone		3–83	0.35–0.45	7–35
Sandstone	22.8–23.6	3–42	0.20–0.45	28–138
Shale	15.7–22.0	3–21	0.25–0.45	7–40
Concrete	15.7–23.6	*Variable*	0.15	15–40

[a] Depends heavily on confining pressure and how determined; E = tangent modulus at approximately 50% of ultimate compressive strength.

It may be possible to use building code values for estimating the allowable bearing capacity of rock; however, geology, type and quality (*RQD*) of rock are significant parameters, which should be used together with the recommended code values. In estimating allowable bearing capacity of a rock, it is common to use a relatively large safety factor, from 5 to 10, which depends to some degree on the *RQD* value of the particular rock. The higher values of safety factor are usually used for *RQD* values less than about 0.75. Table 4.13 may be used as a guide to estimate bearing capacity from code values or

Table 4.14 Typical friction angles Ø for clean fractures in rock (after Wyllie, 1991).

Rock type	Ø°-Values
Schists with high mica content	20–27
Shale	
Marl	
Sandstone	27–34
Siltstone	
Chalk	
Gneiss	
Slate	
Basalt	34–40
Granite	
Limestone	
Conglomerate	

Table 4.15 Approximate relationships of *c* and Ø parameters to *RQD* of rock mass.

	Rock mass properties		
RQD (%)	q_c	c	Ø°
0–70	0.33 q_{uc}	0.1 q_{uc}	30
70–100	0.33–0.80 q_{uc}	0.1 q_{uc}	30–60

to obtain trial values for elastic parameters. In all cases, the upper limit on allowable bearing capacity of rock is taken as the *ultimate compressive strength*, f_c' of the base concrete.

In a rock mass with wide open joints (Figure 4.29a), the ultimate bearing capacity of the foundation is given by the unconfined compression strength of the intact rock, which is measured using representative rock samples and standard test procedure.

The ultimate bearing capacity of a shallow foundation on a rock mass with closed joints (Figure 4.29b) may be obtained using bearing capacity equations of the form given by Terzaghi, as presented in Table 4.2, which requires using bearing capacity factors N_c, N_q and N_γ different from those used for soils. The required cohesion c and angle of internal friction Ø may be obtained from results of high-pressure triaxial tests on large samples of the jointed rock. However, the samples are difficult to obtain and the tests are expensive. According to Tomlinson (2001), for ordinary foundation design it may be satisfactory to use typical values of Ø such as those given by Wyllie (1991); shown in Table 4.14.

Kulhawy and Goodman (1987, 1980) have shown that the c and Ø parameters can be related to *RQD* of the rock mass and they have suggested the approximate relationships shown in Table 4.15.

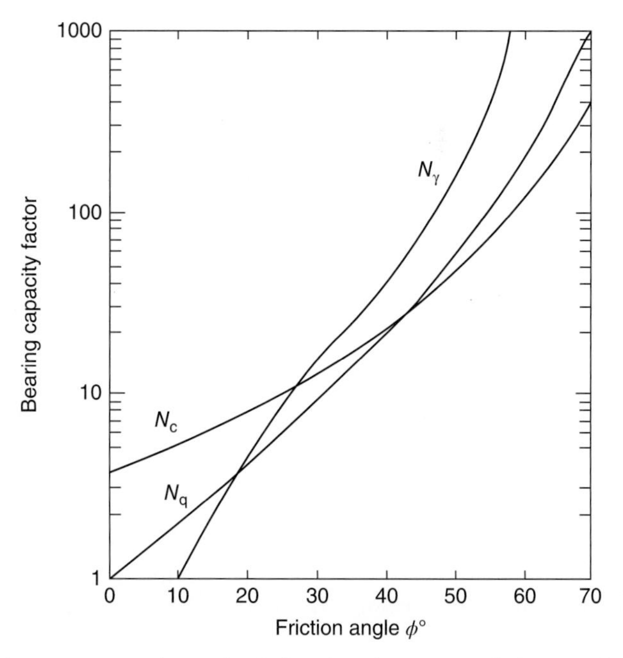

Figure 4.30 Wedge bearing capacity factors for shallow foundations on rock (from Tomlinson, 2001).

Here, q_c and q_{uc} are the unconfined compression strength of the rock mass and intact rock, respectively.

According to Tomlinson (2001), the bearing capacity factors N_c, N_q and N_γ shown in Figure 4.30 are appropriate to the wedge failure conditions shown in Figure 4.29b. They are related to the friction angle Ø of the jointed rock mass.

According to Stagg and Zienkiewicz (1968), the bearing capacity factors for sound rock are approximately

$$N_q = \tan^6\left(45° + \frac{Ø}{2}\right) \quad N_c = 5\tan^4\left(45° + \frac{Ø}{2}\right) \quad N_\gamma = N_q + 1 \tag{4.89}$$

The bearing capacity factors and the appropriate Terzaghi shape factors are used in the Terzaghi bearing capacity equation of Table 4.2 so that the ultimate bearing capacity may be determined.

For rocks the magnitude of the cohesion intercept, c', can be obtained from

$$q_u = 2c' \tan\left(45° + \frac{Ø'}{2}\right) \tag{4.90}$$

The unconfined compression strength q_u and angle of friction $Ø'$ values of rocks can vary widely. Table 4.16 gives a general range of q_u for various types of rocks.

According to Bowles (2001): it is possible to estimate Ø = 45° for most rock except limestone or shale where values between 38° and 45° should be used. Similarly in most cases, it is possible to estimate s_u or $c_u = 5$ MPa as a conservative value. Also, in order to account for the effect of discontinuities of rocks, the q_{ult} may be reduced to q'_{ult} according to *RQD* of the particular rock as follows:

$$q'_{\text{ult}} = q_{\text{ult}}\,(RQD)^2 \tag{4.91}$$

Table 4.16 Ranges of unconfined compression strength of various types of rocks.

Rock type	q_u, MN/m^2	Ø′, degree
Granite	65–250	45–55
Limestone	30–150	35–45
Sandstone	25–130	30–40
Shale	5–40	15–30

Problem Solving

Problem 4.1

A mat foundation 22.5 × 60.0 m supports a silo which completely covers the mat area. The dead weight W of the complete structure, unloaded, is assumed to be 200 MN. The foundation level is located at 3 m below ground surface. The soil profile consists of a uniform saturated clay deposit with average undrained shear strength of 75 kPa and unit weight of 17.5 kN/m^3. The water table is located at the ground surface. If the safety factor SF against bearing capacity failure is to be not less than three and neglecting soil adhesion on the walls of the silo:

(a) Determine the maximum vertical load V which the silo may carry, considering bearing capacity failure only and using Terzaghi bearing capacity equation.
(b) What will be the effect on the SF if the water level rises to 1.5 m above the ground surface in times of flooding?

Solution:
(a) Gross effective foundation pressure is

$$\text{Gross}\, q' = \text{gross}\, q - \text{uplift pressure}$$

$$\text{Gross}\, q' = \left(\frac{V+W}{A}\right) - h\gamma_w = \frac{V + 200\times 10^3}{22.5\times 60} - 3\times 10 \ \ (\text{kPa})$$

Equation (4.2):
$$net\; q' = gross\; q' - \sigma'_o$$

$$net\; q' = \frac{V + 200\times 10^3}{22.5\times 60} - 3\times 10 - 3(17.5-10)$$

$$= (7.41\times 10^{-4})V + 95.7 \ \ (\text{kPa})$$

Table 4.2:
$$gross\; q_{\text{ult}} = cN_c s_c + \gamma' D_f N_q + \frac{1}{2}\gamma' B N_\gamma s_\gamma$$

For saturated clay, $Ø = 0$ (undrained condition); hence, from Table 4.3, $N_c = 5.7$, $N_q = 1$, $N_\gamma = 0$ and from Section 4.6: $s_c = 1 + 0.3\frac{B}{L}$. $Gross\; q_{\text{ult}} = 75\times 5.7\left(1 + 0.3\times\frac{22.5}{60}\right) + \gamma' D_f$ (kPa)

Equation (4.3):
$$net\; q_{\text{ult}} = gross\; q_{\text{ult}} - \sigma'_o$$

$$net\; q_{\text{ult}} = 75\times 5.7\left(1 + 0.3\times\frac{22.5}{60}\right) + \gamma' D_f - \gamma' D_f = 476 \ \text{kPa}$$

Equation (4.4): $net\ q_s = \dfrac{net\ q_{ult}}{SF} = \dfrac{476}{3} = 158.7\,\text{kPa}$

Let $net\ q' = net\ q_s$. Hence, $(7.41 \times 10^{-4})V + 95.7 = 158.7$

$$V = 85 \times 10^3\ kN$$

(b) The reduced $net\ q' = net\ q' -$ the increase in the uplift pressure

$$= [(7.41 \times 10^{-4})(85 \times 10^3) + 95.7] - 1.5 \times 10$$

$$= 143.7\,\text{kPa}$$

$SF = \dfrac{net\ q_{ult}}{reduced\ net\ q'} = \dfrac{476}{143.7} = 3.31$. There will be a slight increase in *SF*.

Problem 4.2

A strip footing 3.5 m wide is to be placed 3 m below ground surface on sandy clay soil having $\gamma = 20.5\,\text{kN/m}^3$. Undrained shear box tests give shear strengths of 35, 47 and 59 kN/m^2 for normal stresses of 70, 140 and 210kN/m^2, respectively. Find the apparent cohesion c and angle of shearing resistance Ø, and then calculate the gross ultimate load per metre run of foundation immediately after construction. Consider bearing capacity failure only. Use Terzaghi bearing capacity equation.

Solution:

Obtain c and Ø values either by calculation ($\tau = c + \sigma \tan Ø$) or from the plot of shear stress τ versus normal stress σ (drawn to the same scale), as shown below.

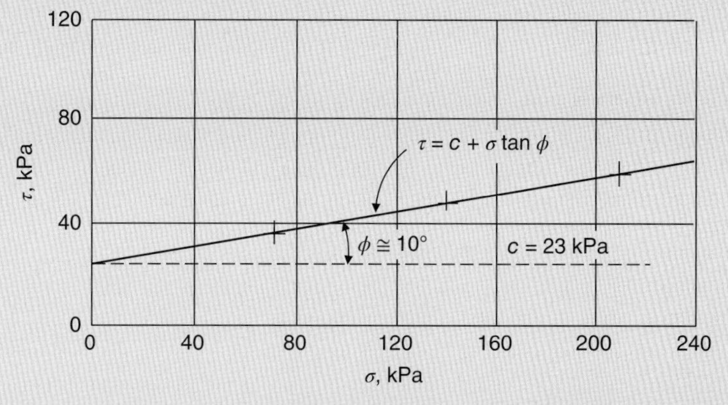

Scheme 4.4

From the plot: $c = 23\,kPa$ and Ø $\cong 10°$

Table 4.2: $gross\ q_{ult} = cN_c s_c + \gamma' D_f N_q + \frac{1}{2}\gamma' B N_\gamma s_\gamma$

For strip footings $s_c = s_\gamma = 1$

Table 4.3: Ø $= 10°$; $N_c = 9.6$; $N_q = 2.7$; $N_\gamma = 1.2$

Equation (4.20): $r_\gamma = 1 - 0.25 \log\left(\dfrac{B}{K}\right)$ for $B > 2$ m

(*Continued*)

$$r_\gamma = 1 - 0.25\log\frac{3.5}{2} = 0.94$$

$$gross\ q_{ult} = 23 \times 9.6 \times 1 + 20.5 \times 3 \times 2.7 + \frac{1}{2} \times 20.5 \times 3.5 \times 1.2 \times 1 \times 0.94$$
$$= 220.80 + 166.05 + 40.47 = 427.32 \text{ kPa}$$

Gross ultimate load = 427.32 × B = 427.32 × 3.5 = 1496 $kN/m\ run$.

Problem 4.3

A square footing is required to carry a gross load of1500 kN. The base of the footing is to be 4.5 m below the ground level. The ground water table will rise to the ground surface. The soil is saturated clay having the following properties: apparent cohesion c = 57.5 kPa, Ø = 0° and γ = 19.2 kN/m^3. Using the Terzaghi bearing capacity equation with a safety factor SF = 3.0, find a suitable size for the footing.

The clay extends to a depth of 8.1 m below the ground surface and below this level there is a stratum of relatively incompressible material. Assume that the mean vertical pressure in the 3.6 m clay layer under the footing is 0.56 of the net contact pressure. Consolidation tests on the clay under similar conditions of pressure distribution and drainage show that an increment of pressure equals 10 kP a causes a settlement of 1.5 % of the sample thickness. Find the probable settlement of the footing.

Solution:

Table 4.2: $$gross\ q_{ult} = cN_c s_c + \gamma' D_f N_q + \frac{1}{2}\gamma' B N_\gamma s_\gamma$$

Table 4.3: $$Ø = 0°;\ N_c = 5.7;\ N_q = 1;\ N_\gamma = 0;$$

For square footings, the Terzaghi $s_c = 1.3$

$$gross\ q_{ult} = cN_c s_c + \gamma' D_f$$

Equation (4.3): $$net\ q_{ult} = gross\ q_{ult} - \sigma'_o = cN_c s_c + \gamma' D_f - \gamma' D_f = cN_c s_c$$

Equation (4.5): $$gross\ q_s = \frac{net\ q_{ult}}{SF} + \sigma'_o = \frac{57.5 \times 5.7 \times 1.3}{3} + (19.2 - 10)(4.5)$$
$$= 142.03 + 41.4 = 183.43 \text{ kPa}$$

Equation (4.4): $$net\ q_s = \frac{net\ q_{ult}}{SF} = \frac{57.5 \times 5.7 \times 1.3}{3} = 142.03 \text{ kPa}$$

Gross effective foundation pressure is

$$\text{Gross } q' = \text{gross } q - \text{uplift pressure}$$
$$\text{Gross } q' = \left(\frac{1500}{A}\right) - h\gamma_w = \left(\frac{1500}{A}\right) - 4.5 \times 10 = \left(\frac{1500}{A}\right) - 45 \text{ (kPa)}$$

Equation (4.2): $$net\ q' = gross\ q' - \sigma'_o$$
$$= \left(\frac{1500}{A}\right) - 45 - 4.5(19.2 - 10)$$
$$= \left(\frac{1500}{A}\right) - 45 - 41.4 \text{ (kPa)}$$

On the basis of gross pressures:

$$\text{Gross } q' = gross\ q_s$$

$$\left(\frac{1500}{A}\right) - 45 = 183.43 \rightarrow A = \frac{1500}{228.43} = 6.57\ \text{m}^2$$

On the basis of net pressures:

$$net\ q' = net\ q_s$$

$$\left(\frac{1500}{A}\right) - 45 - 41.4 = 142.03 \rightarrow A = \frac{1500}{228.43} = 6.57\ \text{m}^2$$

The footing size $= 2.6 \times 2.6\,m$

Settlement of the clay stratum under the footing= $S_c = m_v \times \sigma'_z \times H$

$$\sigma'_z = 0.56\left(\frac{1500}{2.6 \times 2.6} - 45 - 41.4\right) = 75.88\ \text{kPa}$$

Settlement of the clay sample $= 0.015\,h = m_v \times 10 \times h$, where h = thickness of the sample. Hence,

$$\frac{S_c}{0.015\,h} = \frac{m_v \times 75.88 \times 3.6}{m_v \times 10 \times h}$$

The probable settlement of the footing $= S_c = \dfrac{0.015 \times 75.88 \times 3.6}{10} = \mathit{0.41\ m}$

Problem 4.4

A strip footing is to be designed to carry a gross foundation load equals 800 kN/m run at a depth of 0.7 m in a gravelly sand stratum. Appropriate shear strength parameters are $c' = 0$ and $\varnothing = 40°$. Assume that the water table exists at the foundation level. The sand unit weights above and below the water table are 17 and 20 kN/m^3, respectively. Using Terzaghi bearing capacity equation with a safety factor of three, considering shear failure only, determine the width of the footing.

Solution:

Equation (4.3):

$$net\ q_{\text{ult}} = gross\ q_{\text{ult}} - \sigma'_o$$

$$= cN_c s_c + \gamma' D_f N_q + \frac{1}{2}\gamma' B N_\gamma s_\gamma - \gamma' D_f$$

$$= \gamma' D_f (N_q - 1) + \frac{1}{2}\gamma' B N_\gamma s_\gamma$$

Table 4.3: $\varnothing = 40°;\ N_q = 81.3;\ N_\gamma = 100.4$

For strip footings all shape factors areone. Hence,

$$net\ q_{\text{ult}} = 17 \times 0.7(81.3 - 1) + 0.5 \times (20 - 10)(B)(100.4)$$

$$= 955.6 + 502\,B\ \ (\text{kPa})$$

Equation(4.4):

$$net\ q_s = \frac{net\ q_{\text{ult}}}{SF} = \frac{955.6 + 502\,B}{3}\ \ (\text{kPa})$$

Equation (4.1):

$$net\ q = gross\ q - \sigma_o = \frac{800}{B} - 17 \times 0.7\ \ (\text{kPa})$$

(Continued)

Let *net* q = *net* q_s. Hence,

$$\frac{800}{B}-11.9=\frac{955.6+502\,B}{3}$$

$$2400-35.7\,B=955.6\,B+502\,B^2$$

$$B^2+1.97\,B-4.78=0$$

$$B=\frac{-1.97\pm\sqrt{1.97^2-4\times1(-4.78)}}{2\times1}=\frac{-1.97+4.8}{2}=\frac{2.83}{2}=1.415\,\text{m}$$

Use width of the footing = 1.45 *m*

Problem 4.5

A footing 2 m square is located at a depth of 4 m in stiff clay of unit weight 21 kN/m³. The undrained strength of clay at a depth of 4 m is given by the parameters c_u = 120 kPa and $\varnothing_u$ = 0°. For *SF* = 3 with respect to bearing capacity failure, determine the gross foundation vertical load that the footing can carry using: (a) the Vesic method, (b) the Skempton method.

Solution:

(a) The Vesic method. For saturated clay with $\varnothing_u$ = 0°:

Table 4.4: $N_c=5.14,\ N_q=1,\ N_\gamma=0$

Table 4.2: $gross\ q_{\text{ult}}=c\,N_c s_c d_c i_c g_c b_c+\bar{q}N_q s_q d_q i_q g_q b_q$

Table 4.8: for the conditions of this problem all *i*, *g* and *b* factors are one.

Table 4.6: $$s_c=1+\frac{N_q}{N_c}\left(\frac{B}{L}\right)=1+\frac{1}{5.14}\left(\frac{2}{2}\right)=1.2$$

$$d_c=1+0.4\tan^{-1}\frac{D}{B}=1+0.4\tan^{-1}\frac{4}{2}=1.44$$

$$s_q=d_q=1$$

$$gross\ q_{\text{ult}}=120\times5.14\times1.2\times1.44+4\times21=1065.8\ \text{kPa}$$

Equation (4.3): $$net\ q_{\text{ult}}=gross\ q_{\text{ult}}-\sigma'_o=1065.8-4\times21=982\ \text{kPa}$$

Equation (4.4): $$net\ q_s=\frac{net\ q_{\text{ult}}}{SF}=\frac{982}{3}=327\,\text{kPa}$$

Equation (4.5): $$gross\ q_s=\frac{net\ q_{\text{ult}}}{SF}+\sigma'_o=327+4\times21=411\,\text{kPa}$$

Let *gross* q = *gross* q_s. Hence,

The gross load the footing can carry (gross safe foundation load) is 411 × *A* = 411 × 2² = 1644 *kN*

(b) The Skempton method.

Figure 4.4: $$\frac{D}{B}=2;\ N_c=8.35$$

$$gross\ q_{\text{ult}}=c\,N_c+\bar{q}$$

$$net\ q_{\text{ult}}=c\,N_c=120\times8.35=1002\ \text{kPa}$$

Equation (4.5): $$gross\, q_s = \frac{net\, q_{\text{ult}}}{SF} + \sigma'_o = \frac{1002}{3} + 4 \times 21 = 418\,\text{kPa}$$

Let $gross\, q = gross\, q_s$. Hence,

The gross load the footing can carry (gross safe foundation load) is 418 × *A* = 418 × 2^2 = 1672 *kN*

Problem 4.6

The footing of a long retaining wall is 3 m wide and is 1 m below the ground surface in front of the wall. The water table is well below the foundation level. The vertical *V* and horizontal *H* components of the base load are 282 and 102 kN/m run, respectively. The eccentricity of the base load is 0.36 m, as shown in the scheme below. The appropriate shear strength parameters for the foundation soil are $c' = 0$ and $\varnothing' = 35°$ and the unit weight of the soil is 18 kN/m^3. Determine the safety factor against shear failure (a) using Meyerhof equations and effective area method, (b) using Hansen equations and effective area method.

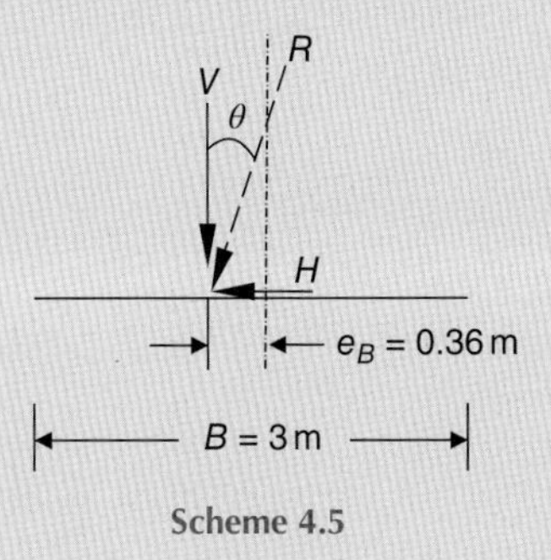

Scheme 4.5

Solution:

(a) Using Meyerhof equations and effective area method.

$$B' = B - 2e_b = 3 - 2 \times 0.36 = 2.28 \text{ m and } L' = L$$

Table 4.5: $$\theta = \tan^{-1}\left(\frac{102}{282}\right) = 20°;\ i_q = \left(1 - \frac{\theta}{90}\right)^2 = \left(1 - \frac{20}{90}\right)^2 = 0.61;$$

$$i_\gamma = \left(1 - \frac{\theta}{\varnothing}\right)^2 = \left(1 - \frac{20}{35}\right)^2 = 0.18;\ K_P = \tan^2\left(45 + \frac{\varnothing}{2}\right) = 3.69;$$

$$d_q = d_\gamma = 1 + 0.1\sqrt{K_P}\,\frac{D}{B} = 1 + 0.1 \times 1.92 \times \frac{1}{3} = 1.064$$

Table 4.4: $\varnothing = 35°;\ N_q = 33.6;\ N_\gamma = 37.8$

Table 4.2: $gross\, q_{\text{ult}} = \bar{q} N_q d_q i_q + 0.5\gamma' B' N_\gamma d_\gamma i_\gamma$

Equation (4.20): $$r_\gamma = 1 - 0.25 \log\left(\frac{B}{K}\right) \quad \text{for } B > 2\,\text{m}$$

$$r_\gamma = 1 - 0.25 \log\frac{2.28}{2} = 0.96$$

$$gross\, q_{\text{ult}} = 1 \times 18 \times 33.6 \times 1.064 \times 0.61$$
$$+ 0.5 \times 18 \times 2.28 \times 37.8 \times 1.064 \times 0.18 \times 0.96 = 535 \text{ kPa}$$

Equation (4.3): $$net\, q_{\text{ult}} = gross\, q_{\text{ult}} - \sigma'_o = 535 - 1 \times 18 = 517 \text{ kPa}$$

(*Continued*)

Equation (4.2): $$net\ q' = gross\ q' - \sigma'_o = \frac{V}{1 \times B'} - \sigma'_o = \frac{282}{2.28} - 18 = 106\ \text{kPa}$$

The safety factor $SF = \frac{net\ q_{ult}}{net\ q'} = \frac{517}{106} = 4.88$; and on the basis of gross pressures, the $SF = \frac{gross\ q_{ult}}{gross\ q'} = \frac{535}{282/(1 \times B')} = \frac{535}{282/2.28} = 4.33$

(b) Using Hansen equations and effective area method.

Table 4.2: $$gross\ q_{ult} = 0 + \bar{q}N_q s_q d_q i_q g_q b_q + 0.5\gamma' B' N_\gamma s_\gamma d_\gamma i_\gamma g_\gamma b_\gamma$$

Table 4.4: $$\varnothing = 35°,\ N_q = 33.6,\ N_\gamma = 34.4$$

Table 4.7: for the conditions of this problem all *g* and *b* factors areone and for long or strip footings all *s* factors are one.

$$i_q = \left[1 - \frac{0.5H}{V + A_f c_a \cot \varnothing}\right]^{\alpha_1},\ \text{using } \alpha_1 = 2.5 \text{ and } c_a = 0:$$

$$i_q = \left[1 - \frac{0.5 \times 102}{282}\right]^{2.5} = 0.61$$

$$i_\gamma = \left[1 - \frac{0.7H}{V + A_f c_a \cot \varnothing}\right]^{\alpha_2},\ \text{using } \alpha_2 = 3.5 \text{ and } c_a = 0:$$

$$i_\gamma = \left[1 - \frac{0.7 \times 102}{282}\right]^{3.5} = 0.36$$

Tables 4.4 and 4.6: $$d_q = 1 + 2\tan \varnothing\,(1 - \sin \varnothing)^2 k = 1.09;\ d_\gamma = 1$$

Equation (4.20): $$r_\gamma = 1 - 0.25\log\frac{2.28}{2} = 0.96$$

$$gross\ q_{ult} = 1 \times 18 \times 33.6 \times 1.09 \times 0.61 + 0.5 \times 18 \times 2.28 \times 34.4 \times 1 \times 0.36 \times 0.96$$
$$= 646\ \text{kPa}$$

Equation (4.3): $$net\ q_{ult} = gross\ q_{ult} - \sigma'_o = 646 - 1 \times 18 = 628\ \text{kPa}$$

Equation (4.2): $$net\ q' = gross\ q' - \sigma'_o = \frac{V}{1 \times B'} - \sigma'_o = \frac{282}{2.28} - 18 = 106\ \text{kPa}$$

The safety factor $SF = \frac{net\ q_{ult}}{net\ q'} = \frac{628}{106} = 5.92$; and on the basis of gross pressures, the $SF = \frac{gross\ q_{ult}}{gross\ q'} = \frac{646}{282/(1 \times B')} = \frac{646}{282/2.28} = 5.22$

Problem 4.7

A load test was carried out on a rectangular footing shown in the scheme below.

The test gave the following results:

The ultimate horizontal load in L direction, $H_{L,\ ult} = 382\ \text{kN}$
The ultimate gross vertical compression load, $V_{ult} = 1060\ \text{kN}$

(a) Find the safety factor against sliding of the footing under the ultimate loads. Neglect the resisting passive earth pressure and assume $\delta = 0.8\ \varnothing$.

(b) Find the gross ultimate bearing capacity by the Hansen method.

(c) Find the gross ultimate bearing capacity by the Vesic method.
(d) Find the gross ultimate bearing capacity by the Meyerhof method.

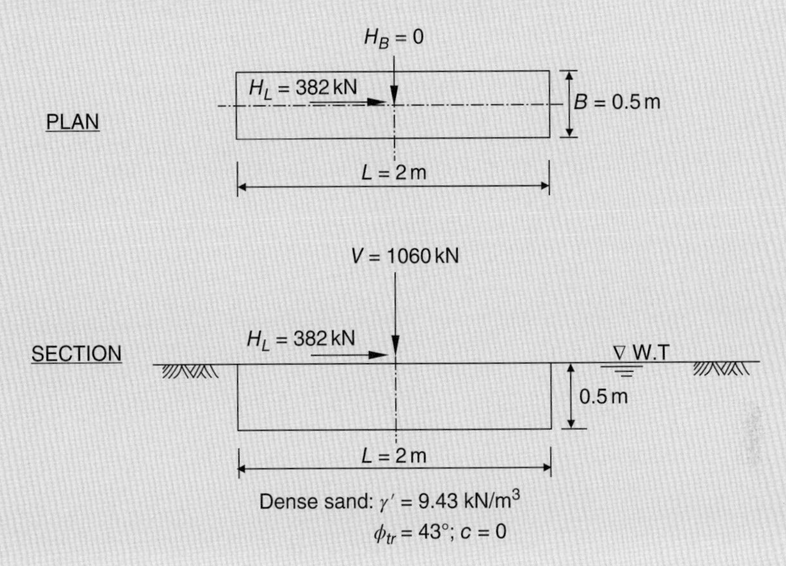

Scheme 4.6

Solution:

(a) Maximum resisting force $= H_{\max} = V\tan\delta + c_a A_f$

$c_a = 0$, since $c = 0$. Hence,

$$H_{\max} = V_{\text{ult}}\tan\delta° = 1060 \times \tan(0.8 \times 43) = 726\,\text{kN}$$

The sliding force $= H_{L,\text{ult}} = 382$ kN

$$\text{Safety factor against sliding of the footing} = SF = \frac{H_{\max}}{H_{L,\text{ult}}} = \frac{726}{382} = 1.9$$

(b) Hansen method.

$\frac{L'}{B'} = \frac{L}{B} = \frac{2}{0.5} = 4 > 2$, and $\text{Ø}_{\text{tr}} = 43° > 34°$. Hence according to Hansen

$$\text{Ø}_{\text{ps}} = 1.5\,\text{Ø}_{\text{tr}} - 17° = 1.5 \times 43 - 17 = 47.5°$$

According to Meyerhof:

$$\text{Ø}_{\text{ps}} = 1.1\,\text{Ø}_{\text{tr}} = 1.1 \times 43 = 47.3°$$

Use $\text{Ø}_{ps} = 47°$

Table 4.2:

$$N_q = e^{\pi\tan\text{Ø}}\tan^2\left(45 + \frac{\text{Ø}}{2}\right)$$

$$= e^{\pi\tan 47}\tan^2\left(45 + \frac{47}{2}\right) = 187.21$$

$$N_\gamma = 1.5(N_q - 1)\tan\text{Ø}$$

$$= 1.5(187.21 - 1)\tan 47 = 299.53$$

Table 4.4:

$$2\tan\text{Ø}\,(1 - \sin\text{Ø})^2 = 2\tan 47(1 - \sin 47)^2 = 0.155$$

(*Continued*)

Because the horizontal component is in L direction, both Equations (4.21) and (4.22) should be used in order to find the smaller gross q_{ult}.

Refer to Tables 4.6 and 4.7 and Section 4.7:

$$d_{q,B} = 1 + 2\tan Ø\,(1 - \sin Ø)^2 D/B = 1 + 0.155\left(\frac{0.5}{0.5}\right) = 1.16$$

$$d_{q,L} = 1 + 2\tan Ø\,(1 - \sin Ø)^2 D/L = 1 + 0.155\left(\frac{0.5}{2}\right) = 1.04$$

$$d_{\gamma,B} = d_{\gamma,L} = 1$$

$$i_{q,\,B} = \left[1 - \frac{0.5H_B}{V + A_f c_a \cot Ø}\right]^{\alpha_1 = 2.5} = 1 \quad (\text{using } \alpha_1 = 2.5 \text{ and } H_B = 0)$$

$$i_{\gamma,B} = \left[1 - \frac{0.7H_B}{V + A_f c_a \cot Ø}\right]^{\alpha_2 = 3.5} = 1 \quad (\text{using } \alpha_2 = 3.5,\ H_B = 0 \text{ and } \eta^\circ = 0)$$

$$c_a = 0 \text{ since } c = 0$$

$$i_{q,\,L} = \left[1 - \frac{0.5H_L}{V + A_f c_a \cot Ø}\right]^{2.5} = \left[1 - \frac{0.5 \times 382}{1060 + 0}\right]^{2.5} = 0.608$$

$$i_{\gamma,L} = \left[1 - \frac{0.7H_L}{V + A_f c_a \cot Ø}\right]^{\alpha_2 = 3.5} = \left[1 - \frac{0.7 \times 382}{1060 + 0}\right]^{\alpha_2 = 3.5} = 0.361$$

$$s_{q,\,B} = 1 + \sin Ø\,(Bi_{q,\,B}/L) = 1 + (\sin 47)\,(0.5 \times 1/2) = 1.18$$

$$s_{q,\,L} = 1 + \sin Ø\,(Li_{q,L}/B) = 1 + (\sin 47)\,(2 \times 0.608/0.5) = 2.78$$

$$s_\gamma \geq 0.6$$

$$s_{\gamma,\,B} = 1 - 0.4\left(\frac{Bi_{\gamma,\,B}}{Li_{\gamma,\,L}}\right) = 1 - 0.4\left(\frac{0.5 \times 1}{2 \times 0.361}\right) = 0.723 > 0.6$$

$$s_{\gamma,\,L} = 1 - 0.4\left(\frac{Li_{\gamma,\,L}}{Bi_{\gamma,\,B}}\right) = 1 - 0.4\left(\frac{2 \times 0.361}{0.5 \times 1}\right) = 0.422 < 0.6$$

Use $s_{\gamma,\,L} = 0.6$

All ground g factors and base b factors are one, since both the ground and base are horizontal.

Equation (4.21):

$$\begin{aligned} q_{\text{ult}} &= \bar{q} N_q s_{q,B} d_{q,B} i_{q,B} g_{q,B} b_{q,B} \\ &= 0.5 \times 9.43 \times 187 \times 1.18 \times 1.16 \times 1 \\ &\quad + 0.5 \times 9.43 \times 0.5 \times 300 \times 0.732 \times 1 \times 1 = 1718 \text{ kPa} \end{aligned}$$

Equation (4.22):

$$\begin{aligned} q_{\text{ult}} &= \bar{q} N_q s_{q,L} d_{q,L} i_{q,L} g_{q,L} b_{q,L} + 0.5\gamma L' N_\gamma s_{\gamma,L} d_{\gamma,L} i_{\gamma,L} g_{\gamma,L} b_{\gamma,L} \\ &= 0.5 \times 9.43 \times 187 \times 2.78 \times 1.04 \times 0.608 \\ &\quad + 0.5 \times 9.43 \times 2 \times 300 \times 0.6 \times 1 \times 0.361 = 2163 \text{ kPa} \end{aligned}$$

Use the smaller value of $q_{\text{ult}} = 1718$ *kPa*

This result is much greater than the *1060 kPa* of the load test.

(c) Vesic method.

Table 4.2:

$$N_q = e^{\pi \tan Ø} \tan^2\left(45 + \frac{Ø}{2}\right) = e^{\pi \tan 47} \tan^2\left(45 + \frac{47}{2}\right) = 187.21$$

$$N_\gamma = 2(N_q + 1)\tan Ø = 2(187.21 + 1)\tan 47 = 403.66$$

Table 4.4: $2\tan Ø\,(1 - \sin Ø)^2 = 2\tan 47(1 - \sin 47)^2 = 0.155$

Refer to Tables 4.6 and 4.8:

$$\frac{B}{L} = \frac{B'}{L'} = \frac{0.5}{2} = 0.25;\ \frac{D}{B'} = \frac{D}{B} = \frac{0.5}{0.5} = 1;\ k = \frac{D}{B} = \frac{0.5}{0.5} = 1;\ \frac{L}{B} = \frac{2}{0.5} = 4$$

$$s_q = 1 + \frac{B}{L}\tan Ø = 1 + \frac{0.5}{2}\tan 47 = 1.27$$

$$s_\gamma = 1 - 0.4\left(\frac{B}{L}\right) = 1 - 0.4(0.25) = 0.9 > 0.6$$

$$d_q = 1 + \frac{2\tan Ø\,(1 - \sin Ø)^2 D}{B} = 1 + 0.155(1) = 1.16;\ d_\gamma = 1$$

$$m = m_L = \frac{2 + \frac{L}{B}}{1 + \frac{L}{B}} = \frac{2 + 4}{1 + 4} = 1.2$$

$$i_{q,\,L} = \left[1 - \frac{H_L}{V + A_f c_a \cot Ø}\right]^m = \left[1 - \frac{382}{1060 + 0}\right]^{1.2} = 0.585$$

$$i_{\gamma,L} = \left[1 - \frac{H_L}{V + A_f c_a \cot Ø}\right]^{m+1} = \left[1 - \frac{382}{1060 + 0}\right]^{1.2+1} = 0.374$$

All ground *g* factors and base *b* factors are one, since both the ground and base are horizontal.

$$gross\ q_{\text{ult}} = \bar{q} N_q s_q d_q i_q + 0.5\gamma' B' N_\gamma s_\gamma d_\gamma i_\gamma$$

$$= 0.5 \times 9.43 \times 187 \times 1.27 \times 1.16 \times 0.585$$

$$+ 0.5 \times 9.43 \times 0.5 \times 404 \times 0.9 \times 1 \times 0.374 = 1081\ kPa$$

This result is practically very close to the 1060 kPa of the load test.

(d) Meyerhof method.

Table 4.5:

$$\theta = \tan^{-1}\left(\frac{382}{1060}\right) \approx 20^o;\ i_q = \left(1 - \frac{\theta}{90}\right)^2 = \left(1 - \frac{20}{90}\right)^2 = 0.61$$

$$i_\gamma = \left(1 - \frac{\theta}{Ø}\right)^2 = \left(1 - \frac{20}{47}\right)^2 = 0.33;\ K_P = \tan^2\left(45 + \frac{Ø}{2}\right) = 6.44$$

$$d_q = d_\gamma = 1 + 0.1\sqrt{K_P}\frac{D}{B} = 1 + 0.1 \times 2.54 \times \frac{0.5}{0.5} = 1.254$$

Table 4.2: $N_q = 187.21$ (the same as that of Hansen)

$$N_\gamma = (N_q - 1)\tan(1.4\,Ø) = (187.21 - 1)\tan(1.4 \times 47)$$

$$= 414.34$$

(Continued)

Table 4.2: $gross\, q_{ult} = \bar{q}N_q d_q i_q + 0.5\gamma' B' N_\gamma d_\gamma i_\gamma$

$$gross\, q_{ult} = 0.5 \times 9.43 \times 187 \times 1.254 \times 0.61$$
$$+ 0.5 \times 9.43 \times 0.5 \times 414 \times 1{,}254 \times 0.33 = 1078 \text{ kPa}$$

This result is practically very close to the 1060 *kPa* of the load test.

Problem 4.8

The scheme below shows a 2 × 2 m square footing with a tilted base, rests on a $c-\varnothing$ soil of the given properties. The ground surface is level and horizontal. The foundation depth equals 0.4 m. The design requires a minimum $SF = 2.5$ against bearing capacity failure and 1.5 against sliding under the given loading conditions.

(a) Check whether the footing is safe against sliding. Assume: $\delta = \varnothing$; $c_a = c$; resisting passive earth pressure is neglected.

(b) Considering the bearing capacity failure only, check the adequacy of the footing dimensions. Use both Hansen and Vesic methods.

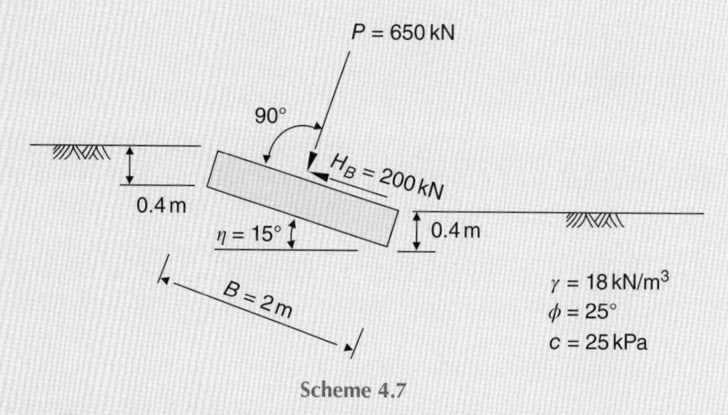

Scheme 4.7

Solution:

(a) Safety against sliding of the footing.
Maximum resisting force $= H_{max} = P \tan\delta + c_a A_f$

$$H_{max} = 650 \times \tan 25 + 25 \times 2 \times 2 = 403 \text{ kN}$$

$$\text{The sliding force} = H_B = 200 \text{ kN}$$

$$\text{Safety factor} = SF = \frac{H_{max}}{H_B} = \frac{403}{200} = 2.02 > 1.5 \text{ (required)}$$

Therefore, the footing is *safe against sliding.*

(b) Adequacy of the footing dimensions.

(i) Hansen method:

Table 4.4: $\varnothing = 25°; N_c = 20.7; N_q = 10.7; N_\gamma = 6.8; \dfrac{N_q}{N_c} = 0.514$

$$2\tan\varnothing\,(1-\sin\varnothing)^2 = 0.311; \frac{D}{B} = \frac{0.4}{2} = 0.2 = k; \frac{B'}{L'} = \frac{B}{L} = \frac{2}{2} = 1$$

Table 4.7: $\beta = 0°$; hence, all g factors are one.

Equation (4.21):

$$gross\ q_{ult} = c\,N_c s_{c,B} d_{c,B} i_{c,B} b_{c,B} + \bar{q} N_q s_{q,B} d_{q,B} i_{q,B} b_{q,B} + 0.5\gamma' B' N_\gamma s_{\gamma,B} d_{\gamma,B} i_{\gamma,B} b_{\gamma,B}$$

Refer to Tables 4.6 and 4.7 and Section 4.7:

$$\eta = 15° = 0.262 \text{ radians}$$

$$b_{c,B} = 1 - \frac{\eta°}{147°} = 1 - \frac{15}{147°} = 0.9$$

$$b_{q,B} = \exp(-2\eta \tan Ø) = \exp[-2(0.262)(\tan 25)] = 0.783$$

$$b_{\gamma,B} = \exp(-2.7\eta \tan Ø) = \exp[-2.7(0.262)(\tan 25)] = 0.719$$

$$d_{c,B} = 1 + 0.4k = 1 + 0.4 \times 0.2 = 1.08$$

$$d_{q,B} = 1 + 2\tan Ø\,(1 - \sin Ø)^2 k = 1 + 0.311 \times 0.2 = 1.062$$

$$d_{\gamma,B} = 1$$

$V = P = 650$ kN. $H_B = 200$ kN. Assume $\alpha_1 = 3$ and $\alpha_2 = 4$

$$V + A_f c_a \cot Ø = 650 + 2 \times 2 \times 25 \times \cot 25 = 864.5$$

$$i_{q,B} = \left[1 - \frac{0.5H_B}{V + A_f c_a \cot Ø}\right]^{\alpha_1} = \left[1 - \frac{0.5 \times 200}{864.5}\right]^3 = 0.692$$

$$i_{\gamma,B} = \left[1 - \frac{[(0.7) - \eta°/450°]H_B}{V + A_f c_a \cot Ø}\right]^{\alpha_2} = \left[1 - \frac{0.678 \times 200}{864.5}\right]^4 = 0.506$$

$$i_{c,B} = i_q - \frac{1 - i_q}{N_q - 1} = 0.692 - \frac{1 - 0.692}{10.7 - 1} = 0.66$$

$$s_{c,B} = 1 + \frac{N_q}{N_c} \times \frac{B i_{c,B}}{L} = 1 + 0.514 \times \frac{2 \times 0.66}{2} = 1.34$$

$$s_{q,B} = 1 + \sin Ø\,(B i_{q,B}/L) = 1 + (\sin 25)\,(2 \times 0.692/2) = 1.29$$

$s_{\gamma,B} = 1 - 0.4\left(\dfrac{B i_{\gamma,B}}{L i_{\gamma,L}}\right)$. The value of $i_{\gamma,L} = 1$, since $H_L = 0$; hence,

$$s_{\gamma,B} = 1 - 0.4\left(\frac{2 \times 0.506}{2 \times 1}\right) = 0.798 > 0.6$$

$$\begin{aligned} gross\ q_{ult} &= 25 \times 20.7 \times 1.34 \times 1.08 \times 0.66 \times 0.9 \\ &\quad + 0.4 \times 18 \times 10.7 \times 1.29 \times 1.062 \times 0.692 \times 0.783 \\ &\quad + 0.5 \times 18 \times 2 \times 6.8 \times 0.798 \times 1 \times 0.506 \times 0.719 \\ &= 444.9 + 57.2 + 35.5 = 537.6\,\text{kPa} \end{aligned}$$

$$\text{gross contact pressure, gross } q = \frac{P}{A} = \frac{650}{2 \times 2} = 162.5\,\text{kPa}$$

(*Continued*)

Safety factor $SF = \frac{gross\ q_{\text{ult}}}{\text{gross}\ q} = \frac{537.6}{162.5} = 3.3 > 2.5$ (required); hence,

The footing dimensions are adequate according to the Hansen method.

(ii) Vesic method:

Table 4.4: $\quad Ø = 25°;\ N_c = 20.7;\ N_q = 10.7;\ N_\gamma = 10.9;\ \frac{N_q}{N_c} = 0.514$

$$2\tan Ø\,(1 - \sin Ø\,)^2 = 0.311;\ \frac{B}{L} = \frac{2}{2} = 1.$$

Refer to Tables 4.6 and 4.8:

$$\eta = 15° = 0.262 \text{ radians};\ \beta = 0°$$

$$b_c = 1 - \frac{2\beta}{5.14\tan Ø} = 1 \quad (\text{since ground slope } \beta = 0°)$$

$$b_q = b_\gamma = (1 - \eta\tan Ø\,)^2 = [1 - (0.262)(\tan 25)]^2 = 0.771$$

$d_c = 1.080;\ d_q = 1.062;\ d_\gamma = 1$ (The same as Hansen depth factors)

$$s_c = 1 + \frac{N_q}{N_c} \times \frac{B}{L} = 1 + 0.514 \times \frac{2}{2} = 1.514$$

$$s_q = 1 + (\tan Ø\,)\left(\frac{B}{L}\right) = 1 + (\tan 25)\left(\frac{2}{2}\right) = 1.466$$

$$s_\gamma = 1 - 0.4\left(\frac{B}{L}\right) = 1 - 0.4 \times \frac{2}{2} = 0.6$$

$$m = \frac{2 + \frac{L}{B}}{1 + \frac{L}{B}} = \frac{2+1}{1+1} = 1.5$$

$$V = P = 650 \text{ kN}.\ H = 200 \text{ kN}.$$

$$V + A_f c_a \cot Ø = 650 + 2 \times 2 \times 25 \times \cot 25 = 864.5$$

$$i_q = \left[1 - \frac{H}{V + A_f c_a \cot Ø}\right]^m = \left(1 - \frac{200}{864.5}\right)^{1.5} = 0.674$$

$$i_\gamma = \left[1 - \frac{H}{V + A_f c_a \cot Ø}\right]^{m+1} = \left(1 - \frac{200}{864.5}\right)^{2.5} = 0.518$$

$$i_c = i_q - \frac{1 - i_q}{N_q - 1} = i_{c,B} = 0.674 - \frac{1 - 0.674}{10.7 - 1} = 0.64$$

$$g_c = i_q - \frac{1 - i_q}{5.14\tan Ø} = 0.674 - \frac{1 - 0.674}{5.14\tan 25} = 0.538$$

$$g_q = g_\gamma = (1 - \tan\beta)^2 = 1$$

$$gross\ q_{\text{ult}} = c N_c s_c d_c i_c b_c g_c + q N_q s_q d_q i_q b_q g_q + 0.5\gamma' B' N_\gamma s_\gamma d_\gamma i_\gamma b_\gamma g_\gamma$$

$$gross\, q_{ult} = 25 \times 20.7 \times 1.514 \times 1.08 \times 0.64 \times 1 \times 0.538$$
$$+ 0.4 \times 18 \times 10.7 \times 1.466 \times 1.062 \times 0.674 \times 0.771 \times 1$$
$$+ 0.5 \times 18 \times 2 \times 10.9 \times 0.6 \times 1 \times 0.518 \times 0.771 \times 1$$
$$= 291.4 + 62.3 + 47.0 = 400.7 \text{ kPa}$$

$$\text{gross contact pressure, gross } q = \frac{P}{A} = \frac{650}{2 \times 2} = 162.5 \text{ kPa}$$

Safety factor $= SF = \dfrac{gross\, q_{ult}}{\text{gross } q} = \dfrac{400.7}{162.5} = 2.5 = 2.5$ (required); hence:

The footing dimensions are probably adequate according to the Vesic method.

The large difference between the two results is mainly due to the g_c factor of the predominating cohesion term; it is considered as unity in the Hansen equation, whereas in the Vesic equation it is about half (0.538) of this value, which reduces the cohesion term by about 50%.

Problem 4.9

Solve Problem 4.6 (a) using the Saran and Agarwal analysis and (b) find whether the Purkaystha and Char reduction factor method, which is originally used for eccentric *vertical* loads, gives reasonable results for eccentric *inclined* loads. Use the Hansen bearing capacity, depth and inclination factors.

Solution:

(a) Saran and Agarwal analysis

Equation (4.35) $$Q_{ult} = B\left[c'N_{c(ei)} + qN_{q(ei)} + \frac{1}{2}\gamma BN_{\gamma(ei)}\right]$$

$$\beta = \tan^{-1}\left(\frac{102}{282}\right) = 20°;\ \varnothing = 35°;\ B = 3\text{ m};\ e = 0.36\text{ m};\ D_f = 1\text{ m}$$

$$\frac{e}{B} = \frac{0.36}{3} = 0.12;\ \gamma = 18 \text{ kN/m}^3;\ q = \gamma D_f;\ c' = 0$$

Figure 4.9: $N_{q(ei)} = 11.5$

Figure 4.11: $N_{\gamma(ei)} = 14$

$$Q_{ult} = 3\left[0 + 18 \times 1 \times 11.5 + \frac{1}{2} \times 18 \times 3 \times 14\right] = 1755\text{ kN}$$

The vertical load $V = 282$ kN

Safety factor $SF = \dfrac{Q_{ult}}{V} = \dfrac{1755}{282} = 6.2$

(b) Purkaystha and Char reduction factor method.

Equation (4.30): $$q_{ult(e)} = q_{ult(c)}\left[1 - a\left(\frac{e}{B}\right)^k\right]$$

Equation (4.31): $$q_{ult(c)} = qN_qF_{qd} + \frac{1}{2}\gamma BN_\gamma F_{\gamma d}$$

Table 4.9: for $\dfrac{D_f}{B} = \dfrac{1}{3}$ obtain $a = 1.793$, $k = 0.79$ (by interpolation)

Table 4.4: for $\varnothing = 35°$: $N_q = 33.6$ and $N_\gamma = 34.4$ (by interpolation)

Table 4.6:
$$F_{qd} = 1 + 2\tan\varnothing\,(1 - \sin\varnothing)^2 k$$
$$= 1 + (2\tan 35)\,(1 - \sin 35)^2\left(\frac{1}{3}\right) = 1.085$$
$$F_{\gamma d} = 1$$

(*Continued*)

Use Hansen i_q and i_γ factors to account for the effect of inclined loading condition. These factors are already computed before (see Solution of Problem 4.6):
$i_q = 0.61$; $i_\gamma = 0.36$

$$q_{\text{ult(e)}} = \begin{bmatrix} 1\times 18\times 33.6\times 1.085\times 0.61 \\ +\dfrac{1}{2}\times 18\times 3\times 34.4\times 1\times 0.36 \end{bmatrix} \left[1-1.793\left(\frac{0.36}{3}\right)^{0.79}\right]$$

$$= (400.3+334.4)(0.664) = 487.8 \text{ kPa}$$

$$Q_{\text{ult}} = Bq_{\text{ult(e)}} = 3\times 487.8 = 1463.4 \text{ kN}$$

Safety factor $SF = \dfrac{Q_{\text{ult}}}{V} = \dfrac{1463.4}{282} = 5.2$
This result represents the average of the three results obtained before. Therefore, one may consider that the Purkaystha and Char reduction factor method gives reasonable results for eccentric *inclined* loading conditions provided that the Hansen bearing capacity, depth and inclination factors are used in Equation (4.30).

Problem 4.10

A mat foundation 9 × 27 m is to be placed at a depth of 3 m in a deep stratum of soft, saturated clay of $\gamma = 16.5 \text{ kN/m}^3$. The water table is at 2.5 m below the ground surface. The strength parameters of the soil, obtained from unconsolidated, undrained tests are $c_u = 25$ kPa, $\varnothing_u = 0$, whereas consolidated, drained tests give $c_d = 5$ kPa, $\varnothing_d = 23°$. The modulus of deformation of the soil in the undrained condition obtained equals to $E_u = 2593$ kPa. The modulus of confined compression (in drained conditions) increases with pressure q (or σ_z) according to $M_v = 12.6\,q'$. Taking the effect of compressibility of the soil into consideration and using the Vesic equations, find the ultimate bearing capacity in the following conditions:

(a) Assume the rate of application of dead and live loads (vertical loads) is fast in comparison with rate of dissipation of excess pore-water pressures caused by loads, so that undrained conditions prevail at failure.
(b) Assume, as the other extreme, that the rate of vertical loading is slow enough so that no excess pore-water pressures are introduced in the foundation soil.

Solution:
(a) *Undrained condition.*
Assume incompressibility soil condition; hence,

$$gross\, q_{\text{ult}} = c N_c s_c d_c i_c b_c g_c + \bar{q} N_q s_q d_q i_q b_q g_q + 0.5\gamma' B' N_\gamma s_\gamma d_\gamma i_\gamma b_\gamma g_\gamma$$

There is no horizontal load component, the base is not tilted and the foundation is not on slope, therefore all i_i, b_i and g_i factors are 1. Hence,

$$gross\, q_{\text{ult}} = c_u N_c s_c d_c + \bar{q} N_q s_q d_q + 0.5\gamma' B' N_\gamma s_\gamma d_\gamma$$

For saturated clay, under undrained conditions, $\varnothing = 0$; hence, from Table 4.4: $N_c = 5.14$; $N_q = 1$ and $N_\gamma = 0$. According to Vesic, from

Table 4.6: $$s_c = 1 + \frac{N_q}{N_c}\times\frac{B}{L} = 1 + \frac{1}{5.14}\times\frac{9}{27} = 1.065$$

$$s_q = 1 + \frac{B}{L}\tan \emptyset = 1$$

$$d_c = 1 + 0.4k;\ \frac{D}{B} = \frac{3}{9} = 0.33 < 1;\ d_c = 1 + 0.4 \times \frac{3}{9} = 1.133$$

$$d_q = 1 + 2\tan \emptyset\,(1 - \sin \emptyset\,)^2 k = 1$$

$$\bar{q} = 2.5 \times 16.5 + 0.5(16.5 - 10) = 44.5\,\text{kPa}$$

$$gross\ q_{\text{ult}} = 25\ \times 5.14 \times 1.065 \times 1.133 + 44.5 \times 1 \times 1 \times 1 + 0$$

$$= 200\ \ \text{kPa}$$

Equation (4.48):

$$I_{r(cr)} = \frac{1}{2}\left\{exp\left[\left(3.30 - 0.45\frac{B}{L}\right)\cot\left(45 - \frac{\emptyset}{2}\right)\right]\right\}$$

$$= \frac{1}{2}\left\{exp\left[\left(3.30 - 0.45 \times \frac{9}{27}\right)(1)\right]\right\} = 11.7$$

Equation (4.47):

$$I_r = \frac{G_s}{c + q'\tan \emptyset}$$

Equation (1.61): shear modulus of the clay $= G_s = \dfrac{E_u}{2(1+\mu)}$

Table 3.10: assume Poisson's ratio of the saturated clay $\mu = 0.5$

$$G_s = \frac{2593}{2(1+0.5)} = 864.33\ \ \text{kPa}$$

$$c = c_u = 25\,\text{kPa; and}\ \emptyset = 0\ \ \text{(undrained condition)}$$

$$I_r = \frac{864.33}{25 + 0} = 35 > 11.7.$$

Therefore, the assumption of soil incompressibility is justified. The computed value of ultimate bearing capacity can be used without reduction. Use *gross* q_{ult} = 200 *kPa*

(b) *Drained condition.*

Assume incompressibility soil condition; hence,

$$gross\ q_{\text{ult}} = c_d\,N_c s_c d_c + \bar{q}N_q s_q d_q + 0.5\gamma' B' N_\gamma s_\gamma d_\gamma$$

For $\emptyset = 23°$, $N_c = 18.05$; $N_q = 8.66$; $N_\gamma = 8.20$ (these values are computed from equations of Table 4.2).
Table 4.6:

$$s_c = 1 + \frac{8.66}{18.05} \times \frac{9}{27} = 1.16$$

$$s_q = 1 + \frac{9}{27}\tan 23 = 1.14$$

$$s_\gamma = 1 - 0.4 \times \frac{B}{L} = 1 - 0.4 \times \frac{9}{27} = 0.87 > 0.6,\ \text{use}\ 0.87$$

$$d_c = 1 + 0.4k = 1 + 0.4 \times \frac{3}{9} = 1.133$$

$$d_q = 1 + 2\tan \emptyset\,(1 - \sin \emptyset\,)^2 k$$

$$= 1 + 2\tan 23(1 - \sin 23)^2(3/9) = 1.105$$

$$d_\gamma = 1$$

(*Continued*)

Equation (4.20):

$$r_\gamma = 1 - 0.25\log\left(\frac{B}{K}\right) \quad \text{for } B \geq 2 \text{ m}$$

$$= 1 - 0.25\log\left(\frac{9}{2}\right) = 0.84$$

$$gross\, q_{\text{ult}} = 5 \times 18.05 \times 1.16 \times 1.133 + 44.5 \times 8.66 \times 1.14 \times 1.105$$
$$+ 0.5(16.5 - 10)(9)(8.2 \times 0.87 \times 1)(0.84)$$

$$gross\, q_{\text{ult}} = 118.61 + 485.45 + 175.28 = 779 \text{ kPa}$$

The average effective overburden pressure in the influence zone, which is taken as pressure at the depth $B/2$ below the foundation level, is

$$q' = 2.5 \times 16.5 + (0.5 + 4.5)(16.5 - 10) = 73.75 \text{ kPa}$$

The modulus of confined compression $M_v = 12.6\,q' = 12.6 \times 73.75 = 929.25$ kPa

For drained condition, Poisson's ratio may be calculated from the following equation (Vesic, 1973):

$$\mu = \frac{1 - \sin 1.2\,\varnothing_d}{2 - \sin 1.2\,\varnothing_d} = \frac{1 - \sin(1.2 \times 23)}{2 - \sin(1.2 \times 23)} = 0.35$$

Drained modulus of deformation E_d may be calculated from the following equation (Vesic, 1973):

$$E_d = M_v\left[\frac{(1+\mu)(1-2\mu)}{(1-\mu)}\right] = (929.25)\frac{(1+0.35)(1-2\times 0.35)}{(1-0.35)} = 579 \text{ kPa}$$

The actual rigidity index is

Equation (4.47):

$$I_r = \frac{G_s}{c + q'\tan\varnothing}$$

$$= \frac{E_d}{2(1+\mu)} \times \frac{1}{c_d + q'\tan\varnothing_d} = \frac{579}{2(1+0.35)(5+73.75\tan 23)} = 5.9$$

Equation (4.48):

$$I_{r(cr)} = \frac{1}{2}\left\{exp\left[\left(3.30 - 0.45\frac{B}{L}\right)\cot\left(45 - \frac{\varnothing}{2}\right)\right]\right\}$$

$$= \frac{1}{2}\left\{exp\left[\left(3.30 - 0.45 \times \frac{9}{27}\right)\cot\left(45 - \frac{23}{2}\right)\right]\right\} = 59$$

$I_r < I_{r(cr)}$. Thus, the assumption of soil incompressibility is not justified. The compressibility factors of Equations (4.44) and (4.45) need to be used in the bearing capacity equation.

Equation (4.45):

$$\zeta_{qc} = \exp\left\{\left(-4.4 + 0.6\frac{B}{L}\right)\tan\varnothing + \left[\frac{(3.07\sin\varnothing)(\log 2I_r)}{1 + \sin\varnothing}\right]\right\}$$

$$= \exp\left\{\left(-4.4 + 0.6 \times \frac{9}{27}\right)\tan 23 + \left[\frac{(3.07\sin 23)(\log 11.8)}{1 + \sin 23}\right]\right\}$$

$$= 0.424$$

$$\zeta_{\gamma c} = \zeta_{qc} = 0.424$$

Equation (4.44):
$$\zeta_{cc} = \zeta_{qc} - \frac{1-\zeta_{qc}}{N_c \tan \varnothing} \quad \text{for } \varnothing' > 0$$
$$= 0.424 - \frac{1-0.424}{18.05 \times \tan 23} = 0.35$$
$$gross\, q_{\text{ult}} = 118.61(0.35) + 485.45(0.424) + 175.28(0.424)$$
$$= 322\ kPa$$

Problem 4.11

Solve Problem 4.10 if the soil consists of a deep stratum of medium-dense sand. Assume the following additional data:

For the sand, the saturated unit weight = 19 kN/m^3 and an average moist unit weight above the water table = 16 kN/m^3.

Drained triaxial tests on sand samples show that the angle of shearing resistance of sand Ø varies with mean normal stress σ according to

$\varnothing = \varnothing_1 - (5.5^\circ)\log\frac{\sigma}{\sigma_1}$, in which $\varnothing_1 = 38^\circ$ = the angle of shearing resistance of the sand at mean normal stress $\sigma_1 = 100$ kPa. According to De Beer (1965), the average mean normal stress along the slip surface is

$$\sigma = \frac{1}{4}(q_{ult} + 3\bar{q})(1 - \sin \varnothing),$$

in which $\bar{q}$ is the effective overburden pressure at the fundation level. In the low and elevated pressure range, the modulus of deformation E of sand increases with the mean normal stress according to the relationship $E = E_1\sqrt{\frac{\sigma}{\sigma_1}}$ in which E_1 is the modulus at the mean normal stress σ_1.

Also, the following equations may be needed in solving the Problem:

$$K_o = 1 - \sin(1.2\varnothing) = \text{coefficient of earth pressure at rest}$$
$$\mu = \frac{K_o}{1+K_o} = \text{Poisson'sratio}$$
$$\sigma = \left(\frac{1+2K_o}{3}\right)(\bar{q}) = \text{initial mean normal stress at depth z.}$$

Solution:

In sands, the drained condition prevails. Assume incompressibility soil condition.

A preliminary estimate of q_{ult} is required so that the mean normal stress along the slip surface can be found. For this preliminary analysis, assume that Ø = 34°, and Table 4.4 gives $N_q = 29.4$; $N_\gamma = 41$.

Table 4.6:

$$s_q = 1 + \frac{9}{27}\tan 34^\circ = 1.22$$
$$s_\gamma = 1 - 0.4 \times \frac{B}{L} = 1 - 0.4 \times \frac{9}{27} = 0.87$$
$$d_q = 1 + 2\tan\varnothing\,(1-\sin\varnothing)^2 k$$
$$= 1 + 2\tan 34(1-\sin 34)^2\left(\frac{3}{9}\right) = 1.09$$
$$d_\gamma = 1$$

(Continued)

Equation (4.20):

$$r_\gamma = 1 - 0.25\log\left(\frac{B}{K}\right) \quad \text{for } B \geq 2\,\text{m}$$

$$= 1 - 0.25\log\left(\frac{9}{2}\right) = 0.84$$

$$\bar{q} = 2.5\times16 + 0.5(19-10) = 44.5\,\text{kPa}$$

$$gross\; q_{\text{ult}} = 44.5\times29.4\times1.22\times1.09$$
$$+0.5(19-10)(9)(41\times0.87\times1)(0.84)$$
$$= 1739.78 + 1213.49 = 2953\,\text{kPa}$$

The mean normal stress along the slip surface is

$$\sigma = \frac{1}{4}(q_{\text{ult}} + 3\bar{q})(1-\sin\varnothing) = \frac{1}{4}(2953 + 3\times44.5)(1-\sin34^\circ)$$
$$= 340\,\text{kPa}$$

The representative angle of shearing resistance is $\varnothing = \varnothing_1 - (5.5^\circ)\log\frac{\sigma}{\sigma_1} = 38^\circ - (5.5^\circ)\log\frac{340}{100} \cong 35^\circ$. The difference between this and the assumed value of $\varnothing$ is small; however, to improve clarity the analysis is now repeated with $\varnothing = 35^\circ$. Thus, from Table 4.2:

$$N_q = e^{\pi\tan\varnothing}\tan^2\left(45+\frac{\varnothing}{2}\right) = e^{\pi\tan35}\tan^2\left(45+\frac{35}{2}\right) = 33.4$$

$$N_\gamma = 2(N_q+1)\tan\varnothing = 2(33.4+1)\tan35 = 48.1$$

$$s_q = 1 + \frac{9}{27}\tan35 = 1.23;\; s_\gamma = 1 - 0.4\times\frac{B}{L} = 1 - 0.4\times\frac{9}{27} = 0.87$$

$$d_q = 1 + 2\tan\varnothing(1-\sin\varnothing)^2 k$$
$$= 1 + 2\tan35(1-\sin35)^2\left(\frac{3}{9}\right) = 1.08$$

$$d_\gamma = 1;\; r_\gamma = 0.84$$

$$gross\; q_{\text{ult}} = 44.5\times33.4\times1.23\times1.08$$
$$+0.5(19-10)(9)(48.1\times0.87\times1)(0.84)$$
$$= 1974.4 + 1423.6 = 3398\;\text{kPa}$$

The mean normal stress along the slip surface is

$$\sigma = \frac{1}{4}(q_{\text{ult}} + 3\bar{q})(1-\sin\varnothing)$$
$$= \frac{1}{4}(3398 + 3\times44.5)(1-\sin35) = 376.5\,\text{kPa}$$

The representative angle of shearing resistance is

$$\varnothing = \varnothing_1 - (5.5^\circ)\log\frac{\sigma}{\sigma_1} = 38 - (5.5^\circ)\log\frac{376.5}{100} = 34.8^\circ.$$

The analysis can be repeated again for more accuracy; however, in view of small change in mean normal stress from the previously found value, a value for *gross* q_{ult} using $\emptyset = 34.8°$ would be retained for incompressible soil. For this value of $\emptyset$, the equations of Tables 4.2 and 4.6 give

$$N_q = 32.8; N_\gamma = 47; s_q = 1 + \frac{9}{27}\tan 34.7 = 1.23; d_\gamma = 1;$$

$$s_\gamma = 1 - 0.4 \times \frac{9}{27} = 0.87; d_q = 1 + 2\tan 34.8(1 - \sin 34.8)^2\left(\frac{3}{9}\right) = 1.09.$$

$$\begin{aligned} gross\ q_{\text{ult}} &= 44.5 \times 32.8 \times 1.23 \times 1.09 \\ &\quad + 0.5(19-10)(9)(47 \times 0.87 \times 1)(0.84) \\ &= 1956.9 + 1391.1 = 3348 \text{ kPa} \end{aligned}$$

To check whether the assumption of incompressibility is justified, the following computations may be necessary.

The mean normal stress in the expansion zone, needed in estimating the elastic parameters E and μ of the sand soil, is taken as the *initial* mean normal stress at a depth $= B/2$ below the foundation level. With $\emptyset = 38°$ for the sand in *elastic* zone the coefficient of earth pressure at rest is

$$K_o = 1 - \sin(1.2\ \emptyset) = 1 - \sin(1.2 \times 38) = 0.29$$

The mean normal stress at a depth of 7.5 m, below ground surface, is

$$\sigma = \left(\frac{1 + 2K_o}{3}\right)(\bar{q}) = \left(\frac{1 + 2 \times 0.29}{3}\right)(2.5 \times 16 + 5 \times 9) = 44.77 \text{ kPa}$$

the modulus of deformation is

$$E = E_1\sqrt{\frac{\sigma}{\sigma_1}} = 36\ 400\sqrt{\frac{44.77}{100}} = 24\ 355 \text{ kPa}$$

Poison's ratio is $\mu = \dfrac{K_o}{1 + K_o} = \dfrac{0.29}{1 + 0.29} = 0.225$

The representative angle of shearing resistance for the *plastic* zone is taken again to be 34.8°.

The critical rigidity index is

Equation (4.48):

$$\begin{aligned} I_{r(cr)} &= \frac{1}{2}\left\{\exp\left[\left(3.30 - 0.45\frac{B}{L}\right)\cot\left(45 - \frac{\emptyset}{2}\right)\right]\right\} \\ &= \frac{1}{2}\left\{\exp\left[\left(3.30 - 0.45 \times \frac{9}{27}\right)\cot\left(45 - \frac{34.8}{2}\right)\right]\right\} = 208 \end{aligned}$$

$$I_r = \frac{G_s}{c + \bar{q}\tan\emptyset} = \frac{E_d}{2(1+\mu)} \times \frac{1}{\bar{q}\tan\emptyset_d} = \frac{24355}{2(1 + 0.225)[(2.5 \times 16 + 5 \times 9)\tan 34.8]} = 168$$

$I_r < I_{r(cr)}$. Thus, the assumption of soil incompressibility is not justified. The compressibility factors of Equation (4.45) need to be used in the bearing capacity equation.

(Continued)

Equation (4.45):

$$\zeta_{qc} = \exp\left\{\left(-4.4+0.6\frac{B}{L}\right)\tan\varnothing + \left[\frac{(3.07\sin\varnothing)(\log 2I_r)}{1+\sin\varnothing}\right]\right\}$$

$$= \exp\left\{\begin{array}{l}\left(-4.4+0.6\times\dfrac{9}{27}\right)\tan 34.8 \\ +\left[\dfrac{(3.07\sin 34.8)(\log 338)}{1+\sin 34.8}\right]\end{array}\right\} = 0.9$$

$$\zeta_{\gamma c} = \zeta_{qc} = 0.9$$

$$gross\ q_{\text{ult}} = (44.5\times 32.8\times 1.23\times 1.09)(0.9)$$

$$+[0.5(19-10)(9)(47\times 0.87\times 1)](0.84)(0.9)$$

$$= 3013\ kPa.$$

(This value is considered too high; however, the allowable soil pressure may be controlled by maximum tolerable settlement for the structure in question).

Problem 4.12

For the design of a shallow foundation, the following data are given:

Foundation: $L = 1.5$ m, $B = 1.0$ m, $D_f = 1.0$ m

Soil: $\varnothing = 25°$

$$c = 50\,\text{kPa}$$

$$\gamma = 17\,\text{kN/m}^3$$

Modulus of deformation $E_s = 1020$ kPa

Poisson's ratio $\mu_s = 0.35$

Calculate the gross q_{ult} of the soil. Take the soil compressibility into consideration.

Solution:

Assume incompressibility soil condition; hence,

$$gross\ q_{\text{ult}} = c_d\, N_c s_c d_c + \bar{q} N_q s_q d_q + 0.5\gamma' B' N_\gamma s_\gamma d_\gamma$$

For $\varnothing = 25°$, $N_c = 20.72$; $N_q = 10.66$; $N_\gamma = 10.88$ (these values are computed from the Vesic equations of Table 4.2).

Table 4.6:

$$s_c = 1 + \frac{10.66}{20.72}\times\frac{1}{1.5} = 1.343$$

$$s_q = 1 + \frac{1}{1.5}\tan 25 = 1.311$$

$$s_\gamma = 1 - 0.4\times\frac{1}{1.5} = 0.733 > 0.6,\ \text{use } 0.733$$

$$\frac{D}{B}=\frac{1}{1}=1\rightarrow k=\frac{D}{B}=1$$

$$d_c=1+0.4k=1+0.4\times 1=1.4$$

$$d_q=1+2\tan Ø\,(1-\sin Ø\,)^2k$$

$$=1+2\tan 25(1-\sin 25)^2(1/1)=1.311$$

$$d_\gamma=1$$

$$\begin{aligned} gross\, q_{\text{ult}} &= 50\ \times 20.72\times 1.343\times 1.4 \\ &\quad +1\times 17\times 10.66\times 1.311\times 1.311 \\ &\quad +0.5\times 17\times 1\times 10.88\times 0.733\times 1 \\ &=1947.89+311.47+67.79=2327\ \text{kPa} \end{aligned}$$

The average overburden pressure in the influence zone, which is taken as pressure at the depth $B/2$ below the foundation level, is

$$\bar{q}=1.5\times 17=25.5\,\text{kPa}$$

Equation (4.47):

$$I_r=\frac{G_s}{c+\bar{q}\tan Ø}=\frac{E_s}{2(1+\mu)}\times\frac{1}{c+\bar{q}\tan Ø}$$

$$=\frac{1020}{2(1+0.35)(50+25.5\tan 25)}=6.1$$

Equation (4.48):

$$I_{r(cr)}=\frac{1}{2}\left\{exp\left[\left(3.30-0.45\frac{B}{L}\right)\cot\left(45-\frac{Ø}{2}\right)\right]\right\}$$

$$=\frac{1}{2}\left\{exp\left[\left(3.30-0.45\times\frac{1}{1.5}\right)\cot\left(45-\frac{25}{2}\right)\right]\right\}=55.53$$

$I_r<I_{r(cr)}$. Thus, the assumption of soil incompressibility is not justified. The compressibility factors of Equations (4.44) and (4.45) need to be used in the bearing capacity equation.

Equation (4.45):

$$\zeta_{qc}=\exp\left\{\left(-4.4+0.6\frac{B}{L}\right)\tan Ø\;+\left[\frac{(3.07\sin Ø\,)(\log 2I_r)}{1+\sin Ø}\right]\right\}$$

$$=\exp\left\{\begin{array}{l}\left(-4.4+\dfrac{0.6\times 1}{1.5}\right)\tan 25\\ +\left[\dfrac{(3.07\sin 25)(\log 12.2)}{1+\sin 25}\right]\end{array}\right\}=0.417$$

Equation (4.44):

$$\zeta_{cc}=\zeta_{qc}-\frac{1-\zeta_{qc}}{N_c\tan Ø}\quad\text{for } Ø>0$$

$$\zeta_{cc}=0.417-\frac{1-0.417}{20.72\times\ \tan 25}=0.357$$

$$\zeta_{\gamma c}=\zeta_{qc}=0.417$$

$$gross\, q_{\text{ult}}=1947.89(0.357)+311.47(0.417)+67.79(0.417)$$

$$=854\ kPa$$

Problem 4.13

The column footing shown in the scheme below measures 1.2 × 1.8 m and is subjected to a centric vertical load Q and two moments M_B and M_L. If $e_B = 0.12$ m, $e_L = 0.26$ m and the depth of foundation $D_f = 1$ m, determine the allowable load the foundation can carry using: (a) Terzaghi equations and effective area method, (b) the Meyerhof equations and effective area method, (c) thePrakash and Saran method. Use the safety factor against bearing capacity failure equals three.

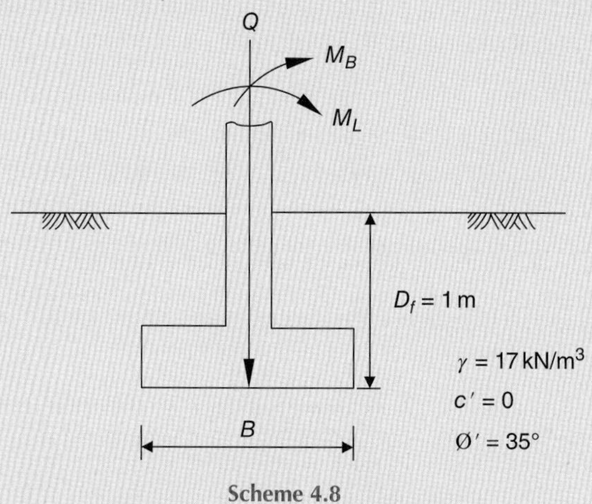

Scheme 4.8

Solution:

(a) Terzaghi equations and effective area method

Table 4.2: $$gross\,q_{\text{ult}} = cN_c s_c + \gamma' D_f N_q + \frac{1}{2}\gamma' B N_\gamma s_\gamma$$

The N_c term is 0, since the cohesion $c = 0$. $B = B'$

Table 4.3: $\varnothing = 35°$; $N_q = 41.4$; $N_\gamma = 42.4$

For rectangular footings, the Terzaghi $s_\gamma = 1 - 0.2\dfrac{B}{L}$ (see Section 4.6)

$$B' = B - 2e_B = 1.2 - 2 \times 0.12 = 0.96 \text{ m}$$

$$L' = L - 2e_L = 1.8 - 2 \times 0.26 = 1.28 \text{ m}$$

$$gross\,q_{ult} = D_f \gamma' N_q + \frac{1}{2}\gamma' B' N_\gamma s_\gamma$$

$$= (1 \times 17)(41.4) + (0.5 \times 17)(0.96)(42.4)\left(1 - 0.2 \times \frac{0.96}{1.08}\right)$$

$$= 703.8 + 284.4 = 988.2 \text{ kPa}$$

$$\text{Gross ultimate load} = gross\,q_{ult} \times B'L' = 988.2 \times 0.96 \times 1.28$$

$$= 1214.3 \text{ kN}$$

The allowable load (safe load) the footing can carry $= \dfrac{1214.3}{SF = 3} = 405\ kN$

(b) Meyerhof equations and effective area method

Table 4.2: $$gross\,q_{\text{ult}} = cN_c s_c d_c + \gamma' D_f N_q s_q d_q + \frac{1}{2}\gamma' B' N_\gamma s_\gamma d_\gamma$$

The N_c term is 0, since the cohesion $c = 0$. $B' = B$

Table 4.5:
$$K_P = \tan^2\left(45 + \frac{\text{Ø}}{2}\right) = 3.69$$

$$d_q = d_\gamma = 1 + 0.1\sqrt{K_P}\,\frac{D}{B} = 1 + 0.1 \times 1.92 \times \frac{1}{1.2} = 1.16$$

$$s_q = s_\gamma = 1 + 0.1K_P\,\frac{D}{B'} = 1 + 0.1 \times 3.69 \times \frac{1}{0.96} = 1.38$$

Table 4.4: Ø $= 35°$; $N_q = 33.6$; $N_\gamma = 37.8$ (by interpolation)

$$gross\, q_{\text{ult}} = (1 \times 17)(33.6)(1.38)(1.16)$$

$$+ (0.5 \times 17)(0.96)(37.8)(1.38)(1.16)$$

$$= 914.4 + 493.8 = 1408.2 \text{ kPa}$$

$$\text{Gross ultimate load} = gross\, q_{\text{ult}} \times B'L' = 1408.2 \times 0.96 \times 1.28$$

$$= 1730.4 \text{ kN}$$

The allowable load (safe load) the footing can carry $= \dfrac{1730.4}{SF = 3} = 577\ kN$

(c) Prakash and Saran method.

Equation (4.34):
$$Q_{\text{ult}} = BL\left[c'N_{c(e)}F_{cs(e)} + qN_{q(e)}F_{qs(e)} + \frac{1}{2}\gamma BN_{\gamma(e)}F_{\gamma s(e)}\right]$$

The N_c term is 0, since the cohesion $c = 0$

For more safety, use $\left(\dfrac{e_B}{B} = \dfrac{0.12}{1.2} = 0.1\right)$ or $\left(\dfrac{e_L}{L} = \dfrac{0.26}{1.8} = 0.14\right)$ whichever is greater; hence, use 0.14.

Figure 4.6: for Ø $= 35°$ and $\dfrac{e}{B} = 0.14$ estimate $N_{\gamma(e)} = 20$

Figure 4.7: for Ø $= 35°$ and $\dfrac{e}{B} = 0.14$ estimate $N_{q(e)} = 29$

$$F_{qs(e)} = 1$$

For the purpose of computing $F_{\gamma s(e)}$ assume the soil is dense sand; hence,

$$F_{\gamma s(e)} = 1.0 + \left(\frac{2e}{B} - 0.68\right)\frac{B}{L} + \left[0.43 - \left(\frac{3}{2}\right)\left(\frac{e}{B}\right)\right]\left(\frac{B}{L}\right)^2$$

$$= 1.0 + (2 \times 0.2 - 0.68)\frac{1.2}{1.8} + \left[0.43 - \left(\frac{3}{2}\right)(0.2)\right]\left(\frac{1.2}{1.8}\right)^2 = 0.873$$

$$Q_{\text{ult}} = 1.2 \times 1.8\left[1 \times 17 \times 29 \times 1 + \frac{1}{2} \times 17 \times 1.2 \times 20 \times 0.873\right]$$

$$= 1450 \text{ kN}$$

The allowable load (safe load) the footing can carry $= \dfrac{1450}{SF = 3} = 483\ kN$

Problem 4.14

A footing 2.5 m square carries a gross foundation pressure of 400 kPa at a depth of 1 m in a sand soil. The saturated and moist unit weights of the sand are 20 and 17 kN/m^3, respectively. The shear strength parameters are $c' = 0$ and $\varnothing' = 40°$. Determine the factor of safety with respect to shear failure for the following cases:

(a) The water table is 5 m below ground level.
(b) The water table is 1 m below ground level.
(c) The water table is 2 m below ground level.
(d) The water table is at ground level and there is seepage vertically upwards under hydraulic gradient $i = 0.2$.

Use the Terzaghi bearing capacity equation and assume the bearing capacity factors N_q and N_γ equal 65 and 95, respectively.

Solution:

(a) Table 4.2: $$gross\, q_{\text{ult}} = c\,N_c s_c + \bar{q}N_q + \frac{1}{2}\gamma' B N_\gamma s_\gamma$$

Equation (4.40): $\gamma' = \gamma_{\text{wet}}$ and $q' = \gamma_{\text{wet}}\, D_f = \bar{q}$ (N_q term)

Equation (4.39): $\gamma' = \gamma_{\text{wet}}$ (N_γ term)

The cohesion term is zero, since $c = 0$.

For square footings, $s_\gamma = 0.8$

Equation (4.20): $$r_\gamma = 1 - 0.25\log\left(\frac{B}{K}\right) \quad \text{for } B > 2 \text{ m}$$

$$r_\gamma = 1 - 0.25\log\frac{2.5}{2} = 0.98$$

$$gross\, q_{\text{ult}} = 1 \times 17 \times 65 + 0.5 \times 17 \times 2.5 \times 95 \times 0.8 \times 0.98$$

$$= 2688 \text{ kPa}$$

Equation (4.3): $$net\, q_{\text{ult}} = gross\, q_{\text{ult}} - \sigma'_o = 2688 - 1 \times 17 = 2671 \text{ kPa}$$

$$gross\, q' = gross\, q - h\gamma_w = 400 - (0)\gamma_w = 400 \text{ kPa}$$

Equation (4.2): $$net\, q' = gross\, q' - \sigma'_o = 400 - 1 \times 17 = 383 \text{ kPa}$$

Safety factor $SF = \dfrac{net\, q_{\text{ult}}}{net\, q'} = \dfrac{2671}{383} = 6.97$

(b) Table 4.2: $$gross\, q_{ult} = c\,N_c s_c + \bar{q}N_q + \frac{1}{2}\gamma' B N_\gamma s_\gamma$$

Equation (4.40): $\gamma' = \gamma_{\text{wet}}$ and $q' = \gamma_{\text{wet}}\, D_f = \bar{q}$ (N_q term)

Equation (4.36): $\gamma' = \gamma_{\text{b}} = \gamma_{\text{sat}} - \gamma_{\text{w}}$ (N_γ term)

The cohesion term is zero, since $c = 0$.

For square footings, $s_\gamma = 0.8$

Equation (4.20): $$r_\gamma = 1 - 0.25\log\left(\frac{B}{K}\right) \quad \text{for } B > 2 \text{ m}$$

$$r_\gamma = 1 - 0.25\log\frac{2.5}{2} = 0.98$$

$$gross\, q_{ult} = 1\times 17\times 65 + 0.5\times(20-10)\times 2.5\times 95\times 0.8\times 0.98$$
$$= 2036\ \text{kPa}$$

Equation (4.3): $$net\, q_{ult} = gross\, q_{ult} - \sigma'_o = 2036 - 1\times 17 = 2019\,\text{kPa}$$

$$gross\, q' = gross\, q - h\gamma_w = 400 - (0)\gamma_w = 400\,\text{kPa}$$

Equation (4.2): $$net\, q' = gross\, q' - \sigma'_o = 400 - 1\times 17 = 383\,\text{kPa}$$

Safety factor $SF = \dfrac{net\, q_{ult}}{net\, q'} = \dfrac{2019}{383} = 5.27$

(c) Table 4.2: $$gross\, q_{ult} = c\,N_c s_c + \bar{q}N_q + \frac{1}{2}\gamma' B N_\gamma s_\gamma$$

Equation (4.40): $$\gamma' = \gamma_{wet} \text{ and } q' = \gamma_{wet}\, D_f = \bar{q}\ (N_q \text{ term})$$

Equation (4.37): $$\gamma' = (2H - z_w)\frac{z_w}{H^2}\gamma_{wet} + \frac{\gamma_b}{H^2}(H - z_w)^2\ (N_\gamma \text{ term})$$

$$H = 0.5\,B\tan\left(45 + \frac{\emptyset}{2}\right) = 0.5\times 2.5\times\tan\left(45 + \frac{40}{2}\right) = 2.68\,\text{m}$$

$$\gamma' = (2\times 2.68 - 1)\left[\frac{1}{(2.68)^2}\right](17) + \frac{10}{(2.68)^2}(2.68-1)^2 = 14.25\,\text{kN/m}^3$$

or

Equation (4.38):
$$\gamma' = \gamma_b + \left(\frac{z_w}{B}\right)(\gamma_{wet} - \gamma_b)$$
$$= 10 + \left(\frac{1}{2.5}\right)(17-10) = 12.8\,\text{kN/m}^3$$

Equation (4.37) gives a more exact result than Equation (4.38) does. However, it would be more conservative to use Equation (4.38). Use $\gamma' = 13\,\text{kN/m}^3$.

The cohesion term is zero, since $c = 0$.

For square footings, $s_\gamma = 0.8$

Equation (4.20):
$$r_\gamma = 1 - 0.25\log\left(\frac{B}{K}\right)\quad \text{for}\ B > 2\,\text{m}$$
$$r_\gamma = 1 - 0.25\log\frac{2.5}{2} = 0.98$$

$$gross\, q_{ult} = 1\times 17\times 65 + 0.5\times 13\times 2.5\times 95\times 0.8\times 0.98$$
$$= 2315\,\text{kPa}$$

Equation (4.3): $$net\, q_{ult} = gross\, q_{ult} - \sigma'_o = 2315 - 1\times 17 = 2298\,\text{kPa}$$

$$gross\, q' = gross\, q - h\gamma_w = 400 - (0)\gamma_w = 400\,\text{kPa}$$

Equation (4.2): $$net\, q' = gross\, q' - \sigma'_o = 400 - 1\times 17 = 383\,\text{kPa}$$

Safety factor $SF = \dfrac{net\, q_{ult}}{net\, q'} = \dfrac{2298}{383} = 6$

(d) Table 4.2: $$gross\, q_{ult} = c\,N_c s_c + \bar{q}N_q + \frac{1}{2}\gamma' B N_\gamma s_\gamma$$

The cohesion term is zero, since $c = 0$

(Continued)

For square footings, $s_\gamma = 0.8$

Equation (4.20):

$$r_\gamma = 1 - 0.25\log\left(\frac{B}{K}\right) \quad \text{for } B > 2 \text{ m}$$

$$r_\gamma = 1 - 0.25\log\frac{2.5}{2} = 0.98$$

Equation (4.42): $\gamma' = \gamma_b = \gamma_{sat} - \gamma_w$ and $q' = \gamma_b D_f = \bar{q}$ (N_q term)

Equation (4.36): $\gamma' = \gamma_b = \gamma_{sat} - \gamma_w$ (N_γ term)

However, the force due to vertically upwards seepage in a soil reduces the effective stress. Therefore, the effective unit weight of the soil is reduced by the same magnitude of seepage force per unit volume.

Seepage force $= i\gamma_w V = 0.2 \times 10 \times V = 2\,V$

Seepage force per unit volume $= \dfrac{2V}{V} = 2\ \text{kN/m}^3$

The reduced $\gamma' = \gamma_b - 2 = 20 - 10 - 2 = 8\,\text{kN/m}^3$

$$gross\ q_{ult} = (1)(8)(65) + (0.5)(8)(2.5)(95)(0.8)(0.98)$$
$$= 1265\,\text{kPa}$$

Equation (4.3):

$$net\ q_{ult} = gross\ q_{ult} - \sigma_o' = 1265 - 1 \times 8 = 1257\,\text{kPa}$$

$$gross\ q' = gross\ q - h\gamma_w = 400 - 1 \times 10 = 390\,\text{kPa}$$

Equation (4.2):

$$net\ q' = gross\ q' - \sigma_o' = 290 - 1 \times 8 = 382\,\text{kPa}$$

Safety factor $SF = \dfrac{net\ q_{ult}}{net\ q'} = \dfrac{1257}{383} = 3.29$

Problem 4.15

A footing foundation 6 m square is located at a depth of 2 m below ground level, the water table being at ground level. A 2 m silty sand layer, below the ground level, overlies a stiff to very stiff saturated clay layer 15 m thick and a firm stratum lies immediately below the clay. The saturated unit weight of the silty sand soil is 20 kN/m^3 and that of the clay is 21 kN/m^3. Assume the following parameters for the clay:

$$c_u = 150\ \text{kPa};\ \text{Ø}_u = 0;\ m_v = 0.06\,\text{m}^2/\text{MN};\ E_u = 60\,\text{MN}/m^2;\ A = 0.3.$$

The design requires a factor of safety with respect to shear failure not to be less than three and the maximum final consolidation settlement not to exceed 35 mm. Determine the net allowable bearing capacity. Use the Hansen bearing capacity equations.

Solution:

Table 4.2:

$$gross\ q_{ult} = 5.14\,s_u\left(1 + s_c' + d_c'\right) + \bar{q} \quad (\text{Ø} = 0)$$

Equation (4.3):

$$net\ q_{ult} = \left[5.14\,s_u\left(1 + s_c' + d_c'\right) + \bar{q}\right] - \gamma' D_f$$
$$= 5.14\,s_u\left(1 + s_c' + d_c'\right)$$

$$s_u = c_u = 150\,\text{kPa};\ \frac{B'}{L'} = \frac{B}{L} = \frac{6}{6} = 1;\ \frac{D}{B} = \frac{2}{6} = 0.33 < 1;\ k = \frac{D}{B} = 0.33$$

Table 4.6:

$$s_c' = 0.2\frac{B'}{L'} = 0.2;\ d_c' = 0.4k = 0.4 \times 0.33 = 0.133$$

$$net\, q_{ult} = 5.14(150)(1 + 0.2 + 0.133) = 1027.7\,\text{kPa}$$

Equation (4.4):
$$net\, q_s = \frac{net\, q_{ult}}{SF} = \frac{1027.7}{3} = 343\,\text{kPa}$$

Calculation of final consolidation settlement may be carried out considering the net effective foundation pressure of 343 kPa. However; experience indicates that this pressure is likely to cause settlement larger than the specified allowable settlement of 35 mm. Therefore, as the first trial use a net design soil pressure of 150 kPa.

The clay layer is relatively thick; there will be significant lateral strain (resulting in immediate settlement in the clay) and it is appropriate to use the Skempton–Bjerrum method. Accordingly, the final settlement $S = S_i + S_c$, may be calculated as follows.

In order to compute the consolidation settlement more accurately, assume the clay is divided into five layers of equal thicknesses, as shown in the scheme below.

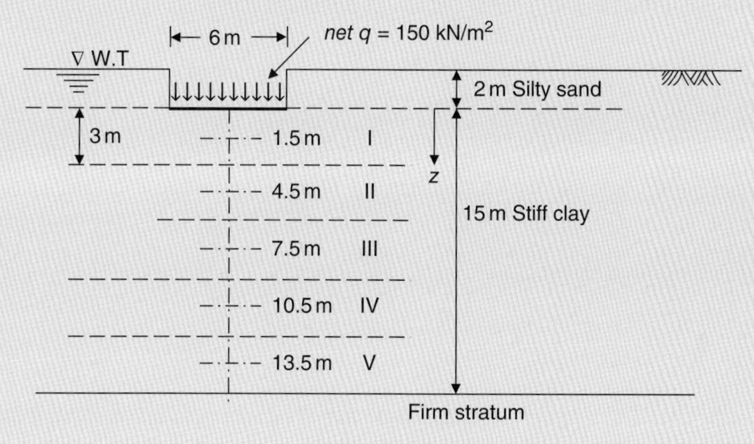

Scheme 4.9

Calculation of the immediate settlement S_i:

Equation (3.33):
$$S_i = \mu_o \mu_1 \frac{qB}{E_u}$$

$\frac{H}{B} = \frac{15}{6} = 2.5$; $\frac{D}{B} = \frac{2}{6} = 0.33$; $\frac{L}{B} = 1$; hence, from figure 3.9:

$$\mu_o = 0.95 \text{ and } \mu_1 = 0.55$$

$$S_i = 0.95 \times 0.55 \times \frac{150 \times 6}{60 \times 1000} = 0.008\,\text{m} = 8\,\text{mm}$$

Calculation of consolidation settlement S_c:

$$S_c = K \sum S_{oed}$$

For each layer:

$$S_{oed} = m_v \times \sigma_z' \times H = \frac{0.06}{1000} \times \sigma_z' \times 3 \times 1000 = 0.18\,\sigma_z'\ (\text{mm})$$

$\sigma_z' = q(4I)$; I is calculated from Table 2.3 or Figure 2.32.

(*Continued*)

Layer	z (m)	I	σ'_z (kPa)	S_{oed} (mm)
I	1.5	0.233	140	25.2
II	4.5	0.121	73	13.1
III	7.5	0.060	36	6.5
IV	10.5	0.033	20	3.6
V	13.5	0.021	13	2.3

$\sum S_{oed} = 50.7\,\text{mm}$

Scheme 4.10

The equivalent diameter of the base area $= B = \sqrt{\frac{4 \times 36}{\pi}} = 6.77\,\text{m}$

$\frac{H}{B} = \frac{15}{6.77} = 2.2,\ A = 0.3$; from Figure 3.15 obtain $K = 0.51$

$$S_c = 0.51 \times 50.7 = 26\,\text{mm}$$

Final settlement $S = S_i + S_c = 8 + 26 = 34\,\text{mm} < 35\,\text{mm}$

As it is clear this result is less than the specified settlement, but too close. Therefore, settlement criterion controls the design. Also, as the result indicates, a second trial computation would not be necessary.

Use: *net* $q_a = 150\,kPa$

Problem 4.16

A long braced excavation in soft clay is 4 m wide and 8 m deep. The saturated unit weight of the clay is 20 kN/m^3 and the undrained shear strength adjacent to the bottom of the excavation is given by $c_u = 40$ kPa, $\varnothing_u = 0$. Determine the safety factor against base failure of the excavation.

Solution:

According to Bjerrum and Eide (1956), the Skempton's values of N_c (Figure 4.4) could also be used in the analysis of base stability in temporary braced excavations in saturated clay (see the scheme below).

Table 4.2: *gross* $q_{\text{ult}} = c N_c s_c + \gamma D_f$ ($\varnothing = 0 \rightarrow N_q = 1$ and $N_\gamma = 0$)

Base failure occurs when *gross* $q_{\text{ult}} = 0$. Therefore, the critical depth $D_c = \frac{c_u N_c(1)}{\gamma}$. In this case, c_u should represent the undrained shear strength of the soil immediately below and adjacent to the excavation base. In general the safety factor against base failure in an excavation of depth D, in clay, is given by

$$SF = \frac{c_u N_c}{\gamma D}$$

Figure 4.4: for $\frac{D}{B} = \frac{8}{4} = 2$; $N_c = 7.1$

$$SF = \frac{c_u N_c}{\gamma D} = \frac{40 \times 7.1}{20 \times 8} = 1.8$$

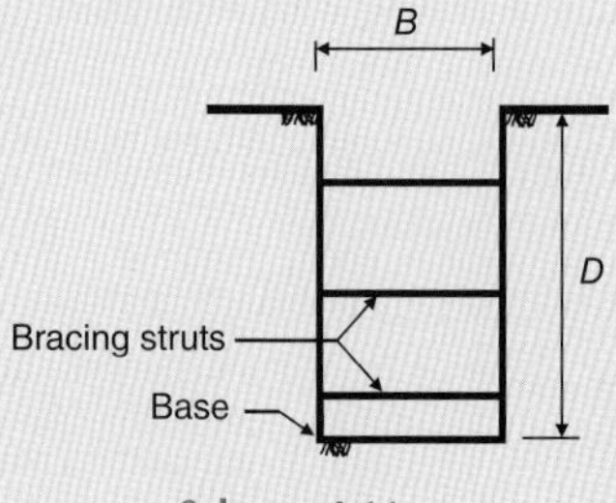

Scheme 4.11

Problem 4.17

A footing foundation 3.5 m square is to be constructed at a depth of 2 m in the deep sand deposit of Problem 1.6. All the data given in that problem are also to be used in solution of this problem without change. If the settlement is not to exceed 30 mm, determine the net allowable bearing capacity using (a) the bearing capacity Equation (3.12), (b) the Meyerhof bearing capacity equation with a safety factor against shear failure $SF = 3$.

Solution:

(a) Refer to the solution of Problem 1.6. In general, in design of shallow foundations, the zone of interest starts from a depth of about one-half footing width (B) above the estimated foundation level to a depth of about ($2B$) below. Therefore, this zone is located between depths 0.25 m and 9 m below the ground surface. All the given SPT N-values should be considered in selecting a design N-value, which is the weighted average of the corrected N-values. In the solution of Problem 1.6, the average N'_{60} selected equalled 10, that is the design $N'_{60} = 10$.

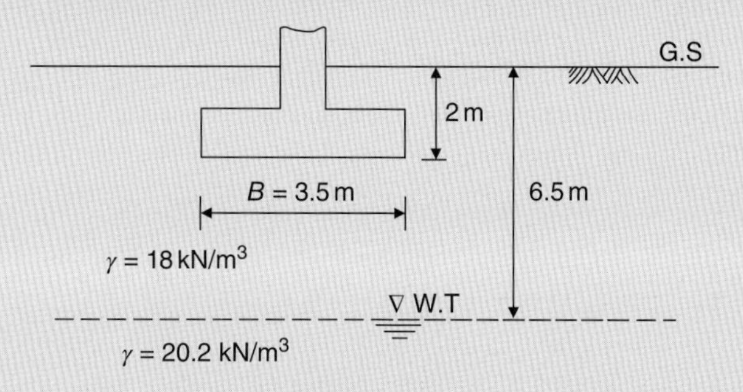

Scheme 4.12

Net allowable bearing capacity for 25 mm settlement:

$$net\ q_a = \frac{N'_{55}}{F_2}\left(\frac{B+F_3}{B}\right)^2 K_d W_r \quad B > F_4$$

Equation (3.12): where: $F_2 = 0.08; F_3 = 0.3; F_4 = 1.2$

$$N'_{55} = N'_{60}\left(\frac{60}{55}\right) = 10 \times \frac{60}{55} = 11$$

The water table is located at 6.5 m depth; hence,

$$z_w = 6.5 - 2 = 4.5\text{ m} > B = 3.5\text{ m}.$$

For $z_w \geq B$: $W_r \cong 1$ (see Section 3.3.2)

$$K_d = 1 + 0.33\left(\frac{D}{B}\right) = 1 + 0.33\left(\frac{2}{3.5}\right) = 1.19 < 1.33$$

$$net\ q_a = \frac{11}{0.08}\left(\frac{3.5+0.3}{3.5}\right)^2 (1.19)(1) = 193\text{ kPa}.$$

Equation (3.14): $S_i = S_o \times \dfrac{netq_{(S_i)}}{netq_{a(S_o)}}$. Hence,

$$netq_a = S_i \times \frac{netq_{(S_o)}}{S_o} = 30 \times \frac{210.4}{25} = 232\text{ kPa}$$

(Continued)

Table 4.2: $$gross\ q_{ult} = cN_c s_c d_c + \gamma' D_f N_q s_q d_q + \frac{1}{2}\gamma' B N_\gamma s_\gamma d_\gamma$$

For the sand soil the N_c term is 0, since c is zero. Hence,

$$gross\ q_{ult} = \gamma' D_f N_q s_q d_q + \frac{1}{2}\gamma' B N_\gamma s_\gamma d_\gamma$$

Because there is no value for Ø given in the Problem data, it is necessary to select a suitable value using the various approximate correlations between Ø, D_r and N given in the discussion of Problem 1.6.

Equation (1.30):

$$\text{Ø}' = 27.1 + 0.3N'_{60} - 0.00054\left(N'_{60}\right)^2$$
$$= 27.1 + 0.3(10) - 0.00054(10)^2 = 30°$$

Equation (1.32): $$\text{Ø}' = \sqrt{20N'_{60}} + 20 = \sqrt{20 \times 11} + 20 = 34°$$

The average denseness of the sand may be classified as loose. For loose sands Table 1.7 gives an average value for Ø $= 31°$. It would be appropriate use the average of these three values and select Ø $= 32°$.

Table 4.5: $$K_P = \tan^2\left(45 + \frac{\text{Ø}}{2}\right) = 3.25$$

$$d_q = d_\gamma = 1 + 0.1\sqrt{K_P}\,\frac{D}{B} = 1 + 0.1 \times 1.8 \times \frac{2}{3.5} = 1.1$$

$$s_q = s_\gamma = 1 + 0.1K_P\,\frac{D}{B} = 1 + 0.1 \times 3.25 \times \frac{2}{3.5} = 1.19$$

$$r_\gamma = 1 - 0.25\log\left(\frac{B}{k}\right) = 1 - 0.25\log\left(\frac{3.5}{2}\right) = 0.94$$

Table 4.4: Ø $= 32°$; $N_q = 23.2$; $N_\gamma = 22.0$

$$gross\ q_{ult} = (2 \times 18)(23.2)(1.19)(1.1)$$
$$+ (0.5 \times 18)(3.5)(22)(1.19)(1.1)(0.94)$$
$$= 1093.3 + 852.7 = 1946\,\text{kPa}$$

Equation (4.3): $$net\ q_{ult} = gross\ q_{ult} - \sigma'_o = 1946 - 18 \times 2 = 1910\,\text{kPa}$$

$$net\ q_s = \frac{net\ q_{ult}}{SF} = \frac{1910}{3} = 637\,\text{kPa}$$

Check settlement:

Equation (3.5): $$S_i = q \times B \times \frac{1-\mu^2}{E_s} \times m \times I_S \times I_F$$

Because there are no given values for the elastic parameters E_s and μ, it is necessary to select suitable values for these parameters using the tables and various approximate correlations between E_s and N given in Section 3.3.1 and in the discussion of Problem 1.6.

E_s value:

Equation (1.33): $$\frac{E_s}{P_a} = \alpha N_{60}$$

$$P_a = \text{atmospheric pressure} = 100\,\text{kPa}$$

Assume $\alpha = 10$ (for clean normally consolidated sand) and $N_{60} \cong N'_{60}$

$$E_s = 100 \times 10 \times 10 = 11\ 000\,\text{kPa}$$

Table 3.8: for normally consolidated sand the following values are suggested:

$$E_s = 7000\sqrt{N_{55}} = 7000\sqrt{11} = 23\ 216\,\text{kPa}$$
$$E_s = 500(N_{55} + 15) = 500(11 + 15) = 13\ 000\,\text{kPa}$$
$$E_s = 2600N_{55} = 2600 \times 11 = 28\ 600\,\text{kPa}$$

Table 3.9: for loose sand E_s ranges 10–25 MPa, with average $E_s = 17.5$ MPa = 17 500 kPa
Use the average of these values, which gives $E_s = 20\ 579$ kPa.
μ value:
Table 3.10a: commonly used value for μ ranges $0.3-0.4$
Table 3.10b: for loose to medium dense sand μ ranges $0.2-0.35$
Assume $\mu = 0.31$

I_s value:

$$I_S = I_1 + \frac{1-2\mu}{1-\mu} I_2$$

$$M = \frac{L/2}{B/2} = 1;\ N = \frac{H}{B/2} = \frac{4B}{B/2} = \frac{4 \times 3.5}{3.5/2} = 8$$

Table 3.7: $I_1 = 0.482$; $I_2 = 0.02$. Hence,

$$I_S = 0.482 + \frac{1-2\times 0.31}{1-0.31} \times 0.02 = 0.493$$

I_F value:

Table 3.6: $\frac{D}{B} = \frac{2}{3.5} = 0.57;\ \frac{B}{L} = \frac{3.5}{3.5} = 1 \rightarrow I_F = 0.75$
For settlement under centre of the loaded area, $m = 4$

$$q = net\,q_s = 637\,\text{kPa}$$

$$S_i = 637 \times \frac{3.5}{2} \times \frac{1-(0.31)^2}{20579} \times 4 \times 0.493 \times 0.75 = 0.0724\,\text{m}$$

Equation (3.6): $S_{i(\text{rigid})} = 0.93 S_{i(\text{flexible})}$

$$S_i = 0.93 \times 0.0724 = 0.067\,\text{m}$$

$S_i = 67$ mm > 30 mm. Hence, settlement criterion controls the design.

Equation (3.14): $S_i = S_o \times \frac{netq_{(S_i)}}{netq_{a(S_o)}}$. Hence,

$$netq_a = S_i \times \frac{netq_{(S_o)}}{S_o} = 30 \times \frac{637}{67} = 285\ kPa$$

Problem 4.18

A footing 3 m square carries a net foundation pressure of 130 kPa at a depth of 1.2 m in a deep deposit of sand. The water table is at a depth of 3 m and the unit weights of the sand above and below the water table are 16 and 19 kN/m^3, respectively. The variation of cone penetration resistance (q_c) with depth (z) is as follows:

z (m)	1.2	1.6	2.0	2.4	2.6	3.0	3.4	3.8	4.2
q_c (MN/m^2)	3.2	2.1	2.8	2.3	6.1	5.0	6.6	4.5	5.5
z (m)	4.6	5.0	5.5	5.8	6.2	6.6	7.0	7.4	8.0
q_c (MN/m^2)	4.5	5.4	10.4	8.9	9.9	9.0	15.1	12.9	14.8

Scheme 4.13

The allowable settlement is specified not to exceed 25 mm and the safety factor against bearing capacity failure should not be less than three. Check whether these two requirements are satisfied. Use the Vesic, Buisman–DeBeer and Schmertmann methods as applicable.

Solution:

The CPT results show that the sand deposit below the foundation level may be divided into four layers as shown in Table 4.17.

Table 4.17 Layers of the sand deposit below the foundation level.

Layer	Depth (m)	Thickness (m)	Average q_c (MN/m^2)
I	1.2–2.5	1.3	2.60
II	2.5–5.2	2.7	5.37
III	5.2–6.8	1.6	9.55
IV	6.8–8.0	1.2	14.27

Check safety factor against bearing capacity failure.

For each layer, select a suitable value for Ø using the various approximate correlations between Ø and q_c given in the discussion of Problem 1.13. If more than one correlation is considered, the average of the results will be taken as a design value for Ø; in this solution the following two correlations are considered:

Equation (1.53):

$$\text{Ø}'_1 = \tan^{-1}\left[0.1 + 0.38\log\left(\frac{q_c}{\sigma'_o}\right)\right]$$

An approximate correlation (Bowles, 2001) for Ø′ is

$\text{Ø}' = 29° + \sqrt{q_c} + 5°$ for gravel; $-5°$ for silty sand, where q_c is in MPa. As a second correlation, assume $\text{Ø}'_2 = 29° + \sqrt{q_c}$ (Table 4.18).

Table 4.18 Values of $\text{Ø}'_1$ and $\text{Ø}'_2$ obtained using correlations.

Layer	Ave. q_c (MN/m^2)	Depth to center of layer (m)	σ'_o (kN/m^2)	$\text{Ø}'_1$ (degree)	$\text{Ø}'_2$ (degree)	Ave. Ø′ (degree)
I	2.60	1.85	29.60	40.0	30.6	35
II	5.37	3.85	55.65	40.5	31.3	36
III	9.55	6.00	75.00	42.0	32.1	37
IV	14.27	7.40	87.60	43.2	32.8	38

In bearing capacity analysis, the influence depth for square footings is about $2B$ below the foundation level and therefore, the $\varnothing'$ values of the first three layers need to be taken into consideration only. Accordingly, the design $\varnothing'$ is

$$\varnothing' = \frac{35+36+37}{3} = 36^\circ$$

Table 4.2:

$$gross\, q_{ult} = c\,N_c s_c d_c i_c g_c b_c + qN_q s_q d_q i_q g_q b_q + \frac{1}{2}\gamma' BN_\gamma s_\gamma d_\gamma i_\gamma g_\gamma b_\gamma$$

The N_c term is 0, since the cohesion $c = 0$

Table 4.8: for the conditions of this problem all $i_c g_c b_c$ and $i_q g_q b_q$ are 1.

$$gross\, q_{ult} = \bar{q}N_q s_q d_q + \frac{1}{2}\gamma' BN_\gamma s_\gamma d_\gamma$$

Table 4.4: for $\varnothing = 36^\circ$: $N_q = 37.7$, $N_\gamma = 56.2$

$$s_q = 1 + \frac{B}{L}\tan\varnothing = 1.73; \; s_\gamma = 1 - 0.4\frac{B}{L} = 0.6$$

Table 4.6:

$$d_q = 1 + 2\tan\varnothing\,(1-\sin\varnothing)^2\left(\frac{D}{B}\right)$$

$$= 1 + 2\tan 36(1-\sin 36)^2(0.4) = 1.1$$

$$= 1 + 2\tan 36(1-\sin 36)^2(0.4) = 1.1$$

$$d_\gamma = 1$$

Equation (4.38):

$$\gamma' = \gamma_b + \left(\frac{z_w}{B}\right)(\gamma_{wet} - \gamma_b) = (19-10) + \left(\frac{1.8}{3}\right)(16-9)$$

$$= 13.2 \text{ kN/m}^3$$

$$gross\, q_{ult} = 16 \times 1.2 \times 37.7 \times 1.73 \times 1.1 + 0.5 \times 13.2 \times 3 \times 56.2 \times 0.6 \times 1$$

$$gross\, q_{ult} = 1252 + 668 = 1920 \text{ kPa}$$

Equation (4.3): $net\, q_{ult} = gross\, q_{ult} - \sigma'_o = 1920 - 1.2 \times 16 = 1901$ Pa

The safety factor against bearing capacity failure is

$$SF = \frac{net\, q_{ult}}{\text{net foundation pressure}} = \frac{1901}{130} = 14.6 \gg 3$$

The required safety against bearing capacity failure is satisfied.

Check allowable settlement using the Buisman–DeBeer method:

$$C = 1.9\frac{q_c}{\sigma'_o} \text{ (Recommended by Meyerhof, 1965)}$$

$$S_i = \frac{H}{C}\ln\frac{\sigma'_o + \sigma_z}{\sigma'_o} = \frac{2.3\,H\,\sigma'_o}{1.9q_c}\log\frac{\sigma'_o + \sigma_z}{\sigma'_o} \text{ or } \Sigma S_i = \sum_0^H \frac{2.3\sigma'_0}{1.9q_c}\Delta z \log\frac{\sigma'_o + \sigma_z}{\sigma'_o}$$

The stress increment σ_z at the centre of each layer will be calculated using either Figure 2.32 or Table 2.3. The above settlement equations are applied to obtain the total settlement, as in Table 4.19.

$$\Sigma S_i = 24.5 \text{ mm} \cong 25 \text{ mm}$$

(*Continued*)

Table 4.19 Total settlement obtained using settlement equations.

Layer	Δ z (m)	σ'_o (kPa)	σ_z (kPa)	$\log\frac{\sigma'_o+\sigma_z}{\sigma'_o}$	q_c (kPa)	$\frac{2.3\sigma'_o}{1.9q_c}\Delta Z$ (mm)	S_i (mm)
I	1.3	29.60	122.2	0.71	2600	17.9	12.7
II	2.7	55.65	52.0	0.29	5370	33.8	9.8
III	1.6	75.00	20.3	0.10	9550	15.2	1.5
IV	1.2	87.60	14.0	0.06	14270	8.9	0.5
							$\sum S_i = 24.5$

The required safety against excessive settlement is satisfied.

Check the allowable settlement using the Schmertmann method:

Equation (3.24):

$$S_i = C_1 C_2 C_3\, q_{net} \sum_{z=0}^{z=z_2} \frac{I_z}{E_s}\Delta z$$

According to Schmertmann, for sand soils supporting square footings:

$E_s = 2.5\, q_c$ (see Section 3.3.3)

Also, the influence depth under square footings is $2B$ (Figure 3.7). Hence, in this case, only the first three layers need to be considered in the settlement computations. The thickness of layer III should be increased by 0.4 m so that the required $2B$ depth is maintained.

Considering square footings:

Equation (3.26)

$$\text{For } z=0 \text{ to } \frac{B}{2}:\ I_z = 0.1 + \left(\frac{z}{B}\right)\left(2I_{zp}-0.2\right)$$

Equation(3.27)

$$\text{For } z=\frac{B}{2} \text{ to } 2B:\ I_z = 0.667 I_{zp}\left(2-\frac{z}{B}\right)$$

Equation (3.25)

$$I_{zp} = 0.5 + 0.1\sqrt{\frac{q_{\text{net}}}{\sigma'_{zp}}}$$

$$q_{\text{net}} = 130 \text{ kPa}$$

For square footings, σ'_{zp} is calculated at depth $B/2$ below foundation level; hence, $\sigma'_{zp} = 2.7\times 16 = 43.2$ kPa

$$I_{zp} = 0.5 + 0.1\sqrt{\frac{130}{43.2}} = 0.673$$

Determine the value of $(I_z\Delta z/E_s)$ at the midpoint of each layer, as shown in Table 4.20.

$$C_1 = 1 - \frac{0.5\sigma'_o}{q_{net}} = 1 - \frac{0.5(1.2\times 16)}{130} = 0.926$$

Table 4.20 Value of $(I_z\Delta z/E_s)$ at the midpoint of each layer.

Layer	E_s (kN/m²)	z (m)	I_z	Δz (m)	$I_Z\Delta z/E_s$
I	6500	0.65	0.348	1.3	6.96×10^{-5}
II	13425	2.65	0.501	2.7	10.08×10^{-5}
III	23875	4.80	0.180	2.0	1.51×10^{-5}
					$\sum(I_z\Delta z/E_s) = 18.55\times 10^{-5}$

$C_2 = 1 + 0.2\log\frac{t}{0.1}$. Assume $t = 0.1$ year; hence, $C_2 = 1$

$C_3 = 1.0$ (for square footings)

$$S_i = C_1 C_2 C_3 q_{\text{net}} \sum_{z=0}^{z=z_2} \frac{I_z}{E_s}\Delta z = 0.926 \times 1 \times 1 \times 130 \times 18.55 \times 10^{-5}$$

$$= 0.0223 \text{ m} = 22.3 \text{ mm} < 25\, mm$$

The required safety against excessive settlement is satisfied.

Problem 4.19

A rectangular footing 0.92 ×1.22 m is shown in the scheme below. Find the gross ultimate load that the footing can carry.

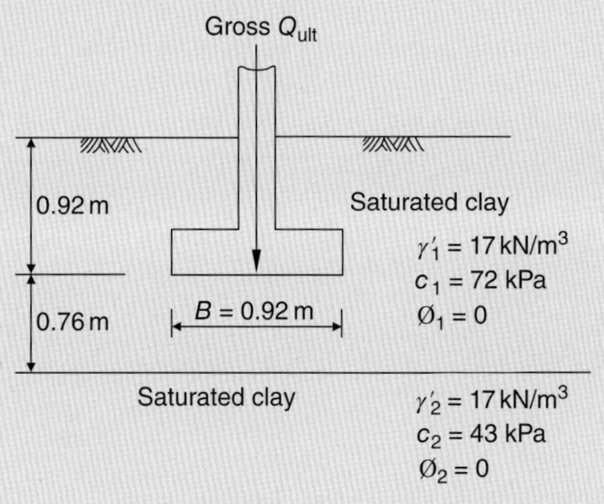

Scheme 4.14

Solution:

This is a case of layered soils where *stronger saturated clay* ($Ø_1 = 0$) is *overlying weaker saturated clay* ($Ø_2 = 0$).

Equation (4.70):

$$q_{\text{ult}} = \left[\left(1 + 0.2\frac{B}{L}\right)5.14c_2 + \left(1 + \frac{B}{L}\right)\left(\frac{2c_a H}{B}\right) + \gamma_1' D_f\right]$$

$$\le \left(1 + 0.2\frac{B}{L}\right)5.14c_1 + \gamma_1' D_f$$

$$H = 0.76 \text{ m};\; D_f = 0.92 \text{ m};\; B = 0.92 \text{ m};\; L = 1.22 \text{ m}$$

Figure 4.22: for $\frac{q_2}{q_1} = \frac{c_2}{c_1} = \frac{43}{72} = 0.6$, find $\frac{c_a}{c_1} \cong 0.97$

$$c_a = 0.97 \times 72 = 69.84 \text{ kPa}$$

$$q_{\text{ult}} = \left[\begin{array}{c}\left(1 + 0.2 \times \frac{0.92}{1.22}\right)(5.14 \times 43) + \left(1 + \frac{0.92}{1.22}\right)\left(\frac{2 \times 69.84 \times 0.76}{0.92}\right) \\ + 17 \times 0.92\end{array}\right]$$

$$= 221.02 + 33.33 + 115.39 + 87.01 + 15.64$$

$$= 472.39 \text{ kPa}$$

(Continued)

As a check, we have:

Equation (4.69):

$$q_{\mathrm{ult},t} = \left(1 + 0.2\frac{B}{L}\right)5.14c_1 + \gamma_1' D_f$$

$$= \left(1 + 0.2 \times \frac{0.92}{1.22}\right)(5.14 \times 72) + 17 \times 0.92$$

$$= 441.54\,\mathrm{kPa} < q_{\mathrm{ult}}$$

Therefore, $q_{\mathrm{ult}} = q_{\mathrm{ult},t}$

The gross ultimate load is

$$Q_{\mathrm{ult}} = 441.54\,(0.92 \times 1.22) = 496\ kN$$

Problem 4.20

A square footing on layered sand is shown in the scheme below. Determine the net allowable load that the foundation can support, considering shear failure only. Use the Meyerhof bearing capacity and shape factors. The safety factor against bearing capacity failure should not be less than four.

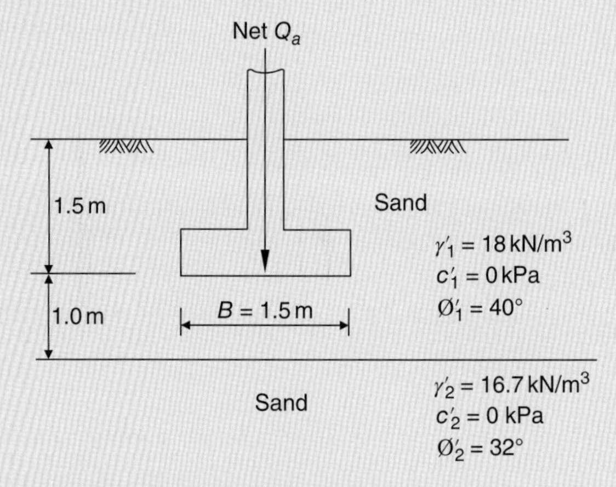

Scheme 4.15

Solution:

This is a case of layered soils where *Stronger sand* $(c_1 = 0)$ is *overlying weaker sand* $(c_2 = 0)$.

Equation (4.67):

$$q_{\mathrm{ult}} = \left[\gamma_1'(D_f + H)N_{q(2)}F_{qs(2)} + \frac{1}{2}\gamma_2' BN_{\gamma(2)}F_{\gamma s(2)} + \gamma_1' H^2\left(1 + \frac{B}{L}\right)\left(1 + \frac{2D_f}{H}\right)\left(\frac{K_s \tan Ø_1'}{B}\right) - \gamma_1' H\right]$$

$$\leq (\gamma_1' Df)N_{q(1)}F_{qs(1)} + \frac{1}{2}\gamma_1' BN_{\gamma(1)}F_{\gamma s(1)}$$

$$H = 1.0\ \mathrm{m};\ D_f = 1.5\ \mathrm{m};\ B = L = 1.5\ \mathrm{m}$$

Table 4.4:

$$Ø_1' = 40°;\ N_{q(1)} = 64.1;\ N_{\gamma(1)} = 93.69$$

$$Ø_2' = 32°;\ N_{q(2)} = 23.2;\ N_{\gamma(2)} = 22.02$$

$$F_{qs(1)} = F_{\gamma s(1)} = 1 + 0.1\frac{B}{L}\tan^2\left(45 + \frac{Ø_1'}{2}\right) = 1 + 0.1 \times \frac{1.5}{1.5} \times 4.6 = 1.5$$

$$F_{qs(2)} = F_{\gamma s(2)} = 1 + 0.1\frac{B}{L}\tan^2\left(45 + \frac{Ø_2'}{2}\right) = 1 + 0.1 \times \frac{1.5}{1.5} \times 3.3 = 1.3$$

Figure 4.21: for $\frac{q_2}{q_1} = \frac{\gamma_2' B N_{\gamma(2)}}{\gamma_1' B N_{\gamma(1)}} = \frac{16.7 \times 22.02}{18 \times 93.69} = 0.22$ and $\varnothing_1' = 40°$,

find $K_s \cong 4.5$

$$q_{\text{ult}} = \begin{bmatrix} (18)(1.5+1)(23.2)(1.3) + (0.5)(16.7)(1.5)(22.02)(1.3) \\ +(18)(1)^2(1+1)\left(1+\frac{2\times1.5}{1}\right)\left(\frac{4.5\times0.84}{1.5}\right) - 18\times1 \end{bmatrix}$$

$$= 1357.2 + 358.5 + 362.9 - 18 = 2060.6\,\text{kPa}$$

As a check, we have: Equation (4.66)-(b)

Equation(4.66)-(b):

$$q_{\text{ult},t} = (\gamma_1' D_f) N_{q(1)} F_{qs(1)} + \frac{1}{2}\gamma_1' B N_{\gamma(1)} F_{\gamma s(1)}$$

$$= (18\times1.5)(64.1\times1.5)$$

$$+0.5\times18\times1.5\times93.69\times1.5$$

$$= 2596.1 + 1897.2 = 4493.3\,\text{kPa} \gg q_{\text{ult}}$$

Use *gross* $q_{\text{ult}} = 2060.6\,\text{kPa}$

Equation (4.3): $\quad net\ q_{\text{ult}} = gross\ q_{\text{ult}} - \sigma_o'$

$$net\ q_{\text{ult}} = 2060.6 - 18\times1.5 = 2033.6\,\text{kPa}$$

Considering shear failure only, $net\ q_a = \frac{net\ q_{\text{ult}}}{SF} = \frac{2033.6}{4} = 508.4\,\text{kPa}$

The net allowable load is

$$\text{Net } Q_a = 508.4\,(1.5)^2 = 1144\ kN$$

Problem 4.21

Find the gross q_{ult} that the 1.22 × 1.83 m footing, shown in the scheme, can carry. The soil profile consists of two layers of saturated clay.

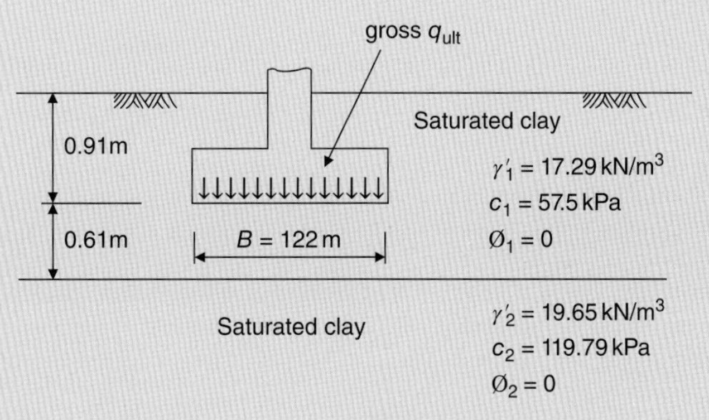

Scheme 4.16

Solution:

This is a case of layered soils where *weaker saturated clay* ($\varnothing_1 = 0$) is *overlying stronger saturated clay* ($\varnothing_2 = 0$).

Equation (4.72): $\quad q_{\text{ult}} = q_{\text{ult},t} + (q_{\text{ult},b} - q_{\text{ult},t})\left(\frac{H}{D}\right)^2 \geq q_{\text{ult},t}$

(Continued)

Equation (4.73):

$$q_{ult,t} = c_1 N_{c(1)} F_{cs(1)} + (\gamma'_1 D_f) N_{q(1)} F_{qs(1)} + \frac{1}{2}\gamma'_1 B N_{\gamma(1)} F_{\gamma s(1)}$$

$$= c_1 N_{c(1)} F_{cs(1)} + (\gamma'_1 D_f) F_{qs(1)}$$

$$F_{qs(1)} = 1 + 0.1\frac{B}{L}\tan^2\left(45 + \frac{Ø'_1}{2}\right) = 1 + 0.1 \times \frac{1.22}{1.83} \times 1 = 1.067$$

$$F_{cs(1)} = 1 + 0.2\frac{B}{L}\tan^2\left(45 + \frac{Ø'_1}{2}\right) = 1 + 0.2 \times \frac{1.22}{1.83} \times 1 = 1.133$$

For $Ø = 0 : N_{c(1)} = N_{c(2)} = 5.14$

$$gross\ q_{ult,t} = 57.5 \times 5.14 \times 1.133$$

$$+ 17.29 \times 0.91 \times 1.067 = 351.74 \text{ kPa}$$

Equation (4.74):

$$gross\ q_{ult,b} = c_2 N_{c(2)} F_{cs(2)} + (\gamma'_2 D_f) N_{q(2)} F_{qs(2)} + \frac{1}{2}\gamma'_2 B N_{\gamma(2)} F_{\gamma s(2)}$$

$$= c_2 N_{c(2)} F_{cs(2)} + (\gamma'_2 D_f) F_{qs(2)}$$

$$F_{qs(2)} = F_{qs(1)} = 1.067; \quad F_{cs(2)} = F_{cs(1)} = 1.133$$

$$gross\ q_{ult,b} = 119.79 \times 5.14 \times 1.133$$

$$+ 19.65 \times 0.91 \times 1.067 = 716.9 \text{ kPa}$$

Equation (4.72):

$$gross\ q_{ult} = q_{ult,t} + (q_{ult,b} - q_{ult,t})\left(\frac{H}{D}\right)^2 \geq q_{ult,t}$$

$$= 351.74 + (716.9 - 351.74)\left(\frac{0.61}{1.22}\right)^2$$

$$gross\ q_{ult} = 351.74 + 91.29 = 443 \text{ kPa} > gross\ q_{ult,t}$$

$$< gross\ q_{ult,b}$$

Use $gross\ q_{ult} = 443\ kPa$

Problem 4.22

Assume locations of the two sand layers of Problem 4.20 are reversed, that is the weaker layer at top and the stronger layer at bottom. Determine the net allowable load that the foundation can support, considering shear failure only. Use the Meyerhof bearing capacity and shape factors. The safety factor against bearing capacity failure should not be less than four.

Solution:

This is a case of layered soils where *weaker sand* $(c_1 = 0)$ is *overlying stronger sand* $(c_2 = 0)$.

Equation (4.72): $$q_{ult} = q_{ult,t} + (q_{ult,b} - q_{ult,t})\left(\frac{H}{D}\right)^2 \geq q_{ult,t}$$

Equation (4.73): $$q_{ult,t} = c_1 N_{c(1)} F_{cs(1)} + (\gamma'_1 D_f) N_{q(1)} F_{qs(1)} + \frac{1}{2}\gamma'_1 B N_{\gamma(1)} F_{\gamma s(1)}$$

$$gross\, q_{ult,t} = (\gamma'_1 D_f) N_{q(1)} F_{qs(1)} + \frac{1}{2}\gamma'_1 B N_{\gamma(1)} F_{\gamma s(1)}$$

$$F_{qs(1)} = F_{\gamma s(1)} = 1 + 0.1\frac{B}{L}\tan^2\left(45 + \frac{Ø'_1}{2}\right) = 1 + 0.1 \times \frac{1.5}{1.5} \times 3.3 = 1.3$$

Table 4.4: for $Ø'_1 = 32°$: $N_{q(1)} = 23.2; N_{\gamma(1)} = 22.02$

$$gross\, q_{ult,t} = (16.7)(1.5)(23.2)(1.3) + (0.5)(16.7)(1.5)(22.02)(1.3)$$
$$= 1114\,\text{kPa}$$

Equation (4.74):

$$gross\, q_{ult,b} = c_2 N_{c(2)} F_{cs(2)} + (\gamma'_2 D_f) N_{q(2)} F_{qs(2)} + \frac{1}{2}\gamma'_2 B N_{\gamma(2)} F_{\gamma s(2)}$$
$$= (\gamma'_2 D_f) N_{q(2)} F_{qs(2)} + \frac{1}{2}\gamma'_2 B N_{\gamma(2)} F_{\gamma s(2)}$$

$$F_{qs(2)} = F_{\gamma s(2)} = 1 + 0.1\frac{B}{L}\tan^2\left(45 + \frac{Ø'_2}{2}\right) = 1 + 0.1 \times \frac{1.5}{1.5} \times 4.6 = 1.5$$

Table 4.4: for $Ø'_2 = 40°$: $N_{q(2)} = 64.1; N_{\gamma(2)} = 93.69$

$$gross\, q_{ult,b} = (18 \times 1.5)(64.1 \times 1.5) + 0.5 \times 18 \times 1.5 \times 93.69 \times 1.5$$
$$= 2596.1 + 1897.2 = 4493.3\ \text{kPa}$$
$$gross\, q_{ult} = 1114 + (4493.3 - 1114)\left(\frac{1}{1.5}\right)^2 = 2616\ \text{kPa}$$
$$> gross\, q_{ult,t}$$
$$< gross\, q_{ult,b}$$

Use $gross\, q_{ult} = 2616\ \text{kPa}$

Equation (4.3): $$net\, q_{ult} = gross\, q_{ult} - \sigma'_o$$

$$net\, q_{ult} = 2616 - 16.7 \times 1.5 = 2591\,\text{kPa}$$

$$net\, q_a = net\, q_s = \frac{net\, q_{ult}}{SF}$$

Considering shear failure only, $$= \frac{2591}{4} = 647.8\,\text{kPa}$$

$$net\, Q_a = 647.8\,(1.5)^2 = \mathit{1457\ kN}$$

Problem 4.23

The scheme below shows a continuous foundation with $B = 2$ m. Determine the gross ultimate load per unit length of the foundation. Use the Meyerhof bearing capacity and shape factors.

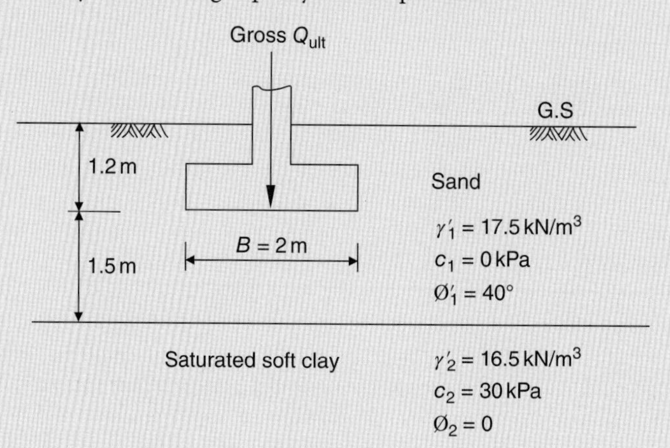

Scheme 4.17

Solution:

This is a case of layered soils where *sand* ($c_1 = 0$) is *overlying saturated soft clay* ($Ø_2 = 0$). For this case, Equation (4.64) will apply.

Equation (4.64):

$$q_{ult} = \left[\begin{matrix}\left(1+0.2\frac{B}{L}\right)5.14c_2 + \gamma'_1 H^2\left(1+\frac{B}{L}\right)\left(1+\frac{2D_f}{H}\right)\left(\frac{K_s \tan Ø'_1}{B}\right) \\ +\gamma'_1 D_f\end{matrix}\right]$$

$$\leq (\gamma'_1 D_f)N_{q(1)}F_{qs(1)} + \frac{1}{2}\gamma'_1 B N_{\gamma(1)}F_{\gamma s(1)}$$

Table 4.4: $Ø'_1 = 40°;\ N_{q(1)} = 64.1;\ N_{\gamma(1)} = 93.69$

For continuous footings, all shape factors are one.

Equation (4.65): $$\frac{q_2}{q_1} = \frac{c_2 N_{c(2)}}{\frac{1}{2}\gamma'_1 B N_{\gamma(1)}} = \frac{5.14c_2}{0.5\gamma'_1 B N_{\gamma(1)}} = \frac{5.14\times 30}{0.5\times 17.5\times 2\times 93.69} = 0.094$$

Figure 4.21: $Ø'_1 = 40°;\ \frac{q_2}{q_1} = 0.094;\ K_s \cong 3$

$$gross\ q_{ult} = (1+0.2\times 0)(5.14\times 30) + 17.5(1.5)^2(1+0)\left(1+\frac{2\times 1.2}{1.5}\right)\left(\frac{3\times 0.84}{2}\right) + 17.5\times 1.2$$
$$= 304.2\,kPa$$

Equation (4.66)-(b): $$gross\ q_{ult,t} = (\gamma'_1 D_f)N_{q(1)}F_{qs(1)} + \frac{1}{2}\gamma'_1 B N_{\gamma(1)}F_{\gamma s(1)}$$

$$gross\ q_{\text{ult},t} = (17.5\times1.2)(64.1)(1) + 0.5(17.5)(2)(93.69)(1)$$

$$= 2986\ \text{kPa} > gross\ q_{\text{ult}}$$

Use $gross\ q_{\text{ult}} = 304.2\ \text{kP}$

$$\text{Gross ultimate load per meter length of the foundation} = (304.2)(B)$$

$$= (304.2)(2)$$

$$= 608\ kN$$

Problem 4.24

Assume locations of the sand and clay layers of Problem 4.23 are reversed, that is the saturated soft clay layer at top and the stronger sand layer at bottom. Determine the net allowable bearing capacity, considering shear failure only. Use the Meyerhof bearing capacity and shape factors. Use a safety factor $SF = 3$.

Solution:

This is a case of layered soils where *weaker saturated clay* ($\varnothing_1 = 0$) *overlying stronger sand layer* ($c_2 = 0$).

Equation (4.72): $$q_{\text{ult}} = q_{\text{ult},t} + (q_{\text{ult},b} - q_{\text{ult},t})\left(\frac{H}{D}\right)^2 \geq q_{\text{ult},t}$$

Equation (4.73): $$q_{\text{ult},t} = c_1 N_{c(1)} F_{cs(1)} + (\gamma_1' D_f) N_{q(1)} F_{qs(1)} + \frac{1}{2}\gamma_1' B N_{\gamma(1)} F_{\gamma s(1)}$$

$$= c_1 N_{c(1)} F_{cs(1)} + (\gamma_1' D_f) F_{qs(1)}$$

Table 4.4: $$\varnothing_2' = 40°;\ N_{q(2)} = 64.1;\ N_{\gamma(2)} = 93.69$$

$$\varnothing = 0;\ N_{c(1)} = 5.14;\ N_{q(1)} = 1;\ N_{\gamma(1)} = 0$$

For continuous footings, all shape factors are one.

$$gross\ q_{\text{ult},t} = 30\times5.14\times1 + 16.5\times1.2\times1 = 174\ \text{kPa}$$

Equation (4.74):

$$gross\ q_{\text{ult},b} = c_2 N_{c(2)} F_{cs(2)} + (\gamma_2' D_f) N_{q(2)} F_{qs(2)} + \frac{1}{2}\gamma_2' B N_{\gamma(2)} F_{\gamma s(2)}$$

$$= (\gamma_2' D_f) N_{q(2)} F_{qs(2)} + \frac{1}{2}\gamma_2' B N_{\gamma(2)} F_{\gamma s(2)}$$

$$= 17.5\times1.2\times64.1 + 0.5\times17.5\times2\times93.69$$

$$= 2986\ \text{kPa}$$

$$H = 1.5\ \text{m};\ D \approx B = 2\ \text{m}$$

$$gross\ q_{\text{ult}} = 174 + (2986 - 174)\left(\frac{1.5}{2}\right)^2 = 1756\ \text{kPa} > gross\ q_{\text{ult},t}$$

$$net\ q_{\text{ult}} = gross\ q_{\text{ult}} - \gamma_1' D_f = 1756 - 16.5\times1.2 = 1736.2\ \text{kPa}$$

(Continued)

Considering shear failure only:

$$net\ q_a = net\ q_s = \frac{net\ q_{ult}}{SF} = \frac{1736.2}{3} = 575\ kPa.$$

(This result is considered too high!)

In the author's opinion, this method overestimates *net* q_{ult} of soft clays which overly strong sands. However, in many of such circumstances the allowable soil pressure is based on allowable settlement rather than *net* q_s.

Problem 4.25

A rectangular footing of 3 × 6 m is to be placed on a two-layer clay deposit, as shown in the scheme below. Estimate the gross ultimate bearing capacity. Use the Meyerhof bearing capacity and shape factors.

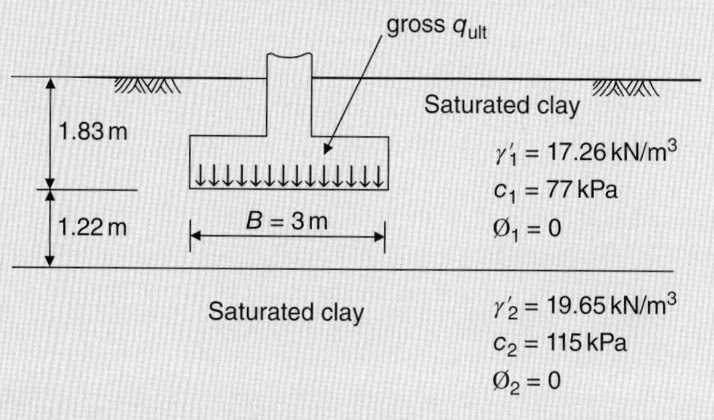

Scheme 4.18

Solution:

This is a case of layered soils where *weaker saturated clay* ($Ø_1 = 0$) is *overlying stronger saturated clay* ($Ø_2 = 0$).

Equation (4.72):

$$q_{ult} = q_{ult,t} + (q_{ult,b} - q_{ult,t})\left(\frac{H}{D}\right)^2 \geq q_{ult,t}$$

Equation (4.73):

$$q_{ult,t} = c_1 N_{c(1)} F_{cs(1)} + (\gamma'_1 D_f) N_{q(1)} F_{qs(1)} + \frac{1}{2}\gamma'_1 B N_{\gamma(1)} F_{\gamma s(1)}$$

$$= c_1 N_{c(1)} F_{cs(1)} + (\gamma'_1 D_f) F_{qs(1)}$$

$$F_{qs(1)} = 1 + 0.1\frac{B}{L}\tan^2\left(45 + \frac{Ø'_1}{2}\right) = 1 + 0.1 \times \frac{3}{6} \times 1 = 1.05$$

$$F_{cs(1)} = 1 + 0.2\frac{B}{L}\tan^2\left(45 + \frac{Ø'_1}{2}\right) = 1 + 0.2 \times \frac{3}{6} \times 1 = 1.1$$

Table 4.4: $Ø = 0;\ N_{c(1)} = N_{c(2)} = 5.14$

$$gross\ q_{ult,t} = 77 \times 5.14 \times 1.1 + 17.26 \times 1.83 \times 1.05 = 468.52\ \text{kPa}$$

Equation (4.74):

$$gross\ q_{ult,b} = c_2 N_{c(2)} F_{cs(2)} + (\gamma'_2 D_f) N_{q(2)} F_{qs(2)} + \frac{1}{2}\gamma'_2 B N_{\gamma(2)} F_{\gamma s(2)}$$

$$= c_2 N_{c(2)} F_{cs(2)} + (\gamma'_2 D_f) F_{qs(2)}$$

$$F_{qs(2)} = F_{qs(1)} = 1.05;\ F_{cs(2)} = F_{cs(1)} = 1.1$$

$$gross\ q_{ult,b} = 115 \times 5.14 \times 1.1 + 17.26 \times 1.83 \times 1.05 = 683.38\ \text{kPa}$$

Equation (4.72):

$$q_{ult} = q_{ult,t} + (q_{ult,b} - q_{ult,t})\left(\frac{H}{D}\right)^2 \geq q_{ult,t}$$

$$D \approx B = 3\ \text{m}$$

Equation (4.72):

$$gross\ q_{ult} = 468.52 + (683.38 - 468.52)\left(\frac{1.22}{3}\right)^2$$

$$gross\ q_{ult} = 504\ \text{kPa} > gross\ q_{ult,t}$$

$$< gross\ q_{ult,b}$$

$$gross\ q_{ult} = 504\ kPa$$

Problem 4.26

A 2 × 2 m square footing is to be placed on a sand layer overlying a clay layer, as shown in the scheme below. Determine the *net* q_a, considering shear failure only. Use the Meyerhof bearing capacity and shape factors. The safety factor against bearing capacity failure should not be less than three.

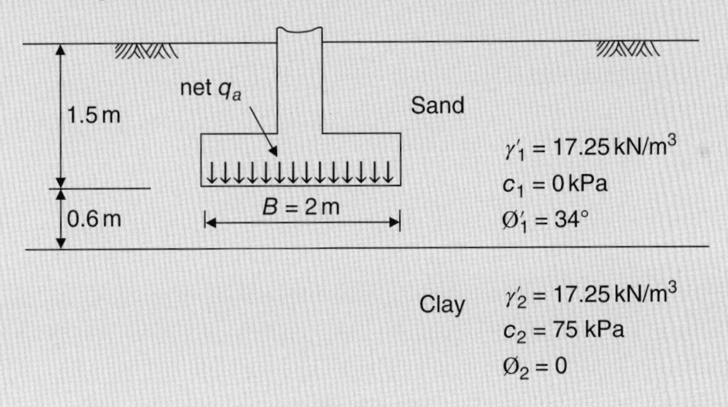

Scheme 4.19

Solution:
In this case it may not be very clear which soil is stronger or weaker than the other. In such a case, it may be reasonable to try the two possibilities then choosing the smaller result.

(a) Assume the sand is stronger than the clay:
Equation (4.64):

$$q_{ult} = \left[\left(1 + 0.2\frac{B}{L}\right)5.14c_2 + \gamma'_1 H^2\left(1 + \frac{B}{L}\right)\left(1 + \frac{2D_f}{H}\right)\left(\frac{K_s \tan Ø'_1}{B}\right) + \gamma'_1 D_f\right]$$

$$\leq (\gamma'_1 D_f)N_{q(1)}F_{qs(1)} + \frac{1}{2}\gamma'_1 B N_{\gamma(1)} F_{\gamma s(1)}$$

Table 4.4: for $Ø'_1 = 34°$: $N_{q(1)} = 29.4$; $N_{\gamma(1)} = 31.1$

(Continued)

$$F_{qs(1)} = F_{\gamma s(1)}$$

Table 4.5:

$$= 1 + 0.1\frac{B}{L}\tan^2\left(45 + \frac{\text{Ø}'_1}{2}\right) = 1 + 0.1 \times \frac{2}{2} \times 3.5 = 1.35$$

Equation (4.65):

$$\frac{q_2}{q_1} = \frac{c_2 N_{c(2)}}{\frac{1}{2}\gamma'_1 B N_{\gamma(1)}} = \frac{5.14c_2}{0.5\gamma'_1 B N_{\gamma(1)}} = \frac{5.14 \times 75}{0.5 \times 17.25 \times 2 \times 31.1} = 0.72$$

Figure 4.21:

$$\text{Ø}'_1 = 34°;\ \frac{q_2}{q_1} = 0.72;\ K_s \cong 6$$

$$gross\ q_{\text{ult}} = (1 + 0.2 \times 1)(5.14 \times 75)$$
$$+ 17.25(0.6)^2(1+1)\left(1 + \frac{2 \times 1.5}{0.6}\right)\left(\frac{6 \times 0.67}{2}\right)$$
$$+ 17.25 \times 1.5 = 638.3\ \text{kPa}$$

$$(\gamma'_1 D_f)N_{q(1)}F_{qs(1)} + \frac{1}{2}\gamma'_1 B N_{\gamma(1)}F_{\gamma s(1)} = (17.25 \times 1.5)(29.4)(1.35)$$
$$+ 0.5(17.25)(2)(31.1)(1.35)$$
$$= 1751.2\ \text{kPa} > gross\ q_{\text{ult}}$$

$$gross\ q_{\text{ult}} = 638.3\ \text{kPa}$$

(b) Assume the sand is weaker than the clay:

$$gross\ q_{\text{ult}} = q_{\text{ult},t} + (q_{\text{ult},b} - q_{\text{ult},t})\left(\frac{H}{D}\right)^2 \geq q_{\text{ult},t}$$

Equation (4.73):

$$q_{\text{ult},t} = c_1 N_{c(1)}F_{cs(1)} + (\gamma'_1 D_f)N_{q(1)}F_{qs(1)} + \frac{1}{2}\gamma'_1 B N_{\gamma(1)}F_{\gamma s(1)}$$
$$= (\gamma'_1 D_f)N_{q(1)}F_{qs(1)} + \frac{1}{2}\gamma'_1 B N_{\gamma(1)}F_{\gamma s(1)}$$

$$F_{qs(1)} = F_{\gamma s(1)} = 1 + 0.1\frac{B}{L}\tan^2\left(45 + \frac{\text{Ø}'_1}{2}\right) = 1 + 0.1 \times \frac{2}{2} \times 3.5 = 1.35$$

Table 4.4: for $\text{Ø}'_1 = 34°$: $N_{q(1)} = 29.4$; $N_{\gamma(1)} = 31.1$

$$gross\ q_{\text{ult},t} = (17.25)(1.5)(29.4)(1.35)$$
$$+ (0.5)(17.25)(2)(31.1)(1.35) = 1751.2\ \text{kPa}$$

Equation (4.74):

$$gross\ q_{\text{ult},b} = c_2 N_{c(2)}F_{cs(2)} + (\gamma'_2 D_f)N_{q(2)}F_{qs(2)} + \frac{1}{2}\gamma'_2 B N_{\gamma(2)}F_{\gamma s(2)}$$
$$= c_2 N_{c(2)}F_{cs(2)} + (\gamma'_2 D_f)N_{q(2)}F_{qs(2)}$$

$$F_{qs(2)} = 1 + 0.1\frac{B}{L}\tan^2\left(45 + \frac{\text{Ø}'_2}{2}\right) = 1 + 0.1 \times \frac{2}{2} \times 1 = 1.1$$

$$F_{cs(2)} = 1 + 0.2\frac{B}{L}\tan^2\left(45 + \frac{\text{Ø}'_2}{2}\right) = 1 + 0.2 \times \frac{2}{2} \times 1 = 1.2$$

Table 4.4: for $\text{Ø}'_2 = 0: N_{c(2)} = 5.14;\ N_{q(2)} = 1;\ N_{\gamma(2)} = 0$

$$gross\ q_{\text{ult},b} = 75 \times 5.14 \times 1.2 + 17.25 \times 1.5 \times 1 \times 1.1$$
$$= 2596.1 + 1897.2 = 488.5\ \text{kPa}$$

$$H = 0.6\ \text{m};\ D \approx B = 2\ \text{m}$$

$$gross\ q_{\text{ult}} = 1751.2 + (488.5 - 1751.2)\left(\frac{0.6}{2}\right)^2$$
$$= 1637.6\ \text{kPa} < gross\ q_{\text{ult},t}.$$

Hence, $gross\ q_{\text{ult}} = 1751.2\ \text{kPa}$

Now, the result of (a) is smaller than that of (b); therefore use:

$$gross\ q_{\text{ult}} = 638.3\ \text{kPa}$$

$$net\ q_{\text{ult}} = gross\ q_{\text{ult}} - \gamma'_1 D_f = 638.3 - 17.25 \times 1.5 = 612.42\ \text{kPa}$$

Considering shear failure only, net allowable bearing capacity is

$$net\ q_a = net\ q_s = \frac{net\ q_{\text{ult}}}{SF} = \frac{612.42}{3} = 204\ kPa$$

Problem 4.27

A 1 m wide strip footing is located on the top of a compacted (not very dense) cohesionless sand slope, as shown in the scheme below. The other available data are: $\beta = 26.5°$; $\text{Ø}_{tr} = 36°$; $\gamma' = 14.85\ \ \text{kN/m}^3$.

Determine the gross ultimate bearing capacity using the Bowles method and the Hansen bearing capacity equation and shape factors.

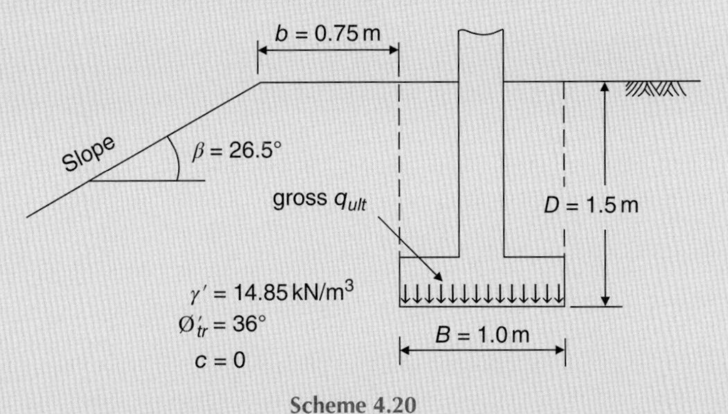

Scheme 4.20

Solution:

Equation (4.53): $gross\ q_{\text{ult}} = c\,N'_c s_c i_c + \bar{q} N'_q s_q i_q + 0.5\gamma' B N'_\gamma s_\gamma i_\gamma$

The cohesion term is zero, since $c = 0$. Hence,

$$gross\ q_{\text{ult}} = \bar{q} N'_q s_q i_q + 0.5\gamma B N'_\gamma s_\gamma i_\gamma$$

For footings on slopes, it is recommended not to adjust $\text{Ø}'_{tr}$ to $\text{Ø}'_{ps}$, as there are considerable uncertainties in the stress state when there is loss of soil support on one side of the base, even for strip (or long) bases. Therefore, use $\text{Ø}'_{ps} = \text{Ø}'_{tr} = 36°$.

(Continued)

For strip footings, all shape (s) factors are one and in this case, since the foundation load is vertical ($H_i = 0$), all load inclination (i) factors are also one.

$$\frac{b}{B} = \frac{0.75}{1} = 0.75 < 2.$$

Hence, the N_γ factor should be adjusted as follows:

Determine the Coulomb passive pressure coefficients using $+\beta$ for $K_{P,\max}$ and $-\beta$ for $K_{P,\min}$, using

$$K_P = \frac{\sin^2(\alpha - \emptyset)}{sin^2\alpha \sin(\alpha+\delta)\left[1 - \sqrt{\dfrac{\sin(\emptyset+\delta)\sin(\emptyset+\beta)}{\sin(\alpha+\delta)\sin(\alpha+\beta)}}\right]^2}$$

This method uses $\alpha = 90°$ and $\delta = \emptyset$.
For $\emptyset = 36°$; $\delta = 36°$; $\alpha = 90°$; $\beta = +26.5°$: $K_{P,\max} = 128.2$
For $\emptyset = 36°$; $\delta = 36°$; $\alpha = 90°$; $\beta = -26.5°$: $K_{P,\min} = 2.8$

$$R = \frac{K_{P,min}}{K_{P,max}} = \frac{2.8}{128.2} = 0.022$$

Table 4.4: $\emptyset = 36°$; $N_\gamma = 40$

$$N'_\gamma = \frac{N_\gamma}{2} + \frac{N_\gamma}{2}\left[R + \frac{b}{2B}(1-R)\right] = \frac{40}{2} + \frac{40}{2}\left[0.022 + \frac{0.75}{2\times 1}(1-0.022)\right]$$

$$N'_\gamma = 28$$

Table 4.11: $\dfrac{b}{B} = 0.75$; $\dfrac{D}{B} = \dfrac{1.5}{1} = 1.5$; $\beta = 26.5°$; $\emptyset = \delta = 36°$; obtain (by interpolation) $N'_q \approx 27$.

$$gross\; q_{\text{ult}} = 14.85 \times 1.5 \times 27 + 0.5 \times 14.85 \times 1 \times 28 = 809\ kPa$$

Problem 4.28

Solve Problem 4.27 using (a) $D = 0$ and $b = 1.5$, (b) $D = b = 0$.

Solution:

(a)

$$gross\; q_{\text{ult}} = \bar{q}N'_q s_q i_q + 0.5\gamma B N'_\gamma s_\gamma i_\gamma$$

$\dfrac{b}{B} = \dfrac{1.5}{1} = 1.5 < 2$. Hence, the N_γ factor should be adjusted.

The $K_{P,\max}$, $K_{P,\min}$, N_γ and R values remain unchanged.

$$N'_\gamma = \frac{N_\gamma}{2} + \frac{N_\gamma}{2}\left[R + \frac{b}{2B}(1-R)\right] = \frac{40}{2} + \frac{40}{2}\left[0.022 + \frac{1.5}{2\times 1}(1-0.022)\right]$$

$$N'_\gamma = 35.11$$

$\bar{q}=0$, since $D=0$. Hence, the N_q term will be 0. All shape (s) and load inclination (i) factors are one for the same reasons stated earlier.

$$gross\ q_{\text{ult}} = 0.5\gamma BN'_{\gamma}s_{\gamma}i_{\gamma} = 0.5\times 14.85\times 1\times 35.11 = 261\ kPa$$

(b) $\frac{b}{B}=\frac{0}{1}=0<2$. Hence, the N_γ factor should be adjusted.

$$N'_{\gamma}=\frac{N_{\gamma}}{2}+\frac{N_{\gamma}}{2}\left[R+\frac{b}{2B}(1-R)\right]=\frac{40}{2}+\frac{40}{2}\left[0.022+\frac{0}{2\times 1}(1-0.022)\right]$$

$$N'_{\gamma}=20.44$$

$\bar{q}=0$, since D = 0. Hence, the N_q term will be 0.All shape (s) and load inclination (i) factorsare one for the same reasons stated earlier.

$$gross\ q_{\text{ult}} = 0.5\gamma BN'_{\gamma}s_{\gamma}i_{\gamma} = 0.5\times 14.85\times 1\times 20.44 = 152\ kPa$$

Problem 4.29

A shallow continuous foundation in a clay soil is shown in the scheme below. Determine the gross ultimate bearing capacity using: (a) the Meyerhof method, (b) the Bowles method.

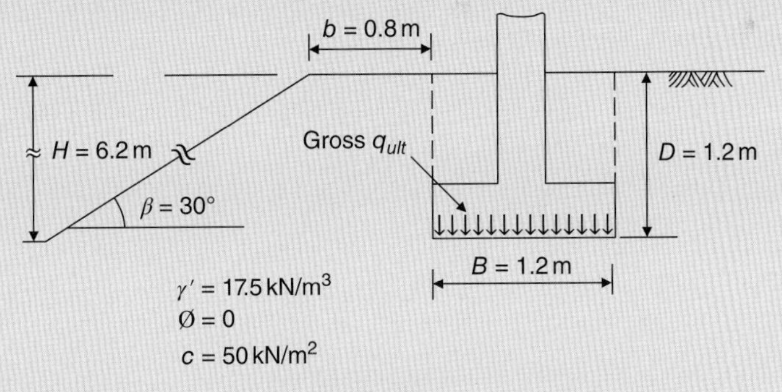

Scheme 4.21

Solution:

(a) Meyerhof method
Equation (4.51): $gross\ q_{\text{ult}} = cN_{cq}$

Assume the stability number N_s = 0, since $B < H$.

Figure 4.14a: $\frac{b}{B}=\frac{0.8}{1.2}=\frac{3}{4};\ \frac{D_f}{B}=\frac{1.2}{1.2}=1;\ \beta=30°; N_s=0; N_{cq}\approx 6.3$

$$gross\ q_{\text{ult}} = 50\times 6.3 = 315\ \text{kN/m}^2$$

(Continued)

(b) Bowles method
Equation (4.53):

$$gross\, q_{ult} = cN'_c s_c i_c + \bar{q}N'_q s_q i_q + 0.5\gamma' BN'_\gamma s_\gamma i_\gamma$$

Since Ø = 0, the N_γ term is zero. For strip footings, all shape (s) factors are one and in this case, since the foundation load is vertical ($H_i = 0$), all load inclination (i) factors are also one.

$$gross\, q_{ult} = cN'_c + \bar{q}N'_q$$

Table 4.11: $\frac{b}{B} = \frac{3}{4}$; $\frac{D_f}{B} = \frac{1.2}{1.2} = 1$; $\beta = 30°$; Ø $= \delta = 0°$; obtain

$N'_c = 5.14$; $N'_q = 1.03$ (by interpolation)

$$gross\, q_{ult} = 50 \times 5.14 + 1.2 \times 1.03 = 258.2\ kPa$$

Problem 4.30

Refer to Problem 4.29. Assume the soil is cohesionless compacted sand which has Ø $= 40°$ and $\gamma' = 20\,\text{kN/m}^3$. Determine the gross ultimate bearing capacity using (a) the Bowles method, (b) the Meyerhof method, (c) the stress characteristic solution method.

Solution:

(a) Bowles method.
Equation (4.53):

$$gross\, q_{ult} = cN'_c s_c i_c + \bar{q}N'_q s_q i_q + 0.5\gamma' BN'_\gamma s_\gamma i_\gamma$$

The cohesion term is zero, since $c = 0$. Hence,

$$gross\, q_{ult} = \bar{q}N'_q s_q i_q + 0.5\gamma' BN'_\gamma s_\gamma i_\gamma$$

For strip footings, all shape (s) factors are one and in this case, since the foundation load is vertical ($H_i = 0$), all load inclination (i) factors are also one.

Table 4.11: $\frac{b}{B} = \frac{0.8}{1.2} = 0.75$; $\frac{D}{B} = \frac{1.2}{1.2} = 1$; $\beta = 30°$; Ø $= \delta = 40°$; $N'_q = 39.2$ (by interpolation).

$\frac{b}{B} = 0.75 < 2$. Hence, the N_γ factor should be adjusted as follows.

Determine the Coulomb passive pressure coefficients using $+\beta$ for $K_{P,\max}$ and $-\beta$ for $K_{P,\min}$

$$K_P = \frac{\sin^2(\alpha - Ø)}{\sin^2\alpha \sin(\alpha + \delta)\left[1 - \sqrt{\frac{\sin(Ø + \delta)\sin(Ø + \beta)}{\sin(\alpha + \delta)\sin(\alpha + \beta)}}\right]^2}$$

This method uses $\alpha = 90°$ and $\delta = Ø$.

For Ø $= 40°$; $\delta = 40°$; $\alpha = 90°$; $\beta = +30°$: $K_{P,\max} = 23.38$

For Ø $= 40°$; $\delta = 40°$; $\alpha = 90°$; $\beta = -30°$: $K_{P,\min} = 3.16$

$$R = \frac{K_{\min}}{K_{\max}} = \frac{3.16}{23.38} = 0.135$$

Table 4.4: $\varnothing = 40°$; $N_\gamma = 93.6$ (according to Meyerhof)

$$N'_\gamma = \frac{N_\gamma}{2} + \frac{N_\gamma}{2}\left[R + \frac{b}{2B}(1-R)\right] = \frac{93.6}{2} + \frac{93.6}{2}\left[0.135 + \frac{0.8}{2\times 1.2}(1-0.135)\right]$$

$$N'_\gamma = 66.61$$

$$gross\ q_{\text{ult}} = 20\times 1.2\times 39.2 + 0.5\times 20\times 1.2\times 66.61 = 1740\ \text{kPa}$$

(b) Meyerhof method.

Equation (4.50): $$gross\ q_{\text{ult}} = \frac{1}{2}\gamma B N_{\gamma q}$$

Figure 4.14b: $\frac{b}{B} = \frac{0.8}{1.2} = 0.75$; $\frac{D}{B} = \frac{1.2}{1.2} = 1$; $\beta = 30°$; $\varnothing = 40°$; $N_{\gamma q} = 113$.

$$gross\ q_{\text{ult}} = \frac{1}{2}\gamma B N_{\gamma q} = 0.5\times 20\times 1.2\times 113 = 1356\ \text{kPa}$$

(c) Stress characteristic solution method.

Equation (4.50): $$gross\ q_{\text{ult}} = \frac{1}{2}\gamma B N_{\gamma q}$$

Figure 4.18a, b: $\frac{b}{B} = \frac{0.8}{1.2} = 0.75$; $\frac{D}{B} = \frac{1.2}{1.2} = 1$; $\beta = 30°$; $\varnothing = 40°$:

$N_{\gamma q} = 170$ (by interpolation).

$$gross\ q_{\text{ult}} = \frac{1}{2}\gamma B N_{\gamma q} = 0.5\times 20\times 1.2\times 170 = \mathit{2040\ kPa}$$

It may be useful to note that the Bowels solution gave a result which is nearly equal to the average of the other two results.

Problem 4.31

Refer to Problem 4.29. Assume a $c-\varnothing$ soil with $c = 25$ kPa, $\varnothing = 30°$ and $\delta = 20°$ for computing K_P. Use 1.2×1.2 m square footing; other data remain unchanged. Determine the gross ultimate bearing capacity using the Bowles method. Use the Hansen bearing capacity and shape factors.

Solution:

Equation (4.53): $$gross\ q_{\text{ult}} = cN'_c s_c i_c + \bar{q}N'_q s_q i_q + 0.5\gamma' BN'_\gamma s_\gamma i_\gamma$$

All load-inclination (i) factors are one, since the foundation load is vertical ($H_i = 0$). Hence,

$$\text{gross}\ q_{\text{ult}} = cN'_c s_c + \bar{q}N'_q s_q + 0.5\gamma' BN'_\gamma s_\gamma$$

Table 4.4: $\varnothing = 30°$; $N_c = 30.13$; $N_q = 18.4$; $N_\gamma = 15.1$

$$s_c = 1 + \frac{N_q}{N_c}\left(\frac{B'}{L'}\right) = 1 + \frac{18.4}{30.13}\left(\frac{1.2}{1.2}\right) = 1.61$$

Table 4.6: $$s_q = 1 + \left(\frac{B'}{L'}\right)\sin\varnothing = 1 + \left(\frac{1.2}{1.2}\right)\sin 30 = 1.5$$

$$s_\gamma = 1 - 0.4\left(\frac{B'}{L'}\right) = 1 - 0.4\left(\frac{1.2}{1.2}\right) = 0.6$$

(Continued)

Table 4.11: $\frac{b}{B} = \frac{0.8}{1.2} = 0.75$; $\frac{D}{B} = \frac{1.2}{1.2} = 1$; $\beta = 30°$; $\varnothing = 30°$; $\delta = 20°$;

$N'_c = 30.14$; $N'_q = 15.03$ (by interpolation).

$\frac{b}{B} = 0.75 < 2$. Hence, the N_γ factor should be adjusted as follows:

Determine the passive earth pressure coefficients using $+\beta$ for $K_{P,max}$ and $-\beta$ for $K_{P,min}$.

$$K_P = \frac{\sin^2(\alpha - \varnothing)}{\sin^2\alpha \sin(\alpha+\delta)\left[1 - \sqrt{\frac{\sin(\varnothing+\delta)\sin(\varnothing+\beta)}{\sin(\alpha+\delta)\sin(\alpha+\beta)}}\right]^2}$$

$$\alpha = 90°;\ \delta = 20°$$

For $\varnothing = 30°$; $\delta = 20°$; $\alpha = 90°$; $\beta = +30°$: $K_{P,max} = 84.65$
For $\varnothing = 30°$; $\delta = 20°$; $\alpha = 90°$; $\beta = -30°$: $K_{P,min} = 0.8$

$$R = \frac{K_{min}}{K_{max}} = \frac{0.8}{84.65} = 0.0095$$

$$N'_\gamma = \frac{N_\gamma}{2} + \frac{N_\gamma}{2}\left[R + \frac{b}{2B}(1-R)\right] = \frac{15.1}{2} + \frac{15.1}{2}\left[0.0095 + \frac{0.8}{2\times1.2}(1-0.0095)\right] = 10.11$$

$$gross\ q_{ult} = 25\times30.14\times1.61 + 17.5\times1.2\times15.03\times1.5$$
$$+0.5\times17.5\times1.2\times10.11\times0.6 = \mathit{1750\ kPa}$$

Problem 4.32

A strip footing is located on the slope shown in the scheme below. What is the net safe bearing capacity using the Bowles method with a safety factor $SF = 3.5$? Use the Hansen bearing capacity and shape factors. Assume $\delta = 20°$ for computing K_P.

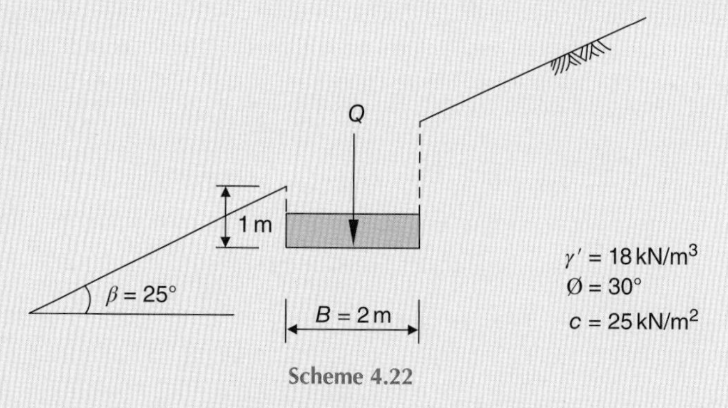

Scheme 4.22

Solution:
Equation (4.53): $$gross\ q_{ult} = cN'_c s_c i_c + \bar{q}N'_q s_q i_q + 0.5\gamma' B N'_\gamma s_\gamma i_\gamma$$

For strip footings, all shape (s) factors are one and in this case since the foundation load is vertical ($H_i = 0$), all load inclination (i) factors are also one.

$$gross\ q_{ult} = cN'_c + \bar{q}N'_q + 0.5\gamma' B N'_\gamma$$

Table 4.11: $\frac{b}{B}=\frac{0}{2}=0;\ \frac{D}{B}=\frac{1}{2}=0.5;\ \beta=25^\circ;\ \varnothing=30^\circ;$
$N'_c=26.63;\ N'_q=12.13.$ (by interpolation)

$\frac{b}{B}=0<2$. Hence, the N_γ factor should be adjusted as follows.

Determine the Coulomb passive earth pressure coefficients using $+\beta$ for $K_{P,\max}$ and $-\beta$ for $K_{P,\min}$.

$$K_P=\frac{\sin^2(\alpha-\varnothing)}{\sin^2\alpha\sin(\alpha+\delta)\left[1-\sqrt{\frac{\sin(\varnothing+\delta)\sin(\varnothing+\beta)}{\sin(\alpha+\delta)\sin(\alpha+\beta)}}\right]^2}$$

$\alpha=90^\circ$ and $\delta=20^\circ$ (assumed).

For $\varnothing=30^\circ;\ \delta=20^\circ;\ \alpha=90^\circ;\ \beta=+25^\circ$: $K_{P,\max}=39.79$
For $\varnothing=30^\circ;\ \delta=20^\circ;\ \alpha=90^\circ;\ \beta=-25^\circ$: $K_{P,\min}=1.54$

$$R=\frac{K_{\min}}{K_{\max}}=\frac{1.54}{39.79}=0.039$$

Table 4.4: $\varnothing=30^\circ;\ N_\gamma=15.1$

$$N'_\gamma=\frac{N_\gamma}{2}+\frac{N_\gamma}{2}\left[R+\frac{b}{2B}(1-R)\right]$$

$$=\frac{15.1}{2}+\frac{15.1}{2}\left[0.039+\frac{1}{2\times2}(1-0.039)\right]=9.66$$

$$gross\ q_{ult}=25\times26.63+18\times1\times12.13+0.5\times18\times2\times9.66$$

$$gross\ q_{ult}=1058\text{ kPa}$$

$$net\ q_{ult}=gross\ q_{ult}-\gamma' D_f=1058-18\times1=1040\text{ kPa}$$

$$net\ q_s=\frac{net\ q_{ult}}{SF}=\frac{1040}{3.5}=297\ kPa$$

Problem 4.33

A square foundation in a sand deposit measures 1.22 × 1.22 m in plan. Given: $D_f=1.52$ m, soil friction angle $\varnothing=35^\circ$ and soil unit weight $\gamma=17.6\ \text{kN/m}^3$. Estimate the ultimate uplift capacity of the foundation.

Solution:

Table 4.12: $\varnothing=35^\circ;\ \left(\frac{D_f}{B}\right)_{cr}=5,\ m=0.25,\ K_u=0.936$

$\frac{D_f}{B}=\frac{1.52}{1.22}=1.25<\left(\frac{D_f}{B}\right)_{cr}$. Hence, it is a shallow foundation.

Equation (4.81):

$$F_q=1+2\left[1+m\left(\frac{D_f}{B}\right)\right]\left(\frac{D_f}{B}\right)K_u\tan\varnothing'$$

$$=1+2[1+0.25(1.25)](1.25)(0.936\times\tan35)=3.15$$

(Continued)

Equation (4.80): $F_q = \frac{Q_u}{A \gamma D_f}$

The ultimate uplift capacity of the foundation is

$$Q_u = F_q(A \gamma D_f) = (3.15)(1.22 \times 1.22 \times 17.6 \times 1.52) = 125.4\ kN$$

Problem 4.34

A circular foundation of 1.5 m diameter is constructed in a sand deposit. Given: $D_f = 1.5\,\text{m}$, soil friction angle $\text{Ø} = 35°$ and soil unit weight $\gamma = 17.4\,\text{kN/m}^3$. Estimate the ultimate uplift capacity of the foundation.

Solution:

Table 4.12: $\text{Ø} = 35°;\ \left(\frac{D_f}{B}\right)_{cr} = 5,\ m = 0.25,\ K_u = 0.936$

$\frac{D_f}{B} = \frac{1.5}{1.5} = 1 < \left(\frac{D_f}{B}\right)_{cr}$. Hence, it is a shallow foundation

Equation (4.81):

$$F_q = 1 + 2\left[1 + m\left(\frac{D_f}{B}\right)\right]\left(\frac{D_f}{B}\right)K_u \tan \text{Ø}'$$

$$= 1 + 2[1 + 0.25(1)](1)(0.936 \times \tan 35) = 2.64$$

Equation (4.80):

$$F_q = \frac{Q_u}{A \gamma D_f}$$

The ultimate uplift capacity of the foundation is

$$Q_u = F_q(A \gamma D_f) = (2.64)\left(\frac{\pi}{4} \times 1.5^2 \times 17.4 \times 1.5\right) = 121.8\ kN$$

Problem 4.35

A rectangular foundation in a saturated clay deposit measures 1.5 m × 3.0 m in plan. The data given: $D_f = 1.8\,\text{m}$, $c_u = 52\,\text{kPa}$ and soil unit weight $\gamma = 18.9\,\text{kN/m}^3$. Estimate the ultimate uplift capacity of the foundation.

Solution:

Equation (4.85):

$$\left(\frac{D_f}{B}\right)_{cr} = 0.107\, c_u + 2.5 \le 7 \quad \text{(for square footings)}$$

$$= 0.107 \times 52 + 2.5 = 8.1 > 7. \text{ Hence, use } 7$$

Equation (4.86):

$$\left(\frac{D_f}{B}\right)_{cr} = \left(\frac{D_f}{B}\right)_{cr,\ sq}\left[0.73 + 0.27\left(\frac{L}{B}\right)\right] \le 1.55\left(\frac{D_f}{B}\right)_{cr,\ sq} \quad \text{(for rectangular footings)}$$

$$= 7\left[0.73 + 0.27\left(\frac{3}{1.5}\right)\right] = 8.89$$

Check:

$$1.55\left(\frac{D_f}{B}\right)_{\mathrm{cr,\ sq}} = 1.55\times 7 = 10.85 > \left(\frac{D_f}{B}\right)_{\mathrm{cr}}. \text{ Hence, } \left(\frac{D_f}{B}\right)_{\mathrm{cr}} = 8.89$$

$$\frac{D_f}{B} = \frac{1.8}{1.5} = 1.2 < \left[\left(\frac{D_f}{B}\right)_{cr} = 8.89\right]. \text{ Hence, it is a shallow foundation.}$$

$$\alpha' = \frac{D_f/B}{(D_f/B)_{\mathrm{cr}}} = \frac{1.2}{8.89} = 0.135$$

From the average curve of Figure 4.28, for $\alpha' = 0.135$ obtain $\beta' = 0.2$
Equation (4.88):

$$Q_u = A\left\{B'\left[7.56 + 1.44\left(\frac{B}{L}\right)\right]c_u + \gamma D_f\right\}$$

$$= (1.5\times 3)\left\{0.2\left[7.56 + 1.44\left(\frac{1.5}{3}\right)\right](52) + (18.9\times 1.8)\right\}$$

$$= 540.6\,\mathrm{kN}$$

The ultimate uplift capacity of the foundation $= Q_u = 540.6\,kN$

Problem 4.36

A foundation measuring 1.2 × 2.4 m in plan is constructed in a saturated clay deposit. The data given: $D_f = 2$ m, $c_u = 74\,\mathrm{kPa}$ and soil unit weight $\gamma = 18\,\mathrm{kN/m^3}$. Estimate the ultimate uplift capacity of the foundation.

Solution:

Equation (4.85):

$$\left(\frac{D_f}{B}\right)_{\mathrm{cr}} = 0.107\,c_u + 2.5 \le 7 \quad (\text{for square footings})$$

$$= 0.107\times 74 + 2.5 = 10.4 > 7. \text{ Hence, use 7}$$

For rectangular footings, Equation (4.86) gives

$$\left(\frac{D_f}{B}\right)_{\mathrm{cr}} = \left(\frac{D_f}{B}\right)_{\mathrm{cr,\ sq}}\left[0.73 + 0.27\left(\frac{L}{B}\right)\right] \le 1.55\left(\frac{D_f}{B}\right)_{\mathrm{cr,\ sq}}$$

$$= 7\left[0.73 + 0.27\left(\frac{2.4}{1.2}\right)\right] = 8.89$$

Check:

$$1.55\left(\frac{D_f}{B}\right)_{\mathrm{cr,\ sq}} = 1.55\times 7 = 10.85 > \left(\frac{D_f}{B}\right)_{\mathrm{cr}}. \text{ Hence, use } \left(\frac{D_f}{B}\right)_{\mathrm{cr}} = 8.89$$

$$\frac{D_f}{B} = \frac{2}{1.2} = 1.67 < \left[\left(\frac{D_f}{B}\right)_{cr} = 8.89\right]. \text{ Hence, it is a shallow foundation.}$$

$$\alpha' = \frac{D_f/B}{(D_f/B)_{\mathrm{cr}}} = \frac{1.67}{8.89} = 0.188$$

(*Continued*)

From the average curve of Figure 4.28, for $\alpha' = 0.188$ obtain $\beta' \approx 0.28$
Equation (4.88):

$$Q18_u = A\left\{B'\left[7.56 + 1.44\left(\frac{B}{L}\right)\right]c_u + \gamma D_f\right\}$$

$$= (1.2\times 2.4)\left\{0.28\left[7.56 + 1.44\left(\frac{1.2}{2.4}\right)\right](74) + (18\times 2)\right\} = 598 \text{ kN}$$

The ultimate uplift capacity of the foundation $= Q_u = 598\,\text{kN}$

Problem 4.37

A composite of the several site borings gave the average soil profile shown below. A square mat, estimated to be in the order of 15 m in plan dimensions, is to be located at $D_f \cong 1.5$ m. Estimate the net allowable bearing capacity which will be recommended for design of the mat foundation.

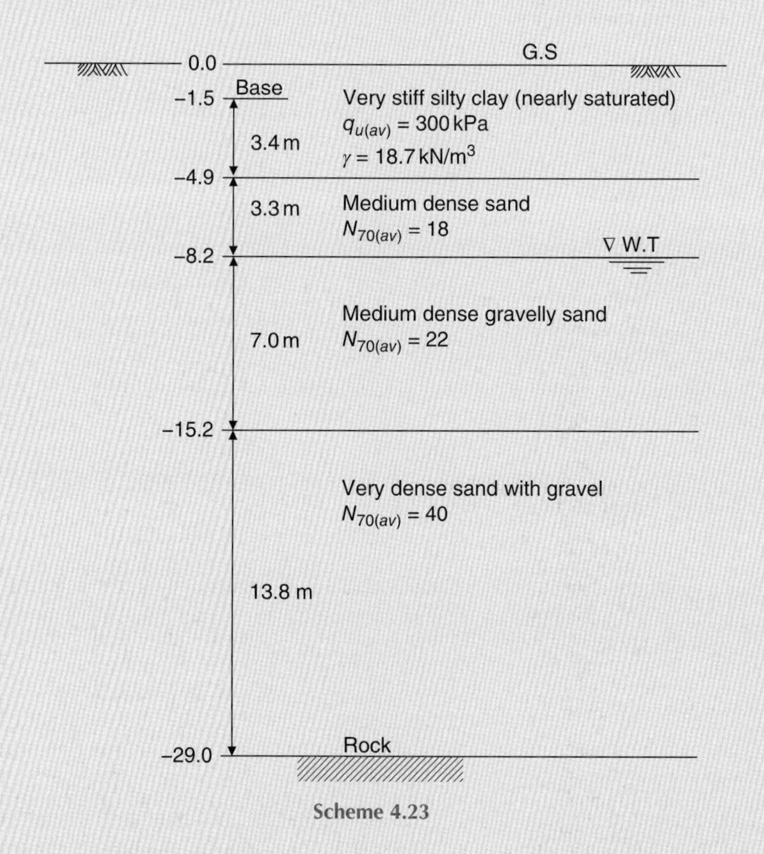

Scheme 4.23

Solution:
It is noteworthy that the very limited data given in this problem are basically the type a geotechnical consultant would have on which to make a design soil pressure recommendation.

Usually, in the analysis of such design problems, a net safe bearing capacity (*net* q_s) is found first and then the final settlement due to a net effective foundation pressure, using *net* q' = *net* q_s, is computed and compared to the desired maximum allowable settlement (not specified in this problem). If the computed settlement is considered acceptable, the net allowable bearing capacity (*net* q_a) will be recommended equal to *net* q_s; otherwise, a lower value for *net* q_a should be selected so that excessive settlement will not occur.

(A) Bearing capacity analysis:
Find *net* q_s with a safety factor $=3$ for the supporting saturated silty clay soil. Assume undrained condition, $\varnothing = 0°$ and use the Hansen bearing capacity equations.

Table 4.2: $$gross\ q_{\text{ult}} = 5.14\, s_u \left(1 + s_c' + d_c' - i_c' - b_c' - g_c'\right) + q'$$

Table 4.7: for the conditions of this problem all i, g and b factors arezero.

Equation (4.3): $$net\ q_{\text{ult}} = gross\ q_{\text{ult}} - \gamma' D_f$$
$$= 5.14\, s_u\left(1 + s_c' + d_c'\right)$$

Table 4.6:
$$s_c' = 0.2\frac{B'}{L'} = 0.2 \times \frac{15}{15} = 0.2$$
$$d_c' = 0.4k;\ k = \frac{D}{B} \text{ for } \frac{D}{B} \le 1;\ \frac{D}{B} = \frac{1.5}{15} = 0.1;\ \text{hence, } d_c' = 0.04$$

$$s_u = \frac{q_u}{2} = \frac{300}{2} = 150\,\text{kPa} \quad (\varnothing = 0°)$$
$$net\ q_{\text{ult}} = 5.14\,(150)(1 + 0.2 + 0.04) = 956.04\ \text{kPa}$$
$$net\ q_s = \frac{net\ q_{\text{ult}}}{SF} = \frac{956.04}{3} = 319\ \text{kPa}$$

(B) Settlement analysis:
Find the final settlement due to a net effective foundation pressure $(net\ q') = net\ q_s = 319$ kPa.

(a) The very stiff saturated silty clay layer:
Assume the secondary compression settlement is neglected (which is common to do so).

Equation (3.3): $$S_T = S_i + S_c$$

In this Problem, consolidation data for the clay is not given. Therefore, some assumptions and approximations may be necessary to be made.

Assume, for the very stiff saturated clay, $S_c = S_i = 0.5\, S_T$, approximately (see Section 3.4.2) and Poisson's $\mu = 0.5$ (see Table 3.10). Also, from Table 3.8, select $E_s = (500 - 1500)\, s_u$ and assume $E_s = 500\, s_u$. Hence, $E_s = 500\, s_u = 500 \times 150/1000 = 75$ MPa (it is the same average value given in Table 3.9).

Equation (3.5): $$S_i = q \times B \times \frac{1-\mu^2}{E_s} \times m \times I_S \times I_F$$

For settlement below the *centre* of the mat:

$$m = 4;\ B = 0.5(\text{width}) = \frac{15}{2} = 7.5\,\text{m};\ L = 0.5(\text{length}) = \frac{15}{2} = 7.5\ \text{m};$$
$$M = \frac{0.5(\text{width})}{0.5(\text{length})} = \frac{7.5}{7.5} = 1;\ N = \frac{H}{0.5(\text{width})} = \frac{3.4}{7.5} = 0.453$$

Table 3.7: $M = 1$; $N = 0.453$; $I_1 = 0.042$; $I_2 = 0.066$

$$I_S = I_1 + \frac{1-2\mu}{1-\mu} I_2 = 0.042 + \frac{1 - 2 \times 0.5}{1 - 0.5} \times 0.066 = 0.042$$

Figure 3.2: $\frac{D}{B} = \frac{1.5}{15} = 0.1;\ \frac{B}{L} = \frac{15}{15} = 1;\ \mu = 0.5;\ I_F = 0.89$

(Continued)

Use the average q at middle of the clay layer below the base = *net* q_s, since the ratios $\frac{B}{z}$ and $\frac{L}{z}$ are so large that the value of $(4I)$ is nearly one (see Figure 2.32 or Table 2.3).

$$S_i = 319 \times 7.5 \times \frac{1-(0.5)^2}{75000} \times 4 \times 0.042 \times 0.89 = 0.0036 \text{ m} = 3.6\,\text{mm}$$

$$S_{T,\,clay} = 2 \times 3.6 = 7.2\,\text{mm}$$

(b) The sand layers:
The depth to the rock is less than $2B$; therefore, all the three sand layers should be included in the settlement computations. It may be sufficient if the weighted average of the estimated E_s values is obtained and used in Equation (3.5) only once. For each sand layer, assume $E_s = 500(N_{55} + 15)$. This equation is the most conservative equation given in Table 3.8.
Determine N_{55} and compute the *weighted average* value of E_s, as shown in Table 4.21.

Table 4.21 Values of N_{55} and E_s.

Layer	N_{70}	$N_{55} = N_{70}$ (70/55)	E_s, kPa	H, m	E_sH
I	18	23	19 000	3.3	62 700
II	22	28	21 500	7.0	150 500
III	40	51	33 000	13.8	455 400
			Σ =	24.1	668 600

$$E_{s,(\text{av})} = \frac{\sum E_s H}{\sum H} = \frac{668\ 600}{24.1} = 27\ 743\,\text{kPa}$$

Determine $q_{(\text{av})}$ at the middle of the three sand layers together due to the 319 kPa pressure acting at top of the first sand layer (the same net contact pressure is, conservatively, used here also because thickness of the clay layer is small compared to the base dimensions).

Figure 2.32: $z = \frac{24.1}{2} = 12.05\,\text{m}$; $\frac{B/2}{z} = \frac{L/2}{z} = \frac{15/2}{12.05} = 0.622$; $I = 0.115$

Therefore, $q_{(\text{av})} = 319(4I) = 319 \times 4 \times 0.115 = 147\,\text{kPa}$

For settlement below the *centre* of the mat:

$$m = 4;\ B = 0.5(\text{width}) = \frac{15}{2} = 7.5\,\text{m};\ L = 0.5(\text{length}) = \frac{15}{2} = 7.5\,\text{m};$$

$$M = \frac{0.5(\text{width})}{0.5(\text{length})} = \frac{7.5}{7.5} = 1;\ N = \frac{H}{0.5(\text{width})} = \frac{24.1}{7.5} = 3.213$$

Table 3.7: $M = 1$; $N = 3.213$; $I_1 = 0.373$; $I_2 = 0.046$

Estimate $\mu = 0.3$ for all sand layers (see Table 3.10). Hence,

$$I_S = I_1 + \frac{1-2\mu}{1-\mu} I_2 = 0.373 + \frac{1-2\times 0.3}{1-0.3} \times 0.046 = 0.399$$

Figure 3.2: $\frac{D}{B} = \frac{4.9}{15} = 0.33$; $\frac{B}{L} = \frac{15}{15} = 1$; $\mu = 0.3$; $I_F = 0.78$

$$S_i = 147 \times 7.5 \times \frac{1-(0.3)^2}{27743} \times 4 \times 0.399 \times 0.78 = 0.045\,\text{m} = 45\,\text{mm}$$

The common value of the allowable total settlement for mat or raft foundations on clay is on the order of about 50 mm (Tables 3.1 and 3.2). The final settlement $= 7.2 + 45 = 52.2\,\text{mm} > 50.0\,\text{mm}$. However, the difference is so small that it can be neglected and therefore the geotechnical engineer may recommend a design net soil pressure (net allowable bearing capacity) $= 300\,kPa < (net\ q_s = 319\,\text{kPa})$. Other geotechnical engineers may be in favour of recommending an even smaller value, such as *250 kPa* for example, on the basis that the available data are very limited and the selected values for E_s and μ are not sufficiently accurate.

Problem 4.38

The retaining wall shown in the scheme below provides lateral support for an existing car park adjacent to a new pavement in a road-widening project. The soil beneath the footing is 3 m clay with average $s_u = 120\,\text{kPa}$ and $m_v = 12 \times 10^{-4}\,\text{m}^2/\text{kN}$, underlain by a very dense silty sand deposit. It is required that the safety factor against shear failure not to be less than $SF = 3$ and the maximum settlement not to exceed 50 mm. Check the safety of the foundation with respect to these two geotechnical design criteria.

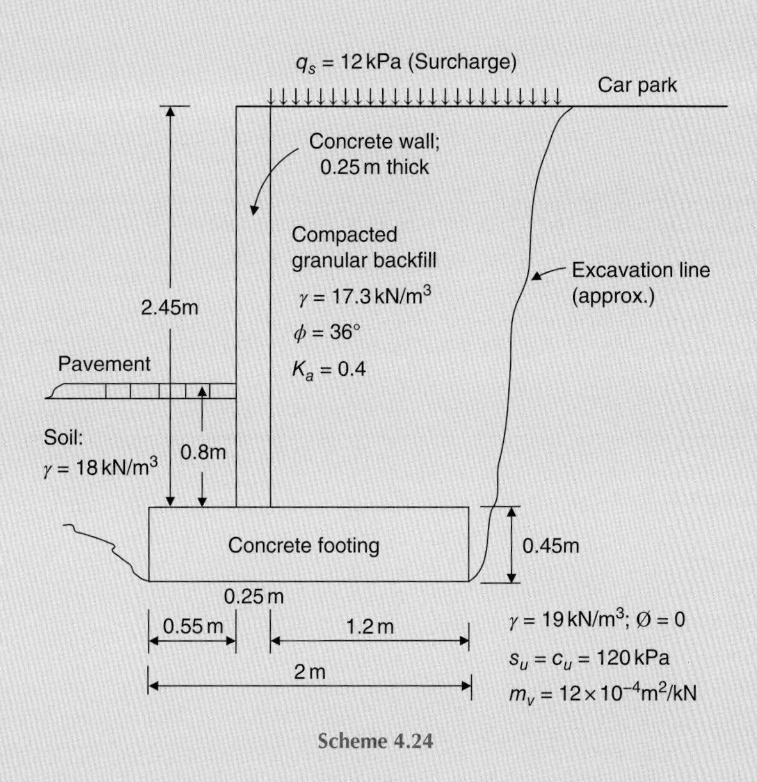

Scheme 4.24

Solution:

(1) Safety against bearing capacity failure.

(a) Determine the horizontal force P_a and its overturning moment M_o due to the *active lateral earth pressures*.

$$q_z = (q_s + \gamma z)K_a = (12 + 17.3z)(0.4)$$

(Continued)

At top: $q_z = (12 + 17.3 \times 0)(0.4) = 4.8\,\text{kPa}$

At base: $q_z = [12 + 17.3(2.45 + 0.45)](0.4) = 24.87\,\text{kPa}$

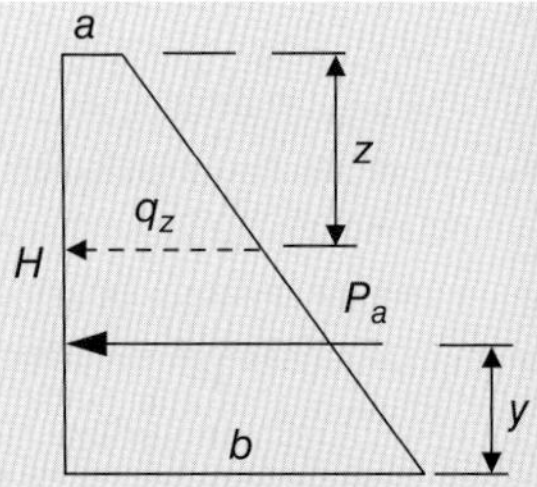

$$P_a = \frac{4.8 + 24.87}{2} \times 2.9 = 43.02\,\text{kN/m run}$$

For trapezoid area: $y = \frac{H}{3}\left(\frac{2a + b}{a + b}\right) = \frac{2.9}{3}\left(\frac{2 \times 4.8 + 24.87}{4.8 + 24.87}\right) = 1.123\ \text{m}$

Scheme 4.25

$$M_o = y \times P_a = 1.123 \times 43.02 = 48.31\,\text{kN.m/m run}$$

(b) Determine all vertical forces and their moments (resisting moment, M_r) about the toe of the base as in Table 4.22.

Table 4.22 Vertical forces and moments (resisting moment, M_r) about the toe of the base.

Part	Weight (kN)	Arm (m)	M_r (kN.m)
Soil above toe	0.55 (18 × 0.8) = 7.92	0.275	2.18
Soil above heel	1.2(17.3 × 2.45 + 12) = 65.26	1.400	91.37
Stem	24(0.25 × 2.45) = 14.70	0.675	9.92
Base slab	24 (0.45 × 2) = 21.60	1.000	21.60
	$R = \sum W = 109.48$	$\sum M_r = 125.07$	

(c) Determine eccentricity e and effective width B'.

The net moment $M_{net} = \Sigma M_r - M_o = 125.07 - 48.31 = 76.76\,\text{kN.m}$

$$\bar{x} = \frac{M_{net}}{R} = \frac{76.76}{109.48} = 0.70\ \text{m; hence, } e = \frac{B}{2} - \bar{x} = \frac{2}{2} - 0.7 = 0.3\ \text{m} < \frac{B}{6}$$

The effective width $B' = B - 2e_B = 2 - 2 \times 0.3 = 1.4\ \text{m}$

(d) Determine *net* q_{ult} and *SF* against bearing capacity failure.

Assume using the Hansen bearing capacity equation for the condition $\varnothing = 0$:

Table 4.2: $gross\ q_{\text{ult}} = 5.14\, s_u \left(1 + s'_c + d'_c - i'_c - b'_c - g'_c\right) + q'$

Table 4.6: $s'_c = 0.2\frac{B'}{L'} = 0$, since $\frac{B'}{L'} \approx 0$ for continuous foundations.

Table 4.7: $b'_c = g'_c = 0$, since $\eta = 0$ and $\beta = 0$

$$d'_c = 0.4k;\ k = \frac{D}{B} \text{ for } \frac{D}{B} \le 1;\ \frac{D}{B} = \frac{1.25}{2} = 0.625;\ \text{hence, } d'_c = 0.25$$

$$i'_c = 0.5 - 0.5\sqrt{1 - \frac{H}{A_f c_a}}$$

$A_f = B' \times 1; H = P_a$. Assume $c_a = 0.7c_u = 0.7 \times 120 = 84\,\text{kPa}$

$$i'_c = 0.5 - 0.5\sqrt{1 - \frac{43.02}{1.4 \times 1 \times 84}} = 0.102$$

$$gross\ q_{ult} = 5.14\, s_u\,(1 + d'_c - i'_c) + \bar{q}$$

$$= 5.14\,(120)\,(1 + 0.25 - 0.102) + 18 \times 1.25$$

$$= 730.59\,\text{kPa}$$

Equation (4.3):

$$net\ q_{ult} = gross\ q_{ult} - \gamma' D_f$$

$$= 730.59 - 18 \times 1.25 = 708.09\,\text{kPa}$$

Net uniform foundation pressure, $net\ q = \frac{R}{B' \times 1} - \bar{q} = \frac{109.48}{1.4 \times 1} - 18 \times 1.25$

$$= 55.7\,\text{kPa}$$

$$SF = \frac{net\ q_{ult}}{net\ q} = \frac{708.9}{55.7} = 12.7 \gg 3$$

Check the maximum foundation pressure, q_{max}:

$$q_{max} = \frac{R}{A} + \frac{(R \times e)\left(\frac{B}{2}\right)}{(1)(B)^3(1/12)} = \frac{109.48}{1 \times 2} + \frac{(109.48 \times 0.3)\left(\frac{2}{2}\right)}{(1)(2)^3(1/12)} = 104.01\,\text{kPa}$$

$$gross\ q_{safe} = \frac{gross\ q_{ult}}{SF} = \frac{730.59}{3} = 243.53\,\text{kPa} \gg 104.01\,\text{kPa}$$

The foundation is very safe against bearing capacity failure.

(2) Safety against excessive settlement:

Assume average uniform net contact pressure $= \frac{R}{A} - \bar{q}$

$$= \frac{109.48}{1 \times 2} - 18 \times 1.25$$

$$= 32.24\ \text{kPa}$$

$\frac{B/2}{z} = \frac{1/2}{1.5} = 0.33,\ \frac{L/2}{z} = \frac{2/2}{1.5} = 0.67$. From Figure 2.32 obtain $I \approx 0.075$

At centre of the clay layer

$$\sigma'_z = 32.24(4I)$$

$$= 32.24 \times 4 \times 0.07 = 10\,\text{kPa}$$

Equation (3.34):

$$S = (h)(m_v)(\sigma'_z) = (3)(12 \times 10^{-4})(10)$$

$$= 0.036\,\text{m} = 36\,\text{mm} < 50\,mm$$

The foundation is safe against excessive settlement.

Problem 4.39

Refer to Problem 3.56. Assume the footing dimensions are not known and the column carries a total net load of 5000 kN. Unit weight γ of the rock is 22 kN/m^3. The foundation design requires a minimum safety factor against shear failure = 3 and a maximum tolerable settlement equals 30 mm. Determine suitable dimensions for the column footing.

Solution:

Assume a square footing with a trial width $B = 1.5$ m.

Refer to the discussions of Article 4.18 and Tables 4.13, 4.14 and 4.15. For the given type of the jointed rock, estimate a suitable value for $\varnothing = 30°$. Also, for $\varnothing = 30°$ and *RQD* range of 0–70, the cohesion *c* may be estimated equals $0.1\,q_c = 0.175$ MPa.

Equation (4.89):

$$N_q = \tan^6\left(45° + \frac{\varnothing}{2}\right) = \tan^6\left(45° + \frac{30}{2}\right) = 27$$

Figure 4.30: $\varnothing = 30°$; $N_q = 9$

Use $N_{q,\text{ave}} = \dfrac{27+9}{2} = 18$

Equation (4.89):

$$N_c = 5\tan^4\left(45° + \frac{\varnothing}{2}\right) = 5\tan^4\left(45° + \frac{30}{2}\right) = 45$$

Figure 4.30: $\varnothing = 30°$; $N_c = 14$

Use $N_{c,\text{ave}} = \dfrac{45+14}{2} = 29$

Equation (4.89):

$$N_\gamma = N_q + 1 = 27 + 1 = 28$$

Figure 4.30: $\varnothing = 30°$; $N_\gamma = 15$

Use $N_{\gamma,\text{ave}} = \dfrac{28+15}{2} = 21$

Table 4.2:

$$gross\ q_{\text{ult}} = c N_c s_c + \bar{q} N_q + \frac{1}{2}\gamma' B N_\gamma s_\gamma$$

$$net\ q_{\text{ult}} = c\ N_c s_c + \bar{q}(N_q - 1) + \frac{1}{2}\gamma' B N_\gamma s_\gamma$$

$$= 0.175 \times 1000 \times 29 \times 1.3 + (22 \times 0.6)(18 - 1) + \frac{1}{2} \times 22 \times 1.5 \times 21 \times 0.8$$

$$= 7099 \text{ kPa}$$

Net foundation pressure $= net\ q = \dfrac{5000}{1.5 \times 1.5} = 2222$ kPa

$$SF = \frac{net\ q_{\text{ult}}}{net\ q} = \frac{7099}{2222} = 3.2 > 3$$

Therefore, try 1.5 × 1.5 m square footing.

Check settlement as follows:

$$\text{Net foundation pressure} = 2222 \ \text{kPa} = 2.222 \ \text{MPa}$$

For the 250 mm joint spacing, the fracture frequency per metre (or joint spacing per metre) is $(1000/250) = 4$. From Table 3.15, for a fracture frequency of four per metre, a suitable value for the mass factor (j) is 0.6. Also, from Table 3.16, for poorly cemented sandstone the modulus ratio (M_r) is 150. Let q_{uc} of the intact rock equals the uniaxial compression strength which is 1.75 MPa. The mass deformation modulus is given by Equation (3.66):

$$E_m = j\,M_r\,q_{uc} = 0.6 \times 150 \times 1.75 = 157.5 \ \text{MPa}$$

Assume the influence depth of the rock below the foundation level is $H = 4B = 4 \times 1.5 = 6$ m.
Divide the footing area into four equal 0.75 × 0.75 m squares.

Figure 3.25: $\frac{L}{B} = \frac{0.75}{0.75} = 1$; $\frac{H}{B} = \frac{6}{0.75} = 8$; $k = 0$ (using $E_m = E_d = E_f$), obtain $I'_p \approx 0.46$

Figure 3.26a: $\frac{H}{B} = \frac{6}{1.5} = 4$; $F_B \approx 1$

Figure 3.26b: $\frac{D}{B} = \frac{0.6}{1.5} = 0.4$; $F_D = 0.925$

Immediate settlement at the corner of each flexible square is given by

Equation (3.63):

$$\rho_i = q\left(\frac{B}{E_f}\right) I'_p F_B F_D$$

$$= (2.222)\left(\frac{0.75}{157.5}\right)(0.46)(1)(0.925) = 4.5 \times 10^{-3}\,\text{m}$$

Immediate settlement at the centre of the flexible raft is

$$4\left(4.5 \times 10^{-3}\right) = 0.018\,\text{m}$$

Equation (3.6):

$$S_{i(\text{rigid})} = 0.93 S_{i(\text{flexible})}$$

$$= 0.93 \times 0.018 = 0.017 \ \text{m} = 17\,\text{mm}$$

Table 3.12: gives rigidity factor = 0.85. Hence,

$$S_{i(\text{rigid})} = 0.85 \times 0.018 = 0.015 \ \text{m} = 15\,\text{mm}$$

Average of the two settlement values = 16 mm < 30 *mm*; O.K.
However, creep might increase the settlement to about the maximum tolerable settlement of 30 mm in the long term.
Use 1.5 m × 1.5 m square footing.

Problem 4.40

A column footing 1.6 × 1.6 m is required to be founded in a moderately strong rock mass of limestone. An embedment depth of 1.5 m is required to get through the top soil and weathered rock zone. Tests on rock cores taken below the foundation level showed representative unconfined compression strength $q_{uc} = 24$ MPa; other tests gave average $RQD = 0.5$; friction angle $\varnothing = 38°$ and rock unit weight $\gamma = 24.5$ kN/m^3. Neglect the relatively thin top soil layer which overlies the rock. For the footing concrete use $f'_c = 28$ MPa. Assume a safety factor against bearing capacity failure = 4. Estimate the gross safe bearing capacity for the foundation, taking into consideration the effect of discontinuities of the rock.

Solution:

Table 4.2: $gross\ q_{\text{ult}} = c\,N_c s_c + \bar{q} N_q + \frac{1}{2}\gamma' B N_\gamma s_\gamma$

(*Continued*)

Equation (4.89):

$$N_c = 5\tan^4\left(45^\circ + \frac{38}{2}\right) = 88.36$$

$$N_q = \tan^6\left(45^\circ + \frac{38}{2}\right) = 74.29$$

$$N_\gamma = N_q + 1 = 74.29 + 1 = 75.29$$

Equation (4.90):

$$q_{uc} = 2c' \tan\left(45^\circ + \frac{\emptyset}{2}\right)$$

$$c = c' = \frac{q_{uc}}{2\tan\left(45^\circ + \frac{\emptyset}{2}\right)} = \frac{24}{2\tan\left(45^\circ + \frac{38}{2}\right)} = 5.85\ \text{MPa} = 5850\ \text{kPa}$$

$$s_c = 1.3;\ s_\gamma = 0.8;\ B = 1.6\ \text{m};\ \bar{q} = 1.5 \times 24.5 = 36.75\ \text{kPa}$$

$$\begin{aligned} gross\ q_{\text{ult}} &= 5850 \times 88.36 \times 1.3 + 36.75 \times 74.29 \\ &+ \frac{1}{2} \times 24.5 \times 1.6 \times 75.29 \times 0.8 \\ &= 671978 + 2730 + 1181 \\ &= 675889\ \text{kPa} = 675.889\ \text{MPa} \end{aligned}$$

Equation (4.91):

$$\begin{aligned} q'_{\text{ult}} &= q_{\text{ult}}\,(RQD)^2 \\ &= (675.889)\,(0.5)^2 = 169\ \text{MPa} \end{aligned}$$

$$\text{gross safe bearing capacity} = \frac{q'_{ult}}{SF} = \frac{169}{4} = 42\ \text{MPa} > (f'_c = 28\ \text{MPa})$$

gross safe bearing capacity of the foundation = *28 MPa*

References

ASCE (1996), *Minimum Design Loads for Buildings and Other Structures,* ANSI/ASCE 7-95, ASCE.

Balla, A. (1961), "Resistance to Breaking Out of Mushroom Foundations for Pylons", *Proceedings of the 5th ICSMFE*, Vol. **1**, pp. 569–576.

Bjerrum, L. and Eide, O. (1956), "Stability of Strutted Excavations in Clay", *Geotechnique*, **6**, pp. 32–47.

Bowles, J. E. (1982), *Foundation Analysis and Design*, 2nd edn, McGraw-Hill, New York.

Bowles, J. E. (2001), *Foundation Analysis and Design*, 5th edn, McGraw-Hill, New York.

Coduto, Donald P. (2001), *Foundation Design: Principles and Practices*, 2nd edn, Prentice-Hall, New Jersey.

Craig, R. F. (2004), *Soil Mechanics*, 7th edn, Chapman and Hall, London.

Das, B. M. (1978), "Model Tests for Uplift Capacity of Foundations in Clay", *Soils and Foundations*, Vol. **18**, No. 2, pp. 17–24.

Das, B. M. (2011), *Principles of Foundation Engineering*, 7th edn, CENGAGE Learning, United States.

Das, Braja M. and Seeley, G. R. (1975), "Breakout Resistance of Horizontal Anchors", *Journal of Geotechnical Engineering Division*, ASCE, Vol. **101**, No. 9, pp. 999–1003.

Das, B. M. (1980), "A Procedure for Estimation of Uplift Capacity of Foundations in Clay", *Soils and Foundations*, Vol. **20**, No.1, pp. 77–82.

Das, B. M. and Jones, A. D. (1982), "Uplift Capacity of Foundations in Sand", *Transaction Research Record* 884, National Research Council, Washington, D.C., pp. 54–58.

DeBeer, E. E. (1965), "The Scale Effect on the Phenomenon of Progressive Rupture in Cohesionless Soils", *Proceedings of the 6th ICSMFE*, Vol. **2**, pp. 13–17.

Graham, J., Andrews, M. and Shields, D. H. (1988), "Stress Characteristics for Shallow Footing on Cohesionless Slopes", *Canadian Geotechnical Journal*, Ottawa, Vol. **25**, No. 2, pp. 238–249.

Hansen, J. B. (1970), "A Revised and Extended Formula for Bearing Capacity", *Danish Geotechnical Institute*, Copenhagen, Bull. No. 28, 21, pp. (successor to Bull. No. 11) and Code of Practice for Foundation Engineering, *Danish Geotechnical Institute* Bull. No. 32(1978).

Hansen, J. B. (1961), "A General Formula for Bearing Capacity", *Danish Geotechnical Institute*, Copenhagen, Bull. No. 11, pp. 38–46.

Highter, W. H. and Anders, J. C. (1985), "Dimensioning Footings Subjected to Eccentric Loads", *Journal of Geotechnical Engineering Division*, ASCE, Vol. **111**, No. GT5, pp. 659–665.

ICBO (1997), *Uniform Building Code*, International Conference of Building Officials, Whittier, CA.

ICC (2000), *International Building Code*, International Code Council.

Knappett, J. A. and Craig, R. F. (2012), *Soil Mechanics*, 8th edn, Spon Press, Abingdon, Oxon, United Kingdom.

Kulhawy, F. H. and Goodman, R. E. (1987), "Foundations in Rock", *Chapter 5 of Ground Engineering Reference Book*, ed. F. G. Bell, Butterworths, London.

Kulhawy, F. H. and Goodman, R. E. (1980), "Design of Foundations on Discontinuous Rock", *Structural Foundations on Rock*, Balkema, Rotterdam, pp. 209–220.

Merifield, R. S., Lyamin, A. and Sloan, S. W. (2003), "Three Dimensional Lower Bound Solution for the Stability of Plate Anchors in Clay", *Journal of Geotechnical and Geoenvironmental Engineering*, ASCE, Vol. **129**, No. 3, pp. 243–253.

Meyerhof, G. G. (1951), "The Ultimate Bearing Capacity of Foundations", *Geotechnique*, Vol. **2**, No. 4, pp. 301–331.

Meyerhof, G. G. (1953), "The Bearing Capacity of Foundations under Eccentric and Inclined Loads", *Third ICSMFE*, Vol. **1**, pp. 440–445.

Meyerhof, G. G. (1955), "Influence of Roughness of Base and Ground-Water Conditions on the Ultimate Bearing Capacity of Foundations", *Geotechnique*, Vol. **5**, pp. 227–242 (reprinted in Meyerhof, 1982).

Meyerhof, G. G. (1956), "Penetration Tests and Bearing Capacity of Cohesionless Soils", *JSMFD, ASCE*, Vol. **82**, SM 1, pp. 1–19.

Meyerhof, G. G. (1957), "The Ultimate Bearing Capacity of Foundations on Slopes", *Fourth ICSMFE*, Vol. **1**, pp. 384–387.

Meyerhof, G. G. (1963), "Some Recent Research on the Bearing Capacity of Foundations", *Canadian Geotechnical Journal*, Ottawa, Vol. 1, No. 1, Sept, pp. 16–26.

Meyerhof, G. G. (1965), "Shallow Foundations", *Proceedings ASCE*, **91**, No. SM2, pp. 21–31.

Meyerhof, G. G. and Adams, J. I. (1968), "The Ultimate Uplift Capacity of Foundations", *Canadian Geotechnical Journal*, Ottawa, Vol. **5**, No. 4, Nov, pp. 225–244.

Meyerhof, G. G. (1974), "Ultimate Bearing Capacity of Footings on Sand layer Overlying Clay", *Canadian Geotechnical Journal*, Ottawa, Vol. **11**, No. 2, May, pp. 223–229.

Meyerhof, G. G. and Hanna, A. M. (1978), "Ultimate Bearing Capacity of Foundations on Layered Soil under Inclined Load", *Canadian Geotechnical Journal*, Ottawa, Vol. **15**, No. 4, pp. 565–572.

Meyerhof, G. G. (1982), *Bearing Capacity and Settlement of Foundations*, Tech-Press, Technical University of Nova Scotia, Halifax.

Prakash, S. and Saran, S. (1971), "Bearing Capacity of Eccentrically Loaded Footings", *Journal of the Soil Mechanics and Foundations Division*, ASCE, Vol. **97**, No. SM1, pp. 95–117.

Prandtl, M. (1921), "Űber die Eindringungsfestigkeit (Hа̀rte) plastischer Baustoffe und die Festigkeit von Schneiden", [On the penetrating strengths (Hardness) of plastic construction materials and the strength of cutting edges], *Zeitchrift für angewandte Mathematik und Mechanik*, Vol. **1**, No.1, pp. 15–20.

Purkayastha, R. D. and Char, R. A. N. (1977), "Stability Analysis of Eccentrically Loaded Footings", *Journal of Geotechnical Engineering Division*, ASCE, Vol. **103**, No. GT6, pp. 647–651.

Reissner, H., (1924), Zum Erddruckproblem", (Concerning the earth-pressure problem), *Proceedings of the 1st International Congress of Applied Mechanics,* Delft, Netherlands, pp. 295–311.

Saran, S. and Agarwal, R. B. (1991), "Bearing Capacity of Eccentrically Obliquely Loaded Footing", *Journal of Geotechnical Engineering Division*, ASCE, Vol. **117**, No. 11, pp. 1669–1690.

Schmertmann, J. H. (1978), *Guidelines for Cone Penetration Test: Performance and Design,* Report FHWA-TS-78-209, Federal Highway Administration, Washington, D.C.

Skempton, A. W. (1951), "The Bearing Capacity of Clays", *Proceedings Building Research Congress*, **1**, pp. 180–189. London.

Stuart, J. G. (1962), "Interference between Foundations with Special Reference to Surface Footings in Sand", *Geotechnique*, London, England, Vol. **12**, No. 1, pp.15–22.

Stagg, K. G. and O. C. Zienkiewicz (1968), *Rock Mechanics in Engineering Practice* (with 12 contributing authors), John Wiley & Sons, Inc., New York, 442 pp.

Terzaghi, K. (1943), *Theoretical Soil Mechanics*, John Wiley & Sons, Inc., New York, 510 pp.

Terzaghi, K., Peck, R. B. (1967), *Soil Mechanics in Engineering Practice*, 2nd edn, John Wiley & Sons, Inc., New York, 729 pp.

Tomlinson, M. J. (1995), *Foundation Design and Construction*, 6th edn, Longman Scientific and Technical, Essex.

Tomlinson, M. J. (2001), *Foundation Design and Construction*, 7th edn, Pearson Education Ltd., Harlow, Essex.

Vesič, A. S. (1969), "Effects of Scale and Compressibility on Bearing Capacity of Surface Foundations", discussion, *Proceedings, 7th ICSMFE, Mexico City, Mexico*, Vol. **3**, pp. 270–272.

Vesič, A. S. (1973), "Analysis of Ultimate Loads of Shallow Foundations", *Journal of the Soil Mechanics and Foundations Division*, ASCE, Vol. **99**, No. SM1, Jan, pp. 45–73.

Vesič, A. S. (1975), "Bearing Capacity of Shallow Foundations", *Foundation Engineering Handbook*, 1st edn, pp. 121–147, Winterkorn, H. F. and Fang, Hsai-Yang, eds., Van Nostrand Reinhold, New York.

West, J. M. and Stuart, J. G. (1965), "Oblique Loading Resulting from Interference between Surface Footings on Sand", *Proceedings, 6th ICSMFE, Montreal, Canada*, Vol. **2**, pp. 214–217.

Wyllie, D. C. (1991), *Foundations in Rock*, Spon, London.

CHAPTER 5

Shallow Foundations – Structural Design

5.1 General

One of the most important matters the foundation performance requirements concern is the *strength* of the foundation. As discussed in Section 2.4, one type of this strength is the *structural strength.* It is important to realise that foundations which are loaded beyond their structural strength capacity will, in principle, fail catastrophically. Therefore, foundations are designed to avoid structural failures, similar to the other structural analyses.

Foundation units such as footings transfer the loads from the structure to the soil or rock supporting the structure. As the soil is generally much weaker than the concrete columns and walls that must be supported, the contact area between the soil and the foundation base is much larger than that between the supported member and the base. The plan dimensions of the base contact area and the embedment depth (foundation depth) are primarily geotechnical concerns, as discussed in Section 4.17. Soon after these design requirements have been set, the next step is to develop a structural design that gives the foundation enough strength and integrity to safely transmit the design loads from the structure to the ground.

In general, the structural design of a reinforced concrete base includes:

- Selecting the more significant combination of loads that the particular foundation must support, and computing the design factored loads.
- Selecting an appropriate value for the compressive strength, f_c', of concrete.
- Selecting an appropriate grade of the reinforcing steel which specifies the yield strength, f_y, of reinforcement.
- Determining the required thickness of the concrete base.
- Determining the size, number, spacing and location of the flexural and temperature reinforcing bars.
- Designing the connection between the supported member (column, pedestal or wall) and the structural base.

Virtually all foundations support a compressive vertical load and, unlike the geotechnical design, it is computed without including the weight of the structural base because this weight is evenly distributed and thus does not produce shear or moment in the base. Sometimes foundations are required to

Shallow Foundations: Discussions and Problem Solving, First Edition. Tharwat M. Baban.

support shear and moment loads also. Again, unlike the geotechnical design, in structural design all the loads supported by the foundation must be factored loads.

In structural design of shallow foundations there are two modes of failure considered: *shear* and *flexure*. A structural base may fail in shear as a wide beam, referred to as *one-way shear* or *beam shear*, or as a result of punching, referred to as *two-way shear* or *punching shear*. Punching shear failure occurs when part of the base is just about to come out of the bottom, which is actually a combination of tension and shear on inclined failure surfaces. These shear stresses are resisted by providing an adequate base thickness; without using *shear reinforcement*. The flexural (bending) stresses are resisted by providing steel reinforcement similar to any other reinforced concrete structural member.

The design and construction of buildings is regulated by municipal bylaws called building codes (MacGregor et al., 2005). Unlike the geotechnical design, the structural design aspects are more strictly codified. Any structural design must satisfy requirements of a specified building code. The most notable, widely used, code is the *Building Code Requirements for Structural Concrete*, generally referred to as the *ACI 318* or *ACI Code*, published by the American Concrete Institute (ACI). The code is explained in a *Commentary*; they are bound together in one volume. Chapter 5 of this book refers extensively to the ACI 318M-08 code. Chapter 15 of the code concerns the structural design of footings. It is recommended that the reader have a copy available.

5.2 Design Loads

Section 9.2 of ACI 318M-08 defines the design factored load, for use in the LRFD method, as the largest of those computed from the following equations:

$$\text{ACI Equation 9-1: } U = 1.4(D+F) \tag{5.1}$$

$$\text{ACI Equation 9-2: } U = 1.2(D+F+T) + 1.6(L+H) + 0.5(L_r \text{ or } S \text{ or } R) \tag{5.2}$$

$$\text{ACI Equation 9-3: } U = 1.2D + 1.6(L_r \text{ or } S \text{ or } R) + (1.0\,L \text{ or } 0.8W) \tag{5.3}$$

$$\text{ACI Equation 9-4: } U = 1.2D + 1.6W + 1.0L + 0.5(L_r \text{ or } S \text{ or } R) \tag{5.4}$$

$$\text{ACI Equation 9-5: } U = 1.2D + 1.0E + 1.0L + 0.2S \tag{5.5}$$

$$\text{ACI Equation 9-6: } U = 0.9D + 1.6W + 1.6H \tag{5.6}$$

$$\text{ACI Equation 9-7: } U = 0.9D + 1.0E + 1.6H \tag{5.7}$$

Each load designation such as *D*, *F*, *T* and so on is defined in Section 2.1 of the ACI Code and in Section 4.17 of this book.

The reader should be aware of the exceptions and limitations regarding some of the factored loads in the above equations, as stated in ACI Section 9.2.1. Also, it is important to remember that the above-mentioned design factored load combinations must only be used in conjunction with *strength-reduction factors* of ACI Section 9.3.2.

As an alternative to the requirements of ACI Sections 9.2.1 and 9.3.2, ACI 318M-08, Appendix C, Section C.9.1 permits the design of reinforced structural concrete members using the load combinations defined in Section C.9.2 and the strength-reduction factors of Section C.9.3.

5.3 Selection of Materials

In the design of shallow foundations the structural engineer (the designer) must select appropriate values for concrete compressive strength, f'_c, and reinforcing steel yield strength, f_y. It is easy for a structural engineer to do so, whereas a geotechnical engineer has little or no control over the engineer properties of the supporting natural soil deposits.

In design of spread footings, selection of *high strength concrete* may not be justified unless the footings carry relatively large loads. The plan dimensions of footings are governed by bearing capacity and settlement requirements; it is the footing thickness only which is governed by the concrete strength. Moreover, high strength concrete requires additional materials and inspection costs. Spread footings are usually designed using f_c' of only 20–25 MPa. For footings that carry relatively large loads, high strength concrete might be justified to keep the footing thickness within reasonable limits, perhaps using f_c' as high as 35 MPa.

Usually, the flexural stresses in footings are small compared to those in the other structural members of the superstructure. Grade-300 steel is usually adequate to resist flexural stresses in spread footings, whereas, Grade-420 steel is, most probably, required on the remaining reinforced concrete members. However, due to practical and economic reasons, usually, designers like to use one grade of steel as much as possible. Therefore, for foundations built of reinforced concrete, designers often use Grade-420 steel, the same as for the other structural members of the structure.

5.4 Structural Action of Vertically and Centrically Loaded Isolated and Continuous (Strip) Footings

5.4.1 General

The structural analysis and design methodology of reinforced concrete footings that engineers now use have been developed, standardised and codified as a result of full-scale tests conducted by many researchers at various times (Talbot, 1913; Richart, 1948; Whitney, 1957; Moe, 1961). Scientific societies, associations and organisations made important contributions (ACI-ASCE, 1962). The structural design of a footing must essentially consider flexure (bending), shear, development of reinforcement, and the transfer of load from the column or wall to the footing. Each of these design aspects will be considered separately in the following Sections.

5.4.2 Flexure

An individual column footing is shown in Figure 5.1. Contact soil pressures cause bending of the footing about axes *A–A* and *B–B*, as shown in Figure 5.1a. Soil pressures acting under the cross-hatched portion of the footing in Figure 5.1b cause moments, M_u, about axis *A–A* at the face of the column. The footing behaves as if it is an inverted cantilever beam. The same argument is true with respect to a continuous (strip) wall footing, in which soil pressures cause bending of the footing about axis *A–A* along the wall length only. *Tensile stresses* due to these bending moments must be resisted by *tensile reinforcement* placed near the footing bottom, in both directions for column footing (Figure 5.1c), and in transverse direction for wall footing. The total factored moment at the critical section (Figure 5.1b) is:

$$M_u = \left(lbq_{factored}\right)(l/2) \tag{5.8}$$

The *critical sections* for maximum factored *moment*, M_u, in the footing are located as follows (ACI Section 15.4.2):

- At face of column, pedestal, or wall, for footings supporting a concrete column, pedestal, or wall;
- Halfway between middle and edge of wall, for footings supporting a masonry wall;
- Halfway between face of column and edge of steel base plate, for footings supporting a column with steel base plate.

According to ACI Section 15.3, for location of critical sections for moment, shear, and development of reinforcement in footings, it shall be permitted to treat *circular* or *regular polygon-shaped* concrete columns or pedestals as square members with the same area.

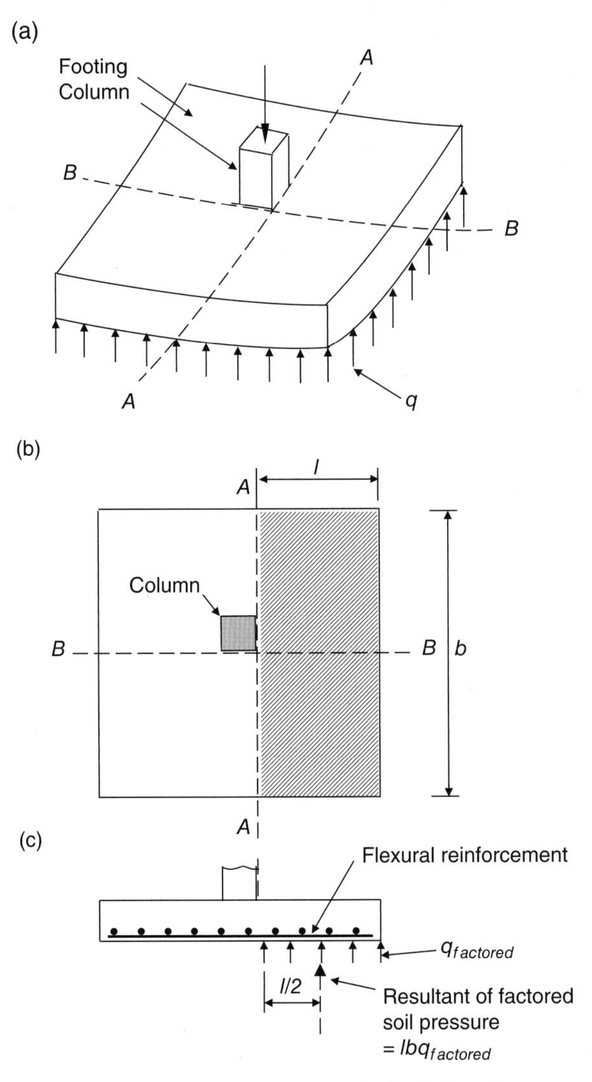

Figure 5.1 Flexural action of a column footing. (a) Footing under load. (b) Tributary area for moment at section *A–A*. (c) Moment about section *A–A*.

As it is clear from Figure 5.1c the moments per unit length (or unit width) vary along lines *A–A* and *B–B*, with the maximum occurring adjacent to the column. However, to simplify reinforcement placing, ACI Section 15.4.3 states that for two-way *square* footings reinforcement shall be distributed uniformly across entire width of footing. In two-way *rectangular* footings, reinforcement shall be distributed in accordance with ACI Sections 15.4.4.1 and 15.4.4.2.

The design equation for flexure is

$$\emptyset M_n \geq M_u \tag{5.9}$$

M_u = factored moment or required ultimate moment at section
M_n = nominal strength or nominal moment capacity at section
$\emptyset M_n$ = design strength or factored moment resistance at section
$\emptyset$ = strength-reduction factor for bending (ACI Section 9.3.2)

Two requirements are satisfied throughout the analysis and design of a reinforced concrete member, (1) *stress and strain compatibility*, which requires *stress* at any point in the beam must correspond to the *strain* at that point, (2) *equilibrium*, which requires the internal forces must balance the external load effects.

Equations forM_n – Tension steel yielding:
ACI Section 10.2.7 permits the use of the equivalent rectangular compressive stress distribution, as shown below:

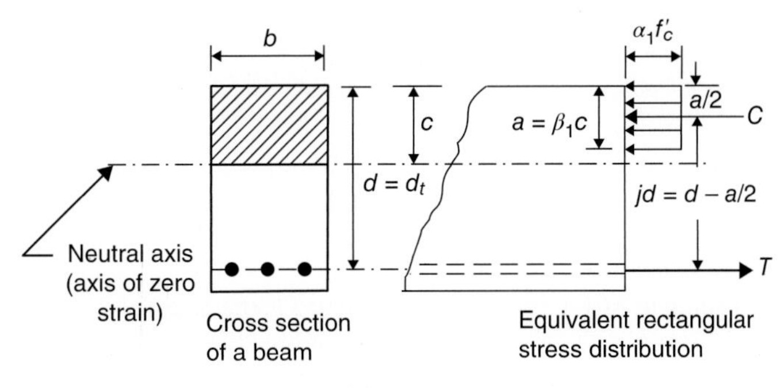

Scheme 5.1

$$\alpha_1 f'_c = 0.85 f'_c \text{ (ACI Section 10.2.7.1)}$$

β_1 = factor relating depth a of the equivalent rectangular compressive stress block to the neutral axis depth c.

For f'_c between 17 and 28 MPa, β_1 shall be taken as 0.85. For f'_c above 28 MPa, β_1 shall be reduced linearly at a rate of 0.05 for each 7 MPa of strength in excess of 28 MPa, but β_1 shall not be taken less than 0.65 (ACI Section 10.2.7.3).

The compressive force in the concrete is

$$C = \text{volume of the compressive stress block} = \left(0.85 f'_c\right) ba$$

The tension force in the steel is

$$T = A_s f_s$$

For equilibrium, $C = T$, and therefore, the depth of the equivalent stress block is

$$a = \frac{A_s f_s}{0.85 f'_c b}$$

If it is assumed that $f_s = f_y$, then

$$a = \frac{A_s f_y}{0.85 f'_c b} \tag{5.10}$$

$M_n = T \times jd$ or $M_n = C \times jd$, and $jd = d - \left(\frac{a}{2}\right)$. Hence,

$M_n = A_s f_y \left(d - \frac{a}{2}\right)$, and

$$\text{Ø}\, M_n = \text{Ø}\left[A_s f_y \left(d - \frac{a}{2}\right)\right] \tag{5.11}$$

Also,

$$M_n = C \times jd = \left(0.85 f_c'\right) ba \left(d - \frac{a}{2}\right). \text{ Hence,}$$

$$\text{Ø}\, M_n = \text{Ø}\left[\left(0.85 f_c'\right) ba \left(d - \frac{a}{2}\right)\right] \tag{5.12}$$

The flexural design requires that the *tension steel* yields before the concrete crushes. The tensile force T equals the yield stress f_y times the area of the tension steel A_s. In other words; it is assumed that $f_s = f_y$. Therefore, it is necessary to check this assumption. This checking may be done by using *strain compatibility* as follows:

According to ACI Sections 10.3.3 and R10.3.3, the nominal *flexural strength* of a member is reached when the strain in the extreme concrete compression fiber ε_{cu} reaches the assumed strain limit 0.003 as shown in thisscheme.

By similar triangles,

$$\varepsilon_t = 0.003\left(\frac{d_t - c}{c}\right)$$

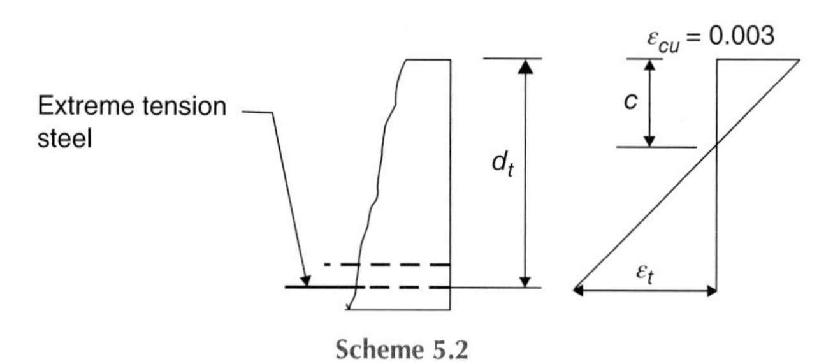

Scheme 5.2

The strain ε_t is the net tensile strain in extreme layer of longitudinal tension steel at nominal strength, excluding strains due to effective pre-stress, creep, shrinkage and temperature.

$$\varepsilon_y = \frac{f_y}{E_s}$$

If $\varepsilon_t > \varepsilon_y$, $f_s = f_y$

Strength-reduction factor, Ø:

In order to restrict the flexural reinforcement ratio or to set the upper limit on this reinforcement in a member, ACI Section 10.3.5 requires that the *net tensile strain* ε_t in the extreme-tension steel shall not be less than 0.004.

Although a footing is not exactly a beam, it is desirable that it be ductile in flexure by limiting ε_t, as just mentioned above.

ACI Section 10.3.3 states that, sections are *compression-controlled* if the net tensile strain in the extreme tension steel ε_t is equal to or less than the compression-controlled strain limit when the concrete in compression reaches its assumed strain limit of 0.003. The compression-controlled strain limit is the net tensile strain in the reinforcement at balanced strain conditions. For Grade-420 reinforcement, and for all pre-stressed reinforcement, it shall be permitted to set the compression-controlled strain limit equal to 0.002.

ACI Section 10.3.4 states that, sections are *tension-controlled* if the net tensile strain in the extreme tension steel ε_t is equal or greater than 0.005 when the concrete in compression reaches its assumed strain limit of 0.003. Sections with ε_t between the compression-controlled strains limit and 0.005 constitute a transition region between compression-controlled and tension-controlled sections.

For tension-controlled sections: Ø = 0.9 (ACI Section 9.3.2.1):

For compression-controlled sections (flexural members other than spiral columns): Ø = 0.65 [ACI Section 9.3.2.2(b)]

For sections in the transition region: 0.90 > Ø > 0.65

According to ACI Section 9.3.2.2, for sections in the transition region, Ø shall be permitted to be linearly increased from that for compression-controlled sections to 0.9 as ε_t increases from the compression-controlled strain limit to 0.005.

For Grade-420 steel, for which the compression-controlled strain limit is 0.002, the variation of Ø with net tensile strain in extreme tension steel ε_t is shown in ACI Figure R9.3.2. Accordingly, Ø is determined by linear proportionality as follows:

$$\text{Ø} = 0.65 + \left(\frac{\varepsilon_t - 0.002}{0.005 - 0.002}\right)(0.9 - 0.65)$$
$$\text{Ø} = 0.65 + (\varepsilon_t - 0.002)(250/3) \tag{5.13}$$

As indicated in ACI Figure R9.3.2, the factor Ø is also determined from the following equation:

$$\text{Ø} = 0.65 + 0.25\left(\frac{1}{c/d_t} - \frac{5}{3}\right) \tag{5.14}$$

Equations (5.13) and (5.14) give

$$\varepsilon_t = \frac{d_t - c}{c} \times 0.003 \tag{5.15}$$

where

c = distance from extreme compression fiber to neutral axis
d_t = the distance from the extreme compression fiber to the extreme tension steel

Similarly, for Grade-280 steel, for which the compression-controlled strain limit is $\left(\frac{f_y}{E_s} = 0.0014\right)$, the factor Ø is determined from the following equations:

$$\text{Ø} = 0.55 + 69.4\varepsilon_t \tag{5.16}$$

or

$$\text{Ø} = 0.34 + \frac{0.21}{c/d_t} \tag{5.17}$$

According to ACI Section 10.5.4, for footings of uniform thickness, the minimum area of flexural tensile reinforcement $A_{s,\ min}$ shall be the same as that required for *shrinkage and temperature reinforcement* in accordance with ACI Section 7.12.2.1. Therefore, for deformed bars of Grade-280 or 530 steel, $A_{s,\ min} = 0.002\ bh$; for deformed bars or welded wire reinforcement of Grade-420 steel, $A_{s,min} = 0.0018\ bh$. Also, ACI Section 10.5.4 requires that, the maximum spacing of this minimum reinforcement shall not exceed three times the footing thickness, or 450 mm whichever is smaller.

In case the required flexural reinforcement exceeds the minimum flexural reinforcement, it shall be adequate to use the same maximum spacing of reinforcement for slabs which is two times the slab thickness or 450 mm, whichever is smaller, as specified by ACI Section 13.3.2.

According to ACI Section 7.12.2.2, shrinkage and temperature reinforcement shall be spaced not farther apart than five times the slab thickness, nor farther apart than 450 mm.

ACI Section 7.7.1-(a) states that, cover for reinforcement in concrete cast against and permanently exposed to earth, shall not be less than 75 mm.

ACI Section 15.7 states that, depth of footing above bottom reinforcement shall not be less than 150 mm for footings on soil, or less than 300 mm for footings on piles.

5.4.3 Shear

As mentioned in Section 5.1, a footing may fail in shear as a *wide beam*, referred to as *one-way shear* or *beam shear* (Figure 5.2a), or as a result of punching, referred to as *two-way shear* or *punching shear* (Figure 5.2b).

Shear reinforcement is very seldom used in spread footings or in mats, due to difficulty in placing it, and due to the fact that it is usually cheaper and easier to deepen the footings than to provide stirrups. ACI Section 11.4.6.1 excludes footings from the minimum *shear reinforcement* requirements. For these reasons the factored shear force, V_u, at any critical section shall be resisted only by the concrete *shear strength* alone.

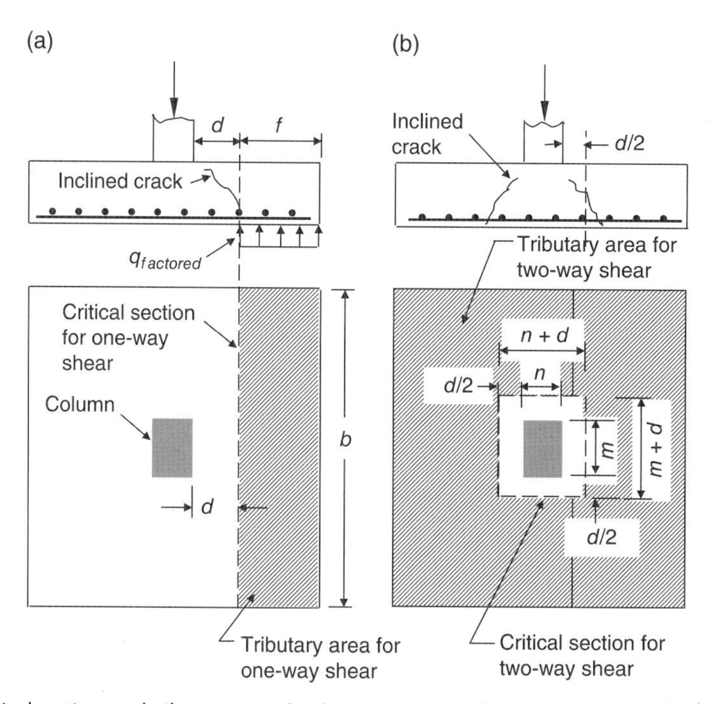

Figure 5.2 Critical sections and tributary areas for shear in a column footing (see text for details).

One-way shear:
According to ACI Sections 15.5.1, 15.5.2, 11.11.1.1 and 11.1.3.1, critical section for *one-way shear* extends in a plane across the entire width of the footing, located at distance *d* from face of the column, pedestal, or wall, as shown in Figure 5.2a. For footings supporting a column or pedestal with steel base plates, the critical section shall be located at distance *d* from a line halfway between the face of the column and the edge of the base plate.

From Figure 5.2a, the factored shear force

$$V_u = bfq_{factored} \tag{5.18}$$

ACI Section 11.1.1 requires that

$$\text{Ø}\, V_c \geq V_u \quad \text{(taking reinforcement shear strength, } V_s = 0) \tag{5.19}$$

where

Ø = strength-reduction factor
  = 0.75 for shear (ACI Section 9.3.2.3)
V_c = nominal shear strength provided by concrete

According to ACI Sections 11.11.1.1 and 11.2.1.1, V_c shall be computed as

$$V_c = 0.17\lambda \sqrt{f_c'}\, b_w d \tag{5.20}$$

where

λ = modification factor (ACI Section 8.6.1)
$\sqrt{f_c'}$ = square root of specified compressive strength of concrete in MPA.
b_w = footing width or length of critical section
d = distance from extreme compression fiber to centroid of longitudinal tension reinforcement

Two-way shear:
In footings subject to *two-way action,* failure may occur by punching along a truncated cone or pyramid around a column. For square or rectangular columns, ACI Sections 11.11.1.2 and 11.11.1.3 allows the use of a critical section with four straight sides drawn parallel to and at a distance *d*/2 from the faces of the column (or the edges of the loaded area), as shown by the dashed lines in Figure 5.2b. The tributary area, assumed critical for design purposes, is shown cross-hatched. For footings supporting a column or pedestal with steel base plates, the critical section shall be located at distance *d*/2 from a line halfway between the face of the column and the edge of the base plate ACI Section 15.5.2.

The applied net factored shear force is:

V_u = factored column loads – factored soil reaction on the shear block

$$V_u = A_f q_{factored} - \left[(m+d)(n+d)q_{factored}\right] \tag{5.21}$$

where A_f is the footing area
Since shear reinforcement is not used in a footing, Ø $V_c \geq V_u$. According to ACI Section 11.11.2.1, V_c shall be the smallest of (a), (b), and (c):

(a)

$$V_c = 0.17\left(1 + \frac{2}{\beta}\right)\lambda\sqrt{f_c'}\,b_o d \tag{5.22}$$

(b)

$$V_c = 0.083\left(\frac{\alpha_s d}{b_o} + 2\right)\lambda\sqrt{f_c'}\,b_o d \tag{5.23}$$

(c)

$$V_c = 0.33\,\lambda\sqrt{f_c'}\,b_o d \tag{5.24}$$

where:

β = the ratio of long side to short side of the column, concentrated load or reaction area.
b_o = the perimeter of the critical section for two-way shear.
α_s = a constant equals 40 for interior columns, 30 for edge columns and 20 for corner columns.

ACI Section 15.5.3 requires that, footing supported on piles (i.e. piles cap) shall satisfy ACI Sections 11.11 and 15.5.4.

ACI Section 15.5.4 states that, computations of shear on any section through a footing supported on piles shall be in accordance with 15.5.4.1, 15.5.4.2, and 15.5.4.3.

5.4.4 Development of Reinforcement

The calculated reinforcing bar stresses at any section must be developed by extending each bar on each side of that section a sufficient distance l_d (embedment length), or by hook (for bars in tension only) at the outer ends.

ACI Section 15.6.2 states that calculated tension or compression in reinforcement at each section shall be developed on each side of that section by embedment length, hook (tension only) or mechanical device or a combination thereof. ACI Section 15.6.1 requires this development of reinforcement in footings be accomplished in accordance with Chapter 12 of the Code. According to ACI Section 15.6.3, critical sections for development of reinforcement shall be assumed at the same locations as defined in ACI Section 15.4.2 for maximum factored moment and at all other vertical planes where changes of section or reinforcement occur.

5.4.5 Transfer of Force at Base of Column, Wall or Pedestal

Connections through which forces and moments are transferred must be designed carefully, since they are often the weak links in structures. There are different types of connections available; each is intended for particular loading conditions, and construction materials of both the supported and supporting members. For example; the methods of connecting concrete, steel and wood columns to concrete footings are different. This Section will discuss concrete column, wall or pedestal connected to concrete footing, and steel column with base plate connected to concrete footing or pedestal.

Concrete column, wall or pedestal connected to concrete footing:
ACI Section 15.8.1 states that, forces and moments at base of column, wall, or pedestal shall be transferred to supporting pedestal or footing by bearing on concrete and by reinforcement, dowels and mechanical connectors.

According to ACI Section 15.8.1.1, bearing stress on concrete at contact surface between concrete column or wall (supported member) and concrete footing or pedestal (supporting member) must not exceed concrete bearing strength for either surface as given by ACI Section 10.14.

According to ACI Section 10.14.1, the maximum design bearing strength of concrete is $\emptyset\left(0.85 f_c' A_1\right)$ except when the supporting surface is wider on all sides than the loaded area; then

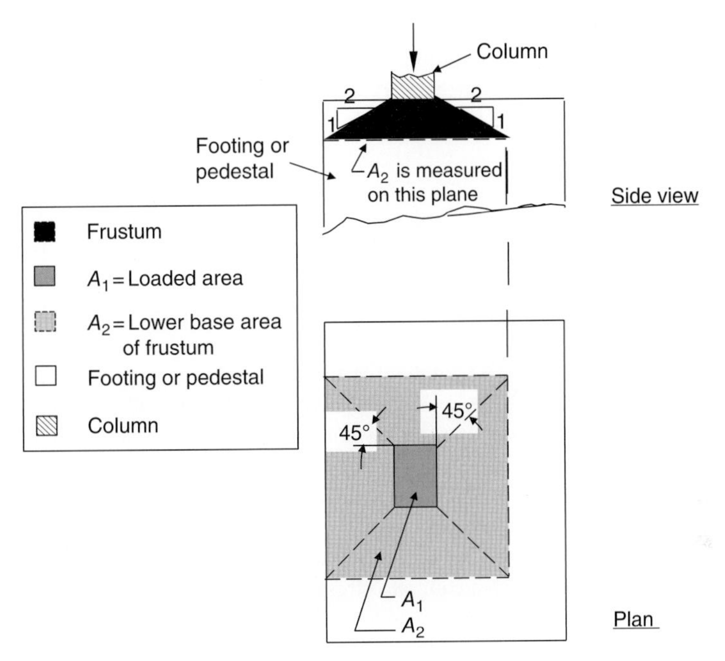

Figure 5.3 Application of frustum to find A_2 in footing or pedestal.

the design bearing strength of the loaded area shall be permitted to be multiplied by $\sqrt{A_2/A_1}$ but not more than two (i.e. the design bearing strength = $\emptyset\left(0.85f_c'\ A_1\right)\sqrt{A_2/A_1}$, and $\sqrt{A_2/A_1} \leq 2$). For the load combinations of ACI Section 9.2.1, ACI Section 9.3.2.4 gives the strength-reduction factor $\emptyset$ = 0.65 for bearing.

Area A_1 is the loaded area (area of the contact surface) but not greater than the bearing cross-sectional area or the bearing plate area. Area A_2 is area of the lower base of larger frustum of a pyramid, cone, or tapered wedge contained wholly within the support and having for its upper base the loaded area (A_1), and having side slopes of 1 vertical to 2 horizontal, as shown in Figure 5.3.

ACI Section 15.8.2 states that, for a castinplace construction, the reinforcement required to satisfying 15.8.1 shall be provided either by extending *longitudinal bars* (reinforcement) into supporting pedestal or footing, or by *dowels*. For cast-in-place columns and pedestals, area of these bars across interface shall be not less than 0.005 A_g, where A_g is the gross area of the supported member (ACI Section 15.8.2.1). For cast-in-place walls, ACI Section 15.8.2.2 requires area of reinforcement across interface shall be not less than minimum vertical reinforcement given in ACI Section 14.3.2.

In practice, generally, the column bars stop at the top of the footing or pedestal, and dowels are used to transfer loads across the interface. This is because it is unwieldy to embed the column steel in the footing or pedestal, due to its unsupported height above the supporting member and the difficulty in locating it properly.

For castinplace columns and pedestals, normally, the number of dowels is equal to the number of *longitudinal bars* in the column or pedestal. However, at least four dowels should be provided across interface.

According to ACI Section 15.8.2.3:

- Dowels shall not be larger than No. 36 bar.
- At footings, it is permitted to lap splice longitudinal bars of sizes as large as No. 43 and No. 57, in compression only, with dowels.

- Dowels shall extend into supported member a distance not less than the compression development length l_{dc} of the longitudinal bars (larger than dowels) and compression *lap splice* length of the dowels, whichever is greater (ACI Sections 12.3.2, 12.3.3, 12.16.1, 12.16.2).
- Dowels shall extend into the footing a distance not less than l_{dc} of the dowels (ACI Sections 12.3.2 and 12.3.3).

Usually, dowels have a 90° hook at the bottom used to facilitate fastening them tightly in place with the footing bottom bars. Figure 5.4a shows a column-footing joint with dowel bars extending through the interface, and Figure 5.4b shows a wall-footing joint.

The preceding discussions concern the condition in which the full section at the interface is under compressive stresses. In other words, no moments are transferred or the eccentricity falls within the kern of the section. The total force transferred by bearing is then calculated as $(A_g - A_{sd})$ times the smaller of the bearing stresses allowed on the supported member (column, pedestal or wall) or the supporting member (footing), where A_{sd} is area of the longitudinal bars or dowels crossing the interface. Any additional load must be transferred by dowels [ACI Section 15.8.1.2-(a)].

For the condition in which moments are transferred to the supporting footing or pedestal, usually, compressive stresses will exist over part, but not all, of the cross-section. The number of dowels required can be obtained by considering the cross-sectional area as an eccentrically loaded column with a maximum compressive concrete stress equal to the smaller of the bearing stresses allowed on the supported member (column, pedestal or wall) or the supporting member (footing). Sufficient reinforcement must cross the interface to provide the necessary axial load and moment capacity. In general, this requires that all the column bars or dowels of the same steel area must cross the interface. According to ACI Section 15.8.1.3, these longitudinal bars or dowels should satisfy ACI Section 12.17. Also, the minimum embedment length of these bars in the footing should not be smaller than the compression development length l_{dc} or tension development length l_{dh}, whichever is greater.

Since the footing and column are poured separately, there is a weak shear plane along the cold joint. For the condition in which lateral forces are imparted to the supporting footing or pedestal, ACI Section 15.8.1.4 requires that, these forces shall be transferred in accordance with shear-friction provisions of ACI Section 11.6, or by other appropriate means.

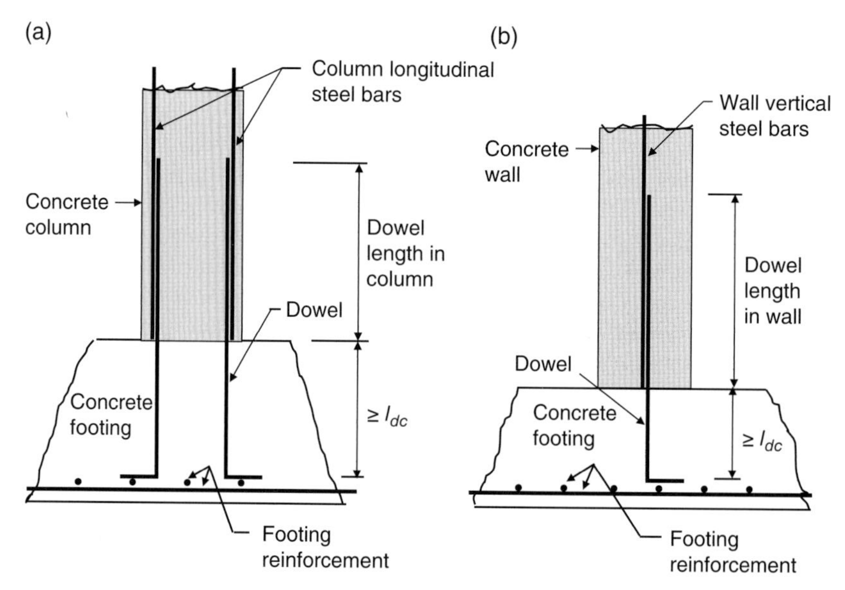

Figure 5.4 (a) Column-footing joint. (b) Wall-footing joint.

Steel column with base plate connected to concrete footing or pedestal:
A steel column requires a base plate either shop-welded or field-bolted to its bottom, to spread the very high stresses in the small column contact area to a value that the supporting footing or pedestal can safely carry. Steel columns are connected to their foundations using *anchor bolts* or *rods* of specified type, grade and ultimate tensile strength (Table 5.1). Slightly oversized holes, 2 to 5 mm larger in diameter than the bolts, are shop punched in the base plate for later attachment to the supporting member. The anchor bolts are usually set in nearly exact position in the wet concrete and become fixed in place. The slightly oversized holes allow a small amount of bolt misalignment when placing the base plate into position; thus, erection of the column onto the foundation is simplified. Figure 5.5 shows a base plate and anchor bolts used to connect a metal column to its foundation. More often, four anchor bolts are used for each column. If possible, it is best to place the bolts in a square pattern to simplify construction and leave less opportunity for mistakes. Of course, more bolts and other patterns also may be used, if necessary.

It is necessary the base plate is carefully aligned horizontally and to elevation, since top surface of the concrete footing or pedestal is usually rough and not necessarily level. Therefore, the contractor must use special construction methods to provide adequate support for the plate and to make the column plumb. One method is to use shims (small and thin steel wedges), which are driven between the base plate and the supporting member. The remaining space is grouted, using non-shrink grout which swells slightly when it cures (as compared to normal grout, which shrinks).

Table 5.1 Ultimate tensile strength, T_u, of selected A307 bolts*.

Bolt diameter and pitch, mm	Net tensile stress area, A_t, mm^2	Tensile force T_u, kN	
		Grade A	Grade B
16P2 **	157	63	108
20P2.5	245	98	169
24P3	353	141	244
30P3.5	561	224	387
36P4	817	327	564
42P4.5	1120	448	773
48P5	1470	588	1014
56P5.5	2030	812	1401
64P6	2680	1072	1849
72P6	3460	1384	2387
80P6	4340	1736	2995
90P6	5590	2236	3857
100P6	6990	2796	4823

* From American National Standards Institute (ANSI) SR 17 (also, ASTM STP 587 dated 1975).
** 16P2 is a nominal bolt diameter of 16 mm with a thread pitch
P = 2 mm.
Grade A, f_{ult} = 400 MPa; f_y = 250 MPa
Grade B, f_{ult} = 690 MPa; f_y = 400 MPa
$A_t = 0.7854\ (\text{Diam.} - 0.9382P)^2$
For 16P2: $A_t = 0.7854\ (16 - 0.9382 \times 2)^2 = 157\ \text{mm}^2$

$$T_u = \frac{400 \times 1000}{1000^2} \times 157 = 63\ \text{kN}$$

P = thread pitch = distance between corresponding points on adjacent thread forms in mm. A pitch of 2 means there are 2 mm between points.

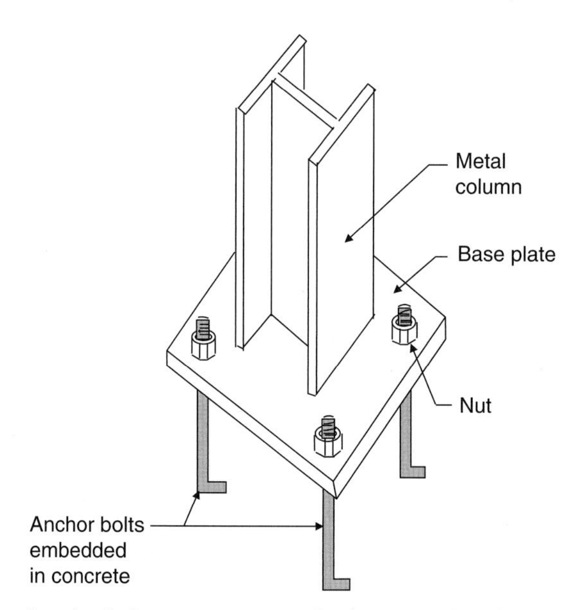

Figure 5.5 Base plate and anchor bolts to connect a metal column to its foundation.

The AISC (1989) manual provides general guidance in the design of base plates. The design of base plates is beyond the scope of this book, but it is covered in most steel design texts and scientific documents such as "Column Base Plates," AISC, by DeWolf and Ricker (1990).

Anchor bolts may be cast-in or post-installed in the supporting member. Cast-in anchor is a headed bolt, headed stud, or hooked bolt installed before placing concrete. Post-installed anchor, such as expansion anchor and undercut anchor, is an anchor installed in hardened concrete. These types of anchor bolts are shown in *Figure RD.1*– Appendix D of ACI 318M-08.

In general, the modes of failure associated with anchor bolts are (i) rupture or fracture of the bolt itself, due to the structural tension and shear loads transmitted from the column to the supporting member, (ii) loss of anchorage in the concrete, such as bolt pullout, concrete breakout, concrete splitting and concrete side-face blowout, due to tension loads; or, such as concrete *spall*, concrete *pryout* and concrete breakout, due to shear loads. All these modes of failure are illustrated in Figure RD.4.1– Appendix D of ACI 318M-08.

Steel, of which anchor bolts are made, is more ductile than concrete. For this reason, it is preferred that anchors are designed so the critical mode of failure is tension or shear of the bolt itself rather than failure of the anchorage.

Compression load transfer:

If the structural design loads between the column and the supporting member consist solely of compression, then theoretically no anchorage will be required. However, anchor bolts are still required to resist erection loads, accidental collisions during erection, and unanticipated shear or tensile loads. The engineer might attempt to estimate these loads and design accordingly, or simply select the bolts using engineering judgment. For example, an engineer might arbitrarily select enough bolts of specified type and diameter to carry 10% of the total compressive load (unfactored) in shear.

Tensile load transfer:

For the design of anchors (except when anchor design includes earthquake forces), each anchor bolt must satisfy the following design criterion (ACI Appendix D Section D.4.1.1):

$$\emptyset N_n \geq N_{ua} \tag{5.25}$$

where

$\O N_n$ = the lowest design strength in tension determined from all appropriate failure modes (ACI Section D. 4.1.2).
$\O$ = strength-reduction factor for anchors in concrete (ACI Section D. 4.4).
N_n = factored tensile force applied to anchor or group of anchors.

Shear load transfer:
For the design of anchors (except when anchor design includes earthquake forces), each anchor bolt must satisfy the following design criterion (ACI Appendix D Section D.4.1.1):

$$\O V_n \geq V_{ua} \tag{5.26}$$

where

$\O V_n$ = the lowest design strength in shear determined from all appropriate failure modes (ACI Section D. 4.1.2).
$\O$ = strength-reduction factor for anchors in concrete (ACI Section D. 4.4).
V_n = nominal shear strength for any mode of failure (ACI Sections D. 6.1, D.6.2, and D.6.3).
V_{ua} = factored shear force applied to a single anchor or group of anchors.

Combined tensile and shear load transfer:
When both tensile force N_{ua} and shear force V_{ua} are present, interaction effects shall be considered in design (ACI Section D. 4.1.3). According to ACI Section D.4.3, this requirement shall be considered satisfied by ACI Section D.7 which specifies that, anchors or groups of anchors shall be designed to satisfy the requirements of ACI Sections D.7.1–D.7.3, as follows:

- If $V_{ua} \leq 0.2\,\O\,V_n$, then full strength in tension shall be permitted:

$$\O N_n \geq N_{ua} \text{ (ACI Sections D.7.1)}$$

- If $N_{ua} \leq 0.2\,\O\,N_n$, then full strength in shear shall be permitted:

$$\O V_n \geq V_{ua} \text{ (ACI Sections D.7.2)}$$

- If $V_{ua} > 0.2\,\O\,V_n$ and $N_{ua} > 0.2\,\O\,N_n$, then

$$\frac{N_{ua}}{\O N_n} + \frac{V_{ua}}{\O V_n} \leq 1.2 \text{ (ACI Section D.7.2)} \tag{5.27}$$

where values of $\O N_n$ and $\O V_n$ shall be as given in Equations (5.25) and (5.26), respectively.

Anchorage:
The designer must determine the required depth of embedment of the anchor bolts into the concrete to provide the necessary anchorage. This required embedment depth is not specifically indicated in most (including ACI) building codes. For an expansion or undercut post-installed anchors only, the ACI Section D.8.5 requires the value of h_{ef} (effective embedment depth of anchor) shall not exceed the greater of two-thirds of the member (footing or pedestal) thickness and the member thickness minus 100 mm. However, it is understood that the h_{ef} must be so determined that the requirements of concrete breakout strength (ACI Section D.5.2) and pullout strength (ACI Section D.5.3) of the anchor in tension are

Table 5.2 Anchorage requirements for bolts and threaded rods (Shipp and Haninger, 1983; © AISC).

Steel grade	Minimum embedment depth	Minimum edge distance
A307, A36	12*d*	5*d* or 100 mm, whichever is greater
A325, A449	17*d*	7*d* or 100 mm, whichever is greater

d = Nominal bolt diameter

satisfied. In addition, the anchorage design must satisfy the requirements of ACI Section D.9 regarding minimum spacing and edge distances for anchors and minimum thickness of members.

According to Coduto (2001), Table 5.2 presents conservative design values for embedment depth and edge distance.

5.5 Eccentrically Loaded Spread Footings

On occasions, footings have overturning moments as well as axial loads (eccentric loading condition), as shown in Figures 2.22 and 4.5. Structural design of eccentrically loaded footings can proceed in the same manner as that for centrically loaded footings. However, the contact soil pressure will not be uniform, as discussed in Section 2.6.2 and shown in Figure 2.22. Therefore, for square footings as well as rectangular footings, it will be necessary to consider both the one-way shear and two-way shear in determining the footing thickness, and to determine the amount of steel required in each direction separately.

A footing is also considered eccentrically loaded when the column, with or without moment, is offset from the footing centroid. If space permits, it will be possible to place the column away from the centre so that the resulting soil pressure is uniform, i.e. the structural load resultant coincides with the soil pressure resultant at centre of the footing, as shown in Figure 5.6. This solution is obviously valid only for moments which always act in the direction shown for that footing configuration. This is not a valid solution for wind moments, since reversals can occur.

In conventional analysis of rigid footings the soil pressure can be computed using Equation (2.7), repeated here for convenience:

$$q = \frac{R}{BL}\left(1 \pm \frac{6e_x}{B} \pm \frac{6e_y}{L}\right) \quad \text{Equation (2.7)}$$

where B and L are in x and y directions. respectively.

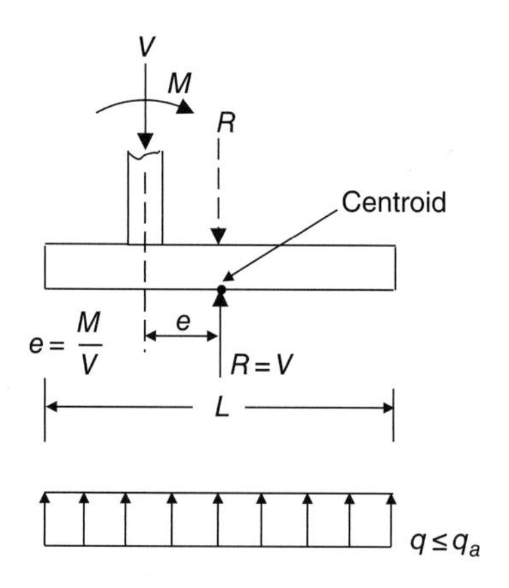

Figure 5.6 Eccentrically loaded spread footing with uniform soil pressure.

According to Bowles (2001), spread footings (assumed somewhat less than rigid) should be designed consistent with the procedure for obtaining the bearing capacity, considering that the soil analogy is almost identical to the Strength Design method of concrete. According to this method of analysis, a *uniform* soil pressure under the effective area $B'L'$ is used to compute design moments and shear; hence, the design is more easily done. In this analysis the soil pressure resultant passes through centroid of the effective area $B'L'$, as shown in Figure 5.7a. The eccentric distances e_x and e_y are computed as $e_x = \dfrac{M_y}{P}$ and $e_y = \dfrac{M_x}{P}$ It is required that:

$$B_{min} = 4e_y + w_y; \;\; L_{min} = 4e_x + w_x$$
$$B'_{min} = 2e_y + w_y; \;\; L'_{min} = 2e_x + w_x$$

Where w_y and w_x are the column cross-section dimensions in y (or B) and x (or L) directions, respectively, as shown in Figure 5.7b.

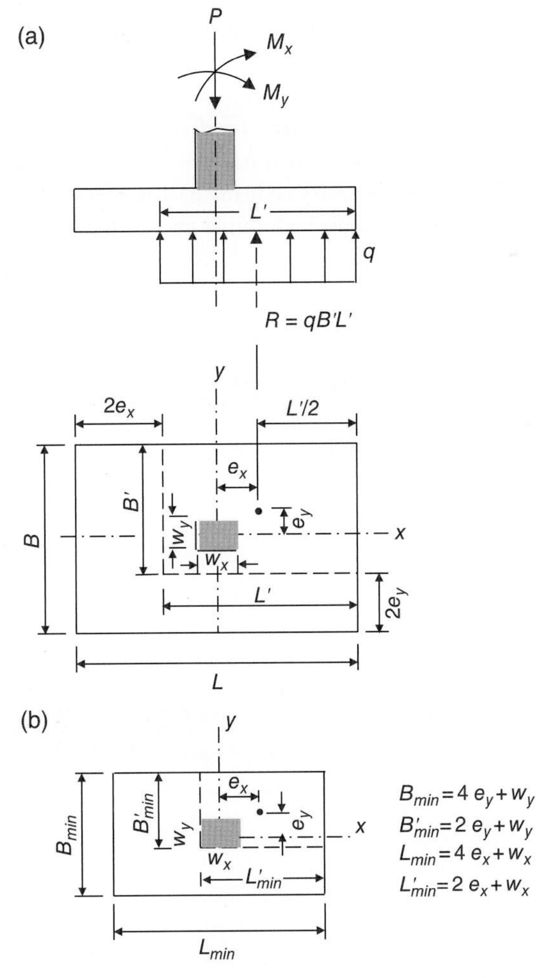

Figure 5.7 A spread footing with overturning moments, considering uniform soil pressure under the effective area $B'L'$ (see text for details).

5.6 Pedestals

Pedestal is a member with the ratio of height-to-least lateral dimension less than or equal to three used primarily to support axial compressive load. For a tapered member, the least lateral dimension is the average of the top and bottom dimensions of the smaller side (ACI Section 2.2). The ACI code allows both reinforced and unreinforced concrete pedestals (ACI Section 22.8). The height-thickness limitation ratio for plain concrete pedestals applies only to the unsupported height; does not apply for portions of pedestals embedded in soil capable of providing lateral restraint (ACI Sections 22.8.2 and R22.8). Usually, a pedestal is used to carry the loads from metal columns through the floor and soil to the footing when the footing is at some depth in the ground, as shown in Figure 5.8. One purpose for using such pedestal is to avoid possible corrosion of the metal from the soil.

Usually, concrete pedestals are reinforced with minimum column steel of $0.01A_g$ but not more than $0.08A_g$ (ACI Section 10.9.1) even when they are designed as unreinforced members. However, if moment or uplift load exists, the vertical steel should always be designed to carry any tension stresses. Steel ties should be liberally added at the top to avoid concrete *spalls* and to keep the edges from cracking (Figure 5.8).

According to ACI Sections 22.8.3 and 22.5.5, maximum factored axial compressive load, P_u, applied to plain concrete pedestals shall not exceed design bearing strength, $\varnothing B_n$, where B_n is nominal bearing strength of loaded area A_1 using

$$B_n = 0.85 f_c' A_1 \tag{5.28}$$

In a case where the supporting surface is wider on all sides than the loaded area, then B_n shall be multiplied by $\sqrt{A_2/A_1}$ but not more than two.

For a reinforced concrete pedestal being designed as a *simply supported* column element (a rather common condition); the following formula may be used:

$$P_u \le \varnothing\left(0.85 f_c' A_c + A_s f_y\right) \tag{5.29}$$

where

A_c = net area of concrete in pedestal (A_g–A_s)
A_s = area of reinforcing steel
f_y = specified – yield strength of reinforcement
f_c' = specified – compressive strength of concrete

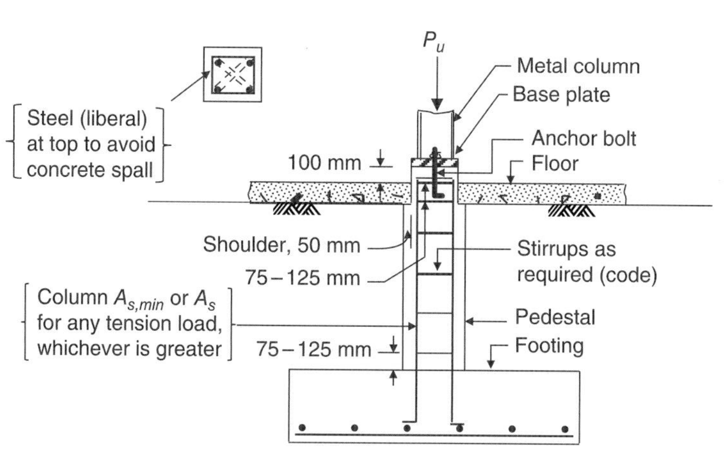

Figure 5.8 Pedestal details (approximate).

5.7 Pile Caps

A *pile cap* or a *cap* is a reinforced concrete element which connects a column, pedestal or wall with a group of piles. The function and structural design of pile caps are very similar to those of spread footings; both elements must distribute loads from the supported member across the bottom of footing or cap. The ACI code calls a cap "a footing supported on piles" (ACI Section 15.5.4). The conventional structural design of rigid pile caps should satisfy the following requirements:

(1) The depth of cap above bottom reinforcement (the effective depth) shall not be less than 300 mm (ACI Section 15.7).
(2) Bending moments, development of reinforcement and shear are taken at the same critical sections as for reinforced-concrete footings (defined in ACI Sections 15.4.2, 15.6.3 and 15.5.2), as shown in Figure 5.9. Computation of shear on any section shall be in accordance with ACI Section 15.5.4. Computations for moments and shear shall be based on the assumption that the reaction from any pile is concentrated at pile centre (ACI Section 15.2.3).
(3) Where necessary, shear around individual piles (punching shear) may be investigated in accordance with ACI Section 11.11.1.2. If shear perimeters overlap, the modified critical perimeter b_o should be taken as described in ACI Section R15.5.2 and illustrated in ACI *Figure R15.5.*
(4) Bearing stress on concrete at contact surface between supported and supporting member, including contact surface between cap and individual concrete pile, shall not exceed concrete bearing strength for either surface as given by ACI Section 10.14 (ACI Section 15.8.1.1).
(5) Piles should extend at least 150 mm into the cap. Some building authorities may allow as little as 75 mm of pile embedment into the cap.
(6) Ordinarily the pile heads are assumed hinged to the pile cap. When pile heads are assumed fixed to the cap, they should extend at least 300 mm into the cap.
(7) The cap bottom reinforcement should be 75 mm above the pile top to control concrete cracking around the pile head.
(8) When piles extend into the cap more than 150 mm, the cap bottom bars should loop around the pile to avoid splitting a part of the cap from pile head.
(9) Reinforcement should be placed so there is a minimum cover of 75 mm for concrete adjacent to the soil (ACI Section 7.7.1).
(10) Pile caps should extend at least 200 mm beyond the outside face of exterior piles.

The conventional pile cap design assumes:

(1) Each pile in a group carries an equal amount of the total concentric axial load Q on the cap. This assumption is practically valid when Q is applied at the centre of the pile group, the piles are all

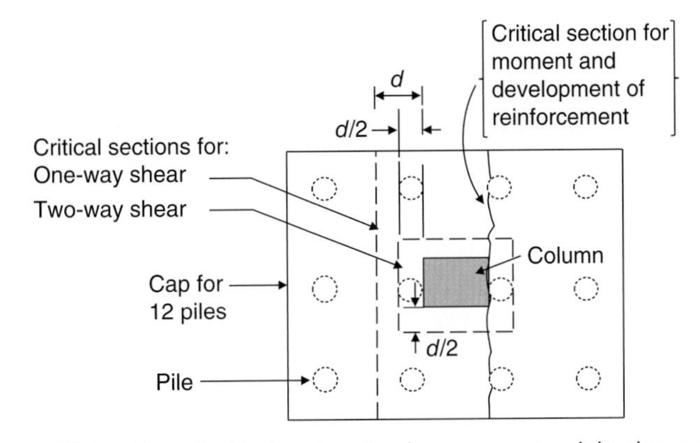

Figure 5.9 A pile cap with locations of critical sections for shear, moment and development of reinforcement.

vertical, the pile group is symmetrical, the cap is rigid and in contact with the ground. The design considers structural axial loads of the supported member, weight of the cap and weight of any soil overlying the cap (if it is below the ground surface). For a group of n piles, the load P_p per pile is

$$P_p = \frac{Q}{n} \tag{5.30}$$

(2) For a pile cap eccentrically loaded with Q or loaded with Q at centre and moments, the combined stress equation [Equation (2.6)] is assumed valid (assuming planar stress distribution). The equation is repeated here for convenience:

$$q = \frac{R}{A} \pm \frac{M_y}{I_y} x \pm \frac{M_x}{I_x} y \quad \text{Equation (2.6)}$$

Where

$$q = \frac{P_p}{(\text{pile cross section area})} = \frac{P_p}{A_p};\ A = nA_p;\ R = Q$$

M_y = Moment about y axis
M_x = Moment about x axis
y = Distance from x axis at pile centre
x = Distance from y axis at pile centre
I_y = Moment of inertia of the pile group about y axis
$= \Sigma$ Moment of inertia of A_p about y axis
$= \Sigma\left(I_o + A_p x^2\right)$
$= A_p \Sigma x^2$ (Considering ΣI_o has a negligible value)
I_x = Moment of inertia of the pile group about x axis
$= \Sigma$ Moment of inertia of A_p about x axis
$= \Sigma\left(I_o + A_p y^2\right)$
$= A_p \Sigma y^2$ (Considering ΣI_o has a negligible value)

Therefore, the combined stress equation can be written in the form

$$\frac{P_p}{A_p} = \frac{Q}{nA_p} \pm \frac{M_y}{A_p \Sigma x^2} x \pm \frac{M_x}{A_p \Sigma y^2} y;\ \text{hence,}$$

$$P_p = \frac{Q}{n} \pm \frac{M_y}{\Sigma x^2} x \pm \frac{M_x}{\Sigma y^2} y \tag{5.31}$$

5.8 Plain Concrete Spread Footings

Structural plain concrete is concrete with no reinforcement or with less reinforcement than the minimum amount specified for reinforced concrete (ACI Section 2.2). Chapter 22 of ACI 318M-08 building code concerns design requirements of structural plain concrete members. However, according to ACI Section 22.1.2, unless in conflict with the provisions of Chapter 22, the provisions of ACI Sections 1.1–7.5, 7.6.1, 7.6.2, 7.6.4, 7.7, 9.1.3, 9.2 and 9.3.5 shall also apply to structural plain concrete. Also, provisions of ACI Chapter 20, ACI Section 21.12.2.5, Sections C.9.2 and C.9.3.5 of ACI Appendix C, and ACI Appendix D are applied.

According to ACI Section 22.2.1, use of structural plain concrete shall be limited to: (a) members that are continuously supported by soil or supported by other structural members capable of providing continuous vertical support; (b) members for which arch action provides compression under all conditions of loading; (c) walls and pedestals. Therefore, it is clear that structural plain concrete can be used for construction of spread footings. However, plain concrete shall not be used for footings on piles (ACI Section 22.7.3). Generally, plain concrete footings are only practical and economical for small column or wall loads.

The structural design of plain concrete spread footings should satisfy the following requirements:

(1) Thickness of structural plain concrete footings$h$ shall not be less than 200 mm (ACI Section 22.7.4).

(2) When computing strength in flexure, combined flexure and axial load, and shear, the thickness$h$ shall be taken as 50 mm less than actual thickness (ACI Section 22.4.7).

(3) When load factor combinations of ACI Section 9.2.1 are used, the *strength-reduction factor* Ø shall be 0.6 for flexure, compression, shear, and bearing (ACI Section 9.3.5). When load factor combinations of ACI Section C.9.2.1 are used, the strength-reduction factor Ø shall be 0.65 for flexure, compression, shear, and bearing (ACI Section C.9.3.5).

(4) According to ACI Section 22.5.1, design of cross-sections subject to flexure shall be based on

$$Ø\, M_n \geq M_u \quad \text{[the same as Equation (5.9)]} \tag{5.32}$$

If tension controls

$$M_n = 0.42\lambda\sqrt{f_c'}\; S_m \tag{5.33}$$

If compression controls

$$M_n = 0.85 f_c'\; S_m \tag{5.34}$$

S_m = The corresponding elastic section modulus.

(5) Maximum factored moment shall be taken at the critical sections defined in ACI Section 22.7.5 (taken at the same critical sections as for reinforced-concrete footings).

(6) According to ACI Section 22.5.4, design of rectangular cross-sections subject to shear shall be based on

$$Ø\, V_n \geq V_u \tag{5.35}$$

where V_n is computed by

$$V_n = 0.11\lambda\sqrt{f_c'}\; b_w h \tag{5.36}$$

for one-way shear, and by

$$V_n = 0.11\left[1 + \frac{2}{\beta}\right]\lambda\sqrt{f_c'}\; b_o h \tag{5.37}$$

for two-way shear, but not greater than$0.22\lambda\sqrt{f_c'}\; b_o h$
where β corresponds to ratio of long side to short side of concentrated load or reaction area.

(7) Location of critical section for shear shall be measured from face of column, pedestal, or wall for footing supporting a column, pedestal, or wall. For footing supporting a column with steel base plates, the critical section shall be measured from location of critical section for moment (ACI Section 22.7.6.1).

(8) Critical section across the entire footing width for one-way shear is located at a distance h from face of concentrated load or reaction area (ACI Section 22.7.6.2). For this condition, the footing shall be designed in accordance with Equation (5.36).

(9) Critical section for two-way shear is located so that the perimeter b_o is a minimum, but need not approach closer than $h/2$ to perimeter of concentrated load or reaction area (ACI Section 22.7.6.2). For this condition, the footing shall be designed in accordance with Equation (5.37).

(10) Circular or regular polygon-shaped concrete columns or pedestals shall be permitted to be treated as square members with the same area for location of critical sections for moment and shear (ACI Section 22.7.7).

(11) According to ACI Sections 22.7.8 and 22.5.5, factored bearing load, B_u, on concrete at contact surface between supporting and supported member shall not exceed design bearing strength, $\varnothing B_n$, for either surface as given in Equation (5.28). In case where the supporting surface is wider on all sides than the loaded area, then B_n shall be multiplied by $\sqrt{A_2/A_1}$ but not more than 2.

(12) Plain concrete shall not be used for footings on piles (ACI Section 22.7.3).

5.9 Combined Footings

5.9.1 General

Generally, a combined footing is a rectangular or trapezoidal reinforced concrete slab supporting more than one column in a line. A combined footing supporting two columns only is usually known as *ordinary combined footing* (Figure 2.3). A combined footing supporting more than two columns is usually called *continuous combined footing* or *strip footing*. Another form of combined footing is *strap* (or *cantilever*) footing which consists of two or more spread footings connected by a rigid beam called *strap*. All these footings are more complicated foundation members than isolated footings due to complexity of their loading and geometry. Therefore, it is appropriate to conduct a more rigorous structural analysis. The available methods of analysis are the *rigid method* (also known as the *conventional method*) and the *flexible method* (also known as the *beam on elastic foundation method*), as reported by ACI Committee 336 (1993).

The rigid method assumes that the footing is much more rigid than the underlying soils, which means any distortions in the footing are too small to significantly impact the distribution of soil bearing pressure. Consequently, the magnitude and distribution of bearing pressure depends only on the applied loads and footing geometry, which is assumed either uniform across the bottom of the footing (if the normal resultant load acts through the centroid) or varies linearly (if the loading is eccentric); thus simplifying the design computations. This method of design, in spite of its substantial approximations and inaccuracies, can be considered appropriate for most combined footings.

For large or heavily loaded combined footings such as continuous strip footings or grid foundations, it may be more appropriate to conduct structural design based on a beam on elastic foundation analysis, as described in Section 5.11. This analysis produces more reliable estimates of the shears, moments and deformations in the footing. However, unfortunately, it is more difficult to implement as it requires consideration of soil-structure interaction which is not as simple. The analysis requires a computer program for maximum design efficiency, such as the program B-5 (FADBEMLP) given in Bowles (2001).

Structural designs will be based on the Load and Resistance Factor Design (LRFD) method. Provisions of Chapter 15 of the ACI code, where applicable, shall apply for design of combined footings and mats also (ACI Section 15.1.1). The direct design method of Chapter 13 of the ACI code shall not be used for design of combined footings and mats (ACI Section 15.10.2).

The following Sections will describe a design procedure, based on the rigid method of analysis, for each of the ordinary rectangular combined footing, trapezoidal combined footing and strap footing.

5.9.2 Rectangular Combined Footings

According to the conventional method of analysis, the design is conducted on the assumption that the footing acts as a rigid body, and the contact pressure is assumed to follow a linear distribution, as discussed earlier. Also, the footing base area and its B and L dimensions should be established such that the maximum contact pressure (unfactored) at no place exceeds the given allowable soil pressure. The structural design steps may be summarised as follows:

(1) Determine the appropriate load factor combination, as discussed in Section 5.2. Then, determine the resultant (R) of all column service (working) loads and moments and the factored resultant (R_{fact}) of all column factored loads and moments.

(2) Determine the factored net contact pressure ($net\ q_{fact}$) using

$$A = \frac{R}{net\ q_a} = \frac{R_{fact}}{net\ q_{fact}}.\text{Hence,}$$

$$net\ q_{fact} = (net\ q_a)\left(\frac{R_{fact}}{R}\right)$$

(3) Find footing dimensions B and L. First find the location of R_{fact} by taking moments about the centre of column 1 (exterior column). Let this location represent the centre of the footing base area A and the centre of contact pressure instantaneously; hence, $L/2$ equals the distance between the centre and the given property line (exterior edge of the footing). Find $B = R/L(net\ q_a)$.

(4) Draw factored load, shear and moment diagrams considering the footing as a reinforced concrete *beam*. The diagrams should give shear and moment values at the critical sections and other necessary locations. For convenience, take each column load as a concentrated load acting at the column centre.

(5) Find the thickness of the footing by calculating d based on analysis for both wide-beam and two-way shear at the most critical sections.

(6) Find steel for bending in long direction required by the maximum (+) and (−) bending moments. Steel bars for (+) bending shall be placed in the bottom of the footing near the columns, and those for (−) bending in the top near or in the centre portion between columns. When it is found that trying to cut the (−) bars in accordance with ACI code requirements is not worth the extra engineering and bar placing effort, it would be preferable to extend the bars full length of the footing. Also, enough number of the (+) bottom bars are needed full length of the footing so that the transverse bottom bars (bars in short direction) can be supported. The design should provide number, size, spacing and length of steel bars for each type of reinforcement, in accordance with ACI code requirements.

(7) Check the development of the top and bottom steel bars in long direction in accordance with the requirements of ACI Section 12.2.

(8) Find steel for bending in short direction required at bottom of the footing. For this purpose (Bowles, 2001), the footing is divided into three zones or strips of the defined widths shown in Figure 5.10. Zones I and II, usually known as *effective zones*, should be analysed as beams; the provided steel should not be less than that required for bending or $A_{s(min)}$, whichever is greater. For zone III (the remaining portions), the provided steel should satisfy $A_{s(min)}$ requirement only.

(9) Check the development of the bottom steel bars in the short direction in accordance with the requirements of ACI Section 12.2.

(10) Check column bearing on the footing at the location of each column, and design the column-to-footing dowels. Follow the same procedure used for isolated spread footings, as discussed in Section 5.4.4.

5.9.3 Trapezoidal Combined Footings

A trapezoidal combined footing may be necessary if a rectangular combined footing cannot provide the assumed uniform contact pressure. This condition will exist if the column which has too limited space for a spread footing carries much larger load. In this case the resultant of all column loads (including moments) will be much closer to the heavier column, and doubling the centroid distance as done for the rectangular combined footing will not provide sufficient footing length to reach the other column. In order to obtain uniform contact pressure, the location of centroid can be adjusted to agree with location of the resultant load by using a trapezoid-shaped base area with dimensions L, B_1 and B_2, where B_2 is the larger width near or adjacent to the heavier column, as shown in Figure 5.11. Using the known values of the base area A, footing length L and centroid distance x', the unique values of widths B_1 and B_2 are computed by solving the following two equations simultaneously:

$$A = \frac{B_1 + B_2}{2} L \tag{5.38}$$

$$x' = \frac{L}{3}\left(\frac{2B_1 + B_2}{B_1 + B_2}\right) \tag{5.39}$$

Equation (5.39) reveals that the solution for $B_1 = 0$ is a triangle and for $B_1 = B_2$ is a rectangle. Therefore, a trapezoid solution exists only for $(L/3) < x' < (L/2)$ with the minimum value of L as out-to-out of the column faces.

When the values of B_1 and B_2 are computed, the footing is treated similarly to the rectangular combined footing (as a reinforced concrete beam); following the same design steps mentioned in

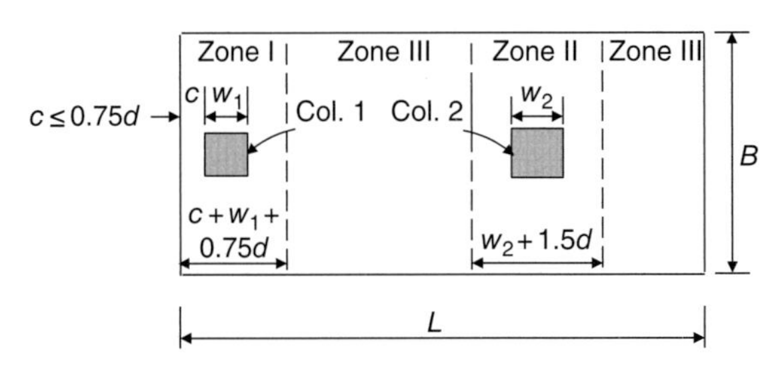

Figure 5.10 Zones (transverse strips) for steel bars at bottom of a rectangular combined footing in short direction.

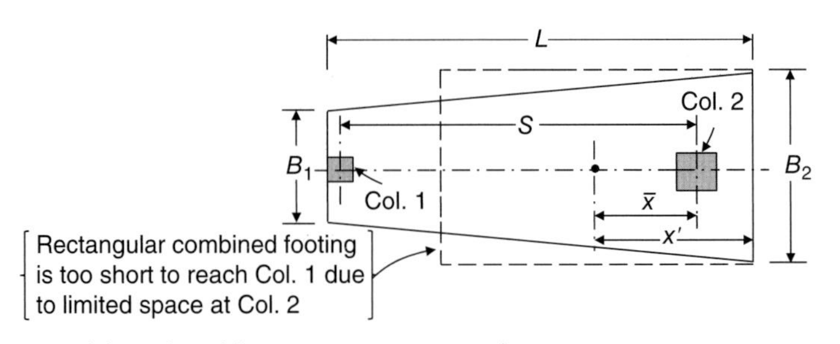

Figure 5.11 A trapezoidal combined footing supporting two columns.

Section 5.9.2. However, it must be realised that the contact pressure per unit length of the footing will vary linearly due to the varying width from B_1 to B_2. Therefore, the shear diagram will be a second-degree curve and the moment diagram will be a third-degree curve.

In most cases a trapezoidal combined footing would be used with only two columns as shown in Figure 5.11, but the solution proceeds similarly for more than two columns.

5.9.4 Strap (or Cantilever) Footings

A strap (or cantilever) footing is a combined footing designed as two spread footings connected by a rigid beam called *strap*. Strap footings may have a number of arrangements or configurations (see Figure 2.5); however, the one which produce the high degree of rigidity is more desirable (Figure 5.12). The main purpose in using the strap is to transmit the moment of the eccentrically loaded exterior footing to the interior column footing so that a uniform contact pressure is assumed beneath each footing. A strap footing may be used in the case when the distance of the resultant of all column loads (including moments) from the exterior face of one of the columns (i.e. x') is less than $L/3$, where L is the footing length out-to-out of the column faces. In this case neither rectangular nor trapezoid combined footing can be used. Also, if the distance between columns is relatively large and/or the design soil pressure is relatively high, a strap footing will be preferable and more economical than rectangular or trapezoid combined footings. However, due to extra labor and forming costs, a strap footing should be considered only after a careful analysis shows that the other footing types including isolated footings (even if oversize) will not be satisfactory.

Refer to Figure 5.12. The distance between columns S, column widths w_1 and w_2, loads P_1 and P_2, and footing projection c have given values. The contact pressure is assumed uniformly distributed beneath each footing and their resultants R_1 andR_2 is acting at the footings centre. These resultants are computed only when the distance S' is known. Therefore, it will be necessary to assume an appropriate value for either the eccentricity e or length L_1, since $S' = S - e$ or $S' = S - [(L_1/2) - (w_1/2) - c]$. When the resultants R_1 and R_2 are found, the footing dimensions are proportioned using the equation $R_i = BLq_a$, where q_a is the design soil pressure.

In conventional design of a strap footing the following points should be considered:

(1) The strap dimensions must provide adequate rigidity so that any rotation of the footings is avoided. According to Bowles (2001), it may be necessary that the design provide $(I_{strap}/I_{footing}) > 2$.
(2) The strap should not be subjected to any direct soil pressure from below (Figure 5.12). Usually, the strap weight is neglected in the design.
(3) Width of the strap should be at least equal to the smallest column width. In case the strap depth is restricted, it would be necessary to increase the width to obtain the necessary rigidity.

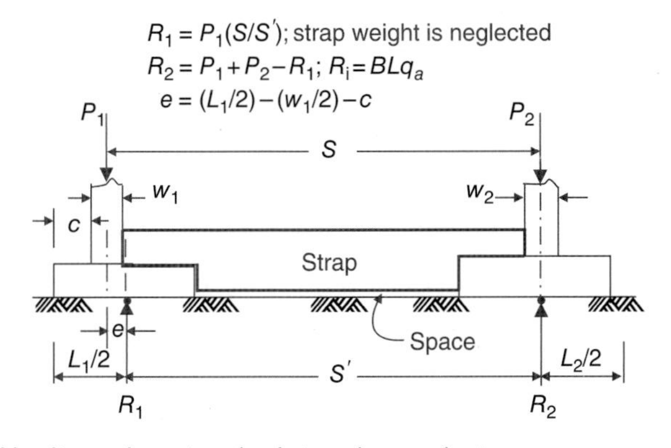

Figure 5.12 Assumed loading and reactions for design of a strap footing.

(4) The strap shall be designed as a reinforced concrete beam. However, it may be desirable not to use shear reinforcement in either the strap or the two footings so that structural rigidity of the system is increased. It is necessary to check depth to span (clear span between footings) to see if it is a *deep* beam (ACI Section 10.7).

(5) The strap must be properly connected to the footing and column by sufficient dowels so that the system acts as a unit.

(6) An appropriate value for either the eccentricity$e$ or length L_1 must be selected so that the footing widths B_1 and B_2 will not be greatly different. This is necessary to control or reduce footing differential settlement.

(7) Footing thicknesses are designed for the worst case of two-way action or wide-beam shear. The design wide-beam shear values are obtained from the factored shear diagram.

(8) Design footing reinforcing as a spread footing for both directions.

(9) It is not possible to obtain a unique design, since the footing dimensions are dependent upon the designer's arbitrarily selected value of the eccentricity e or length L_1.

5.10 Modulus of Subgrade Reaction

All flexible methods of structural analysis require the relationship between deflection (settlement) and soil pressure be defined. This relationship is known as *modulus* (or *coefficient*) *of subgrade reaction*, K_s, expressed as:

$$K_s = \frac{q}{\delta} \tag{5.40}$$

where q = bearing pressure; δ deflection (settlement)

It is important to realise that the modulus of subgrade reaction is not a constant for a given soil, but rather depends on several factors, such as load intensity, size and shape of the foundation base and foundation embedment depth. Furthermore, it is different for different points at the base of the same foundation. Also, it should be clear that the load-settlement relationship of a soil is non-linear, so the K_s value in the above equation must represent some assumed linear function, such as an initial tangent or secant modulus, as shown in Figure 5.13.

The modulus of subgrade reaction may be estimated on the basis of field experiments, such as plate-load tests, which yield load-deflection relationships, or on the basis of known soil characteristics. For small foundations of width B less than about 1.5 m, K_s may be estimated from results of plate-load tests

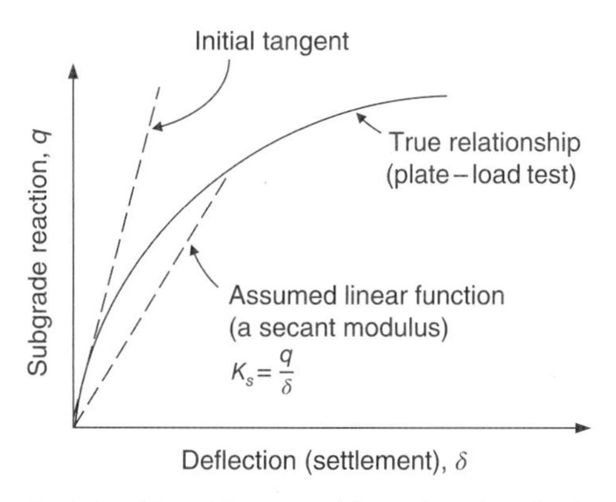

Figure 5.13 A non-linear $q - \delta$ relationship, with assumed linear functions for the modulus of subgrade reaction, K_s.

as suggested by Terzaghi (1955). He proposed certain equations for estimating K_s for full-sized footings on clay and sand soils (see Das, 2011; Bowles, 2001). However, even for small-sized footings, the evaluation of K_s involves many uncertainties, and the customary procedure for determining K_s on the basis of small-scale load tests is subject to all limitations of the plate-load test described in the discussion of Problem 1.9. For mat foundations, K_s cannot be reliably estimated on the basis of field plate-load tests because the scale effects are too severe.

According to Vesic (1961), the modulus of subgrade reaction could be computed as

$$K_s = \frac{K_s'}{B} = \frac{1}{B}\left[0.65\sqrt[12]{\frac{E_s B^4}{E_f I_f}}\,\frac{E_s}{1-\mu_s^2}\right] \tag{5.41}$$

where

K_s' = modulus of subgrade reaction, in units of E_s
K_s = modulus of subgrade reaction, in units of unit weight
E_s = deformation modulus of soil
E_f = modulus of elasticity of foundation material
μ_s = Poisson's ratio of soil
I_f = moment of inertia of the cross-section of the foundation
B = foundation width

For all practical purposes, value of $\left(0.65\sqrt[12]{\frac{E_s B^4}{E_f I_f}}\right)$ can be approximated as one; hence, the above Vesic equation reduces to

$$K_s = \frac{E_s}{B\left(1-\mu_s^2\right)} \tag{5.42}$$

Equation (5.42) shows a direct relationship between K_s and E_s. According to Bowles (2001), one may rearrange the general equation of immediate settlement, Equation (3.5), using $E_s' = \left(1-\mu_s^2\right)/E_s$, and obtain

$$S_i = \delta = q \times B \times E_s' \times m \times I_S \times I_F.$$

Hence,

$$K_s = \frac{q}{\delta} = \frac{1}{B \times E_s' \times m \times I_S \times I_F} \tag{5.43}$$

Carefully note in using Equation (5.43) that its basis is in the settlement equation of Chapter 3, Equation (3.5), and use B, m, I_S and I_F as defined there.

Also, Bowles has suggested the following equation for approximating K_S from the allowable bearing capacity q_a provided by the geotechnical engineer:

$$K_s = 40\,(SF)\,q_a \tag{5.44}$$

where

K_S = modulus of subgrade reaction in kN/m^3.
SF = safety factor
q_a = allowable bearing capacity in kPa

Table 5.3 Values of modulus of subgrade reaction, K_s (from Bowles, 2001).

Soil	K_s, kN/m^3
Loose sand	4800–16 000
Medium dense sand	9600–80 000
Dense sand	64 000–128 000
Clayey medium dense sand	32 000–80 000
Silty medium dense sand	24 000–48 000
Clayey soil:	
$q_a \leq 200$ kPa	12 000–24 000
$200 < q_a \leq 800$ kPa	24 000–48 000
$q_a > 800$ kPa	>48 000

Table 5.4 Typical values of modulus of subgrade reaction, K_s (from Das, 2001).

Soil	K_s, MN/m^3
Dry or moist sand:	
Loose	8–25
Medium	25–125
Dense	125–375
Saturated sand:	
Loose	10–15
Medium	35–40
Dense	130–150
Clay:	
Stiff	10–25
Very stiff	25–50
Hard	>50

Tables 5.3 and 5.4 provide approximate ranges of K_S values for sandy and clayey soils which may be used as a guide and for comparison when using the approximate equations.

The K_S values of Table 5.4 were obtained from results of load tests using square plates of size 0.3 × 0.3 m.

Equation (5.44) is based on relationships: $q_a = \frac{q_{ult}}{SF}$; $K_s = \frac{q_{ult}}{\delta}$; and q_{ult} is at a settlement $\delta =$ 0.0254 m (or 25 mm). According to Bowles, for $\delta = 6,\ 12,\ 20$ mm and so on, the factor 40 can be adjusted to 165, 83, 50 and so on. Factor 40 is reasonably conservative but smaller assumed displacements can always be used.

5.11 Beams on Elastic Foundations

When flexural rigidity of a continuous strip footing (either isolated or taken from a mat) is taken into account, the foundation is considered as an elastic (or flexible) member interacting with an elastic soil; analysed on the basis of some form of a *beam on elastic foundation*. The analysis may be based on: (1) *classical solutions*, (2) *discrete-element formulations*.

(1) *Classical solutions.*

A classical solution (also known as classical Winkler solution) is an approximate flexible method of analysis in which the soil is considered to be equivalent to a bed of infinite number of coil springs (Figure 5.14) used to compute the shears, moments, and deformations in the supported structural base. This equivalent foundation is sometimes referred to as the *Winkler foundation* because the earliest use of these "springs" to represent the interaction between soil and foundations has been attributed to Winkler (1867). The elastic stiffness of these assumed springs represents the modulus of subgrade reaction, K_S.

Refer to Figure 5.15. The beam has an *infinite* length and a width B; rests on an elastic soil equivalent to an infinite number of elastic springs. Using deflection $\delta = y$, soil reaction per unit length of the beam $= q$, and $K_s' = K_s B$, then from Equation (5.40)

$$q = -y K_s' \tag{5.45}$$

From the mechanic of materials

$$M = E_f I_f \frac{d^2y}{dx^2}; \quad V = \frac{dM}{dx}; \quad q = \frac{dV}{dx}$$

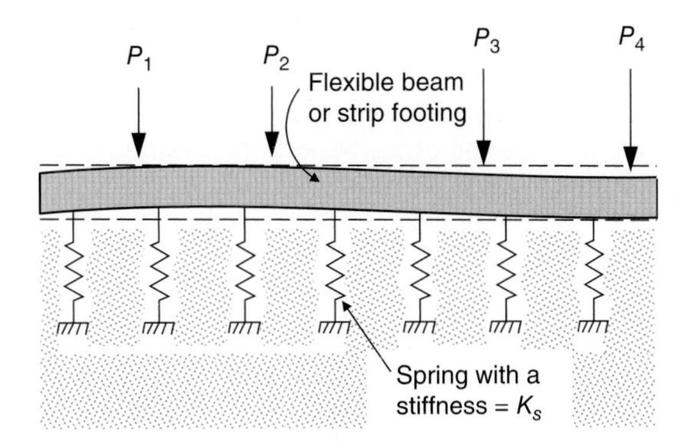

Figure 5.14 A flexible foundation on a "bed of springs" (Winkler concept).

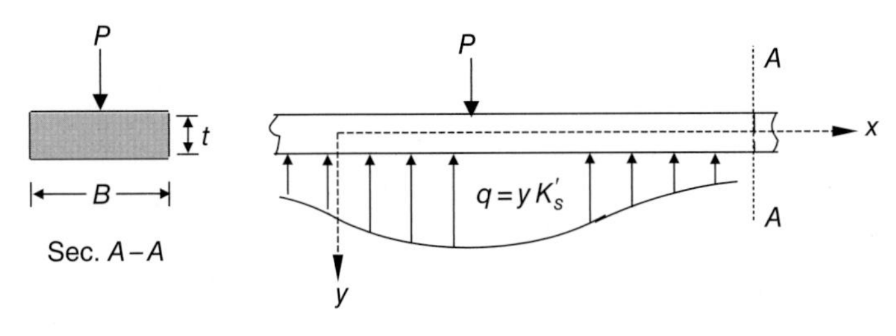

Figure 5.15 Beam of infinite length on elastic foundation.

hence, $q = E_f I_f \frac{d^4y}{dx^4}$, and from Equation (5.45)

$$E_f I_f \frac{d^4y}{dx^4} = -yK_s' \tag{5.46}$$

where

M = moment at any section
V = shear force at any section
E_f = modulus of elasticity of the material of the beam
I_f = moment of inertia of the cross-section of the beam

The solution of Equation (5.46) is

$$y = e^{\lambda x}(C_1 \cos\lambda x + C_2 \sin\lambda x) + e^{-\lambda x}(C_3 \cos\lambda x + C_4 \sin\lambda x) \tag{5.47}$$

where factor $\lambda = \sqrt[4]{\frac{K_s'}{4E_f I_f}} = \sqrt[4]{\frac{K_s B}{4E_f I_f}}$

The factor λ is impotant in determining whether a foundation should be analysed on the basis of the conventional rigid procedure or as a beam on an elastic foundation (Section 5.12.1). It has been suggested that rigid members of length L should have $\lambda L < (\pi/4)$, flexible members have $\lambda L > \pi$, and members of intermediate flexibility have $(\pi/4) < \lambda L < \pi$. However, these criteria are of limited application because of the number of loads and their locations along the member.

The closed-form solution of the basic differential equations concerning the deflections, slopes, moments and shears in the beam of Figure 5.15, for several loadings, is given in Table 5.5.

Deflections, moments and shears in a beam of *finite* length L, rests on an elastic foundation and carries a concentrated load P at any point (Figure 5.16), are given in Table 5.6. The equations were developed by Hetenyi (1946) using the *Winkler concept*. The distance x to use in the equations is measured from the end of the beam to the point for which the deflection, moment, or shear is desired. For $x < a$, use the equations as given, and measure$x$ from E (Figure 5.16). For $x > a$, replace a with b in the equations, and measure x from F. Applying the boundary conditions, as illustrated in Figure 5.16, the equations of Table 5.6 can be expressed as

$$y = \frac{P\lambda}{K_s'}A' \quad M = \frac{P}{2\lambda}B' \quad V = PC'$$

where the coefficients A', B' and C' are the values for the hyperbolic and trigonometric remainder of the equations. These coefficients are so complicated that a computer program will be required to carry out the design computations.

The classical Winkler solutions, being of closed form, suffer from many problems and difficulties. According to the classical approach each spring is linear and acts independently from the others, and that all of the springs have the same K_S. In other words, it is difficult to allow for change in subgrade reaction along footing. In reality, the load-deformation behavior of soil is nonlinear, and a load at one point on a footing induces deflection both at that point and in the adjacent parts of the footing. Another problem with the classical solution it is difficult to account for boundary conditions of known rotation or deflection at desired points. Also, it is difficult to apply multiple types of loads to a footing. The classical solution assumes weightless footing, but weight will be a factor when footing tends to separate from the soil. Due to all these shortcomings, the classical approach is rarely a better model than a discrete-element analysis.

Table 5.5 Closed-form solution of a beam of infinite length on an elastic foundation (from Bowles, 1982).

Concentrated load at end	**Moment at end**
$y=\dfrac{2V_1\lambda}{k'_s}D_{\lambda x}$	$y=\dfrac{-2M_1\lambda^2}{k'_s}C_{\lambda x}$
$\theta=\dfrac{-2V_1\lambda^2}{k'_s}A_{\lambda x}$	$\theta=\dfrac{4M_1\lambda^3}{k'_s}D_{\lambda x}$
$M=\dfrac{-V_1}{\lambda}B_{\lambda x}$	$M = M_1A_{\lambda x}$
$Q = -V_1C_{\lambda x}$	$Q = -2M_1\lambda B_{\lambda x}$
Concentrated load at centre, + ↓	**Moment at centre, + ↷**
$y=\dfrac{P\lambda}{2k'_s}A_{\lambda x}$	$y=\dfrac{M_0\lambda^2}{k'_x}B_{\lambda x}$ deflection
$\theta=\dfrac{-P\lambda^2}{k^r_i}B_{\lambda x}$	$\theta=\dfrac{\mathrm{M}_0\lambda^3}{k'_s}C_{\lambda x}$ slope
$M=\dfrac{P}{4\lambda}C_{\lambda x}$	$M=\dfrac{M_0}{2}D_{\lambda x}$ moment
$Q=\dfrac{-P}{2}D_{\lambda x}$	$Q=\dfrac{-M_0\lambda}{2}A_{\lambda x}$ shear

The A, B, C, and D coefficients (use only $+\ x$) are as follows:

$$A_{\lambda x} = e^{-\lambda x}(\cos\lambda x + \sin\lambda x)$$

$$B_{\lambda x} = e^{-\lambda x}\sin\lambda x$$

$$C_{\lambda x} = e^{-\lambda x}(\cos\lambda x - \sin\lambda x)$$

$$D_{\lambda x} = e^{-\lambda x}\cos\lambda x$$

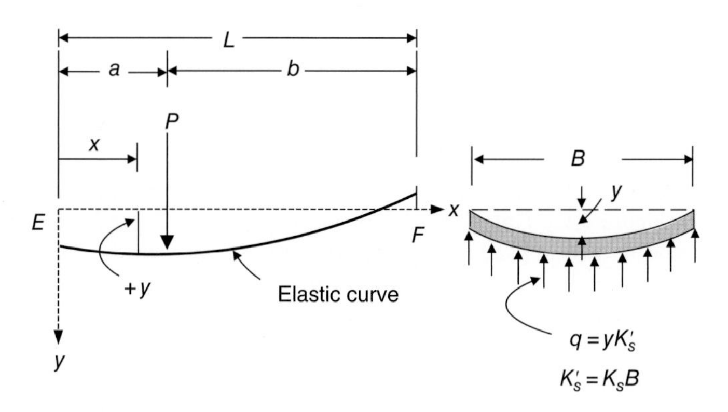

Figure 5.16 Beam of finite length on elastic foundation.

Table 5.6 Closed-form solution of a beam of finite length on an elastic foundation (from Bowles, 2001).

Deflection	$y=\dfrac{P\lambda}{k_s'(\sinh^2\lambda L-\sin^2\lambda L)}\{2\cosh\lambda x\cos\lambda x(\sinh\lambda L\cos\lambda a\cosh\lambda b-\sin\lambda L\cosh\lambda a\cos\lambda b)$ $+(\cosh\lambda x\sin\lambda x+\sinh\lambda x\cos\lambda x)[\sinh\lambda L(\sin\lambda a\cosh\lambda b-\cos\lambda a\sinh\lambda b)$ $+\sin\lambda L(\sinh\lambda a\cos\lambda b-\cosh\lambda a\sin\lambda b)]\}$
Moment	$M=\dfrac{P}{2\lambda(\sinh^2\lambda L-\sin^2\lambda L)}\{2\sinh\lambda x\sin\lambda x(\sinh\lambda L\cos\lambda a\cosh\lambda b-\sin\lambda L\cosh\lambda a\cos\lambda b)$ $+(\cosh\lambda x\sin\lambda x-\sinh\lambda x\cos\lambda x)$ $\times[\sinh\lambda L(\sin\lambda a\cosh\lambda b-\cos\lambda a\sinh\lambda b)+\sin\lambda L(\sinh\lambda a\cos\lambda b-\cosh\lambda a\sin\lambda b)]\}$
Shear	$Q=\dfrac{P}{\sinh^2\lambda L-\sin^2\lambda L}\{(\cosh\lambda x\sin\lambda x+\sinh\lambda x\cos\lambda x)$ $\times(\sinh\lambda L\cos\lambda a\cos\lambda b-\sin\lambda L\cosh\lambda a\cos\lambda b)$ $+\sinh\lambda x\sin\lambda x\,[\sinh\lambda L(\sin\lambda a\cosh\lambda b-\cos\lambda a\sinh\lambda b)$ $+\sin\lambda L(\sinh\lambda a\cos\lambda b-\cosh\lambda a\sin\lambda b)]\}$

(2) *Discrete-element formulations.*
Among the discrete-element methods of analysis, the finite-element method (FEM) is the most efficient and reliable means for developing solutions to problems of beams on elastic foundations [based on Equation (5.46)]. With this method it becomes easy to overcome most of the above-mentioned shortcomings the classical solutions suffer from. However, the FEM requires a digital (or personal) computer. It is practical only when written into a computer program, because there are usually too many equations for hand solving. Compared to the finite-difference method (FDM), the FEM is more capable and adaptable because an equation model for one element can be used for all the other elements in the beam model. The FDM requires all the elements have the same length and cross-section, otherwise great difficulty will arise. Also, it requires a different equation formulation for end elements than for the interior ones, and modeling boundary conditions is not easy. However, the finite difference method may provide good results for the approximations used, and it does not require much computer memory.

The FEM divides the beam into a number of elements as desired. Each element has certain defined dimensions, a specified stiffness and strength, and is connected to the adjacent elements at certain points or *nodes* in a specified way. The beam elements and their nodal points are numbered, and connected to the ground through a series of "springs", which are defined using the modulus of subgrade reaction. Typically, one spring is required at each nodal point, as shown in Figure 5.17. Thus, a *beam-finite element* model is produced. The loads on the beam include the externally applied column loads and the weight of the beam itself. These loads cause downward movement of the beam elements (or the nodes) which is resisted by the supporting springs. These resisting forces, along with the stiffness of the beam, can be evaluated simultaneously using matrix algebra, which allows us to compute the induced deflections, moments and shears in the beam. The fundamental components (developed equations and element matrices) of FEM are found in Bowles (2001), and if more background is required the reader is referred to references, such as Zienkiewicz (1977), Cook (1974), Bowles (1974a) or Wang (1970). There are computer programs, such as the program B-5 (FADBEMLP) given in Bowles (2001), which can be used to illustrate the procedure.

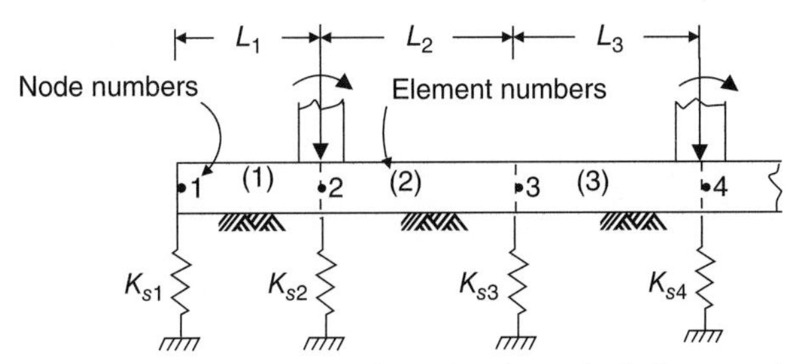

Figure 5.17 A continuous beam or strip footing on elastic foundation (bed of springs) divided into a number of beam finite elements.

5.12 Mat Foundations

5.12.1 General

Mat, raft and floating foundations were described in some detail in Section 2.2. Usually, a mat foundation is a reinforced concrete slab, relatively thick and with or without depressions or openings, supporting an array of columns in several lines both ways. Most mat foundations are supported directly on the underlying soils. However, mats may be supported on piles, or, partially on the underlying soil, and use piles to carry the balance of the load. These foundation options may be necessary when the contact pressure is too high or the underlying supporting soil is too compressible; that a soil-supported mat may experience large excessive settlement. This Section concerns structural design of mat foundations supported on soils. Depending upon the flexural rigidity of foundation, the methods of design of mat foundations can be classified into the following two categories:

(1) *Rigid* (or *conventional*) *methods.*
As mentioned earlier (Section 5.9.1), the conventional method assumes that the base (mat) is perfectly rigid and that the contact pressure follows a planar distribution; its centroid coincides with the line of action of the resultant force of all loads acting on the base. This method does not consider redistribution of contact pressure beneath the mat, and it assumes that the individual columns will not settle differentially. These simplifying assumptions make it easy to compute the shears, moments, and deflections using the principles of structural mechanics. However, although the conventional analysis is appropriate for isolated and most of combined footings, it does not reliably model mat foundations because the rigidity assumption, in most of cases, is no longer valid. A continuous strip footing (or a mat which consists of continuous strip footings in both directions) having adjacent column loads and column spacing vary by not more than 20% of the greater value, and the average of two adjacent spans is less than $1.75/\lambda$, can be considered rigid and the variation of soil pressure determined on the basis of simple statics (ACI Committee 336). The conventional design method may be adequate where these rigidity requirements are satisfied. The factor λ is

$$\lambda = \sqrt[4]{\frac{K_s B}{4E_c I}} \tag{5.48}$$

where

K_S = coefficient (or modulus) of vertical subgrade reaction
E_C = modulus of elasticity of concrete
I = moment of inertia of the beam (strip) section

The geotechnical engineer should furnish the designer K_S values even when a simplified design method is used. The design procedure will be described in Section 5.12.2.

(2) *Flexible methods.*
For general cases falling outside the rigidity limitations given in method (1), it is recommended to design the mat as a flexible plate supported by an elastic foundation (the soil). Flexible methods include the following:
(A) *Approximate flexible method.* It is primarily based on the theory of plates on elastic foundation, using the Winkler's concept (i.e., the plate or mat rests on a bed of "springs" which have the same K_S). This method was proposed by ACI Committee 336 (1993) to calculate moments, shears, and deflections at all points in a mat foundation with the help of charts. The design procedure will be briefly described in Section 5.12.3.
(B) *Discrete (finite) element methods.* In these methods the mat is divided into a number of discrete elements using grid lines. These methods use modulus of subgrade reaction K_S as the soil contribution to the structural model. There are three general discrete element formulations which may be used:
- Finite difference method (FDM)
- Finite element method (FEM)
- Finite grid method (FGM).

5.12.2 Design Procedure for the Conventional (Rigid) Method

The procedure for design a mat foundation, using the conventional method, consists of the following steps:

(1) All the columns and walls are numbered and their total unfactored (working) axial loads, moments and any overturning moment due to wind or other causes, are calculated separately. The mat self-weight, however, may not be considered because it is taken directly by the supporting soil. It may be useful if a suitable table is used for this purpose. The table should also contain summations of all the vertical loads and moments, that is Σ (column loads), ΣM_x and ΣM_y.
(2) The line of action of the resultant R of all the axial loads and moments is determined using statics; summing moments about two of the adjacent mat edges, and computing the $\bar{x}$ and $\bar{y}$ moment arms of the resultant force. Then, the eccentricities e_x and e_y are computed (Figure 5.18a).
(3) Determine the maximum contact pressure (unfactored), q_{max}, below one of the mat corners using Equation (2.7), repeated here for convenience.

$$q = \frac{R}{BL}\left(1 \pm \frac{6e_x}{B} \pm \frac{6e_y}{L}\right) \quad \text{Equation (2.7)}$$

In this equation, the B and L dimensions are in x and y directions, respectively.
(4) Compare the computed q_{max} with the allowable soil pressure q_a furnished by the geotechnical engineer to check if $q_{max} \leq q_a$.
(5) Compute the contact pressure (unfactored) at selected points beneath the mat. These selected points are corners of continuous beam strips (or combined footings with multiple columns) to which the mat is divided in both x and y directions, as shown in Figure 5.18b. The contact pressure q at any point below the mat is computed using Equation (2.6), repeated here:

$$q_{(x,\,y)} = \frac{R}{A} \pm \frac{M_y}{I_y}x \pm \frac{M_x}{I_x}y \quad \text{Equation (2.6)}$$

(6) Check static equilibrium of each individual beam strip, and modify the column loads and contact pressures accordingly. Since a mat transfers load horizontally, any given strip may not satisfy a vertical load summation (vertical equilibrium) unless consideration is given to the shear transfer

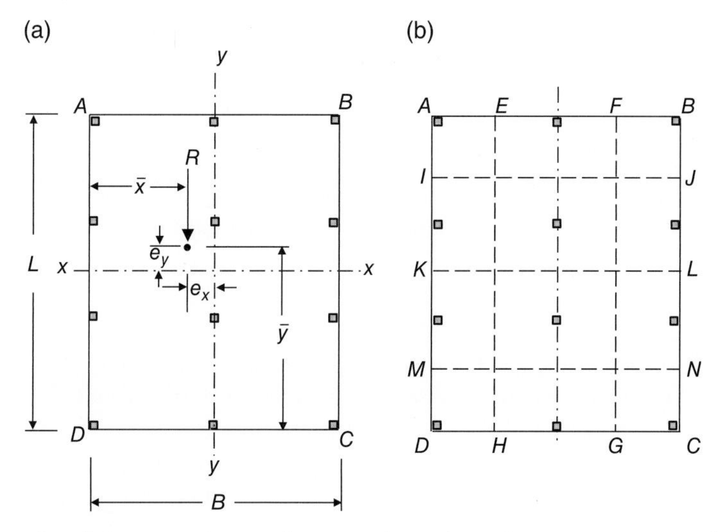

Figure 5.18 A mat foundation supporting 12 columns. (a) The resultant *R* has eccentricities e_x and e_y. (b) The mat is divided into beam strips in both directions.

between strips. However, each strip acts independently as assumed without considering any shear transfer. Therefore, vertical equilibrium may not be satisfied. As Bowles (1982) suggests, we may average the column loads and soil pressures so a strip is in vertical equilibrium. For example, take strip AEHD (Figure 5.18b):

Let the average soil (contact) pressure on the strip be q_{av}, and the strip area be *A*; hence, the upward soil reaction is Aq_{av}. Let the downward column loads on the strip be Σ column loads. Therefore, the average load on the strip is

$$Q_{av} = \frac{1}{2}(\Sigma \text{ column loads} + Aq_{av}) \tag{5.49}$$

The column load modification factor is

$$MF_{(col.)} = \frac{Q_{av}}{\Sigma \text{ column loads}} \tag{5.50}$$

Each column load on the strip should be multiplied by the factor $MF_{(col.)}$. The contact pressure modification factor is

$$MF_{(soil)} = \frac{Q_{av}}{Aq_{av}} \tag{5.51a}$$

Assume $\bar{q}_{av}$ represents the modified average soil pressure, then

$$\bar{q}_{av} = q_{av}\left(\frac{Q_{av}}{Aq_{av}}\right) = \frac{\left(MF_{(col.)}\right)\Sigma \text{ column loads}}{A} \tag{5.51b}$$

Note if the resultant of the modified column loads (i.e. $\sum$ column loads) does not fall at the centre of strip, a non-uniform soil pressure diagram must be used; the soil pressure at any point is computed using Equation (2.6).

(7) Change each modified column load (dead load D + live load L) to factored load. For example, a factored modified column load may be computed as

$$1.2\left(MF_{(col.)}\right)(D)+1.6\left(MF_{(col.)}\right)(L)$$

However, in order to avoid repetition, this computation can be done directly as soon as the factor $MF_{(col.)}$ is determined; using the factored modified load as

$$\left(MF_{(col.)}\right)(1.2D+1.6L)$$

Compute modified factored contact pressure, modified $q_{factored}$, as follows:

If the resultant of the factored modified loads (modified $R_{factored}$) passes through the centre of strip, the modified $q_{factored}$ is simply given by

$$\text{modified } q_{factored}=\frac{\text{modified } R_{factored}}{A} \quad (5.52)$$

If the modified $R_{factored}$ has an eccentricity e_L (in x or y direction), modified $q_{factored}$ at any point below the strip is given by

$$\text{modified } q_{factored}=\frac{\text{modified } R_{factored}}{A}\pm\frac{(e_L)\left(\text{modified } R_{factored}\right)}{I_B}l \quad (5.53)$$

where

l = distance of point from the strip centre in L direction
I_B = moment of inertia of the strip area about an axis passes through the strip centre in B direction

(8) Draw factored load, shear and moment diagrams for the continuous beam strips in both directions.

(9) Determine the minimum mat thickness considering punching shear (two-way or diagonal tension shear) at critical columns, based on factored column load and shear perimeter, similarly as for a spread footing. Note that the factored actual column load should be used and not the factored modified column load. Columns adjacent to a mat edge often control the mat depth d, and may require investigation of a two-sided (corner column) and three-sided (side column) diagonal tension shear perimeter. Also, the condition of unbalanced moment transfer should be investigated, and the design must satisfy the requirements of ACI Sections 11.11.7.1.

It is common practice not to use shear reinforcement so that depth is a maximum. This increases the flexural stiffness (rigidity) and increases the reliability of using Equation (2.6) or Equation (2.7).

(10) Check the computed mat depth d considering beam shear (one-way shear) at the most critical section.For this purpose, the maximum factored shear value may be obtained from the factored shear diagram of the most critical strip. Furthermore, some designers also perform this checking computations at most critical section in a given direction taking the mat as a whole.

(11) Obtain the maximum factored positive and negative moments per unit width using the factored moment diagrams of the strips in each direction.

(12) Determine the positive and negative reinforcement required in each direction in accordance with the requirements of ACI 318M-08.

(13) Check development of the reinforcement in both directions in accordance with the requirements of ACI code-Chapter 12.

(14) Check column bearing on the mat at critical column locations and design the column-to-mat dowels. Follow the same procedure used for isolated spread footings, as discussed in Section 5.4.4.

5.12.3 Design Procedure for the Approximate Flexible Method

The procedure for design a mat foundation, using the approximate flexible method, consists of the following major steps:

(1) Find the required total thickness t of the mat. The computations proceed exactly in the same manner as that described in step 9 for the conventional rigid method.

(2) Determine the flexural rigidity D of the mat as

$$D = \frac{E_f t^3}{12\left(1 - \mu_f^2\right)} \tag{5.54}$$

where

E_f = modulus of elasticity of foundation material
μ_f = Poisson's ratio of foundation material.

(3) Determine the radius of effective stiffness L' as

$$L' = \sqrt[4]{\frac{D}{K_s}} \tag{5.55}$$

where K_S = coefficient of subgrade reaction.

The radius of influence of any column load is on the order of $3L'$ to $4L'$.

(4) Determine the radial moment M_r and tangential moment M_t at any point (in polar coordinates) caused by a column load P using the following equations:

$$M_r = -\frac{P}{4}\left[Z_4 - \left(\frac{1-\mu_f}{x}\right) Z_3'\right] \tag{5.56}$$

$$M_t = -\frac{P}{4}\left[\mu_f Z_4 - \left(\frac{1-\mu_f}{x}\right) Z_3'\right] \tag{5.57}$$

Where

Z_i = factors depend on x, obtained from Figure 5.19
x = distance ration $\left(\frac{r}{L'}\right)$ shown in Figure 5.19

In rectangular coordinates the above moment equations can be written as

$$M_x = M_r \cos^2\theta + M_t \sin^2\theta \tag{5.58}$$

$$M_y = M_r \sin^2\theta + M_t \cos^2\theta \tag{5.59}$$

where θ is the angle which the radius r makes with x – axis (Figure 5.19).

(5) Determine the shear force V per unit width of the mat caused by a column load as

$$V = -\frac{P}{4L'} Z_4' \tag{5.60}$$

(6) Determine the deflection δ caused by a column load as

$$\delta = \frac{P(L')^2}{8D} \quad \text{(at location of a column load)} \tag{5.61}$$

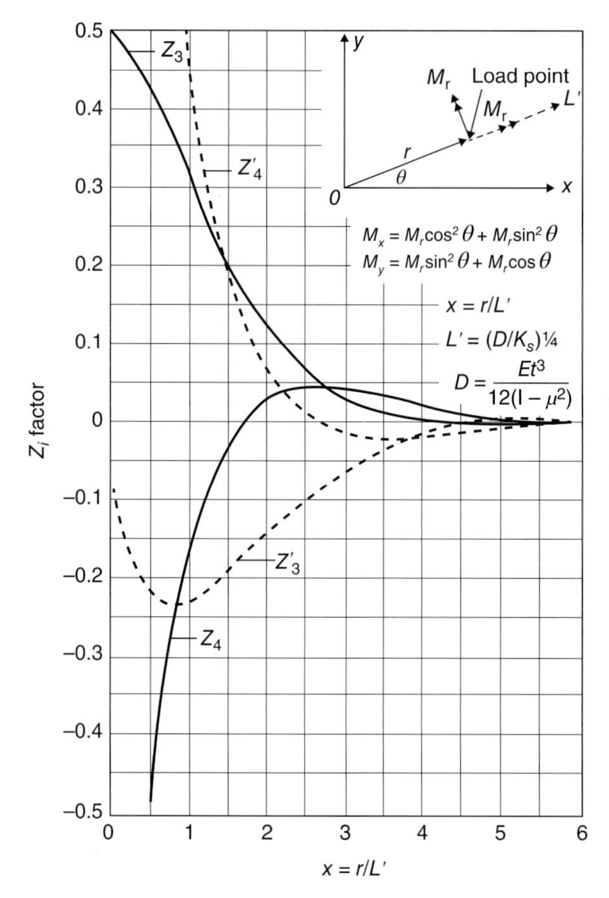

Figure 5.19 Z_i factors for computing moments, shears, and deflections in a flexible mat (reproduced from Bowles, 2001; after Hetenyi, 1946).

$$\delta = \frac{P(L')^2}{4D} Z_3 \quad \text{(at any point a distance } r \text{ from a column load)} \tag{5.62}$$

(7) When the zones of influence of two or more columns overlap, the method of superposition can be used to obtain the net values of responses (moment, shear and deflection) at any desired point.

(8) If the free edge of the mat is at a distance less than the radius of influence from an individual column load, a correction should be applied to the calculated response net values, as follows:

(a) Responses at the edge points of the mat due to column loads within the radius of influence should be calculated by Equations (5.56–5.62).

(b) Assuming strips of 1 m width as semi-infinite beams, shear and moments equal and opposite to those obtained in (a) should be applied as edge loads and their effects at various points superimposed on the net value of the respective response.

(c) Moment, shear, and deflection in a semi-infinite beam are

$$M = M_1 A_{\lambda x} - \frac{P_1}{\lambda} B_{\lambda x} \qquad V = -2 M_1 \lambda B_{\lambda x} - P_1 C_{\lambda x}$$

$$\delta = -\frac{2 M_1 \lambda^2}{K_s} C_{\lambda x} + \frac{2 P_1 \lambda}{K_s} D_{\lambda x}$$

Where M_1, P_1 = moment and shear, respectively, from (a)

$$\lambda = \left(\frac{K_s b}{4E_c I_b}\right)^{0.25}$$

b = strip width (1 m)
I_b = moment of inertia of the strip
$A_{\lambda x}$; $B_{\lambda x}$; $C_{\lambda x}$; $D_{\lambda x}$ = coefficients obtained from tables or figures (see Hetenyi, 1946).

5.12.4 Finite Difference Method for the Design of Mat Foundations

The following fourth-order differential equation, which corresponds to Equation (5.46), concerns deflection of a mat foundation, considered as a flexible plate:

$$\frac{\partial^4 w}{\partial x^4} + 2\frac{\partial^4 w}{\partial x^2 \partial y^2} + \frac{\partial^4 w}{\partial y^4} = \frac{q}{D} + \frac{P}{D(\partial x\, \partial y)} \tag{5.63}$$

where

w = deflection (contact settlement)
q = subgrade reaction per unit area of mat (contact pressure)
D = flexural rigidity of mat
P = concentrated load at a given point

The mat rests on an assumed bed of uniformly distributed coil springs representing the supporting soil. The modulus of subgrade reaction K_S at a point is represented by the elastic stiffness of a spring at that point. It is common to use the concept of modulus of subgrade reaction in the solution of this type of problem. Because the subgrade reaction becomes converted to an equivalent spring applied at the node of interest, the analytical methods for mat foundations using this concept are considerably simplified over other methods of computations.

When a mat is divided into a grid of elements of ($rh \times h$) dimension (Figure 5.20), Equation (5.63) can be reproduced in a finite- difference equation for deflection at an interior point O, as follows:

$$\begin{aligned}&\left(\frac{6}{r^4} + \frac{8}{r^2} + 6\right) w_O + \left(-\frac{4}{r^4} - \frac{4}{r^2}\right)(w_L + w_R) + \left(-\frac{4}{r^2} - 4\right)(w_T + w_B)\\ &+ \frac{2}{r^2}(w_{TL} + w_{TR} + w_{BL} + w_{BR}) + w_{TT} + w_{BB} + \frac{1}{r^4}(w_{LL} + w_{RR})\\ &= \frac{qh^4}{D} + \frac{Ph^2}{rD}\end{aligned} \tag{5.64}$$

when $r = 1$, that is using a grid of square elements, Equation (5.64) becomes

$$\begin{aligned}&20w_O - 8(w_L + w_R + w_T + w_B) + 2(w_{TL} + w_{TR} + w_{BL} + w_{BR})\\ &+ (w_{TT} + w_{BB} + w_{LL} + w_{RR}) = \frac{qh^4}{D} + \frac{Ph^2}{D}\end{aligned} \tag{5.65}$$

Since $q = -K_s w_O$, Equation (5.65) becomes

$$\begin{aligned}&\left(20 + \frac{K_s h^4}{D}\right) w_O - 8(w_L + w_R + w_T + w_B)\\ &+ 2(w_{TL} + w_{TR} + w_{BL} + w_{BR}) + (w_{TT} + w_{BB} + w_{LL} + w_{RR}) = \frac{Ph^2}{D}\end{aligned} \tag{5.66}$$

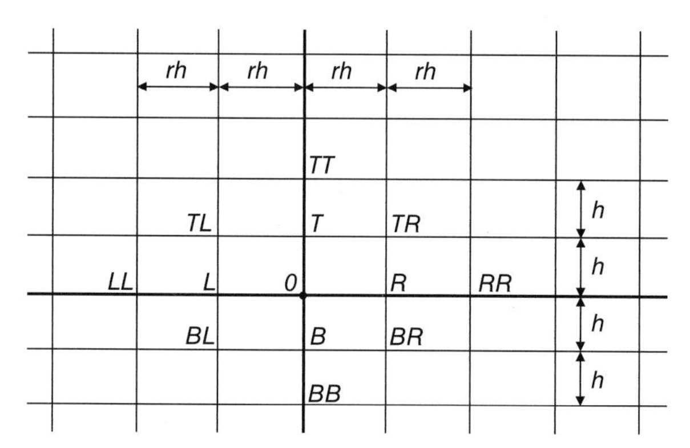

Figure 5.20 A finite-difference grid of elements of $rh \times h$ dimension shows points of deflection, w_i, included in Equation (5.64).

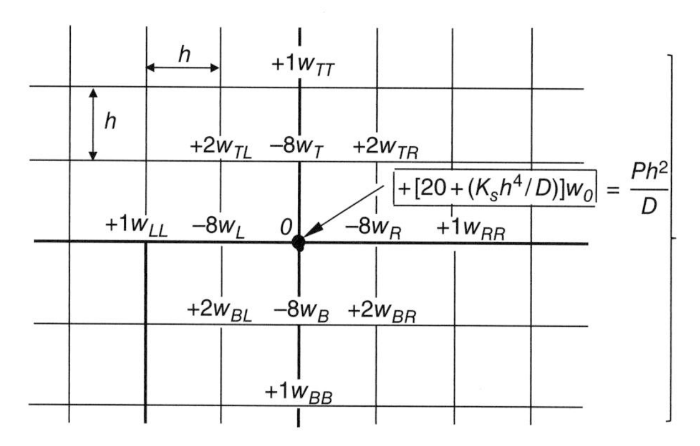

Figure 5.21 Diagrammatical representation of the finite-difference equation, Equation (5.66), for deflection at an interior node, using a grid of square elements.

The notations w_O, w_T, w_R, ..., represent deflection at points O, T, R, ... Suffixes L, T, R and B are respectively stand for left, top, right and bottom, as shown in Figure 5.20. Equation (5.66) is diagrammatically represented by Figure 5.21. For a given mat foundation, one difference equation can be written for each point (node) of the network. For points at or near free edges, the difference equations are modified to account for boundary conditions. By solving these simultaneous equations, the deflections at all points are computed. The computations can be carried out rapidly using a digital computer.

By using finite-difference operators which relate moments to deflections, bending moments per unit width in x or y direction can be determined for any point in the network. For example, the total moment per unit width in $L-R$ direction for an interior node O (Figure 5.21) is determined as

$$M_{L-R} = -\frac{D}{h^2}[(w_L - 2w_O + w_R) + \mu(w_T - 2w_O + w_B)] \tag{5.67}$$

The finite-difference method was extensively used in the past, but is sometimes used as a check on alternative methods where it is practical. It is reliable if the mat can be modeled using a finite-difference grid. It does not require massive computer resources, since the input data are minimal compared with any other discrete method. However, it is very difficult for the method to model boundary conditions for column fixity, to allow for holes, notches, or re-entrant corners. Also, it is difficult to account for moments applied at nodes (such as column moments) since the difference model uses moment per unit of width.

There are FDM computer programs, such as program FADMATFD, B19, which can be used to illustrate the procedure. Bowles (2001) has given an example problem (Example 10-4) which illustrates typical input and output from this FDM program. All the necessary equations for various nodes and a FDM program for their solutions are given in Bowles 1974a.

5.12.5 Modulus of Subgrade Reaction and Node Coupling of Soil Effects for Mats

It is very common to use the concept of modulus of subgrade reaction in the discrete element analyses for mat foundations. As it has already been mentioned in Section 5.10.4, the use of K_S in analysing mats is rather widespread because of the greater convenience of this parameter. These analyses use K_S to compute the node springs. At any node, a node spring (or spring constant) K is simply equals to the product of the node K_S and the contributory area from any element, expressed as

$$K_i = K_s \times \text{Contributory area} \quad (\text{in units of kN/m}) \tag{5.68}$$

Since the result of this product has units of a "spring" it is commonly called a *node spring.*

Figure 5.22 shows grid lines dividing an irregular-shaped mat into discrete elements. The *spring constant* K at various nodal points is computed as follows:

Assume the area of any element such as A; B; C; and so on equals A_A; A_B; A_C; and so on. Also, assume the modulus of subgrade reaction within any element A; B; C; and so on equals K_A; K_B; K_C; and so on. Then:

$$K_1 = \frac{K_A A_A}{3} + \frac{K_B A_B}{4}, \quad K_2 = \frac{K_B A_B}{4} + \frac{K_C A_C}{4}, \quad K_3 = \frac{K_C A_C}{4} + \frac{K_D A_D}{4}, \quad K_4 = \frac{K_D A_D}{4},$$

$$K_5 = \frac{K_A A_A}{3} + \frac{K_E A_E}{3} + \frac{K_F A_F}{4}, \quad K_6 = \frac{K_A A_A}{3} + \frac{K_B A_B}{4} + \frac{K_F A_F}{4} + \frac{K_G A_G}{4},$$

$$K_7 = \frac{K_B A_B}{4} + \frac{K_C A_C}{4} + \frac{K_G A_G}{4} + \frac{K_H A_H}{4}, \quad K_8 = \frac{K_C A_C}{4} + \frac{K_D A_D}{4} + \frac{K_H A_H}{4} + \frac{K_I A_I}{4},$$

$$K_9 = \frac{K_D A_D}{4} + \frac{K_I A_I}{4}$$

When the coil springs with constant K_S are acting independent of each other, the bed of springs supporting the mat is termed a "Winkler" foundation (see Section 5.11). Springs of this foundation are

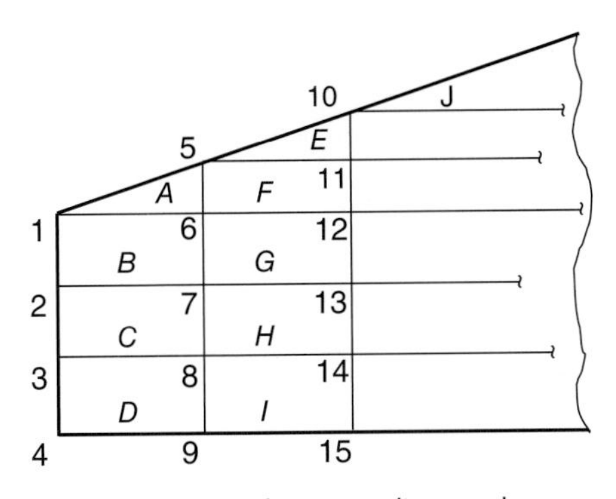

Figure 5.22 Grid lines dividing an irregular shaped mat into discrete elements.

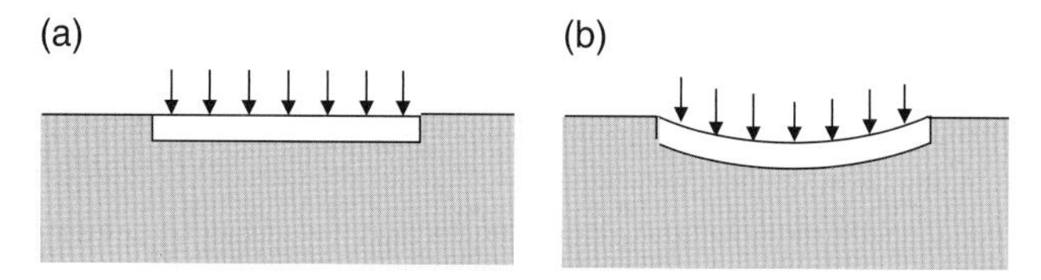

Figure 5.23 Settlement of a uniformly loaded flexible mat on a uniform soil: (a) constant settlement, Winkler concept; (b) dishing settlement, actual.

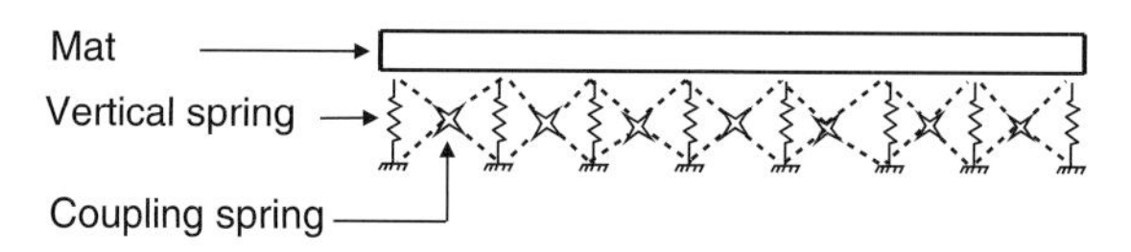

Figure 5.24 Soil–structure interaction using coupled springs.

uncoupled; hence, the deflection of any spring is not influenced by adjacent springs. If we compute node springs based on contributing node area, as just outlined above, analysis of a uniformly loaded mat (base of an oil tank) will produce a constant settlement profile across the slab with uncoupled springs and a dishing settlement profile with coupled springs, as shown in Figure 5.23. Boussinesq theory indicates the dishing profile is correct. Because of this major shortcoming associated with uncoupled springs, some designers do not like to use the concept of K_S, preferring instead to use a discrete element method of the elastic foundation bed with E_S and μ_S as elastic parameters. This choice does somewhat couple the effects; however, the computations are extensive and only as good as one's estimate of E_S and μ_S. It was already shown in Section 5.10 that there is a direct relationship between these elastic parameters and K_S. Since the elastic parameters E_S and K_S usually increase with depth from overburden and preconsolidation it appears this may be lessen the effect of ignoring coupling (Christian, 1976).

More accurate analysis will be achieved when the node springs are *coupled*, as shown in Figure 5.24. Accordingly, the vertical springs no longer act independently, and the uniformly loaded mat of Figure 5.23 exhibits the desired dish shape. In principle, this approach is more realistic than the Winkler analysis, but it is difficult to select appropriate K_S values for the coupling springs, and the equations to be programmed become much more complicated. Additionally, fractions of the springs K_i appear in off-diagonal terms of the stiffness matrix, making it difficult to perform any kind of nonlinear analysis (soil-base separation or excessive deformation).

In order to avoid the difficulties of true coupling and at the same time overcome the lack of coupling in the Winkler method, several ways have been proposed (Bowles, 1986; Liao, 1991; Horvath, 1993; ACI Committee 336) which indirectly allow for coupling (approximately). These proposed ways are known as the *pseudo-coupled method*. This method uses "springs" that act independently, but have different K_S values depending on their location on the mat. In order to produce the actual dish-shaped deformation (settlement) in a uniformly loaded mat resting on a uniform soil, the method requires the mat area be concentrically zoned using softer springs in the innermost (central) zone and transitioning to the outer (exterior) zone. Some authorities recommend using K_S values along the mat perimeter be about twice those in the centre. In case concentrated loads, such as column loads, also are present, the resulting deformations are automatically superimposed on the dishing deformation profile.

There are different methods to assign a K_S value to each zone so that approximate coupling can be done. The following two methods may be used:

(1) Compute values of K_S at 1/4 or 1/8 points, depending on the mat size and L/B value, along centre line of the mat area in L or B direction as applicable. Subsequently, the mat plan is zoned with

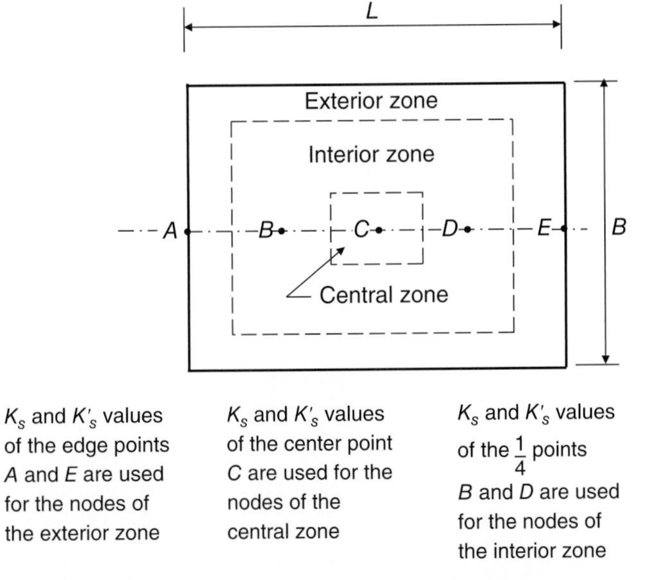

Figure 5.25 A rectangular mat foundation divided into three zones for the pseudo-coupled analysis. The coefficients of subgrade reaction K_s and K_s' are assigned for the nodes of each zone.

different values of K_S, as shown in Figure 5.25. Value of K_S at any point may be computed from Equation (5.43), repeated here for convenience, as follows:

$$K_s = \frac{q}{\delta} = \frac{1}{B \times E_s' \times m \times I_S \times I_F} \quad \left[E_s' = \frac{(1-\mu_s^2)}{E_s};\ \delta = S_i\right]$$

Carefully note in using Equation (5.43) that its basis is Equation (3.5) of Chapter 3, and use B, m, I_S, and I_F as defined there.

Consolidation settlement S_c, which is a time-dependent effect, can be incorporated into the mat analyses in an approximate manner using a revised modulus of subgrade reaction K_s'. This parameter can be computed from the basic definition of K_S as follows:

$$K_s' = \frac{\text{base contact pressure (remains constant)}}{\text{total settlement}} = \frac{q}{S_T} = \frac{q}{S_i + S_c} \tag{5.69}$$

$K_s = \dfrac{q}{S_i}$, as indicated above. Hence,

$$K_s' = \frac{K_s S_i}{S_i + S_c} \tag{5.70}$$

As before, the mat plan is zoned with different values of K_s' (Figure 5.25). If the computer output shows that the contact pressure in the zone of interest is much different from the consolidation pressure q (which may change as computations progress), a new value of S_c would have to be estimated and the problem recycled.

(2) This method also requires that the mat plan be divided into three or more concentric zones. The innermost (central) zone should be about half as wide and half as long as the mat, as shown in Figure 5.26. Then, a K_S value shall be assigned to each zone using softer springs in the innermost zone and transitioning to the outermost (exterior) zone. Usually, the outermost zone will have a K_S about twice as large as that of the innermost zone. For example, assume the central zone (Figure 5.26) has K_S equals to $K_{s,\ A}$. Exterior and interior zones may have modulus of subgrade

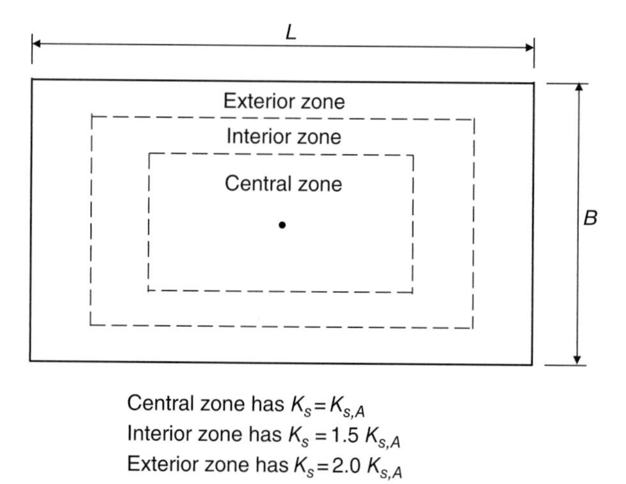

Figure 5.26 A rectangular mat foundation divided into three zones for the pseudo-coupled analysis. Modulus of subgrade reaction, K_S, progressively increases from the central zone to the exterior zone.

reaction equals to $2K_{s,A}$ and 1.5 $K_{s,A}$, respectively. Summation of product of each zone area and its K_S should equal to the product of the mat area and $K_{s,av}$. Average modulus of subgrade reaction $K_{s,av}$ is usually furnished by the geotechnical consultant. Thus, K_S for each zone will be computed.

According to ACI Committee 336 (1993), the pseudo-coupled method produced computed moments 18 to 25 percent higher than those determined from the Winkler method, which is an indication of how unconservative Winkler method can be.

5.12.6 Finite Element Method for the Design of Mat Foundations

Finite element method (FEM) is a computer-based technique which can model soil–mat interaction with good realism over a wide range of practical conditions. Use of finite element method is a necessity for the detailed analysis of those cases where it is unrealistic to assume a mat as being either infinitely rigid or infinitely flexible. The usual manner of carrying out an analysis using FEM involves use of plate bending finite elements to model the mat. These elements, rectangular and/or triangular in shape, are defined by a two-dimensional mesh with specific node points. The surface of the soil under the mat is defined by an equivalent mesh and node points. In order to include soil contribution to the structural model, the mat elements are connected to the ground through a series of "springs," which are defined using the modulus of subgrade reaction K_S (Section 5.10.5). Typically, one spring is located at each corner of each element. The finite element method is the most efficient means for analysing mats with curved boundaries or notches with re-entrant corners as shown in Figure 5.27a. Generally, finite element models require gridding that produces large number of elements, nodes, and equations. For this reason, the FEM is computationally intensive. The simple gridding of Figure 5.27b produces 70 elements, 82 nodes and 246 equations.

Finite element programs use displacement functions to produce conforming inter-element compatibility at nodes and along element boundaries. The displacement function for a plate finite element is

$$u = a_1 + a_2X + a_3Y + a_4X^2 + a_5XY + a_6Y^2 + a_7X^3 + a_8X^2Y + a_9XY^2 + a_{10}Y^3 + a_{11}X^4 + a_{12}X^3Y + a_{13}X^2Y^2 + a_{14}XY^3 + a_{15}Y^4 \tag{5.71}$$

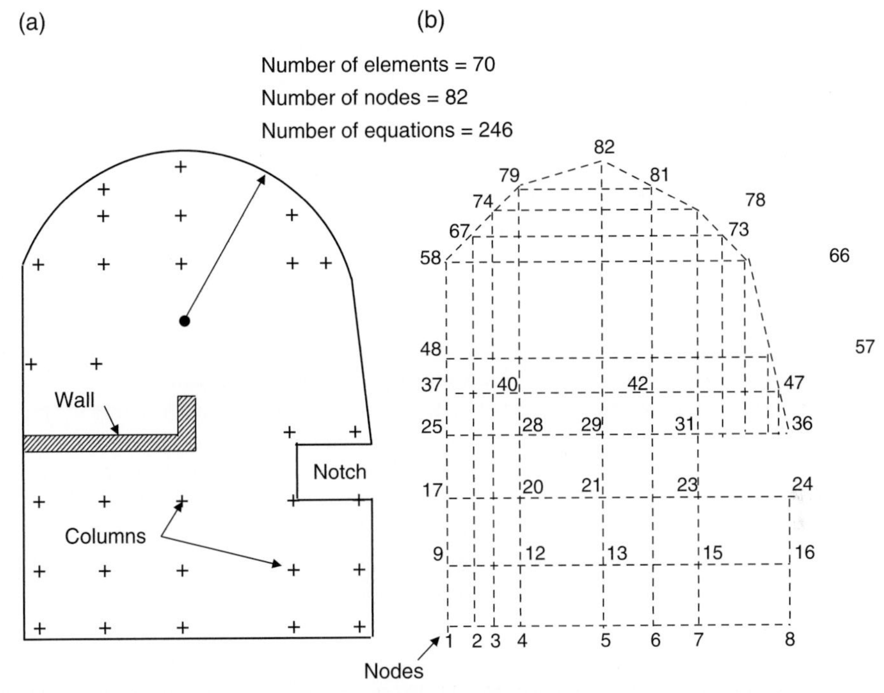

Figure 5.27 Mat gridding for finite element or finite grid method. (a) Mat with wall, notch with re-entrant corners and irregular shape, including a curved area. (b) Finite element model for the mat shown in (a); the curved boundary has been replaced with straight segments and triangles are utilised.

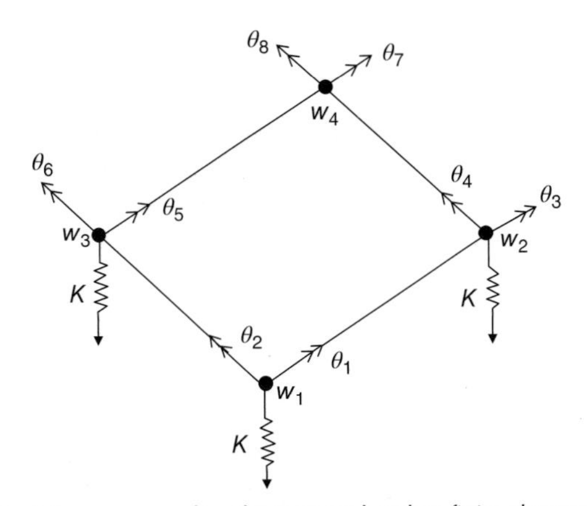

Figure 5.28 Displacements at the corner nodes of a rectangular plate finite element.

This general displacement equation consists of 15 unknown terms. For bending, the vertical displacement and slopes (rotations) in the X- and Y-directions are required at each node. Therefore, with a rectangular plate element and these three general displacements (three degrees of freedom) at each corner node only 12 unknowns (four translations and eight rotations) are required, as shown in Figure 5.28. For this reason one must reduce Equation (5.71) to one with 12 terms instead of 15 or add a node and use the 15-term displacement function. There are computer programs which delete terms, combined terms and add nodes. These programs will give about the same computed output so the preferred program is that one most familiar to the user.

An important advance in FEM is using an *iso-parametric element* approach. An element is of this type if the same function can be used to describe both shape and displacement. Iso-parametric formulation can allow a given element have more nodes than an adjacent one, which produces some mesh refinement without the large increase in grid lines of the other methods. It was found that adequate mat analysis can be carried out through implementing computer programs based on this methodology. However, the method is heavily computation-intensive.

The FEM technique is mathematically efficient, can model boundary condition displacements effectively and utilise an iso-parametric approach. All these are considered as major advantages of the method. However, the followings may be considered as disadvantages of FEM:

- Identification of the incorrect output is not easy since the methodology uses advanced mathematical concepts which are not commonly known to many geotechnical and structural engineers.
- The formulation of the stiffness matrix is computationally intensive
- The methodology gives output node moments per unit width, whereas the input moments are concentrated at the nodes. This unit incompatibility makes direct moment summation and nodal statics check difficult. Similarly a vertical force summation is not easy since element node shears are difficult to compute with the element moments obtained on a unit width basis
- FEM is particularly sensitive to aspect ratios of rectangular elements and intersection angles of triangles. To control these factors, the designer needs to increase the grid lines and number of nodes. This solution increases user input and causes rapid increase in the size of the stiffness matrix

In addition to the references mentioned in Section 5.11, the finite element necessary equations and matrix formulations can be found in many other publications and textbooks (for example Bowles, 1976; Ghali and Neville, 1972). Also, a variety of FEM programs have been implemented into readily-available software packages. Commercial computerised finite element programs, such as SAP, SAFE, NASTRAN, ANSYS and so on, may be used to run a mat analysis and design.

5.12.7 Finite Grid Method for the Design of Mat Foundations

The finite grid method (FGM) is similar to the beam-finite element described in Section 5.11 but extended to a beam-column finite element, used for a plate with bending and torsional resistance, as shown in Figure 5.29. The torsional resistance is used to incorporate the plate twist using the shear modulus G. The FGM produces nonconforming elements, that is inter-element compatibility is insured only at the nodes.

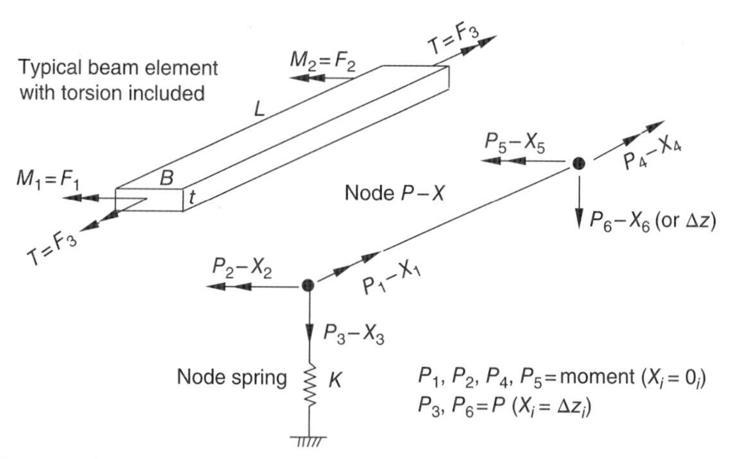

Figure 5.29 Element coding for nodes and element forces for the finite grid method.

The analysis is begun by drawing the mat plan to a suitable scale with all column and wall locations. Then, a grid is laid on this plan such that the grid nodes occur at any points of zero rotations or displacements, i.e. at column faces, wall edges, fixed edges and similar (Figure 5.27). If no nodes have unknown rotations or displacements, any convenient gridding will be used. It is not necessary the grid elements have the same size, but best results are obtained if very small members are not adjacent to large ones. For pinned columns between nodes the grid can be at convenient divisions. Suitable orientation of node numbers helps in reducing size of stiffness matrix. It is recommended the orientation be such that the origin of node numbers located at the upper left corner of the grid, a minimum number of nodes are horizontal and coding starts first across and then down. Since the elements input data are enormous, a data generator (e.g. program B-18; Bowles, 2001) to produce all element data is a necessity.

A comprehensive theoretical development of FGM specifically for mats is found in Bowles (2001). This reference also presents several examples which are used to illustrate mat analyses using the FGM. Computer program FADMAT (B-6) with necessary example data sets (EXAM? DATA), given in the same reference, are used to obtain the particular example output.

The FGM is particularly well-suited for use for the analysis of mats and plates. Usually, this method of analysis requires large number of data entries and large matrices (even using band-matrix solution methods) to solve. Some designers consider these problems as disadvantages of the method. However, the FGM has the following distinct advantages (Bowles, 2001):

- It is easy to input concentrated column moments directly.
- The output is easy to interpret since beam-column type elements that have only bending and torsion are used. The moment per unit width is simply the node moment (from a node summation) divided by the element width.
- It is easy to model notches or slots, holes, or re-entrant corners as with the FEM.
- Boundary conditions are as easily modeled as with the FEM.
- It is easy to obtain design shears at the ends of the elements. The shear is simply the sum of the element end moments divided by element length. Then one divides the total element shear by the element width to get the shear per unit width.
- It is relatively simple to extend the three degrees of freedom (d.o.f.) nodes of this method to use six d.o.f. nodes that are required for pile-cap analysis (Bowles, 1983).
- The method can be fixed (reprogrammed) to solve circular mats or plates.

Problem Solving

Problem 5.1

A reinforced concrete circular column is supported by a square isolated footing (spread footing). The footing is of plain concrete, and concentrically loaded. Design the footing in accordance with the requirements of ACI 318M-08 and using the following available data:

Net allowable soil pressure: *net* $q_a = 150$ kPa.3
Column loads (unfactored): $D = 150$ kN $L = 100$ kN
Column diameter = 300 mm
Column steel: six No. 25 bars; $f_y = 420$ MPa, $E_{steel} = 200\ 000$ MPa
Column concrete: $f_c' = 28$ MPa
Footing concrete: $f_c' = 21$ MPa

(*Continued*)

Solution:

Step 1. Find footing base dimensions (ACI Section 22.7.2).

Refer to the set of load combinations of Section 4.17. The greatest combination of the given loads is $D + L = 150 + 100 = 250$ kN.

$$\text{Footing base area} = \frac{D+L}{net\ q_a} = \frac{250}{150} = 1.67\,\text{m}^2$$

The footing base area is square; hence, $B = L = \sqrt{1.67} = 1.29\,\text{m}$.

Try square footing 1.3 m × 1.3 m, and check $net\ q_a$:

Equation (4.11): $net\ q' \leq net\ q_a$

Equation (4.10): $net\ q' = \dfrac{V}{A} = \dfrac{D+L}{A} = \dfrac{250}{1.3 \times 1.3} = 148\,\text{kPa} < net\ q_a\,(\text{OK.})$

Use 1.3 × 1.3 m square footing

Step 2. Compute the design factored net load and factored net soil pressure

Because the statement of the problem mentioned only dead and live loads, we will assume these as the only applicable loads. This reduces the set of factored load combinations of Section 5.2 to the following:

$$U = 1.4(D) = 1.4 \times 150 = 210\,\text{kN}$$

$$U = 1.2(D) + 1.6(L) = 1.2 \times 150 + 1.6 \times 100 = 340\ \text{kN}$$

The design factored net load is the greater of these two values. Use:

Design factored net load = *340 kN.*

$$\text{Design factored net soil pressure} = net\ q_{factored} = \frac{\text{design factored net load}}{\text{area}}$$

$$= \frac{340}{1.3 \times 1.3} = 201.2\,kPa$$

Note: This factored net soil pressure, strictly, is the net effective foundation pressure $net\ q'$, defined in Section 4.2, but factored.

Step 3. Find footing thickness h.

(a) Considering bending. The critical section for the maximum bending moment is located at face of the equivalent square column (ACI Sections 22.7.5 and 22.7.7), as shown in the scheme below:

The side of an equivalent square column is $w' = \sqrt{\dfrac{\pi \times 0.3^2}{4}} = 0.26\,\text{m}$

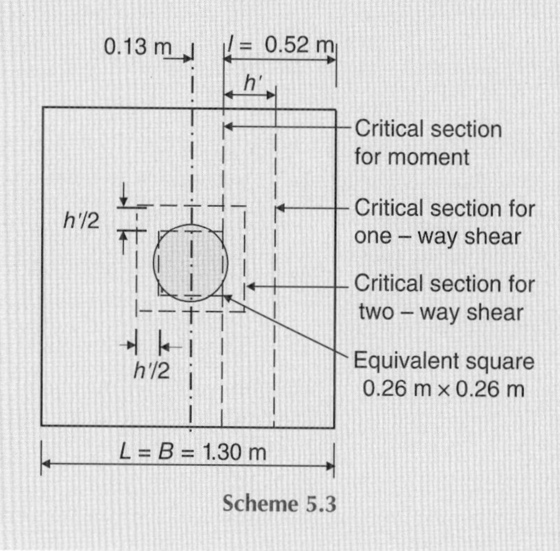

Scheme 5.3

Use effective footing thickness = $h' = h - 0.05$ m (ACI Section 22.4.7)
The applied factored bending moment per meter of the critical section is

$$M_u = \frac{(net\ q_{factored})(l)^2}{2} = \frac{201.2 \times 0.52^2}{2} = 27.20\,\text{kN.m}$$

Equation (5.32): $\varnothing M_n \geq M_u$
When load factor combinations of ACI Section 9.2.1 are used, the strength-reduction factor Ø shall be 0.6 for flexure, compression, shear, and bearing (ACI Section 9.3.5).
Equation (5.33): $M_n = 0.42\lambda\sqrt{f_c'}\ S_m$
where

λ = modification factor (ACI Section 8.6.1)
= 1.0 for normal-weight concrete
S_m = the corresponding elastic section modulus
$= \frac{b(h')^2}{6}$ (for rectangular section)

Let $\varnothing M_n = M_u$; hence,

$$(0.6)(0.42)(1)\left(\sqrt{21}\right)(1000)\left[\frac{(1)(h')^2}{6}\right] = 27.2$$

$$h' = 0.376\,\text{m} = 376\,\text{mm; hence,}$$

$$h = 376 + 50 = 426\,\text{mm}$$

$$> (h_{min} = 200\text{mm; ACI Section 22.7.4}) \text{ (OK)}.$$

(b) Considering two-way shear. The critical section for two-way shear is located a distance $h'/2$ from faces of the equivalent square column (ACI Sections 22.7.6.2 and 22.7.7), as shown in the scheme above.

Equation (5.35): $\varnothing V_n \geq V_u$

$$V_u = A_f(net\ q_{factored}) - \text{soil reaction on the shear block}$$

For columns of small cross-section, such as the column of this problem, the soil reaction on the shear block is small; it is usually neglected. Hence, use

$$V_u = A_f(net\ q_{factored}) = 1.3 \times 1.3 \times 201.2 = 340\,\text{kN}$$

Equation (5.37): $$V_n = 0.11\left[1 + \frac{2}{\beta}\right]\lambda\sqrt{f_c'}\ b_o h'$$

$$V_n \leq 0.22\lambda\sqrt{f_c'}\, b_o h' \text{ (ACI Section 22.5.4)}$$

where β corresponds to ratio of long side to short side of concentrated load or reaction area.
Use $V_n = 0.22\lambda\sqrt{f_c'}\ b_o h'$, since $\beta = \frac{0.26}{0.26} = 1$
Let $\varnothing V_n = V_u$

(Continued)

$$(0.6)(0.22)(1)\left(\sqrt{21}\right)(1000)[4(0.26+h')(h')]=340$$

$$2420(h')^2+629h'-340=0$$

$$h'=\frac{-629\pm\sqrt{629^2-(4)(2420)(-340)}}{2\times2420}$$

$$=\frac{-629+1920}{4840}=0.267\text{ m}=267\text{ mm; hence,}$$

$h=267+50=317$ mm $< h$ required by bending moment

(c) Considering one-way shear. The critical section for one-way shear is located a distance h' from face of the equivalent square column (ACI Sections 22.7.6.2 and 22.7.7), as shown in the scheme above.
Equation (5.35): $\varnothing V_n \geq V_u$

$$V_u=[B(0.52-h')](net\ q_{factored})=(1.3)(0.52-h')(201.2)$$
$$=(136-261.56h')\text{ kN}$$

Equation (5.36):

$$V_n=0.11\lambda\sqrt{f_c'}\,b_w h'$$
$$=(0.11)(1)\left(\sqrt{21}\right)(1000)(1.3)(h')=(655.3h')\text{ kN}$$

Let $\varnothing V_n = V_u$

$$0.6\times655.3h'=136-261.56h'$$

$$h'=\frac{136}{654.74}=0.208\text{ m}=208\text{ mm; hence,}$$

$$h=208+50=258\text{ mm}<(h=426\text{ mm})$$

Use h as required by bending moment.

Use footing thickness h = 450 mm

Step 4. Check column bearing on the footing (ACI Sections 22.7.8 and 22.5.5).

$$\varnothing B_n \geq B_u$$

$$B_u=\text{factored bearing load}=1.2D+1.6L=340\text{ kN}$$

(a) Top surface of the footing. The supporting surface is wider on all sides than the loaded area A_1. Diameter of the lower base area A_2 of the frustum, shown in the scheme below, is

$$0.3+2(0.5)=1.3\text{ m}$$

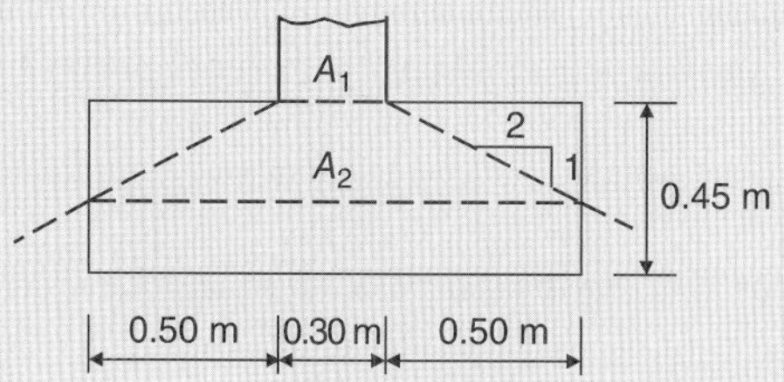

Scheme 5.4

$$A_2 = \frac{\pi(1.3)^2}{4} = 1.33\text{m}^2;\ A_1 = \frac{\pi(0.3)^2}{4} = 0.071\text{m}^2;\ \sqrt{\frac{A_2}{A_1}} = 4.3 > 2;\text{ hence, use } \sqrt{\frac{A_2}{A_1}} = 2.$$

$$\varnothing B_n = \varnothing\left(0.85 f_c' A_1\right)\sqrt{\frac{A_2}{A_1}} \quad \text{(ACI Section 22.5.5)}$$

$$\varnothing B_n = (0.6)(0.85 \times 21 \times 0.071)(2)$$
$$= 1.521\ \text{MN} = 1521\ kN \gg B_u\ \text{(OK.)}$$

Therefore, vertical compression reinforcement or dowels through the supporting surface at the interface is theoretically not required.

(b) Base of the column. The allowable bearing strength of the column base at the interface is

$$\varnothing B_n = \varnothing\left(0.85 f_c' A_1\right)$$
$$= (0.6)(0.85 \times 28 \times 0.071) = 1.014\ \text{MN} = 1014\ kN \gg B_u \quad \text{(OK.)}$$

Therefore, dowels through the column base at the interface are theoretically not required.

Step 5. Design dowels to satisfy the minimum area of reinforcement across interface (ACI Section 15.8.2.1).

$$A_{s,min} = 0.005\, A_g$$

where A_g is the gross area of the supported member

$$A_{s,min} = 0.005\left(\frac{\pi(0.3)^2}{4}\right) = 0.005\,(0.071) = 0.355 \times 10^{-3}\ \text{m}^2$$

Assume using (arbitrarily) six No. 20 dowel bars:

$$A_{s,provided} = 6\left(\frac{\pi(0.020)^2}{4}\right) = 1.88 \times 10^{-3}\ \text{m}^2 \gg A_{s,min} \quad \text{(OK.)}$$

Usesix No. 20 dowels (f_y = 420 MPa).

Step 6. Find the embedment length of dowels in both the footing and the column.

As mentioned in Section 5.4.4, dowels shall not be larger than a No. 36 bar and shall extend into supported member a distance not less than the larger of l_{dc}, of the longitudinal bars (No. 57 and smaller but larger than dowels) and compression lap splice length of the dowels, whichever is greater, and into the footing a distance not less than l_{dc} of the dowels. Also, according to ACI Section 12.16.2, when bars of different size are lap spliced in compression, splice length shall be the larger of l_{dc} of larger bar and compression lap splice length of smaller bar.

(a) Embedment length of dowels in the footing:

According to ACI Section 12.3.2,

$$l_{dc} = \frac{0.24 f_y d_b}{\lambda\sqrt{f_c'}} \text{ or } l_{dc} = (0.043 f_y)\,d_b, \text{ whichever is the larger}$$

$$l_{dc} = \frac{0.24(420)(20)}{(1)\sqrt{21}} = 440\ \text{mm},$$

(Continued)

or,

$$l_{dc} = \left(\frac{0.043 \times 10^{-6}}{10^{-6}}\right)(420)(20) = 361\ \text{mm}$$

$$l_{dc,\ min} = 200\ \text{mm} \quad (\text{ACI Section } 12.3.1)$$

$$l_{dc} = 440\ \text{mm controls.}$$

$$h = 450\ \text{mm} > (l_{dc} = 440\ \text{mm}) \quad (\text{OK.})$$

Use the embedment length of dowels in the footing = 440 mm.

Note: One may use a smaller length as permitted by ACI Section 12.3.3, but not less than 200 mm.

(b) Embedment length of dowels in the column:

$$l_{dc} = \frac{0.24(420)(25)}{(1)\sqrt{28}} = 476\ \text{mm, or,}$$

$$l_{dc} = \left(\frac{0.043 \times 10^{-6}}{10^{-6}}\right)(420)(25) = 452\ \text{mm}$$

Use $l_{dc} = 476$ mm.

According to ACI Section 12.16.I, for $f_y \le 420$ MPa and $f_c' \ge 21$ MPa, compression lap splice length shall be $0.071 f_y d_b$, but not less than 300 mm. Lap splice length = 0.071 × 420 × 20 = 596 mm > (l_{dc} = 476 mm)

Use the embedment length of dowels in the column = 600 mm

Note: One (arbitrarily) may use a larger length.

Step 7. Decide whether shrinkage and temperature (S and T) steel is required for this *plain* concrete footing.

For footings, this decision depends mainly on the designer's judgment of effects of shrinkage and temperature cracks, since the ACI Code, strictly, is not clear on this point. The ACI Section 2.2 defines plain concrete as structural concrete with no reinforcement or with less reinforcement than the minimum amount specified for reinforced concrete. This definition allows designers to use some amount of reinforcement as S and T steel in plain concrete footings. However, some authorities are of the opinion that concrete placed in the ground does not require S and T steel since the temperature differentials are not large. In any case a more conservative solution is obtained by using shrinkage and temperature reinforcement in both directions. For this problem, if S and T steel is desired, the required amount of steel shall be computed in accordance with ACI Section 7.12.2.1, as follows:

$$A_s = 0.0018\,bh = 0.0018 \times 1.30 \times 0.45 = 1.053 \times 10^{-3}\ \text{m}^2$$
$$= 1053\ \text{mm}^2.$$

An amount of $A_s = 1000\ \text{mm}^2$ will be sufficient.

Try five No. 16 bars:

$$A_{s,provided} = 5 \times 199 \cong 1000\ \text{mm}^2 \quad (\text{OK.})$$

Using 75 mm minimum concrete cover (ACI Section 7.7.1) at each side, centre to centre bar spacing will be 283 mm. Check concrete cover provided at each side:

$$\frac{1300 - (4 \times 283 + 16)}{2} = 76\ \text{mm} > 75\ \text{mm} \quad (\text{OK.})$$

The centre to centre bar spacing of 283 mm also satisfies the maximum spacing requirement of ACI Section 7.12.2.2.

Use five No. 16 bars @ 283 mm c.c. both ways, distributed near the top of the footing.

Step 8. Decide whether the footing should be sloped or stepped so that some economy may be achieved.

Isolated footings may be of constant thickness or either sloped or stepped. Sloped or stepped plain concrete footings are most commonly used to reduce the quantity of concrete away from the column where the bending moments are small. If labor and material costs permit, these alternative footings may be economical. However, when labor costs are high relative to material, it is usually more economical to use constant-thickness reinforced footings.

Assume a sloped footing is desired. One can check the footing thickness and determine the material savings as follows:

Let the footing thickness at its edges = 200 mm, which is the minimum thickness required by ACI Section 22.7.4. Leave a 100 mm around the column to obtain a shoulder of square perimeter, as shown in the scheme below, since the column must be formed after the footing has been poured. In sloped or stepped footings, angle of slope or depth and location of steps shall be such that design requirements are satisfied at every section (ACI Section 15.9.1). With this slope, shown in the scheme below, the depth furnished at critical section for two-way shear, that is at location $(h'/2) = 0.2$ m from face of the equivalent square column, is

$$h = 0.2 + \frac{0.32}{0.42} \times 0.25 = 0.390\ \text{m} > 0.317\ m \quad \text{(OK.)}$$

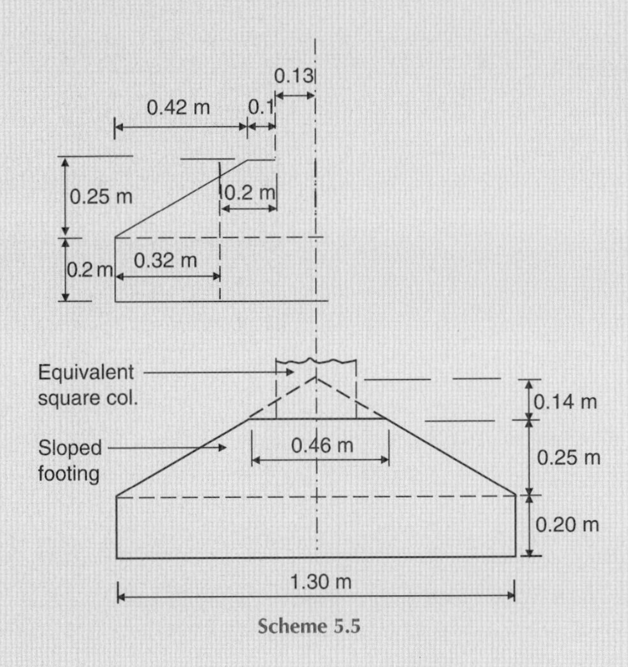

Scheme 5.5

The applied factored bending moment at this location is

$$M_u = \frac{(net\ q_{\text{factored}})(l)^2}{2} = \frac{201.2 \times 0.32^2}{2} = 10.3\ \text{kN.m/m}$$

(Continued)

Let $\emptyset M_n = M_u$

$$\emptyset M_n = (\emptyset)0.42\lambda\sqrt{f_c'}\,S_m$$

$$= (0.6)(0.42)(1)(\sqrt{21})(1000)\left[\frac{(1)(h')^2}{6}\right] = 10.3$$

$h' = 0.23$ m; hence, the depth required for bending is

$$h = 0.23 + 0.05 = 0.28\text{ m} < 0.39\text{ m} \quad \text{(OK.)}$$

The sloped footing is satisfactory.

The material savings is calculated (approximately) as follows:

$$\text{Volume of right pyramid} = \frac{1}{3} \times \text{base area} \times \text{height to apex}$$

$$\text{Volume of the sloped footing} = \frac{1}{3}\left[1.3^2(0.25 + 0.14) - 0.46^2 \times 0.14\right]$$

$$+ 1.3^2 \times 0.2 = 0.21 + 0.338$$

$$= 0.548\text{ m}^3$$

$$\text{Volume of the footing with constant thickness} = 1.3^2 \times 0.45$$

$$= 0.761\text{ m}^3$$

$$\text{Material savings} = \frac{0.761 - 0.548}{0.761} \times 100 = 28\%$$

Step 9. Draw a final design sketch.

Problem 5.2

Design a reinforced concrete square footing using the same data given in Problem 5.1. Use the same f_y for the column and footing steel.

Solution:

Step 1. Find footing base dimensions (ACI Section15.2.2).

Since the footing shape, design loads and $netq_a$ are remained unchanged, the design and computations proceed exactly in the same manner as those for the plain concrete footing; presented inSolution of Problem 5.1, *Step1*. Therefore,

Use 1.3 × 1.3 m square footing.

Step 2. Compute the design factored net load and factored net soil pressure.

For the same reasons mentioned in *Step 1*, and since the footing base area is remained unchanged, the design loads will remain unchanged too. Therefore,

Design factored net load, $U = 1.2(D) + 1.6(L) = 340\ kN$

Design factored net soil pressure, $net\ q_{\text{factored}} = 201.2\ kPa$

Step 3. Find footing thickness h.

Since the footing is *square* and centrally loaded by the column axial load, the required thickness shall be based on two-way shear only. The critical section for two-way shear is located a distance $d/2$ from faces of the equivalent square column (ACI Sections 11.11.1.2 and 15.3), as shown in the scheme below. The side of the equivalent square column is

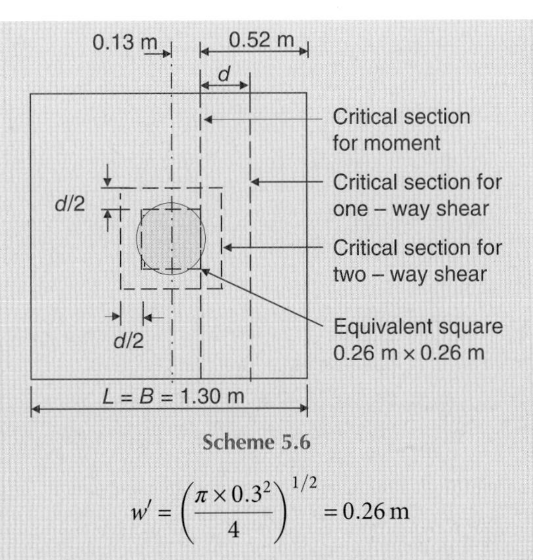

Scheme 5.6

$$w' = \left(\frac{\pi \times 0.3^2}{4}\right)^{1/2} = 0.26\,\text{m}$$

ACI Section 11.1.1 requires that

$$\text{Ø}\, V_c \geq V_u \quad (\text{taking reinforcement shear strength}, V_s = 0)$$

where

Ø = strength-reduction factor (ACI Section 9.3.2.3)
=0.75 for shear
$V_u = A_f(net q_{factored})$ – soil reaction on the shear block

For columns of small cross-section, such as the column of this problem, the soil reaction on the shear block is usually too small and would be safer if neglected. Hence,

$$V_u = A_f(net\, q_{\text{factored}}) = 1.3 \times 1.3 \times 201.2 = 340\,\text{kN}$$

The shear strength of concrete V_c shall be the smallest of (a), (b) and (c):

(a)

$$V_c = 0.17\left(1 + \frac{2}{\beta}\right)\lambda\sqrt{f_c'}\, b_o d \quad \text{Equation (5.22)}$$

$$= (0.17)\left(1 + \frac{2}{1}\right)(1)\sqrt{f_c'}\, b_o d = (0.51)\sqrt{f_c'}\, b_o d$$

(b)

$$V_c = 0.083\left(\frac{\alpha_s d}{b_o} + 2\right)\lambda\sqrt{f_c'}\, b_o d \quad \text{Equation (5.23)}$$

$$= (0.083)\left(\frac{40d}{b_o} + 2\right)(1)\sqrt{f_c'}\, b_o d = (0.083)\left(\frac{40d}{b_o} + 2\right)\sqrt{f_c'}\, b_o d$$

(c)

$$V_c = 0.33\lambda\sqrt{f_c'}\, b_o d \quad \text{Equation (5.24)}$$

$$= 0.33(1)\sqrt{f_c'}\, b_o d = (0.33)\sqrt{f_c'}\, b_o d$$

(Continued)

Use $V_c = (0.33)\sqrt{f_c'}\ b_o d$. Calculate d, then check V_c of Equation (5.23).

Let $\text{Ø}\ V_c = V_u$

$$\text{Ø}\ V_c = (0.75)(0.33)\sqrt{21}(1000)[4(0.26+d)(d)] = 340$$

$$4537(d)^2 + 1180d - 340 = 0$$

$$d = \frac{-1180 \pm \sqrt{1180^2 - (4)(4537)(-340)}}{2 \times 4537}$$

$$= \frac{-1180 + 2750}{9074} = 0.173\ \text{m}$$

Check V_c of Equation (5.23):

$$V_c = (0.083)\left(\frac{40 \times 0.173}{4 \times 0.26 + 4 \times 0.173} + 2\right)\sqrt{f_c'}\ b_o d$$

$$= (0.498)\sqrt{f_c'} b_o d > (0.33)\sqrt{f_c'}\ b_o d \quad \text{(OK.)}$$

Use $d = 0.2$ m $= 200$ mm.

The distance d will be taken to the intersection of the steel bars running each way at bottom of the *square* footing.

Assume using No. 19 bars; hence, depth of footing above reinforcement $\cong$ 180 mm > (d_{min}150 mm); ACI Section 15.7 (OK.)

Use minimum concrete cover for reinforcement = 75 mm; ACI Section 7.7.1. Hence, the overall footing thickness is

$$h = 200 + 75 + 1\ \text{bar diameter} = 275 + 19.1 = 294.1\ \text{mm}$$

Use $h = 300$ mm.

Step 4. Design the flexural reinforcement.

Equation (5.9):

$$\text{Ø}\ M_n \geq M_u$$

Equation (5.8):

$$M_u = (lbq_{factored})(l/2) = \frac{(net\ q_{factored})(l)^2}{2}/m$$

$$= \frac{201.2 \times 0.52^2}{2} = 27.20\ \text{kN.m/m}$$

Assume tension-controlled section, Ø = 0.9, and $f_s = f_y$.

Equation (5.12):

$$\text{Ø}\ M_n = \text{Ø}\left[(0.85 f_c')ba\left(d - \frac{a}{2}\right)\right]$$

$$= (0.9)\left[(0.85 \times 21 \times 1000)(1)\left(0.2a - \frac{a}{2}\right)\right]$$

$$= 3213a - 8033a^2 \quad (\text{kN.m/m})$$

Let $\text{Ø}\ M_n = M_u$:

$$8033a^2 - 3213a + 27.2 = 0$$

$$a = \frac{-(-3213) \pm \sqrt{3213^2 - (4)(8033)(27.2)}}{2 \times 8033} = \frac{139}{16\ 066} = 8.65 \times 10^{-3}\,\text{m}$$

Equation (5.10):

$$a = \frac{A_s f_y}{0.85 f_c' \ b}$$

$$A_s = \frac{0.85 f_c' \ b(a)}{f_y} = \frac{0.85 \times 21 \times 1000 \times 1 \times 8.65 \times 10^{-3}}{420 \times 1000}$$

$$= 3.7 \times 10^{-4}\,\text{m}^2 = 368\,\text{mm}^2/\text{m}$$

$A_{s,min} = 0.0018\ bh$ (ACI Sections 10.5.4 and 7.12.2.1)

$$= 0.0018 \times 1 \times 0.3 = 5.4 \times 10^{-4}\,\text{m}^2$$

$$= 540\ \text{mm}^2/\text{m} > A_s \text{ required by analysis}$$

Hence, use $A_s = A_{s,min}$.
Use three No. 16; $A_{s,provided} = 3 \times 199 = 597\ \text{mm}^2/\text{m}$
Compute a for $A_s = 597\text{mm}^2/\text{m}$, and check if $f_s = f_y$ and whether the section is tension-controlled:

$$a = \frac{A_s f_y}{0.85 f_c' b} = \frac{5.97 \times 10^{-4} \times 420}{0.85 \times 21 \times 1} = 0.014\,\text{m} = 14\,\text{mm}$$

$$c = \frac{a}{\beta_1}$$

For f_c' between 17 and 28 MPa, β_1 shall be taken as 0.85. Hence,

$$c = \frac{14}{0.85} = 16.47$$

$$d_t = d + (16/2) = 200 + 8 = 208\,\text{mm}$$

$$\varepsilon_t = 0.003\left(\frac{d_t - c}{c}\right) = 0.003\left(\frac{208 - 16.47}{16.47}\right) = 0.035 > 0.005 \text{Equation (5.15)}$$

Hence, the section is tension-controlled, and Ø = 0.9 (ACI Sections 10.3.4 and 9.3.2.1).

$$\varepsilon_y = \frac{f_y}{E_s} = \frac{420}{200\ 000} = 0.0021.$$

Therefore, $\varepsilon_t > \varepsilon_y$, and $f_s = f_y$

The assumptions made are satisfied. Note that this checking may not be necessary when $A_{s,min}$ governs.

Total number of bars required each way $= \frac{1.3}{1} \times 3 = 3.9$; use four bars.

Note: The reinforcement and thickness results indicate clearly that it may be more economical and effective to use reinforced concrete footing than plain concrete footing with S and T steel provided.

Using 75 mm concrete cover at each side, centre to centre bars spacing will be 378 mm. ACI Section 10.5.4 requires maximum spacing shall not exceed three times the slab or footing thickness, or 450 mm, whichever is smaller. Therefore, the 378 mm spacing is adequate.

Try four No. 16 bars @ 378 mm c.c. each way.

Step 5. Check the development of reinforcement.

In this case, the bars are in tension, the provided bar size is smaller than No. 19, the clear spacing of the bars exceeds $2d_b$, and the clear cover exceeds d_b. Therefore, the development length l_d of bars at each side of the

(*Continued*)

critical section (which is the same critical section for moment) shall be determined from the following equation, but not less than 300 mm:

$$l_d = \left(\frac{f_y \psi_t \psi_e}{2.1\lambda\sqrt{f_c'}}\right) d_b \quad \text{(ACI Sections 12.2.1, 12.2.2 and 12.2.4)}$$

where the factors ψ_t, ψ_e and λ are 1 (ACI Sections 12.2.4 and 8.6.1)

$$l_d = \left(\frac{420 \times 1 \times 1}{2.1 \times 1 \times \sqrt{21}}\right)\left(\frac{15.9}{1000}\right) = 0.694\ \text{m} \cong 0.7\ \text{m} = 700\ \text{mm} > 300\ \text{mm}.$$

The required $l_{d,min} = 700$ mm

The bar extension past the critical section (i.e. the available length) is

$$520\ \text{mm} - 75\ \text{mm cover} = 445\ \text{mm} < 700\ \text{mm} \quad \text{(Not OK.)}$$

We must consider smaller bars or use bars terminating in a standard hook.

(a) Assume considering smaller bars. Try 10 No. 10 bars each way:

$$A_{s,provided} = 10 \times 71 = 710\ \text{mm}^2 > \left(\frac{1.3}{1} \times 540 = 702\ \text{mm}^2\right) \quad \text{(OK.)}$$

$$l_{d,min} = \left(\frac{420 \times 1 \times 1}{2.1 \times 1 \times \sqrt{21}}\right)\left(\frac{9.5}{1000}\right) = 0.415\ \text{m} = 415\ \text{mm} < 445\ mm \quad \text{(OK.)}$$

Using 75 mm concrete cover at each side, centre to centre bar spacing will be 126.6 mm. Therefore, steel bars shall be distributed at 126 mm centre to centre.

However, such bar spacing may not be desirable, since overcrowded reinforcement may cause inadequate concrete placement and increase labour costs.

(b) Assume using bars terminating in a 180 degree standard hook, as shown below (ACI Section 12.5.1 and Figure R12.5).

Try four No. 16 hooked bars each way.

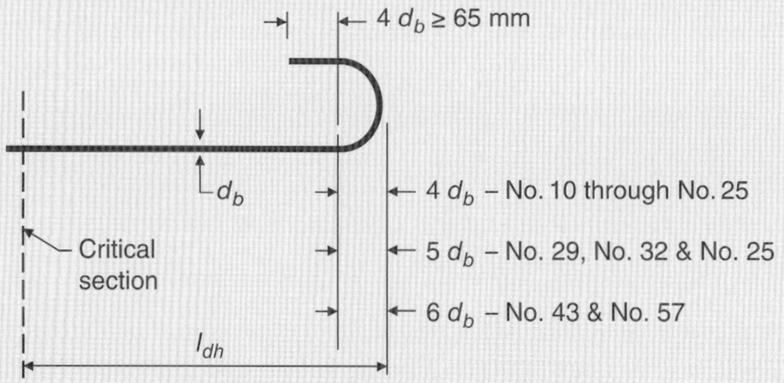

4 d_b ≥ 65 mm
d_b
Critical section
4 d_b – No. 10 through No. 25
5 d_b – No. 29, No. 32 & No. 25
6 d_b – No. 43 & No. 57
l_{dh}

Scheme 5.7

$$l_{dh} = \left(\frac{0.24\psi_e f_y}{\lambda\sqrt{f_c'}}\right) d_b. \quad \text{(ACI Sections 12.5.2)}$$

where the factors ψ_e and λ shall be taken as 1.

$$l_{dh} = \left(\frac{0.24 \times 1 \times 420}{1 \times \sqrt{21}}\right)\left(\frac{15.9}{1000}\right) = 0.35\ \text{m} = 350\ \text{mm} < 445\ \text{mm}. \quad \text{(OK.)}$$

Assume the designer prefers using the specified hooked bars.

Use four No. 16 bars @ 378 mm c.c. each way at bottom of the footing. The bars must have 180 degree standard hooks at each end.

Step 6. Check column bearing on the footing (ACI Section 10.14).

The factored load at the base of the column = 1.2*D* + 1.6*L* = 340 kN

(a) Top surface of the footing. From *Step* 4 – (a) of the solution of Problem 5.1, the maximum bearing load on the top of the footing may be taken equal to $[\text{Ø}\,(0.85f_c'\,A_1)(2)]$. (ACI Section 10.14.1)

ACI Section 9.3.2.4 gives Ø = 0.65 for bearing. Hence,

The maximum bearing factored load = 0.65(0.85)(21)(0.071)(2)

$$= 1.648\ \text{MN} = 1648\ \text{kN} \gg 340\ kN$$

Therefore, vertical compression reinforcements or dowels through the supporting surface at the interface are, theoretically, not required.

(b) Base of the column. The allowable bearing strength of the column base at the interface is $\text{Ø}(0.85f_c'\,A_1)$. Hence,

The allowable bearing factored load = (0.65)(0.85 × 28 × 0.071)

$$= 1.098\ \text{MN} = 1098\ kN \gg 340\ kN$$

Therefore, dowels through the column base at the interface are, theoretically, not required.

Step 7. Design dowels to satisfy the minimum area of reinforcement across interface required by ACI Section 15.8.2.1.

From *Step 5* of the Solution of Problem 5.1, $A_{s,min} = 0.36 \times 10^{-3}\ \text{m}^2$

Assume, arbitrarily, usingsix No 20 dowel bars:

$$A_{s,provided} = 6\left(\frac{\pi(0.020)^2}{4}\right) = 1.88 \times 10^{-3}\ \text{m}^2 \gg A_{s,min}$$

Try six No. 20 dowels (f_y = 420 MPa).

Step 8. Find the embedment length of dowels in both the footing and the column.

(a) Embedment length of dowels in the footing:

Refer to ACI Section 12.3.

ACI Section 12.3.2: $l_{dc} = \dfrac{0.24 f_y d_b}{\lambda\sqrt{f_c'}}$ or $l_{dc} = (0.043 f_y) d_b$, whichever is larger, but shall not be less than 200 mm.

$$l_{dc} = \frac{0.24(420)(20)}{(1)\sqrt{21}} = 440\ \text{mm, or}$$

$$l_{dc} = \left(\frac{0.043 \times 10^{-6}}{10^{-6}}\right)(420)(20) = 361\ \text{mm}$$

$$l_{dc,min} = 200\ \text{mm}\quad (\text{ACI Section 12.3.1})$$

$$l_{dc} = 440\ \text{mm} > h = 300\ \text{mm}\quad (\text{Not OK.})$$

According to ACI Section 12.3.3, for reinforcement in excess of that required by analysis, l_{dc} is permitted to be reduced as follows:

Required $l_{dc} = l_{dc}$ (A_s required/A_s provided)

$$= 440\left(0.36 \times \frac{10^{-3}}{1.88} \times 10^{-3}\right) = 84\ \text{mm} < l_{dc,min}\quad (\text{OK.})$$

(Continued)

Use the embedment length of dowels in the footing = 200 mm.

(b) Embedment length of dowels in the column:

The necessary computations are exactly the same as those presented in *Step* 6(b) of the Solution of Problem 5.1. Therefore,

Use the embedment length of dowels in the column = 600 mm.

Note: One, arbitrarily, may use a larger length.

Step 9. Draw a final design sketch as shown in the scheme below.

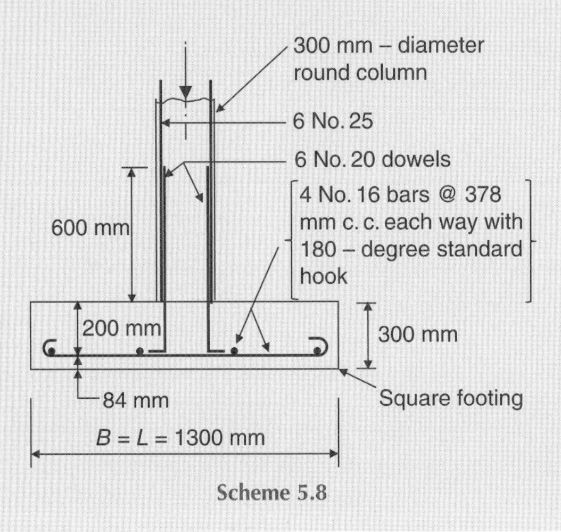

Scheme 5.8

Problem 5.3

A rectangular reinforced concrete footing is required to support an interior concrete column carrying a service (working) dead load of 2000 kN and a service live load of 1350 kN, at centre of the footing. The column cross-section is 450 × 450 mm. The column is built of 35-MPa concrete and has eight No. 29 longitudinal steel bars with $f_y = 420$ MPa and $E_{steel} = 2 \times 10^5$ MPa. The gross allowable soil pressure (*gross* q_a) at an expected foundation level of 1.0–1.5 m depth has been recommended by the geotechnical consultant equals 250 kPa. The maximum width of the footing is limited to 3 m. The top of the footing will be covered with 0.15 m fill with a unit weight of 20 kN/m^3 and a 0.15-m concrete basement floor ($\gamma_c = 24$ kN/m^3) with a uniform live loadof 3.75 kPa. Using $f_c' = 21$ MPa and $f_y = 420$ MPa, design the footing in accordance with the requirements of ACI 318M-08.

Solution:

Step 1. Find footing base dimensions (ACI Section15.2.2).

Assume the foundation depth $D_f = 1.2$ m.

Hence, footing thickness (assumed) $h = 1.2 - (0.15 + 0.15) = 0.9$ m

The greatest combination of the given loads is

$$D + L = [2000 + A(0.15 \times 24 + 0.15 \times 20 + 0.9 \times 24)]$$
$$+ 1350 + 3.75A$$
$$= 3350 + 31.95A \quad (\text{kN})$$

The gross effective foundation pressure = *gross* $q' = D + L$

$$= 3350 + 31.95A$$

Footing base area $= A = \dfrac{gross\, q'}{gross\, q_a} = \dfrac{3350 + 31.95\,A}{250}$; hence,

$$A = \frac{3350}{218.05} = 15.36\,\text{m}^2$$

Use the maximum footing width $B = 3$ m. Therefore, the footing length is

$$L = \frac{A}{B} = \frac{15.36}{3} = 5.12\,\text{m}$$

Use:

3.0 × 5.2 mrectangular footing [*gross* $q' <$ *gross* q_a, OK.]

Step 2. Compute the design factored net load and factored net soil pressure.

Because the statement of the problem mentioned only dead and live loads, we will assume these as the only applicable loads. This reduces the set of foundation net factored load combinations of Section 5.2 to the following:

$$U = 1.4(D)$$
$$U = 1.2(D) + 1.6(L)$$

The factored soil pressure is usually computed without including the weights of footing, backfill material, floor and the floor uniform live load because these loads are evenly distributed and supported; thus they do not produce shear or moment in the footing.

$$U = 1.4(D) = 1.4 \times 2000 = 2800\,\text{kN}; \text{ or,}$$
$$U = 1.2(D) + 1.6(L) = 1.2(2000) + 1.6(1350) = 4560\,\text{kN}$$

Use design factored net load = 4560 kN.

Design factored net soil pressure $= net\; q_{\text{factored}} = \dfrac{4560}{3 \times 5.2} = 292.31\,\text{kPa}$

Step 3. Find footing thickness *h*.

Since the footing is *rectangular* in shape, determine the required thickness based on both one-way shear and two-way shear analyses.

(a) Considering one-way shear:

The critical section for one-way shear is located at a distance*d* from face of the column across the entire footing width (ACI Sections 11.11.1.1 and 11.1.3.1), as shown in the scheme below.

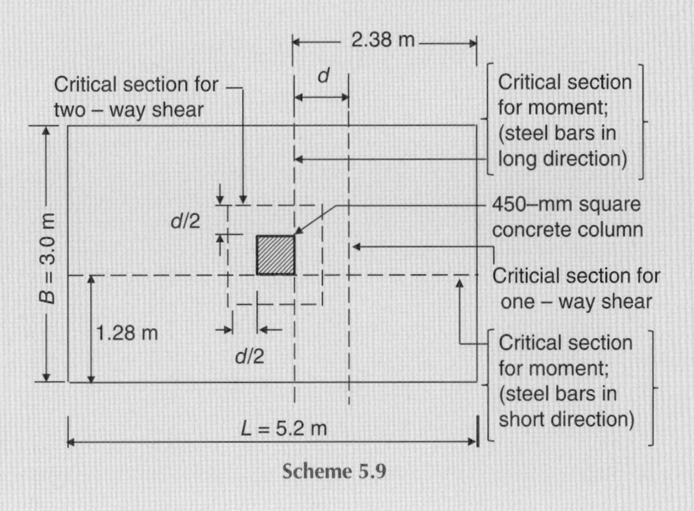

Scheme 5.9

(*Continued*)

Equation (5.19): $\varnothing V_c \geq V_u$ (taking reinforcement shear strength, $V_S = 0$)

where
$\varnothing$ = strength-reduction factor (ACI Section 9.3.2.3)
= 0.75 for shear

$$V_u = [B(2.38-d)](net\ q_{factored}) = (3)(2.38-d)(292.31)$$

$$= (2087.09 - 876.93\,d)\ \text{kN}$$

Equation (5.20): $$V_c = 0.17\lambda\sqrt{f_c'}\,b_w d$$

Let $\varnothing V_c = V_u$:

$$(0.75\times 0.17\times 1)\left(\sqrt{21}\right)(1000)(3)(d) = (2087.09 - 876.93\,d)$$

$$d = \frac{2087.09}{2629.77} = 0.794\ \text{m} = 794\ \text{mm}$$

(b) Considering two-way shear:

$$V_u = A_f(net\ q_{\text{factored}}) - \text{soil reaction on the shear block}$$

$$= (3\times 5.20)(292.31) - (0.45+d)^2(292.31)$$

$$= 4560 - 59.19 - 263.08\,d - 292.31\,d^2$$

$$= 4500.81 - 263.08\,d - 292.31\,d^2$$

The shear strength of concrete V_c shall be the smallest of (i), (ii) and (iii):

(i)

$$V_c = 0.17\left(1+\frac{2}{\beta}\right)\lambda\sqrt{f_c'}\,b_o d \quad \text{Equation (5.22)}$$

$$= (0.17)\left(1+\frac{2}{1}\right)(1)\sqrt{f_c'}\,b_o d = (0.51)\sqrt{f_c'}\,b_o d$$

(ii)

$$V_c = 0.083\left(\frac{\alpha_s d}{b_o}+2\right)\lambda\sqrt{f_c'}\,b_o d \quad \text{Equation (5.23)}$$

$$= (0.083)\left(\frac{40d}{b_o}+2\right)(1)\sqrt{f_c'}\,b_o d = (0.083)\left(\frac{40d}{b_o}+2\right)\sqrt{f_c'}\,b_o d$$

(iii)

$$V_c = 0.33\,\lambda\sqrt{f_c'}\,b_o d \quad \text{Equation (5.24)}$$

$$= 0.33\,(1)\sqrt{f_c'}\,b_o d = (0.33)\sqrt{f_c'}\,b_o d$$

Use $V_c = (0.33)\sqrt{f'_c}b_o d$. Calculate d, then check V_c of Equation (5.23).

$$\varnothing V_c = (0.75)(0.33)\sqrt{21}(1000)[4(0.45+d)(d)] = 4537(d)^2 + 2042\,d$$

Let $\varnothing V_n = V_u$:

$$4537(d)^2 + 2042\,d = 4500.81 - 263.08\,d - 292.31\,d^2$$

$$4829\,d^2 + 2305\,d - 4501 = 0$$

$$d = \frac{-2305 \pm \sqrt{2305^2 - (4)(4829)(-4501)}}{2 \times 4829} = \frac{-2305 + 9605}{9658} = 0.756\,\text{m} = 756\,\text{mm}$$

Check V_c of Equation (5.23):

$$V_c = (0.083)\left(\frac{40 \times 0.756}{4(0.45+0.756)} + 2\right)\sqrt{f'_c}\;b_o d$$

$$= (0.686)\sqrt{f'_c}\;b_o d > V_c = (0.33)\sqrt{f'_c}\;b_o d \quad \text{(OK.)}$$

$d = 756$ mm < ($d = 794$ mm, required by one way shear)

Therefore, the one-way shear controls:

Use $d = 0.8\ m = 800\ mm.$

The distance d will be taken to the centre of the steel bars in long direction.

Assume using No. 25 bars, and 75 mm minimum concrete cover (ACI Section 7.7.1). Hence, the overall footing thickness is

$$h = 800 + 75 + 1/2\,\text{bar diameter} = 875 + 12.7 = 887.7\,\text{mm}.$$

Use $h = 900$ mm.

Step 4. Design the flexural reinforcement.

(a) Reinforcement in long direction. The critical section is shown in the figure of *Step* 3.

Equation (5.9):
$$\varnothing M_n \geq M_u$$

Equation (5.8):
$$M_u = (lbq_{factored})(l/2) = \frac{(net\ q_{factored})(l)^2}{2} \Big/ m$$

$$= \frac{(net\ q_{factored})(l)^2}{2} = \frac{292.31 \times 2.38^2}{2} = 828\,\text{kN.m/m}$$

Assume tension-controlled section, $\varnothing = 0.9$, and $f_s = f_y$.

$$a = \frac{A_s f_y}{0.85 f'_c b} = \frac{A_s \times 420}{0.85 \times 21 \times 1} = 23.53 A_s$$

$$\varnothing M_n = \varnothing\left[A_s f_y\left(d - \frac{a}{2}\right)\right] \quad \text{Equation (5.11)}$$

$$= (0.9)\left[(A_s \times 420 \times 1000)\left(0.8 - \frac{23.53 A_s}{2}\right)\right]$$

$$= 302\,400 A_s - 4\,447\,170 A_s^2 \quad \text{(kN.m/m)}$$

(*Continued*)

Let $\varnothing M_n = M_u$:

$$302\ 400A_s - 4\ 447\ 170A_s^2 = 828$$

$$4\ 447\ 170A_s^2 - 302\ 400A_s + 828 = 0$$

$$A_s = \frac{-(-302\ 400) \pm \sqrt{(-302\ 400)^2 - (4)(4\ 447\ 170)(828)}}{2 \times 4\ 447\ 170} = \frac{302\ 400 - 276\ 978}{8\ 894\ 340}$$

$$= 2858 \times 10^{-6}\,\text{m}^2/\text{m} = 2858\,\text{mm}^2/\text{m}$$

$A_{s,min} = 0.0018\,bh$ (ACI Sections 10.5.4 and 7.12.2.1)

$$= 0.0018 \times 1 \times 0.9 = 1.62 \times 10^{-3}\,\text{m}^2/\text{m}$$

$$= 1620\,\text{mm}^2/\text{m} < A_s \text{ required by analysis}$$

Try six No. 25; $A_{s,provided} = 6 \times 510 = 3060\ \text{mm}^2/\text{m} > A_s$ (OK.)

Compute a for $A_s = 3060$ mm^2/m, and check if $f_s = f_y$ and whether the section is tension-controlled:

$$a = 23.53A_s = 23.53 \times 3060 \times 10^{-6} = 0.072\,\text{m} = 72\,\text{mm}$$

$c = \dfrac{a}{\beta_1}$; For f_c' between 17 and 28 MPa, β_1 shall be taken as 0.85.

$$c = \frac{72}{0.85} = 84.71\,\text{mm}$$

$$d_t = d = 800\,\text{mm}$$

$$\varepsilon_t = 0.003\left(\frac{d_t - c}{c}\right) \quad \text{Equation (5.23)}$$

$$= 0.003\left(\frac{800 - 84.71}{84.71}\right) = 0.025 > 0.005.$$

Hence, the section is tension-controlled, and $\varnothing = 0.9$ (ACI Sections 10.3.4 and 9.3.2.1).

$$\varepsilon_y = \frac{f_y}{E_s} = \frac{420}{200\ 000} = 0.0021.$$

Hence,

$$\varepsilon_t > \varepsilon_y,$$

and

$$f_s = f_y$$

Therefore the assumptions made are satisfied.

Total number of bars required in long direction = 3 × 6 = 18

Using 75 mm minimum concrete cover at each side, centre to centre bar spacing will be 166 mm. Check concrete cover provided at each side:

$$\frac{3000 - (17 \times 166 + 25.4)}{2} = 76.3\,\text{mm} > 75\ \text{mm} \quad \text{(OK.)}$$

In case the required flexural reinforcement exceeds the minimum flexural reinforcement, it shall be adequate to use the same maximum spacing of reinforcement for slabs which is two times the slab thickness, or 450 mm whichever is smaller, as specified by ACI Section 13.3.2. Therefore, the 166 mm spacing is adequate.

Try 18 No. 25 @ 166 mm c.c. across the footing width.

Check the development of reinforcement:

In this case, the bars are in tension, the provided bar size is larger than No. 19, the clear spacing of the bars exceeds $2d_b$, and the clear cover exceeds d_b. Therefore, the development length l_d of bars at each side of the critical section (which is the same critical section for moment) shall be determined from the following equation, but not less than 300 mm:

$$l_d = \left(\frac{f_y \psi_t \psi_e}{1.7\lambda\sqrt{f_c'}}\right) d_b \text{ (ACI Sections 12.2.1, 12.2.2 and 12.2.4)}$$

where the factors ψ_t, ψ_e and λ are 1. (ACI Sections 12.2.4 and 8.6.1)

$$l_d = \left(\frac{420 \times 1 \times 1}{1.7 \times 1 \times \sqrt{21}}\right)\left(\frac{25.4}{1000}\right) = 1.37\,\text{m} = 1370\,\text{mm} > 300\,\text{mm}$$

Therefore, the required $l_d = 1370$ mm.

The bar extension past the critical section (i.e. the available length) is

$$2380\,\text{mm} - 75\,\text{mm cover} = 2305\,\text{mm} > 1370\,\text{mm} \quad \text{(OK.)}$$

Use 18 No. 25 @ 166 mm c.c. across the footing width, at bottom in long direction.

(b) Reinforcement in short direction. The critical section is shown in the scheme in *Step 3*.

Bars in short direction are placed on bars in long direction. Therefore,

$$d = 0.8 - \left(\begin{array}{l}\frac{1}{2}\text{ diameter of bar in long direction} + \\ \frac{1}{2}\text{ diameter of bar in short direction}\end{array}\right)$$

Assume using No. 19 bars in short direction. Hence,

$$d = 0.8 - \left(\frac{1}{2} \times 0.0254 + \frac{1}{2} \times 0.0191\right) = 0.778\,\text{m}$$

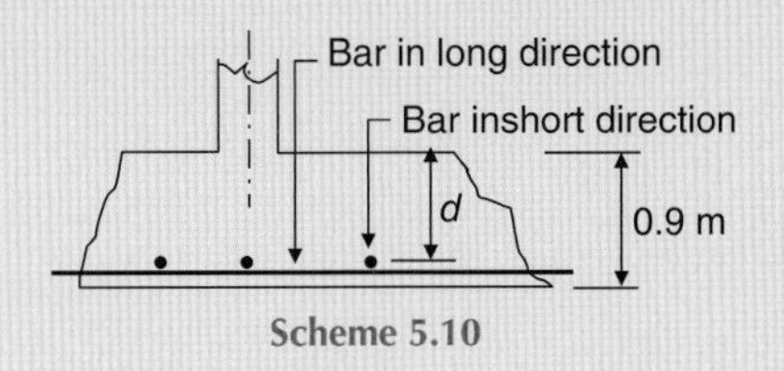

Scheme 5.10

Equation (5.9):

$$\varnothing M_n \geq M_u$$

(Continued)

$$M_u = \frac{(net\ q_{factored})(l)^2}{2} = \frac{292.31 \times 1.28^2}{2} = 239.5\,\text{kN.m/m}$$

Assume tension-controlled section, $\varnothing = 0.9$, and $f_s = f_y$.

$$a = \frac{A_s f_y}{0.85 f_c' b} = \frac{A_s \times 420}{0.85 \times 21 \times 1} = 23.53 A_s$$

$$\varnothing M_n = \varnothing \left[A_s f_y \left(d - \frac{a}{2} \right) \right] \quad \text{Equation (5.11)}$$

$$= (0.9)\left[(A_s \times 420 \times 1000)\left(0.778 - \frac{23.53 A_s}{2} \right) \right]$$

$$= 294084 A_s - 4447170 A_s^2 \quad (\text{kN.m/m})$$

Let $\varnothing M_n = M_u$:

$$294\,084 A_s - 4\,447\,170 A_s^2 = 239.5$$

$$4\,447\,170 A_s^2 - 294\,084 A_s + 239.5 = 0$$

$$A_s = \frac{-(-294\,084) \pm \sqrt{(-294\,084)^2 - (4)(4\,447\,170)(239.5)}}{2 \times 4\,447\,170} = \frac{294\,084 - 286\,749}{8\,894\,340}$$

$$= 824.7 \times 10^{-6}\,\text{m}^2/\text{m} = 825\,\text{mm}^2/\text{m}$$

$A_{s,min} = 0.0018\, bh$ (ACI Sections 10.5.4 and 7.12.2.1)

$$= 0.0018 \times 1 \times 0.9 = 1.62 \times 10^{-3}\,\text{m}^2/\text{m}$$

$$= 1620\,\text{mm}^2/\text{m} > A_s \text{ required by analysis}$$

Therefore, use A_s required $= A_{s,min} = 1620$ mm^2/m

The assumptions made are satisfied, since A_s required $= A_{s,min}$.

Total A_s required $= 1620 \times 5.2 = 8424$ mm^2

Try 30 No. 19:

$$A_{s,provided} = 30 \times 284 = 8520\,\text{mm}^2 > 8424\,\text{mm}^2 \quad \text{(OK.)}$$

Check the development of reinforcement:

In this case, the bars are in tension, the provided bar size is No. 19, the clear spacing of the bars exceeds $2d_b$, and the clear cover exceeds d_b. Therefore, the development length l_d of bars at each side of the critical section (which is the same critical section for moment) shall be determined from the following equation, but not less than 300 mm:

$$l_d = \left(\frac{f_y \psi_t \psi_e}{2.1 \lambda \sqrt{f_c'}} \right) d_b \quad \text{(ACI Sections 12.2.1, 12.2.2 and 12.2.4)}$$

where the factors ψ_t, ψ_e and λ are 1. (ACI Sections 12.2.4 and 8.6.1)

$$l_d = \left(\frac{420 \times 1 \times 1}{2.1 \times 1 \times \sqrt{21}} \right) \left(\frac{19.1}{1000} \right) = 0.833\,\text{m} = 833\,\text{mm} > 300\,\text{mm}.$$

Therefore, the required $l_d = 833$ mm.

The bar extension past the critical section (i.e. the available length) is

$$1280\,\text{mm} - 75\,\text{mm cover} = 1205\,\text{mm} > 833\,\text{mm} \quad (\text{OK.})$$

According to ACI Section 15.4.4.2, for reinforcement in short direction, a portion of the total reinforcement ($\gamma_s A_s$) shall be distributed uniformly over a band width (centred on centreline of column or pedestal) equal to the length of short side of footing. Remainder of reinforcement ın short direction, $(1 - \gamma_s)A_s$, shall be distributed uniformly outside centre band width of footing.

$$\gamma_s = \frac{2}{(\beta + 1)}$$

where β is ratio of long to short sides of footing.

In this case: $A_s = 8520\ \text{mm}^2$ or 30 No. 19 bars;

$$\gamma_s = \frac{2}{\left(\frac{5.2}{3} + 1\right)} = 0.732$$

$$0.732\,(30\,\text{bars}) = 21.96\,\text{bars}$$

Provide 22 bars uniformly distributed in the middle strip of 3 m width.

When $A_{s,min}$ controls, ACI Section 10.5.4 requires maximum spacing shall not exceed three times the slab or footing thickness, or 450 mm, whichever is smaller.

Provide 22 No. 19 bars @ 142 mm c.c. in the middle strip in short direction, placed on top of bars in long direction.

Provide 4 No. 19 bars @ 255 mm c.c. in each strip outside the middle strip in short direction, placed on top of bars in long direction.

Step 5. Check column bearing on the footing (ACI Section 10.14).

The factored load at the base of the column = $1.2D + 1.6L = 1.2 \times 2000 + 1.6 \times 1350 = 4560$ kN

(a) Top surface of the footing. The supporting surface is wider than the loaded area A_1 on all sides. Base area of the frustum, shown in the scheme below, is

$$A_2 = 3 \times 3 = 9\,\text{m}^2$$

$$A_1 = 0.45 \times 0.45 = 0.203\,\text{m}^2$$

$$\sqrt{\frac{A_2}{A_1}} = \sqrt{\frac{9}{0.203}} = 6.6 > 2$$

0.45 m × 0.45 m concrete column

1.275 m

1.275 m

A_1

2

1

A_2

0.9 m

3.0 m

Scheme 5.11

Use $\sqrt{\frac{A_2}{A_1}} = 2.$

The maximum bearing load on the top of the footing may be taken as

$$\text{Ø}\left(0.85 f_c'\, A_1\right)(2) \quad (\text{ACI Section 10.14.1})$$

Where $\quad \text{Ø} = 0.65 \quad (\text{ACI Section 9.3.2.4})$

The maximum bearing factored load = 0.65(0.85)(21)(0.203)(2)

$$= 4.711\,\text{MN} = 4711\,\text{kN}$$

(Continued)

The factored load at the base of the column is less than the maximum bearing factored load. Therefore, vertical compression reinforcement or dowels through the supporting surface at the interface is theoretically not required.

(b) Base of the column. The allowable bearing strength of the column base at the interface is $\varnothing\left(0.85f_c' A_1\right)$.

The allowable bearing factored load $= (0.65)(0.85 \times 35 \times 0.203)$

$$= 3.925\,\text{MN} = 3925\,\text{kN}$$

$$< 4560\,\text{kN}$$

Therefore, dowels through the column base at the interface are needed to transfer the excess load.

Area of dowels required $= A_d = \dfrac{4560 - 3925}{\varnothing f_y}$

$$\varnothing = 0.65 \quad (\text{ACI Sections } 9.3.2.4 \text{ and } 9.3.2.2(\text{b}))$$

$$A_d = \frac{4560 - 3925}{0.65 \times 420 \times 1000} = 2.326 \times 10^{-3}\,\text{m}^2 = 2326\,\text{mm}^2$$

The area of dowels must also satisfy ACI Section 15.8.2.1, which requires

$$A_d \geq 0.005\,A_g$$

$$A_d = 0.005 \times 0.45^2 = 1.013 \times 10^{-3}\,\text{m}^2 = 1013\,\text{mm}^2 < 2326\,\text{mm}^2$$

Trysix No. 22 dowels. $A_{s,provided} = 6 \times 387 = 2322\ \text{mm}^2$ (acceptable).

Provide six No. 22 dowels (f_y = 420 MPa); dowel each corner bar and two other (opposite) bars.

Step 6. Find the embedment length of dowels in both the footing and the column.

(a) Embedment length of dowels in the footing:

According to ACI Sections 15.8.2.3 and 12.3.2:

$$l_{dc} = \frac{0.24 f_y d_b}{\lambda\sqrt{f_c'}},\ \text{or},\ l_{dc} = (0.043 f_y)\,d_b,\ \text{whichever is the larger}$$

$$l_{dc} = \frac{0.24(420)(22.2)}{(1)\sqrt{21}} = 488\,\text{mm},\ \text{or},\ l_{dc} = (0.043)(420)(22.2)$$

$$= 401\,\text{mm}$$

$$l_{dc,\,min} = 200\,\text{mm} \quad (\text{ACI Section } 12.3.1)$$

Therefore, the required l_{dc} = 488 mm < d

The dowel bars will be extended down to the level of the main footing steel and hooked 90° (ACI standard 90° hook). The hooks will be tied (wired) to the main steel to hold the dowels in place.

Use the embedment length of dowels in the footing ≅ 768 mm.

(b) Embedment length of dowels in the column:

According to ACI Sections 15.8.2.3, 12.3.2 and 12.16.1:

$$l_{dc} = \frac{0.24(420)(28.7)}{(1)\sqrt{35}} = 489\,\text{mm},\ \text{or},\ l_{dc} = (0.043)(420)(28.7)$$

$$= 518\,\text{mm}$$

$$l_{dc} = \left(\frac{0.043 \times 10^{-6}}{10^{-6}}\right)(420)(28.7) = 518\,\text{mm}$$

Therefore, the required $l_{dc} = 518$ mm

For $f_y \leq 420$ MPa and $f'_c \geq 21$ MPa: compression lap splice length shall be $0.071\, f_y d_b$, but not less than 300 mm.

Lap splice length $= 0.071 \times 420 \times 22.2 = 662$ mm $> (l_{dc} = 518$ mm)

Therefore, the minimum embedment length = lap splice length = 662 mm

Use the embedment length of dowels in the column = 700 mm.

Note: One, arbitrarily, may use a larger length.

Step 7. Develop the final design sketch as shown in the scheme below.

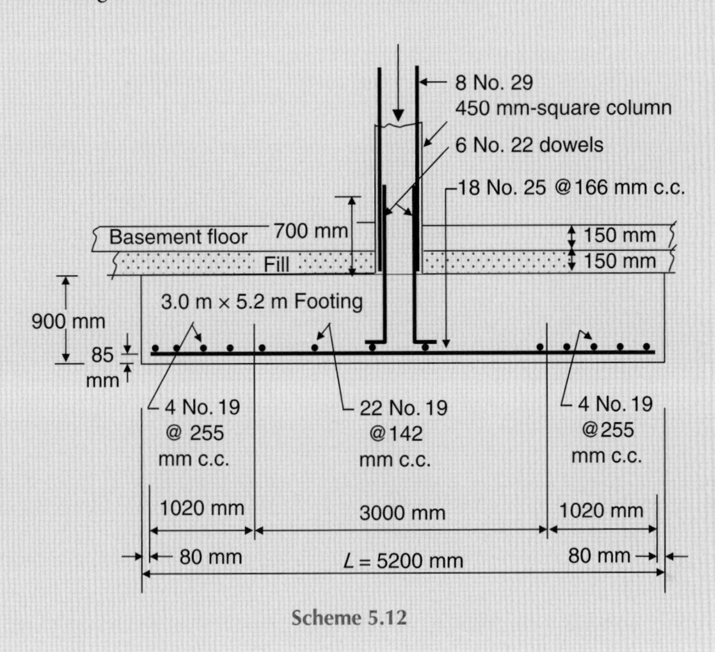

Scheme 5.12

Problem 5.4

A single column footing is loaded with an axial column load at centre, a moment and a horizontal load, as shown in the scheme below. Design the footing for the following given design data:

Loads:

$P = D + L$	$D = 420$ kN	$L = 535$ kN
$M_y = M_{y,D} + M_{y,L}$	$M_{y,D} = 228$ kN.m	$M_{y,L} = 250$ kN.m
$H = H_D + H_L$	$H_D = 42$ kN	$H_L = 53$ kN

Column:

Reinforced concrete 500 mm × 500 mm 8 No. 19 bars

$f_y = 420$ MPa $E_{steel} = 2 \times 10^5$ MPa. $f'_c = 28$ MPa

Footing:

Reinforced concrete footing

$f_y = 420$ MPa $E_{steel} = 2 \times 10^5$ MPa. $f'_c = 21$ MPa

Foundation soil:

Net allowable soil pressure = *net* $q_a = 150$ kPa

(*Continued*)

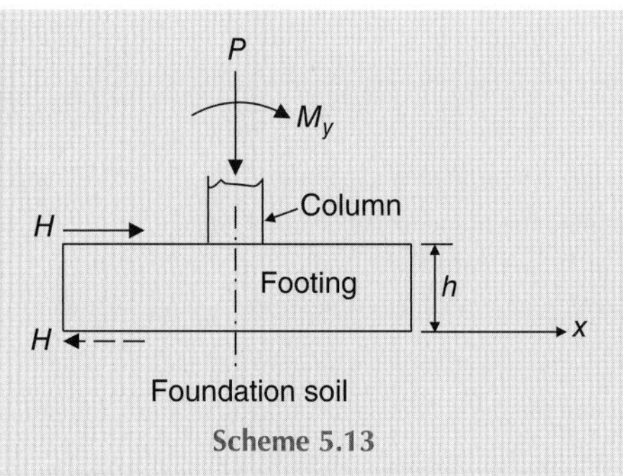

Scheme 5.13

Solution:

Step 1. Find footing base dimensions (ACI Section 15.2.2).

$$P = D + L = 420 + 535 = 955\,\text{kN}$$

$$M_{y,D} + M_{y,L} = 228 + 250 = 478\,\text{kN.m}$$

$$H_D + H_L = 42 + 53 = 95\,\text{kN}$$

$$M_{y,H} = h\,(H_D + H_L)$$

Assume $h = 0.5$ m. Hence,

$$M_{y,H} = 95 \times 0.5 = 47.5\,\text{kN.m}$$

$$M_{y,\,\text{total}} = 478 + 47.5 = 525.5\,\text{kN.m}$$

$$\text{Eccentricity}, e_x = \frac{M_{y,\,\text{total}}}{P} = \frac{525.5}{955} = 0.55\,\text{m}$$

Since eccentrically loading condition prevails, it will be more economical to use rectangular footing with its length$L$ parallel to e_x.

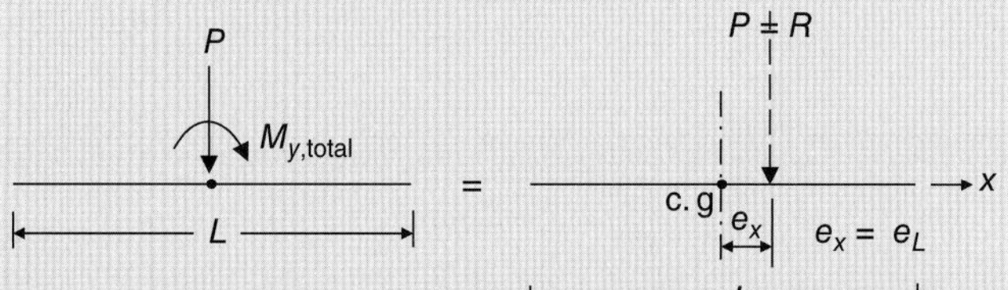

Scheme 5.14

$$q = \frac{R}{BL} \pm \frac{M_y(x)}{I_y} \quad \text{or} \quad q = \frac{R}{BL}\left(1 \pm \frac{6e_L}{L}\right)$$

$$Net\ q_{max} = \frac{P}{BL}\left(1 + \frac{6e_L}{L}\right) = \frac{955}{BL}\left(1 + \frac{6 \times 0.55}{L}\right)$$

Let net q_a = net q_{max}:

$$150 = \frac{955}{BL}\left(1 + \frac{6 \times 0.55}{L}\right) = \frac{955}{BL} + \frac{3151.5}{BL^2}$$

$$B = \frac{6.37}{L} + \frac{21.01}{L^2}$$

Consider $e_L \leq (L/6)$: $L \geq (6\,e_L = 3.3\ m)$:

Compute B *and* A *using suitable values for* L, *as shown in the table below:*

L, m	B, m	A, m^2
3.5	3.54 > L?	12.39
4.0	2.91	11.64
4.5	2.45	11.03
5.0	2.11	10.55

These results indicate that when L increases, area decreases providing more economical design. However, it is also required not to use a large value for L in order to avoid the appearance of the column on a beam. Therefore, it is necessary to take these requirements into consideration when L and B dimensions are selected.

Use rectangular footing 2.5 × 4.5 m.

Check $net\ q_a$: $$net\ q_{max} = \frac{P}{BL}\left(1+\frac{6e_L}{L}\right) = \frac{955}{2.5\times 4.5}\left(1+\frac{6\times 0.55}{4.5}\right) = 84.89 + 62.25$$

$$= 147.14\ \text{kPa} < net\ q_a\ \ \text{(OK.)}$$

Step 2. Compute the design factored net loads, moments and soil pressures.

Factored $P = P_{ult} = 1.2 \times 420 + 1.6 \times 535 = 1360$ kN

Factored $M_y = M_{y,\ ult} = 1.2(228 + 42 \times 0.5) + 1.6(250 + 53 \times 0.5)$

$$= 298.8 + 442.4 = 741.2\ kN.m$$

$$e_L = \frac{M_{y,ult}}{P_{ult}} = \frac{741.2}{1360} = 0.55\ \text{m} < \left(\frac{L}{6} = 0.75\ \text{m}\right)\ \ \text{(OK.)}$$

$$\text{Factored}\ net\ q_{max} = \frac{P_{ult}}{BL}\left(1+\frac{6e_L}{L}\right) = \frac{1360}{2.5\times 4.5}\left(1+\frac{6\times 0.55}{4.5}\right)$$

$$= 120.89 + 88.65 = 209.54\ \text{kPa}$$

$$\text{Factored}\ net\ q_{min} = \frac{P_{ult}}{BL}\left(1-\frac{6e_L}{L}\right) = 120.89 - 88.65 = 32.24\ \text{kPa}$$

Step 3. Draw the factored soil pressure diagram and locate the necessary critical sections.

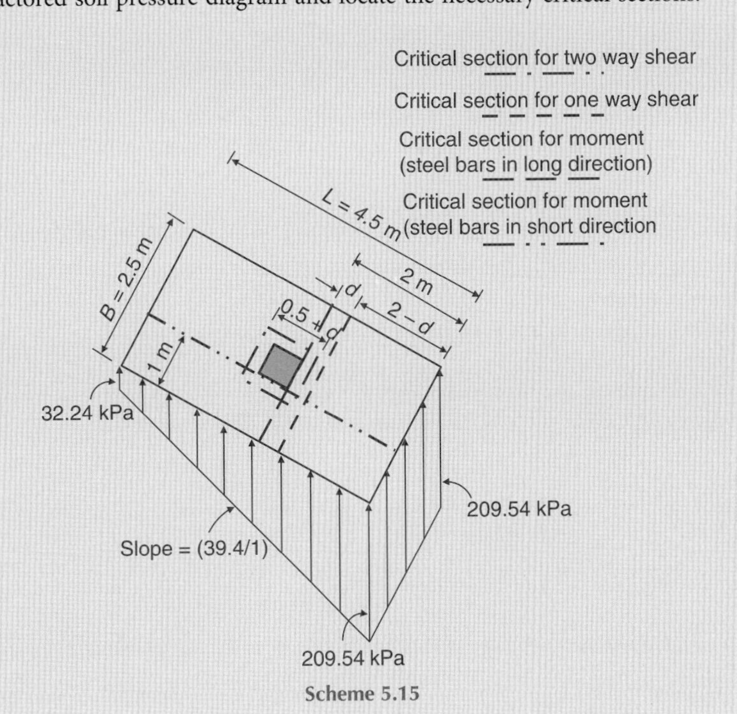

Scheme 5.15

(Continued)

Step 4. Find footing thickness h.

Since the footing is eccentrically loaded and *rectangular* in shape, the required thickness shall be determined on the basis of both one-way and two-way shear analyses.

(a) Considering one-way shear:

The critical section for one-way shear is located a distance d from face of the column across the entire footing width (ACI Sections 11.11.1.1 and 11.1.3.1), as shown on the factored soil pressure diagram of *Step 3*.

$$\text{Slope of the soil pressure line} = \frac{209.54 - 32.24}{4.5} = \frac{39.4}{1}$$

$$\text{Factored } net\ q \text{ at the critical section} = 32.24 + [39.4(4.5 - 2 + d)]$$

$$= 130.74 + 39.4\,d$$

$$V_u = \left[\frac{209.54 + (130.74 + 39.4\,d)}{2}\right](2 - d)(2.5)$$

$$= (170.14 + 19.7\,d)(5 - 2.5\,d)$$

$$= 850.7 + 98.5\,d - 425.35\,d - 49.25\,d^2$$

$$= 850.7 - 326.85\,d - 49.25\,d^2 \text{ kN}$$

Equation (5.19): $\varnothing V_c \geq V_u$ (taking reinforcement shear strength, $V_S = 0$)

where $\varnothing$ = strength-reduction factor (ACI Section 9.3.2.3) = 0.75 for shear

Equation (5.20): $V_c = 0.17\lambda\sqrt{f_c'}\,b_w d$

Let $\varnothing V_c = V_u$

$$(0.75 \times 0.17 \times 1)(\sqrt{21})(1000)(2.5)(d) = 850.7 - 326.85\,d - 49.25\,d^2$$

$$1460.7\,d - 850.7 + 326.85\,d + 49.25\,d^2 = 0$$

$$d^2 + 36.3\,d - 17.3 = 0$$

$$d = \frac{-36.3 \pm \sqrt{36.3^2 - (4)(1)(-17.3)}}{2 \times 1} = \frac{-36.3 + 37.24}{2} = 0.47\text{ m} = 470\text{ mm}$$

(b) Considering two-way shear:

V_u = volume of the factored pressure diagram or P_{ult}
- soil reaction on the shear block
=1360 – soil reaction on the shear block (kN)

The shear strength of concrete V_c shall be the smallest of (a), (b), and (c):

(a)

$$V_c = 0.17\left(1 + \frac{2}{\beta}\right)\lambda\sqrt{f_c'}\,b_o d \quad \text{Equation (5.22)}$$

$$= (0.17)\left(1 + \frac{2}{1}\right)(1)\sqrt{f_c'}\,b_o d = (0.51)\sqrt{f_c'}\,b_o d$$

(b)

$$V_c = 0.083\left(\frac{\alpha_s d}{b_o} + 2\right)\lambda\sqrt{f_c'}\,b_o d \quad \text{Equation (5.23)}$$

$$= (0.083)\left(\frac{40d}{b_o} + 2\right)(1)\sqrt{f_c'}\,b_o d = (0.083)\left(\frac{40d}{b_o} + 2\right)\sqrt{f_c'}\,b_o d$$

(c)

$$V_c = 0.33\lambda\sqrt{f_c'}\,b_o d \quad \text{Equation (5.24)}$$

$$= 0.33(1)\sqrt{f_c'}\,b_o d = (0.33)\sqrt{f_c'}\,b_o d$$

Use $V_c = (0.33)\sqrt{f_c'}\,b_o d$. Calculate d, then check V_c of Equation (5.23).

Assume $d = 0.47$ m (as required by one-way shear)

$$\text{Ø}\,V_c = (0.75)(0.33)\sqrt{21}\,(1000)[4(0.5 + 0.47)(0.47)] = 2068.3\,\text{kN}$$

$\text{Ø}\,V_n > V_u$. Therefore, one-way shear controls.

Check V_c of Equation (5.23) using $d = 0.47$ m:

$$V_c = (0.083)\left(\frac{40 \times 0.47}{4(0.5 + 0.47)} + 2\right)\sqrt{f_c'}\,b_o d = (0.568)\sqrt{f_c'}\,b_o d$$

$$V_c = (0.33)\sqrt{f_c'}\,b_o d < (0.568)\sqrt{f_c'}\,b_o d \quad \text{(OK.)}$$

Use d = 0.47 m = 470 mm.

The distance d will be taken to the centre of the steel bars in the long direction.

Assume using No. 25 bars, and 75 mm minimum concrete cover (ACI Section 7.7.1). Hence, the overall footing thickness is

$$h = 470 + 75 + 1/2\,\text{bar diameter} = 545 + 12.7 = 557.7\,\text{mm}.$$

Use h = 600 mm.

Step 5. Designthe flexural reinforcement.

(a) Reinforcement in long direction. The most critical section is located at the right face (or the left face) of the column, as shown on the factored soil pressure diagram.

Equation (5.9): $\text{Ø}\,M_n \geq M_u$

Factored *net q* at the critical section = 209.54 – 39.4 × 2 = 130.74 kPa

The maximum factored moment/m is

$$M_u = \frac{(130.74)(2)^2}{2} + \frac{(209.54 - 130.74)(2)}{2}\left(\frac{2}{3} \times 2\right)$$

$$= 366.55\,\text{kN.m/m}$$

(Continued)

Assume tension-controlled section,

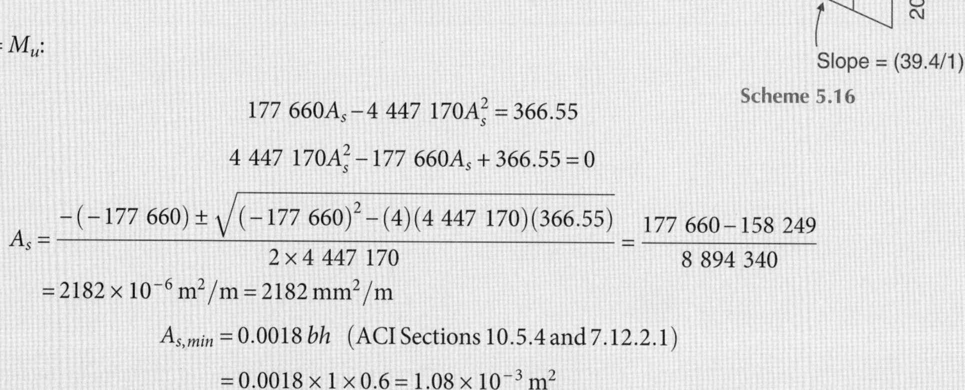

Scheme 5.16

$$\text{Ø} = 0.9;\ f_s = f_y$$

$$a = \frac{A_s f_y}{0.85 f_c' b} = \frac{A_s \times 420}{0.85 \times 21 \times 1} = 23.53 A_s$$

$$\text{Ø} M_n = \text{Ø}\left[A_s f_y\left(d - \frac{a}{2}\right)\right] \text{ Equation (5.11)}$$

$$= (0.9)\left[(A_s \times 420 \times 1000)\left(0.47 - \frac{23.53 A_s}{2}\right)\right]$$

$$= 177\,660\,A_s - 4\,447\,170 A_s^2$$

Let $\text{Ø} M_n = M_u$:

$$177\,660 A_s - 4\,447\,170 A_s^2 = 366.55$$

$$4\,447\,170 A_s^2 - 177\,660 A_s + 366.55 = 0$$

$$A_s = \frac{-(-177\,660) \pm \sqrt{(-177\,660)^2 - (4)(4\,447\,170)(366.55)}}{2 \times 4\,447\,170} = \frac{177\,660 - 158\,249}{8\,894\,340}$$

$$= 2182 \times 10^{-6}\ \text{m}^2/\text{m} = 2182\ \text{mm}^2/\text{m}$$

$$A_{s,min} = 0.0018\,bh \quad (\text{ACI Sections 10.5.4 and 7.12.2.1})$$

$$= 0.0018 \times 1 \times 0.6 = 1.08 \times 10^{-3}\ \text{m}^2$$

$$= 1080\ \text{mm}^2/\text{m} < A_s \text{ required by analysis}$$

Try six No. 22; $A_{s,provided} = 6 \times 387 = 2322\ \text{mm}^2/\text{m}$

Compute a for $A_s = 2322\ \text{mm}^2/\text{m}$, and check if $f_s = f_y$ and whether the section is tension-controlled:

$$a = 23.53 A_s = 23.53 \times 2322 \times 10^{-6} = 0.055\ \text{m} = 55\ \text{mm}$$

$c = \frac{a}{\beta_1}$; For f_c' between 17 and 28 MPa, β_1 shall be taken as 0.85

$$c = \frac{55}{0.85} = 64.71\ \text{mm}$$

$$d_t = d = 470\ \text{mm}$$

$$\varepsilon_t = 0.003\left(\frac{d_t - c}{c}\right) \text{ Equation (5.15)}$$

$\varepsilon_t = 0.003\left(\frac{470 - 64.71}{64.71}\right) = 0.019 > 0.005$. Hence, the section is tension-controlled, and $\text{Ø} = 0.9$ (ACI Sections 10.3.4 and 9.3.2.1).

$$\varepsilon_y = \frac{f_y}{E_s} = \frac{420}{200\,000} = 0.0021.$$

Hence, $\varepsilon_t > \varepsilon_y$ and $f_s = f_y$.

Therefore the assumptions are satisfied.

Total number of bars required in long direction = 2.5 × 6 = 15

Using 75 mm minimum concrete cover at each side, centre to centre bar spacing will be 166 mm. Check concrete cover provided at each side:

$$\frac{2500-(14\times 166+22.2)}{2}=76.9\,\text{mm}>75\,\text{mm}\quad (\text{OK.})$$

In case the required flexural reinforcement exceeds the minimum flexural reinforcement, it shall be adequate to use the same maximum spacing of reinforcement for slabs which is two times the slab thickness, or 450 mm whichever is smaller, as specified by ACI Section 13.3.2.

Try 15 No. 22 @ 166 mm c.c. in long direction at bottom of the footing.

Check the development of reinforcement:

In this case, the bars are in tension, the provided bar size is larger than No. 19, the clear spacing of the bars exceeds $2d_b$, and the clear cover exceeds d_b. Therefore, the development length l_d of bars at each side of the critical section (which is the same critical section for moment) shall be determined from the following equation, but not less than 300 mm:

$$l_d=\left(\frac{f_y\psi_t\psi_e}{1.7\lambda\sqrt{f_c'}}\right)d_b\quad (\text{ACI Sections 12.2.1, 12.2.2 and 12.2.4})$$

where the factors ψ_t, ψ_e and λ are 1. (ACI Sections 12.2.4 and 8.6.1)

$$l_d=\left(\frac{420\times 1\times 1}{1.7\times 1\times \sqrt{21}}\right)\left(\frac{22.2}{1000}\right)=1.20\,\text{m}=1200\,\text{mm}>300\ \text{mm}.$$

Therefore, the required l_d = 1200 mm

The bar extension past the critical section (i.e. the available length) is

$$2000\,\text{mm}-75\,\text{mm cover}=1925\,\text{mm}>1200\,\text{mm}\ (\text{OK.})$$

Use 15 No. 22 @ 166 mm c.c. across the footing width, at bottom in long direction.

(b) Reinforcement in short direction. The critical section is located at the face of the column, as shown on the factored soil pressure diagram.

Bars in short direction are placed on bars in long direction. Therefore,

$$d=0.47-\left(\begin{array}{l}\frac{1}{2}\text{ diameter of bar in long direction }+\\ \frac{1}{2}\text{ diameter of bar in short direction}\end{array}\right)$$

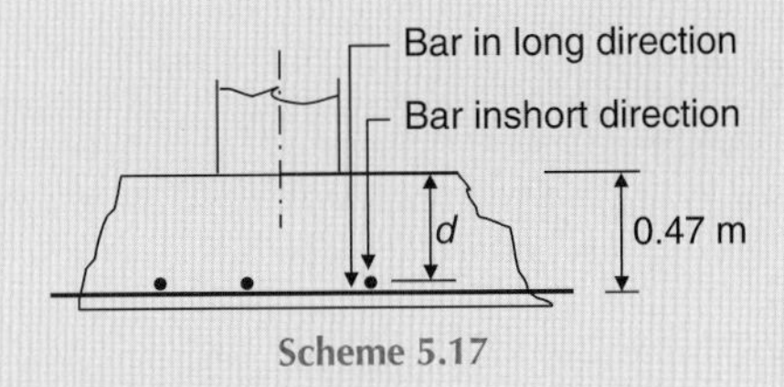

Scheme 5.17

Assume using No. 19 bars in short direction. Hence,

$$d=0.47-\left(\frac{1}{2}\times 0.0222+\frac{1}{2}\times 0.0191\right)=0.449\,\text{m}$$

(Continued)

$$\emptyset M_n \geq M_u$$

$$M_u = \left[\frac{(32.24 + 209.54)(4.5)}{2}(1)\left(\frac{1}{2}\right)\right] / 4.5$$

$$= 60.45 \text{ kN.m/m}$$

Assume tension-controlled section, $\emptyset = 0.9$, and $f_s = f_y$.

$$a = \frac{A_s f_y}{0.85 f_c' b} = \frac{A_s \times 420}{0.85 \times 21 \times 1} = 23.53 A_s$$

$$\emptyset M_n = \emptyset \left[A_s f_y \left(d - \frac{a}{2}\right)\right] \quad \text{Equation (5.11)}$$

$$= (0.9)\left[(A_s \times 420 \times 1000)\left(0.449 - \frac{23.53 A_s}{2}\right)\right]$$

$$= 1\ 697\ 224\, A_s - 4\ 447\ 170\, A_s^2$$

Let $\emptyset M_n = M_u$

$$169\ 722\, A_s - 4\ 447\ 170\, A_s^2 = 60.45$$

$$4\ 447\ 170\, A_s^2 - 169\ 722\, A_s + 60.45 = 0$$

$$A_s = \frac{-(-169\ 722) \pm \sqrt{(-169\ 722)^2 - (4)(4\ 447\ 170)(60.45)}}{2 \times 4\ 447\ 170} = \frac{169\ 722 - 166\ 524}{8\ 894\ 340}$$

$$= 360 \times 10^{-6}\ \text{m}^2/\text{m} = 360\ \text{mm}^2/\text{m}$$

$$A_{s,min} = 0.0018\, bh = 0.0018 \times 1 \times 0.6 = 1.08 \times 10^{-3}\ \text{m}^2/\text{m}$$

$$= 1080\ \text{mm}^2/\text{m} > A_s \text{ required by analysis}$$

Use A_s required = $A_{s,min}$ = 1080 mm^2/m

The assumptions made are satisfied, since A_s required = $A_{s,min}$.

Total A_s required = 1080 × 4.5 = 4860 mm^2

Try 18 No. 19: $A_{s,provided}$ = 18 × 284 = 5112 mm^2 > 4860 mm^2 (OK.)

Check the development of reinforcement:

In this case, the bars are in tension, the provided bar size is No. 19, the clear spacing of the bars exceeds $2d_b$, and the clear cover exceeds d_b. Therefore, the development length l_d of bars at each side of the critical section (which is the same critical section for moment) shall be determined from the following equation, but not less than 300 mm:

$$l_d = \left(\frac{f_y \psi_t \psi_e}{2.1 \lambda \sqrt{f_c'}}\right) d_b \quad \text{(ACI Sections 12.2.1, 12.2.2 and 12.2.4)}$$

Where the factors ψ_t, ψ_e and λ are 1. (ACI Sections 12.2.4 and 8.6.1)

$$l_d = \left(\frac{420 \times 1 \times 1}{2.1 \times 1 \times \sqrt{21}}\right)\left(\frac{19.1}{1000}\right) = 0.833\ \text{m} = 833\ \text{mm} > 300\ \text{mm}.$$

Therefore, the required l_d = 833 mm

The bar extension past the critical section (i.e. the available length) is

$$1000\ \text{mm} - 75\ \text{mm cover} = 925\ \text{mm} > 833\ \text{mm} \quad \text{(OK.)}$$

According to ACI Section 15.4.4.2, for reinforcement in short direction, a portion of the total reinforcement, $\gamma_s A_s$, shall be distributed uniformly over a band width (centred on centreline of column or pedestal) equal to the length of short side of footing. Remainder of reinforcement in short direction $(1 - \gamma_s)A_s$, shall be distributed uniformly outside centre band width of footing.

$$\gamma_s = \frac{2}{(\beta + 1)}$$

where β is ratio of long to short sides of footing.

In this case: $A_s = 5112\ \text{mm}^2$ or 18 No.19 bars;

$$\gamma_s = \frac{2}{\left(\frac{4.5}{2.5} + 1\right)} = 0.714$$

$$0.714\,(18\,\text{bars}) = 12.85\ \text{bars}$$

Provide 13 bars uniformly distributed in the middle strip of 2.5 m width.

When $A_{s,min}$ controls, ACI Section 10.5.4 requires maximum spacing shall not exceed three times the slab or footing thickness, or 450 mm, whichever is smaller.

Provide 13 No. 19 @ 207 mm c.c. in the middle strip in short direction, placed on top of bars in long direction.

Provide 3 No. 19 @ 300 mm c.c. in each strip outside the middle strip in short direction, placed on top of bars in long direction.

Step 6. Check the column bearing on the footing (ACI Section 10.14).

The factored load at the base of the column $= 1.2D + 1.6L$

$$= 1.2 \times 420 + 1.6 \times 535$$

$$= 1360\,\text{kN}.$$

(a) Top surface of the footing. The supporting surface is wider than the loaded area A_1 on all sides.

Base area of the frustum shown in the scheme below, is

$$A_2 = 2.5 \times 2.5 = 6.25\,\text{m}^2$$

$$A_1 = 0.5 \times 0.5 = 0.25\,\text{m}^2$$

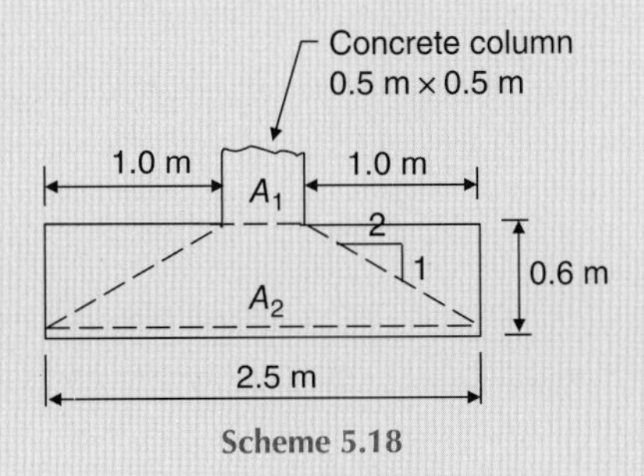

Scheme 5.18

$$\sqrt{\frac{A_2}{A_1}} = \sqrt{\frac{6.25}{0.25}} = 5 > 2$$

Use $\sqrt{\frac{A_2}{A_1}} = 2.$

(*Continued*)

The maximum bearing load on the top of the footing may be taken as

$$\text{Ø}\left(0.85f_c'\ A_1\right)(2)\quad \text{(ACI Section 10.14.1)}$$

ACI Section 9.3.2.4 gives Ø = 0.65 for bearing. Hence,
The maximum bearing factored load = 0.65(0.85)(21)(0.25)(2)

$$= 5.801\ \text{MN}$$

$$= 5801\ \text{kN} \gg 1360\ \text{kN}$$

Therefore, vertical compression reinforcement or dowels through the supporting surface at the interface are theoretically not required.

(b) Base of the column. The allowable bearing strength of the column base at the interface is $\text{Ø}\left(0.85f_c'A_1\right)$. Hence,
The allowable bearing factored load $= (0.65)(0.85 \times 28 \times 0.25)$

$$= 3868\ \text{MN} = 3868\ \text{kN} \gg 1360\ \text{kN}$$

Therefore, dowels through the column base at the interface are theoretically not required.

It should be realised that the preceding analysis concerns the condition in which the full section at the interface is under compressive stresses. In other words, no moments are transferred to the section or the eccentricity falls within the kern of the section. For the condition in which moments are transferred to the supporting footing or pedestal, usually, compressive stresses will exist over part, but not all, of the section at the interface, which is the same condition of this Problem. The number of dowels required can be obtained by considering the cross-sectional area as an eccentrically loaded column with a maximum compressive concrete stress equal to the smaller of the bearing stresses allowed on the supported member (column, pedestal or wall) or the supporting member (footing). Sufficient reinforcement must cross the interface to provide the necessary axial load and moment capacity.

In general, this requires that all the column bars or dowels of the same steel area must cross the interface. These longitudinal bars or dowels should satisfy ACI Sections 15.8.1.3, 12.17 and 15.8.2.3. Also, the minimum embedment length of these bars in the footing should not be smaller than the compression development length l_{dc} or tension development length l_{dh}, whichever is greater.

Step 7. Design dowels to satisfy the requirements of moment transfer and the minimum reinforcement area across interface (ACI Section 15.8.2.1).

Assume using dowels of the same number and size of the column bars. Hence, use eight No. 19 dowels.

$$A_{s,min} = 0.005\,A_g = 0.005 \times 0.5^2 = 1.25^{-3}\,m^2 = 1250\ \text{mm}^2$$

$$A_{s,provided} = 8 \times 284 = 2272\ \text{mm}^2 > A_{s,min}\quad \text{(OK.)}$$

Provide eight No.19 dowels (f_y = 420 MPa).

Step 8. Find the embedment length of dowels in both the footing and the column.

(a) Embedment length of dowels in the footing.

According to ACI Sections 15.8.2.3 and 12.3.2:

$$l_{dc} = \frac{0.24 f_y d_b}{\lambda\sqrt{f_c'}}\ \text{or}\ l_{dc} = \left(0.043 f_y\right) d_b,$$

whichever is the larger

$$l_{dc} = \frac{0.24(420)(19.1)}{(1)\sqrt{21}} = 420\,\text{mm or}$$

$$l_{dc} = \left(\frac{0.043 \times 10^{-6}}{10^{-6}}\right)(420)(19.1) = 345\,\text{mm}$$

$$l_{dc,min} = 200\,\text{mm} \quad (\text{ACI Section 12.3.1})$$

Therefore, the required $l_{dc} = 420$ mm $< d = 470$ mm

According to ACI Sections 12.5.1 and 12.5.2:

$$l_{dh} = \frac{0.24 f_y d_b}{\lambda\sqrt{f'_c}} \text{ or } 8\,d_b \text{ or } 150\,\text{mm},$$

which ever is the larger.

Therefore, it is clear that the minimum embedment length of dowels in the footing is 420 mm.

The dowel bars will be extended down to the level of the main footing steel and hooked 90° (ACI standard 90° hook). The hooks will be tied (wired) to the main steel to hold the dowels in place.

Use the embedment length of dowels in the footing $\cong$ *450 mm.*

(b) Embedment length of dowels in the column.

According to ACI Sections 15.8.2.3, 12.3.2 and 12.16.1:

$$l_{dc} = \frac{0.24 f_y d_b}{\lambda\sqrt{f'_c}} \text{ or } l_{dc} = (0.043 f_y)\,d_b,$$

whichever is the larger

$$l_{dc} = \frac{0.24(420)(19.1)}{(1)\sqrt{28}} = 364\,\text{mm (or)}$$

$$l_{dc} = 0.043 \times 420 \times 19.1 = 345\,\text{mm}$$

$$l_{dc,min} = 200\,\text{mm (ACI Section 12.3.1)}$$

Therefore, the required $l_{dc} = 364$ mm

For $f_y \le 420$ MPa and $f'_c \ge 21$ MPa: compression lap splice length shall be 0.071 $f_y d_b$, but not less than 300 mm.

Lap splice length = 0.071 × 420 × 19.1 = 570 mm > 364 mm.

According to ACI Sections 12.17.2.2 and 12.15:

Minimum length of lap for tension lap splices shall be for class B splice, but not less than 300 mm.

$$\text{Class B-splice length } = 1.3\,l_d = 1.3\left[\left(\frac{f_y \psi_t \psi_e}{2.1\lambda\sqrt{f'_c}}\right) d_b\right]$$

where the factors ψ_t, ψ_e and λ are 1. (ACI Sections 12.2.4 and 8.6.1)

$$1.3\,l_d = 1.3\left[\left(\frac{420 \times 1 \times 1}{2.1 \times 1\sqrt{28}}\right)\frac{19.1}{1000}\right] = 0.94\,\text{m} = 940\,\text{mm} > 570\,\text{mm}.$$

Use the embedment length of dowels in the column = 1000 *mm.*

(Continued)

Step 9. Develop the final design sketch as shown in the scheme below.

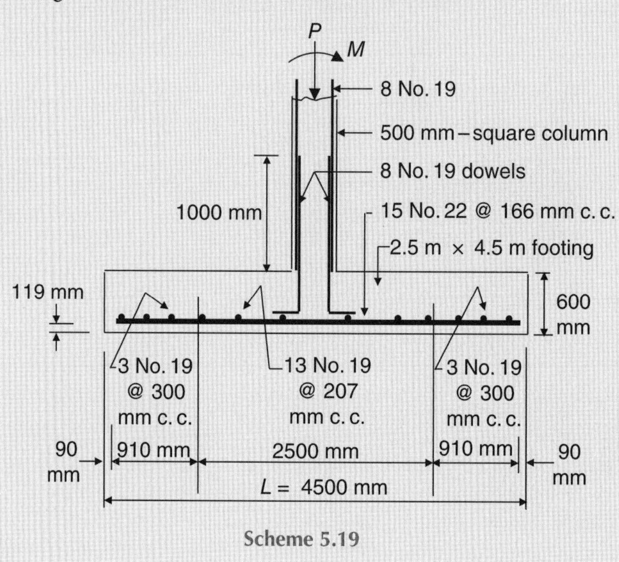

Scheme 5.19

Problem 5.5

Given: Same data as that of Problem 5.4 *except* we have the same moment about both axes, the load $H = 0$, and instead the given *net* q_a the following geotechnical data is available:

- Angle of internal friction $\text{Ø} = 0°$
- Undrained shear strength $s_u = 100$ kPa
- Depth of foundation $D_f = 1$ m
- Unit weight of soil $\gamma = 17.5\ \text{kN/m}^3$

Required: Design the footing using uniform soil pressure distribution as suggested by Bowles (2001). Assume bearing capacity failure controls the design soil pressure. Use Hansen's bearing capacity equation with $SF = 3$.

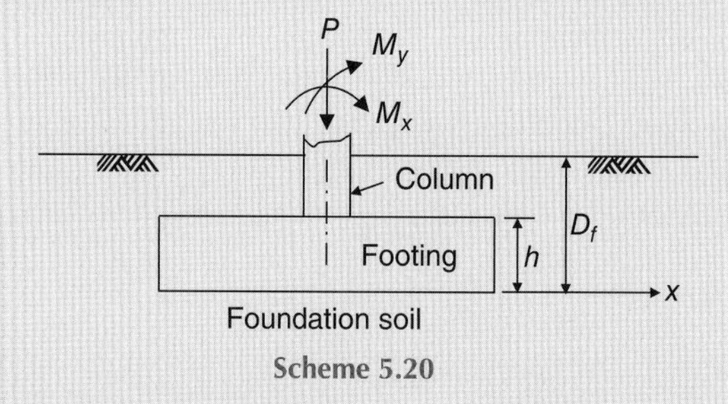

Scheme 5.20

Solution:

Step 1. Determine the design *net* q_a using the Hansen's bearing capacity equation.

We realise that with equal moments about both axes the optimum footing shape will be *square* with $e_x = e_y$, $L = B$ and $L' = B'$.

Table 4.2 $gross\ q_{ult} = 5.14\ s_u\left(1 + s'_c + d'_c\right) + \bar{q} \quad (\text{Ø} = 0°)$

Equation (4.3):

$$net\ q_{ult} = gross\ q_{ult} - \sigma'_o$$

$$\sigma'_o = \bar{q} = \gamma' D_f$$

$$net\ q_{ult} = 5.14\ s_u\left(1 + s'_c + d'_c\right)$$

Assume $B = 3$ m. Hence, $\frac{D}{B} = \frac{1}{3} = 0.33$

Table 4.6: $s'_c = 0.2\frac{B'}{L'} = 0.2; k = \frac{D}{B}$ for $\frac{D}{B} \le 1$

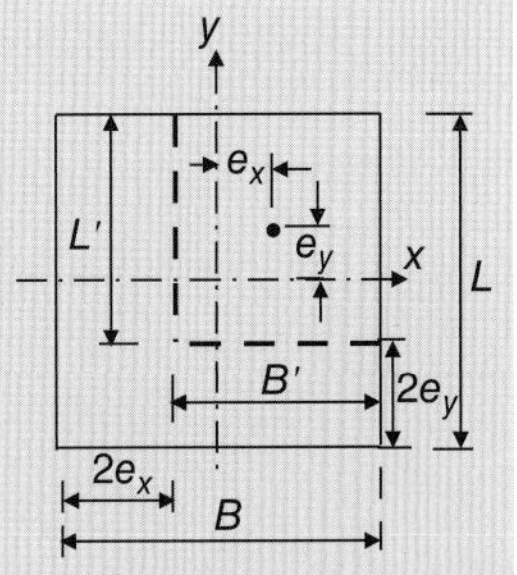

$$d'_c = 0.4k = 0.4 \times 0.33 = 0.13; s_u = 100 \text{ kPa}$$

$$net\ q_{ult} = 5.14(100)(1 + 0.2 + 0.13)$$
$$= 683.6 \text{ kPa}$$

$$net\ q_s = \frac{net\ q_{ult}}{SF} = \frac{683.6}{3} = 227.87 \text{ kPa}$$

Scheme 5.21

Let the design soil pressure $net\ q_a = net\ q_s = 228\ kPa$ since, in this case, bearing capacity failure controls the design soil pressure.

Step 2. Find footing base dimensions (ACI Section15.2.2).

$$\text{Net foundation unfactored load } P = 420 + 535 = 955 \text{ kN}$$

$$\text{Net allowable unfactored load} = (net\ q_a)(L'B')$$

Let $(net\ q_a)(L'B') = P$. Hence,

$L'B' = \frac{955}{228} = 4.2 \text{ m}^2$, or, $B' = \sqrt{4.2} = 2.05$ m, since $L' = B'$.

$$e_x = \frac{M_y}{P} = \frac{228 + 250}{955} = 0.50 \text{ m} = e_y, \text{ since } M_x = M_y$$

$$B = B' + 2e_x = 2.05 + 2 \times 0.5 = 3.05 \text{ m}$$

Try $L = B = 3.1$ m

$$d'_c = 0.4k = 0.4 \times \frac{1}{3.1} = 0.129 \cong 0.13 \quad \text{(OK.)}$$

Check $\quad e_x \le \frac{B}{6}$: $e_x = 0.5$ m; $\frac{B}{6} = \frac{3.1}{6} = 0.52 \text{ m} > e_x$ (OK.)

Check $\quad e_y \le \frac{L}{6}$: $e_y = 0.5$ m; $\frac{L}{6} = \frac{3.1}{6} = 0.52 \text{ m} > e_y$ (OK.)

Check B_{min}: $\quad B_{min} = 4(e_B \text{ or } e_x) + w_x$; where w_x = col. dimension $\|$ B

$$= 4 \times 0.5 + 0.5 = 2.5 \text{ m} < (B = 3.1 \text{ m}) \quad \text{(OK.)}$$

Check L_{min}: $\quad L_{min} = 4(e_L \text{ or } e_y) + w_y$; where w_y = col. dimension $\|$ L

$$= 4 \times 0.5 + 0.5 = 2.5 \text{ m} < (L = 3.1 \text{ m}) \quad \text{(OK.)}$$

Use 3.1 × 3.1 m reinforced concrete square footing.

(*Continued*)

Step 3. Compute the design factored net loads, moments and soil pressures.

$$\text{Factored } P = P_{ult} = 1.2 \times 420 + 1.6 \times 535 = 1360\,\text{kN}$$

$$\text{Factored } M_x = M_{x,ult} = 1.2(228) + 1.6(250) = 673.6\,\text{kN.m}$$

$$\text{Factored } M_y = M_{y,\,ult} = 1.2(228) + 1.6(250) = 673.6\,\text{kN.m}$$

$$e_x = e_y = \frac{673.6}{1360} \cong 0.50\,\text{m} < \left(\frac{3.1}{6} = 0.52\,\text{m}\right) \quad \text{(OK.)}$$

$$B' = L' = 3.1 - 2 \times 0.5 = 2.1\,\text{m}$$

$$\text{Factored } net\ q = \frac{P_{ult}}{B'L'} = \frac{1360}{2.1 \times 2.1} = 308.4\,\text{kP}a$$

Step 4. Draw the factored soil pressure diagram and locate the necessary critical sections.

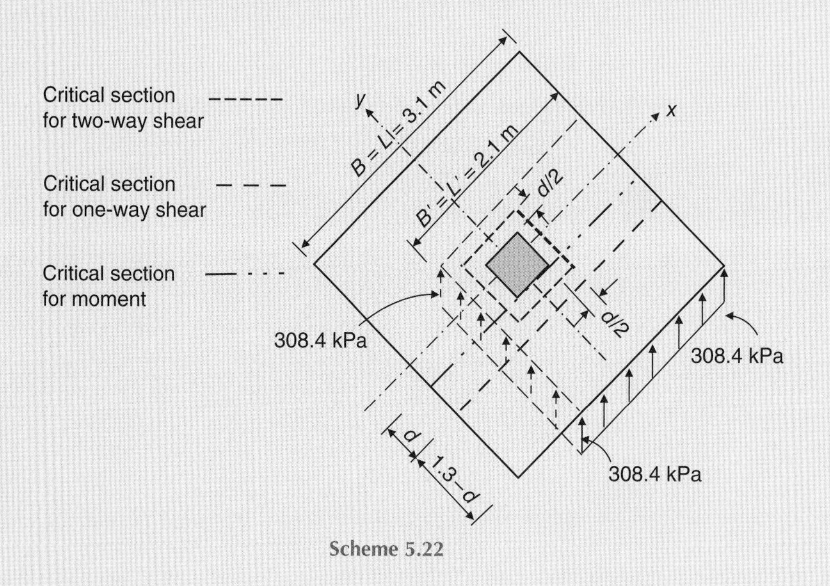

Scheme 5.22

Step 5. Find footing thickness *h*.

Since the footing is eccentrically loaded, the required thickness shall be based on both one-way shear and two-way shear analyses. The critical sections for shear are shown on the scheme of *Step 4*.

(a) Consider one-way shear: The critical section for one-way shear is located at distance *d* from face of the column.

$$V_u = [B(1.3-d)]\left(net\ q_{factored}\right) = (1)(1.3-d)(308.4)$$

$$= (400.92 - 308.4\,d)\text{kN/m}$$

Equation (5.19):

$$\varnothing\, V_c \geq V_u \,(\text{taking reinforcement shear strength, } V_s = 0)$$

$$\text{where } \varnothing = \text{strength reduction factor (ACI Section 9.3.2.3)}$$

$$= 0.75 \text{ for shear}$$

Equation (5.20): $V_c = 0.17\lambda\sqrt{f'_c}b_w d$

Let $\text{Ø}\, V_c = V_u$

$(0.75 \times 0.17 \times 1)(\sqrt{21})(1000)(1)(d) = (400.92 - 308.4\,d)$

$d = \dfrac{400.92}{892.7} = 0.45\text{ m} = 450\text{ mm}$

(b) Consider two-way shear:

$$V_u = P_{ult} - \text{soil reaction on the shear block}$$
$$= 1360 - \text{soil reaction on the shear block}\quad(\text{kN})$$

The shear strength of concrete V_c shall be the smallest of (a), (b) and (c):

(a)

$$V_c = 0.17\left(1 + \frac{2}{\beta}\right)\lambda\sqrt{f'_c}\,b_o d$$
$$= (0.17)\left(1 + \frac{2}{1}\right)(1)\sqrt{f'_c}b_o d = (0.51)\sqrt{f'_c}\,b_o d \qquad \text{Equation (5.22)}$$

(b)

$$V_c = 0.083\left(\frac{\alpha_s d}{b_o} + 2\right)\lambda\sqrt{f'_c}\,b_o d$$
$$= (0.083)\left(\frac{40d}{b_o} + 2\right)(1)\sqrt{f'_c}\,b_o d = (0.083)\left(\frac{40d}{b_o} + 2\right)\sqrt{f'_c}\,b_o d \qquad \text{Equation (5.23)}$$

(c)

$$V_c = 0.33\lambda\sqrt{f'_c}\,b_o d$$
$$= 0.33(1)\sqrt{f'_c}\,b_o d = (0.33)\sqrt{f'_c}\,b_o d \qquad \text{Equation (5.24)}$$

Use $V_c = (0.33)\sqrt{f'_c}\,b_o d$. Calculate d, then check V_c of Equation (5.23).
Assume $d = 0.45$ m

$$\text{Ø}\, V_c = (0.75)(0.33)\sqrt{21}(1000)[4(0.5 + 0.45)(0.45)] = 1939.5\text{ kN}$$

$\text{Ø}V_c > V_u$. Therefore, one-way shear controls.
Check V_c of Equation (5.23):

$$V_c = (0.083)\left(\frac{40 \times 0.45}{4(0.5 + 0.45)} + 2\right)\sqrt{f'_c}b_o d = (0.56)\sqrt{f'_c}\,b_o d$$
$$V_c = (0.33)\sqrt{f'_c}b_o d < (0.568)\sqrt{f'_c}\,b_o d(\text{OK.})$$

Use d = 0.45 m = 450 mm.

The distance d will be taken to the intersection of the steel bars running each way at bottom of the footing.
Assume using No. 22 bars; hence, depth of footing above reinforcement = (450 – 22.2) mm > 150 mm minimum required (ACI Section 15.7).
Minimum concrete cover for reinforcement (ACI Section 7.7.1) = 75 mm Hence, the overall footing thickness is

$$h = 450 + 75 + 22.2 = 275 + 19.1 = 547.2\text{ mm}.$$

(Continued)

Use h = 550 *mm.*

Step 6. Design the flexural reinforcement.

The most critical section is located at the face of the column, shown in the schemes 5.22 and 5.23.

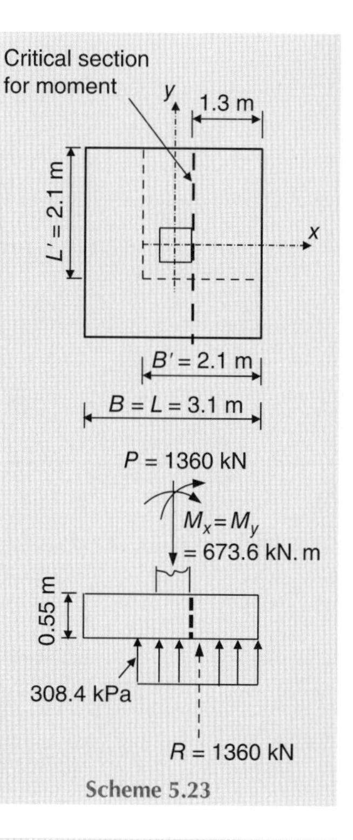

Scheme 5.23

Equation (5.9):

$$\emptyset M_n \geq M_u$$

$$net\ q_{factored} = 308.4\,\text{kN}$$

The maximum factored moment is

$$M_u = \frac{net\ q_{factored} \times l^2}{2}$$

$$= \frac{(308.4)(1.3)^2}{2}$$

$$= 260.6\,\text{kN.m/m}$$

The rest of the design computations proceed in the same manner as that for the reinforcement of the square footing of Problem 5.2.

Also, for the rest of the design steps concerning the column bearing on footing, the necessary dowels and their embedment lengths in both the footing and the column, the design proceeds in the same manner as that for the footing of Problem 5.4.

The necessary computations are left for the reader.

Problem 5.6

Using thesame data given in Problem 5.4, design an eccentrically loaded footing such that the soil pressure will be approximately uniform. Assume there will be no space limitation in layout of the footing and the moments always act in the directions shown.

Solution:

Step 1. Compute theeccentricity *e* and place centre of the footing away from the column centre a distance equals *e*.

If space permits, it will be possible to place the footing centre away from the column so that the resulting soil pressure is uniform, that is the resultant of structural loads coincides with the soil pressure resultant at centre of the footing (see Figure 5.6). This solution is obviously valid only for moments which always act in the given direction. This is not a valid solution for wind moments, since reversals can occur.

The greatest combinations of the given loads are

$$P = D + L = 420 + 535 = 955\,\text{kN}$$

$$M_{y,D} + M_{y,L} = 228 + 250 = 478\,\text{kN.m}$$

$$H_D + H_L = 42 + 53 = 95\,\text{kN}$$

$$M_{y,H} = h\,(H_D + H_L)$$

Assume $h = 0.5$ m. Hence, $M_{y,H} = 95 \times 0.5 = 47.5$ kN. m

$$M_{y,\text{total}} = 478 + 47.5 = 525.5\,\text{kN.m}$$

$$\text{Resultant of the structural loads} = R = P = 955\,\text{kN}$$

$$\text{Eccentricity of } P = e_x = \frac{M_{y,\text{total}}}{P} = \frac{525.5}{955} = 0.55\,\text{m}$$

Place the column centre at 0.55 m to the left of the footing centre; thus, the resultant of structural loads coincides with the soil pressure resultant at centre of the footing, as shown below:

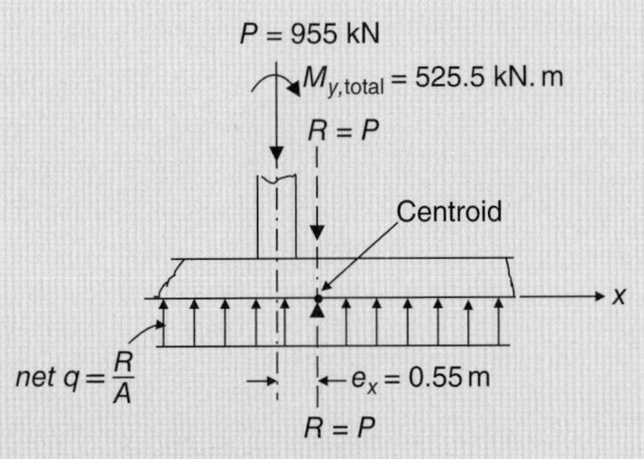

Scheme 5.24

Step 2. Find footing base dimensions (ACI Section 15.2.2).

Since the soil pressure is uniformly distributed, a square footing may be used.

Let $net\ q = net\ q_a$:

$$A = \frac{R}{net\ q_a} = \frac{955}{150} = 6.4\,\text{m}^2.\ \text{Hence, } B = L = \sqrt{6.4} = 2.53\,\text{m}$$

Use 2.6 × 2.6 m *square footing.*

Step 3. Compute the design factored net load, moment and soil pressure.

$$\text{Factored } P = P_{ult} = 1.2 \times 420 + 1.6 \times 535 = 1360\,\text{kN}$$

$$\text{Factored } M_y = M_{y,ult} = 1.2(228 + 42 \times 0.5) + 1.6(250 + 53 \times 0.5)$$

$$= 298.8 + 442.4 = 741.2\ \text{kN.m}$$

$$\text{Factored}\ net\ q = \frac{P_{ult}}{BL} = \frac{1360}{2.6 \times 2.6} = 201.2\ \text{kP}a$$

Check the eccentricity of the factored *P*:

$$e_x = \frac{\text{Factored } M_y}{\text{Factored P}} = \frac{741.2}{1360} = 0.545\,\text{m} \cong 0.55\,\text{m}\quad \text{(OK.)}$$

Step 4. Find footing thickness *h*.

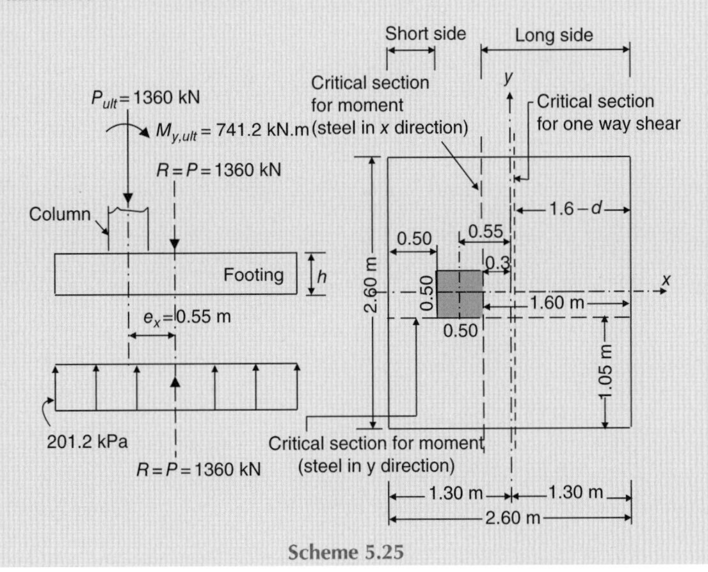

Scheme 5.25

(*Continued*)

Determine the required thickness based on both one-way shear and two-way shear analyses.

(a) Considering one-way shear: The most critical section for one-way shear is located a distance *d* from the right face of the column, as shown in the scheme above.

$$V_u = [(2.60)(1.6-d)](net\ q_{factored}) = (2.60)(1.60-d)(201.2)$$
$$= (836.99 - 523.12\,d)\text{kN}$$

Equation (5.20):

$$V_c = 0.17\lambda\sqrt{f_c'}b_w d$$

Let $\text{Ø}\,V_c = V_u$

$$(0.75\times0.17\times1)(\sqrt{21})(1000)(2.60)(d) = (836.99 - 523.12\,d)$$

$$d = \frac{836.99}{2042.24} = 0.41\text{ m} = 410\text{ mm}$$

(b) Considering two-way shear:

The critical section for two-way shear is located a distance *d*/2 from the column faces (ACI Sections 11.11.1.2 and 15.3), as shown in the scheme.

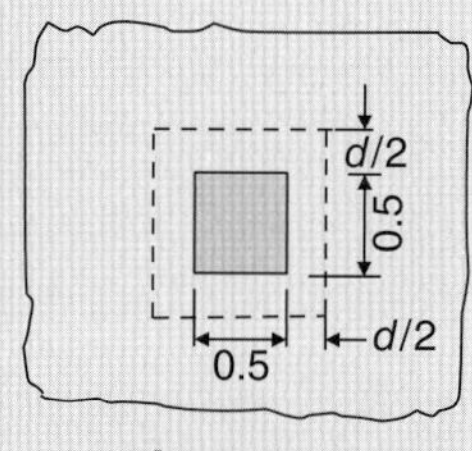

Scheme 5.26

Assume $d = 0.41$ m as required by one-way shear.

$$V_u = A_f(net\ q_{factored}) - \text{soil reaction on the shear block}$$
$$= (2.6\times2.6)(201.2) - \text{soil reaction on the shear block}$$
$$= 1360.11 - \text{soil reaction on the shear block(kN)}$$

The shear strength of concrete V_c shall be the smallest of (a), (b) and (c):

(a)

$$V_c = 0.17\left(1+\frac{2}{\beta}\right)\lambda\sqrt{f_c'}b_o d$$

Equation (5.22)

$$= (0.17)\left(1+\frac{2}{1}\right)(1)\sqrt{f_c'}b_o d = (0.51)\sqrt{f_c'}b_o d$$

(b)

$$V_c = 0.083\left(\frac{\alpha_s d}{b_o}+2\right)\lambda\sqrt{f_c'}b_o d$$

$$= (0.083)\left(\frac{40d}{b_o}+2\right)(1)\sqrt{f_c'}b_o d$$

Equation (5.23)

$$= (0.083)\left(\frac{40(0.41)}{4(0.5+0.41)}+2\right)\sqrt{f_c'}b_o d = (0.54)\sqrt{f_c'}b_o d$$

(c)

$$V_c = 0.33\lambda\sqrt{f_c'}b_o d$$

Equation (5.24)

$$= 0.33(1)\sqrt{f_c'}b_o d = (0.33)\sqrt{f_c'}b_o d$$

Use $V_c = (0.33)\sqrt{f_c'}b_o d$.

$$\varnothing V_c = (0.75)(0.33)\sqrt{21}(1000)[4(0.5+0.41)(0.41)]$$

$$= 1692.7 \text{ kN} > V_u$$

Therefore one-way shear controls.

Use d = 0.41 m = 410 mm.

This depth d will be taken to the centre of the steel bars in the x-direction.

Using No. 22 bars with 75 mm concrete cover (ACI Section 7.7.1), the overall footing thickness is

$$h = 410 + 75 + 11.1 = 875 + 12.7 = 496.1 \text{ mm}$$

Use h = 500 mm.

Step 5. Design the flexural reinforcement.

(a) Reinforcement in the x-direction. The critical section is shown in the scheme 5.25.

Equation (5.9):

$$\varnothing M_n \geq M_u$$

The maximum factored moment is

$$M_u = \frac{(net\, q_{\text{factored}})(l)^2}{2} = \frac{201.2 \times 1.6^2}{2} = 258 \text{ kN.m/m}$$

Assume tension-controlled section, $\varnothing = 0.9$, and $f_s = f_y$.

$$a = \frac{A_s f_y}{0.85 f_c' b} = \frac{A_s \times 420}{0.85 \times 21 \times 1} = 23.53\, A_s$$

$$\varnothing M_n = \varnothing \left[A_s f_y \left(d - \frac{a}{2}\right)\right]$$

$$= (0.9)\left[(A_s \times 420 \times 1000)\left(0.41 - \frac{23.53\, A_s}{2}\right)\right] \text{ Equation (5.11)}$$

$$= 154980\, A_s - 4447170\, A_s^2$$

Let $\varnothing M_n = M_u$

$$154980\, A_s - 4447170\, A_s^2 = 258$$

$$4447170\, A_s^2 - 154980\, A_s + 258 = 0$$

$$A_s = \frac{-(-154\,980) \pm \sqrt{(-154\,980)^2 - (4)(4\,447\,170)(258)}}{2 \times 4\,447\,170} = \frac{154980 - 139389}{8894340}$$

$$= 1753 \times 10^{-6} \text{ m}^2/\text{m} = 1753 \text{ mm}^2/\text{m}$$

$$A_{s,min} = 0.0018\, bh = 0.0018 \times 1 \times 0.5 = 0.9 \times 10^{-3} \text{ m}^2/\text{m}$$

$$= 900 \text{ mm}^2/\text{m} < A_s \text{ required by analysis}$$

Try five No.22 bars: $A_{s,provided} = 5 \times 387 = 1935 \text{ mm}^2/\text{m} > A_s$ (OK.)

Compute a for $A_s = 1935$ mm²/m and check if $f_s = f_y$ and whether the section is tension-controlled:

$$a = 23.53 A_s = 23.53 \times 1935 \times 10^{-6} = 0.0455 \text{ m} = 45.5 \text{ mm}$$

(Continued)

$c = \frac{a}{\beta_1}$; For f'_c between 17 and 28 MPa, β_1 shall be taken as 0.85

$$c = \frac{45.5}{0.85} = 53.53\,\text{mm}$$

$$d_t = d = 410\,\text{mm}$$

$$\varepsilon_t = 0.003\left(\frac{d_t - c}{c}\right) \quad \text{Equation (5.15)}$$

$$= 0.003\left(\frac{410 - 53.53}{53.53}\right) = 0.020 > 0.005.$$

Hence, the section is tension-controlled, and Ø = 0.9 (ACI Sections 10.3.4 and 9.3.2.1).

$$\varepsilon_y = \frac{f_y}{E_s} = \frac{420}{200000} = 0.0021.$$

Hence, $\varepsilon_t > \varepsilon_y$ and $f_s = f_y$.

Therefore the assumptions made are satisfied.

Total number of bars required in the x-direction = 2.6 × 5 = 13

Using 75 mm minimum concrete cover at each side, centre to centre bar spacing will be 202 mm. Check concrete cover provided at each side:

$$\frac{2600 - (12 \times 202 + 22.2)}{2} = 76.9\,\text{mm} > 75\ \text{mm(OK.)}$$

In case the required flexural reinforcement exceeds the minimum flexural reinforcement, it shall be adequate to use the same maximum spacing of reinforcement for slabs which is two times the slab thickness, or 450 mm, whichever is smaller, as specified by ACI Section 13.3.2.

Try 13 No.22 bars @ 202 mm *c.c. in the x -direction at bottom of the footing.*

Check the development of reinforcement:

In this case, the bars are in tension, the provided bar size is larger than No.19, the clear spacing of the bars exceeds $2d_b$, and the clear cover exceeds d_b. Therefore, the development length l_d of bars at each side of the critical section (which is the same critical section for moment) shall be determined from the following equation, but not less than 300 mm:

$$l_d = \left(\frac{f_y \psi_t \psi_e}{1.7\lambda\sqrt{f'_c}}\right) d_b \quad \text{(ACI Sections 12.2.1, 12.2.2 and 12.2.4)}$$

where the factors ψ_t, ψ_e and λ are 1. (ACI Sections 12.2.4 and 8.6.1)

$$l_d = \left(\frac{420 \times 1 \times 1}{1.7 \times 1 \times \sqrt{21}}\right)\left(\frac{22.2}{1000}\right) = 1.2\,\text{m} = 1200\,\text{mm} > (l_{d,min} = 300\,\text{mm})$$

Therefore, the required l_d = 1200 mm

The smaller bar extension past the critical section (i.e. the available length) is

$$1000\,\text{mm} - 75\,\text{mm cover} = 925\,\text{mm} < 1200\,\text{mm} \quad \text{(Not OK.)}$$

We must consider smaller bars or use bars terminating in a standard hook.

Assume the designer prefers using smaller bars:

Tryseven No.19 bars; $A_{s,provided} = 7 \times 284 = 1988\ \text{mm}^2/\text{m} > A_s$ required by analysis (OK.)

Total number of bars required in the x-direction $= 2.6 \times 7 = 19$

$$l_d = \left(\frac{420 \times 1 \times 1}{2.1 \times 1 \times \sqrt{21}}\right)\left(\frac{19.1}{1000}\right) = 0.834\,\text{m} = 834\,\text{mm} < 925\,\text{mm} \quad \text{(OK.)}$$

Using 75 mm concrete cover at each side, centre to centre bar spacing will be 135 mm. Check concrete cover provided at each side:

$$\frac{2600 - (18 \times 135 + 19.1)}{2} = 75.5\,\text{mm} > 75\,\text{mm, O.K.}$$

Use 19 No.19 bars @ 135 mm *c.c. in the* $\mathbf{x}$ *-direction at bottom of the footing.*

(b) Reinforcement in the y-direction. The critical section is shown in the scheme of *Step 4.*

$$\emptyset M_n \geq M_u$$

The maximum factored moment is

$$M_u = \frac{(net\ q_{factored})(l)^2}{2} = \frac{201.2 \times 1.05^2}{2} = 111\,\text{kN.m/m}$$

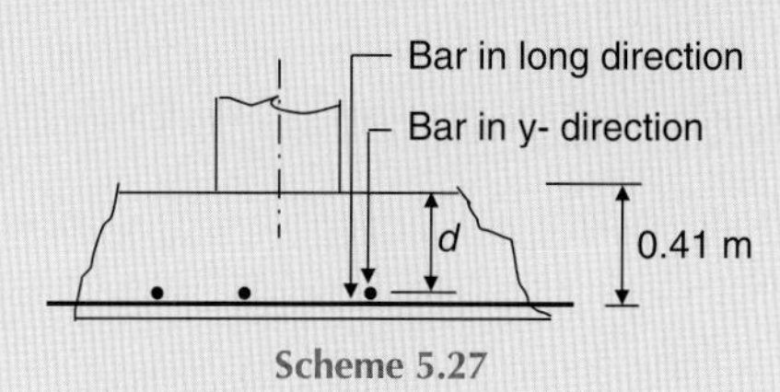

Scheme 5.27

Bars in short direction are placed on bars in long direction. Therefore,

$$d = 0.41 - \left(\begin{array}{l}\frac{1}{2}\text{diameter of bar in } x-\text{direction} + \\ \frac{1}{2}\text{diameter of bar in } y-\text{direction}\end{array}\right)$$

Assume using No. 16 bars in y-direction. Hence,

$$d = 0.41 - \left(\frac{1}{2} \times 0.0191 + \frac{1}{2} \times 0.0159\right) = 0.39\,\text{m}$$

Assume tension-controlled section, $\emptyset = 0.9$, and $f_s = f_y$.

$$a = \frac{A_s f_y}{0.85 f_c' b} = \frac{A_s \times 420}{0.85 \times 21 \times 1} = 23.53\,A_s$$

$$\emptyset M_n = \emptyset\left[A_s f_y\left(d - \frac{a}{2}\right)\right]$$

$$= (0.9)\left[(A_s \times 420 \times 1000)\left(0.39 - \frac{23.53\,A_s}{2}\right)\right] \quad \text{Equation (5.11)}$$

$$= 147420\,A_s - 4447170\,A_s^2$$

(Continued)

Let $\emptyset M_n = M_u$

$$147420\,A_s - 4447170\,A_s^2 = 111$$

$$4447170\,A_s^2 - 147420 + 111 = 0$$

$$A_s = \frac{-(-147\,420) \pm \sqrt{(-147\,420)^2 - (4)(4\,447\,170)(111)}}{2 \times 4447170} = \frac{147420 - 140563}{8\,894\,340}$$

$$= 771 \times 10^{-6}\ \text{m}^2/\text{m} = 771\ \text{mm}^2/\text{m}$$

$$A_{s,\min} = 0.0018\,bh = 0.0018 \times 1 \times 0.5 = 0.9 \times 10^{-3}\ \text{m}^2$$

$$= 900\ \text{mm}^2/\text{m} > A_s\ \text{required by analysis}$$

Therefore, use $A_s = A_{s,min} = 900\ \text{mm}^2/\text{m}$

Try five No.16 bars; $A_{s,provided} = 5 \times 199 = 995\ \text{mm}^2/\text{m} > A_s$ (OK.)

The assumptions made are satisfied, since A_s required $= A_{s,min}$.

Total number of bars required in the y -direction $= 2.6 \times 5 = 13$.

It may be more appropriate to place eight bars (about 60% of the total bars) within the half of the footing area that contains the column and the other five bars in the other half, uniformly distributed.

Check the development of reinforcement:

In this case, the bars are in tension, the provided bar size is smaller than No. 19, the clear spacing of the bars exceeds $2d_b$, and the clear cover exceeds d_b. Therefore, the development length l_d of bars at each side of the critical section (which is the same critical section for moment) shall be determined from the following equation, but not less than 300 mm:

$$l_d = \left(\frac{f_y \psi_t \psi_e}{2.1\lambda\sqrt{f_c'}}\right) d_b \quad \text{(ACI Sections 12.2.1, 12.2.2 and 12.2.4)}$$

where the factors ψ_t, ψ_e and λ are 1. (ACI Sections 12.2.4 and 8.6.1)

$$l_d = \left(\frac{420 \times 1 \times 1}{2.1 \times 1 \times \sqrt{21}}\right)\left(\frac{15.9}{1000}\right) = 0.694\ \text{m} = 694\ \text{mm} > (l_{d,min} = 300\ \text{mm})$$

Therefore, the required $l_d = 694$ mm

The bar extension past the critical section (i.e. the available length) is

$$1050\ \text{mm} - 75\ \text{mm cover} = 975\ \text{mm} > 694\ \text{mm} \quad \text{(OK.)}$$

When $A_{s,min}$ controls, ACI Section 10.5.4 requires maximum spacing shall not exceed three times the slab or footing thickness, or 450 mm, whichever is smaller.

Using 75 mm minimum concrete cover at each side, the reinforcement distribution and centre to centre bar spacing shall be as follow:

Provide eight No.16 bars @ 173 mm *c.c.in the half of the footing area that contains the column, placed on top of bars in x-direction.*

Provide five No.13 bars @ 243 mm *c.c. in the other half of the footing area, placed on top of bars in x-direction.*

Step 6. Check the column bearing on the footing (ACI Section 10.14).

For this step and the other design steps concerning the necessary dowels and their embedment lengths in both the footing and the column, the design proceeds in the same manner as that for the footing of Problem 5.4. The necessary computations are left for the reader.

Problem 5.7

The bearing walls of a one-story industrial building will be built of concrete block 200 × 300 × 400 mm. Design a wall footing for the following data:

$$\text{Wall loads including wall, floor, and roof contribution:}\quad D = 70\ \text{kN/m}$$
$$L = 30\ \text{kN/m}$$
$$D + L = 100\ \text{kN/m}$$

Wall thickness = 300 mm

Footing: $f_c' = 21\,\text{MPa}$ $f_y = 420\ \text{MPa}$ $E_{steel} = 2 \times 10^5\,\text{MPa}$

Foundation soil: $net\ q_a = 150\,\text{kPa}$

Solution:

Step 1. Find the footıng width and the factored net soil pressure.

Neglect any weight increase from displacing the lighter soil with heavier concrete, since it is usually too small for lightly loaded footings.

$$\text{Footing width } = B = \text{area per meter length of wall} = \frac{\text{wall load}}{net\ q_a}$$

$$B = \frac{100}{150} = 0.67\,\text{m}$$

Let the footing projects 200 mm on each side of the wall to facilitate placing the foundation wall. Use $B = 0.7$ m = 700 mm.

$$\text{Factored net soil pressure} = net\ q_{\text{factored}} = \frac{\text{factored net load}}{\text{area}} = \frac{1.2D + 1.6L}{\text{area}}$$

$$= \frac{1.2 \times 70 + 1.6 \times 30}{0.7 \times 1} = 188.6\ \text{kPa}$$

Step 2. Find footing thickness h.

One-way shear only is significant in wall footings. Critical section for one-way shear is located a distance d from face of the wall (ACI Sections 11.11.1.1 and 11.1.3.1). However, in this case, the critical section may fall out the footing projection, and to account for the most severe condition, we consider the section located at face of the wall, as shown in the scheme below.

$$V_u = (1 \times 0.2)\left(net\ q_{factored}\right) = (1 \times 0.2)(188.6) = 37.72\ \text{kN/m}$$

Equation (5.19): $\varnothing V_c \geq V_u$ (taking reinforcement shear strength, $V_s = 0$)

where $\varnothing$ = strength reduction factor (ACI Section 9.3.2.3)

= 0.75 for shear

(Continued)

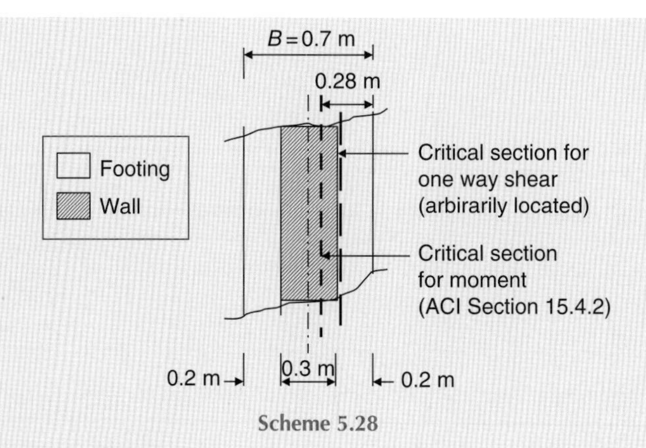

Scheme 5.28

Equation(5.20): $V_c = 0.17\lambda\sqrt{f_c'}b_w d$

Let $\varnothing V_c = V_u$:

$$(0.75 \times 0.17 \times 1)(\sqrt{21})(1000)(1)(d) = 37.72$$

$$d = \frac{37.72}{584.28} = 0.065\,\text{m} = 65\,\text{mm} \qquad \text{(ACI Section 15.7)}$$

$$d_{min} = 150\,\text{mm}$$

Choose arbitrarily d = 250 mm.

The depth d will be taken to the centre of the transverse steel bars.

$$\text{Minimum concrete cover} = 75\,\text{mm} \quad \text{(ACI Section 7.7.1)}$$

Assume using No.13 bars. Hence, the overall footing thickness is

$$h = 250 + 75 + 6.4 = 331.4\,\text{mm}.$$

Use h = 350 mm.

Step 3. Design the flexural reinforcement.

The critical section for transverse bending is shown in the scheme of *Step 2.*

Equation (5.9): $\varnothing M_n \geq M_u$

The maximum factored moment is

$$M_u = \frac{(net\ q_{factored})(l)^2}{2} = \frac{188.6 \times 0.28^2}{2} = 7.4\,\text{kN.m/m}$$

Assume tension-controlled section, $\varnothing = 0.9$, and $f_s = f_y$

$$a = \frac{A_s f_y}{0.85 f_c' b} = \frac{A_s \times 420}{0.85 \times 21 \times 1} = 23.53\,A_s$$

$$\varnothing M_n = \varnothing\left[A_s f_y\left(d - \frac{a}{2}\right)\right]$$

$$= (0.9)\left[(A_s \times 420 \times 1000)\left(0.25 - \frac{23.53 A_s}{2}\right)\right] \quad \text{Equation (5.11)}$$

$$= 94\ 500\ A_s - 4\ 447170\ A_s^2$$

Let $\varnothing M_n = M_u$

$$94\ 500 A_s - 4\ 447\ 170 A_s^2 = 7.4$$

$$4\ 447\ 170 A_s^2 - 94\ 500 A_s + 7.4 = 0$$

$$A_s = \frac{-(-94\ 500) \pm \sqrt{(-94\ 500)^2 - (4)(4\ 447\ 170)(7.4)}}{2 \times 4\ 447\ 170} = \frac{94\ 500 - 93\ 801}{8\ 894\ 340}$$

$$= 78.6 \times 10^{-6} \text{m}^2/\text{m} = 78.6 \text{ mm}^2/\text{m}$$

$$A_{s,\min} = 0.0018\ bh$$

$$= 0.0018 \times 1 \times 0.35 = 0.63 \times 10^{-3} \text{ m}^2/\text{m} \quad \text{(ACI Sections 10.5.4 and 7.12.2.1)}$$

$$= 630 \text{ mm}^2/\text{m} \gg A_s \text{ required by analysis}$$

Therefore, use A_s required = $A_{s,\min}$ = 630 mm²/m

The assumptions made are satisfied, since A_s required = $A_{s,\min}$.

Try No.13 bars @ 250 mm c.c., that is five No.13 bars per metre length of the footing. $A_{s,\text{provided}} = 5 \times 129 =$ 645 mm²/m > $A_{s,\min}$ (OK.)

Check the development of reinforcement:

In this case, the bars are in tension, the provided bar size is smaller than No. 19, the clear spacing of the bars exceeds $2d_b$, and the clear cover exceeds d_b. Therefore, the development length l_d of bars at each side of the critical section (which is the same critical section for moment) shall be determined from the following equation, but not less than 300 mm:

$$l_d = \left(\frac{f_y \psi_t \psi_e}{2.1 \lambda \sqrt{f_c'}}\right) d_b \quad \text{(ACI Sections 12.2.1, 12.2.2 and 12.2.4)}$$

where the factors ψ_t, ψ_e and λ are 1. (ACI Sections 12.2.4 and 8.6.1)

$$l_d = \left(\frac{420 \times 1 \times 1}{2.1 \times 1 \times \sqrt{21}}\right)\left(\frac{12.7}{1000}\right) = 0.554 \text{ m} = 554 \text{ mm} > 300 \text{ mm}.$$

Therefore, the required l_d = 554 mm.

The bar extension past the critical section (i.e. the available length) is

$$280 \text{ mm} - 75 \text{ mm cover} = 205 \text{ mm} < 554 \text{ mm} \quad \text{(Not OK.)}$$

Use bars terminating in a 180-degree standard hook (ACI Section 12.5.1 and ACI Figure R12.5). Hooked bar details for development of 180-degree standard hooks are shown in the scheme in Solution of Problem 5.2, *Step 5*.

$$l_{dh} = \left(\frac{0.24 \psi_e f_y}{\lambda \sqrt{f_c'}}\right) d_b \quad \text{(ACI Sections 12.5.2)}$$

where the factors ψ_e and λ shall be taken as 1.

$$l_{dh} = \left(\frac{0.24 \times 1 \times 420}{1 \times \sqrt{21}}\right)\left(\frac{12.7}{1000}\right) = 0.28 \text{ m} = 280 \text{ mm} > 150 \text{ mm} > 8\ d_b$$

Therefore, the required l_{dh} = 280 mm.

(Continued)

$$l_{dh} = 280\ \text{mm} > 205\ \text{mm available. (Not OK.)}$$

However, l_{dh} can be reduced by the applicable modification factor of ACI Section 12.5.3(a), as follows:

$$\text{Reduced } l_{dh} = 0.7 \times l_{dh} = 0.7 \times 280 = 196\ \text{mm} < 205\ \text{mm}$$
$$> 150\ \text{mm (OK.)}$$

Therefore, the bars shall be terminating in a 180-degree standard hook.
Use No.13 bars @ 250 mm c.c. placed transversely at bottom of the footing. The bars must have 180-degree standard hook at each end.

Step 4. Select longitudinal steel based on the minimum shrinkage and temperature reinforcement (ACI Sections 7.12.1 and 7.12.2.1).

$$A_s = 0.0018\,bh = 0.0018 \times 0.7 \times 0.35 = 4.41 \times 10^{-4}\ \text{m}^2 = 441\ \text{mm}^2$$

Try seven No. 10 bars: $A_{s,provided} = 7 \times 71 = 497\ \text{mm}^2 > 441\ \text{mm}^2$ (OK.)
Longitudinal steel will, in general, be more effective in the top of the footing than in the bottom; it could control cracks when the foundation settles. According to ACI Section 7.12.2.2, shrinkage and temperature reinforcement shall be spaced not farther apart than five times the slab thickness, nor farther apart than 450 mm. Provide four No. 10 bars at top and three No. 10 bars at bottom, which will also provide support for the transverse flexural reinforcement.

Provide four No.10 bars @ 180 mm c.c. at top of the footing in long direction.

Provide three No.10 bars @ 250 mm c.c. at bottom of the footing in long direction as support for the transverse bars.

Step 5. Draw a final design sketch as shown in the scheme below.

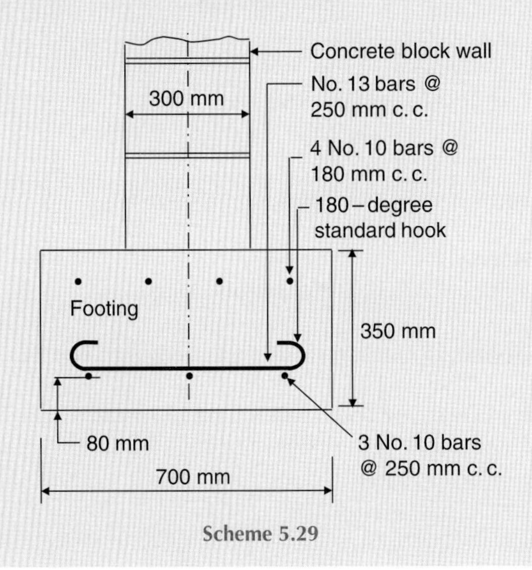

Scheme 5.29

Problem 5.8

A 300 mm thick reinforced concrete wall carries a service dead load of 150 kN/m and live load of 185 kN/m. The wall is reinforced with No. 13 vertical steel bars in two layers parallel with faces of the wall at 350 mm centre to centre. The supporting soil has *gross* $q_a = 240$ kPa at the foundation level, which is 1.5 m below the final ground surface. The soil unit weight is 19 kN/m^3. Design the wall footing using $f_c' = 21$ MPa, $f_y = 420$ MPa and $E_{steel} = 2 \times 10^5$ MPa.

Solution:

Step 1. Find footing width B and the factored net soil pressure.

Equation 4.8b: $$net\ q = \frac{V}{A} - \gamma_w h + D_c(\gamma_c - \gamma) \leq net\ q_a$$

where $\gamma_w h$ = uplift water pressure = 0 (in this case) Hence,

$$net\ q = \frac{V}{A} + D_c(\gamma_c - \gamma)$$

Guess a trial value for D_c between 1 and 1.5 times the wall thickness.

Assume $D_c = 0.4$ m

$$net\ q = \frac{150 + 185}{A} + (0.4)(24 - 19) = \frac{335}{A} + 2\ (\text{kPa})$$

$$net\ q_a = gross\ q_a - \sigma_o' = 240 - 19 \times 1.5 = 211.5\ \text{kPa}$$

Let $net\ q = net\ q_a$

$$\frac{335}{A} + 2 = 211.5$$

$$A = \frac{335}{209.5} = 1.6\ \text{m}^2$$

$A = B \times 1$. Hence, B = 1.6 m.

Another method to find B is as follows:

$$\text{Gross foundation load} = (150 + 185) + (B \times 1)(24 \times 0.4 + 19 \times 1.1)$$

$$= 335 + 30.5\ B\ \ (\text{kN})$$

$$A = B \times 1 = \frac{\text{gross foundation load}}{\text{gross } q_a} = \frac{335 + 30.5\ B}{240}$$

$$B = \frac{335}{209.5} = 1.6\ \text{m}^2$$

The factored soil pressure is computed without including the weights of footing and backfill material because these loads are evenly distributed and thus do not produce shear or moment in the footing.

$$\text{Foundation factored net load} = 1.2(150) + 1.6 \times 185 = 476\ \text{kN}$$

$$\text{Factored net soil pressure} = net\ q_{factored} = \frac{476}{\text{A}} = \frac{476}{1.6 \times 1} \cong 300\ \text{kPa.}$$

Step 2. Find footing thickness h. One-way shear only is significant in strip footings.

Critical section for one-way shear is located a distance d from face of the wall (ACI Sections 11.11.1.1 and 11.1.3.1), as shown in the scheme below.

(*Continued*)

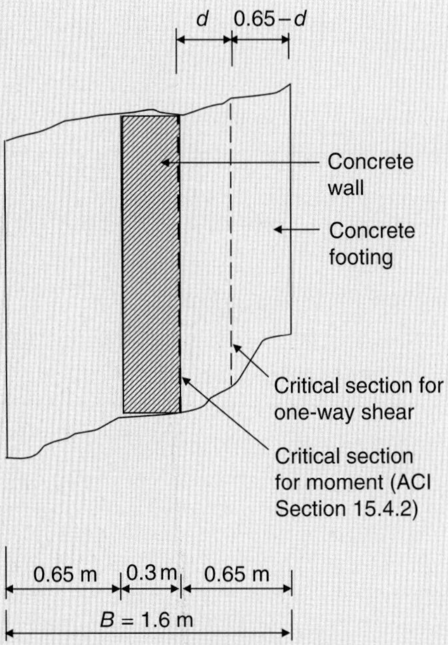

Scheme 5.30

Equation (5.19): $\varnothing V_c \geq V_u$ (taking reinforcement shear strength, $V_s = 0$)

where $\varnothing$ = strength reduction factor (ACI Section 9.3.2.3)

= 0.75 for shear

$$V_u = (1)(0.65 - d)(net\ q_{factored}) = (0.65 - d)(300) = 195 - 300\ d$$

Equation (5.20):

$$V_c = 0.17\lambda\sqrt{f'_c}b_w d$$

$$\varnothing V_c = (0.75 \times 0.17 \times 1)(\sqrt{21})(1000)(1)(d)$$

Let $\varnothing V_c = V_u$:

$$(0.75 \times 0.17 \times 1)\left(\sqrt{21}\right)(1000)(1)(d) = 195 - 300\,d$$

$$d = \frac{195}{884.28} = 0.221\ \text{m} = 221\ \text{mm} > (d_{min} = 150\ \text{mm})\ \text{ACI Section 15.7}$$

Try $d = 221$ mm.

The depth d will be taken to the centre of the transverse steel bars. Assume using No. 13 bars, and 75 mm minimum concrete cover (ACI Section 7.7.1). Hence, the overall footing thickness is

$$h = 221 + 75 + 6.4 = 302.4\ \text{mm}$$

Try $h = 310$ mm.

Check gross q_a:

$$\text{gross } q = \frac{\text{gross foundation load}}{A} = \frac{(150 + 185) + (1.6 \times 1)(24 \times 0.31 + 19 \times 1.19)}{1.6 \times 1}$$

$$= 239.4\ \text{kPa} < (\text{gross } q_a = 240\ \text{kPa}) \qquad \text{(OK.)}$$

Use: $d = 221$ mm
$h = 310$ mm

Step 3. Design the flexural reinforcement.
The critical section for transverse bending is shown in the scheme of *Step 2.*

Equation (5.9):
$$\varnothing M_n \geq M_u$$

$$M_u = \frac{(net\ q_{\text{factored}})(l)^2}{2} = \frac{300 \times 0.65^2}{2} = 63.4\ \text{kN.m/m}$$

Equation (5.11):
$$\varnothing M_n = \varnothing\left[A_s f_y\left(d - \frac{a}{2}\right)\right]$$

Assume tension-controlled section, $\varnothing = 0.9$, and $f_s = f_y$.

$$a = \frac{A_s f_y}{0.85 f_c' b} = \frac{A_s \times 420}{0.85 \times 21 \times 1} = 23.53\, A_s$$

$$\varnothing M_n = (0.9)\left[(A_s \times 420 \times 1000)\left(0.221 - \frac{23.53 A_s}{2}\right)\right]$$

$$= 83\ 538\, A_s - 4\ 447\ 170\, A_s^2$$

Let $\varnothing M_n = M_u$

$$83\ 538\, A_s - 4\ 447\ 170\, A_s^2 = 63.4$$
$$4\ 447\ 170\, A_s^2 - 83\ 538\, A_s + 63.4 = 0$$

$$A_s = \frac{-(-83\ 538) \pm \sqrt{(-83\ 538)^2 - (4)(4\ 447\ 170)(63.4)}}{2 \times 4\ 447\ 170} = \frac{83\ 538 - 76\ 490}{8\ 894\ 340}$$

$$= 792 \times 10^{-6}\ \text{m}^2/\text{m} = 792\ \text{mm}^2/\text{m}$$

$$A_{s,min} = 0.0018\, bh = 0.0018 \times 1 \times 0.31 = 5.58 \times 10^{-4}\ \text{m}^2$$
$$= 558\ \text{mm}^2/\text{m} < A_s\ \text{required by analysis}$$

In case the required flexural reinforcement exceeds the minimum flexural reinforcement, it shall be adequate to use the same maximum spacing of reinforcement for slabs which is two times the slab thickness, or 450 mm whichever is smaller, as specified by ACI Section 13.3.2.

Try No. 13 @ 160 mm c.c.

$$A_{s,provided} = \frac{1000}{160} \times 129 = 806\ \text{mm}^2/\text{m} > A_s\ \ (\text{OK.})$$

Compute a for $A_s = 806$ mm^2/m, and check if $f_s = f_y$ and whether the section is tension-controlled:

$$a = 23.53 A_s = 23.53 \times 806 \times 10^{-6} = 0.019\ \text{m} = 19\ \text{mm}$$

(*Continued*)

$c = \dfrac{a}{\beta_1}$; For f_c' between 17 and 28 MPa, β_1 shall be taken as 0.85

$$c = \frac{19}{0.85} = 22.4\,\text{mm}$$

$$d_t = d = 221\ \text{mm}$$

$$\varepsilon_t = 0.003\left(\frac{d_t - c}{c}\right)$$

$$= 0.003\left(\frac{221 - 22.4}{22.4}\right) = 0.027 > 0.005. \quad \text{Equation (5.15)}$$

Hence, the section is tension-controlled, and Ø = 0.9 (ACI Sections 10.3.4 and 9.3.2.1).

$$\varepsilon_y = \frac{f_y}{E_s} = \frac{420}{200\,000} = 0.0021.$$

Hence, $\varepsilon_t > \varepsilon_y$ and $f_s = f_y$.

Therefore the assumptions made are satisfied.

Check the development of reinforcement:

In this case, the bars are in tension, the provided bar size is smaller than No. 19, the clear spacing of the bars exceeds $2d_b$, and the clear cover exceeds d_b. Therefore, the development length l_d of bars at each side of the critical section (which is the same critical section for moment) shall be determined from the following equation, but not less than 300 mm:

$$l_d = \left(\frac{f_y \psi_t \psi_e}{2.1\lambda\sqrt{f_c'}}\right) d_b \quad \text{(ACI Sections 12.2.1, 12.2.2 and 12.2.4)}$$

where the factors ψ_t, ψ_e and λ are 1. (ACI Sections 12.2.4 and 8.6.1)

$$l_d = \left(\frac{420\times 1\times 1}{2.1\times 1\times \sqrt{21}}\right)\left(\frac{12.7}{1000}\right) = 0.554\,\text{m} = 554\,\text{mm} > 300\ \text{mm}.$$

Use $l_d = 0.554$ m = 554 mm.

The bar extension past the critical section (i.e. the available length) is

$$650\,\text{mm} - 75\,\text{mm cover} = 575\,\text{mm} > l_d\ \text{(OK.)}$$

Use No. 13 bars @ 160 mm c.c. placed at bottom of the footing in the transverse direction.

Step 4. Select longitudinal steel based on the minimum shrinkage and temperature reinforcement (ACI Sections 7.12.1 and 7.12.2.1).

$$A_s = 0.0018\,bh = 0.0018 \times 1.6 \times 0.31$$

$$= 8.928 \times 10^{-4}\,\text{m}^2 = 893\,\text{mm}^2$$

Try 13 No. 10 bars: $A_{s,provided} = 13 \times 71 = 923\,\text{mm}^2 > A_s$ (OK.)

Longitudinal steel will, in general, be more effective in the top of the footing than in the bottom; it could control cracks when the foundatıon settles.

According to ACI Section 7.12.2.2, shrinkage and temperature reinforcement shall be spaced not farther apart than five times the slab thickness, nor farther apart than 450 mm.

Provide eight No. 10 @ 205 mm c.c. at top of the footing in long direction.

Provide five No. 10 @ 360 mm c.c. at bottom of the footing in long direction as support for the transverse bars.

Step 5. Check wall bearing on the footing (ACI Section 10.14).

$$\begin{aligned}\text{The factored load at the base of the wall} &= 1.2D + 1.6L\\ &= 1.2\times 150 + 1.6\times 185\\ &= 476\ \text{kN}\end{aligned}$$

(a) Base of the wall. The allowable bearing strength of concrete at the wall base is

$$\begin{aligned}\emptyset\left(0.85f_c'A_1\right) &= 0.65[0.85\times 21(0.3\times 1)]\\ &= 3.48\ \text{MN} = 3480\ \text{kN} \gg (1.2D+1.6L)\quad \text{(OK.)}\end{aligned}$$

(b) Top surface of the footing. The design bearing strength on the top surface of the footing shall not exceed $\emptyset\left(0.85f_c'A_1\right)$, except when the supporting surface is wider on all sides than the loaded area; then the design bearing strength of the loaded area shall be permitted to be multiplied by $\sqrt{A_2/A_1}$ but not more than 2, (ACI Section 10.14.1).

Since both the footing and wall concrete have the same f_c', it is clear that the design bearing strength on the top surface of the footing is also much greater than $(1.2D + 1.6L)$.

Therefore, vertical compression reinforcement or dowels through the supporting surface at the interface are theoretically not required. However, ACI Section 15.8.2.2 requires area of reinforcement across interface shall be not less than the minimum vertical reinforcement given in ACI Section 14.3.2.

Step 6. Design dowels to satisfy the minimum area of reinforcement across interface required by ACI Sections 15.8.2.2 and14.3.2.

ACI Section 14.3.2 states that minimum ratio of vertical reinforcement area to gross concrete area, ρ_e, shall be:

(a) 0.0012 for deformed bars not larger than No. 13 with f_y not less than 420 MPa; or
(b) 0.0015 for other deformed bars; or
(c) 0.0012 for welded wire reinforcement not larger than MW200 or MD200.

Assume using No. 10 dowels with $f_y = 420$ MPa. Hence, $\rho_e = 0.0012$.

$$A_{s,\min} = 0.0012\times 1\times 0.3 = 3.6\times 10^{-4}\ \text{m}^2/\text{m} = 360\ \text{mm}^2/\text{m}$$

Try No. 10 @ 350 mm c.c. (the same spacing of the wall vertical bars) in two layers parallel with faces of the wall:

$$A_{s,\text{provided}} = 2\left(\frac{1000}{350}\times 71\right) = 406\ \text{mm}^2/\text{m} > A_{s,\min}\quad \text{(OK.)}$$

Provide No. 10 @ 350 mm c.c. in two layers parallel with faces of the wall.

Step 7. Find the embedment length of dowels in both the footing and the wall.

For this step, the design proceeds in the same manner as that of Solution of Problem 5.3, *Step 6*.

Step 8. Develop the final design sketch.

Problem 5.9

A metal column with steel base bearing plate is supported by a concrete pedestal rests on a reinforced concrete footing. The pedestal should have shoulder to provide bearing for the floor slab which will be placed 1.5 m above the footing, as shown in the scheme below. Design the pedestal and bearing plate for the following given data:

$$\text{Loads: } P = D + L \quad D = 600\,\text{kN} \quad L = 450\,\text{kN}$$

Column: **W** 310×107 metal (A-36 steel) column; $f_y = 250$ MPa
Bearing plate: metal (A-36 steel) plate; $f_y = 250$ MPa
Anchor bolts: A307-grade A (A-36 steel); $f_{ult} = 400$ MPa; $f_y = 250$ MPa
Concrete pedestal: $f_c' = 24\,\text{MPa}$
Steel bars: $f_y = 420$ MPa; $E_{steel} = 2 \times 10^5$ MPa

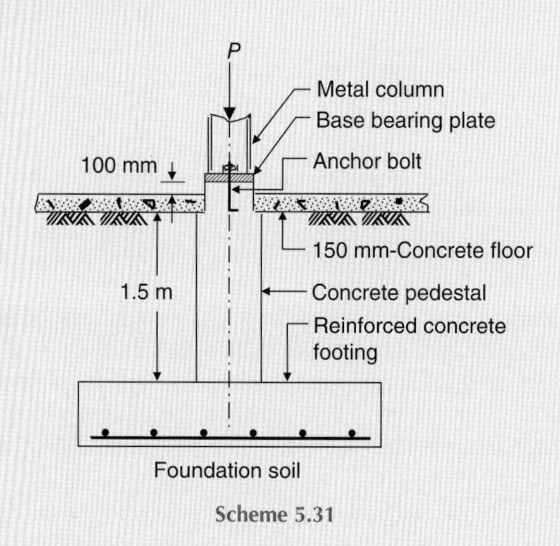

Scheme 5.31

Solution:

Step 1. Find area of the base plate and the cross-section dimensions of the pedestal.

Usually, the cross-sectional area of the pedestal is set for the base plate but increased by at least 50 mm to provide bearing for the floor slab.

(a) Area of the base plate, A_1. Design according to the AISC specifications.

$$A_1 = \frac{1}{A_2}\left(\frac{P}{0.35 f_c'}\right)^2, \quad \text{or,} \quad A_1 = \frac{P}{0.7 f_c'}, \text{ whichever is greater.}$$

$A_2 =$ Surface area of the supporting member

$$A_2 = \frac{P}{0.175 f_c'}$$

$$A_2 = \frac{P}{0.175 f_c'} = \frac{600 + 450}{0.175 \times 24 \times 1000} = 0.25\,\text{m}^2$$

$$A_1 = \frac{1}{0.25}\left(\frac{1050}{0.35 \times 24 \times 1000}\right)^2 = 0.0625\,\text{m}^2, \quad \text{or,}$$

$$A_1 = \frac{1050}{0.7 \times 24 \times 1000} = 0.0625\,\text{m}^2$$

Use a plate area $A_1 \geq 0.0625\,\text{m}^2$.

The plate plan dimensions should be larger than the column cross-section dimensions on all sides by about 12 mm in order that the plate can be fillet-welded to the bottom of the column.

Column cross-section dimensions:

For **W** 31 × 107 metal, AISC Tables give:

$$d = 311\,\text{mm}$$

$$b_f = 306\,\text{mm}$$

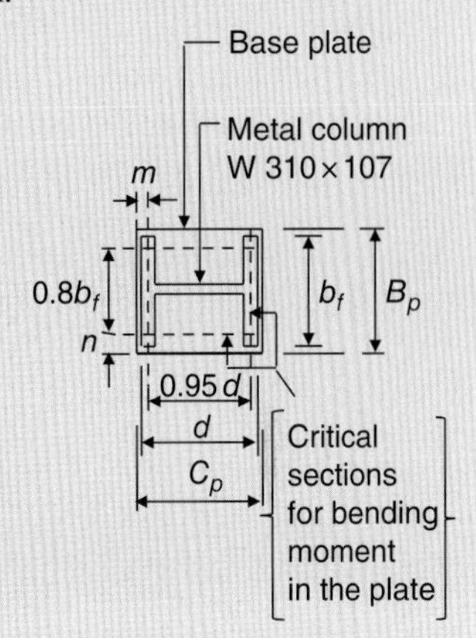

Scheme 5.32

Try a plate with dimensions:

$$B_p = b_f + 2 \times 12 = 306 + 24 = 330\,\text{mm}$$

$$C_p = d + 2 \times 12 = 311 + 24 = 335\,\text{mm}$$

Check A_1:

The furnished area is

$$0.330 \times 0.335 = 0.116\,\text{m}^2 > \left(A_{1,min} = 0.0625\,\text{m}^2\right) \quad \text{(OK.)}$$

Use 330 mm × 335 mm base bearing plate.

(b) Cross-section dimensions of the pedestal.

Assume using a pedestal with cross-section dimensions as follows:

For the height between the bottom levels of the base plate and floor slab, the cross-section shall be *0.5 × 0.5 m*.

For the height between the bottom level of the floor slab and top level of the footing, the cross-section shall be *0.6 × 0.6 m*.

Thus, the pedestal will have a *50 mm* shoulder on all sides as bearing for the floor slab.

Check A_2:

$$\text{The furnished area} = 0.5 \times 0.5 = 0.25\,\text{m}^2$$
$$A_2 = 0.25\,\text{m}^2 \quad \text{(OK.)}$$

Check the ratio $\frac{L_u}{B}$:

$$L_u = \text{Unsupported height of a pedestal}$$
$$B = \text{Least lateral dimension of a pedestal}$$

Assume neglecting lateral restraint due to backfill. Hence,

$$\frac{L_u}{B} = \frac{1.5}{0.6} = 2.5 < 3.$$

Therefore, the supporting member is considered pedestal (ACI Section 2.2).

(Continued)

Step 2. Check plate bearing on pedestal.

$$\text{Maximum allowable bearing load should be} \leq \left(0.7 f_c' A_1\right)$$
$$\geq (D+L)$$
$$0.7 f_c' A_1 = 0.7 \times 24 \times 0.33 \times 0.36 = 1.996\ \text{MN}$$
$$D + L = 600 + 450 = 1050\ \text{kN}$$

According to AISC, for the case $A_2 > A_1$, the maximum allowable bearing load on the top of the pedestal is $0.35 f_c' A_1 \sqrt{A_2/A_1}$, on the condition $\sqrt{A_2/A_1} \leq 2$.

$$\sqrt{A_2/A_1} = \sqrt{0.25/(0.33 \times 0.36)} = 1.45 < 2$$
$$\text{Maximum allowable bearing load} = (0.35)(24)(0.33 \times 0.36)(1.45)$$
$$= 1.447\ \text{MN} = 1447\ kN$$
$$< \left(0.7 f_c' A_1\right)$$
$$> (D+L) \qquad \text{(OK.)}$$

Step 3. Find the plate thickness t_p.

Actually, computation of t_p is beyond the scope of this text. However, for completeness, we may compute t_p using the AISC specifications.

Refer to the scheme of *Step 1*:

$$t_p = 2\nu \sqrt{\frac{f_p}{f_y}}$$

where f_y = tensile yield stress of the plate steel

f_p = service load $P/(B_p \times C_p)$

$\nu = (m)$ or (n) or $(\lambda n')$, whichever is the greatest, mm

$m = (C_p - 0.95\, d)/2$

$n = (B_p - 0.80\, b_f)/2$

$n' = 0.25 \sqrt{d \times b_f}$

$\lambda = \dfrac{2\sqrt{X}}{1 + \sqrt{1 - X}}$ or 1.0 whichever is smaller. If $X > 1$, use $\lambda = 1$

$X = \dfrac{4 P_o}{L^2 F_b}$. Use $P_o = P$ and $F_b = 0.35 f_c' \sqrt{A_2/A_1}$

$L = d + b_f$

$L = 311 + 306 = 617\ \text{mm} = 0.617\ \text{m}$

$$m = \frac{(335 - 0.95 \times 311)}{2} = 19.8\ \text{mm};\quad n = \frac{(330 - 0.80 \times 306)}{2} = 42.6\ \text{mm} > \text{m}$$

$$n' = 0.25\sqrt{311 \times 306} = 77.1\ \text{mm};\ F_b = 0.35(24)\,(1.45) = 12.18\ \text{MPa}$$

$$X = \frac{4 \times 1050}{0.617^2\,(12.18 \times 1000)} = 0.906$$

$$\lambda = \frac{2 \times \sqrt{0.906}}{1 + \sqrt{1 - 0.906}} = 1.457 > 1. \text{ Hence, use } \lambda = 1$$

$$\lambda n' = 1 \times 77.1 = 77.1 \text{ mm} > n. \text{ Hence, } \nu = 77.1 \text{ mm}$$

$$f_p = \frac{1050}{(0.330 \times 0.335)} = 9498 \text{ kPa} = 9.498 \text{ MPa}; \ f_y = 250 \text{ MPa}$$

$$t_p = 2\,(77.1)\sqrt{\frac{9.498}{250}} = 30.06 \text{ mm}$$

Use plate thickness t_p = 30 mm.

Step 4. Design pedestal reinforcement.

Longitudinal reinforcement:

As mentioned in Section 5.6, usually, concrete pedestals are reinforced with minimum column steel of 0.01 A_g but not more than 0.08 A_g (ACI Section 10.9.1) even when they are designed as unreinforced members.

The top surface area of the pedestal $A_2 = 0.25 \text{ m}^2 = 250\ 000 \text{ mm}^2$

$A_{s,\,min} = 0.01\,A_g = 0.01 \times 250\ 000 = 2500 \text{ mm}^2$

Try 8 No. 22 bars. $A_{s,provided} = 8 \times 387 = 3096 \text{ mm}^2$

$$> A_{s,\,min}$$

$$< 0.08A_g \quad \text{(OK.)}$$

Provide eight No. 22 bars.

The bars should terminate at about 100 mm below the top surface of the pedestal in order to minimise point loading on the base plate. The bars should be extended through the pedestal-footing interface into the footing, hooked (90-degree standard hook) and fastened tightly with the footing bottom bars.

Ties:

Lateral ties should be at least No. 10 in size (ACI Section 7.10.5.1).

Provide No. 10 square-shaped ties.

$$\text{The longitudinal bar clear spacing} = \frac{[500 - (2 \times 40 + 2 \times 9.5 + 3 \times 22.2)]}{2}$$

$$= 167.2 \text{ mm} > 150 \text{ mm}$$

Therefore, diamond-shaped ties should be provided also (ACI Section 7.10.5.3). (See the final design sketches of *Step 7*).

Vertical spacing of ties shall not exceed 16 longitudinal bar diameters, 48 tie bar or wire diameters, or least dimension of the compression member (ACI Section 7.10.5.2):

16 longitudinal bar diameters = $16 \times 22.2 = 355.2$ mm

48 tie bar diameters = $48 \times 9.5 = 456$ mm

Least dimension of the compression member = 0.5 m = 500 mm

Use maximum tie spacing = 355 mm.

(Continued)

Place a tie not more than one-half tie spacing above the top of the footing (ACI Section 7.10.5.4). The ties shall be located vertically as follows:

Locate the first tie at 100 mm below the top of the pedestal (at top end of the longitudinal bars). Locate the second tie at 100 mm below the first one. Use 355 mm tie spacing for the rest of the required ties. The lowest tie will be placed at 130 mm above the footing, which is less than one-half tie spacing. With this arrangement the required number of ties will be 6 square-shaped and 6 diamond-shaped ties.

Step 5. Design anchor bolts.

If the structural design loads between the column and the supporting member consist solely of compression (as in this Problem), then theoretically no anchorage will be required. However, anchor bolts are still required to resist erection loads, accidental collisions during erection, and unanticipated shear or tensile loads. The engineer might attempt to estimate these loads and design accordingly, or simply select the bolts using engineering judgment. For example, an engineer might arbitrarily select enough bolts of specified type and diameter to carry 10 % of the total compressive load in shear. Assume we consider the latter option as follows:

Equation (5.26): $\varnothing V_n \geq V_{ua}$

$$V_{ua} = \text{total factored shear force} = 0.1 \times \text{factored } P$$

$$= (0.1)(1.2 \times 600 + 1.6 \times 450) = 144\,\text{kN}$$

$$\varnothing V_n = \text{the lowest of } \varnothing V_{sa},\ \varnothing V_{cb} \text{ and } \varnothing V_{cp} \text{ (ACI Section D.4.1.2)}$$

$$\varnothing = \text{strength reduction factor for anchors in concrete}$$

$$= 0.65 \text{ for anchors of ductile steel in shear (ACI Section D.4.4)}$$

Since this analysis is based on an arbitrary choice or option and the anchor bolts will be installed in a deep member (the pedestal) relatively far from the edges, the concrete breakout and pryout effects may be neglected.

Assume using A307 bolt material grade A (A-36 steel of $f_{ult} = 400$ MPa and $f_y = 250$ MPa). From Table 5.1 select a bolt diameter and pitch equal to 20P2.5. For threaded bolts, ANSI/ASME B1.1 uses the same effective cross-sectional area of a bolt in tension or in shear (see ACI Sections RD.5.1.2 and RD.6.1.2). For the 20P2.5 bolt, Table 5.1 gives $A_t = 245\ \text{mm}^2$.

For cast-in headed bolt and hooked bolt anchors (ACI Section D.6.1.2):

$$\varnothing V_{sa} = \varnothing\,(n\,0.6\,A_{se,\,V}\,f_{uta})$$

where n is the number of anchors in the group, and f_{uta} shall not be taken greater than the smaller of $1.9\,f_{ya}$ and 860 Mpa.

$$f_{ya} = f_y \text{ of bolt steel.}$$

$$\varnothing V_{sa} = (0.65)\left(0.6 \times 245 \times 1.9 \times 250/10^6\right)(n) = 0.0454\,(n)$$

Let $\varnothing V_{sa} = \varnothing V_n = V_{ua}$:

$$0.0454(n) = 144/1000$$

$$n = \frac{144}{45.4} = 3.17$$

Provide four 20P2.5 A307 Grade A cast-in headed bolt or hooked bolt anchors.

Step 6. Determine the effective embedment depth h_{ef}, spacing and edge distances of the anchor bolts.

Effective embedment depth h_{ef}:

The required anchor bolt embedment depth h_{ef} is not directly and specifically indicated in most (including ACI) building codes. We may, however, use the minimum depth of Table 5.2, compression development length l_{dc} (ACI Section 12.3.2) and tension development length l_{dh} (ACI Section 12.5.2), whichever is the greatest.

$$\text{Table 5.2 gives minimum effective embedment depth} = 12\ d_a = 12 \times 20 = 240\ \text{mm}$$

$$l_{dc} = \frac{0.24 f_y d_b}{\lambda \sqrt{f_c'}} \text{ or } l_{dc} = \left(0.043 f_y\right) d_b, \textit{whichever is the larger}$$

$$\mathrm{l_{dc}} = \frac{0.24(250)(20)}{(1)\sqrt{21}} = 262\ \text{mm, or}$$

$$l_{dc} = \left(\frac{0.043 \times 10^{-6}}{10^{-6}}(250)\right)(20) = 215\ \text{mm}$$

$$l_{dc,min} = 200\ \text{mm} \quad (\text{ACI Section 12.3.1})$$

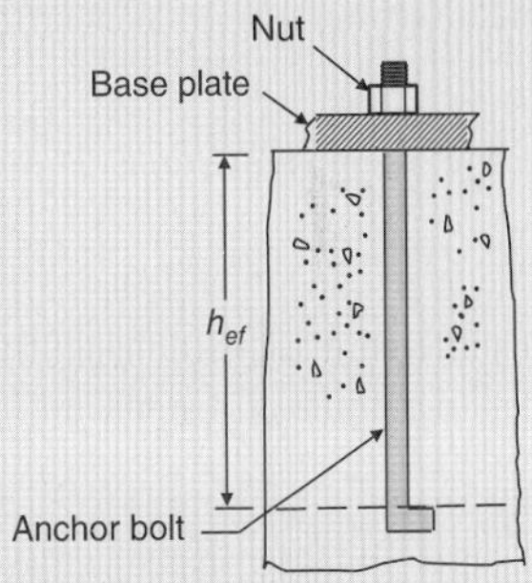

Nut

Base plate

h_{ef}

Anchor bolt

Scheme 5.33

Required $l_{dc} = 262$ mm

$$l_{dh} = \frac{0.24 f_y d_b}{\lambda \sqrt{f_c'}} \text{ or } 8\ d_b \text{ or } 150\ \text{mm, whichever is the larger.}$$

$$l_{dh} = \frac{0.24(250)(20)}{(1)\sqrt{21}} = 262\ \text{mm} > 8\ d_b > 150\ \text{mm}$$

Required $l_{dh} = 262$ mm

Therefore, use $h_{ef} = 262\ mm$

Also, additional lengths due to the plate and nut thicknesses, grout bed, bend and so on are to be specified.

Anchor bolt spacing:

ACI Section D.8.1 requires minimum centre-to-centre spacing of anchors shall be $4d_a$ for untorqued cast-in anchors and $6d_a$ for torqued cast-in anchors and post-installed anchors.

In this case, it is not exactly clear whether the anchors will be torqued or untorqued. Therefore, it may be better to use anchor spacing not less than $6d_a$.

Use minimum anchor spacing = $6d_a = 6 \times 20 = 120$ mm.

Edge distances for anchor bolts:

Table 5.2 gives minimum edge distances for A307 (A-36 steel) anchor bolts = $5d_a$ or 100 mm, whichever is greater.

ACI Section D.8.2 requires minimum edge distances for cast-in headed anchors that will not be torqued shall be based on concrete cover for reinforcement specified in ACI Section 7.7. For cast-in headed anchors that will be torqued, the minimum edge distances shall be $6d_a$.

(Continued)

As mentioned above, here also it may be better to use minimum edge distance not less than $6d_a$, which is larger than $5d_a$ or 100 mm.

Use minimum edge distance = $6d_a = 6 \times 20 = 120$ mm

In order to satisfy the above anchor spacing and edge distance requirements, the anchor bolts may be placed in the pattern shown in the scheme below.

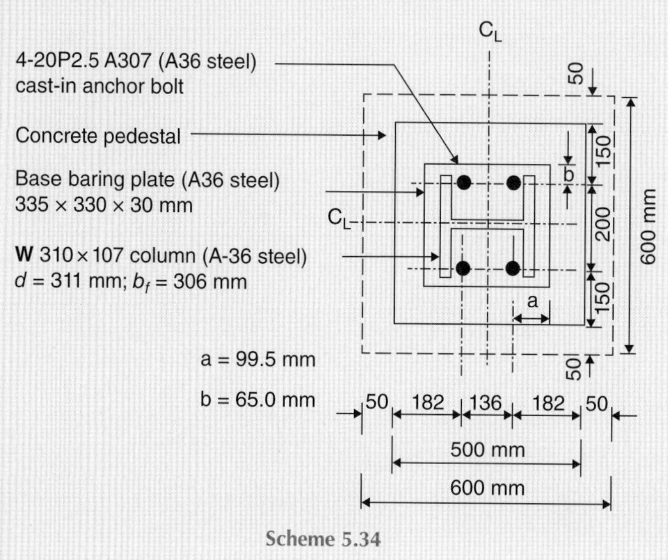

Scheme 5.34

Step 7. Develop the final design sketch.

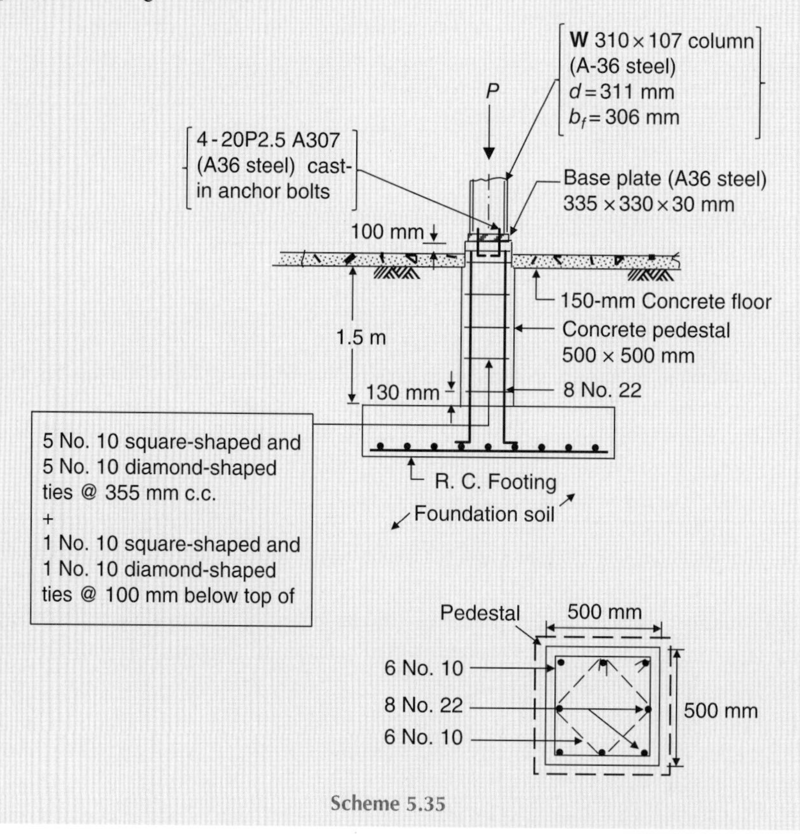

Scheme 5.35

Problem 5.10

A rigid pile cap is required to support at its centre an interior metal column of type HP 360 × 174 with 660 × 660 mm base plate. The column will carry a service dead load of 1000 kN and live load of 1400 kN. Each pile working load and diameter should not exceed 300 kN and 400 mm, respectively. Top of the cap will be covered with 0.15 m compacted fill with a unit weight of 20 kN/m^3 and a 0.15 m concrete floor ($\gamma_c = 24$ kN/m^3) with a uniform live load of 3.75 kPa. Design the pile cap using $f_c' = 28$ MPa for concrete, $f_y = 420$ MPa for reinforcing steel and minimum centre-to-centre pile spacing = two pile diameters but not less than 760 mm.

Solution:

Step 1. Determine number of piles required, pile-group pattern, pile spacing and plan dimensions of the pile cap.

Equation (5.30): $P_p = \dfrac{Q}{n}$

$$n = \frac{Q}{P_p}$$

A trial estimate of the cap dimensions may be necessary in order to compute the total load Q. Since the cap will be concentrically loaded (the column is at centre without moments), it may be appropriate to use a square cap and pile-group pattern.

Try a square cap with dimensions 2.8 × 28 × 0.85 m.

$$Q = 1000 + 1400 + (2.8 \times 2.8)(0.85 \times 24 + 0.15 \times 20 + 0.15 \times 24 + 3.75)$$

$$= 2641.08\,\text{kN}$$

$$n = \frac{2641.08}{300} = 8.8\,\text{piles}$$

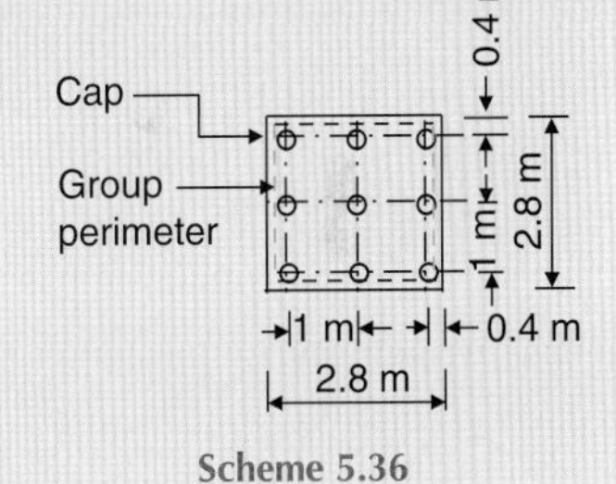

Scheme 5.36

Use nine piles distributed uniformly in a square pattern.

Assume pile diameter = 400 mm

Minimum centre to centre pile spacing

$= 2 \times 400$

$= 800$ mm

Use centre-to-centre pile spacing = 1000 mm.

Plan dimensions of the pile cap shall be larger than that of its pile-group pattern by at least 200 mm on all sides (see Section 5.7).

Use the cap plan dimensions = 2.8 × 2.8 m.

Step 2. Locate the necessary critical sections according to ACI Sections 15.4, 15.5 and 15.6.

AISC Tables give the dimensions for the HP 360 174 as shown in Scheme 5.37.

Step 3. Find cap thickness h.

Determine the required thickness based on both two-way shear and one-way shear analyses.

ACI Section 15.5.3 states that where the distance between the axis of any pile and the axis of the column is more than two times the distance between the top of the pile cap and the top of the pile, the pile cap shall satisfy 11.11 and 15.5.4. Other pile caps shall satisfy either Appendix A or 11.11 and 15.5.4.

In this case, we design the cap so that the requirements of ACI Sections 11.11 and 15.5.4 are satisfied and, therefore, Appendix A will not be considered.

(a) Two-way shear.

Two-way shear at location of the column:

The critical section is located at distance $d/2$ from a line halfway between the face of the column and the edge of the base plate (ACI Section 15.5.2), as shown in the scheme 5.37.

(Continued)

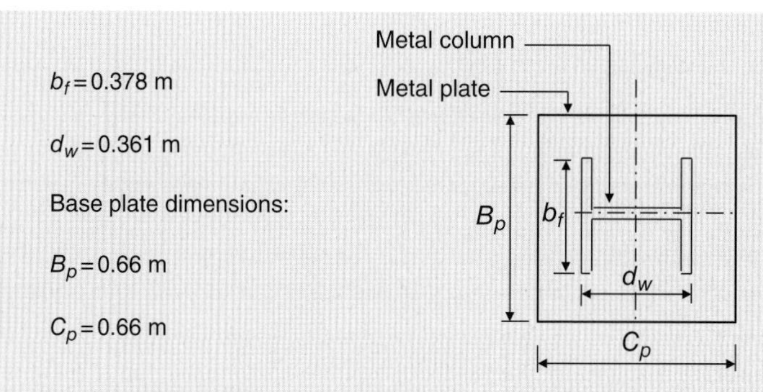

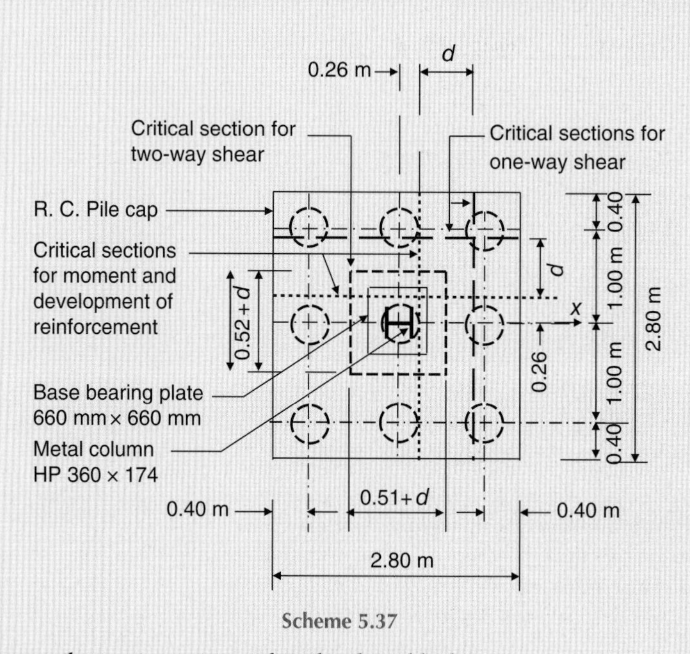

Scheme 5.37

Assume only one pile reaction exists within the shear block.

$$\text{Factored } Q = 1.2\left[1000 + (2.8 \times 2.8)\begin{pmatrix} 0.85 \times 24 + 0.15 \times 20 \\ + 0.15 \times 24 \end{pmatrix}\right]$$

$$+ 1.6(1400 + 2.8 \times 2.8 \times 3.75)$$

$$= 1454.02 + 2287.04 = 3741.06\,\text{kN}$$

$$\text{Factored } P_P = \frac{\text{factored } Q}{n} = \frac{3741.06}{9} = 415.67\,\text{kN}$$

Equation (5.19):

$$\emptyset\, V_c \geq V_u \,(\text{taking reinforcement shear strength, } V_s = 0)$$

$$\text{where } \emptyset = \text{strength reduction factor (ACI Section 9.3.2.3)}$$

$$= 0.75 \text{ for shear}$$

$$V_u = \text{factored Q} - \text{one pile reaction on the shear block}$$

$$= 3741.06 - 415.67 = 3325.39\,\text{kN}$$

The shear strength of concrete V_c shall be the smallest of (i), (ii) and (iii):

(i)

$$V_c = 0.17\left(1+\frac{2}{\beta}\right)\lambda\sqrt{f_c'}b_o d$$

$$= (0.17)\left(1+\frac{2}{1}\right)(1)\sqrt{f_c'}b_o d = (0.51)\sqrt{f_c'}b_o d \qquad \text{Equation (5.22)}$$

(ii)

$$V_c = 0.083\left(\frac{\alpha_s d}{b_o}+2\right)\lambda\sqrt{f_c'}b_o d$$

$$= (0.083)\left(\frac{40d}{b_o}+2\right)(1)\sqrt{f_c'}b_o d = (0.083)\left(\frac{40d}{b_o}+2\right)\sqrt{f_c'}b_o d \qquad \text{Equation (5.23)}$$

(iii)

$$V_c = 0.33\,\lambda\sqrt{f_c'}b_o d$$

$$= 0.33\,(1)\sqrt{f_c'}b_o d = (0.33)\sqrt{f_c'}b_o d \qquad \text{Equation (5.24)}$$

Use $V_c = (0.33)\sqrt{f_c'}b_o d$. Calculate d, then check V_c of Equation (5.23).

Perimeter of the critical section is $2(0.51 + d + 0.52 + d)$ m

$$\varnothing V_c = (0.75)(0.33)\sqrt{28}(1000)[2(0.51+d+0.52+d)(d)]$$

$$= 5239(d)^2 + 2698\,d(\text{kN})$$

Let $\varnothing V_c = V_u$

$$5239(d)^2 + 2698\,d = 3325.39$$

$$5239(d)^2 + 2698\,d - 3325.39 = 0$$

$$d = \frac{-2698 \pm \sqrt{2698^2-(4)(5239)(-3325.39)}}{2\times 5239} = \frac{-2698+8773}{10478} = 0.58\text{ m} = 580\text{ mm}$$

Check V_c of Equation (5.23):

$$V_c = (0.083)\left(\frac{40\times 0.580}{2(0.51+0.580+0.52+0.580)}+2\right)\sqrt{f_c'}b_o d = (0.606)\sqrt{f_c'}b_o d$$

$$V_c = (0.33)\sqrt{f_c'}b_o d < (0.606)\sqrt{f_c'}b_o d \qquad \text{(OK.)}$$

Try d = 0.6 m = 600 mm and h = 850 mm.

Check whether only one pile reaction exists within the shear block:

Face of the nearest pile is located at 0.8 m from the cap centre, whereas, the critical section is at 0.56 m Hence, there is only one pile reaction located within the shear block area, as assumed. (OK.)

Two-way shear around individual piles:

Where necessary, shear around individual piles may be investigated in accordance with ACI Sections 11.11.1.2. In this case, by inspection, it is clear that shear around individual piles is not critical. However, for completeness, it may be performed as follows:

Check first whether there is any overlapping of critical shear perimeters from the adjacent piles.

Since the centre-to-centre pile spacing (i.e. 1000 mm) is not less than two times the radius of shear perimeter [i.e. 2(300 + 200) = 1000 mm)], there is no overlapping of critical shear perimeters. Therefore, the modified critical perimeter for shear defined in ACI Section R 15.5.2 and Figure R 15.5 is not applicable.

It will be sufficient to investigate two-way shears around one of the piles only, since the pile cap has a constant d and the piles have the same P_p and diameter.

(*Continued*)

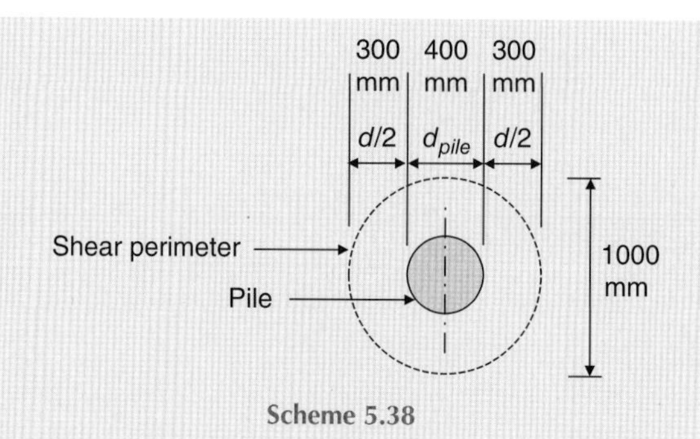

Scheme 5.38

$$\varnothing V_c = \varnothing (0.33)\sqrt{f_c'}b_o d$$
$$= (0.75)(0.33)\sqrt{28}(1000)(\pi \times 1)(0.6) = 2470\,\text{kN}$$

$$V_u = \text{factored } (P_p) - \text{factored weights of cap, fill and floor loads}$$
$$= 415.67 - \text{factored weights of cap, fill and floor loads}$$

$V_u < \varnothing V_c$. Therefore the cap is safe against two-way shear around any individual pile.

(b) One-way shear:

The critical section is located at distance d from a line halfway between the face of the column and the edge of the plate, as shown in Scheme 5.37.

Assume $d = 0.6$ m. The critical section is located at $(0.26 + 0.6) =$ 0.86 m from the cap centre, whereas face of the nearest exterior pile is located at 0.8 m. Hence, the critical section passes through the exterior piles, as shown below. Accordıng to ACI Section 15.5.4.3, the portion of the full pile reaction contributes to the applied shear load equals

$$(0.34/0.40)(\text{full pile reaction})$$
$$= (0.85)(415.67)$$
$$= 353.32 \text{ kN}$$

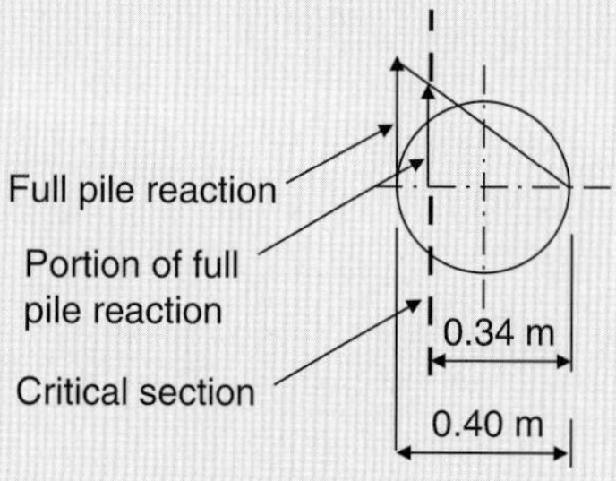

Scheme 5.39

V_u = portion of full pile reaction from three piles – (weights of cap, filland floor loads covering the cross-hatched area shown in Scheme 5.40)

$$= 3 \times 353.32 - (2.8 \times 0.54)[(1.2)(0.85 \times 24 + 0.15 \times 20 + 0.15 \times 24) + 1.6 \times 3.75]$$
$$= 1060.0 - 58.0 = 1002\,\text{kN}$$

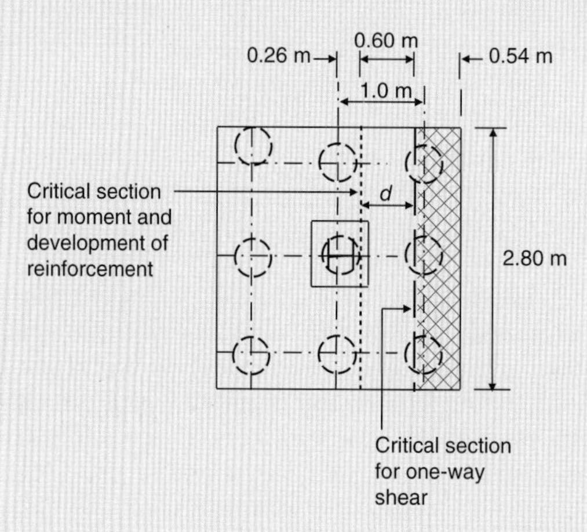

Scheme 5.40

Equation (5.20):

$$\varnothing V_c = (\varnothing)\left(0.17\lambda\sqrt{f_c'}b_w d\right)$$
$$= (0.75)(0.17 \times 1)\left(\sqrt{28}\right)(1000)(2.80)(0.6)$$
$$= 1133\,\text{kN} > V_u$$

Therefore, the two-way shear controls.

Use $d = 0.60$ m $= 600$ mm.

The depth d will be taken to the intersection of the steel bars running each way at bottom of the cap.

Check the cap thickness h:

$$h = d + \text{one bar diameter (say 25 mm)}$$
$$+ 75\,\text{mm space between steel bars and top of piles}$$
$$+ 150\,\text{mm pile embedment into the cap}$$
$$= 600 + 25 + 75 + 150 = 850\,\text{mm} \qquad \text{(OK.)}$$

Use the cap thickness $h = 850$ *mm.*

Step 4. Design the flexural reinforcement.

The cap is square and centrically loaded, the pile pattern is symmetrical about both x and y axes and the critical sections for moment in both directions are approximately located at the same distance (0.260 m or 0.255 m) from the cap centre. Therefore, practically, the same bottom reinforcement will be required for both directions.

Equation (5.9): $\varnothing M_n \geq M_u$

The maximum factored moment is

$$M_u = (0.74)(3 \times \text{factored}\, P_P) - (0.57)[\text{factored}\,(W + \text{floor load})]$$
$$= (0.74)(3 \times 415.67) -$$
$$(0.57)\begin{bmatrix} 1.2(2.8 \times 1.14)(0.15 \times 24 + 0.15 \times 20 + 0.85 \times 24) \\ + 1.6(2.8 \times 1.14)(3.75) \end{bmatrix}$$
$$= 923 - 70 = 853\,\text{kN.m}$$

$$M_u \text{ Per metre width} = \frac{853}{2.8} = 305\,\text{kN.m/m}$$

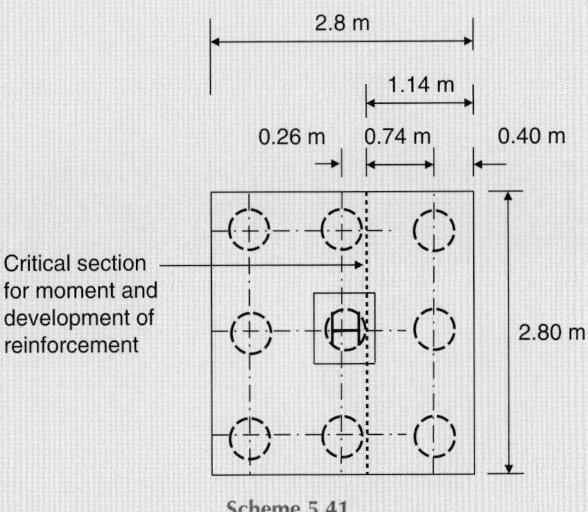

Scheme 5.41

(Continued)

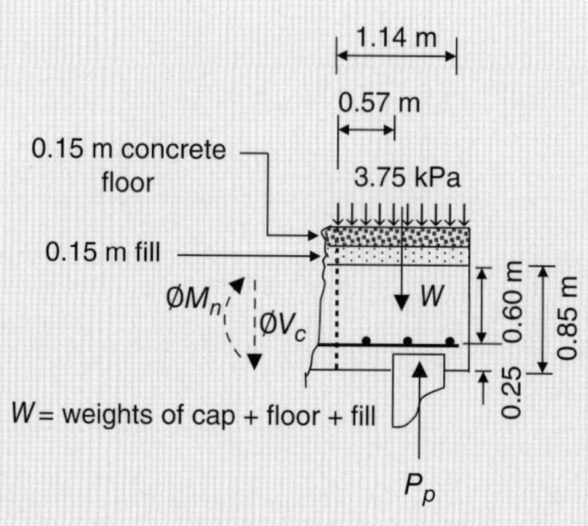

Scheme 5.42

Assume tension-controlled section, $\text{Ø} = 0.9$, and $f_s = f_y$.

$$a = \frac{A_s f_y}{0.85 f_c' b} = \frac{A_s \times 420}{0.85 \times 28 \times 1} = 17.65\, A_s$$

$$d = 0.6\,\text{m}$$

Equation (5.11):

$$\text{Ø}\, M_n = \text{Ø}\left[A_s f_y \left(d - \frac{a}{2}\right)\right]$$

$$= (0.9)\left[(A_s \times 420 \times 1000)\left(0.6 - \frac{17.65 A_s}{2}\right)\right]$$

$$= 226\ 800\, A_s - 3\ 335\ 850\, A_s^2$$

Let $\text{Ø}\, M_n = M_u$

$$226\ 800\, A_s - 3\ 335\ 850\, A_s^2 = 305$$

$$3\ 335\ 850\, A_s^2 - 226\ 800\, A_s + 305 = 0$$

$$A_s = \frac{-(-226\ 800) \pm \sqrt{(-226\ 800)^2 - (4)(3\ 335\ 850)(305)}}{2 \times 3\ 335\ 850} = \frac{226\ 800 - 217\ 643}{6\ 671\ 700}$$

$$= 1.373 \times 10^{-3}\,\text{m}^2/\text{m} = 1373\,\text{mm}^2/\text{m},$$

required by analysis.

In order to provide a more strong and rigid pile cap it may be more appropriate to use $A_{s,\min}$ as required for flexural members (ACI Section 10.5.1) rather than $A_{s,\min}$ based on shrinkage and temperature reinforcement (ACI Sections 10.5.1 and 7.12.2.1). Accordingly, ACI Section 10.5.1 requires

$$A_{s,min} = \frac{0.25\sqrt{f_c'}}{f_y} b_w d$$

$$= \frac{0.25\sqrt{28}}{420} \times 1 \times 0.6$$

$$= 1.89 \times 10^{-3}\,\text{m}^2/\text{m} = 1890\,\text{mm}^2/\text{m}$$

and not less than 1.4

$$\frac{b_w d}{f_y} = \frac{1.4 \times 1 \times 0.6}{420}$$

$$= 2 \times 10^{-3}\,\text{m}^2/\text{m} = 2000\,\text{mm}^2/\text{m}$$

Therefore, $A_{s,\text{min}} = 2000\ \text{mm}^2/\text{m}$ controls. However, according to ACI Section 10.5.3, this $A_{s,\text{min}}$ need not be used if $A_{s,\text{provided}}$ is at least one-third greater than that required by analysis.

$$1.33A_s = 1.33 \times 1373 = 1826\,\text{mm}^2/\text{m}$$

$$A_{s,\text{total}} = 2.8 \times 1826 = 5113\,\text{mm}^2$$

Try 18 No. 19 bars:

$$A_{s,\text{provided}} = 18 \times 284$$

$$= 5112\,\text{mm}^2 > 1.33\,A_s\ (\text{OK.})$$

The assumptions made will be satisfied, since $A_{s,\text{min}}$ controls.

Provide 18 No. 19 bars each way

Check the development of reinforcement:

In this case, the bars are in tension, the provided bar size is No. 19, the clear spacing of the bars exceeds $2d_b$, and the clear cover exceeds d_b. Therefore, the development length l_d of bars at each side of the critical section (which is the same critical section for moment) shall be determined from the following equation, but not less than 300 mm:

$$l_d = \left(\frac{f_y \psi_t \psi_e}{2.1\lambda\sqrt{f'_c}}\right) d_b\ (\text{ACI Sections 12.2.1, 12.2.2 and 12.2.4})$$

where the factors ψ_t, ψ_e and λ are 1. (ACI Sections 12.2.4 and 8.6.1)

$$l_d = \left(\frac{420 \times 1 \times 1}{2.1 \times 1 \times \sqrt{28}}\right)\left(\frac{19.1}{1000}\right) = 0.722\,\text{m} = 722\,\text{mm} > 300\,\text{mm}.$$

Therefore, the required $l_d = 0.722$.

The bar extension past the critical section (i.e. the available length) is

$$1140\,\text{mm} - 75\,\text{mm cover} = 1065\,\text{mm} > 722\,\text{mm}\quad (\text{OK.})$$

However, we shall hook the bars (90-degree standard hook) at both ends to insure better anchorage and keep the pile cap safe against bursting of the side cover where the pile transfers its load to the cap.

ACI Sections 7.6.5 and 10.5.4 requires maximum spacing shall not exceed three times the slab or footing thickness, or 450 mm, whichever is smaller.

(Continued)

Use centre to centre bar spacing = 154 mm. Check the 75 mm minimum concrete cover at each side:

$$\text{Concrete cover provided} = \frac{(2800 - 17 \times 154 - 19.1)}{2} = 81.5\,\text{mm} \quad (\text{OK.})$$

Provide 18 No. 19 bars @ 154 mm c.c. at bottom of the cap in both directions. All the bars shall be hooked (90-degree standard hook) at both ends. The extreme bottom bars shall be placed 81 mm above top of piles.

Step 5. Check plate bearing on pile cap.

For this step, the design proceeds in the same manner as that of Solution of Problem 5.9, *Step 2.*

However, where necessary, bearing strength of concrete at location of individual piles may be investigated in accordance with ACI Sections 15.8.1.1 and 10.14.

In this case, at location of any pile, the pile cap is the supported member and the pile is the supporting member. The supporting contact surface area is the loaded area A_1 equals to the pile cross-section area. The maximum design bearing strength of the cap is

$$\varnothing\left(0.85 f_c' A_1\right) = (0.65)(0.85)(28 \times 1000)\left(0.4^2 \pi/4\right) = 1945\,\text{kN}$$

$$\text{Factored } P_p = 415.67\,\text{kN} < \varnothing\left(0.85 f_c' A_1\right) \quad (\text{OK.})$$

Step 6. Design anchor bolts.

Design of the anchor bolts proceeds in the same manner as that of Solution of Problem 5.9, *Step 5.*

Step 7. Determine the effective embedment depth h_{ef}, spacing and edge distances of the anchor bolts.

The design proceeds in the same manner as that of Solution of Problem 5.9, *Step 6.*

Step 8. Develop the final design sketch.

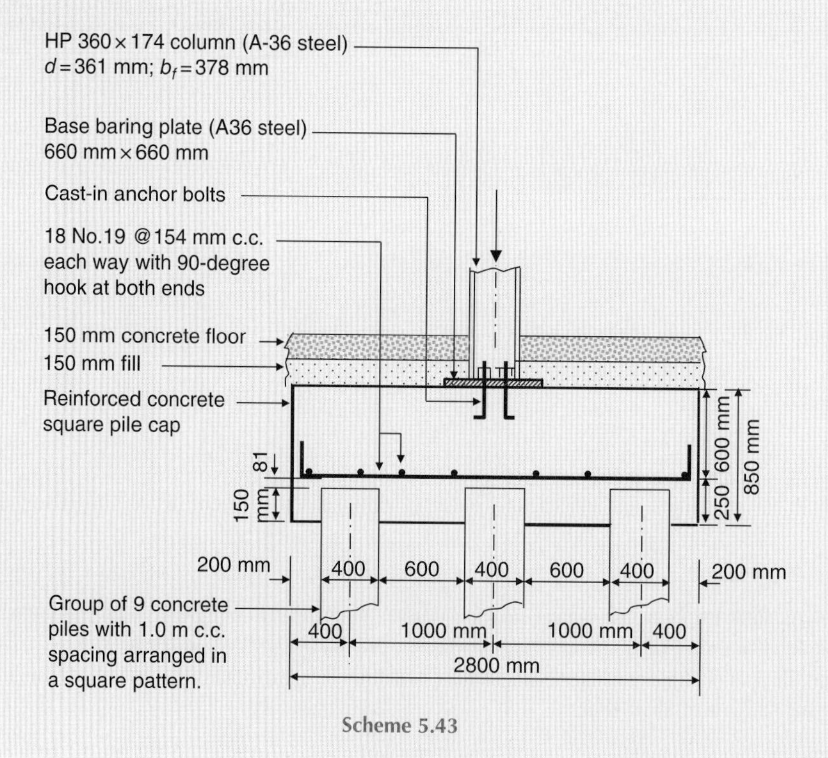

Scheme 5.43

Problem 5.11

Given: same data as that of Problem 5.10 except the column, in addition to its axial loads, carries the same moment M about both x and y axes, and each pile working load should not exceed 435kN.

$$M = M_D + M_L \quad M_D = 55\,\text{kN.m} \quad M_L = 65\ \text{kN.m}$$

Required: Design the pile cap.

Solution:

Step 1. Determine number of piles required, pile-group pattern, pile spacing and plan dimensions of the pile cap.

The greatest combinations of the given loads are

$$Q = P = D + L$$

$$M_y = M_x = M = M_D + M_L = 55 + 65 = 120\,\text{kN.m}$$

Since the column is at centre of the pile cap and carries the same moment M about both x and y axes, it may be appropriate to use a *square* cap.

Try a square cap with trial dimensions: 3.0 m × 3.0 m × 0.85 m

For the purpose of estimating a trial number of piles let:

$$n = \frac{Q}{\text{allowable pile working load}}$$

$$Q = 1000 + 1400 + (3\times 3)(0.85\times 24 + 0.15\times 20 + 0.15\times 24 + 3.75)$$

$$= 2677\,\text{kN}$$

$$n = \frac{2677}{435} = 6.2\,\text{piles}$$

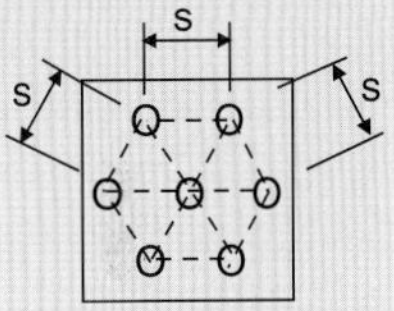

Scheme 5.44

Try seven piles

Pile diameter = 0.4 m

A square pile cap can accommodate seven piles distributed in a stable hexagonal pattern as shown. The pile group pattern will be symmetrical about both x and y axes.

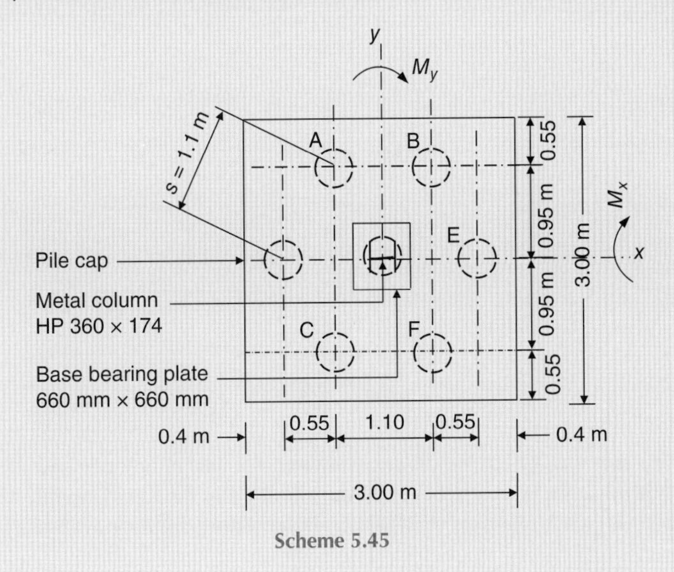

Scheme 5.45

(*Continued*)

Pile caps should extend at least 200 mm beyond the outside face of exterior piles (see Section 5.7).
Centre-to-centre spacing of adjacent piles is

$$s = [3 - 2\,(\text{cap projection at each side}) - 2\,(\text{half pile diameter})]/2$$
$$= (3 - 2\times 0.2 - 2\times 0.2)\,/2 = 1.1 \text{ m}$$

Equation (5.31): $$P_p = \frac{Q}{n} \pm \frac{M_y}{\Sigma x^2}x \pm \frac{M_x}{\Sigma y^2}y$$

It is clear that $P_{p,\,max}$ is located at centre of pile B. Compute $P_{p,\,max}$ and check the 435 kN pile working load.

$$P_{p,\,max} = \frac{Q}{n} + \frac{M_y}{\Sigma x^2}x + \frac{M_x}{\Sigma y^2}y = \frac{2677}{7} + \frac{120}{4\times 0.55^2 + 2\times 1.1^2}\times 0.55 + \frac{120}{4\times 0.95^2}\times 0.95$$
$$= 382.4 + 18.2 + 31.6 = 432\,\text{kN} < 435\,\text{kN} \qquad \text{(OK.)}$$

Also, it is clear that $P_{p,min}$ is located at pile C.

$$P_{p,min} = \frac{Q}{n} - \frac{M_y}{\Sigma x^2}x - \frac{M_x}{\Sigma y^2}y = \frac{2699}{7} - \frac{120}{4\times 0.55^2 + 2\times 1.1^2}\times 1.1 - \frac{120}{4\times 0.95^2}\times 0.95$$
$$= 382.4 - 18.2 - 31.6 = 333\,\text{kN}$$

Use seven piles arranged in a hexagonal pattern.
Use centretocentre pile spacing = 1100 mm.
Use the cap plan dimensions = 3000 × 3000 mm.

Step 2. Locate the necessary critical sections according to ACI Sections 15.4, 15.5 and 15.6.
AISC Tables give the dimensions for the HP 360 174 as shown in Scheme 5.46.

Step 3. Find cap thickness h.
Determine the required thickness based on both two-way shear and one-way shear analyses.

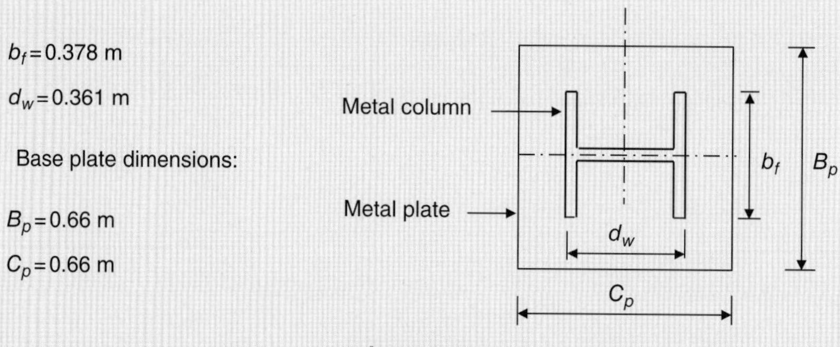

Scheme 5.46

ACI Section 15.5.3 states that where the distance between the axis of any pile and the axis of the column is more than two times the distance between the top of the pile cap and the top of the pile, the pile cap shall satisfy Sections 11.11 and 15.5.4. Other pile caps shall satisfy either Appendix A or Sections 11.11 and 15.5.4.

In this case, we design the cap so that the requirements of ACI Sections 11.11 and 15.5.4 are satisfied, and therefore, Appendix A will not be considered.

(a) Two-way shear:

Two-way shear at location of the column:

The critical section is located at distance $d/2$ from a line halfway between face of the column and edge of the base plate, as shown in the scheme below.

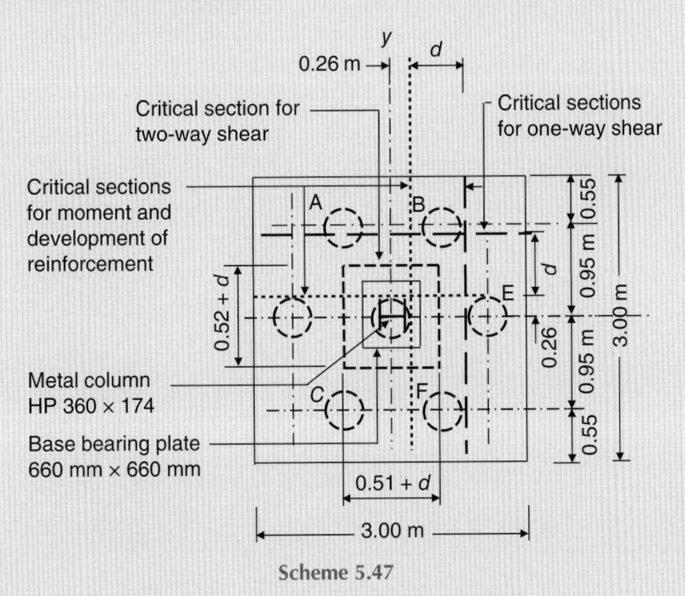

Scheme 5.47

Assume only one pile (centre pile) reaction exists within the shear block.

$$\text{Factored } Q = 1.2\left[1000 + (3\times 3)\begin{pmatrix} 0.85\times 24 + 0.15\times 20 \\ +0.15\times 24 \end{pmatrix}\right]$$

$$+1.6(1400 + 3\times 3\times 3.75)$$

$$= 1491.6 + 2294.0 = 3785.6\,\text{kN}$$

$$\text{Factored } P_P \text{ (center pile)} = P_p = \frac{Q}{n} \pm \frac{M_y}{\Sigma x^2}x \pm \frac{M_x}{\Sigma y^2}y$$

$$= \frac{\text{factored } Q}{n} + 0 + 0 = \frac{3785.6}{7} = 540.8\,\text{kN}$$

Equation (5.19): $\varnothing V_c \geq V_u$ (taking reinforcement shear strength, $V_s = 0$).

where $\varnothing$ = strength reduction factor (ACI Section 9.3.2.3)

$= 0.75$ for shear

V_u = factored Q − one pile reaction on the shear block

$= 3785.6 - 540.8 = 3245\,\text{kN}$

(Continued)

The shear strength of concrete V_c shall be the smallest of (i), (ii) and (iii):

(i)

$$V_c = 0.17\left(1+\frac{2}{\beta}\right)\lambda\sqrt{f_c'}b_o d$$

Equation (5.22)

$$= (0.17)\left(1+\frac{2}{1}\right)(1)\sqrt{f_c'}b_o d = (0.51)\sqrt{f_c'}b_o d$$

(ii)

$$V_c = 0.083\left(\frac{\alpha_s d}{b_o}+2\right)\lambda\sqrt{f_c'}b_o d$$

Equation (5.23)

$$= (0.083)\left(\frac{40d}{b_o}+2\right)(1)\sqrt{f_c'}b_o d = (0.083)\left(\frac{40d}{b_o}+2\right)\sqrt{f_c'}b_o d$$

(iii)

$$V_c = 0.33\lambda\sqrt{f_c'}b_o d$$

Equation (5.24)

$$= 0.33(1)\sqrt{f_c'}b_o d = (0.33)\sqrt{f_c'}b_o d$$

Use $V_c = (0.33)\sqrt{f_c'}b_o d$. Calculate d, then check V_c of Equation (5.23).

$$\varnothing V_c = (0.75)(0.33)\sqrt{28}(1000)[2(0.51+d+0.52+d)(d)]$$

$$= 5239(d)^2 + 2698d$$

Let $\varnothing V_n = V_u$:

$$5239(d)^2 + 2698d = 3245$$

$$5239(d)^2 + 2698d - 3245 = 0$$

$$d = \frac{-2698 \pm \sqrt{2698^2-(4)(5239)(-3245)}}{2\times 5239} = \frac{-2698+8246}{10\,478} = 0.53\text{ m} = 530\text{ mm}$$

Check V_c of Equation (5.23):

$$V_c = (0.083)\left(\frac{40\times 0.53}{2(0.51+0.53+0.52+0.53)}+2\right)\sqrt{f_c'}b_o d = (0.60)\sqrt{f_c'}b_o d$$

$$V_c = (0.33)\sqrt{f_c'}b_o d < (0.60)\sqrt{f_c'}b_o d \quad \text{(OK.)}$$

Try $d = 0.6$ m = 600 mm and $h = 850$ mm

Check whether only one pile reaction exists within the shear block:

Face of the nearest pile to the critical section is located (0.95 – half pile diameter = 0.95 – 0.2 = 0.75 m) from the cap centre.

$$\frac{0.52+d}{2} = \frac{0.52+0.6}{2} = 0.56\text{ m} < 0.75\text{ m}.$$

Hence, there is only one pile (centre pile) reaction located within the shear block area. (OK.)

Two-way shear around individual piles:

Where necessary, shear around individual piles may be investigated in accordance with ACI Section 11.11.1.2. In this case, by inspection, it is clear that shear around individual piles is not critical. However, for completeness, it may be performed as follows:

Check first whether there is any overlapping of critical shear perimeters from the adjacent piles.

Since the centretocentre pile spacing (i.e.1100 mm) is more than two times the radius of shear perimeter [i.e. 2(300 + 200) = 1000 mm)], there is no overlapping of critical shear perimeters. Therefore, the modified critical perimeter for shear defined in ACI Section R 15.5.2 and Figure R 15.5 is not applicable.

It will be sufficient to investigate two-way shears around the pile of maximum P_P, which is pile B.

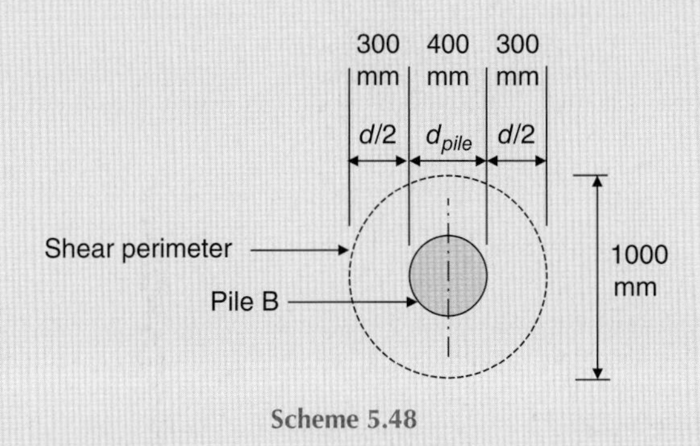

Scheme 5.48

$$\varnothing V_c = \varnothing\,(0.33)\sqrt{f_c'}\,b_o d$$

$$= (0.75)(0.33)\sqrt{28}\,(1000)(\pi \times 1)(0.6) = 2470\,\text{kN}$$

V_u = factored $P_{p,max}$ –factored weights of cap, fill and floor loads

$$\text{Factored M} = 1.2 \times 55 + 1.6 \times 65 = 170\,\text{kN.m}$$

$$\text{Factored}\,P_{p,max} = \frac{Q}{n} + \frac{M_y}{\Sigma x^2}x + \frac{M_x}{\Sigma y^2}y$$

$$= \frac{3785.6}{7} + \frac{170}{4 \times 0.55^2 + 2 \times 1.1^2} \times 0.55 + \frac{170}{4 \times 0.95^2} \times 0.95$$

$$= 540.8 + 25.76 + 44.74 = 611.3\,\text{kN}$$

V_u = 611.3 –factored weights of cap, fill and floor loads.

$$V_u \ll \varnothing V_c.$$

Therefore the cap is safe against two-way shear around any individual pile.

(b) One-way shear:

The critical section is located at distance d from a line halfway between face of the column and edge of the base plate, as shown in Schemes 5.47, 5.49 and 5.50.

$$\text{Assume}\; d = 0.6\,\text{m}.$$

(Continued)

The critical section in each direction is located at (0.26 + 0.6) = 0.86 m from the cap centre.

(1) Consider the critical section in y direction:

The inner face of pile E (the outer pile) is located at 0.90 m from the cap centre, whereas the exterior faces of piles B and F are located at 0.75 m. Therefore, only factored full reaction of pile E contributes to shear at the critical section. Full reaction of pile E is

$$P_p = \frac{3785.6}{7} + \frac{170}{4 \times 0.55^2 + 2 \times 1.1^2} \times 1.1 + \frac{170}{4 \times 0.95^2} \times 0 = 540.8 + 51.52 + 0$$
$$= 592.32 \text{ Kn}$$

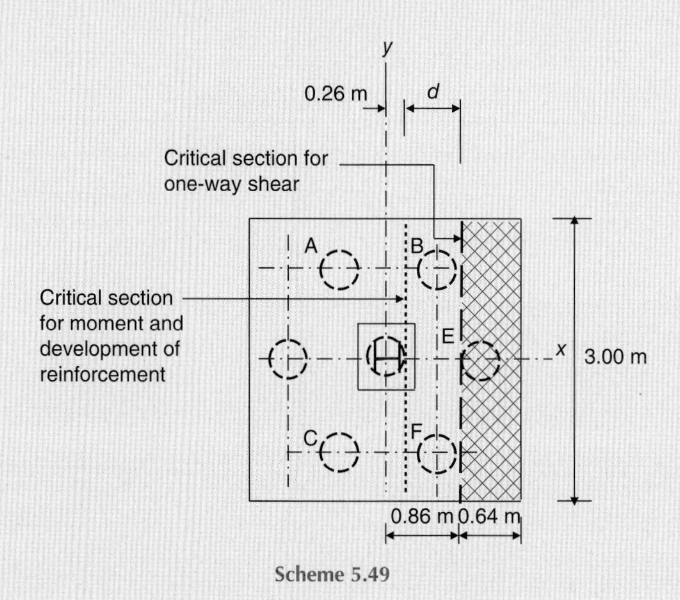

Scheme 5.49

V_u = (factored full reaction of pile E) – (weights of cap, fill and floor loads covering the diamond outlined area shown in the scheme above)

$$= 592.32 - (3 \times 0.64)[(1.2)(0.85 \times 24 + 0.15 \times 20 + 0.15 \times 24 + 1.6 \times 3.75$$

$$= 592.32 - 73.73 = 519 \text{ kN}$$

Equation (5.20):

$$\text{Ø}\, V_c = (\text{Ø})\left(0.17\lambda\sqrt{f_c'}b_w d\right)$$

$$\text{Ø}\, V_c = (\text{Ø})\left(0.17\lambda\sqrt{f_c'}b_w d\right) = (0.75)(0.17 \times 1)\left(\sqrt{28}\right)(1000)(3.0)(0.6)$$

$$= 1214 \text{kN} \gg V_u$$

Therefore, the two-way shear controls.

(2) Consider the critical section in *x* direction:

The inner face of each of the upper two piles, that is piles A and B, is at 0.75 m from the cap centre. Therefore, the critical section passes through these two piles as shown below. According to ACI Section 15.5.4.3, for each of these two piles only a portion of the full pile reaction contributes to the applied shear load.

$$\text{Portion of full pile reaction} = \left(\frac{0.29}{0.40}\right)(\text{full pile reaction})$$

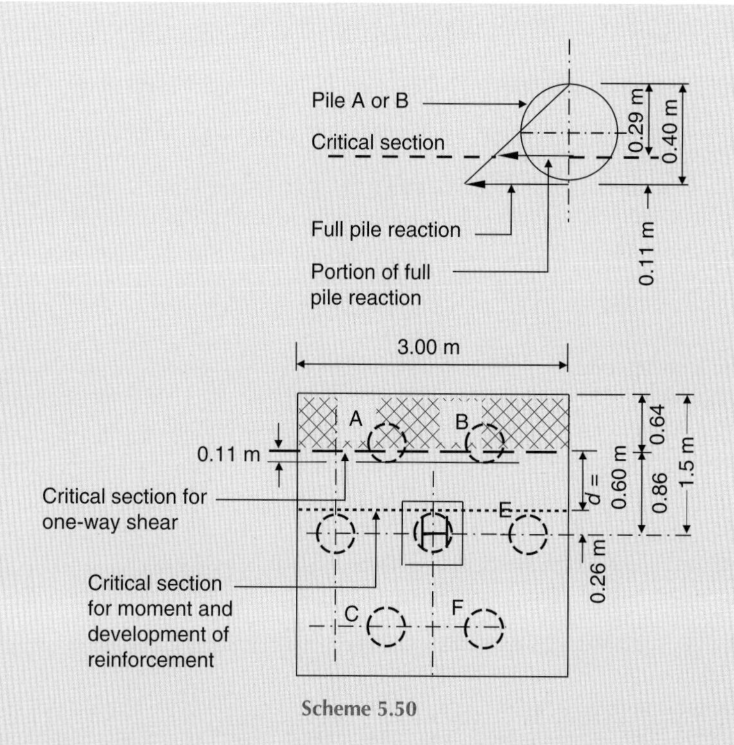

Scheme 5.50

The portion of full pile reaction for pile B is

$$\left(\frac{0.29}{0.40}\right)P_p = \left(\frac{0.29}{0.40}\right)\left(\frac{Q}{n} + \frac{M_y}{\Sigma x^2}x + \frac{M_x}{\Sigma y^2}y\right)$$

$$= \left(\frac{0.29}{0.40}\right)\left(\frac{3785.6}{7} + \frac{170}{4\times 0.55^2 + 2\times 1.1^2}\times 0.55 + \frac{170}{4\times 0.95^2}\times 0.95\right)$$

$$= (0.725)(540.8 + 25.76 + 44.74) = 0.725\times 611.3$$

$$= 443.2\,\text{kN}$$

The portion of full pile reaction for pile A is

$$\left(\frac{0.29}{0.40}\right)P_p = \left(\frac{0.29}{0.40}\right)\left(\frac{Q}{n} - \frac{M_y}{\Sigma x^2}x + \frac{M_x}{\Sigma y^2}y\right)$$

$$= \left(\frac{0.29}{0.40}\right)\left(\frac{3785.6}{7} - \frac{170}{4\times 0.55^2 + 2\times 1.1^2}\times 0.55 + \frac{170}{4\times 0.95^2}\times 0.95\right)$$

$$= (0.725)(540.8 - 25.76 + 44.74) = 0.725\times 559.8$$

$$= 405.9\,\text{kN}$$

(*Continued*)

V_u = pile reactions – (weights of cap, fill and floor loads covering the cross-hatched area shown in the scheme)

$$= (443.2 + 405.9) -$$
$$(3 \times 0.64)\left[(1.2)(0.9 \times 24 + 0.15 \times 20 + 0.15 \times 24) + 1.6 \times 3.75\right]$$
$$= 849.1 - 76.49 = 772.61 \text{ kN}$$

$$\text{Ø}\, V_c = (\text{Ø})\left(0.17\lambda \sqrt{f_c'}\, b_w d\right) = (0.75)(0.17 \times 1)\left(\sqrt{28}\right)(1000)(3.0)(0.6)$$
$$= 1214\,\text{kN} > V_u$$

Therefore, the two-way shear controls.

Use $d = 0.60$ *m* $= 600$ *mm.*

The depth d will be taken to the centre of the steel bars placed at bottom of the cap.

Check the assumed cap thickness h:

$$h = d + \text{one bar diameter (say 25 mm)}$$
$$+ 75 \text{ mm space between steel bars and top of piles}$$
$$+ 150 \text{ mm pile embedment into the cap}$$
$$= 600 + 25 + 75 + 150 = 850 \text{ mm} \qquad \text{(OK.)}$$

Use the cap thickness $h = 850$ mm

Check the pile working load or the number of piles required, using the designed cap dimensions:

Actually, this checking is unnecessary, since the provided cap dimensions are exactly the same assumed dimensions which were used in the design computations.

$$Q = 1000 + 1400 + (3 \times 3)(0.85 \times 24 + 0.15 \times 20 + 0.15 \times 24 + 3.75)$$
$$= 2677 \text{ kN}$$

$$P_{p,max} = \frac{Q}{n} + \frac{M_y}{\Sigma x^2}x + \frac{M_x}{\Sigma y^2}y$$
$$= \frac{2677}{7} + \frac{120}{4 \times 0.55^2 + 2 \times 1.1^2} \times 0.55 + \frac{120}{4 \times 0.95^2} \times 0.95$$
$$= 382.43 + 18.2 + 31.6 = 432.2\,\text{kN} < 435\,\text{kN} \qquad \text{(OK.)}$$

Step 4. Design the flexural reinforcement.

(a) Reinforcement in x direction. The critical section is located at 0.26 m to the right of y axis, as shown in the scheme below. Use $d = 0.60$ m.

From *Step 3*:

$$\text{Factored } P_{p,(\text{E})} = 592.32\,\text{kN and } P_{p,(\text{B})} = 611.3\,\text{kN}$$

$$\text{Factored } P_{p,(\text{F})} = \frac{Q}{n} + \frac{M_y}{\Sigma x^2}x - \frac{M_x}{\Sigma y^2}y$$
$$= \frac{3785.6}{7} + \frac{170}{4 \times 0.55^2 + 2 \times 1.1^2} \times 0.55 - \frac{170}{4 \times 0.95^2} \times 0.95$$
$$= 540.8 + 25.76 - 44.74 = 521.82\,\text{kN}$$

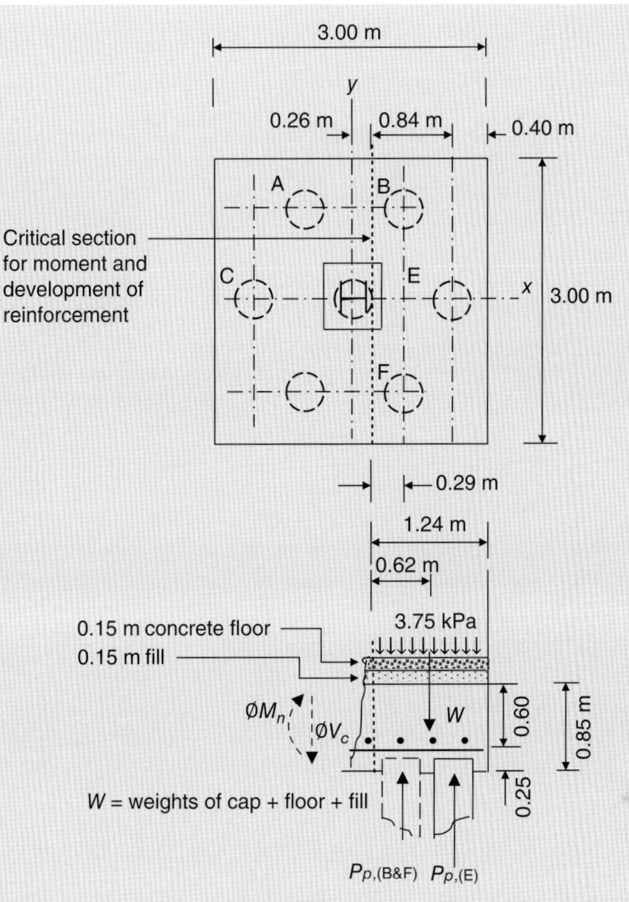

Scheme 5.51

The maximum factored moment is

$$M_u = (0.29)\left(P_{P,(B)} + P_{P,(F)}\right) + (0.84)\left(P_{P,(E)}\right) - (0.62)[\textit{Factored } (W + \text{floor load})]$$

$$= (0.29)(611.3 + 521.82) + (0.84)(592.32)$$

$$-(0.62)\begin{bmatrix} 1.2(3 \times 1.24)(0.15 \times 24 + 0.15 \times 20 + 0.85 \times 24) \\ + (1.6)(3 \times 1.24)(3.75) \end{bmatrix}$$

$$= 826.15 - 88.57 = 737.58\,\text{kN.m}$$

Equation (5.9):

$$\varnothing M_n \geq M_u$$

Assume tension-controlled section, $\varnothing = 0.9$, and $f_s = f_y$.

$$a = \frac{A_s f_y}{0.85 f_c' b} = \frac{A_s \times 420}{0.85 \times 28 \times 1} = 17.65\, A_s$$

(Continued)

$$\varnothing M_n = \varnothing \left[A_s f_y \left(d - \frac{a}{2}\right)\right]$$

$$= (0.9)\left[(A_s \times 420 \times 1000)\left(0.6 - \frac{17.65A_s}{2}\right)\right] \text{ Equation (5.11)}$$

$$= 226\ 800\,A_s - 3335\ 850\,A_s^{\,2}$$

Let $\varnothing M_n = M_u$

$$226\ 800\,A_s - 3\ 335\ 850\,A_s^{\,2} = 246$$

$$3\ 335\ 850\,A_s^{\,2} - 226\ 800\,A_s + 246 = 0$$

$$A_s = \frac{-(-226\ 800) \pm \sqrt{(-226\ 800)^2 - (4)(3\ 335\ 850)(246)}}{2 \times 3\ 335\ 850} = \frac{226\ 800 - 219\ 444}{6\ 671\ 700}$$

$$= 1.108 \times 10^{-3}\ \text{m}^2/\text{m} = 1108\ \text{mm}^2/\text{m}$$

In order to provide a more strong and rigid pile cap it may be more appropriate to use $A_{s,\text{min}}$ as required for flexural members (ACI Section 10.5.1) rather than $A_{s,\text{min}}$ based on shrinkage and temperature reinforcement (ACI Sections 10.5.1 and 7.12.2.1). Accordingly, ACI Section 10.5.1 requires

$$A_{s,\text{min}} = \frac{0.25\sqrt{f'_c}}{f_y} b_w d = \frac{0.25\sqrt{28}}{420} \times 1 \times 0.6$$

$$= 1.89 \times 10^{-3}\ \text{m}^2/\text{m} = 1890\ \text{mm}^2/\text{m}$$

and not less than 1.4

$$\frac{b_w d}{f_y} = \frac{1.4 \times 1 \times 0.6}{420}$$

$$= 2 \times 10^{-3}\ \text{m}^2/\text{m} = 2000\ \text{mm}^2/\text{m}$$

Therefore, $A_{s,\text{min}}$ = 2000 mm^2/m controls. However, according to ACI Section 10.5.3, this $A_{s,\text{min}}$ need not be used if $A_{s,\text{provided}}$ is at least one-third greater than that required by analysis.

1.33 A_s required by analysis = 1.33 × 1108 = 1474 mm^2/m

$$A_{s,\text{total}} = 3 \times 1474 = 4422\ \text{mm}^2$$

Try 16 No. 19 bars:

$$A_{s,\text{provided}} = 16 \times 284 = 4544\ \text{mm}^2 > 1.33\,A_s \quad \text{(OK.)}$$

The assumptions made are satisfied, since $A_{s,\text{min}}$ controls.

Provide 16 No. 19 bars.

Check the development of reinforcement:

In this case, the bars are in tension, the provided bar size is No. 19, the clear spacing of the bars exceeds $2d_b$, and the clear cover exceeds d_b. Therefore, the development length l_d of bars at each side of the critical section (which is the same critical section for moment) shall be determined from the following equation, but not less than 300 mm:

$$l_d = \left(\frac{f_y \psi_t \psi_e}{2.1\lambda\sqrt{f'_c}}\right) d_b \quad \text{(ACI Sections 12.2.1, 12.2.2 and 12.2.4)}$$

where the factors ψ_t, ψ_e and λ are 1 (ACI Sections 12.2.4 and 8.6.1) $l_d = \left(\frac{420 \times 1 \times 1}{2.1 \times 1 \times \sqrt{28}}\right)\left(\frac{19.1}{1000}\right) = 0.722\ \text{m} = 722\ \text{mm} > 300\ \text{mm}$. Therefore, the required $l_d = 722$ mm.

The bar extension past the critical section (i.e. the available length) is

$$1240\ \text{mm} - 75\ \text{mm cover} = 1165\ \text{mm} > 722\ \text{mm} \quad \text{(OK.)}$$

However, we shall hook the bars (90-degree standard hook) at both ends to insure better anchorage and keep the pile cap safe against bursting of the side cover where the pile transfers its load to the cap.

ACI Sections 7.6.5 and 10.5.4 require maximum spacing shall not exceed three times the slab or footing thickness, or 450 mm, whichever is smaller.

Use centre to centre bar spacing = 188 mm.

Check the 75 mm minimum concrete cover at each side:

$$\text{Concrete cover} = \frac{(3000 - 15 \times 188 - 19.1)}{2} = 80.5\ \text{mm} > 75\ \text{mm, O.K.}$$

Provide 16 No.19 bars @ 188 mm c.c. at bottom of the pile cap in x direction, placed 90 mm above top of piles and hooked (90-degree standard hook) at both ends.

(b) Reinforcement in y direction. The critical section is located at 0.26 m above the x axis, as shown in the scheme below.

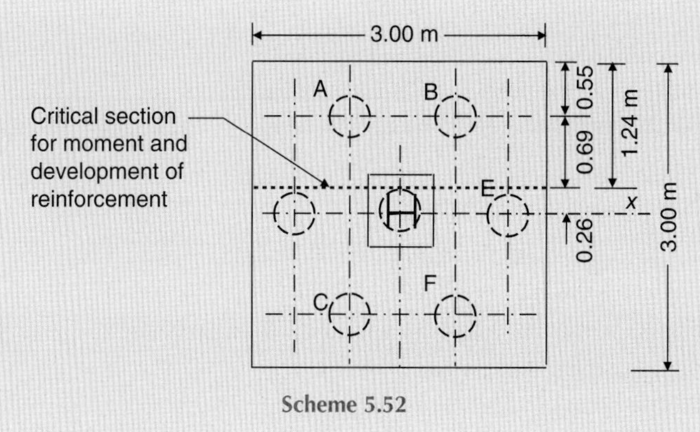

Scheme 5.52

From *Step 3*:

Factored $P_{p,(A)} = 559.8$ kN and $P_{p,(B)} = 611.3$ kN

The maximum factored moment is

$$\begin{aligned} M_u &= (0.69)\left(P_{P,(A)} + P_{P,(B)}\right) - (0.62)[\text{factored}\,(W + \text{floor load})] \\ &= (0.69)(559.8 + 611.3) \\ &\quad - (0.62)\begin{bmatrix} 1.2(3 \times 1.24)(0.15 \times 24 + 0.15 \times 20 + 0.85 \times 24) \\ + (1.6)(3 \times 1.24)(3.75) \end{bmatrix} \\ &= 808.06 - 88.57 = 719.5\ \text{kN.m} \end{aligned}$$

$$M_u = \frac{719.5}{3} = 240\ \text{kN.m/m}$$

This moment is very close to the M_u just calculated in (a) above. Therefore, it is very clear here again $A_{s,min}$ controls. Consequently, it is appropriate to use reinforcement in y direction equals to that in x direction. Also, the same

(Continued)

size, number and spacing of the steel bars will be used for both directions, since the cap is square, h is constant and the critical sections are located at the same distance from the cap centre.

Provide 16 No.19 bars @ 188 mm c.c. at bottom of the pile cap in y direction, placed on top of bars in x direction, and hooked (90-degree standard hook) at both ends.

Step 5. Check plate bearing on pile cap.

For this step, the design proceeds in the same manner as that of Solution of Problem 5.9, *Step 2* and Solution of Problem 5.10, *Step 5.*

Step 6. Design the anchor bolts and determine their effective embedment depth h_{ef}, spacing and edge distances.

The design computations proceed in the same manner as those presented in Solution of Problem 5.9, *Steps 5 and 6.*

Step 7. Develop the final design sketch.

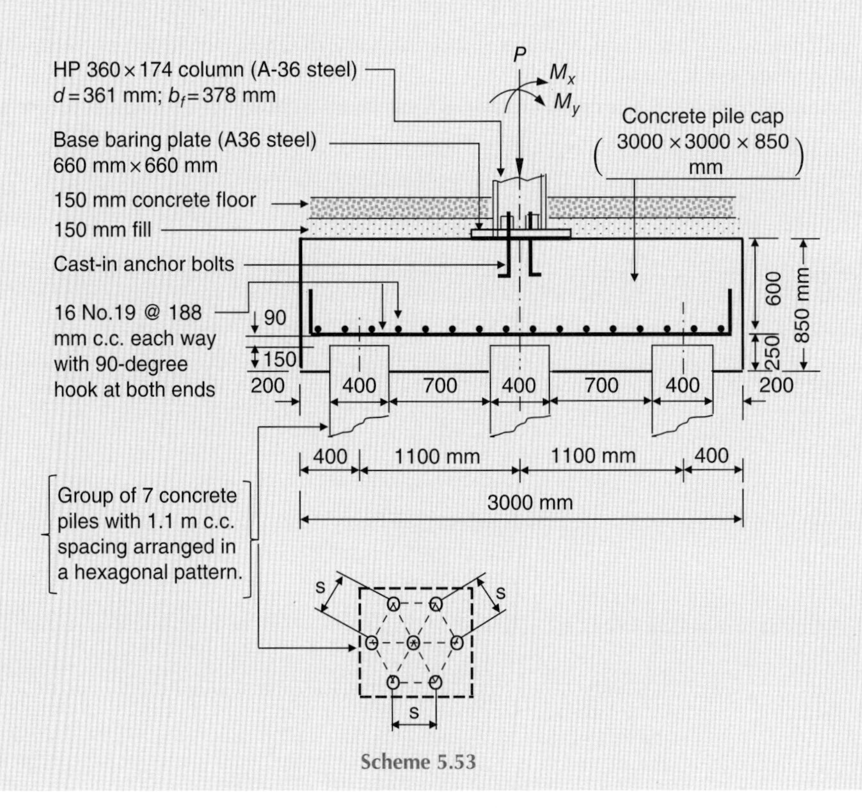

Scheme 5.53

Problem 5.12

The following design data belong to a rectangular ordinary combined footing which supports two columns. The distance between the columns is 6 m, centre to centre. The exterior face of the exterior column (Col. 1) is located right on the property line. The gross allowable soil pressure (*gross* q_a) is 250 kPa at a depth of 1.3 m below the finished basement floor. The basement concrete floor is 0.15 m thick and supports a live load of 5 kPa. The density of the fill above the footing is $\gamma'_s = 20\ \text{kN/m}^3$. Design the footing using the conventional method.Footing: concrete strength $f'_c = 21$ MPa; reinforcing steel $f_y = 420$ MPa.

Column No.	Size, m	Working loads				
		D,kN	L,kN	M_D,kN.m	M_L,kN.m	M,kN.m
1	0.6 × 0.4	900	675	90	70	160 ↗
2	0.6 × 0.6	1350	1000	130	100	230 ↗

Scheme 5.54

Solution:

Step 1. Estimate the net factored soil pressure and footing dimensions.

Let resultant of all moments and axial loads passes through the centroid of the footing base area to achieve uniform soil pressure. The gross load of this resultant equals to the dead and live loads of the two columns plus the floor load and the weights of footing, backfill material and floor. For a first trial guess the footing thickness h equals 1 m.

$$D_f = \text{foundation depth} = \text{thicknesses of footing + fill + floor} = 1.3\,\text{m}$$

$$\text{Therefore, thickness of the fill above the footing} = 1.3 - 1 - 0.15 = 0.15\ \text{m}$$

The net resultant load is

$$net\,R = gross\,R - A_f D_f \gamma'_s$$

$$net\,q = \frac{net\,R}{A_f} = \frac{gross\,R - A_f D_f \gamma'_s}{A_f} = \frac{gross\,R}{A_f} - D_f \gamma'_s$$

$$= \frac{(900 + 675 + 1350 + 1000) + A_f(1 \times 24 + 0.15 \times 24 + 0.15 \times 20 + 5)}{A_f} - 1.3 \times 20$$

$$= \frac{3925}{A_f} + 35.6 - 26.0 = \frac{3925}{A_f} + 9.6 \quad \text{(kPa)}$$

$$net\,q_a = gross\,q_a - D_f \gamma'_s = 250 - 1.3 \times 20 = 224\,\text{kPa}$$

Let $net\ q = net\ q_a$:

$$\frac{3925}{A_f} + 9.6 = 224$$

$$A_f = \frac{3925}{224 - 9.6} = \frac{3925}{214.4} = 18.31\,\text{m}^2$$

As for the concentrically loaded spread footings, the net factored soil pressure ($net\ q_{\text{factored}}$) is computed without including the uniform floor load and the weights of footing, backfill material and floor because these loads are evenly distributed and directly transferred to the soil below and, thus, do not affect the footing shear forces and moments. Therefore, the factored net soil pressure is

$$net\,q_{\text{factored}} = \frac{\text{factored column loads}}{A_f} = \frac{1.2(900 + 1350) + 1.6(675 + 1000)}{18.31}$$

$$= \frac{5380}{18.31} = 293.83\ \text{kPa}$$

Compute factored moments, factored column loads and factored resultant.

Column No. 1: factored moment $M_1 = 1.2 \times 90 + 1.6 \times 70 = 220$ kN. m

(*Continued*)

Column No. 2: factored moment $M_2 = 1.2 \times 130 + 1.6 \times 100$

$$= 316\,\text{kN.m}$$

Factored $\Sigma M = 536$ kN. m

Column No. 1: factored load $P_1 = 1.2 \times 900 + 1.6 \times 675 = 2160$ kN

Column No. 2: factored load $P_2 = 1.2 \times 1350 + 1.6 \times 1000$

$$= 3220\ kN$$

$$\text{Factored resultant } R = \text{Factored } \Sigma P = 5380\ kN$$

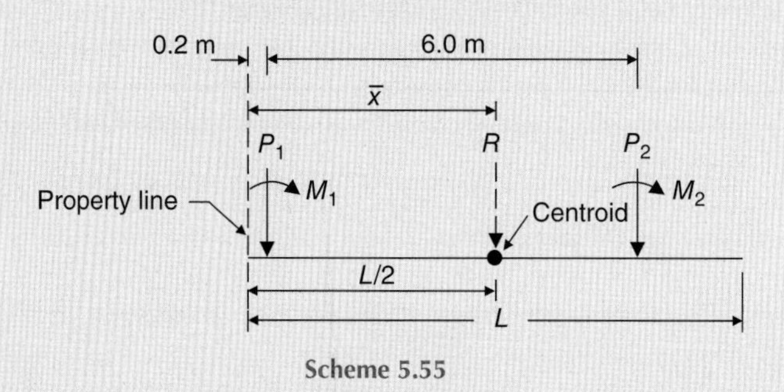

Scheme 5.55

$$\bar{X} = \frac{0.2 \times P_1 + (0.2 + 6.0)(P_2) + M}{R} = \frac{0.2 \times 2160 + (0.2 + 6.0)(3220) + 536}{5380} = 3.891\,\text{m}$$

Let $\frac{L}{2} = \bar{X}$:

$$L = 2 \times 3.891 = 7.782\,\text{m}$$

$$B = \frac{A}{L} = \frac{18.31}{7.782} = 2.353\,\text{m}$$

Use rectangular combined footing 2.4 m × 7.8 m

Check eccentricity of unfactored R:

$$\bar{X} = \frac{0.2(900 + 675) + (0.2 + 6.0)(1350 + 1000) + (160 + 230)}{(900 + 675) + (1350 + 1000)} = \frac{15275}{3925} = 3.892\,\text{m}$$

$$\text{eccentricity } e = \bar{X} - L/2 = 3.892 - 3.891 = 0.001\,\text{m} \cong 0 \qquad \text{(OK.)}$$

Step 2. Obtain data for shear-force and bending-moment diagrams considering the combined footing as a reinforced concrete beam.

Any convenient method, for example integral calculus, may be used for calculating the shear and moment values at different locations necessary to obtain informable diagrams. Usually, the design calculations involve tedious busywork; therefore, it may be preferable to use any available computer program concerns the conventional design of ordinary combined footings.

In the design computations it may be necessary to use the exact B and L values (i.e. $B = 2.353$ m and $L = 7.782$ m) so that the shear and moment diagrams will close.

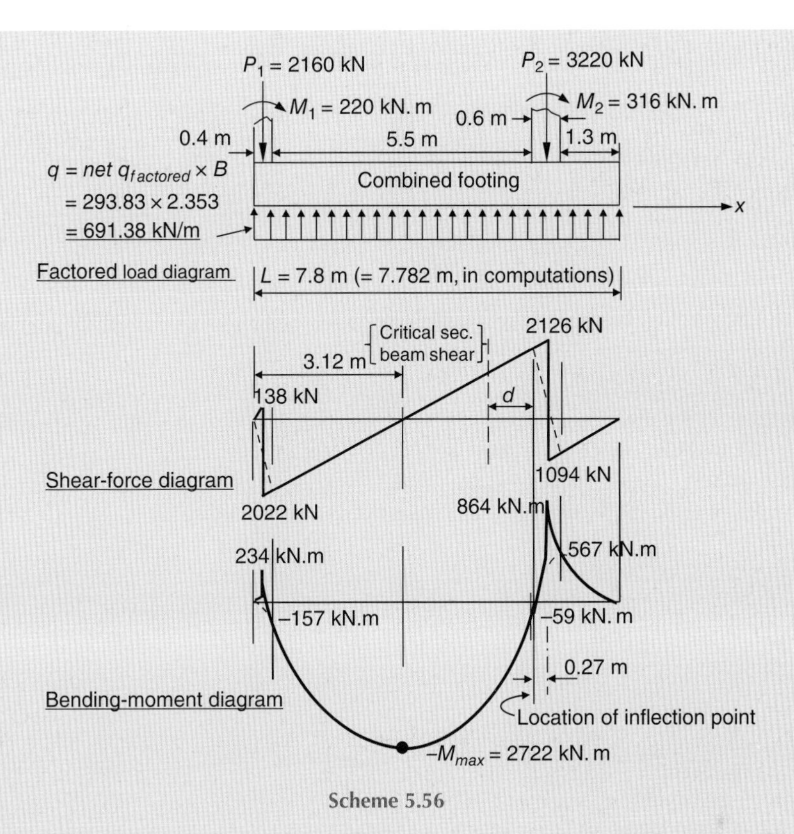

Scheme 5.56

It is preferable to draw the complete load diagram first, and then the shear and moment diagrams, as shown above. The factored distributed loads on the footing, shear forces and bending moments are computed for the full 2.4-m width of the footing.

Step 3. Find footing thickness h.

Determine the required thickness based on both one-way shear and two-way shear analyses.

(a) One-way shear. It is critical at distance d from the interior face of the interior column (Col. No. 2), as indicated on the shear-force diagram.

Equation (5.19): $$\varnothing V_c \geq V_u \text{ (taking reinforcement shear strength, } V_s = 0)$$

$$\text{where } \varnothing = \text{strength reduction factor (ACI Section 9.3.2.3)}$$

$$= 0.75 \text{ for shear}$$

$$V_u = 2126 - 691.38(d + 0.3) = 1919 - 691.38\,d \quad \text{(kN)}$$

Equation (5.20): $$V_c = 0.17\lambda\sqrt{f_c'}b_w d$$

Let $\varnothing V_c = V_u$: $$= (0.75 \times 0.17 \times 1)\left(\sqrt{21}\right)(1000)(2.353)(d)$$

$$(0.75 \times 0.17 \times 1)\left(\sqrt{21}\right)(1000)(2.353)(d) = 1919 - 691.38\,d$$

$$d = \frac{1919}{2066.19} = 0.929 \text{ m} = 929 \text{ mm}.$$

Try $d = 0.93$ m.

(b) Two-way shear. This shear action is most critical at one of the two columns. Therefore, both locations should be considered. The critical sections are shown in the scheme below.

(Continued)

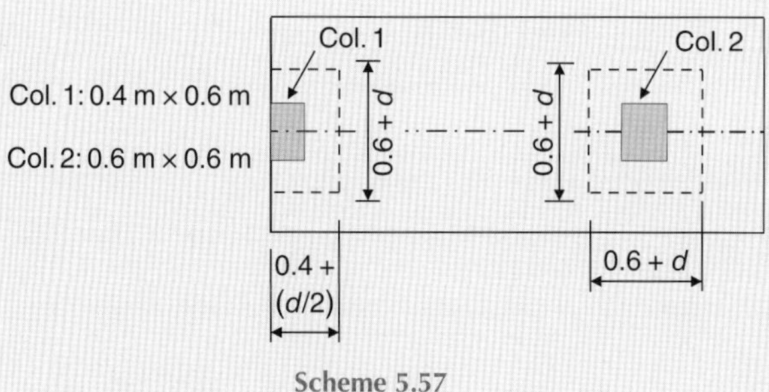

Scheme 5.57

Two-way shear at the exterior column:
The shear perimeter has three sides only.
Assume $d = 0.93$ m.
Shear strength of concrete V_c shall be the smallest of (a), (b) and (c):

(a)

$$V_c = 0.17\left(1+\frac{2}{\beta}\right)\lambda\sqrt{f_c'}b_o d$$

Equation (5.22)

$$= (0.17)\left(1+\frac{2}{1.5}\right)(1)\sqrt{f_c'}b_o d = (0.397)\sqrt{f_c'}b_o d$$

(b)

$$V_c = 0.083\left(\frac{\alpha_s d}{b_o}+2\right)\lambda\sqrt{f_c'}b_o d$$

$$= (0.083)\left(\frac{30d}{b_o}+2\right)(1)\sqrt{f_c'}b_o d = (0.083)\left(\frac{30d}{b_o}+2\right)\sqrt{f_c'}b_o d$$

Equation (5.23)

$$= (0.083)\left(\frac{30\times 0.93}{2\left(0.4+\frac{0.93}{2}\right)+1.53}+2\right)\sqrt{f_c'}b_o d = (0.88)\sqrt{f_c'}b_o d$$

(c)

$$V_c = 0.33\lambda\sqrt{f_c'}b_o d$$

Equation (5.24)

$$= 0.33(1)\sqrt{f_c'}b_o d = (0.33)\sqrt{f_c'}b_o d$$

Use $V_c = (0.33)\sqrt{f_c'}b_o d$

$$\text{Ø}\,V_c = (0.75)(0.33)\sqrt{21}(1000)[2(0.4+0.465)+(0.6+0.93)]0.93$$
$$= 3439\text{ kN}$$

For this three-sided shear perimeter, the condition of *unbalanced moment transfer* exists, as shown in the scheme below. According to ACI Section 11.11.7.1, where unbalanced moment M_u exists, the fraction $\gamma_f M_u$ shall be transferred by flexure in accordance with ACI Section 13.5.3, and the remainder fraction of the unbalanced moment, $\gamma_v M_u$, shall be considered to be transferred by eccentricity of shear about the centroid of the critical shear perimeter defined in ACI Section 11.11.1.2. Also, ACI Section 11.11.7.2 requires that the shear stress resulting from moment transfer by eccentricity of shear shall be assumed to vary linearly about the centroid of the critical shear perimeter, and the maximum shear stress due to V_u and M_u shall not exceed Ø $V_c/(b_o d)$. Factors γ_f and γ_v are used to determine the unbalanced moment transferred by flexure and by eccentricity of shear, respectively.

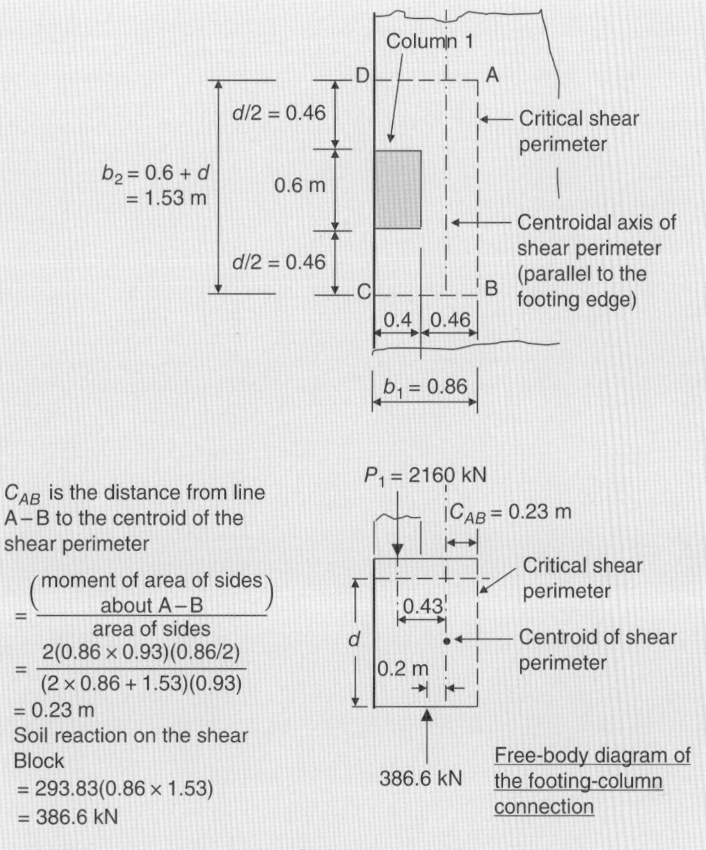

Scheme 5.58

Summation of moments about the centroid of the shear perimeter equals the unbalanced moment. Hence,

$$M_u = 2160 \times 0.43 - 386.6 \times 0.2 = 851.5\,\text{kN.m}$$

$$\gamma_v = \left(1 - \gamma_f\right) \quad \text{(ACI Section 11.11.7.1)}$$

$$\gamma_f = \frac{1}{1 + (2/3)\sqrt{b_1/b_2}} \quad \text{(ACI Section 13.5.3.2)}$$

$$= \frac{1}{1 + (2/3)\sqrt{0.86/1.53}} = 0.667$$

$$\gamma_v = 0.667 = 0333$$

The shear stresses due to the direct shear and the shear due to moment transfer will add at points D and C, giving the largest shear stresses on the critical shear perimeter. Hence,

$$v_{u(DC)} = \frac{V_u}{b_o d} + \frac{\gamma_v M_u C_{DC}}{J_c} \quad \text{(ACI Section R11.11.7.2)}$$

$J_c = J_z$ = Property of the shear perimeter analogous to *polar moment of inertia*

(Continued)

$$J_c = 2\left(\frac{b_1 d^3}{12}\right) + 2\left(\frac{d b_1^3}{12}\right) + 2(b_1 d)\left(\frac{b_1}{2} - C_{AB}\right)^2 + (b_2 d) C_{AB}{}^2$$

$$= 2\left(\frac{0.86 \times 0.93^3}{12}\right) + 2\left(\frac{0.93 \times 0.86^3}{12}\right) + 2(0.86 \times 0.93)\left(\frac{0.86}{2} - 0.23\right)^2 +$$

$$(1.53 \times 0.93)(0.23^2)$$

$$= 0.115 + 0.099 + 0.064 + 0.075 = 0.353\ \text{m}^4$$

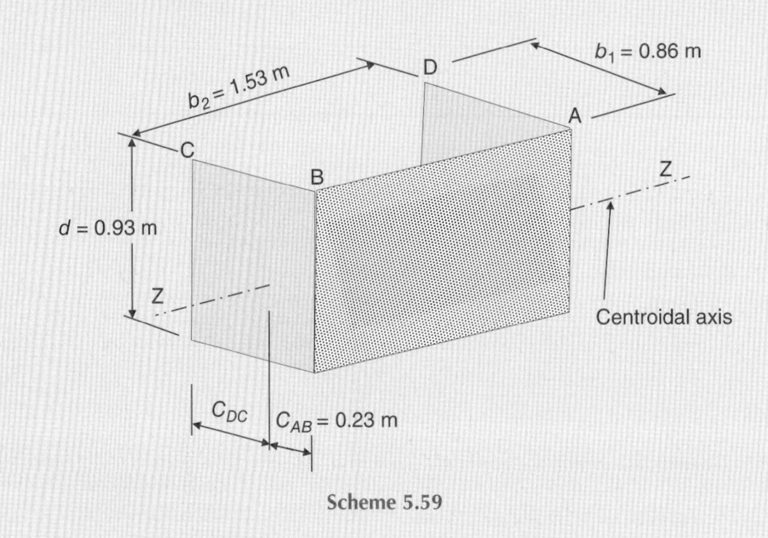

Scheme 5.59

$$v_{u(DC)} = \frac{2160 - 386.6}{(2 \times 0.86 + 1.53)(0.93)} + \frac{0.333 \times 851.5(0.86 - 0.23)}{0.353} = 586.73 + 506.05$$

$$= 1093\ \text{kN/m}^2$$

$$\varnothing\, v_c = \frac{\varnothing\, V_c}{b_o d} = \frac{3439}{[2(0.4 + 0.465) + (0.6 + 0.93)]0.93} = 1134\ \text{kN/m}^2 > v_{u(DC)} \quad \text{(OK.)}$$

Therefore, the one-way shear controls.

Two-way shear at the interior column:

$$\varnothing\, V_c \geq V_u$$

The shear perimeter has four sides.

Assumed $d = 0.93$ m.

The shear strength of concrete V_c shall be the smallest of (a), (b) and (c):

(a)

$$V_c = 0.17\left(1 + \frac{2}{\beta}\right)\lambda \sqrt{f_c'}\, b_o d \qquad \text{Equation (5.22)}$$

$$= (0.17)\left(1 + \frac{2}{1}\right)(1)\sqrt{f_c'}\, b_o d = (0.51)\sqrt{f_c'}\, b_o d$$

(b)

$$V_c = 0.083\left(\frac{\alpha_s d}{b_o} + 2\right)\lambda\sqrt{f_c'}b_o d$$

$$= (0.083)\left(\frac{40d}{b_o} + 2\right)(1)\sqrt{f_c'}b_o d \qquad \text{Equation (5.23)}$$

$$= (0.083)\left(\frac{40 \times 0.93}{4(0.6 + 0.93)} + 2\right)\sqrt{f_c'}b_o d = (0.67)\sqrt{f_c'}b_o d$$

(c)

$$V_c = 0.33\lambda\sqrt{f_c'}b_o d$$

$$= 0.33(1)\sqrt{f_c'}b_o d = (0.33)\sqrt{f_c'}b_o d \qquad \text{Equation (5.24)}$$

Use $V_c = (0.33)\sqrt{f_c'}b_o d$.

$$\varnothing V_c = (0.75)(0.33)\sqrt{21}(1000)[4(0.6 + 0.93)]0.93 = 6455\,\text{kN}$$

$V_u = P_2$ –soil reaction on the shear block
$V_u = 3220$ –soil reaction on the shear block (kN)
$V_u < \varnothing V_c$. Hence, the one-way shear controls.
Use $d = 0.93\ m = 930\ mm$.

The depth d will be taken to the centre of the steel bars at top or at bottom of the footing in long direction.
Assume using No. 25 bars, and 75 mm minimum concrete cover (ACI Section 7.7.1). Hence, the overall footing thickness is

$$h = 930 + 75 + 1/2\,\text{bar diameter} = 1005 + 12.7 = 1017.7\,\text{mm}.$$

This value of h is considered very close to the assumed 1000 mm. For more safety, one may use a slightly larger value.
Use $h = 1050\ mm$.

Step 4. Design the flexural reinforcement in the long direction.
(a) Midspan negative reinforcement at top of the footing.
Refer to the factored bending moment diagram of *Step* 2.

$$-M_{max} = 2722\,\text{kN.m}$$

Equation (5.9):

$$\varnothing M_n \geq M_u$$

$$M_u = \frac{2722}{B} = \frac{2722}{2.4} = 1134.2\,\text{kN.m/m}$$

Assume tension-controlled section, $\varnothing = 0.9$, and $f_s = f_y$.
Equation (5.11):

$$\varnothing M_n = \varnothing\left[A_s f_y\left(d - \frac{a}{2}\right)\right]$$

$$a = \frac{A_s f_y}{0.85 f_c' b} = \frac{A_s \times 420}{0.85 \times 21 \times 1} = 23.53 A_s$$

(*Continued*)

$$\emptyset M_n = \emptyset\left[A_s f_y\left(d-\frac{a}{2}\right)\right]$$

$$=(0.9)\left[(A_s \times 420 \times 1000)\left(0.93-\frac{23.53A_s}{2}\right)\right]$$

$$=351\ 540\,A_s - 4\ 447\ 170A_s^2$$

Let $\emptyset M_n = M_u$:

$$351\ 540\,A_s - 4\ 447\ 170A_s^2 = 1134.2$$

$$4\ 447\ 170A_s^2 - 351\ 540A_s + 1134.2 = 0$$

$$A_s = \frac{-(-351\ 540) \pm \sqrt{(-351\ 540)^2 - (4)(4\ 447\ 170)(1134.2)}}{2 \times 4\ 447\ 170} = \frac{351\ 540 - 321\ 566}{8\ 894\ 340}$$

$$= 3370 \times 10^{-6}\ \text{m}^2/\text{m} = 3370\ \text{mm}^2/\text{m}$$

ACI Section 10.5.1:

$$A_{s,\min} = \left(\frac{0.25\ \sqrt{f_c'}}{f_y} b_w d\right), \text{ and not less than } \left(1.4\frac{b_w d}{f_y}\right)$$

$$A_{s,\min} = \frac{0.25\ \sqrt{f_c'}}{f_y} b_w d = \frac{0.25\sqrt{21}}{420} \times 1 \times 0.93 = 2.537 \times 10^{-3}\ \text{m}^2/\text{m}$$

$$= 2537\ \text{mm}^2/\text{m}$$

$$1.4\frac{b_w d}{f_y} = \frac{1.4 \times 1 \times 0.93}{420} = 3.1 \times 10^{-3}\frac{\text{m}^2}{\text{m}} = 3100\ \text{mm}^2/\text{m}$$

Therefore, use $A_s = 3370\ \text{mm}^2/\text{m}$

$$A_{s,\text{total}} = 3370 \times B = 3370 \times 2.4 = 8088\ \text{mm}^2$$

Try 16 No. 25 bars: $A_{s,\text{provided}} = 16 \times 510 = 8160\ \text{mm}^2$ (OK.)

Compute a for $A_s = \dfrac{8160}{2.4} = 3400\ \text{mm}^2/\text{m}$, and check if $f_s = f_y$ and whether the section is tension-controlled:

$$a = 23.53A_s = 23.53 \times 3400 \times 10^{-6} = 0.080\ \text{m} = 80\ \text{mm}$$

$c = \dfrac{a}{\beta_1}$; For f_c' between 17 and 28 MPa, β_1 shall be taken as 0.85

$$c = \frac{80}{0.85} = 94.12\ \text{mm}$$

$$d_t = d = 930\ \text{mm}$$

$$\varepsilon_t = 0.003\left(\frac{d_t - c}{c}\right) \qquad \text{Equation (5.15)}$$

$$= 0.003\left(\frac{930 - 94.12}{94.12}\right) = 0.027 > 0.005.$$

Hence, the section is tension-controlled, and Ø = 0.9 (ACI Sections 10.3.4 and 9.3.2.1).

$\varepsilon_y = \frac{f_y}{E_s} = \frac{420}{200\ 000} = 0.0021$. Hence, $\varepsilon_t > \varepsilon_y$ and $f_s = f_y$.

Therefore, the assumptions made are satisfied.

In case the required flexural reinforcement exceeds the minimum flexural reinforcement, it shall be adequate to use the same maximum spacing of reinforcement for slabs which is two times the slab thickness or 450 mm, whichever is smaller, as specified by ACI Section 13.3.2.

Use centre to centre bar spacing = 148 mm. Check the 75 mm minimum concrete cover at each side:

$$\frac{2400-(15\times 148+25.4)}{2} = 77.3\,\text{mm} > 75\,\text{mm} \quad \text{(OK.)}$$

Try 16 No. 25 bars @ 148 mm c. c. at top, across the footing width.

Check the development of reinforcement:

In this case, the bars are in tension, the provided bar size is No. 25, the clear spacing of the bars exceeds $2d_b$ and the clear cover exceeds d_b. Therefore, the development length l_d of bars at each side of the critical section (which is located where the maximum factored negative moment exists) shall be determined from the following equation, but not less than 300 mm:

$$l_d = \left(\frac{f_y \psi_t \psi_e}{1.7\lambda\sqrt{f_c'}}\right) d_b \quad \text{(ACI Sections 12.2.1, 12.2.2 and 12.2.4)}$$

where the factors ψ_e and λ are 1 (ACI Sections 12.2.4 and 8.6.1) .

The factor $\psi_t = 1.3$ because the reinforcement is placed such that more than 300 mm of fresh concrete exists below the top bars (ACI Section 12.2.4).

$$l_d = \left(\frac{420\times 1.3\times 1}{1.7\times 1\times\sqrt{21}}\right)\left(\frac{25.4}{1000}\right) = 1.78\,\text{m} = 1780\,\text{mm} > 300\ \text{mm}.$$

Therefore, the required $l_d = 1780$ mm.

The bar extension past the critical section (i.e. the available length) is

$$3120\,\text{mm} - 75\,\text{mm cover} = 3045\,\text{mm} > 1780\,\text{mm} \quad \text{(OK.)}$$

According to MacGregor and Wight (2005), since the loading, the supports and the shape of the moment diagram are all inverted from those found in a normally loaded beam, the necessary check of the ACI Section 12.11.3 requirement which concerns positive moment reinforcement shall be made for the negative moment reinforcement.

It may be appropriate to extend all the bars into the column regions at both ends. At the point of inflection, 0.27 m from the centre of the interior column, V_u = 1940 kN and by ACI Section 12.11.3,

$$l_d \le \frac{M_n}{V_u} + l_a$$

$$M_n = \text{nominal flexural strength of the cross section} = A_s f_y\left(d-\frac{a}{2}\right)$$

$$a = \frac{A_s f_y}{0.85 f_c' b} = \frac{(8160/10^6)(420)}{0.85\times 21\times 2.4} = 0.08$$

$$M_n = (8160/10^6)(420\times 1000)\left(0.93-\frac{0.08}{2}\right) = 3050\,\text{kN.m}$$

(Continued)

Let $l_d = \frac{M_n}{V_u} + l_a$. Hence, $l_{a,\,min} = l_d - \frac{M_n}{V_u} = 1.78 - \frac{3050}{1940} = 0.21$ m. The distance l_a at a support shall be the embedment length beyond centre of support.

Accordingly, extend the bars to the exterior face of the interior column. This will give

$$l_a = 0.30\,\text{m} > l_{a,min}.$$

However, in order to satisfy the requirements of ACI Sections 12.12.3 and 12.10.3, the negative moment reinforcement shall have an embedment length beyond the point of inflection not less than d, $12d_b$ or $\ell_n/16$, whichever is greater. In this case, d controls. Therefore, the top bars should be extended not less than 0.93 m beyond the point of inflection at $x = 5.93$ m. However, since the remaining distance to the footing edge is only 0.94 m, it may be more practical to extend the top bars to the right edge of the footing (less 75 mm minimum concrete cover).

At centre of the exterior column, $V_u = 2022$ kN; hence,

$$l_{a,min} = l_d - \frac{M_n}{V_u} = 1.78 - \frac{3050}{2022} = 0.27\,\text{m}$$

The available space beyond the column centre is 0.2 m less concrete cover, which is less than $l_{a,min}$. Therefore, the top bars will all have to be hooked at the exterior end. This is also necessary to anchor the bars transferring the unbalanced moment from the column to the footing.

Provide 16 No.25 bars @ 148 mm c.c. at top, full length of the footing (less concrete cover at the ends). All the bars should be hooked at the exterior end (at the exterior face of the exterior column).

(b) Positive reinforcement at bottom of the footing.

Let us design, conservatively, for $+M_{max} = 864$ kN. m instead the 567 kN.m positive moment. However, in most of cases $A_{s,min}$ controls and it makes no difference which moment is used.

Equation (5.9): $\varnothing M_n \geq M_u$

$$M_u = \frac{864}{B} = \frac{864}{2.4} = 360\,\text{kN.m/m}$$

Assume tension-controlled section, $\varnothing = 0.9$, and $f_s = f_y$.

$$a = \frac{A_s f_y}{0.85 f_c' b} = \frac{A_s \times 420}{0.85 \times 21 \times 1} = 23.53 A_s$$

$$\varnothing M_n = \varnothing \left[A_s f_y \left(d - \frac{a}{2}\right)\right]$$

$$= (0.9)\left[(A_s \times 420 \times 1000)\left(0.93 - \frac{23.53 A_s}{2}\right)\right] \text{ Equation (5.11)}$$

$$= 351\,540\,A_s - 4\,447\,170 A_s^2$$

Let $\varnothing M_n = M_u$:

$$351\,540\,A_s - 4\,447\,170 A_s^2 = 360$$

$$4\,447\,170 A_s^2 - 351\,540 A_s + 360 = 0$$

$$A_s = \frac{-(-351\,540) \pm \sqrt{(-351\,540)^2 - (4)(4\,447\,170)(360)}}{2 \times 4\,447\,170} = \frac{351\,540 - 342\,310}{8\,894\,340}$$

$$= 1038 \times \frac{10^{-6}\,\text{m}^2}{\text{m}} = 1038\,\text{mm}^2/\text{m} < (A_{s,min} = 3100\,\text{mm}^2/\text{m})$$

Therefore, $A_{s,\min}$ controls.

$$A_{s,\text{total}} = 3100 \times B = 3100 \times 2.4 = 7440\ \text{mm}^2.$$

Try 15 No. 25 bars: $A_{s,\text{provided}} = 15 \times 510 = 7650\ \text{mm}^2/\text{m}$
ACI Sections 7.6.5 and 10.5.4 requires maximum spacing shall not exceed three times the slab or footing thickness or 450 mm, whichever is smaller.

Use centretocentre bar spacing = 158 mm. Check the 75 mm minimum concrete cover at each side:

$$\text{Concrete cover} = \frac{2400 - (14 \times 158 + 25.4)}{2} = 81.3\ \text{mm} > 75\ \text{mm} \quad (\text{OK.})$$

Try 15 No. 25 bars @ 158 mm c.c. at bottom, across the footing width.

Check the development of reinforcement:

In this case, the bars are in tension, the provided bar size is No. 25, the clear spacing of the bars exceeds $2d_b$, and the clear cover exceeds d_b. Therefore, the development length l_d of bars at each side of the critical section (which is located where the maximum factored positive moment exists) shall be determined from the following equation, but not less than 300 mm:

$$l_d = \left(\frac{f_y \psi_t \psi_e}{1.7\lambda\sqrt{f_c'}}\right) d_b \quad (\text{ACI Sections 12.2.1, 12.2.2 and 12.2.4})$$

where the factors ψ_t, ψ_e and λ are 1 (ACI Sections 12.2.4 and 8.6.1).

$$l_d = \left(\frac{420 \times 1 \times 1}{1.7 \times 1 \times \sqrt{21}}\right)\left(\frac{25.4}{1000}\right) = 1.37\ \text{m} = 1370\ \text{mm} > 300\ \text{mm}.$$

Therefore, the required $l_d = 1370$ mm.

The bar extension past the critical section (i.e. the available length) is

$$1600\ \text{mm} - 75\ \text{mm cover} = 1525\ \text{mm} > 1370\ \text{mm} \quad (\text{OK.})$$

In order to satisfy the requirements of ACI Sections 12.10.3 and even 12.12.3 when we consider the footing as an inverted normally loaded beam (the same as we did for the top bars), it is necessary to extend the bottom bars a minimum distance d (greater than $12d_b$) past the point of inflection. Therefore, cut off eight bottom bars (alternately spaced) at 1.4 m from centre of the interior column toward the exterior column. The other seven bars shall be extended full length (less 75 mm cover at each end) of the footing and hooked at the exterior end (at the exterior column). These seven bars will resist the tensile stresses due to the relatively small positive moment at the exterior column, and provide supports to the transverse bottom bars.

Provide 15 No.25 bars @ 158 mm c.c. at bottom of the footing. Cut off eight bars (alternately spaced) at 1.40 m from the centre of the interior column. The other seven bars shall be extended full length of the footing and hooked at the exterior end (at the exterior column).

Step 5. Design the transverse reinforcement at bottom of the footing.

The footing is divided into three zones or strips of the defined widths shown in the scheme below. Zones I and II, usually known as effective zones, should be analyzed as beams; the provided steel should not be less than that required for bending or $A_{s,\min}$, whichever is greater. For zone III (the remaining portions), the provided steel should satisfy $A_{s,\min}$ requirement only. All the transverse steel bars should be placed on top of bars in long direction at the bottom of the footing.

(*Continued*)

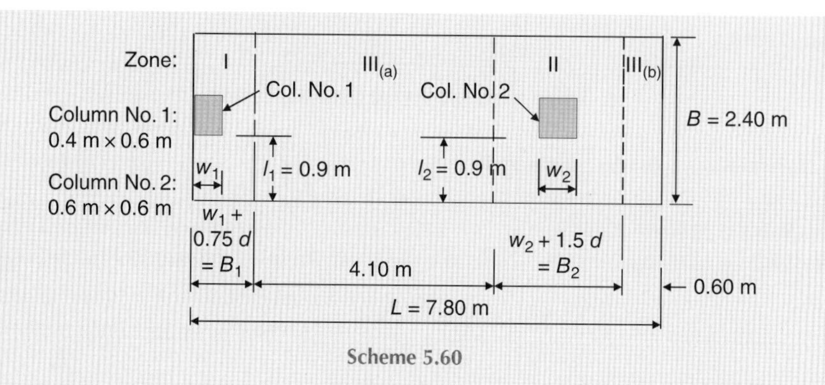

Scheme 5.60

Zone I: 1.1 m × 2.4 m
Zone II: 2.0 m × 2.4 m
Zone III$_{(a)}$: 4.1 m × 2.4 m (between zones I and II)
Zone III$_{(b)}$: 0.6 m × 2.4 m (the portion of the cantilever part)

Transverse reinforcement for zone I:

$$net\ q_{\text{factored}} = \frac{P_1}{B_1 \times B} = \frac{2160}{1.1 \times 2.4} = 818.18\,\text{kPa}$$

$$M_u = \frac{net\ q_{\text{factored}} \times l_1^2}{2} = \frac{818.18 \times 0.9^2}{2} = 331.4\,\text{kN.m/m}$$

$$d = 0.93\,\text{m} - \text{one bar diameter} = 0.93 - 0.0254 = 0.9046\,\text{m}$$

Use d = 0.9 m.

Assume tension-controlled section, $\varnothing = 0.9$ and $f_s = f_y$.

$$a = \frac{A_s f_y}{0.85 f_c' b} = \frac{A_s \times 420}{0.85 \times 21 \times 1} = 23.53 A_s$$

$$\varnothing M_n = \varnothing \left[A_s f_y \left(d - \frac{a}{2}\right)\right]$$

$$= (0.9)\left[(A_s \times 420 \times 1000)\left(0.9 - \frac{23.53 A_s}{2}\right)\right] \text{ Equation (5.11)}$$

$$= 340\ 200\,A_s - 4\ 447\ 170 A_s^{\,2}$$

Let $\varnothing M_n = M_u$:

$$4\ 447\ 170 A_s^{\,2} - 340\ 200 A_s + 331.4 = 0$$

$$A_s = \frac{-(-340\ 200) \pm \sqrt{(-340200)^2 - (4)(4\ 447\ 170)(331.4)}}{2 \times 4\ 447\ 170} = \frac{340\ 200 - 331\ 422}{8\ 894\ 340}$$

$$= 987 \times 10^{-6}\,\text{m}^2/\text{m} = 987\,\text{mm}^2/\text{m}$$

$$A_{s,\min} = \frac{0.25\sqrt{f_c'}}{f_y} b_w d = \frac{0.25\sqrt{21}}{420} \times 1 \times 0.9 = 2.455 \times 10^{-3}\,\text{m}^2/\text{m}$$

$$= 2455\,\text{mm}^2/\text{m}$$

and not less than 1.4

$$\frac{b_w d}{f_y} = \frac{1.4 \times 1 \times 0.9}{420} = 3.0 \times \frac{10^{-3}\text{m}^2}{\text{m}}$$
$$= 3000\,\text{mm}^2/\text{m}$$

$A_{s,min} > A_s$ required by analysis. Use $A_s = A_{s,min} = 3000$ mm^2/m

Transverse reinforcement for zone II:

$$net\ q_{\text{factored}} = \frac{P_2}{B_2 \times B} = \frac{3220}{2 \times 2.4} = 670.83\,\text{kPa}$$

$$M_u = \frac{net\ q_{\text{factored}} \times l_1^2}{2} = \frac{670.83 \times 0.9^2}{2} = 271.7\,\text{kN.m/m} \ < M_u \text{ of zone I.}$$

Therefore, $A_{s,min}$ controls. Use $A_s = 3000$ mm^2/m

Transverse reinforcement for zone III:

Use $A_s = A_{s,\min} = 3000\,\text{mm}^2/\text{m}$

Thus, in this case, for all the three zones the same amount of reinforcement will be required.

$$A_{s,\text{total}} = 3000 \times L = 3000 \times 7.8 = 23\ 400\,\text{mm}^2$$

Try 46 No. 25 bars: $A_{s,\text{provided}} = 46 \times 510 = 23\ 460$ mm^2.

ACI Sections 7.6.5 and 10.5.4 require maximum spacing shall not exceed three times the slab or footing thickness or 450 mm, whichever is smaller.

Use centre to centre bar spacing = 169 mm. Check the 75 mm minimum concrete cover at each side:

$$\text{Concrete cover} = \frac{7800 - (45 \times 169 + 25.4)}{2} = 84.8\,\text{mm} > 75\,\text{mm} \quad \text{(OK.)}$$

Thus, theoretically, the bars will be uniformly distributed with 169 mm centretocentre.

However, it may be more realistic if the amount of steel in the effective zones I and II are somewhat increased and that in zone III decreased. We shall do this by arbitrarily decreasing and increasing the bar spacing in the three zones but without decreasing the required total steel amount, as follows:

Zone I: Eight No.25 bars; seven spacings @ 142 mm c.c.
Zone II: 15 No.25 bars;14 spacings @ 141 mm c.c.
Zone III$_{(a)}$: 19 No.25 bars;18 spacings @ 226 mm c.c.
Zone III$_{(b)}$: Four No.25 bars; three spacings @ 166 mm c.c.

Check the development of reinforcement:

In this case, the bars are in tension, the provided bar size is No. 25, the clear spacing of the bars exceeds $2d_b$, and the clear cover exceeds d_b. Therefore, the development length l_d of bars at each side of the critical section (which is located where the maximum factored positive moment exists) shall be determined from the following equation, but not less than 300 mm:

$$l_d = \left(\frac{f_y \psi_t \psi_e}{1.7\lambda\sqrt{f'_c}}\right) d_b \quad \text{(ACI Sections 12.2.1, 12.2.2 and 12.2.4)}$$

(*Continued*)

where the factors ψ_t, ψ_e and λ are 1 (ACI Sections 12.2.4 and 8.6.1)

$$l_d = \left(\frac{420 \times 1 \times 1}{1.7 \times 1 \times \sqrt{21}}\right)\left(\frac{25.4}{1000}\right) = 1.37\ \text{m} = 1370\ \text{mm} > 300\ \text{mm}.$$

Therefore, the required $l_d = 1370$ mm.

The bar extension past the critical section (i.e. the available length) is

$$l_1\ (\text{or}\ l_2) - 75\ \text{mm cover} = 900 - 75 = 825\ \text{mm} < l_{d,min}\quad (\text{Not OK.})$$

The bottom transverse bars will all have to be hooked at both ends.

Because *two-way shear cracks* would extend roughly the entire width of the footing, the hooked transverse bars provide adequate anchorage outside the inclined cracks (MacGregor and Wight, 2005).

Step 6. Check columns bearing on the footing (ACI Section 10.14).

The design proceeds in the same manner as that of Solution of Problem 5.4, *Step 6.*

Step 7. Design dowels to satisfy the requirements of moment transfer and minimum reinforcement area across interface (ACI Section 15.8.2.1).

The design proceeds in the same manner as that of Solution of Problem 5.4, *Step 7.*

Step 8. Find the embedment length of dowels in both the footing and the column.

The design proceeds in the same manner as that of Solution of Problem 5.4, *Step 8.*

Step 9. Develop the final design sketch as shown in the scheme below.

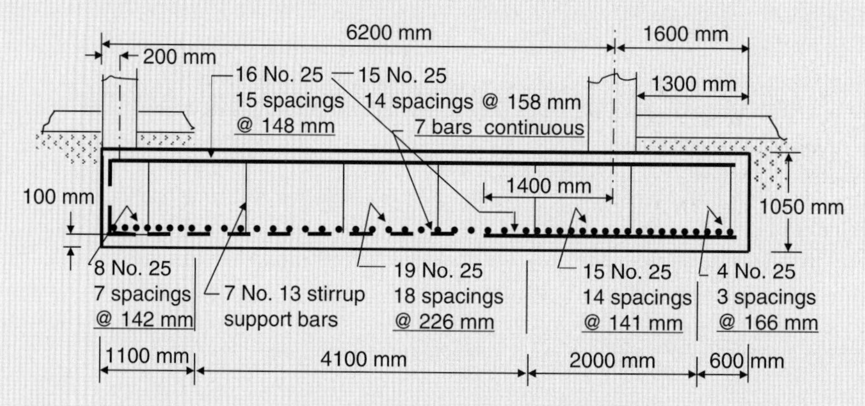

Scheme 5.61

Problem 5.13

The following design data belong to a combined footing which supports two columns. The distance between the columns is 6 m, centre to centre. The exterior face of the exterior column (Col. 1) is located right on the property line. The interior column (Col. 2) has too limited space for a centrally loaded spread footing. The net allowable soil pressure (*net* q_a) is 151 kPa at the expected foundation depth. Seeking a uniform soil pressure distribution under the footing, select either a rectangular or a trapezoid combined footing, and then design the footing using the conventional method.

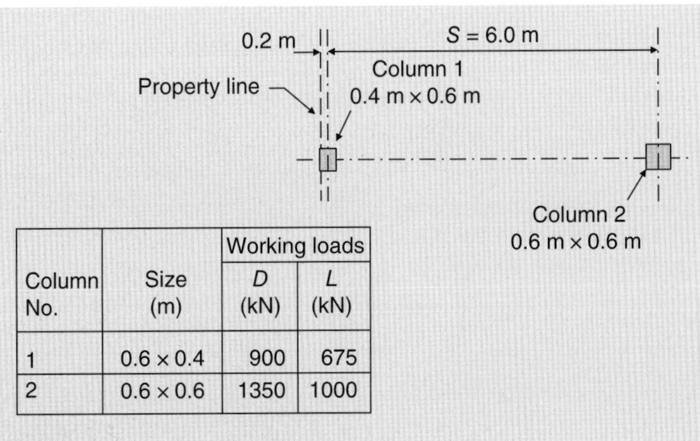

Column No.	Size (m)	Working loads	
		D (kN)	L (kN)
1	0.6 × 0.4	900	675
2	0.6 × 0.6	1350	1000

Footing:

Concrete strength $f'_c = 30$ Mpa
Reinforcing steel $f_y = 420$ Mpa

Scheme 5.62

Solution:

Step 1. Locate the resultant of the loads and find the distance x' shown in the figure below.

Because the space is too limited at column 2, it is desirable not to extend the footing beyond the column. Therefore,

$$L = 6.0 + 0.2 + 0.3 = 6.5 \text{ m}$$

$R = 3925$ kN
$P_1 = 1575$ kN
$P_2 = 2350$ kN
$S = 6.0$ m
$L = 6.5$ m

L; 0.2 m; 0.3 m; x′; P1; R; P2; Column 1; Column 2; S

Scheme 5.63

The loads resultant R must pass through the centroid of the footing base area which is located at distance x' from the right face of the interior column.

$$x' = \frac{(P_1)(6+0.3) + (P_2)(0.3)}{R} = \frac{1575 \times 6.3 + 2350 \times 0.3}{3925} = 2.71 \text{ m}$$

$L/3 < x' < L/2$. This means that the resultant R is much closer to the heavier column, and doubling the centroid distance x', as it was done for the rectangular combined footing of Problem 5.12, will not provide sufficient footing length to reach the other column. Therefore, in this case, a trapezoid combined footing is required.

Select a trapezoid combined footing.

Step 2. Find footing dimensions B_1 and B_2 and compute *net* q_{factored}.

In order to obtain uniform contact pressure, locations of the centroid and R must coincide, as shown in the scheme below.

(*Continued*)

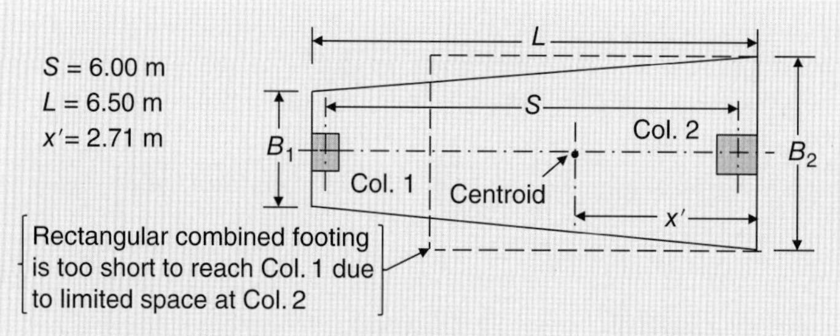

Scheme 5.64

Assume neglecting any weight increase from displacing the lighter soil with heavier concrete, since it is usually too small compared to the foundation loads.

$$\text{The foundation net load} = R = 3925\,\text{kN}$$

$$\text{The footing base area} = A = \frac{R}{net\ q_a} = \frac{3925}{151} = 26.0\,\text{m}^2$$

Equation (5.38):

$$A = \frac{B_1 + B_2}{2}L$$

$$26 = \frac{B_1 + B_2}{2}(6.5) = (3.25)(B_1 + B_2)$$

$$B_2 = 8 - B_1$$

Equation (5.39):

$$x' = \frac{L}{3}\left(\frac{2B_1 + B_2}{B_1 + B_2}\right)$$

$$\frac{6.5}{3}\left(\frac{2B_1 + B_2}{B_1 + B_2}\right) = \frac{6.5}{3}\left(\frac{2B_1 + 8 - B_1}{B_1 + 8 - B_1}\right) = \frac{2.167B_1 + 17.333}{8} = 2.71$$

$$B_1 = \frac{4.347}{2.167} = 2.0\ m$$

$$B_2 = 8 - 2 = 6.0\ m$$

Factored $P_1 = 1.2 \times 900 + 1.6 \times 675 = 2160\ \text{kN}$
Factored $P_2 = 1.2 \times 1350 + 1.6 \times 1000 = 3220\ \text{kN}$
Factored $R = 5380$ kN

$$net\ q_{\text{factored}} = \frac{\text{Factored}\,R}{A} = \frac{5380}{26} = 206.92\,\text{kPa}.$$

Use $net\ q_{\text{factored}} = 207\ kPa$
Check x' using factored loads:

$$x' = \frac{(2160)(6 + 0.3) + (3220)(0.3)}{5380} = \frac{14574}{5380} = 2.71\,\text{m} \quad \text{(OK)}$$

Step 3. Obtain data for shear-force and bending-moment diagrams considering the combined footing as a reinforced concrete beam.

It is preferable to draw the complete load diagram first, and then the shear and moment diagrams, as shown below. The factored distributed loads on the footing, shear forces and bending moments are computed for full width (variable) of the footing at necessary locations. Obviously, the contact pressure per unit length of the footing will vary linearly due to the varying width from B_1 to B_2. Therefore, the shear diagram will be a second-degree curve and the moment diagram will be a third-degree curve.

Any convenient method, for example integral calculus (with attention to values at the limits), may be used for calculating the shear and moment values at different locations necessary to obtain informable diagrams. Usually, the design calculations involve an enormous amount of busywork, and more tedious than that with a rectangular combined footing. Therefore, it may be preferable to use any available computer program concerns the conventional design of combined footings.

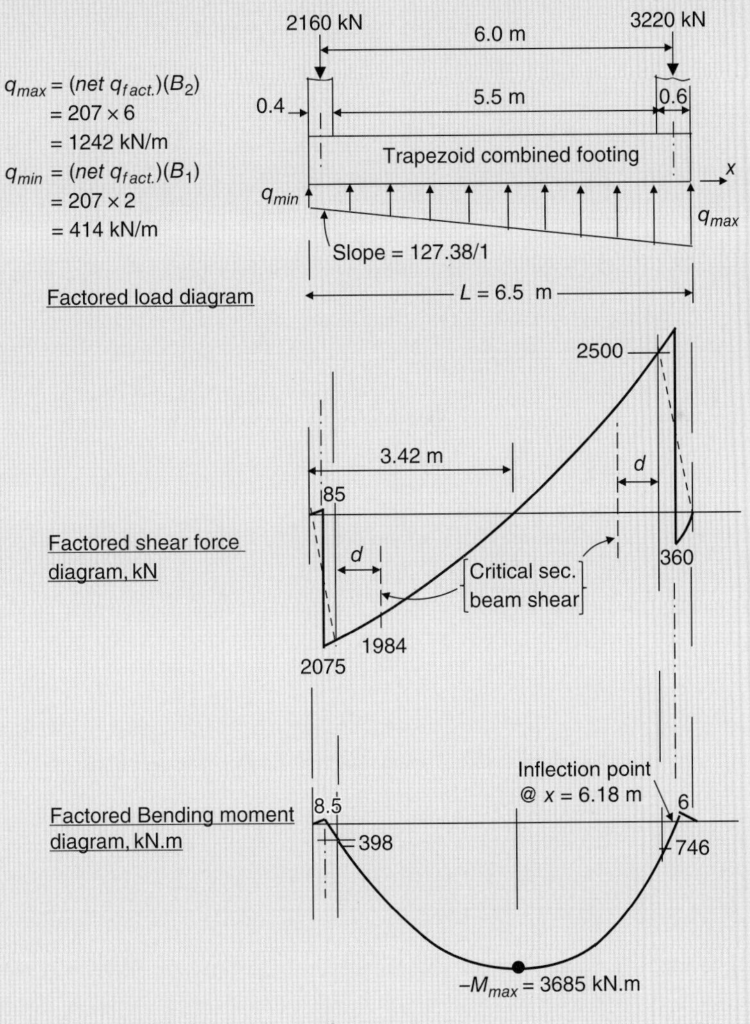

Scheme 5.65

Step 4. Find footing thickness h.

Determine the required thickness based on both one-way shear and two-way shear analyses.

(*Continued*)

(a) One-way shear. Since the footing width varies, we should consider shear at two critical sections; one at distance d from the right face of the exterior column, and the other at distance d from the left face of the interior column, as indicated on the shear-force diagram. However, in this case, one may detect location of the most critical shear by comparing the B ratio with the ratio of shear forces at the column faces, as follows:

The width B ratio (approximately) $= \frac{6}{2} = 3$

The shear-force ratio $= \frac{2500}{1984} = 1.26$

Since the width ratio is much larger than the shear-force ratio, the footing depth d will probably be based on shear at distance d from the interior face of the exterior column. Based on this reasoning, d is calculated as follows:

$$V_u = 414(0.4+d) + 127.38(0.4+d)^2(0.5) - 2160$$

$$= 165.6 + 414d + 10.19 + 50.95d + 63.69d^2 - 2160$$

$$= 464.95d + 63.69d^2 - 1984.21 \quad \text{(kN)}$$

It may be noticed from the shear diagram that at this critical section V_u is negative. Therefore, it would be necessary to use it positive when equated to the shear strength.

Equation (5.19):

$$\varnothing\, V_c \geq V_u \,(\text{taking reinforcement shear strength, } V_s = 0)$$

$$\text{where } \varnothing = \text{strength reduction factor (ACI Section 9.3.2.3)}$$

$$= 0.75 \text{ for shear}$$

Equation (5.20):

$$V_c = 0.17\lambda\sqrt{f_c'}\, b_w d$$

$$b_w = 2 + \frac{0.4+d}{6.5} \times 4 = 2 + 0.25 + 0.62d = 2.25 + 0.62d$$

$$\varnothing\, V_c = (0.75 \times 0.17 \times 1)(\sqrt{30})(1000)(2.25 + 0.62d)(d)$$

$$= 1571\, d + 433\, d^2$$

Let $\varnothing\, V_c = V_u$:

$$1571\, d + 433\, d^2 = -(464.95d + 63.69d^2 - 1984.21)$$

$$497\, d^2 + 2036\, d - 1984.21 = 0$$

$$d^2 + 4d - 4 = 0$$

$$d = \frac{-4 \pm \sqrt{4^2 - 4 \times 1(-4)}}{2 \times 1} = \frac{-4 + 5.66}{2} = 0.83 \text{ m}$$

Since there will be moment transfer at the column regions, the two-way shear may require d larger than 0.83 m.

Try $d = 1.0$ m $= 1000$ mm

(b) Two-way shear at location of each column. If it is not very clear at which column this shear is most critical, both locations should be considered. The shear perimeter at each column has three sides only, as shown in the scheme below. One may detect location of the most critical shear by comparing the shear- perimeter ratio with the shear-force ratio.

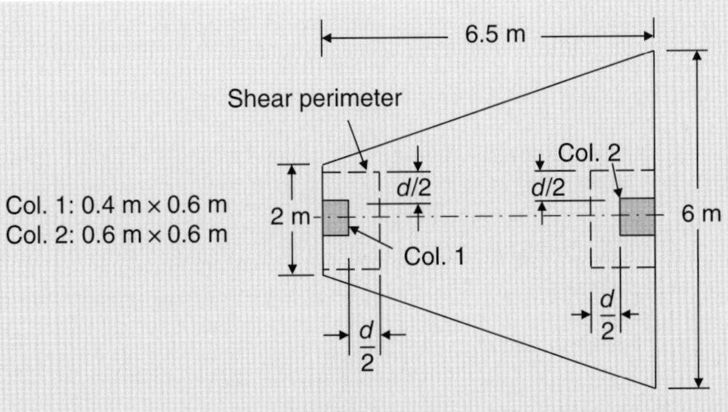

Scheme 5.66

Assume $d = 1.0$ m.

$$\text{The shear-perimeter ratio } = \frac{2(0.6+0.5)+(0.6+1)}{2(0.4+0.5)+(0.6+1)} = 1.12$$

$$\text{The shear-force ratio } = \frac{3220-207(0.6+0.5)(0.6+1)}{2160-207(0.4+0.5)(0.6+1)} = \frac{2855.68}{1861.92} = 1.53$$

Since the shear-perimeter ratio is much smaller than the shear-force ratio, the most critical two-way shear will be at the interior column.

Shear strength of concrete V_c shall be the smallest of (a), (b) and (c):

(a)

$$V_c = 0.17\left(1+\frac{2}{\beta}\right)\lambda\sqrt{f_c'}b_o d \qquad \text{Equation (5.22)}$$

$$= (0.17)\left(1+\frac{2}{1.5}\right)(1)\sqrt{f_c'}b_o d = (0.397)\sqrt{f_c'}b_o d$$

(b)

$$V_c = 0.083\left(\frac{\alpha_s d}{b_o}+2\right)\lambda\sqrt{f_c'}b_o d$$

$$= (0.083)\left(\frac{30d}{b_o}+2\right)(1)\sqrt{f_c'}b_o d \qquad \text{Equation (5.23)}$$

$$= (0.083)\left[\frac{30\times 1}{2(0.6+0.5)+(0.6+1)}+2\right]\sqrt{f_c'}b_o d = (0.821)\sqrt{f_c'}b_o d$$

(c)

$$V_c = 0.33\lambda\sqrt{f_c'}b_o d \qquad \text{Equation (5.24)}$$

$$= 0.33(1)\sqrt{f_c'}b_o d = (0.33)\sqrt{f_c'}b_o d$$

$$V_c = (0.33)\sqrt{f_c'}b_o d.$$

$$\text{Use } \varnothing V_c = (0.75)(0.33)\sqrt{30}(1000)[2(0.6+0.5)+(0.6+1)](1)$$

$$= 5151.33\,\text{kN}$$

The condition of unbalanced moment transfer exists, and the design must satisfy the requirements of ACI Sections 11.11.7.1, 11.11.7.2 and 13.5.3. The design proceeds in the same manner as that for the three-sided shear perimeter at location of the exterior column of Problem 5.12.

(Continued)

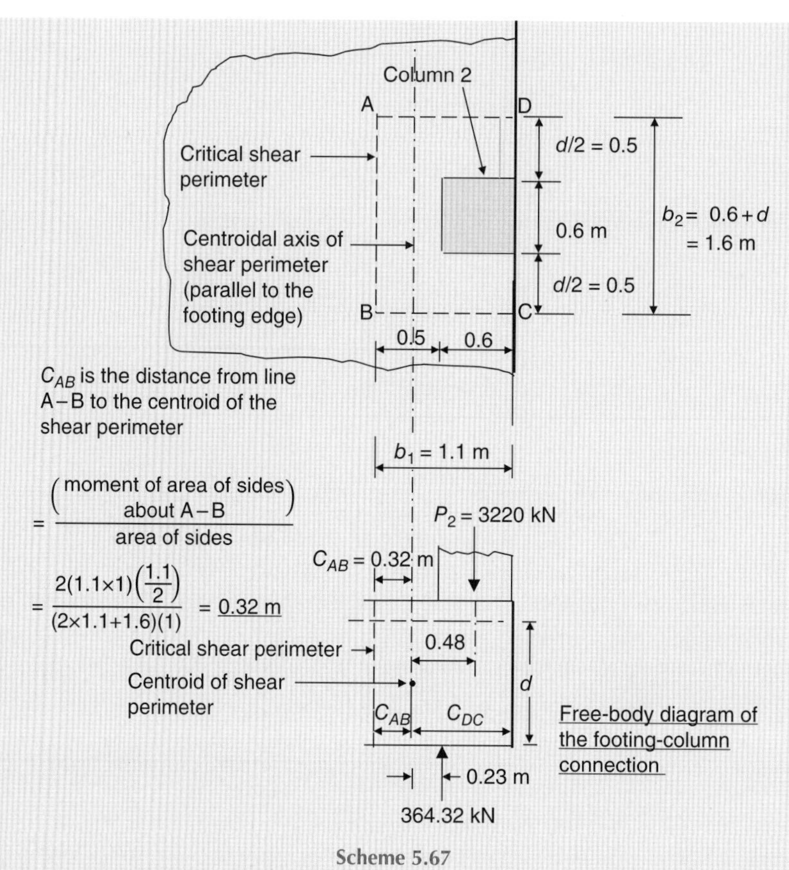

Scheme 5.67

Soil reaction on the shear block = 207(1.1 × 1.6) = 364.32 kN

$$C_{DC} = b_1 - C_{AB} = 1.1 - 0.32 = 0.78\,\text{m}$$

Summation of moments about the centroid of the shear perimeter equals the unbalanced moment. Hence,

$$M_u = 3220 \times 0.48 - 364.32 \times 0.23 = 1461.81\,\text{kN.m}$$

$$\gamma_v = \left(1 - \gamma_f\right) \quad \text{(ACI Section 11.11.7.1)}$$

$$\gamma_f = \frac{1}{1 + (2/3)\sqrt{b_1/b_2}} \quad \text{(ACI Section 13.5.3.2)}$$

$$= \frac{1}{1 + (2/3)\sqrt{1.1/1.6}} = 0.644$$

$$\gamma_v = 1 - 0.644 = 0.356$$

The shear stresses due to the direct shear and the shear due to moment transfer will add at points D and C, giving the largest shear stresses on the critical shear perimeter. Hence,

$$v_{u(DC)} = \frac{V_u}{b_o d} + \frac{\gamma_v M_u C_{DC}}{J_c} \quad \text{(ACI Section R11.11.7.2)}$$

$J_c = J_z$ = Property of the shear perimeter analogous to polar moment of inertia

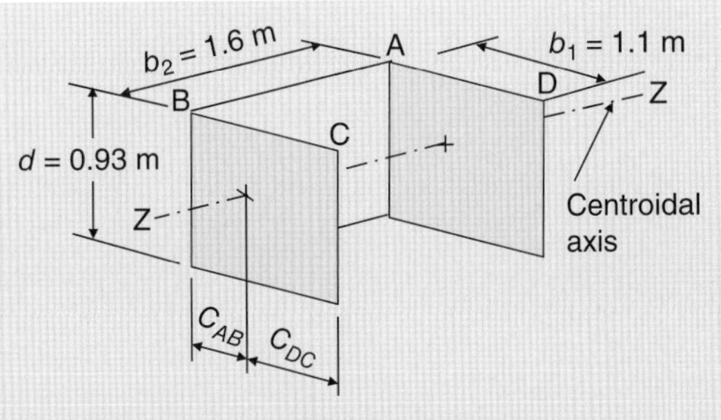

Scheme 5.68

$$J_c = 2\left(\frac{b_1 d^3}{12}\right) + 2\left(\frac{d b_1^3}{12}\right) + 2(b_1 d)\left(\frac{b_1}{2} - C_{AB}\right)^2 + (b_2 d) C_{AB}{}^2$$

$$= 2\left(\frac{1.1 \times 1^3}{12}\right) + 2\left(\frac{1 \times 1.1^3}{12}\right) + 2(1.1 \times 1)\left(\frac{1.1}{2} - 0.32\right)^2 + (1.6 \times 1)(0.32^2)$$

$$= 0.183 + 0.222 + 0.116 + 0.164 = 0.685\,\text{m}^4$$

$$v_{u(DC)} = \frac{3220 - 364.32}{(2 \times 1.1 + 1.6)(1)} + \frac{0.356 \times 1461.81(0.78)}{0.685}$$

$$= 751.49 + 592.58 = 1344.07\,\text{kN/m}^2$$

$$\text{Ø}\,v_c = \frac{\text{Ø}\,V_c}{b_o d} = \frac{5151.33}{[2(0.6+0.5) + (0.6+1)](1)} = 1355.61\,\text{kN/m}^2 > v_{u(DC)} \quad \text{(OK.)}$$

The factored shear stress $v_{u(DC)}$ is too close to Ø v_c (the choice of using d = 1 m was successful), which means that two-way shear requires $d \cong 1$ m, whereas, one-way shear requires $d = 0.83$ m and, therefore, the two-way shear controls the footing depth.

Use d = 1.0 m = 1000 mm.

The depth d will be taken to the centre of the steel bars in long direction.

Assume using No. 29 bars, and 75 mm minimum concrete cover (ACI Section 7.7.1). Hence, the overall footing thickness is

$$h = 1000 + 75 + 1/2\,\text{bar diameter} = 1075 + 14.4 = 1089.4\,\text{mm}.$$

Use h = 1010 mm.

Step 5. Design the flexural reinforcement in the long direction.

The bending-moment diagram requires that the flexural reinforcement in long direction must be placed at top of the footing. The maximum factored negative moment of 3685 kN.m is located at $x = 3.42$ m (from the left face of the exterior column). At this location the footing width is

$$B = 2 + \frac{3.42}{6.5} \times 4 = 4.1\,\text{m}$$

Equation (5.9):

$$\text{Ø}\,M_n \geq M_u$$

$$M_u = \frac{3685}{B} = \frac{3685}{4.1} = 899\ \text{kN.m/m}$$

Assume tension-controlled section, Ø = 0.9, and $f_s = f_y$.

(*Continued*)

$$a = \frac{A_s f_y}{0.85 f_c' b} = \frac{A_s \times 420}{0.85 \times 30 \times 1} = 16.47 A_s$$

$$\varnothing M_n = \varnothing \left[A_s f_y \left(d - \frac{a}{2}\right)\right]$$

$$= (0.9)\left[(A_s \times 420 \times 1000)\left(1 - \frac{16.47 A_s}{2}\right)\right] \text{ Equation (5.11)}$$

$$= 378\ 000\, A_s - 3\ 112\ 830 A_s^2 (\text{kN.m/m})$$

Let $\varnothing M_n = M_u$:

$$378\ 000\, A_s - 3\ 112\ 830 A_s^2 = 899$$

$$A_s^2 - 0.121 A_s + 2.888 \times 10^{-4} = 0$$

$$A_s = \frac{-(-0.121) \pm \sqrt{(-0.121)^2 - (4)(1)(2.888 \times 10^{-4})}}{2 \times 1} = \frac{0.121 - 0.116}{2}$$

$$= 2.5 \times 10^{-3}\ \text{m}^2/\text{m} = 2500\ \text{mm}^2/\text{m}$$

$$A_{s,min} = \frac{0.25\sqrt{f_c'}}{f_y} b_w d = \frac{0.25\sqrt{30}}{420} \times 1 \times 1 = 3.26 \times 10^{-3}\ \text{m}^2/\text{m}$$

$$= 3260\ \text{mm}^2/\text{m}$$

and not less than $1.4\dfrac{b_w d}{f_y} = \dfrac{1.4 \times 1 \times 1}{420}$

$$= 3.333 \times 10^{-3}\ \text{m}^2/\text{m} = 3333\ \text{mm}^2/\text{m}$$

Therefore, use $A_s = A_{s,min} = 3333\ \text{mm}^2/\text{m}$

$$A_{s,total} = 3333 \times B = 3333 \times 4.1 = 13\ 667\ \text{mm}^2$$

Try 22 No. 29 bars: $A_{s,provided} = 22 \times 645 = 14\ 190\ \text{mm}^2/\text{m}$ (OK.)

ACI Sections 7.6.5 and 10.5.4 requires maximum spacing shall not exceed three times the slab or footing thickness, or 450 mm, whichever is smaller.

Use centre to centre bar spacing = 186 mm. Check the 75 mm minimum concrete cover at each side:

$$\text{Concrete cover} = \frac{4100 - (21 \times 186 + 28.7)}{2} = 82.7\ \text{mm} > 75\ mm\ (\text{OK.})$$

Provide 22 No. 29 bars @ 188 mm c.c. in long direction, at top across the footing width where B = 4.1 m.

Since the footing is subjected to negative bending moment for its full length and $A_{s,min}$ controls, the same amount of reinforcement per meter width of the footing shall be provided at top in long direction for the full footing length. Also, since the footing width varies, one should compute total A_s for several locations so that cutting of the steel bars (if justified) can be done properly as required. The necessary computations are left for the reader.

Also, checking the development of reinforcement, at the critical sections, is left for the reader.

Step 6. Design the transverse reinforcement at bottom of the footing.

The footing is divided into three zones or strips of the defined widths shown in the figure below. Zones I and II, usually known as effective zones, should be analysed as beams; the provided steel should not be less than that required for bending or $A_{s,min}$, whichever is greater. For zone III (between zones I and II), the provided steel should satisfy the $A_{s,min}$ requirement only. All the transverse steel bars should be placed on top of the supporting bars in long direction at the bottom of the footing.

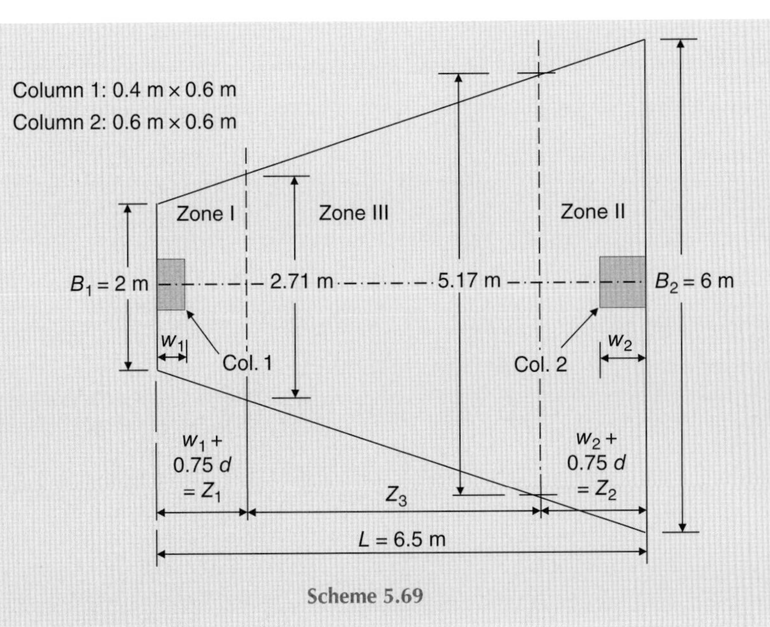

Scheme 5.69

Zone I: $Z_1 = 0.4 + 0.75 \times 1 = 1.15$ m
Zone II: $Z_2 = 0.6 + 0.75 \times 1 = 1.35$ m
Zone III: $Z_3 = 6.5 - (1.15 + 1.35) = 4$ m

$$\text{Zone I: average footing width } = \frac{2 + \left(2 + \dfrac{4}{6.5} \times 1.15\right)}{2} = 2.35\,\text{m}$$

$$\text{Zone II: average footing width } = \frac{6 + \left(2 + \dfrac{4}{6.5} \times 5.15\right)}{2} = 5.58\,\text{m}$$

$$\text{Zone III: average footing width } = \frac{\left(2 + \dfrac{4}{6.5} \times 1.15\right) + \left(2 + \dfrac{4}{6.5} \times 5.15\right)}{2} = 3.94\,\text{m}$$

(a) Transverse reinforcement for Zone I:

$$net\ q_{\text{factored}} = \frac{\text{factored } P_1}{\text{zone area}} = \frac{2160}{Z_1 \times \text{average width}} = \frac{2160}{1.15 \times 2.35} = 799.26\,\text{kPa}$$

$$l_1 = \frac{\text{average width} - 0.6}{2} = \frac{2.35 - 0.6}{2} = 0.875\,\text{m}$$

$$M_u = \frac{net\ q_{\text{factored}} \times l_1^2}{2} = \frac{799.26 \times 0.875^2}{2} = 306\,\text{kN.m/m}$$

Assume tension-controlled section, $\varnothing = 0.9$, and $f_s = f_y$.

$$a = \frac{A_s f_y}{0.85 f_c' b} = \frac{A_s \times 420}{0.85 \times 30 \times 1} = 16.47 A_s$$

$$d = 1\,\text{m} - \text{one bar diameter} = 1 - 0.025 = 0.975\,\text{m}$$

(Continued)

$$\varnothing M_n = \varnothing\left[A_s f_y\left(d-\frac{a}{2}\right)\right]$$

$$= (0.9)\left[(A_s \times 420 \times 1000)\left(0.975-\frac{16.47A_s}{2}\right)\right] \text{ Equation (5.11)}$$

$$= 368550\,A_s - 3112830\,A_s^2 \text{ (kN.m/m)}$$

$$\varnothing M_n \geq M_u$$

Let $\varnothing M_n = M_u$:

$$368550\,A_s - 3112830\,A_s^2 = 306$$

$$A_s^2 - 0.118\,A_s + 9.83 \times 10^{-5} = 0$$

$$A_s = \frac{-(-0.118) \pm \sqrt{(-0.118)^2-(4)(1)(9.83\times10^{-5})}}{2\times1} = \frac{0.118-0.116}{2}$$

$$= 1\times10^{-3}\,\text{m}^2/\text{m} = 1000\,\text{mm}^2/\text{m}$$

$$A_{s,\min} = \frac{0.25\sqrt{f_c'}}{f_y}b_w d = \frac{0.25\sqrt{30}}{420}\times1\times0.975 = 3.18\times10^{-3}\,\text{m}^2/\text{m}$$

and not less than $1.4\dfrac{b_w d}{f_y} = \dfrac{1.4\times1\times0.975}{420} = 3.325\times10^{-3}\,\text{m}^2/\text{m}$

Therefore, use $A_s = A_{s,\min} = 3.325 \times 10^{-3}$ m²/m = 3325 mm²/m

$$A_{s,\text{total}} = 3325 \times Z_1 = 3325 \times 1.15 = 3824\,\text{mm}^2$$

Try eight No. 25 bars: $A_{\text{provided}} = 8 \times 510 = 4080$ mm² (OK.)

Check the development of reinforcement:

In this case, the bars are in tension, the provided bar size is No. 25, the clear spacing of the bars exceeds $2d_b$, and the clear cover exceeds d_b. Therefore, the development length l_d of bars at each side of the critical section (which is located where the maximum factored positive moment exists) shall be determined from the following equation, but not less than 300 mm:

$$l_d = \left(\frac{f_y\psi_t\psi_e}{1.7\lambda\sqrt{f_c'}}\right)d_b \quad \text{(ACI Sections 12.2.1, 12.2.2 and 12.2.4)}$$

Where the factors ψ_t, ψ_e and λ are 1 (ACI Sections 12.2.4 and 8.6.1)

$$l_d = \left(\frac{420\times1\times1}{1.7\times1\times\sqrt{30}}\right)\left(\frac{25.4}{1000}\right) = 1.15\,\text{m} = 1150\,\text{mm} > 300\,mm$$

Therefore, the required l_d = 1150 mm.

The bar extension past the critical section (i.e. the shortest available length) is

$$\frac{B_1-600}{2} - 75\,\text{mm cover} = \frac{2000-600}{2} - 75 = 625\,\text{mm} < l_{d,\min} \quad \text{(Not OK.)}$$

Therefore, thebars will all have to be hooked (90-degree standard hook) at both ends.

Check l_{dh} (ACI Section 12.5.2):

$$l_{dh} = \left(\frac{0.24\psi_e f_y}{\lambda\sqrt{f'_c}}\right)d_b$$

where the factors ψ_e and λ shall be taken as 1.

$$l_{dh} = \left(\frac{0.24\times1\times420}{1\times\sqrt{30}}\right)\left(\frac{25.4}{1000}\right) = 0.47\ \text{m} = 470\ \text{mm} > 8d_b > 150\ mm$$

Therefore, the required $l_{dh} = 470$ mm.

$$l_{dh} < 625\ \text{mm}\quad \text{(OK.)}$$

ACI Sections 7.6.5 and 10.5.4 require maximum spacing shall not exceed three times the slab or footing thickness or 450 mm, whichever is smaller.

Use centre-to-centre bar spacing = 150 mm. Check the 75 mm minimum concrete cover at the small end of the footing:

$$\text{Concrete cover} = 1150 - 7\times150 - \left(\frac{25.4}{2}\right)$$
$$= 87.3\ \text{mm} > 75\ mm \qquad \text{(OK.)}$$

Provide eight No. 25 bars @ 150 mm c.c. in Zone I, placed at bottom of the footing in the transverse direction. The bars will all have to be hooked (90-degree standard hook) at both ends.

(b) Transverse reinforcement for zone II:

$$net\ q_{\text{factored}} = \frac{\text{factored}\ P_2}{\text{zone area}} = \frac{3220}{Z_2\times\text{average width}} = \frac{3220}{1.35\times5.58} = 427.45\ \text{kPa}$$

$$l_2 = \frac{\text{average width} - 0.6}{2} = \frac{5.58-0.6}{2} = 2.49\ \text{m}$$

$$M_u = \frac{net\ q_{\text{factored}}\times l_2^2}{2} = \frac{427.45\times2.49^2}{2} = 1325.12\ \text{kN.m/m}$$

$$d = 1\ \text{m} - \text{one bar diameter} = 1 - 0.025 = 0.975\ \text{m}$$

Assume tension-controlled section, $\text{Ø} = 0.9$, and $f_s = f_y$.

$$a = \frac{A_s f_y}{0.85 f'_c b} = \frac{A_s\times420}{0.85\times30\times1} = 16.47A_s$$

$$\text{Ø}\,M_n = \text{Ø}\left[A_s f_y\left(d - \frac{a}{2}\right)\right]$$

$$= (0.9)\left[(A_s\times420\times1000)\left(0.975 - \frac{16.47A_s}{2}\right)\right]\quad \text{Equation (5.11)}$$

$$= 368\ 550\,A_s - 3\ 112\ 830\,A_s^2\ \text{(kN.m/m)}$$

$$\text{Ø}\,M_n \geq M_u$$

Let $\text{Ø}\,M_n = M_u$:

$$368\ 550\,A_s - 3\ 112\ 830\,A_s^2 = 1325.12$$

(Continued)

$$A_s^2 - 0.118A_s + 4.26\times10^{-4} = 0$$

$$A_s = \frac{-(-0.118)\pm\sqrt{(-0.118)^2-(4)(1)(4.26\times10^{-4})}}{2\times1} = \frac{0.118-0.111}{2}$$

$$= 3.5\times10^{-3}\,\mathrm{m^2/m} = 3500\,\mathrm{mm^2/m}$$

$$A_{s,\min} = \frac{0.25\sqrt{f_c'}}{f_y}b_w d = \frac{0.25\sqrt{30}}{420}\times1\times0.975 = 3.18\times10^{-3}\,\mathrm{m^2/m}$$

and not less than $1.4\dfrac{b_w d}{f_y} = \dfrac{1.4\times1\times0.975}{420} = 3.325\times10^{-3}\,\mathrm{m^2/m} < A_s$

Therefore, use $A_s = 3.5\times10^{-3}$ m²/m = 3500 mm²/m

$$A_{s,\mathrm{total}} = 3500\times Z_2 = 3500\times1.35 = 4725\,\mathrm{mm^2}$$

Try 10 No. 25 bars: $A_{s,provided} = 10\times510 = 5100$ mm² (OK.)

Compute a for $A_s = \dfrac{5100}{1.35} = 3778\,\mathrm{mm^2/m}$, and check if $f_s = f_y$ and whether the section is tension-controlled:

$$a = 16.47A_s = 16.47\times3778\times10^{-6} = 0.062\,\mathrm{m} = 62\,\mathrm{mm}$$

$c = \dfrac{a}{\beta_1}$. For f_c' between 17 and 28 MPa, β_1 shall be taken as 0.85. For f_c' above 28 MPa, β_1 shall be reduced linearly at a rate of 0.05 for each 7 MPa of strength in excess of 28 MPa, but β_1 shall not be taken less than 0.65 (ACI Section 10.2.7.3).

$$\beta_1 = 0.85 - \frac{2}{7}\times0.05 = 0.836$$

$$c = \frac{62}{0.836} = 74.16\,\mathrm{mm}$$

$$d_t = d = 975\,\mathrm{mm}$$

$$\varepsilon_t = 0.003\left(\frac{d_t - c}{c}\right) \qquad \text{Equation (5.15)}$$

$$= 0.003\left(\frac{975-74.16}{74.16}\right) = 0.036 > 0.005.$$

Hence, the section is tension-controlled, and Ø = 0.9 (ACI Sections 10.3.4 and 9.3.2.1).

$$\varepsilon_y = \frac{f_y}{E_s} = \frac{420}{200\,000} = 0.0021$$

Hence, $\varepsilon_t > \varepsilon_y$ and $f_s = f_y$.

Therefore, the assumptions made are satisfied.

Check the development of reinforcement:

$$l_d = \left(\frac{f_y\psi_t\psi_e}{1.7\lambda\sqrt{f_c'}}\right)d_b \quad \text{(ACI Sections 12.2.1, 12.2.2 and 12.2.4)}$$

where the factors ψ_t, ψ_e and λ are 1 (ACI Sections 12.2.4 and 8.6.1)

$$l_d = \left(\frac{420\times1\times1}{1.7\times1\times\sqrt{30}}\right)\left(\frac{25.4}{1000}\right) = 1.15\,\mathrm{m} = 1150\,\mathrm{mm} > 300\,mm$$

Therefore, the required $l_d = 1150$ mm.

The bar extension past the critical section (i.e. the shortest available length) is

$$\frac{5170-600}{2} - 75\,\text{mm cover} = 2210\,\text{mm} \gg l_d \quad \text{(OK.)}$$

In case the required flexural reinforcement exceeds the minimum flexural reinforcement, it shall be adequate to use the same maximum spacing of reinforcement for slabs which is two times the slab thickness, or 450 mm whichever is smaller, as specified by ACI Section 13.3.2.

Use centretocentre bar spacing = 140 mm. Check the 75 mm minimum concrete cover at the large end of the footing:

$$\text{Concrete cover} = 1350 - 9 \times 140 - \left(\frac{25.4}{2}\right)$$

$$= 77.3\,\text{mm} > 75\,\text{mm} \qquad \text{(OK.)}$$

Provide 10 No. 25 bars @ 140 mm c.c. in Zone II, placed at bottom of the footing in the transverse direction.

(c) Transverse reinforcement for zone III:

Use $A_s = A_{s,\min} = 3325\ \text{mm}^2/\text{m}$

$$A_{s,\text{total}} = 3325 \times Z_3 = 3325 \times 4 = 13\ 300\,\text{mm}^2$$

Try 26 No. 25 bars: $A_{s,\text{provided}} = 26 \times 510 = 13260\ \text{mm}^2$ (Acceptable)

Use centretocentre bar spacing $= \dfrac{4000}{25} = 160\,\text{mm}$.

By inspection, the development of reinforcement is not a problem.

Provide 26 No. 25 bars @ 160 mm c.c. in Zone III, placed at bottom of the footing in the transverse direction.

Step 7. Check columns bearing on the footing (ACI Section 10.14), and design the necessary dowels.

The design proceeds in the same manner as that of Solution of Problem 5.3, *Step 5.*

Step 8. Find the embedment length of dowels in both the footing and the columns.

The design proceeds in the same manner as that of Solution of Problem 5.3, *Step 6.*

Step 9. Develop the final design sketches.

Problem 5.14

The design data given in Scheme 5.70 belong to a strap footing which will support two columns. The distance between the columns is 5.5 m, centre to centre. Centre of the exterior column (Col. 1) is located at a distance of 0.4 m from the property line where the footing edge will be located. At location of the interior column (Col. 2), there will be a sufficient space for a centrally loaded spread footing. At the expected foundation depth, the net allowable soil pressure (*net* q_a) is 150 kPa. In this Problem, it is required to design the strap as beam but not "deep" beam, and the minimum stiffness ratio ($I_{\text{strap}}/I_{\text{footing}}$) equals 1.5. Design the strap footing using the conventional (rigid) method.

Solution:

Step 1. Compute working and factored column loads, their resultants, and factored net contact pressure (*net* q_{factored}).

(Continued)

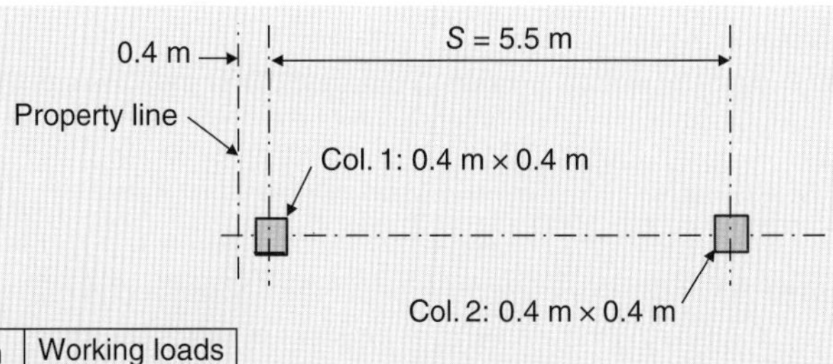

Column	Size, m	Working loads	
		D, kN	L, kN
1	0.4 × 0.4	200	350
2	0.4 × 0.4	400	300

Footing: concrete strength $f'_c = 21$ Mpa.
reinforcing steel $f_y = 420$ Mpa.

Scheme 5.70

Column	Working loads, kN	Factored loads, kN
1	200 + 350 = *550*	$P_1 = 1.2 \times 200 + 1.6 \times 350 = 800$
2	400 + 300 = *700*	$P_2 = 1.2 \times 400 + 1.6 \times 300 = 960$
For Cols 1 + 2	Working R = 1250	Factored R = 1760

$$\text{Total area of footings} = \frac{\text{working}\,R}{net\ q_a} = \frac{\text{factored}\,R}{net\ q_{factored}}$$

$$net\ q_{factored} = \frac{150 \times 1760}{1250} = 211.2\ \text{kN}$$

Step 2. Estimate a trial value for the eccentricity e of the load of Col. 1 or a trial value for the length L_1 of Footing 1, and then compute S', R_1 and R_2 shown below.

Each of the factored soil reactions R_1 and R_2 should act at the centre of its footing so that a uniform soil pressure distribution can be assumed. Also, in order to reduce differential settlement, it is required that the selected value for e or L_1 results in very close footing widths, that is B_1 and B_2 should not be greatly different. In design of strap footings, usually, the strap weight is neglected.

Estimate the eccentricity $e = 0.7$ m

$$S' = S - e = 5.5 - 0.7 = 4.8\ \text{m}$$

Take moments about centre of Col. 2:

$$\Sigma M = 0 - R_1 S' - P_1 S = 0$$

$$\text{Factored}\ R_1 = \frac{800 \times 5.5}{4.8} = 916\ \text{kN}$$

$$\Sigma F_v = 0$$

$$\text{Factored}\ R_2 = P_1 + P_2 - R_1 = 1760 - 916.67 = 843.33\ \text{kN}$$

Scheme 5.71

Step 3. Find the plan dimensions of each footing.
Footing 1:

$$\frac{L_1}{2} = 0.4 + e = 0.4 + 0.7 = 1.1\,\text{m} \rightarrow L_1 = 2 \times 1.1 = 2.2\,\text{m}$$

$$R_1 = B_1\ L_1\ (net\ q_{\text{factored}}) \rightarrow B_1 = \frac{R_1}{L_1\ (net\ q_{\text{factored}})} = \frac{916.67}{2.2 \times 211.2} = \mathit{1.973}\,\text{m}$$

Footing 2:
Use a square column footing. Hence, $B_2 = L_2$

$$R_2 = (B_2)^2 (net\ q_{\text{factored}}) \rightarrow B_2 = \sqrt{\frac{R_2}{(net\ q_{\text{factored}})}} = \sqrt{\frac{843.33}{211.2}} = \mathit{1.998}\,\text{m}$$

Use: Footing 1 *2.0 × 2.2 m*
Footing 2 *2.0 × 2.0 m*

It is expected that settlement of the footings should be nearly equal, since $net\ q_{\text{factored}}$ is the same for both footings and the widths B_1 and B_2 are equal.

In the design computations it may be necessary to use the exact B and L values so that the shear and moment diagrams will close.

Step 4. Obtain data for shear-force and bending-moment diagrams considering the strap footing as a reinforced concrete structural member with variable cross-section dimensions and loadings.

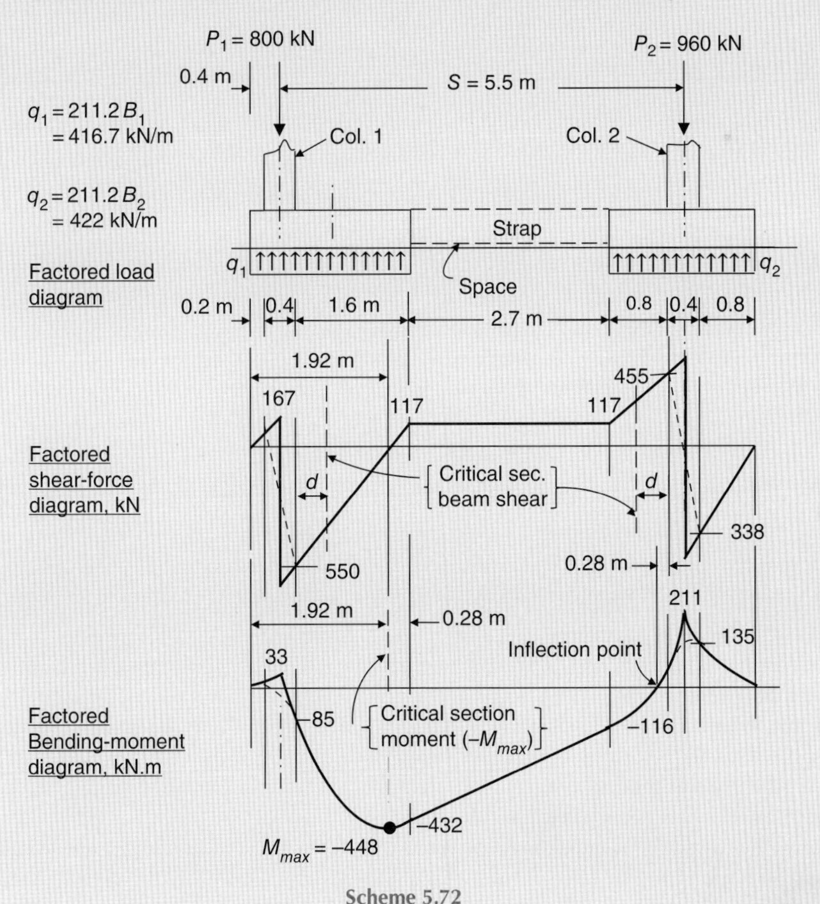

Scheme 5.72

(*Continued*)

It is preferable to draw the complete load diagram first, and then the shear and moment diagrams, as shown above.

The factored distributed loads on the footings (uniform contact pressure), shear forces and bending moments at any location shall be computed for the full width of the member.

Any convenient method, for example integral calculus (with attention to values at the limits), may be used for calculating the shear and moment values at necessary locations so that informable diagrams are obtained.

If the necessary computations involve tedious busywork, it will be preferable to use an appropriate computer program.

Step 5. Find footing thickness h. Determine the required thickness for each footing based on both one-way shear and two-way shear analyses.

(a) One-way shear.

Footing 1 (the exterior footing):

The critical section is located at distance d from the interior face of the exterior column, as indicated on the shear-force diagram.

Equation (5.19):

$$\varnothing V_c \geq V_u \text{ (taking reinforcement shear strength, } V_s = 0)$$

$$\text{where } \varnothing = \text{strength reduction factor (ACI Section 9.3.2.3)}$$

$$= 0.75 \text{ for shear}$$

$$V_u = 550 - 416.7d \qquad \text{(kN)}$$

Equation (5.20):

$$V_c = 0.17\lambda\sqrt{f_c'}b_w d$$

$$\text{Let } \varnothing V_c = V_u:$$

$$(0.75 \times 0.17 \times 1)(\sqrt{21})(1000)(2)(d) = 550 - 416.7d$$

$$d = \frac{550}{1585} = 0.35 \text{ m} = 350 \text{ mm}.$$

Try $d = 0.36$ m = 360 mm.

Footing 2 (the interior footing):

The critical section is located at distance d from the interior face of the interior column, as indicated on the shear-force diagram.

$$V_u = 454 - 422d \text{ (kN)}$$

Let $\varnothing V_c = V_u$:

$$(0.75 \times 0.17 \times 1)(\sqrt{21})(1000)(2)(d) = 454 - 422d$$

$$d = \frac{454}{1591} = 0.29 \text{ m} = 290 \text{ mm}.$$

This depth is close to that of Footing 1 (only 0.06 m difference); it may be more practical, in this case, to use the same d for both footings.

Try $d = 0.36$ m = 360 mm.

(b) Two-way shear.

Footing 1 (the exterior footing):

Consider the shear perimeter has three sides only.

In this case, the condition of unbalanced moment transfer exists, and the design must satisfy the requirements of ACI Sections 11.11.7.1, 11.11.7.2 and 13.5.3. The design proceeds in the same manner as that for the three-sided shear perimeter at location of column1 of Problem 5.12.

Assume $d = 0.36$ m = 360 mm.

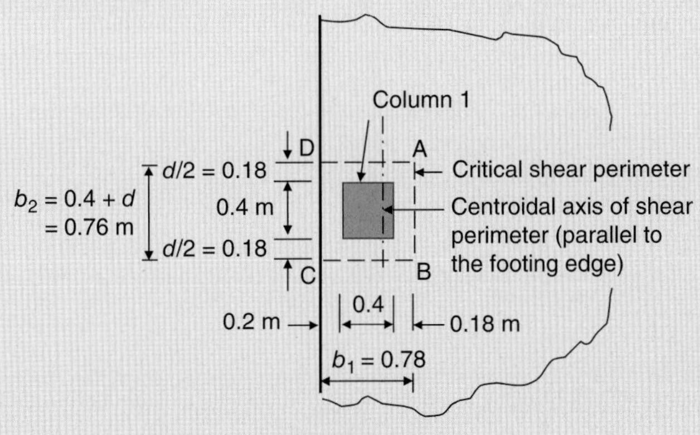

Scheme 5.73

C_{AB} is the distance from line A–B to the centroid of the shear perimeter

$$= \frac{\begin{pmatrix}\text{moment of area of sides}\\ \text{about A-B}\end{pmatrix}}{\text{area of sides}} = \frac{2(0.78 \times 0.36)(0.78/2)}{(2 \times 0.78 + 0.76)(0.36)} = 0.26\,\text{m}$$

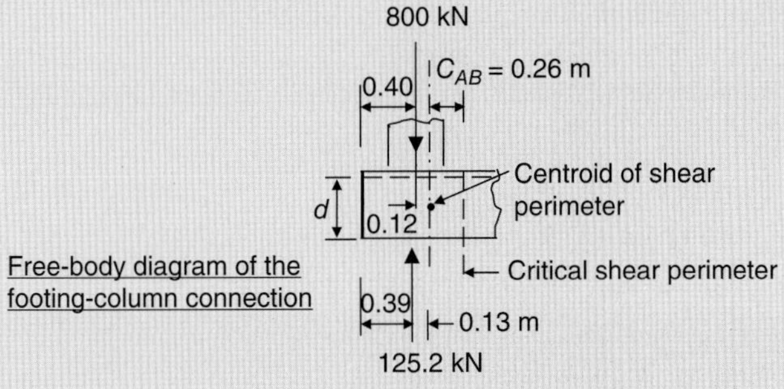

Scheme 5.74

Soil reaction on the shear block = 211.2(0.78 × 0.76) = 125.2 kN
Summation of moments about the centroid of the shear perimeter equals the unbalanced moment M_u. Hence, $M_u = 800 \times 0.12 - 125.2 \times 0.13 = 80$ kN. m

$$\gamma_v = \left(1 - \gamma_f\right) \quad \text{(ACI Section 11.11.7.1)}$$

$$\gamma_f = \frac{1}{1 + (2/3)\sqrt{b_1/b_2}}$$

$$= \frac{1}{1 + (2/3)\sqrt{0.78/0.76}} = 0.60$$

$$\gamma_v = 1 - 0.6 = 0.4$$

The shear stresses due to the direct shear and the shear due to moment transfer will add at points D and C, giving the largest shear stresses on the critical shear perimeter. Hence,

$$v_{u(DC)} = \frac{V_u}{b_o d} + \frac{\gamma_v M_u C_{DC}}{J_c} \quad \text{(ACI Section R11.11.7.2)}$$

(Continued)

$J_c = J_z$ = Property of the shear perimeter analogous to polar moment of inertia

$$J_c = 2\left(\frac{b_1 d^3}{12}\right) + 2\left(\frac{d b_1^3}{12}\right) + 2(b_1 d)\left(\frac{b_1}{2} - C_{AB}\right)^2 + (b_2 d) C_{AB}{}^2$$

$$= 2\left(\frac{0.78 \times 0.36^3}{12}\right) + 2\left(\frac{0.36 \times 0.78^3}{12}\right) + 2(0.78 \times 0.36)\left(\frac{0.78}{2} - 0.26\right)^2 +$$

$$(0.76 \times 0.36)(0.26^2) = 0.0061 + 0.0285 + 0.0095 + 0.0185 = 0.063\ \text{m}^4$$

$V_u = 800 - 125.2 = 674.8\ \text{kN}$

$C_{DC} = b_1 - C_{AB} = 0.78 - 0.26 = 0.52\ \text{m}$

$$v_{u(DC)} = \frac{674.8}{(2 \times 0.78 + 0.76)(0.36)} + \frac{0.4 \times 80 \times 0.52}{0.063} = 807.95 + 264.13$$

$$= 1072.1\ \text{kN/m}^2$$

$$\varnothing\, v_c \geq v_{u(DC)}$$

$$\varnothing\, v_c = \frac{\varnothing\, V_c}{b_o d}$$

Shear strength of concrete V_c shall be the smallest of (i), (ii) and (iii):

(i)

$$V_c = 0.17\left(1 + \frac{2}{\beta}\right)\lambda \sqrt{f_c'}\, b_o d$$

Equation (5.22)

$$= (0.17)\left(1 + \frac{2}{1}\right)(1)\sqrt{f_c'}\, b_o d = (0.51)\sqrt{f_c'}\, b_o d$$

(ii)

$$V_c = 0.083\left(\frac{\alpha_s d}{b_o} + 2\right)\lambda \sqrt{f_c'}\, b_o d$$

Equation (5.23)

$$= (0.083)\left(\frac{30 \times 0.36}{2(0.78) + 0.76} + 2\right)\sqrt{f_c'}\, b_o d = (0.46)\sqrt{f_c'}\, b_o d$$

(iii)

$$V_c = 0.33\,\lambda \sqrt{f_c'}\, b_o d$$

Equation (5.24)

$$= 0.33\,(1)\sqrt{f_c'}\, b_o d = (0.33)\sqrt{f_c'}\, b_o d$$

Use $V_c = (0.33)\sqrt{f_c'}\, b_o d$

$$\varnothing\, V_c = (0.75)(0.33)\sqrt{21}\,(1000)(b_o d) = 1134.2\, b_o d \quad \text{(kN)}$$

$$\varnothing\, v_c = \frac{1134.2\, b_o d}{b_o d} = 1134.2\ \text{kN/m}^2 > v_{u(DC)} \quad \text{(OK.)}$$

Footing 2 (the interior footing):
The footing is concentrically loaded, and the shear perimeter has four sides, as shown below.

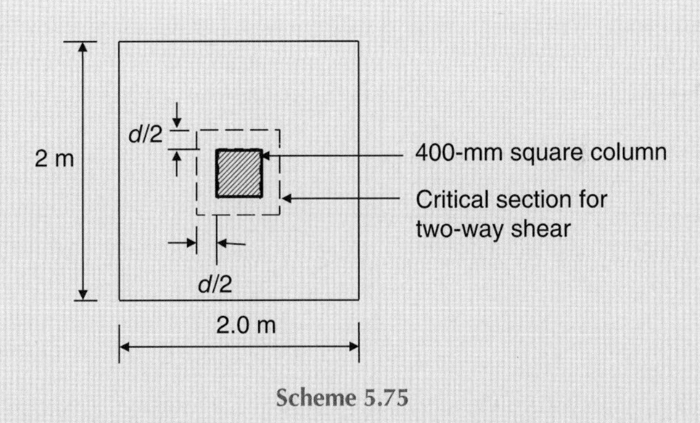

Scheme 5.75

Assume $d = 0.36$ m.
Shear strength of concreter V_c shall be the smallest of (i), (ii) and (iii):

(i)

$$V_c = 0.17\left(1 + \frac{2}{\beta}\right)\lambda\sqrt{f_c'}b_o d$$

$$= (0.17)\left(1 + \frac{2}{1}\right)(1)\sqrt{f_c'}b_o d = (0.51)\sqrt{f_c'}b_o d \quad \text{Equation (5.22)}$$

(ii)

$$V_c = 0.083\left(\frac{\alpha_s d}{b_o} + 2\right)\lambda\sqrt{f_c'}b_o d$$

$$= (0.083)\left(\frac{40d}{b_o} + 2\right)(1)\sqrt{f_c'}b_o d \quad \text{Equation (5.23)}$$

$$= (0.083)\left(\frac{40 \times 0.36}{4(0.4 + 0.36)} + 2\right)\sqrt{f_c'}b_o d$$

$$= (0.56)\sqrt{f_c'}b_o d$$

(iii)

$$V_c = 0.33\lambda\sqrt{f_c'}b_o d$$

$$= 0.33(1)\sqrt{f_c'}b_o d = (0.33)\sqrt{f_c'}b_o d \quad \text{Equation (5.24)}$$

Use $V_c = (0.33)\sqrt{f_c'}b_o d$.

$$\text{Ø}\, V_c = (0.75)(0.33)\sqrt{21}(1000)[4(0.4 + 0.36)]0.36 = 1241\,\text{kN}$$

$V_u = P_2$ –soil reaction on the shear block
$V_u = 960$ –soil reaction on the shear block (kN)

$$V_u < \text{Ø}\, V_c\,(\text{OK.})$$

Therefore, for both footings, the depth d is controlled by one-way shear.

(Continued)

Assume using No. 19 bars, with 75 mm minimum concrete cover (ACI Section 7.7.1). Hence, the overall footing thickness is

$$h = 360 + 75 + \frac{1}{2}\text{bar diameter} = 435 + \left(\frac{19.1}{2}\right) = 444.6\,\text{mm}.$$

Therefore, for both footings, use:

$$d = 0.36\,\text{m} = \mathit{360}\,\text{mm}$$

$$h = 450\ \text{mm}.$$

Step 6. Design theflexural reinforcement for each footing.

(a) Footing 1 (the exterior footing).

It is expected that the strap depth will be larger than that of the footings, since the design requires the strap-footing stiffness ratio not be less than 1.5. In such a case, it is usual practice to extend the strap beam over the footings to the columns. Thus, Footing 1, actually, acts as a wall footing, cantilevering out on the two sides of the strap. Therefore, the main flexural reinforcement at the footing bottom shall be the transverse reinforcement only (i.e. one-way footing). Obviously, shrinkage and temperature reinforcement shall be provided normal to the flexural reinforcement (ACI Sections 7.12.1 and 7.12.2.1). As indicated by the bending-moment diagram, the footing is subjected to negative moments in long direction. This moment will be resisted by the strap top bars, since the strap extends over the footing to Column 1.

Transverse flexural positive reinforcement:

Design the footing reinforcement as a spread footing. Consider the critical section located at the face of Column 1 (not the strap face so that conservative design is achieved), as shown below.

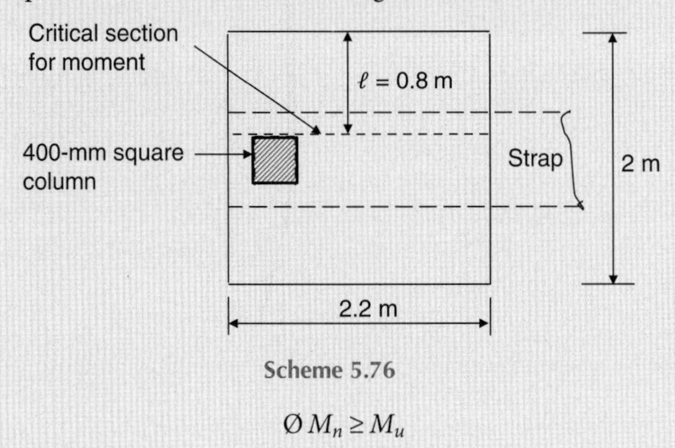

Scheme 5.76

$$\text{Ø}\, M_n \geq M_u$$

The maximum factored moment is

$$M_u = \frac{(net\; q_{\text{factored}})(\ell)^2}{2} = \frac{211.2 \times 0.8^2}{2} = 68\,\text{kN.m/m}$$

Assume tension-controlled section, Ø = 0.9 and $f_s = f_y$.

$$a = \frac{A_s f_y}{0.85 f_c' b} = \frac{A_s \times 420}{0.85 \times 21 \times 1} = 23.53 A_s$$

$$d = 0.36\,\text{m}$$

$$\text{Ø}\, M_n = \text{Ø}\left[A_s f_y \left(d - \frac{a}{2}\right)\right]$$

$$= (0.9)\left[(A_s \times 420 \times 1000)\left(0.36 - \frac{23.53 A_s}{2}\right)\right] \quad \text{Equation (5.11)}$$

$$= 136\ 080 A_s - 4\ 447\ 170 A_s^2 \quad (\text{kN.m/m})$$

Let $\emptyset M_n = M_u$:

$$136\ 080A_s - 4\ 447\ 170A_s^2 = 68$$

$$4\ 447\ 170A_s^2 - 136\ 080A_s + 68 = 0$$

$$A_s^2 - 0.031A_s + 1.529 \times 10^{-5} = 0$$

$$A_s = \frac{-(-0.031) \pm \sqrt{(-0.031)^2 - (4)(1)(1.529 \times 10^{-5})}}{2 \times 1} = \frac{0.0310 - 0.0299}{2}$$

$$= 550 \times 10^{-6}\ \text{m}^2/\text{m} = 550\ \text{mm}^2/\text{m}$$

$$A_{s,\min} = 0.0018\, bh = 0.0018 \times 1 \times 0.45 = 810 \times 10^{-6}\ \text{m}^2/\text{m}$$

$$= 810\ \text{mm}^2/\text{m} > A_s$$

Use $A_s = A_{s,\min} = 810\ \text{mm}^2/\text{m}$

The assumptions made are satisfied, since A_s required = $A_{s,\min}$.

$$A_{s,\text{total}} = A_s \times L_1 = 810 \times 2.2 = 1782\ \text{mm}^2$$

Try nine No.16 bars: $A_{s,\text{provided}} = 9 \times 199 = 1791\ \text{mm}^2$ (OK.)

Check the development of reinforcement:

In this case, the bars are in tension, the provided bar size is smaller than No. 19, the clear spacing of the bars exceeds $2d_b$, and the clear cover exceeds d_b. Therefore, the development length l_d of bars at each side of the critical section (which is the same critical section for moment) shall be determined from the following equation, but not less than 300 mm:

$$l_d = \left(\frac{f_y \psi_t \psi_e}{2.1\lambda \sqrt{f'_c}}\right) d_b \quad \text{(ACI Sections 12.2.1, 12.2.2 and 12.2.4)}$$

where the factors ψ_t, ψ_e and λ are 1. (ACI Sections 12.2.4 and 8.6.1)

$$l_d = \left(\frac{420 \times 1 \times 1}{2.1 \times 1 \times \sqrt{21}}\right)\left(\frac{15.9}{1000}\right) = 0.694\ \text{m} = 694\ \text{mm} > 300\ \text{mm}.$$

Therefore, the required $l_d = 694$ mm.

The bar extension past the critical section (i.e. the available length) is

$$800\ \text{mm} - 75\ \text{mm cover} = 725\ \text{mm} > 694\ \text{mm (OK.)}$$

ACI Sections 7.6.5 and 10.5.4 requires maximum spacing shall not exceed three times the slab or footing thickness or 450 mm, whichever is smaller.

Use centretocentre bar spacing = 253 mm. Check the 75 mm minimum concrete cover at each side:

$$\text{Concrete cover} = \frac{(2200 - 8 \times 253 - 15.9)}{2} = 80\ \text{mm} > 75\ \text{mm (OK.)}$$

Provide nine No. 16 bars @ 253 mm c.c., placed at bottom of the footing in the transverse direction.

(*Continued*)

Shrinkage and temperature reinforcement:

This reinforcement shall be provided, in long direction, normal to the flexural reinforcement (ACI Sections 7.12.1 and 7.12.2.1).

Total

$$A_s = 0.0018\,bh = 0.0018 \times 2 \times 0.45 = 1620 \times 10^{-6}\,\text{m}^2$$
$$= 1620\,\text{mm}^2$$

Try nine No.16 bars: $A_{s,\text{provided}} = 9 \times 199 = 1791\ \text{mm}^2$ (OK.)

According to ACI Section 7.12.2.2, shrinkage and temperature reinforcement shall be spaced not farther apart than five times the slab thickness, nor farther apart than 450 mm.

Use centre-to-centre bar spacing = 229 mm. Check the 75 mm minimum concrete cover at each side:

$$\text{Concrete cover} = \frac{(2000 - 8 \times 229 - 15.9)}{2} = 76\,\text{mm} > 75\,\text{mm} \quad \text{(OK.)}$$

Provide nine No. 16 @ 229 mm c.c. in the long direction, placed on top of the transverse reinforcement at bottom of the footing.

Longitudinal flexural negative reinforcement:

As mentioned earlier, this reinforcement will be provided by the strap top bars (see design of the strap later).

(b) Footing 2 (the interior footing).

Dimension of Footing 2 in both directions is the same as the transverse dimension of Footing 1(i.e. $B = 2$ m); the design contact pressure for both footings is the same (i.e. *net* $q_{\text{factored}} = 211.2$ kPa); the columns have the same cross-section dimensions (i.e. 0.4 m × 0.4 m); the same d, f_c' and f_y are used in design of both footings. For these reasons, the same amount of flexural positive reinforcement which was computed for Footing 1 shall also be used for Footing 2 and in both directions.

Use centre-to-centre bar spacing = 229 mm. Check the 75 mm minimum concrete cover at each side:

$$\text{Concrete cover} = \frac{(2000 - 8 \times 229 - 15.9)}{2} = 76\,\text{mm} > 75\,\text{mm} \quad \text{(OK.)}$$

As indicated by the bending-moment diagram, a small portion of the footing is subjected to a small negative moment. This moment will be resisted by the strap top bars, since the strap extends over the footing to Column 2.

Provide nine No. 16 @ 229 mm c.c.,placed at bottom of the footing in both directions.

Step 7. Check column bearing on each footing (ACI Section 10.14), and design the necessary dowels.

Proceed in the same manner as that of Solution of Problem 5.3, *Step 5.*

Step 8. Find the embedment length of dowels in both the footing and the column for each column-footing joint.

Proceed in the same manner as that of Solution of Problem 5.3, *Step 6.*

Step 9. Design the strap as a reinforced concrete structural beam.

This design step involves findingb, d and A_s. Bending-moment and shear-force diagrams indicate that the strap must be designed for a negative moment of 432 kN.m and a shear force of 117 kN. However, this design moment is too close to the maximum negative moment (448 kN.m) and it would be conservative if the latter is used for design. In order to compute the cross-section dimensions b and d, it may be necessary to select a trial steel ratio ρ. For Grade-420 reinforcement and f_c' values between 21 and 35 MPa, it may be appropriate to start a beam design by assuming that $\rho \cong 0.01$. This choice or assumption of ρ value, generally, is economical, gives desirable level of structural ductility and assures easy placement of reinforcement.

The required strap-footing minimum stiffness ratio of 1.5 will be a factor in selecting a minimum value for the strap height h. Also, if the designer wants to avoid deflection calculations, ACI Section 9.5.2.1 and ACI Table 9.5(a) require $h \geq \frac{\ell}{16}$. Furthermore, it is required the strap shall be designed as beam but not "deep" beam, which requires $h < \frac{\ell_n}{4}$ (ACI Section 10.7).Taking all these factors into consideration, the strap may be designed as follows:

Assume tension-controlled section, $\varnothing = 0.9$, and $f_s = f_y$.

Equation (5.10):

$$a = \frac{A_s f_y}{0.85 f_c' b}$$

If we substitute $A_s = \rho bd$ into Equation (5.10), we get

$$a = \frac{\rho f_y}{f_c'}\left(\frac{d}{0.85}\right)$$

Equation (5.12):

$$\varnothing M_n = \varnothing\left[(0.85 f_c')ba\left(d - \frac{a}{2}\right)\right]$$

If we substitute $a = \frac{\rho f_y}{f_c'}\left(\frac{d}{0.85}\right)$ into Equation (5.12), we get

$$\varnothing M_n = \varnothing\left\{(0.85 f_c')(b)\left[\frac{\rho f_y}{f_c'}\left(\frac{d}{0.85}\right)\right]\left[d - \frac{\rho f_y}{2 f_c'}\left(\frac{d}{0.85}\right)\right]\right\}.$$

Hence, for $f_c' = 21$ MPa, $f_y = 420$ MPa, $\rho = 0.01$

$$\varnothing M_n = \varnothing\left\{(0.85 \times 21)(b)\left(\frac{0.01 \times 420d}{21 \times 0.85}\right)\left[d - \frac{0.01 \times 420d}{2 \times 21 \times 0.85}\right]\right\}$$

$$= \varnothing bd^2(4.2)\left(1 - \frac{4.2}{35.7}\right) = 3.706\,\varnothing bd^2 \quad \text{(MN.m)}$$

$$= 3706\,\varnothing bd^2 \quad \text{(kN.m)}$$

$$M_u = 448 \text{ kN.m}$$

$$\varnothing M_n \geq M_u$$

Let $\varnothing M_n = M_u$:

$$3706\,\varnothing bd^2 = 448$$

$$bd^2 = \frac{448}{3706 \times 0.9} = 0.134 \text{ m}^3;\ b = \frac{0.134}{d^2} \text{ (m)}$$

The strap width b should be at least equal to the smaller width of the columns. Therefore, use $b \geq 0.4$ m, or, $b_{min} = 0.40$ m.

Assume $b = 0.40$ m. Hence, $d = \sqrt{\frac{0.134}{0.4}} = 0.52$ m.

According to ACI Section10.7, deep beams have $\ell_n \leq 4\,h$.

The strap clear span $\ell_n = 2.7$ m.

Let $\ell_n = 4\,h$:

$$h = \frac{2.7}{4} = 0.675 \text{ m}$$

(*Continued*)

It is required to design the strap as beam but not "deep" beam. Therefore, the overall strap depth or height h must be smaller than 0.675 m.

Assume: h = 0.67 m; concrete cover = 0.08 m. Hence,

$$d = 0.67 - 0.08 = 0.59\,\text{m} > (d = 0.52\,\text{m})$$

Try: $h = 0.67\,\text{m}$ $d = 0.59\,\text{m}$ $b = 0.40\,\text{m}$

Check the minimum stiffness ratio:

$$\left(\frac{I_{\text{strap}}}{I_{\text{footing}}}\right) = \frac{\left(\frac{bh^3}{12}\right)_{\text{strap}}}{(bh^3/12)_{\text{footing}}} = \frac{0.4 \times \frac{0.67^3}{12}}{2 \times (0.45^3/12)} = 0.66 \ll 1.5 \quad \text{(Not OK.)}$$

Increase the strap width and check the stiffness ratioagain.

Try $b = 0.9$ m:

$$\frac{\left(\frac{bd^3}{12}\right)_{\text{strap}}}{(bd^3/12)_{\text{footing}}} = \frac{0.9 \times \frac{0.67^3}{12}}{2 \times (0.45^3/12)} = 1.49 \cong 1.5\,\text{(Acceptable)}$$

Check deflection of the strap beam:

According to ACI Section 9.5.2.1 and ACI Table 9.5(a), for a simple beam constructed with structural normal-weight concrete and Grade 420 reinforcement, not supporting or attached to partitions or other construction likely to be damaged by large deflections, the minimum overall depth or h to avoid deflection calculations is $\frac{\ell}{16}$, where ℓ is the span length of the beam. In this case, the individual footings are supporting the strap, and it may be appropriate to let ℓ represent the distance between the footings, centre to centre, which is 4.8 m.

Therefore,

$\frac{\ell}{16} = \frac{4.8}{16} = 0.3\,\text{m} < (h = 0.67\,\text{m})$. Therefore, deflections should not be a problem.

Use a strap beam with b = 0.90 m, *d* = 0.59 m and *h* = 0.67 m.

Flexural reinforcement at the top of the strap:

$$\varnothing M_n \geq M_u$$

$$M_u = 448\,\text{kN.m}$$

Assume tension-controlled section, $\varnothing = 0.9$, and $f_s = f_y$.

$$a = \frac{A_s f_y}{0.85 f_c' b} = \frac{A_s \times 420}{0.85 \times 21 \times 0.9} = 26.14\,A_s \quad \text{(m)}$$

$$\varnothing M_n = \varnothing\left[(0.85 f_c')\,ba\left(d - \frac{a}{2}\right)\right]$$

$$= 0.9\left[(0.85 \times 21 \times 1000)(0.9 \times 26.14 A_s)\left(0.59 - \frac{26.14 A_s}{2}\right)\right]$$

$$= 222\,988\,A_s - 4\,939\,744 A_s^2 \quad \text{(kN.m)}$$

Let $\varnothing M_n = M_u$:

$$222\,988 A_s - 4939\,744 A_s^2 = 448$$

$$A_s^2 - 45.14 \times 10^{-3} A_s + 9.07 \times 10^{-5} = 0$$

$$A_s = \frac{-(-45.14\times10^{-3}) \pm \sqrt{(-45.14\times10^{-3})^2 - 4\times1\times9.07\times10^{-5}}}{2\times1} = = \frac{0.04514-0.04092}{2}$$

$$= 2.11\times10^{-3}\ \text{m}^2 = 2110\ \text{mm}^2$$

$$A_{s,min} = \frac{0.25\ \sqrt{f_c'}}{f_y} b_w d = \frac{0.25\ \sqrt{21}}{420} \times 0.9\times0.59 = 1.448\times10^{-3}\ \text{m}^2$$

$$= 1448\ \text{mm}^2$$

and not less than $1.4\ \dfrac{b_w\,d}{f_y} = \dfrac{1.4\times0.9\times0.59}{420} = 1.77\times10^{-3}\ \text{m}^2 = 1770\ \text{mm}^2$

Use $A_s = 2110\ \text{mm}^2$

Try six No. 22 bars: $A_{s,provided} = 6\times387 = 2322\ \text{mm}^2$ (OK.)

Compute a for $A_s = 2322\ \text{mm}^2$ and check if $f_s = f_y$ and whether the section is tension-controlled:

$$a = 26.14A_s = 26.14\times2322\times10^{-6} = 0.061\ \text{m} = 61\ \text{mm}$$

$c = \dfrac{a}{\beta_1}$; For f_c' between 17 and 28 MPa, β_1 shall be taken as 0.85

$$c = \frac{61}{0.85} = 71.8\ \text{mm}$$

$$d_t = d = 590\ \text{mm}$$

$$\varepsilon_t = 0.003\left(\frac{d_t - c}{c}\right)$$

$$= 0.003\left(\frac{590-71.8}{71.8}\right) = 0.022 > 0.005.\ \text{Equation (5.15)}$$

Hence, the section is tension-controlled, and Ø = 0.9 (ACI Sections 10.3.4 and 9.3.2.1).

$\varepsilon_y = \dfrac{f_y}{E_s} = \dfrac{420}{200000} = 0.0021$. Hence, $\varepsilon_t > \varepsilon_y$ and $f_s = f_y$.

Therefore the assumptions made are satisfied.

Because $\varepsilon_t = 0.022$ exceeds 0.004, the strap section satisfies the definition of beam in ACI Section 10.3.5.

It may be a good practice to compute Ø M_n to check whether the provided A_s is adequate. This checking guards against errors in computations. Thus,

$$\text{Ø}\,M_n = \text{Ø}\left[(0.85f_c')ba\left(d - \frac{a}{2}\right)\right]$$

$$= 0.9\times0.85\times21\times1000\times0.9\times0.061\left(0.59 - \frac{0.061}{2}\right)$$

$$= 493\ \text{kN.m} > (M_u = 448\ \text{kN.m}) \qquad \text{(OK.)}$$

Provide six No. 22 bars at the top of the strap in one layer.

Check the development of reinforcement:

In this case, the bars are in tension, the provided bar size is No. 22, the clear spacing of the bars exceeds $2d_b$, and the clear cover exceeds d_b. Therefore, the development length l_d of bars at each side of the critical section (which is located where the maximum factored negative moment exists) shall be determined from the following equation, but not less than

$l_d = \left(\dfrac{f_y \psi_t \psi_e}{1.7\lambda\sqrt{f_c'}}\right) d_b$ (ACI Sections 12.2.1, 12.2.2 and 12.2.4)

where the factors ψ_e and λ are 1 (ACI Sections 12.2.4 and 8.6.1). *(Continued)*

The factor ψ_t is 1.3 because the reinforcement is placed such that more than 300 mm of fresh concrete exists below the top bars (ACI Section 12.2.4).

$$l_d = \left(\frac{420 \times 1.3 \times 1}{1.7 \times 1 \times \sqrt{21}}\right)\left(\frac{22.2}{1000}\right) = 1.556\ \text{m} = 1556\ \text{mm} > 300\ \text{mm}.$$

Therefore, the required $l_d = 1556$ mm.

Since top of the strap will be about 0.3 m above the top of each footing, the left end of the strap shall be extended to the edge of footing 1 at the property limit and its right end to a point 0.2 m beyond column 2. Thus, the bar extension past the critical section (i.e. the shortest available length) is

$$2200\ \text{mm} - 75\ \text{mm cover} = 2125\ \text{mm} > 1556\ \text{mm} \quad \text{(OK.)}$$

The point of inflection is located at 0.28 m from the interior face of column 2 as shown on the bending-moment diagram. The available length of bars beyond this point is

$$0.28 + 0.40 + 0.20\ (\text{beyond Col. 2}) = 0.88\ \text{m} > (d = 0.59\ \text{m}) > (12d_b)$$

Therefore, the requirements of ACI Section 12.10.3 are satisfied.

Check the shear strength, $\varnothing V_c$:

Equation (5.19):

$$\varnothing V_c \geq V_u\ (\text{taking reinforcement shear strength, } V_s = 0)$$

$$\text{where } \varnothing = \text{shear strength reduction factor (ACI Section 9.3.2.3)}$$

$$= 0.75$$

Equation (5.20):

$$V_c = 0.17\lambda\sqrt{f'_c}\,b_w d$$

$$\varnothing V_c = (0.75 \times 0.17 \times 1)(\sqrt{21})(1000)(0.9)(0.59) = 311\ \text{kN}$$

$$V_u = 117\ \text{kN} < \varnothing V_c\ \text{(OK.)}$$

ACI Section 11.4.6.1 requires that a minimum area of shear reinforcement, $A_{v,\min}$, shall be provided in all reinforced concrete flexural members (prestressed and non-prestressed) where V_u exceeds $0.5\,\varnothing\,V_c$.

$$0.5\,\varnothing\,V_c = 0.5 \times 311 = 158\ \text{kN} > (V_u = 117\ \text{kN}).$$

Therefore, the $A_{v,min}$ will not be required. However, a sufficient number of stirrup support bars shall be provided to hold and support the longitudinal negative reinforcement properly.

Step 10. Develop the final design sketches as shown in the schemes below.

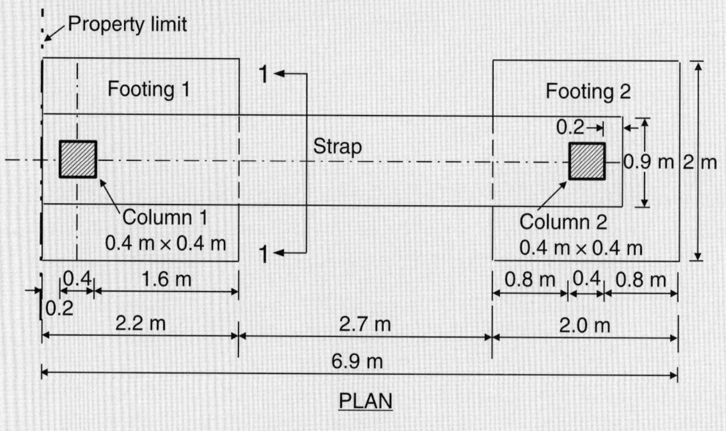

Scheme 5.77

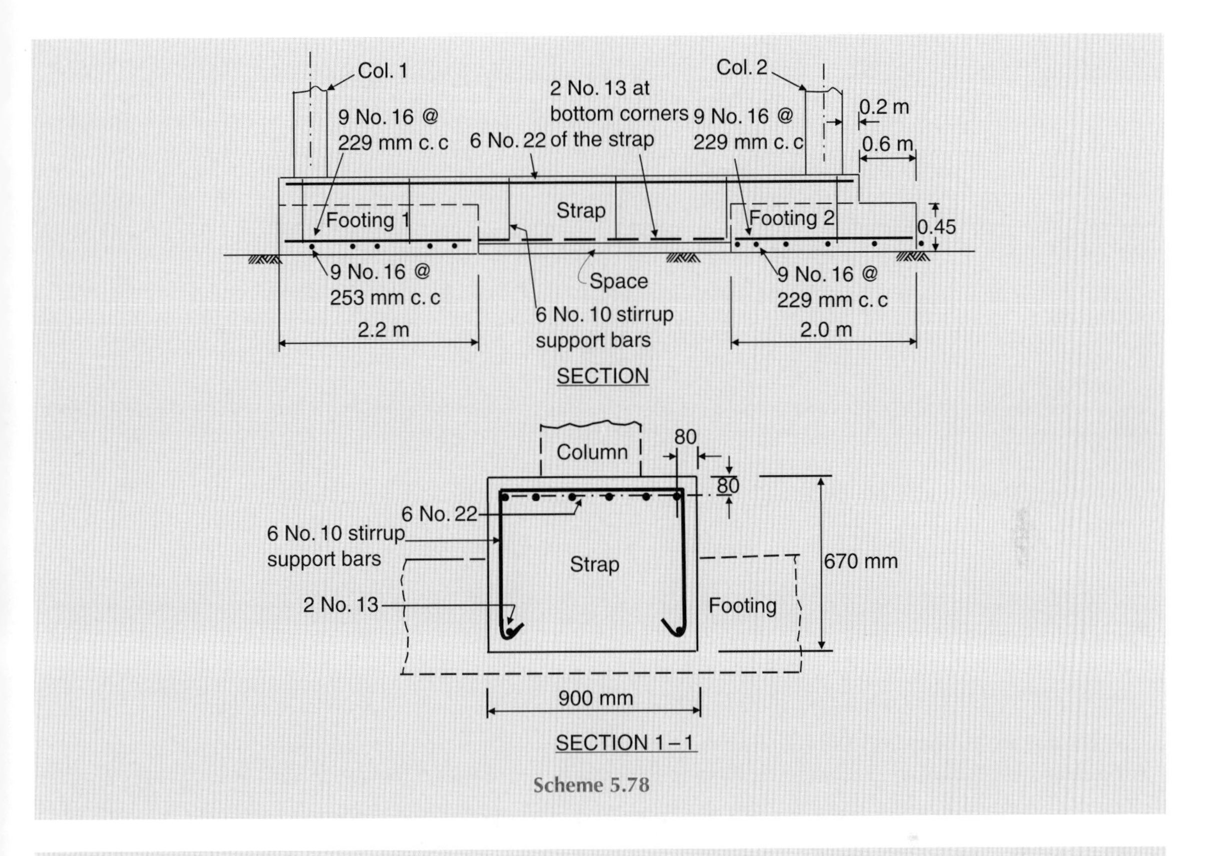

Scheme 5.78

Problem 5.15

Design the strap of Problem 5.14 using the minimum stiffness ratio ($I_{strap}/I_{footing}$) equals 1.0 instead of the 1.5.

Solution:

Equation (5.10):

$$a = \frac{A_s f_y}{0.85 f_c' b}$$

If we substitute $A_s = \rho bd$ into Equation (5.10), we get

$$a = \frac{\rho f_y}{f_c'}\left(\frac{d}{0.85}\right)$$

Equation (5.12):

$$\varnothing M_n = \varnothing\left[(0.85 f_c')ba\left(d - \frac{a}{2}\right)\right]$$

If we substitute $a = \frac{\rho f_y}{f_c'}\left(\frac{d}{0.85}\right)$ into Eq. (5.12) we get

$$\varnothing M_n = \varnothing\left\{(0.85 f_c')(b)\left[\frac{\rho f_y}{f_c'}\left(\frac{d}{0.85}\right)\right]\left[d - \frac{\rho f_y}{2 f_c'}\left(\frac{d}{0.85}\right)\right]\right\}$$

For Grade-420 reinforcement and f_c' values between 21 and 35 MPa, it may be appropriate to start a beam design by assuming that $\rho \cong 0.01$. This choice or assumption of ρ value, generally, is economical, gives desirable level of structural ductility, and assures easy placement of reinforcement.

For $f_c' = 21$ MPa, $f_y = 420$ MPa, assume $\rho = 0.01$

(Continued)

$$\emptyset M_n = \emptyset\left\{(0.85\times 21)(b)\left(\frac{0.01\times 420d}{21\times 0.85}\right)\left[d-\frac{0.01\times 420d}{2\times 21\times 0.85}\right]\right\}$$

$$= \emptyset\, bd^2(4.2)\left(1-\frac{4.2}{35.7}\right) = 3.706\,\emptyset\, bd^2 \quad (\text{MN.m})$$

$$= 3706\,\emptyset\, bd^2 \quad (\text{kN.m})$$

Mu = 448 kN.m (see the B.M.D. of Solution of Problem 5.14, *Step 4*)

$$\emptyset M_n \geq M_u$$

Let $\emptyset M_n = M_u$:

$$3706\,\emptyset\, bd^2 = 448$$

Assume tension-controlled section, $\emptyset = 0.9$.

$$bd^2 = \frac{448}{3706\times 0.9} = 0.134\,\text{m}^3$$

The strap width b should be at least equal to the smaller width of the columns. Therefore, use $b \geq 0.4$ m or $b_{min} = 0.40$ m. Assume $b = 0.40$ m. Hence,

$$d = \sqrt{\frac{0.134}{0.4}} = 0.52\,\text{m}$$

According to ACI Section10.7, deep beams have $\ell_n \leq 4\,h$.
Strap clear span $\ell_n = 2.7$ m, and the overall strap depth or height $h = d + 0.080$ (assumed).
Let $\ell_n = 4\,h$. Hence, $h = \frac{2.7}{4} = 0.675\,\text{m}$ If a deep strap is not required, the strap height h should be decreased.
Assume $h_{max} = 0.67$ m. Hence,

$$d_{max} = 0.67 - 0.08 = 0.59\,\text{m} > (d = 0.52\,\text{m})$$

Try $h = 0.67$ m, $d = 0.59$ m and $b = 0.40$ m. Check the required strap-footing stiffness ratio, $(I_{\text{strap}}/I_{\text{footing}}) \geq 1$.

$$\left(\frac{I_{\text{strap}}}{I_{\text{footing}}}\right) = \frac{\left(\frac{bh^3}{12}\right)_{\text{strap}}}{(bh^3/12)_{\text{footing}}} = \frac{0.4\times\frac{0.67^3}{12}}{2\times(0.45^3/12)} = 0.66 < 1 \text{ (Not OK.)}$$

Increase the strap width and check the stiffness ratioagain.
Try $b = 0.6$ m:

$$\frac{\left(\frac{bd^3}{12}\right)_{\text{strap}}}{(bd^3/12)_{\text{footing}}} = \frac{0.6\times\frac{0.67^3}{12}}{2\times(0.45^3/12)} = 0.99 \cong 1.0 \ (\textit{Acceptable})$$

Check deflection of the strap beam:
According to ACI Section 9.5.2.1 and ACI Table 9.5(a), for a simple beam constructed with structural normal-weight concrete and Grade 420 reinforcement, not supporting or attached to partitions or other construction likely to be damaged by large deflections, the minimum overall depth or h to avoid deflection calculations is $\frac{\ell}{16}$, where ℓ is the span length of the beam. In this case, the individual footings are supporting the strap, and it may be appropriate to let ℓ represent the distance between the footings, centre to centre, which is 4.8 m,
$\frac{\ell}{16} = \frac{4.8}{16} = 0.3\,\text{m} < (h = 0.67\,\text{m})$. Therefore, deflections should not be a problem.

Use $h = 0.67$ m, $d = 0.59$ m and $b = 0.6$ m.
Compute the area of reinforcement, A_s:
Assume tension-controlled section, $\varnothing = 0.9$ and $f_s = f_y$.

$$a = \frac{A_s f_y}{0.85 f_c' b} = \frac{A_s \times 420}{0.85 \times 21 \times 0.6} = 39.22\,A_s \quad \text{(m)}$$

$$\varnothing M_n = \varnothing\left[(0.85 f_c')ba\left(d - \frac{a}{2}\right)\right]$$

$$= 0.9\left[(0.85 \times 21 \times 1000)(0.6 \times 39.22 A_s)\left(0.59 - \frac{39.22 A_s}{2}\right)\right]$$

$$= 223\,045 A_s - 7\,413\,395 A_s^2 \quad \text{(kN.m)}$$

$$M_u = 448\,\text{kN.m}$$

$$\varnothing M_n \geq M_u$$

Let $\varnothing M_n = M_u$:

$$223\,045 A_s - 7\,413\,395 A_s^2 = 448$$

$$A_s^2 - 0.03 A_s + 6.04 \times 10^{-5} = 0$$

$$A_s = \frac{-(-0.03) \pm \sqrt{(-0.03)^2 - 4 \times 1 \times 6.04 \times 10^{-5}}}{2 \times 1} = \frac{0.03 - 0.02857}{2}$$

$$= 2.17 \times 10^{-3}\,\text{m}^2 = 2170\,\text{mm}^2$$

$$A_{s,min} = \frac{0.25\sqrt{f_c'}}{f_y} b_w d = \frac{0.25\sqrt{21}}{420} \times 0.6 \times 0.59 = 9.66 \times 10^{-4}\,\text{m}^2$$

$$= 966\,\text{mm}^2$$

and not less than:

$$1.4\frac{b_w d}{f_y} = \frac{1.4 \times 0.6 \times 0.59}{420} = 1.18 \times 10^{-3}\,\text{m}^2$$

$$= 1180\,\text{mm}^2$$

Use $A_s = 2170$ mm^2
Try six No. 22 bars: $A_{s,provided} = 6 \times 387 = 2322$ mm^2 (OK.)
Compute a for $A_s = 2322$ mm^2 and check if $f_s = f_y$ and whether the section is tension-controlled:

$$a = 39.22 A_s = 39.22 \times 2322 \times 10^{-6} = 0.091\,\text{m} = 91\,\text{mm}$$

$c = \frac{a}{\beta_1}$; For f_c' between 17 and 28 MPa, β_1 shall be taken as 0.85

$$c = \frac{91}{0.85} = 107.06\,\text{mm};\ d_t = d = 590\,\text{mm}$$

$$\varepsilon_t = 0.003\left(\frac{d_t - c}{c}\right) \qquad \text{Equation (5.15)}$$

$$= 0.003\left(\frac{590 - 107.06}{107.06}\right) = 0.014 > 0.005.$$

Hence, the section is tension-controlled, and $\varnothing = 0.9$ (ACI Sections 10.3.4 and 9.3.2.1).

$\varepsilon_y = \frac{f_y}{E_s} = \frac{420}{200000} = 0.0021$. Hence, $\varepsilon_t > \varepsilon_y$ and $f_s = f_y$.

(Continued)

Therefore the assumptions made are satisfied.

Because $\varepsilon_t = 0.014$ exceeds 0.004, the strap section satisfies the definition of beam in ACI Section 10.3.5. It may be a good practice to compute $\emptyset M_n$ to check whether the provided A_s is adequate. This checking guards against errors in computations. Thus,

$$\emptyset M_n = \emptyset \left[(0.85 f_c') ba \left(d - \frac{a}{2} \right) \right]$$

$$= 0.9 \times 0.85 \times 21 \times 1000 \times 0.6 \times 0.091 \left(0.59 - \frac{0.091}{2} \right) \quad \text{(OK.)}$$

$$= 478 \text{ kN.m} > (M_u = 448 \text{ kN.m})$$

Provide six No. 22 bars at the top of the strap in one layer.

Check the development of reinforcement:
The check proceeds exactly in the same manner as that for the strap of Problem 5.14; its repetition is unnecessary.
Check the shear strength, $\emptyset V_c$:
Equation (5.19):

$$\emptyset V_c \geq V_u \text{ (taking reinforcement shear strength, } V_s = 0)$$

where $\emptyset$ = shear strength reduction factor (ACI Section 9.3.2.3)

$= 0.75$

Equation (5.20):

$$V_c = 0.17\lambda \sqrt{f_c'} b_w d$$

$$\emptyset V_c = (0.75 \times 0.17 \times 1)(\sqrt{21})(1000)(0.6)(0.59) = 206.8 \text{ kN}$$

$$V_u = 117 \text{ kN} < \emptyset V_c \quad \text{(OK.)}$$

However, ACI Section 11.4.6.1 requires that a minimum area of shear reinforcement, $A_{v,\min}$, shall be provided in all reinforced concrete flexural members (prestressed and non-prestressed) where V_u exceeds $0.5\,\emptyset V_c$.
$0.5\,\emptyset V_c = 0.5 \times 206.8 = 103.4$ kN < ($V_u = 117$ kN).

Therefore, $A_{v,\min}$ is required.
Try No.10 double-leg stirrups ($f_{yt} = 300$ MPa) with standard 135° stirrup hooks.

$$A_v = 2 \times 71 = 142 \text{ mm}^2$$

Anchorage of stirrups:

ACI Sections 7.1.3 and 12.13.2.1 allow No. 25 and smaller stirrups to be anchored by standard90° or 135° stirrup hooks. The stirrup legs should be hooked around longitudinal bars.

Provide two No. 13 bars in the lower corners of the strap cross-section to anchor the stirrups properly.
Stirrups spacing, s.

(a) Based on the strap depth d:
According to ACI Section 11.4.5.1, spacing of shear reinforcement perpendicular to axis of member shall not exceed the smaller of $d/2$ of the member and 600 mm.

$$\frac{d}{2}=\frac{590}{2}=295\text{ mm}<600\text{ mm}$$

Try $s_{max}=295$ mm
$\varnothing V_n \geq V_u$(ACI Section 11.1.1)

$$V_n=V_c+V_s$$

$$V_s=\text{Reinforcement shear strength}$$

According to ACI Section 11.4.5.3, where V_s exceeds $0.33\sqrt{f_c'}b_w d$, the above s_{max} shall be reduced by one-half. Thus, s_{max} must be cut in half if V_n exceeds $\left(V_c+0.33\sqrt{f_c'}b_w d\right)$.

$$V_c=0.33\sqrt{f_c'}b_w d$$

$$V_c+0.33\sqrt{f_c'}b_w d=3V_c=(3)(0.17)\sqrt{21}\times 0.6\times 0.59\times 1000$$

$$=827\text{ kN}$$

Let $\varnothing V_n \geq V_u$:

$$V_n=\frac{V_u}{\varnothing}=\frac{117}{0.75}=156\text{ kN}<3V_c$$

Therefore, s_{max} based on the strap depth is 295 mm.

(b) Based on $A_{v,min}$:

According to ACI Section 11.4.6.3, $A_{v,\min}=0.062\sqrt{f_c'}\dfrac{b_w s}{f_{yt}}$

$$\geq \frac{(0.35\,b_w s)}{f_{yt}}$$

Let $A_{v,min}=A_v$:

$$s=\frac{A_v\times f_{yt}}{b_w\times 0.062\sqrt{f_c'}}=\frac{142\times 10^{-6}\times 300}{0.6\times 0.062\sqrt{21}}=0.25\text{ m}=250\text{ mm}\quad\text{or}$$

$$s=\frac{A_v\times f_{yt}}{0.35\times b_w}=\frac{2\times 71\times 10^{-6}\times 300}{0.35\times 0.6}=0.203\text{ m}\leq 295\text{ mm}$$

Maximum spacing based on $A_{v,min}$ governs.
Try s_{max} = 203 mm (conservatively).
Since the factored shear force V_u is constant along the strap span, uniform stirrup spacing may be used. Place the first and last stirrups at $\frac{s}{2}$ from the interior edges of footings, since each stirrup is assumed to reinforce a length of strap web extending $\frac{s}{2}$ on each side of the stirrup, as shown in the scheme below.

$$\text{Number of stirrups}=\frac{2700}{203}=13.3$$

Use 15 stirrups.
Provide 15 No. 10 U-stirrups @ 180 mm c.c. Place the first and last stirrups at 90 mm from the interior edge of footings.
Make the necessary sketches.

(Continued)

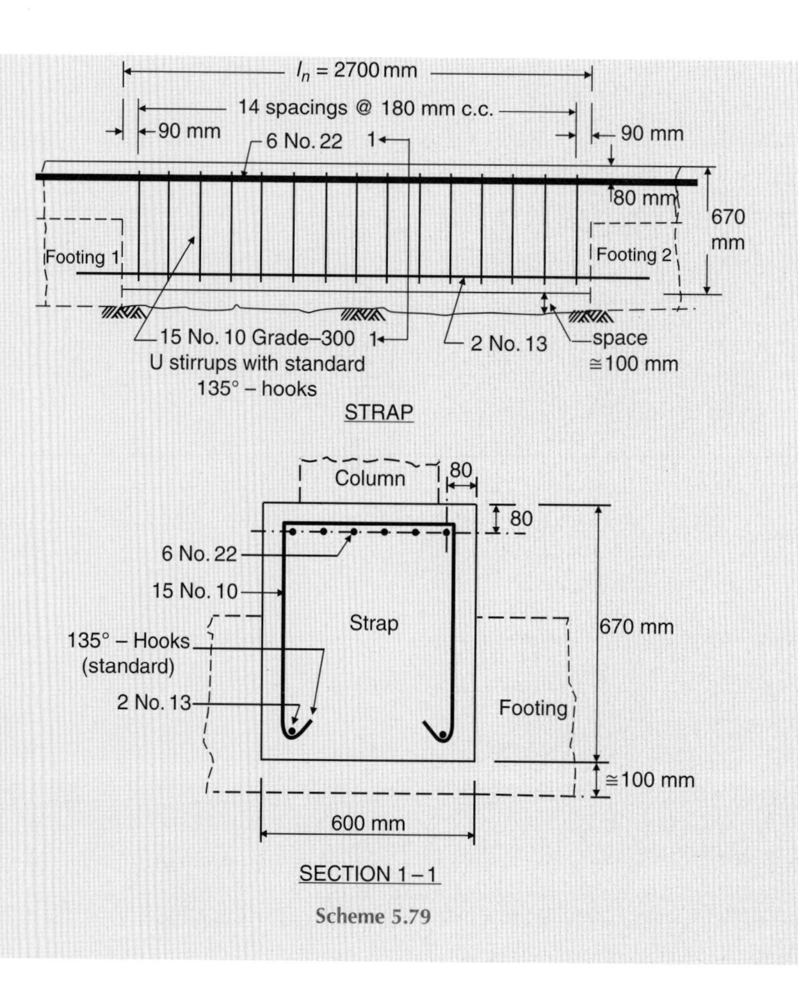

Scheme 5.79

Problem 5.16

(A) Design a rectangular combined footing using the same data given in Problem 5.14.
(B) Make a rough quantity estimate of concrete and steel materials required for construction of the rectangular combined footing designed in (A).
(C) Make a rough quantity estimate of concrete and steel materials required for construction of the strap footing of Problem 5.14 but using the strap of Problem 5.15.

Solution:
(A) Design of the rectangular combined footing.
Step 1. Compute working and factored column loads and factored resultant.

Column	Working loads, kN	Factored loads, kN
1	200 + 350 = *550*	$P_1 = 1.2 \times 200 + 1.6 \times 350 = 800$
2	400 + 300 = *700*	$P_2 = 1.2 \times 400 + 1.6 \times 300 = 960$
		$R = 1760$

Step 2. Estimate the footing dimensions and *net* q_{factored}.

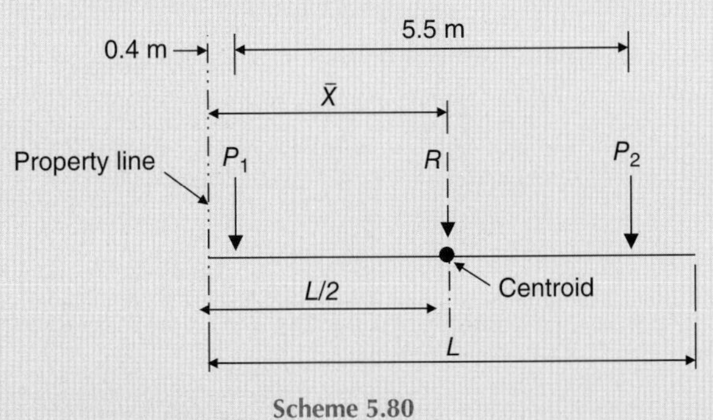

Scheme 5.80

The factored resultant R is located at

$$\bar{X} = \frac{0.4 \times P_1 + (0.4 + 5.5)(P_2)}{R} = \frac{0.4 \times 800 + (5.9)(960)}{1760} = 3.4\,\text{m}$$

Let $\frac{L}{2} = \bar{X}$

$$L = 2 \times 3.4 = 6.8\,\text{m}$$

$$\text{Area of footing } A_f = \frac{\text{working } R}{net\ q_a}$$

$$= \frac{550 + 700}{150} = 8.33\,\text{m}^2$$

$$B = \frac{A_f}{L} = \frac{8.33}{6.8} = 1.225\,\text{m}$$

With these B and L dimensions the footing appears like a beam supporting columns, which is, usually, undesirable. Therefore, it may be more adequate if a wider footing is provided.

Let us, arbitrarily, use $B = 2.0$ m:

$$A_f = 2 \times 6.8 = 13.6\,\text{m}^2 > 8.33\,\text{m}^2.$$

$$net\ q = \frac{\text{working } R}{A_f} = \frac{1250}{13.6} = 92\,\text{kPa} \ll net\ q_a\ (OK.)$$

Check eccentricity of unfactored R:

$$\bar{X} = \frac{0.4(550) + (5.9)(700)}{1250} = \frac{4350}{1250} = 3.48\,\text{m}$$

$$\text{eccentricity } e = \bar{X} - \left(\frac{L}{2}\right)$$

$$= 3.48 - 3.40 = 0.08\,\text{m}\ \ (\text{Acceptable})$$

$$\text{Average } net\ q_{\text{factored}} = \frac{\text{factored } R}{A_f} = \frac{1760}{13.6} = 129.41\ kPa.$$

Use *2.0 × 6.8 m* rectangular combined footing.

Step 3. Obtain data for shear-force and bending-moment diagrams considering the combined footing as a reinforced concrete beam.

(*Continued*)

It is preferable to draw the complete load diagram first, and then the shear and moment diagrams, as shown below. The factored distributed loads on the footing, shear forces and bending moments shall be computed for the full 2.0-m width of the footing.

$$q = net\ q_{\text{factored}} \times B$$

$$= 129.41 \times 2 = 258.82\ kN/m$$

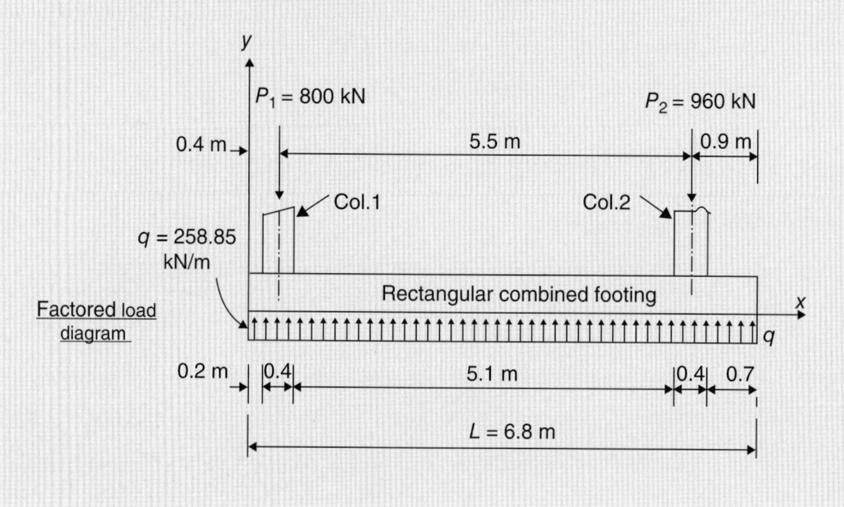

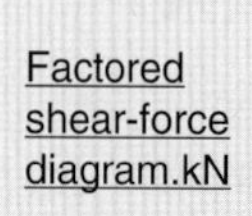

Factored shear-force diagram.kN

Factored Bending-moment diagram,kN.m

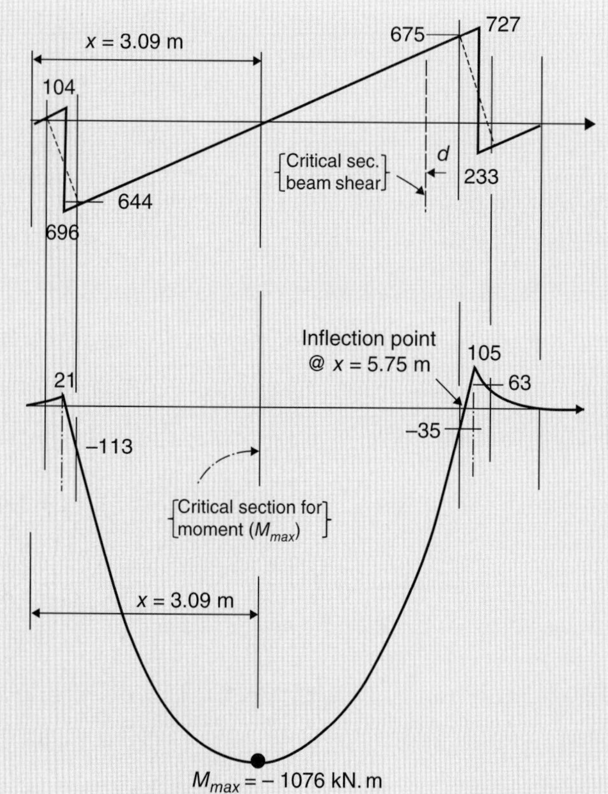

Scheme 5.81

Step 4. Find footing thickness h.

Determine the required thickness based on both one-way shear and two-way shear analyses.

(a) One-way shear. It is critical at distance d from the interior face of the interior column (Col. No. 2), as indicated on the shear-force diagram.

$$V_u = 675 - 258.82\,d \quad (\text{kN})$$

Equation (5.19):

$$\varnothing\, V_c \geq V_u \; (\text{taking reinforcement shear strength, } V_s = 0)$$

$$\text{where } \varnothing = \text{strength reduction factor (ACI Section 9.3.2.3)}$$
$$= 0.75 \text{ for shear}$$

Equation (5.20): $V_c = 0.17\lambda\sqrt{f_c'}\,b_w d$

Let $\varnothing\, V_c = V_u$:

$$(0.75 \times 0.17 \times 1)(\sqrt{21})(1000)(2)(d) = 675 - 258.82\,d$$

$$d = \frac{675}{1427.38} = 0.473\text{ m} = 473\text{ mm}$$

Try $d = 0.5$ m $= 500$ mm

(b) Two-way shear at location of each column. If it is not very clear at which column this shear is most critical, both locations should be considered. One may detect location of the most critical shear by comparing the shear-perimeter ratio with the shear-force ratio.

The shear perimeter at Column 1 (the exterior column) has three sides only, whereas at Column 2 (the interior column) it has four sides, as shown in the scheme below.

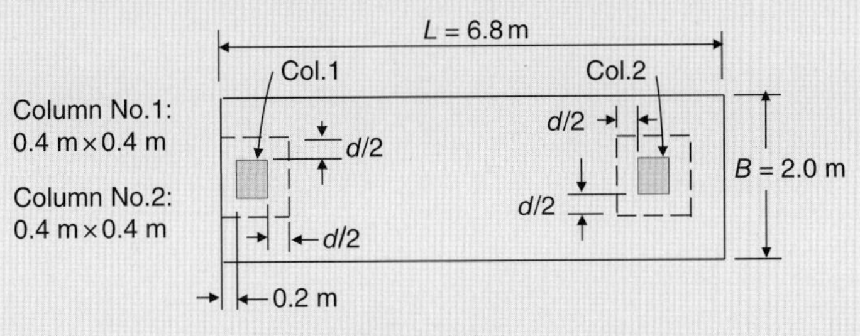

Scheme 5.82

Assume using $d = 0.5$ m, as required by one-way shear.

$$\text{The shear-perimeter ratio} = \frac{2(0.2 + 0.4 + 0.25) + (0.4 + 0.5)}{4(0.4 + 0.5)} = 0.72$$

$$\text{The shear-force ratio} = \frac{800 - 129.41(0.2 + 0.4 + 0.25)(0.4 + 0.5)}{960 - 129.41(0.4 + 0.5)(0.4 + 0.5)} = \frac{701}{855.18} = 0.82$$

Since the shear-perimeter ratio is smaller than the shear-force ratio, the most critical shear will probably be at column 1. Furthermore, there will be an increase in applied shear stress at Column 1 due to transfer of unbalanced moment. However, for completeness, we will check two-way shear at location of each column.

(Continued)

Two-way shear at location of Column 1:

In this case, we should realise that the condition of unbalanced moment transfer exists, and the design must satisfy the requirements of ACI Sections 11.11.7.1, 11.11.7.2 and 13.5.3. The analysis proceeds in the same manner as that for the three-sided shear perimeter at location of column 1 of Problem 5.14.

Area of the shear block = (0.2 + 0.4 + 0.25) × (0.4 + 5) = 0.765 m^2

Soil reaction on the shear block = 129.41 × 0.765 = 99 kN

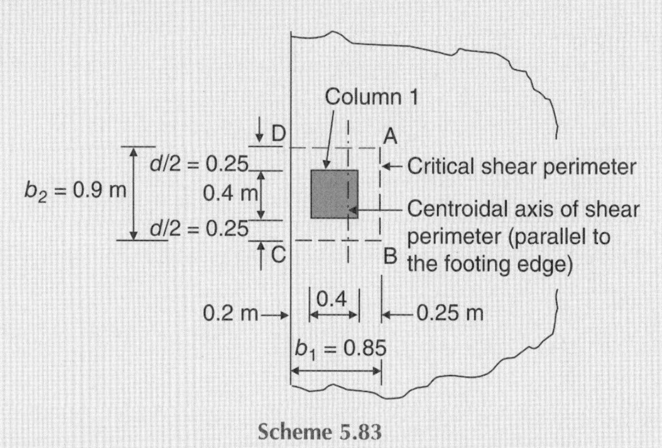

Scheme 5.83

C_{AB} is the distance from line A–B to the centroid of the shear perimeter.

$$= \frac{\left(\begin{array}{c}\text{moment of area of sides}\\ \text{about A-B}\end{array}\right)}{\text{area of sides}} = \frac{2(0.85 \times 0.5)(0.85/2)}{(2 \times 0.85 + 0.9)(0.5)} = 0.28\,\text{m}$$

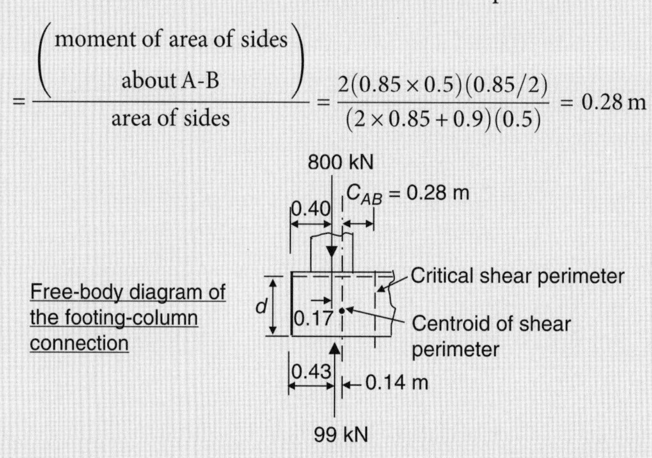

Scheme 5.84

Summation of moments about the centroid of the shear perimeter equals the unbalanced moment, M_u. Hence,

$$M_u = 800 \times 0.17 - 99 \times 0.14 = 122.14\,\text{kN.m}$$

$$\gamma_v = \left(1 - \gamma_f\right) \qquad \text{(ACI Section 11.11.7.1)}$$

$$\gamma_f = \frac{1}{1 + (2/3)\sqrt{b_1/b_2}} \qquad \text{(ACI Section 13.5.3.2)}$$

$$= \frac{1}{1 + (2/3)\sqrt{0.85/0.0.9}} = 0.607$$

$$\gamma_v = 1 - 0.607 = 0.393$$

The shear stresses due to the direct shear and the shear due to moment transfer will add at points D and C, giving the largest shear stresses on the critical shear perimeter. Hence,

$$v_{u(\mathrm{DC})} = \frac{V_u}{b_o d} + \frac{\gamma_v M_u C_{\mathrm{DC}}}{J_c} \quad \text{(ACI Section } R11.11.7.2\text{)}$$

$J_c = J_z =$ Property of the shear perimeter analogous to polar moment of inertia

$$J_c = 2\left(\frac{b_1 d^3}{12}\right) + 2\left(\frac{d b_1^3}{12}\right) + 2(b_1 d)\left(\frac{b_1}{2} - C_{\mathrm{AB}}\right)^2 + (b_2 d) C_{\mathrm{AB}}{}^2$$

$$= 0.0177 + 0.0512 + 0.0179 + 0.0353 = 0.122\,\mathrm{m}^4$$

$$V_u = 800 - 99 = 701\,\mathrm{kN}$$

$$C_{\mathrm{DC}} = b_1 - C_{\mathrm{AB}} = 0.85 - 0.28 = 0.57\,\mathrm{m}$$

$$v_{u(\mathrm{DC})} = \frac{701}{(2 \times 0.85 + 0.9)(0.5)} + \frac{0.393 \times 122.14 \times 0.57}{0.122}$$

$$= 539.23 + 224.27$$

$$= 763.5\ \mathrm{kN/m^2}$$

$$\varnothing\, v_c \geq v_{u(DC)}$$

$$\varnothing\, v_c = \frac{\varnothing\, V_c}{b_o d}$$

Shear strength of concrete V_c shall be the smallest of (i), (ii) and (iii):

(i)

$$V_c = 0.17\left(1 + \frac{2}{\beta}\right)\lambda \sqrt{f_c'}\, b_o d \qquad \text{Equation (5.22)}$$

$$= (0.17)\left(1 + \frac{2}{1}\right)(1)\sqrt{f_c'}\, b_o d = (0.51)\sqrt{f_c'}\, b_o d$$

(ii)

$$V_c = 0.083\left(\frac{\alpha_s d}{b_o} + 2\right)\lambda \sqrt{f_c'}\, b_o d \qquad \text{Equation (5.23)}$$

$$= (0.083)\left(\frac{30 \times 0.5}{2(0.85) + 0.9} + 2\right)\sqrt{f_c'}\, b_o d = (0.64)\sqrt{f_c'}\, b_o d$$

(iii)

$$V_c = 0.33\,\lambda \sqrt{f_c'}\, b_o d \qquad \text{Equation (5.24)}$$

$$= 0.33\,(1)\sqrt{f_c'}\, b_o d = (0.33)\sqrt{f_c'}\, b_o d$$

Use $V_c = (0.33)\sqrt{f_c'}\, b_o d$

(Continued)

$$\varnothing V_c = (0.75)(0.33)\sqrt{21}(1000)(b_o d)$$
$$= 1134.2\, b_o d \ \ (\text{kN})$$
$$\varnothing v_c = \frac{1134.2\, b_o d}{b_o d} = 1134.2\,\text{kN/m}^2 > v_{u(\text{DC})}$$

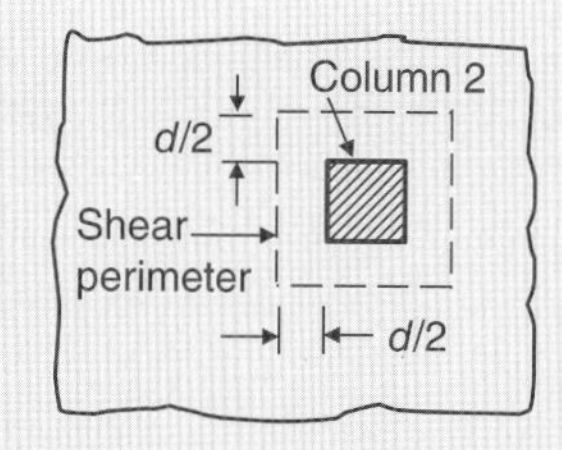

Scheme 5.85

Therefore, one-way shear controls.

Two-way shear at location of Column 2:

The shear perimeter has four sides, as shown.

It is clear that shear strength of concrete $V_c = (0.33)\sqrt{f_c'}b_o d$ controls. Therefore,

$$\varnothing V_c = (0.75)(0.33)\sqrt{21}(1000)[4(0.4+0.5)]0.5 = 2041\,\text{kN}$$

$V_u = P_2$ −soil reaction on the shear block

$V_u = 960$ −soil reaction on the shear block (kN)

$V_u < \varnothing V_c$

Therefore, one-way shear controls.

Assume using No. 25 bars, with 75 mm minimum concrete cover (ACI Section 7.7.1). Hence, the overall footing thickness is

$$h = 500 + 75 + \frac{1}{2}\text{bar diameter} = 575 + \left(\frac{25.4}{2}\right) = 588\,\text{mm}.$$

Use: d = 0.5 *m* = 500 *mm*; *h* = 600 *mm*.

The depth d will be taken to the centre of the steel bars in long direction.

Step 5. Design the flexural reinforcement in the long direction.

(a) Midspan negative reinforcement at top of the footing.

Design negative factored moment = 1076 kN. m

Equation (5.9):

$$\varnothing M_n \geq M_u$$

$$M_u = \frac{1076}{B} = \frac{1076}{2} = 538\,\text{kN.m/m}$$

Assume tension-controlled section, $\varnothing = 0.9$, and $f_s = f_y$.

Equation (5.10):

$$a = \frac{A_s f_y}{0.85 f_c' b}$$
$$= \frac{A_s \times 420}{0.85 \times 21 \times 1} = 23.53\, A_s$$

Equation (5.11):

$$\varnothing M_n = \varnothing\left[A_s f_y\left(d - \frac{a}{2}\right)\right]$$
$$= (0.9)\left[(A_s \times 420 \times 1000)\left(0.5 - \frac{23.53\, A_s}{2}\right)\right]$$
$$= 189\,000\, A_s - 4\,447\,170\, A_s^2$$

Let $\varnothing M_n = M_u$:

$$189\,000 A_s - 4\,447\,170 A_s^2 = 538$$

$$A_s^2 - 42.5\times10^{-3} A_s + 1.21\times10^{-4} = 0$$

$$A_s = \frac{-(-42.5\times10^{-3}) \pm \sqrt{(-42.5\times10^{-3})^2 - (4)(1)(1.21\times10^{-4})}}{2\times1} = \frac{0.0425 - 0.0364}{2}$$

$$= 3.05\times10^{-3}\,\text{m}^2/\text{m} = 3050\,\text{mm}^2/\text{m}$$

ACI Section 10.5.1:

$$A_{s,\min} = \frac{0.25\sqrt{f_c'}}{f_y} b_w d = \frac{0.25\sqrt{21}}{420}\times1\times0.5 = 1.364\times10^{-3}\,\text{m}^2/\text{m}$$

$$= 1364\,\text{mm}^2/\text{m}$$

and not less than $1.4\dfrac{b_w d}{f_y} = \dfrac{1.4\times1\times0.5}{420} = 1.667\times10^{-3}\dfrac{\text{m}^2}{\text{m}} = 1667\,\text{mm}^2/\text{m}$

Therefore, use $A_s = 3050\ \text{mm}^2/\text{m}$

$$\text{Total}\, A_s = 3050\times B = 3050\times2 = 6100\,\text{mm}^2$$

Try 12 No. 25 bars. $A_{s,\text{provided}} = 12\times510 = 6120\ \text{mm}^2$ (OK.)

Compute a for $A_s = \dfrac{6120}{2} = 3060\,\text{mm}^2/\text{m}$, and check if $f_s = f_y$ and whether the section is tension-controlled:

$$a = 23.53\,A_s = 23.53\times3060\times10^{-6} = 0.072\,\text{m} = 72\,\text{mm}$$

$c = \dfrac{a}{\beta_1}$. For f_c' between 17 and 28 MPa, β_1 shall be taken as 0.85. For f_c' above 28 MPa, β_1 shall be reduced linearly at a rate of 0.05 for each 7 MPa of strength in excess of 28 MPa, but β_1 shall not be taken less than 0.65 (ACI Section 10.2.7.3).

$$c = \frac{72}{0.85} = 84.71\,\text{mm}$$

$$d_t = d = 500\,\text{mm}$$

$$\varepsilon_t = 0.003\left(\frac{d_t - c}{c}\right) \qquad \text{Equation (5.15)}$$

$$= 0.003\left(\frac{500 - 84.71}{84.71}\right) = 0.015 > 0.005.$$

Hence, the section is tension-controlled, and $\varnothing = 0.9$ (ACI Sections 10.3.4 and 9.3.2.1).

$\varepsilon_y = \dfrac{f_y}{E_s} = \dfrac{420}{200\,000} = 0.0021$. Hence, $\varepsilon_t > \varepsilon_y$ and $f_s = f_y$

Therefore the assumptions made are satisfied.

In case the required flexural reinforcement exceeds the minimum flexural reinforcement, it shall be adequate to use the same maximum spacing of reinforcement for slabs which is two times the slab thickness or 450 mm, whichever is smaller, as specified by ACI Section 13.3.2.

(*Continued*)

Use centre-to-centre bar spacing = 165 mm. Check the 75 mm minimum concrete cover at the large end of the footing:

$$\text{Concrete cover} = \frac{2000 - (11 \times 165 + 25.4)}{2} = 79.8\,\text{mm} > 75\,mm \quad (\text{OK.})$$

Try 12 No. 25 bars@ 165 mm c.c. at top, across the footing width.

Check the development of reinforcement:
In this case, the bars are in tension, the provided bar size is No. 25, the clear spacing of the bars exceeds $2d_b$, and the clear cover exceeds d_b. Therefore, the development length l_d of bars at each side of the critical section (which is located where the maximum factored negative moment exists) shall be determined from the following equation, but not less than 300 mm:

$$l_d = \left(\frac{f_y \psi_t \psi_e}{1.7\lambda\sqrt{f_c'}}\right) d_b \quad (\text{ACI Sections 12.2.1, 12.2.2 and 12.2.4})$$

Where the factors ψ_e and λ are 1 (ACI Sections 12.2.4 and 8.6.1).

The factor $\psi_t = 1.3$ because the reinforcement is placed such that more than 300 mm of fresh concrete exists below the top bars (ACI Section 12.2.4).

$$l_d = \left(\frac{420 \times 1.3 \times 1}{1.7 \times 1 \times \sqrt{21}}\right)\left(\frac{25.4}{1000}\right) = 1.78\,\text{m} = 1780\,\text{mm} > 300\,mm.$$

Therefore, the required $l_d = 1780$ mm.

The bar extension past the critical section (i.e. the available length) is

$$3090\,\text{mm} - 75\,\text{mm cover} = 3015\,\text{mm} > 1780\,\text{mm} \quad (\text{OK.})$$

According to MacGregor and Wight (2005), since the loading, the supports and the shape of the moment diagram are all inverted from those found in a normally loaded beam, the necessary check of the ACI Section 12.11.3 requirement which concerns positive moment reinforcement shall be made for the negative moment reinforcement.

It may be appropriate to extend all the bars into the column regions. At the point of inflection, 0.15 m from the centre of the interior column,

$$V_u = 727 - 0.15 \times 258.82 = 688.2\,\text{kN}.$$

ACI Section 12.11.3 requires $l_d \le \dfrac{M_n}{V_u} + l_a$

$$M_n = \text{nominal flexural strength of the cross section} = A_s f_y\left(d - \frac{a}{2}\right)$$

$$a = \frac{A_s f_y}{0.85 f_c' b} = \frac{(6120/10^6)(420)}{0.85 \times 21 \times 2} = 0.072$$

$$M_n = (6120/10^6)(420 \times 1000)\left(0.5 - \frac{0.072}{2}\right) = 1193\,\text{kN.m}$$

Let $l_d = \dfrac{M_n}{V_u} + l_a$:

$$l_{a,\min} = l_d - \frac{M_n}{V_u} = 1.78 - \frac{1193}{688.2} = 1.780 - 1.734 = 0.046\ \text{m}$$

l_a at a support shall be the embedment length beyond centre of support.
Accordingly, extend the bars to the exterior face (right face) of the interior column. This will give $l_a = 0.20\ \text{m} \gg l_{a,\min}$.

However, in order to satisfy the requirements of ACI Sections 12.12.3 and 12.10.3, the negative moment reinforcement shall have an embedment length beyond the point of inflection not less than d, $12d_b$ or $\ell_n/16$, whichever is greater. In this case, d controls. Therefore, the top bars should be extended not less than 0.5 m beyond the point of inflection at $x = 5.75$ m. However, since the remaining distance to the footing edge is only 0.55 m, it may be more practical to extend the top bars to the right edge of the footing (less 75 mm minimum concrete cover).

At the region of the exterior column a similar condition prevails. However, the available space beyond the column centre is only 0.4 m less concrete cover, which is smaller than the minimum required embedment length d or 0.5 m. Therefore, the top bars will all have to be hooked (90-degree standard hook) at the footing left edge. This is also necessary to anchor the bars transferring the unbalanced moment from the column to the footing.

Provide 12 No. 25 bars @ 165 mm c.c. at top, full length of the footing (less concrete cover at the ends). All the bars should be hooked (90-degree standard hook) at the footing left edge.

(b) Positive reinforcement at bottom of the footing.

Design positive factored moment = 105 kN.m (more conservative than using 63 kN.m). However, the moments are so small that $A_{s,\min}$ controls and it makes no difference which moment is used.

$$M_u = \frac{105}{B} = \frac{105}{2} = 53\ \text{kN.m/m}$$

Assume tension-controlled section $\emptyset = 0.9$ and $f_s = f_y$.

$$a = \frac{A_s f_y}{0.85 f_c' b} = \frac{A_s \times 420}{0.85 \times 21 \times 1} = 23.53 A_s$$

Equation (5.11):

$$\emptyset M_n = \emptyset \left[A_s f_y \left(d - \frac{a}{2}\right)\right]$$

$$= (0.9)\left[(A_s \times 420 \times 1000)\left(0.5 - \frac{23.53 A_s}{2}\right)\right]$$

$$= 189\,000\,A_s - 4\,447\,170 A_s^2\ (\text{kN.m/m})$$

$\emptyset M_n \geq M_u$. Let $\emptyset M_n = M_u$

$$189\,000\,A_s - 4\,447\,170 A_s^2 = 53$$

$$A_s^2 - 42.5 \times 10^{-3} A_s + 1.19 \times 10^{-5} = 0$$

$$A_s = \frac{-(-42.5 \times 10^{-3}) \pm \sqrt{(-42.5 \times 10^{-3})^2 - (4)(1)(1.19 \times 10^{-5})}}{2 \times 1} = \frac{0.0425 - 0.0419}{2}$$

$$= 6 \times \frac{10^{-4}\ \text{m}^2}{\text{m}} = 600\ \text{mm}^2/\text{m, required by analysis.}$$

(Continued)

ACI Section 10.5.1:

$$A_{s,\min} = \frac{0.25\sqrt{f_c'}}{f_y} b_w d = \frac{0.25\sqrt{21}}{420} \times 1 \times 0.5 = 1.364 \times 10^{-3}\ \text{m}^2/\text{m}$$

$$= 1364\ \text{mm}^2/\text{m}$$

and not less than:

$$1.4\frac{b_w d}{f_y} = \frac{1.4 \times 1 \times 0.5}{420} = 1.667 \times 10^{-3}\ \text{m}^2/\text{m} = 1667\ \text{mm}^2/\text{m}$$

Therefore, $A_{s,\min}$ controls. However, ACI Section 10.5.3 states that the requirements of ACI Section 10.5.1 need not be applied if, at every section, A_s provided is at least one-third greater than that required by analysis.

Accordingly, total $A_{s,\min} = (1.33 \times 600)(B) = 800 \times 2 = 1600\ \text{mm}^2$.

Try 6 No. 19 bars: $A_{s,\text{provided}} = 6 \times 284 = 1704\ \text{mm}^2$ (OK.)

ACI Sections 7.6.5 and 10.5.4 requires maximum spacing shall not exceed three times the slab or footing thickness, or 450 mm, whichever is smaller.

Use centre to centre bar spacing = 365 mm. Check the 75 mm minimum concrete cover at each side:

$$\text{Concrete cover} = \frac{2000 - (5 \times 365 + 19.1)}{2} = 78\ \text{mm} > 75\ mm\quad (OK.)$$

Try six No. 19 bars @ 364 mm c.c. at bottom, across the footing width.

Check the development of reinforcement:

In this case, the bars are in tension, the provided bar size is No. 19, the clear spacing of the bars exceeds $2d_b$, and the clear cover exceeds d_b. Therefore, the development length l_d of bars at each side of the critical section (which is located where the maximum factored positive moment exists) shall be determined from the following equation, but not less than 300 mm:

$$l_d = \left(\frac{f_y \psi_t \psi_e}{2.1\lambda\sqrt{f_c'}}\right) d_b \quad \text{(ACI Sections 12.2.1, 12.2.2 and 12.2.4)}$$

Where the factors ψ_t ψ_e, λ and are 1 (ACI Sections 12.2.4 and 8.6.1).

$$l_d = \left(\frac{420 \times 1 \times 1}{2.1 \times 1 \times \sqrt{21}}\right)\left(\frac{19.1}{1000}\right) = 0.83\ \text{m} = 830\ \text{mm} > 300\ mm.$$

Therefore, the required $l_d = 830$ mm.

The bar extension past the critical section (i.e. the available length) is

$$900\ \text{mm} - 75\ \text{mm cover} = 825\ \text{mm} < 830\ \text{mm}\quad \text{(NotOK.)}$$

Therefore, all the bottom bars in long direction have to be hooked (90-degree standard hook) at the footing right edge.

In order to satisfy the requirements of ACI Sections 12.10.3 and 12.11.3, it is necessary to extend the bottom bars a minimum distance d past the point of inflection. Therefore, cut off two bottom bars at 1.0 m from the center of the interior column toward the exterior column. The otherfour bars shall be extended full length of the footing (less75 mm cover at each end). These four bars will resist the tensile stresses due to the relatively small positive moment at the exterior column and provide support to the transverse bottom bars.

Provide six No. 19 bars @ 365 mm c.c. at bottom of the footing in long direction. Cut off 2 bars at **1.0 m** *from the centre of the interior column. The other four bars shall be extended full length of the footing. All the bars have to be hooked (90-degree standard hook) at the footing right edge.*

Step 6. Design the transverse reinforcement at bottom of the footing.

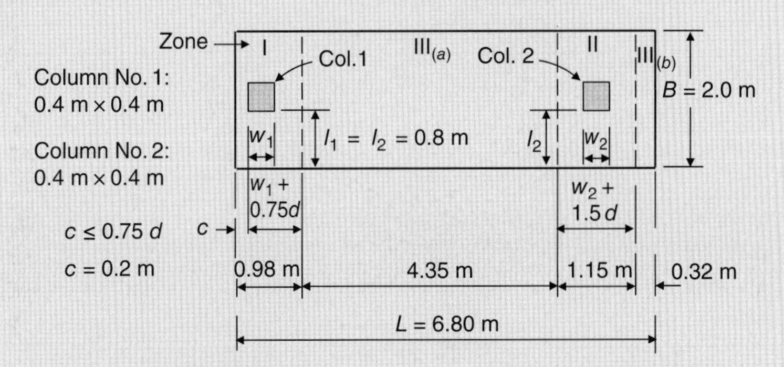

Scheme 5.86

Transverse reinforcement for zone I:

$$net\ q_{\text{factored}} = \frac{P_1}{\text{area}} = \frac{800}{0.98 \times 2.0} = 408.2\,\text{kPa}$$

$$M_u = \frac{net\ q_{\text{factored}} \times l_1^2}{2} = \frac{408.2 \times 0.8^2}{2} = 131\,\text{kN.m/m}$$

$$d = 0.5\,\text{m} - \text{one bar diameter} = 0.5 - 0.0191 = 0.4809\,\text{m}$$

Use $d = 0.48$ m.

Assume tension-controlled section, $\varnothing = 0.9$, and $f_s = f_y$.

$$a = \frac{A_s f_y}{0.85 f_c' b} = \frac{A_s \times 420}{0.85 \times 21 \times 1} = 23.53\ A_s$$

Equation (5.11):

$$\varnothing M_n = \varnothing\left[A_s f_y \left(d - \frac{a}{2}\right)\right]$$

$$= (0.9)\left[(A_s \times 420 \times 1000)\left(0.48 - \frac{23.53\,A_s}{2}\right)\right]$$

$$= 181\,440\,A_s - 4\,447\,170\,A_s^2$$

Let $\varnothing M_n = M_u$:

$$181\,440\,A_s - 4\,447\,170 A_s^2 = 131$$

$$A_s^2 - 0.041\,A_s + 2.95 \times 10^{-5} = 0$$

$$A_s = \frac{-(-0.041) \pm \sqrt{(-0.041)^2 - (4)(1)(2.95 \times 10^{-5})}}{2 \times 1} = \frac{0.0410 - 0.0395}{2}$$

$$= 1.5 \times 10^{-3}\,\text{m}^2/\text{m} = 1500\,\text{mm}^2/\text{m}$$

(*Continued*)

$$A_{s,\min} = \frac{0.25\sqrt{f_c'}}{f_y} b_w d = \frac{0.25\sqrt{21}}{420} \times 1 \times 0.48 = 1.31 \times 10^{-3}\ \text{m}^2/\text{m}$$

$$= 1310\ \text{mm}^2/\text{m}$$

and not less than: $$1.4\frac{b_w d}{f_y} = \frac{1.4 \times 1 \times 0.48}{420}$$

$$= 1.6 \times 10^{-3}\frac{\text{m}^2}{\text{m}} = 1600\ \text{mm}^2/\text{m}$$

$A_{s,\min} > A_s$ required by analysis. Therefore, $A_{s,\min}$ controls.
Use $A_s = A_{s,\min} = 1600\ \text{mm}^2/\text{m}$.

Transverse reinforcement for zone II:

$$net\ q_{\text{factored}} = \frac{P_2}{\text{area}} = \frac{960}{1.15 \times 2} = 417.4\ \text{kPa}$$

$$M_u = \frac{net\ q_{\text{factored}} \times l_2^2}{2} = \frac{417.4 \times 0.8^2}{2} = 134\ \text{kN.m/m}$$

$$\varnothing M_n = 181\,440\,A_s - 4\,447\,170 A_s^2 \quad \text{(as before)}$$

Let $\varnothing M_n = M_u$:

$$181\,440\,A_s - 4\,447\,170 A_s^2 = 134$$

$$A_s^2 - 0.041\,A_s + 3.01 \times 10^{-5} = 0$$

$$A_s = \frac{-(-0.041) \pm \sqrt{(-0.041)^2 - (4)(1)(3.01 \times 10^{-5})}}{2 \times 1} = \frac{0.0410 - 0.0395}{2}$$

$$= 1.5 \times 10^{-3}\ \text{m}^2/\text{m} = 1500\ \text{mm}^2/\text{m} < A_{s,\min}$$

Therefore, $A_{s,\min}$ controls. Use $A_s = A_{s,\min} = 1600\ \text{mm}^2/\text{m}$.
Transverse reinforcement for zone III:
Use $A_s = A_{s,\min} = 1600\ \text{mm}^2/\text{m}$.
Thus, in this case, the required amount of reinforcement for each zone is 1600 mm^2 per meter length of the footing.

$$\text{Total } A_s = 1600 \times L = 1600 \times 6.8 = 10\,880\ \text{mm}^2.$$

Try 39 No. 19 bars. $A_{s,\text{provided}} = 39 \times 284 = 11\,076\ \text{mm}^2$ (OK.)
ACI Sections 7.6.5 and 10.5.4 requires maximum spacing shall not exceed three times the slab or footing thickness or 450 mm, whichever is smaller.
Use centre to centre bar spacing = 174 mm. Check the 75 mm minimum concrete cover at each side:

$$\text{Concrete cover} = \frac{6800 - (38 \times 174 + 19.1)}{2} = 84\ \text{mm} > 75\ mm \quad \text{(OK.)}$$

Thus, theoretically, the bars will be uniformly distributed with 174 mm centretocentre spacing. However, it may be more realistic if the amount of steel in the effective zones I and II are somewhat increased and that in zone III decreased. We shall do this by arbitrarily decreasing and increasing the bar spacing in the three zones but without decreasing the required total steel amount, as follows:

Zone I: seven No.19 bars; six spacings @ 147 mm c.c.
Zones II+ Zone III$_{(b)}$: 11 No.19 bars;10 spacings @ 137 mm c.c.
Zone III$_{(a)}$: 21 No.19 bars;20 spacings @ 216 mm c.c.

Check the development of reinforcement:
In this case, the bars are in tension, the provided bar size is No. 19, the clear spacing of the bars exceeds $2d_b$, and the clear cover exceeds d_b. Therefore, the development length l_d of bars at each side of the critical section (which is located where the maximum factored positive moment exists) shall be determined from the following equation, but not less than 300 mm.

$$l_d = \left(\frac{f_y \psi_t \psi_e}{2.1\lambda\sqrt{f_c'}}\right) d_b \quad \text{(ACI Sections 12.2.1, 12.2.2 and 12.2.4)}$$

where the factors ψ_t, ψ_e and λ are 1 (ACI Sections 12.2.4 and 8.6.1)

$$l_d = \left(\frac{420 \times 1 \times 1}{2.1 \times 1 \times \sqrt{21}}\right)\left(\frac{19.1}{1000}\right) = 0.83\ \text{m} = 830\ \text{mm} > 300\ mm.$$

Therefore, the required $l_d = 830$ mm.
The bar extension past the critical section (i.e. the available length) is

$$800\ \text{mm} - 75\ \text{mm cover} = 725\ \text{mm} < 830\ \text{mm (Not OK.)}$$

Therefore, all the bottom transverse bars have to be hooked at both ends.
The bottom transverse bars will all have to be hooked (90-degree standard hook) at both ends.
Because two-way shear cracks would extend roughly the entire width of the footing, the hooked transverse bars provide adequate anchorage outside the inclined cracks (MacGregor and Wight, 2005).
*Step*7. Check columns bearing on the footing (ACI Section 10.14), and design the necessary dowels.
The design proceeds in the same manner as that of Solution of Problem 5.3, *Step 5.*
*Step*8. Find the embedment length of dowels in both the footing and the columns.
The design proceeds in the same manner as that ofSolution of Problem 5.3, *Step 6.*
Step 9. Develop the final design sketches.
(B) and **(C)** Rough quantity estimate of concrete and steel materials.

Table 5.7 Quantity estimates of concrete and steel materials.

Material	Rectangular combined footing	Strap footing		Total
		Footing 1 and Footing 2	Strap	
Concrete, m^3 ($f_c' = 21$ MPa)	0.6 × 2.0 × 6.8 = *8.16*	0.45(2 × 2.2 + 2 × 2) = 3.78	0.67 × 0.6 × 2.7 + 0.32 × 0.6 × 3.6 = 1.7	*5.48*
Steel, kg (f_y = 420 MPa)	(12)(6.8 − 0.15 + 12 × 0.0254)(4) = 339 (2)(1.9 − 0.075 + 12 × 0.0191)(2.24) = 9 (4)(6.8 − 0.15 + 12 × 0.0191)(2.24) = 62 (39)(2 − 0.15 + 24 × 0.0191)(2.24) = 202 *Total* (339 + 9 + 62 + 202) = *612*	(3 × 9)(2.0 − 0.15)(1.55) = 77 (9)(2.2 − 0.15)(1.55) = 29 *Total* (77 + 29) = 106	(6)(6.2 − 0.15)(3.04) = 110 (20)(0.56)(2 × 0.57 + 1 × 0.5 + 24 × 0.01) = 21 (2)(6.2 − 0.15)(0.994) = 12 *Total* (110 + 21 + 12) = 143	*249*

Problem 5.17

Design the reinforced concrete cantilever retaining wall in the scheme shown below to provide lateral support for a recreational park. In the required length of the wall, the largest wall height above the ground level (in front the wall) is 2 m. Factors such as soil bearing capacity, depth of frost penetration and seasonal volume change require the base (footing) be placed at a foundation depth $D = 1.2$ m, as shown in the scheme. It is required the wall shall have a front batter not less than 1H : 48V and thickness at top = 0.3 m. Other given data are:

Surcharge: $q_s = 10$ kPa.

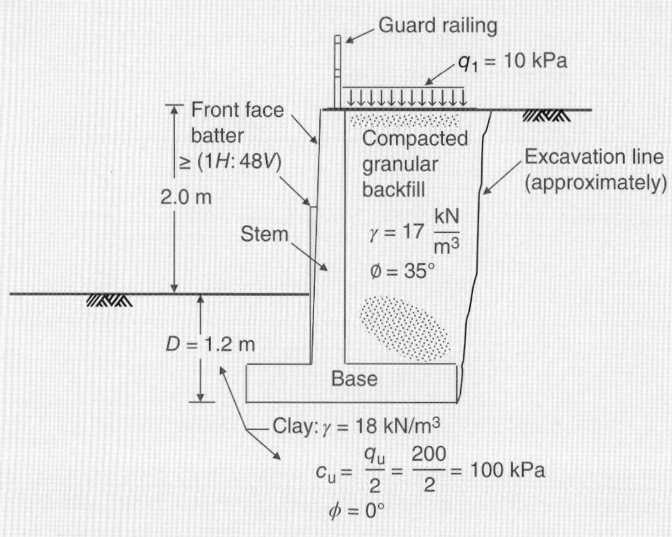

Scheme 5.87

Backfill material: compacted granular backfill in the limited zone over the heel, which will be compacted to $\gamma = 17$ kN/m^3 and an estimated Ø = 35°.

Supporting soil: clay of $\gamma = 18$ kN/m^3, $q_u = 200$ kPa, $c_u = 100$ kPa and Ø = 0°.

Wall stability : *SF* against overturning should not be less than 2.0.

SF against sliding should not be less than 1.5.

Concrete: $f_c' = 21$ MPa

Reinforcing steel: $f_y = 420$ MPa

Solution:

Step 1. Select tentative values for the base dimensions, stem thickness and toe distance. For this purpose, it is common to use the following approximate relationships as a guide (based on experience accumulated with stable walls):

Base width $B = (0.4 \text{ to } 0.7)H$, where H is the total height of the retaining wall including the footing. In this case, $H = 2.0 + 1.2 = 3.2\ m$

Base thickness $h = \left(\frac{1}{12} \text{ to } \frac{1}{10}\right)H$

Stem bottom thickness $b \cong h$

Toe distance $l \cong B/3$

Assume the following dimensions:

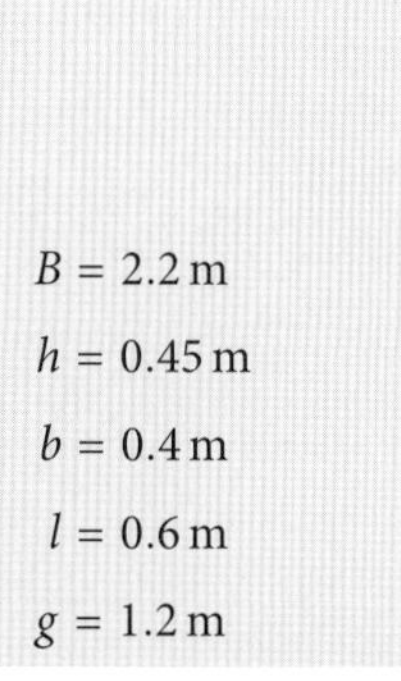

$B = 2.2$ m

$h = 0.45$ m

$b = 0.4$ m

$l = 0.6$ m

$g = 1.2$ m

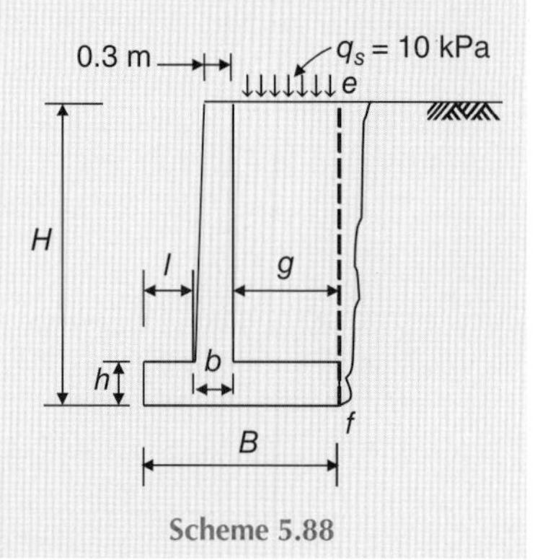

Scheme 5.88

Check the front batter:

$$\text{Slope of the front face} = \frac{b-0.3}{H-h} = \frac{0.4-0.3}{3.20-0.45} = \frac{0.10}{2.75} = \frac{1}{27.5} > \frac{1}{48} \quad \text{(OK.)}$$

Step 2. Using earth pressure principles, obtain the profiles of the active lateral pressure on both the stem and vertical line *ef* at the heel (the "virtual" back, shown in the scheme of *Step 1*).

Lateral earth pressure due to the vertical surcharge pressure q_s and the vertical soil pressure γz, at any depth z, is

$$q_{hz} = (q_s + \gamma z)K_a$$

where K_a = Rankine active earth pressure coefficient.

The coefficient K_a depends on the backfill slope angle β and friction angle Ø can be obtained from tables or may be calculated from the equation:

$$K_a = \cos\beta \frac{\cos\beta - \sqrt{\cos^2\beta - \cos^2 \text{Ø}}}{\cos\beta + \sqrt{\cos^2\beta - \cos^2 \text{Ø}}}$$

$$\text{For } \beta = 0^o \text{ and } \text{Ø} = 35^o : K_a = (1)\frac{(1) - \sqrt{(1) - \cos^2 35}}{(1) + \sqrt{(1) - \cos^2 35}} = \frac{1 - \sin 35}{1 + \sin 35} = 0.271$$

However, there will be a relatively large lateral pressure induced by compaction of the backfill, which may be accounted for by using a larger value for K_a. Also, the probability that a Rankine active wedge will not form in the limited backfill zone should be realised. Moreover, in this case, the guard rail would occasionally have a lateral load from persons leaning against it, which increases lateral pressure on the wall. In order to account for all these effects, it may be advisable (Bowles, 2001) to use K_a equals to the at rest lateral earth pressure coefficient K_o. This gives

$$K_a = K_o = 1 - \sin \text{Ø} = 1 - \sin 35 = 0.426$$

$$q_{hz} = (10 + 17z)(0.426)$$

At top: $q_{hz} = (10 + 17 \times 0.0)(0.426)$ $= 4.26$ kPa

At bottom of stem: $q_{hz} = [(10) + 17(3.20 - 0.45)](0.426)$ $= 24.18$ kPa

At bottom of base: $q_{hz} = [(10) + 17(3.20)](0.426) = 27.43$ kPa

(Continued)

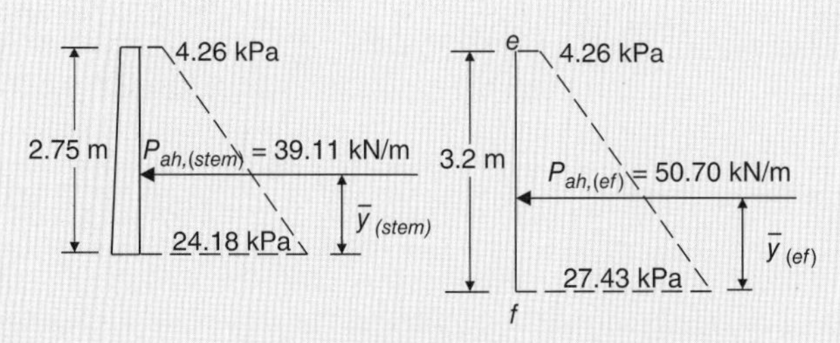

Scheme 5.89

Total lateral force per metre length of wall is P_{ah} = area of pressure diagram

$$P_{ah,(\text{stem})} = \frac{4.26 + 24.18}{2} \times 2.75 = 39.11\ kN/m$$

$$\bar{y}_{(\text{stem})} = \frac{L}{3}\left(\frac{2a+b}{a+b}\right) = \frac{2.75}{3}\left(\frac{2 \times 4.26 + 24.18}{4.26 + 24.18}\right) = 1.05\ \text{m}$$

$$P_{ah,(ef)} = \frac{4.26 + 27.43}{2} \times 3.2 = 50.7\ kN/m$$

$$\bar{y}_{(ef)} = \frac{L}{3}\left(\frac{2a+b}{a+b}\right) = \frac{3.2}{3}\left(\frac{2 \times 4.26 + 27.43}{4.26 + 27.43}\right) = 1.21\ \text{m}$$

Step 3. Check wall stability for overturning and sliding and that the resultant R is in the middle third of base width B.

Before these computations are started, the designer must make a decision on whether to use passive pressure from the soil in front of the toe and whether the adjacent soil directly in front of the stem (covering the base top at the toe) will be available for resisting overturning moments and sliding. Most times toe soil is neglected for a conservative solution, espacially, when the foundation depth D is small. In this case, assume we neglect the soil covering the base in front of the stem, and neglect the *passive resistance* unless necessary.

(a) Compute the vertical resultant force R and its moment M_r about the base toe at point A.

For these computations refer to the scheme of *Step 1* It is convenient to use Table 5.8, where the load sources are labelled, weights computed, moment arms given and so on. Include the surcharge q_s in the backfill soil weight W_s, take concrete unit weight $\gamma_c = 24\ \text{kN/m}^3$ and ignore the weight of the guard rail since its mass per metre length is negligible. Ignore the heel friction $P'_{av,(ef)}$ in order to accomplish more conservative stability analyses.

Table 5.8 Load sources, loads, moment arms and moments for wall stability, *Step 3*.

Load source	Load (weight), kN/m	Arm, m	Moment (M_A), kN.m/m
Backfill soil + surcharge	$1.2(10 + 17 \times 2.75) = 68.1$	1.6	108.96
Stem	$24\left(\frac{0.3+0.4}{2} \times 2.75\right) = 23.1$	0.828	19.13
Base slab	$24(0.45 \times 2.2) = 23.8$	1.1	26.18
$P_{av,(ef)} = 0 \quad (\beta = 0°)$	0.0	—	0.00
$P'_{av,(ef)} = P_{ah,(ef)} \tan(0.8\,\emptyset)$	(ignored)	—	—
	$\mathbf{R = 115.0}$		$\mathbf{\Sigma M_A = M_r = 154.27}$

(b) Check the overturning stability.

Overturning moment $M_o = P_{ah,(ef)} \times \bar{y}_{(ef)}$

$$= 50.7 \times 1.21 = 61.35 \text{ kN.m/m}$$

Resisting moment (ignoring the resisting moment due to the heel friction $P'_{av,(ef)}$, as shown in Table 5.8) = $M_r = 154.27$ kN.m/m

$$\text{Safety factor against overturning}: SF = \frac{M_r}{M_o} = \frac{154.27}{61.35}$$

$$= 2.5 > 2.0 \text{ (OK)}$$

(c) Check that the location of the resultant R is inside the middle third of the base width B.

$$\text{The net overturning moment } M_{o,net} = M_r - M_o = 154.27 - 61.35$$

$$= 92.92 \text{ kN.m/m}$$

$$\bar{x} = \frac{M_{o,net}}{R} = \frac{92.92}{115} = 0.81 \text{ m (from the toe)}$$

$$\text{Eccentricity} = e = \frac{B}{2} - \bar{x} = \frac{2.2}{2} - 0.81 = 0.29 \text{ m}$$

$$\frac{B}{6} = \frac{2.2}{6} = 0.37 \text{ m} > e \text{ (OK.)}$$

The resultant R is inside the middle third of the base width B.

(d) Check the sliding stability.

Resistance against sliding is

$$F_r = R\tan\delta + c_a A_f + P_P$$

where A_f = base effective area $= B' \times 1 = (B - 2e)(1)$

$= 2.2 - 2 \times 0.29 = 1.62 \text{ m}^2/\text{m}$

c_a = base adhesion (0.6 to 1.0c). Assume $c_a = 0.6c_u$

$= 0.6(100) = 60$ kPa

δ = friction angle between base and soil (0.5 to 1.0 Ø)

$= 0$ for Ø $= 0$

P_P = passive resistance

In this case, assume P_P is neglected.

$$F_r = 0 + 60 \times 1.62 + 0 = 97.2 \text{ kN/m}$$

$$\text{Safety factor against sliding } SF = \frac{F_r}{P_{ah,(ef)}} = \frac{97.2}{50.7} = \mathbf{1.9} > 1.5 \text{ (OK.)}$$

Step 4. Compute safe bearing capacity of the supporting soil. Assume using Hansen bearing capacity equation for the Ø = 0° condition.

(Continued)

Table 4.2 $gross\ q_{ult} = 5.14\ s_u\left(1 + s_c' + d_c' - i_c' - b_c' - g_c'\right) + \bar{q}$
Since the toe soil is neglected, the overburden pressure $\bar{q} = 0$. Hence, the safe bearing capacity is

$$gross\ q_{safe} = 5.14\ s_u\left(1 + d_c' - i_c'\right)/SF$$

$$s_u = c_u = 100\ \text{kPa}$$

Table 4.6: $s_c' = 0.2\frac{B'}{L'} = 0$, since $\frac{B'}{L'} \approx 0$ for continuous foundations.
Table 4.7: $b_c' = g_c' = 0$, since $\eta = 0$ and $\beta = 0$

$$d_c' = 0.4k;\ \ k = \frac{D}{B}\ \text{for}\ \frac{D}{B} \le 1;\ \frac{D}{B} = \frac{1.2}{2.2} = 0.545;\ \ \text{hence,}\ d_c' = 0.218$$

$$i_c' = 0.5 - 0.5\sqrt{1 - \frac{H}{A_f c_a}}.\ \text{Use}\ c_a = 60\ \text{kPa};\ A_f = 1.62\ \text{m}^2\ \text{as in}\ \textit{Step 3 (d)}.$$

$$i_c' = 0.5 - 0.5\sqrt{1 - \frac{50.7}{1.62 \times 60}} = 0.154$$

$$gross\ q_{safe} = 5.14\,(100)\,(1 + 0.218 - 0.154)/3$$
$$= 182.3\ kPa$$

$$\text{Gross foundation pressure,}\ gross\ q = \frac{R}{B' \times 1} = \frac{115}{1.62}$$
$$= 71.0\ \text{kPa} \ll gross\ q_{safe}\ \text{(OK.)}$$

Check the maximum gross foundation pressure, $gross\ q_{max}$:

$$gross\ q_{max} = \frac{R}{A} + \frac{(R \times e)\left(\frac{B}{2}\right)}{(1)(B)^3\left(\frac{1}{12}\right)}\quad \text{Check}$$

$$= \frac{115}{1 \times 2.2} + \frac{(115 \times 0.29)\left(\frac{2.2}{2}\right)}{(1)(2.2)^3(1/12)} = 93.62\ \text{kPa} < gross\ q_{safe}\ \text{(OK.)}$$

Step 5. Apply the appropriate load factors (ACI Section 9.2) to all the horizontal and vertical forces, compute eccentricity e of the factored resultant force and calculate the base factored contact pressures.

In this case, the only applicable loads are D (dead loads, or related internal moments and forces) and H (loads due to weight and pressure of soil, or related internal moments and forces). The applicable load combinations of the ACI Section 9.2.1 are reduced to:

$$U = 1.2\,D + 1.6\,(H)$$

where H represents loads due to soil and surcharge (Table 5.9).

$$\text{Factored overturning moment}\ M_o = 1.6\left[P_{ah,(ef)} \times \bar{y}_{(ef)}\right]$$
$$= 1.6 \times 61.35 = 98.16\ \text{kN.m/m}$$

Table 5.9 Factored loads and moments required for calculating factored contact pressures, *Step 5.*

Load source	Factored load (weight), kN/m	Arm, m	Factored moment (M_A), Kn.m/m
Backfill soil + surcharge	$1.6[1.2(10+17\times 2.75)]=109.0$	1.6	174.4
Stem	$1.2\left[24\left(\frac{0.3+0.4}{2}\times 2.75\right)\right]=27.7$	0.83	23.0
Base slab	$1.2[24(0.45\times 2.2)]=23.5$	1.1	25.9
$P_{av,(ef)}=0\ (\beta=0°)$	0.0	—	0.0
$P'_{av,(ef)}=1.6\left[P_{ah,(ef)}\tan(0.8\,\varnothing)\right]$	(ignored)	—	—
	$R=160.2\ kN/m$		$\Sigma M_A=M_r=223.3\ kN.m/m$

$$\text{Factored net overturning moment } M_{o,\text{net}}=M_r-M_o=223.3-98.16=125.14\,\text{kN.m/m}$$

$$\bar{x}=\frac{M_{o,\text{net}}}{R}=\frac{125.14}{160.2}=0.78\text{ m (from the toe)}$$

$$\text{Eccentricity } e=\frac{B}{2}-\bar{x}=\frac{2.2}{2}-0.78=0.32\ m$$

$$\frac{B}{6}=\frac{2.2}{6}=0.37\text{ m}>e\ \ \text{(OK.)}$$

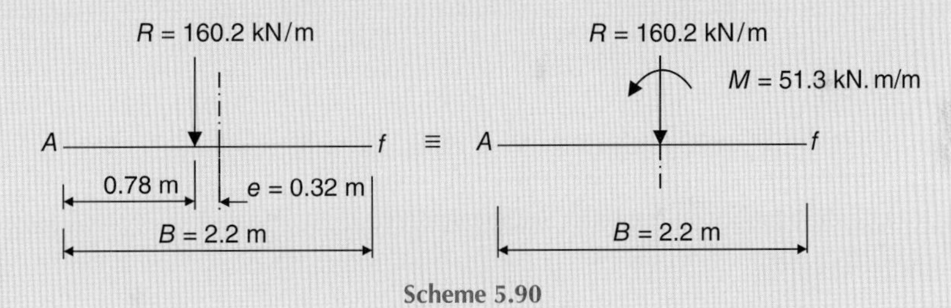

Scheme 5.90

$$\text{Factored } q_{A,(max)}=\frac{R}{A}+\frac{(M)\left(\frac{B}{2}\right)}{(1)(B)^3\left(\frac{1}{12}\right)}=\frac{160.2}{2.2\times 1}+\frac{(51.3)\left(\frac{2.2}{2}\right)}{(1)(2.2)^3\left(\frac{1}{12}\right)}$$

$$=72.82+63.60=136.42\ kPa$$

$$\text{Factored } q_{f,(min)}=\frac{R}{A}-\frac{(M)\left(\frac{B}{2}\right)}{(1)(B)^3\left(\frac{1}{12}\right)}=72.82-63.60=9.22\ kPa$$

Step 6. Check the assumed footing thickness h using (a) uniform soil pressure distribution as suggested by Bowles (2001), (b) linear non-uniform soil pressure distribution.

(a) Using uniform soil pressure distribution. Uniform soil pressure is

$$q=\frac{R}{B'\times 1}=\frac{160.2}{(2.2-2\times 0.32)(1)}=\frac{160.2}{1.56}=103.0\,\text{kPa}$$

(Continued)

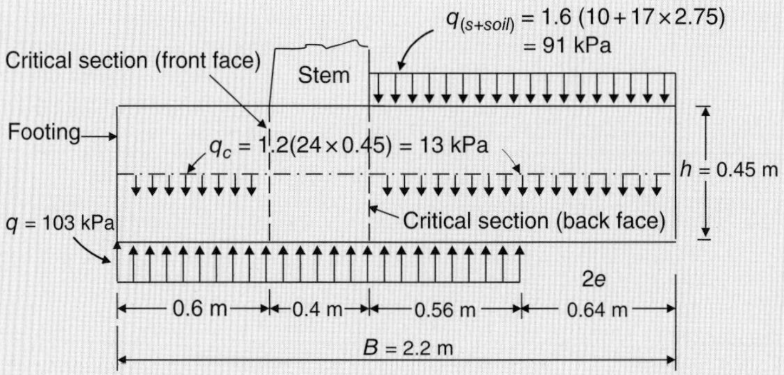

$q_{(s+soil)} = 1.6\,(10 + 17 \times 2.75)$
$= 91$ kPa
Critical section (front face)
Stem
Footing
$q_c = 1.2(24 \times 0.45) = 13$ kPa
$h = 0.45$ m
$q = 103$ kPa
Critical section (back face)
2e
0.6 m 0.4 m 0.56 m 0.64 m
$B = 2.2$ m

Scheme 5.91

One-way shear is only significant in a continuous-strip footing supporting a wall. To consider the most severe condition we use, conservatively, the critical sections located at the faces of the wall as shown.

Shear at the stem front face:

Neglect backfill soil over the toe as being conservative.

$$q_{net} = q - q_c = 103 - 13 = 90\,\text{kPa}$$

$$V_u = q_{net} \times A = (90)(1 \times 0.6) = 54\,\text{kN}$$

Shear at the stem back face:

$$V_u = (91 + 13)(1 \times 0.64) + (91 + 13 - 103)(1 \times 0.56) = 67.12\,\text{kN}$$

Use $V_u = 67.12$ kN/m

Equation (5.19): $ØV_c \geq V_u$ (taking reinforcement shear strength, $V_s = 0$)

$$\text{where } Ø = \text{strength reduction factor (ACI Section 9.3.2.3)}$$
$$= 0.75 \text{ for shear}$$

Equation (5.20): $$V_c = 0.17\lambda\sqrt{f'_c}\,b_w d$$

$$\text{Estimate } d = h - \text{concrete cover} - 0.5\,\text{bar diameter (assume No. 25)}$$
$$= 0.450 - 0.075 - 0.013 = 0.362\text{ m}$$
$$Ø\,V_c = (0.75 \times 0.17 \times 1)\left(\sqrt{21}\right)(1000)(1)(0.362)$$
$$= 212\,\text{kN/m} \gg V_u \quad \text{(OK.)}$$

(b) Using linear non-uniform soil pressure distribution.

Shear at the stem front face:

$$V_u = \left(\frac{q_{\max} + q_1}{2}\right)(0.6)(1) - q_c(0.6)(1)$$
$$= \left(\frac{136.42 + 101.73}{2}\right)(0.6)(1) - (13)(0.6)(1) = 64\,\text{kN/m}$$
$$Ø\,V_c = 212\,\text{kN/m} \gg V_u \quad \text{(OK.)}$$

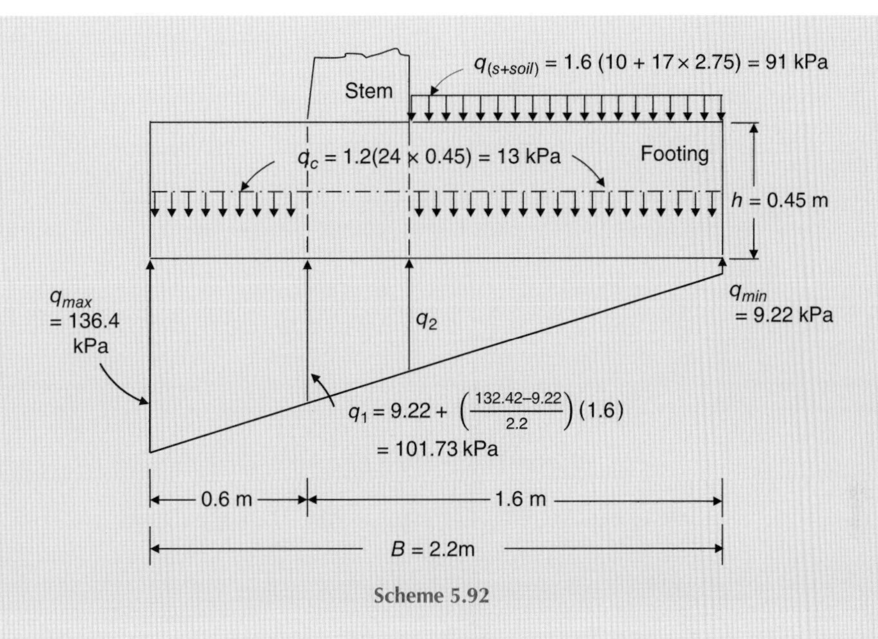

Scheme 5.92

Shear at the stem back face:

$$q_2 = 9.22 + \left(\frac{136.42 - 9.22}{2.2}\right)(1.2) = 78.6\,\text{kPa}$$

$$V_u = (91 + 13)(1.2)(1) - \left(\frac{78.6 + 9.22}{2}\right)(1.2)(1) = 72.11\,\text{kN/m}$$

$$\text{Ø}\, V_c = 212\,\text{kN/m} \gg V_u\,(\text{OK.})$$

Use footing thickness h = 0.45 m.

Step 7. Find the flexural reinforcement needed for the toe and heel.

(a) Toe flexural reinforcement. The shear computations of *Step 6* indicate that the ultimate moment M_u based on uniform soil pressure (Bowles method) will be smaller than that based on non-uniform pressure. The latter method, furnish more conservative design but not as economical as the Bowles method. Since the provided base thickness h is more than adequate, it may be appropriate to use the more economical or Bowles method.

The critical section for bending is at the front face of the stem where the maximum factored positive moment is

$$M_u = \frac{(q_{\text{net}})(l)^2}{2} = \frac{90 \times 0.6^2}{2} = 16.2\ \text{kN.m/m}$$

Equation (5.9): $$\text{Ø}\, M_n \geq M_u$$

Assume tension-controlled section, Ø = 0.9, and $f_s = f_y$.

Equation (5.11): $$\text{Ø}\, M_n = \text{Ø}\left[A_s f_y\left(d - \frac{a}{2}\right)\right]$$

(*Continued*)

Equation (5.10):
$$a = \frac{A_s f_y}{0.85 f'_c b}$$
$$= \frac{A_s \times 420}{0.85 \times 21 \times 1} = 23.53\ A_s$$
$$\varnothing M_n = (0.9)\left[(A_s \times 420 \times 1000)\left(0.362 - \frac{23.53\ A_s}{2}\right)\right]$$
$$= 136\ 836\, A_s - 4\ 447\ 170\ A_s^2\ \ (\text{kN.m/m})$$

Let $\varnothing M_n = M_u$:

$$136\ 836\, A_s - 4\ 447\ 170\ A_s^2 = 16.2$$
$$A_s^2 - 0.031\, A_s + 3.64 \times 10^{-6} = 0$$
$$A_s = \frac{-(-0.031) \pm \sqrt{(-0.031)^2 - (4)(1)(3.64 \times 10^{-6})}}{2 \times 1} = \frac{0.031 - 0.03076}{2}$$
$$= 1.2 \times 10^{-4} \text{m}^2/\text{m} = 120\ \text{mm}^2/\text{m, required by analysis.}$$
$$A_{s,\min} = 0.0018\, bh\ \ (\text{ACI Sections 10.5.4 and 7.12.2.1})$$
$$= 0.0018 \times 1 \times 0.45 = 8.1 \times 10^{-4}\ \text{m}^2 = 810\ \text{mm}^2/\text{m}$$

$A_{s,\min} > A_s$. Therefore, use $A_s = A_{s,\min} = 810\ \text{mm}^2/\text{m}$
The assumptions made are satisfied, since $A_{s,\min}$ controls.

Try four No. 16 bars. $A_{s,provided} = 4 \times 199 = 796\ \text{mm}^2/\text{m}$. It is less than $A_{s,\min}$ but much greater than 1.33 A_s allowed by ACI Section 10.5.3. Because the difference is too small and the base thickness h is somewhat overdesigned, the provided A_s is considered adequate.

ACI Sections 10.5.4 and 7.6.5 requires maximum spacing shall not exceed three times the slab or footing thickness, or 450 mm, whichever is smaller.

Try No. 16 bars @ 250 mm c.c.

Check the development of reinforcement:
In this case, the bars are in tension, the provided bar size is No. 16, the clear spacing of the bars exceeds $2d_b$, and the clear cover exceeds d_b. Therefore, the development length l_d of bars at *each side* of the critical section (which is located where the maximum factored positive moment exists) shall be determined from the following equation, but not less than 300 mm:

$$l_d = \left(\frac{f_y \psi_t \psi_e}{2.1 \lambda \sqrt{f'_c}}\right) d_b\ \ (\text{ACI Sections 12.2.1, 12.2.2 and 12.2.4})$$

where the factors ψ_t, ψ_e and λ are 1 (ACI Sections 12.2.4 and 8.6.1).

$$l_d = \left(\frac{420 \times 1 \times 1}{2.1 \times 1 \times \sqrt{21}}\right)\left(\frac{15.9}{1000}\right) = 0.694\ \text{m} = 694\ \text{mm} > 300\ \text{mm.}$$

Therefore, the required $l_d = 694$ mm.
The bar extension past the critical section (i.e. the available length) is

$$600\ \text{mm} - 75\ \text{mm cover} = 525\ \text{mm} < l_d\quad (\text{Not OK.})$$

Therefore, the bottom bars in the transverse direction will all have to be hooked (90-degree standard hook) at the toe end.

Check l_{dh} (ACI Section 12.5.2):

$$l_{dh} = \left(\frac{0.24\psi_e f_y}{\lambda\sqrt{f_c'}}\right) d_b$$

where the factors ψ_e and λ shall be taken as 1.

$$l_{dh} = \left(\frac{0.24 \times 1 \times 420}{1 \times \sqrt{21}}\right)\left(\frac{15.9}{1000}\right) = 0.35\,\text{m} = 350\,\text{mm} > 8d_b > 150\ mm$$

Therefore, the required $l_{dh} = 350$ mm.

The available length = 525 mm > ($l_{dh} = 350$ mm)

These bars may be cut at a minimum distance = 0.6 + 0.694 = 1.3 m from the toe end. However, since full-width bar length is only 2.05 m, it may not worth the effort to cut the bars. In this case, assume we prefer to cut some of the bars as indicated below.

Provide No. 16 bars @ 250 mm c.c. placed at bottom of the footing in the transverse direction. All the bars have to be hooked (90-degree standard hook) at the toe end. Two out of each four bars shall be extended full width of the footing and the other two bars cut at 1.3 m from the toe end.

(b) Heel flexural reinforcement. The critical section for bending is at the back face of the stem.

The maximum factored negative moment is

$$M_u = \frac{103 \times 0.56^2}{2} - \frac{(91+13)(1.2)^2}{2} = -59\,\text{kN.m/m}$$

$$\text{Ø}\,M_n \geq M_u$$

Assume tension-controlled section, Ø = 0.9, and $f_s = f_y$.

$$\text{Ø}\,M_n = \text{Ø}\left[A_s f_y\left(d - \frac{a}{2}\right)\right]$$

$$a = \frac{A_s f_y}{0.85 f_c' b} = \frac{A_s \times 420}{0.85 \times 21 \times 1} = 23.53\ A_s\ \text{Equation (5.11)}$$

$$\text{Ø}\,M_n = (0.9)\left[(A_s \times 420 \times 1000)\left(0.362 - \frac{23.53 A_s}{2}\right)\right]$$

$$= 136\,836\,A_s - 4447170\ A_s^2\ \ (\text{kN.m/m})$$

Let $\text{Ø}M_n = M_u$:

$$136\,836\,A_s - 4447170 A_s^2 = 59$$

$$A_s^2 - 0.031\,A_s + 1.33 \times 10^{-5} = 0$$

$$A_s = \frac{-(-0.031) \pm \sqrt{(-0.031)^2 - (4)(1)(1.33 \times 10^{-5})}}{2 \times 1} = \frac{0.031 - 0.0301}{2}$$

$$= 4.5 \times 10^{-4} = 450\,\text{mm}^2/\text{m},\ \text{required by analysis.}$$

$$A_{s,\min} = 0.0018\,bh\ \ (\text{ACI Sections 10.5.4 and 7.12.2.1})$$

$$= 0.0018 \times 1 \times 0.45 = 8.1 \times 10^{-4}\ \text{m}^2 = 810\ \text{mm}^2/\text{m}$$

(Continued)

$A_{s,\min} > A_s$. Therefore, use $A_s = A_{s,\min} = 810$ mm^2/m

The assumptions made are satisfied, since $A_{s,\min}$ controls.

Try four No. 16 bars. $A_{s,provided} = 4 \times 199 = 796$ mm^2/m. It is considered adequate for the same reasons just mentioned in (a).

ACI Sections 10.5.4 and 7.6.5 requires maximum spacing shall not exceed three times the slab or footing thickness, or 450 mm, whichever is smaller.

Try No. 16 bars @ 250 mm c.c.

Check the development of reinforcement:

In this case, the bars are in tension, the provided bar size is smaller than No. 19, the clear spacing of the bars exceeds $2d_b$, and the clear cover exceeds d_b. Therefore, the development length l_d of bars at *each side* of the critical section (which is located where the maximum factored positive moment exists) shall be determined from the following equation, but not less than 300 mm:

$$l_d = \left(\frac{f_y \psi_t \psi_e}{2.1\lambda\sqrt{f'_c}}\right) d_b \text{ (ACI Sections 12.2.1, 12.2.2 and 12.2.4)}$$

where the factors ψ_e and λ are 1 (ACI Sections 12.2.4 and 8.6.1).

The factor $\psi_t = 1.3$ because the reinforcement is placed such that more than 300 mm of fresh concrete exists below the top bars (ACI Section 12.2.4).

$$l_d = \left(\frac{420 \times 1.3 \times 1}{2.1 \times 1 \times \sqrt{21}}\right)\left(\frac{15.9}{1000}\right) = 0.902\text{ m} = 902\text{ mm} > 300\text{ mm}.$$

Therefore, the required $l_d = 902$ mm.

The bar extension past the critical section (i.e. the available length) is

$$1000\text{ mm} - 75\text{ mm cover} = 925\text{ mm} > 902\text{ mm (OK.)}$$

Provide No. 16 bars @ 250 mm c.c. placed at top of the footing, full width, in the transverse direction.

Step 8. Provide longitudinal shrinkage and temperature reinforcement.

ACI Chapter 15 (Footings) does not give any provision or requirement concerning shrinkage and temperature reinforcement. The provisions of ACI Section 7.12 are intended for structural slabs only; they are not intended for slabs on ground (ACI R7.12.1). Some authorities (e.g. Bowles, 1996) do not use this reinforcement in one-way footings, whereas others (MacGregor and Wight) use it to satisfy ACI Section 7.12.2.1. In this case, assume we also prefer to provide minimum shrinkage and temperature reinforcement perpendicular to the toe and heel transverse reinforcement.

Select longitudinal steel based on the minimum shrinkage and temperature reinforcement (ACI Sections 7.12.1 and 7.12.2.1).

$$A_s = 0.0018\,bh = 0.0018 \times 2.2 \times 0.45 = 1.782 \times 10^{-3}\text{ m}^2$$
$$= 1782\text{ mm}^2$$

Try 14 No. 13 bars: $A_{s,\text{provided}} = 14 \times 129 = 1806$ mm$^2 > A_s$ (OK.)

Longitudinal steel will, in general, be more effective in the top of the footing than in the bottom; it could control cracks when the foundation settles.

According to ACI Section 7.12.2.2, shrinkage and temperature reinforcement shall be spaced not farther apart than five times the slab thickness, nor farther apart than 450 mm.

Use eight No. 13 bars at top and six No. 13 bars at bottom, which will also provide support for the transverse flexural reinforcement.

Provide eight No. 13 @ 288 mm c.c. at top of the footing in long direction, as support for the transverse bars.

Provide six No. 13 @ 405 mm c.c. at bottom of the footing in long direction.

Step 9. Check thickness b at the stem bottom.

Refer to the profiles of the active lateral pressure of *Step 2* and the scheme shown below.

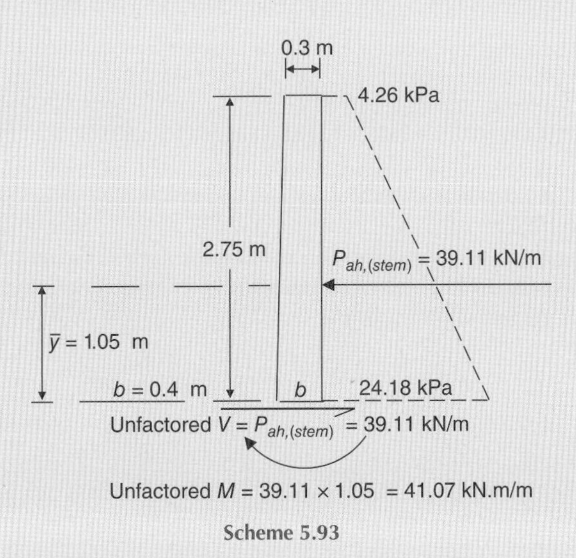

Scheme 5.93

Unfactored shear force at the stem bottom is

$$V = P_{ah,\text{stem}} = 39.11 \ \text{kN/m}$$

$$V_u = 1.6 \times 39.11 = 63 \,\text{kN/m}$$

Equation (5.19): $ØV_c \geq V_u$ (taking reinforcement shear strength, $V_s = 0$)

where Ø = shear strength reduction factor (ACI Section 9.3.2.3)
= 0.75

Equation (5.20):

$$V_c = 0.17\lambda\sqrt{f_c'}b_w d$$

$$b_w = 1\,\text{m}$$

$$d = 0.315\ \text{m}\left[\text{allowing 85 mm for concrete cover} + \left(\frac{1}{2}\right)\text{bar diameter}\right]$$

$$Ø\,V_c = (0.75 \times 0.17 \times 1)(\sqrt{21})(1000)(1)(0.315)$$
$$= 184\ \text{kN/m} > (V_u = 63\ \text{kN/m}) \qquad \text{(OK.)}$$

Check *shear friction* (ACI Section 11.6) since the stem is built after the base has been poured and partially cured.

According to ACI Section 11.6.5, the nominal shear strength V_n shall not exceed the smaller of $0.2f_c'A_c$ or $5.5\,A_c$, where A_c is area of concrete section resisting shear transfer.

$$V_n = V_c \ \text{(taking reinforcement shear strength, } V_s = 0)$$
$$= 0.2f_c'A_c = 0.2 \times 21 \times 0.4 \times 1 \times 1000 = 1680\,\text{kN/m}$$

(*Continued*)

or,

$$V_n = V_c = 5.5\,A_c = 5.5 \times 0.4 \times 1 \times 1000 = 2200 \text{ kN/m} > \textit{1680 kN/m}$$

Use $V_c = 1680$ kN/m

$$\emptyset\, V_c = 0.75 \times 1680 = 1260\,\text{kN/m} \gg (V_u = 63\,\text{kN/m})$$

From this computation it would appear that shear friction seldom controls except possibly for a very high wall with a thin stem (Bowles, 1996).

Step 10. Determine bending moment at the stem bottom and moments at other necessary points of the stem height.

Since the stem height is only 2.75 m, it may be necessary to compute moments at the stem bottom and at about the mid-point of its height, as shown below.

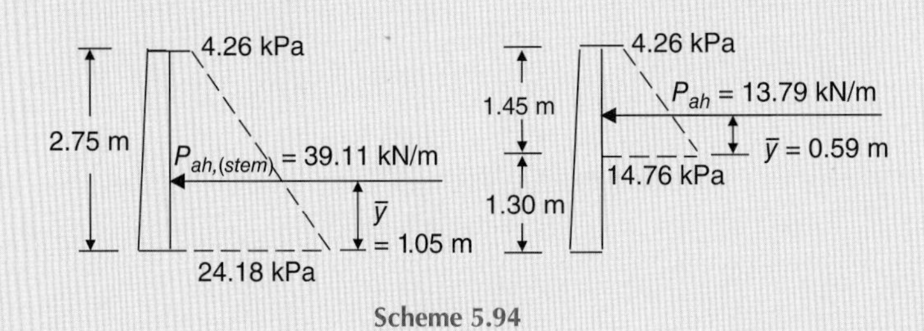

Scheme 5.94

At 1.3 m height:

$$q_{hz} = [(10) + 17(2.75 - 1.3)](0.426) = 14.76\,\text{kPa}$$

$$P_{ah} = \frac{4.26 + 14.76}{2} \times 1.45 = 13.79\,\text{kN/m}$$

$$\bar{y} = \frac{L}{3}\left(\frac{2a+b}{a+b}\right) = \frac{1.45}{3}\left(\frac{2 \times 4.26 + 14.76}{4.26 + 14.76}\right) = 0.59\,\text{m}$$

Bending moment at the level of the stem bottom:

$$M_{u,\max} = 1.6 \times 39.11 \times 1.05 = 65.7 \text{ kN.m/m}$$

Bending moment at the 1.3 m height level:

$$M_u = 1.6 \times 13.79 \times 0.59 = 13.02\,\text{kN.m/m}$$

Step 11. Determine the stem vertical and horizontal reinforcement.

ACI Section 14.1.2 states that cantilever retaining walls are designed according to flexural design provisions of Chapter 10 with minimum horizontal reinforcement according to 14.3.3.

ACI Section 15.8.2.2 states that for cast-in-place walls, area of reinforcement across interface shall be not less than minimum vertical reinforcement given in 14.3.2.

Vertical reinforcement at the stem bottom level:

Equation (5.9): $$\emptyset\, M_n \geq M_u$$

Assume tension-controlled section, $\emptyset = 0.9$, and $f_s = f_y$.

Equation (5.11): $$\emptyset M_n = \emptyset\left[A_s f_y\left(d - \frac{a}{2}\right)\right]$$

Equation (5.10): $$a = \frac{A_s f_y}{0.85 f_c' b}$$

$$= \frac{A_s \times 420}{0.85 \times 21 \times 1} = 23.53\ A_s$$

$$\emptyset M_n = (0.9)\left[(A_s \times 420 \times 1000)\left(0.315 - \frac{23.53\ A_s}{2}\right)\right]$$

$$= 119\ 070\ A_s - 4\ 447\ 170\ A_s^2 \quad (\text{kN.m/m})$$

Let $\emptyset M_n = M_u$:

$$119\ 070\ A_s - 4\ 447\ 170 A_s^2 = 65.7$$

$$A_s^2 - 0.027\ A_s + 1.48 \times 10^{-5} = 0$$

$$A_s = \frac{-(-0.027) \pm \sqrt{(-0.027)^2 - (4)(1)(1.48 \times 10^{-5})}}{2 \times 1} = \frac{0.0270 - 0.0259}{2}$$

$$= 5.6 \times 10^{-4} = 560\ \text{mm}^2/\text{m}$$

Try three No. 16 bars:

$$A_{s,\text{provided}} = 3 \times 199 = 597\ \text{mm}^2/\text{m} \quad (\text{OK.})$$

Minimum ratio of vertical reinforcement area to gross concrete area perpendicular to that reinforcement, ρ_ℓ, shall be 0.0012 for deformed bars not larger than No. 16 with f_y not less than 420 MPa (ACI Section 14.3.2).

$$A_{s,\min} = 0.0012\ b\ell = 0.0012 \times 0.4 \times 1 = 4.8 \times 10^{-4}\ \text{m}^2 = 480\ \text{mm}^2/\text{m}$$

$$A_{s,\min} < A_s \quad (\text{OK.})$$

Compute a for $A_s = 597$ mm^2/m, and check if $f_s = f_y$ and whether the section is tension-controlled:

$$a = 23.53\ A_s = 23.53 \times 597 \times 10^{-6} = 0.014\ \text{m} = 14\ \text{mm}$$

$c = \dfrac{a}{\beta_1}$; For f_c' between 17 and 28 MPa, β_1 shall be taken as 0.85.

$$c = \frac{14}{0.85} = 16.47\ \text{mm};$$

$$d_t = d = 315\ \text{mm}$$

$$\varepsilon_t = 0.003\left(\frac{d_t - c}{c}\right) \quad \text{Equation (5.15)}$$

(Continued)

$\varepsilon_t = 0.003\left(\dfrac{315-16.47}{16.47}\right) = 0.054 > 0.005$. Hence, the section is tension-controlled, and $\varnothing = 0.9$ (ACI Sections 10.3.4 and 9.3.2.1).

$$\varepsilon_y = \frac{f_y}{E_s} = \frac{420}{200\ 000} = 0.0021.$$

Hence, $\varepsilon_t > \varepsilon_y$ and $f_s = f_y$.

The assumptions made are satisfied.

Vertical and horizontal reinforcement shall not be spaced farther apart than three times the wall thickness or farther apart than 450 mm, whichever is smaller (ACI Section 14.3.5).

Provide No. 16 bars @ 333 mm c.c. at bottom level of the stem.

Vertical reinforcement at the 1.3 m stem height level:

Stem thickness $b = 0.3 + \dfrac{0.4-0.3}{2.75} \times 1.45 = 0.305\text{ m}$

$$d = 0.305 - 0.085 = 0.22\text{ m}$$

$$\varnothing M_n = \varnothing\left[A_s f_y\left(d - \frac{a}{2}\right)\right] = (0.9)\left[(A_s \times 420 \times 1000)\left(0.22 - \frac{23.53\,A_s}{2}\right)\right]$$
$$= 83160\,A_s - 4447170\,A_s^2 \quad (\text{kN.m/m})$$

Let $\varnothing M_n = M_u$:

$$83160\,A_s - 4447170\,A_s^2 = 13.02$$

$$A_s^2 - 0.0187\,A_s + 2.928 \times 10^{-6} = 0$$

$$A_s = \frac{-(-0.0187) \pm \sqrt{(-0.0187)^2 - (4)(1)(2.928 \times 10^{-6})}}{2 \times 1} = \frac{0.0187 - 0.0184}{2}$$
$$= 1.5 \times 10^{-4} = 150\text{ mm}^2/\text{m}$$

$$A_{s,\min} = 0.0012\,b\ell = 0.0012 \times 0.305 \times 1$$
$$= 3.66 \times 10^{-4}\text{ m}^2 = 366\text{ mm}^2/\text{m} > A_s$$

Therefore, $A_{s,\min}$ controls.

This reinforcement will be provided by extending two out of each three bars of the stem reinforcement, at the bottom level, full height of the stem. The third bar shall be cut at d or $12d_b$, whichever is greater, above the 1.3 m height level (ACI Section 12.10.3). Accordingly, the cut point will be located at 1.52 m height level.

All the vertical bars shall be extended through the interface into the footing, hooked (90-degree standard hook) and wired to the bars at the toe bottom.Thus, the requirement of ACI Section 15.8.2.2 will be satisfied.

Provide No. 16 bars @ 333 mm c.c. placed vertically and extended through the interface into the footing. All the bars shall be hooked (90-degree standard hook) at the lower end only and wired to the bars at the toe bottom. Two out of each three bars shall be extended full height of the stem and the third bar shall be cut at 1.52 m stem height.

Horizontal reinforcement (ACI Section 14.3.3):
Minimum ratio of horizontal reinforcement area to gross concrete area perpendicular to that reinforcement ρ_t, shall be 0.0020 for deformed bars not larger than No. 16 with f_y not less than 420 MPa.

$$A_{s,\min} = 0.0020\,b_{ave}h = 0.002 \times \frac{0.4+0.3}{2} \times 1$$
$$= 7 \times 10^{-4}\text{m}^2 = 700\text{ mm}^2 \text{ per metre height of the stem}$$

Try four No. 16 bars: $A_{s,provided} = 4 \times 199 = 796\ \text{mm}^2/\text{m}$ (OK.)

Provide No. 16 bars @ 250 mm c.c. placed horizontally for the full stem height and wired to the vertical reinforcement.

Step 12. Develop the final design sketches.

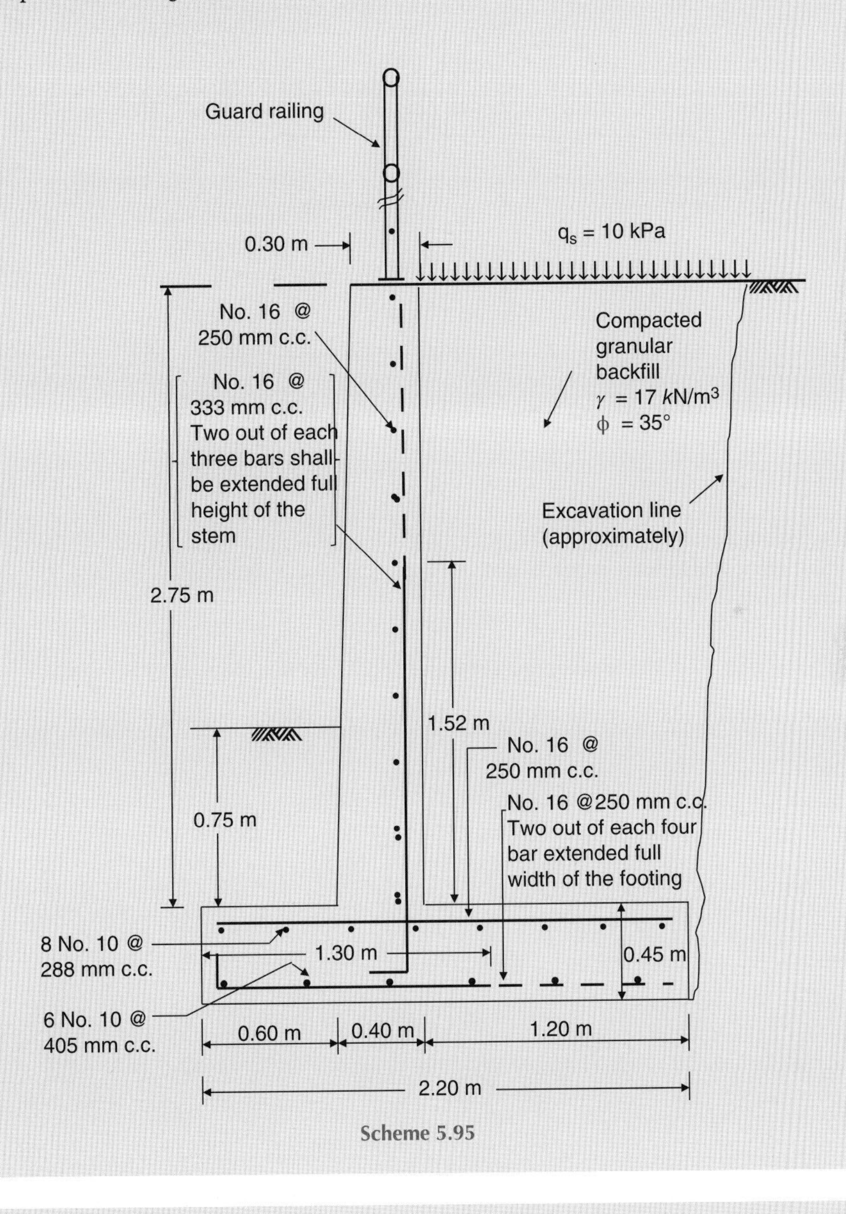

Scheme 5.95

Problem 5.18

Refer to Problem 5.17. Find the flexural reinforcement needed for the toe and heel based on linear non-uniform soil pressure distribution.

(Continued)

Solution:
From Problem 5.17, *Step 6*(b):

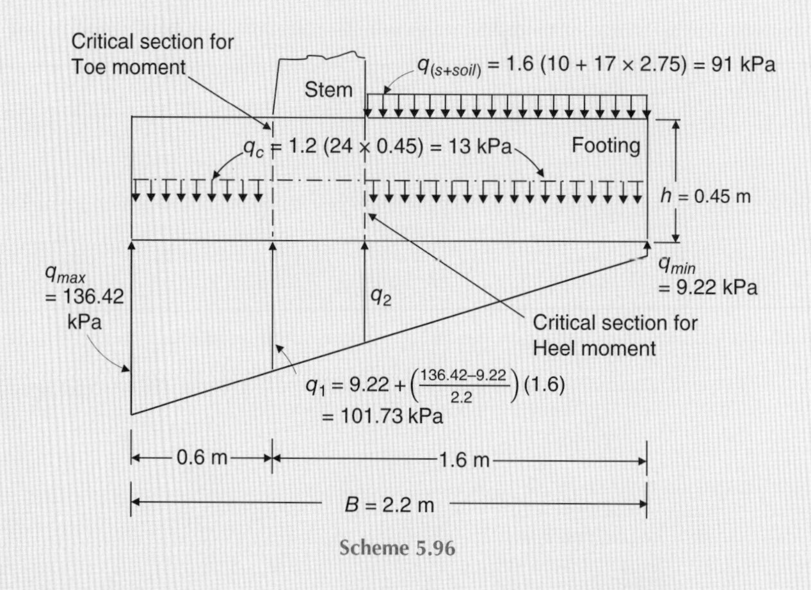

Scheme 5.96

(a) Toe flexural reinforcement.
The critical section is located at the stem front face as shown in the scheme above. Maximum factored bending moment is

$$M_u = (\text{B.M due to contact pressure}) - (\text{B.M due to toe weight})$$

$$= \frac{101.73 \times 0.6^2}{2} + \frac{136.42 - 101.73}{2} \times 0.6 \times \frac{2}{3} \times 0.6 - \frac{13 \times 0.6^2}{2}$$

$$= 18.314.16 - 2.34 = 20.13\,\text{kN.m/m}$$

Equation (5.9): $$\varnothing M_n \geq M_u$$

Assume tension-controlled section, $\varnothing = 0.9$ and $f_s = f_y$.

Equation (5.11): $$\varnothing M_n = \varnothing \left[A_s f_y \left(d - \frac{a}{2}\right)\right]$$

Equation (5.10): $$a = \frac{A_s f_y}{0.85 f_c' b}$$

$$= \frac{A_s \times 420}{0.85 \times 21 \times 1} = 23.53\,A_s$$

Estimate

$$d = h - \text{concrete cover} - 0.5\,\text{bar diameter (assume No. 25)}$$

$$= 0.450 - 0.075 - 0.013 = 0.362\,\text{m}$$

$$\varnothing M_n = (0.9)\left[(A_s \times 420 \times 1000)\left(0.362 - \frac{23.53\ A_s}{2}\right)\right]$$

$$= 136\ 836\, A_s - 4\ 447\ 170\ A_s^2 \quad (\text{kN.m/m})$$

Let $\varnothing M_n = M_u$:

$$136\ 836\, A_s - 4\ 447\ 170 A_s^2 = 20.13$$

$$A_s^2 - 0.031\, A_s + 4.53 \times 10^{-6} = 0$$

$$A_s = \frac{-(-0.031) \pm \sqrt{(-0.031)^2 - (4)(1)(4.53 \times 10^{-6})}}{2 \times 1} = \frac{0.031 - 0.03071}{2}$$

$$= 1.45 \times 10^{-4} \text{m}^2/\text{m} = 145\,\text{mm}^2/\text{m, required by analysis.}$$

$$A_{s,\min} = 0.0018\, bh \quad (\text{ACI Sections 10.5.4 and 7.12.2.1})$$

$$= 0.0018 \times 1 \times 0.45 = 8.1 \times 10^{-4}\ \text{m}^2 = 810\ \text{mm}^2/\text{m}$$

$A_{s,\min} > A_s$ by analysis. Therefore, use $A_s = A_{s,\min} = 810\ \text{mm}^2/\text{m}$

The assumptions made are satisfied, since $A_{s,\min}$ controls.

Try four No.16 bars: $A_{s,\text{provided}} = 4 \times 199 = 796\ \text{mm}^2/\text{m}$. It is less than $A_{s,\min}$ but much greater than 1.33 A_s required by analysis. Since the difference is too small and the base thickness h is somewhat overdesigned, it is considered acceptable.

ACI Sections 10.5.4 and 7.6.5 requires maximum spacing shall not exceed three times the slab or footing thickness, or 450 mm, whichever is smaller.

Try No. 16 bars @ 250 mm c.c.

Check the development of reinforcement:

The necessary computations is exactly the same as that carried out in Problem 5.17, *Step 7*(a). Since the reinforcement, concrete and footing dimensions are all remained unchanged, the same results will be obtained.

Provide No. 16 bars @ 250 mm c.c. placed at bottom of the footing in the transverse direction. All the bars have to be hooked (90-degree standard hook) at the toe end. Two out of each four bars shall be extended full width of the footing and the other two bars cut at 1.3 m from the toe end.

(b) Heel flexural reinforcement.

The critical section for bending is located at the back face of the stem, as shown in the scheme above.

$$q_2 = 9.22 + \left(\frac{136.42 - 9.22}{2.2}\right)(1.0) = 67.04\,\text{kPa}$$

M_u = (B.M due to weight of backfill soil and load) + (B.M due to heel weight) – (B.M due to contact pressure)

$$M_u = \left(\frac{91 \times 1.0^2}{2}\right) + \left(\frac{13 \times 1.0^2}{2}\right) - \left(\frac{9.22 \times 1.0^2}{2} + \frac{67.04 - 9.22}{2} \times 1.0 \times \frac{1}{3} \times 1.0\right)$$

$$= 45.5 + 6.5 - 4.61 - 9.64 = 37.75\,\text{kN.m/m}$$

Equation (5.9): $\varnothing M_n \geq M_u$

Assume tension-controlled section, $\varnothing = 0.9$ and $f_s = f_y$.

Equation (5.11): $\varnothing M_n = \varnothing\left[A_s f_y\left(d - \frac{a}{2}\right)\right]$

(Continued)

$$a = \frac{A_s f_y}{0.85 f_c' \, b}$$

Equation (5.10):

$$= \frac{A_s \times 420}{0.85 \times 21 \times 1} = 23.53\, A_s$$

$$\varnothing M_n = (0.9)\left[(A_s \times 420 \times 1000)\left(0.362 - \frac{23.53\, A_s}{2}\right)\right]$$

$$= 136836\, A_s - 4447170\, A_s^2 (\text{kN.m/m})$$

Let $\varnothing M_n = M_u$:

$$136836\, A_s - 4447170\, A_s^2 = 37.75$$

$$A_s^2 - 0.031\, A_s + 8.49 \times 10^{-6} = 0$$

$$A_s = \frac{-(-0.031) \pm \sqrt{(-0.031)^2 - (4)(1)(8.49 \times 10^{-6})}}{2 \times 1} = \frac{0.031 - 0.0304}{2}$$

$$= 3 \times 10^{-4} \text{m}^2/\text{m} = 300\ \text{mm}^2/\text{m},\ \text{required by analysis.}$$

$$A_{s,\min} = 0.0018\, bh \quad (\text{ACI Sections 10.5.4 and 7.12.2.1})$$

$$= 0.0018 \times 1 \times 0.45 = 8.1 \times 10^{-4}\ \text{m}^2 = 810\ \text{mm}^2/\text{m}$$

$A_{s,\min} > A_s$ required by analysis.

Therefore, use $A_s = A_{s,\min} = 810$ mm^2/m.

The assumptions made are satisfied, since $A_{s,\min}$ controls.

Try four No.16 bars: $A_{s,\text{provided}} = 4 \times 199 = 796$ mm^2/m. It is acceptable for the same reasons mentioned in (a) above.

ACI Sections 10.5.4 and 7.6.5 requires maximum spacing shall not exceed three times the slab or footing thickness, or 450 mm, whichever is smaller.

Try No. 16 bars @ 250 mm c.c.

Check the development of reinforcement:

The necessary computations is exactly the same as that carried out in Problem 5.17, *Step 7*(a). Since the reinforcement, concrete and footing dimensions are all remained unchanged, the same results will be obtained.

Provide No. 16 bars @ 250 mm c.c. placed at top, extended full width of the footing in the transverse direction.

Thus, in this case, we find that the required transverse flexural reinforcement based on the linear non-uniform soil pressure distribution is the same as that based on the uniform soil pressure distribution since $A_{s,min}$ governs in both cases.

Problem 5.19

Refer to Problem 5.17. Assume the base soil and that in front the wall is gravelly sand of $\varnothing = 33°$ and $\gamma = 18$ kN/m^3, overlying a soft-clay deposit at 1.0 m depth below the base level, as shown in the scheme below. The ground water table is also located at the same depth. The soft clay has an average $c = 20$ kPa, $\varnothing = 0°$ and $\gamma = 17.5$ kN/m^3. Determine the safety factor (*SF*) against a deep-seated shear failure (or rotational instability) for a trial cylindrical slip surface through the heel shown in Scheme 5.97. Assume radius of the cylindrical slip surface is 4.33 m, and its centre *O* located at 2.87 m above the foundation level. Use the conventional *method of slices* (the Fellenius or Swedish solution).

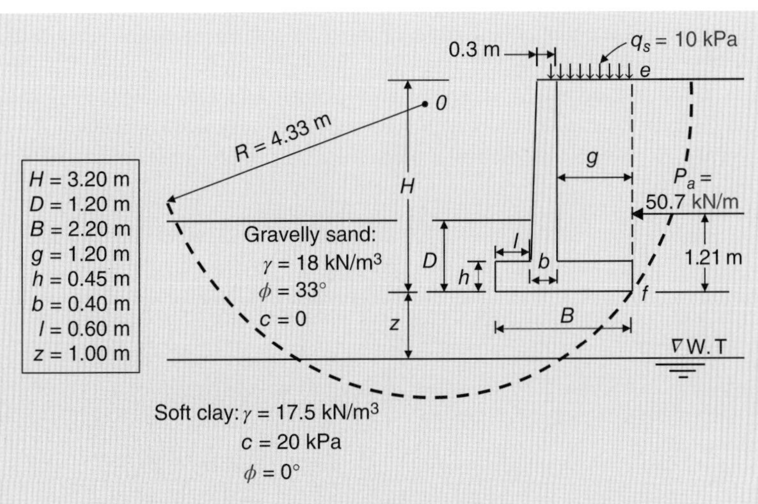

Scheme 5.97

Solution

Normally, the safety factor (*SF*) against a deep-seated shear failure (or rotational instability) should be at least two. For this analysis, several trial failure surfaces (circles) should be drawn, the safety factor computed, and the minimum value taken. It may be useful to mention that the trial failure circles passing through the heel apply for ground or backfill slope angles of $\beta = 0$ to perhaps $\beta = 10°$; for larger values of β, irregular surfaces or surcharges, the assumed failure circles may not pass through the heel, and other failure locations should be investigated.

In this case, we try to compute the safety factor for the circle indicated in the given scheme of the Problem as an example. Due to the repetitive nature of the calculations and the need to select an adequate number of trial surfaces, the method of slices is particularly suitable for solution by computer, especially, where more complex slope geometry and different soil strata are required to be introduced.

Step 1. Draw the wall-soil system to a convenient scale, as shown below.

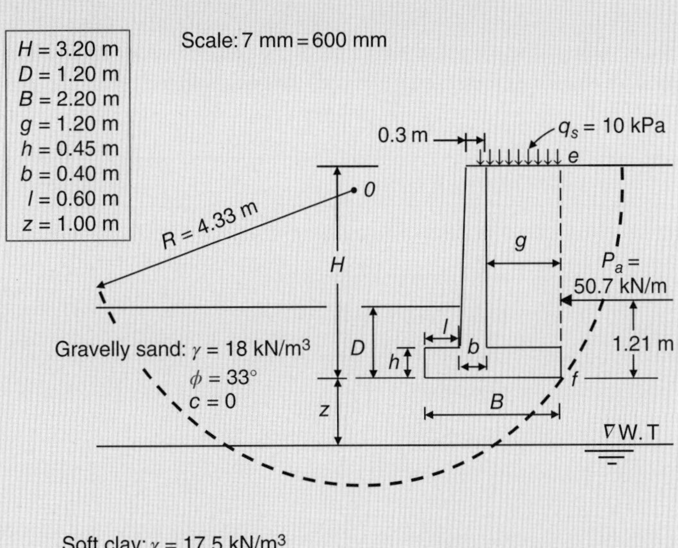

Scheme 5.98

(Continued)

Step 2. Divide the circular segment *amnf* and the material above by vertical planes into a series of *slices*, as shown in the scheme below. The base b of each slice is assumed to be a straight line. For any slice the inclination of the base to the horizontal is α and the height, measured on the centre line, is h. Weight W (including surcharge) of each slice is resolved into its normal component $N = W\cos\alpha$ and tangential component $T = W\sin\alpha$, which are determined graphically, as shown. Alternatively, the value of α can be measured or calculated.

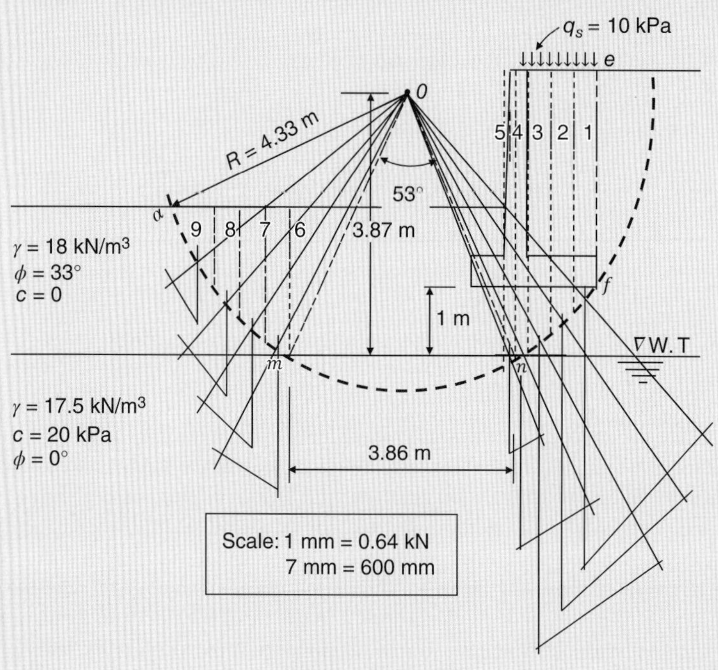

Scheme 5.99

Note: Since the centre of rotation O is centered with respect to the circular arc *mn* (located in the soft clay), it is not necessary to divide this arc and the material above into slices and find N and T; their resultant effects will be zero (neglecting the difference in weights of a portion of the concrete footing and displaced soil). However, some correction was applied to include the weight of a portion of the concrete stem in slice 5, as shown in the scheme.

Step 3. Determine values of N and T per metre length of the retaining wall for each slice, as indicated in Table 5.10.

Table 5.10 Values of N and T per metre length for each slice of the retaining wall.

Slice number	Weight of slice, W (kN)	N (kN)	T (kN)
1	$0.4\,(10 + 2.92 \times 18) + 0.4 \times 0.45 \times 24 = 29.3$	21.2	20.3
2	$0.4\,(10 + 3.26 \times 18) + 0.4 \times 0.45 \times 24 = 31.8$	25.0	19.7
3	$0.4\,(10 + 3.56 \times 18) + 0.4 \times 0.45 \times 24 = 34.0$	28.8	18.1
4	$0.2\,(10 + 1.00 \times 18) + 0.2 \times 3.20 \times 24 = 21.0$	18.6	9.7
5	$0.15 \times 2.75 \times 24 = 9.9$	9.0	4.5
6	$0.43 \times 2.06 \times 18 = 15.9$	14.1	-7.5
7	$0.43 \times 1.84 \times 18 = 13.8$	10.2	-9.3
8	$0.43 \times 1.37 \times 18 = 10.6$	7.2	-7.7
9	$0.5 \times 0.63 \times 1.11 \times 18 = 6.3$	3.8	-5.0
		$\Sigma N = 137.8$. $\Sigma T = 42.8$	

Step 4. Determine the driving and resisting moments about centre of rotation at point *O*, then compute the safety factor *SF*.

(a) Driving moment M_d.

This moment consists of moment of the total active thrust P_a plus moment of $\Sigma\ T$, computed as

$$M_d = \bar{y} P_a + R \sum T$$

The thrust P_a is located at 1.21 m above the foundation level, whereas, point *O* is located at 2.87 m above the same level. Therefore,

$$\bar{y} = 2.87 - 1.21 = 1.66\,\text{m}$$

$$M_d = 1.66 \times 50.7 + 4.33 \times 42.8 = 269.5\ kN.m/m$$

(b) Resisting moment M_r.

This moment equals to moment of the total shear strength T_f along the circular arc *amnf*, computed as

$$M_r = R \sum T_f$$

$$T_f = \text{cohesion force} + \text{friction force} = (\text{arc length} \times c) + (N \tan Ø°)$$

Gravelly sand:

The base of slice 5 is located in soft clay. Therefore, *N* for this slice should not be considered.

$$\sum T_f = 0 + (\sum N - N \text{ of slice } 5) \tan Ø° = (137.8 - 9) \tan 33$$
$$= 83.6\,\text{kN/m}$$

Soft clay:

$$\sum T_f = (\text{length of arc } mn \times c) + 0 = \frac{53}{360}(2\pi R)(c)$$

$$= \frac{53}{360} \times 2 \times \frac{22}{7} \times 4.33 \times 20 = 80.1\,\text{kN/m}$$
$$M_r = 4.33(83.6 + 80.1) = 708.8\ kN.m/m$$

Safety factor is

$$SF = \frac{M_r}{M_d} = \frac{708.8}{269.5} = 2.63 > 2\ \ (\text{OK.})$$

Problem 5.20

A gravity concrete retaining wall is shown in the scheme below. Determine the maximum and minimum pressures under the base of the wall, and the safety factors against overturning and sliding (ignore any material which may exist in front the wall). Use Coulomb's active earth pressure theory and wall friction angle $\delta = (2/3)Ø$.

(Continued)

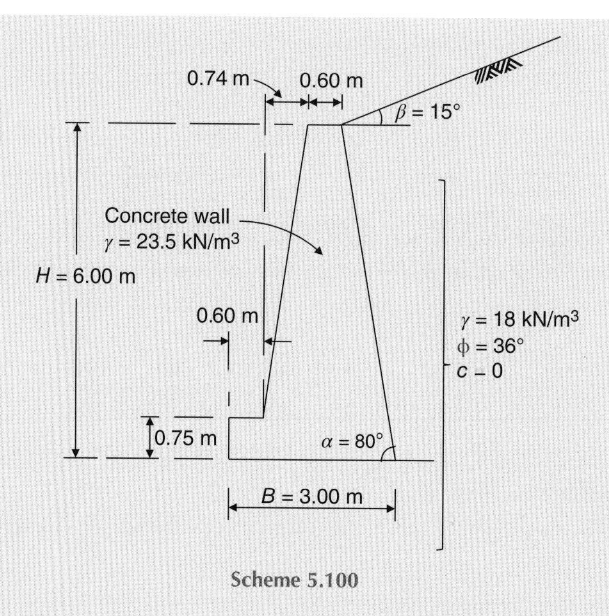

Scheme 5.100

Solution:

Step 1. Determine the total active earth pressure thrust P_a and its components P_{ah} and P_{av}.

$$P_a = \frac{\gamma H^2}{2} K_a$$

Where K_a = Coulomb's coefficient of lateral active earth pressure.
For cohesionless soils, K_a may be obtained from tables or calculated as

$$K_a = \frac{\sin^2(\alpha + \varnothing)}{\sin^2\alpha \sin(\alpha - \delta)\left[1 + \sqrt{\dfrac{\sin(\varnothing + \delta)\sin(\varnothing - \beta)}{\sin(\alpha - \delta)\sin(\alpha + \beta)}}\right]^2}$$

Angles:

$$\varnothing = 36^\circ$$

$$\beta = 15^\circ$$

$$\alpha = 80^\circ$$

$$\delta = \frac{2\varnothing^\circ}{3} = \frac{2}{3} \times 36 = 24^\circ$$

$$K_a = \frac{\sin^2 116}{\sin^2 80 \sin 56\left[1 + \sqrt{\dfrac{\sin 60 \sin 21}{\sin 56 \sin 95}}\right]^2}$$

$$K_a = \frac{0.808}{0.804\left[1 + \sqrt{\dfrac{0.310}{0.826}}\right]^2} = \frac{0.808}{0.84[1 + 0.613]^2}$$

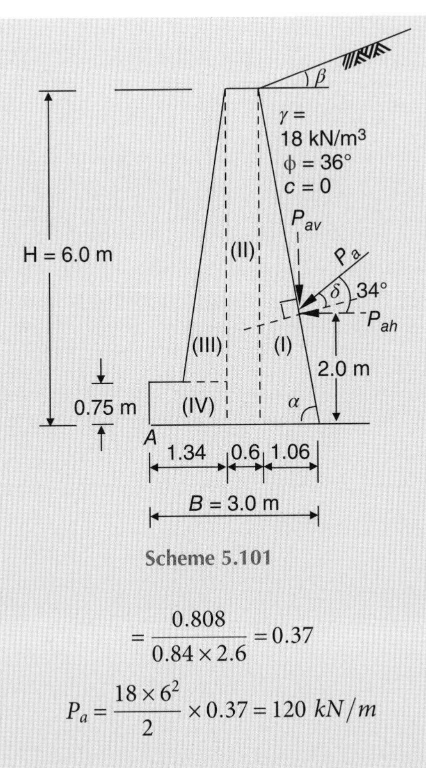

Scheme 5.101

$$= \frac{0.808}{0.84 \times 2.6} = 0.37$$

$$P_a = \frac{18 \times 6^2}{2} \times 0.37 = 120\ kN/m$$

The active thrust P_a acts on the wall back face at a height of $\frac{1}{3}H$ and at δ above the normal, or at 34° above the horizontal as shown.

$$P_{ah} = P_a \cos 34^\circ = 120 \times 0.83 = 99.6\ kN/m$$

$$P_{av} = P_a \sin 34^\circ = 120 \times 0.56 = 67.2\ kN/m$$

Step 2. Compute the vertical resultant force R, net moment M_{net}, eccentricity e, overturning moment M_o about base toe at point A and the resisting moment M_r.

For these computations refer to the scheme in *Step 1*. It is convenient to tabulate the load sources, loads, moment arms and moments, as in Table 5.11.

Table 5.11 Load sources, loads, moment arms and moments.

Load source	Load (weight) kN/m	Arm m	Moment M_A kN.m/m + ↶
(I)	0.5 × 6 × 1.06 × 23.5 = 74.73	2.29	171.13
(II)	0.6 × 6 × 23.5 = 84.60	1.64	138.74
(III)	0.5 × 5.25 × 0.74 × 23.5 = 45.65	1.09	49.76
(IV)	0.75 × 1.34 × 23.5 = 23.62	0.67	15.83
P_{av}	= 67.20	2.65	178.08
P_{ah} = 99.6 kN/m		2.00	−199.20
	R = **295.80**		M_{net} = **354.34**

(*Continued*)

$$\bar{x}R = M_{net}$$

$$\bar{x} = \frac{M_{net}}{R} = \frac{354.34}{295.80} = 1.2\,\text{m}$$

$$e = \frac{B}{2} - \bar{x} = \frac{3}{2} - 1.2 = 0.3\ m < \left(\frac{B}{6} = 0.5\,\text{m}\right)$$

Overturning moment $M_o = P_{ah} \times 2 = 99.6 \times 2 = 199.2$ kN.m/m
Resisting moment $M_r = M_{net} - M_o = 354.34 - (-199.20) = 553.54$ kN.m/m

Step 3. Determine the safety factor against overturning.
Safety factor against overturning is

$$SF = \frac{M_r}{M_o} = \frac{553.54}{199.2} = 2.78$$

Step 4. Determine the safety factor against sliding.
Resisting force against sliding is

$$F_r = R\tan\delta + c_a A_f + P_P$$

In this case, c_a is zero since the base soil cohesion $c = 0$. Also, since any material which may exist in front the wall is ignored, P_P is zero.

$$F_r = R\tan\delta = 295.8 \times \tan 24 = 131.7\,\text{kN/m}$$

Sliding force $= P_{ah} = 99.6$ kN/m
Safety factor against sliding is

$$SF = \frac{F_r}{P_{ah}} = \frac{131.7}{99.6} = 1.32$$

Step 5. Determine the maximum and minimum pressures under the base.

$$q = \frac{R}{BL}\left(1 \pm \frac{6e_x}{B} \pm \frac{6e_y}{L}\right) \quad \text{Equation (2.7)}$$

Maximum pressure under the base is

$$q_{max} = \frac{R}{BL}\left(1 + \frac{6e_x}{B}\right) = \frac{295.8}{3 \times 1}\left(1 + \frac{6 \times 0.3}{3}\right)$$
$$= 98.60 + 59.16 = 157.76\ kPa$$

Minimum pressure under the base is

$$q_{min} = \frac{R}{BL}\left(1 - \frac{6e_x}{B}\right) = \frac{295.8}{3 \times 1}\left(1 - \frac{6 \times 0.3}{3}\right) = 98.6 - 59.16 = 39.44\ kPa$$

Problem 5.21

Solve Problem 5.20 using Rankine's active earth pressure theory.

Solution:

Step 1. Determine the total active earth pressure thrust P_a and its components P_{ah} and P_{av}.

$$P_a = \frac{\gamma (H')^2}{2} K_a$$

where K_a = Rankine's coefficient of lateral active earth pressure.

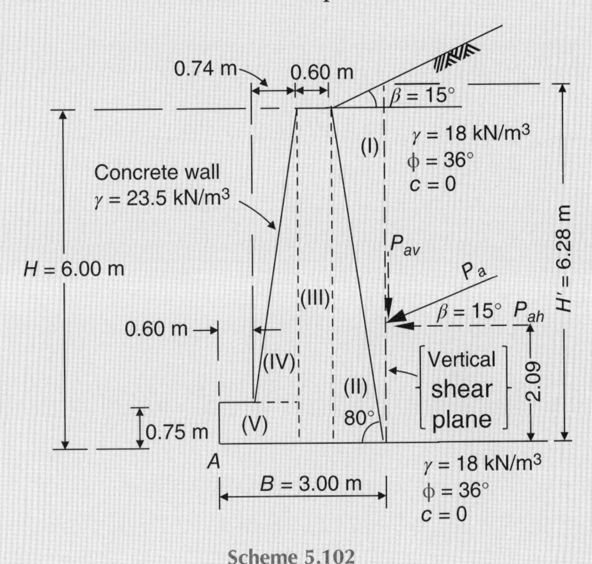

Scheme 5.102

The active thrust P_a acts on the vertical shear plane at a height of $\frac{1}{3}H'$ and at β above the horizontal as shown.

For cohesionless soils, the coefficient K_a depends on the backfill slope angle β and friction angle $\varnothing$. It can be obtained from tables or may be calculated from the following equation

$$K_a = \cos\beta \frac{\cos\beta - \sqrt{\cos^2\beta - \cos^2\varnothing}}{\cos\beta + \sqrt{\cos^2\beta - \cos^2\varnothing}}$$

For $\beta = 15^\circ$ and $\varnothing = 36^\circ$:

$$K_a = 0.966 \times \frac{0.966 - \sqrt{0.966^2 - 0.809^2}}{0.966 + \sqrt{0.966^2 - 0.809^2}} = 0.966 \times \frac{0.966 - 0.528}{0.966 + 0.528} = \frac{0.423}{1.494} = 0.283$$

$$P_a = \frac{18 \times 6.28^2}{2} \times 0.283 = 100.45\ kN/m$$

$$P_{ah} = P_a \cos 15^\circ = 100.45 \times 0.966 = 97.03\ kN/m$$

$$P_{av} = P_a \sin 15^\circ = 100.45 \times 0.259 = 26.0\ kN/m$$

Step 2. Compute the vertical resultant force R, net moment M_{net}, eccentricity e, overturning moment M_o about base toe at point A and the resisting moment M_r.

(Continued)

For these computations refer to the scheme of *Step* 1. It is convenient to tabulate the load sources, loads, moment arms and moments in a table such as Table 5.12.

Table 5.12 Load sources, loads, moment arms and moments.

Load source	Load (weight) kN/m	Arm m	Moment M_A kN. m/m + ↷
(I)	0.5 × 6.28 × 1.06 × 18.0 = 59.91	2.65	158.56
(II)	0.5 × 6.0 × 1.06 × 23.5 = 74.73	2.29	171.13
(III)	0.6 × 6 × 23.5 = 84.60	1.64	138.74
(IV)	0.5 × 5.25 × 0.74 × 23.5 = 45.65	1.09	49.76
(V)	0.75 × 1.34 × 23.5 = 23.62	0.67	15.83
P_{av}	= 26.00	3.00	78.00
P_{ah} = 97.03 kN/m		2.09	−202.79
	R = 314.51		**M_{net} = 409.23**

$$\bar{x}R = M_{\text{net}}$$

$$\bar{x} = \frac{M_R}{R} = \frac{409.23}{314.51} = 1.3\,\text{m}$$

$$e = \frac{B}{2} - \bar{x} = \frac{3}{2} - 1.3 = 0.2\ m < \left(\frac{B}{6} = 0.5\,\text{m}\right)$$

Overturning moment

$$M_o = P_{ah} \times 2.09 = 97.03 \times 2.09 = 202.79\,\text{kN.m/m}$$

Resisting moment

$$M_r = M_{\text{net}} - M_o = 409.23 - (-202.79) = 612.02\,\text{kN.m/m}$$

Step 3. Determine the maximum and minimum pressures under the base.

$$q = \frac{R}{BL}\left(1 \pm \frac{6e_x}{B} \pm \frac{6e_y}{L}\right) \text{ Equation (2.7)}$$

Maximum pressure under the base is

$$q_{max} = \frac{R}{BL}\left(1 + \frac{6e_x}{B}\right) = \frac{314.51}{3 \times 1}\left(1 + \frac{6 \times 0.2}{3}\right) = 104.84 + 41.93 = 146.77\ kPa$$

Minimum pressure under the base is

$$q_{min} = \frac{R}{BL}\left(1 - \frac{6e_x}{B}\right) = \frac{314.51}{3 \times 1}\left(1 - \frac{6 \times 0.2}{3}\right) = 104.84 - 41.93 = 62.91\ kPa$$

Step 4. Determine the safety factor against overturning.
Safety factor against overturning is

$$SF = \frac{M_r}{M_o} = \frac{612.02}{202.79} = 3.0$$

Step 5. Determine the safety factor against sliding.
Resisting force against sliding is

$$F_r = R\tan\delta + c_a A_f + P_P$$

In this case, c_a is zero since the base soil cohesion $c = 0$. Also, since any material which may exist in front the wall is ignored, P_P is zero.

$$F_r = R\tan\delta = 314.51 \times \tan 24 = 140.0\,\text{kN/m}$$

Sliding force $= P_{ah} = 97.03$ kN/m
Safety factor against sliding is

$$SF = \frac{F_r}{P_{ah}} = \frac{140.0}{97.03} = 1.44$$

Problem 5.22

A trapezoidal-shaped solid gravity concrete wall is required to retain a 5.0-m embankment which has a slope angle $\beta = 15°$. A stiff silty clay will support the wall structure at a foundation depth of about 1.0 m below the lowest ground level. Safety factors against bearing capacity failure, overturning and sliding should not be less than 3.0, 2.0 and 1.5, respectively. The following tentative dimensions for a gravity retaining wall may be used as a guide:
Base width $B: 0.5$ to $0.7\,H$

where H = total height of wall structure

Base thickness $h: \dfrac{H}{8}$ to $\dfrac{H}{6}$

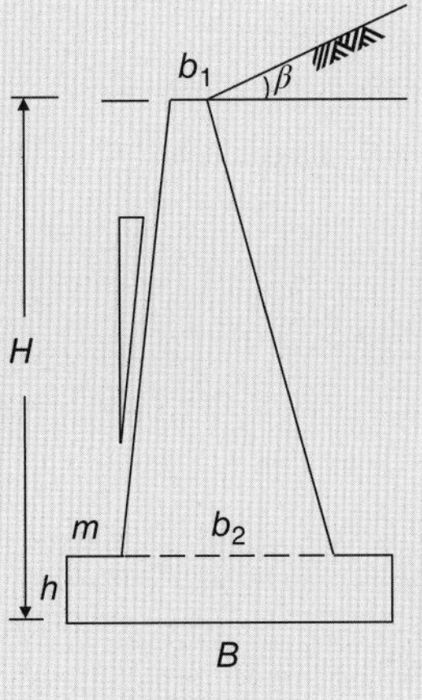

Scheme 5.103

(*Continued*)

Toe projection m: 0.5 h to h

Stem thickness b_1 (at top) : $0.30\,\text{m to}\,\dfrac{H}{12}$

Minimum front face batter: 1 : 48

Design the retaining wall using the following soil and concrete data:

Base soil	Backfill	Concrete
$\gamma = 18\ \text{kN/m}^3$	$\gamma = 17\ \text{kN/m}^3$	$\gamma = 23.5\ \text{kN/m}^3$
$c_u = \dfrac{q_u}{2} = 105\,\text{kPa}$	$c_u = 0$	$f_c' = 17\,\text{MPa}$
$\varnothing = 0$	$\varnothing = 34°$	—

Solution:

Step 1. Select initial values for the base and stem cross-section dimensions.

Assume the following tentative dimensions:

Total height $H = 6.00$ m

Base thickness $h = 0.75$ m

Stem height $Z = H - h = 6.00 - 0.75 = 5.25$ m

Base width $B = 3.50$ m

Toe projection $m = 0.60$ m

Heel projection $n = 0.60$ m

Stem thickness at top $b_1 = 0.40$ m

Stem thickness at bottom $b_2 = 2.30$ m

Front face batter = 1 : 10.5

Angle $\alpha = (75.1)°$

Angle $\beta = 15°$

Foundation depth $D = 1.00$ m

Height of the vertical shear plane at heel is

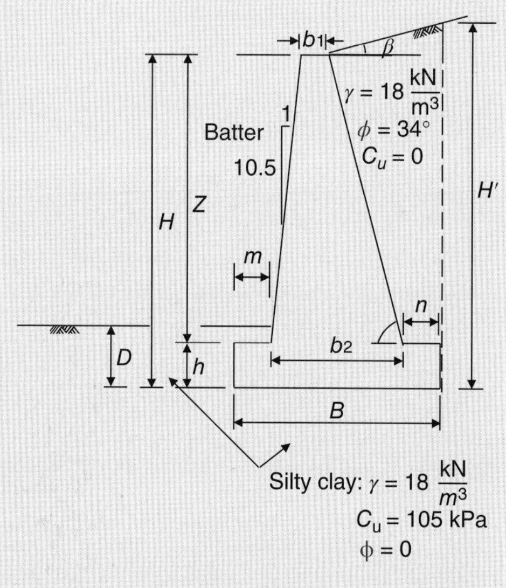

Scheme 5.104

$$H' = H + 2\tan\beta = H + 2\tan 15^{\circ}$$
$$= 6 + 2 \times 0.268 = 6.54\,\text{m}$$

Step 2. Determine the total active earth pressure thrust P_a and its components P_{ah} and P_{av}, using Rankine's lateral earth pressure theory.

$$P_a = \frac{\gamma H'^2}{2} K_a$$

Where K_a = Rankine's coefficient of lateral active earth pressure.

The active thrust P_a acts on the vertical shear plane at a height of $\frac{1}{3}H'$ and at β above the horizontal as shown in the scheme above.

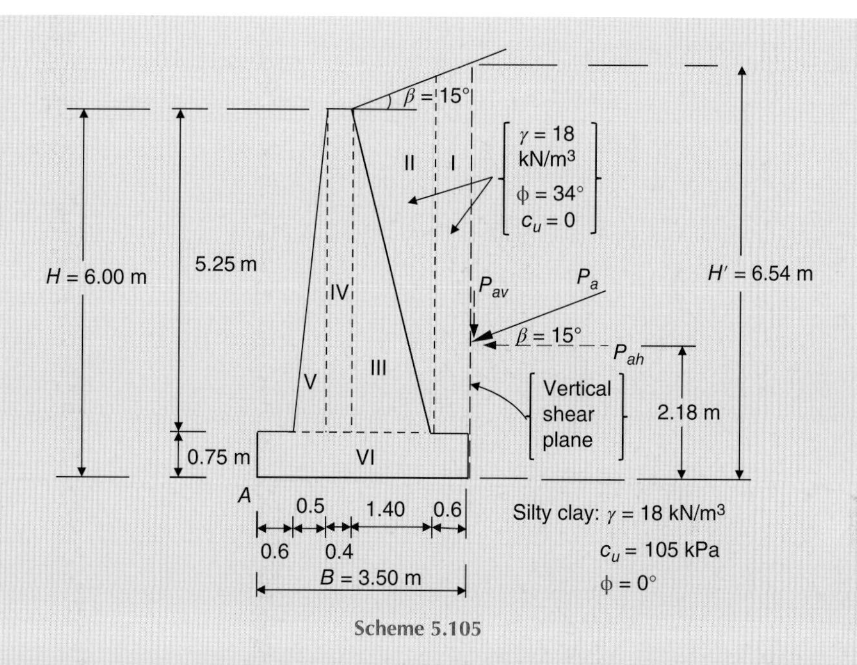

Scheme 5.105

For cohesionless soils, the coefficient K_a depends on the backfill slope angle β and friction angle Ø, can be obtained from tables or may be calculated from the equation

$$K_a = \cos\beta \frac{\cos\beta - \sqrt{\cos^2\beta - \cos^2 Ø}}{\cos\beta + \sqrt{\cos^2\beta - \cos^2 Ø}}$$

For $\beta = 15^\circ$ and $Ø = 34^\circ$:

$$K_a = 0.966 \times \frac{0.966 - \sqrt{0.966^2 - 0.829^2}}{0.966 + \sqrt{0.966^2 - 0.829^2}} = 0.966 \times \frac{0.966 - 0.496}{0.966 + 0.496} = \frac{0.454}{1.462} = 0.311$$

$$P_a = \frac{18 \times 6.54^2}{2} \times 0.311 = \mathit{119.72\ kN/m}$$

$$P_{ah} = P_a \cos 15^\circ = 119.72 \times 0.966 = \mathit{115.65\ kN/m}$$

$$P_{av} = P_a \sin 15^\circ = 119.72 \times 0.259 = \mathit{31.01\ kN/m}$$

Step 3. Compute the vertical resultant force R, net moment M_{net}, eccentricity e, overturning moment M_o about base toe at point A, and the resisting moment M_r.

For these computations refer to the scheme in *Step 2*. It is convenient to tabulate the load sources, loads, moment arms and moments, as in Table 5.13. Assume, conservatively, we neglect the toe soil and that covering the base in front of the wall stem.

(Continued)

Table 5.13 Load sources, loads, moment arms and moments for *Step 3*.

Load source	Load kN/m	Arm m	Moment M_A, kN. m/m, +↷
I	$(0.6)\left(5.25+\frac{2.0+1.4}{2}\tan 15^\circ\right)(18.0)=61.62$	3.21	197.80
II	$(1.4/2)(5.25 + 1.4 \tan 15^\circ)(18.0) = 70.88$	2.43	172.47
III	$0.5 \times 1.4 \times 5.25 \times 23.5 = 86.36$	1.97	169.84
IV	$0.4 \times 5.25 \times 23.5 = 49.35$	1.30	64.16
V	$0.5 \times 0.5 \times 5.25 \times 23.5 = 30.84$	0.93	28.68
VI	$0.75 \times 3.5 \times 23.5 = 61.69$	1.75	107.96
P_{av}	= 31.01	3.50	108.54
P_{ah} = 115.65 kN/m		2.18	−252.12
	R = **391.75**		M_{net} = **597.33**

$$\bar{x}\, R = M_{\text{net}}$$

$$\bar{x} = \frac{M_{\text{net}}}{R} = \frac{597.33}{391.75} = 1.52 \text{ m}$$

$$e = \frac{B}{2} - \bar{x} = \frac{3.5}{2} - 1.52 = 0.23\ m < \left(\frac{B}{6} = 0.58 \text{ m}\right)$$

Step 4. Check wall stability for overturning and sliding.

(a) Overturning stability.

Overturning moment:

$$M_o = P_{ah} \times 2.18 = 115.65 \times 2.18$$
$$= 252.12\,\text{kN.m/m}$$

Resisting moment:

$$M_r = M_{net} - M_o = 597.33 - (-252.12)$$
$$= 849.45\,\text{kN.m/m}$$

Safety factor against overturning is

$$SF = \frac{M_r}{M_o} = \frac{849.45}{252.12} = 3.4 > 2 \quad (\text{OK.})$$

(b) Sliding stability.

Resisting force against sliding (neglecting any passive resistance force) is

$$F_r = R \tan\delta + c_a A_f$$

In this case, δ is zero since $\varnothing = 0$, and therefore $F_r = c_a A_f$.

Assume $c_a = 0.6\ c_u = 0.6 \times 105 = 63$ kPa

$$A_f = \text{base effective area} = B' \times 1 = (B - 2e)(1)$$
$$= (3.5 - 2 \times 0.23)(1) = 3.04\ \text{m}^2/\text{m}$$
$$F_r = 63 \times 3.04 = 191.52\ \text{kN/m}$$

Sliding force $= P_{ah} = 115.65$ kN/m
Safety factor against sliding is

$$SF = \frac{F_r}{P_{ah}} = \frac{191.52}{115.65} = 1.66 > 1.50\ \text{(OK.)}$$

Step 5. Check the foundation stability against bearing capacity failure.

Compute safe bearing capacity of the supporting soil using Hansen bearing capacity equation for the Ø = 0° condition.

Table 4.2: $gross\ q_{\text{ult}} = 5.14\, s_u\left(1 + s'_c + d'_c - i'_c - b'_c - g'_c\right) + \bar{q}$

$$s_u = c_u = 105\ \text{kPa}$$

Table 4.6: $s'_c = 0.2\frac{B'}{L'} = 0$, since $\frac{B'}{L'} \approx 0$ for continuous foundations.

Table 4.7: $b'_c = 0$ since $\eta^\circ = 0$; $g'_c = \frac{\beta^\circ}{147^\circ} = \frac{15}{147} = 0.102$

$$d'_c = 0.4k;\ k = \frac{D}{B}\ \text{for}\ \frac{D}{B} \le 1;\ \frac{D}{B} = \frac{1.0}{3.5} = 0.286;\ \text{hence,}\ d'_c = 0.114$$

$$i'_c = 0.5 - 0.5\sqrt{1 - \frac{H}{A_f c_a}}.$$

Use $c_a = 63$ kPa; $A_f = 3.04\ \text{m}^2$, as in *Step* 4(b).

$$i'_c = 0.5 - 0.5\sqrt{1 - \frac{115.65}{3.04 \times 63}} = 0.184$$
$$gross\ q_{\text{ult}} = 5.14\, s_u\left(1 + d'_c - i'_c - g'_c\right) + \bar{q}$$

Since the toe soil is neglected, the overburden pressure $\bar{q} = 0$. Hence, the safe bearing capacity is

$$gross\ q_{\text{safe}} = 5.14\, s_u\left(1 + d'_c - i'_c - g'_c\right)/SF$$
$$= 5.14(105)(1 + 0.114 - 0.184 - 0.102)/3$$
$$= 464.87/3 = \mathit{155\ kPa}$$

Gross foundation pressure is

$$gross\ q = \frac{R}{B' \times 1} = \frac{391.75}{3.04 \times 1} = \mathit{128.87\ kPa} < gross\ q_{\text{safe}}\ \text{(OK.)} \tag{5.73}$$

Check the maximum gross foundation pressure, $gross\ q_{\text{max}}$:

$$gross\ q_{\text{max}} = \frac{R}{BL}\left(1 + \frac{6e_x}{B}\right)$$
$$= \frac{391.75}{3.5 \times 1}\left(1 + \frac{6 \times 0.23}{3.5}\right) = \mathit{156\ kPa} \cong gross\ q_{\text{safe}}\ \text{(OK.)}$$

(*Continued*)

Step 6. Apply the appropriate load factors (ACI Section 9.2) to all the horizontal and vertical forces, compute eccentricity e of the factored resultant force and calculate the base factored contact pressures.

In this case, the applicable load combinations of the ACI Section 9.2.1 are reduced to:

$U = 1.2\ D + 1.6(H)$, where H represents weight of soil.

$U = 1.6(H)$, where H represents load of lateral pressures

Refer to Table 5.13 in *Step 3* to produce Table 5.14.

Table 5.14 Load sources, loads, moment arms and moments for *Step 6*.

Load source	Factored Load kN/m	Arm, m	Factored Moment M_A + kN. m/m
I	(1.6)(61.62) = 98.59	3.21	316.47
II	(1.6)(70.88) = 113.41	2.43	275.59
III	1.2 × 86.36 = 103.63	1.97	204.15
IV	1.2 × 49.35 = 59.22	1.30	76.99
V	1.2 × 30.84 = 37.01	0.93	34.42
VI	1.2 × 61.69 = 74.03	1.75	129.55
P_{av}	1.6 × 31.01 = 49.62	3.50	173.67
P_{ah} = 1.6 × 115.65 kN/m		2.18	−403.39
	R = **535.51**		M_{net} = **807.45**

$$\bar{x}\, R = M_{\text{net}}$$

$$\bar{x} = \frac{M_{\text{net}}}{R} = \frac{807.45}{535.51} = 1.51\text{ m}$$

$$e = \frac{B}{2} - \bar{x} = \frac{3.5}{2} - 1.51 = 0.24\ m < \left(\frac{B}{6} = 0.58\text{ m}\right)\quad \text{(OK.)}$$

$$q_{\max} = \frac{R}{BL}\left(1 + \frac{6e_x}{B}\right) = \frac{535.51}{3.5 \times 1}\left(1 + \frac{6 \times 0.24}{3.5}\right)$$

$$= 153.00 + 62.95 = 215.95\ kPa\quad \text{(At the toe)}$$

$$q_{\min} = \frac{R}{BL}\left(1 - \frac{6e_x}{B}\right) = \frac{535.51}{3.5 \times 1}\left(1 - \frac{6 \times 0.24}{3.5}\right)\quad \text{(At the heel)}$$

$$= 153.00 - 62.95 = 90.05\ kPa$$

Step 7. Check the assumed footing thickness h considering shear and bending in accordance with ACI Sections 22.7.6 and 22.5.1, respectively.

(a) Considering shear.

In this case, only one-way shear is significant. To consider the most severe condition we use the critical sections located at the faces of the wall. It is clear that the most critical section is in the toe at the wall front face, 0.6 m from the footing edge, as shown below.

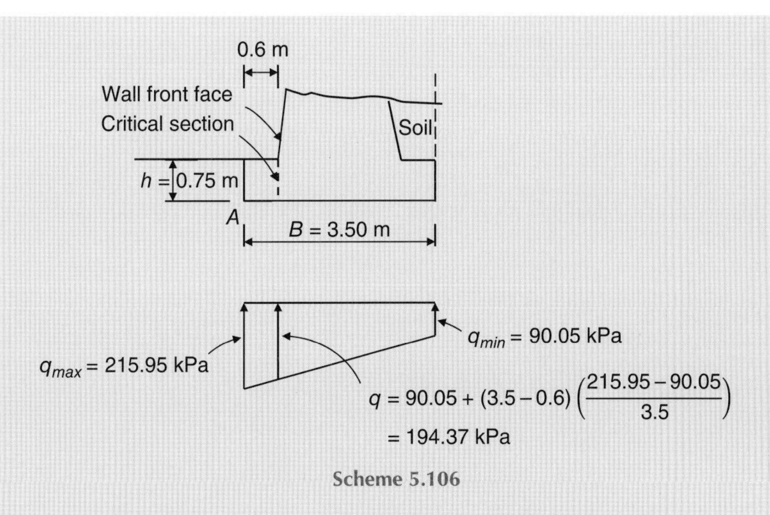

Scheme 5.106

Use effective footing thickness $= h' = h - 0.05$ m (ACI Section 22.4.7)

$$V_u = \left(\frac{q_{\max} + q}{2}\right)(0.6)(1) - \text{factored weight of concrete}$$
$$= \left(\frac{215.95 + 194.37}{2}\right)(0.6)(1) - (1.2)(h')(0.6)(1)(23.5)$$
$$= 123.1 - 16.92\,h'$$

$$\text{Ø}\, V_n \geq V_u$$

When the load factor combinations of ACI Section 9.2.1 are used, the strength reduction factor Ø shall be 0.6 for flexure, compression, shear and bearing of structural plain concrete (ACI Section 9.3.5).

$$V_n = 0.11\lambda\sqrt{f_c'}b_w h' \text{ Equation (5.36)}$$

where λ = modification factor (ACI Section 8.6.1)
= 1.0 for normal-weight concrete

$$\text{Ø}\, V_n = (0.6)(0.11)(1)\left(\sqrt{17}\right)(1000)(1.0)(h') = 272.12\,h'$$

Let Ø $V_n = V_u$:

$$272.12\,h' = 123.1 - 16.92\,h'$$
$$h' = \frac{123.1}{289.04} = 0.43 \text{ m}$$
$$h = 0.43 + 0.05 = 0.48 \text{ m} < h = 0.75\text{m (provided)} \quad \text{(OK.)}$$

Therefore, the section is safe against shear.

(b) Considering bending.
The critical section for the maximum bending moment is located at the wall front face, as shown in the scheme above.

(Continued)

Assume $h' = 0.75 - 0.05 = 0.7$ m (as provided)
The applied factored bending moment per metre length of the wall is

$$M_u = \frac{q \times 0.6^2}{2} + \frac{(q_{max} - q)(0.6)}{2} \times \frac{2 \times 0.6}{3} - \frac{(1.2)(0.7)(0.6^2)(23.5)}{2}$$

$$= 0.18 \times 194.37 + 0.12(215.95 - 194.37) - 3.55 = 34.03 \text{ kN.m}$$

$$\varnothing M_n \geq M_u \qquad \text{Equation (5.32)}$$

$$M_n = 0.42\,\lambda\sqrt{f_c'}S_m \qquad \text{Equation (5.33)}$$

$$S_m = \text{the corresponding elastic section modulus} = \frac{b(h')^2}{6}$$

$$= \frac{(1)(0.7^2)}{6} = 0.082\,m^3$$

$$\varnothing M_n = (\varnothing)(0.42\lambda\sqrt{f_c'})(S_m) = (0.6)(0.42)(1)(\sqrt{17} \times 1000)(0.082)$$

$$= 85.2 \text{ kN.m} > M_u$$

Therefore, the section is safe against bending.

The assumed footing thickness ($h = 0.75$ m) is safe against shear and bending.

Step 8. Check the wall thickness b at about half of its height; say 3 m from top, considering (a) shear and (b) combined bending and axial load in compression.

(a) Considering shear.

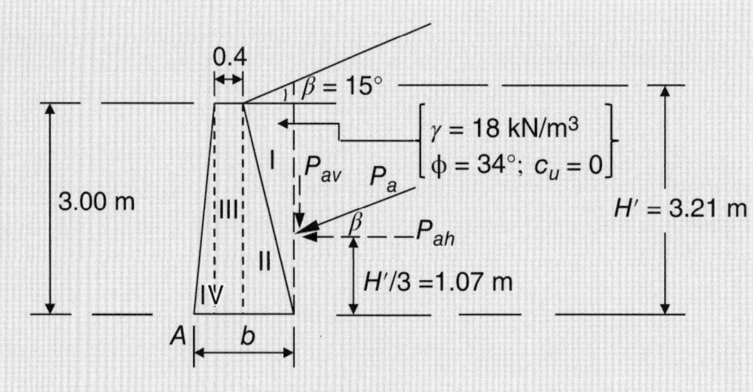

Scheme 5.107

Refer to the scheme in *Step 2* and the figure above:

$$H' = 3 + \left(\frac{1.4}{5.25} \times 3\right)\tan 15^\circ = 3.21 \text{ m}$$

$$b = \frac{0.5}{5.25} \times 3 + 0.4 + \frac{1.4}{5.25} \times 3 = 1.49 \text{ m}; \; b' = b - 0.05 = 1.44 \text{ m}$$

$$P_a = \frac{\gamma H'^2}{2} K_a = \frac{18 \times 3.21^2}{2} \times 0.311 = 28.84 \text{ kN/m}$$

$$P_{ah} = P_a \cos 15^\circ = 28.84 \times 0.966 = 27.86 \text{ kN/m}$$

$$P_{av} = P_a \sin 15^\circ = 28.84 \times 0.259 = 7.47 \text{ kN/m}$$

$$V_u = \text{factored } P_{ah} = 1.6 \times 27.86 = 44.58 \text{ kN/m}$$

$$\varnothing V_n \geq V_u$$

$$V_n = 0.11\lambda\sqrt{f_c'}b_w\, b'$$

$$\varnothing\, V_n = (0.6)(0.11)(1)\left(\sqrt{17}\right)(1000)(1.0)(1.44)$$

$$= 391.86\,\text{kN/m} \gg V_u$$

Therefore,the section is safe against shear.

(b) Considering combined bending and axial load in compression.
Determine factored resultant R of all loads and its eccentricity e. For these computations refer to the figure shown above and that of *Step 2*. It is convenient to use Table 5.15, where the load sources are labeled, weights computed, moment arms given and so on.

Table 5.15 Load sources, loads, moment arms and moments for *Step 8* (b).

Load source	**Factored Load kN/m**	**Arm, m**	**Moment M_A, + kN.m/m, + ↷**
I	$(1.6)\left(\frac{3.21}{2} \times \frac{1.4\times 3}{5.25}\right)(18.0) = 36.98$	1.22	45.12
II	$(1.2)\left(\frac{3}{2} \times \frac{1.4\times 3}{5.25}\right)(23.5) = 33.84$	0.96	32.49
III	1.2 × 0.4 × 3 × 23.5 = 33.84	0.49	16.58
IV	$(1.2)\left(\frac{3}{2} \times \frac{0.5\times 3}{5.25}\right)(23.5) = 12.09$	0.19	2.30
P_{av}	1.6 × 7.47 = 11.95	1.49	17.81
P_{ah} = 1.6 × 27.86 = 44.58 kN/m		1.07	−47.70
	R = **128.70**		M_{net} = **66.60**

$$\bar{x}\, R = M_{\text{net}}$$

$$\bar{x} = \frac{M_{\text{net}}}{R} = \frac{66.6}{128.7} = 0.52\,\text{m}$$

$$e = \frac{b}{2} - \bar{x} = \frac{1.49}{2} - 0.52 = 0.23\ m < \left(\frac{b}{6} = 0.25\,\text{m}\right)$$

Therefore, R is located within the middle-third of the wall thickness and the whole section is in compression.

Maximum compressive stress $f_{\max} = \frac{R}{b'L}\left(1 + \frac{6e_x}{b'}\right) = \frac{128.70}{1.44\times 1}\left(1 + \frac{6\times 0.23}{1.44}\right)$

$$= 89.38 + 85.65 = 175.03\,\text{kPa}$$

(Continued)

Allowable compressive stress of plain concrete, in bearing, is

$$f_c = \emptyset\,(0.85 f_c') = 0.6 \times 0.85 \times 17 \times 1000 = 8670\,\text{kPa} \gg f_{max}$$

According to ACI Section 22.6.5, if the resultant of all factored loads is located within the middle-third of the wall thickness, structural plain concrete walls of solid rectangular cross section shall be permitted to be designed by

$$\emptyset P_n \geq P_u$$

In this case $P_u = R = 128.7$ kN/m, and P_n is nominal axial strength calculated by

$$P_n = 0.45 f_c' A_g \left[1 - \left(\frac{\ell_c}{32b}\right)^2\right]$$

The term $\left(\frac{\ell_c}{32b}\right)^2$ is neglected since it is too small.

$$\emptyset P_n = 0.6 \times 0.45 \times 17 \times 1000 \times 1.44 \times 1$$
$$= 6609.6\,\text{kN/m} \gg P_u$$

Therefore, the section is safe against the applied factored compressive stress.
Step 9. Develop the final design sketches.

Problem 5.23

A rectangular footing 2 × 3 m is to be placed at a depth of 1 m in a deep stratum of soft to medium saturated clay. Average elastic parameters E_s and μ of the clay are 8 MPa and 0.5, respectively. Estimate the average modulus of-subgrade reaction K_s using:

(a) Equation (5.42).
(b) Equation (5.43).
(c) Equation (5.44). Assume $q_a = 80$ kPa and $SF = 3$

Solution:

(a) Equation (5.42):

$$K_s = \frac{E_s}{B(1-\mu_s^2)}$$

$$K_s = \frac{E_s}{B(1-\mu_s^2)} = \frac{8 \times 1000}{2(1-0.5^2)} = 5333\ kN/m^3$$

(b) Equation (5.43):

$$K_s = \frac{1}{B \times E_s' \times m \times I_S \times I_F}$$

$$E_s' = \frac{(1-\mu_s^2)}{E_s} = \frac{1-0.5^2}{8 \times 1000} = 9.38 \times 10^{-5}\,\text{m}^2/\text{kN}$$

For the *centre* of the loaded area:

Use: $\frac{B}{2}$ instead of B, $\frac{L}{2}$ instead of L, m = 4 and $H = 4B$

$M = L/B$, $N = H/B$, H = Thickness of the soil layer, in units of B

$$M = \frac{1.5}{1} = 1.5, N = 4 \times \frac{2}{1} = 8$$

From Table 3.7 obtain $I_1 = 0.561$ and $I_2 = 0.029$

$$I_S = I_1 + \frac{1-2\mu}{1-\mu} I_2 = 0.561 + \frac{1-2\times0.5}{1-0.5} \times 0.029 = 0.561$$

For actual values of $\frac{B}{L} = 0.667$; $\frac{D}{B} = 0.5$; $\mu = 0.5$, from Table 3.6, obtain $I_F = 0.88$.

$$K_s = \frac{1}{1\times9.38\times10^{-5}\times4\times0.561\times0.88} = 5399\,\text{kN/m}^3$$

For a *corner* of the loaded area:

Use: actual B and L dimensions; m = 1; $H = 4B$;

$$M = \frac{3}{2} = 1.5;\ N = 4 \times \frac{2}{2} = 4$$

From Table 3.7 obtain $I_1 = 0.455$ and $I_2 = 0.054$

$$I_S = I_1 + \frac{1-2\mu}{1-\mu} I_2 = 0.455 + \frac{1-2\times0.5}{1-0.5} \times 0.054 = 0.455$$

For actual values of $\frac{B}{L} = 0.667$, $\frac{D}{B} = 0.5$, $\mu = 0.5$ Table 3.6 gives

$$I_F = 0.88\ (\text{as before})$$

$$K_s = \frac{1}{2\times9.38\times10^{-5}\times1\times0.455\times0.88} = 13313\,\text{kN/m}^3$$

Compute weighted average K_s using four centre K_s values and one corner value, as follows:

$$\text{Average } K_s = \frac{4\times5399+13313}{5} = \mathit{6982\ kN/m^3}$$

(c) Equation (5.44):

$$K_s = 40\,(SF)\,q_a$$

$$K_s = 40\,(SF)\,q_a = 40\times3\times80 = \mathit{9600\ kN/m^3}$$

Problem 5.24

For the 15 × 15 m mat foundation and soil data of Problem 4.37 recommend a single value of K_s that may be considered as an average value for the base.

(Continued)

Solution:

Step 1. Zone the area beneath the mat. In this case, it would be sufficient to compute K_s for three zones, namely, centre (point C), intermediate or one-quarter point (point B) and edge (point A), as shown below.

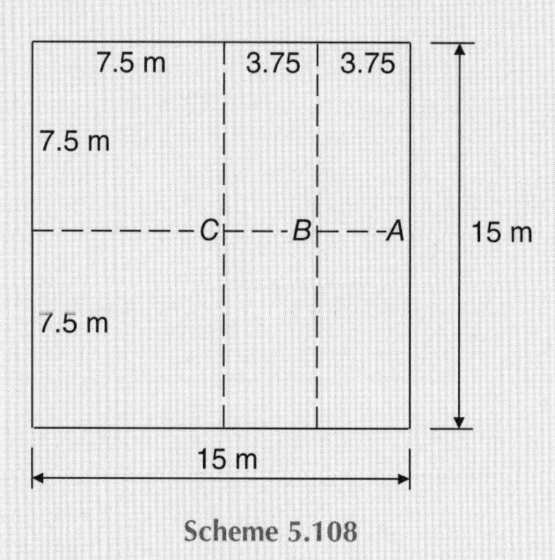

Scheme 5.108

Step 2. Compute $K_{s,\,C}$ for the centre zone (point C).

$$K_{s,\,C} = \frac{1}{B \times E_s' \times m \times I_S \times I_F}$$

$$E_s' = \frac{(1-\mu_s^2)}{E_s}$$

Refer to Solution of Problem 4.37. Estimate weighted average E_s of soils beneath the mat, using the estimated E_s of the stiff silty clay and sand soils, as follows:

$$E_{s,(av)} = \frac{3.4 \times 75\ 000 + 668\ 600}{3.4 + 24.1} = 33\ 585\ \text{kPa}$$

Also, estimate weighted average $\mu = \dfrac{3.4 \times 0.5 + 24.1 \times 0.3}{3.4 + 24.1} = 0.32$

$$E_s' = \frac{(1-0.32^2)}{33585} = 2.67 \times 10^{-5}\ \text{m}^2/\text{kN}$$

$$I_S = I_1 + \frac{1-2\mu}{1-\mu} I_2;\ M = \frac{L/2}{B/2};\ N = \frac{H}{B/2}$$

$$H = \text{total thickness of the soil layers} = 27.5\ \text{m} < 4B$$

$$M = \frac{7.5}{7.5} = 1;\ N = \frac{27.5}{7.5} = 3.67$$

From Table 3.7 obtain $I_1 = 0.393$ and $I_2 = 0.041$

$$I_S = 0.393 + \frac{1-2 \times 0.32}{1-0.32} \times 0.041 = 0.415$$

For actual values of $\dfrac{B}{L} = 1,\ \dfrac{D}{B} = \dfrac{1.5}{15} = 0.1$ and $\mu = 0.32$, from Table 3.6 obtain $I_F = 0.95$ (approximately).

$$K_{s,C} = \frac{1}{B \times E_s' \times m \times I_S \times I_F} = \frac{1}{7.5 \times 2.67 \times 10^{-5} \times 4 \times 0.415 \times 0.95} = 3167\,\text{kN/m}^3$$

Step 3. Compute $K_{s,B}$ for the intermediate zone at the one-quarter point (point B).
There are two sets of rectangles contribute at point B, as shown below:

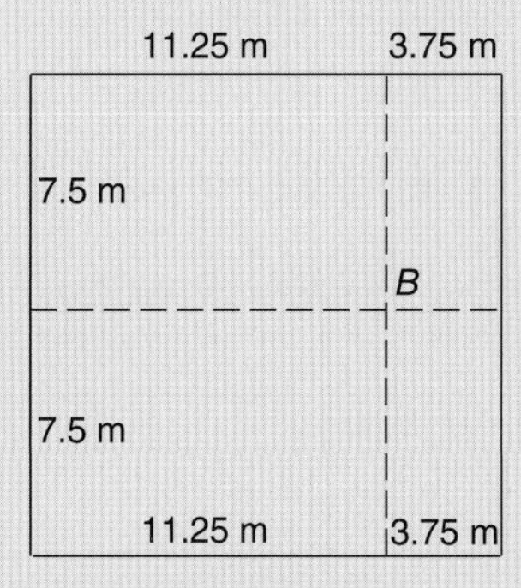

Scheme 5.109

Set 1: two rectangles of $B = 7.5$ m and $L = 11.25$ m

$$M = \frac{L}{B} = 1.5;\; N = \frac{H}{B} = 3.67$$

$$I_1 = 0.41;\; I_2 = 0.065;\; I_S = 0.444$$

$$\frac{B}{L} = 0.67;\; \frac{D}{B} = \frac{1.5}{7.5} = 0.2;\; I_F = 0.926$$

$$S_i = q \times B \times \frac{(1-\mu_s^2)}{E_s} \times m \times I_S \times I_F$$

$$S_i = q \times B \times E_s' \times m \times I_S \times I_F$$

$$\frac{S_i}{q} = 7.5 \times 2.67 \times 10^{-5} \times 2 \times 0.444 \times 0.926 = 1.647 \times 10^{-4}\,\text{m}^3/\text{kN}$$

Set 2: two rectangles of $B = 3.75$ m and $L = 7.5$ m

$$M = \frac{L}{B} = 2;\; N = \frac{H}{B} = 7.33;\; I_1 = 0.597;\; I_2 = 0.041;\; I_S = 0.619$$

$$\frac{B}{L} = 0.5;\; \frac{D}{B} = \frac{1.5}{3.75} = 0.4;\; I_F = 0.866$$

$$S_i = q \times B \times E_s' \times m \times I_S \times I_F$$

$$\frac{S_i}{q} = 3.75 \times 2.67 \times 10^{-5} \times 2 \times 0.619 \times 0.86 = 1.073 \times 10^{-4}\,\text{m}^3/\text{kN}$$

$$\frac{1}{K_{s,B}} = \sum \frac{S_i}{q} = 1.647 \times 10^{-4} + 1.073 \times 10^{-4} = 2.72 \times 10^{-4}\,\text{m}^3/\text{kN}$$

$$K_{s,B} = \frac{1}{2.72 \times 10^{-4}} = 3676\,\text{kN/m}^3$$

(Continued)

Step 4. Compute $K_{s,A}$ for the edge zone (point A).

There is one set of two rectangles contributing at point A, as shown below:

$B = 7.5$ m and $L = 15.0$ m

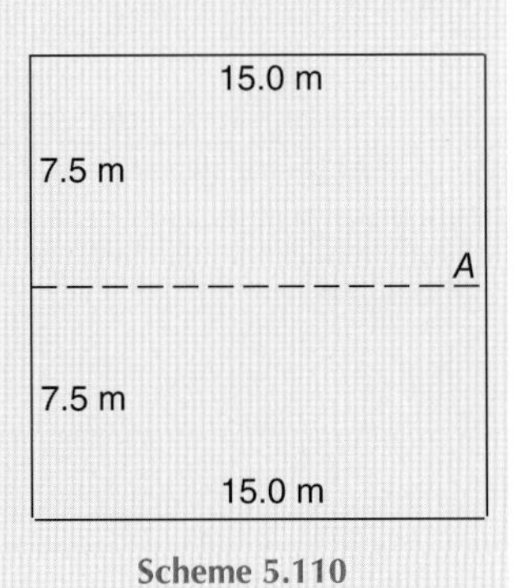

Scheme 5.110

$$M = \frac{L}{B} = 2; N = \frac{H}{B} = \frac{27.5}{7.5} = 3.67$$

$$I_1 = 0.452; I_2 = 0.074; I_S = 0.491$$

$$\frac{B}{L} = 0.5; \frac{D}{B} = \frac{1.5}{7.5} = 0.2; I_F = 0.936$$

$$S_i = q \times B \times E_s' \times m \times I_S \times I_F$$

$$\frac{1}{K_{s,A}} = \frac{S_i}{q} = B \times E_s' \times m \times I_S \times I_F$$

$$K_{s,A} = \frac{1}{7.5 \times 2.67 \times 10^{-5} \times 2 \times 0.491 \times 0.936}$$

$$= 5433\ kN/m^3$$

Step 5. Recommend a single value of K_S that may be considered as an average value for the base.

If a single value of K_S is to be provided, one might simply use

$$K_s = \frac{K_{s,C} + K_{s,B} + K_{s,A}}{3} = \frac{3167 + 3676 + 5433}{3} = 4092\ \text{kN/m}^3$$

or (weighting $K_{s,C}$),

$$K_s = \frac{4K_{s,C} + K_{s,A}}{5} = \frac{4 \times 3167 + 5433}{5} = 3620\ \text{kN/m}^3$$

or (weighting $K_{s,B}$ and using *Simpson's rule*),

$$K_s = \frac{K_{s,C} + 4K_{s,B} + K_{s,A}}{6} = \frac{3167 + 4 \times 3676 + 5433}{6} = 3884\ \text{kN/m}^3$$

Another method may be by equating summation of product of each zone area and its K_s to the product of the mat area and $K_{s,av}$, as follows:

Divide the mat area into three zones, as shown in the scheme below, and compute the area of each zone.

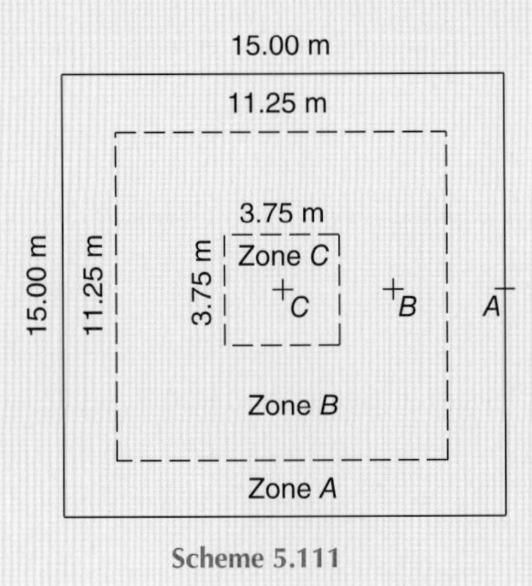

Scheme 5.111

$$\text{Area of zone } C \text{ (central)}: A_C = 3.75 \times 3.75 = 14.063 \text{ m}^2$$

$$\text{Area of zone } B \text{ (interior)}: A_B = 11.25 \times 11.25 - 14.063 = 112.5 \text{ m}^2$$

$$\text{Area of zone } A \text{ (exterior)}: A_A = 15 \times 15 - 112.5 = 98.437 \text{ m}^2$$

$$A_A(K_{s,A}) + A_B(K_{s,B}) + A_C(K_{s,C}) = (A_A + A_B + A_C)\, K_{s,av}$$

$$98.437(5433) + 112.5(3676) + 14.063(3167) = (15^2)\, K_{s,av}$$

$$K_{s,av} = \frac{992896}{225} = 4413 \text{ kN/m}^3$$

It may be appropriate to recommend the average of all the computed values as a single value for the required modulus of subgrade reaction, which is

$$K_s = \frac{4092 + 3620 + 3884 + 4413}{4} = 4002.25 \text{ kN/m}^3$$

Recommend $K_s = 4000\ kN/m^3$

Problem 5.25

A structure is to be supported on a 15-m square mat foundation. The average contact pressure is 100 kPa.

A settlement analysis conducted by the geotechnical engineer gave an average settlement $S = 25$ mm.

Recommend design values for the modulus of subgrade reaction K_S which may be used in a pseudo-coupled analysis.

Solution:

Step 1. Estimate $K_{s,av}$.

Equation (5.40): $K_s = \frac{q}{\delta}$

Where

$$q = \text{bearing pressure}$$
$$\delta = \text{deflection (settlement)}$$

$$\text{Average modulus of subgrade reaction} = K_{s,av} = \frac{100}{0.025} = 4000 \text{ kN/m}^3$$

Step 2. Divide the mat area into three zones and compute the area of each zone. Note that the innermost (central) zone shall be half as wide and half as long as the mat.

$$\text{Area of zone } A \text{ (central)}: A_A = 7.5 \times 7.5 = 56.25 \text{ m}^2$$

$$\text{Area of zone } B \text{ (interior)}: A_B = 11.25 \times 11.25 - 7.5 \times 7.5 = 70.313 \text{ m}^2$$

$$\text{Area of zone } C \text{ (exterior)}: A_C = 15 \times 15 - 11.25 \times 11.25 = 98.438 \text{ m}^2$$

*Step 3.*Assign a K_S value to each zone.

The K_S values should progressively increase from the centre such that the outermost zone (exterior zone) has a K_S about twice as large as the innermost zone (central zone). In this case, let us use:

(*Continued*)

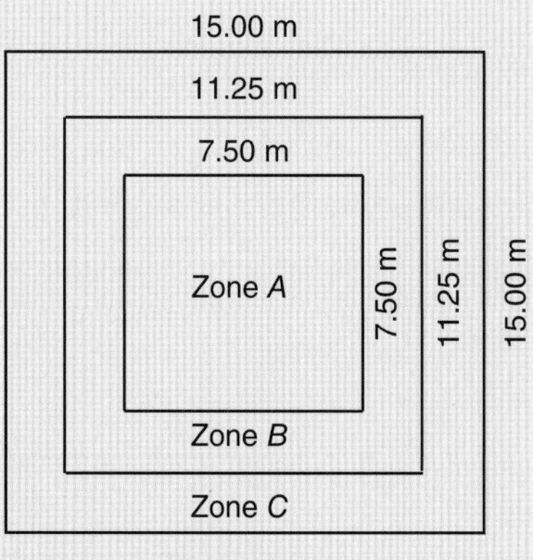

Scheme 5.112

$$K_{s,C} = 2\,K_{s,A};\; K_{s,B} = 1.5\,K_{s,A}$$

Step 4. Compute the design K_S values.

Summation of the product of each zone area and its K_S should equal to the product of the mat area and $K_{s,av}$.

$$A_A(K_{s,A}) + A_B(K_{s,B}) + A_C(K_{s,C}) = (A_A + A_B + A_C)\,K_{s,av}$$

$$56.25(K_{s,A}) + 70.313(1.5)(K_{s,A}) + 98.438(2)(K_{s,A}) = (15^2)\,K_{s,av}$$

$$K_{s,A} = \frac{225}{358.6} = 0.627\,K_{s,av}$$

$$K_{s,A} = 0.627 \times 4000 = \mathit{2508\ kN/m^3}$$

$$K_{s,B} = 1.5\,K_{s,A} = 1.5 \times 2508 = \mathit{3762\ kN/m^3}$$

$$K_{s,C} = 2\,K_{s,A} = 2 \times 2508 = \mathit{5016\ kN/m^3}$$

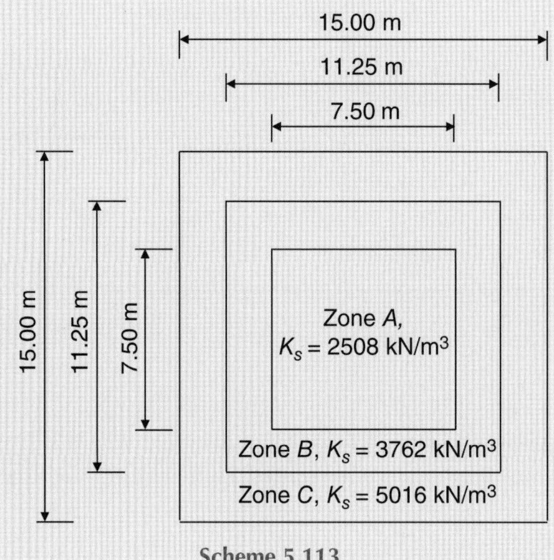

Scheme 5.113

Problem 5.26

The plan of a simple mat foundation and the tabulated column working loads are shown in the scheme below. All the columns are 0.4 × 0.4 m and carry no moment of significant value.

The design soil pressure or *net* q_a and average modulus of subgrade reaction $K_{s,av}$, recommended by the geotechnical engineer, are 60 kPa and 7200 kN/m^3, respectively.

Design the mat by the *conventional rigid method*, and then check if this design method could be considered appropriate. Use:

$$f_c' = 24 \text{ MPa}$$
$$f_y = 420 \text{ MPa}$$

Col.No.	Axial col.loads D, kN	Axial col.loads L, kN
1	80	160
2	120	240
3	100	200
4	320	640
5	320	640
6	270	540
7	320	640
8	320	640
9	270	540
10	80	160
11	120	240
12	90	180

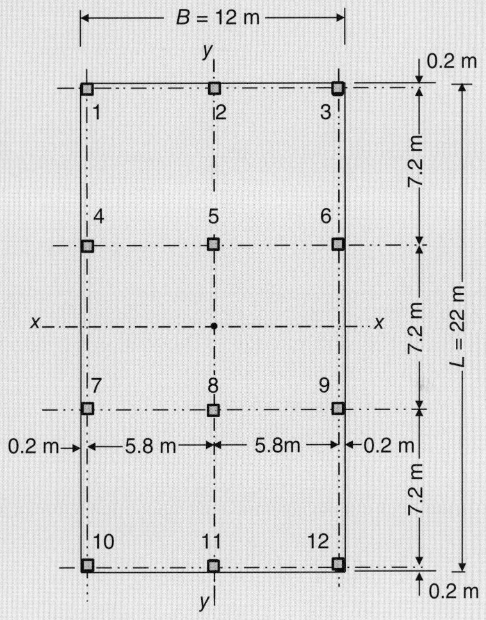

Scheme 5.114

Solution:

Step 1. Compute resultant R of all working (unfactored) loads, as shown in Table 5.16.

Step 2. Find location of R and determine the eccentricities e_x and e_y.

(a) Take moments about the left edge of the mat and compute $\bar{x}$ as follows:

$$R\bar{x} = \Sigma \text{ moment of column } (D+L) \pm \Sigma M_y$$

$$\bar{x} = \frac{\begin{array}{c} 2(0.2)(240) + 2(0.2)(960) + 2(6.0)(360) + 2(6.0)(960) \\ + (11.8)(300 + 810 + 810 + 270) \pm (0) \end{array}}{7230} = 5.83 \text{ m}$$

$$e_x = \frac{12}{2} - 5.83 = 0.17 \ m < \frac{B}{6}$$

(*Continued*)

Table 5.16 Computation of resultant R of all working (unfactored) loads.

	Axial loads			Moments	
Col. No.	***D*, kN**	***L*, kN**	**(*D* + *L*), kN**	$M_{x,(D+L)}$	$M_{y,(D+L)}$
1	80	160	240	– –	– –
2	120	240	360	– –	– –
3	100	200	300	– –	– –
4	320	640	960	– –	– –
5	320	640	960	– –	– –
6	270	540	810	– –	– –
7	320	640	960	– –	– –
8	320	640	960	– –	– –
9	270	540	810	– –	– –
10	80	160	240	– –	– –
11	120	240	360	– –	– –
12	90	180	270	– –	– –
			$R = \Sigma\,(D + L) = \mathbf{7230\ kN}$	$\Sigma\,M_x = 0$	$\Sigma\,M_y = 0$

(b) Take moments about the bottom edge of the mat and compute $\bar{y}$ as follows:

$$R\bar{y} = \Sigma\,\text{moment of column}\,(\text{D} + \text{L}) \pm \Sigma\,M_x$$

$$\bar{y} = \frac{(21.8)(240 + 360 + 300) + (14.6)(960 + 960 + 810) + (7.4)(960 + 960 + 810) + (0.2)(240 + 360 + 270) \pm (0)}{7230} = 11.04\ \text{m}$$

$$e_y = 11.04 - \frac{22}{2} = 0.04\ m < \frac{L}{6}$$

Step 3. Determine the maximum net contact pressure (unfactored) using Equation (2.7) and compare it with the given *net* q_a.

$$q = \frac{R}{BL}\left(1 \pm \frac{6e_x}{B} \pm \frac{6e_y}{L}\right)$$

$$net\ q_{\max} = \frac{R}{BL}\left(1 + \frac{6e_x}{B} + \frac{6e_y}{L}\right) \qquad \text{Equation (2.7)}$$

$$= \frac{7230}{12 \times 22}\left(1 + \frac{6 \times 0.17}{12} + \frac{6 \times 0.04}{22}\right) = 30.01\ kPa$$

$$net\ q_a = 60\,\text{kPa} > net\ q_{\max} \quad (OK.)$$

Step 4. Compute unfactored net contact pressure q at selected points beneath the mat using Equation (2.6). These selected points are corners of continuous beam strips (or combined footings with multiple columns) to which the mat is divided in both x and y directions, as shown in the scheme below.

Point	*R/A* kPa	± 0.39 *x* kPa	± 0.03 *y* kPa	*q* kPa
A	+ 27.39	+ 2.34	+ 0.33	+ 30.06*
E	+ 27.39	+ 1.13	+ 0.33	+ 28.85
F	+ 27.39	− 1.13	+ 0.33	+ 26.59
B	+ 27.39	− 2.34	+ 0.33	+ 25.38
I	+ 27.39	+ 2.34	+ 0.22	+ 29.95
J	+ 27.39	− 2.34	+ 0.22	+ 25.27
K	+ 27.39	+ 2.34	0	+ 29.73
L	+ 27.39	− 2.34	0	+ 25.05
M	+ 27.39	+ 2.34	− 0.22	+ 29.51
N	+ 27.39	− 2.34	− 0.22	+ 24.83
C	+ 27.39	+ 2.34	− 0.33	+ 29.40
G	+ 27.39	+ 1.13	− 0.33	+ 28.19
H	+ 27.39	− 1.13	− 0.33	+ 25.93
D	+ 27.39	− 2.34	− 0.33	+ 24.72

*The + sign indicates that the soil is in "compression"

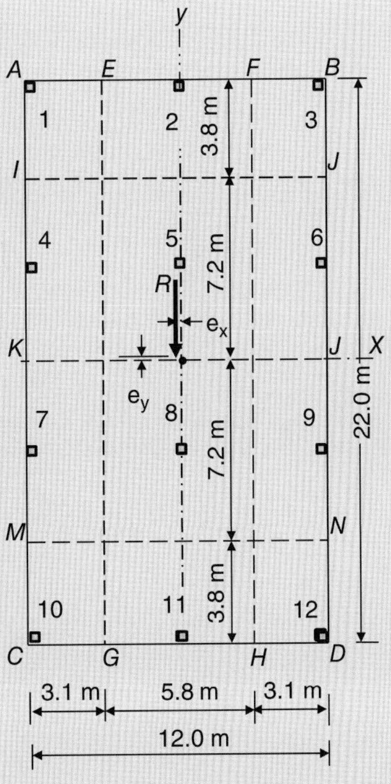

Scheme 5.115

$$q_{(x,\,y)} = \frac{R}{A} \pm \frac{M_y}{I_y}x \pm \frac{M_x}{I_x}y \quad \text{Equation (2.6)}$$

$$M_y = R \times e_x = 7230 \times 0.17 = 1229.1 \text{ kN.m}$$

$$M_x = R \times e_y = 7230 \times 0.04 = 289.2 \text{ kN.m}$$

$$I_y = \frac{LB^3}{12} = \frac{22 \times 12^3}{12} = 3168 \text{ m}^3;\; I_x = \frac{BL^3}{12} = \frac{12 \times 22^3}{12} = 10\,648 \text{ m}^3$$

$$q_{(x,y)} = \frac{7230}{12 \times 22} \pm \frac{1229.1}{3168}x \pm \frac{289.2}{10\,648}y = 27.39 \pm 0.39\,x \pm 0.03\,y$$

Step 5. Check static equilibrium of each individual beam strip and modify the column loads and contact pressures accordingly.

(A) Beam strips in y – or L – direction.

Strip AEGC (3.1 × 22 m):

This strip is symmetrically loaded by columns No. 1, 4, 7 and 10.

$$\sum \text{column loads} = 240 + 960 + 960 + 240 = 2400 \text{ kN}$$

$$\text{Average soil reaction} = (A)(q_{av}) = (3.1 \times 22)\left(\frac{30.06 + 28.85 + 28.19 + 29.40}{4}\right)$$

$$= 1986.33 \text{ kN} \neq \sum \text{column loads}$$

Therefore, *static equilibrium is not satisfied* ($\sum F_v \neq 0$); the column loads and contact pressures need to be modified.

(Continued)

$$Q_{av} = \frac{1}{2}(\Sigma \text{ column loads} + Aq_{av})$$

$$Q_{av} = \frac{1}{2}(2400 + 1986.33) = 2193.17 \text{ kN} \qquad \text{Equation (5.49)}$$

The reducing factor for column loads of strip *AEGC* is

$$MF_{(\text{col.})} = \frac{Q_{av}}{\Sigma \text{column loads}} = \frac{2193.17}{2400} = 0.914 \qquad \text{Equation (5.50)}$$

The increasing factor for contact pressures is

$$MF_{(\text{soil})} = \frac{Q_{av}}{Aq_{av}} = \frac{2193.17}{1986.33} = 1.104 \qquad \text{Equation (5.51)-(a)}$$

For each column load, the applicable factored load combinations of the ACI Section 9.2.1 are reduced to:

$$U = 1.2D + 1.6L$$

Each column factored modified load (Table 5.17) = $[MF_{(\text{col.})}](1.2\ D + 1.6\ L)$.

Table 5.17 Column factored modified loads.

Column No.	Factored modified column loads
1	(0.914)(1.2 × 80 + 1.6 × 160) = **322 kN**
4	(0.914)(1.2 × 320 + 1.6 × 640) = **1287 kN**
7	(0.914)(1.2 × 320 + 1.6 × 640) = **1287 kN**
10	(0.914)(1.2 × 80 + 1.6 × 160) = **322 kN**
	***R* = 3218 kN**

Since the strip is symmetrically loaded, the factored *R* falls at the centre and the factored contact pressure q_{factored} is assumed uniformly distributed.

$$q_{\text{factored}} = \frac{R_{\text{factored}}}{A} = \frac{3218}{3.1 \times 22} = 47.185 \text{ kPa} \qquad \text{Equation (5.52)}$$

The factored contact pressure per metre length of the strip is

$$q_{\text{factored}} = 47.185 \times 3.1 = 146.27 \text{ kN/m length}$$

Strip EFHG (5.8 × 22 m):
This strip is also symmetrically loaded by columns No. 2, 5, 8 and 11.

$$\Sigma \text{ column loads} = 360 + 960 + 960 + 360 = 2640\,\text{kN}$$

$$\text{Average soil reaction} = (A)(q_{av}) = (5.8 \times 22)\left(\frac{28.85 + 26.59 + 25.93 + 28.19}{4}\right)$$

$$= 3494.96\,\text{kN} \neq \Sigma \text{ column loads}$$

Therefore, *static equilibrium is not satisfied* ($\Sigma F_v \neq 0$); the column loads and contact pressures need to be modified.

$$Q_{av} = \frac{1}{2}(2640 + 3494.96) = 3067.48\ \text{kN}$$

$MF_{(\text{col.})} = \dfrac{3067.48}{2640} = 1.162$. It is an increasing factor for the columns of strip *EFHG*.
The contact pressure modification factor is

$$MF_{(\text{soil})} = \frac{Q_{av}}{Aq_{av}} = \frac{3067.48}{3494.96} = 0.878.$$

It is a reducing factor in this case.
Each column factored modified load = $[MF_{(\text{col.})}](1.2\ D + 1.6\ L)$, computed as shown in Table 5.18.

Table 5.18 Column factored modified loads.

Column No.	Factored modified column loads
2	(1.162)(1.2 × 120 + 1.6 × 240) = **614 kN**
5	(1.162)(1.2 × 320 + 1.6 × 640) = **1636 kN**
8	(1.162)(1.2 × 320 + 1.6 × 640) = **1636 kN**
11	(1.162)(1.2 × 120 + 1.6 × 240) = **614 kN**
	***R* = 4500 kN**

$$q_{\text{factored}} = \frac{4500}{5.8 \times 22} = 35.27\,\text{kPa}$$

The factored contact pressure per metre length of the strip is

$$q_{\text{factored}} = 35.27 \times 5.8 = \mathit{204.57\ kN/m\ length}$$

Strip FBDH (3.1 × 22 m):
This strip is unsymmetrically loaded by columns No. 3, 6, 9 and 12.

$$\Sigma \text{ column loads} = 300 + 810 + 810 + 270 = 2190\,\text{kN}$$

$$\text{Average soil reaction} = (A)(q_{av}) = (3.1 \times 22)\left(\frac{26.59 + 25.38 + 24.72 + 25.93}{4}\right)$$

$$= 1749.67\,\text{kN} \neq \Sigma \text{ column loads}$$

(*Continued*)

Therefore, *static equilibrium is not satisfied* ($\Sigma F_v \neq 0$); the column loads and contact pressures need to be modified.

$$Q_{av} = \frac{1}{2}(2190 + 1749.67) = 1969.84 \text{ kN}$$

$MF_{(\text{col.})} = \dfrac{1969.84}{2190} = 0.899$. This is a reducing factor for columns of strip *FBDH*.

Each column factored modified load $= [MF_{(\text{col.})}](1.2\ D + 1.6\ L)$, computed as shown in Table 5.19.

Table 5.19 Column factored modified loads.

Column No.	Factored modified column loads
3	(0.899)(1.2 × 100 + 1.6 × 200) = **396 kN**
6	(0.899)(1.2 × 270 + 1.6 × 540) = **1068 kN**
9	(0.899)(1.2 × 270 + 1.6 × 540) = **1068 kN**
12	(0.899)(1.2 × 90 + 1.6 × 180) = **356 kN**
	***R* = 2888 kN**

Since the strip is unsymmetrically loaded, the factored *R* does not fall at the centre; it has an eccentricity e_L or e_y, calculated as follows:

$$R\bar{y} = \sum \text{moment of factored column loads}$$

$$\bar{y} = \frac{396 \times 21.8 + 1068 \times 14.6 + 1068 \times 7.4 + 356 \times 0.2}{2888} = 11.15 \text{ m}$$

$$e_L = 11.15 - \frac{22}{2} = 0.15 \text{ m} < \frac{L}{6}$$

The factored contact pressures are calculated, using Equation (5.53), as follows:

$$q_{\text{factored}} = \frac{R_{\text{factored}}}{A} \pm \frac{e_L R_{\text{factored}}}{I_B} l$$

$$q_{\max} = \frac{2888}{3.1 \times 22} + \frac{0.15 \times 2888}{3.1 \times 22^3/12} \times 11 = 42.346 + 1.732 = 44.078 \text{ kPa}$$

$$q_{\min} = 42.346 - 1.732 = 40.614 \text{ kPa}$$

$$q_{\max} = 44.078 \times 3.1 = \mathit{136.64\ kN/m\ length}$$

$$q_{\min} = 40.614 \times 3.1 = \mathit{125.9\ kN/m\ length}$$

(B) Beam strips in x – or B – direction.

Strip ABJI (3.8 × 12 m):

This strip is unsymmetrically loaded by columns No. 1, 2 and 3.

$$\sum \text{column loads} = 240 + 360 + 300 = 900 \text{ kN}$$

$$\text{Average soil reaction} = (A)(q_{av}) = (3.8 \times 12)\left(\frac{30.06 + 25.38 + 25.27 + 29.95}{4}\right)$$

$$= 1261.52 \text{ kN} \neq \Sigma \text{ column loads}$$

Therefore, *static equilibrium is not satisfied* ($\Sigma F_v \neq 0$); the column loads and contact pressures need to be modified.

$$Q_{av} = \frac{1}{2}(900 + 1261.52) = 1080.76\,\text{kN}$$

$MF_{(\text{col.})} = \dfrac{1080.76}{900} = 1.2$ It is an increasing factor for columns of strip *ABJI*

Each column factored modified load = $[MF_{(\text{col.})}](1.2\,D + 1.6\,L)$, computed as shown in Table 5.20.

Table 5.20 Column factored modified loads.

Column No.	Factored modified column loads
1	(1.2)(1.2 × 80 + 1.6 × 160) = **422 kN**
2	(1.2)(1.2 × 120 + 1.6 × 240) = **634 kN**
3	(1.2)(1.2 × 100 + 1.6 × 200) = **528 kN**
	R **= 1584 kN**

Since the strip is unsymmetrically loaded, the factored R does not fall at the centre; it has an eccentricity e_x calculated as follows:

$$R\bar{x} = \Sigma\,\text{moment of factored column loads}$$

$$\bar{x} = \frac{422 \times 0.2 + 634 \times 6.0 + 528 \times 11.8}{1584} = 6.388\text{ m}$$

$$e_x = 6.388 - \frac{12}{2} = 0.388\text{ m} < \frac{12}{6}$$

The factored contact pressures are calculated, using Equation (5.53), as follows:

$$q_{\text{factored}} = \frac{R_{\text{factored}}}{A} \pm \frac{e_x R_{\text{factored}}}{I_y}\,l$$

$$q_{\max} = \frac{1584}{3.8 \times 12} + \frac{0.388 \times 1584}{3.8 \times 12^3/12} \times 6 = 34.737 + 6.739 = 41.476\ \text{kPa}$$

$$q_{\min} = 34.737 - 6.739 = 27.998\text{ kPa}$$

$$q_{\max} = 41.476 \times 3.8 = \mathit{157.61\ kN/m\ length}$$

$$q_{\min} = 27.998 \times 3.8 = \mathit{106.39\ kN/m\ length}$$

Strip IJLK (7.2 × 12 m):
This strip is unsymmetrically loaded by columns No. 4, 5 and 6.

$$\sum \text{column loads} = 960 + 960 + 810 = 2730\,\text{kN}$$

$$\text{Average soil reaction} = (A)(q_{av}) = (7.2 \times 12)\left(\frac{29.95 + 25.27 + 25.05 + 29.73}{4}\right)$$

$$= 2376\,\text{kN} \neq \sum \text{column loads}$$

(*Continued*)

Therefore, *static equilibrium is not satisfied* ($\Sigma F_v \neq 0$); the column loads and contact pressures need to be modified.

$$Q_{av} = \frac{1}{2}(2730 + 2376) = 2553\ \text{kN}.$$

$MF_{(\text{col.})} = \dfrac{2553}{2730} = 0.935$. This is a reducing factor for columns of strip *IJLK*.

Each column factored modified load = $[MF_{(\text{col.})}](1.2\ D + 1.6\ L)$, computed as shown in Table 5.21.

Table 5.21 Column factored modified loads.

Column No.	Factored modified column loads
4	(0.935)(1.2 × 320 + 1.6 × 640) = **1316 kN**
5	(0.935)(1.2 × 320 + 1.6 × 640) = **1316 kN**
6	(0.935)(1.2 × 270 + 1.6 × 540) = **1111 kN**
	***R* = 3743 kN**

Since the strip is unsymmetrically loaded, the factored R does not fall at the centre; it has an eccentricity e_x calculated as follows:

$$R\bar{x} = \Sigma\,\text{moment of factored column loads}$$

$$\bar{x} = \frac{1316 \times 0.2 + 1316 \times 6.0 + 1111 \times 11.8}{3743} = 5.682\ \text{m}$$

$$e_x = \frac{12}{2} - 5.682 = 0.318\ \text{m} < \frac{12}{6}$$

The factored contact pressures are calculated, using Equation (5.53), as follows:

$$q_{\text{factored}} = \frac{R_{\text{factored}}}{A} \pm \frac{e_x R_{\text{factored}}}{I_y} l$$

$$q_{\max} = \frac{3743}{7.2 \times 12} + \frac{0.318 \times 3743}{7.2 \times 12^3/12} \times 6 = 43.322 + 6.888 = 50.210\ \text{kPa}$$

$$q_{\min} = 43.322 - 6.888 = 36.434\ \text{kPa}$$

$$q_{\max} = 50.210 \times 7.2 = \mathit{361.51\ kN/m\ length}$$

$$q_{\min} = 36.434 \times 7.2 = \mathit{262.32\ kN/m\ length}$$

Strip KLNM (7.2 × 12 m):

This strip is the same as strip *IJLK*. However, there is a very small difference in their average soil reactions; computed equals 0.8% only, which may be considered negligible. For design purposes, assume both strips identical.

Strip MNDC (3.8 × 12 m):

This strip is unsymmetrically loaded by columns No. 10, 11 and 12.

$$\Sigma \text{ column loads} = 240 + 360 + 270 = 870 \text{ kN}$$

$$\text{Average soil reaction} = (A)(q_{av}) = (3.8 \times 12)\left(\frac{29.51 + 24.83 + 24.72 + 29.40}{4}\right)$$

$$= 1236.44 \text{ kN} \neq \Sigma \text{ column loads}$$

Therefore, *static equilibrium is not satisfied* ($\Sigma F_v \neq 0$); the column loads and contact pressures need to be modified.

$$Q_{av} = \frac{1}{2}(870 + 1236.44) = 1053.22 \text{ kN}$$

$MF_{(col.)} = \dfrac{1053.22}{870} = 1.211$. This is an increasing factor for columns of strip *MNDC*.

Each column factored modified load = $[MF_{(col.)}](1.2\,D + 1.6\,L)$, computed as shown in Table 5.22:

Table 5.22 Column factored modified loads.

Column No.	Factored modified column loads
10	(1.211)(1.2 × 80 + 1.6 × 160) = **426 kN**
11	(1.211)(1.2 × 120 + 1.6 × 240) = **639 kN**
12	(1.211)(1.2 × 90 + 1.6 × 180) = **480 kN**
	***R* = 1545 kN**

Since the strip is unsymmetrically loaded, the factored R does not fall at the centre; it has an eccentricity e_x calculated as follows:

$$R\bar{x} = \Sigma \text{ moment of factored column loads}$$

$$\bar{x} = \frac{426 \times 0.2 + 639 \times 6.0 + 480 \times 11.8}{1545} = 6.203 \text{ m}$$

$$e_x = 6.203 - \frac{12}{2} = 0.203 \text{ m} < \frac{12}{6}$$

The factored contact pressures are calculated, using Equation (5.53), as follows:

$$q_{\text{factored}} = \frac{R_{\text{factored}}}{A} \pm \frac{e_x R_{\text{factored}}}{I_y} l$$

$$q_{\max} = \frac{1545}{3.8 \times 12} + \frac{0.203 \times 1545}{3.8 \times 12^3/12} \times 6 = 33.882 + 3.439 = 37.321 \text{ kPa}$$

$$q_{\min} = 33.882 - 3.439 = 30.443 \text{ kPa}$$

$$q_{\max} = 37.321 \times 3.8 = \mathit{141.82\ kN/m\ length}$$

$$q_{\min} = 30.443 \times 3.8 = \mathit{115.68\ kN/m\ length}$$

Step 6. Draw factored load, shear and moment diagrams for the continuous beam strips in both directions.

Note that some of the strips are unsymmetrically loaded; so shear and moment computations become more tedious and troublesome unless a programmable calculator is available (use calculus to obtain shear and moment values).

(*Continued*)

(A) Beam strips in $y-$ or $L-$ direction.
Strip AEGC (3.1 × 22.0 m):

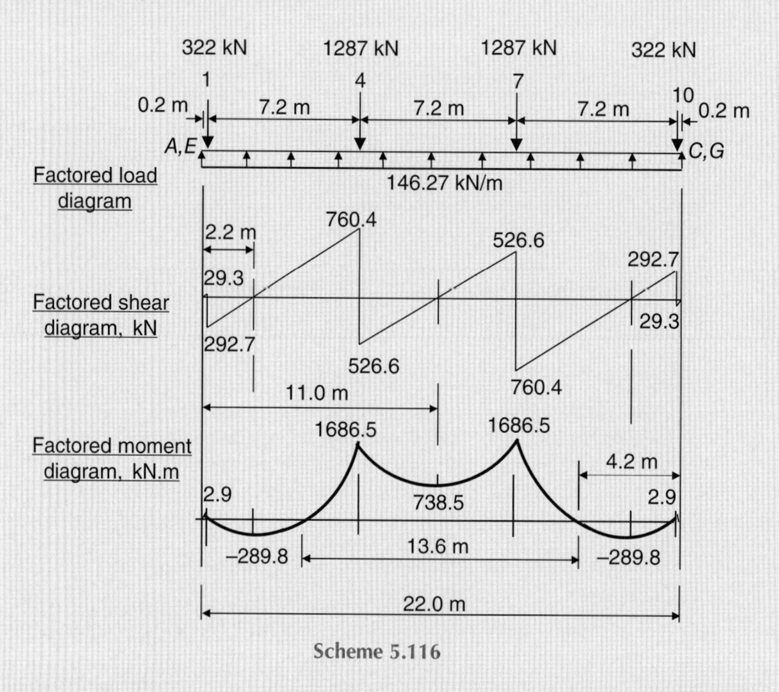

Scheme 5.116

Strip EFHG (5.8 × 22.0 m):

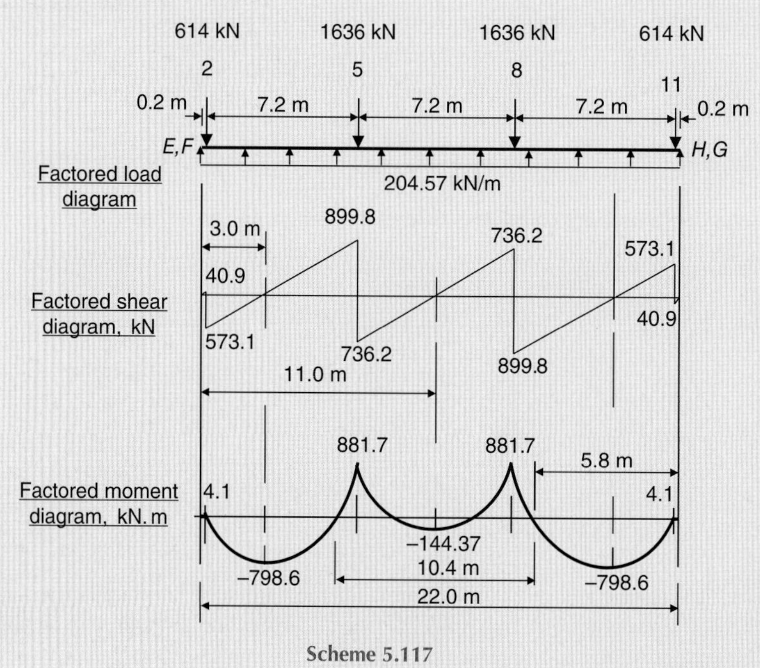

Scheme 5.117

Strip FBDH (3.1 × 22.0 m):

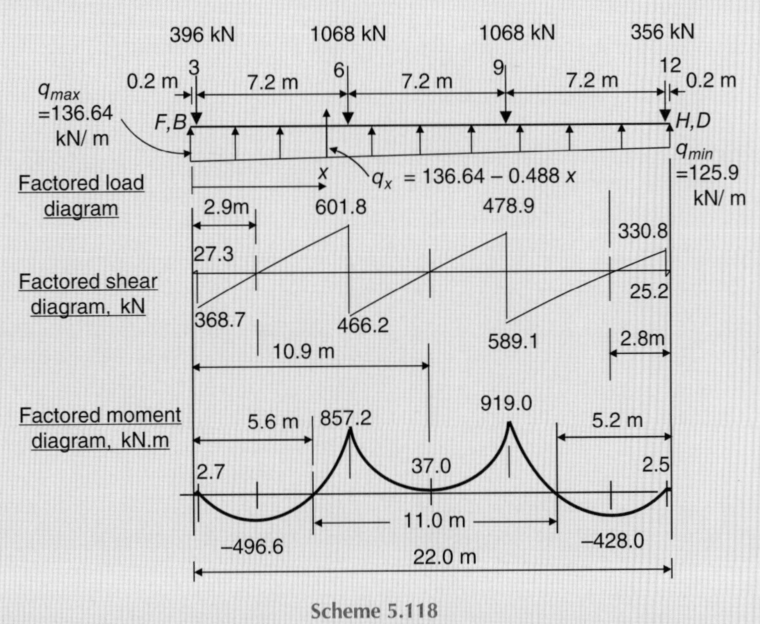

Scheme 5.118

(B) Beam strips in x – or B – direction.
Strip ABJI (3.8 × 12.0 m):

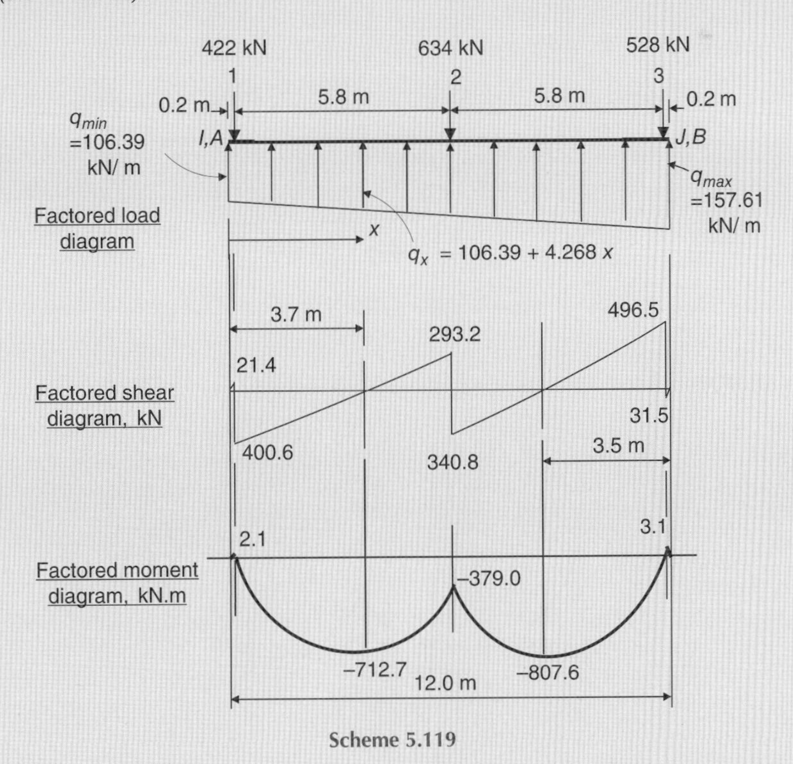

Scheme 5.119

(*Continued*)

Strip IJLK (7.2 × 12.0 m):

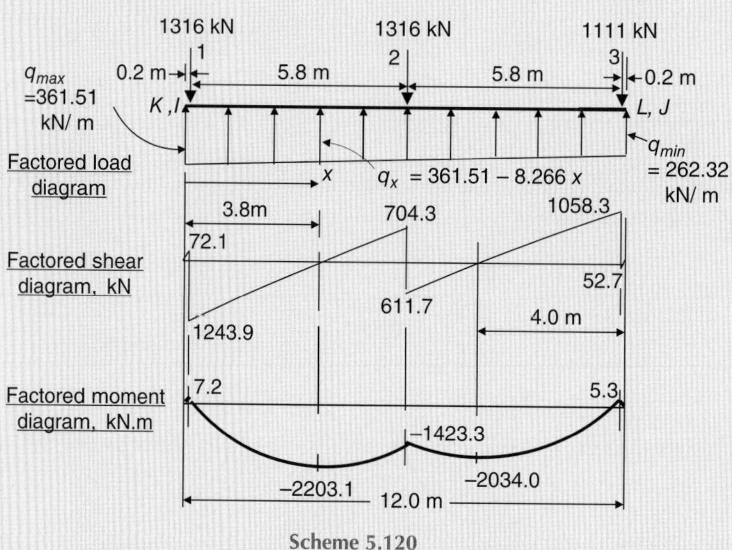

Scheme 5.120

Strip KLNM (7.2 × 12.0 m):

The same load, shear and moment diagrams which belong to strip *IJLK* are also used for strip *KLNM*, since these two strips are assumed identical, as mentioned earlier.

Strip MNDC (3.8 × 12.0 m):

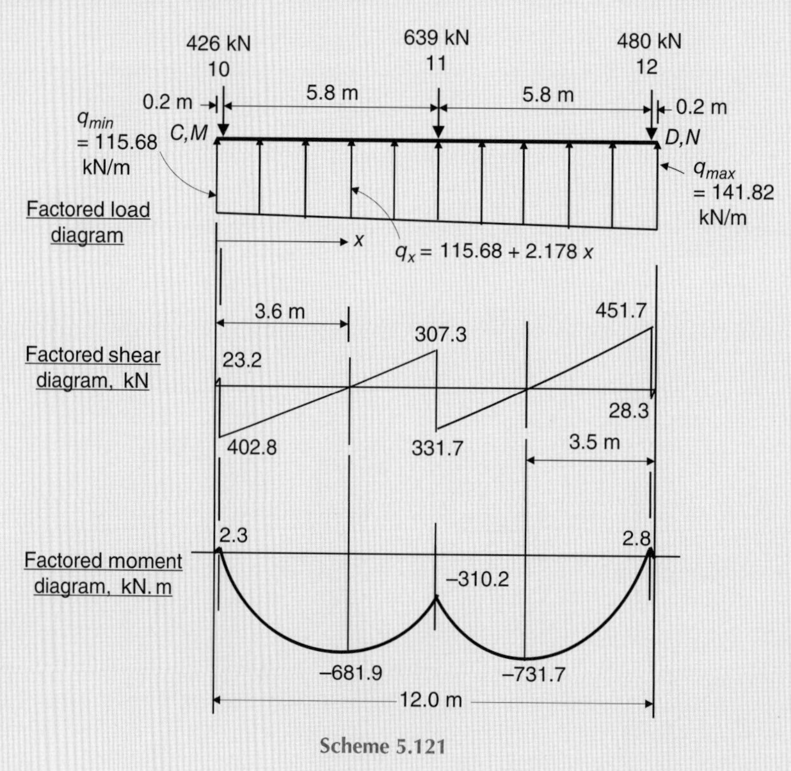

Scheme 5.121

A summary of the maximum positive and negative bending moments per metre width of each strip is given in Table 5.23.

Table 5.23 Summary of the maximum positive and negative bending moments per metre width of each strip.

Moment (kN.m/m)	Strips in y- or L-direction			Strips in x- or B-direction			
	AEGC	*EFHG*	*FBDH*	*ABJI*	*IJLK*	*KLNM*	*MNDC*
$+M_{max}$	544	152	297	1	1	1	1
$-M_{min}$	94	138	160	213	306	306	1

Step 7. Determine the minimum mat thickness considering punching shear (two-way or diagonal tension shear) at critical columns.

Considering the factors which contribute to two-way shear (column load, soil reaction, column size and length of shear perimeter at each column), by inspection, it appears that the maximum shear occurs at column No. 4 (or column No. 7) with a three-side shear perimeter or at column No. 3 with a two-side shear perimeter. Let us check first at which of these two columns the two-way shear is more critical:

Assume soil reaction on the shear block under each column is neglected (for the purpose of this checking only) since it is too small compared to the column load.

$$\text{Column No. 4: factored load} = 1.2\,D + 1.6\,L = 1.2\,(320) + 1.6(640) = 1408 \text{ kN}$$

$$\text{Column No. 3: factored load} = 1.2\,D + 1.6\,L = 1.2\,(100) + 1.6(200) = 440 \text{ kN}$$

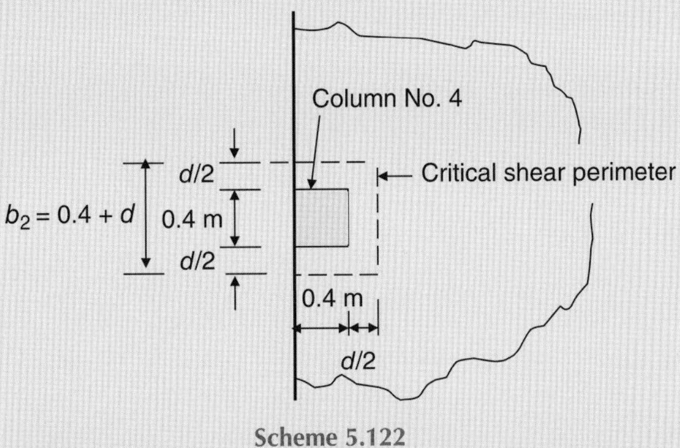

Scheme 5.122

$$\text{The shear–force ratio} = \frac{1408}{440} = 3.2$$

Assume a range for the mat thickness d between 0.5 and 1.0 m.

$$\text{The shear-perimeter ratio} = \frac{2(0.4 + 0.5d) + (0.4 + d)}{2(0.4 + 0.5d)} = 1.72 \quad \text{for } d = 0.5\text{m}$$

$$= 1.78 \quad \text{for } d = 1.0\text{m}$$

Since the shear–force ratio is much larger than the shear–perimeter ratios shear at column No. 4 is more critical.

(Continued)

Consider two-way shear at column No. 4:
Equation (5.19): $\varnothing V_c \geq V_u$ (taking reinforcement shear strength, $V_s = 0$)
or

$$\varnothing v_c \geq v_u$$

where $\varnothing$ = shear strength reduction factor = 0.75 (ACI Section 9.3.2.3)

Compute $\varnothing v_c$:
Assume the mat thickness $d = 0.6$ m.
Shear strength of concrete V_c shall be the smallest of (a), (b) and (c):

(a)

$$V_c = 0.17\left(1 + \frac{2}{\beta}\right)\lambda\sqrt{f_c'}b_o d$$

$$= (0.17)\left(1 + \frac{2}{1.0}\right)(1)\sqrt{f_c'}b_o d = (0.51)\sqrt{f_c'}b_o d \quad \text{Equation (5.22)}$$

(b)

$$V_c = 0.083\left(\frac{\alpha_s d}{b_o} + 2\right)\lambda\sqrt{f_c'}b_o d$$

$$= (0.083)\left(\frac{30d}{b_o} + 2\right)(1)\sqrt{f_c'}b_o d \quad \text{Equation (5.23)}$$

$$= (0.083)\left(\frac{30 \times 0.6}{2(0.4 + 0.3) + (0.4 + 0.6)} + 2\right)\sqrt{f_c'}b_o d = (0.789)\sqrt{f_c'}b_o d$$

(c)

$$V_c = 0.33\lambda\sqrt{f_c'}b_o d$$

$$= 0.33(1)\sqrt{f_c'}b_o d = (0.33)\sqrt{f_c'}b_o d \quad \text{Equation (5.24)}$$

Use: $V_c = (0.33)\sqrt{f_c'}\, b_o d; v_c = \frac{V_c}{b_o d} = (0.33)\sqrt{f_c'}$

$$\varnothing v_c = (0.75)(0.33)\sqrt{24}(1000) = 1212.5 \text{ kN/m}^2$$

Compute the factored shear stress, v_u:
Column No. 4 is a side column with a three-side shear perimeter. The condition of unbalanced moment transfer exists; and the design must satisfy the requirements of ACI Sections 11.11.7.1, 11.11.7.2 and 13.5.3. Therefore, we should consider factored shear stresses due to both the direct shear V_u and the moment transfer M_u, as given by the following equation:

$$v_u = \frac{V_u}{b_o d} \pm \frac{\gamma_v M_u C}{J_c} \quad \text{(ACI Section R11.11.7.2)}$$

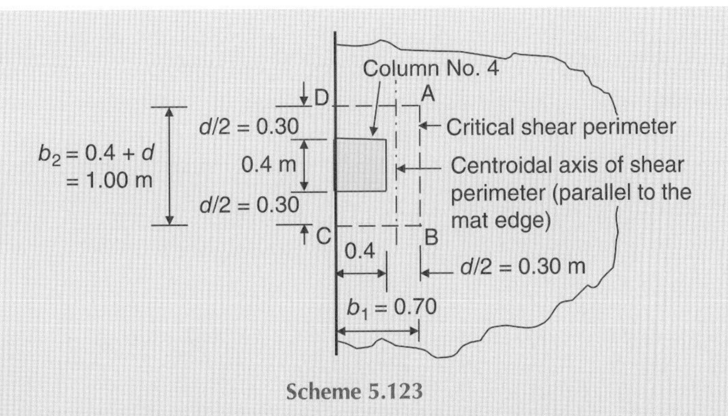

Scheme 5.123

C_{AB} = distance from line A – B to the centroid of the shear perimeter

$$= \frac{\left(\begin{array}{c} \text{moment of area of sides} \\ \text{about } A-B \end{array}\right)}{\text{area of sides}} = \frac{2(0.7 \times d)\left(\dfrac{0.7}{2}\right)}{(2 \times 0.70 + 1.0)(d)}$$

$$= \frac{2(0.7 \times 0.6)(0.7/2)}{(2 \times 0.70 + 1.0)(0.6)} = 0.204\,\text{m}$$

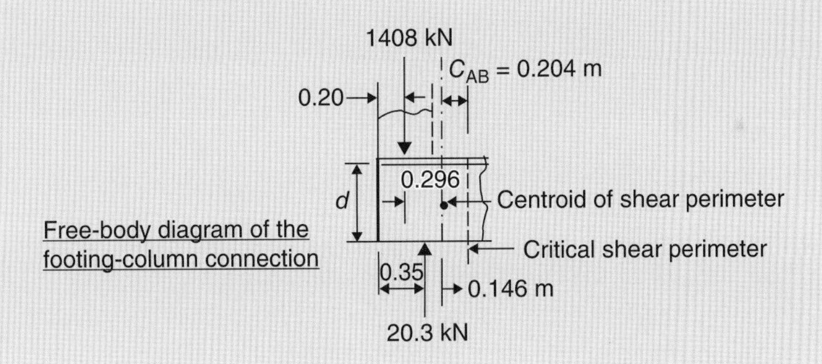

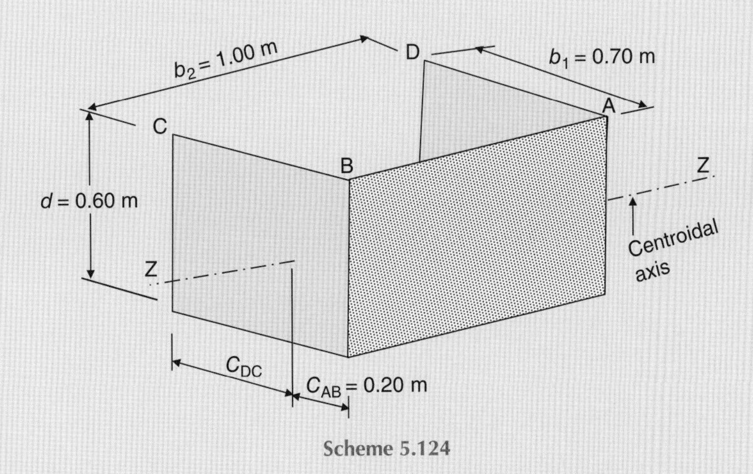

Scheme 5.124

(Continued)

Estimate the average soil pressure on the shear block = 29 kPa. (*Note:* it would be more correct if the factored value is used. However, since its effect is nearly negligible and in order to obtain a more conservative design the un-factored value is considered.)

Soil reaction on the shear block = 29 × 1 × 0.7 = 20.3 kN

Summation of moments about the centroid of the shear perimeter equals the unbalanced moment, M_u. Hence,

$$M_u = 1408 \times 0.296 - 20.3 \times 0.146 = 413.8\,\text{kN.m}$$

$$V_u = 1408 - 20.3 = 1387.7\,\text{kN}$$

$$\gamma_v = \left(1 - \gamma_f\right)$$

$$\gamma_f = \frac{1}{1 + (2/3)\sqrt{b_1/b_2}} \quad \text{(ACI Section 11.11.7.1)}$$

$$= \frac{1}{1 + (2/3)\sqrt{0.7/1.0}} = 0.642 \quad \text{(ACI Section 13.5.3.2)}$$

$$\gamma_v = 1 - 0.642 = 0.358$$

$J_c = J_z$ = Property of the shear perimeter analogous to polar moment of inertia

$$J_c = 2\left(\frac{b_1 d^3}{12}\right) + 2\left(\frac{d b_1^3}{12}\right) + 2(b_1 d)\left(\frac{b_1}{2} - C_{AB}\right)^2 + (b_2 d) C_{AB}{}^2$$

$$= 2\left(\frac{0.7 \times 0.6^3}{12}\right) + 2\left(\frac{0.6 \times 0.7^3}{12}\right) + 2(0.7 \times 0.6)\left(\frac{0.7}{2} - 0.204\right)^2 + (1 \times 0.6)(0.204^2)$$

$$= 0.0252 + 0.0343 + 0.0179 + 0.0250 = 0.1024\,\text{m}^4$$

$$C = C_{DC} = b_1 - C_{AB} = 0.70 - 0.204 = 0.496\,\text{m}$$

In this case, the shear stresses due to direct shear and shear due to moment transfer will add at points D and C, giving the largest shear stresses on the critical shear perimeter. Hence,

$$v_{u(DC)} = \frac{V_u}{b_o d} + \frac{\gamma_v M_u C_{DC}}{J_c} \quad \text{(ACI Section R11.11.7.2)}$$

$$v_{u(DC)} = \frac{1387.7}{(2 \times 0.7 + 1)(0.6)} + \frac{0.358 \times 413.8 \times 0.496}{0.1024} = 963.68 + 717.56$$

$$= 1681.24\,\text{kPa} > (\varnothing v_c = 1212.5\ \text{kPa}) \quad \text{(Not OK.)}$$

Assume another value for d and check $\varnothing v_c$ again.

Try $d = 0.76$ m:

$$b_1 = 0.4 + (d/2) = 0.78\,\text{m};\ b_2 = 0.4 + (d) = 1.16\,\text{m}$$

$$C_{AB} = \frac{\begin{pmatrix}\textit{moment of area of sides} \\ \textit{about A} - B\end{pmatrix}}{\text{area of sides}} = \frac{2(0.78 \times d)(0.78/2)}{(2 \times 0.78 + 1.16)(d)}$$

$$\frac{2(0.78 \times 0.76)(0.39)}{(2 \times 0.78 + 1.16)(0.76)} = 0.224\ \text{m}$$

$$C = C_{DC} = b_1 - C_{AB} = 0.78 - 0.224 = 0.56\ \text{m}$$

$$\text{Soil reaction on the shear block} = 29 \times 1.16 \times 0.78 = 26.24\,\text{kN}$$

$$\text{Moment arm of soil reaction} = \left(\frac{b_1}{2}\right) - C_{AB}$$

$$= \left(\frac{0.78}{2}\right) - 0.224 = 0.17\,\text{m}$$

$$\text{Moment arm of column load} = (b_1 - 0.2) - C_{AB}$$

$$= 0.58 - 0.22 = 0.36\,\text{m}$$

$$M_u = 1408 \times 0.36 - 26.24 \times 0.17 = 502.42\,\text{kN.m}$$

$$V_u = 1408 - 26.24 = 1381.76\,\text{kN}$$

$$\gamma_f = \frac{1}{1 + (2/3)\sqrt{b_1/b_2}}$$

$$= \frac{1}{1 + (2/3)\sqrt{0.78/1.16}} = 0.646; \gamma_v = 1 - 0.646 = 0.354$$

$$J_c = 2\left(\frac{b_1 d^3}{12}\right) + 2\left(\frac{d b_1^3}{12}\right) + 2(b_1 d)\left(\frac{b_1}{2} - C_{AB}\right)^2 + (b_2 d){C_{AB}}^2$$

$$= 2\left(\frac{0.78 \times 0.76^3}{12}\right) + 2\left(\frac{0.76 \times 0.78^3}{12}\right) + 2(0.78 \times 0.76)\left(\frac{0.78}{2} - 0.224\right)^2 + (1.16 \times 0.76)(0.224^2)$$

$$= 0.0571 + 0.0601 + 0.0327 + 0.0442 = 0.194\,\text{m}^4$$

$$v_{u(DC)} = \frac{1381.76}{(2 \times 0.78 + 1.16)(0.76)} + \frac{0.354 \times 502.42 \times 0.56}{0.194} = 668.42 + 513.40$$

$$= 1181.82\,\text{kPa} < (\text{Ø}\, v_c = 1212.5\,\text{kPa}) \quad \text{(OK.)}$$

Use d = 0.76 m; h = 0.85 m

Step 8. Check the computed mat depth d considering beam shear (one-way shear) at the most critical section.

Refer to the shear diagrams in *Step 6*. It is clear that the most critical section for shear per metre width is located in strip *AEGC* at distance d from either the left face of column No. 4 or the right face of column No. 7. The applied beam shear at any of these two locations is

$$V_u = \frac{760.4 - q(d + 0.2)}{3.1} = \frac{760.4 - 146.27(0.76 + 0.2)}{3.1} = 200\,\text{kN/m width}$$

Equation (5.19):

$$\text{Ø}\, V_c \geq V_u \quad \text{(taking reinforcement shear strength, } V_s = 0)$$

Equation (5.20):

$$V_c = 0.17\lambda\sqrt{f_c'}\,b_w d$$

$$\text{Ø}\, V_c = (0.75 \times 0.17 \times 1)(\sqrt{24})(1000)(1)(0.76)$$

$$= 474.7\,\text{kN/m} \gg (V_u = 200\,\text{kN/m}) \quad \text{(OK.)}$$

Step 9. Obtain the design factored positive and negative moments per metre width using the factored moment diagrams in *Step 6*.

Refer to Table 5.23 at the end of *Step 6*. Obtain the following factored maximum moments per metre width:

(A) Moments required for design of reinforcement in y – or L – direction.

(Continued)

$$+M_{max} = 544\text{ kN.m/m (for steel in midspans, at bottom of the mat)}$$
$$-M_{max} = 160\text{ kN.m/m (for steel in exterior spans, at top)}$$

(B) Moments required for design of reinforcement in $x-$ or $B-$ direction.
$+\ M_{max} \cong 1$ kN. m/m. For all practical purposes, it is considered negligible.
$-\ M_{max} = 306$ kN. m/m (for steel in both spans, at top of the mat)

Step 10. Determine the positive and negative flexural reinforcement required in each direction.

(A) Flexural reinforcement required in $y-$ or $L-$ direction.
(a) Negative reinforcement at top of the mat:
Equation (5.9): $\emptyset M_n \geq M_u$

$$M_u = 160\text{ kN.m/m}$$

Assume tension-controlled section, $\emptyset = 0.9$ and $f_s = f_y$.
Equation (5.10):

$$a = \frac{A_s f_y}{0.85 f_c' b}$$
$$= \frac{A_s \times 420}{0.85 \times 24 \times 1} = 20.59 A_s \text{ (m)}$$

Equation (5.11):

$$\emptyset M_n = \emptyset \left[A_s f_y \left(d - \frac{a}{2}\right)\right]$$
$$= (0.9)\left[(A_s \times 420 \times 1000)\left(0.76 - \frac{20.59\,A_s}{2}\right)\right]$$
$$= 287\,280\,A_s - 3\,891\,510\,A_s^2$$

Let $\emptyset M_n = M_u$:

$$287\,280\,A_s - 3\,891\,510\,A_s^2 = 160$$
$$A_s^2 - 0.074\,A_s + 4.11 \times 10^{-5} = 0$$
$$A_s = \frac{-(-0.074) \pm \sqrt{(-0.074)^2 - (4)(1)(4.11 \times 10^{-5})}}{2 \times 1} = \frac{0.074 - 0.073}{2}$$
$$= 0.5 \times \frac{10^{-3}\text{ m}^2}{\text{m}} = 500\text{ mm}^2/\text{m (required by analysis)}$$

ACI Section 10.5.1:

$$A_{s,\min} = \frac{0.25\sqrt{f_c'}}{f_y} b_w d \text{ and not less than } 1.4\frac{b_w d}{f_y}$$
$$A_{s,\min} = \frac{0.25\sqrt{24}}{420} \times 1 \times 0.76 = 2.216 \times 10^{-3}\text{ m}^2/\text{m} = 2216\text{ mm}^2/\text{m}$$
$$1.4\frac{b_w d}{f_y} = \frac{1.4 \times 1 \times 0.76}{420} = 2.533 \times 10^{-3}\text{ m}^2/\text{m} = 2533\text{ mm}^2/\text{m}$$

$A_{s,\min} > A_s$ by analysis. However, the requirements of ACI Section 10.5.1 need not be applied if, at every section, A_s provided is at least one-third greater than that required by analysis (According to ACI Section 10.5.3).

$$(1.333)(A_s \text{ by analysis}) = 1.333 \times 500 = 667 \text{ mm}^2/\text{m}$$

ACI Section 10.5.4:

$$A_{s,\min} = 0.0018\,bh = (0.0018)(1)(0.85) = 1.53 \times 10^{-3} \text{ m}^2/\text{m}$$
$$= 1530 \text{ mm}^2/\text{m} > (1.333 \times A_s \text{ by analysis})$$

Use $A_{s,\min} = 1530 \text{ mm}^2/\text{m}$.

The assumptions made are satisfied, since required $A_s = A_{s,\min}$.

$$A_{s,\text{total}} = 1530 \times B = 1530 \times 12 = 18\,360 \text{ mm}^2$$

Try 36 No. 25 bars, $A_{s,\text{provided}} = 36 \times 510 = 18\,360 \text{ mm}^2$ (OK.)

ACI Sections 7.6.5 and 10.5.4 requires maximum spacing shall not exceed three times the slab or footing thickness or 450 mm, whichever is smaller.

Use centre to centre bar spacing = 337 mm. Check the 75 mm minimum concrete cover at each side:

$$\text{Concrete cover} = \frac{12\,000 - (35 \times 337 + 25.4)}{2} = 89.8 \text{ mm} > 75\ mm \text{ (OK.)}$$

Check the development of reinforcement:

Refer to the moment diagrams of strips *AEGC*, *EFHG* and *FBDH*. Negative moments exist in the exterior spans of all the three strips except strip *EFHG* where a very small negative moment of 25 kN.m/m exists in the middle span also. Therefore, for all practical purposes there are no negative reinforcement required in the mid spans; they are needed in the exterior spans only. However, we shall extend one out of each three bars full length (less 75 mm concrete cover at the ends) of the mat. The other bars shall be extended beyond the inflection points and into the column regions of the exterior spans.

In this case, the bars are in tension, the provided bar size is No. 25, the clear spacing of the bars exceeds $2d_b$ and the clear cover exceeds d_b. Therefore, the development length l_d of bars at each side of the critical section (which is located where the maximum factored negative moment exists) shall be determined from the following equation, but not less than 300 mm:

$l_d = \left(\frac{f_y \psi_t \psi_e}{1.7\lambda\sqrt{f'_c}}\right) d_b$ (ACI Sections 12.2.1, 12.2.2 and 12.2.4)

where the factors ψ_e and λ are 1 (ACI Sections 12.2.4 and 8.6.1).

The factor $\psi_t = 1.3$ because the reinforcement is placed such that more than 300 mm of fresh concrete exists below the top bars (ACI Section 12.2.4).

$$l_d = \left(\frac{420 \times 1.3 \times 1}{1.7 \times 1 \times \sqrt{24}}\right)\left(\frac{25.4}{1000}\right) = 1.665 \text{ m} = 1665 \text{ mm} > 300\ mm.$$

Therefore, the required $l_d = 1665$ mm.

The moment diagrams show that the smallest bar extension past the critical section (i.e. the available length) is

$$2200 \text{ mm} - 75 \text{ mm cover} = 2125 \text{ mm} > 1665 \text{ mm} \quad \text{(OK.)}$$

MacGregor and Wight (2005) suggest that, in conventional design of combined footings (the mat strips are continuous combined footings), the computed l_d of the *negative* tension reinforcement (top steel bars) shall satisfy Equation (12.5) of ACI Section 12.11.3. This is due to the fact that in the case of a combined footing (designed as a reinforced concrete beam), the loading, the supports, and the shape of the moment diagram are all inverted from those found in a normally loaded beam carrying gravity loads (loaded on the top surface and supported on the bottom surface), where the ACI equation concerns the positive tension reinforcement.

(*Continued*)

Check ACI Section 12.11.3:

$$l_d \leq \frac{M_n}{V_u} + l_a \text{ [ACI Equation (12.5)]}$$

M_n = nominal flexural strength of the cross section. It is calculated assuming all reinforcement at the section to be stressed to f_y

V_u = shear at the section where the inflection point is located.

l_a = additional embedment length beyond centerline of support or point of inflection l_a at a support shall be the embedment length beyond center of support; at a point of inflection shall be limited to d or 12 d_b, whichever is greater.

$$M_n = A_s f_y \left[d - \frac{a}{2}\right]$$

$$a = \frac{A_s f_y}{0.85 f_c' b} = \frac{\left(\frac{1000}{337} \times \frac{510}{10^6}\right)(420)}{(0.85)(24)(1)} = 0.031 \text{ m}$$

$$M_n = \left(\frac{1000}{337} \times \frac{510}{10^6}\right)(420 \times 1000)\left[0.76 - \frac{0.031}{2}\right] = 473.2 \text{ kN.m/m}$$

As mentioned before, we shall extend two-thirds of all the steel bars beyond the inflection points and into the column regions of the exterior spans.

In y – or L – direction, the exterior supports are columns No. 1, 2, 3, 10, 11 and 12; and the interior supports are columns No. 4, 5, 6, 7, 8 and 9. From shear diagrams of the strips in long direction, at points of inflection, maximum shear values are:

At all the exterior supports of the three strips, it is clear that maximum shear per metre width is:

$$V_u = \frac{368.7}{3.1} = 119 \text{ kN/m}$$

At the interior supports, maximum shear per metre width at inflection points in each of the three strips are:

Strip *AEGC*: $V_u = \frac{[760.4 - (7.4 - 4.2)(146.27)]}{3.1} = 94.3 \text{ kN/m}$

Strip *EFHG*: $V_u = \frac{[899.8 - (7.4 - 5.8)(204.57)]}{5.8} = 98.7 \text{ kN/m}$

Strip *FBDH*:

$$\text{at } x = 5.6 \text{ m} \quad q = 136.64 - 0.488 \times 5.6 = 133.91 \text{ kN/m}$$

$$\text{at } x = 7.4 \text{ m} \quad q = 136.64 - 0.488 \times 7.4 = 133.03 \text{ kN/m}$$

$$\text{at } x = 5.6 \text{ m} \quad V_u = \frac{\left[601.8 - \frac{(7.4 - 5.6)(133.91 + 133.03)}{2}\right]}{3.1} = 116.6 \text{ kN/m (approximately)}$$

$$\text{at } x = 14.6 \text{ m} \quad q = 136.64 - 0.488 \times 14.6 = 129.52 \text{ kN/m}$$

$$\text{at } x = 16.8 \text{ m} \quad q = 136.64 - 0.488 \times 16.8 = 128.44 \text{ kN/m}$$

$$\text{at } x = 16.8 \text{ m} \quad V_u = \frac{\left[589.1 - \frac{(16.8 - 14.6)(129.52 + 128.44)}{2}\right]}{3.1} = 98.5 \text{ kN/m (approximately)}$$

For strip *FBDH*, maximum $V_u = 116.6$ kN/m (approximately)
Therefore, use maximum $V_u = 116.6$ kN/m (approximately)
At the exterior supports, the available distance beyond the centreline of each column is

$$l_a = 0.20\text{ m} - 0.075\text{ m cover} = 0.125\text{ m}$$

$$\frac{M_n}{V_u} + l_a = \frac{473.2}{119.0} + 0.125 = 4.1\text{ m} \gg (l_d = 1.665\text{ m}) \quad \text{(OK.)}$$

At the interior supports, extend the top bars to the interior face (away from exterior column) of the interior columns. The available distance beyond the centreline of each column is

$$l_a = 0.20\text{ m}$$

$$\frac{M_n}{V_u} + l_a = \frac{473.2}{116.6} + 0.2 = 4.3\text{ m} \gg (l_d = 1.665\text{ m}) \quad \text{(OK.)}$$

Provide 36 No. 25 bars @ 337 mm c.c., placed at top of the mat in L-direction. Extend one out of each three bars full length (less 75 mm concrete cover at the ends) of the mat. The other bars shall be cut at the interior face of the interior columns to provide reinforcement for the exterior spans.
(b) Positive reinforcement at bottom of the mat:

$$M_u = 544\text{ kN.m/m}$$

$$\emptyset M_n \geq M_u$$

Assume tension-controlled section, $\emptyset = 0.9$ and $f_s = f_y$.

$$\emptyset M_n = \emptyset \left[A_s f_y \left(d - \frac{a}{2}\right)\right] \quad \text{Equation (5.11)}$$

Equation (5.10):

$$a = \frac{A_s f_y}{0.85 f'_c b}$$

$$= \frac{A_s \times 420}{0.85 \times 24 \times 1} = 20.59 A_s \quad \text{(m)}$$

$$\emptyset M_n = (0.9)\left[(A_s \times 420 \times 1000)\left(0.76 - \frac{20.59\, A_s}{2}\right)\right]$$

$$= 287280\, A_s - 3891510\, A_s^2 \quad \text{(kN.m/m)}$$

Let $\emptyset M_n = M_u$:

$$287280\, A_s - 3891510\, A_s^2 = 544$$

$$A_s^2 - 0.074\, A_s + 1.4 \times 10^{-4} = 0$$

$$A_s = \frac{-(-0.074) \pm \sqrt{(-0.074)^2 - (4)(1)(1.4 \times 10^{-4})}}{2 \times 1} = \frac{0.074 - 0.070}{2}$$

$$= 2 \times \frac{10^{-3}\text{ m}^2}{\text{m}} = 2000\text{ mm}^2/\text{m (required by analysis)}.$$

(Continued)

ACI Section 10.5.1:

$$A_{s,\min} = \frac{0.25\sqrt{f'_c}}{f_y} b_w d \text{ and not less than } 1.4\frac{b_w d}{f_y}$$

$$A_{s,\min} = \frac{0.25\sqrt{24}}{420} \times 1 \times 0.76 = 2.216 \times 10^{-3}\ \text{m}^2/\text{m} = 2216\ \text{mm}^2/\text{m}$$

$$1.4\frac{b_w d}{f_y} = \frac{1.4 \times 1 \times 0.76}{420} = 2.533 \times 10^{-3}\ \text{m}^2/\text{m} = 2533\ \text{mm}^2/\text{m}$$

$A_{s,\min} > A_s$ by analysis Therefore, use $A_s = A_{s,\min} = 2533\ \text{mm}^2/\text{m}$

The assumptions made are satisfied, since A_s required $= A_{s,\min}$.

$$A_{s,\text{total}} = 2533 \times B = 2533 \times 12 = 30\,396\ \text{mm}^2$$

Try 60 No. 25 bars, $A_{s,\text{provided}} = 60 \times 510 = 30\,600\ \text{mm}^2$ (OK.)

ACI Sections 7.6.5 and 10.5.4 requires maximum spacing shall not exceed three times the slab or footing thickness, or 450 mm, whichever is smaller.

Use centre to centre bar spacing = 200 mm. Check the 75 mm minimum concrete cover at each side:

$$\text{Concrete cover} = \frac{12\,000 - (59 \times 200 + 25.4)}{2} = 87.3\ \text{mm} > 75\ mm \quad \text{(OK.)}$$

Check the development of reinforcement:

We shall extend one out of each four bars full length (less 75 mm concrete cover at the ends) of the mat. The other bars shall be extended beyond the inflection points a distance equal to d or $12d_b$, whichever is greater (ACI Sections 12.10.3). In this case, it is clear that d controls. Maximum distance between any two inflection points in all the three strips is 13.6 m (strip *AEGH*). Therefore, each cut off point shall be located beyond the column centre at a distance equal to

$$\frac{13.6 - 7.2}{2} + 0.76 = 3.96\ \text{m; say, } 4\ \text{m.}$$

In this case, the bars are in tension, the provided bar size is No. 25, the clear spacing of the bars exceeds $2d_b$ and the clear cover exceeds d_b. Therefore, the development length l_d of bars at each side of the critical section (which is located where the maximum factored positive moment exists) shall be determined from the following equation, but not less than 300 mm:

$$l_d = \left(\frac{f_y \psi_t \psi_e}{1.7\lambda\sqrt{f'_c}}\right) d_b \text{ (ACI Sections 12.2.1, 12.2.2 and 12.2.4)}$$

Where the factors ψ_t, ψ_e and λ are 1 (ACI Sections 12.2.4 and 8.6.1).

$$l_d = \left(\frac{420 \times 1 \times 1}{1.7 \times 1 \times \sqrt{24}}\right)\left(\frac{25.4}{1000}\right) = 1.28\ \text{m} = 1280\ \text{mm} > 300\ mm.$$

Therefore, the required $l_d = 1280$ mm.

The bar extension past the critical section (at column centre) is

$$4\ \text{m} = 4000\ \text{mm} \gg 1280\ \text{mm} \quad \text{(OK.)}$$

Provide 60 No. 25 bars @ 200 mm c.c., placed at bottom of the mat in L-direction. Extend one out of each four bars full length (less 75 mm concrete cover at the ends) of the mat. The other bars (short bars of 15.2 m length) shall be extended 4 m beyond the centre line of each interior column toward the exterior column.

(B) Flexural reinforcement required in x – or B – direction.

(a) Negative reinforcement at top of the mat:

$$M_u = 306 \text{ kN.m/m}$$

$$\text{Ø}\, M_n \geq M_u$$

Assume tension-controlled section, $\text{Ø} = 0.9$ and $f_s = f_y$.

$$\text{Ø}\, M_n = \text{Ø}\left[A_s f_y\left(d - \frac{a}{2}\right)\right] \qquad \text{Equation (5.11)}$$

Equation (5.10):

$$a = \frac{A_s f_y}{0.85 f_c' b}$$

$$= \frac{A_s \times 420}{0.85 \times 24 \times 1} = 20.59 A_s \text{ (m)}$$

$$d = 0.7600 - 0.0254 = 0.735 \text{ m}$$

$$\text{Ø}\, M_n = (0.9)\left[(A_s \times 420 \times 1000)\left(0.735 - \frac{20.59\, A_s}{2}\right)\right]$$

$$= 277\ 830\, A_s - 3\ 891\ 510\, A_s^2 \text{ (kN.m/m)}$$

Let $\text{Ø}\, M_n = M_u$:

$$277\ 830 A_s - 3\ 891\ 510\, A_s^2 = 306$$

$$A_s^2 - 0.0714\, A_s + 4.11 \times 10^{-5} = 0$$

$$A_s = \frac{-(-0.0714) \pm \sqrt{(-0.0714)^2 - (4)(1)(7.86 \times 10^{-5})}}{2 \times 1} = \frac{0.0714 - 0.0692}{2}$$

$$= 1.1 \times 10^{-3} \text{ m}^2/\text{m} = 1100 \text{ mm}^2/\text{m (required by analysis)}.$$

ACI Section 10.5.1:

$$A_{s,\min} = \frac{0.25\sqrt{f_c'}}{f_y} b_w d \text{ and not less than } 1.4\frac{b_w d}{f_y}$$

$$A_{s,\min} = \frac{0.25\sqrt{24}}{420} \times 1 \times 0.735 = 2.143 \times 10^{-3} \text{ m}^2/\text{m} = 2143 \text{ mm}^2/\text{m}$$

$$1.4\frac{b_w d}{f_y} = \frac{1.4 \times 1 \times 0.735}{420} = 2.450 \times 10^{-3} \text{ m}^2/\text{m} = 2450 \text{ mm}^2/\text{m}$$

(*Continued*)

$A_{s,\min} > A_s$ by analysis. However, the requirements of ACI Section 10.5.1 need not be applied if, at every section, A_s provided is at least one-third greater than that required by analysis (According to ACI Section 10.5.3).

$$(1.333)(A_s \text{ by analysis}) = 1.333 \times 1100 = 1463\ \text{mm}^2/\text{m}$$

ACI Section 10.5.4:

$$A_{s,\min} = 0.0018\,bh = (0.0018)(1)(0.85) = 1.53 \times 10^{-3}\ \text{m}^2/\text{m}$$
$$= 1530\ \text{mm}^2/\text{m} > (1.333 \times A_s \text{ by analysis})$$

Use $A_s = A_{s,\min} = 1530\ \text{mm}^2/\text{m}$

The assumptions made are satisfied, since required $A_s = A_{s,min}$.

$$A_{s,\text{total}} = 1530 \times L = 1530 \times 22 = 33\,660\ \text{mm}^2$$

Try 66 No. 25 bars, $A_{s,\text{provided}} = 66 \times 510 = 33\ \ 660$ (OK.)

ACI Sections 7.6.5 and 10.5.4 requires maximum spacing shall not exceed three times the slab or footing thickness, or 450 mm, whichever is smaller.

Use centre to centre bar spacing = 335 mm. Check the 75 mm minimum concrete cover at each side:

$$\text{Concrete cover} = \frac{22\,000 - (65 \times 335 + 25.4)}{2} = 99.8\ \text{mm} > 75\ mm \quad (\text{OK.})$$

Check the development of reinforcement:

We shall extend all the bars full width (less 75 mm concrete cover at the ends) of the mat.

In this case, the bars are in tension, the provided bar size is No. 25, the clear spacing of the bars exceeds $2d_b$, and the clear cover exceeds d_b Therefore, the development length l_d of bars at each side of the critical section (which is located where the maximum factored negative moment exists) shall be determined from the following equation, but not less than 300 mm:

$$l_d = \left(\frac{f_y \psi_t \psi_e}{1.7\lambda\sqrt{f_c'}}\right) d_b \text{ (ACI Sections 12.2.1, 12.2.2 and 12.2.4)}$$

where the factors ψ_e and λ are 1 (ACI Sections 12.2.4 and 8.6.1).

The factor $\psi_t = 1.3$ because the reinforcement is placed such that more than 300 mm of fresh concrete exists below the top bars (ACI Section 12.2.4).

$$l_d = \left(\frac{420 \times 1.3 \times 1}{1.7 \times 1 \times \sqrt{24}}\right)\left(\frac{25.4}{1000}\right) = 1.665\ \text{m} = 1665\ \text{mm} > 300\ mm.$$

Therefore, the required $l_d = 1665$ mm.

The moment diagrams show that the smallest bar extension past the critical section (i.e. the available length) is

$$3500\ \text{mm} - 75\ \text{mm cover} = 3425\ \text{mm} > 1665\ \text{mm} \quad (\text{OK.})$$

Check ACI Section 12.11.3:

In this case, it is required to check ACI Equation (12.5) at the exterior supports only, since all the bars are extended full width of the mat.

$$l_d \le \frac{M_n}{V_u} + l_a\ [\text{ACI Equation (12.5)}]$$

$$M_n = A_s f_y \left[d - \frac{a}{2} \right]$$

$$a = \frac{A_s f_y}{0.85 f_c' b} = \frac{\left(\frac{1000}{335} \times \frac{510}{10^6}\right)(420)}{(0.85)(24)(1)} = 0.0313 \text{ m}$$

$$M_n = \left(\frac{1000}{335} \times \frac{510}{10^6}\right)(420 \times 1000)\left[0.735 - \frac{0.0313}{2}\right]$$

$$V_u = \frac{1243.9}{7.2} = 172.8 \text{ kN/m}$$

At all the exterior supports of the three strips, it is clear that maximum shear per metre length is

$$V_u = \frac{1243.9}{7.2} = 172.8 \text{ kN/m}$$

At the exterior supports, the available distance beyond the centreline of each column is

$$l_a = 0.20 \text{ m} - 0.075 \text{ m cover} = 0.125 \text{ m}$$

$$\frac{M_n}{V_u} + l_a = \frac{460.0}{172.8} + 0.125 = 2.8 \text{ m} > (l_d = 1.665 \text{ m}) \quad \text{(OK.)}$$

Provide 66 No. 25 bars @ 335 mm c.c., placed at top of the mat in B – direction and directly below the bars in L-direction. Extend all the bars full width (less 75 mm concrete cover at the ends) of the mat.

(b) Positive reinforcement at bottom of the mat:

As mentioned in *Step 9*, for all practical purposes there are no (+) moments in any span; so (+) moment steel in x – or B – direction is not required by analysis. However, enough supportive steel bars shall be provided and wired to the bottom bars in L – direction so that the latter bars can be positioned and held properly.

Provide 50-No. 13 bars @ 450 mm c.c., at bottom of the mat in B – direction, wired to the bars in L – direction. Extend all the bars full width (less 75 mm concrete cover at the ends) of the mat.

Step 11. Check column bearing on the mat at critical column locations and design the column to mat dowels.

The design computations proceed in the same manner as that presented in the design *Steps 5 and 6* for the column footing of Problem 5.3. The necessary computations are left for the reader.

Step 12. Check if the conventional rigid design method could be considered appropriate.

The rigid design method is considered appropriate if the mat strips, both ways, can be considered or treated as a rigid body. As mentioned in Section 5.10.1, strips (continuous footings or combined footings with multiple columns) can be considered rigid if the following requirements or criteria are met:

(a) Variation in adjacent column loads and spacing is not over 20% of the greater value.

(b) Average of two adjacent spans is $< (1.75/\lambda)$.

Equation (5.48):

$$\lambda = \sqrt[4]{\frac{K_s B}{4 E_c I}}$$

Where

K_s = coefficient (or modulus) of vertical subgrade reaction

E_c = modulus of elasticity of concrete

I = moment of inertia of the beam (strip) section

B = width of the strip

(Continued)

Criterion (a):

All strips (both ways) have no variation in the adjacent column spacing. The problem is with the column loads. All strips except strips *IJLK* and *KLNM* have adjacent column loads varying by more than 20%.

Therefore criterion (a) is not fully satisfied. One may estimate that this requirement of rigidity is only 50–60% met.

Criterion (b):

$E_c = 4700\sqrt{f_c'}$ (ACI Section 8.5.1)

$$E_c = 4700 \times \sqrt{24} = 23025.2 \text{ MPa} = 5.2 \times 10^7 \text{ kPa}$$

$$K_{s,\,av} = 7200\,\text{kN/m}^3 \quad \text{(given)}$$

$$I = \frac{bh^3}{12} = \frac{(B)(0.85^3)}{12} = (0.051)\,(B)\,\text{m}^4$$

$$\lambda = \sqrt[4]{\frac{K_s B}{4E_c I}}$$

$$= \sqrt[4]{\frac{(7200)(B)}{(4)(5.2\times 10^7)(0.051)(B)}} = 0.161$$

$$\frac{1.75}{\lambda} = \frac{1.75}{0.161} = 10.87\,\text{m}$$

Maximum span length in all the strips (both ways) is 7.2 m < (1.75/ λ)

Therefore, the rigidity requirement of criterion (b) is fully satisfied.

However, since both rigidity criteria were not met simultaneously, one may consider the conventional rigid method is unsuitable or not appropriate for design of the given mat.

Step 13. Draw the final design sketches showing top and bottom steel bars, as shown below.

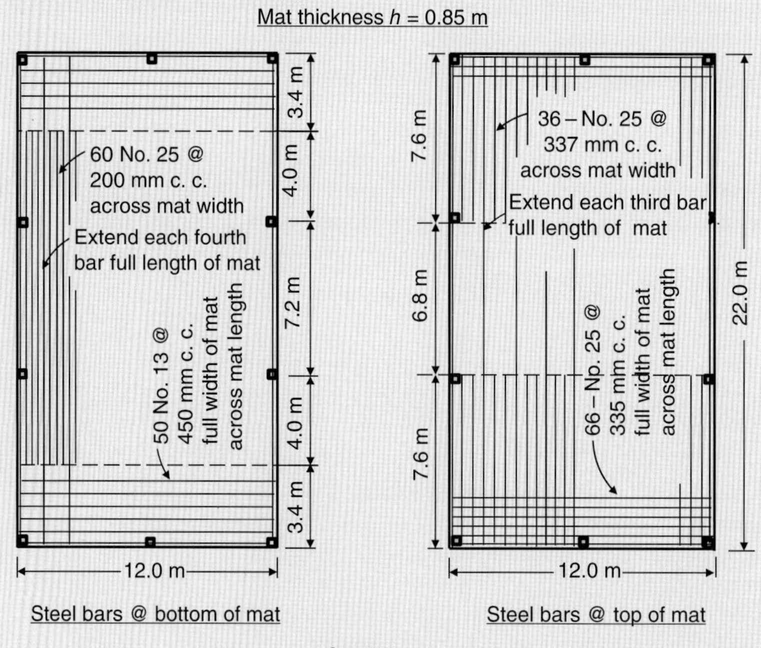

Scheme 5.125

Problem 5.27

The plan of a mat foundation with nine reinforced concrete columns is shown in the scheme below. The scheme also shows the working dead load D and live load L for each column. All the columns are 0.5 × 0.5 m in cross-section and are spaced at 8.0 m each way. The edge distance of the mat from the centreline of the exterior columns is 1.0 m. All the columns carry no moment of significant value. The design soil pressure or *net* q_a and average modulus of subgrade reaction $K_{s,av}$, recommended by the geotechnical engineer, are 100 kPa and 13 000 kN/m^3, respectively. Design the mat by the *approximate flexible method.* Use: $f_c' = 24$ MPa; $f_y = 420$ MPa.

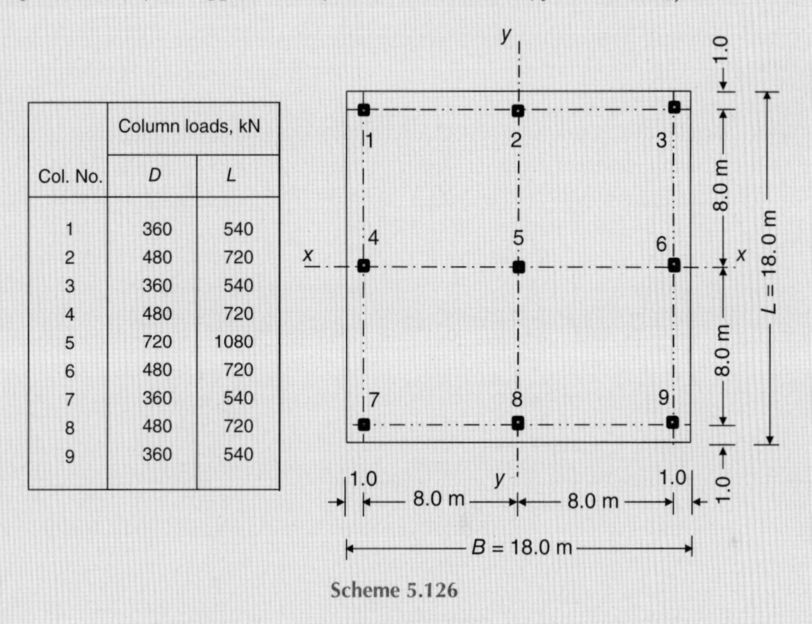

Scheme 5.126

Solution:

Step 1. Compute the working and factored $(D + L)$ column loads and their resultants; then check *net* q_a.

All the column loads and resultants are computed and tabulated, as shown in Table 5.24.

Table 5.24 Column loads and resultants.

Col. No.	Working loads D, kN	Working loads L, kN	Unfactored col. loads (D + L), kN	Factored col. loads (1.2 D + 1.6 L), kN
1	360	540	900	1296
2	480	720	1200	1728
3	360	540	900	1296
4	480	720	1200	1728
5	720	1080	1800	2592
6	480	720	1200	1728
7	360	540	900	1296
8	480	720	1200	1728
9	360	540	900	1296
			Unfactored R = **10200 kN**	Factored R = **14688 kN**

(Continued)

$$\text{Unfactored } net\ q = \frac{\text{unfactored } R}{A} = \frac{10\ 200}{18 \times 18}$$
$$= 31.5\,\text{kPa} \ll (net\ q_a = 100\,\text{kPa}) \quad \text{(OK.)}$$

Step 2. Determine the minimum mat thickness t considering punching shear.

In this problem, all the columns are of the same size with the same four-side shear perimeter. Therefore, punching shear is most critical at the heaviest column which is the interior column (Col. No. 5). Since it is common practice not to use shear reinforcement, we conservatively neglect the resisting factored soil reaction (which is too small compared to the factored column load) on the shear block. Thus, the factored applied shear force on the shear perimeter is

$$V_u = 2592 \text{ kN}$$

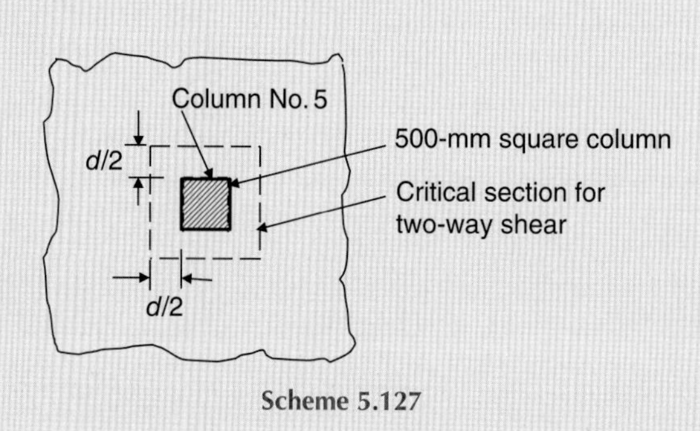

Scheme 5.127

Equation (5.19):

$$\varnothing\, V_c \geq V_u \text{ (taking reinforcement shear strength, } V_s = 0)$$

$$\text{where } \varnothing = \text{strength reduction factor (ACI Section 9.3.2.3)}$$
$$= 0.75 \text{ for shear}$$

Shear strength of concrete V_c shall be the smallest of (a), (b) and (c):

(a)

$$V_c = 0.17\left(1 + \frac{2}{\beta}\right)\lambda\sqrt{f_c'}b_o d$$
$$= (0.17)\left(1 + \frac{2}{1.0}\right)(1)\sqrt{f_c'}b_o d = (0.51)\sqrt{f_c'}b_o d \qquad \text{Equation (5.22)}$$

(b)

$$V_c = 0.083\left(\frac{\alpha_s d}{b_o} + 2\right)\lambda\sqrt{f_c'}b_o d$$
$$= (0.083)\left(\frac{40d}{b_o} + 2\right)(1)\sqrt{f_c'}b_o d \qquad \text{Equation (5.23)}$$

(c)

$$V_c = 0.33\,\lambda\sqrt{f_c'}b_o d$$
$$= 0.33\,(1)\sqrt{f_c'}\ b_o d = (0.33)\sqrt{f_c'}b_o d \qquad \text{Equation (5.24)}$$

Use $V_c = (0.33)\sqrt{f_c'}\, b_o d$. Calculate d then check V_c of Equation (5.23).

Let $\emptyset\, V_c = V_u$:

$$\emptyset\, V_c = (0.75)(0.33)\sqrt{24}(1000)[4(0.5+d)(d)] = 4850\,d^2 + 2425\,d$$

$$4850\,d^2 + 2425\,d = 2592$$

$$d^2 + 0.50\,d - 0.534 = 0$$

$$d = \frac{(-0.5) \pm \sqrt{(0.5)^2 - 4(1)(-0.534)}}{2} = \frac{-0.5 + 1.545}{2} = 0.523 \text{ m}$$

Check V_c of Equation (5.23):

$$V_c = (0.083)\left(\frac{40 \times 0.523}{4(0.5+0.523)} + 2\right)\sqrt{f_c'}b_o d = (0.59)\sqrt{f_c'}b_o d$$

$$V_c = (0.33)\sqrt{f_c'}b_o d < (0.59)\sqrt{f_c'}b_o d \quad \text{(OK.)}$$

Assume using No. 22 bars and 75 mm clear concrete cover. Therefore, $t = 523 + 11 + 75 = 609$ mm

Try $t = 0.65\ m = 650\ mm$.

Step 3. Determine the flexural rigidity D of the mat.

Equation (5.54):
$$D = \frac{E_f t^3}{12\left(1-\mu_f^2\right)}$$

$$E_f = E_c = 4700\sqrt{f_c'} \quad \text{(ACI Section 8.5.1)}$$

$$E_c = 4700 \times \sqrt{24} = 23025.2 \text{ MPa} = 23.03 \times 10^6 \text{ kPa}$$

$\mu_f = \mu_c$. For concrete, assume $\mu_c = 0.15$.

$$D = \frac{23.03 \times 10^6 \times 0.65^3}{12(1-0.15^2)} = 5.395 \times 10^5\ kN.m$$

Step 4. Determine the radius of effective stiffness L' and define a value for the radius of influence on the order of $3L'$ to $4L'$

Equation (5.55):
$$L' = \sqrt[4]{\frac{D}{K_s}}$$

$$L' = \sqrt[4]{\frac{5.392 \times 10^5}{13000}} = 2.54\ m$$

$$3L' = 3 \times 2.54 = 7.62 \text{ m};\ 4L' = 4 \times 2.567 = 10.16 \text{ m}$$

Assume we recommend: *radius of influence < 10 m.*

Step 5. Determine moments (M_r, M_t, M_x, M_y), shear force V and deflection δ at points of interest so that a complete analysis of the given mat may be furnished.

The mat and its loads are symmetrical in both x and y directions. Therefore, it is clear that only one-eighth of the mat area needs to be analysed. Thus, M, V, and δ values at Points α through i will complete the analysis, as shown in the scheme below.

The foundation responses, namely, moments, shear and deflection, per unit width of the mat are expressed by Equations (5.56) to (5.62). Values of Z for the desired values of the ratio r/L' are obtained from Figure 5.19, as shown in Table 5.25.

(*Continued*)

Table 5.26 gives the calculated factored moments, shear and deflection at the points of interest. It is noteworthy that, theoretically, as observed from Figure 5.19 (Z_4 and Z'_4 are ∞ for zero r/L'), the moments and shear due to a point load at the load itself will be infinite. Therefore, it will be necessary to determine these responses at the column faces and, for design purpose, assume their maximum values uniform under the column. In the present Problem with square columns, M_x and M_y (moments in x and y directions) at the column faces will have the same maximum values. Also, M_x and M_y, calculated from Equations (5.58) and (5.59), are moments about the $y-$ and $x-$ axis, respectively.

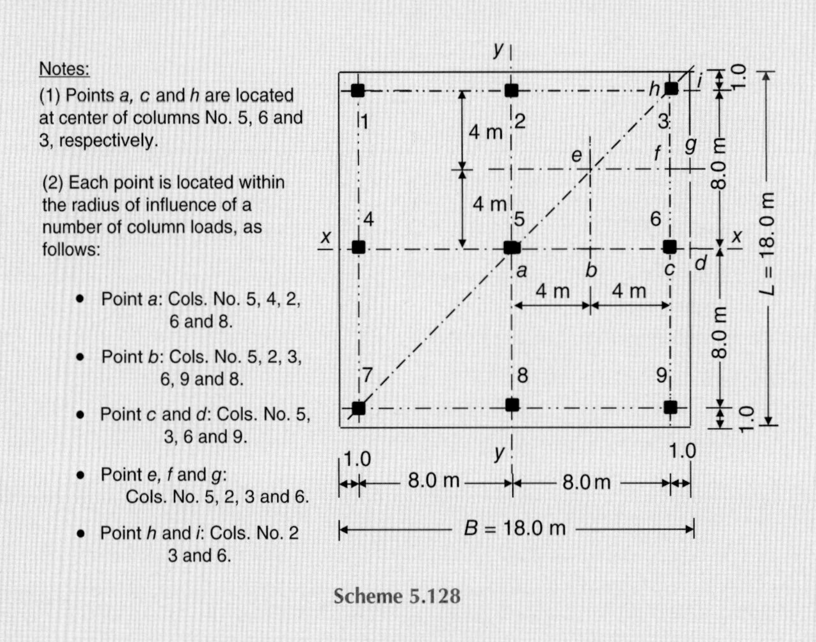

Scheme 5.128

Table 5.25 Values of Z for the desired values of the ratio r/L'.

r, m	r/L'	Z_3	Z'_3	Z_4	Z'_4
0.25	0.10	0.490	−0.100	−0.850	1.550
1.00	0.39	0.440	−0.210	−0.575	1.200
4.00	1.57	0.190	−0.180	−0.025	0.170
4.12	1.62	0.180	−0.170	−0.020	0.165
5.66	2.23	0.100	−0.125	0.040	0.040
8.00	3.15	0.025	−0.060	0.040	−0.020
8.06	3.17	0.024	−0.058	0.038	−0.022
8.94	3.52	0.015	−0.040	0.035	−0.025
9.00	3.54	0.013	−0.035	0.034	−0.023
9.85	3.88	0.002	−0.020	0.025	−0.020

Table 5.26 Calculated factored moments, shear and deflection.

(1) Point a:

P (kN)	r (m)	M_r (kN, m)	M_t (kN, m)	θ (deg)	$\sin^2\theta$	$\cos^2\theta$	M_x (kN, m)	M_y (kN, m)	V (kN)	δ mm
2592	0.25	0	633.4	0, 90	0,1	1,0	633.4	633.4	−395.4	3.9
1728	8.00	−24.3	4.4	0	0	1	−24.3	4.4	3.4	0.1
1728	8.00	−24.3	4.4	−90	1	0	4.4	−24.3	3.4	0.1
1728	8.00	−24.3	4.4	180	0	1	−24.3	4.4	3.4	0.1
1728	8.00	−24.3	4.4	90	1	0	4.4	−24.3	3.4	0.1
						Σ =	593.6	593.6	−381.8	4.3

(2) Point b:

P (kN)	r (m)	M_r (kN, m)	M_t (kN, m)	θ (deg)	$\sin^2\theta$	$\cos^2\theta$	M_x (kN, m)	M_y (kN, m)	V (kN)	δ mm
2592	4.00	−46.9	65.6	0	0	1	−46.9	65.6	−43.4	1.5
1728	8.94	−19.3	1.9	−63.5	0.8	0.2	−2.3	−15.1	4.3	0.1
1296	8.94	−14.5	1.4	243.5	0.8	0.2	−1.8	−11.3	3.2	0.1
1728	4.00	−31.3	43.7	180	0	1	−31.3	43.7	−28.9	1.0
1296	8.94	−14.5	1.4	116.5	0.8	0.2	−1.8	−11.3	3.2	0.1
1728	8.94	−19.3	1.9	63.5	0.8	0.2	−2.3	−15.1	4.3	0.1
						Σ =	−86.4	56.5	−57.3	2.9

(3) Point c:

P (kN)	r (m)	M_r (kN, m)	M_t (kN, m)	θ (deg)	$\sin^2\theta$	$\cos^2\theta$	M_x (kN, m)	M_y (kN, m)	V (kN)	δ mm
2592	8.00	−36.4	6.6	0	0	1	−36.4	6.6	5.1	0.2
1296	8.00	−18.2	3.3	−90	1	0	3.3	−18.2	2.6	0.1
1728	0.25	0	422.3	0, 90	0, 1	1, 0	422.3	422.3	−263.6	2.6
1296	8.00	−18.2	3.3	90	1	0	3.3	−18.2	2.6	0.1
						Σ =	392.5	392.5	−253.3	3.0

(4) Point d:

P (kN)	r (m)	M_r (kN, m)	M_t (kN, m)	θ (deg)	$\sin^2\theta$	$\cos^2\theta$	M_x (kN, m)	M_y (kN, m)	V (kN)	δ mm
2592	9.00	−27.5	2.1	0	0	1	−27.5	2.1	5.9	0.1
1296	8.06	−17.4	3.2	−82.9	0.98	0.02	2.8	−17.0	2.8	0.1
1728	1.00	50.7	235.0	0	0	1	50.7	235.0	−204.1	2.3
1296	8.06	−17.4	3.2	82.9	0.98	0.02	2.8	−17.0	2.8	0.1
						Σ =	28.8	203.1	−192.6	2.6

(Continued)

Table 5.26 (*Continued*)

(5) Point *e*:

P (kN)	r (m)	M_r (kN, m)	M_t (kN, m)	θ (deg)	$\sin^2\theta$	$\cos^2\theta$	M_x (kN, m)	M_y (kN, m)	V (kN)	δ mm
2592	5.66	−56.8	27.0	45	0.5	0.5	−14.9	−14.9	−10.2	0.8
1728	5.66	−37.9	18.0	−45	0.5	0.5	−10.0	−10.0	−6.8	0.5
1296	5.66	−28.4	13.5	225	0.5	0.5	−7.5	−7.5	−5.1	0.4
1728	5.66	−37.9	18.0	135	0.5	0.5	−10.0	−10.0	−6.8	0.5
						Σ =	−42.4	−42.4	−28.9	2.2

(5) Point *f*:

P (kN)	r (m)	M_r (kN, m)	M_t (kN, m)	θ (deg)	$\sin^2\theta$	$\cos^2\theta$	M_x (kN, m)	M_y (kN, m)	V (kN)	δ mm
2592	9.85	−28.9	2.9	26.6	0.2	0.8	−22.5	−3.5	6.4	0.1
1728	9.85	−19.3	1.9	−26.6	0.2	0.8	−15.1	−2.3	4.3	0.1
1296	4.00	−23.5	32.8	−90	1	0	32.8	−23.5	−21.7	0.7
1728	4.00	−31.3	43.7	90	1	0	43.7	−31.3	−28.9	1.0
						Σ =	38.9	−60.6	−39.9	1.9

(6) Point *g*:

P (kN)	r (m)	M_r (kN, m)	M_t (kN, m)	θ (deg)	$\sin^2\theta$	$\cos^2\theta$	M_x (kN, m)	M_y (kN, m)	V (kN)	δ mm
2592	9.85	−19.0	0.4	24	0.17	0.83	−15.7	−2.9	5.1	0.1
1728	9.85	−12.7	0.3	−24	0.17	0.83	−10.5	−1.9	3.4	0
1296	4.12	−22.4	29.9	−76	0.94	0.06	26.8	−19.3	−21.0	0.7
1728	4.12	−29.9	39.8	76	0.94	0.06	35.6	−25.7	−28.1	0.9
						Σ =	36.2	−49.8	−40.6	1.7

(7) Point *h*:

P (kN)	r (m)	M_r (kN, m)	M_t (kN, m)	θ (deg)	$\sin^2\theta$	$\cos^2\theta$	M_x (kN, m)	M_y (kN, m)	V (kN)	δ mm
1728	8.00	−24.3	4.4	0	0	1	−24.3	4.4	3.4	0.1
1296	0.25	0	316.7	0,90	0,1	1,0	316.7	316.7	−197.7	2.0
1728	8.00	−24.3	4.4	90	1	0	4.4	−24.3	3.4	0.1
						Σ =	296.8	296.8	−190.9	2.2

(8) Point *i*:

P (kN)	r (m)	M_r (kN, m)	M_t (kN, m)	θ (deg)	$\sin^2\theta$	$\cos^2\theta$	M_x (kN, m)	M_y (kN, m)	V (kN)	δ mm
1728	9.00	−18.3	1.4	0	0	1	−18.3	1.4	3.9	0.1
1296	1.00	38.0	176.3	0	0	1	38.0	176.3	−153.1	1.7
1728	8.06	−23.2	4.3	82.9	0.98	0.02	3.8	−22.7	3.7	0.1
						Σ =	23.5	155.0	−145.5	1.9

Step 6. Apply moment and shear equal and opposite to those given in the tables of *Step 5* for the edge points *d*, *g* and *i* as edge redundant loads, and superimpose their effects (moment, shear and deflection) on the respective values at the points located less than 10 m (radius of influence) from the particular edge point. For this purpose, use the moment, shear and deflection equations given in Section 5.10.3, repeated here for convenience:

$$M = M_1\,A_{\lambda x} - \frac{P_1}{\lambda}\,B_{\lambda x} \quad V = -2\,M_1\,\lambda\,B_{\lambda x} - P_1 C_{\lambda x}$$

$$\delta = -\frac{2\,M_1\,\lambda^2}{K_s}\,C_{\lambda x} + \frac{2\,P_1\,\lambda}{K_s}\,D_{\lambda x}$$

where M_1, P_1 = the applied edge moment and shear, respectively.

$$\lambda = \left(\frac{K_s b}{4E_c I_b}\right)^{0.25} \quad \text{Equation (5.48)}$$

where b = strip width (1 m); I_b = moment of inertia of the strip.

$A_{\lambda x}$, $B_{\lambda x}$, $C_{\lambda x}$ and $D_{\lambda x}$ are coefficients obtained from tables or figures. For example, see Table 9-3 of *Foundation Analysis and Design* (Bowles, 1982).

In these equations, P_1 and M_1 are equal and opposite to the calculated redundant loads V and M_x (actually should be zero), respectively, of a point at the east or west edge. They are applied at the edge point; the effects M, V and δ calculated and superimposed on the respective previous values at all the points within the radius of influence on the E–W line which contains the edge point. Similarly, at the north or south edge the redundant loads are V and M_y; the necessary corrections should be applied to respective previous values at all the points within the radius of influence on the N–S line which contains the edge point.

As mentioned earlier, in the present convention M_x is the moment about the y-axis and M_y is the moment about the x-axis. It may be noted that, due to symmetry of the mat of the present Problem, M_x and M_y at the east edge points are equal to M_y and M_x, respectively, at the corresponding north edge points.

$$\lambda = \left(\frac{K_s b}{4E_c I_b}\right)^{0.25} = \left(\frac{13\ 000 \times 1}{4 \times 23.03 \times 10^6 (1 \times 0.65^3/12)}\right)^{0.25} = 0.28\ \text{m}^{-1}$$

The calculated correction values of M, V and δ are given in the Table 5.27.

Final net results (after superposition of the respective values) of moment, shear and deflection values at points *a*, *b*, *c*, *d*, *e*, *f* and *h* on the E–W line are given in Table 5.28. Due to symmetry of the given mat, the corresponding points on the N–W line have the same moment, shear and deflection values. The moment and shear values are per unit width of the mat each way. Thus, these values will complete the analysis.

Step 7. Check the mat thickness (t = 650 mm) considering beam shear (one-way shear) using maximum shear force V_u = 359.4 kN/m width.

Equation (5.19): $\quad \varnothing\, V_c \geq V_u$ (taking reinforcement shear strength, $V_s = 0$)

Equation (5.20): $\quad V_c = 0.17\lambda\,\sqrt{f_c'}\,b_w d$

$$d = t - \text{concrete cover} - \text{half bar diameter}$$

$$= 650 - 75 - 11 = 564\ \text{mm}$$

$$\varnothing\, V_c = (0.75 \times 0.17 \times 1)(\sqrt{24})(1000)(1)(0.564)$$

$$= 352.3\ \text{kN/m} < (V_u = 359.4\ \text{kN/m}) \quad (\textit{Not OK.})$$

(Continued)

Table 5.27 Calculated correction values of M, V and δ.

(1) Edge point d, ($E - W$ line): $M_1 = -28.8$ kN.m; $P_1 = 192.6$ kN

Point	x (m)	λx	$A_{\lambda x}$	$B_{\lambda x}$	$C_{\lambda x}$	$D_{\lambda x}$	M (kN.m)	V (kN)	δ (mm)
a	9	2.52	−0.0175	0.0475	−0.1124	−0.0649	−32.2	22.4	0.0
b	5	1.40	0.2849	0.2430	−0.2011	0.0419	−175.4	42.7	0.0
c	1	0.28	0.9416	0.2078	0.5190	0.7266	−170.1	−96.6	0.0

(2) Edge point g, ($E - W$ line): $M_1 = -36.2$ kN.m; $P_1 = 40.6$ kN

Point	x (m)	λx	$A_{\lambda x}$	$B_{\lambda x}$	$C_{\lambda x}$	$D_{\lambda x}$	M (kN.m)	V (kN)	δ (mm)
e	5	1.40	0.2849	0.2430	−0.2011	0.0419	−45.5	13.1	0.0
f	1	0.28	0.9416	0.2078	0.5190	0.7266	−64.2	−16.9	0.0

(3) Edge point i, ($E - W$ line): $M_1 = -23.5$ kN.m; $P_1 = 145.5$ kN

Point	x (m)	λx	$A_{\lambda x}$	$B_{\lambda x}$	$C_{\lambda x}$	$D_{\lambda x}$	M (kN.m)	V (kN)	δ (mm)
h	1	0.28	0.9416	0.2078	0.5190	0.7266	−130.1	−72.8	0.0

Table 5.28 Moment, shear and deflection values.

Point	M (kN. m)	V (kN)	δ (mm)
a	561.4	−359.4	4.3
b	−261.8	−14.6	2.9
c	222.4	−349.9	3.0
e	−87.9	−15.8	2.2
f	−25.3	−56.8	1.9
h	166.7	−263.7	2.2

However, the difference (1.7%) is so small that the 0.65-m thickness may be considered acceptable or increased by 15 mm (i.e t = 665 mm). This slight increase in the mat thickness would not require any analysis revision.

Use mat thickness t = 665 mm.

Step 8. Determine the positive and negative flexural reinforcement required in E–W and N–S directions, that is x and y directions, respectively.

Due to the symmetry of the given mat and its loads, the same positive and negative flexural reinforcement shall be required in both x and y directions. The design maximum positive and negative moments per metre width of the mat are: $+M = 561.4$ kN.m; $-M = 261.8$ kN. m. The design of flexural reinforcement proceeds in the same manner as that for the mat of the Solution of Problem 5.26, *Step 10*, with attention to the moment locations. The remainder of the solution is left for the reader.

Step 9. Check column bearing on the mat at critical column locations and design the column to mat dowels.

The design computations proceed in the same manner as that presented in the design *Steps 5 and 6* for the column footing of Problem 5.3. The necessary computations are left for the reader.

Step 10. Develop the final design sketches.

Problem 5.28

Consider the mat foundation of Problem 5.27.

(a) Check the rigidity of the mat. Use $t = 665$ mm, $E_c = 23.03 \times 10^6$ kPa and $K_s = 13000$ kN/m^3.

(b) Using the conventional (rigid) design method, determine the minimum mat thickness t and the design maximum positive and negative moments. Compare the results with those obtained in the Solution of Problem 5.27.

Solution:

(a) A strip taken from a mat may be considered rigid if the following criteria are met:

(1) Variation in adjacent column loads and spacing is not over 20% of the greater value.

(2) Average of two adjacent spans is $< (1.75/\lambda)$. The factor λ is

$$\lambda = \sqrt[4]{\frac{K_s B}{4E_c I}} \quad \text{Equation (5.48)}$$

Criterion (1) is not fully satisfied since variation in the adjacent column loads is over 20% of the greater value (25 and 33%).

$$= \sqrt[4]{\frac{K_s B}{4E_c I}} = \sqrt[4]{\frac{K_s B}{4E_c\left(\frac{Bt^3}{12}\right)}} = \sqrt[4]{\frac{13\,000 \times 12}{4 \times 23.03 \times 10^6 \times 0.665^3}} = 0.275\,\text{m}^{-1}; \quad \frac{1.75}{\lambda} = \frac{1.75}{0.275} = 6.36\,\text{m}$$

The given span length = 8 m. Therefore, criterion (2) is also not satisfied.

The given mat cannot be considered rigid.

(b) Refer to Solution of Problem 5.27, *Step 7*. The provided mat thickness t considering two-way shear is *665 mm*. However, this thickness should be checked for one-way shear soon after shear diagrams are obtained.

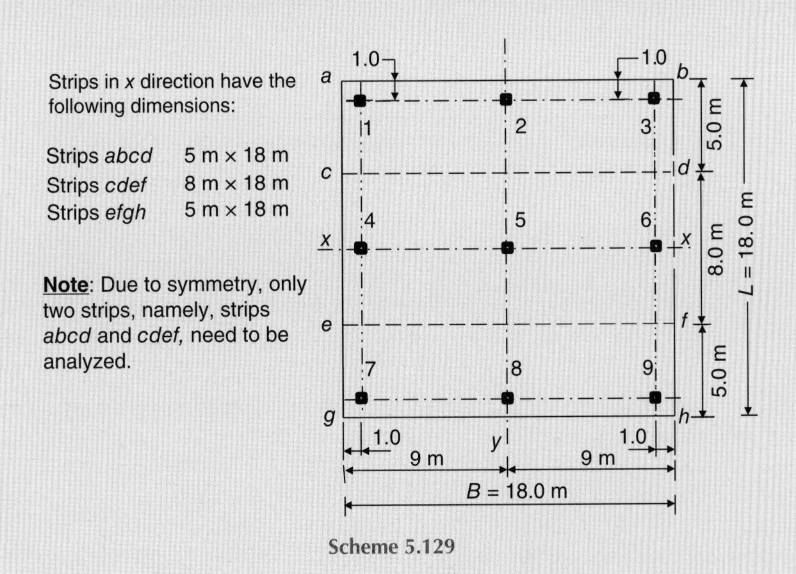

Scheme 5.129

(*Continued*)

Divide the mat into three continuous beam strips (or combined footings with multiple columns), as shown above. Due to symmetry of the mat and its loads, it makes no difference in which direction the mat is divided.

The net contact pressure, *net q*, is assumed uniformly distributed since the mat and its loads are symmetrical about both the x and y axes ($e_x = e_y = 0$).

$$\text{Unfactored } net\, q = \frac{\text{unfactored } R}{A} = \frac{10\ 200}{18 \times 18} = 31.48\,\text{kPa}.$$

Check the static equilibrium of strips *abcd* and *efgh* and modify the column loads and contact pressures accordingly.

Strip *abcd*:

$$\Sigma\,\text{column unfactored loads} = 900 + 1200 + 900 = 3000\,\text{kN}$$

$$\text{Average soil reaction} = (A)(\text{Unfactored } net\, q) = (5 \times 18)(31.48)$$
$$= 2833.2\,\text{kN} \neq \Sigma\,\text{column loads}$$

Therefore, *static equilibrium is not satisfied* ($\Sigma F_v \neq 0$); the column loads and contact pressures need to be modified.

$$Q_{av} = \frac{1}{2}(\Sigma\,\text{column loads} + \text{average soil reaction})$$
$$= \frac{1}{2}(3000 + 2833.2) = 2916.6\,\text{kN} \qquad \text{Equation (5.49)}$$

$$MF_{(\text{col.})} = \frac{Q_{av}}{\Sigma\,\text{column loads}} = \frac{2916.6}{3000} = 0.972. \qquad \text{Equation (5.50)}$$

This is a reducing factor for columns of strip *abcd*.

Each column load on the strip should be multiplied by the factor $MF_{(\text{col.})}$.

Each column factored modified load = $(MF_{(\text{col.})})(1.2\,D + 1.6\,L)$. The computed load values are as shown in Table 5.29 (for the column factored loads refer to Table 5.24 in Solution of Problem 5.27, *Step 1*).

Table 5.29 Computed load values.

Column No.	Factored modified column loads
1	(0.972)(1296) = 1258 kN
2	(0.972)(1728) = 1680 kN
3	(0.972)(1296) = 1258 kN
	R = 4196 kN

Since the strip is symmetrically loaded, the factored R falls at the centre and the factored contact pressure q_{factored} is assumed uniformly distributed.

$$q_{\text{factored}} = \frac{R_{\text{factored}}}{A} = \frac{4196}{5 \times 18} = 46.622\ \text{kPa} \qquad \text{Equation (5.52)}$$

The factored contact pressure per metre length of the strip is
$q_{factored} = 46.622 \times 5 = 233.11\ kN/m\ \ length$
Strip *cdef*:

$$\Sigma\,\text{column unfactored loads} = 1200 + 1800 + 1200 = 4200\,\text{kN}$$

$$\text{Average soil reaction} = (A)(\text{unfactored } net\ q) = (8 \times 18)(31.48)$$
$$= 4533.12\ \text{kN} \neq \Sigma\,\text{column loads}$$

Therefore, *static equilibrium is not satisfied* ($\Sigma F_v \neq 0$); the column loads and contact pressures need to be modified.

$$Q_{av} = \frac{1}{2}(\Sigma\,\text{column loads} + \text{average soil reaction})$$

$$= \frac{1}{2}(4200 + 4533.12) = 4366.56\ \text{kN}$$

$$MF_{(\text{col.})} = \frac{Q_{av}}{\Sigma\,\text{column loads}} = \frac{4366.56}{4200}$$
$$= 1.04.$$

It is an increasing factor for columns of strip *cdef*.
Each column load on the strip should be multiplied by the factor $MF_{(\text{col.})}$.
Each column factored modified load = $[MF_{(\text{col.})}](1.2\ D + 1.6\ L)$. The computed load values are as shown in Table 5.30 (for the column factored loads refer to Table 5.24 in Solution of Problem 5.27, *Step 1*).

Table 5.30 Computed load values.

Column No.	Factored modified column loads
4	(1.04)(1728) = *1797 kN*
5	(1.04)(2592) = *2696 kN*
6	(1.04)(1728) = *1797 kN*
	R = *6290 kN*

Since the strip is symmetrically loaded, the factored *R* falls at the centre and the factored contact pressure $q_{factored}$ is assumed uniformly distributed.

$$q_{factored} = \frac{R_{factored}}{A}$$

$$= \frac{6290}{8 \times 18} = 43.681\ \text{kPa}$$

(*Continued*)

The factored contact pressure per metre length of the strip is

$$q_{\text{factored}} = 43.681 \times 8 = \mathit{349.45\ kN/m\ length}$$

Draw factored load, shear and moment diagrams for the continuous beam strips in both directions.

Strip *abcd*:

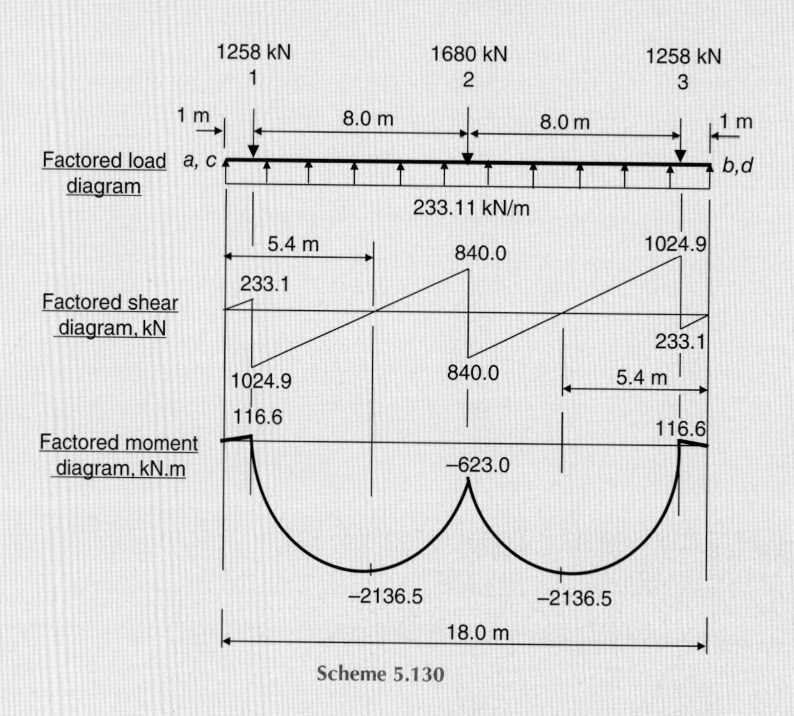

Scheme 5.130

Strip *cdef*:

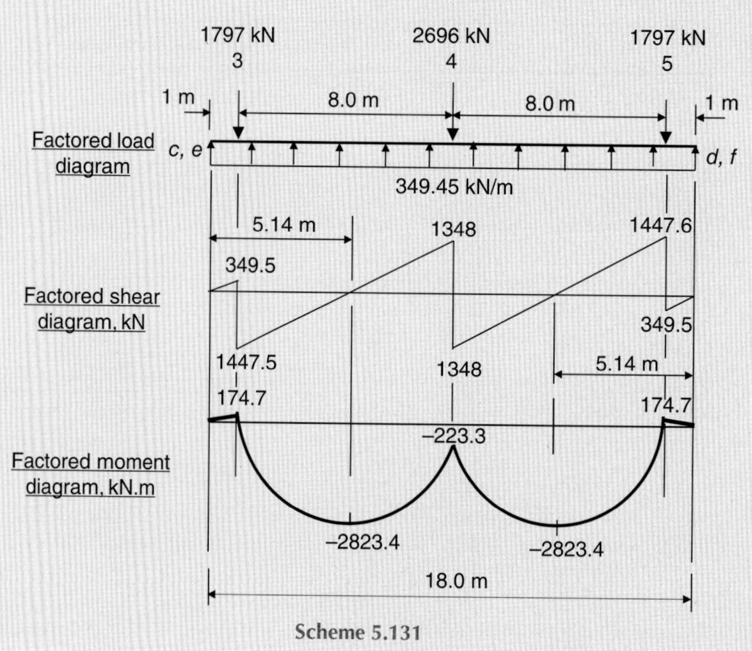

Scheme 5.131

The provided mat thickness $t = 665$ mm

$$d = t - \text{concrete cover} - \text{half bar diameter}$$

$$= 665 - 75 - 11 = 579 \text{ mm.}$$

Check the designed mat depth d considering beam shear (one-way shear) at the most critical section: Refer to the shear diagrams. It is clear that the most critical section for shear per metre width is located in strip *abcd*, at distance d from either the left face of column No. 3 or the right face of column No. 1. The applied beam shear at any of these two locations is

$$V_u = \frac{1024.9 - q(d + 0.2)}{5} = \frac{1024.9 - 233.11(0.579 + 0.200)}{5} = 168.7 \text{ kN/m width}$$

Equation (5.19): $\quad \varnothing V_c \geq V_u$ (taking reinforcement shear strength, $V_s = 0$)

Equation (5.20): $\quad V_c = 0.17\lambda\sqrt{f_c'}b_w d$

$$\varnothing V_c = (0.75 \times 0.17 \times 1)(\sqrt{24})(1000)(1)(0.579)$$

$$= 352.3\frac{\text{kN}}{\text{m}} \gg (V_u = 168.7 \text{ kN/m width}) \quad \text{(OK.)}$$

Table 5.31 elucidates comparison of results obtained using the approximate flexible and rigid methods of analysis:

Table 5.31 A comparison of results obtained using the approximate flexible and rigid methods of analysis.

	Design method	
Item	**Approximate flexible method**	**Conventional (rigid) method**
Thickness	665 mm	665 mm
Maximum one-way shear	359.4 kN/m width	$\left(\frac{1024.9}{5}\right) = 205.0 \text{kN/m width}$
Maximum two-way shear	2592 kN	2592 kN
Maximum positive moment	561.4 kN.m/m width	23.3 kN.m/m width
Maximum Negative moment	261.8 kN.m/m width	427.3 kN.m/m width

Note: For a flexible mat such as the one given in Problems 5.27 and 5.28, one may conclude that the approximate flexible method gives more realistic results than the conventional (rigid) method.

References

ACI 318 (2008), Building Code Requirements for Structural Concrete *(ACI 318M-08) and* Commentary, *reported by ACI Committee 318, an ACI Standard,* American Concrete Institute, Farmington Hills, MI, U.S.A., 473 pp.

ACI-ASCE (1962), "Shear and Diagonal Tension", Part 3, "Slabs and Footings", reported by ACI-ASCE Committee 326, *Proceedings, ACI Journal,* March, Vol. **59**, No. 3, pp. 353–396.

ACI Committee 336 (1993), "Suggested Analysis and Design Procedures for Combined Footings and Mats, ACI 336.2R-88 (Reapproved 1993)", *ACI Manual of Concrete Practice*, American Concrete Institute, Farmington Hills, MI, pp. 336.2R-1 to 336.2R-21 plus discussion.

AISC (1989), *Manual of Steel Construction, Allowable Stress Design*, 9th edn, American Institute of Steel Construction, Chicago.

Bowles, J. E. (2001), *Foundation Analysis and Design*, 5th edn, McGraw-Hill, New York.

Bowles, J. E. (1982), *Foundation Analysis and Design*, 3rd edn, McGraw-Hill, New York.

Bowles, J. E. (1974a), *Analytical and Computer Methods in Foundation Engineering*, McGraw-Hill, New York, 519 pp.

Bowles, J. E. (1976), "Mat Foundations" and "Computer Analysis of Mat Foundations", *Proceedings*, Short Course-Seminar on Analysis and Design of Building Foundations (Lehigh University), Envo Press, Lehigh Valley, pp. 209–232 and 233–256.

Bowles, J. E. (1983), "Pile Cap Analysis," *Proceedings 8th Conference on Electronic Computation*, ASCE, pp. 102–113.

Bowles, J. E. (1986), "Mat Design", *Journal of American Concrete Institute*, Vol. **83**, No. 6, Nov/Dec, pp. 1010–1017.

Christian, J. T. (1976), *"Soil-Foundation-Structure Interaction"*, Proceedings, *Short Course-Seminar on Analysis and Design of Building Foundations (Lehigh University)*, Envo Press, Lehigh Valley, pp. 149–179.

Coduto, Donald P. (2001), *Foundation Design: Principles and Practices*, 2nd edn, Prentice-Hall, Inc., New Jersey.

Cook, R. D. (1974), *Concepts and Applications of Finite Element Analysis*, John Wiley & Sons, Inc., New York, 402 pp.

Das, Braja M. (2011), *Principles of Foundation Engineering*, 7th edn, CENGAGE Learning, United States.

DeWolf, J. T., and Ricker, D. T. (1990), *Column Base Plates*, American Institute of Steel Construction, Chicago, IL.

Ghali, A. and Neville, A. M. (1972), *Structural Analysis: A Unified Classical and Matrix Approach*, Intext Educational Publishers (Now Harper and Row), Chapters 17–20.

Hetenyi, M. (1946), *Beams on Elastic Foundations*, The University of Michigan Press, Ann Arbor, MI, 255 pp.

Horvath, J. S. (1993), *Subgrade Modeling for Soil-Structure Interaction Analysis of Horizontal Foundation Elements*, Manhattan College Research Report No. CE/GE-93-1, Manhattan College, New York.

Liao, S. S. C. (1991), *Estimating the Coefficient of Subgrade Reaction for Tunnel Design*, Internal Research Report, Parsons Brinkerhoff, Inc., New York.

MacGregor, J. G., and Wight J. K. (2005), *Reinforced Concrete: Mechanics and Design*, 4th edn, Pearson Prentice Hall, Upper Saddle River, New Jersey, 1132 pp.

Moe, J. (1961), *Shearing Strength of Reinforced Concrete Slabs and Footings under Concentrated Loads*, Portland cement Association Bulletin D47, Skokie, Illinois.

Richart, F. E. (1948), "Reinforced Concrete Wall and Column Footings", *Journal of the American Concrete Institute*, Vol. **20**, No. 2, pp. 97–127, and Vol. **20**, No. 3, pp. 237–260.

Shipp, J. G., and Haninger, E. R. (1983), "Design of Headed Anchor Bolts", *Engineering Journal*, Vol. **20**, No. 2, pp. 58–69, American Institute of Steel Construction, Chicago, IL.

Talbot, A. N. (1913), *Reinforced Concrete Wall Footings and Column Footings*, Bulletin No. 67, University of Illinois Engineering Experiment Station, Urbana.

Terzaghi, K. (1955), "Evaluation of Coefficients of Subgrade Reaction", *Geotechnique*, Vol. **5**, No. 4, pp. 297–326.

Vesič, A. S. (1973), "Bending of Beams Resting on Isotropic Elastic Solid", *Journal of Engineering Mechanics Division*, ASCE, Vol. **87**, No. EM2, pp. 35–53.

Wang, C. K. (1970), *Matrix Methods of Structural Analysis*, 2nd edn, Intext Educational Publishers, Scranton, PA, 406 pp.

Whitney, Charles S. (1957), "Ultimate Shear Strength of Reinforced Concrete Flat Slabs, Footings, Beams, and Frame Members without Shear Reinforcement", *Journal of the American Concrete Institute*, Vol. **29**, No. 4, pp. 265–298.

Winkler, E. (1867), *Die Lehre von Elastizität und Festigkeit* (On Elasticity and Fixity), H. Dominicus, Prague.

CHAPTER 6

Eurocode Standards and the Design of Spread Foundations

6.1 General

In line with the European Union's (EU's) strategy of smart, sustainable and inclusive growth, *Standardisation* plays an important part in supporting the industrial policy for the globalisation era. All 33 national members of the *European Committee for Standardisation* (CEN; French: *Comité Européen de Normalisation*, founded in 1961) work together to develop *European Standards* (ENs) in various sectors in order to build a European market for goods and services and to position Europe in the global economy. CEN is officially recognised as a European standards body by the EU.

The construction sector is of strategic importance to the EU as it delivers the buildings and infrastructure needed. It is the largest single economic activity and it is the biggest industrial employer in Europe. These facts urged CEN to produce what is known as *Eurocode Standards*, also called *Structural Eurocodes* or *Eurocodes*.They are a set of harmonised technical rules and requirements developed by the CEN for the structural design of construction works in the European Union. The development of the Eurocodes started in 1975; since then they have evolved significantly and are now claimed to be among the most technically advanced structural codes in the world. With the publication of all the 58 Eurocodes Parts in 2007, the implementation of the Eurocode Standards has been extended to all the European Union countries and there are firm steps toward their adoption internationally. The purposes of the Eurocodes may be summarised as follows:

- A means to prove compliance with the requirements for mechanical strength and stability and safety in case of fire established by European Union law.
- A basis for construction and engineering contract specifications.
- A framework for creating harmonised technical specifications for building products.

There are 10 Structural Eurocodes,each published as a separate *European Standard* and each having a number of parts (so far 58 parts have been produced), covering all the main structural materials (Figure 6.1).

It has been a legal requirement from March 2010 that all European public-sector clients base their planning and building control applications on structural designs that meet the requirements of the Eurocode. To comply with this, changes have been necessary to the building regulations. The Eurocodes therefore replace the existing national building codes published by national standard bodies (e.g. BS 5950), although many countries had a period of co-existence. Each country is required to

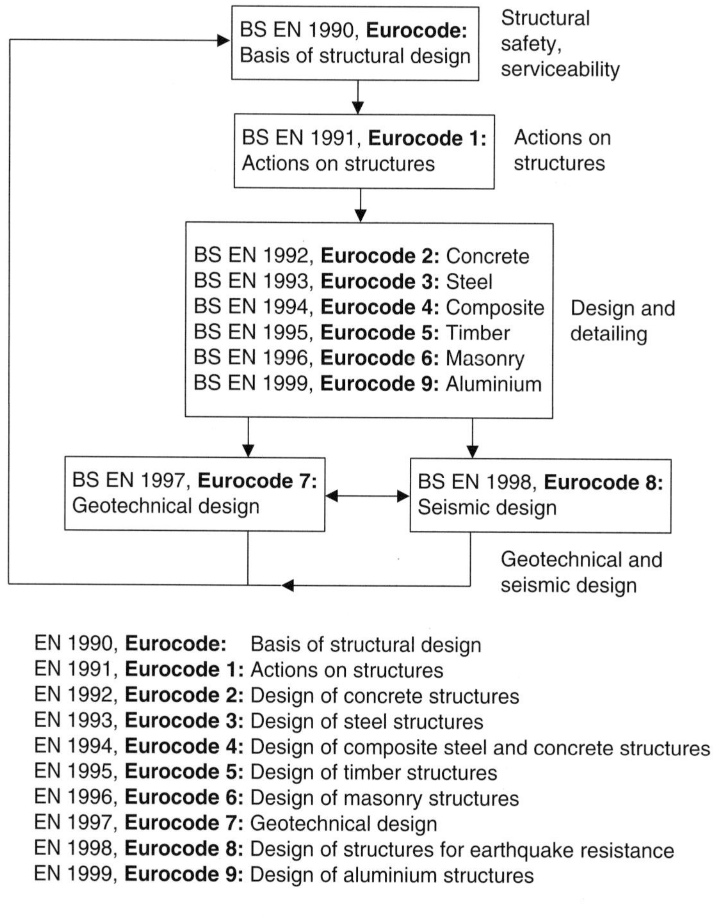

Figure 6.1 The Eurocodes and their relationships.

publish a Eurocode with a national title page and forward but the original text of the Eurocode (including any annexes) must appear as produced by CEN as the main body of the document. A National Annex can be included at the back of the document (Figure 6.2). Throughout Chapter 6 the relevant UK National Annexes will be used along with the Eurocodes.

There are two categories of Annexes to the Structural Eurocodes. One type is labelled "I" and is Informative (i.e. for information and not as a mandatory part of the code). The second type is labelled "N" and is Normative (i.e. a mandatory part of the code). In their National Annex a country can choose to make an Informative annex Normative if they so whish. In other words, a National annex may only contain information on those parameters which are left open in the Eurocode for national choice, known as Nationally Determined Parameters (NDP), to be used for the design of buildings and civil engineering works in the country concerned, that is:

- Values and/or classes where alternatives are given in the Eurocode,
- Values to be used where a symbol only is given in the Eurocode,
- Country-specific data (geographical, climatic, etc.), for example a snow map,
- A procedure to be used where an alternative procedure is given in the Eurocode,
- Decisions on the application of informative annexes,
- References to non-contradictory complementary information so that the Eurocode may easily be used without complications.

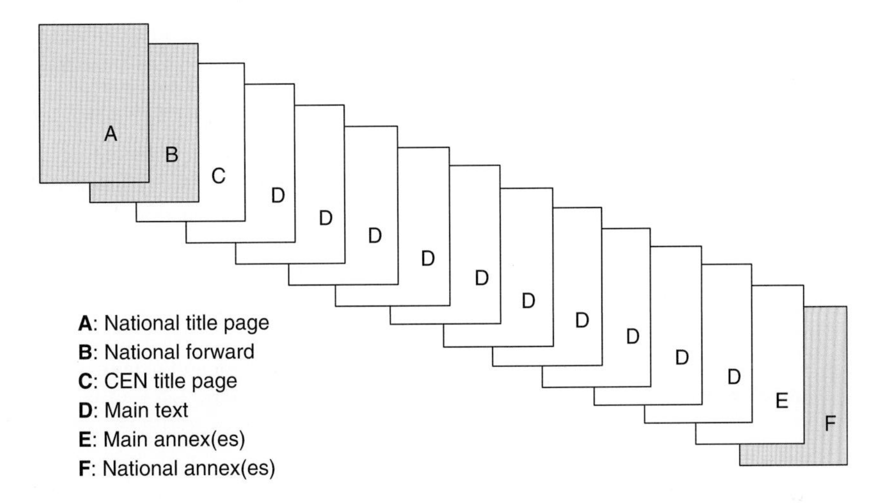

Figure 6.2 A typical Eurocode layout.

Finally, it may be useful to mention some of the following significant benefits of using the new Eurocodes:

(1) Use of Eurocodes provides more opportunity for marketing and use of structural components in EU Member States.
(2) Eurocodes facilitate the marketing and use of materials and constituent products, the properties of which enter into design calculations.
(3) Eurocodes facilitate the acquisition of European contracts as well as public sector contracts.
(4) Use of the Eurocodes provides more opportunity for designers to work throughout Europe.
(5) In Europe all public works must allow the Eurocodes to be used.
(6) Eurocodes provide a design framework and detailed implementation rules which are valid across Europe and likely to find significant usage worldwide.
(7) Eurocodes are among the most advanced technical views prepared by the best informed groups of experts in their fields across Europe.
(8) Eurocodes are considered as the most comprehensive treatment of subjects, with many aspects not codified now being covered by agreed procedures.
(9) Eurocodes provide common design criteria and methods of meeting necessary requirements for mechanical strength (resistance), serviceability and durability.
(10) Use of Eurocodes enables the preparation of common design aids and software.
(11) Use of Eurocodes increases competitiveness of European civil engineering firms, contractors, designers and manufacturers in their global activities.
(12) Eurocodes provide a common understanding regarding the design of structures between owners, designers, contractors and manufacturers of construction products.

6.2 Basis of Design Irrespective of the Material of Construction

6.2.1 Introduction

The basis for the design and verification of structures irrespective of the material of construction can be found in the EN 1990 Eurocode. This code is a fully operative material-independent code, establishes principles and requirements for safety, serviceability and durability of structures and gives guidelines for related aspects of structural reliability. It uses a statistical approach to determine realistic values for actions that occur in combination with each other. It is based on the limit state concept and used in conjunction with the partial factor method.

The EN 1990 Eurocode is intended to be used in conjunction with the other Eurocodes for the structural design of buildings and civil engineering works, including geotechnical aspects, structural fire design and situations involving earthquakes, execution and temporary structures. Also, it is applicable for: design of structures where other materials or actions outside the scope of the EN 1991 to EN 1999 Eurocodes are involved, structural appraisal of existing construction, developing the design of repairs and alterations or in assessing changes of use.

As a fully operative material-independent key code, EN 1990 is new to the European design engineer. It needs to be fully understood as it is a key code to designing structures that have an acceptable level of safety and economy, with opportunities for innovation.

As mentioned above and shown in Figure 6.1, for the design of new structures, EN 1990 will be used together with:

- EN 1991 (Eurocode 1: *Actions on structures*);
- EN 1992 to EN 1999 (design Eurocodes 2–9).

Gulvanessian et al. (2012) provide a comprehensive description, background and commentary to EN 1990 Eurocode.

6.2.2 Terms and Definitions

To assist familiarity, the followings are definitions of some important Eurocode terminology, symbols and subscripts, which will be used throughout Chapter 6.

Common terms and special terms relating to design in general:

(1) *Structure*: organised combination of connected parts designed to carry loads and provide adequate rigidity.

(2) *Structural member*: physically distinguishable part of a structure, for example a column, a beam, a foundation pile.

(3) *Structural system*: load-bearing members of a building or civil engineering works and the way in which these members function together.

(4) *Structural model*: idealisation of the structural system used for the purposes of analysis, design and verification.

(5) *Design criteria*: quantitative formulations that describe for each limit state the conditions to be fulfilled.

(6) *Design situations*: sets of physical conditions representing the real conditions occurring during a certain time interval for which the design will demonstrate that relevant limit states are not exceeded.

(7) *Persistent design situation*: design situation that is relevant during a period of the same order as the design working life of the structure.

(8) *Transient design situation*: design situation that is relevant during a period much shorter than the design working life of the structure and which has a high probability of occurrence.

(9) *Seismic design situation*: design situation involving exceptional conditions of the structure when subjected to a seismic event.

(10) *Accidental design situation*: design situation involving exceptional conditions of the structure or its exposure, including fire, explosion, impact or local failure.

(11) *Fire design*: design of a structure to fulfill the required performance in case of fire.

(12) *Hazard*: for the purpose of EN 1991 to EN 1999, an unusual and severe event, for example an abnormal action or environmental influence, insufficient strength or resistance, or excessive deviation from intended dimensions.

(13) *Design working life*: assumed period for which a structure or part of it is to be used for its intended purpose with anticipated maintenance but without major repair being necessary.
(14) *Limit states*: states beyond which the structure no longer fulfills the relevant design criteria.
(15) *Ultimate limit states*: states associated with collapse or with other similar forms of structural failure.
(16) *Serviceability limit states*: states that correspond to conditions beyond which specified service requirements for a structure or structural member are no longer met.
(17) *Reversible serviceability limit states*: serviceability limit states where no consequences of actions exceeding the specified service requirements will remain when the actions are removed.
(18) *Irreversible serviceability limit states*: serviceability limit states where some consequences of actions exceeding the specified service requirements will remain when the actions are removed.
(19) *Serviceability criterion*: design criterion for a serviceability limit state.
(20) *Resistance* (R): capacity of a member or component, or a cross-section of a member or component of a structure, to withstand actions without mechanical failure, for example bending resistance, buckling resistance, tension resistance
(21) *Strength*: mechanical property of a material indicating its ability to resist actions, usually given in units of stress.
(22) *Basic variable*: part of a specified set of variables representing physical quantities which characterise actions and environmental influences, geometrical quantities and material properties including soil properties.
(23) *Nominal value*: value fixed on non-statistical bases, for instance on acquired experience or on physical conditions.
(24) *Characteristic value*: a value that may be derived statistically with a probability of not being exceeded during a reference period. The value corresponds to a specified fractile (that point below which a stated fraction of the values lie) for a particular property of material or product. The characteristic values are denoted by subscript "k" (e.g. Q_k etc.). It is the principal representative value from which other representative values may be derived.
(25) *Representative value*: value used for verification of a limit state. It may be the characteristic value or an accompanying value, for example combination, frequent or quasi-permanent.
(26) *Reference period*: chosen period of time that is used as a basis for assessing statistically variable actions, and possibly for accidental actions.
(27) *Combination of actions*: set of design values used for the verification of the structural stability for a limit state under the simultaneous influence of different and statistically independent actions.

Terms relating to actions:

(28) *Action* (F): (a) set of forces (loads) applied to the structure (direct action), (b) set of imposed deformations or accelerations caused for example, by temperature changes, moisture variation, uneven settlement or earthquakes (indirect actions).
(29) *Effect of action* (E): effect of actions (or action effect) on structural members, (e.g. internal force, moment, stress, strain) or on the whole structure (e.g. deflection, rotation).
(30) *Permanent action* (G): action that is likely to act throughout a given reference period and for which the variation in magnitude with time is negligible, or for which the variation is always in the same direction (monotonic) until the action attains a certain limit value.
(31) *Variable action* (Q): action for which the variation in magnitude with time is neither negligible nor monotonic.
(32) *Accidental action* (A): action, usually of short duration but of significant magnitude; that is unlikely to occur on a given structure during the design working life.
(33) *Seismic action* (A_E): action that arises due to earthquake ground motion.
(34) *Fixed action*: action that have a fixed distribution and position over the structure or structural member such that the magnitude and direction of the action are determined unambiguously for the whole structure or structural member if this magnitude and direction are determined at one point on the structure or structural member.

(35) *Free action*: action that may have various spatial distributions over the structure.

(36) *Single action*: action that can be assumed to be statically independent in time and space of any other action.

(37) *Static action*: action that does not cause significant acceleration of the structure or structural members.

(38) *Dynamic action*: action that cause significant acceleration of the structure or structural members.

(39) *Quasi-static action*: dynamic action represented by an equivalent static action in a static model.

(40) *Characteristic value of an action* (F_k): principal representative value of an action (see definition 24.).

(41) *Combination value of a variable action* ($\Psi_0 Q_k$): value chosen – in so far as it can be fixed on statistical bases – so that the probability of the effects caused by the combination is approximately the same as by the characteristic value of an individual action. It may be expressed as a determined part of the characteristic value by using a factor $\Psi_0 \leq 1$.

(42) *Frequent value of a variable action* ($\Psi_1 Q_k$): value determined – in so far as it can be fixed on statistical bases – so that either the total time, within the reference period, during which it is exceeded is only a small given part of the reference period, or the frequency of it being exceeded is limited to a given value. It may be expressed as a determined part of the characteristic value by using a factor $\Psi_1 \leq 1$.

(43) *Quasi-permanent value of a variable action* ($\Psi_2 Q_k$): value determined so that the total period of time for which it will be exceeded is a large fraction of the reference period. It may be expressed as a determined part of the characteristic value by using a factor $\Psi_2 \leq 1$.

(44) *Accompanying value of a variable action* (ΨQ_k): value of a variable action that accompanies the leading action in a combination. It may be the combination value, the frequent value or the quasi-permanent value.

(45) *Representative value of an action* (F_{rep}): value used for the verification of a limit state. A representative value may be the characteristic value (F_k) or an accompanying value ΨF_k).

(46) *Design value of an action* (F_d): value obtained by multiplying the representative value by the partial factor γ_f.

(47) *Load arrangement*: identification of the position, magnitude and direction of a free action.

Terms relating to material and product properties:

(48) *Characteristic value (X_k or R_k)*: value of a material or product property having a prescribed probability of not being attained in a hypothetical unlimited test series. This value generally corresponds to a specified fractile of the assumed statistical distribution of the particular property of the material or product. A nominal value is used as the characteristic value in some circumstances.

(49) *Design value of a material or product property (X_d or R_d)*: value obtained by dividing the characteristic value by a partial factor γ_m, or γ_M or, in special circumstances, by direct determination.

(50) *Nominal value of a material or product property (X_{nom} or R_{nom})*: value normally used as a characteristic value and established from an appropriate document such as a European Standard or Prestandard.

Terms relating to geometrical data:

(51) *Characteristic value of a geometrical property* (α_k): value usually corresponding to the dimensions specified in the design. Where relevant, values of geometrical quantities may correspond to some prescribed fractile of the statistical distribution.

(52) *Design value of a geometrical property* (α_d): generally a nominal value. Where relevant, values of geometrical quantities may correspond to some prescribed fractile of the statistical distribution.

Note: The design value of a geometrical property is generally equal to the characteristic value. However, it may be treated differently in cases where the limit state under consideration is very sensitive to the value of the geometrical property, for example when considering the effect of geometrical imperfections on buckling. In such cases, the design value will normally be established as a value specified directly, for example in an appropriate European Standard or Prestandard.

Alternatively, it can be established from a statistical basis, with a value corresponding to a more appropriate fractile (e.g. a rarer value) than applies to the characteristic value.

Symbols and subscripts
So far as we may notice, most of the significant symbols and subscripts have been presented. The following are some others:

A_{Ed} – Design value of seismic action $A_{Ed} = \gamma_I A_{Ed}$.
A_{Ek} – Characteristic value of seismic action.
C_d – Nominal value or a function of certain properties of materials.
$E_{d,dst}$ – Design value of effect of destabilising section.
$E_{d,stb}$ – Design value of effect of stabilising section.
F_w – Wind force (general symbol).
F_{wk} – Characteristic value of the Wind force.
P – Relevant representative value of a prestressing action.
Q_{Sn} – Characteristic value of snow load.
T – Thermal climatic action (general symbol).
X – Material property.
a_d – Design values of geometrical data.
a_d – Characteristic values of geometrical data.
a_{nom} –Nominal value of geometrical data.
d_{set} – Difference in settlement of an individual foundation or part of a foundation compared to a reference level.
u – Horizontal displacement of a structure or structural member.
w – Vertical deflection of a structural member.
γ – Partial factor (safety or serviceability).
γ_G – Partial factor for permanent action.
γ_Q – Partial factor for variable action.
γ_M – Partial factor for a material property, also accounting for model uncertainties and dimensional variations.
γ_F – Partial factor for actions, also accounting for model uncertainties and dimensional variations.
$\gamma_{G,set}$ – Partial factor for permanent actions due to settlements, also accounting for model uncertainties.
ψ – Factor for converting a characteristic value to a representative value.
ψ_0 – Factor for combination value of a variable action.
ψ_1 – Factor for frequent value of a variable action.
ψ_2 – Factor for quasi-permanent value of a variable action.
ξ – Reduction factor for permanent action.

Note: Other symbols will be defined when they appear for the first time.

6.2.3 Requirements

The followings are the requirements that must be adhered to by all the Eurocode suite and construction product standards so that safety, serviceability and durability of structures are achieved in a proper manner:

- Basic requirements,
- Reliability management and differentiation,
- Design working life,
- Durability.

These requirements are thoroughly covered by EN 1990 Section 2, and may be summarised as in the following paragraphs.

Basic requirements

According to EN 1990 Section 2, Clause 2.1, the basic requirements stipulate that:

A structure and structural members should be designed and executed in such a way that it will, during its intended life, with appropriate degrees of reliability and in an economic way:

- Sustain all actions and influences likely to occur during execution and use (safety requirement relating to the ultimate limit state); and remain fit for use under all expected conditions (serviceability requirement relating to the serviceability limit state).
- Have satisfactory structural resistance, serviceability and durability.
- In the case of fire, provide satisfactory structural resistance for the required period of time.
- Not be damaged by events such as explosions, impact or consequences of human errors, to an extent disproportionate to the original cause (robustness requirement).

EN 1990 Section 2, Clause 2.1 provides methods of avoiding or limiting potential damage. Also, it provides methods of achieving basic requirements satisfied.

Reliability management and differentiation

EN 1990 Section 2, Clause 2.2 provides: methods of achieving and adopting reliability, factors to be considered in selecting reliability levels, methods of specifying reliability levels that apply to a particular structure, measures of achieving reliability levels relating to structural resistance and serviceability.

Reliability is ability of a structure or a structural member to fulfill the specified requirements, including the design working life, for which it has been designed. It covers safety, serviceability and durability of a structure or a structural member. Reliability is usually expressed in probabilistic terms. Design and execution according to the suite of the Eurocodes, together with appropriative quality control measures, will ensure an appropriate degree of reliability for the majority of structures.

Reliability differentiation means measures intended for the socio-economic optimisation of the resources to be used to build construction works, taking into account all the expected consequences of failures and the cost of construction works. It is a concept which is not covered by BSI codes. EN 1990 provides guidance for adopting reliability differentiation. It gives further guidance in an Informative Annex "Management of Structural Reliability for Construction Works" (2005). Calgaro et al. (2001) describe the management of structural reliability in EN 1990, where the concept of the risk background of the Eurocodes is described more comprehensively.

Design working life

The design working life should be specified. Table 6.1, taken from the UK National Annex for EN 1990, gives indicative design working life classifications. BSI codes do not have a design working life requirement for buildings. The given values may also be used for time-dependent performance (e.g. fatigue-related calculations).

Buildings subject to Building Regulations, hospitals, schools and so on will be in Category 4 (Table 6.1).

The design working life requirement is useful for:

- The selection of design actions,
- Consideration of material property deterioration,
- Life cycle costing,
- Evolving maintenance strategies.

Durability

The durability of a structure is its ability to remain fit for use during the design working life given appropriate maintenance. EN 1990 stipulates that the structure needs to be designed so that deterioration over its design working life does not impair the performance of the structure. The Factors that should be considered to ensure adequately durable structures are listed in EN 1990 Section 2, Clause 2.4.

Table 6.1 Design working life classification.

Design working life category	Indicative design working life (years)	Examples
1	10	Temporary structures[a]
2	10 to 25	Replaceable structural parts, e.g. gantry girders, bearings
3	15 to 30	Agricultural and similar structures
4	50	Building structures and other common structures
5	100	Monumental Building structures, bridges, and other civil engineering structures

[a] Structures or parts of structures that can be dismantled with a view to being reused should not be considered as temporary.

6.2.4 Quality Management

EN 1990 Section 2, Clause 2.5 states: In order to provide a structure that corresponds to the requirements and to the assumptions made in the design, appropriate quality management measures should be in place. These measures comprise:

- Definition of the reliability requirements,
- Organisational measures,
- Controls at the stages of design, execution, use and maintenance.

6.2.5 Principles of Limit States Design

EN 1990 considers two different types of limit state, namely *ultimate limit state* (ULS) and *serviceability limit state* (SLS). These limit states shall be related to design situations. Verification of one of these two limit states may be omitted provided that sufficient information is available to prove that it is satisfied by the other. Verification of limit states that are concerned with time dependent effects (e.g. fatigue) should be related to the design working life of the construction. Limit state concept is used in conjunction with the partial safety factor method.

EN 1990 stipulates that the structure needs to be designed so that no limit state is exceeded when relevant design values for actions, material and product properties, and geometrical data are used. This is satisfied by the *partial factor method.* In this method the basic variables (i.e. actions, resistances and geometrical properties) are given design values through the use of partial factors, γ, and reduction coefficients, ψ, of their characteristic values.

Design situations

EN 1990 states: The relevant design situations shall be selected taking into account the circumstances under which the structure is required to fulfill its function.

The selected design situations shall be sufficiently severe and varied so as to encompass all conditions that can reasonably be foreseen to occur during the execution and use of the structure.

EN 1990 considers persistent, transient, accidental and seismic design situations related to the ultimate limit states verification; their definitions have been presented in Section 6.2.2.

The design situations related to the serviceability limit state concern:

- The functioning of the structure or structural members under normal use,
- The comfort of people,
- The appearance of the construction works.

EN 1990 requires that verification of serviceability limit states should be based on criteria concerning the following aspects:

(a) Deformations that affect
- the appearance,
- the comfort of users, or
- the functioning of the structure (including the functioning of machines or services), or that cause damage to finishes or non- structural members;

(b) Vibrations
- that cause discomfort to people, or
- that limit the functional effectiveness of the structure;

(c) Damage that is likely to adversely affect
- the appearance,
- the durability, or
- the functioning of the structure.

Actions

General definitions of different actions have been given in Section 6.2.2. Actions are classified by their variation in time as follows:

- Permanent actions, G, for example self-weight of structures, fixed equipment and road surfacing, and indirect actions caused by shrinkage and uneven settlements;
- Variable actions, Q, for example imposed loads on building floors, beams and roofs, and wind actions or snow loads;
- Accidental actions, A, for example explosions or impact from vehicle;
- Seismic action, A_E, action that arises due to earthquake ground motion.

For each variable action there are four representative values. The principal representative value is the *characteristic* value, Q_k, and this can be determined statistically or, where there is insufficient data, a nominal value may be used. The other representative values are *combination*, *frequent* and *quasi- permanent;* these are obtained by applying to the characteristic value the factors ψ_0, ψ_1 and ψ_2 respectively (Figure 6.3).Table 6.2 gives the representative values of the different actions mentioned above.

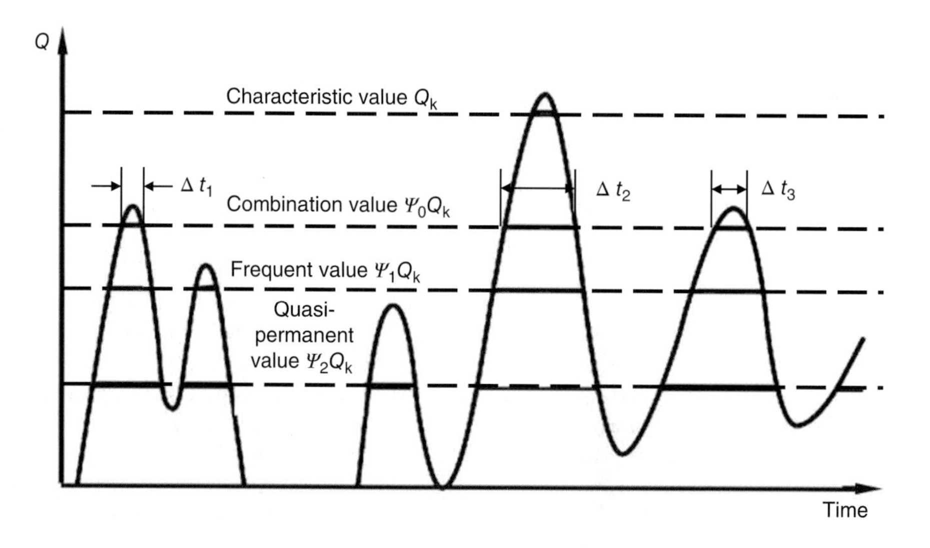

Figure 6.3 Representative values of variable actions.

Table 6.2 Representative values of actions.

	Action			
Value	**Permanent**	**Variable**	**Accidental**	**Seismic**
Character	G_k	Q_k	---	A_{Ek}, or
Nominal	---	---	A_d	$A_{Ed} = \gamma_I A_{Ek}$
Combination	---	$\psi_0 Q_k$	---	---
Frequent	---	$\psi_1 Q_k$	---	---
Quasi-permanent	---	$\psi_2 Q_k$	---	---

Table 6.3 Recommended values of factors for buildings (from UK National Annex A to EN 1990).

Action	ψ_0	ψ_1	ψ_2
Imposed loads in buildings (see BS EN 1991 – 1 – 1)			
• Category A: domestic, residential areas	0.7	0.5	0.3
• Category B: office areas	0.7	0.5	0.3
• Category C: congregation areas	0.7	0.7	0.6
• Category D: shopping areas	0.7	0.7	0.6
• Category E: storage areas	1.0	0.9	0.8
• Category F: traffic area, vehicle weight < 30 kN	0.7	0.7	0.6
• Category G: traffic area, 30 kN < vehicle weight < 160 kN	0.7	0.5	0.3
• Category H: roofs [a]	0.7	0.0	0.0
Snow loads on buildings (see BS EN 1991–3)			
• For sites located at altitude H > 1000 m above see level	0.7	0.5	0.2
• For sites located at altitude H < 1000 m above see level	0.5	0.2	0.0
Wind loads on buildings (see BS EN 1991 – 1 – 4)	0.5[b]	0.2	0.0
Temperature (non – fire) in buildings (see BS EN 1991 – 1 – 5)	0.6	0.5	0.0

[a] See also EN 1991-1-1: Clause 3.3.2 where $\psi_0 = 0$.
[b] 0.6 in the EN 1991-1-1

A semi-probabilistic method is used to derive the ψ factors, which vary depending on the type of imposed load (see Table 6.3). Further information on derivation of the ψ factors can be found in Appendix C of the Eurocode.

The combination value ($\psi_0 Q_k$) of an action is associated with the combination of actions for ultimate and irreversible serviceability limit states (e.g. functionality of fittings with brittle behaviour) in order to take account of the reduced probability of simultaneous occurrence of the most unfavourable values of two or more independent actions (i.e. applied to the characteristic value of all accompanying actions).

The frequent value ($\psi_1 Q_k$) is primarily associated with the frequent combination in the serviceability limit states, but it is also used for verification of the accidental design situation of the ultimate limit states (e.g. every day office use). In both cases, the factor ψ_1 is applied as a multiplier of the leading variable action. In accordance with EN 1990, the frequent value $\psi_1 Q_k$ of a variable action Q is determined so that the total time, within a chosen period of time, during which $Q > \psi_1 Q_k$ is only a specified (small) part of the period, or the frequency of the event $Q > \psi_1 Q_k$ is limited to a given value. The total time for which $\psi_1 Q_k$ is exceeded is equal to sum of time periods, shown in Figure 6.3 by continuous sections of the horizontal line that belongs to the frequent value $\psi_1 Q_k$.

The quasi-permanent values ($\psi_2 Q_k$) are mainly used in the assessment of long-term effects, (e.g. cosmetic cracking of a slab). They are also used for the representation of variable actions in accidental and seismic combinations of actions (ultimate limit states) and for verification of frequent and quasi-permanent combinations (long-term effects) of serviceability limit states. In accordance with EN 1990, the quasi-permanent values ($\psi_2 Q_k$) is defined so that the total time, within a chosen period during which it is exceeded, that is when $Q > \psi_2 Q_k$ is a considerable part (0.5) of the chosen period of time. The value may also be determined as the value averaged over the chosen period of time. The total time of $\psi_2 Q_k$ being exceeded is equal to the sum of periods, shown in Figure 6.3 by continuous sections of the horizontal line that belongs to the frequent value $\psi_2 Q_k$.

The representative values $\psi_0 Q_k$, $\psi_1 Q_k$ and $\psi_2 Q_k$ and the characteristic values are used to define the design values of the actions and the combinations of the actions as explained in the following paragraphs.

Combinations of actions

The term "combinations of actions", as defined in Section 6.2.2, should not be confused with "load cases", which are concerned with the arrangement of the variable actions to give the most unfavourable conditions.

The following process can be used to determine the value of actions used for analysis:

(1) Identify the design situation (e.g. persistent, transient, etc.).
(2) Identify all realistic actions.
(3) Determine the partial factors for each applicable combination of actions.
(4) Arrange the actions to produce the most critical conditions.

Where there is only one variable action (e.g. imposed load) in a combination, the magnitude of the actions can be obtained by multiplying them by the appropriate partial factors. Where there is more than one variable action in a combination, it is necessary to identify the leading action ($Q_{k,1}$) and other accompanying actions ($Q_{k,i}$).

For the persistent and transient design situations for ultimate limit states and for the characteristic (rare) combinations of serviceability limit states, only the non-leading variable actions may be reduced using the ψ_0 factors. In other cases (for accidental design situation and combinations of serviceability limit states), the leading as well as accompanying actions may be reduced using the appropriate ψ factors (see Table 6.4).

Verification by the partial factor method

(A) *Ultimate limit states*

(a) *Ultimate limit state categories and verifications.*

EN 1990 stipulates that the effects of design actions do not exceed the design resistance of the structure at the ultimate limit state (ULS); and the following four categories to which ULS are divided needs to be verified, where relevant:

Table 6.4 Application of factors ψ_0, ψ_1 and ψ_2 for leading and non-leading variable actions at ultimate and serviceability limit states.

Limit state	Design situation or Combination	Combination value ψ_0	Frequent value ψ_1	Qusi-permanent value ψ_2
Ultimate	Persistent and transient	non-leading	×	×
	Accidental	×	leading	leading and non-leading
	Seismic	×	×	all variable actions
Serviceability	Characteristic	non-leading	×	×
	Frequent	×	leading	non-leading
× means not applied	Quasi-permanent	×	×	all variable actions

(1) EQU. Loss of static equilibrium of the structure or any part of it considered as a rigid body, where:

- Minor variations in the value or the spatial distribution of actions from a single source are significant;
- The strengths of construction materials or ground are generally not governing.

The verification shall be based on:

$$E_{d,dst} \le E_{d,stb} \tag{6.1}$$

where:
$E_{d,dst}$ = design value of the effect of destabilizing action,
$E_{d,stb}$ = design value of the effect of stabilizing actions.

(2) STR. Internal failure or excessive deformation of the structure or structural members, including footings, piles and basement walls and so on, where the strength of construction materials of the structure governs.
The verification shall be based on:

$$E_d \le R_d \tag{6.2}$$

where:
E_d = design value of the effect of actions
R_d = design value of the corresponding resistance.

(3) GEO. Failure or excessive deformation of the ground where the strengths of soil or rock are significant in providing resistant.
The verification shall be based on Equation (6.2).

(4) FAT. Fatigue failure of the structure or structural members. The combinations apply:

- Persistent or transient design situation (fundamental combination);
- Accidental design situation;
- Seismic design situation.

Specific rules for FAT limit states are given in the design Eurocodes EN 1992 to EN 1999.

(b) *Load combination expressions in EN 1990 for the verification of ULS for the persistent and transient design situations.*
EN 1990 Clause 6.4.3.2 defines three alternative sets of expressions for the verification of ultimate limit states for the *persistent* and *transient* design situations as follows:

(1) EN 1990 Expression (6.10)

$$\sum\nolimits_{j\ge 1}\gamma_{G,j}G_{k,j}\text{"} + \text{"}\gamma_P P\text{"} + \text{"}\gamma_{Q,1}Q_{k,1}\text{"} + \text{"}\sum\nolimits_{i>1}\gamma_{Q,i}\psi_{0,i}Q_{k,i}$$

(2) Less favourable of EN 1990 Expressions (6.10a) and (6.10b)
EN 1990 Expressions (6.10a)

$$\sum_{j\geq 1}\gamma_{G,j}G_{k,j}\text{"}+\text{"}\gamma_P P\text{"}+\text{"}\gamma_{Q,1}\psi_{0,1}Q_{k,1}\text{"}+\text{"}\sum_{i>1}\gamma_{Q,i}\psi_{0,i}Q_{k,i}$$

EN 1990 Expressions (6.10b)

$$\sum_{j\geq 1}\xi_j\gamma_{G,j}G_{k,j}\text{"}+\text{"}\gamma_P P\text{"}+\text{"}\gamma_{Q,1}Q_{k,1}\text{"}+\text{"}\sum_{i>1}\gamma_{Q,i}\psi_{0,i}Q_{k,i}$$

(3) Less favourable of EN 1990 Expression (6.10a, modified) to include permanent actions (i.e. self-weight) only and EN 1990 Expression (6.10b).
EN 1990 Expression (6.10a, modified)

$$\sum_{j\geq 1}\gamma_{G,j}G_{k,j}\text{"}+\text{"}\gamma_P P\text{"}$$

EN 1990 Expressions (6.10b)

$$\sum_{j\geq 1}\xi_j\gamma_{G,j}G_{k,j}\text{"}+\text{"}\gamma_P P\text{"}+\text{"}\gamma_{Q,1}Q_{k,1}\text{"}+\text{"}\sum_{i>1}\gamma_{Q,i}\psi_{0,i}Q_{k,i}$$

In the preceding Expressions:

+implies "to be combined with"
Σ implies "the combined effect of"
ξ is a reduction factor for unfavourable permanent actions G

In Expression (6.10) the combination of actions is governed by a leading variable action $Q_{k,1}$ represented by its characteristic value and multiplied by its appropriate safety factor γ_Q. Other variable action $Q_{k,i}$ for $i > 1$ which may act simultaneously with the leading variable action $Q_{k,1}$ are taken into account as accompanying variable actions and are represented by their combination value, that is their characteristic value reduced by the relevant combination factor ψ_0, and are multiplied by the appropriate safety factor to obtain the design values. The permanent actions are taken into account with their characteristic values, and are multiplied by the load factor γ_G. Depending on whether the permanent actions are favourably or unfavourably they have different design values.
In Expression (6.10a) all the variable actions are taken into account with their combination value, that is there is no leading variable action. The permanent actions are taken into account as in Expression (6.10). All the actions are multiplied by the appropriate safety factors, γ_G or γ_Q.

In Expression (6.10b) the combination of actions is governed by a leading variable action represented by its characteristic value as in Expression (6.10) with the other variable actions being taken into account as accompanying variable actions and are represented by their combination value, that is their characteristic value is reduced by the appropriate combination factor of a variable action ψ_0. The unfavourable permanent actions are taken into account with a characteristic value reduced by a reduction factor, ξ, which may be considered as a combination factor. All the actions are multiplied by the appropriate factors, γ_G or γ_Q, as mentioned before.

The third Expression set is very similar to the second set except that the Expression (6.10a, modified) includes only permanent actions.

Using Expression (6.10), the following comparison between BSI structural code and EN 1990 with regard to combination of the effects of actions may be made:

For one variable action (imposed or wind)

- BSI $1.4G_k + (1.4 \text{ or } 1.6)\,Q_k$
- EN 1990 $1.35G_k + 1.5\,Q_k$

For two or more variable actions (imposed + wind)

- BSI $1.2G_k + 1.2\,Q_{k,1} + 1.2\,Q_{k,2}$
- EN 1990 $1.35G_k + 1.5\,Q_{k,1} + 0.75\,Q_{k,2}$

According to Gulvanessian et al. (2012), the following combinations have been adopted in the UK national annex for EN 1990 for buildings:

- Expression (6.10) using $\gamma_G = 1.35$ and $\gamma_Q = 1.5$
- Expression (6.10a) and (6.10b) using $\gamma_G = 1.35$, $\gamma_Q = 1.5$ and $\xi = 0.925$

The recommended values for all γ and ψ factors (except ψ_0 for wind actions, where in the UK national annex $\psi_0 = 0.5$) have been adopted by the UK national annex for EN 1990, and are generally being adopted by most CEN Member States.

For UK buildings, Expression (6.10) is always equal to or more conservative than the less favourable of Expressions (6.10a) and (6.10b). Expression (6.10b) will normally apply when the permanent actions are not greater than 4.5 times the variable actions [except for storage imposed load (category E, Table 6.3) where Expression (6.10a) always applies]. Therefore, for a typical concrete frame building, Expression (6.10b) will give the most structurally economical combination of actions.

For bridges only the use of Expression (6.10) is permitted.

(c) *Load combination expressions in EN 1990 for the verification of ULS for the accidental design situation.*

The expressions for the accidental design situation given in EN 1990 basically use the same concept as BSI codes for the accidental action but accompanying loads are treated as in (b) above.

EN 1990 Clause 6.4.3.3 requires the following combination expression to be investigated: EN 1990 Expression (6.11b)

$$\sum_{j\geq 1} G_{k,j}\text{"} + \text{"}P\text{"} + \text{"}A_d\text{"} + \text{"}(\psi_{1,1} \text{ or } \psi_{2,1})Q_{k,1}\text{"} + \text{"}\sum_{i>1} \psi_{2,i}Q_{k,i}$$

The choice between $\psi_{1,1}Q_{k,1}$ or $\psi_{2,1}Q_{k,1}$ should be related to the relevant accidental design situation (impact, fire or survival after an accidental event or situation). In the UK national annex to EN 1990, $\psi_{1,1}Q_{k,1}$ is chosen.

The combinations of actions for accidental design situations should either (i) involve explicit an accidental actions A (fire or impact), or (ii) refer to a situation after an accidental event $(A = 0)$. For fire situations, apart from the temperature effect on the material properties, A_d should represent the design value of the direct thermal action due to fire.

The expression for the accidental design situation specifies factor of safety of unity both for the self-weight and the accidental action A_d and a frequent or quasi-permanent value for the leading variable action. The philosophy behind this is the recognition that an accident on a building or construction works is a very rare event (although when it does occur the consequences may be severing) and hence EN 1990 provides an economic solution.

(d) *Load combination expressions in EN 1990 for the verification of ULS for the seismic design situation.*

EN 1990 Clause 6.4.3.4 requires the following combination expression to be investigated:

EN 1990 Expression (6.12b)

$$\sum_{j\geq 1} G_{k,j}\text{"} + \text{"}P\text{"} + \text{"}A_{Ed}\text{"} + \text{"}\sum_{i\geq 1} \psi_{2,i}Q_{k,i}$$

Notes:

(1) The values of γ and ψ factors for actions should be obtained from EN 1991 and from Annex A.

(2) The partial factors for properties of materials and products should be obtained from EN 1992 to EN 1999.

(B) *Serviceability limit states*

As mentioned earlier, EN 1990 gives guidance on the following serviceability limit state (SLS) verifications:

- The functioning of the structure or structural members under normal use,
- The comfort of people,
- The appearance of the construction works.

This is different from the concept of BSI codes.

In the verification of serviceability limit states in EN 1990, separate load combination expressions are used depending on the design situation being considered. For each of the particular design situation an appropriate representative value for an action is used.

For the SLS verification, EN 1990 Clause 6.5.1 stipulates that

$$E_d \le C_d \tag{6.3}$$

where:

C_d = limmiting design value of the relevant serviceability criterion

E_d = design value of the effects of actions specified in the serviceability criterion, determined on the basis of the relevant combination.

EN 1990 Clause 6.5.3 requires that the combinations of actions to be taken into account in the relevant design situations should be appropriate for the serviceability requirements and performance criteria being verified.

For the serviceability limit states verification, EN 1990 gives three expressions, namely, *characteristic, frequent* and *quasi-permanent.* Care should be taken not to confuse these SLS three expressions with the representative values that have the same titles. For design of concrete structures, EN 1992 indicates which combination should be used for which phenomenon (e.g. deflection is checked using the quasi-permanent combination). It is assumed, in the three expressions, that all partial factors are equal to 1. See Annex A and EN 1991 to EN 1999. Also, for SLS the partial factors γ_M for the properties of materials should be taken as 1 except if differently specified in EN 1992 to EN 1999.The Eurocode three expressions are:

(1) EN 1990 Expression (6.14b): the characteristic (rare) combination is used mainly in those cases when exceedance of a limit state causes a permanent (irreversible) local damage or permanent unacceptable deformation.

$$\sum_{j\ge 1} G_{k,j}\text{"} + \text{"}P\text{"} + \text{"}Q_{k,1}\text{"} + \text{"}\sum_{i>1} \psi_{0,i} Q_{k,i}$$

(2) EN 1990 Expression (6.15b): the frequent combination is used mainly in those cases when exceedance of a limit state causes local damage, large deformation or vibrations which are temporary (reversible).

$$\sum_{j\ge 1} G_{k,j}\text{"} + \text{"}P\text{"} + \text{"}\psi_{1,1} Q_{k,1}\text{"} + \text{"}\sum_{i>1} \psi_{2,i} Q_{k,i}$$

(3) EN 1990 Expression (6.16b): the quasi permanent combination is used mainly when long term effects are of importance.

$$\sum_{j\ge 1} G_{k,j}\text{"} + \text{"}P\text{"} + \text{"}\sum_{i>1} \psi_{2,i} Q_{k,i}$$

6.3 Design of Spread Foundations

6.3.1 Introduction

All foundations should be designed so that the underlying materials(soil and/ or rock) safely resist the actions applied to the structure. The design of any foundation consists of two components; the *geotechnical design* and the *structural design* of the foundation itself. However, for some foundations (e.g. flexible mats or rafts) the effect of the interaction between the soil and structure (i.e. soil-structure interaction) may be critical and must also be considered.

The design of "Spread foundations" is covered by EN 1997 (Eurocode 7). There are two parts to EN 1997, Part 1 (or EN 1997-1): *General rules* and Part 2 (or EN 1997-2): *Ground investigation and testing.* Section 6 of EN 1997-1 "Spread foundations" applies to pad (e.g. footing of a column), strip (e.g. footing of a wall or a long pedestal) and raft foundations, and some provisions may be applied to deep foundations, such as caissons. Section 9 of EN 1997-1 "Retaining structures" applies to gravity walls (walls of stone or plain or reinforced concrete), embedded walls and composite retaining structures (see Subsections 9.1.2.1, 9.1.2.2 and 9.1.2.3 of EN 1997-1).

EN 1997-1 §6.8 gives principles and application rules related to "Structural design of spread foundations". However, it provides no guidance on the procedures for assessing the required amount or detailing of reinforcement in the concrete – this is dealt with by EN 1992.

National annex for EN 1997-1 gives alternative procedures and recommended values with notes indicating where national choices may have to be made. Therefore the National Standard implementing EN 1997-1 should have a National annex containing all Nationally Determined Parameters (NDP) to be used for the design of buildings and civil engineering works to be constructed in the relevant country.

The Eurocodes adopt, for all civil and building engineering materials and structures, a common design philosophy based on the use of separate limit states and partial factors, rather than global safety factors; this is a significant change in the traditional geotechnical design practice as embodied in BS Codes such as the superseded BS 8004. Moreover, EN 1997-1 provides one, unified methodology for all geotechnical design problems. An advantage of EN 1997-1 is that its design methodology is largely identical with that for all of the structural Eurocodes, making the integration of geotechnical design with structural design more rational.

6.3.2 Geotechnical Categories

EN 1997-1 Section 2.1 introduces three Geotechnical Categories to assist in establishing the geotechnical design requirements for a structure, as shown in Table 6.5.

It is expected that structural engineers will take responsibility for the geotechnical design of category 1 structures, and that geotechnical engineers will take responsibility for category 3 structures. The geotechnical design of category 2 structures may be undertaken by members of either profession; this decision will very much depend on individual circumstances.

Geotechnical design differs from design in other structural materials in that both the design actions and the design resistances are functions of the effect of the actions, material properties and dimensions of the problem. For example, when assessing sliding in a retaining wall design the horizontal forces due

Table 6.5 Geotechnical categories of structures.

Category	Description	Risk of geotechnical failure	Examples from EN 1997
1	Small and relatively simple structures	Negligible	None given
2	Conventional types of structure and foundation with no difficult ground or loading conditions	No exceptional risk	Spread foundation
3	All other structures	Abnormal risks	Large or unusual structures Exceptional ground conditions

to the earth (design effect of the actions) are derived from the material properties of the soil and the dimensions of the problem; similarly, the resistance to sliding is also derived from the actions, material properties and the dimensions. Therefore, geotechnical calculations have a greater level of complexity than design in other structural materials where actions are not normally a function of the material properties and resistances not a function of the actions.

6.3.3 Limit States

The following limit states should be satisfied for geotechnical design:

(a) *Ultimate limit states*
Ultimate limit states (ULS) are those that will lead to failure of the ground and/or the associated structure.

EN 1997-1 identifies the following five ultimate limit states. They should be satisfied for geotechnical design; each has its combination of actions (For an explanation of Eurocode terminology, please refer to the Section 6.2.2 of this chapter):
EQU – Loss of static equilibrium of the structure.
STR – Internal failure or excessive deformation of the structure.
GEO – Failure or excessive deformation in the ground.
UPL – Loss of equilibrium or excessive deformation due to uplift.
HYD – Failure due to hydraulic heave, piping and erosion.

(b) *Serviceability limit states*
Serviceability limit states (SLS) are those that result in unacceptable levels of deformation (e.g. excessive settlement or heave), vibration and noise. They should be satisfied for geotechnical design.

The designer of a geotechnical structure tries first to identify the possible ultimate and serviceability limit states that are likely to affect the structure. Usually, it will be clear that one of the limit states will govern the design and therefore it will not be necessary to carry out verifications for all of them, although it is good practice to record that they have all been considered. Verifications of serviceability limit states is a key requirement of a EN 1997 design, which does not explicitly feature in more traditional approaches. In traditional design it is assumed in many calculation procedures that the lumped safety factor provides not only safety against failure but also will limit deformations to tolerable levels. For example, it is common to adopt safety factors against bearing capacity failure in the foundation soil in excess of 3.0. This is far greater than needed to ensure sufficient reserve against failure. EN 1997 generally requires a specific verification that serviceability limits are met and thus serviceability requirements are more likely to be the governing factor for many settlement sensitive projects.

Traditional geotechnical design methods, adopting lumped safety factors, have proved satisfactory over many decades and much experience has been built on such methods. However, the use of a single factor to account for all uncertainties in the analysis does not provide an adequate control of different levels of doubt in various parts of the design process. The limit state approach forces designers to think more rigorously about possible modes of failure and those parts of the calculation process where there is most uncertainty. The partial factors in EN 1997 have been chosen to give similar designs to those obtained using lumped factors; thus, ensuring that the wealth of previous experience is not lost by the introduction of a radically different design methodology. The Eurocodes present a unified approach to all structural materials and should lead to less confusion and fewer errors when considering soil–structure interaction.

EN 1997-1 §6.2 lists eight different types of limit states (repeated below for convenience) that should be considered in geotechnical design of spread foundations including pads, strips and rafts. The limit states are:

- Loss of overall stability;
- Bearing resistance failure, punching failure, squeezing;
- Failure by sliding;
- Combined failure in the ground and in the structure;
- Structural failure due to foundation movement;
- Excessive settlements;
- Excessive heave due to swelling, frost and other causes;
- Unacceptable vibrations.

We may notice that the first five limit states are related to ultimate limit state design; their details are covered in EN 1997-1 §6.5. The remaining three limit states are related to serviceability limit state design; their details are covered in EN 1997-1 §6.6.

6.3.4 Geotechnical Design

(A) *Design situations*

EN 1997-1 Section 6.3 requires that design situations shall be selected in accordance with EN 1997-1 Section 2.2 which states: both short-term and long-term design situations shall be considered. This consideration is to reflect the sometimes vastly different resistances obtained from drained and undrained soils. Examples of these situations may be given as follows:

- For persistent design situation: long-term is considered where structures are founded on course soils and fully-drained fine soils; and short-term is considered where there is partially-drained fine soils (with design working life less than 25 years).
- For transient design situation: long-term is considered where there are temporary works in coarse soils; and short-term is considered where there are temporary works in fine soils
- For accidental design situation: long-term is considered where structures are founded on course soils and quick-draining fine soils; and short-term is considered where structures are founded on slow-draining fine soils.
- For seismic design situation: short-term is considered where structures are founded on slow-draining fine soils.

EN 1997-1 Section 2.2 provides a list of nine different items that the detailed specifications of design situations should include.

(B) *Actions*

EN 1997-1 Section 6.3 requires that the actions listed in EN 1997-1 Section 2.4.2(4) should be considered when selecting the limit states for calculation. This section provides a list of 20 types of actions that should be included in geotechnical design.

In geotechnical design it is necessary to identify which action is *favourable* and which action is *unfavourable*. The Eurocodes make an important distinction between favourable (or stabilising) and unfavourable(or destabilising) actions, which is reflected in the values of the partial factors γ_F applied to each type of action. Unfavourable actions are typically increased by the partial factor (i.e. $\gamma_F > 1$) and favourable actions are decreased or left unchanged (i.e. $\gamma_F \leq 1$). For example, to provide sufficient reliability against bearing capacity failure, the self-weight of a wall and its strip footing should be considered as unfavourable actions, since they increase the effective stress beneath the footing. However, the same actions should be considered as favourable for sliding, since they increase resistance.

The designer should be aware of not treating an action as both favourable and unfavourable in the *same* calculation since it is illogical to do so. EN 1997 deals with this issue in what has been known as the *Single-Source Principle*: "unfavourable (or destabilising) and favourable (or stabilising) permanent actions may in some situations be considered as coming from a single source. If they are considered so, a single partial factor may be applied to the some of these actions or to the some of their effects". [EN 1997-1 Section 2.4.2(9) P Note]. In the example just given if the water table is located above the foundation level, the upward thrust on the footing bottom owing to water pressure will be considered as favourable and the horizontal thrust on the wall as unfavourable for bearing (both thrusts coming from a single source). The Note allows these thrusts to be treated in the same way; either both unfavourable or both favourable, whichever represents the more critical design condition.

(C) *Basic requirements*
Considering the ULS, the design effect of destabilising actions $E_{d,dst}$ and design effect of actions E_d must be less than or equal to the design effect of stabilising actions $E_{d,st}$ and design resistance R_d, respectively, as indicated in Equations (6.1) and (6.2). Considering the SLS, the design effect of the actions E_d must be less than or equal to the limiting design value of the relevant serviceability criterion C_d, as indicated in Equation (6.3).

(D) *Geometrical data*
According to EN 1997-1 Section 2.4.4, the level and slope of ground surface, water levels, levels of interfaces between strata, excavation levels and the dimensions of the geotechnical structure shall be treated as geometrical data.

(E) *Design approaches and combinations*
There has not been a consensus amongst geotechnical engineers over the application of limit state principles to geotechnical design. EN 1997-1, to allow for these differences of opinion, presents three different Design Approaches (DAs) for carrying out ultimate limit state analysis for the GEO and STR limit states. The decision on which approach to use for a particular country is given in its National Annex. The design approaches are:

- DA1 – Combination 1: *A1* "+" *M1* "+" *R1* (EN 1997-1§2.4.7.3.4.2)
  – Combination 2: *A2* "+" *M2* "+" *R1*
- DA2: *A1* "+" *M1* "+" *R2* (EN 1997-1§2.4.7.3.4.3)
- DA3: (*A1* or *A2*) "+" *M2* "+" *R3* (EN 1997-1§2.4.7.3.4.4)
  "+"implies "to be combined with" (EN 1997-1§2.4.7.3.4.2)

These approaches apply the partial factor sets in different combinations in order to provide reliability in the design. The method of applying the partial factors reflects the differing opinions regarding geotechnical design held across the European Union. Actually, Design Approach 1 provides reliability by applying different partial factor sets to two variables (actions and ground strength parameters or ground resistance) in two separate calculations (Combinations 1 and 2), whereas Design Approaches 2 and 3 apply factor sets to two variables simultaneously, in a single calculation. Further clarification of the Design approaches is provided in Annex B of EN 1997-1.

In Annex A of EN 1997-1 the partial factors are grouped in sets denoted by *A* (for actions or effects of actions), M (for soil parameters) and *R* (for resistances).

In the UK, Design Approach 1 is specified in the National Annex. For this Design Approach (excluding design of axially loaded piles and anchors) there are two sets of combinations to use for the STR and GEO ultimate limit states, as indicated above. The values for the partial factors to be applied to the actions for these combinations are given in Table 6.6 and the partial factors for the geotechnical material properties are given in Table 6.7. Combination 1 will generally govern the structural resistance, and Combination 2 will generally govern the size of the foundations.

The partial factors to be applied to the geotechnical material properties and actions at the EQU limit state are given in Table 6.7 and Table 6.8, respectively.

For the SLS, EN 1997-1 does not give any advice on whether the characteristic, frequent or quasi-permanent combination should be used. Where the *prescriptive* design method is used for spread foundations (will be discussed in the following paragraphs) then the characteristic combination

Table 6.6 Design values of actions derived for UK design, STR and GEO ultimate limit states– persistent and transient design situations.

Combination Exp. reference BS EN 1990	Permanent actions		Leading variable action	Accompanying variable actions	
	Unfavorable	Favorable		Main (if any)	Others
Combination 1{Application of combination 1(BS EN 1997) to set B (BS EN 1990)}					
Exp. (6.10)	$1.35\,G_k^a$	$1.0\,G_k^a$	$1.5^b Q_k$	–	$1.5^b \psi_{0,i}^c Q_{k,i}$
Exp. (6.10a)	$1.35\,G_k^a$	$1.0\,G_k^a$	–	$1.5\,\psi_{0,1}^c Q_k$	$1.5^b \psi_{0,i}^c Q_{k,i}$
Exp. (6.10b)	$0.925^d \times 1.35\,G_k^a$	$1.0\,G_k^a$	$1.5^b Q_k$	–	$1.5^b \psi_{0,i}^c Q_{k,i}$
Combination 2{Application of combination 2(BS EN 1997) to set C (BS EN 1990)}					
Exp. (6.10)	$1.0\,G_k^a$	$1.0\,G_k^a$	$1.3^b Q_{k,1}$	–	$1.3^b \psi_{0,i}^c Q_{k,i}$

Key:

[a] Where the variation in permanent action is not considered significant $G_{k,jsup}$ and $G_{k,jinf}$ may be taken as G_k; $G_{k,jsup}$ & $G_{k,jinf}$ are upper & lower characteristic values of permanent action j.

[b] Where the action is favorable, $\gamma_{Q,i} = 0$ and the variable actions should be ignored.

[c] The value of ψ_0 can be obtained from NA A1.1 of the UK NA to BS EN 1990 or from Table 6.3.

[d] The value of ξ in the UK NA to BS EN 1990 is 0.925

Table 6.7 Partial factors for geotechnical material properties.

Symbol	Angle of shearing resistance (apply to tan φ) γ_φ	Effective cohesion γ_c	Undrained shear strength γ_{cu}	Unconfined strength γ_{qu}	Bulk density γ_γ
Combination 1	1.0	1.0	1.0	1.0	1.0
Combination 2	1.25	1.25	1.4	1.4	1.0
EQU	1.1	1.1	1.2	1.2	1.0

Table 6.8 Design values of actions derived for UK design, EQU ultimate limit states– persistent and transient design situations.

Combination Exp.reference BS EN 1990	Permanent actions		Leading variable action	Accompanying variable actions	
	Unfavorable	Favorable		Main (if any)	Others
Exp.(6.10)	$1.1\,G_k^a$	$0.90\,G_k^a$	$1.5^b Q_k$	–	$1.5^b \psi_{0,i}^c Q_{k,i}$

Key:

[a] Where the variation in permanent action is not considered significant $G_{k,jsup}$ and $G_{k,jinf}$ may be taken as G_k; $G_{k,jsup}$ & $G_{k,jinf}$ are upper & lower characteristic values of permanent action j.

[b] Where the action is favorable, $\gamma_{Q,i} = 0$ and the variable actions should be ignored.

[c] The value of ψ_0 can be obtained from NA A1.1 of the UK NA to BS EN 1990 or from Table 6.3.

should be adopted. For *direct* design method the frequent combination can be used for sizing of foundations and the quasi-permanent combination can be used for settlement calculations.

(F) *Design methods*

Section 6 of EN 1997-1 requires spread foundations (e.g. pad and strip) to be designed using the following methods:

(1) *Direct method* – calculation is carried out for each limit state. At the ULS, the bearing resistance of the soil should be checked using partial factors on the soil properties as well as on the actions. At the SLS, the settlement of the foundations should be calculated and checked against permissible limits. As mentioned earlier, the frequent combination can be used for sizing of foundations and the quasi-permanent combination can be used for settlement calculations.

(2) *Indirect method* – experience and testing used to determine serviceability limit state parameters that also satisfy all relevant limit states. This method is used predominantly for Geotechnical Category 1 structures (see Table 6.5), where there is negligible risk in terms of overall stability or ground movements, which are known from comparable local experience to be sufficiently straightforward. In these cases the procedures may consist of routine methods for foundation design and construction. The procedures should be used only if there is no excavation below the water table or if comparable local experience indicates that a proposed excavation below the water table will be straightforward. Indirect methods may also be used for higher risk structures where it is difficult to predict the structural performance with sufficient accuracy from analytical solutions. In these cases, reliance is placed on the observational method, and identification of potential behaviour. Then the final, suitably conservative, design of the foundation can be decided.

(3) *Prescriptive method* – a presumed bearing resistance is used. This method may be used where calculation of the soil properties is not possible or necessary and can be used provided that conservative rules of design are used. Tables (for example Tables in the Building Regulations) of presumed bearing values may be used to determine presumed (allowable) bearing pressures for Geotechnical Category 1 structures where ground conditions are well known, and in preliminary calculations for Geotechnical Category 2 structures. Alternatively, the presumed bearing pressure to allow for settlement can be calculated by the geotechnical designer and included in the geotechnical design report. Unlike British standard BS 8004 – which was giving allowable bearing pressures for rocks, non-cohesive soils, cohesive soils, peat and organic soils, made ground, fill, high porosity chalk and the Mercia Mudstone – Annex G of EN 1997-1-1 only provides values of presumed bearing resistance for rock (which was also appeared in BS 8004).

This chapter does not attempt to provide complete guidance on the design of spread foundations, for which the reader should refer to any well-established text on the subject.

(G) *Footings subject to vertical actions*

(a) *Eccentricity* $e = 0$

EN 1997-1 requires the design vertical action V_d acting on the foundation to be less than or equal to the design bearing resistance R_d of the ground beneath it:

$$V_d \leq R_d \tag{6.4}$$

The action V_d should include the self-weight of the foundation and any backfill on it. Equation (6.4) is simply a re-statement of Equation (6.2). Designers more commonly consider pressures and stresses rather than work in terms of forces, so we will re-write Equation (6.4) as:

$$q_{Ed} \leq q_{Rd} \tag{6.5}$$

The pressure q_{Ed} is the design bearing pressure on the ground (an action effect); q_{Rd} is the corresponding design resistance.

Design baring pressure:

Figure 6.4 shows a footing carrying characteristic vertical actions V_{Gk} (permanent) and V_{Qk} (variable) imposed on it by the super-structure. The characteristic self-weights of the footing and of the backfill upon it are both permanent actions (W_{Gk})

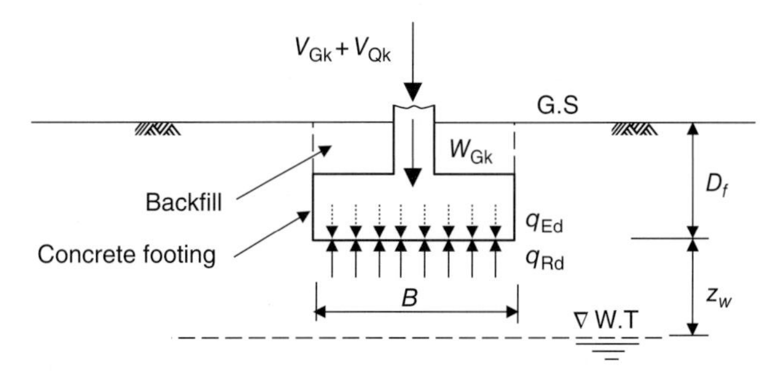

Figure 6.4 Vertical actions on a spread foundation.

Refer to Figure 6.4:
The characteristic bearing pressure q_{Ek} is given by

$$q_{Ek} = \frac{\sum V_{rep}}{A'} = \frac{\left(V_{Gk} + \sum_i \psi_i V_{Qk,i}\right) + W_{Gk}}{A'} \tag{6.6}$$

The action V_{rep} is a representative vertical action, and A' is the footing's effective area. The factor ψ_i is the combination factor applicable to the variable action (see Table 6.3).

If we assume that only one variable action is applied to the footing, Equation (6.6) simplifies to

$$q_{Ek} = \frac{(V_{Gk} + V_{Qk,1}) + W_{Gk}}{A'} \tag{6.7}$$

This is because $\psi = 1.0$ for the leading variable action $(i = 1)$.
The design bearing pressure q_{Ed} beneath the footing is then

$$q_{Ed} = \frac{\sum V_d}{A'} = \frac{\gamma_G(V_{Gk} + W_{Gk}) + \gamma_Q V_{Qk,1}}{A'} \tag{6.8}$$

where γ_G and γ_Q are partial factors on permanent and variable actions, respectively (see the UK national annex for EN 1990).

Design baring resistance:

The design resistance q_{Rd} may be calculated using analytical or semi-empirical formulae. Annex D (informative) of BS EN 1997-1-1 provides widely-recognised formulae for bearing resistance, considering both the drained and undrained conditions. (Also, see the relevant formulations in Chapter 4). These apply for homogeneous ground conditions. We should notice that for drained conditions, water pressures must be included as actions. The question then arises, which partial factors should be applied to the weight of a submerged structure? Since the water pressure acts to reduce the value of the design vertical action V_d, it may be considered as favourable action, while the total weight is unfavourable. Physically however, the soil has to sustain the submerged weight.

For the design of structural members, water pressure may be considered as unfavourable action (Scarpelli and Orr, 2013).

(b) *Eccentricity* $e \neq 0$

It may be realised that the ability of a spread foundation to carry forces reduces dramatically when those forces are applied eccentrically from the centre of the foundation.

Refer to Figure 6.5. To prevent contact with the ground being lost at the footing's edges, it is customary to keep the total action R within the foundation's "middle third". This requires that $e_B \leq (B/6)$ and $e_L \leq (L/6)$, where B and L are the footing's breadth and length; and e_B and e_L are eccentricities in the direction of B and L. (see Section 4.8 of Chapter 4).

A method to take account of the effect of eccentric loading on bearing capacity calculations is by assuming that the load acts at centre of a smaller foundation, as shown in Figure 6.5. The actual foundation area $(A = B \times L)$ is therefore reduced to an *effective area* A' (the shaded part of area A) where

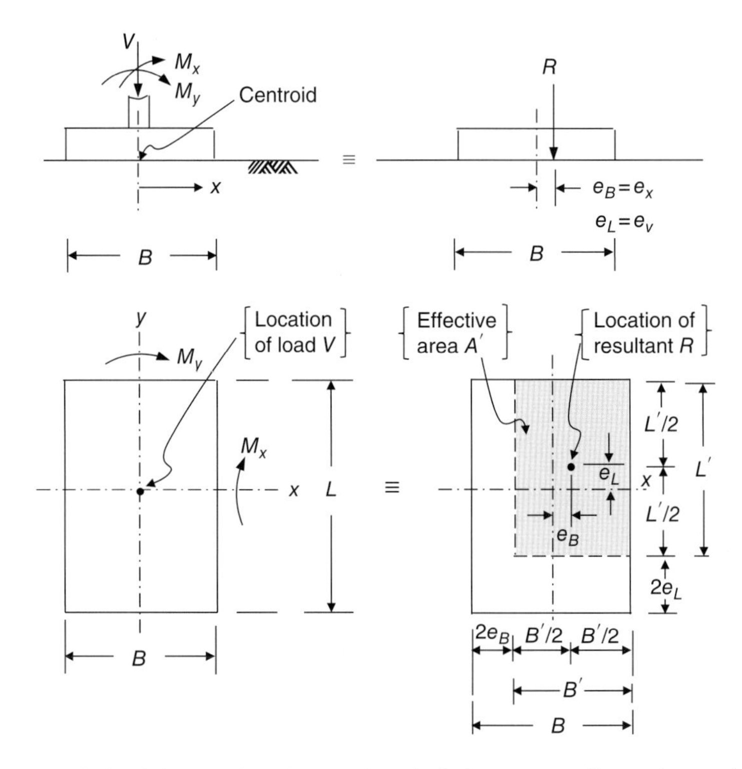

Figure 6.5 Eccentrically loaded spread foundation and method of computing effective footing dimensions.

$$A' = B' \times L' = (B - 2e_B) \times (L - 2e_L) \tag{6.9}$$

The dimensions B' and L' are the footing's effective width and length.

It may be useful to mention that there is no guidance in the Eurocode as to whether the eccentricity should be calculated for the characteristic or design values of the actions. In the author's and many others' opinion it would be best to base the calculation on design actions. In this case, the eccentricities e_B and e_L may be denoted by e'_B and e'_L.

EN 1997-1 §6.5.4 requires that special precautions be taken when the eccentricity of the loading exceeds one-third of the width or length of a rectangular footing or 0.6 of the radius of a circular footing. It should be noted that this is not the middle-third rule, but rather a "middle two-thirds" rule. Such precautions include: careful review of the design values of actions; designing the location of the foundation edge by taking into account the magnitude of construction tolerances up to 0.1 m (i.e. tolerances up to 0.1 m in the dimensions of the foundation). However, it may be appropriate that foundations continue to be designed using the middle-third rule until the implications of EN 1997's more relaxed Principle have been thoroughly tested in practice (Bond and Harris, 2010).

According to Scarpelli and Orr (2013), for the eccentric loading condition shown in Figure 6.6, it may be necessary to analyse different load combinations, by considering the permanent vertical load as both favourable and unfavourable and by changing the leading variable load.

$$V_{\text{unfavourable}} \text{ and } H_{\text{unfavourable}}$$

$$V_d = \gamma_G G_k + \gamma_{Qv} \psi_0 Q_{vk} \qquad H_d = \gamma_{Qh} Q_{hk}$$

$$\gamma_G = 1.35 \quad \gamma_{Qv} = 1.5 \quad \gamma_{Qh} = 1.5$$

$$V_{\text{unfavourable}} \text{ and } H_{\text{unfavourable}}$$

$$V_d = \gamma_G G_k + \gamma_{Qv} Q_{vk} \qquad H_d = \gamma_{Qh} \psi_0 Q_{hk}$$

$$\gamma_G = 1.35 \quad \gamma_{Qv} = 1.5 \quad \gamma_{Qh} = 1.5$$

$V_{\text{favourable}}$ and $H_{\text{unfavourable}}$

$$V_d = \gamma_G\, G_k + \gamma_{Qv}\, Q_{vk} \qquad H_d = \gamma_{Qh}\, Q_{hk}$$

$$\gamma_G = 1.0 \quad \gamma_{Qv} = 0.0 \quad \gamma_{Qh} = 1.5$$

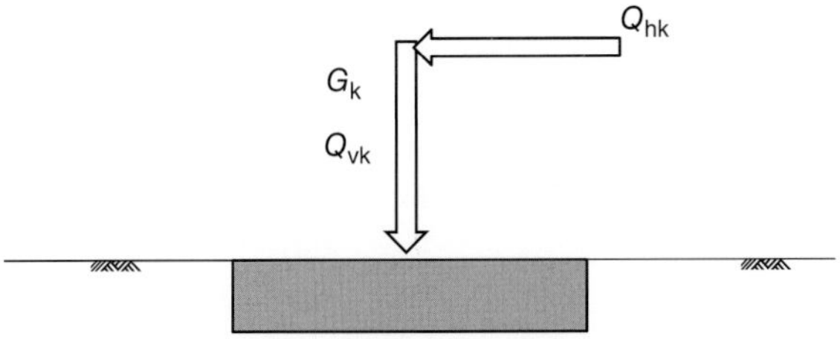

Q_{hk}

G_k

Q_{vk}

Figure 6.6 Spread footing under vertical and horizontal loads (eccentric loading condition).

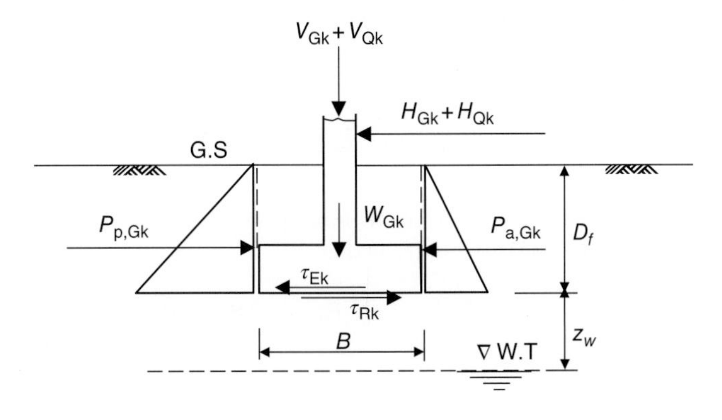

Figure 6.7 Horizontal actions on a spread foundation.

(H) *Footings subject to overturning*

When there is a moment applied to the foundation, the EQU limit state should also be checked. Assuming the potential overturning of the base is due to the variable action from the wind ($Q_{k,w}$), the following combination should be used (the variable imposed action is not considered to contribute to the stability of the structure):

$$0.9\, G_k + 1.5\, Q_{k,w} \quad \text{(EQU combination)}$$

The action G_k is the stabilising characteristic permanent action. (Use 1.1 G_k for a destabilising permanent action)

The action $Q_{k,w}$ is the destabilising characteristic variable action.

(I) *Footings subject to horizontal actions and sliding*

Figure 6.7 shows the footing from Figure 6.4 subject to characteristic horizontal actions H_{Gk} (permanent) and H_{Qk} (variable), in addition to characteristic vertical actions V_{Gk} (permanent), V_{Qk} (variable), and W_{Gk} (permanent).

EN 1997-1 requires the design horizontal action H_d acting on the foundation to be less than or equal to the sum of the design resistance R_d from the ground beneath the footing and any design passive thrust R_{pd} on the side of the foundation:

$$H_d \le R_d + R_{pd} \tag{6.10}$$

This equation is merely a re-statement of Equation (6.2)

Designers more commonly consider shear stresses rather than work in terms of shear forces, so we will re-write Equation (6.10) as:

$$\tau_{Ed} \leq \tau_{Rd} \tag{6.11}$$

The stress τ_{Ed} is the design shear stresses acting across the base of the footing (an action effect); τ_{Rd} is the design resistance to that shear stress.

Design shear stresses

The characteristic shear stress τ_{Ek} shown in Figure 6.7 is given by

$$\tau_{Ek} = \frac{\sum H_{rep}}{A'} = \frac{\left(H_{Gk} + \sum_i \psi_i H_{Qk,i}\right) + P_{a,Gk}}{A'} \tag{6.12}$$

The action H_{rep} is a representative horizontal action; $P_{a,Gk}$ is the characteristic thrust due to active earth pressures on the side of the foundation (a permanent action); A' is the footing's effective area. The factor ψ_i is the combination factor applicable to the variable action (see Table 6.3).

If we assume that only one variable horizontal action is applied to the footing, Equation (6.12) simplifies to

$$\tau_{Ek} = \frac{\left(H_{Gk} + H_{Qk,1}\right) + P_{a,Gk}}{A'} \tag{6.13}$$

This is because $\psi = 1.0$ for the leading variable action $(i = 1)$.

The design shear stress τ_{Ed} is then

$$\tau_{Ed} = \frac{\sum H_d}{A'} = \frac{\gamma_G\left(H_{Gk} + P_{a,Gk}\right) + \gamma_Q H_{Qk,1}}{A'} \tag{6.14}$$

where γ_G and γ_Q are partial factors on permanent and variable actions, respectively.

Sliding resistance

(a) *Drained sliding resistance*

For drained conditions, the characteristic shear resistance τ_{Rk} shown in Figure 6.7, excluding the passive thrust in front of the foundation, is given by:

$$\tau_{Rk} = \frac{V'_{Gk} \tan\delta_k}{A'} = \frac{(V_{Gk} - U_{Gk})\tan\delta_k}{A'} \tag{6.15}$$

where V_{Gk} and V'_{Gk} represent the characteristic total and effective permanent vertical actions on the footing, respectively; U_{Gk} is the characteristic uplift owing to pore water pressures acting on the underside of the base (also a permanent action); and δ_k is the characteristic angle of friction between the base and the ground.

Variable vertical actions have been excluded from Equation (6.15), since they are favourable. Also, the equation conservatively ignores any effective adhesion between the underside of the base and the ground, as suggested by EN 1997 §6.5.3(10).

The design shear resistance τ_{Rd} (ignoring passive pressures) is then given by:

$$\tau_{Rd} = \frac{V'_{Gd} \tan\delta_d}{\gamma_{Rh} A'} = \frac{(V_{Gd} - U_{Gd})\tan\delta_d}{\gamma_{Rh} A'} \tag{6.16}$$

where γ_{Rh} is a partial factor on horizontal sliding resistance and the subscript, d, denote design values.

Normally it is assumed that the soil at the interface with concrete is remolded. So the design friction angle δ_d may be assumed equal to the design value of the effective critical state angle of shearing resistance $\varphi'_{cv,d}$ for cast in situ concrete foundations and equal to 2/3 $\varphi'_{cv,d}$ for smooth precast foundations [EN 1997 §6.5.3(10)]. The design effective critical state angle is

$$\varphi'_{cv,d} = \tan^{-1}\left(\frac{\tan\varphi'_{cv,k}}{\gamma_\varphi}\right) \tag{6.17}$$

where γ_φ is the partial factor on shearing resistance.

The vertical action V_{Gk} is *favourable*, since an increase in its value increases the shear resistance; whereas U_{Gd} is *unfavourable* action, since an increase in its value decreases the resistance. Introducing into Equation (6.15) partial factors on *favourable* and *unfavourable* permanent actions ($\gamma_{G,fav}$ and γ_G) results in:

$$\begin{aligned}\tau_{Rd} &= \frac{(\gamma_{G,fav} V_{Gk} - \gamma_G U_{Gk})\tan\delta_k}{\gamma_{Rh}\gamma_\varphi A'} \\ &= \left[\left(\frac{\gamma_{G,fav}}{\gamma_{Rh}\times\gamma_\varphi}\right)V_{Gk} - \left(\frac{\gamma_G}{\gamma_{Rh}\times\gamma_\varphi}\right)U_{Gk}\right]\times\left(\frac{\tan\delta_k}{A'}\right)\end{aligned} \tag{6.18}$$

If, however, partial factors are applied to the net effects of actions rather than to the actions themselves, then Equation (6.18) will become

$$\tau_{Rd} = \frac{\gamma_{G,fav}(V_{Gk} - U_{Gk})\tan\delta_k}{\gamma_{Rh}\gamma_\varphi A'} = \left(\frac{\gamma_{G,fav}}{\gamma_{Rh}\times\gamma_\varphi}\right)(V_{Gk} - U_{Gk})\times\left(\frac{\tan\delta_k}{A'}\right) \tag{6.19}$$

Table 6.9 summarises the values of these partial factors for each of the EN 1997's three Design Approaches.

As far as an ultimate limit state of sliding is concerned, the horizontal component of an inclined load on the footing is considered unfavourable, whereas the vertical component is favourable, although they have the same source. Except in Combination 2 of Design Approach 1, a favourable permanent action attracts a partial factor $\gamma_{G,fav} = 1$ and an unfavourable permanent action a partial

Table 6.9 Partial factors for each of the EN 1997's three Design Approaches.

Individual Partial Factor or Partial Factor "Grouping"	Design Approach 1: Combination 1	Design Approach 1: Combination 2	2	3
γ_G	1.35	1.0	1.35	1.35/1
$\gamma_{G,fav}$	1.0	1.0	1.0	1.0
γ_φ	1.0	1.25	1.0	1.0
γ_{cu}	1.0	1.4	1.0	1.4
γ_{Rh}	1.0	1.0	1.1	1.0
$\gamma_{G,fav}/(\gamma_{Rh}\times\gamma_\varphi)$	1.0	0.8	0.91	0.8
$\gamma_G/(\gamma_{Rh}\times\gamma_\varphi)$	1.35	0.8	1.23	1.08/0.8
$1/(\gamma_{Rh}\times\gamma_{cu})$	1.0	0.71	0.91	0.71

factor $\gamma_G = 1.35$. In case the components of the inclined action are treated differently, then the inclination of that action will change. This is a prime example of where the Single Source Principle [discussed in Section 6.3.4 (B)] should be invoked and the whole action should be treated either as unfavourable or as favourable, whichever represents the more critical design condition.

(b) *Undrained sliding resistance*

For undrained conditions, the characteristic shear resistance τ_{Rk} shown in Figure 6.7, excluding the passive thrust in front of the foundation, is given by:

$$\tau_{Rk} = C_{uk} \quad (6.20)$$

where C_{uk} represents the characteristic undrained shear strength of the soil.

The design shear resistance τ_{Rd} (ignoring passive pressures) is then given by:

$$\tau_{Rd} = \frac{C_{ud}}{\gamma_{Rh}} = \frac{C_{uk}}{\gamma_{cu} \times \gamma_{Rh}} \quad (6.21)$$

where γ_{cu} is the partial factor on undrained shear strength (see Table 6.9).

(J) *SLS verification*

When design is carried out by direct methods, settlement calculations are required to check SLS. For soft clays, irrespective of Geotechnical categories, settlement calculations shall be carried out. For spread foundations on stiff and firm clays in Geotechnical categories 2 and 3, vertical displacement should usually be calculated [EN 1997 §6.6.1(3, 4)].

As mentioned in Sections 6.2.5(B) and 6.3.3(b), EN 1997 requires the design movements E_d of a foundation to be less than or equal to the limiting movement C_d specified for the project; see Equation (6.3).

The following three components of settlement should be considered for partially or fully saturated soils [EN 1997-1 §6.6.2(2)]:

- s_0: immediate settlement; for fully-saturated soil due to shear deformation at constant volume, and for partially-saturated soil due to both shear deformation and volume reduction;
- s_1: settlement caused by consolidation;
- s_2: settlement caused by creep.

So, Equation (6.3) can be re-written for settlement of foundations as follows:

$$s_{Ed} = s_0 + s_1 + s_2 \leq s_{Cd} \quad (6.22)$$

where s_{Ed} is the total settlement (an action effect), and s_{Cd} is the limiting value of that settlement.

The components of foundation movement, which should be considered include: settlement, relative (or differential) settlement, rotation, tilt, relative deflection, relative rotation, horizontal displacement and vibration amplitude. Definitions and limiting values of some terms for foundation movement and deformation are given in Annex H to EN1997-1. For example, for normal structures with isolated foundations, total settlements up to 50 mm are often acceptable. Larger settlements may be acceptable provided the relative rotations remain within acceptable limits and provided the total settlements do not cause problems with the services entering the structure, or cause tilting and so on.

In verifications of SLS, partial factors are normally set to 1. The combination factors ψ applied to accompanying variable actions are those specified for the characteristic, frequent, or quasi-permanent combinations, which are the ψ_2 values from EN 1990.

Annex F of EN 1997-1 presents two methods to evaluate settlement. Other methods (from the in situ tests) are given in the Annexes to EN 1997-2.

EN 1997 emphasises the fact that settlement calculations should not be regarded as accurate. They merely provide an approximate indication [EN 1997-1 §6.6.1(6)].

Where the depth of compressible layers is large, it is normal to limit the analysis to depths where the increase in effective vertical stress is greater than 20% of the effective overburden stress [EN 1997-1 §6.6.2(6)].

For conventional structures founded on clays, settlements should be calculated explicitly when the ratio of the characteristic bearing resistance R_k to the applied serviceability loads E_k is less than three. If this ratio is less than two, those calculations should take account of the ground's non-linear stiffness [EN 1997-1 §6.6.2(16)].

Therefore, the serviceability limit state may be deemed to have been verified if:

$$E_k \leq \frac{R_k}{\gamma_{R,SLS}} \tag{6.23}$$

where E_k characteristic effects of actions, $R_k =$ characteristic resistance to those actions, and $\gamma_{R,SLS} =$ a partial resistance factor ≥ 3.

Concluding remarks

The geotechnical design of spread foundations to EN 1997 involves checking that the ground has sufficient bearing resistance to withstand vertical actions, sufficient sliding resistance to withstand horizontal and inclined actions and sufficient stiffness to prevent unacceptable settlement. The first two requirements concern ultimate limit states (ULS) and the last concerns a serviceability limit state (SLS).

Verification of ULS is carried out by satisfying Equations (6.4) and (6.10), repeated here for convenience:

$$V_d \leq R_d \text{ and } H_d \leq R_d + R_{pd}$$

These two equations are merely specific forms of Equation (6.2):

$$E_d \leq R_d$$

Verification of SLS is carried out by satisfying Equation (6.22), repeated here for convenience:

$$s_{Ed} = s_0 + s_1 + s_2 \leq s_{Cd}$$

This equation is merely a specific form of Equation (6.3):

$$E_d \leq C_d$$

Alternatively, serviceability limit states may be verified by satisfying Equation (6.23), repeated here for convenience:

$$E_k \leq \frac{R_k}{\gamma_{R,SLS}}$$

In this equation the partial factor $\gamma_{R,SLS}$ is equal or greater than 3.

6.3.5 Structural Design

(A) *General*

As mentioned earlier in Section 6.3.1, the principles and application rules related to "Structural design of spread foundations" is given in EN 1997-1 §6.8. It requires that structural failure of a spread foundation shall be prevented in accordance with EN 1997-1 §2.4.6.4, which states: The design strength properties of structural materials and the design resistances of structural elements shall be calculated in accordance with EN 1992 to EN 1996 and EN 1999. Therefore, it is clear that all the structural reinforced concrete elements of a spread foundation shall be designed in accordance with EN 1992.

For stiff footings, EN 1997-1 §6.8 recommends a linear distribution of bearing pressures may be used to calculate bending moments and shear stresses in the structural elements. A more detailed analysis of soil-structure interaction may be used to justify a more economic design.

The distribution of bearing pressures beneath flexible foundations may be derived by modeling the foundation as a beam or raft resting on a deforming continuum or series of springs, with appropriate stiffness and strength.

(B) *Selected symbols and their definitions*

A_c Cross-sectional area of concrete $= bh$
A_s Area of tension steel
$A_{s,prov}$ Area of tension steel provided
$A_{s,req'd}$ Area of tension steel required
d Effective depth
d_{eff} Average effective depth $= (d_y + d_z)/2$
f_{cd} Design value of concrete compressive strength $= a_{cc}f_{ck}/\gamma_c$
f_{ck} Characteristic cylinder strength of concrete
f_{ctm} Mean of axial tensile strength $= 0.3{f_{ck}}^{2/3}$ for $f_{ck} \leq$ C50/60 (From Table 3.1, EN 1992)
G_k Characteristic value of a permanent action
h Overall depth of the section
l_{eff} Effective span of member
M Design moment at the ULS
Q_k Characteristic value of a variable action
Q_{kw} Characteristic value of a variable wind action
V_{Ed} Design value of applied shear force
v_{Ed} Design value of applied shear stress
$V_{Rd,c}$ Design value of the punching shear resistance without punching shear reinforcement
$v_{Rd,c}$ Design value of the punching shear stress resistance without punching shear reinforcement
$v_{Rd,max}$ Design value of the maximum punching shear stress resistance along the control section considered
x Depth to neutral axis $= (d - z)/0.4$
x_{max} Limiting value for depth to neutral axis $= (\delta - 0.4)d$, where $\delta \leq 1$
z Lever arm
a_{cc} Coefficient taking account of long term effects on compressive strength and of unfavourable effects resulting from the way load is applied (from UK National Annex) $= 0.85$ for flexure and axial loads, 1.0 for other phenomena
β Factor for determining punching shear stress
δ Ratio of the redistributed moment to the elastic bending moment
γ_m Partial factor for material properties
ρ_0 Reference reinforcement ratio $= f_{ck}/1000$
ρ_l Required tension on reinforcement at mid-span to resist the moment due to the design loads (or at support for cantilevers) $= A_s/bd$
ψ_0 Factor for combination value of a variable action
ψ_1 Factor for frequent value of a variable action
ψ_2 Factor for quasi-permanent value of a variable action

(C) *Material properties*

(a) *Concrete*

The concrete compressive strength classes are based on the characteristic cylinder strength f_{ck} determined at 28 days with a maximum value of C_{max}. The value of C_{max} for use in a Country may be found in its National Annex. EN 1992-1-1 §3.1.2(2) recommends C90/105. Details of

Table 6.10 Selected concrete properties based on Table 3.1 of EN 1992, Part1-1.

Symbol	Properties										
f_{ck} (MPa)	12	16	20	25	30	35	40	45	50	28[a]	32[a]
$f_{ck,cube}$ (MPa)	15	20	25	30	37	45	50	55	60	35[a]	40[a]
f_{ctm} (MPa)	1.6	1.9	2.2	2.6	2.9	3.2	3.5	3.8	4.1	2.8[a]	3.0[a]
E_{cm} [b](GPa)	27	29	30	31	33	34	35	36	37	32[a]	34[a]

Key

f_{ck} = Characteristic cylinder strength

$f_{ck,\ cube}$ = Characteristic cube strength

f_{ctm} = Mean tensile strength

E_{cm} = Mean secant modulus of elasticity

[a] Concrete properties not cited in Table 3.1 of EN 1992, Part1-1.

[b] Mean secant modulus of elasticity at 28 days for concrete with quartzite aggregates. For concrete with other aggregates refer to CI 3.1.3 (2)

the required concrete properties such as strength, elastic deformation and creep and shrinkage, are given in EN 1992-1-1 §3.1.

Typical concrete properties are given in Table 6.10.

In EN 1992 the design of reinforced concrete is based on the characteristic cylinder strength f_{ck} rather than cube strength and should be specified to BS 8500: Concrete-complementary British Standard to BS EN 206-1.

A key change in BS 8500 was the introduction of dual classification system where the concrete cylinder strength is given alongside the equivalent concrete cube strength (e.g. for class C28/35 concrete the cylinder strength is 28 MPa, whereas the cube strength is 35 MPa).

As mentioned above, EN 1992 allows the designer to use high strength concrete, up to a class C90/105 (or class C70/85 for bridges). Above a class C50/60, the engineer will find that there are restrictions placed in EN 1992. For example, there are lower strain limits, additional requirements for fire resistance and in the UK the resistance to shear should be limited to that of a class C50/60 concrete.

(b) *Reinforcing steel*

According to EN 1992-1-1 §3.2.2(1), the behaviour of reinforcing steel is specified by the following properties:

- Yield strength (f_{yk} or $f_{0,2k}$)
- Maximum actual yield strength ($f_{y,max}$),
- Tensile strength (f_t),
- Ductility (ε_{uk} and f_t/f_{yk})
- Bond characteristics (f_R: See Annex C),
- Section sizes and tolerances,
- Fatigue strength,
- Weldability,
- Shear and weld strength for welded fabric and lattice girders.

Details of all these required reinforcing steel properties are given in EN 1992-1-1 §3.2.

The application rules for design and detailing in EN 1992-1-1 are valid for a specified yield strength range f_{yk} = 400 – 600 MPa. The upper limit of f_{yk} within this range for use within a Country may be found in its National Annex. The recommended value for partial factor γ_s is 1.15 for ULS, persistent and transient design situations [EN 1992-1-1 §2.4.2.4(1)].

The characteristic strength of reinforcing steel supplied in the UK is 500 MPa; previously the minimum strength was 460 MPa. In order to ensure that there was no confusion with older steel grades the 500 grade steel is designated with an "H". The properties of reinforcing steel in UK for use with EN 1992 are given in BS 4449 (2005): "Specification for carbon steel bars for the reinforcement of concrete".

There are three classes of reinforcement: A, B and C, which provide increasing ductility, as shown in Table 6.11. Class A is not suitable where redistribution of 20% and above has been assumed in the design.

There is no provision for the use of plain bar or mild steel reinforcement, but guidance is given in the background paper to the UK National Annex.

(D) *Actions*

EN 1991: *Actions on structures* consists of 10 parts giving details of a wide variety of actions. A relatively brief description of different types of actions has been given in Section 6.2.5 of this Chapter.

EN 1991, Part 1-1: *General Actions-Densities, self-weight, imposed loads for buildings* gives the densities and self-weights of building material. The draft National Annex to this Eurocode gives the imposed loads for UK buildings. Table 6.12 gives bulk density of three selected materials. A selection of imposed loads is presented in Table 6.13.

(E) *Combination of actions*

As defined earlier, the term combination of actions refers to a set of values of actions to be used when a limit state is under the influence of different actions. The ULS and SLS combinations of actions are covered in EN 1990 and in Section 6.2.5 of this Chapter. Also, the values of partial factors for these limit state combinations can be obtained by referring to same sources.

(F) *Reinforced concrete pads (column footings)*

Successful structural design of a reinforced concrete pad foundation must ensure:

- Beam shear strength,
- Punching shear strength,
- Sufficient reinforcement to resist bending moments.

The moments and shear forces should be assessed using one of the three alternative sets of expressions given in Section 6.2.5 and the design values of actions of Table 6.6. For example, the moments and shear forces may be assessed using Expression (6.10) and STR combination 1:

Table 6.11 Characteristic tensile properties of reinforcing steel.

Property	Reinforcement class		
	A	*B*	*C*
Characteristic yield strength f_{yk} or $f_{0,k}$ (MPa)	500	500	500
Minimum value of $k = (f_t/f_y)_k$	≥ 1.05	≥ 1.08	≥ 1.15 < 1.35
Characteristic strain at maximum force ε_{uk}(%)	≥ 2.5	≥ 5.0	≥ 7.5

Notes
[a] Table derived from BS EN 1992–1–1 Annex C, BS 4449: 2005 and BS EN 10080
[b] The nomenclature used in BS 4449: 2005 differs from that used in BS EN 1992–1–1 Annex C and used here.
[c] In accordance with BS 8666, class H may be specified, in which case class A, B or C may be supplied.

Table 6.12 Selected bulk density of material (from EN 1991–1–1).

Material	Bulk density (kN/m^3)
Normal weight concrete	24.0
Reinforced normal weight concrete	25.0
Wet Normal weight reinforced concrete	26.0

Table 6.13 Selected imposed loads for buildings (from draft UK National Annex to EN 1991-1-1).

Category	Example use	q_k (kN/m^2)	Q_k (kN)
A1	All uses within self-contained dwelling units	1.5	2.0
A2	Bedrooms and dormitories	1.5	2.0
A3	Bedrooms in hotels and motels, hospital wards and toilets	2.0	2.0
A5	Balconies in single family dwelling units	2.5	2.0
A7	Balconies in hotels and motels	4.0 min.	2.0 at outer edge
B1	Offices for general use	2.5	2.7
C5	Assembly area without fixed seating, concert halls, bars, places of worship	5.0	3.6
D1/2	Shopping areas	4.0	3.6
E12	General storage	2.4 per m height	7.0
E17	Dense mobile stacking in warehouses	4.8 per m height	7.0
F	Gross vehicle weight ≤ 30 kN	2.5	10.0

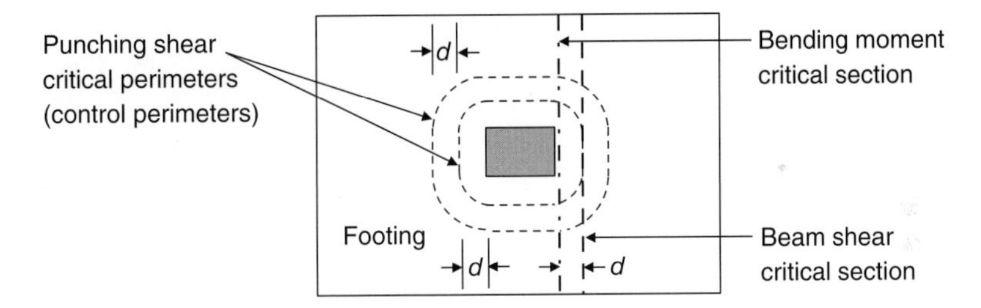

Figure 6.8 Location of critical section for beam shear, punching shear and bending moment in reinforced concrete pad foundations.

For one variable action: $1.35G_k + 1.5\,Q_k$

However, there may be economies to make from using the other two Expressions (6.10a) and (6.10b).

The design applied moment, beam shear and punching shear should be checked at their critical sections located as shown in Figure 6.8.

Treating the reinforced pad as a beam in bending, the critical bending moments for design of bottom reinforcement are calculated at the column faces, as shown in Figure 6.8. A procedure for determining flexural reinforcement for pad foundations is shown in Figure 6.9 (and Table 6.14). Particular rules and detailing of column footings are given in EN 1992-1-1 §9.8.2.

EN 1992 provides specific guidance on the design of foundations for punching shear, and this varies from that given for slabs. The critical shear perimeter has rounded corners and the forces directly resisted by the ground should be deducted (to avoid unnecessary conservative designs). The critical peripheral section should be found iteratively, but it is generally acceptable to check at d and $2d$ [EN 1992-1-1 §6.4.2 (1, 2)]. A procedure for determining punching shear capacity for pad foundations as suggested by The Concrete Centre (TCC), UK, is shown in Figure 6.10. It is not usual for a pad foundation to contain shear reinforcement; therefore it is only necessary that the concrete shear stress

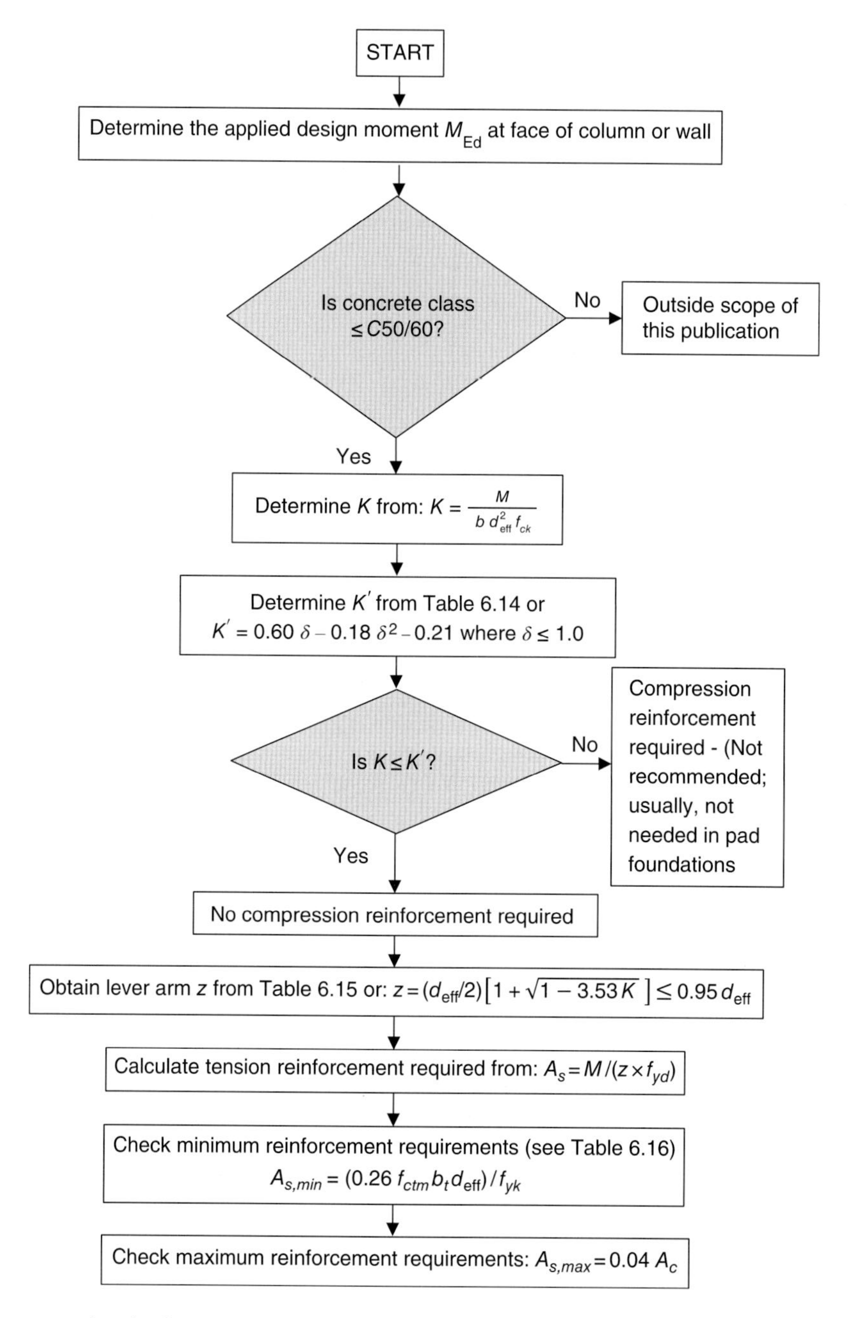

Figure 6.9 Procedure for determining flexural reinforcement for pad foundations.

capacity without shear reinforcement ($v_{Rd,c}$; see Table 6.17) is greater than the applied shear stress [$v_{Ed} = V_{Ed}/(bd)$]. If the basic shear stress is exceeded, the designer may increase the thickness of the pad. Alternatively, the amount of main reinforcement could be increased or, less desirably, shear links could be provided. EN 1992-1-1 §6.4.4(2) gives equations for calculation of applied and resisting shear stresses.

Table 6.14 Values for K'.

% Redistribution	δ (Redistribution ratio)	K'
0	1.00	0.208[a]
10	0.90	0.182[a]
15	0.85	0.168
20	0.80	0.153
25	0.75	0.137
30	0.70	0.120

Key

[a] It is often recommended in the UK that K' should be limited to 0.168 to ensure ductile failure.

Table 6.15 Values of z/d_{eff} for singly reinforced rectangular sections.

K	z/d_{eff}	K	z/d_{eff}
≤ 0.05	0.950[a]	0.13	0.868
0.06	0.944	0.14	0.856
0.07	0.934	0.15	0.843
0.08	0.924	0.16	0.830
0.09	0.913	0.17	0.816
0.10	0.902	0.18	0.802
0.11	0.891	0.19	0.787
0.12	0.880	0.20	0.771

Key

[a] Limiting z to 0.95 d_{eff} is not a requirement of EN 1992, but is considered to be good practice.

Table 6.16 Minimum percentage of reinforcement required.

f_{ck}	f_{ctm}	Minimum % ($0.26\, f_{ctm}/f_{yk}$[a])
25	2.6	0.13%
28	2.8	0.14%
30	2.9	0.15%
32	3.0	0.16%
35	3.2	0.17%
40	3.5	0.18%
45	3.8	0.20%
50	4.1	0.21%

Key

[a] Where $f_{yk} = 500$ MPa; $(0.26 f_{ctm}/f_{yk})\% \geq 0.13\,\%$

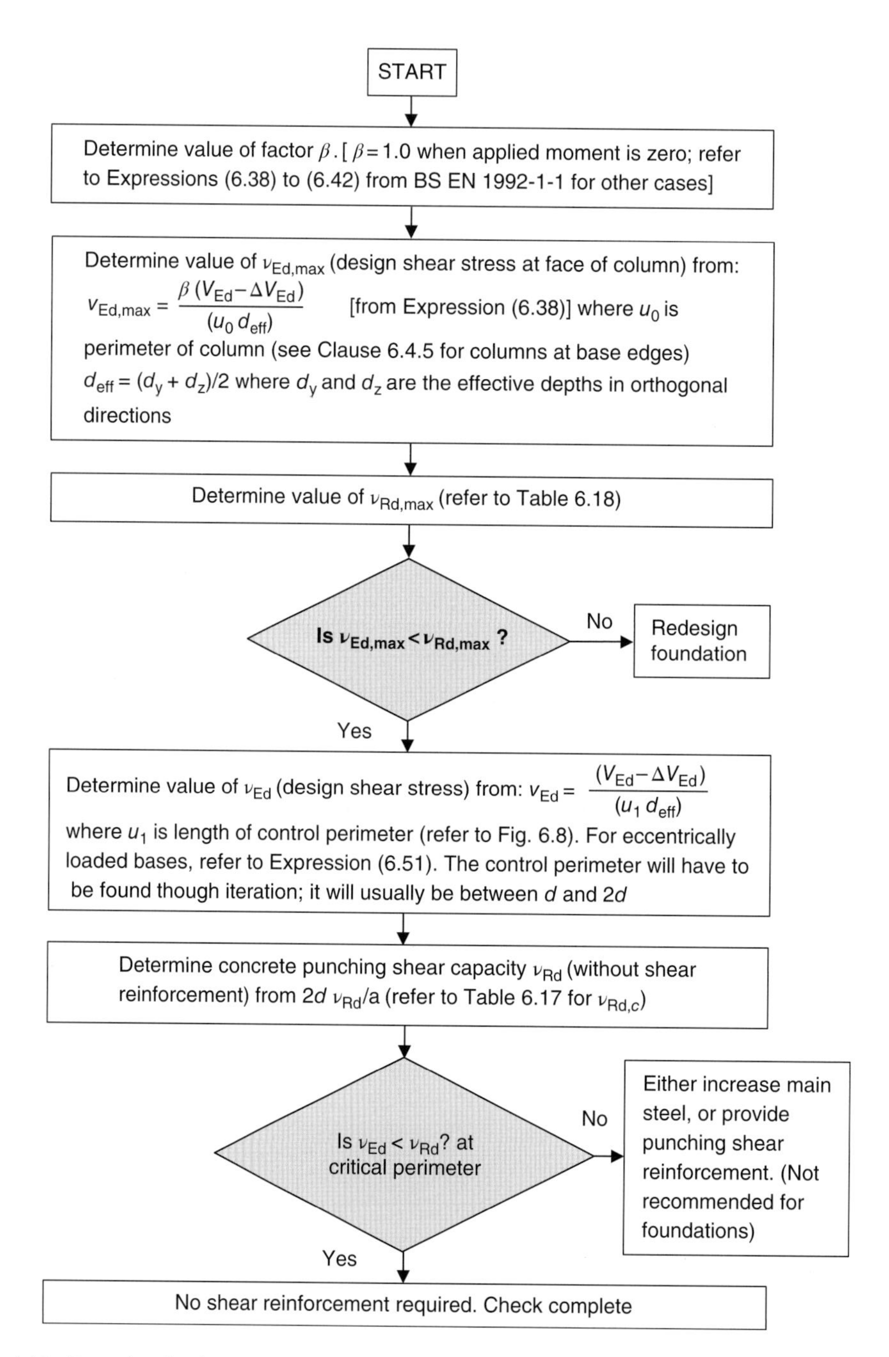

Figure 6.10 Procedure for determining punching shear capacity for pad foundations.

(G) *Reinforced concrete strip foundations*

Where the supporting soil is weak and the wall load is heavy a wide strip footing may be required. For this case, a reinforced concrete strip foundation, at a comparatively shallow depth, is likely to show an advantage in safety and cost over unreinforced concrete. In addition to the main transverse reinforcement, *longitudinal* bars is also desirable in strip foundations on highly variable soils when the foundation is enabled to bridge over local weak or hard spots in the soil at the foundation level, or when there is an abrupt change in loading.

Table 6.17 Resistance $v_{Rd,c}$ of members without shear reinforcement, MPa.

ρ_l	Effective depth, d (mm)							
	300	400	500	600	700	800	900	1000[a]
0.25%	0.47	0.43	0.40	0.38	0.36	0.35	0.35	0.34
0.50%	0.54	0.51	0.48	0.47	0.45	0.44	0.44	0.43
0.75%	0.62	0.58	0.55	0.53	0.52	0.51	0.50	0.49
1.00%	0.68	0.64	0.61	0.59	0.57	0.56	0.55	0.54
1.25%	0.73	0.69	0.66	0.63	0.62	0.60	0.59	0.58
1.50%	0.78	0.73	0.70	0.67	0.65	0.64	0.63	0.62
1.75%	0.82	0.77	0.73	0.71	0.69	0.67	0.66	0.65
≥ 2.00%	0.85	0.80	0.77	0.74	0.72	0.70	0.69	0.68
K	1.816	1.707	1.632	1.577	1.535	1.500	1.471	1.447
f_{ck}		25	28	32	35	40	45	50
Factor		0.94	0.98	1.02	1.05	1.10	1.14	1.19

Key

[a] For depths greater than 1000 mm calculate $v_{Rd,c}$ directly.

Notes

(1) Table derived from: $v_{Rd,c} = 0.12k\,(100\rho_l f_{ck})^{(1/3)} \geq 0.035k^{1.5} f_{ck}{}^{0.5}$ where $k = 1 + \sqrt{(200/d)} \leq 2$ and $\rho_l = \sqrt{\left(\rho_{ly} \times \rho_{lz}\right)} \leq 0.02$, $\rho_{ly} = A_{sy}/(bd)$ and $\rho_{lz} = A_{sz}/(bd)$

(2) This Table has been prepared for $f_{ck} = 30$; where ρ_l exceed 0.40 % the following factors may be used:

f_{ck}	25	28	32	35	40	45	50
Factor	0.94	0.98	1.02	1.05	1.10	1.14	1.19

The critical bending moments for design of bottom transverse reinforcement are calculated at the wall faces on the assumption that the footing projections behave as cantilevers. The slab must also be designed to withstand shear and bond stresses. It is necessary to consider only the beam shear stress, which is usually checked at the critical sections, in long direction, located a distance d from the wall face. Particular rules and detailing of wall footings are given in EN 1992-1-1 §9.8.2.

(H) *Pile caps*

The function and structural design of pile caps are very similar to those of reinforced concrete pads; both elements must distribute loads from the supported member across their bottoms.

Here also, a pile cap may be treated as a beam in bending, where the critical bending moments for the design of the bottom reinforcement are located at the column faces. Alternatively, a truss analogy may be used; this is covered in EN 1992-1-1 §5.6.4 and 6.5.

Both beam shear and punching shear should be checked as shown in Figure 6.11a. For calculating the design values of shear resistances, Table 6.17 may be used. Again here also, if the basic shear stress is exceeded, the designer may increase the thickness of the cap. Alternatively, the amount of main reinforcement could be increased or, less desirably, shear links could be provided. Care should be taken that main reinforcing bars are fully anchored. The compression caused by the support reaction from the pile may be assumed to spread at 45 degree angles from the edge of the pile (Figure 6.11b).This compression may be taken into account when calculating the anchorage length. Also, when assessing the shear capacity in a pile cap, only the tension steel placed within the stress zone should be considered as contributing to the shear capacity.

Particular rules and detailing of column footings on piles are given in EN 1992-1-1 §9.8.1.

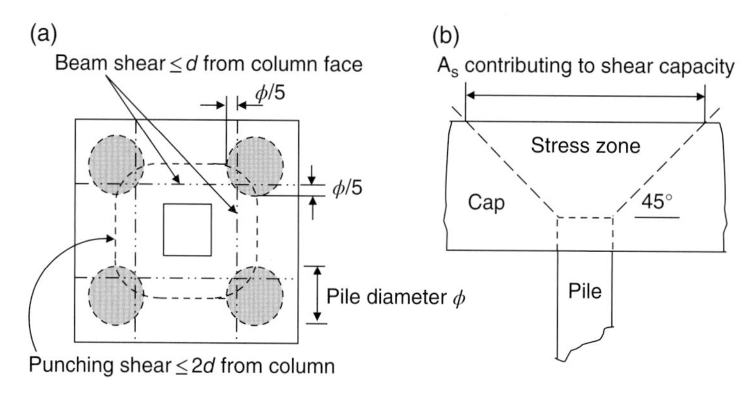

Figure 6.11 A pile cap: (a) critical shear perimeters, (b) compressed zone.

Table 6.18 Values for $v_{Rd,max}$, MPa.

f_{ck}	20	25	28	30	32	35	40	45	50
$v_{Rd,max}$	3.68	4.50	4.97	5.28	5.58	6.02	6.72	7.38	8.00

Table 6.19 Minimum percentage of reinforcement required.

f_{ck}	25	28	30	32	35	40	45	50
f_{ctm}	2.6	2.8	2.9	3.0	3.2	3.5	3.8	4.1
Minimum % ($0.26\, f_{ctm}/f_{yk}$); $f_{yk} = 500$ MPa	0.13	0.14	0.15	0.16	0.17	0.18	0.20	0.21

(I) *Raft foundations*

The basic design processes for rafts are similar to those for isolated pad foundations or pile caps. The only difference in approach lies in the selection of an appropriate method for analysing the interaction between the raft and he ground so as to achieve a reasonable representation of their behaviour (TCC, 2006).

For stiffer rafts (i.e. span to thickness smaller than 10) with a fairly regular layout, simplified approaches such as yield line or the flat slab equivalent frame method may be employed, once an estimation of the variations in bearing pressure has been obtained from a geotechnical specialist. Whatever simplifications are made, individual elastic raft reactions should equate to the applied column loads.

For thinner, more flexible rafts or for those with a complex layout, the application of a finite element or grillage analysis may be required. For rafts bearing on cohesionless subgrades or when contiguous walls or diaphragm perimeter walls are present, the ground may be modeled as a bed of Winkler springs (see Section 5.11 of Chapter 5).For rafts bearing on cohesive subgrades, this approach is unlikely to be valid, and specialist software will be required (TCC, 2006).

(J) *Spacing and quantity of reinforcement*

The minimum clear spacing of reinforcing bars should be the greatest of (see EN 1992-1-1 § 8.2):

- Bar diameter
- Aggregate size plus 5 mm
- 20 mm

The minimum area of principal reinforcement is $A_{s,min} = 0.26 \frac{f_{ctm} b_t d}{f_{yk}}$, but not less than $0.0013\, b_t d$ (see Table 6.19; EN 1992-1-1 §9.2.1.1).

Except at lap locations, the maximum area of tension or compression reinforcement, should not exceed $A_{s,max} = 0.04\,A_c$ (EN 1992-1-1 §9.2.1.1).

(K) *Bar bend and anchorage length*

The minimum diameter to which a bar is bent shall be such as to avoid bending cracks in the bar, and to avoid failure of the concrete inside the bend of the bar. In order to avoid these damages, EN 1992-1-1 §8.3(2) specifies minimum mandrel diameter $\emptyset_{m,min}$ (diameter to which the bar is bent) to be 4∅ and 7∅ for bar diameters $\emptyset \le 16$ mm and $\emptyset > 16$ mm, respectively. EN 1992-1-1 §8.3(3) give further guidance and provisions regarding the mandrel diameter.

Reinforcing bars shall be so anchored that the bond forces are safely transmitted to the concrete avoiding longitudinal cracking or spalling. Transverse reinforcement shall be provided if necessary. The calculation of the required anchorage length shall take into consideration the type of steel and bond properties of the bar. EN 1992-1-1 §8.4.3give guidance and an equation for calculation of the basic required anchorage length $l_{b,rqd}$. Also, EN 1992-1-1 §8.4.4 give guidance and an equation for calculation of the design anchorage length l_{bd}. For bent bars the lengths $l_{b,rqd}$ and l_{bd} should be measured along the centre line of the bar [EN 1992-1-1 §8.4.3(3)]. Bends and hooks do not contribute to compression anchorages [EN 1992-1-1 §8.4.1(3)].

Table 6.20 gives anchorage length l_{bd} for reinforcing straight bars, and length $l_{b,rqd}$ for standard bends, hooks and loops, in footings, based on the simplified procedure suggested in EN 1992-1-1 §8.4.4(2). The given length values account for good or poor bond conditions (related to concreting) as well as reinforcement function (i.e. tension or compression reinforcement). Also, for standard bends or hooks, the table data are based on the assumption that minimum bar spacing $s_{min} \ge 2c_{nom}$, using the nominal concrete cover to reinforcement $c_{nom} = 40$ mm.

(L) *Plain concrete foundations*

EN 1992-1-1 §12.9.3(1) states: In the absence of more detailed data, axially loaded strip and pad footings may be designed and constructed as plain concrete provided that:

$$\frac{0.85 \times h_F}{a} \ge \sqrt{\left(3\sigma_{gd}/f_{ctd,pl}\right)}$$

Where:

h_F = The foundation depth (footing thickness; see Figure 6.12)
a = The projection from the column face
σ_{gd} = The design value of the ground bearing pressure

Table 6.20 Anchorage lengths for reinforcing bars in footings (C25/30).

Diameter ∅ (mm)	Straight bar l_{bd} (mm)				Bend, hook and loop (mm)	
	Tension		Compression		Tension	
	Good	Poor	Good	Poor	Good	Poor
8	226	323	323	461	226	323
10	283	404	404	577	283	404
12	339	484	484	692	339	484
14	408	582	565	807	565	807
16	500	715	646	922	646	922
20	686	980	807	1153	807	1153
25	918	1312	1009	1441	1009	1441

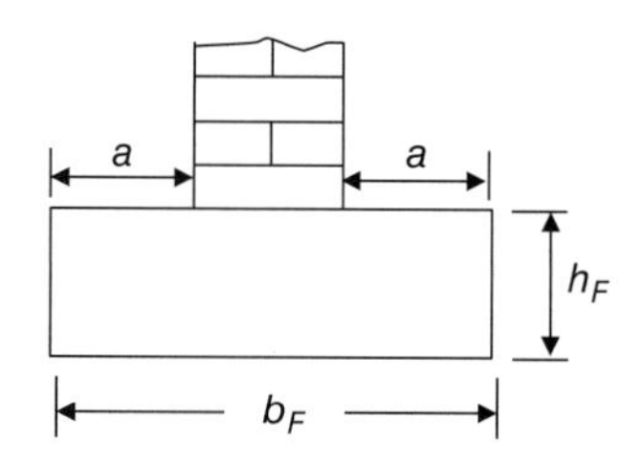

Figure 6.12 Unreinforced strip and pad footings; notations.

$f_{ctd,pl}$ = The design concrete tensile strength from Expression (3.16).

As a simplification the relation $(h_F/a) \geq 2$ may be used.

For plain concrete in compression: the value of a_{cc} should be taken as 0.6 instead of the 0.85 usually used for reinforced concrete. The coefficient a_{cc} is applied to the design compressive strength to take account of long-term effects.

The possibility of splitting forces, especially, for footings on rock, as advised in EN 1992-1-1 §9.8.4, may need to be considered.

EN 1992-1-1 §12.1(4) allows plain concrete to contain reinforcement needed to satisfy serviceability and/or durability requirements, for example, to control cracking.

Problem Solving

Problem 6.1

A building consists of $n = 3$ stories with plan dimensions $L = 48$ m and $B = 15$ m; divided into $N_L = 8$ bays in the L direction and $N_B = 2$ bays in the B direction, as shown in the scheme below. The height of each story is $h = 3.2$ m. The floors of the building are $d_{floor} = 250$ mm thick.

Shear walls, intended to resist overturning, are located at both ends of the building and are $t = 300$ mm thick by $b_w = 4$ m wide on plan.

A water tank, $d_{tank} = 2$ m deep by $l_{tank} = 5$ m long by $b_{tank} = 5$ m wide, sits over the shear wall at one end of the building.

The shear walls are supported by strip foundations of length $l_{fdn} = 6.5$ m, width $b_{fdn} = 2$ m and $d_{fdn} = 1.5$ m. The following characteristic imposed/wind actions act on the building:

- roof loading $q_{rf,k} = 0.6\,\text{kPa}$
- office floor loading $q_{off,k} = 2.5\,\text{kPa}$
- partition loading $q_{par,k} = 0.8\,\text{kPa}$
- wind (horizontal) $q_{w,k} = 1.15\,\text{kPa}$

The characteristic unit weight of reinforced concrete is $\gamma_{c,k} = 25\,\text{kN/m}^3$ and of water $\gamma_{w,k} = 10\,\text{kN/m}^3$.

Required: the way in which actions should be combined according to Eurocode (EN 1990), in a way that is suitable for geotechnical design of foundations.

Solution:

Geometry

The total plan area of building is

$$A_{total} = L \times B = 48 \times 15 = 720\,\text{m}^2$$

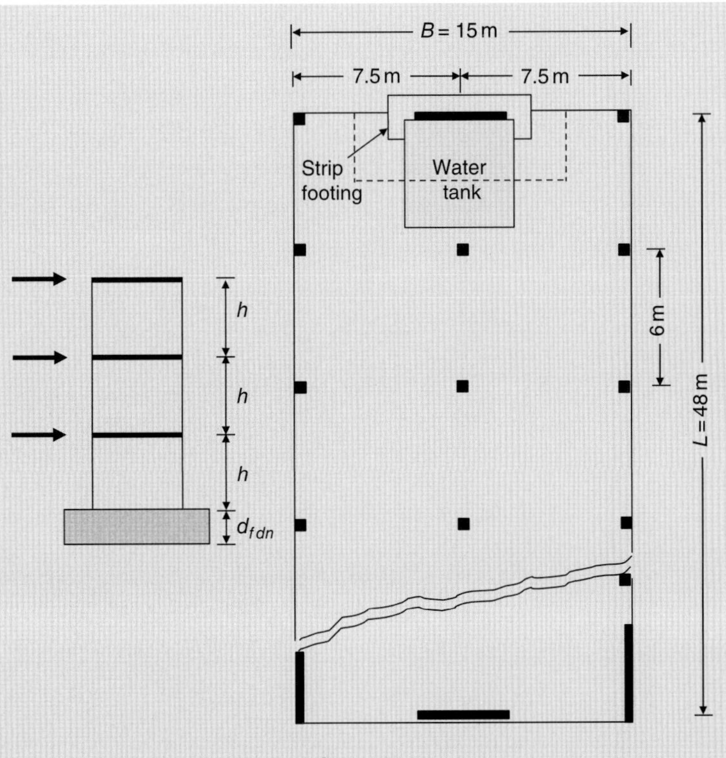

Scheme 6.1

The tributary area above the stability wall is

$$A = \left(\frac{B + b_w}{2}\right) \times \frac{1}{2}\left(\frac{L}{N_L}\right) = \left(\frac{15 + 4}{2}\right) \times \frac{1}{2}\left(\frac{48}{8}\right) = 28.5\,\text{m}^2$$

Characteristic ***permanent*** *actions*
Self-weight of slabs:

- Floor $g_{fl,Gk} = \gamma_{c,\,k} \times d_{floor} = 25 \times 0.25 = 6.25\,\text{kPa}$
- Screed on roof $g_{scr,Gk} = 1.5\,\text{kPa}$
- Raised floor $g_{r,-fl,Gk} = 0.5\,\text{kPa}$ (removable)

Self-weight of water tank on roof-only half total weight is carried by the core wall:

$$W_{tank,Gk} = \frac{1}{2}\gamma_{w,\,k} \times d_{tank} \times l_{tank} \times b_{tank} = \frac{1}{2} \times 10 \times 2 \times 5 \times 5$$
$$= 250\text{ kN} \qquad \text{(removable)}$$

Self-weight of core wall:

$$W_{wall,Gk} = \gamma_{c,\,k} t_w b_w (nh) = 25 \times 0.3 \times 4 \times 3 \times 3.2 = 288\,\text{kN}$$

Self-weight of strip foundation:

$$W_{fdn,Gk} = \gamma_{c,\,k} d_{fdn} b_{fdn} l_{fdn} = 25 \times 1.5 \times 2 \times 6.5 = 488\,\text{kN}$$

(*Continued*)

Total self-weight of *non-removable* members (normal to ground):

$$N_{Gk1} = A\left(ng_{fl,Gk} + g_{scr,Gk}\right) + W_{wall,Gk} + W_{fdn,Gk}$$
$$= 28.5(3 \times 6.25 + 1.5) + 288 + 488 = 1353\, kN$$

Total self-weight of *removable* members (normal to ground):

$$N_{Gk2} = A(n-1)g_{r,-fl,Gk} + W_{tank,\,Gk}$$
$$= 28.5(3-1) \times 0.5 + 250 = 279\, kN$$

Characteristic ***variable*** *actions*
Imposed actions (normal to ground):

- On roof $N_{rf,Qk} = q_{rf,k}A = 0.6 \times 28.5 = 171\, kN$
- On floor $N_{fl,Qk} = (n-1)\left(q_{off,k} + q_{par,k}\right)A$ $= (3-1)(2.5+0.8) \times 28.5 = 188.1\, kN$

Wind actions (horizontal direction):

- On roof $Q_{w,rf,\,Qk} = q_{w,\,k}\left(\frac{h}{2}\right)\left(\frac{L}{2}\right) = 1.15\left(\frac{3.2}{2}\right)\left(\frac{48}{2}\right) = 44.2\,\text{kN}$
- On each floor $Q_{w,fl,\,Qk} = q_{w,\,k}(h)\left(\frac{L}{2}\right)$ $= 1.15(3.2)\left(\frac{48}{2}\right) = 88.3\,\text{kN}$

Total wind action (normal to ground):

$$N_{w,\,Qk} = 0\, kN$$

Moment effect of wind action (on ground):

- First floor $M_{w,\,Qk1} = Q_{w,fl,\,Qk}\left[(n-2)h + d_{fdn}\right]$ $= 88.3[(3-2)(3.2) + 1.5] = 415\,\text{kN.m}$
- Second floor $M_{w,\,Qk2} = Q_{w,fl,\,Qk}\left[(n-1)h + d_{fdn}\right]$ $= 88.3[(3-1)(3.2) + 1.5] = 698\,\text{kN.m}$
- Roof $M_{w,\,Qk3} = Q_{w,rf,\,Qk}\left[(n-0)h + d_{fdn}\right]$ $= 44.2[(3-0)(3.2) + 1.5] = 490\,\text{kN.m}$
- total $M_{w,\,Qk} = \sum \text{Moment effect of wind action}$ $= 415 + 698 + 490 = 1603\, kN.m$

Combinations of actions for persistent and transient design situations – ULS verification
Combination 1 – wind as leading variable action, vertical actions unfavourable, partial factors from Set B in Table 6.6 or Section 6.3.4(G).

- Partial factors:
  - On permanent actions $\gamma_G = 1.35$
  - On variable actions (wind) $\gamma_{Q,w} = 1.5$
  - On variable actions (imposed loads) $\gamma_{Q,i} = 1.5$
- Combination factors (Table 6.3):
  - For wind (leading variable action) $\psi_w = 1.0$
  - For imposed load in office areas (Category B) $\psi_{fl} = \psi_{0,i,B} = 0.7$
  - For imposed load on roof (Category H) $\psi_{rf} = \psi_{0,i,H} = 0.0$ (also see EN 1991-1-1: Clause 3.3.2)
- Design value of normal action effect:

$$N_{Ed} = \gamma_G(N_{Gk1} + N_{Gk2}) + \gamma_{Q,w}\psi_w N_{w,\,Qk} + \gamma_{Q,I}\left(\psi_{fl}N_{fl,\,Qk} + \psi_{rf}N_{rf,\,Qk}\right)$$
$$= 1.35(1353 + 279) + 0 + 1.5(0.7 \times 188.1 + 0^*) = 2401\ kN$$

$*$ See EN 1991-1-1 §3.3.2(1)

- Design value of moment effect:

$$M_{Ed} = \gamma_{Q,w}\psi_w M_{w,\,Qk} = 1.5 \times 1 \times 1603 = 2405\ kN.m$$

- Maximum bearing pressure ($P_{\max,\,Ed}$) on underside of strip foundation:
  - Check eccentricity e_l:

$$e_l = (M_{Ed}/N_{Ed}) = (2405/2401) = 1.00\ \text{m} < \frac{l_{fdn}}{6}\ (OK.)$$

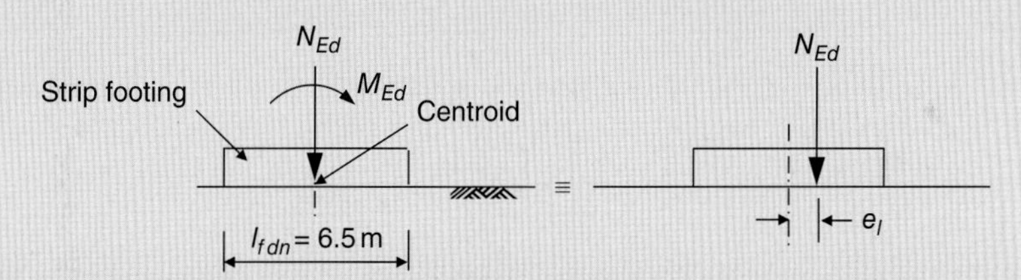

Scheme 6.2

$$P_{\max,\,Ed} = \frac{N_{Ed}}{A}\left(1 + \frac{6\,e_l}{l_{fdn}}\right) \qquad \text{[Equation(2.7)] or}$$

$$P_{\max,\,Ed} = \left(\frac{N_{Ed}}{b_{fdn}l_{fdn}}\right) + \left(\frac{6\,M_{Ed}}{b_{fdn}l_{fdn}^2}\right)$$
$$= \left(\frac{2401}{2 \times 6.5}\right) + \left(\frac{6 \times 2405}{2 \times 6.5^2}\right) = 184.69 + 170.77 = 355.5\ kPa$$

Combination 2 – wind as leading variable action, vertical actions favourable ($\gamma_G = 1$ and $\gamma_{Q,i} = 0$), partial factors from Set B in Table 6.6 or Subsection 6.3.4(G).

- Design value of normal action effect:

$$N_{Ed} = \gamma_{G,\,fav}(N_{Gk1} + N_{Gk2}) = (1)(1353 + 279) = 1632\ kN$$

- Design value of moment effect:

$$M_{Ed} = \gamma_{Q,w}\psi_w M_{w,\,Qk} = 1.5 \times 1 \times 1603 = 2405\ kN.m$$

(Continued)

- Maximum bearing pressure ($P_{max,Ed}$) on underside of strip foundation:
  - Check eccentricity e_l:

$$e_l = (M_{Ed}/N_{Ed}) = (2405/1632) = 1.47\,\text{m} > \frac{l_{fdn}}{6} \quad \text{(Not OK.)}$$

Therefore, Equation (2.7) is not applicable.

Since tension cannot occur between ground and the underside of the footing, the triangular bearing pressure distribution shown below may be considered (see Section 2.6.2 of Chapter 2).

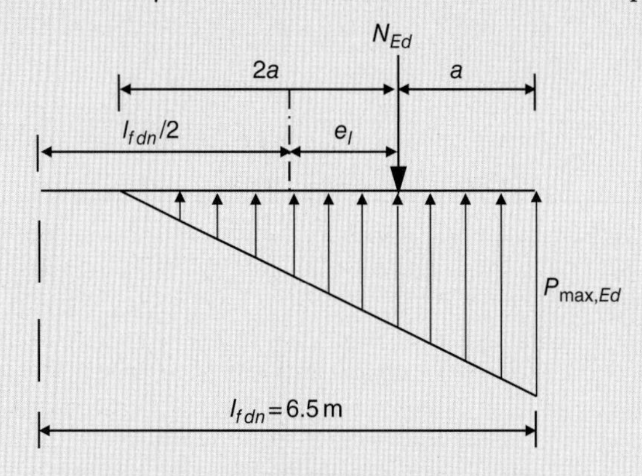

Scheme 6.3

$$a = \frac{l_{fdn}}{2} - e_l = \left(\frac{1}{2}\right)(l_{fdn} - 2e_l)$$

$$N_{Ed} = \left(\frac{P_{max,\,Ed} \times 3a}{2}\right)(b_{fdn}) = (P_{max,\,Ed})\left(\frac{3}{4}\right)(b_{fdn})(l_{fdn} - 2e_l)$$

$$P_{max,\,Ed} = \frac{4N_{Ed}}{3b_{fdn}(l_{fdn} - 2e_l)} = \frac{4 \times 1632}{3 \times 2(6.5 - 2 \times 1.47)} = 306\,kPa$$

However, TCC (2006) recommends a rectangular ULS pressure distribution diagram to be considered and not triangular, as shown below.

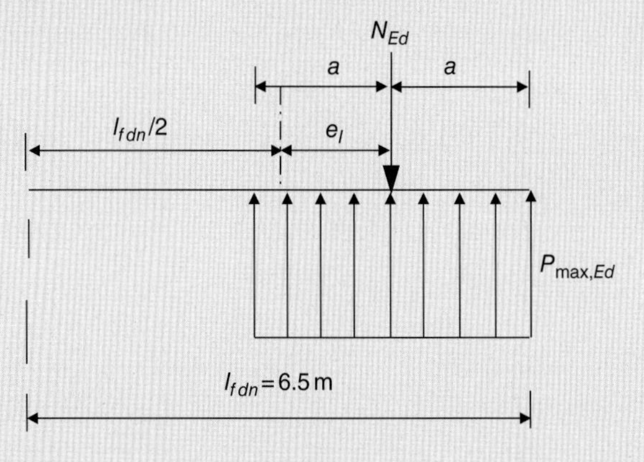

Scheme 6.4

Accordingly, the maximum bearing pressure ($P_{max,\,Ed}$) on the underside of the strip foundation is

$$P_{max,\,Ed} = \frac{N_{Ed}}{b_{fdn} \times 2a} = \frac{1632}{2 \times 2\left(\frac{1}{2}\right)(l_{fdn} - 2e_l)} = \frac{1632}{2(6.5 - 2 \times 1.47)} = 229\,\text{kPa}$$

Combination 3 – imposed loads as leading variable action, vertical actions unfavourable, partial factors from Set B in Table 6.6 or Section 6.3.4(G).

- Partial factors:
  - On permanent actions $\gamma_G = 1.35$
  - On variable actions (imposed loads) $\gamma_{Q,i} = 1.5$
  - On variable actions (imposed loads) (wind) $\gamma_{Q,w} = 1.5$
- Combination factors (Table 6.3):
  - For imposed load in office areas (Category B) $\psi_{fl} = 1.0$
  - For imposed load on roof (Category H) $\psi_{rf} = 1.0$
  - For wind $\psi_w = \psi_{0,w} = 0.6$
- Design value of normal action effect:

$$N_{Ed} = \gamma_G(N_{Gk1} + N_{Gk2}) + \gamma_{Q,i}\left(\psi_{fl}N_{fl,Qk} + \psi_{rf}N_{rf,Qk}\right) + \gamma_{Q,w}\psi_{0,w}N_{w,Qk}$$

$$= 1.35(1353 + 279) + 1.5(1 \times 188.1 + 1 \times 17.1) + 0$$

$$= \mathit{2511\,kN}$$

- Design value of moment effect:

$$M_{Ed} = \gamma_{Q,w}\psi_{0,w}M_{w,Qk} = 1.5 \times 0.6 \times 1603 = \mathit{1443\,kN.m}$$

- Maximum bearing pressure ($P_{max,Ed}$) on underside of strip foundation:
  - Check eccentricity e_l:

$$e_l = (M_{Ed}/N_{Ed}) = (1443/2511) = 0.57\,\text{m} < \frac{l_{fdn}}{6}\,(\text{OK}).$$

$$P_{max,Ed} = \left(\frac{N_{Ed}}{b_{fdn}l_{fdn}}\right) + \left(\frac{6M_{Ed}}{b_{fdn}l_{fdn}^2}\right)$$

$$= \left(\frac{2511}{2 \times 6.5}\right) + \left(\frac{6 \times 1443}{2 \times 6.5^2}\right) = 193.15 + 102.46 = \mathit{296\,kPa}$$

Combination 4 – wind as leading variable action, vertical actions unfavourable, partial factors from Set C in Table 6.6 or Section 6.3.4(G).

- Partial factors:
  - On permanent actions $\gamma_G = 1.0$
  - On variable actions (wind) $\gamma_{Q,w} = 1.3$
  - On variable actions (imposed loads) $\gamma_{Q,i} = 1.3$
- Combination factors (Table 6.3):
  - For wind (leading variable action) $\psi_w = 1.0$
  - For imposed load in office areas (Category B) $\psi_{fl} = \psi_{0,i,B} = 0.7$
  - For imposed load on roof (Category H) $\psi_{rf} = \psi_{0,i,H} = 0.0$ (also see EN 1991-1-1: Clause 3.3.2)
- Design value of normal action effect:

$$N_{Ed} = \gamma_G(N_{Gk1} + N_{Gk2}) + \gamma_{Q,w}\psi_wN_{w,Qk} + \gamma_{Q,I}\left(\psi_{fl}N_{fl,Qk} + \psi_{rf}N_{rf,Qk}\right)$$

$$= 1.0(1353 + 279) + 0 + 1.3(0.7 \times 188.1 + 0^*) = \mathit{1803\,kN}$$

(Continued)

∗ See EN 1991-1-1 §3.3.2(1)

- Design value of moment effect:

$$M_{Ed} = \gamma_{Q,w}\psi_w M_{w,\,Qk} = 1.3 \times 1 \times 1603 = 2084\,kN.m$$

- Maximum bearing pressure ($P_{\max,\,Ed}$) on underside of strip foundation.
  The calculations proceed in the same manner as those for the preceding combinations; left for the reader.

  Combination 5– wind as leading variable action, vertical actions favourable ($\gamma_G = 1$ and $\gamma_{Q,i} = 0$), partial factors from Set C in Table 6.6 or Section 6.3.4(G).
- Design value of normal action effect:

$$N_{Ed} = \gamma_{G,\,fav}(N_{Gk1} + N_{Gk2}) = (1)(1353 + 279) = 1632\,kN$$

- Design value of moment effect:

$$M_{Ed} = \gamma_{Q,w}\psi_w M_{w,\,Qk} = 1.3 \times 1 \times 1603 = 2084\,kN.m$$

- Maximum bearing pressure ($P_{\max,\,Ed}$) on underside of strip foundation:
  The calculations proceed in the same manner as those for the preceding combinations; left for the reader.
  Combination 6 – imposed loads as leading variable action, vertical actions unfavourable, partial factors from Set C in Table 6.6 or Section 6.3.4(G).
- Partial factors:
  - On permanent actions $\gamma_G = 1.0$
  - On variable actions (imposed loads) $\gamma_{Q,i} = 1.3$
  - On variable actions (imposed loads) (wind) $\gamma_{Q,w} = 1.3$
- Combination factors (Table 6.3):
  - For imposed load in office areas (Category B) $\psi_{fl} = 1.0$
  - For imposed load on roof (Category H) $\psi_{rf} = 1.0$
  - For wind $\psi_w = \psi_{0,\,w} = 0.6$
- Design value of normal action effect:

$$\begin{aligned} N_{Ed} &= \gamma_G(N_{Gk1} + N_{Gk2}) + \gamma_{Q,i}\left(\psi_{fl}N_{fl,\,Qk} + \psi_{rf}N_{rf,\,Qk}\right) + \gamma_{Q,w}\psi_{0,w}N_{w,\,Qk} \\ &= 1.0(1353 + 279) + 1.3(1 \times 188.1 + 1 \times 17.1) + 0 \\ &= 1899\,kN \end{aligned}$$

- Design value of moment effect:

$$M_{Ed} = \gamma_{Q,w}\psi_{0,w} M_{w,\,Qk} = 1.3 \times 0.6 \times 1603 = 1250\,kN.m$$

- Maximum bearing pressure ($P_{\max,\,Ed}$) on underside of strip foundation.
  The calculations proceed in the same manner as those for the preceding combinations; left for the reader.
  Combinations of actions for quasi-persistent design situations – SLS verifications
  Combination 1 – wind as leading variable action, vertical actions unfavourable, partial factors for SLS.
- Partial factors [see Section 6.3.4(J)]:
  - On permanent actions $\gamma_G = \gamma_{G,\,SLS} = 1.0$
  - On variable actions (wind) $\gamma_{Q,w} = \gamma_{Q,w,\,SLS} = 1.0$
  - On variable actions (imposed loads) $\gamma_{Q,i} = \gamma_{Q,i,\,SLS} = 1.0$
- Combination factors [see Section 6.3.4(J) and Table 6.3]:
  - For wind $\psi_w = \psi_{2,\,w} = 0.0$
  - For imposed load in office areas (Category B) $\psi_{fl} = \psi_{2,i} = 0.3$
  - For imposed load on roof (Category H) $\psi_{rf} = \psi_{2,i} = 0.0$

- Design value of normal action effect:

$$N_{Ed} = \gamma_G(N_{Gk1} + N_{Gk2}) + \gamma_{Q,w}\psi_w N_{w,\,Qk} + \gamma_{Q,i}\left(\psi_{fl} N_{fl,\,Qk} + \psi_{rf} N_{rf,\,Qk}\right)$$
$$= 1.0(1353 + 279) + 0 + 1.0(0.3 \times 188.1 + 0) = \mathit{1688\,kN}$$

- Design value of moment effect:

$$M_{Ed} = \gamma_{Q,w}\psi_w M_{w,\,Qk} = 1.0 \times 0 \times 1603 = \mathit{0.0\,kN.m}$$

Combination 2 – wind as leading variable action, vertical actions favourable, partial factors for SLS.

- Design value of normal action effect:

$$N_{Ed} = \gamma_{G,\,\text{fav}}(N_{Gk1} + N_{Gk2}) = 1.0(1353 + 279) = \mathit{1632\,kN}$$

- Design value of moment effect:

$$M_{Ed} = \gamma_{Q,w}\psi_w M_{w,\,Qk} = 1.0 \times 0 \times 1603 = \mathit{0.0\,kN.m}$$

Combination 3 – imposed loads as leading variable action, vertical actions unfavourable, partial factors for SLS.

- Combination factors [see Section 6.3.4(J) and Table 6.3]:
  - For wind $\psi_w = \psi_{2,\,w} = 0.0$
  - For imposed load in office areas (Category B) $\psi_{fl} = \psi_{2,i} = 0.3$
  - For imposed load on roof (Category H) $\psi_{rf} = \psi_{2,i} = 0.0$
- Design value of normal action effect:

$$N_{Ed} = \gamma_G(N_{Gk1} + N_{Gk2}) + \gamma_{Q,i}\left(\psi_{fl} N_{fl,\,Qk} + \psi_{rf} N_{rf,\,Qk}\right) + \gamma_{Q,w}\psi_w N_{w,\,Qk}$$
$$= 1.0(1353 + 279) + 1.0(0.3 \times 188.1 + 0) + 0$$
$$= \mathit{1688\,kN}$$

- Design value of moment effect:

$$M_{Ed} = \gamma_{Q,w}\psi_w M_{w,\,Qk} = 1.0 \times 0 \times 1603 = \mathit{0.0\,kN.m}$$

Problem 6.2

A centrally loaded column footing (pad) of length $L = 2.5$ m, width $B = 1.5$ m and depth (thickness) $d = 0.5$ m is required to carry a vertical imposed permanent action $V_{Gk} = 800$ kN and a vertical imposed variable action $V_{Qk} = 450$ kN. The footing base is horizontal (i.e. $\alpha = 0°$) and located at a depth of 0.5 m below ground surface (i.e. $D = 0.5$ m). The weight density (unit weight) of the reinforced concrete is $\gamma_{c,\,k} = 25\,\text{kN/m}^3$ (see Table 6.12). The footing is founded on dry sand which has the following characteristic parameters:

- Angle of shearing resistance $\phi_k = 35°$
- Effective cohesion $c'_k = 0.0$ kPa
- weight density $\gamma_k = 18\,\text{kN/m}^3$

It is required to perform verification of strength (GEO ultimate limit state), that is verification of bearing resistance, using all the three Design Approaches. *Note:* In order to concentrate on the EN 1997 rather than the geotechnical related issues a relatively simple problem has been selected which excludes the effects of groundwater.

(Continued)

Solution:

(A) *Design Approach 1*

DA1 – Combination 1: $A1$ "+"$M1$ "+"$R1$

– Combination 2: $A2$" + "M2" + "$R1$

Actions and effects:

$$\text{Characteristic self-weight of footing } W_{Gk} = \gamma_{c,k} \times L \times B \times d$$
$$= 25 \times 2.5 \times 1.5 \times 0.5$$
$$= 46.9\,\text{kN}$$

Area of base $A_b = L \times B = 2.5 \times 1.5 = 3.75\,\text{m}^2$

Partial factors from Table 6.6 for $\begin{pmatrix}\text{Combination } 1\\ \text{Combination } 2\end{pmatrix}$:

$$\gamma_G = \begin{pmatrix}1.35\\1\end{pmatrix};\ \gamma_Q = \begin{pmatrix}1.5\\1.3\end{pmatrix}$$

Design vertical action: $V_d = \gamma_G(W_{Gk} + V_{Gk}) + \gamma_Q V_{Qk}$

$$= \begin{pmatrix}1.35(46.9+800) + 1.5 \times 450\\ 1(46.9+800) + 1.3 \times 450\end{pmatrix} = \begin{pmatrix}1818.3\\1431.9\end{pmatrix}\text{kN}$$

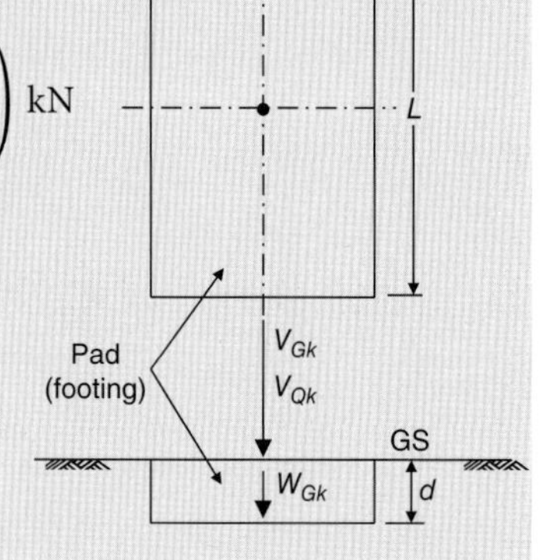

Scheme 6.5

Design bearing pressure [Equation (6.8)]:

$$q_{Ed} = \frac{V_d}{A_b} = \begin{pmatrix}\dfrac{1818.3}{3.75}\\ \dfrac{1431.9}{3.75}\end{pmatrix} = \begin{pmatrix}484.9\\381.8\end{pmatrix}\text{kPa}$$

Materials properties and resistance:

Partial factors from Table 6.7 for $\begin{pmatrix}\text{Combination } 1\\ \text{Combination } 2\end{pmatrix}$:

$$\gamma_\varnothing = \begin{pmatrix}1\\1.25\end{pmatrix};\gamma_c = \begin{pmatrix}1\\1.25\end{pmatrix}$$

Design angle of shearing resistance

$$\phi_d = \tan^{-1}\left(\frac{\tan\phi_k}{\gamma_\varnothing}\right) = \begin{bmatrix}\tan^{-1}\left(\dfrac{\tan 35}{1}\right)\\ \tan^{-1}\left(\dfrac{\tan 35}{1.25}\right)\end{bmatrix}$$

$$= \begin{pmatrix}35\\29.3\end{pmatrix}^\circ$$

Design effective cohesion $c'_d = \frac{c'_k}{\gamma_c} = \begin{pmatrix} 0 \\ 0 \end{pmatrix}$ kPa

Bearing capacity factors from Annex D of EN 1997-1:

For effect of overburden, $N_q = e^{\pi \tan \varnothing_d} \tan^2\left(45^o + \frac{\varnothing_d}{2}\right) = \begin{pmatrix} 33.3 \\ 17.0 \end{pmatrix}$

For effect of cohesion, $N_c = (N_q - 1)\cot\phi_d = \begin{pmatrix} 46.1 \\ 28.4 \end{pmatrix}$

For effect of self-weight, $N_\gamma = 2(N_q - 1)\tan\phi_d = \begin{pmatrix} 45.2 \\ 17.8 \end{pmatrix}$

Shape factors:

For effect of overburden, $s_q = 1 + \frac{B}{L}\sin\phi_d = \begin{pmatrix} 1.34 \\ 1.29 \end{pmatrix}$

For effect of cohesion, $s_c = \frac{(s_q N_q - 1)}{(N_q - 1)} = \begin{pmatrix} 1.35 \\ 1.31 \end{pmatrix}$

For effect of self-weight, $s_\gamma = 1 - 0.3\frac{B}{L} = \begin{pmatrix} 0.82 \\ 0.82 \end{pmatrix}$

Depth factors:

The suggested method in Annex D does not include depth factors which are present in other formulations of the extended bearing capacity formula (see Section 4.5 of Chapter 4). There has been concern in using these depth factors as their influence can be significant and the reliance on the additional capacity provided by its inclusion is not conservative.

Load-inclination factors, caused by a horizontal load H:

Since $H = 0$, each of the load-inclination factors i_c, i_q and i_γ is 1.0.

Base-inclination factors:

Since the foundation base is horizontal, that is $\alpha = 0$, each of the base-inclination factors b_c, b_q and b_γ is 1.0.

Ground-inclination factors:

The suggested method in Annex D omits the ground-inclination factors which are present in other formulations of the extended bearing capacity formula. However, neglecting these factors is unsafe.

Effective overburden pressure q' (or $\sigma'_{vk,b}$):

$$q' = \gamma_k \times D = 18 \times 0.5 = 9\,\text{kN/m}^2$$

Gross ultimate bearing capacity (*Gross* q_{ult}) or gross bearing resistance (R/A'):

Annex D of EN 1997-1 suggests that bearing resistance may be calculated from

$$R/A' = \underbrace{c' N_c\, b_c\, s_c\, i_c}_{\text{Cohesion}} + \underbrace{q' N_q\, b_q\, s_q\, i_q}_{\text{Overburden}} + \underbrace{0.5\gamma' B' N_\gamma\, b_\gamma s_\gamma\, i_\gamma}_{\text{Self-weight}}$$

- From cohesion: $(Gross\, q_{ult})_1 = \begin{pmatrix} 0 \\ 0 \end{pmatrix}\begin{pmatrix} 46.1 \\ 28.4 \end{pmatrix}\begin{pmatrix} 1.35 \\ 1.31 \end{pmatrix} = \begin{pmatrix} 0 \\ 0 \end{pmatrix}$ kPa

(*Continued*)

- From overburden: $(Gross\, q_{ult})_2 = \begin{pmatrix} 9 \\ 9 \end{pmatrix} \begin{pmatrix} 33.3 \\ 17.0 \end{pmatrix} \begin{pmatrix} 1.34 \\ 1.29 \end{pmatrix} = \begin{pmatrix} 402 \\ 197 \end{pmatrix}$ kPa

- From self-weight: $(Gross\, q_{ult})_3 = \begin{pmatrix} 0.5 \\ 0.5 \end{pmatrix} \begin{pmatrix} 18 \\ 18 \end{pmatrix} \begin{pmatrix} 1.5 \\ 1.5 \end{pmatrix} \begin{pmatrix} 45.2 \\ 17.8 \end{pmatrix} \begin{pmatrix} 0.82 \\ 0.82 \end{pmatrix}$

$$= \begin{pmatrix} 500 \\ 197 \end{pmatrix} \text{kPa}$$

$$\text{Gross bearing resistance } R/A' = Gross\, q_{ult} = \sum_{i=1}^{3} (Gross\, q_{ult})_i$$

$$= \begin{pmatrix} 902 \\ 394 \end{pmatrix} \text{kPa}$$

From Set $R1$ in Table A.5 of Annex A to EN 1997-1: the partial factor on bearing resistance is

$$\gamma_{R;v} = \begin{pmatrix} 1 \\ 1 \end{pmatrix}$$

$$\textit{Design bearing resistance: } q_{Rd} = \frac{R/A'}{\gamma_{R;v}} = \frac{\begin{pmatrix} 902 \\ 394 \end{pmatrix}}{\begin{pmatrix} 1 \\ 1 \end{pmatrix}} = \begin{pmatrix} 902 \\ 394 \end{pmatrix} kPa$$

Verification of bearing resistance

$$\text{Utilization factor } \Lambda_{GEO} = \frac{q_{Ed}}{q_{Rd}} = \begin{pmatrix} 484.9/902 \\ 381.8/394 \end{pmatrix} = \begin{pmatrix} 54\% \\ 97\% \end{pmatrix}$$

Since design is unacceptable if *utilisation factor* is greater than 100%, combination 2 of the Design Approach 1(i.e. DA1-2) is critical with a utilisation factor of 97%, implying that the requirements of the Eurocode are only just met.

(B) *Design Approach 2 A1 "+" M1 "+" R2*
Actions and effects:

$$\text{Characteristic self-weight of footing } W_{Gk} = \gamma_{c,k} \times L \times B \times d$$

$$= 25 \times 2.5 \times 1.5 \times 0.5$$

$$= 46.9\, \text{kN}$$

$$\text{Area of base } A_b = L \times B = 2.5 \times 1.5 = 3.75\, \text{m}^2$$

Partial factors from Set $A1$ in Table A.3 of Annex A to EN 1997-1:

$$\gamma_G = 1.35;\ \gamma_Q = 1.5$$

Design vertical action: $V_d = \gamma_G(W_{Gk} + V_{Gk}) + \gamma_Q V_{Qk}$

$$= 1.35(46.9 + 800) + 1.5 \times 450 = 1818.3 \text{ kN}$$

Design bearing pressure [Equation (6.8)]: $q_{Ed} = \frac{V_d}{A_b} = \frac{1818.3}{3.75} = 484.9\,kPa$

Materials properties and resistance:
Partial factors from Set *M*1 in Table A.4 of Annex A to EN 1997-1:

$$\gamma_{\varnothing} = 1;\ \gamma_c = 1$$

Design angle of shearing resistance $\phi_d = \tan^{-1}\left(\frac{\tan\phi_k}{\gamma_{\varnothing}}\right) = \tan^{-1}\left(\frac{\tan 35}{1}\right)$

$$= 35^\circ$$

Design effective cohesion $c'_d = \frac{c'_k}{\gamma_c} = \frac{0}{1} = 0\,\text{kPa}$

Bearing capacity factors from Annex D of EN 1997-1:

For effect of overburden, $N_q = e^{\pi \tan \varnothing_d} \tan^2\left(45^\circ + \frac{\varnothing_d}{2}\right) = 33.3$

For effect of cohesion, $N_c = (N_q - 1)\cot\phi_d = 46.1$

For effect of self-weight, $N_\gamma = 2(N_q - 1)\tan\phi_d = 45.2$

Shape factors:

For effect of overburden, $s_q = 1 + \frac{B}{L}\sin\phi_d = 1.34$

For effect of cohesion, $s_c = \frac{(s_q N_q - 1)}{(N_q - 1)} = 1.35$

For effect of self-weight, $s_\gamma = 1 - 0.3\frac{B}{L} = 0.82$

Depth factors:[see the comments given in (A) *Design Approach 1*]

Load-inclination factors, caused by a horizontal load *H*:

Since $H = 0$, each of the load-inclination factors i_c, i_q and i_γ is 1.0.

Base-inclination factors:

Since the foundation base is horizontal, that is $\alpha = 0$, each of the base-inclination factors b_c, b_q and b_γ is 1.0.

Ground-inclination factors: [see the comments given in (A) *Design Approach 1*]

Effective overburden pressure q' (or $\sigma'_{vk,b}$):

$$q' = \gamma_k \times D = 18 \times 0.5 = 9\,\text{kN/m}^2$$

Gross ultimate bearing capacity (*Gross* q_{ult}) or gross bearing resistance (R/A'):

Annex D of EN 1997-1 suggests that bearing resistance may be calculated from

$$R/A' = \underbrace{c' N_c\, b_c\, s_c\, i_c}_{\text{Cohesion}} + \underbrace{q' N_q\, b_q\, s_q\, i_q}_{\text{Overburden}} + \underbrace{0.5\gamma' B' N_\gamma\, b_\gamma s_\gamma\, i_\gamma}_{\text{Self-weight}}$$

(*Continued*)

- From cohesion: $(Gross\, q_{ult})_1 = (0)(46.1)(1.35) = 0\,\text{kPa}$
- From overburden: $(Gross\, q_{ult})_2 = (9)(33.3)(1.34) = 402\,\text{kPa}$
- From self-weight: $(Gross\, q_{ult})_3 = (0.5)(18)(1.5)(45.2)(0.82)$

$$= 500\,\text{kPa}$$

$$\text{Gross bearing resistance } R/A' = Gross\, q_{ult} = \sum_{i=1}^{3} (Gross\, q_{ult})_i$$

$$= 902\,\text{kPa}$$

From Set *R*2 in Table A.5 of Annex A to EN 1997-1: the partial factor on bearing resistance is

$$\gamma_{R;v} = 1.4$$

As it may be noticed, for Design Approach 2 the uncertainty in the calculation is covered through partial factors on the actions and on overall factor on the calculated resistance.

Design bearing resistance: $q_{\text{Rd}} = \dfrac{R/A'}{\gamma_{R;v}} = \dfrac{902}{1.4} = 644.3\,kPa$

Verification of bearing resistance

$$\text{Utilization factor } A_{\text{GEO}} = \frac{q_{\text{Ed}}}{q_{\text{Rd}}} = \frac{484.9}{644.3} = 75\%$$

Design is unacceptable if utilisation factor is greater than 100%.

The calculated utilisation factor (75%) would indicate that, for the design situation of this problem, according to DA2 the footing is potentially over-designed.

(C) *Design Approach 3 A1 " + "M2 " + "R3*

Actions and effects:

$$\text{Characteristic self-weight of footing } W_{Gk} = \gamma_{c,k} \times L \times B \times d$$

$$= 25 \times 2.5 \times 1.5 \times 0.5$$

$$= 46.9\,\text{kN}$$

Area of base $A_b = L \times B = 2.5 \times 1.5 = 3.75\,\text{m}^2$

Partial factors on structural actions from Set *A*1 in Table A.3 of Annex A to EN 1997-1:

$$\gamma_G = 1.35;\ \gamma_Q = 1.5$$

Design vertical action:

$$V_d = \gamma_G (W_{Gk} + V_{Gk}) + \gamma_Q V_{Qk}$$

$$= 1.35(46.9 + 800) + 1.5 \times 450 = 1818.3\,\text{kN}$$

Design bearing pressure [° (6.8)]: $q_{Ed} = \dfrac{V_d}{A_b} = \dfrac{1818.3}{3.75} = 484.9\,kPa$

Materials properties and resistance:

Partial factors from Set $M2$ in Table A.4 of Annex A to EN 1997-1:

$$\gamma_{\emptyset} = 1.25; \gamma_c = 1.25$$

Design angle of shearing resistance $\phi_d = \tan^{-1}\left(\frac{\tan\phi_k}{\gamma_{\emptyset}}\right) = \tan^{-1}\left(\frac{\tan 35}{1.25}\right)$

$$= 29.3^{\circ}$$

Design effective cohesion $c'_d = \frac{c'_k}{\gamma_c} = \frac{0}{1.25} = 0 \text{ kPa}$

Bearing capacity factors from Annex D of EN 1997-1:

For effect of overburden, $N_q = e^{\pi \tan \emptyset_d} \tan^2\left(45^o + \frac{\emptyset_d}{2}\right) = 17$

For effect of cohesion, $N_c = (N_q - 1)\cot\phi_d = 28.4$

For effect of self-weight, $N_\gamma = 2(N_q - 1)\tan\phi_d = 17.8$

Shape factors:

For effect of overburden, $s_q = 1 + \frac{B}{L}\sin\phi_d = 1.29$

For effect of cohesion, $s_c = \frac{(s_q N_q - 1)}{(N_q - 1)} = 1.31$

For effect of self-weight, $s_\gamma = 1 - 0.3\frac{B}{L} = 0.82$

Depth factors: [see the comments given in (A) *Design Approach 1*]

Load-inclination factors, caused by a horizontal load H:

Since $H = 0$, each of the load-inclination factors i_c, i_q and i_γ is 1.0.

Base-inclination factors:

Since the foundation base is horizontal, that is $\alpha = 0$, each of the base-inclination factors b_c, b_q and b_γ is 1.0.

Ground-inclination factors: [see the comments given in (A) *Design Approach 1*]

Effective overburden pressure q' (or $\sigma'_{vk,b}$):

$$q' = \gamma_k \times D = 18 \times 0.5 = 9 \text{ kN/m}^2$$

Gross ultimate bearing capacity (*Gross* q_{ult}) or gross bearing resistance (R/A'):

Annex D of EN 1997-1 suggests that bearing resistance may be calculated from

$$R/A' = \underbrace{c' N_c\, b_c\, s_c\, i_c}_{\text{Cohesion}} + \underbrace{q' N_q\, b_q\, s_q\, i_q}_{\text{Overburden}} + \underbrace{0.5\gamma' B' N_\gamma\, b_\gamma s_\gamma\, i_\gamma}_{\text{Self-weight}}$$

- From cohesion: $(Gross\, q_{ult})_1 = (0)(28.4)(1.31) = 0$ kPa
- From overburden: $(Gross\, q_{ult})_2 = (9)(17)(1.29) = 197$ kPa
- From self-weight: $(Gross\, q_{ult})_3 = (0.5)(18)(1.5)(17.8)(0.82)$

$$= 197 \text{ kPa}$$

Gross bearing resistance $R/A' = Gross\, q_{ult} = \sum_{i=1}^{3} (Gross\, q_{ult})_i$

$$= 394 \text{ kPa}$$

(Continued)

From Set $R3$ in Table A.5 of Annex A to EN 1997-1: the partial factor on bearing resistance is

$$\gamma_{R;v} = 1.0$$

As it may be noticed, Design Approach 3 applies partial factors to both actions and material properties at the same time.

Design bearing resistance: $$q_{\mathrm{Rd}} = \frac{R/A'}{\gamma_{R;v}} = \frac{394}{1.0} = 394\,kPa$$

Verification of bearing resistance

$$\text{Utilisation factor } A_{\mathrm{GEO}} = \frac{q_{\mathrm{Ed}}}{q_{\mathrm{Rd}}} = \frac{484.9}{394} = 123\%$$

The design is unacceptable since the utilisation factor is greater than 100%. Thus the DA3 calculation suggests the design is unsafe and re-design would be required.

Finally, one may comment on the three DAs calculations as follows:

The three design approaches gives different evaluation of the suitability of the proposed foundation for the given design situation and loading. Of the three approaches, DA1 suggests the footing is only just satisfactory whilst DA3 suggests redesign would be required and DA2 may indicate that the footing is overdesigned!

It may not be so easy to decide on which approach is the most appropriate or which one to be neglected or considered. However, it would appear that DA3 is unnecessary conservative as providing significant partial factors on both actions and material properties, and therefore, the footing redesign would not be necessary. At the end, considering both safety and economy requirements, one may decide on the DA1-2 calculations; knowing that this approach generally governs the size of the foundations.

Problem 6.3

Solve Problem 2 assuming that the rectangular pad footing is eccentrically loaded by the imposed actions. The eccentricities e_B and e_L are 0.075 and 0.100 m, respectively.

Solution:

(A) *Design Approach 1* DA1 – Combination 1: $A1$ "+" $M1$ "+" $R1$
– Combination 2: A1 "+" M2 "+" R1

Effective footing plan dimensions:

The footing self-weight and partial factors for actions are the same as those calculated in the Solution of Problem 6.2.

As mentioned earlier in Section 6.3.4(G), it would be best to base calculation of the eccentricities on design actions. The self-weight of the footing still acts through the centre of the footing. Accordingly, we can write:
Eccentricity of total vertical action

$$e'_B = \frac{e_B\left(\gamma_G V_{Gk} + \gamma_Q V_{Qk}\right)}{\gamma_G\left(W_{Gk} + V_{Gk}\right) + \gamma_Q V_{Qk}}$$

$$= \begin{pmatrix} \dfrac{0.075(1.35 \times 800 + 1.5 \times 450)}{1.35(46.9 + 800) + 1.5 \times 450} \\ \\ \dfrac{0.075(1 \times 800 + 1.3 \times 450)}{1(46.9 + 800) + 1.3 \times 450} \end{pmatrix} = \begin{pmatrix} 0.0724 \\ 0.0725 \end{pmatrix} \mathrm{m}$$

Eccentricity of total vertical action

$$e'_L = \frac{e_L(\gamma_G V_{Gk} + \gamma_Q V_{Qk})}{\gamma_G(W_{Gk} + V_{Gk}) + \gamma_Q V_{Qk}}$$

$$= \begin{pmatrix} \dfrac{0.100(1.35 \times 800 + 1.5 \times 450)}{1.35(46.9 + 800) + 1.5 \times 450} \\ \dfrac{0.100(1 \times 800 + 1.3 \times 450)}{1(46.9 + 800) + 1.3 \times 450} \end{pmatrix} = \begin{pmatrix} 0.0965 \\ 0.0967 \end{pmatrix} \text{m}$$

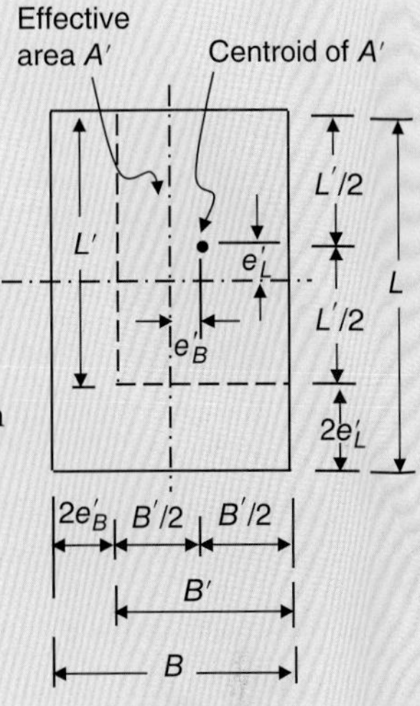

Scheme 6.6

Load is within the middle third since $e'_B < \left(\dfrac{B}{6} = 0.25\,\text{m}\right)$ and

$$e'_L < \left(\frac{L}{6} = 0.42\,\text{m}\right)$$

Effective width $B' = B - 2e'_B = 1.5 - 2\begin{pmatrix} 0.0724 \\ 0.0725 \end{pmatrix} = \begin{pmatrix} 1.36 \\ 1.35 \end{pmatrix}$ m

Effective length $L' = L - 2e'_L = 2.5 - 2\begin{pmatrix} 0.0965 \\ 0.0967 \end{pmatrix} = \begin{pmatrix} 2.31 \\ 2.31 \end{pmatrix}$ m

Effective area $A' = (L' \times B') = \begin{pmatrix} 2.31 \times 1.36 \\ 2.31 \times 1.35 \end{pmatrix} = \begin{pmatrix} 3.13 \\ 3.13 \end{pmatrix} \text{m}^2$

Actions and effects:

From previous calculation (Solution of Problem 6.2.), $V_d = \begin{pmatrix} 1818.3 \\ 1431.9 \end{pmatrix}$ kN

Design bearing pressure [Equation (6.8)]: $q_{Ed} = \dfrac{V_d}{A'_b} = \begin{pmatrix} \dfrac{1818.3}{3.13} \\ \dfrac{1431.9}{3.13} \end{pmatrix} = \begin{pmatrix} 581.6 \\ 458.2 \end{pmatrix}$ *kPa*

Materials properties and resistance:

From previous calculation (Solution of Problem 6.2.), $\phi_d = \begin{pmatrix} 35 \\ 29.3 \end{pmatrix}^{\circ}$ and

$$c'_d = \begin{pmatrix} 0 \\ 0 \end{pmatrix} \text{kPa}$$

Also, bearing capacity factors $N_q = \begin{pmatrix} 33.3 \\ 17.0 \end{pmatrix}$, $N_c = \begin{pmatrix} 46.1 \\ 28.4 \end{pmatrix}$ and $N_\gamma = \begin{pmatrix} 45.2 \\ 17.8 \end{pmatrix}$

(*Continued*)

Shape factors:

For effect of overburden, $s_q = 1 + \frac{B'}{L'}\sin\phi_d = \begin{pmatrix} 1.34 \\ 1.29 \end{pmatrix}$

For effect of cohesion, $s_c = \frac{(s_q N_q - 1)}{(N_q - 1)} = \begin{pmatrix} 1.35 \\ 1.31 \end{pmatrix}$

For effect of self-weight, $s_\gamma = 1 - 0.3\frac{B'}{L'} = \begin{pmatrix} 0.82 \\ 0.82 \end{pmatrix}$

Depth factors: [see the comments given in Solution of Problem 6.2- (A) *Design Approach 1*]

Load-inclination factors, caused by a horizontal load H:

Since $H = 0$, each of the load-inclination factors i_c, i_q and i_γ is 1.0.

Base-inclination factors:

Since the foundation base is horizontal, that is $\alpha = 0$, each of the base-inclination factors b_c, b_q and b_γ is 1.0.

Ground-inclination factors: [see the comments given in Solution of Problem 6.2- (A) *Design Approach 1*]

Effective overburden pressure q' (or $\sigma'_{vk,b}$):

$$q' = \gamma_k \times D = 18 \times 0.5 = 9\,\text{kN/m}^2$$

Gross ultimate bearing capacity (*Gross* q_{ult}) or gross bearing resistance (R/A'):

Annex D of EN 1997-1 suggests that bearing resistance may be calculated from

$$R/A' = \underbrace{c' N_c b_c s_c i_c}_{\text{Cohesion}} + \underbrace{q' N_q b_q s_q i_q}_{\text{Overburden}} + \underbrace{0.5\gamma' B' N_\gamma b_\gamma s_\gamma i_\gamma}_{\text{Self-weight}}$$

- From cohesion: $(Gross\,q_{ult})_1 = \begin{pmatrix} 0 \\ 0 \end{pmatrix}\begin{pmatrix} 46.1 \\ 28.4 \end{pmatrix}\begin{pmatrix} 1.35 \\ 1.31 \end{pmatrix} = \begin{pmatrix} 0 \\ 0 \end{pmatrix}$ kPa

- From overburden: $(Gross\,q_{ult})_2 = \begin{pmatrix} 9 \\ 9 \end{pmatrix}\begin{pmatrix} 33.3 \\ 17.0 \end{pmatrix}\begin{pmatrix} 1.34 \\ 1.29 \end{pmatrix} = \begin{pmatrix} 402 \\ 197 \end{pmatrix}$ kPa

- From self-weight: $(Gross\,q_{ult})_3 = \begin{pmatrix} 0.5 \\ 0.5 \end{pmatrix}\begin{pmatrix} 18 \\ 18 \end{pmatrix}\begin{pmatrix} 1.36 \\ 1.35 \end{pmatrix}\begin{pmatrix} 45.2 \\ 17.8 \end{pmatrix}\begin{pmatrix} 0.82 \\ 0.82 \end{pmatrix}$

  $= \begin{pmatrix} 454 \\ 177 \end{pmatrix}$ kPa

$$\text{Gross bearing resistance } R/A' = Gross\ q_{ult} = \sum_{i=1}^{3} (Gross\,q_{ult})_i$$

$$= \begin{pmatrix} 856 \\ 374 \end{pmatrix}\ \text{kPa}$$

From Set *R*1 in Table A.5 of Annex A to EN 1997-1: the partial factor on bearing resistance is

$$\gamma_{R;v} = \begin{pmatrix} 1 \\ 1 \end{pmatrix}$$

$$\textit{Design bearing resistance: } q_{\text{Rd}} = \frac{R/A'}{\gamma_{R;v}} = \frac{\begin{pmatrix} 856 \\ 374 \end{pmatrix}}{\begin{pmatrix} 1 \\ 1 \end{pmatrix}} = \begin{pmatrix} 856 \\ 374 \end{pmatrix} \textbf{ kPa}$$

Verification of bearing resistance

$$\text{Utilization factor } A_{\text{GEO}} = \frac{q_{\text{Ed}}}{q_{\text{Rd}}} = \begin{pmatrix} 581.6/856 \\ 458.2/374 \end{pmatrix} = \begin{pmatrix} 68\% \\ 122\% \end{pmatrix}$$

Design is unacceptable if utilisation factor is greater than 100%.

(B) *Design Approach* 2 *A*1" + "*M*1" + "*R*2
Effective footing plan dimensions:
The footing self-weight and partial factors for actions are the same as those calculated in the Solution of Problem 6.2.

$$e'_B = \frac{e_B(\gamma_G V_{Gk} + \gamma_Q V_{Qk})}{\gamma_G(W_{Gk} + V_{Gk}) + \gamma_Q V_{Qk}} = 0.0724\text{ m (as before)}$$

Load is within middle-third since $e'_B < \left(\frac{B}{6} = 0.25\text{ m}\right)$

$$e'_L = \frac{e_L(\gamma_G V_{Gk} + \gamma_Q V_{Qk})}{\gamma_G(W_{Gk} + V_{Gk}) + \gamma_Q V_{Qk}} = 0.0965\text{ m}$$

Load is within middle-third since $e'_L < \left(\frac{L}{6} = 0.42\text{ m}\right)$
Effective width $B' = B - 2e'_B = 1.5 - 2 \times 0.0724 = 1.36\text{ m}$
Effective length $L' = L - 2e'_L = 2.5 - 2 \times 0.0965 = 2.31\text{ m}$
Effective area $A' = (L' \times B') = 2.31 \times 1.36 = 3.14\text{ m}^2$
Actions and effects:
From previous calculation (Solution of Problem 6.2.), $V_d = 1818.3\text{ kN}$
Design bearing pressure [Equation (6.8)]: $q_{Ed} = \frac{V_d}{A'_b} = \frac{1818.3}{3.14} = 579.1\ kPa$
Materials properties and resistance:
From previous calculation (Solution of Problem 6.2.), $\phi_d = 35^o$ and $c'_d = 0\text{ kPa}$
Also, bearing capacity factors $N_q = 33.3$, $N_c = 46.1$ and $N_\gamma = 45.2$
Shape factors:
For effect of overburden, $s_q = 1 + \frac{B'}{L'}\sin\phi_d = 1.34$
For effect of cohesion, $s_c = \frac{(s_q N_q - 1)}{(N_q - 1)} = 1.35$

(*Continued*)

For effect of self-weight, $s_\gamma = 1 - 0.3\dfrac{B'}{L'} = 0.82$

Depth factors: [see the comments given in Solution of Problem 6.2]

Load-inclination factors, caused by a horizontal load H:

Since $H = 0$, each of the load-inclination factors i_c, i_q and i_γ is 1.0.

Base-inclination factors:

Since the foundation base is horizontal, that is $\alpha = 0$, each of the base-inclination factors b_c, b_q and b_γ is 1.0.

Ground-inclination factors: [see the comments given in Solution of Problem 6.2]

Effective overburden pressure q' (or $\sigma'_{vk,b}$):

$$q' = \gamma_k \times D = 18 \times 0.5 = 9\,\text{kN/m}^2$$

Gross ultimate bearing capacity (*Gross* q_{ult}) or gross bearing resistance (R/A'):

Annex D of EN 1997-1 suggests that bearing resistance may be calculated from

$$R/A' = \underbrace{c'N_c\, b_c\, s_c\, i_c}_{\text{Cohesion}} + \underbrace{q'N_q\, b_q\, s_q\, i_q}_{\text{Overburden}} + \underbrace{0.5\gamma' B' N_\gamma\, b_\gamma s_\gamma\, i_\gamma}_{\text{Self-weight}}$$

- From cohesion: $(Gross\, q_{ult})_1 = (0)(46.1)(1.35) = (0)\,\text{kPa}$
- From overburden: $(Gross\, q_{ult})_2 = (9)(33.3)(1.34) = 402\,\text{kPa}$
- From self-weight: $(Gross\, q_{ult})_3 = (0.5)(18)(1.36)(45.2)(0.82)$

$$= 454\,\text{kPa}$$

$$\text{Gross bearing resistance } R/A' = Gross\, q_{ult} = \sum_{i=1}^{3} (Gross\, q_{ult})_i$$

$$= 856\,\text{kPa}$$

From Set $R2$ in Table A.5 of Annex A to EN 1997-1: the partial factor on bearing resistance is

$$\gamma_{R;v} = 1.4$$

Design bearing resistance: $q_{\text{Rd}} = \dfrac{R/A'}{\gamma_{R;v}} = \dfrac{856}{1.4} = 611\,kPa$

Verification of bearing resistance

$$\text{Utilization factor } A_{\text{GEO}} = \frac{q_{\text{Ed}}}{q_{\text{Rd}}} = \frac{597.1}{611} = 98\%$$

Design is unacceptable if utilisation factor is greater than 100%.

(C) *Design Approach 3 A1" + "M2" + "R3*

Effective footing plan dimensions:

The footing self-weight and partial factors for actions are the same as those calculated in the Solution of Problem 6.2.

$$e'_B = \frac{e_B(\gamma_G V_{Gk} + \gamma_Q V_{Qk})}{\gamma_G(W_{Gk} + V_{Gk}) + \gamma_Q V_{Qk}} = 0.0724\,\text{m (as before)}$$

Load is within middle-third since $e'_B < \left(\dfrac{B}{6} = 0.25\,\text{m}\right)$

$$e'_L = \frac{e_L(\gamma_G V_{Gk} + \gamma_Q V_{Qk})}{\gamma_G(W_{Gk} + V_{Gk}) + \gamma_Q V_{Qk}} = 0.0965\,\text{m}$$

Load is within middle-third since $e'_L < \left(\frac{L}{6} = 0.42\,\text{m}\right)$

Effective width $B' = B - 2e'_B = 1.5 - 2 \times 0.0724 = 1.36\,\text{m}$

Effective length $L' = L - 2e'_L = 2.5 - 2 \times 0.0965 = 2.31\,\text{m}$

Effective area $A' = (L' \times B') = 2.31 \times 1.36 = 3.14\,\text{m}^2$

Actions and effects:

From previous calculation (Solution of Problem 6.2.), $V_d = 1818.3\,\text{kN}$

Design bearing pressure [Equation (6.8)]: $q_{Ed} = \frac{V_d}{A'_b} = \frac{1818.3}{3.14} = 579.1\,kPa$

Materials properties and resistance:

From previous calculation (Solution of Problem 6.2.), $\phi_d = 29.3°$ and $c'_d = 0$ kPa

Also, bearing capacity factors $N_q = 16.9$, $N_c = 28.4$ and $N_\gamma = 17.8$

Shape factors:

For effect of overburden, $s_q = 1 + \frac{B'}{L'}\sin\phi_d = 1.29$

For effect of cohesion, $s_c = \frac{(s_q N_q - 1)}{(N_q - 1)} = 1.31$

For effect of self-weight, $s_\gamma = 1 - 0.3\frac{B'}{L'} = 0.82$

Depth factors (see the comments given in Solution of Problem 6.2):

Load-inclination factors, caused by a horizontal load H:

Since $H = 0$, each of the load-inclination factors i_c, i_q and i_γ is 1.0.

Base-inclination factors:

Since the foundation base is horizontal, that is $\alpha = 0$, each of the base-inclination factors b_c, b_q and b_γ is 1.0.

Ground-inclination factors: [see the comments given in Solution of Problem 6.2]

Effective overburden pressure q' (or $\sigma'_{vk,b}$):

$$q' = \gamma_k \times D = 18 \times 0.5 = 9\,\text{kN/m}^2$$

Gross ultimate bearing capacity (*Gross* q_{ult}) or gross bearing resistance (R/A'):

Annex D of EN 1997-1 suggests that bearing resistance may be calculated from

$$R/A' = \underbrace{c' N_c\, b_c\, s_c\, i_c}_{\text{Cohesion}} + \underbrace{q' N_q\, b_q\, s_q\, i_q}_{\text{Overburden}} + \underbrace{0.5\gamma' B' N_\gamma\, b_\gamma s_\gamma\, i_\gamma}_{\text{Self-weight}}$$

- From cohesion: $(Gross\, q_{ult})_1 = (0)(28.4)(1.31) = (0)\,\text{kPa}$
- From overburden: $(Gross\, q_{ult})_2 = (9)(16.9)(1.29) = 196.2\,\text{kPa}$
- From self-weight: $(Gross\, q_{ult})_3 = (0.5)(18)(1.36)(17.8)(0.82)$

$$= 178.7\,\text{kPa}$$

(*Continued*)

$$\text{Gross bearing resistance } R/A' = Gross\; q_{ult} = \sum_{i=1}^{3} (Gross\; q_{ult})_i$$

$$= 374.9\,\text{kPa}$$

From Set *R*3 in Table A.5 of Annex A to EN 1997-1: the partial factor on bearing resistance is

$$\gamma_{R;v} = 1.0$$

Design bearing resistance: $q_{\text{Rd}} = \dfrac{R/A'}{\gamma_{R;v}} = \dfrac{374.9}{1.0} = 374.9\,kPa$

Verification of bearing resistance:

$$\text{Utilization factor } A_{\text{GEO}} = \frac{q_{\text{Ed}}}{q_{\text{Rd}}} = \frac{597.1}{374.9} = 154\%$$

Design is unacceptable if utilisation factor is greater than 100%.

Finally, one may comment on the three DAs calculations as follows:

The introduction of eccentricity into Problem 6.3 results in the foundation being inadequate for Design Approach 1. Thus the footing would need to be re-designed in order to satisfy EN 1997 requirements. Also, the footing does not satisfy Design Approach 3 and needs to be re-designed. Design Approach 2 suggests the footing is only just satisfactory. These results might necessitate redesigning the footing (making the footing larger) or, if possible, repositioning the source of the applied loads.

Problem 6.4

A long strip footing of width $B = 2.5$ m and depth (thickness) $d = 1.5$ m, is required to carry a vertical imposed permanent action $V_{Gk} = 250\,\text{kN/m}$ and a vertical imposed variable action $V_{Qk} = 110\,\text{kN/m}$. The footing base is horizontal (i.e. $\alpha = 0°$) and located at a depth of 1.5 m below ground surface (i.e. $D = 1.5$ m). The weight density (unit weight) of the reinforced concrete is $\gamma_{c,\,k} = 25\,\text{kN/m}^3$. The water table currently exists at a depth $d_w = 1$ m (see the scheme below). Assume the weight density of groundwater $\gamma_w = 10\,\text{kN/m}^3$. The footing is founded on a medium strength clay layer which has the following characteristic parameters:

- Undrained strength $c_{uk} = 45\,\text{kPa}$
- Angle of shearing resistance $\phi_k = 25^o$
- Effective cohesion $c'_k = 5\,\text{kPa}$
- Saturated weight density (saturated unit weight) $\gamma_k = 21\,\text{kN/m}^3$

It is required to perform verification of strength (GEO ultimate limit state), that is verification of bearing resistance, using Design Approaches 1.

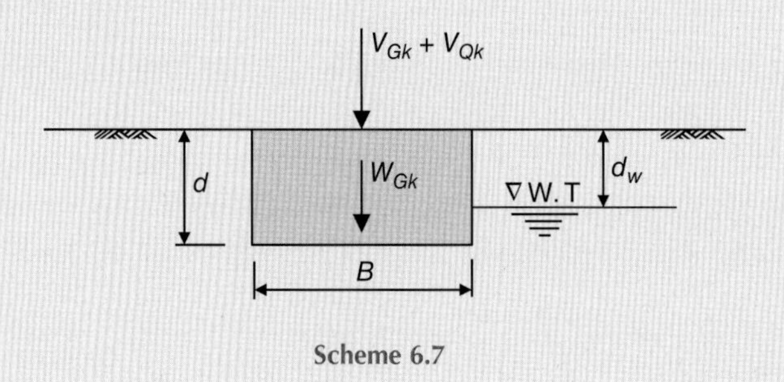

Scheme 6.7

Solution:

DA1 – Combination 1: $A1“+”M1“+”R1$

– Combination 2: $A2“+”M2“+”R1$

Geometrical parameters

In an ultimate limit state, the design water level should represent the most onerous that could occur during the design working life of the structure. Therefore, it would be appropriate to take the ground water level at the ground surface.

Use the design depth of water table $d_w = 0.0$ m

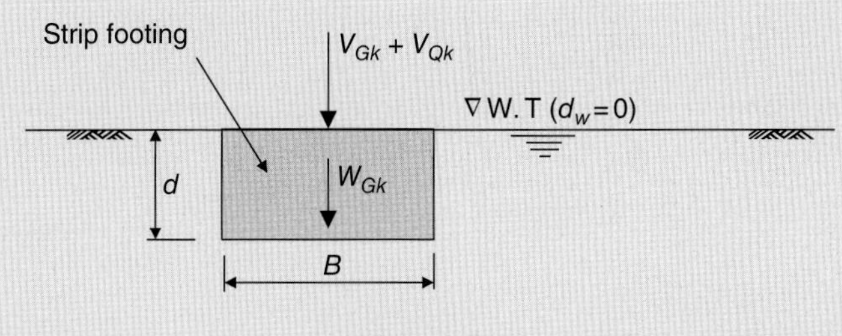

Scheme 6.8

Actions and effects

$$\text{Characteristic self-weight of footing } W_{Gk} = \gamma_{c,k} \times B \times d = 25 \times 2.5 \times 1.5 = 93.8\,\text{kN/m}$$

Area of base $A_b = L \times B = 1 \times 2.5 = 2.5\ \text{m}^2/\text{m}$

Partial factors from Table 6.6 for $\begin{pmatrix}\text{Combination }1\\ \text{Combination }2\end{pmatrix}$:

$$\gamma_G = \begin{pmatrix}1.35\\1\end{pmatrix};\ \gamma_{G,\text{fav}} = \begin{pmatrix}1\\1\end{pmatrix};\ \gamma_Q = \begin{pmatrix}1.5\\1.3\end{pmatrix}$$

Design vertical action:

$$V_d = \gamma_G(W_{Gk} + V_{Gk}) + \gamma_Q V_{Qk} = \begin{pmatrix}1.35(93.8+250)+1.5\times110\\1(93.8+250)+1.3\times110\end{pmatrix} = \begin{pmatrix}629.1\\486.8\end{pmatrix}\text{kN/m}$$

Design bearing pressure (total stress) $q_{Ed} = \dfrac{V_d}{B} = \begin{pmatrix}251.6\\194.7\end{pmatrix}$ *kPa*

Characteristic pore pressure (uplift pressure) underneath footing is

$$u_{k,\text{base}} = \gamma_w(d - d_w) = 10(1.5-0) = 15\,\text{kPa}$$

Design uplift pressure (favourable) is

$$u_d = \gamma_{G,\text{fav}} \times u_{k,\text{base}} = \begin{pmatrix}1\times15\\1\times15\end{pmatrix} = \begin{pmatrix}15\\15\end{pmatrix}\text{kPa}$$

(*Continued*)

$$\textit{Design bearing pressure (effective stress) } q'_{Ed} = q_{Ed} - u_d = \begin{pmatrix} 236.6 \\ 179.7 \end{pmatrix} kPa$$

Materials properties and resistance:

Partial factors from Table 6.7 for $\begin{pmatrix} \text{Combination 1} \\ \text{Combination 2} \end{pmatrix}$:

$$\gamma_{\emptyset} = \begin{pmatrix} 1 \\ 1.25 \end{pmatrix}; \gamma_c = \begin{pmatrix} 1 \\ 1.25 \end{pmatrix}; \gamma_{cu} = \begin{pmatrix} 1 \\ 1.4 \end{pmatrix}$$

$$\text{Design angle of shearing resistance } \phi_d = \tan^{-1}\left(\frac{\tan\phi_k}{\gamma_{\emptyset}}\right) = \begin{bmatrix} \tan^{-1}\left(\frac{\tan 25}{1}\right) \\ \tan^{-1}\left(\frac{\tan 25}{1.25}\right) \end{bmatrix}$$

$$= \begin{pmatrix} 25 \\ 20.5 \end{pmatrix}^{\circ}$$

$$\text{Design effective cohesion } c'_d = \frac{c'_k}{\gamma_c} = \begin{pmatrix} 5/1 \\ 5/1.25 \end{pmatrix} = \begin{pmatrix} 5 \\ 4 \end{pmatrix} \text{kPa}$$

$$\text{Design undrained strength } c_{u,\,d} = \frac{c_{u,k}}{\gamma_{cu}} = \begin{pmatrix} 45/1 \\ 45/1.4 \end{pmatrix} = \begin{pmatrix} 45 \\ 32.1 \end{pmatrix} \text{kPa}$$

Drained bearing capacity factors from Annex D of EN 1997-1:

$$\text{For effect of overburden, } N_q = e^{\pi \tan \emptyset_d} \tan^2\left(45^o + \frac{\emptyset_d}{2}\right) = \begin{pmatrix} 10.7 \\ 6.7 \end{pmatrix}$$

$$\text{For effect of cohesion, } N_c = (N_q - 1)\cot\phi_d = \begin{pmatrix} 20.7 \\ 15.3 \end{pmatrix}$$

$$\text{For effect of self-weight, } N_\gamma = 2(N_q - 1)\tan\phi_d = \begin{pmatrix} 9 \\ 4.3 \end{pmatrix}$$

Shape factors:

Drained and undrained shape factors for strip footing are taken as 1.0.

Depth factors:

The suggested method in Annex D does not include depth factors whatsoever, which are present in other formulations of the extended bearing capacity formula (see Section 4.5 of Chapter 4). There has been concern in using these depth factors as their influence can be significant and the reliance on the additional capacity provided by its inclusion is not conservative.

Load-inclination factors, caused by a horizontal load H:

Since $H = 0$, all the load-inclination factors, for both drained and undrained conditions, i_c, i_q and i_γ are 1.0.

Base-inclination factors:

Since the foundation base is horizontal, that is $\alpha = 0$, all the base-inclination factors, for both drained and undrained conditions, b_c, b_q and b_γ are 1.0.

Ground-inclination factors:

The suggested method in Annex D omits the ground-inclination factors which are present in other formulations of the extended bearing capacity formula. However, neglecting these factors is unsafe.

Undrained bearing resistance:

Total overburden pressure at foundation level q_k (or $\sigma_{vk,b}$):

$$q = \gamma_k \times D = 21 \times 1.5 = 31.5 \text{ kPa}$$

Gross ultimate bearing capacity (*Gross* q_{ult}) or gross bearing resistance (R/A'):

Annex D of EN 1997-1 suggests that undrained bearing resistance may be calculated from

$$\frac{R}{A'} = (\pi + 2)c_{u,d}b_c s_c i_c + q$$

$$= 5.14\begin{pmatrix} 45 \\ 32.1 \end{pmatrix}(1)(1)(1) + 31.5 = \begin{pmatrix} 262.8 \\ 196.5 \end{pmatrix} \text{ kPa}$$

From Set $R1$ in Table A.5 of Annex A to EN 1997-1: the partial factor on bearing resistance is

$$\gamma_{R;v} = \begin{pmatrix} 1 \\ 1 \end{pmatrix}$$

Design undrained bearing resistance: $q_{\text{Rd}} = \dfrac{R/A'}{\gamma_{R;v}} = \dfrac{\begin{pmatrix} 262.8 \\ 196.5 \end{pmatrix}}{\begin{pmatrix} 1 \\ 1 \end{pmatrix}} = \begin{pmatrix} 262.8 \\ 196.5 \end{pmatrix}$ *kPa*

Verification of undrained bearing resistance:

$$\text{Utilization factor } A_{\text{GEO}} = \frac{q_{\text{Ed}}}{q_{\text{Rd}}} = \begin{pmatrix} 251.6/262.8 \\ 194.7/196.5 \end{pmatrix} = \begin{pmatrix} 96\% \\ 99\% \end{pmatrix}$$

Design is unacceptable if utilisation factor is greater than 100%.

Drained bearing resistance:

Effective overburden pressure at foundation level q'_k (or $\sigma'_{vk,b}$):

$$q'_k = q - u_{k,\text{base}} = 31.5 - 15 = 16.5 \text{ kPa}$$

Annex D of EN 1997-1 suggests that drained bearing resistance may be calculated from

$$R/A' = \underbrace{c'N_c\, b_c\, s_c\, i_c}_{\text{Cohesion}} + \underbrace{q'N_q\, b_q\, s_q\, i_q}_{\text{Overburden}} + \underbrace{0.5\gamma' B' N_\gamma\, b_\gamma s_\gamma\, i_\gamma}_{\text{Self-weight}}$$

- From cohesion: $\left(Gross\, q'_{ult}\right)_1 = c'_d N_c = \begin{pmatrix} 5 \\ 4 \end{pmatrix}\begin{pmatrix} 20.7 \\ 15.3 \end{pmatrix} = \begin{pmatrix} 104 \\ 61 \end{pmatrix}$ kPa

(Continued)

- From overburden: $(Gross\, q'_{ult})_2 = q'N_q = 16.5\begin{pmatrix}10.7\\6.7\end{pmatrix}$

$$= \begin{pmatrix}177\\111\end{pmatrix} \text{kPa}$$

- From self-weight: $(Gross\, q'_{ult})_3 = 0.5\gamma' B' N_\gamma$

$$= \begin{pmatrix}0.5\\0.5\end{pmatrix}\begin{pmatrix}11\\11\end{pmatrix}\begin{pmatrix}2.5\\2.5\end{pmatrix}\begin{pmatrix}9\\4.3\end{pmatrix} = \begin{pmatrix}124\\59\end{pmatrix} \text{kPa}$$

Gross drained bearing resistance $R/A' = Gross\, q_{ult}$

$$= \begin{pmatrix}405\\231\end{pmatrix} \text{kPa}$$

From Set $R1$ in Table A.5 of Annex A to EN 1997-1: the partial factor on bearing resistance is

$$\gamma_{R;v} = \begin{pmatrix}1\\1\end{pmatrix}$$

Design drained bearing resistance: $q'_{\text{Rd}} = \dfrac{R/A'}{\gamma_{R;v}} = \dfrac{\begin{pmatrix}405\\231\end{pmatrix}}{\begin{pmatrix}1\\1\end{pmatrix}} = \begin{pmatrix}405\\231\end{pmatrix}$ *kPa*

Verification of drained bearing resistance

$$\text{Utilization factor } A'_{\text{GEO}} = \frac{q'_{\text{Ed}}}{q'_{\text{Rd}}} = \begin{pmatrix}236.6/405\\179.7/231\end{pmatrix} = \begin{pmatrix}58\%\\78\%\end{pmatrix}$$

Design is unacceptable if utilisation factor is greater than 100%.

Finally, one may comment on the Design Approach 1 calculation as follows:

The calculation indicates that the undrained (short-term) situation is more critical than the drained (long-term). The main reason for this may be due to the favourable depth and shape factors which are ignored in both situations. However, finite element studies indicate that these factors could be more significant when undrained condition prevails, and therefore, it may not be realistic to assume that short-term situation is always critical.

Combination 2 governs in both cases and is verified, since the utilisation factors in each case is less than 100%, although it is too close in the undrained condition.

Problem 6.5

Consider the same long strip footing of Problem 6.4 with the following available additional data:

- The clay layer overlies a rigid layer at a depth $d_R = 4.5$ m below the ground surface, as shown in the scheme below.
- The clay's characteristic coefficient of compressibility $m_{v,k} = 0.12\,\text{m}^2/\text{MN}$ and its undrained Young's modulus is assumed to be $E_{u,k} = 600\, c_{u,k}$.
- The limiting value of total settlement $s_{Cd} = 50$ mm.

It is required to perform verification of serviceability of the strip footing (a) implicitly (through an ultimate limit state, ULS), (b) explicitly (using a serviceability limit state, SLS, calculation).

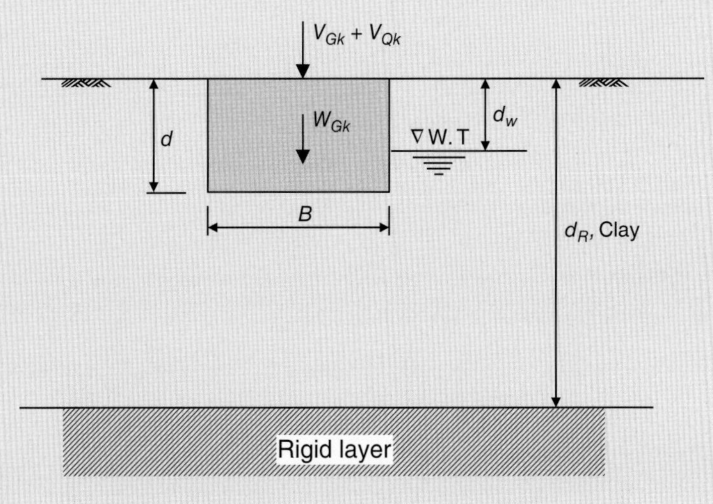

Scheme 6.9

Solution:

(a) Implicit verification of serviceability (based on ULS check)

Geometrical parameters

For serviceability limit states, the design depth of the water table is the most adverse level that could occur in normal circumstances. Therefore, it would be more appropriate not to raise the water table to ground surface. Hence, use the current depth of water table as the design depth.

Design depth of water table $d_{w,d} = d_w = 1$ m.

Actions and effects

$$\text{Characteristic self-weight of footing } W_{Gk} = \gamma_{c,k} \times B \times d$$
$$= 25 \times 2.5 \times 1.5$$
$$= 93.8\,\text{kN/m}$$

Imposed permanent action $V_{Gk} = 250\,\text{kN/m}$

Imposed variable action $V_{Qk} = 110\,\text{kN/m}$

Characteristic uplift pore pressure on the underside of the footing is

$$u_{k,\text{base}} = \gamma_w(d - d_{w,d}) = 10(1.5 - 1) = 5\,\text{kPa}$$

Partial load factors for SLS (see Section 6.2.5):

$$\gamma_G = 1, \gamma_{G,\text{fav}} = 1, \gamma_Q = 1$$

Design vertical action: $V_d = \gamma_G(W_{Gk} + V_{Gk}) + \gamma_Q V_{Qk}$

$$= 1(93.8 + 250) + 1 \times 110 = 453.8\,\text{kN/m}$$

(Continued)

Design bearing pressure (total stress) $q_{Ed} = \frac{V_d}{B} = \frac{453.8}{2.5} = 181.5\,kPa$

Design uplift pressure (favourable) is

$$u_d = \gamma_{G,\,fav} \times u_{k,base} = 1 \times 5 = 5\,\text{kPa}$$

Design bearing pressure (effective stress) $q'_{Ed} = q_{Ed} - u_d = 176.5\,kPa$

Materials properties and resistance:

Characteristic material properties are:

- undrained shear strength $c_{uk} = 45\,\text{kPa}$
- angle of shearing resistance $\phi_k = 25°$
- cohesion $c'_k = 5\,\text{kPa}$

Partial material factors for SLS (see Section 6.2.5):

$$\gamma_{cu} = 1;\ \gamma_{\phi} = 1;\ \gamma_c = 1$$

Design angle of shearing resistance $\phi_d = \tan^{-1}\left(\frac{\tan\phi_k}{\gamma_\varnothing}\right) = 25°$

Design effective cohesion $c'_d = \frac{c'_k}{\gamma_c} = 5\,\text{kPa}$

Design undrained strength $c_{u,\,d} = \frac{c_{u,k}}{\gamma_{cu}} = 45\,\text{kPa}$

Drained bearing capacity factors from Annex D of EN 1997-1:

For effect of overburden, $N_q = e^{\pi \tan\varnothing_d}\tan^2\left(45° + \frac{\varnothing_d}{2}\right) = 10.7$

For effect of cohesion, $N_c = (N_q - 1)\cot\phi_d = 20.7$

For effect of self-weight, $N_\gamma = 2(N_q - 1)\tan\phi_d = 9$

Shape factors:

Drained and undrained shape factors for strip footing are normally taken as 1.0.

Depth factors:

The suggested method in Annex D does not include depth factors whatsoever, which are present in other formulations of the extended bearing capacity formula (see Section 4.5 of Chapter 4). There has been concern in using these depth factors as their influence can be significant and the reliance on the additional capacity provided by its inclusion is not conservative.

Load-inclination factors, caused by a horizontal load H:

Since $H = 0$, all the load-inclination factors, for both drained and undrained conditions, i_c, i_q and i_γ are 1.0.

Base-inclination factors:

Since the foundation base is horizontal, that is $\alpha = 0$, all the base-inclination factors, for both drained and undrained conditions, b_c, b_q and b_γ are 1.0.

Ground-inclination factors:

The suggested method in Annex D omits the ground-inclination factors which are present in other formulations of the extended bearing capacity formula. However, neglecting these factors is unsafe.

Undrained bearing resistance:

Total overburden pressure at foundation level q_k (or $\sigma_{vk,b}$) assume γ_k clay above W.T. is $\cong$ 21 kN/M^3:

$$q = \gamma_k \times D = 21 \times 1.5 = 31.5\,\text{kPa}$$

Gross ultimate bearing capacity (*Gross* q_{ult}) or gross bearing resistance (R/A'):

Annex D of EN 1997-1 suggests that undrained bearing resistance may be calculated from

$$\frac{R}{A'} = (\pi + 2)c_{u,d}b_c s_c i_c + q$$

$$= 5.14(45)(1)(1)(1) + 31.5 = 262.8\,\text{kPa}$$

Partial resistance factor for SLS is $\gamma_{Rv,\,SLS} = 3.0$

Design undrained bearing resistance: $q_{\text{Rd}} = \dfrac{R/A'}{\gamma_{Rv,\,SLS}} = \dfrac{262.8}{3.0} = 87.6\,kPa$

Verification of undrained bearing resistance:

$$\text{Utilization factor } A_{\text{SLS}} = \frac{q_{\text{Ed}}}{q_{\text{Rd}}} = \frac{181.5}{87.6} = 207\%$$

Design is unacceptable if utilisation factor is greater than 100%.

Drained bearing resistance:

Effective overburden pressure at foundation level q'_k (or $\sigma'_{vk,b}$):

$$q'_k = q - u_{k,\text{base}} = 31.5 - 5 = 26.5\,\text{kPa}$$

Annex D of EN 1997-1 suggests that drained bearing resistance may be calculated from

$$R/A' = \underbrace{c'N_c b_c s_c i_c}_{\text{Cohesion}} + \underbrace{q'N_q b_q s_q i_q}_{\text{Overburden}} + \underbrace{0.5\gamma' B' N_\gamma b_\gamma s_\gamma i_\gamma}_{\text{Self-weight}}$$

- From cohesion: $(Gross\, q'_{ult})_1 = c'_d N_c = (5)(20.7) = 103.5\,\text{kPa}$
- From overburden: $(Gross\, q'_{ult})_2 = q' N_q = 26.5(10.7)$
  $= 283.6\,\text{kPa}$
- From self-weight: $(Gross\, q'_{ult})_3 = 0.5\gamma' B' N_\gamma$
  $= (0.5)(21 - 10)(2.5)(9)$
  $= 123.8\,\text{kPa}$

$$\text{Gross drained bearing resistance } R/A' = Gross\, q_{ult} = \sum_{i=1}^{3} (Gross\, q'_{ult})_i$$

$$= 510.9\,\text{kPa}$$

Design drained bearing resistance: $q'_{\text{Rd}} = \dfrac{R/A'}{\gamma_{Rv,\,SLS}} = \dfrac{510.9}{3.0} = 170.3\,kPa$

Verification of drained bearing resistance

$$\text{Utilization factor } A'_{\text{GEO}} = \frac{q'_{\text{Ed}}}{q'_{\text{Rd}}} = \frac{176.5}{170.3} = 104\%$$

Design is unacceptable if utilisation factor is greater than 100%.

As it is clear, the calculation based on a resistance factor ($\gamma_{Rv,\,SLS}$) of 3.0 does not work for both the undrained and drained conditions and therefore an explicit settlement calculation is required.

(*Continued*)

(b) Explicit verification of serviceability

Actions and effects

The net bearing pressure at the foundation level ($z = 0$ m) is

$$q_{net,\,d} = q_{Ed} - \sigma_{vk,b} = 181.5 - 31.5 = 150 \text{ kPa}$$

Immediate settlement s_o (or s_i)

For evaluation of settlement Annex F to EN 1997-1 suggests using an equation of the form:

$$s = p \times B \times f / E_m$$

The symbols are defined in the Annex F.

The following settlement equation [Equation (3.33)] is one of many that are available and follows the guidance given in the Annex F, as indicated above.

$$s_o = \mu_o \mu_1 \frac{qB}{E_m}$$

where the coefficient μ_o depends on the depth of foundation and μ_1 depends on the layer thickness and the shape of the loaded area (see Figure 3.9). The width B is in m, q in kPa and E_m in MPa.

The immediate settlement is considered to be the short-term component of the total settlement, which occurs without drainage.

The values adopted for the stiffness parameters (such as E_m and Poisson's ratio) should in this case represent the undrained behaviour.

Refer to Figure 3.9:

$\frac{D}{B} = \frac{d}{B} = \frac{1.5}{2.5} = 0.6$. For this value of $\frac{d}{B}$ obtain $\mu_o \cong 0.93$

$\frac{H}{B} = \frac{d_R - d}{B} = \frac{4.5 - 1.5}{2.5} = 1.2$. For this value of $\frac{d_R - d}{B}$ obtain $\mu_1 \cong 0.4$

$$q = q_{net,\,d} = 150 \text{ kPa}$$

$$E_m = E_{u,k} = 600\, c_{u,k} = 600 \times 45 = 27\ 000 \text{ kPa} = 27 \text{ MPa}$$

$$s_o = \mu_o \mu_1 \frac{q_{net,\,d}\, B}{E_{u,k}} = 0.93 \times 0.4 \times \frac{150 \times 2.5}{27} = 5.2 \text{ mm}$$

This settlement equation gives average vertical displacement under a flexible uniformly loaded area. It does not include a factor to account for rigidity. It was found that if the footing is rigid the settlement will be uniform and reduced by about 7%, as Equation (3.6) indicates. If we assume that the given strip footing is rigid, then

$$s_o \cong 0.93 \times 5.2 = 4.8 \text{ mm}$$

Consolidation settlement s_1 *(or* s_c*)*

In order to compute the consolidation settlement more accurately, assume the 3 m clay layer below the foundation level is divided into five sub-layers, each 0.6 m thick, as shown in the scheme below.

Consolidation settlement in each sub-layer is

$$s_{c,\,i} = m_{vk} \times \sigma'_{z,i} \times h$$

$m_{v,k} = 0.12 \text{ m}^2/\text{MN}$

$h = 0.6\,\text{m}$ = thickness of sublayer i
$\sigma'_{z,i}$ = net increase in pressure at middle of sublayer $i = q_{net,\,d}\,(4I)$
I = stress influence factor obtained from Table 2.3 or Figure 2.32

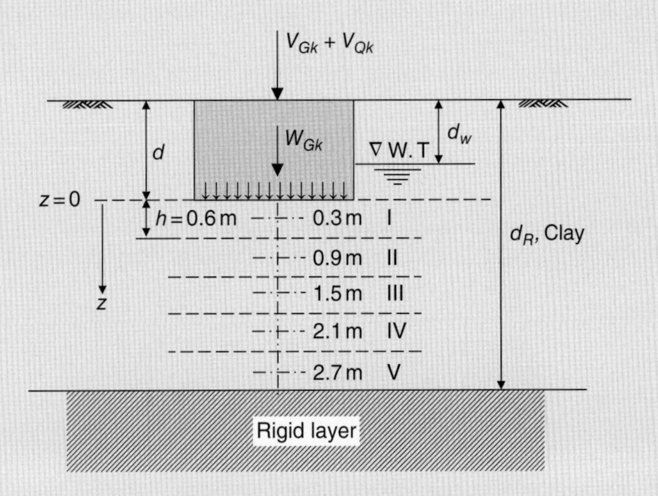

Scheme 6.10

Results of the necessary calculations are presented in the table below:

Layer (m)	z	I (kPa)	$q_{net,\,d}$ (kPa)	$\sigma'_{z,i}$ (m²/Mpa)	$m_{v,k}$ (m)	h (mm)	s_c
I	0.3	0.24	150	144	0.12	0.6	10.4
II	0.3	0.22	150	132	0.12	0.6	9.5
III	0.3	0.18	150	108	0.12	0.6	7.8
IV	0.3	0.15	150	90	0.12	0.6	6.5
V	0.3	0.12	150	72	0.12	0.6	5.2
							$s_1 = \sum s_c = 39.4$

The total consolidation settlement $s_1 = 39.4$ mm

Total settlement

In this solution only immediate and consolidation settlements have been considered. The creep component s_2 (secondary compression) is considered negligible.

Sum of settlements is $s = s_o + s_1 = 5.2 + 39.4 = 44.6\,\text{mm}$

Design effect of actions is $s_{Ed} = s = 45\,mm$

Verification of settlement (SLS)

Equation (6.22):

$$s_{\text{Ed}} = s_0 + s_1 + s_2 \leq s_{\text{Cd}}$$

$$s_{\text{Cd}} = 50\,\text{mm (given)}$$

Utilisation factor $A_{\text{SLS}} = \dfrac{s_{\text{Ed}}}{s_{\text{Cd}}} = \dfrac{45}{50} = 90\,\%$

Design is unacceptable if utilisation factor is greater than 100%.

Serviceability is satisfied by the explicit calculation since $A_{\text{SLS}} = 90\,\%$

Problem 6.6

The scheme shown below represents geometry of a heavily loaded column footing which has a total depth (thickness) $h = 0.8$ m. The footing is centrally loaded by the vertical imposed actions. The gross design soil pressure q_{Ed} on the underside of the footing was calculated equals 1445 kPa. The design requires using class C25/30 concrete and type B500B steel.

It is required to design the bottom reinforcement using the approach described in EN 1992-1-1§9.8.2.2.

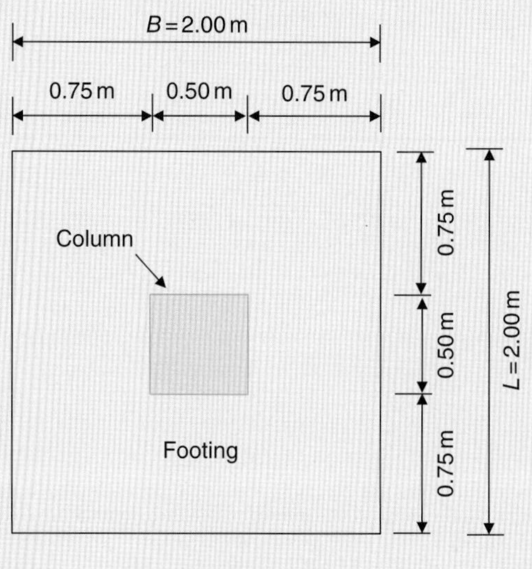

Scheme 6.11

Solution:

The EN 1992-1-1 § 9.8.2.2 design approach for anchorage of steel bars provides the maximum force in the reinforcement. Hence, the required bottom reinforcement may be designed.

Actions and effects:

Characteristic self-weight of footing $W_{Gk} = \gamma_{c,k} \times L \times B \times h$

$$= 25 \times 2.0 \times 2.0 \times 0.8$$

$$= 80 \text{ kN}$$

Table 6.6 gives partial factor $\gamma_G = 1.35$. Hence,

$$W_d = 1.35 \times 80 = 108 \text{ kN}$$

The effective design soil pressure on the underside of the footing is

$$q'_{Ed} = q_{Ed} - \left(\frac{W_d}{BL}\right) = 1445 - \left(\frac{108}{2 \times 2}\right) = 1418 \, kPa$$

The tensile force to be anchored is given by:

$$F_s = R_d \times \frac{Z_e}{Z_i}$$

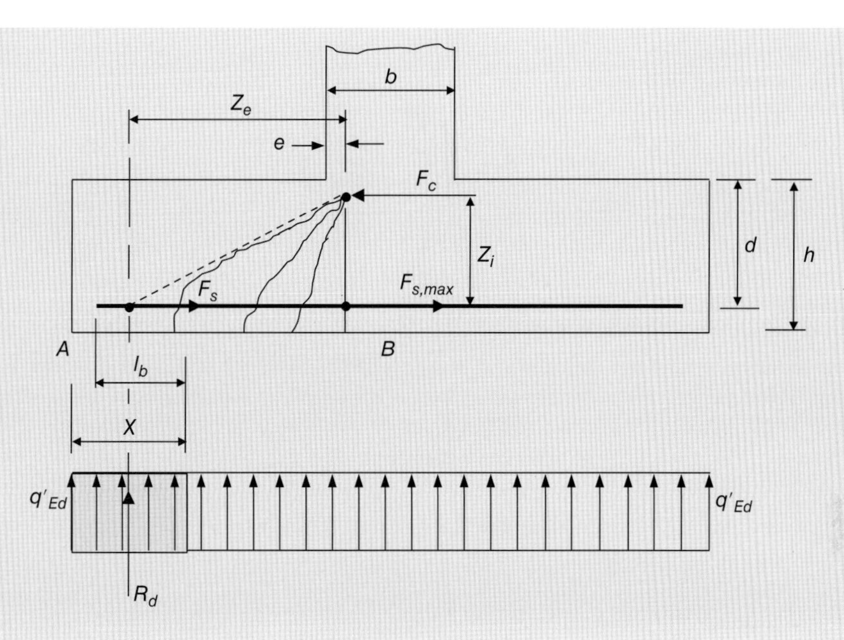

Scheme 6.12

where:
R_d = design resultant of ground pressure within distance X
Z_e = External lever arm
Z_i = Internal lever arm

EN 1992-1-1 §9.8.2.2(3) suggests, as simplifications, Z_e may be determined assuming $e = 0.15\,b$ and Z_i may be taken as 0.9 d.

Assume using bar diameter $\varnothing = 16$ mm, and nominal concrete cover to the reinforcing bars $c_{nom} = K_2 = 75$ mm [EN 1992-1-1 §4.4.1.3(4)]

$$e = 0.15 \times 0.5 = 0.075\text{ m}$$

$$Z_e = \frac{B}{2} - \frac{X}{2} - \frac{b}{2} + e$$

Maximum tensile force $F_{s,\,max}$

Assuming $X = \frac{B}{2} - 0.35\,b$ (the distance between sections at A and B), the maximum tensile force $F_{s,max}$ on the reinforcement is obtained using

$$F_{s,max} = R_d \times \frac{Z_e}{Z_i}$$

$$X = \frac{B}{2} - 0.35\,b = \frac{2}{2} - 0.35 \times 0.5 = 0.825\text{ m}$$

$$d = h - c_{nom} - \left(\frac{\varnothing}{2}\right) = 0.800 - 0.075 - 0.008 = 0.717\text{ m}$$

$$Z_i = 0.9\,d = 0.9 \times 0.717 = 0.645\text{ m}$$

$$Z_e = \frac{B}{2} - \frac{X}{2} - \frac{b}{2} + e = \frac{2}{2} - \frac{0.825}{2} - \frac{0.5}{2} + 0.075 = 0.413\text{ m}$$

$$R_d = q'_{Ed} \times L \times X = 1418 \times 2 \times 0.825 = 2340\text{ kN}$$

$$F_{s,max} = 2340 \times \frac{0.413}{0.645} = \mathit{1498\ kN}$$

(*Continued*)

Required reinforcement area A_s
Partial factor for reinforcing steel $\gamma_s = 1.15$

Required reinforcement area

$$A_s = \frac{F_{s,\,max}}{f_{yk}/\gamma_s} = \frac{1498/1000}{500/1.15}$$

$$= 3.445 \times 10^{-3}\,\text{m}^2 = 3445\,mm^2$$

***Provide** 18 Ø 16 both ways with 91 mm clear bar spacing.*
Verifications and reinforcement arrangement

- The recommended minimum bar diameter is 8 mm [EN 1992-1-1 §9.8.2.1(1)]. The provided bar diameter is 16 mm > $Ø_{min}$ (OK.)

$$A_{s,\,provided} = 18 \times 200 = 3600\,\text{mm}^2 > A_{s,required} \quad (OK.)$$

- The minimum clear spacing of reinforcing bars $s_{min} = 25$ mm (EN 1992-1-1 §8.2). The provided clear bar spacing is 91 mm > s_{min} (OK.)

$$A_{s,\,max} = 0.04\,A_c \text{ (EN 1992-1-1 §9.2.1.1)}$$
$$= 0.04 \times 800 \times 2000 = 64000\,\text{mm}^2 \gg A_{s,provided} \quad (OK.)$$

- From Table 6.16: minimum percentage of reinforcement is

$$A_{s,min} = 0.13\% = 0.0013 \times 717 \times 2000$$
$$= 1864\,\text{mm}^2 < A_{s,\,provided} \quad (OK.)$$

- The provided concrete cover to the external reinforcing bars is

$$c_{provided} = \frac{2000 - 18 \times 16 - 17 \times 91}{2} = 82.5\,\text{mm} > 75\,mm \quad (OK.)$$

- Verification of straight bar anchorage:
  For straight bars without end anchorage the minimum value of X is the most critical. As a simplification, $X_{min} = h/2$ may be assumed [EN 1992-1-1 §9.8.2.2(5)]

$$X_{min} = \frac{h}{2} = \frac{0.8}{2} = 0.4\,\text{m}$$

$$Z_e = \frac{B}{2} - \frac{X_{min}}{2} - \frac{b}{2} + e = \frac{2}{2} - \frac{0.4}{2} - \frac{0.5}{2} + 0.075 = 0.625\,\text{m}$$

$$R_d = q'_{Ed} \times L \times X_{min} = 1418 \times 2 \times 0.4 = 1134.4\text{ kN}$$

$$F_s = 1134.4 \times \frac{0.625}{0.645} = 1099.2\,kN$$

  for the purpose of this verification assume $c_{nom} = 0.040$ m.
  From Table 6.20: the design anchorage length in case of tensile force and good bond conditions is $l_{bd} = 500\,\text{mm} = 0.5$ m.

$$l_b = \frac{F_s}{A_s f_{yd}} \times l_{bd} = \frac{1099.2}{\frac{3600}{10^6} \times \frac{500 \times 1000}{1.15}} \times 0.5 = 0.351\,\text{m}$$

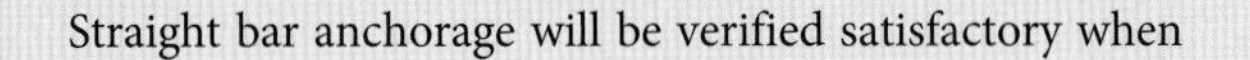

$$l_b + c_{nom} < X_{min}$$

$$0.351 + 0.040 = 0.391\ \text{m} < (X_{min} = 0.4\ \text{m; OK.})$$

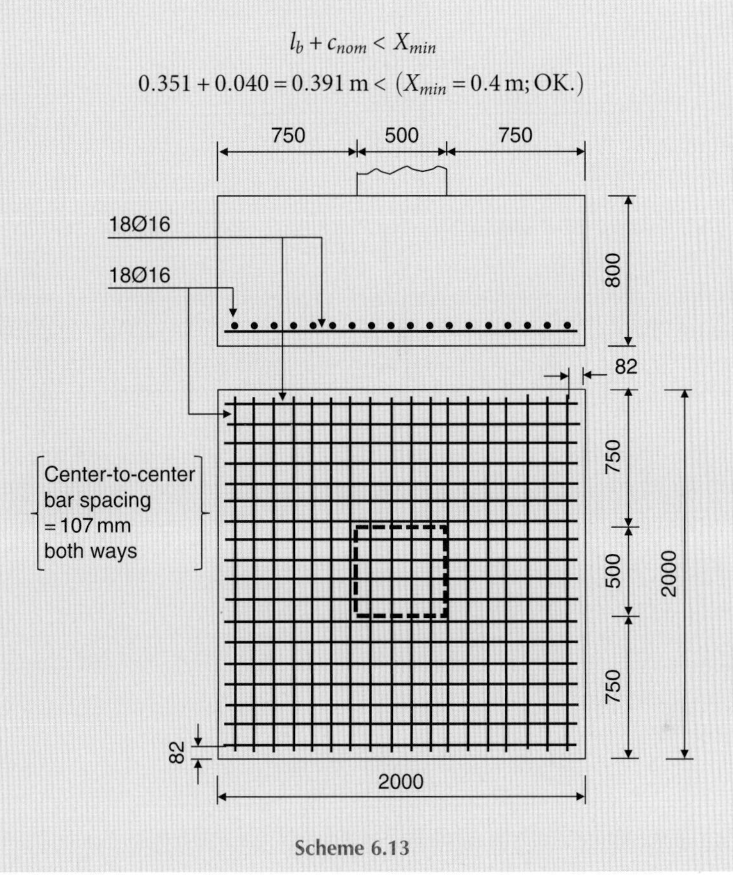

Scheme 6.13

Problem 6.7

The tensile reinforcing steel bars at bottom of a column footing are of the type H16 and Grade B500B. The footing depth h is 500 mm. Assume the concrete strength class is C25/30, nominal concrete cover c_{nom} is 40 mm and bar spacing is 200 mm. Calculate:

(a) Bond stress, f_{bd}
(b) Basic anchorage length, $l_{b,req}$
(c) Design tension anchorage length l_{bd} for straight bars
(d) Design tension anchorage length l_{bd} for 90°-bent bars
(e) Design lap length in tension, l_0

Solution:

(a) Bond stress, f_{bd}

$f_{bd} = 2.25\eta_1\eta_2 f_{ctd}$ [EN 1992-1-1 § 8.4.2 (2)]

- $\eta_1 = 1.0$ for "Good" bond conditions
- $\eta_2 = 1.0$ for bar $\varnothing \leq 32$ mm

f_{ctd} = design value of concrete tensile strength according to EN 1992-1-1 §3.1.6(2)

(Continued)

$$= \frac{\alpha_{ct} f_{ctk,\,0.05}}{\gamma_c}, \text{ where } \alpha_{ct} = 1 \text{ and } \gamma_c = 1.5$$

$f_{ctk,\,0.05} = 0.7 f_{ctm}$ (Table 3.1 of EN 1992-1-1)

$f_{ctm} = 0.3 \times f_{ck}^{2/3} \leq C50/60$ (Table 3.1 of EN 1992-1-1)

$f_{ctk,\,0.05} = 0.7 \times 0.3 \times f_{ck}^{2/3} = 0.21 \times 25^{2/3} = 1.8\,\text{MPa}$

$$f_{ctd} = \frac{1 \times 1.8}{1.5} = 1.2\,\text{MPa}$$

$$f_{bd} = 2.25 \times 1 \times 1 \times 1.2 = 2.7\,MPa$$

(b) Basic anchorage length, $l_{b,req}$

The $l_{b,req}$ for anchoring the force $\sigma_{sd} A_s$ in a straight bar assuming constant bond stress f_{bd} is

$l_{b,req} = (Ø/4)(\sigma_{sd}/f_{bd})$ [EN 1992-1-1 §8.4.3(2)]

Where σ_{sd} is the design stress of the bar at the position the anchorage is measured from.

For bent bars, the length $l_{b,req}$ and design length l_{bd} should be measured along the centre-line of the bar [EN 1992-1-1 §8.4.3(3)].

Maximum stress in the bar $\sigma_{sd,max} = \frac{f_{yk}}{\gamma_s} = \frac{500}{1.15} = 435\,\text{MPa}$.

$l_{b,\,req} = (Ø/4)(435/2.7) = 40.3\,Ø$ (For concrete class C25/30)

(c) Design tension anchorage length l_{bd} for straight bars

$l_{bd} = \alpha_1 \alpha_2 \alpha_3 \alpha_4 \alpha_5\, l_{b,\,req} \geq l_{b,\,min}$ [EN 1992-1-1 §8.4.4(1)]

Where:

α_1, α_2, α_3, α_4 and α_5 are coefficients given in Table 8.2 of EN 1992-1-1

$\alpha_1 = 1.0$

$\alpha_2 = 1 - 0.15(c_d - Ø)/Ø$

≥ 0.7

≤ 1.0

C1

a

C

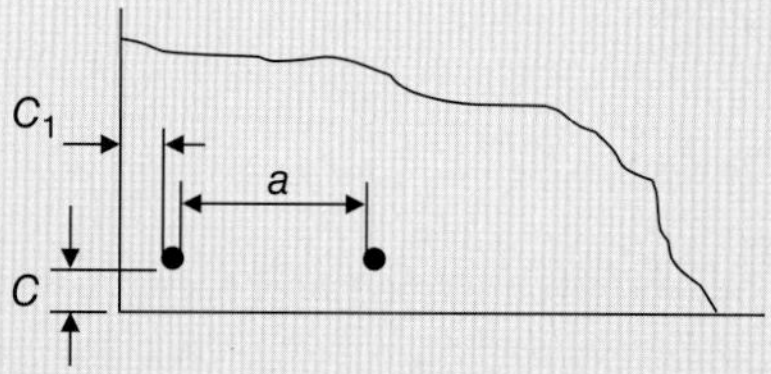

Scheme 6.14

Let $c = c_1 = c_{nom} = 40\,\text{mm}$

$a = 200\,\text{mm};\ Ø = 16\,\text{mm}$

$c_d = \min(a/2, c_1, c)$ [EN 1992-1-1 §8.4.4(1)]

$$\alpha_2 = 1 - \frac{0.15(40 - 16)}{16} = 0.775$$

$\alpha_3 = 1 - K\lambda$

≥ 0.7

≤ 1.0

Assume using conservative value for α_3, that is $K = 0$. Hence,

$$\alpha_3 = 1$$

Also, assume using conservative values for both α_4 and α_5, but not greater than 1.0. Hence,

$$\alpha_4 = \alpha_5 = 1$$

$$l_{bd} = 1 \times 0.775 \times 1 \times 1 \times 1 \times 40.3\ \varnothing = 31.23 \times 16 = 500\ \text{mm}$$

$l_{b,\,min} \geq \max\left(0.3 l_{b,\,req},\ 10\ \varnothing,\ 100\ \text{mm}\right)$ [EN 1992-1-1 §8.4.4(1)]

$$l_{b,\,min} = 0.3 l_{b,\,req} = 0.3 \times 40.3 \times 16 = 193.4\ \text{mm} < (l_{bd} = 500\ \text{mm})$$

$l_{bd} = 500\ mm$ (For concrete class C25/30)

(d) Design tension anchorage length l_{bd} for 90°-bent bars

$l_{bd} = \alpha_1\alpha_2\alpha_3\alpha_4\alpha_5\, l_{b,\,req} \geq l_{b,\,min}$ [EN 1992-1-1 §8.4.4(1)]

$\alpha_1 = 0.7$ if $c_d > 3\varnothing$; otherwise $\alpha_1 = 1.0$ (Table 8.2 of EN 1992-1-1)

$c_d = \min(a/2,\ c_1)$ (Figure 8.6 of EN 1992-1-1)

Let $c_1 = c_{nom} = 40$ mm.

$a = 200$ mm; $\varnothing = 16$ mm

$c_d = 40\ \text{mm} < 3\varnothing$. Hence,

$$\alpha_1 = 1.0$$

$\alpha_2 = 1 - 0.15(c_d - 3\varnothing)/\varnothing$ (Table 8.2 of EN 1992-1-1)

$$\geq 0.7$$

$$\leq 1.0$$

$$\alpha_2 = 1 - \frac{0.15(40 - 3 \times 16)}{16} = 1.075 > 1.0.\ \text{Hence,}$$

$$\alpha_2 = 1$$

$$\alpha_3 = 1 - K\lambda$$

$$\geq 0.7$$

$$\leq 1.0$$

Assume using conservative value for, α_3, that is $K = 0$. Hence,

$$\alpha_3 = 1$$

Also, assume using conservative values for both α_4 and α_5, but not greater than1.0. Hence,

$$\alpha_4 = \alpha_5 = 1$$

$$l_{bd} = 1 \times 1 \times 1 \times 1 \times 1 \times 40.3\,\varnothing = 40.3 \times 16 = 645\ \text{mm}$$

$l_{b,\,min} \geq \max\left(0.3 l_{b,\,req},\ 10\varnothing,\ 100\ \text{mm}\right)$ [EN 1992-1-1 §8.4.4(1)]

$$l_{b,\,min} = 0.3 l_{b,\,req} = 0.3 \times 40.3 \times 16$$

$$= 193.4\ \text{mm} < (l_{bd} = 500\ \text{mm})$$

$l_{bd} = 645\ mm$ (For concrete class C25/30)

(e) Design lap length in tension, l_0

$l_0 = \alpha_1\alpha_2\alpha_3\alpha_5\alpha_6\, l_{b,\,req} \geq l_{0,\,min}$ [EN 1992-1-1 §8.7.3(1)]

(Continued)

Values of α_1, α_2, α_3 and α_5 may be taken from Table 8.2 of EN 1992-1-1; however, for the calculation of α_3, $\Sigma A_{st,\ min}$ should be taken as $1.0A_s\left(\frac{\sigma_{sd}}{f_{yd}}\right)$, with A_s = area of one lapped bar.

From Table 8.2 of EN 1992-1-1:

$\alpha_1 = 1.0$

$\alpha_2 = 0.775$ [As calculated in (c)]
$\alpha_3 = 1.0$ [Conservative value as calculated in (c)]
$\alpha_5 = 1.0$ [Conservative value]
$\alpha_6 = \left(\frac{\rho_1}{25}\right)^{0.5}$ with $1.5 \geq a_6 \geq 1.0$. The ratio ρ_1 is the percentage of reinforcement lapped within 0.65 l_0 from centre of the lap length considered (see Figure 8.8 of EN 1992-1-1). Values of α_6 are given in Table 8.3 of EN 1992-1-1.

Percentage of reinforcement lapped within a bar spacing of 200 mm is conservatively calculated as $\rho_1 = \frac{A_{st}}{A_c} = \frac{2(\pi Ø^2/4)}{A_c} = \frac{2(\pi \times 16^2/4)}{200 \times 500} = 0.004 = 0.4\% < 25\%$. Hence,
Table 8.3 of EN 1992-1-1 gives $\alpha_6 = 1.0$
$l_{0,\ min} \geq \max\{0.3\,\alpha_6 l_{b,\ req}; 15Ø; 200\ \text{mm}\}$ [EN 1992-1-1 §8.7.3(1)]

$\geq \max\{0.3 \times 1 \times 40.3 \times 16; 15 \times 16; 200\ \text{mm}\}$

$\geq \max\{193.4\ \text{mm}; 240\ \text{mm}; 200\ \text{mm}\}$

$\geq 240\ \text{mm}$

$l_0 = 1 \times 0.775 \times 1 \times 1 \times 1 \times 40.3 \times 16 = 500\ \text{mm} > l_{0,\ min}$. Hence,
$l_0 = 500\ mm$ (For concrete class C25/30)

Problem 6.8

Fairly loose becoming medium-dense medium sand will support the widely spaced individual column footings of a structure. One of the centrally loaded interior column footings is required to carry a vertical imposed permanent action $V_{Gk} = 1100$ kN and a vertical imposed variable action $V_{Qk} = 700$ kN. The column is 600 mm square. The geotechnical designer recommended using an isolated footing 4 m square, and a foundation depth = 1.2 m. Also, he reported that the differential settlement between columns founded on loosest and densest soils will be about 2 mm, which is negligible.

The structural design requires using class C25/30 concrete ($f_{ck} = 25$ MPa) and type B500B steel ($f_{yk} = 500$ MPa). Using DA1-Combination 1, perform the necessary design calculations for flexure and shear (i.e. determine the footing depth (thickness) h and design the reinforcement).

Solution:

Actions and effects:

Partial factors from Annex A to EN 1990, or from Table 6.6 of this chapter:

$$\gamma_G = 1.35;\ \gamma_Q = 1.5$$

Design vertical action:

$$V_d = \gamma_G V_{Gk} + \gamma_Q V_{Qk}$$
$$= 1.35 \times 1100 + 1.5 \times 700 = 2535\ \text{kN}$$

Design bearing pressure [Equation (6.8)]: $q_{Ed} = \frac{V_d}{A_b} = \frac{2535}{4 \times 4} = 158.4\ kPa$

Design assumptions:

- A trial value for $h = 750\,\text{mm} = 0.75\,\text{m}$
- Concrete cover to the tensile reinforcement at the footing bottom: $c_{nom} = K_2 = 75\,\text{mm}$ [EN 1992-1-1 §4.4.1.3(4)]
- There will be no shear reinforcement provided
- Steel bar $\varnothing = 16\,\text{mm}$

Design applied moment M_{Ed} *at the column face*

$$M_{Ed} = q_{Ed} \times \frac{l^2}{2} \times B = 158.4 \times \frac{1.7^2}{2} \times 4 = 915.6\ \text{kN.m}$$

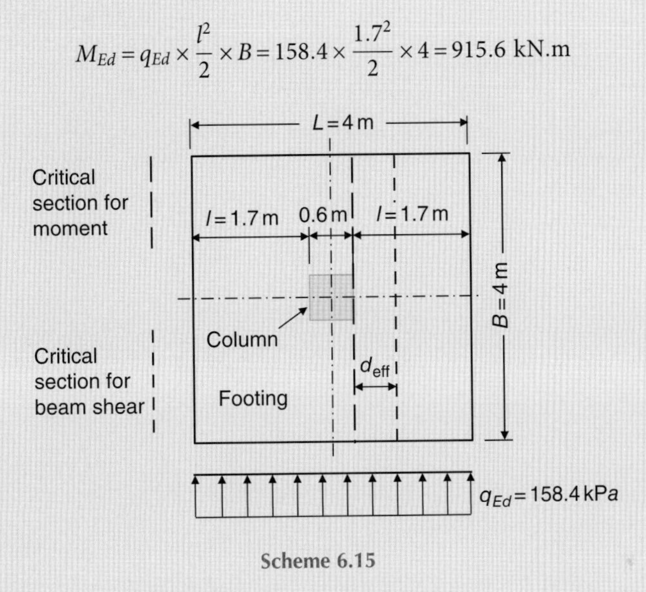

Scheme 6.15

Reinforcement

Since the square footing is symmetrically loaded and the depth h is constant, it would be appropriate to provide the same reinforcement both ways. Hence, it would also be appropriate to take the effective depth d_{eff} to the intersection of the steel bars running each way. Thus, the successive stages of the design calculations will be minimised.

$$d_{eff} = h - c_{nom} - \varnothing = 0.75 - 0.075 - 0.016 = 0.659\ \text{m}$$

$$K = \frac{M_{Ed}}{b\, d_{eff}^2 f_{ck}} = \frac{915.6}{4 \times 0.659^2 \times 25 \times 1000} = 0.021$$

$$K' = 0.60\delta - 0.18\delta^2 - 0.21 \text{ where } \delta \le 1$$

It may also be obtained from Table 6.14. However, it is often recommended in the UK that K' should be limited to 0.168 to ensure ductile failure.

Let $K' = 0.168 > K$ (OK.). Compression reinforcement is not required.

Lever arm z may be obtained from tables, such as Table 6.15, or calculated using

$$z = \frac{d_{eff}}{2}\left[1 + \sqrt{1 - 3.53\,K}\right] \le 0.95 d_{eff}$$

$$= \frac{d_{eff}}{2}\left[1 + \sqrt{1 - 3.53 \times 0.021}\right] = 0.981 d_{eff} > 0.95 d_{eff}. \text{Hence,}$$

$$z = 0.95 d_{eff} = 0.95 \times 0.659 = 0.626\ \text{m}$$

(Continued)

The required tension steel area A_s may be calculated from

$$A_s = \frac{M_{Ed}}{\frac{f_{yk}}{\gamma_s} \times z}$$

$\gamma_s = 1.15$ [EN 1992-1-1 §2.4.2.4(1)]

$$A_s = \frac{915.6}{\frac{500 \times 1000}{1.15} \times 0.626} = 3.364 \times 10^{-3}\ \text{m}^2 = 3364\ mm^2$$

Try 18Ø16 : $A_{s,\ provided} = 3618\ \text{mm}^2$
Check minimum reinforcement requirements using Table 6.16 or by calculating $A_{s,\ min}$ from:

$$A_{s,\ min} = 0.26 \frac{f_{ctm} b_t d_{\text{eff}}}{f_{yk}}, \text{but not less than } 0.0013\ b_t d_{\text{eff}}$$

$$\text{where } f_{ck} \geq 25\ \text{MPa}$$

$$A_{s,\ min} = 0.26 \times \frac{2.6 b_t d_{\text{eff}}}{500} = 0.0013\ b_t d_{\text{eff}}$$

Minimum percentage of reinforcement required $= 0.13\%$
Percentage of reinforcement provided $= \frac{3618}{4000 \times 659} \times 100 = 0.14\%$ (OK.)
Check maximum reinforcement requirements:

$$A_{s,\ max} = 0.04\,A_c = 0.04 \times 4000 \times 659 = 105440\ \text{mm}^2 \gg A_{s,\ provided}\ (\text{OK}).$$

Provide 18 Ø 16 steel bars both ways at the bottom of the footing
Bar clear spacing and concrete cover:
For footings on soil, minimum concrete cover is
$c_{nom} = K_2 = 75\ \text{mm}$ [EN 1992-1-1 §4.4.1.3(4)]
Bar clear spacing $s = \frac{4000 - 2 \times 75 - 18 \times 16}{17} = 209.5\ \text{mm}$. Use:
$s_{provided} = 209\ mm$. Hence,
$c_{provided} = \frac{4000 - 17 \times 209 - 18 \times 16}{2} = 79.5\ mm > c_{nom}$ (OK).
Centre to centre bar spacing $= 209 + 16 = 225\ mm$

Beam Shear (one-way shear)

Design beam shear at critical section is

$$V_{Ed} = q_{Ed}(l - d_{\text{eff}})(B) = 158.4(1.7 - 0.659)(4) = 660\ \text{kN}$$

Design shear stress at critical section is

$$v_{Ed} = \frac{V_{Ed}}{(B d_{\text{eff}})} = \frac{660}{4 \times 0.659} = 250.4\ kPa$$

Design concrete resisting shear stress, $v_{Rd,\ c}$:

$$v_{Rd,\ c} = C_{Rd,\ c} k (100 \rho_l f_{ck})^{(1/3)} \geq 0.035 k^{1.5} f_{ck}^{\ 0.5} \text{ [EN 1992-1-1 §6.2.2]}$$

$$k = 1 + \sqrt{\frac{200}{d_{\text{eff}}}} \leq 2; f_{ck} = 25\text{ MPa}; \rho_l = \frac{A_s}{Bd_{\text{eff}}} = 0.0014; C_{Rd,\,c} = \frac{0.18}{\gamma_c}$$

$$k = 1 + \sqrt{\frac{200}{659}} = 1.55; C_{Rd,\,c} = \frac{0.18}{1.5} = 0.12; 0.035k^{1.5}f_{ck}{}^{0.5} = 0.338\text{ MPa}$$

$$v_{Rd,\,c} = 0.12 \times 1.55(100 \times 0.0014 \times 25)^{(1/3)} = 0.282\text{ MPa}$$

$v_{Rd,\,c} < 0.035k^{1.5}f_{ck}{}^{0.5}$. Hence,
$v_{Rd,\,c} = 0.338\text{ MPa} = 338\ kPa > v_{Ed}$ (OK.); no shear reinforcement is required by beam shear in both directions.

Punching shear [EN 1992-1-1 § 6.4.1(4), 6.4.3(2) and 6.4.4]
Refer to the flow chart of Figure 6.10.

(a) Considering punching shear perimeter u_0 at the column face:
The factor $\beta = 1.0$ since there is no external applied moment.
Design maximum shear stress at the face of the column is

$$v_{Ed,max} = \frac{V_{Ed} - \Delta V_{Ed}}{(u_0 d_{\text{eff}})}.$$

u_0 = perimeter of the column = $4 \times 0.6 = 2.4$ m

$\Delta V_{Ed} = \Delta V_d$ = the net upward force within the control perimeter (u_0)

$= 158.4 \times 0.6 \times 0.6 - 0.75 \times 0.6^2 \times 25 = 50.3$ kPa

$V_{Ed} = V_d$ = the applied design shear force = 2535 kN

d_{eff} = the average of the effective depths in orthogonal directions

$$v_{Ed,max} = \frac{2535 - 50.3}{2.4 \times 0.659} = 1567\text{ kPa} = 1.571\ MPa$$

Refer to Table 6.18: for $f_{ck} = 25$ MPa, the maximum resisting shear is

$$v_{Ed,max} = \underline{4.5\ MPa} > v_{Ed,\ max}\ (\text{OK}).$$

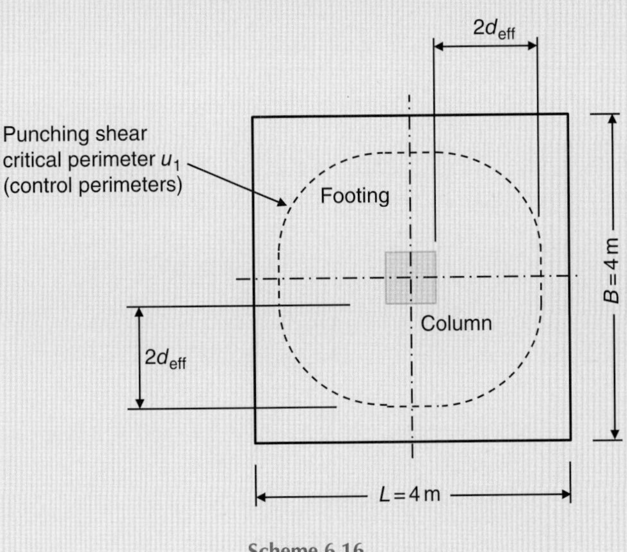

Scheme 6.16

(*Continued*)

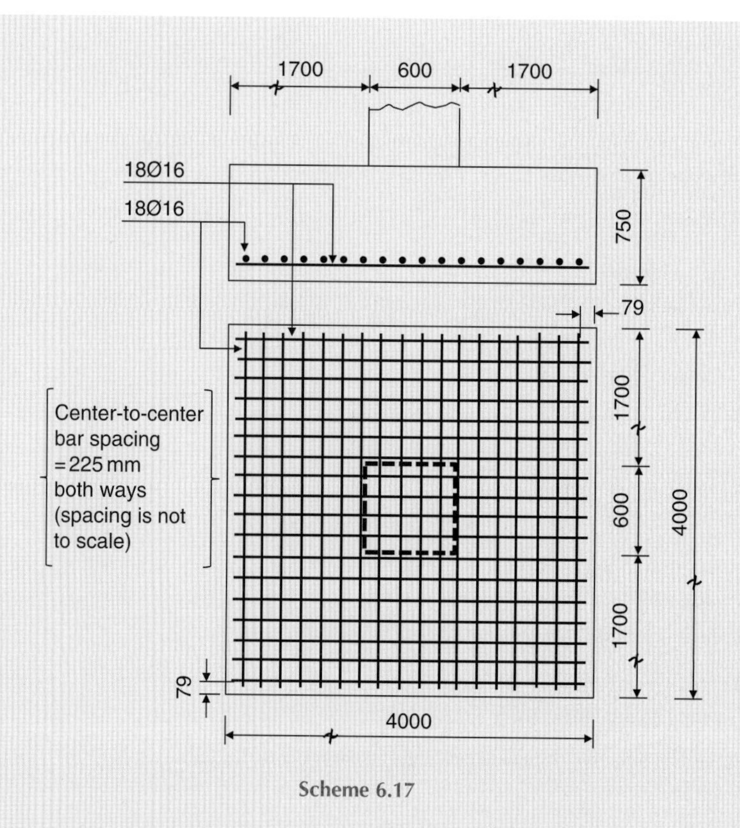

Scheme 6.17

No shear reinforcement is required by punching shear [EN 1992-1-1 §6.4.3(2)].

(b) Considering punching shear perimeter u_1 at distance $2d_{eff}$ from the column face:

Design punching shear stress at distance $2d_{eff}$ from the column face is

$$v_{Ed} = \frac{V_{Ed} - \Delta V_{Ed}}{(u_1 d_{\text{eff}})}$$

$$\Delta V_{Ed} = \Delta V_d = \text{the net upward force within the control perimeter } (u_1)$$

$$= q_{Ed}\left[\frac{\pi(2\times 2d_{\text{eff}})^2}{4} + 4(0.6\times 2d_{\text{eff}}) + 0.6\times 0.6\right]$$

$$-0.75\left[\frac{\pi(2\times 2d_{\text{eff}})^2}{4} + 4(0.6\times 2d_{\text{eff}}) + 0.6\times 0.6\right]\times 25$$

$$= 158.4\left[\frac{\pi(2\times 2\times 0.659)^2}{4} + 4(0.6\times 2\times 0.659) + 0.36\right]$$

$$-0.75\left[\frac{\pi(2\times 2\times 0.659)^2}{4} + 4(0.6\times 2\times 0.659) + 0.36\right]\times 25$$

$$= 1423 - 168 = 1255\,\text{kN}$$

$$u_1 = 4\times 0.6 + 2\times 2\times 0.659\pi = 10.68\,\text{m}$$

$$v_{Ed} = \frac{2535 - 1255}{10.68\times 0.659} = 189\,\text{kPa} = 0.189\,MPa$$

Design punching resisting shear stress, $v_{Rd,\,c}$:
$v_{Rd,\,c} = C_{Rd,\,c}k(100\rho_l f_{ck})^{(1/3)} \geq 0.035k^{1.5}f_{ck}{}^{0.5}$ [EN 1992-1-1 §6.4.4]

$$k = 1 + \sqrt{\frac{200}{d_{\text{eff}}}} \leq 2; f_{ck} = 25\,\text{MPa}; \rho_l = \frac{A_s}{Bd_{\text{eff}}} = 0.0014; C_{Rd,\,c} = \frac{0.18}{\gamma_c}$$

$$k = 1 + \sqrt{\frac{200}{659}} = 1.55; C_{Rd,\,c} = \frac{0.18}{1.5} = 0.12; 0.035k^{1.5}f_{ck}^{0.5} = 0.338\,\text{MPa}$$

$$v_{Rd,\,c} = 0.12 \times 1.55(100 \times 0.0014 \times 25)^{(1/3)} = 0.282\,\text{MPa}$$

$v_{Rd,\,c} < 0.035k^{1.5}f_{ck}^{0.5}$. Hence,
$v_{Rd,\,c} = 0.338\,\text{MPa} = 338\,\text{kPa} = 0.338\,MPa > v_{Ed}$ (OK).
No shear reinforcement is required by punching shear [EN 1992-1-1 § 6.4.3(2)].

Summary of design calculation results

- Footing depth $h = 750\,mm$
- No shear reinforcement is required
- No compression reinforcement is required
- Tension reinforcement is required at the footing bottom: 18 Ø 16 straight bars, both ways
- Centre-to-centre bar spacing $= 225\,mm$
- Concrete cover to the bottom steel bars $= 79\,mm$
- Concrete: $f_{ck} = 25$ MPa; Steel: $f_{yk} = 500$ MPa

References

BCA (2006), *How to Design Concrete Structures using Eurocode 2*, British Cement Association and The Concrete Centre, Blackwater, Camberley.

Bond, A., and Harris, A. (2010), *Decoding Eurocode 7*, 1st edn, Taylor and Francis, 270 Madison Avenue, New York, NY 10016, USA.

BS NA EN 1997-1 (2004), *UK National Annex to Eurocode 7: Geotechnical Design: General Rules*, British Standards Institution, London.

BS NA EN 1990 (2005) *UK National Annex to Eurocode: Basis of Structural Design*, British Standards Institution, London

Burridge, J. (2010), "RC Detailing to Eurocode 2", The Concrete Centre, Riverside House, 4 Meadows Business Park, Station Approach, Blackwater, Camberley, Surrey GU17 9AB, UK.

Calgaro J-A., Tchumi M., and M. Gulvanessian, H. (2009), *Designers' Guide to Eurocode* 1: Actions on Buildings, Thomas Telford Publishing, London.

EN 1990:2002+A1 (2005) *Eurocode –Basis of Structural Design*, European Committee for Standardisation, Brussels

EN 1991-1-1 (2002), *Eurocode 1 –Actions on Structures* – Part 1-1: *General Actions – Densities, Self-weight, Imposed Loads for Buildings*: Committee for Standardisation, Brussels.

EN 1992-1-1 (2004), *Eurocode 2 –Design of Concrete Structures* – Part 1-1: *General Rules and Rules for Buildings*: Committee for Standardisation, Brussels.

EN 1997-1 (2004b), *Eurocode 7 –Geotechnical Design* – Part 1: *General Rules*: Committee for Standardisation, Brussels.

EN 1997-1: Annex F (2004c), *EN Annex F to Eurocode 7 – Geotechnical Design* – Part 1: *General Rules*: *Sample Methods for Settlement Evaluation*: Committee for Standardisation, Brussels.

Gulvanessian, H., Calgaro J-A., and Holicky, M. (2002), *Designers' Guide to EN 1990 Eurocode: Basis of Structural Design*, Thomas Telford Publishing, London.

Scarpelli, G., and Orr, Trevor L.L (2013), "Eurocode 7: Geotechnical Design – Worked Examples", *JRC Scientific and Policy Re ports*, Publications Office of the European Union, Brussels.

(TCC) (2006), *How to Design Concrete Structures using Eurocode 2: The Concrete Centre and British Cement Association*, Blackwater, Camberley, UK.

Index

Shallow Foundations: Discussions and Problem Solving, First Edition. Tharwat M. Baban.